Frequently Used Formulas

Slope of a Line Containing the Points (x_1, y_1) and (x_2, y_2):

$$m = \frac{y_2 - y_1}{x_2 - x_1}, \text{ where } x_2 - x_1 \neq 0.$$

Slope—Intercept Form of a Line: $y = mx + b$

Standard Form of a Line:

$Ax + By = C$ where A, B, and C are integers and A is positive.

Point-Slope Formula: If (x_1, y_1) is a point on line L and m is the slope of line L, then the equation of L is given by

$$y - y_1 = m(x - x_1).$$

Factoring Formulas
1. Difference of Two Squares: $a^2 - b^2 = (a + b)(a - b)$
2. Sum and Difference of Two Cubes:

$$a^3 + b^3 = (a + b)(a^2 - ab + b^2)$$
$$a^3 - b^3 = (a - b)(a^2 + ab + b^2)$$

Distance Formula: The distance, d, between the points (x_1, y_1) and (x_2, y_2) is given by $d = \sqrt{(x_2 - x_1)^2 + (y_2 - y_1)^2}$.

Quadratic Formula: The solutions of any quadratic equation of the form $ax^2 + bx + c = 0$ $(a \neq 0)$ are

$$x = \frac{-b \pm \sqrt{b^2 - 4ac}}{2a}.$$

Standard Form for the Equation of a Parabola That Opens Vertically:

$$y = a(x - h)^2 + k, \text{ where the vertex is at } (h, k)$$

Standard Form for the Equation of a Parabola That Opens Horizontally:

$$x = a(y - k)^2 + h, \text{ where the vertex is at } (h, k)$$

Standard Form for the Equation of a Circle:

$$(x - h)^2 + (y - k)^2 = r^2, \text{ where the center is } (h, k) \text{ and the radius is } r$$

Standard Form for the Equation of an Ellipse

$$\frac{(x - h)^2}{a^2} + \frac{(y - k)^2}{b^2} = 1, \text{ where th}$$

D1511273

Standard Form for the Equation of a Hyperbola

$$\frac{(x - h)^2}{a^2} - \frac{(y - k)^2}{b^2} = 1 \text{ or } \frac{(y - k)^2}{b^2} - \frac{(x - h)^2}{a^2} = 1$$

where the center is (h, k)

About The Cover

After years of teaching, it became clear to Sherri Messersmith that studying math for most students was less about the memorization of facts, but rather a journey of studying and understanding what may seem to be complex topics. Like a cyclist training on a long road, as pictured on the front cover, an athlete typically builds his or her training base with short intervals to successfully work up to longer, more challenging rides. Similarly, students in mathematics courses must put in many hours of effort to attain the skills needed for success in mathematics.

In **"Beginning and Intermediate Algebra, 2e,"** Sherri Messersmith presents a map of the material in "bite-size" pieces, providing the needed rationale for greater student understanding. **Messersmith—*Mapping the Journey to Mathematical Success!*** We hope you enjoy your journey, and wish you all the best in your training!

Beginning and Intermediate Algebra

second edition

Sherri Messersmith

College of DuPage

McGraw-Hill Higher Education

Boston   Burr Ridge, IL   Dubuque, IA   New York   San Francisco   St. Louis
Bangkok   Bogotá   Caracas   Kuala Lumpur   Lisbon   London   Madrid   Mexico City
Milan   Montreal   New Delhi   Santiago   Seoul   Singapore   Sydney   Taipei   Toronto

The McGraw·Hill Companies

McGraw-Hill
Higher Education

BEGINNING AND INTERMEDIATE ALGEBRA, SECOND EDITION

Published by McGraw-Hill, a business unit of The McGraw-Hill Companies, Inc., 1221 Avenue of the
Americas, New York, NY 10020. Copyright © 2009 by The McGraw-Hill Companies, Inc. All rights reserved.
Previous edition © 2007. No part of this publication may be reproduced or distributed in any form or by any
means, or stored in a database or retrieval system, without the prior written consent of The McGraw-Hill
Companies, Inc., including, but not limited to, in any network or other electronic storage or transmission, or
broadcast for distance learning.

Some ancillaries, including electronic and print components, may not be available to customers outside the
United States.

This book is printed on acid-free paper.

3 4 5 6 7 8 9 0 QPV/QPV 0 9

ISBN 978–0–07–304775–1
MHID 0–07–304775–9

ISBN 978–0–07–304776–8 (Annotated Instructor's Edition)
MHID 0–07–304776–7

Editorial Director: *Stewart K. Mattson*
Senior Sponsoring Editor: *Richard Kolasa*
Vice-President New Product Launches: *Michael Lange*
Senior Developmental Editor: *Michelle L. Flomenhoft*
Marketing Manager: *Victoria Anderson*
Lead Project Manager: *Peggy J. Selle*
Senior Production Supervisor: *Sherry L. Kane*
Lead Media Project Manager: *Stacy A. Patch*
Senior Freelance Design Coordinator: *Michelle D. Whitaker*
Cover/Interior Designer: *Pam Verros/pv Design*
(USE) Cover Image: © *Peter Griffith/Masterfile*
Lead Photo Research Coordinator: *Carrie K. Burger*
Project Coordinator: *Melissa M. Leick*
Compositor: *Aptara, Inc.*
Typeface: *10.5/12 Times Roman*
Printer: *Quebecor World Versailles, KY*

The credits section for this book begins on page C-1 and is considered an extension of the copyright page.

Library of Congress Cataloging-in-Publication Data

Messersmith, Sherri.
  Beginning and intermediate algebra / Sherri Messersmith. — 2nd ed.
    p. cm.
  Includes index.
  ISBN 978–0–07–304775–1 — ISBN 0–07–304775–9 (hard copy : alk. paper)
  1. Algebra—Textbooks. I. Title.
QA152.3.M47 2009
512.9—dc22

                    2007042472

www.mhhe.com

Colleagues,

When I first started teaching over 20 years ago, I would teach a course using the methods that were given in the book we were using for the class. As time went on, however, I figured out how my students learned best and developed my own ideas, philosophies, techniques, and worksheets that I used to teach my courses. Eventually, no matter what book we were given for the course, I taught the way I knew worked best. I found that fewer students were dropping the courses, attitudes and grades improved, students' confidence rose, and they were telling me, "I learned more math this semester than I learned in all of high school!" At this point I thought about writing a textbook so that I could share with others what I had learned so that I could, hopefully, help even more students succeed in their classrooms.

I constantly ask, "What else can I do to help my students learn?" One thing my students taught me is they learn best when material is presented in bite-sized pieces. Developmental math students are overwhelmed when too much material is presented at once. Yet, this is how most textbooks are written! My book reflects the way I learned to teach in class: concepts are presented in bite-sized pieces so that students have time to digest fewer concepts at once before moving on and learning more ideas.

I have also asked *why* students have so much trouble learning particular concepts. A good example of this is radicals. Why is it SO hard for students to simplify something like $\sqrt[4]{16x^{12}}$? One reason is that they don't know that $\sqrt[4]{16} = 2$. I have figured out some of the underlying causes, usually based in arithmetic, that are behind what makes it difficult to learn the algebra we are trying to teach. To help improve the skills in which they are weak, I have developed supplemental worksheets. When I began using them in class, my students' understanding of many concepts, including the rules of exponents and radicals, greatly improved. There is a worksheet to accompany every section of the book, and they fall into these three categories: worksheets to strengthen basic skills, those which help teach new concepts, and worksheets that tie multiple concepts together. They are available on the website www.mhhe.com/messersmith.

Also available on *MathZone* is an assessment tool to determine students' strengths and weaknesses at the beginning of the semester. By identifying and targeting the weaknesses, in particular, *MathZone* can provide a way for students to work on improving their basic skills so that they are better prepared to learn the algebra that will be taught in the course.

Beginning and Intermediate Algebra, **2e** has evolved from notes, worksheets, and teaching techniques I have developed throughout my two decades of teaching. This text is a compilation of that as well as what I have learned from faculty nationwide, interaction with the Board of Advisors, the review process, and national conferences and faculty forums I have attended. The sharing of ideas has provided me greater insight not only in presenting topics in this textbook, but also in making me a better teacher.

I am very aware that many students who are in a developmental math course are also enrolled in, or should be enrolled in, a reading course. Without sacrificing mathematical rigor, I have written this text using language that is mathematically sound yet easy for students to understand. Reviewers and students have told us that they find the reading style of this book to be very "friendly" and not as "intimidating" as other books. Breaking down concepts into bite-sized pieces also helps these students more easily learn the mathematics. Plus, I have written interesting, real-world application problems so that, hopefully, the students are more engaged and don't find solving applications to be the nightmare they usually think they are. Ultimately, my goal is to help students **map their journey to mathematical success!**

Thank you to everyone who has helped me develop this textbook. My commitment has been to write the most mathematically sound, readable, student-friendly, and up-to-date text with unparalleled resources available for both students and instructors. I would love to hear from you. If you would like to share your comments, ideas, or suggestions for making the text even better, please contact me at sherri.messersmith@gmail.com.

Sherri Messersmith

About the Author

Sherri Messersmith has been teaching at College of DuPage in Glen Ellyn, Illinois, since 1994. She has over 20 years of experience teaching many different courses from developmental mathematics through calculus. She earned a bachelor of science degree in the teaching of mathematics at the University of Illinois at Urbana-Champaign and went on to teach at the high school level for two years. Sherri returned to UIUC and earned a master of science in applied mathematics and stayed on at the university to teach and coordinate large sections of undergraduate math courses. This is her second textbook, and she has also appeared in videos accompanying several McGraw-Hill texts.

Sherri lives outside of Chicago with her husband, Phil, and their daughters, Alex and Cailen. In her precious free time, she likes to read, play the guitar, and travel—the manuscripts for this and her previous book have accompanied her from Spain to Greece and many points in between.

Table of Contents

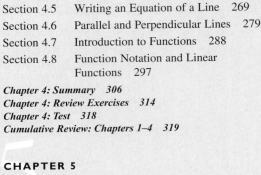

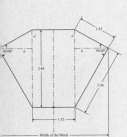

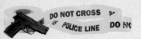

How Does Messersmith Help Students Map the Journey to Mathematical Success?

Beginning and Intermediate Algebra, second edition, centers around helping students to be more successful in mathematics, and all of the revisions made for this edition support this theme. From the start, Messersmith presents the content in "bite-size" pieces, focusing not only on learning mathematical concepts, but also explaining the why behind those concepts. By breaking down the sections into manageable chunks, the author has identified the core places where students traditionally struggle—the book gives students the tools they need to *map their journey to mathematical success.* Messersmith offers three significant tools to help students along the way: applications, pedagogy, and technology.

Messersmith provides an abundant quantity of exercises and **applications** to help students practice and reinforce what they have learned before they chart the next step of their mathematical journey. The traditional, time-tested applications are all available. In addition, there are a significant number of modern applications with pop-culture references to better engage students. Messersmith appreciates the fact that students get turned off by word problems and seeks to combat that by using names and references that students will easily recognize. Once students are able to apply the mathematics, their problem-solving skills should improve—good problem-solving skills are a hallmark of mathematical success.

Several of the **pedagogical** features in Messersmith are designed as checkpoints for students to practice what they have learned before moving forward. One well-received feature of the book are the *You Try* problems that follow nearly every example. These problems provide students with the opportunity to work out a problem similar to the example to make sure they understand the concept being presented. The more students are able to practice, the better their journey will be. Another well-liked feature is the *Mid-Chapter Summary* that appears in several of the chapters. These sections, complete with their own examples and exercises, help students to synthesize important concepts sooner rather than later, which should help in their overall understanding of the material. Messersmith also provides *In-Class Examples* in the margins for instructors only. These examples are matched to the examples in the book to give instructors additional material to present in the classroom. The more examples that students see, the greater their chances are for success.

Finally, Messersmith is available with a number of **technology** solutions to help students succeed on their journey. Through McGraw-Hill's online homework management system, *MathZone,* students can access additional practice exercises, videos demonstrating specific exercises or key concepts, and assessment tools—these assets give students the opportunity for more practice, which should lead to a greater understanding of the important mathematical concepts. Sherri Messersmith presents in all of the exercise *videos* that accompany the book. Because of this, the problem-solving style in the videos directly matches the book, which should make these videos more accessible for students. Also available through MathZone are author-created *worksheets* for each section of the book that fall into three categories: review worksheets/basic skills, worksheets to teach new content, and worksheets to reinforce/pull together different concepts. These worksheets will help the instructor avoid the common stumbling blocks that students encounter in their mathematical journey, like not remembering their multiplication facts when they get to factoring. In addition, McGraw-Hill offers *ALEKS Assessment* tools to assist in diagnosing strengths and weaknesses before students fall behind in their journey.

Key Content Highlights

The content in Messersmith is geared toward students' needs and the pitfalls they may encounter along the way. Chapter 1 includes a review of basic concepts from **geometry.** Throughout beginning and intermediate algebra courses, students need to know these basics, but many do not. Section 1.3 provides the material necessary for students to review the geometry concepts they will need later in the course. Reviewing the geometry early, rather than in an appendix or not at all, removes a common stumbling block for the students. The book also includes geometry applications where appropriate throughout.

"I love the geometry section, most of our students have limited knowledge of geometry and this will really help." Jack L. Haughn, Glendale Community College

Chapter 2 is an entire chapter devoted to the rules of **exponents.** While most texts cover these rules in only two or three sections, Messersmith makes it easier for students to learn and become proficient at applying the rules by breaking down the concepts into bite-size pieces. This chapter includes a Mid-Chapter Summary to help students synthesize what they have learned before moving forward. The author places exponents early because she found that the traditional order of the Real Numbers (Chapter 1 in many books) followed by Solving Linear Equations (Chapter 2 in many books) lulled her students into a false sense of security because they, generally, remembered these topics pretty well from high school. Presenting the rules of exponents in Chapter 2 grabs students' attention because this is not a topic most of them remember well. And covering the rules of exponents early allows for constant review of the topic throughout the course. Students and instructors revisit exponents when they get to the polynomials chapter—Chapter 6. Section 6.1 is a review of the rules of exponents.

"Exponent rules are very important in many later areas to be covered. The mid-chapter summary does an excellent job of pulling together the exponent rules. Section 2.2b is good for an early summary since there are so many of these rules. There are an abundance of practice exercises. Lots of practice is needed to reinforce the exponent rules." Jamie W. McGill, East Tennessee State University

"I think that this introduction to exponents [in Chapter 2] will benefit the student immensely when it comes time to do polynomials. It gets them familiar with how they look—they can look scary." Denise Lujan, University of Texas at El Paso

In response to reviewer feedback, **functions** are now introduced beginning in Chapter 4 and then integrated in subsequent chapters as appropriate. Messersmith recognizes that students need to get familiar with the language of functions earlier in order to be successful later in the course.

"I truly believe that this chapter will help my students in beginning and intermediate algebra to be well prepared for the function concept when they get to the college algebra course." Hamid Attarzadeh, Jefferson Community and Technical College

Also as a result of reviewer feedback, the book now includes a **Beginning Algebra Review** in an appendix to bridge the gap to intermediate algebra for those who need it. It is included as an appendix so that the instructor can use it where it best fits his or her curriculum. This serves as another source of review and reinforcement for the student.

"I like having the basics all within an easy to access spot. This would be great for just a quick refresher for topics that students may have forgotten but just need a reminder." Rhoda Oden, Gadsden State Community College

How Does Messersmith Motivate Students and Help Instructors?

Messersmith integrates a number of features to help both instructors and students navigate through the material in the book. Each feature was reviewed extensively throughout the development process and is meant to help reinforce Messersmith's goal to **map the journey to mathematical success** for the student. We are confident that the entire learning package is of great value to your students as well as you as an instructor. Key features include:

>You Try Problems   After nearly every example, there is a **You Try** problem that mirrors that example. This provides students the opportunity to practice a problem similar to what the instructor has presented before moving on to the next concept. Answers are provided at the end of the section for immediate feedback. Successfully solving the **You Try** problems give students confidence to proceed to the next topic.

"I really like the "You Try" exercises because [they are] built in to facilitate the student that is willing to read through the text and try to learn the material. They also reinforce the idea that a math textbook is not to be just read but requires writing and working problems to understand." Nicole Sifford, Three Rivers Community College

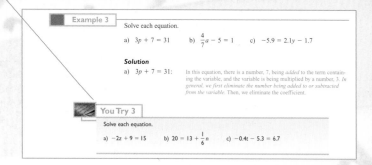

Example 3

Solve each equation.

a) $3p + 7 = 31$ b) $\frac{4}{7}a - 5 = 1$ c) $-5.9 = 2.1y - 1.7$

Solution

a) $3p + 7 = 31$:   In this equation, there is a number, 7, being *added* to the term containing the variable, and the variable is being multiplied by a number, 3. *In general, we first eliminate the number being added to or subtracted from the variable.* Then, we eliminate the coefficient.

You Try 3

Solve each equation.

a) $-2z + 9 = 15$ b) $20 = 13 + \frac{1}{6}n$ c) $-0.4t - 5.3 = 6.7$

>Mid-Chapter Summaries   Several chapters include a **Mid-Chapter Summary** section—this is in keeping with Messersmith's philosophy of breaking sections into manageable chunks to increase student comprehension. These sections help students to synthesize key concepts before moving on to the rest of the chapter.

"I really like the chapter summary material that is [put] in the middle of some of the longer chapters. This allows students to 'catch their breath' in a manner of speaking while working through longer, more difficult chapters." Glenn Jablonski, Triton College

Mid-Chapter Summary

Objectives

1. Learn Strategies for Factoring a Given Polynomial

1. Learn Strategies for Factoring a Given Polynomial

In this chapter, we have discussed several different types of factoring problems:

1) Factoring out a GCF (Section 7.1)
2) Factoring by grouping (Section 7.1)
3) Factoring a trinomial of the form $x^2 + bx + c$ (Section 7.2)
4) Factoring a trinomial of the form $ax^2 + bx + c$ (Section 7.3)
5) Factoring a perfect square trinomial (Section 7.4)
6) Factoring the difference of two squares (Section 7.4)
7) Factoring the sum and difference of two cubes (Section 7.4)

We have practiced the factoring methods separately in each section, but how do we know which factoring method to use given many different types of polynomials together? We will discuss some strategies in this section. First, recall the steps for factoring *any* polynomial:

>Worksheets Supplemental **worksheets** for every section are available online through MathZone. They fall into three categories: review worksheets/basic skills, worksheets to teach new content, and worksheets to reinforce/pull together different concepts. These **worksheets** provide a quick, engaging way for students to work on key concepts. They save instructors from having to create their own supplemental material, and they address potential stumbling blocks. They are also a great resource for standardizing instruction across a mathematics department.

"I love your worksheets!!! My students can never get enough practice on basic skills. One of the main complaints that I hear from my colleagues is that their students do not have the basic skills they need to be successful. At the same time I do not have time to make out extra practice worksheets for my students." Dr. Angela Everett. Chattanooga State Technical Community College

---

Worksheet 1A   Name: _____

Messersmith—Beginning & Intermediate Algebra

1) $8 \cdot 3$ _____	16) $5 \cdot 8$ _____	
2) $4 \cdot 9$ _____	17) $12 \cdot 7$ _____	
3) $12 \cdot 5$ _____	18) $6 \cdot 4$ _____	
4) $7 \cdot 3$ _____	19) $3 \cdot 8$ _____	
5) $4 \cdot 4$ _____	20) $8 \cdot 9$ _____	
6) $10 \cdot 6$ _____	21) $12 \cdot 2$ _____	
7) $11 \cdot 5$ _____	22) $7 \cdot 1$ _____	
8) $6 \cdot 9$ _____	23) $3 \cdot 11$ _____	
9) $4 \cdot 7$ _____	24) $2 \cdot 4$ _____	
10) $8 \cdot 12$ _____	25) $5 \cdot 4$ _____	
11) $5 \cdot 9$ _____	26) $9 \cdot 7$ _____	
12) $9 \cdot 2$ _____	27) $11 \cdot 11$ _____	
13) $7 \cdot 8$ _____	28) $6 \cdot 7$ _____	
14) $3 \cdot 4$ _____	29) $12 \cdot 6$ _____	
15) $8 \cdot 0$ _____	30) $9 \cdot 9$ _____	

>In-Class Examples In order to give instructors additional material to use in the classroom, a matching **In-Class Example** is provided in the margin of the Annotated Instructor's Edition for every example in the book. The more examples a student reviews, the better chance he or she will have to understand the related concept.

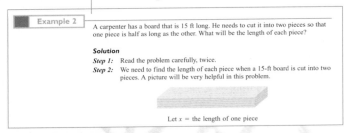

Example 2

A carpenter has a board that is 15 ft long. He needs to cut it into two pieces so that one piece is half as long as the other. What will be the length of each piece?

Solution

Step 1: Read the problem carefully, twice.

Step 2: We need to find the length of each piece when a 15-ft board is cut into two pieces. A picture will be very helpful in this problem.

Let x = the length of one piece

>End-of-Section Exercises The **end-of-section exercises** are presented from the most basic to the most rigorous so that students can see how the concepts work at the simplest level before progressing to more challenging problems. Messersmith incorporates interesting real-world, up-to-date, relevant information that will appeal to students of all backgrounds into the **applications** in the book. Students have identified a number of the problems as interesting and fun in previous use.

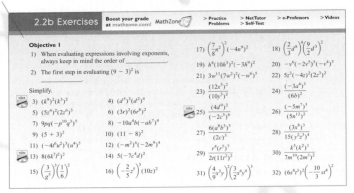

2.2b Exercises

Objective 1

1) When evaluating expressions involving exponents, always keep in mind the order of _____.

2) The first step in evaluating $(9 - 3)^2$ is _____.

Simplify.

3) $(k^9)^2(k^3)^2$
4) $(d^5)^3(d^2)^4$
5) $(5z^4)^2(2z^6)^3$
6) $(3r)^2(6r^8)^2$
7) $9pq(-p^{10}q^3)^5$
8) $-10a^4b(-ab^7)^4$
9) $(5 + 3)^2$
10) $(11 - 8)^2$
11) $(-4t^6u^2)^3(u^4)^5$
12) $(-m^2)^6(-2m^9)^4$
13) $8(6k^7t^2)^2$
14) $5(-7c^4d)^2$
15) $\left(\frac{3}{g}\right)^3\left(\frac{1}{6}\right)^2$
16) $\left(-\frac{2}{5}z^2\right)^3(10z)^2$

17) $\left(\frac{7}{8}n^2\right)^2(-4n^9)^2$
18) $\left(\frac{2}{3}d^8\right)^4\left(\frac{9}{2}d^3\right)^2$
19) $h^4(10h^3)^2(-3h^9)^2$
20) $-v^6(-2v^5)^3(-v^4)^3$
21) $3w^{11}(7w^2)^2(-w^6)^5$
22) $5z^3(-4z)^2(2z^3)^2$
23) $\frac{(12x^3)^2}{(10y^5)^2}$
24) $\frac{(-3a^4)^3}{(6b)^2}$
25) $\frac{(4d^9)^2}{(-2c^5)^6}$
26) $\frac{(-5m^7)^3}{(5n^{12})^2}$
27) $\frac{6(a^8b^3)^5}{(2c)^3}$
28) $\frac{(3x^8)^3}{15(y^2z^3)^4}$
29) $\frac{r^4(r^5)^7}{2t(11t^2)^3}$
30) $\frac{k^5(k^2)^3}{7m^{10}(2m^3)^2}$
31) $\left(\frac{4}{9}x^3y\right)^2\left(\frac{3}{2}x^6y^4\right)^3$
32) $(6s^8t^3)^2\left(-\frac{10}{3}st^4\right)^2$

>Icon Key   Within the exercises, the reader will find several different icons identifying different types of exercises.

Video—exercise has a video with Sherri Messersmith walking through the solution.

Calculator—exercise can be done using a calculator.

Writing—exercise that requires students to explain an idea or answer a question in their own words.

>Chapter-Opening Vignettes   Each chapter opens with a real-world **vignette** to capture the student's attention and engage them in the upcoming material.

>Learning Objectives   The **objectives** are clearly identified at the beginning of each section so that the student knows what key outcomes they will be expected to understand at the end of the section. The objectives then appear within the body of the text, showing when a particular objective is about to be developed. References are also included within the exercise sets to help students quickly locate related material if they need more practice.

CHAPTER 1

The Real Number System and Geometry

Algebra at Work: Landscape Architecture

Jill is a landscape architect and uses multiplication, division, and geometry formulas on a daily basis. Here is an example of the type of landscaping she designs. When Jill is asked to create the landscape for a new house, her first job is to draw the plans.

The ground in front of the house will be dug out into shapes that include rectangles and circles, shrubs and flowers will be planted, and mulch will cover the ground. To determine the volume of mulch that will be needed, Jill must use the formulas for the area of a rectangle and a circle and then multiply by the depth of the mulch. She will calculate the total cost of this landscaping job only after determining the cost of the plants, the mulch, and the labor.

It is important that her numbers are accurate. Her company and her clients must have an accurate estimate of the cost of the job. If the estimate is too high, the customer might choose another, less expensive landscaper to do the job. If the estimate is too low, either the client will have to pay more money at the end or the company will not earn as much profit on the job.

In this chapter, we will review formulas from geometry as well as some concepts from arithmetic.

Section 1.4 Sets of Numbers and Absolute Value

Objectives

1. Identify Numbers and Graph Them on a Number Line
2. Compare Numbers Using Inequality Symbols
3. Find the Additive Inverse and Absolute Value of a Number

1. Identify Numbers and Graph Them on a Number Line

In Section 1.1, we defined the following sets of numbers:

Natural numbers: {1, 2, 3, 4, . . .}
Whole Numbers: {0, 1, 2, 3, 4, . . .}

We will begin this section by discussing other sets of numbers.
On a **number line**, positive numbers are to the right of zero and negative numbers are to the left of zero.

>Be Careful Boxes   These **boxes** contain mistakes that are very common for students to make. They are included to make students aware of these common errors so that, hopefully, they will not make these mistakes themselves.

Can the product rule be applied to $4^3 \cdot 5^2$? **No!** The bases are not the same, so we cannot simply add the exponents. To evaluate $4^3 \cdot 5^2$, we would evaluate $4^3 = 64$ and $5^2 = 25$, then multiply:

$$4^3 \cdot 5^2 = 64 \cdot 25 = 1600$$

>Using Technology Boxes For those instructors who want to make use of graphing calculator–related material, **Using Technology boxes** are included at the ends of sections where relevant. For those instructors who don't want to use this material, they are easily skipped. Answers to these exercises are included after the answers to the You Try problems.

Using Technology

A graphing calculator, like the TI-83 Plus or TI-84, can display numbers in scientific notation. First, the calculator must be set to scientific mode.

Press MODE to see the eight different options that allow you to change calculator settings, depending on what you need. The default settings are shown on the left. To change to scientific mode, use the arrow key, > , to highlight SCI, and press ENTER. This is shown on the right.

>Chapter Summaries
The comprehensive **summaries** at the end of each chapter enable students to review important concepts. A definition or concept is presented, along with a related example and a page reference from the relevant example.

Chapter 2: Summary

Definition/Procedure	Example	Reference
Simplifying Expressions		2.1
Combining Like Terms **Like terms** contain the same variables with the same exponents.	Combine like terms and simplify. $4k^2 - 3k + 1 - 2(6k^2 - 5k + 7)$ $= 4k^2 - 3k + 1 - 12k^2 + 10k - 14$ $= -8k^2 + 7k - 13$	p. 89
Writing Mathematical Expressions	Write a mathematical expression for the following: *sixteen more than twice a number* Let x = the number. Sixteen more than twice a number $16 \quad + \quad 2x$ $16 + 2x$	p. 91
The Product Rule and Power Rules of Exponents		2.2a
Exponential Expression: $a^n = \underbrace{a \cdot a \cdot a \cdot \ldots \cdot a}_{n \text{ factors of } a}$ a is the **base**, n is the exponent	$5^4 = 5 \cdot 5 \cdot 5 \cdot 5$ 5 is the **base**, 4 is the exponent.	p. 93
Product Rule: $a^m \cdot a^n = a^{m+n}$	$x^8 \cdot x^2 = x^{10}$	p. 94
Basic Power Rule: $(a^m)^n = a^{mn}$	$(t^3)^5 = t^{15}$	p. 95
Power Rule for a Product: $(ab)^n = a^n b^n$	$(2c)^4 = 2^4 c^4 = 16c^4$	p. 96
Power Rule for a Quotient: $\left(\dfrac{a}{b}\right)^n = \dfrac{a^n}{b^n}$, where $b \neq 0$.	$\left(\dfrac{w}{5}\right)^3 = \dfrac{w^3}{5^3} = \dfrac{w^3}{125}$	p. 96
Combining the Rules of Exponents		2.2b
Remember to follow the order of operations.	Simplify. $(3y^4)^2(2y^9)^3$ $= 9y^8 \cdot 8y^{27}$ Exponents come before multiplication. $= 72y^{35}$ Use the product rule and multiply coefficients.	p. 100
Integer Exponents with Real Number Bases		2.3a
Zero Exponent: If $a \neq 0$, then $a^0 = 1$.	$(-9)^0 = 1$	p. 102

>Review Exercises and Chapter Tests

These appear in each chapter to provide students with an opportunity to check their progress and to review important concepts, as well as to provide confidence and guidance in preparing for in-class tests or exams.

Chapter 2 Review Exercises

(2.1) Combine like terms and simplify.

1) $15y^2 + 8y - 4 + 2y^2 - 11y + 1$

2) $7t + 10 - 3(2t + 3)$

3) $\frac{3}{2}(5n - 4) + \frac{1}{4}(n + 6)$

4) $1.4(a + 5) - (a + 2)$

Write a mathematical expression for each phrase, and combine like terms if possible. Let x represent the unknown quantity.

5) Ten less than a number

6) The sum of a number and twelve

7) The sum of a number and eight decreased by three

8) Six less than twice a number

(2.2a)

9) Write in exponential form

   a) $8 \cdot 8 \cdot 8 \cdot 8 \cdot 8 \cdot 8$ b) $(-7)(-7)(-7)(-7)$

10) Identify the base and the exponent.

   a) -6^5 b) $(4t)^3$

   c) $4t^3$ d) $-4t^3$

11) Use the rules of exponents to simplify.

   a) $2^3 \cdot 2^2$ b) $\left(\frac{1}{3}\right)^2 \cdot \left(\frac{1}{3}\right)$

   c) $(7^3)^4$ d) $(k^5)^6$

12) Use the rules of exponents to simplify.

   a) $(3^2)^2$ b) $8^3 \cdot 8^7$

   c) $(m^4)^9$ d) $p^9 \cdot p^7$

13) Simplify using the rules of exponents.

   a) $(5y)^3$ b) $(-7m^4)(2m^{12})$

   c) $\left(\frac{a}{b}\right)^6$ d) $6(xy)^2$

   e) $\left(\frac{10}{9}c^4\right)(2c)\left(\frac{15}{4}c^3\right)$

14) Simplify using the rules of exponents.

   a) $\left(\frac{x}{y}\right)^{10}$ b) $(-2z)^5$

   c) $(6t^7)\left(-\frac{5}{8}t^5\right)\left(\frac{2}{3}t^2\right)$

   d) $-3(ab)^4$ e) $(10j^6)(4j)$

(2.2b)

15) Simplify using the rules of exponents.

   a) $(z^3)^2(z^3)^4$ b) $-2(3c^5d^8)^2$

   c) $(9 - 4)^3$ d) $\frac{(10t^3)^2}{(2u^7)^3}$

16) Simplify using the rules of exponents.

   a) $\left(\frac{-20d^4c}{5b^3}\right)^3$ b) $(-2y^8z)^3(3yz^2)^2$

   c) $\frac{x^7 \cdot (x^2)^5}{(2y^3)^4}$ d) $(6 - 8)^2$

(2.3a)

17) Evaluate.

   a) 8^0 b) -3^0

   c) 9^{-1} d) $3^{-2} - 2^{-2}$

   e) $\left(\frac{4}{5}\right)^{-3}$

18) Evaluate.

   a) $(-12)^0$ b) $5^0 + 4^0$

   c) -6^{-2} d) 2^{-4}

   e) $\left(\frac{10}{3}\right)^{-2}$

(2.3b)

19) Rewrite the expression with positive exponents. Assume the variables do not equal zero.

   a) v^{-9} b) $\left(\frac{9}{c}\right)^{-2}$

   c) $\left(\frac{1}{y}\right)^{-8}$ d) $-7k^{-9}$

   e) $\frac{19z^{-4}}{a^{-1}}$

   g) $\left(\frac{2j}{k}\right)^{-5}$

20) Rewrite the ex_____ the variables d_____

   a) $\left(\frac{1}{x}\right)^{-5}$

   c) $a^{-8}b^{-3}$

Chapter 1 Test

1) Find the prime factorization of 210.

2) Write in lowest terms:

   a) $\frac{45}{72}$

   b) $\frac{420}{560}$

Perform the indicated operation(s). Write all answers in lowest terms.

3) $\frac{9}{14} \cdot \frac{7}{24}$

4) $\frac{1}{3} + \frac{4}{15}$

5) $5\frac{1}{4} - 2\frac{1}{6}$

6) $\frac{12}{13} \div 6$

7) $\frac{4}{7} - \frac{5}{6}$

8) $-11 - (-19)$

9) $25 + 15 \div 5$

10) $\frac{9}{10} \cdot \left(-\frac{2}{5}\right)$

11) $-8 \cdot (-6)$

12) $-13.4 + 6.9$

13) $30 - 5[-10 + (2 - 6)^2]$

14) $\frac{(50 - 26) \div 3}{3 \cdot 5 - 7}$

15) Both the highest and lowest points in the continental United States are in California. Badwater, Death Valley, is 282 ft below sea level while, 76 mi away, Mount Whitney reaches an elevation of 14,505 ft. What is the difference between these two elevations?

Evaluate.

16) 2^5

17) -3^4

18) $|-92|$

19) $|2 - 12| - 4|6 - 1|$

20) The complement of 12° is _____.

21) Find the missing angle, and classify the triangle as acute, obtuse, or right.

22) Find the area of this triangle. Include the correct units.

23) Find the area and perimeter of each figure. Include the correct units.

   a)

   b)

24) Find the volume of this cube:

25) The radius of a baseball is approximately 1.5 in. Find the exact value of each in terms of π. Include the correct units.

   a) Find its circumference.

   b) Find the volume of the baseball.

26) Given the set of numbers,

$$\left\{41, -8, 0, 2.\overline{83}, \sqrt{75}, 6.5, 4\frac{5}{8}, 6.37528861\ldots\right\},$$

list the

   a) integers

   b) irrational numbers

   c) natural numbers

>Cumulative Reviews   Cumulative Reviews are included starting with Chapter 2. These reviews help students build on previously covered material and give them an opportunity to reinforce the skills necessary in preparing for midterm and final exams. These reviews assist students with the retention of knowledge throughout the course.

Cumulative Review: Chapters 1–5

Perform the operations and simplify.

1) $\dfrac{3}{10} - \dfrac{7}{15}$

2) $5\dfrac{5}{6} \div 1\dfrac{13}{15}$

3) $(5-8)^3 + 40 \div 10 - 6$

4) Find the area of the triangle.

5) Simplify $-8(3x^2 - x - 7)$

Simplify. The answer should not contain any negative exponents.

6) $(2y^5)^4$

7) $3c^2 \cdot 5c^{-8}$

8) $\dfrac{36a^{-2}b^5}{54ab^{13}}$

9) Write 0.00008319 in scientific notation.

10) Solve $11 - 3(2k - 1) = 2(6 - k)$.

11) Solve $0.04(3p - 2) - 0.02p = 0.1(p + 3)$.

12) Solve. Write the answer in interval notation.
$-47 \le 7t - 5 \le 6$

13) Write an equation and solve.

The number of plastic surgery procedures performed in the United States in 2003 was 293% more than the number performed in 1997. If approximately 8,253,000 cosmetic procedures were performed in 2003, how many took place in 1997? (Source: American Society for Aesthetic Plastic Surgery)

14) The area, A, of a trapezoid is $A = \dfrac{1}{2}h(b_1 + b_2)$
where h = height of the trapezoid,
b_1 = length of one base of the trapezoid, and
b_2 = length of the second base of the trapezoid.

a) Solve the equation for h.

b) Find the height of the trapezoid that has an area of 39 cm^2 and bases of length 8 cm and 5 cm.

15) Graph $2x + 3y = 5$.

16) Find the x- and y-intercepts of the graph of $4x - 5y = 10$.

17) Write the slope-intercept form of the equation of the line containing $(-7, 4)$ and $(1, -3)$.

18) Determine whether the lines are parallel, perpendicular, or neither.

$$10x + 18y = 9$$
$$9x - 5y = 17$$

Solve each system of equations.

19) $9x + 7y = 7$
$3x + 4y = -11$

20) $3(2x - 1) - (y + 10) = 2(2x - 3) - 2y$
$3x + 13 = 4x - 5(y - 3)$

21) $\dfrac{5}{6}x - \dfrac{1}{2}y = \dfrac{2}{3}$
$-\dfrac{5}{4}x + \dfrac{3}{4}y = \dfrac{1}{2}$

Write a system of equations and solve.

22) Dhaval used twice as many 6-ft boards as 4-ft boards when he made a playhouse for his children. If he used a total of 48 boards, how many of each size did he use?

23) Through 2003, Aretha Franklin had won 10 more Grammy Awards than Whitney Houston, and Christina Aguilera had won half as many as Whitney Houston. All together these three singers had won 25 Grammy Awards. How many did each woman win? (Source: www.grammy.com)

Content Changes

Sherri Messersmith and McGraw-Hill would like to thank all of the reviewers for their feedback throughout the development process. The following key content changes were all made based on feedback from instructors across the country.

- **Chapter 4:** Functions have been introduced earlier in the second edition. Previously, they were found in Chapter 12; now they can be found in the new sections, 4.7 and 4.8. Functions now appear throughout the book for those schools that like to approach functions in this way, but these problems can be skipped easily enough for those schools that prefer to teach functions later.
- **Chapter 5:** In Section 5.1 (Solving Systems of Linear Equations by Graphing), exercises involving functions have been added at the end of the exercise set. A new section has been added: Section 5.5 on Systems of Linear Equations in Three Variables. In the preliminary edition, this was found in Chapter 9.
- **Chapter 6:** In Section 6.2 (Addition and Subtraction of Polynomials), an example and exercises on polynomial functions have been added.
- **Chapter 7:** In Section 7.5 (Solving Quadratic Equations by Factoring), exercises involving functions have been added. In Section 7.6 (Applications of Quadratic Equations), some explanations and exercises involving functions have been added.
- **Chapter 8:** In Section 8.1 (Simplifying Rational Expressions), the definition of a rational function is given, two examples that include finding the domain of rational functions have been added, and exercises have been added.
- **Chapter 9:** The section on Systems of Linear Equations in Three Variables has been removed from Chapter 9 and has been placed in Chapter 5.
- **Chapter 12:** Much of this chapter has been rewritten to reflect the changes made when the introduction of functions was moved to Chapter 4. Section 12.1 is still an introduction to functions, but the chapter now contains six sections instead of the eight sections it contained in the preliminary edition.
- **Chapter 14:** Section 14.6 (Second-Degree Inequalities and Systems of Inequalities) has been removed.
- **Chapter 15:** This chapter (Sequences and Series) has been removed from the print version of the book but is available online or can be custom bound on request.
- **Beginning Algebra Review Appendix:** This appendix is intended to bridge the gap to intermediate algebra for those who need it. It is included as an appendix so that the instructor can use it where it best fits his or her curriculum.

Supplements

Multimedia Supplements

www.mathzone.com

MathZone

McGraw-Hill's MathZone is a complete online tutorial and homework management system for mathematics and statistics, designed for greater ease of use than any other system available. Instructors have the flexibility to create and share courses and assignments with colleagues, adjunct faculty, and teaching assistants with only a few clicks of the mouse. All algorithmic exercises, online tutoring, and a variety of video and animations are directly tied to text-specific materials.

MathZone is completely customizable to suit individual instructor and student needs. Exercises can be easily edited, multimedia is assignable, importing additional content is easy, and instructors can even control the level of help available to students while doing their homework. Students have the added benefit of full access to the study tools to individually improve their success without having to be part of a MathZone course.

MathZone has automatic grading and reporting of easy-to-assign algorithmically generated problem types for homework, quizzes, and tests. Grades are readily accessible through a fully integrated grade book that can be exported in one click to Microsoft Excel, WebCT, or BlackBoard.

MathZone offers

- Practice exercises, based on the text's end-of-section material, generated in an unlimited number of variations, for as much practice as needed to master a particular topic.
- Subtitled videos demonstrating text-specific exercises and reinforcing important concepts within a given topic.
- NetTutor™ integrating online whiteboard technology with live personalized tutoring via the Internet.
- Assessment capabilities, which provide students and instructors with the diagnostics to offer a detailed knowledge base through advanced reporting and remediation tools.
- Faculty the ability to create and share courses and assignments with colleagues and adjuncts, or to build a course from one of the provided course libraries.
- An Assignment Builder that provides the ability to select algorithmically generated exercises from any McGraw-Hill math textbook, edit content, as well as assign a variety of MathZone material, including an ALEKS Assessment.
- Accessibility from multiple operating systems and Internet browsers.

Instructors: To access MathZone, request registration information from your McGraw-Hill sales representative.

Worksheets

Available through MathZone, supplemental worksheets are provided for every section. They fall into three categories: review worksheets/basic skills, worksheets to teach new content, and worksheets to reinforce/pull together different concepts. These worksheets are a great way to both enhance instruction and to give the students more tools to be successful in studying a given topic.

Computerized Test Bank (CTB) Online (Instructors Only)

Available through MathZone, this **computerized test bank**, utilizing Brownstone Diploma® algorithm-based testing software, enables users to create customized exams quickly. This user-friendly program enables instructors to search for questions by topic, format, or difficulty level; to edit existing questions or to add new ones; and to scramble questions and answer keys for multiple versions of the same test. Hundreds of text-specific open-ended and multiple-choice questions are included in the question bank. Sample chapter tests in Microsoft Word® and PDF formats are also provided.

Online Instructor's Solutions Manual (Instructors Only)

Available on MathZone, the Instructor's Solutions Manual provides comprehensive, **worked-out solutions** to all exercises in the text. The methods used to solve the problems in the manual are the same as those used to solve the examples in the textbook.

NetTutor

Available through MathZone, NetTutor is a revolutionary system that enables students to interact with a live tutor over the World Wide Web. NetTutor's Web-based, graphical chat capabilities enable students and tutors to use mathematical notation and even to draw graphs as they work through a problem together. Students can also submit questions and receive answers, browse previously answered questions, and view previous live-chat sessions. Tutors are familiar with the textbook's objectives and problem-solving styles.

Video Lectures on Digital Video Disk (DVD)

In the videos, author Sherri Messersmith works through selected exercises from the textbook, following the solution methodology employed in the text. The video series is available on DVD or online as an assignable element of MathZone. The DVDs are closed-captioned for the hearing impaired, subtitled in Spanish, and meet the Americans with Disabilities Act Standards for Accessible Design. Instructors may use them as resources in a learning center, for online courses, and/or to provide extra help for students who require extra practice.

ALEKS

www.aleks.com

ALEKS (**A**ssessment and **LE**arning in **K**nowledge **S**paces) is a dynamic online learning system for mathematics education, available over the Web 24/7. ALEKS assesses students, accurately determines their knowledge, and then guides them to the material that they are most ready to learn. With a variety of reports, Textbook Integration Plus, quizzes, and homework assignment capabilities, ALEKS offers flexibility and ease of use for instructors.

- ALEKS uses artificial intelligence to determine exactly what each student knows and is ready to learn. ALEKS remediates student gaps and provides highly efficient learning and improved learning outcomes
- ALEKS is a comprehensive curriculum that aligns with syllabi or specified textbooks. Used in conjunction with a McGraw-Hill text, students also receive links to text-specific videos, multimedia tutorials, and textbook pages.
- Textbook Integration Plus enables ALEKS to be automatically aligned with syllabi or specified McGraw-Hill textbooks with instructor chosen dates, chapter goals, homework, and quizzes.
- ALEKS with AI-2 gives instructors increased control over the scope and sequence of student learning. Students using ALEKS demonstrate a steadily increasing mastery of the content of the course.
- ALEKS offers a dynamic classroom management system that enables instructors to monitor and direct student progress toward mastery of course objectives.

See www.aleks.com for more information.

Printed Supplements

Annotated Instructor's Edition (Instructors Only)

This ancillary contains answers to exercises in the text, including answers to all section exercises, all *Review Exercises, Chapter Tests,* and *Cumulative Reviews.* These answers are printed in a separate color for ease of use by the instructor and are located on the appropriate pages throughout the text, space permitting. The AIE contains answers to all the exercises in an appendix at the end of the book. The student edition contains answers to selected exercises.

Student's Solutions Manual

The *Student's Solutions Manual* provides comprehensive, **worked-out solutions** to all of the odd-numbered exercises. The steps shown in the solutions match the style of solved examples in the textbook.

Acknowledgments

I would like to thank the reviewers who provided me with thoughtful and valuable feedback.

Board of Advisors

Kay Cornelius, *Sinclair Community College*
Calandra Davis, *Georgia Perimeter College*
Said Fariabi, *San Antonio College*
Corinna Goehring, *Jackson State Community College*
Marc Grether, *University of North Texas*
Glenn Jablonski, *Triton College*
George Johnson, III, *St. Philip's College*
Rhoda Oden, *Gadsden State Community College*

Manuscript Reviewers

David Anderson, *South Suburban College*
Darla Aguilar, *Pima Community College*
Frances Alvarado, *University of Texas–Pan American*
Hamid Attarzadeh, *Jefferson Community & Technical College*
Jon Becker, *Indiana University–Northwest*
Charles Belair, *Medaille College*
Suanne Benowicz, *Lewis & Clark College*
Sylvia Brite, *Jefferson Community College*
Don Brown, *Macon State College*
Susan Caldiero, *Cosumnes River College*
Peggy Clifton, *Redlands Community College*
Deborah Cochener, *Austin Peay State University*
Omri Crewe, *Collin County Community College*
Patrick Cross, *University of Oklahoma*
Joseph Ediger, *Portland State University*
Barbara Elzey-Miller, *Lexington Community College*
Azin Enshai, *American River College*
Angela Everett, *Chattanooga State Technical Community College*
Tonia Faulling, *Tri-County Technical College*
Steven Felzer, *Lenoir Community College*
Donna Flint, *South Dakota State University*
Robert Frank, *Westmoreland County Community College*
Larry Green, *Lake Tahoe Community College*
Charles Groce, *Eastern Kentucky University*
Gloria Guerra, *St. Philip's College*
Nikki Handley, *Odessa College*
Jack Haughn, *Glendale Community College*
Andrea Hendricks, *Georgia Perimeter College*

Celeste Hernandez, *Richland College*
Evan Innerst, *Canada College*
Jo Johansen, *Rutgers University*
Rex Johnson, *University of Texas–El Paso*
Stacy Jurgens, *Mesabi Range Community & Technical College*
Lynette King, *Gadsden State Community College*
David Kedrowski, *Mid Michigan Community College*
Gopala Krishna, *South Carolina State University*
Betty Larson, *South Dakota State University*
Wilene Leach, *Arkansas State University*
Paul Lee, *St. Philip's College*
Christine Lehmann, *Purdue University–North Central*
Yixia Lu, *South Suburban College*
Denise Lujan, *University of Texas–El Paso*
Sarah Luther, *Umpqua Community College*
Charlotte Matthews, *University of South Alabama*
James Mays, *Muscatine Community College*
Jamie McGill, *East Tennessee State University*
Dan McGlasson, *University of Texas–El Paso*
Hazel McKenna, *Utah Valley State College*
Timothy McKenna, *University of Michigan–Dearborn*
Iris McMurtry, *Motlow State Community College*
Patricia Bederman Miller, *Keystone College*
Mary Ann Misko, *Gadsden State Community College*
Dennis Monbrod, *South Suburban College*
Kimberly Morgan, *Rochester College*
Jack Morrell, *Atlanta Metropolitan College*
Ben Moulton, *Utah Valley State College*
Carol Murphy, *San Diego Miramar College*
Mark Naber, *Monroe County Community College*
Anita Ann Nelson, *Tri-County Technical College*
Agashi Nwogbaga, *Wesley College*
Zacchaeus Oguntebi, *Georgia Perimeter College*
Sheila Palmer, *Walla Walla Community College*
Mari Peddycoart, *Kingwood College*
William Peters, *San Diego Mesa College*
James Pierce, *Lincoln Land Community College*
Carolyn Rieffel, *Louisiana Tech College–Fletcher*
Laurie Riggs, *California State Polytechnic University–Pomona*

Sharon Robertson, *University of Tennessee–Martin*

Mary Robinson, *University of New Mexico–Valencia*

Richard Rupp, *Del Mar College*

Haazim Sabree, *Georgia Perimeter College*

Rebecca Schantz, *East Central College*

Vicki Schell, *Pensacola Junior College*

Janet Schlaak, *University of Cincinnati*

Mohsen Shirani, *Tennessee State University*

Mark Shore, *Allegany College of Maryland*

Nicole Sifford, *Three Rivers Community College*

Dave Sobecki, *Miami University–Hamilton*

John Squires, *Cleveland State Community College*

Dan Taylor, *Centralia College*

John Thoo, *Yuba College*

James Vicich, *Scottsdale Community College*

Julien Viera, *University of Texas–El Paso*

John Ward, *Jefferson Community College–Louisville*

Pamela Webster, *Texas A&M University–Commerce*

Mary Jo Westlake, *Itasca Community College*

Diane Williams, *Northern Kentucky University*

Paige Wood, *Kilgore College*

Bella Zamansky, *University of Cincinnati*

Loris Zucca, *Kingwood College*

AMATYC Focus Group Participants

James Carr, *Normandale Community College*

John Collado, *South Suburban College*

Kay Cornelius, *Sinclair Community College*

Paul W. Jones, II, *University of Cincinnati*

Stacy Jurgens, *Mesabi Range Community and Technical College*

Abbas Meigooni, *Lincoln Land Community College*

Carol Schmidt, *Lincoln Land Community College*

Lee Ann Spahr, *Durham Technical Community College*

Additionally I would like to thank my husband, Phil, and my daughters, Alex and Cailen, for their patience, support, and willingness to eat "whatever" for dinner while I have been working on the book. A big high five goes out to Sue Xander for her great friendship and support.

Thank you to all of my colleagues at College of DuPage, especially Jerry Krusinski, Adenuga Atewologun, Chris Picard, Marge Peters, Patrick (Jim) Bradley, Gloria Olsen, Keith Kuchar, and Mary Hill who wrote the Online Computerized Test Bank to accompany the book.

Thanks to Jill McClain-Wardynski, Dorothy Wendel, Pat Steele, Elizabeth Siebenaler, and Geoff Krader for their accuracy checking and proofreading. And thanks to Vicki Schell for her contributions to the Using Technology boxes.

To all of the baristas at the local coffee shop: thanks for having my drink ready before I even get to the register and for letting me sit at the same table for hours on end when I just had to get out of my office.

There are so many people to thank at McGraw-Hill: Jeff Huettman for starting it all with a conversation on a bus out of Bedrock at 5 A.M., Liz Haefele for all of her encouragement and support when we were just getting underway, and Rich Kolasa and Michelle Flomenhoft for all of their hard work when we went through the changing of the guard. I would also like to thank Marty Lange, Stewart Mattson, Torie Anderson, Amber Bettcher, Peggy Selle, Sherry Kane, Michelle Whitaker, Pam Verros, and Stacy Patch for everything they have done on this project.

My greatest thanks go to my assistant, Bill Mulford, for being there from the beginning, every step of the way. Whether he was doing research or trying to survive the madness of meeting deadlines, no one could have done it better. The support he has provided, both technical and moral, has been invaluable, and for that I am extremely grateful. Thank you so much. You are the best.

Sherri Messersmith

Sherri Messersmith
College of DuPage

A Commitment to Accuracy

You have a right to expect an accurate textbook, and McGraw-Hill invests considerable time and effort to make sure that we deliver one. Listed below are the many steps we take to make sure this happens.

Our Accuracy Verification Process

First Round

Step 1: Numerous **college math instructors** review the manuscript and report on any errors that they may find, and the authors make these corrections in their final manuscript.

Second Round

Step 2: Once the manuscript has been typeset, the **authors** check their manuscript against the first page proofs to ensure that all illustrations, graphs, examples, exercises, solutions, and answers have been correctly laid out on the pages, and that all notation is correctly used.

Step 3: An outside, **professional mathematician** works through every example and exercise in the page proofs to verify the accuracy of the answers.

Step 4: A proofreader adds a triple layer of accuracy assurance in the first pages by hunting for errors; then a second, corrected round of page proofs is produced.

Third Round

Step 5: The **author team** reviews the second round of page proofs for two reasons: (1) to make certain that any previous corrections were properly made, and (2) to look for any errors they might have missed on the first round.

Step 6: A **second proofreader** is added to the project to examine the new round of page proofs to double check the author team's work and to lend a fresh, critical eye to the book before the third round of paging.

Fourth Round

Step 7: A **third proofreader** inspects the third round of page proofs to verify that all previous corrections have been properly made and that there are no new or remaining errors.

Step 8: Meanwhile, in partnership with **independent mathematicians,** the text accuracy is verified from of fresh perspectives:

- The **test bank author** checks for consistency and accuracy as they prepare the computerized test item file.
- The **solutions manual author** works every single exercise and verifies their answers, reporting any errors to the publisher.
- A **consulting group of mathematicians,** who write material for the text's MathZone site, notifies the publisher of any errors they encounter in the page proofs.
- A video production company employing **expert math instructors** for the text's videos will alert the publisher of any errors they might find in the page proofs.

Final Round

Step 9: The **project manager,** who has overseen the book from the beginning, performs a **fourth proofread** of the textbook during the printing process, providing a final accuracy review.

⇒ What results is a mathematics textbook that is as accurate and error-free as is humanly possible, and our authors and publishing staff are confident that our many layers of quality assurance have produced textbooks that are the leaders of the industry for their integrity and correctness.

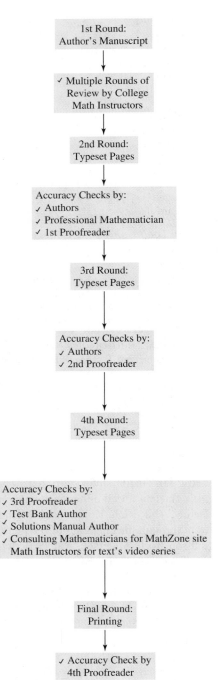

1st Round:
Author's Manuscript

✓ Multiple Rounds of
Review by College
Math Instructors

2nd Round:
Typeset Pages

Accuracy Checks by:
✓ Authors
✓ Professional Mathematician
✓ 1st Proofreader

3rd Round:
Typeset Pages

Accuracy Checks by:
✓ Authors
✓ 2nd Proofreader

4th Round:
Typeset Pages

Accuracy Checks by:
✓ 3rd Proofreader
✓ Test Bank Author
✓ Solutions Manual Author
✓ Consulting Mathematicians for MathZone site
  Math Instructors for text's video series

Final Round:
Printing

✓ Accuracy Check by
  4th Proofreader

Applications Index

CHAPTER 1

The Real Number System and Geometry

Algebra at Work: Landscape Architecture

Jill is a landscape architect and uses multiplication, division, and geometry formulas on a daily basis. Here is an example of the type of landscaping she designs. When Jill is asked to create the landscape for a new house, her first job is to draw the plans.

The ground in front of the house will be dug out into shapes that include rectangles and circles, shrubs and flowers will be planted, and mulch will cover the ground. To determine the volume of mulch that will be needed, Jill must use the formulas for the area of a rectangle and a circle and then multiply by the depth of the mulch. She will calculate the total cost of this landscaping job only after determining the cost of the plants, the mulch, and the labor.

It is important that her numbers are accurate. Her company and her clients must have an accurate estimate of the cost of the job. If the estimate is too high, the customer might choose another, less expensive landscaper to do the job. If the estimate is too low, either the client will have to pay more money at the end or the company will not earn as much profit on the job.

In this chapter, we will review formulas from geometry as well as some concepts from arithmetic.

Section 1.1 Review of Fractions

Objectives

1. Understand What a Fraction Represents

2. Write Fractions in Lowest Terms

3. Multiply and Divide Fractions

4. Add and Subtract Fractions

Why review fractions and arithmetic skills? Because the manipulations done in arithmetic and with fractions are precisely the same skills needed to learn algebra.

Let's begin by defining some numbers used in arithmetic:

Natural numbers: 1, 2, 3, 4, 5, . . .

Whole numbers: 0, 1, 2, 3, 4, 5, . . .

Natural numbers are often thought of as the counting numbers. **Whole numbers** consist of the natural numbers and zero.

Natural and whole numbers are used to represent complete quantities. To represent a part of a quantity we can use a fraction.

1. Understand What a Fraction Represents

What is a fraction?

Definition

A **fraction** is a number in the form $\frac{a}{b}$, where $b \neq 0$. a is called the **numerator**, and b is the **denominator**.

1) A fraction describes a part of a whole quantity.

2) $\frac{a}{b}$ means $a \div b$.

Example 1

What part of the figure is shaded?

Solution

The whole figure is divided into three equal parts. Two of the parts are shaded.

Therefore, the part of the figure that is shaded is $\frac{2}{3}$.

$\dfrac{2}{3}$ $\begin{array}{l} \to \text{number of shaded parts} \\ \to \text{total number of equal parts in the figure.} \end{array}$

 You Try 1

What part of the figure is shaded?

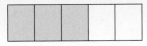

2. Write Fractions in Lowest Terms

A fraction is in **lowest terms** when the numerator and denominator have no common factors except 1. Before discussing how to write a fraction in lowest terms, we need to know about factors.

Consider the number 12.

$$12 \quad = \quad 3 \quad \cdot \quad 4$$
$$\uparrow \qquad\qquad \uparrow \qquad\quad \uparrow$$
$$\text{product} \qquad \text{factor} \qquad \text{factor}$$

3 and *4* are *factors* of 12. (When we use the term **factors**, we mean natural numbers.) Multiplying *3* and *4* results in 12. *12* is the **product**.

Does 12 have any other factors?

Example 2

Find all factors of 12.

Solution

$$12 = 3 \cdot 4 \qquad \text{Factors are 3 and 4}$$
$$12 = 2 \cdot 6 \qquad \text{Factors are 2 and 6}$$
$$12 = 1 \cdot 12 \qquad \text{Factors are 1 and 12}$$

These are all of the ways to write *12* as the product of two factors. The factors of 12 are 1, 2, 3, 4, 6, and 12.

 You Try 2

Find all factors of 30.

We can also write 12 as a product of *prime numbers*.

Definition

A **prime number** is a natural number whose only factors are 1 and itself. (The factors are natural numbers.)

Example 3

Is 7 a prime number?

Solution

Yes. The only way to write 7 as a product of natural numbers is $1 \cdot 7$.

 You Try 3

Is 19 a prime number?

> **Definition**   A **composite number** is a natural number with factors other than 1 and itself. Therefore, if a natural number is not prime, it is composite.

> The number 1 is neither prime nor composite.

You Try 4

   a) What are the first six prime numbers?

   b) What are the first six composite numbers?

To perform various operations in arithmetic and algebra, it is helpful to write a number as the product of its **prime factors**. This is called finding the **prime factorization** of a number. We can use a **factor tree** to help us find the prime factorization of a number.

Example 4

Write 12 as the product of its prime factors.

Solution

Use a factor tree.

$$
\begin{array}{c}
12 \\
/ \ \backslash \\
③ \cdot 4 \qquad \text{Think of } \textit{any} \text{ two natural numbers which multiply to 12.} \\
\quad\; / \ \backslash \\
\quad ② \cdot ② \quad \text{4 is not prime, so break it down into the product of two factors, } 2 \times 2.
\end{array}
$$

When a factor is a prime number, circle it, and that part of the factor tree is complete. When all of the numbers at the end of the tree are primes, you have found the *prime factorization* of the number.

   Therefore, $12 = 2 \cdot 2 \cdot 3$. Write the prime factorization from the smallest factor to the largest.

Example 5

Write 120 as the product of its prime factors.

Solution

$$
\begin{array}{c}
120 \\
/ \quad \backslash \\
10 \quad \cdot \quad 12 \qquad \text{Think of } \textit{any} \text{ two natural numbers that multiply to 120.} \\
/ \ \backslash \quad / \ \backslash \\
② \cdot ⑤ \; ② \cdot 6 \qquad \text{10 and 12 are not prime, so write them as the product of two factors.} \\
\qquad\qquad / \ \backslash \qquad \text{Circle the primes.} \\
\qquad\qquad ② \cdot ③ \qquad \text{6 is not prime, so write it as the product of two factors. The factors are} \\
\qquad\qquad\qquad\qquad \text{primes. Circle them.}
\end{array}
$$

Prime factorization: $120 = 2 \cdot 2 \cdot 2 \cdot 3 \cdot 5$.

You Try 5

Use a factor tree to write each number as the product of its prime factors.

a) 20   b) 36   c) 90

Let's return to writing a fraction in lowest terms.

Example 6

Write each fraction in lowest terms.

a) $\dfrac{4}{6}$ b) $\dfrac{48}{42}$

Solution

a) $\dfrac{4}{6}$ There are two ways to approach this problem:

Method 1
Write 4 and 6 as the product of their primes, and divide out common factors.

$$\frac{4}{6} = \frac{2 \cdot 2}{2 \cdot 3} \qquad \text{Write 4 and 6 as the product of their prime factors.}$$

$$= \frac{\overset{1}{\cancel{2}} \cdot 2}{\underset{1}{\cancel{2}} \cdot 3} \qquad \text{Divide out common factor.}$$

$$= \frac{2}{3} \qquad \begin{array}{l}\text{Since 2 and 3 have no common factors other than 1,}\\\text{the fraction is in lowest terms.}\end{array}$$

Method 2
Divide 4 and 6 by a common factor.

$$\frac{4}{6} = \frac{4 \div 2}{6 \div 2} = \frac{2}{3}$$

$\dfrac{4}{6}$ and $\dfrac{2}{3}$ are **equivalent fractions** since $\dfrac{4}{6}$ simplifies to $\dfrac{2}{3}$.

b) $\dfrac{48}{42}$ $\dfrac{48}{42}$ is an **improper fraction**. A fraction is *improper* if its numerator is greater than or equal to its denominator. We will use two methods to express this fraction in lowest terms.

Method 1
Using a factor tree to get the prime factorizations of 48 and 42 and then dividing out common factors, we have

$$\frac{48}{42} = \frac{\overset{1}{\cancel{2}} \cdot 2 \cdot 2 \cdot 2 \cdot \overset{1}{\cancel{3}}}{\underset{1}{\cancel{2}} \cdot \underset{1}{\cancel{3}} \cdot 7} = \frac{2 \cdot 2 \cdot 2}{7} = \frac{8}{7} \text{ or } 1\frac{1}{7}$$

The answer may be expressed as an improper fraction, $\frac{8}{7}$, or as a **mixed number**, $1\frac{1}{7}$, as long as each is in lowest terms.

Method 2
48 and 42 are each divisible by 6, so we can divide each by 6.

$$\frac{48 \div 6}{42 \div 6} = \frac{8}{7} \text{ or } 1\frac{1}{7}$$

You Try 6

Write each fraction in lowest terms.

a) $\frac{8}{14}$ b) $\frac{63}{36}$

3. Multiply and Divide Fractions

Multiplying Fractions

To multiply fractions, $\frac{a}{b} \cdot \frac{c}{d}$, we multiply the numerators and multiply the denominators. That is,

$\frac{a}{b} \cdot \frac{c}{d} = \frac{a \cdot c}{b \cdot d}$ if $b \neq 0$ and $d \neq 0$.

Example 7

Multiply. Write each answer in lowest terms.

a) $\frac{3}{8} \cdot \frac{7}{4}$ b) $\frac{10}{21} \cdot \frac{21}{25}$ c) $4\frac{2}{5} \cdot 1\frac{7}{8}$

Solution

a) $\frac{3}{8} \cdot \frac{7}{4} = \frac{3 \cdot 7}{8 \cdot 4}$ Multiply numerators; multiply denominators.

$\quad\quad = \frac{21}{32}$ 21 and 32 contain no common factors, so $\frac{21}{32}$ is in lowest terms.

b) $\frac{10}{21} \cdot \frac{21}{25}$

If we follow the procedure in the previous example we get

$$\frac{10}{21} \cdot \frac{21}{25} = \frac{10 \cdot 21}{21 \cdot 25}$$

$$= \frac{210}{525} \quad \frac{210}{525} \text{ is not in lowest terms.}$$

We must reduce $\dfrac{210}{525}$ to lowest terms:

$$\frac{210 \div 5}{525 \div 5} = \frac{42}{105}$$ $\dfrac{42}{105}$ is not in lowest terms. Each number is divisible by 3.

$$= \frac{42 \div 3}{105 \div 3} = \frac{14}{35}$$ 14 and 35 have a common factor of 7.

$$= \frac{14 \div 7}{35 \div 7} = \frac{2}{5}$$ $\dfrac{2}{5}$ is in lowest terms.

Therefore, $\dfrac{10}{21} \cdot \dfrac{21}{25} = \dfrac{2}{5}$.

However, we can take out the common factors before we multiply to avoid all of the reducing in the steps above.

5 is the greatest common factor of 10 and 25. Divide 10 and 25 by **5**.

$$\frac{\overset{2}{\cancel{10}}}{\underset{1}{\cancel{21}}} \times \frac{\overset{1}{\cancel{21}}}{\underset{5}{\cancel{25}}} = \frac{2}{1} \times \frac{1}{5} = \frac{2 \times 1}{1 \times 5} = \boxed{\frac{2}{5}}$$

21 is the greatest common factor of 21 and 21. Divide each 21 by **21**.

> Usually, it is easier to remove the common factors before multiplying rather than after finding the product.

c) $4\dfrac{2}{5} \cdot 1\dfrac{7}{8}$

Before multiplying mixed numbers, they must be changed to improper fractions. Recall that $4\dfrac{2}{5}$ is the same as $4 + \dfrac{2}{5}$. Here is one way to rewrite $4\dfrac{2}{5}$ as an improper fraction:

1) Multiply the denominator and the whole number:

    $5 \cdot 4 = 20.$

2) Add the numerator:

    $20 + 2 = 22.$

3) Put the sum over the denominator:

    $\dfrac{22}{5}$

To summarize, $4\dfrac{2}{5} = \dfrac{(5 \cdot 4) + 2}{5} = \dfrac{20 + 2}{5} = \dfrac{22}{5}$.

Then, $1\dfrac{7}{8} = \dfrac{(8 \cdot 1) + 7}{8} = \dfrac{8 + 7}{8} = \dfrac{15}{8}$.

$$4\dfrac{2}{5} \cdot 1\dfrac{7}{8} = \dfrac{22}{5} \cdot \dfrac{15}{8}$$

$$= \dfrac{\overset{11}{\cancel{22}}}{\underset{1}{\cancel{5}}} \cdot \dfrac{\overset{3}{\cancel{15}}}{\underset{4}{\cancel{8}}} \qquad \text{5 and 15 each divide by \textbf{5}.}$$

$$\qquad\qquad\qquad\qquad \text{8 and 22 each divide by \textbf{2}.}$$

$$= \dfrac{11}{1} \cdot \dfrac{3}{4}$$

$$= \dfrac{33}{4} \text{ or } 8\dfrac{1}{4} \qquad \text{Express the result as an improper fraction or as a mixed number.}$$

You Try 7

Multiply. Write the answer in lowest terms.

a) $\dfrac{1}{5} \cdot \dfrac{4}{9}$ b) $\dfrac{8}{25} \cdot \dfrac{15}{32}$ c) $3\dfrac{3}{4} \cdot 2\dfrac{2}{3}$

Dividing Fractions

To divide fractions, we must define a reciprocal.

> **Definition**
>
> The **reciprocal** of a number, $\dfrac{a}{b}$, is $\dfrac{b}{a}$ since $\dfrac{a}{b} \cdot \dfrac{b}{a} = 1$. That is, a nonzero number times its reciprocal equals 1.

For example, the reciprocal of $\dfrac{5}{9}$ is $\dfrac{9}{5}$ since $\dfrac{\overset{1}{\cancel{5}}}{\underset{1}{\cancel{9}}} \cdot \dfrac{\overset{1}{\cancel{9}}}{\underset{1}{\cancel{5}}} = \dfrac{1}{1} = 1$.

> **Definition**
>
> **Division of fractions:** Let a, b, c, and d represent numbers so that b, c, and d do not equal zero. Then,
>
> $$\dfrac{a}{b} \div \dfrac{c}{d} = \dfrac{a}{b} \cdot \dfrac{d}{c}.$$

To perform division involving fractions, multiply the first fraction by the reciprocal of the second.

Example 8

Divide. Write the answer in lowest terms.

a) $\dfrac{3}{8} \div \dfrac{10}{11}$　　　　b) $\dfrac{3}{2} \div 9$　　　　c) $5\dfrac{1}{4} \div 1\dfrac{1}{13}$

Solution

a) $\dfrac{3}{8} \div \dfrac{10}{11} = \dfrac{3}{8} \cdot \dfrac{11}{10}$　　　Multiply $\dfrac{3}{8}$ by the reciprocal of $\dfrac{10}{11}$.

　　　　$= \dfrac{33}{80}$　　　Multiply.

b) $\dfrac{3}{2} \div 9 = \dfrac{3}{2} \cdot \dfrac{1}{9}$　　　The reciprocal of 9 is $\dfrac{1}{9}$.

　　　　$= \dfrac{\overset{1}{\cancel{3}}}{2} \cdot \dfrac{1}{\underset{3}{\cancel{9}}}$　　　Divide out a common factor of 3.

　　　　$= \dfrac{1}{6}$　　　Multiply.

c) $5\dfrac{1}{4} \div 1\dfrac{1}{13} = \dfrac{21}{4} \div \dfrac{14}{13}$　　　Change the mixed numbers to improper fractions.

　　　　$= \dfrac{21}{4} \cdot \dfrac{13}{14}$　　　Multiply $\dfrac{21}{4}$ by the reciprocal of $\dfrac{14}{13}$.

　　　　$= \dfrac{\overset{3}{\cancel{21}}}{4} \cdot \dfrac{13}{\underset{2}{\cancel{14}}}$　　　Divide out a common factor of 7.

　　　　$= \dfrac{39}{8}$ or $4\dfrac{7}{8}$　　　Multiply. Express the answer as an improper fraction or mixed number.

 You Try 8

Divide. Write the answer in lowest terms.

a) $\dfrac{2}{7} \div \dfrac{3}{5}$　　b) $\dfrac{3}{10} \div \dfrac{9}{16}$　　c) $9\dfrac{1}{6} \div 5$

4. Add and Subtract Fractions

Think about a pizza cut into eight equal slices:

If you eat two pieces and your friend eats three pieces, what fraction of the pizza was eaten?

Five out of the eight pieces were eaten. As a fraction we can say that you and your friend ate $\frac{5}{8}$ of the pizza.

Let's set up this problem as the sum of two fractions.

Fraction you ate + Fraction your friend ate = Fraction of the pizza eaten

$$\frac{2}{8} \qquad + \qquad \frac{3}{8} \qquad = \qquad \frac{5}{8}$$

To add $\frac{2}{8} + \frac{3}{8}$, we added the numerators and kept the denominator the same. Notice, these fractions have the same denominators.

Definition   Let a, b, and c be numbers such that $c \neq 0$.

$$\frac{a}{c} + \frac{b}{c} = \frac{a+b}{c} \quad \text{and} \quad \frac{a}{c} - \frac{b}{c} = \frac{a-b}{c}$$

To add or subtract fractions, the denominators must be the same. (This is called a **common denominator**.) Then, add (or subtract) the numerators and keep the same denominator.

Example 9

Perform the operation and simplify.

a) $\dfrac{3}{11} + \dfrac{5}{11}$ b) $\dfrac{17}{30} - \dfrac{13}{30}$

Solution

a) $\dfrac{3}{11} + \dfrac{5}{11} = \dfrac{3+5}{11}$ Add the numerators and keep the denominators the same.

$\qquad = \dfrac{8}{11}$

b) $\dfrac{17}{30} - \dfrac{13}{30} = \dfrac{17-13}{30}$ Subtract the numerators and keep the denominators the same.

$\qquad = \dfrac{4}{30}$ This is not in lowest terms, so reduce.

$\qquad = \dfrac{2}{15}$ Simplify.

You Try 9

Perform the operation and simplify.

a) $\dfrac{5}{9} + \dfrac{2}{9}$ b) $\dfrac{19}{20} - \dfrac{7}{20}$

When adding or subtracting mixed numbers, either work with them as mixed numbers or change them to improper fractions first.

Example 10

Add $2\dfrac{4}{15} + 1\dfrac{7}{15}$.

Solution

Method 1

To add these numbers while keeping them in mixed number form, add the whole number parts and add the fractional parts.

$$2\frac{4}{15} + 1\frac{7}{15} = (2 + 1) + \left(\frac{4}{15} + \frac{7}{15}\right)$$
$$= 3\frac{11}{15}$$

Method 2

Change each mixed number to an improper fraction, then add.

$$2\frac{4}{15} + 1\frac{7}{15} = \frac{34}{15} + \frac{22}{15}$$
$$= \frac{34 + 22}{15}$$
$$= \frac{56}{15} \text{ or } 3\frac{11}{15}$$

You Try 10

Add $4\dfrac{3}{7} + 5\dfrac{1}{7}$.

The examples given so far contain common denominators. How do we add or subtract fractions that do not have common denominators? We find the least common denominator for the fractions and rewrite each fraction with this denominator.

The **least common denominator (LCD)** of two fractions is the least common multiple of the numbers in the denominators.

Example 11

Find the LCD for $\dfrac{3}{4}$ and $\dfrac{1}{6}$.

Solution

Method 1

List some multiples of 4 and 6.

4:   4, 8, ⬚12⬚, 16, 20, *24*, . . .
6:   6, ⬚12⬚, 18, *24*, 30, . . .

Although 24 is a multiple of 6 and of 4, the *least* common multiple, and therefore the least common denominator, is 12.

Method 2

We can also use the prime factorization of 4 and 6 to find the LCD.

To find the LCD:

1) Find the prime factorization of each number.
2) The least common denominator will include each different factor appearing in the factorizations.
3) If a factor appears more than once in any prime factorization, use it in the LCD the *maximum number of times* it appears in any single factorization. Multiply the factors.

$$4 = 2 \cdot 2$$
$$6 = 2 \cdot 3$$

The least common multiple of 4 and 6 is

$$\underbrace{2 \cdot 2}_{\substack{\text{2 appears at} \\ \text{most twice in} \\ \text{any single} \\ \text{factorization.}}} \quad \cdot \quad \underbrace{3}_{\substack{\text{3 appears} \\ \text{once in a} \\ \text{factorization.}}} \quad = 12$$

The LCD of $\dfrac{3}{4}$ and $\dfrac{1}{6}$ is 12.

 **You Try 11**

Find the LCD for $\dfrac{5}{6}$ and $\dfrac{4}{9}$.

To add or subtract fractions with unlike denominators, begin by identifying the least common denominator. Then, we must rewrite each fraction with this LCD. This will not change the value of the fraction; we will obtain an *equivalent* fraction.

Example 12

Rewrite $\dfrac{3}{4}$ with a denominator of 12.

Solution

We want to find a fraction that is equivalent to $\dfrac{3}{4}$ so that $\dfrac{3}{4} = \dfrac{?}{12}$.

To obtain the new denominator of 12, the "old" denominator, 4, must be multiplied by 3. But, if the denominator is multiplied by 3, the numerator must be multiplied by 3 as well. When we multiply $\dfrac{3}{4}$ by $\dfrac{3}{3}$, we have multiplied by 1 since $\dfrac{3}{3} = 1$. This is why the fractions are equivalent.

$$\frac{3}{4} \cdot \frac{3}{3} = \frac{9}{12}$$

So, $\dfrac{3}{4} = \dfrac{9}{12}$.

Adding or Subtracting Fractions with Unlike Denominators

To add or subtract fractions with unlike denominators:

1) Determine, and write down, the least common denominator (LCD).

2) Rewrite each fraction with the LCD.

3) Add or subtract.

4) Express the answer in lowest terms.

 You Try 12

Rewrite $\dfrac{5}{6}$ with a denominator of 42.

Example 13

Add or subtract.

a) $\dfrac{2}{9} + \dfrac{1}{6}$ b) $6\dfrac{7}{8} - 3\dfrac{1}{2}$

Solution

a) $\dfrac{2}{9} + \dfrac{1}{6}$ LCD = 18   Identify the least common denominator.

$\dfrac{2}{9} \cdot \dfrac{2}{2} = \dfrac{4}{18}$ $\dfrac{1}{6} \cdot \dfrac{3}{3} = \dfrac{3}{18}$ Rewrite each fraction with a denominator of 18.

$\dfrac{2}{9} + \dfrac{1}{6} = \dfrac{4}{18} + \dfrac{3}{18}$

$= \dfrac{7}{18}$

b) $6\dfrac{7}{8} - 3\dfrac{1}{2}$

Method I

Keep the numbers in mixed number form. Subtract the whole number parts and subtract the fractional parts. Get a common denominator for the fractional parts.

LCD = 8   Identify the least common denominator.

$6\dfrac{7}{8}$: $\dfrac{7}{8}$ has the LCD of 8.

$3\dfrac{1}{2}$: $\dfrac{1}{2} \cdot \dfrac{4}{4} = \dfrac{4}{8}$.   So, $3\dfrac{1}{2} = 3\dfrac{4}{8}$.   Rewrite $\dfrac{1}{2}$ with a denominator of 8.

$6\dfrac{7}{8} - 3\dfrac{1}{2} = 6\dfrac{7}{8} - 3\dfrac{4}{8}$

$= 3\dfrac{3}{8}$ Subtract whole number parts and subtract fractional parts.

Method 2

Rewrite each mixed number as an improper fraction, get a common denominator, then subtract.

$6\dfrac{7}{8} - 3\dfrac{1}{2} = \dfrac{55}{8} - \dfrac{7}{2}$ LCD = 8

$\dfrac{55}{8}$ has a denominator of 8.

$\dfrac{7}{2} \cdot \dfrac{4}{4} = \dfrac{28}{8}$ Rewrite $\dfrac{7}{2}$ with a denominator of 8.

$6\dfrac{7}{8} - 3\dfrac{1}{2} = \dfrac{55}{8} - \dfrac{7}{2}$

$= \dfrac{55}{8} - \dfrac{28}{8}$

$= \dfrac{27}{8}$ or $3\dfrac{3}{8}$.

 You Try 13

Perform the operations and simplify.

a) $\dfrac{11}{12} - \dfrac{5}{8}$ b) $\dfrac{1}{3} + \dfrac{5}{6} + \dfrac{3}{4}$ c) $4\dfrac{2}{5} + 1\dfrac{7}{15}$

Answers to You Try Exercises

1) $\frac{3}{5}$ 2) 1, 2, 3, 5, 6, 10, 15, 30   3) yes   4) a) 2, 3, 5, 7, 11, 13   b) 4, 6, 8, 9, 10, 12

5) a) $2 \cdot 2 \cdot 5$ b) $2 \cdot 2 \cdot 3 \cdot 3$ c) $2 \cdot 3 \cdot 3 \cdot 5$ 6) a) $\frac{4}{7}$ b) $\frac{7}{4}$ 7) a) $\frac{4}{45}$ b) $\frac{3}{20}$ c) 10

8) a) $\frac{10}{21}$ b) $\frac{8}{15}$ c) $\frac{11}{6}$ or $1\frac{5}{6}$ 9) a) $\frac{7}{9}$ b) $\frac{3}{5}$ 10) $9\frac{4}{7}$ 11) 18   12) $\frac{35}{42}$

13) a) $\frac{7}{24}$ b) $\frac{23}{12}$ or $1\frac{11}{12}$ c) $\frac{88}{15}$ or $5\frac{13}{15}$

1.1 Exercises

Boost your grade at mathzone.com!   MathZone   > Practice Problems   > NetTutor   > Self-Test   > e-Professors   > Videos

Objective 1

1) What fraction of each figure is shaded? If the fraction is not in lowest terms, reduce it.

a)

b)

c)

2) What fraction of each figure is *not* shaded? If the fraction is not in lowest terms, reduce it.

a)

b)

c)

3) Draw a rectangle divided into 8 equal parts. Shade in $\frac{4}{8}$ of the rectangle. Write another fraction to represent how much of the rectangle is shaded.

4) Draw a rectangle divided into 6 equal parts. Shade in $\frac{2}{6}$ of the rectangle. Write another fraction to represent how much of the rectangle is shaded.

Objective 2

5) Find all factors of each number.
a) 18
b) 40
c) 23

6) Find all factors of each number.
a) 20
b) 17
c) 60

7) Identify each number as prime or composite.
a) 27
b) 34
c) 11

8) Identify each number as prime or composite.
a) 2
b) 57
c) 90

9) Is 3072 prime or composite? Explain your answer.

10) Is 4185 prime or composite: Explain your answer.

11) Use a factor tree to find the prime factorization of each number.

   a) 18

   b) 54

   c) 42

   d) 150

12) Explain, in words, how to use a factor tree to find the prime factorization of 72.

13) Write each fraction in lowest terms.

   a) $\dfrac{9}{12}$

   b) $\dfrac{54}{72}$

   c) $\dfrac{84}{35}$

   d) $\dfrac{120}{280}$

14) Write each fraction in lowest terms.

   a) $\dfrac{21}{35}$

   b) $\dfrac{48}{80}$

   c) $\dfrac{125}{500}$

   d) $\dfrac{900}{450}$

Objective 3

15) Multiply. Write the answer in lowest terms.

   a) $\dfrac{2}{7} \cdot \dfrac{3}{5}$

   b) $\dfrac{15}{26} \cdot \dfrac{4}{9}$

   c) $\dfrac{1}{2} \cdot \dfrac{14}{15}$

   d) $\dfrac{42}{55} \cdot \dfrac{22}{35}$

   e) $4 \cdot \dfrac{1}{8}$

   f) $6\dfrac{1}{8} \cdot \dfrac{2}{7}$

16) Multiply. Write the answer in lowest terms.

   a) $\dfrac{1}{6} \cdot \dfrac{5}{9}$

   b) $\dfrac{9}{20} \cdot \dfrac{6}{7}$

   c) $\dfrac{12}{25} \cdot \dfrac{25}{36}$

   d) $\dfrac{30}{49} \cdot \dfrac{21}{100}$

   e) $\dfrac{7}{15} \cdot 10$

   f) $7\dfrac{5}{7} \cdot 1\dfrac{5}{9}$

17) When Elizabeth multiplies $5\dfrac{1}{2} \cdot 2\dfrac{1}{3}$, she gets $10\dfrac{1}{6}$. What was her mistake? What is the correct answer?

18) Explain how to multiply mixed numbers.

19) Divide. Write the answer in lowest terms.

   a) $\dfrac{1}{42} \div \dfrac{2}{7}$

   b) $\dfrac{3}{11} \div \dfrac{4}{5}$

   c) $\dfrac{18}{35} \div \dfrac{9}{10}$

   d) $\dfrac{14}{15} \div \dfrac{2}{15}$

    e) $6\dfrac{2}{5} \div 1\dfrac{13}{15}$

   f) $\dfrac{4}{7} \div 8$

20) Explain how to divide mixed numbers.

Objective 4

21) Find the least common multiple of 10 and 15.

22) Find the least common multiple of 12 and 9.

23) Find the least common denominator for each group of fractions.

a) $\dfrac{9}{10}, \dfrac{11}{30}$

b) $\dfrac{7}{8}, \dfrac{5}{12}$

c) $\dfrac{4}{9}, \dfrac{1}{6}, \dfrac{3}{4}$

24) Find the least common denominator for each group of fractions.

a) $\dfrac{3}{14}, \dfrac{2}{7}$

b) $\dfrac{17}{25}, \dfrac{3}{10}$

c) $\dfrac{29}{30}, \dfrac{3}{4}, \dfrac{9}{20}$

25) Add or subtract. Write the answer in lowest terms.

a) $\dfrac{6}{11} + \dfrac{2}{11}$

b) $\dfrac{19}{20} - \dfrac{7}{20}$

c) $\dfrac{4}{25} + \dfrac{2}{25} + \dfrac{9}{25}$

d) $\dfrac{2}{9} + \dfrac{1}{6}$

e) $\dfrac{3}{5} + \dfrac{11}{30}$

f) $\dfrac{13}{18} - \dfrac{2}{3}$

g) $\dfrac{4}{7} + \dfrac{5}{9}$

h) $\dfrac{5}{6} - \dfrac{1}{4}$

i) $\dfrac{3}{10} + \dfrac{7}{20} + \dfrac{3}{4}$

j) $\dfrac{1}{6} + \dfrac{2}{9} + \dfrac{10}{27}$

26) Add or subtract. Write the answer in lowest terms.

a) $\dfrac{8}{9} - \dfrac{5}{9}$

b) $\dfrac{14}{15} - \dfrac{2}{15}$

c) $\dfrac{11}{36} + \dfrac{13}{36}$

d) $\dfrac{16}{45} + \dfrac{8}{45} + \dfrac{11}{45}$

e) $\dfrac{15}{16} - \dfrac{3}{4}$

f) $\dfrac{1}{8} + \dfrac{1}{6}$

g) $\dfrac{5}{8} - \dfrac{2}{9}$

h) $\dfrac{23}{30} - \dfrac{19}{90}$

i) $\dfrac{1}{6} + \dfrac{1}{4} + \dfrac{2}{3}$

j) $\dfrac{3}{10} + \dfrac{2}{5} + \dfrac{4}{15}$

27) Add or subtract. Write the answer in lowest terms.

a) $8\dfrac{5}{11} + 6\dfrac{2}{11}$

b) $2\dfrac{1}{10} + 9\dfrac{3}{10}$

c) $7\dfrac{11}{12} - 1\dfrac{5}{12}$

d) $3\dfrac{1}{5} + 2\dfrac{1}{4}$

e) $5\dfrac{2}{3} - 4\dfrac{4}{15}$

f) $9\dfrac{5}{8} - 5\dfrac{3}{10}$

g) $4\dfrac{3}{7} + 6\dfrac{3}{4}$

h) $7\dfrac{13}{20} + 4\dfrac{4}{5}$

28) Add or subtract. Write the answer in lowest terms.

a) $3\dfrac{2}{7} + 1\dfrac{3}{7}$

b) $8\dfrac{5}{16} + 7\dfrac{3}{16}$

c) $5\frac{13}{20} - 3\frac{5}{20}$

d) $10\frac{8}{9} - 2\frac{1}{3}$

e) $1\frac{5}{12} + 2\frac{3}{8}$

f) $4\frac{1}{9} + 7\frac{2}{5}$

g) $1\frac{5}{6} + 4\frac{11}{18}$

h) $3\frac{7}{8} + 4\frac{2}{5}$

Objectives 3 and 4

29) For Valentine's Day, Alex wants to sew teddy bears for her friends. Each bear requires $1\frac{2}{3}$ yd of fabric. If she has 7 yd of material, how many bears can Alex make? How much fabric will be left over?

30) A chocolate chip cookie recipe that makes 24 cookies uses $\frac{3}{4}$ cup of brown sugar. If Raphael wants to make 48 cookies, how much brown sugar does he need?

31) After 8 weeks of the baseball season in 2004, Barry Bonds had been up to bat 99 times. He got a hit $\frac{4}{11}$ of the time. How many hits did Barry Bonds have?
(*Chicago Tribune*, June 1, 2004, Sports, p. 5)

32) When all children are present, Ms. Yamoto has 30 children in her fifth-grade class. One day during flu season, $\frac{3}{5}$ of them were absent. How many children were absent on this day?

33) Mr. Burnett plans to have a picture measuring $18\frac{3}{8}''$ by $12\frac{1}{4}''$ custom framed. The frame he chose is $2\frac{1}{8}''$ wide. What will be the new length and width of the picture plus the frame?

34) Andre is building a table in his workshop. For the legs, he bought wood that is 30 in. long. If the legs are to be $26\frac{3}{4}$ in. tall, how many inches must he cut off to get the desired height?

35) When Rosa opens the kitchen cabinet, she finds three partially filled bags of flour. One contains $\frac{2}{3}$ cup, another contains $1\frac{1}{4}$ cups, and the third contains $1\frac{1}{2}$ cups. How much flour does she have all together?

36) Tamika takes the same route to school every day. (See the figure.) How far does she walk to school?

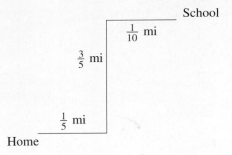

37) The gas tank of Jenny's car holds $11\frac{3}{5}$ gal, while Scott's car holds $16\frac{3}{4}$ gal. How much more gasoline does Scott's car hold?

38) Mr. Johnston is building a brick wall along his driveway. He estimates that one row of brick plus mortar will be $4\frac{1}{4}$ in. high. How many rows will he need to construct a wall that is 34 in. high?

39) For homework, Bill's math teacher assigned 42 problems. He finished $\frac{5}{6}$ of them. How many problems did he do?

40) Clarice's parents tell her that she must deposit $\frac{1}{3}$ of the money she earns from babysitting into her savings account, but she can keep the rest. If she earns \$117 in 1 week during the summer how much does she deposit, and how much does she keep?

41) A welder must construct a beam with a total length of $32\frac{7}{8}$ in. If he has already joined a $14\frac{1}{6}$-in. beam with a $10\frac{3}{4}$-in. beam, find the length of a third beam needed to reach the total length.

42) Telephia, a market research company, surveyed 1500 teenage cell phone owners. The company learned that $\frac{2}{3}$ of them use cell phone data services. How many teenagers surveyed use cell phone data services? (*American Demographics,* May 2004, Vol. 26, Issue 4, p. 10)

43) A study conducted in 2000 indicated that about $\frac{3}{5}$ of the full-time college students surveyed had consumed alcohol sometime during the 30 days preceding the survey. If 400 students were surveyed, how many of them drank alcohol within the 30 days before the survey? (*Alcohol Research & Health,* The Journal of the Nat'l Institute on Alcohol Abuse & Alcoholism, Vol. 27, No. 1, 2003)

Section 1.2 Exponents and Order of Operations

Objectives

1. Use Exponents
2. Use the Order of Operations

1. Use Exponents

In Section 1.1, we discussed the prime factorization of a number. Let's find the prime factorization of 8.

$$
\begin{array}{cc}
8 & 8 = 2 \cdot 2 \cdot 2 \\
\diagup \diagdown & \\
4 \cdot \textcircled{2} & \\
\diagup \diagdown & \\
\textcircled{2} \cdot \textcircled{2} &
\end{array}
$$

We can write $2 \cdot 2 \cdot 2$ another way, by using an *exponent*.

$$2 \cdot 2 \cdot 2 = 2^3 \leftarrow \text{exponent (or power)}$$
$$\uparrow$$
$$\text{base}$$

2 is the *base*. 2 is a *factor* that appears three times. 3 is the *exponent* or *power*. An **exponent** represents repeated multiplication. We read 2^3 as "2 to the third power" or "2 cubed." 2^3 is called an **exponential expression**.

Example 1

Rewrite each product in exponential form.

a) $6 \cdot 6 \cdot 6 \cdot 6 \cdot 6$ b) $3 \cdot 3$

Solution

a) $6 \cdot 6 \cdot 6 \cdot 6 \cdot 6 = 6^5$ 6 is the base. It appears as a factor 5 times. So, 5 is the exponent.

b) $3 \cdot 3 = 3^2$ 3 is the base. 2 is the exponent.
   This is read as "3 squared."

You Try 1

Rewrite each product in exponential form.

a) $4 \cdot 4 \cdot 4 \cdot 4 \cdot 4 \cdot 4$ b) $\dfrac{3}{8} \cdot \dfrac{3}{8} \cdot \dfrac{3}{8} \cdot \dfrac{3}{8}$

We can also evaluate an exponential expression.

Example 2

Evaluate.

a) 2^5 b) 5^3 c) $\left(\dfrac{4}{7}\right)^2$ d) 8^1 e) 1^4

Solution

a) $2^5 = 2 \cdot 2 \cdot 2 \cdot 2 \cdot 2 = 32$ 2 appears as a factor 5 times.

b) $5^3 = 5 \cdot 5 \cdot 5 = 125$ 5 appears as a factor 3 times.

c) $\left(\dfrac{4}{7}\right)^2 = \dfrac{4}{7} \cdot \dfrac{4}{7} = \dfrac{16}{49}$ $\dfrac{4}{7}$ appears as a factor 2 times.

d) $8^1 = 8$ 8 is a factor only once.

e) $1^4 = 1 \cdot 1 \cdot 1 \cdot 1 = 1$ 1 appears as a factor 4 times.

> 1 raised to any natural number power is 1 since 1 multiplied by itself equals 1.

You Try 2

Evaluate.

a) 3^4 b) 7^2 c) $\left(\dfrac{2}{5}\right)^3$

It is generally agreed that there are some skills in arithmetic that everyone should have in order to be able to acquire other math skills. Knowing the basic multiplication facts, for example, is essential for learning how to add, subtract, multiply, and divide fractions as well as how to perform many other operations in arithmetic and algebra. Similarly, memorizing powers of certain bases is necessary for learning how to apply the rules of exponents (Chapter 2) and for working with radicals (Chapter 10). Therefore, the powers listed here must be memorized in order to be successful in the previously mentioned, as well as other, topics. Throughout this book, it is assumed that students know these powers:

Powers to Memorize							
$2^1 = 2$	$3^1 = 3$	$4^1 = 4$	$5^1 = 5$	$6^1 = 6$	$8^1 = 8$	$10^1 = 10$	
$2^2 = 4$	$3^2 = 9$	$4^2 = 16$	$5^2 = 25$	$6^2 = 36$	$8^2 = 64$	$10^2 = 100$	
$2^3 = 8$	$3^3 = 27$	$4^3 = 64$	$5^3 = 125$			$10^3 = 1000$	
$2^4 = 16$	$3^4 = 81$						
$2^5 = 32$				$7^1 = 7$	$9^1 = 9$	$11^1 = 11$	
$2^6 = 64$				$7^2 = 49$	$9^2 = 81$	$11^2 = 121$	
						$12^1 = 12$	
						$12^2 = 144$	
						$13^1 = 13$	
						$13^2 = 169$	

(Hint: Making flashcards might help you learn these facts.)

2. Use the Order of Operations

We will begin this topic with a problem for the student:

You Try 3

Evaluate $36 - 12 \div 4 + (6 - 1)^2$.

What answer did you get? 31? or 58? or 8? Or, did you get another result?

Most likely you obtained one of the three answers just given. Only one is correct, however. If we do not have rules to guide us in evaluating expressions, it is easy to get the incorrect answer.

Therefore, here are the rules we follow. This is called the **order of operations**.

The Order of Operations

Simplify expressions in the following order:

1) If parentheses or other grouping symbols appear in an expression, simplify what is in these grouping symbols first.

2) Simplify expressions with exponents.

3) Perform multiplication and division from left to right.

4) Perform addition and subtraction from left to right.

Think about the "You Try" problem. Did you evaluate it using the order of operations? Let's look at that expression:

Example 3

Evaluate $36 - 12 \div 4 + (6 - 1)^2$.

Solution

$36 - 12 \div 4 + (6 - 1)^2$ First, perform the operation in the parentheses.
$36 - 12 \div 4 + 5^2$
$36 - 12 \div 4 + 25$ Exponents are done before division, addition, and subtraction.
$36 - 3 + 25$ Perform division before addition and subtraction.
$33 + 25$ When an expression contains only addition and subtraction, per-
58 form the operations starting at the left and moving to the right.

You Try 4

Evaluate: $12 \cdot 3 - (2 + 1)^2 \div 9$.

A good way to remember the order of operations is to remember the sentence, "**P**lease **E**xcuse **M**y **D**ear **A**unt **S**ally." (**P**arentheses, **E**xponents, **M**ultiplication, **D**ivision, **A**ddition, **S**ubtraction)

Example 4

Evaluate.

a) $9 + 20 - 5 \cdot 3$

b) $5(8 - 2) + 3^2$

c) $2[20 - (40 \div 5)] + 7$

d) $\dfrac{(7 - 5)^3 \cdot 6}{30 - 9 \cdot 2}$

Solution

a) $9 + 20 - 5 \cdot 3 = 9 + 20 - 15$ Perform multiplication before addition and subtraction.
    $= 29 - 15$ When an expression contains only addition and
         subtraction, work from left to right.
    $= 14$ Subtract.

b) $5(8 - 2) + 3^2 = 5(6) + 3^2$ Parentheses
    $= 5(6) + 9$ Exponent
    $= 30 + 9$ Multiply
    $= 39$ Add

c) $2[20 - (40 \div 5)] + 7$
    This expression contains two sets of grouping symbols: **brackets** [ ] and **parentheses** ( ). Perform the operation in the **innermost** grouping symbol first which is the parentheses in this case.

    $2[20 - (40 \div 5)] + 7 = 2[20 - 8] + 7$ Innermost grouping symbol
    $= 2[12] + 7$ Brackets
    $= 24 + 7$ Perform multiplication before addition.
    $= 31$ Add.

d) $\dfrac{(7-5)^3 \cdot 6}{30 - 9 \cdot 2}$

The fraction bar in this expression acts as a grouping symbol. Therefore, simplify the numerator, simplify the denominator, then simplify the resulting fraction, if possible.

$$\dfrac{(7-5)^3 \cdot 6}{30 - 9 \cdot 2} = \dfrac{2^3 \cdot 6}{30 - 18} \qquad \text{Parentheses} \\ \text{Multiply}$$

$$= \dfrac{8 \cdot 6}{12} \qquad \text{Exponent} \\ \text{Subtract}$$

$$= \dfrac{48}{12} \qquad \text{Multiply}$$

$$= 4$$

 You Try 5

Evaluate:

a) $18 - 3 \cdot 4 + 2$ b) $7 \cdot 6 - (1 + 3)^2 \div 2$

c) $5 + 3[15 - 2(3 + 1)]$ d) $\dfrac{7^2 - 3 \cdot 3}{5(12 - 8)}$

Answers to You Try Exercises

1) a) 4^6 b) $\left(\dfrac{3}{8}\right)^4$ 2) a) 81   b) 49   c) $\dfrac{8}{125}$ 3) 58   4) 35   5) a) 8   b) 34   c) 26   d) 2

1.2 Exercises

Objective 1

1) Identify the base and the exponent.

   a) 6^4

   b) 2^3

   c) $\left(\dfrac{9}{8}\right)^5$

2) Identify the base and the exponent.

   a) 5^1

   b) 1^8

   c) $\left(\dfrac{3}{7}\right)^2$

3) Write in exponential form.

   a) $9 \cdot 9 \cdot 9 \cdot 9$

   b) $2 \cdot 2 \cdot 2 \cdot 2 \cdot 2 \cdot 2 \cdot 2 \cdot 2$

   c) $\dfrac{1}{4} \cdot \dfrac{1}{4} \cdot \dfrac{1}{4}$

4) Explain, in words, why $7 \cdot 7 \cdot 7 \cdot 7 \cdot 7 = 7^5$.

5) Evaluate.

   a) 8^2

   b) 11^2

   c) 2^4

d) 5^3

e) 3^4

f) 12^2

g) 1^2

h) $\left(\dfrac{3}{10}\right)^2$

i) $\left(\dfrac{1}{2}\right)^6$

j) $(0.3)^2$

6) Evaluate.

a) 9^2

b) 13^2

c) 3^3

d) 2^5

e) 4^3

f) 1^4

g) 6^2

h) $\left(\dfrac{7}{5}\right)^2$

i) $\left(\dfrac{2}{3}\right)^4$

j) $(0.02)^2$

7) Evaluate $(0.5)^2$ two different ways.

8) Explain why $1^{200} = 1$.

Objective 2

9) In your own words, summarize the order of operations.

Evaluate.

10) $20 + 12 - 5$

11) $17 - 2 + 4$

12) $35 - 7 + 8 - 3$

13) $50 \div 10 + 15$

14) $6 \cdot 4 - 2$

15) $20 - 3 \cdot 2 + 9$

16) $22 + 10 \div 2 - 1$

17) $8 + 12 \cdot \dfrac{3}{4}$

18) $27 \div \dfrac{9}{5} - 1$

19) $\dfrac{3}{4} \cdot \dfrac{1}{6} + \dfrac{1}{2} \cdot \dfrac{1}{3}$

20) $\dfrac{1}{2} \cdot \dfrac{4}{5} - \dfrac{2}{5} \cdot \dfrac{3}{10}$

21) $2 \cdot \dfrac{3}{4} - \left(\dfrac{2}{3}\right)^2$

22) $\left(\dfrac{3}{2}\right)^2 - \left(\dfrac{5}{4}\right)^2$

23) $25 - 11 \cdot 2 + 1$

24) $2 + 16 + 14 \div 2$

25) $15 - 3(6 - 4)^2$

26) $7 + 2(9 - 5)^2$

27) $60 \div 15 + 5 \cdot 3$

28) $27 \div (10 - 7)^2 + 8 \cdot 3$

29) $6[21 \div (3 + 4)] - 9$

30) $2[23 + (11 - 9)^3] + 3$

31) $4 + 3[(1 + 3)^3 \div (10 - 2)]$

32) $(8 + 2)^2 - 5[9 \times (3 + 1) - 5^2]$

33) $\dfrac{12(5 + 1)}{2 \cdot 5 - 1}$

34) $\dfrac{(14 - 4)^2 - 4^3}{4 \cdot 9 - 3 \cdot 11}$

35) $\dfrac{4(7 - 2)^2}{(12)^2 - 8 \cdot 3}$

36) $\dfrac{6(8 - 6)^2}{10 + 12 \div 2 + 4}$

Section 1.3 Geometry Review

Objectives

1. Understand the Definitions of Different Angles and Parallel and Perpendicular Lines

2. Identify Equilateral, Isosceles, and Scalene Triangles

3. Learn and Apply Area, Perimeter, and Circumference Formulas

4. Learn and Apply Volume Formulas

Thousands of years ago, the Egyptians collected taxes based on how much land a person owned. They developed measuring techniques to accomplish such a task. Later the Greeks formalized the process of measurements such as this into a branch of mathematics we call geometry. "Geometry" comes from the Greek words for "earth measurement." In this section, we will review some basic geometric concepts that we will need in the study of algebra.

Let's begin by looking at angles. An angle can be measured in **degrees**. For example, 45° is read as "45 degrees."

1. Understand the Definitions of Different Angles and Parallel and Perpendicular Lines

Angles

An **acute angle** is an angle whose measure is greater than 0° and less than 90°.

A **right angle** is an angle whose measure is 90°.

An **obtuse angle** is an angle whose measure is greater than 90° and less than 180°.

A **straight angle** is an angle whose measure is 180°.

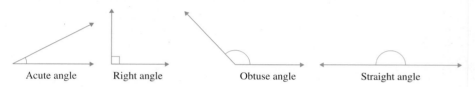

Acute angle   Right angle   Obtuse angle   Straight angle

Two angles are **complementary** if their measures add to 90°.

Two angles are **supplementary** if their measures add to 180°.

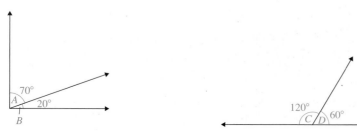

A and B are **complementary angles** since $m\angle A + m\angle B = 70° + 20° = 90°$.

C and D are **supplementary angles** since $m\angle C + m\angle D = 120° + 60° = 180°$.

The measure of angle A is denoted by $m\angle A$.

Example 1

$m\angle A = 37°$. Find its complement.

Solution

$$\text{Complement} = 90° - 37° = 53°$$

Since the sum of two complementary angles is 90°, if one angle measures 37°, its complement has a measure of $90° - 37° = 53°$.

You Try 1

$m\angle A = 75°$. Find its supplement.

Next, we will explore some relationships between lines and angles.

Vertical Angles

When two lines intersect, four angles are formed (see Figure 1.1). The pair of opposite angles are called **vertical angles**. Angles *A* and *C* are *vertical angles,* and angles *B* and *D* are *vertical angles. The measures of vertical angles are equal.* Therefore, $m\angle A = m\angle C$ and $m\angle B = m\angle D$.

Figure 1.1

Parallel and Perpendicular Lines

Parallel lines are lines in the same plane that do not intersect (Figure 1.2). **Perpendicular lines** are lines that intersect at right angles (Figure 1.3).

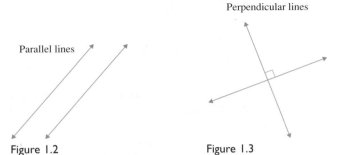

Parallel lines

Figure 1.2

Perpendicular lines

Figure 1.3

2. Identify Equilateral, Isosceles, and Scalene Triangles

We can classify triangles by their angles and by their sides.

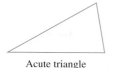 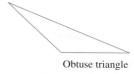 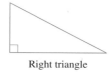

Acute triangle   Obtuse triangle   Right triangle

An **acute triangle** is one in which all three angles are acute.

An **obtuse triangle** contains one obtuse angle.

A **right triangle** contains one right angle.

Property: The sum of the measures of the angles of any triangle is 180°.

Equilateral triangle   Isosceles triangle   Scalene triangle

If a triangle has three sides of equal length, it is an **equilateral triangle**. (Each angle measure of an equilateral triangle is 60°.)

If a triangle has two sides of equal length, it is an **isosceles triangle**. (The angles opposite the equal sides have the same measure.)

If a triangle has no sides of equal length, it is a **scalene triangle**. (No angles have the same measure.)

Example 2

Find the measures of angles *A* and *B* in this isosceles triangle.

Solution

The single hash mark on the two sides of the triangle means that those sides are of equal length.

$$m\angle B = 38°$$ Angle measures opposite sides of equal length are the same.
$$38° + m\angle B = 38° + 38° = 76°.$$

We have found that the sum of two of the angles is 76°. Since all of the angle measures add up to 180°,

$$m\angle A = 180° - 76° = 104°$$
$$m\angle A = 104°$$

You Try 2

Find the measures of angles *A* and *B* in this isosceles triangle.

3. Learn and Apply Area, Perimeter, and Circumference Formulas

The **perimeter** of a figure is the distance around the figure, while the **area** of a figure is the number of square units enclosed within the figure. For some familiar shapes, we have the following formulas:

Figure		Perimeter	Area
Rectangle:		$P = 2l + 2w$	$A = lw$
Square:		$P = 4s$	$A = s^2$
Triangle: h = height		$P = a + b + c$	$A = \dfrac{1}{2}bh$
Parallelogram: h = height		$P = 2a + 2b$	$A = bh$
Trapezoid: h = height		$P = a + c + b_1 + b_2$	$A = \dfrac{1}{2}h(b_1 + b_2)$

The perimeter of a circle is called the **circumference**. The **radius**, *r*, is the distance from the center of the circle to a point on the circle. A line segment that passes through the center of the circle and has its endpoints on the circle is called a **diameter**.

Pi, π, is the ratio of the circumference of any circle to its diameter. $\pi \approx 3.14159265\ldots$, but we will use 3.14 as an approximation for π.

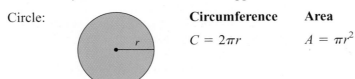

Circle:

Circumference   **Area**

$C = 2\pi r$ $A = \pi r^2$

Example 3

Find the perimeter and area of each figure.

a)

3 in.

5 in.

b)

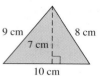

9 cm   8 cm

7 cm

10 cm

Solution

a)   This figure is a rectangle.

Perimeter: $P = 2l + 2w$
$P = 2(5 \text{ in.}) + 2(3 \text{ in.})$
$P = 10 \text{ in.} + 6 \text{ in.}$
$P = 16 \text{ in.}$

Area: $A = lw$
$A = (5 \text{ in.})(3 \text{ in.})$
$A = 15 \text{ in}^2$ or 15 square inches

b)   This figure is a triangle.

Perimeter: $P = a + b + c$
$P = 8 \text{ cm} + 10 \text{ cm} + 9 \text{ cm}$
$P = 27 \text{ cm}$

Area: $A = \dfrac{1}{2}bh$

$A = \dfrac{1}{2}(10 \text{ cm})(7 \text{ cm})$

$A = 35 \text{ cm}^2$ or 35 square centimeters

You Try 3

Find the perimeter and area of the figure.

6 cm

9 cm

Example 4

Find the (a) circumference and (b) area of the circle. Give an exact answer for each and give an approximation using 3.14 for π.

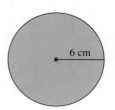

6 cm

Solution

a) The formula for the circumference of a circle is $C = 2\pi r$. The radius of the given circle is 6 cm. Replace r with 6 cm.

$$
\begin{aligned}
C &= 2\pi r \\
&= 2\pi(6 \text{ cm}) \quad \text{Replace } r \text{ with 6 cm.} \\
&= 12\pi \text{ cm} \quad \text{Multiply.}
\end{aligned}
$$

Leaving the answer in terms of π gives us the exact circumference of the circle. The exact circumference is 12π cm.

To find an approximation for the circumference, substitute 3.14 for π and simplify.

$$
\begin{aligned}
C &= 12\pi \text{ cm} \\
&\approx 12(3.14) \text{ cm} = 37.68 \text{ cm}
\end{aligned}
$$

b) The formula for the area of a circle is $A = \pi r^2$. Replace r with 6 cm.

$$
\begin{aligned}
A &= \pi r^2 \\
&= \pi(6 \text{ cm})^2 \quad \text{Replace } r \text{ with 6 cm.} \\
&= 36\pi \text{ cm}^2 \quad 6^2 = 36
\end{aligned}
$$

Leaving the answer in terms of π gives us the exact area of the circle. The exact area is 36π cm^2.

To find an approximation for the area, substitute 3.14 for π and simplify.

$$
\begin{aligned}
A &= 36\pi \text{ cm}^2 \\
&\approx 36(3.14) \text{ cm}^2 \\
&= 113.04 \text{ cm}^2
\end{aligned}
$$

 You Try 4

Find the (a) circumference and (b) area of the circle. Give an exact answer for each and give an approximation using 3.14 for π.

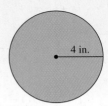

4 in.

A **polygon** is a closed figure consisting of three or more line segments. (See the figure.) We can extend our knowledge of perimeter and area to determine the area and perimeter of a polygon.

Polygons:

Example 5

Find the perimeter and area of the figure shown here.

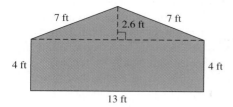

Solution

Perimeter: The perimeter is the distance around the figure.

$$P = 7 \text{ ft} + 7 \text{ ft} + 4 \text{ ft} + 13 \text{ ft} + 4 \text{ ft}$$
$$P = 35 \text{ ft}$$

Area: To find the area of this figure, think of it as two regions: a triangle and a rectangle.

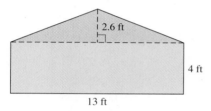

$$\text{Total area} = \text{Area of triangle} + \text{Area of rectangle}$$
$$= \frac{1}{2}bh + lw$$
$$= \frac{1}{2}(13 \text{ ft})(2.6 \text{ ft}) + (13 \text{ ft})(4 \text{ ft})$$
$$= 16.9 \text{ ft}^2 + 52 \text{ ft}^2$$
$$= 68.9 \text{ ft}^2$$

 You Try 5

Find the perimeter and area of the figure.

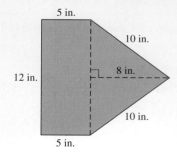

4. Learn and Apply Volume Formulas

The **volume** of a three-dimensional object is the amount of space occupied by the object. Volume is measured in cubic units such as cubic inches (in³), cubic centimeters (cm³), cubic feet (ft³), and so on. Volume also describes the amount of a substance that can be enclosed within a three-dimensional object. Therefore, volume can also be measured in quarts, liters, gallons, and so on. In the figures, l = length, w = width, h = height, s = length of a side, and r = radius.

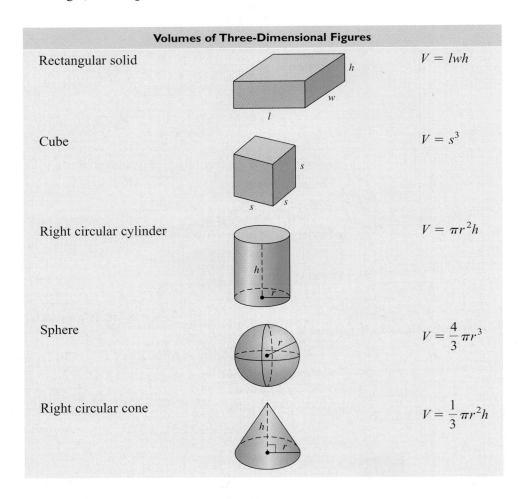

Volumes of Three-Dimensional Figures		
Rectangular solid		$V = lwh$
Cube		$V = s^3$
Right circular cylinder		$V = \pi r^2 h$
Sphere		$V = \dfrac{4}{3}\pi r^3$
Right circular cone		$V = \dfrac{1}{3}\pi r^2 h$

Example 6

Find the volume of each. In (b) give the answer in terms of π.

a)   b)

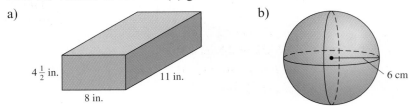

$4\frac{1}{2}$ in.   11 in.   8 in.   6 cm

Solution

a) $V = lwh$ Volume of a rectangular solid

$$= (11 \text{ in.})(8 \text{ in.})\left(4\frac{1}{2} \text{ in.}\right)$$ Substitute values.

$$= (11 \text{ in.})(8 \text{ in.})\left(\frac{9}{2} \text{ in.}\right)$$ Change to an improper fraction.

$$= \left(88 \cdot \frac{9}{2}\right) \text{in}^3$$ Multiply.

$$= 396 \text{ in}^3 \text{ or } 396 \text{ cubic inches}$$

b) $V = \dfrac{4}{3}\pi r^3$ Volume of a sphere

$$= \frac{4}{3}\pi(6 \text{ cm})^3$$ Replace r with 6 cm.

$$= \frac{4}{3}\pi(216 \text{ cm}^3)$$ $6^3 = 216$

$$= 288\pi \text{ cm}^3$$ Multiply.

You Try 6

Find the volume of each figure. In (b) give the answer in terms of π.

a) A box with length = 4 ft, width = 3 ft, and height 2 ft

b) A sphere with radius = 3 in.

Example 7

Application (Oil Drum)

A large truck has a fuel tank in the shape of a right circular cylinder. Its radius is 1 ft, and it is 4 ft long.

a) How many cubic feet of diesel fuel will the tank hold? (Use 3.14 for π.)

b) How many gallons will it hold? Round to the nearest gallon. ($1 \text{ ft}^3 \approx 7.48$ gal)

c) If diesel fuel costs \$1.75 per gallon, how much will it cost to fill the tank?

Solution

a) We're asked to determine how much fuel the tank will hold. We must find the *volume* of the tank.

$$\text{Volume of a cylinder} = \pi r^2 h$$
$$\approx (3.14)(1 \text{ ft})^2(4 \text{ ft})$$
$$= 12.56 \text{ ft}^3$$

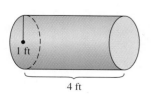

1 ft

4 ft

The tank will hold 12.56 ft^3 of diesel fuel.

b) We must convert 12.56 ft^3 to gallons. Since $1 \text{ ft}^3 \approx 7.48$ gal, we can change units by multiplying:

$$12.56 \text{ ft}^3 \cdot \left(\frac{7.48 \text{ gal}}{1 \text{ ft}^3}\right) = 93.9488 \text{ gal}$$
$$\approx 94 \text{ gal}$$

We can divide out units in fractions the same way we can divide out common factors.

The tank will hold approximately 94 gal.

c)   Diesel fuel costs $1.75 per gallon. We can figure out the total cost of the fuel the same way we did in (b).

$1.75 *per* gallon
↓

$$94 \text{ gal} \cdot \left(\frac{\$1.75}{\text{gal}}\right) = \$164.50 \qquad \text{Divide out the units of gallons.}$$

It will cost about $164.50 to fill the tank.

You Try 7

A large truck has a fuel tank in the shape of a right circular cylinder. Its radius is 1 ft, and it is 3 ft long.

a) How many cubic feet of diesel fuel will the tank hold? (Use 3.14 for π.)

b) How many gallons of fuel will it hold? Round to the nearest gallon. ($1 \text{ ft}^3 \approx 7.48 \text{ gal}$)

c) If diesel fuel costs $1.75 per gallon, how much will it cost to fill the tank?

Answers to You Try Exercises

1) $105°$ 2) $m\angle A = 126°$, $m\angle B = 27°$ 3) $P = 30$ cm; $A = 54$ cm^2

4) a) $C = 8\pi$ in; $C \approx 25.12$ in.   b) $A = 16\pi$ in^2; $A \approx 50.24$ in^2 5) $P = 42$ in; $A = 108$ in^2

6) a) 24 ft^3 b) 36π in^3 7) a) 9.42 ft^3 b) 70 gal   c) 122.50

1.3 Exercises

Boost your grade at mathzone.com!

> Practice Problems   > NetTutor   > e-Professors   > Videos
> Self-Test

Objective 1

1) An angle whose measure is 90° is a(n) _____ angle.

2) An angle whose measure is between 0° and 90° is a(n) _____ angle.

3) An angle whose measure is between 90° and 180° is a(n) _____ angle.

4) An angle whose measure is 180° is a(n) _____ angle.

5) If the sum of two angles is 180°, the angles are _____. If the sum of two angles is 90°, the angles are _____.

6) If two angles are supplementary, can both of them be obtuse? Explain.

Find the complement of each angle.

7) 24°

8) 63°

9) 45°

10) 80°

Find the supplement of each angle.

11) 100°

12) 59°

13) 71°

14) 175°

15) Find the measure of the missing angles.

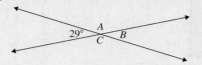

16) Find the measure of the missing angles.

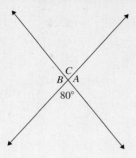

Objective 2

17) The sum of the angles in a triangle is _____ degrees.

Find the missing angle and classify each triangle as acute, obtuse, or right.

18)

19)

20)

21)

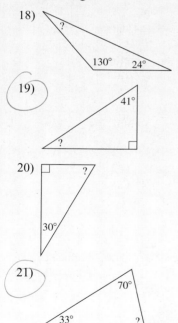

22) Can a triangle contain more than one obtuse angle? Explain.

Classify each triangle as equilateral, isosceles, or scalene.

23)

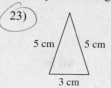

24)

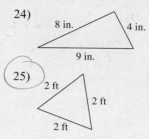

25)

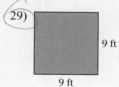

26) What can you say about the measures of the angles in an equilateral triangle?

27) True or False: If a triangle has two sides of equal length, then the angles opposite these sides are equal.

28) True or False: A right triangle can also be isosceles.

Objective 3

Find the area and perimeter of each figure. Include the correct units.

29)

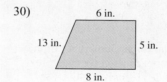

30)

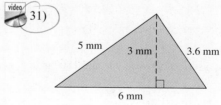

video

31)

32)

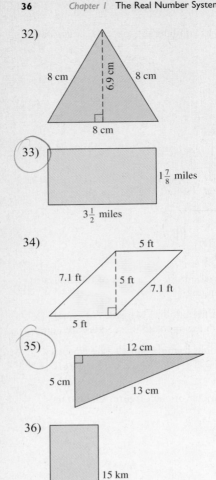

33)

34)

35)

36)

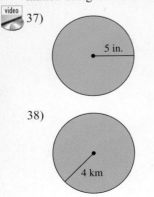

For 37–40, find the (a) area and (b) circumference of the circle. Give an exact answer for each and give an approximation using 3.14 for π. Include the correct units.

video 37)

38)

39)

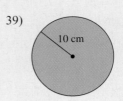

40)

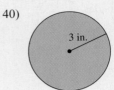

For 41–44, find the exact area and circumference of the circle in terms of π. Include the correct units.

41)

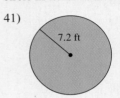

42)

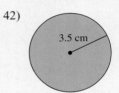

43)

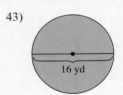

44)

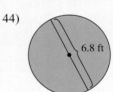

45) Find the area and perimeter of each figure. Include the correct units.

a)

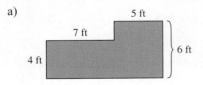

b)

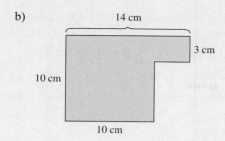

14 cm

3 cm

10 cm

10 cm

c)

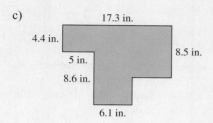

17.3 in.

4.4 in.

5 in.

8.5 in.

8.6 in.

6.1 in.

Find the area of the shaded region. Use 3.14 for π. Include the correct units.

46)

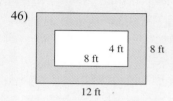

4 ft

8 ft

8 ft

12 ft

47)

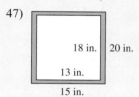

18 in.   20 in.

13 in.

15 in.

48)

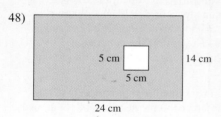

5 cm   14 cm

5 cm

24 cm

49)

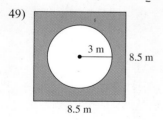

3 m

8.5 m

8.5 m

50)

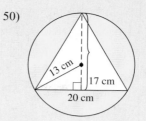

13 cm   17 cm

20 cm

Objective 4

Find the volume of each figure. Where appropriate, give the answer in terms of π. Include the correct units.

51)

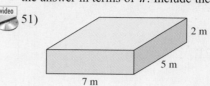

2 m

5 m

7 m

52)

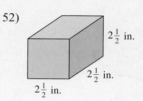

$2\frac{1}{2}$ in.

$2\frac{1}{2}$ in.

$2\frac{1}{2}$ in.

53)

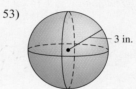

3 in.

54)

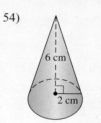

6 cm

2 cm

55)

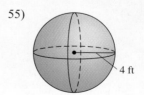

4 ft

56)

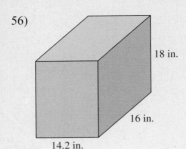

18 in.

16 in.

14.2 in.

57)

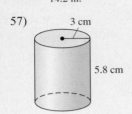

3 cm

5.8 cm

58)

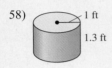

1 ft

1.3 ft

Objectives 3 and 4

Applications of Perimeter, Area, and Volume: Use 3.14 for π and include the correct units.

59) Kyeung wants to carpet her living room which measures 18 ft by 15 ft. She has budgeted $550 for this purchase.

   a)   How much carpet does she need?

   b)   Her favorite carpet costs $2.30/ft². Can she make this purchase and stay within her budget?

60) A rectangular flower box has dimensions 30" × 6" × 9". How much dirt will it hold?

61) The inside of a thermos has a radius of 3 cm and a height of 15 cm. The thermos is in the shape of a right circular cylinder. How much liquid will the thermos hold?

62) Sam's kitchen window is cracked and he wants to replace it himself. The window measures 2 ft by 3 ft. If a sheet of glass costs $11.25/ft², how much will it cost him to fix his window?

63) The medium-sized pizza at Marco's Pizzeria has a 14-in. diameter.

   a)   What is the perimeter of the pizza?

   b)   What is the area of the pizza?

64) Find the perimeter of home plate given the dimensions below.

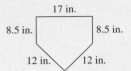

17 in.

8.5 in.   8.5 in.

12 in.   12 in.

65) A rectangular reservoir at a water treatment facility is 60 ft long, 19 ft wide, and 50 ft deep. If $1 \text{ ft}^3 \approx 7.48$ gal, how many gallons of water will this reservoir hold?

66) An above-ground, circular pool has a radius of 12 ft. How many feet of soft padding are needed to cover the sharp edges around the pool's rim?

67) Micah needs to rent office space for his growing business. He can afford to pay at most $1300 per month. He visits an office with the following layout. The owner charges $1.50/ft² for all office space in the building. Can Micah afford this office?

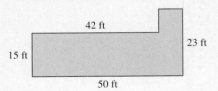

42 ft

15 ft   23 ft

50 ft

68) The Florida Highway Department must order four concrete columns for a bridge. Each column is a right circular cylinder with a radius of 2 ft and a height of 14 ft. The cost of the concrete is $3.10/ft³. Find the cost of the four columns. (Round to the nearest dollar.)

69) The radius of a basketball is approximately 4.7 in. Find its circumference to the nearest tenth of an inch.

70) Arun's Pet Supplies sells two medium-sized rectangular fish tanks. Model A has dimensions of 30" × 10" × 14" while model B has dimensions of 24" × 13" × 16".

    a) How many cubic inches of water will each tank hold?

    b) How many gallons of water will each tank hold? (Hint: Change cubic inches to cubic feet, then use the relationship $1 \text{ ft}^3 \approx 7.48$ gal.)

71) Phil plans to put a fence around his 9 ft × 5 ft rectangular garden to keep out the rabbits. The fencing costs $1.60/ft. How much will the fence cost?

72) A plot of land in the shape of a trapezoid has the dimensions given in the figure. Find the area and perimeter of this property.

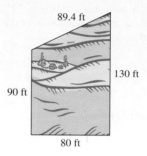

89.4 ft

130 ft

90 ft

80 ft

73) A pile of sand is in the shape of a right circular cone. The radius of the base is 2 ft, and the pile is 6 ft high. Find the volume of sand in the pile.

74) Find the volume of ice cream pictured below. Assume that the right circular cone is completely filled and that the scoop on top is half of a sphere.

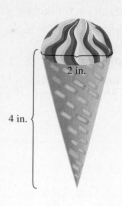

2 in.

4 in.

Section 1.4 Sets of Numbers and Absolute Value

Objectives

1. Identify Numbers and Graph Them on a Number Line

2. Compare Numbers Using Inequality Symbols

3. Find the Additive Inverse and Absolute Value of a Number

1. Identify Numbers and Graph Them on a Number Line

In Section 1.1, we defined the following sets of numbers:

    Natural numbers: {1, 2, 3, 4, . . .}

    Whole Numbers: {0, 1, 2, 3, 4, . . .}

We will begin this section by discussing other sets of numbers.

    On a **number line**, positive numbers are to the right of zero and negative numbers are to the left of zero.

Definition	The set of **integers** includes the set of natural numbers, their negatives, and zero. The set of *integers* is $\{\ldots, -3, -2, -1, 0, 1, 2, 3, \ldots\}$.

Example I

Graph each number on a number line.
$5, 1, -2, 0, -4$

Solution

5 and 1 are to the right of zero since they are positive.
-2 is two units to the left of zero, and -4 is four units to the left of zero.

You Try I

Graph each number on a number line. $3, -1, 6, -5, -3$

Positive and negative numbers are also called **signed numbers**.

Example 2

Given the set of numbers $\left\{-11, 0, 9, -5, -1, \dfrac{2}{3}, 6\right\}$, list the

a)   whole numbers   b)   natural numbers   c)   integers

Solution

a)   whole numbers: 0, 6, 9
b)   natural numbers: 6, 9
c)   integers: $-11, -5, -1, 0, 6, 9$

You Try 2

Given the set of numbers $\left\{3, -2, -9, 4, 0, \dfrac{5}{8}, -\dfrac{1}{3}\right\}$, list the

a) whole numbers   b) natural numbers   c) integers

Notice in Example 2 that $\dfrac{2}{3}$ did not belong to any of these sets. That is because the whole numbers, natural numbers, and integers do not contain any fractional parts. $\dfrac{2}{3}$ is a *rational number*.

Definition

A **rational number** is any number of the form $\frac{p}{q}$, where p and q are integers and $q \neq 0$.

Therefore, a *rational number* is any number that can be written as a fraction where the numerator and denominator are integers and the denominator does not equal zero.

Rational numbers include much more than numbers like $\frac{2}{3}$, which are already in fractional form.

Example 3

Explain why each of the following numbers is rational.

a)   3   b)   0.5   c)   −7

d) $2\frac{1}{8}$ e) $0.\overline{4}$ f) $\sqrt{9}$

Solution

Rational Number	Reason
3	3 can be written as $\frac{3}{1}$.
0.5	0.5 can be written as $\frac{5}{10}$.
−7	−7 can be written as $\frac{-7}{1}$.
$2\frac{1}{8}$	$2\frac{1}{8}$ can be written as $\frac{17}{8}$.
$0.\overline{4}$	$0.\overline{4}$ can be written as $\frac{4}{9}$.
$\sqrt{9}$	$\sqrt{9} = 3$ and $3 = \frac{3}{1}$.

$\sqrt{9}$ is read as "the square root of 9." This means, "What number times itself equals 9?" That number is 3.

 You Try 3

Explain why each of the following numbers is rational.

a) 6   b) 0.3   c) −2   d) $3\frac{2}{5}$ e) $0.\overline{3}$ f) $\sqrt{25}$

To summarize, the set of rational numbers includes

1) Integers, whole numbers, and natural numbers.
2) Repeating decimals.
3) Terminating decimals.
4) Fractions and mixed numbers.

The set of rational numbers does *not* include nonrepeating, nonterminating decimals. These decimals cannot be written as the quotient of two integers. Numbers such as these are called *irrational numbers*.

Definition

The set of numbers that cannot be written as the quotient of two integers is called the set of **irrational numbers**. Written in decimal form, an *irrational number* is a nonrepeating, nonterminating decimal.

Example 4

Explain why each of the following numbers is irrational.
a)   0.21598354...   b) π c) $\sqrt{5}$

Solution

Irrational Number	Reason
0.21598354...	It is a nonrepeating, nonterminating decimal.
π	$\pi \approx 3.14159265...$ It is a nonrepeating, nonterminating decimal.
$\sqrt{5}$	"5" is not a perfect square, and the decimal equivalent of the square root of a nonperfect square is a nonrepeating, nonterminating decimal. Here, $\sqrt{5} \approx 2.236067....$

You Try 4

Explain why each of the following numbers is irrational.
a)   0.7593614...   b) $\sqrt{7}$

If we put together the sets of numbers we have discussed up to this point, we get the *real numbers*.

Definition

The set of **real numbers** consists of the rational and irrational numbers.

We summarize the information next with examples of the different sets of numbers:

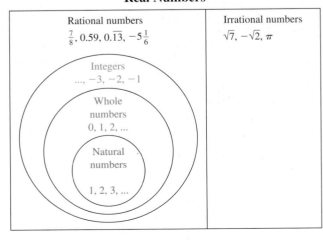

From the figure we can see, for example, that all whole numbers $\{0, 1, 2, 3, \ldots\}$ are integers, but not all integers are whole numbers (-3, for example).

Example 5

Given the set of numbers $\left\{0.\overline{2}, 37, -\dfrac{4}{15}, \sqrt{11}, -19, 8.51, 0, 6.149235\ldots\right\}$, list the

a) integers   b) natural numbers   c) whole numbers

d) rational numbers   e) irrational numbers   f) real numbers

Solution

a) integers: $-19, 0, 37$

b) natural numbers: 37

c) whole numbers: $0, 37$

d) rational numbers: $0.\overline{2}, 37, -\dfrac{4}{15}, -19, 8.51, 0$ Each of these numbers can be written as the quotient of two integers.

e) irrational numbers: $\sqrt{11}, 6.149235\ldots$

f) real numbers: All of the numbers in this set are real.
$$\left\{0.\overline{2}, 37, -\dfrac{4}{15}, \sqrt{11}, -19, 8.51, 0, 6.149235\ldots\right\}$$

You Try 5

Given the set of numbers $\left\{-38, 0, \sqrt{15}, 6, \dfrac{3}{2}, 5.4, 0.\overline{8}, 4.981162\ldots\right\}$, list the

a) whole numbers   b) integers   c) rational numbers   d) irrational numbers

2. Compare Numbers Using Inequality Symbols

Let's review the inequality symbols.

     $<$ less than $\leq$ less than or equal to

     $>$ greater than $\geq$ greater than or equal to

     $\neq$ not equal to $\approx$ approximately equal to

We use these symbols to compare numbers as in $5 > 2$, $6 \leq 17$, $4 \neq 9$, and so on. How do we compare negative numbers?

As we move to the *left* on the number line, the numbers get smaller. As we move to the *right* on the number line, the numbers get larger.

Example 6

Insert $>$ or $<$ to make the statement true. Look at the number line, if necessary.

a) 5 __ 1 b) -4 __ 3 c) -1 __ -5 d) -5 __ -2

Solution

a) $5 \ge 1$ 5 is to the right of 1.
b) $-4 \le 3$ −4 is to the left of 3.
c) $-1 \ge -5$ −1 is to the right of −5.
d) $-5 \le -2$ −5 is to the left of −2.

You Try 6

Insert $>$ or $<$ to make the statement true.

a) 4 __ 9 b) 6 __ -8 c) -3 __ -10

Application of Signed Numbers

Example 7

Use a signed number to represent the change in each situation.

a) After a storm passed through Kansas City, the temperature dropped 18°.

b) Between 1995 and 2000, the number of "Generation Xers" moving to the Seattle, Washington, area increased by 27,201 people.
(*American Demographics,* May 2004, Vol. 26, Issue 4, p. 18)

Solution

a) $-18°$ The negative number represents a decrease in temperature.

b) 27,201   The positive number represents an increase in population.

You Try 7

Use a signed number to represent the change.
After taking his last test, Julio raised his average by 3.5%.

3. Find the Additive Inverse and Absolute Value of a Number

Distance = 2   Distance = 2

Notice that both −2 and 2 are a distance of 2 units from 0 but are on opposite sides of 0. We say the 2 and −2 are *additive inverses.*

Definition Two numbers are **additive inverses** if they are the same distance from 0 on the number line but on the opposite side of 0. Therefore, if a is any real number, then $-a$ is its additive inverse.

Furthermore, $-(-a) = a$. We can see this on the number line.

Example 8

Find $-(-4)$.

Solution

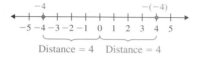

So, beginning with -4, the number on the opposite side of zero and 4 units away from zero is 4. $-(-4) = 4$

 You Try 8

Find $-(-11)$.

This idea of "distance from zero" can be explained in another way: *absolute value*.

The absolute value of a number is the distance between that number and 0 on the number line. It just describes the distance, *not* what side of zero the number is on. Therefore, the absolute value of a number is always positive or zero.

Definition If a is any real number, then the **absolute value of a**, denoted by $|a|$, is

    i) a if $a \geq 0$

    ii) $-a$ if $a < 0$

Remember, $|a|$ is never negative.

Example 9

Evaluate each.

a) $|9|$ b) $|-7|$ c) $|0|$ d) $-|5|$ e) $|11-5|$

Solution

a) $|9| = 9$ 9 is 9 units from 0.

b) $|-7| = 7$ -7 is 7 units from 0.

c) $|0| = 0$

d) $-|5| = -5$ First, evaluate $|5|$. $|5| = 5$. Then, apply the negative symbol to get -5.

e) $|11 - 5| = |6|$ The absolute value symbols work like parentheses. First, evaluate what is
                         inside: $11 - 5 = 6$.

      $= 6$ Find the absolute value.

You Try 9

Evaluate each.

a) $|16|$ b) $|-4|$ c) $-|6|$ d) $|14 - 9|$

Answers to You Try Exercises

1)
A number line marked from -6 to 6 with points at -5, -3, -1, 3, and 6.

2) a) 0, 3, 4   b) 3, 4   c) $-9, -2, 0, 3, 4$

3) a) $6 = \dfrac{6}{1}$ b) $0.3 = \dfrac{3}{10}$ c) $-2 = \dfrac{-2}{1}$ d) $3\dfrac{2}{5} = \dfrac{17}{5}$ e) $0.\overline{3} = \dfrac{1}{3}$ f) $\sqrt{25} = 5$

4) a) It is a nonrepeating, nonterminating decimal.   b) 7 is not a perfect square, so the decimal
equivalent of $\sqrt{7}$ is a nonrepeating, nonterminating decimal.

5) a) 0, 6   b) $-38, 0, 6$ c) $-38, 0, 6, \dfrac{3}{2}, 5.4, 0.\overline{8}$ d) $\sqrt{15}, 4.981162\ldots$

6) a) $<$ b) $>$ c) $>$ 7) 3.5%   8) 11   9) a) 16   b) 4   c) -6 d) 5

1.4 Exercises

Boost your grade at mathzone.com!   MathZone

> Practice Problems   > NetTutor   > e-Professors   > Videos
> Self-Test

Objective 1

1) Given the set of numbers

$$\left\{-14, 6, \frac{2}{5}, \sqrt{19}, 0, 3.\overline{28}, -1\frac{3}{7}, 0.95\right\},$$

list the

a)   whole numbers

b)   integers

c)   irrational numbers

d)   natural numbers

e)   rational numbers

f)   real numbers

2) Given the set of numbers

$$\left\{5.2, 34, -\frac{9}{4}, -18, 0, 0.\overline{7}, \frac{5}{6}, \sqrt{6}, 4.3811275\ldots\right\},$$

list the

a)   integers

b)   natural numbers

c)   rational numbers

d)   whole numbers

e)   irrational numbers

f)   real numbers

Determine if each statement is true or false.

3) Every whole number is an integer.

4) Every rational number is a whole number.

5) Every real number is an integer.

6) Every natural number is a whole number.

7) Every integer is a rational number.

8) Every whole number is a real number.

Graph the numbers on a number line. Label each.

9) $6, -4, \dfrac{3}{4}, 0, -1\dfrac{1}{2}$

10) $5\dfrac{2}{3}, 1, -3, -4\dfrac{5}{6}, 2\dfrac{1}{4}$

11) $-5, 6\dfrac{1}{8}, 2\dfrac{3}{4}, -2\dfrac{5}{7}, 4.3$

12) $1.7, -\dfrac{4}{5}, 3\dfrac{1}{5}, -5, -2\dfrac{1}{2}$

 13) $-6.8, -\dfrac{3}{8}, 0.2, 1\dfrac{8}{9}, -4\dfrac{1}{3}$

14) $-1, 5.9, 1\dfrac{7}{10}, -\dfrac{2}{3}, 0.61$

Objective 3

Evaluate.

15) $|-13|$

16) $|8|$

17) $\left|\dfrac{3}{2}\right|$

18) $|-23|$

19) $-|10|$

20) $-|6|$

21) $-|-19|$

22) $-\left|-1\dfrac{3}{5}\right|$

Find the additive inverse of each.

23) 11

24) 5

25) -7

26) $-\dfrac{1}{2}$

27) -4.2

28) 2.9

Objectives 2 and 3

Write each group of numbers from smallest to largest.

 29) $7, -2, 3.8, -10, 0, \dfrac{9}{10}$

30) $-6, -7, 5.2, 5.9, 6, -1$

31) $-4\dfrac{1}{2}, \dfrac{5}{8}, \dfrac{1}{4}, -0.3, -9, 1$

32) $14, -5, 13.6, -5\dfrac{2}{3}, 1, \dfrac{6}{7}$

Decide if each statement is true or false.

33) $9 \geq -2$

34) $-6 > 3$

35) $-7 \leq -4$

36) $10.8 \geq 10.2$

37) $\dfrac{1}{6} \leq \dfrac{1}{8}$

38) $-8.1 > -8.5$

 39) $-5\dfrac{3}{10} < -5\dfrac{3}{4}$

40) $\dfrac{4}{5} \neq 0.8$

41) $|-9| \geq 9$

42) $-|-31| = 31$

Use a signed number to represent the change in each situation.

43) In 2001, Barry Bonds set the all-time home run record with 73. In 2002, he hit 46. That was a decrease of 27 home runs. (*Total Baseball: The Ultimate Baseball Encyclopedia,* 8th edition by John Thorn, Phil Birnbaum, and Bill Deane, © 2004, SPORT Media Pub., Inc.)

44) In 2004, the median income for females 25 years and older with a bachelor's degree was $31,585, and in 2005 it increased by $1083 to $32,668. (U.S. Census Bureau)

45) In 2001 the U.S. unemployment rate was 4.7%, and in 2002 it increased by 1.1% to 5.8%. (U.S. Bureau of Labor Statistics)

46) During the 2001–2002 NBA season, Kobe Bryant averaged 25.2 points per game. The following season he averaged 30.0 points per game, an increase of 4.8 over the previous year. (*Total Basketball: The Ultimate Basketball Encyclopedia,* by Ken Shouler, Bob Ryan, Sam Smith, Leonard Koppett, and Bob Bellotti, © 2003, SPORT Media Pub., Inc.)

47) In Michael Jordan's last season with the Chicago Bulls (1997–1998) the average attendance at the United Center was 23,988. During the 2002–2003 season, the average attendance fell to 19,617. This was a decrease of 4371 people per game. (*Total Basketball: The Ultimate Basketball Encyclopedia,* by Ken Shouler, Bob Ryan, Sam Smith, Leonard Koppett, and Bob Bellotti, © 2003, SPORT Media Pub., Inc.)

48) The 2002 Indianapolis 500 was won by Helio Castroneve with an average speed of 166.499 mph. The next year Gil de Ferran won it with an average speed of 156.291 mph or 10.208 mph slower than in 2002. (www.indy500.com)

49) According to the 1990 census, the population of North Dakota was 639,000. In 2000 it increased by 3000 to 642,000. (U.S. Census Bureau)

50) The 1999 Hennessey Viper Venom can go from 0 to 60 mph in 3.3 sec. The 2000 model goes from 0 to 60 mph in 2.7 sec, which is a decrease of 0.6 sec to go from 0 to 60 mph. (www.supercars.net)

Section 1.5 Addition and Subtraction of Real Numbers

Objectives

1. Add Integers Using a Number Line
2. Add Real Numbers with the Same Sign
3. Add Real Numbers with Different Signs
4. Subtract Real Numbers
5. Solve Applied Problems Involving Addition and Subtraction of Real Numbers
6. Apply the Order of Operations to Real Numbers
7. Translate English Expressions to Mathematical Expressions

In Section 1.4, we defined real numbers. In this section, we will discuss adding and subtracting real numbers.

1. Add Integers Using a Number Line

Let's use a number line to add numbers.

Example 1

Use a number line to add each pair of numbers.

a) $2 + 5$ b) $-1 + (-4)$ c) $2 + (-5)$ d) $-8 + 12$

Solution

a) $2 + 5$: Start at 2 and move 5 units to the right.

$2 + 5 = 7$

b) $-1 + (-4)$: Start at -1 and move 4 units to the left. (Move to the left when adding a negative.)

$-1 + (-4) = -5$

c) $2 + (-5)$: Start at 2 and move 5 units to the left.

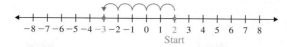

$$2 + (-5) = -3$$

d) $-8 + 12$: Start at -8 and move 12 units to the right.

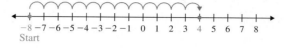

$$-8 + 12 = 4$$

You Try 1

Use a number line to add each pair of numbers.

a) $1 + 3$ b) $-3 + (-2)$ c) $8 + (-6)$ d) $-10 + 7$

2. Add Real Numbers with the Same Sign

We found that

$$2 + 5 = 7, \qquad -1 + (-4) = -5, \qquad 2 + (-5) = -3, \qquad -8 + 12 = 4.$$

Notice that when we add two numbers with the same sign, the result has the same sign as the numbers being added.

> To add numbers with the same sign, find the absolute value of each number and add them. The sum will have the same sign as the numbers being added.

Apply this rule to $-1 + (-4)$.

The result will be negative
↓
$$-1 + (-4) = -\underbrace{(|-1| + |-4|)}_{\substack{\text{Add the} \\ \text{absolute} \\ \text{value of} \\ \text{each number.}}} = -(1 + 4) = -5$$

Example 2

Add.

a) $-8 + (-2)$ b) $-35 + (-71)$

Solution

a) $-8 + (-2) = -(|-8| + |-2|) = -(8 + 2) = -10$

b) $-35 + (-71) = -(|-35| + |-71|) = -(35 + 71) = -106$

You Try 2

Add.

a) $-2 + (-9)$ b) $-48 + (-67)$

3. Add Real Numbers with Different Signs

In Example 1, we found that $2 + (-5) = -3$ and $-8 + 12 = 4$.

> To add two numbers with different signs, find the absolute value of each number. Subtract the smaller absolute value from the larger. The sum will have the sign of the number with the larger absolute value.

Let's apply this to $2 + (-5)$ and $-8 + 12$.

$2 + (-5)$: $|2| = 2$ $|-5| = 5$
Since $2 < 5$, subtract $5 - 2$ to get 3. Since $|-5| > |2|$, the sum will be negative.
$2 + (-5) = -3$

$-8 + 12$: $|-8| = 8$ $|12| = 12$
Subtract $12 - 8$ to get 4. Since $|12| > |-8|$, the sum will be positive.
$-8 + 12 = 4$

Example 3

Add.

a) $-19 + 4$ b) $10.3 + (-4.1)$ c) $\dfrac{1}{4} + \left(-\dfrac{5}{9}\right)$ d) $-7 + 7$

Solution

a) $-19 + 4 = -15$ — The sum will be negative since the number with the larger absolute value, $|-19|$, is negative.

b) $10.3 + (-4.1) = 6.2$ — The sum will be positive since the number with the larger absolute value, $|10.3|$, is positive.

c) $\dfrac{1}{4} + \left(-\dfrac{5}{9}\right) = \dfrac{9}{36} + \left(-\dfrac{20}{36}\right)$ — Get a common denominator.

$\qquad = -\dfrac{11}{36}$ — The sum will be negative since the number with the larger absolute value, $\left|-\dfrac{20}{36}\right|$, is negative.

d) $-7 + 7 = 0$

> The sum of a number and its additive inverse is always 0. That is, if a is a real number, then $a + (-a) = 0$. Notice in part d) of Example 3 that -7 and 7 are additive inverses.

You Try 3

Add.

a) $11 + (-8)$ b) $-17 + (-5)$ c) $-\dfrac{5}{8} + \dfrac{2}{3}$ d) $59 + (-59)$

4. Subtract Real Numbers

Subtraction of numbers can be defined in terms of the additive inverse. We'll begin by looking at a basic subtraction problem.

Represent $6 - 4$ on a number line.

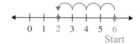

We begin at 6. To subtract 4, move 4 units to the left to get 2.

$$6 - 4 = 2$$

We would use the same process to find $6 + (-4)$. This leads us to a definition of subtraction:

> **Definition**   If a and b are real numbers, then $a - b = a + (-b)$.

The definition tells us that to subtract $a - b$,
1) Change subtraction to addition.
2) Find the additive inverse of b.
3) Add a and the additive inverse of b.

Example 4

Subtract.

a) $5 - 16$ b) $-19 - 4$ c) $20 - 5$ d) $7 - (-11)$

Solution

a) $5 - 16 = 5 + (-16) = -11$

         Change to   Additive inverse
         addition   of 16

b) $-19 - 4 = -19 + (-4) = -23$

              Change to   Additive inverse
              addition   of 4

c) $20 - 5 = 20 + (-5)$
                 $= 15$

d) $7 - (-11) = 7 + 11 = 18$

            Change to   Additive inverse
            addition   of -11

You Try 4

Subtract.

a) $3 - 10$ b) $-6 - 12$ c) $18 - 7$ d) $9 - (-16)$

In part d) of Example 4, $7 - (-11)$ changed to $7 + 11$. So, *subtracting a negative number is equivalent to adding a positive number.* Therefore, $-2 - (-6) = -2 + 6 = 4$.

5. Solve Applied Problems Involving Addition and Subtraction of Real Numbers

Sometimes, we use signed numbers to solve real-life problems.

Example 5

The lowest temperature ever recorded was $-129°F$ in Vostok, Antarctica. The highest temperature on record is $136°F$ in Al'Aziziyah, Libya. What is the difference between these two temperatures? (*Source*: *Encyclopedia Britannica Almanac 2004*)

Solution

$$\text{Difference} = \text{Highest temperature} - \text{Lowest temperature}$$
$$= \qquad 136 \qquad - \qquad (-129)$$
$$= 136 + 129$$
$$= 265$$

The difference between the temperatures is $265°F$.

You Try 5

The best score in a golf tournament was -12, and the worst score was $+17$. What is the difference between these two scores?

6. Apply the Order of Operations to Real Numbers

We discussed the order of operations in Section 1.2. Let's explore it further with the real numbers.

Example 6

Simplify.

a) $(10 - 18) + (-4 + 6)$ b) $-13 - (-21 + 5)$
c) $|-31 - 4| - 7|9 - 4|$

Solution

a) $(10 - 18) + (-4 + 6) = -8 + 2$ First, perform the operations in parentheses.
$$= -6$$ Add.

b) $-13 - (-21 + 5) = -13 - (-16)$ First, perform the operations in parentheses.
$$= -13 + 16$$ Change to addition.
$$= 3$$ Add.

c) $|-31 - 4| - 7|9 - 4| = |-31 + (-4)| - 7|9 - 4|$
$$= |-35| - 7|5|$$ Perform the operations in the absolute values.
$$= 35 - 7(5)$$ Evaluate the absolute values.
$$= 35 - 35$$
$$= 0$$

You Try 6

Simplify.

a) $[12 + (-5)] - [-16 + (-8)]$ b) $-\dfrac{4}{9} + \left(\dfrac{1}{6} - \dfrac{2}{3}\right)$ c) $-|7 - 15| - |4 - 2|$

7. Translate English Expressions to Mathematical Expressions

Knowing how to translate from English expressions to mathematical expressions is a skill students need to learn algebra. Here, we will be discussing how to "translate" from English to mathematics.

    Let's look at some key words and phrases you may encounter.

Phrase	Operation
sum	addition
more than	addition
increased by	addition
difference between	subtraction
less than	subtraction
decreased by	subtraction
subtracted from	subtraction

Here are some examples:

Example 7

Write a mathematical expression for each and simplify.

a) 9 more than -2 b) 10 less than 41   c) -8 decreased by 17
d) the sum of 13 and -4 e) 8 less than the sum of -11 and -3

Solution

a) 9 more than -2

    9 more than a quantity means we *add 9* to the quantity, in this case, -2.

$$-2 + 9 = 7$$

b) 10 less than 41

    10 less than a quantity means we *subtract 10 from* that quantity, in this case, 41.

$$41 - 10 = 31$$

c) −8 decreased by 17

If −8 is being *decreased by 17*, then we subtract 17 *from* −8.

$$-8 - 17 = -8 + (-17)$$
$$= -25$$

d) the sum of 13 and −4

Sum means add.

$$13 + (-4) = 9$$

e) 8 less than the sum of −11 and −3.

8 less than means we are subtracting 8 *from* something. From what? From the sum of −11 and −3.
Sum means add, so we must find the sum of −11 and −3 and subtract 8 from it.

$$[-11 + (-3)] - 8 = -14 - 8 \qquad \text{First, perform the operation in the brackets.}$$
$$= -14 + (-8) \qquad \text{Change to addition.}$$
$$= -22 \qquad \text{Add.}$$

You Try 7

Write a mathematical expression for each and simplify.

a) −14 increased by 6   b) 27 less than 15   c) The sum of 23 and −7 decreased by 5

Answers to You Try Exercises

1) a) 4   b) −5   c) 2   d) −3   2) a) −11   b) −115   3) a) 3   b) −22   c) $\frac{1}{24}$ d) 0

4) a) −7   b) −18   c) 11   d) 25   5) 29   6) a) 31   b) $-\frac{17}{18}$ c) −10   7) a) −14 + 6; −8

b) 15 − 27; −12   c) [23 + (−7)] − 5; 11

1.5 Exercises   **Boost your grade at mathzone.com!**   MathZone   > Practice Problems   > NetTutor   > e-Professors   > Videos
> Self-Test

Objectives 1–4 and 6

1) Explain, in your own words, how to add two negative numbers.

2) Explain, in your own words, how to add a positive and a negative number.

3) Explain, in your own words, how to subtract two negative numbers.

Use a number line to represent each sum or difference.

4) −8 + 5

5) 6 − 11

6) $-1 - 5$

7) $-2 + (-7)$

8) $10 + (-6)$

Add or subtract as indicated.

9) $9 + (-13)$

10) $-7 + (-5)$

11) $-2 - 12$

12) $-4 + 11$

13) $-25 + 38$

14) $10 - (-17)$

15) $-1 - (-19)$

16) $-40 - (-6)$

17) $-794 - 657$

18) $380 + (-192)$

19) $-\dfrac{3}{10} + \dfrac{4}{5}$

20) $\dfrac{2}{9} - \dfrac{5}{6}$

21) $-\dfrac{5}{8} - \dfrac{2}{3}$

22) $\dfrac{3}{7} - \left(-\dfrac{1}{8}\right)$

23) $-\dfrac{11}{12} - \left(-\dfrac{5}{9}\right)$

24) $-\dfrac{3}{4} + \left(-\dfrac{1}{6}\right)$

25) $7.3 - 11.2$

26) $-14.51 + 20.6$

27) $-5.09 - (-12.4)$

28) $8.8 - 19.2$

29) $9 - (5 - 11)$

30) $-2 + (3 - 8)$

31) $-1 + (-6 - 4)$

32) $14 - (-10 - 2)$

33) $(-3 - 1) - (-8 + 6)$

34) $[14 + (-9)] + (1 - 8)$

35) $-16 + 4 + 3 - 10$

36) $8 - 28 + 3 - 7$

37) $5 - (-30) - 14 + 2$

38) $-17 - (-9) + 1 - 10$

 39) $\dfrac{4}{9} - \left(\dfrac{2}{3} + \dfrac{5}{6}\right)$

40) $-\dfrac{1}{2} + \left(\dfrac{3}{5} - \dfrac{3}{10}\right)$

41) $\left(\dfrac{1}{8} - \dfrac{1}{2}\right) + \left(\dfrac{3}{4} - \dfrac{1}{6}\right)$

42) $\dfrac{11}{12} - \left(\dfrac{3}{8} - \dfrac{2}{3}\right)$

43) $(2.7 + 3.8) - (1.4 - 6.9)$

44) $-9.7 - (-5.5 + 1.1)$

 45) $|7 - 11| + |6 + (-13)|$

46) $|8 - (-1)| - |3 + 12|$

47) $-|2 - (-3)| - 2|-5 + 8|$

48) $|-6 + 7| + 5|-20 - (-11)|$

Determine whether each statement is true or false. For any real numbers a and b,

49) $|a - b| = |b - a|$

50) $|a + b| = a + b$

51) $|a| + |b| = a + b$

52) $|a + b| = |a| + |b|$

53) $a + (-a) = 0$

54) $-b - (-b) = 0$

Objective 5

Applications of Signed Numbers: Write an expression for each and simplify. Answer the question with a complete sentence.

 55) The world's tallest mountain, Mt. Everest, reaches an elevation of 29,028 ft. The Mariana Trench in the Pacific Ocean has a maximum depth of 36,201 ft below sea level. What is the difference between these two elevations? *(Encyclopedia Britannica Almanac 2004)*

56) In 2002, the total attendance at Major League Baseball games was 67,859,176. In 2003, that figure was 67,630,052. What was the difference in the number of people who went to ballparks from 2002 to 2003? *(Total Baseball: The Ultimate Baseball Encyclopedia,* 8th edition by John Thorn, Phil Birnbaum, and Bill Deane, © 2004, SPORT Media Pub., Inc.)

57) The median income for a male 25 years or older with a bachelor's degree in 2004 was $51,081. In 2005, the median income was $51,700. What was the difference in the median income from 2004 to 2005? *(U.S. Census Bureau)*

58) Mt. Washington, New Hampshire, rises to an elevation of 6288 ft. New Orleans, Louisiana, lies 6296 ft below this. What is the elevation of New Orleans? (www.infoplease.com)

59) The lowest temperature ever recorded in the United States was −79.8°F in the Endicott Mountains of Alaska. The highest U.S. temperature on record was 213.8°F more than the lowest and was recorded in Death Valley, California. What is the highest temperature on record in the United States? (www.infoplease.com)

60) The lowest temperature ever recorded in Hawaii was 12°F while the lowest temperature in Colorado was 73°F less than that. What was the coldest temperature ever recorded in Colorado? (www.infoplease.com)

61) During one offensive drive in the first quarter of Super Bowl XXXVIII, the New England Patriots ran for 7 yd, gained 4 yd on a pass play, gained 1 yd on a running play, gained another 6 yd on a pass by Tom Brady, then lost 10 yd on a running play. What was the Patriots' net yardage on this offensive drive? (www.superbowl.com)

62) The bar graph shows Dale Earnhardt Jr.'s winnings from races for the years 2000–2003. Use a signed number to represent the change in his winnings over the given years. (www.dalejrpitstop.com)

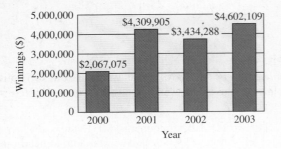

a)   2000–2001

b)   2001–2002

c)   2002–2003

63) The bar graph shows the number of housing starts (in thousands) during five months in 2005 in the northeastern United States. Use a signed number to represent the change over the given months. (www.census.gov)

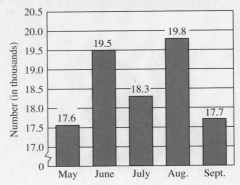

a)   May–June   c)   July–August

b)   June–July   d)   August–September

64) The bar graph shows the estimated population of Oakland, California, from 2000 to 2003. Use a signed number to represent the change in Oakland's population over the given years. (U.S. Census Bureau)

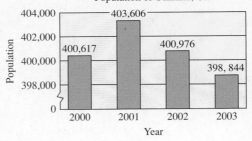

a)   2000–2001

b)   2001–2002

c)   2002–2003

Objective 7

Write a mathematical expression for each and simplify.

65) 7 more than 5

66) 3 more than 11

67) 16 less than 10

68) 15 less than 4

69) −8 less than 9

70) −25 less than −19

71) The sum of −21 and 13

72) The sum of −7 and 20

73) −20 increased by 30

74) −37 increased by 22

75) 23 decreased by 19

76) 8 decreased by 18

77) 18 less than the sum of −5 and 11

78) 35 less than the sum of −17 and 3

Section 1.6 Multiplication and Division of Real Numbers

Objectives

1. Multiply Real Numbers
2. Evaluate Exponential Expressions
3. Divide Real Numbers
4. Apply the Order of Operations to Real Numbers
5. Translate English Expressions to Mathematical Expressions

1. Multiply Real Numbers

What is the meaning of $4 \cdot 5$? It is repeated addition.

$$4 \cdot 5 = 5 + 5 + 5 + 5 = 20.$$

So, what is the meaning of $4 \times (-5)$? It, too, represents repeated addition.

$$4 \cdot (-5) = -5 + (-5) + (-5) + (-5) = -20.$$

Let's make a table of some products:

×	5	4	③	2	1	0	−1	−2	−3	−4	−5
④	20	16	12	8	4	0	−4	−8	−12	−16	−20

$$4 \cdot 3 = 12$$

The bottom row represents the product of 4 and the number above it ($4 \cdot 3 = 12$). Notice that as the numbers in the first row decrease by 1, the numbers in the bottom row decrease by 4. Therefore, once we get to $4 \cdot (-1)$, the product is negative. From the table we can see that,

> The product of a positive number and a negative number is negative.

Example 1

Multiply.

a) $-6 \cdot 9$ b) $\dfrac{3}{8} \cdot (-12)$ c) $-5 \cdot 0$

Solution

a) $-6 \cdot 9 = -54$

b) $\dfrac{3}{8} \cdot (-12) = \dfrac{3}{\overset{}{\underset{2}{8}}} \cdot \left(-\dfrac{\overset{3}{\cancel{12}}}{1} \right) = -\dfrac{9}{2}$

c) $-5 \cdot 0 = 0$ The product of zero and any real number is zero.

You Try 1

Multiply.

a) $-7 \cdot 3$ b) $\dfrac{8}{15} \cdot (-10)$

What is the sign of the product of two negative numbers? Again, we'll make a table.

$\times$	3	2	1	0	-1	-2	-3
-4	-12	-8	-4	0	4	8	12

As we decrease the numbers in the top row by 1, the numbers in the bottom row *increase* by 4. When we reach $-4 \cdot (-1)$, our product is a positive number, 4. The table illustrates that,

> The product of two negative numbers is positive.

We can summarize our findings this way:

Multiplying Real Numbers

1) The product of two positive numbers is positive.

2) The product of two negative numbers is positive.

3) The product of a positive number and a negative number is negative.

4) The product of any real number and zero is zero.

Example 2

Multiply.

a) $-7 \cdot (-3)$

b) $-2.5 \cdot 8$

c) $-\dfrac{4}{5} \cdot \left(-\dfrac{1}{6}\right)$

d) $-3 \cdot (-4) \cdot (-5)$

Solution

a) $-7 \cdot (-3) = 21$ The product of two negative numbers is positive.

b) $-2.5 \cdot 8 = -20$ The product of a negative number and a positive number is negative.

c) $-\dfrac{4}{5} \cdot \left(-\dfrac{1}{6}\right) = -\dfrac{\overset{2}{\cancel{4}}}{5} \cdot \left(-\dfrac{1}{\underset{3}{\cancel{6}}}\right)$

$\qquad\qquad = \dfrac{2}{15}$ The product of two negatives is positive.

d) $\underbrace{-3 \cdot (-4)}_{12} \cdot (-5) = 12 \cdot (-5)$ Order of operations—multiply from left to right.

$\qquad\qquad\qquad\quad = -60$

You Try 2

Multiply.

a) $-2 \cdot 8$

b) $-\dfrac{10}{21} \cdot \dfrac{14}{15}$

c) $-2 \cdot (-3) \cdot (-1) \cdot (-4)$

Note: It is helpful to know that

1)   An **even number** of negative factors in a product gives a positive result.

$$-3 \cdot 1 \cdot (-2) \cdot (-1) \cdot (-4) = 24 \qquad \text{Four negative factors}$$

2)   An **odd number** of negative factors in a product gives a negative result.

$$5 \cdot (-3) \cdot (-1) \cdot (-2) \cdot (3) = -90 \qquad \text{Three negative factors}$$

2. Evaluate Exponential Expressions

In Section 1.2 we discussed exponential expressions. Recall that exponential notation is a shorthand way to represent repeated multiplication:

$$2^4 = 2 \cdot 2 \cdot 2 \cdot 2 = 16$$

Now we will discuss exponents and negative numbers. Consider a base of -2 raised to different powers. (The -2 is in parentheses to indicate that it is the base.)

$$(-2)^1 = -2$$
$$(-2)^2 = -2 \cdot (-2) = 4$$
$$(-2)^3 = -2 \cdot (-2) \cdot (-2) = -8$$
$$(-2)^4 = -2 \cdot (-2) \cdot (-2) \cdot (-2) = 16$$
$$(-2)^5 = -2 \cdot (-2) \cdot (-2) \cdot (-2) \cdot (-2) = -32$$
$$(-2)^6 = -2 \cdot (-2) \cdot (-2) \cdot (-2) \cdot (-2) \cdot (-2) = 64$$

Do you notice that

1) -2 raised to an *odd* power gives a negative result?

and

2) -2 raised to an *even* power gives a positive result?

This will always be true.

> A negative number raised to an *odd* power will give a *negative* result. A negative number raised to an *even* power will give a *positive* result.

Example 3

Evaluate.

a) $(-6)^2$ b) $(-10)^3$

Solution

a) $(-6)^2 = 36$

b) $(-10)^3 = -1000$

You Try 3

Evaluate.

a) $(-9)^2$ b) $(-5)^3$

How do $(-2)^4$ and -2^4 differ? Let's identify their bases and evaluate each.

$(-2)^4$:   Base $= -2$ $(-2)^4 = 16$

-2^4:   Since there are no parentheses,

-2^4 is equivalent to $-1 \cdot 2^4$. Therefore, the base is 2.

$$-2^4 = -1 \cdot 2^4$$
$$= -1 \cdot 2 \cdot 2 \cdot 2 \cdot 2$$
$$= -1 \cdot 16$$
$$= -16$$

So, $(-2)^4 = 16$ and $-2^4 = -16$.

BE CAREFUL   When working with exponential expressions, be able to identify the base.

Example 4

Evaluate.

a) $(-5)^3$ b) -9^2 c) $\left(-\dfrac{1}{7}\right)^2$

Solution

a) $(-5)^3$: Base $= -5$ $(-5)^3 = -5 \cdot (-5) \cdot (-5)$
$$= -125$$

b) -9^2: Base $= 9$ $-9^2 = -1 \cdot 9^2$
$$= -1 \cdot 9 \cdot 9$$
$$= -81$$

c) $\left(-\dfrac{1}{7}\right)^2$: Base $= -\dfrac{1}{7}$ $\left(-\dfrac{1}{7}\right)^2 = -\dfrac{1}{7} \cdot \left(-\dfrac{1}{7}\right)$
$$= \dfrac{1}{49}$$

You Try 4

Evaluate.

a) -3^4 b) $(-11)^2$ c) -8^2 d) $-\left(-\dfrac{2}{3}\right)^3$

3. Divide Real Numbers

We can state these rules for dividing signed numbers:

Dividing Signed Numbers

1)   The quotient of two positive numbers is a positive number.

2)   The quotient of two negative numbers is a positive number.

3)   The quotient of a positive and a negative number is a negative number.

Example 5

Divide.

a) $-48 \div 6$ b) $-\dfrac{1}{12} \div \left(-\dfrac{4}{3}\right)$ c) $\dfrac{-6}{-1}$ d) $\dfrac{-27}{72}$

Solution

a) $-48 \div 6 = -8$

b) $-\dfrac{1}{12} \div \left(-\dfrac{4}{3}\right) = -\dfrac{1}{12} \cdot \left(-\dfrac{3}{4}\right)$ When dividing by a fraction, multiply by the reciprocal.

$\qquad = -\dfrac{1}{\overset{}{\underset{4}{\cancel{12}}}} \cdot \left(-\dfrac{\overset{1}{\cancel{3}}}{4}\right)$

$\qquad = \dfrac{1}{16}$

c) $\dfrac{-6}{-1} = 6$ The quotient of two negative numbers is positive, and $\dfrac{6}{1}$ simplifies to 6.

d) $\dfrac{-27}{72} = -\dfrac{27}{72}$ The quotient of a negative number and a positive number is negative, so reduce $\dfrac{27}{72}$.

$\qquad = -\dfrac{3}{8}$ 27 and 72 each divide by 9.

It is important to note here that there are three ways to write the answer: $-\dfrac{3}{8}, \dfrac{-3}{8},$ or $\dfrac{3}{-8}$. These are equivalent. However, we usually write the negative sign in front of the entire fraction as in $-\dfrac{3}{8}$.

You Try 5

Divide.

a) $-\dfrac{4}{21} \div \left(-\dfrac{2}{7}\right)$ b) $\dfrac{72}{-8}$ c) $\dfrac{-19}{-1}$

4. Apply the Order of Operations to Real Numbers

Example 6

Simplify.

a) $-24 \div 12 - 2^2$ b) $-5(-3) - 4(2 - 3)$

Solution

a) $-24 \div 12 - 2^2 = -24 \div 12 - 4$ Simplify exponent first.

$\qquad = -2 - 4$ Perform division before subtraction.

$\qquad = -6$

b) $-5(-3) - 4(2 - 3) = -5(-3) - 4(-1)$ Simplify the difference in parentheses.
$\qquad\qquad\qquad\quad = 15 - (-4)$ Find the products.
$\qquad\qquad\qquad\quad = 15 + 4$
$\qquad\qquad\qquad\quad = 19$

You Try 6

Simplify.

a) $-13 - 4(-5 + 2)$ b) $(-10)^2 + 2[8 - 5(4)]$

5. Translate English Expressions to Mathematical Expressions

Here are some words and phrases you may encounter and how they would translate to mathematical expressions:

Word or Phrase	Operation
Times	Multiplication
Product of	Multiplication
Divided by	Division
Quotient of	Division

Example 7

Write a mathematical expression for each and simplify.
a) The quotient of -56 and 7
b) The product of 4 and the sum of 15 and -6
c) Twice the difference of -10 and -3
d) Half of the sum of -8 and 3

Solution

a) The quotient of -56 and 7:
Quotient means division with -56 in the numerator and 7 in the denominator.
The expression is $\dfrac{-56}{7} = -8$.

b) The product of 4 and the sum of 15 and -6:
The *sum of 15 and -6* means we must add the two numbers. *Product* means multiply.

$$\overset{\text{Sum of 15}}{\underset{\text{Product of 4}}{4(\overbrace{15 + (-6)}}}) = 4(9)$$
$$\underset{\text{and the sum}}{}$$
$$= 36$$

c)   Twice the difference of -10 and -3:
The *difference of -10 and -3* will be in parentheses with -3 being subtracted from -10. *Twice* means "two times."

$$2(-10 - (-3)) = 2(-10 + 3)$$
$$= 2(-7)$$
$$= -14$$

d)   Half of the sum of -8 and 3:
The *sum of -8 and 3* means that we will add the two numbers. They will be in parentheses. *Half of* means multiply by $\dfrac{1}{2}$.

$$\frac{1}{2}(-8 + 3) = \frac{1}{2}(-5)$$
$$= -\frac{5}{2}$$

You Try 7

Write a mathematical expression for each and simplify.

a) 12 less than the product of -7 and 4

b) Twice the sum of 19 and -11

c) The sum of -41 and -23 divided by the square of -2

Answers to You Try Exercises

1) a) -21 b) $-\dfrac{16}{3}$ 2) a) -16 b) $-\dfrac{4}{9}$ c) 24   3) a) 81   b) -125

4) a) -81 b) 121   c) -64 d) $\dfrac{8}{27}$ 5) a) $\dfrac{2}{3}$ b) -9 c) 19   6) a) -1 b) 76

7) a) $(-7) \cdot 4 - 12;\ -40$ b) $2[19 + (-11)];\ 16$ c) $\dfrac{-41 + (-23)}{(-2)^2};\ -16$

Objective 1

Fill in the blank with *positive* or *negative*.

1) The product of two negative numbers is _____.

2) The product of a positive number and a negative number is _____.

3) The quotient of a negative number and a positive number is _____.

4) The quotient of two negative numbers is _____.

Multiply.

5) $-5 \cdot 9$

6) $3 \cdot (-11)$

7) $-14 \cdot (-3)$

8) $-16 \cdot (-31)$

9) $-2 \cdot 5 \cdot (-3)$

10) $-1 \cdot (-6) \cdot (-7)$

11) $\dfrac{7}{9} \cdot \left(-\dfrac{6}{5}\right)$

12) $-\dfrac{15}{32} \cdot \left(-\dfrac{8}{25}\right)$

13) $(-0.25)(1.2)$

14) $(-3.8)(-7.1)$

15) $8 \cdot (-2) \cdot (-4) \cdot (-1)$

16) $-5 \cdot (3) \cdot (-2) \cdot (-1) \cdot (-4)$

17) $(-8) \cdot (-9) \cdot 0 \cdot \left(-\dfrac{1}{4}\right) \cdot (-2)$

18) $(-6) \cdot \left(-\dfrac{2}{3}\right) \cdot 2 \cdot (-5)$

Objective 2

Fill in the blank with *positive* or *negative*.

19) If a is a positive number, then $-a^6$ is _____.

20) If a is a positive number, then $(-a)^6$ is _____.

21) If a is a negative number, then $-a^5$ is _____.

22) Explain the difference between how you would evaluate -3^4 and $(-3)^4$. Then, evaluate each.

Evaluate.

23) $(-6)^2$

24) -6^2

25) -5^3

26) $(-2)^4$

27) $(-3)^2$

28) $(-1)^5$

 29) -7^2

30) -4^3

31) -2^5

32) $(-12)^2$

Objective 3

Divide.

33) $-42 \div (-6)$

34) $-108 \div 9$

35) $\dfrac{56}{-7}$

36) $\dfrac{-32}{-4}$

37) $\dfrac{-3.6}{0.9}$

38) $\dfrac{12}{-0.5}$

 39) $-\dfrac{12}{13} \div \left(-\dfrac{6}{5}\right)$

40) $-14 \div \left(-\dfrac{10}{3}\right)$

41) $\dfrac{0}{-4}$

42) $-\dfrac{0}{9}$

43) $\dfrac{360}{-280}$

44) $\dfrac{-84}{-210}$

Objective 4

Use the order of operations to simplify.

45) $7 + 8(-5)$

46) $-40 \div 2 - 10$

47) $(9 - 14)^2 - (-3)(6)$

48) $-23 - 6^2 \div 4$

 49) $10 - 2(1 - 4)^3 \div 9$

50) $-7(4) + (-8 + 6)^4 + 5$

51) $\left(-\dfrac{3}{4}\right)(8) - 2[7 - (-3)(-6)]$

52) $-2^5 - (-3)(4) + 5[(-9 + 30) \div 7]$

53) $\dfrac{-46 - 3(-12)}{(-5)(-2)(-4)}$

54) $\dfrac{(8)(-6) + 10 - 7}{(-5 + 1)^2 - 12 + 5}$

Objective 5

Write a mathematical expression for each and simplify.

55) The product of -12 and 6

56) The quotient of -80 and -4

57) 9 more than the product of -7 and -5

58) The product of -10 and 2 increased by 11

59) The quotient of 63 and -9 increased by 7

60) 8 more than the quotient of 54 and -6

 61) 19 less than the product of -4 and -8

62) The product of -16 and -3 decreased by 20

63) The quotient of -100 and 4 decreased by the sum of -7 and 2

64) The quotient of -35 and 5 increased by the product of -11 and -2

65) Twice the sum of 18 and -31

66) Twice the difference of -5 and -15

67) Two-thirds of -27

68) Half of -30

69) The product of 12 and -5 increased by half of 36

70) One third of -18 decreased by half the sum of -21 and -5

Section 1.7 Algebraic Expressions and Properties of Real Numbers

Objectives

1. Identify the Terms and Coefficients in an Algebraic Expression

2. Evaluate Algebraic Expressions

3. Use the Commutative Properties

4. Use the Associative Properties

5. Use the Identity and Inverse Properties

6. Use the Distributive Property

1. Identify the Terms and Coefficients in an Algebraic Expression

Here is an algebraic expression:

$$5x^3 - 9x^2 + \frac{1}{4}x + 7$$

x is the *variable*. A **variable** is a symbol, usually a letter, used to represent an unknown number. The *terms* of this algebraic expression are $5x^3$, $-9x^2$, $\frac{1}{4}x$, and 7. A **term** is a number or a variable or a product or quotient of numbers and variables. 7 is the **constant** or **constant term**. The value of a constant does not change. Each term has a **coefficient**.

Term	Coefficient
$5x^3$	5
$-9x^2$	-9
$\dfrac{1}{4}x$	$\dfrac{1}{4}$
7	7

Definition	An **algebraic expression** is a collection of numbers, variables, and grouping symbols connected by operation symbols such as $+$, $-$, $\times$, and $\div$.

Examples of expressions:

$$10k + 9, \qquad 3(2t^2 + t - 4), \qquad 6a^2b^2 - 13ab - 2a + 5$$

Example 1

List the terms and coefficients of

$$4x^2y + 7xy - x + \frac{y}{9} - 12$$

Solution

Term	Coefficient
$4x^2y$	4
$7xy$	7
$-x$	-1
$\dfrac{y}{9}$	$\dfrac{1}{9}$
-12	-12

The minus sign indicates a negative coefficient.

$\dfrac{y}{9}$ can be rewritten as $\dfrac{1}{9}y$.

-12 is also called the "constant."

You Try 1

List the terms and coefficients of $-15r^3 + r^2 - 4r + 8$.

Next, we will use our knowledge of operations with real numbers to evaluate algebraic expressions.

2. Evaluate Algebraic Expressions

We can **evaluate** an algebraic expression by substituting a value for a variable and simplifying. The value of an algebraic expression changes depending on the value that is substituted.

Example 2

Evaluate $5x - 3$ when (a) $x = 4$ and (b) $x = -2$.

Solution

a) $5x - 3$ when $x = 4$ Substitute 4 for x.
     $= 5(4) - 3$ Use parentheses when substituting a value for a variable.
     $= 20 - 3$ Multiply.
     $= 17$ Subtract.

b) $5x - 3$ when $x = -2$ Substitute -2 for x.
     $= 5(-2) - 3$ Use parentheses when substituting a value for a variable.
     $= -10 - 3$ Multiply.
     $= -13$

You Try 2

Evaluate $10x + 7$ when $x = -4$.

Example 3

Evaluate $4c^2 - 2cd + 1$ when $c = -3$ and $d = 5$.

Solution

$4c^2 - 2cd + 1$ when $c = -3$ and $d = 5$ Substitute -3 for c and 5 for d.
$= 4(-3)^2 - 2(-3)(5) + 1$ Use parentheses when substituting.
$= 4(9) - 2(-15) + 1$ Evaluate exponent and multiply.
$= 36 - (-30) + 1$ Multiply.
$= 36 + 30 + 1$
$= 67$

You Try 3

Evaluate $b^2 + 7ab - 4a - 5$ when $a = \dfrac{1}{2}$ and $b = -6$.

Properties of Real Numbers

Like the order of operations, the properties of real numbers guide us in our work with numbers and variables. We begin with the commutative properties of real numbers.

     True or false?

1) $7 + 3 = 3 + 7$ *True*: $7 + 3 = 10$ and
                             $3 + 7 = 10$

2) $8 - 2 = 2 - 8$ *False*: $8 - 2 = 6$ but
                             $2 - 8 = -6$

3) $(-6)(5) = (5)(-6)$ *True*: $(-6)(5) = -30$ and
                               $(5)(-6) = -30$

3. Use the Commutative Properties

In 1) we see that adding 7 and 3 in any order still equals 10. The third equation shows that multiplying $(-6)(5)$ and $(5)(-6)$ each equal -30. But, 2) illustrates that changing the order in which numbers are subtracted does *not* necessarily give the same result: $8 - 2 \neq 2 - 8$. Therefore, subtraction is **not commutative**, while the addition and multiplication of real numbers **is commutative**. This gives us our first property of real numbers:

Commutative Properties

If a and b are real numbers, then

1) $a + b = b + a$ Commutative property of addition

2) $ab = ba$ Commutative property of multiplication

We have already shown that subtraction is not commutative. Is division commutative? No. For example,

$$20 \div 4 \overset{?}{=} 4 \div 20$$
$$5 \neq \frac{1}{5}$$

Example 4

Use the commutative property to rewrite each expression.

a) $12 + 5$ b) $x \cdot 9$

Solution

a) $12 + 5 = 5 + 12$

b) $x \cdot 9 = 9 \cdot x$ or $9x$

You Try 4

Use the commutative property to rewrite each expression.

a) $9 + 7$ b) $y \cdot 8$

4. Use the Associative Properties

Another important property involves the use of grouping symbols. Let's determine if these two statements are true:

$$(9 + 4) + 2 \overset{?}{=} 9 + (4 + 2)$$
$$13 + 2 \overset{?}{=} 9 + 6$$
$$15 = 15$$
TRUE

and

$$(2 \cdot 3)4 \overset{?}{=} 2(3 \cdot 4)$$
$$(6)4 \overset{?}{=} 2(12)$$
$$24 = 24$$
TRUE

We can generalize and say that when adding or multiplying real numbers, the way in which we group them to evaluate them will not affect the result. Notice that the *order* in which the numbers are written does not change.

Associative Properties

If $a, b,$ and c are real numbers, then

1) $(a + b) + c = a + (b + c)$ Associative property of addition

2) $(ab)c = a(bc)$ Associative property of multiplication

Sometimes, applying the associative property can simplify calculations.

Example 5

Apply the associative property to simplify $\left(7 \cdot \dfrac{2}{5} \right) 5$.

Solution

By the associative property,

$$\left(7 \cdot \frac{2}{5} \right) 5 = 7 \cdot \left(\frac{2}{\cancel{5}} \cdot \cancel{5}^{\,1} \right)$$

$$= 7 \cdot 2$$
$$= 14$$

You Try 5

Apply the associative property to simplify $\left(9 \cdot \dfrac{4}{3} \right) 3$.

Example 6

Use the associative property to simplify each expression.

a) $5 + (2 + n)$ b) $\left(-\dfrac{3}{11} \cdot \dfrac{8}{5} \right) \dfrac{5}{8}$

Solution

a) $5 + (2 + n) = (5 + 2) + n$
$$= 7 + n$$

b) $\left(-\dfrac{3}{11} \cdot \dfrac{8}{5} \right) \dfrac{5}{8} = -\dfrac{3}{11} \left(\dfrac{8}{5} \cdot \dfrac{5}{8} \right)$

$$= -\frac{3}{11} (1) \qquad \text{A number times its reciprocal equals 1.}$$

$$= -\frac{3}{11}$$

You Try 6

Use the associative property to simplify each expression.

a) $(k + 3) + 9$ b) $\left(-\dfrac{5}{12} \cdot \dfrac{4}{7}\right)\dfrac{7}{4}$

The identity properties of addition and multiplication are also ones we need to know.

5. Use the Identity and Inverse Properties

For addition we know that, for example,

$$5 + 0 = 5, \qquad 0 + \frac{2}{3} = \frac{2}{3}, \qquad -14 + 0 = -14$$

When zero is added to a number, the value of the number is unchanged. *Zero is the* **identity element for addition** (also called the **additive identity**).

What is the identity element for multiplication?

$$-4(1) = -4, \qquad 1(3.82) = 3.82, \qquad \frac{9}{2}(1) = \frac{9}{2}$$

When a number is multiplied by 1, the value of the number is unchanged. *One is the* **identity element for multiplication** (also called the **multiplicative identity**).

Identity Properties

If a is a real number, then

1) $a + 0 = 0 + a = a$ Identity property of addition

2) $a \cdot 1 = 1 \cdot a = a$ Identity property of multiplication

---

The next properties we will discuss give us the additive and multiplicative identities as results. In Section 1.4, we introduced an **additive inverse**.

Number	Additive Inverse
3	-3
-11	11
$-\dfrac{7}{9}$	$\dfrac{7}{9}$

Let's add each number and its additive inverse:

$$3 + (-3) = 0, \qquad -11 + 11 = 0, \qquad -\frac{7}{9} + \frac{7}{9} = 0$$

> The sum of a number and its additive inverse is zero (the identity element for addition).

Given a number such as $\frac{3}{5}$, we know that its **reciprocal** (or **multiplicative inverse**) is $\frac{5}{3}$. We have also established the fact that the product of a number and its reciprocal is 1 as in

$$\frac{3}{5} \cdot \frac{5}{3} = 1$$

Therefore, multiplying a number b by its reciprocal (multiplicative inverse) $\frac{1}{b}$ gives us the identity element for multiplication, 1. That is,

$$b \cdot \frac{1}{b} = \frac{1}{b} \cdot b = 1$$

Inverse Properties

If a is any real number and b is a real number not equal to 0, then

1) $a + (-a) = -a + a = 0$ Inverse property of addition

2) $b \cdot \dfrac{1}{b} = \dfrac{1}{b} \cdot b = 1$ Inverse property of multiplication

---

Example 7

Which property is illustrated by each statement?

a) $0 + 9 = 9$ b) $-1.3 + 1.3 = 0$ c) $\dfrac{1}{12} \cdot 12 = 1$ d) $7(1) = 7$

Solution

a) $0 + 9 = 9$ Identity property of addition

b) $-1.3 + 1.3 = 0$ Inverse property of addition

c) $\dfrac{1}{12} \cdot 12 = 1$ Inverse property of multiplication

d) $7(1) = 7$ Identity property of multiplication

You Try 7

Which property is illustrated by each statement?

a) $4 \cdot \dfrac{1}{4} = 1$ b) $-8 + 8 = 0$ c) $15(1) = 15$ d) $7 + 0 = 7$

6. Use the Distributive Property

The last property we will discuss is the **distributive property**. It involves both multiplication and addition or multiplication and subtraction.

Distributive Properties

If a, b, and c are real numbers, then

1) $a(b + c) = ab + ac$ and $(b + c)a = ba + ca$

2) $a(b - c) = ab - ac$ and $(b - c)a = ba - ca$

Example 8

Evaluate using the distributive property.

a) $3(2 + 8)$ b) $-6(7 - 8)$ c) $-(6 + 3)$

Solution

a) $\overparen{3(2 + 8)} = 3 \cdot 2 + 3 \cdot 8$ Apply distributive property.
$\qquad\qquad = 6 + 24$
$\qquad\qquad = 30$

Note: We would get the same result if we would apply the order of operations:

$$3(2 + 8) = 3(10)$$
$$= 30$$

b) $\overparen{-6(7 - 8)} = -6 \cdot 7 - (-6)(8)$ Apply distributive property.
$\qquad\qquad\quad = -42 - (-48)$
$\qquad\qquad\quad = -42 + 48$
$\qquad\qquad\quad = 6$

c) $\overparen{-(6 + 3)} = -1(6 + 3)$
$\qquad\qquad\quad = -1 \cdot 6 + (-1)(3)$ Apply distributive property.
$\qquad\qquad\quad = -6 + (-3)$
$\qquad\qquad\quad = -9$

A negative sign in front of parentheses is the same as multiplying by -1.

You Try 8

Evaluate using the distributive property.

a) $2(11 - 5)$ b) $-5(3 - 7)$ c) $-(4 + 9)$

The distributive property can be applied when there are more than two terms in parentheses and when there are variables.

Example 9

Use the distributive property to rewrite each expression. Simplify if possible.

a) $-2(3 + 8 - 5)$ b) $7(x + 4)$ c) $-(-10m + 3n - 8)$

Solution

a) $-2(3 + 8 - 5) = -2 \cdot 3 + (-2)(8) - (-2)(5)$ Apply distributive property.
$= -6 + (-16) - (-10)$ Multiply.
$= -6 + (-16) + 10$
$= -12$

b) $7(x + 4) = 7x + 7 \cdot 4$ Apply distributive property.
$= 7x + 28$

c) $-(-10m + 3n - 8) = -1(-10m + 3n - 8)$

$= -1(-10m) + (-1)(3n) - (-1)(8)$ Apply distributive property.
$= 10m + (-3n) - (-8)$ Multiply.
$= 10m - 3n + 8$

You Try 9

Use the distributive property to rewrite each expression. Simplify if possible.

a) $6(a + 2)$ b) $3(8x - 5y + 11z)$ c) $-(-r + 4s - 9)$

The properties stated previously are summarized next.

Properties of Real Numbers

If a, b, and c are real numbers, then

Commutative Properties: $a + b = b + a$ and $ab = ba$
Associative Properties: $(a + b) + c = a + (b + c)$ and $(ab)c = a(bc)$
Identity Properties: $a + 0 = 0 + a = a$
  $a \cdot 1 = 1 \cdot a = a$
Inverse Properties: $a + (-a) = -a + a = 0$
  $b \cdot \dfrac{1}{b} = \dfrac{1}{b} \cdot b = 1 \ (b \neq 0)$
Distributive Properties: $a(b + c) = ab + ac$ and $(b + c)a = ba + ca$
  $a(b - c) = ab - ac$ and $(b - c)a = ba - ca$

Answers to You Try Exercises

1)

Term	Coeff.
$-15r^3$	-15
r^2	1
$-4r$	-4
8	8

2) -33 3) 8 4) a) $7 + 9$ b) $8y$ 5) 36

6) a) $k + 12$ b) $-\dfrac{5}{12}$

7) a) inverse property of multiplication b) inverse property of addition c) identity property of multiplication d) identity property of addition 8) a) 12 b) 20 c) -13 9) a) $6a + 12$ b) $24x - 15y + 33z$ c) $r - 4s + 9$

1.7 Exercises

Boost your grade at mathzone.com! MathZone

> Practice Problems > NetTutor > e-Professors > Videos
> Self-Test

Objective 1

For each expression, list the terms and their coefficients. Also, identify the constant.

1) $7p^2 - 6p + 4$

2) $-8z + \dfrac{5}{6}$

3) $x^2 y^2 + 2xy - y + 11$

4) $w^3 - w^2 + 9w - 5$

 5) $-2g^5 + \dfrac{g^4}{5} + 3.8g^2 + g - 1$

6) $121c^2 - d^2$

Objective 2

7) Evaluate $4c + 3$ when

  a) $c = 2$

  b) $c = -5$

8) Evaluate $8m - 5$ when

  a) $m = 3$

  b) $m = -1$

9) Evaluate $2j^2 + 3j - 7$ when

  a) $j = 4$

  b) $j = -5$

10) Evaluate $6 - t^3$ when

  a) $t = -3$

  b) $t = 3$

Evaluate each expression when $x = -2$, $y = 7$, and $z = -3$.

11) $8x + y$

12) $10z - 3x$

13) $x^2 + xy + 10$

14) $2xy + 5xz$

15) $z^3 - x^3$

16) $x^2 - 2z^2 + 4xy$

17) $\dfrac{2x}{y + z}$

18) $\dfrac{x - y}{3z}$

 19) $\dfrac{x^2 - y^2}{2z^2 + y}$

20) $\dfrac{5 + 3(y + 4z)}{x^2 - z^2}$

Objectives 3–6

21) What is the identity element for addition?

22) What is the identity element for multiplication?

23) What is the multiplicative inverse of 6?

24) What is the additive inverse of -9?

Which property of real numbers is illustrated by each example? Choose from the commutative, associative, identity, inverse, or distributive property.

25) $(-11 + 4) + 9 = -11 + (4 + 9)$

26) $5 \cdot 7 = 7 \cdot 5$

27) $20 + 8 = 8 + 20$

28) $16 + (-16) = 0$

29) $3(8 \cdot 4) = (3 \cdot 8) \cdot 4$

30) $5(3 + 7) = 5 \cdot 3 + 5 \cdot 7$

31) $(10 + 2)6 = 10 \cdot 6 + 2 \cdot 6$

32) $\dfrac{3}{4} \cdot 1 = \dfrac{3}{4}$

33) $-24 + 0 = -24$

34) $\left(\dfrac{8}{15}\right)\left(\dfrac{15}{8}\right) = 1$

35) $9(a - b) = 9a - 9b$

36) $-3(c + d) = -3c - 3d$

Rewrite each expression using the indicated property.

37) $7(u + v)$; distributive

38) $12 + (3 + 7)$; associative

39) $k + 4$; commutative

40) $-8(c + 5)$; distributive

41) $m + 0$; identity

42) $-4z + 0$; identity

43) $(-5 + 3) + 6$; associative

44) $9 + 11r$; commutative

45) Is $10c - 3$ equivalent to $3 - 10c$? Why or why not?

46) Is $8 + 5n$ equivalent to $5n + 8$? Why or why not?

Rewrite each expression using the distributive property. Simplify if possible.

47) $5(4 + 3)$

48) $8(1 + 5)$

 49) $-2(5 + 7)$

50) $6(9 - 4)$

51) $4(11 - 3)$

52) $-7(2 - 6)$

53) $-(9 - 5)$

54) $-(6 + 1)$

55) $(8 - 2)4$

56) $(-10 + 3)5$

57) $2(-6 + 5 + 3)$

58) $3(8 + 7 - 2)$

59) $9(g + 6)$

60) $10(p + 7)$

61) $4(t - 5)$

62) $6(n - 8)$

63) $-5(z + 3)$

64) $-2(m + 11)$

65) $-8(u - 4)$

66) $-3(h - 9)$

67) $-(v - 6)$

68) $-(y - 13)$

69) $10(m + 5n - 3)$

70) $12(2a - 3b + c)$

71) $-(-8c + 9d - 14)$

72) $-(x - 4y + 10z)$

Definition/Procedure	Example	Reference
Review of Fractions		1.1
Reducing Fractions A fraction is in **lowest terms** when the numerator and denominator have no common factors other than 1.	Write $\dfrac{36}{48}$ in lowest terms. Divide 36 and 48 by a common factor, 12. Since $36 \div 12 = 3$ and $48 \div 12 = 4$, $\dfrac{36}{48} = \dfrac{3}{4}$.	p. 5
Multiplying Fractions To multiply fractions, multiply the numerators and multiply the denominators. Common factors can be divided out either before or after multiplying.	Multiply $\dfrac{21}{45} \cdot \dfrac{9}{14}$. $\dfrac{\overset{3}{\cancel{21}}}{\underset{5}{\cancel{45}}} \cdot \dfrac{\overset{1}{\cancel{9}}}{\underset{2}{\cancel{14}}}$ ← 9 and 45 each divide by 9. ← 21 and 14 each divide by 7. $= \dfrac{3}{5} \cdot \dfrac{1}{2}$ $= \dfrac{3}{10}$ Multiply numerators and multiply denominators.	p. 6
Dividing Fractions To divide fractions, multiply the first fraction by the reciprocal of the second.	Divide $\dfrac{7}{5} \div \dfrac{4}{3}$. $\dfrac{7}{5} \div \dfrac{4}{3} = \dfrac{7}{5} \cdot \dfrac{3}{4} = \dfrac{21}{20}$	p. 8
Adding and Subtracting Fractions To add or subtract fractions, 1) Identify the least common denominator (LCD). 2) Write each fraction as an equivalent fraction using the LCD. 3) Add or subtract. 4) Express the answer in lowest terms.	Add $\dfrac{5}{11} + \dfrac{2}{11}$. $\dfrac{5}{11} + \dfrac{2}{11} = \dfrac{7}{11}$ Subtract $\dfrac{8}{9} - \dfrac{3}{4}$. $\dfrac{8}{9} - \dfrac{3}{4} = \dfrac{32}{36} - \dfrac{27}{36}$ $= \dfrac{5}{36}$	p. 9
Exponents and Order of Operations		1.2
Exponents An **exponent** represents repeated multiplication.	Write $8 \cdot 8 \cdot 8 \cdot 8 \cdot 8$ in exponential form. $8 \cdot 8 \cdot 8 \cdot 8 \cdot 8 = 8^5$ Evaluate 3^4. $3^4 = 3 \cdot 3 \cdot 3 \cdot 3 = 81$	p. 19
Order of Operations **P**arentheses, **E**xponents, **M**ultiplication, **D**ivision, **A**ddition, **S**ubtraction	Evaluate $10 + (2 + 3)^2 - 8 \cdot 4$. $10 + (2 + 3)^2 - 8 \cdot 4$ $= 10 + 5^2 - 8 \cdot 4$ Parentheses $= 10 + 25 - 8 \cdot 4$ Exponents $= 10 + 25 - 32$ Multiply $= 35 - 32$ Add $= 3$ Subtract	p. 21

Definition/Procedure	Example	Reference

Important Angles

The definitions for an acute angle, an obtuse angle, and a right angle can be found on p. 25.

Two angles are **complementary** if the sum of their angles is 90°.

Two angles are **supplementary** if the sum of their angles is 180°.

The measure of an angle is 62°. Find the measure of its complement and its supplement.

The measure of its complement is 28° since $90° - 62° = 28°$.

The measure of its supplement is 118° since $180° - 62° = 118°$.

p. 25

Triangle Properties

The sum of the measures of the angles of any triangle is 180°.

An **equilateral triangle** has three sides of equal length. Each angle measures 60°.

An **isosceles triangle** has two sides of equal length. The angles opposite the sides have the same measure.

A **scalene triangle** has no sides of equal length. No angles have the same measure.

Find the measure of $\angle A$.

$m\angle B + m\angle C = 86° + 71° = 157°$
$m\angle A = 180° - 157° = 23°$

p. 27

Perimeter and Area

The formulas for the perimeter and area of a rectangle, square, triangle, parallelogram, and trapezoid can be found on p. 28.

Find the area and perimeter of this triangle.

$$\text{Area} = \frac{1}{2}(\text{base})(\text{height})$$
$$= \frac{1}{2}(12 \text{ in.})(6 \text{ in.})$$
$$= 36 \text{ in}^2$$
$$\text{Perimeter} = 12 \text{ in.} + 10 \text{ in.} + 7.2 \text{ in.}$$
$$= 29.2 \text{ in.}$$

p. 28

Definition/Procedure	Example	Reference
Volume The formulas for the volume of a rectangular solid, cube, right circular cylinder, sphere, and right circular cone can be found on p. 32.	Find the volume of the cylinder pictured here. 12 cm 4 cm Give an exact answer and give an approximation using 3.14 for π. $V = \pi r^2 h \qquad\qquad V = 192\pi \text{ cm}^3$ $ = \pi(4 \text{ cm})^2(12 \text{ cm}) \quad \approx 192(3.14) \text{ cm}^3$ $ = \pi(16 \text{ cm}^2)(12 \text{ cm}) \quad = 602.88 \text{ cm}^3$ $ = 192\pi \text{ cm}^3$	p. 32
Sets of Numbers and Absolute Value		1.4
Natural numbers: $\{1, 2, 3, 4, \ldots\}$ **Whole numbers:** $\{0, 1, 2, 3, 4, \ldots\}$ **Integers:** $\{\ldots, -3, -2, -1, 0, 1, 2, 3, \ldots\}$ A **rational number** is any number of the form $\dfrac{p}{q}$, where p and q are integers and $q \neq 0$.	The following numbers are rational: $-1, 2, \dfrac{3}{4}, 3.\overline{6}, 4.5$	p. 39
An **irrational number** cannot be written as the quotient of two integers.	The following numbers are irrational: $\sqrt{7}, 5.1948\ldots$	p. 42
The set of **real numbers** includes the rational and irrational numbers.	Any number that can be represented on the number line is a real number.	p. 42
The **additive inverse** of a is $-a$.	The additive inverse of 13 is -13.	p. 45
Absolute Value $|a|$ is the distance of a from zero.	$|-9| = 9$	p. 45
Addition and Subtraction of Real Numbers		1.5
Adding Real Numbers To add numbers with the **same sign**, add the absolute value of each number. The sum will have the same sign as the numbers being added.	$-7 + (-4) = -11$	pp. 49–50
To add two numbers with **different signs**, subtract the smaller absolute value from the larger. The sum will have the sign of the number with the largest absolute value.	$-16 + 10 = -6$	

Definition/Procedure	Example	Reference
Subtracting Real Numbers To subtract $a - b$, change subtraction to addition and add the additive inverse of b: $a - b = a + (-b)$.	$5 - 9 = 5 + (-9) = -4$ $-14 - (-6) = -14 + 6 = -8$ $11 - 4 = 11 + (-4) = 7$	**p. 51**
Multiplication and Division of Real Numbers		**1.6**
Multiplying Real Numbers The product of two real numbers with the **same** sign is positive. The product of a positive number and a negative number is **negative**. An **even number** of negative factors in a product gives a **positive** result. An **odd number** of negative factors in a product gives a **negative** result.	$9 \cdot 4 = 36 \qquad -6 \cdot (-5) = 30$ $-3 \cdot 7 = -21 \qquad 8 \cdot (-1) = -8$ $\underbrace{(-2)(-1)(3)(-4)(-5)}_{\text{4 negative factors}} = 120$ $\underbrace{(4)(-3)(-2)(-1)(3)}_{\text{3 negative factors}} = -72$	**p. 57**
Evaluating Exponential Expressions	Evaluate $(-2)^4$. The base is -2. $\quad (-2)^4 = (-2)(-2)(-2)(-2) = 16$ Evaluate -2^4. The base is 2. $-2^4 = -1 \cdot 2^4 = -1 \cdot 2 \cdot 2 \cdot 2 \cdot 2 = -16$	**p. 59**
Dividing real numbers The quotient of two numbers with the **same** sign is positive. The quotient of two numbers with **different** signs is negative.	$\dfrac{100}{4} = 25 \qquad -63 \div (-7) = 9$ $\dfrac{-20}{4} = -5 \qquad 32 \div (-8) = -4$	**p. 61**
Algebraic Expressions and Properties of Real Numbers		**1.7**
An **algebraic expression** is a collection of numbers, variables, and grouping symbols connected by operation symbols such as $+$, $-$, $\times$, and $\div$.	$4y^2 - 7y + \dfrac{3}{5}$	**p. 66**

Important terms
    Variable   Constant
    Term   Coefficient
 	
We can evaluate expressions for different values of the variables.
	Evaluate $2xy - 5y + 1$ when $x = -3$ and $y = 4$.
	
	Substitute -3 for x and 4 for y and simplify.
	
	$2xy - 5y + 1 = 2(-3)(4) - 5(4) + 1$
	$\qquad\qquad\qquad = -24 - 20 + 1$
	$\qquad\qquad\qquad = -24 + (-20) + 1$
	$\qquad\qquad\qquad = -43$
		p. 67

Definition/Procedure	Example	Reference

Properties of Real Numbers

If a, b, and c are real numbers, then the following properties hold.

p. 68

Commutative Properties:

$a + b = b + a$

$\quad ab = ba$

$9 + 2 = 2 + 9$

$(-4)(7) = (7)(-4)$

Associative Properties:

$(a + b) + c = a + (b + c)$

$\quad (ab)c = a(bc)$

$(4 + 1) + 7 = 4 + (1 + 7)$

$(2 \cdot 3)10 = 2(3 \cdot 10)$

Identity Properties:

$a + 0 = 0 + a = a$

$a \cdot 1 = 1 \cdot a = a$

$\dfrac{3}{4} + 0 = \dfrac{3}{4} \qquad 5 \cdot 1 = 5$

Inverse Properties:

$a + (-a) = -a + a = 0$

$b \cdot \dfrac{1}{b} = \dfrac{1}{b} \cdot b = 1$

$6 + (-6) = 0 \qquad 8 \cdot \dfrac{1}{8} = 1$

Distributive Properties:

$a(b + c) = ab + ac$ and $(b + c)a = ba + ca$

$a(b - c) = ab - ac$ and $(b - c)a = ba - ca$

$$9(3 + 4) = 9 \cdot 3 + 9 \cdot 4$$
$$= 27 + 36$$
$$= 63$$

$$4(n - 7) = 4n - 4 \cdot 7$$
$$= 4n - 28$$

(1.1) Find all factors of each number.

1) 16

2) 37

Find the prime factorization of each number.

3) 28

4) 66

Write each fraction in lowest terms.

5) $\dfrac{12}{30}$

6) $\dfrac{414}{702}$

Perform the indicated operation. Write the answer in lowest terms.

7) $\dfrac{4}{11} \cdot \dfrac{3}{5}$

8) $\dfrac{45}{64} \cdot \dfrac{32}{75}$

9) $\dfrac{5}{8} \div \dfrac{3}{10}$

10) $35 \div \dfrac{7}{8}$

11) $4\dfrac{2}{3} \cdot 1\dfrac{1}{8}$

12) $\dfrac{30}{49} \div 2\dfrac{6}{7}$

13) $\dfrac{2}{9} + \dfrac{4}{9}$

14) $\dfrac{2}{3} + \dfrac{1}{4}$

15) $\dfrac{9}{40} + \dfrac{7}{16}$

16) $\dfrac{1}{5} + \dfrac{1}{3} + \dfrac{1}{6}$

17) $\dfrac{21}{25} - \dfrac{11}{25}$

18) $\dfrac{5}{8} - \dfrac{2}{7}$

19) $3\dfrac{2}{9} + 5\dfrac{3}{8}$

20) $9\dfrac{3}{8} - 2\dfrac{5}{6}$

21) A pattern for a skirt calls for $1\dfrac{7}{8}$ yd of fabric. If Mary Kate wants to make one skirt for herself and one for her twin, how much fabric will she need?

(1.2) Evaluate.

22) 3^4

23) 2^6

24) $\left(\dfrac{3}{4}\right)^3$

25) $(0.6)^2$

26) $13 - 7 + 4$

27) $8 \cdot 3 + 20 \div 4$

28) $\dfrac{12 - 56 \div 8}{(1 + 5)^2 - 2^4}$

(1.3)

29) The complement of 51° is _____.

30) The supplement of 78° is _____.

31) Is this triangle acute, obtuse, or right? Find the missing angle.

Find the area and perimeter of each figure. Include the correct units.

32)

6.5 mi

6.5 mi

33)

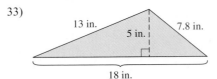

13 in.   7.8 in.

5 in.

18 in.

34)

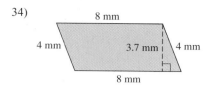

35)

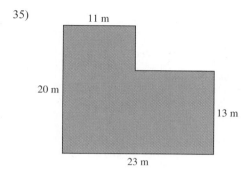

Find the a) area and b) circumference of each circle. Give an exact answer for each and give an approximation using 3.14 for π. Include the correct units.

36)

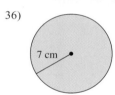

37)

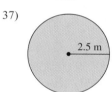

Find the area of the shaded region. Use 3.14 for π. Include the correct units.

38)

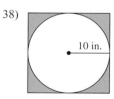

Find the volume of each figure. Where appropriate, give the answer in terms of π. Include the correct units.

39)

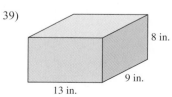

40)

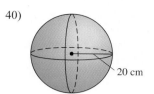

41)

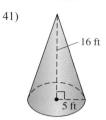

42)

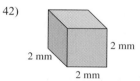

43) Find the volume of a hockey puck given that it is one inch thick and three inches in diameter. (Hint: A hockey puck is a right circular cylinder.)

(1.4)

44) Given the set of numbers,

$$\left\{ \sqrt{23},\ -6,\ 14.38,\ \frac{3}{11},\ 2,\ 5.\overline{7},\ 0,\ 9.21743819... \right\},$$

list the

a) whole numbers

b) natural numbers

c) integers

d) rational numbers

e) irrational numbers

45) Graph and label these numbers on a number line.

$$-2,\ 5\frac{1}{3},\ 0.8,\ -4.5,\ 3,\ -\frac{3}{4}$$

46) Evaluate $|-10|$.

(1.5) Add or subtract as indicated.

47) $-18 + 4$

48) $60 - (-15)$

49) $-\dfrac{5}{8} + \left(-\dfrac{2}{3}\right)$

50) $0.8 - 5.9$

51) The lowest temperature on record in the state of Wyoming is $-66°F$. Georgia's record low is $49°$ higher than Wyoming's. What is the lowest temperature ever recorded in Georgia? (www.infoplease.com)

(1.6) Multiply or divide as indicated.

52) $(-10)(-7)$

53) $\left(-\dfrac{2}{3}\right)(15)$

54) $(3.7)(-2.1)$

55) $(-3)(-5)(-2)$

56) $(-1)(6)(-4)\left(-\dfrac{1}{2}\right)(-5)$

57) $-54 \div 6$

58) $\dfrac{-24}{-12}$

59) $\dfrac{38}{-44}$

60) $-\dfrac{20}{27} \div \dfrac{8}{15}$

61) $-\dfrac{8}{9} \div (-4)$

Evaluate.

62) -5^2

63) $(-5)^2$

64) $(-3)^4$

65) $(-1)^9$

66) -2^6

Use the order of operations to simplify.

67) $64 \div (-8) + 6$

68) $15 - (3 - 7)^3$

69) $-11 - 3 \cdot 9 + (-2)^1$

70) $\dfrac{6 - 2(5 - 1)}{(-3)(-4) + 7 - 3}$

Write a mathematical expression for each and simplify.

71) The quotient of -120 and -3

72) Twice the sum of 22 and -10

73) 15 less than the product of -4 and 7

74) 11 more than half of -18

(1.7)

75) List the terms and coefficients of $c^4 + 12c^3 - c^2 - 3.8c + 11$.

76) Evaluate $-3m + 7n$ when $m = 6$ and $n = -2$.

77) Evaluate $\dfrac{t - 6s}{s^2 - t^2}$ when $s = -4$ and $t = 5$.

Which property of real numbers is illustrated by each example? Choose from the commutative, associative, identity, inverse, or distributive property.

78) $0 + 12 = 12$

79) $(8 + 1) + 5 = 8 + (1 + 5)$

80) $\left(\dfrac{4}{7}\right)\left(\dfrac{7}{4}\right) = 1$

81) $35 + 16 = 16 + 35$

82) $-6(3 + 8) = (-6)(3) + (-6)(8)$

Rewrite each expression using the distributive property. Simplify if possible.

83) $3(10 - 6)$

84) $(3 + 9)2$

85) $-(12 + 5)$

86) $-7(2c - d + 4)$

1) Find the prime factorization of 210.

2) Write in lowest terms:

   a) $\dfrac{45}{72}$

   b) $\dfrac{420}{560}$

Perform the indicated operation(s). Write all answers in lowest terms.

3) $\dfrac{9}{14} \cdot \dfrac{7}{24}$

4) $\dfrac{1}{3} + \dfrac{4}{15}$

5) $5\dfrac{1}{4} - 2\dfrac{1}{6}$

6) $\dfrac{12}{13} \div 6$

7) $\dfrac{4}{7} - \dfrac{5}{6}$

8) $-11 - (-19)$

9) $25 + 15 \div 5$

10) $\dfrac{9}{10} \cdot \left(-\dfrac{2}{5}\right)$

11) $-8 \cdot (-6)$

12) $-13.4 + 6.9$

13) $30 - 5[-10 + (2 - 6)^2]$

14) $\dfrac{(50 - 26) \div 3}{3 \cdot 5 - 7}$

15) Both the highest and lowest points in the continental United States are in California. Badwater, Death Valley, is 282 ft below sea level while, 76 mi away, Mount Whitney reaches an elevation of 14,505 ft. What is the difference between these two elevations?

Evaluate.

16) 2^5

17) -3^4

18) $|-92|$

19) $|2 - 12| - 4|6 - 1|$

20) The complement of 12° is _____.

21) Find the missing angle, and classify the triangle as acute, obtuse, or right.

22) Find the area of this triangle. Include the correct units.

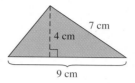

23) Find the area and perimeter of each figure. Include the correct units.

   a)

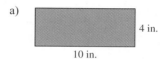

   b)

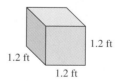

24) Find the volume of this cube:

25) The radius of a baseball is approximately 1.5 in. Find the exact value of each in terms of π. Include the correct units.

   a) Find its circumference.

   b) Find the volume of the baseball.

26) Given the set of numbers,
   $$\left\{ 41, -8, 0, 2.\overline{83}, \sqrt{75}, 6.5, 4\dfrac{5}{8}, 6.37528861\ldots \right\},$$
   list the

   a) integers

   b) irrational numbers

   c) natural numbers

d) rational numbers

e) whole numbers

27) Graph the numbers on a number line. Label each.

$6, \dfrac{7}{8}, -4, -1.2, 4\dfrac{3}{4}, -\dfrac{2}{3}$

28) Write a mathematical expression for each and simplify.

a) The sum of -4 and 27

b) The product of 5 and -6 subtracted from 17

29) List the terms and coefficients of

$5a^2b^2 + 2a^2b + \dfrac{8}{9}ab + a - \dfrac{b}{7} - 14$

30) Evaluate $9g^2 + 3g - 6$ when $g = -1$.

31) Which property of real numbers is illustrated by each example? Choose from the commutative, associative, identity, inverse, or distributive property.

a) $9(5 - 7) = 9 \cdot 5 - 9 \cdot 7$

b) $-6 + 6 = 0$

c) $8 \cdot 3 = 3 \cdot 8$

32) Rewrite each expression using the distributive property. Simplify if possible.

a) $-2(5 + 3)$

b) $5(t + 9u + 1)$

Variables and Exponents

Algebra at Work: Custom Motorcycle Shop

The people who build custom motorcycles use a lot of mathematics to do their jobs. Mark is building a chopper frame and needs to make the supports for the axle. He has to punch holes in the plates that will be welded to the frame.

Mark has to punch holes with a diameter of 1 in. in mild steel that is $\frac{3}{8}$-in. thick. The press punches two holes at a time. He must determine how much power is needed to do this job, so he uses a formula containing an exponent, 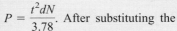 $P = \dfrac{t^2 dN}{3.78}$. After substituting the numbers into the expression, he calculates that the power needed to punch these holes is 0.07 hp.

In this chapter, we will learn more about working with expressions containing exponents.

Section 2.1 Simplifying Expressions

Objectives

1. Identify Like Terms
2. Combine Like Terms
3. Translate English Expressions to Mathematical Expressions

What do we mean when we say, "Simplify this expression"? Consider something we already know:

$$\text{Simplify } 15 + 11 - 8 + 3.$$

We use the order of operations and properties of arithmetic to combine terms:

$$
\begin{aligned}
15 + 11 - 8 + 3 &= 26 - 8 + 3 \\
&= 18 + 3 \\
&= 21
\end{aligned}
$$

But, what if we were asked to "Simplify this expression"?

$$15a + 11a - 8a + 3a$$

How would we combine terms? This is the type of problem we will discuss in this section.

1. Identify Like Terms

In the expression $15a + 11a - 8a + 3a$ there are four **terms**: $15a$, $11a$, $-8a$, $3a$. In fact, they are **like terms**. *Like terms contain the same variables with the same exponents.*

Example 1

Determine if the following groups of terms are like terms.

a) $4y^2, -9y^2, \dfrac{2}{3}y^2$ b) $-5x^6, 0.8x^9, 3x^4$

c) $6a^2b^3, a^2b^3, -\dfrac{5}{8}a^2b^3$ d) $9c, 4d$

Solution

a) $4y^2, -9y^2, \dfrac{2}{3}y^2$

Yes. Each contains the variable y with an exponent of 2. They are y^2-terms.

b) $-5x^6, 0.8x^9, 3x^4$

No. Although each contains the variable x, the exponents are not the same.

c) $6a^2b^3, a^2b^3, -\dfrac{5}{8}a^2b^3$

Yes. Each contains a^2 and b^3.

d) $9c, 4d$

No. The terms contain different variables.

 **You Try 1**

Determine if the following groups of terms are like terms.

a) $2k^2, -9k^2, \frac{1}{5}k^2$ b) $-xy^2, 8xy^2, 7xy^2$ c) $3r^3s^2, -10r^2s^3$

2. Combine Like Terms

Let's return to our problem, "Simplify this expression."

$$15a + 11a - 8a + 3a$$

To simplify, we combine like terms using the distributive property.

$$
\begin{aligned}
15a + 11a - 8a + 3a &= (15 + 11 - 8 + 3)a && \text{Distributive property} \\
&= (26 - 8 + 3)a && \text{Order of operations} \\
&= (18 + 3)a && \text{Order of operations} \\
&= 21a
\end{aligned}
$$

Remember, we can add and subtract only those terms that are like terms.

Example 2

Combine like terms.

a) $-9k + 2k$ b) $n + 8 - 4n + 3$ c) $\frac{3}{5}t^2 + \frac{1}{4}t^2$

d) $10x^2 + 6x - 2x^2 + 5x$

Solution

a) We can use the distributive property to combine like terms.

$$-9k + 2k = (-9 + 2)k = -7k$$

Notice that using the distributive property to combine like terms is the same as combining the coefficients of the terms and leaving the variable and its exponent the same.

b) $n + 8 - 4n + 3 = n - 4n + 8 + 3$ Rewrite like terms together.
$\qquad\qquad\qquad\quad = -3n + 11$ Remember, n is the same as $1n$.

c) $\frac{3}{5}t^2 + \frac{1}{4}t^2 = \frac{12}{20}t^2 + \frac{5}{20}t^2$ Get a common denominator.

$\qquad\qquad = \frac{17}{20}t^2$

d) $10x^2 + 6x - 2x^2 + 5x = 10x^2 - 2x^2 + 6x + 5x$ Rewrite like terms together.
$\qquad\qquad\qquad\qquad\quad = 8x^2 + 11x$

$8x^2 + 11x$ cannot be simplified more because the terms are *not* like terms.

You Try 2

Combine like terms.

a) $6z + 5z$ b) $q - 9 - 4q + 11$ c) $\dfrac{5}{6}c^2 - \dfrac{2}{3}c^2$ d) $2y^2 + 8y + y^2 - 3y$

If an expression contains parentheses, we use the distributive property to clear the parentheses, and then combine like terms.

Example 3

Combine like terms.

a) $5(2c + 3) - 3c + 4$ b) $3(2n + 1) - (6n - 11)$

c) $\dfrac{3}{8}(8 - 4p) + \dfrac{5}{6}(2p - 6)$

Solution

a) $5(2c + 3) - 3c + 4 = 10c + 15 - 3c + 4$ Distributive property
$= 10c - 3c + 15 + 4$ Rewrite like terms together.
$= 7c + 19$

b) $3(2n + 1) - (6n - 11) = 3(2n + 1) - 1(6n - 11)$ Remember, $-(6n - 11)$ is the same as $-1(6n - 11)$.

$= 6n + 3 - 6n + 11$ Distributive property
$= 6n - 6n + 3 + 11$ Rewrite like terms together.
$= 0n + 14$ $0n = 0$
$= 14$

c) $\dfrac{3}{8}(8 - 4p) + \dfrac{5}{6}(2p - 6) = \dfrac{3}{8}(8) - \dfrac{3}{8}(4p) + \dfrac{5}{6}(2p) - \dfrac{5}{6}(6)$ Distributive property

$= 3 - \dfrac{3}{2}p + \dfrac{5}{3}p - 5$ Multiply.

$= -\dfrac{3}{2}p + \dfrac{5}{3}p + 3 - 5$ Rewrite like terms together.

$= -\dfrac{9}{6}p + \dfrac{10}{6}p + 3 - 5$ Get a common denominator.

$= \dfrac{1}{6}p - 2$ Combine like terms.

You Try 3

Combine like terms.

a) $9d^2 - 7 + 2d^2 + 3$ b) $10 - 3(2k + 5) + k - 6$

3. Translate English Expressions to Mathematical Expressions

Translating from English to a mathematical expression is a skill that is necessary to solve applied problems. We will practice writing mathematical expressions.

Read the phrase carefully, choose a variable to represent the unknown quantity, then translate the phrase to a mathematical expression.

Example 4

Write a mathematical expression for each and simplify. Define the unknown with a variable.

a)   Seven more than twice a number

b)   The sum of a number and four times the same number

Solution

a)   Seven more than twice a number

   i) **Define the unknown.** This means that you should clearly state on your paper what the variable represents.

$$\text{Let } x = \text{the number.}$$

   ii) **Slowly, break down the phrase.** How do you write an expression for "seven more than" something?

$$7+$$

   iii) **What does "twice a number" mean?** It means two times the number. Since our number is represented by x, "twice a number" is $2x$.

   iv) **Put the information together:**

$$\text{Seven more than twice a number}$$
$$7 \qquad + \qquad 2x$$

   The expression is $7 + 2x$ or $2x + 7$.

b)   The sum of a number and four times the same number

   i) **Define the unknown.**

$$\text{Let } y = \text{the number.}$$

   ii) **Slowly, break down the phrase.** What does *sum* mean? **Add.** So, we have to add a number and four times the same number:

$$\text{number} + 4(\text{number})$$

   iii) Since y represents the number, *four times the number* is $4y$.

   iv) Therefore, to translate from English to a mathematical expression we know that we must add the number, y, to four times the number, $4y$. Our expression is $y + 4y$. It simplifies to $5y$.

You Try 4

Write a mathematical expression for each and simplify. Let x equal the unknown number.

a) Five less than twice a number

b) The sum of a number and two times the same number

Answers to You Try Exercises

1) a) yes   b) yes   c) no   2) a) $11z$ b) $-3q + 2$ c) $\frac{1}{6}c^2$ d) $3y^2 + 5y$

3) a) $11d^2 - 4$ b) $-5k - 11$ 4) a) $2x - 5$ b) $x + 2x; 3x$

2.1 Exercises

Boost your grade at mathzone.com! MathZone

> Practice Problems   > NetTutor   > e-Professors   > Videos
> Self-Test

Objective 1

1) Are $16t^2$ and $-3t$ "like" terms? Why or why not?

2) Are $-\frac{1}{2}a^3$ and $6a^3$ "like" terms? Why or why not?

3) Are x^4y^3 and $5x^4y^3$ "like" terms? Why or why not?

4) Are $8c$ and $-13d$ "like" terms? Why or why not?

Objective 2

Combine like terms and simplify.

5) $10p + 9 + 14p - 2$

6) $11 - k^2 + 12k^2 - 3 + 6k^2$

7) $-18y^2 - 2y^2 + 19 + y^2 - 2 + 13$

8) $-7x - 3x - 1 + 9x + 6 - 2x$

9) $\frac{4}{9} + 3r - \frac{2}{3} + \frac{1}{5}r$

10) $6a - \frac{3}{8}a + 2 + \frac{1}{4} - \frac{3}{4}a$

11) $2(3w + 5) + w$

12) $-8d^2 + 6(d^2 - 3) + 7$

13) $9 - 4(3 - x) - 4x + 3$

14) $m + 11 + 3(2m - 5) + 1$

15) $3g - (8g + 3) + 5$

16) $-6 + 4(10b - 11) - 8(5b + 2)$

17) $-5(t - 2) - (10 - 2t)$

18) $11 + 8(3u - 4) - 2(u + 6) + 9$

19) $3[2(5x + 7) - 11] + 4(7 - x)$

20) $22 - [6 + 5(2w - 3)] - (7w + 16)$

21) $\frac{4}{5}(2z + 10) - \frac{1}{2}(z + 3)$

22) $\frac{2}{3}(6c - 7) + \frac{5}{12}(2c + 5)$

23) $1 + \frac{3}{4}(10t - 3) + \frac{5}{8}\left(t + \frac{1}{10}\right)$

24) $\frac{7}{15} - \frac{9}{10}(2y + 1) - \frac{2}{5}(4y - 3)$

25) $2.5(x - 4) - 1.2(3x + 8)$

26) $9.4 - 3.8(2a + 5) + 0.6 + 1.9a$

Objective 3

Write a mathematical expression for each phrase, and combine like terms if possible. Let x represent the unknown quantity.

27) Eighteen more than a number

28) Eleven more than a number

29) Six subtracted from a number

30) Eight subtracted from a number

31) Three less than a number

32) Fourteen less than a number

33) The sum of twelve and twice a number

34) Five added to the sum of a number and six

35) Seven less than the sum of three and twice a number

36) Two more than the sum of a number and nine

37) The sum of a number and fifteen decreased by five

38) The sum of -8 and twice a number increased by three

Section 2.2a The Product Rule and Power Rules of Exponents

Objectives

1. Identify the Base and Exponent in an Exponential Expression and Evaluate the Expression

2. Use the Product Rule for Exponents

3. Use the Power Rule $(a^m)^n = a^{mn}$

4. Use the Power Rule $(ab)^n = a^n b^n$

5. Use the Power Rule $\left(\dfrac{a}{b}\right)^n = \dfrac{a^n}{b^n}$, Where $b \neq 0$

1. Identify the Base and Exponent in an Exponential Expression and Evaluate the Expression

Recall from Chapter 1 that exponential notation is used as a shorthand way to represent a multiplication problem.

For example, $3 \cdot 3 \cdot 3 \cdot 3 \cdot 3$ can be written as 3^5.

> **Definition**
>
> An **exponential expression** of the form a^n is $a^n = \underbrace{a \cdot a \cdot a \cdot \cdots \cdot a}_{n \text{ factors of } a}$, where a is any real number and n is a positive integer. a is the **base**, and n is the **exponent**.

We can also evaluate an exponential expression.

Example 1

Identify the base and the exponent in each expression and evaluate.

a) 2^4 b) $(-2)^4$ c) -2^4

Solution

a) 2^4 *2* is the base, *4* is the exponent. Therefore, $2^4 = 2 \cdot 2 \cdot 2 \cdot 2 = 16$.

b) $(-2)^4$ -2 is the base, *4* is the exponent. Therefore, $(-2)^4 = (-2) \cdot (-2) \cdot (-2) \cdot (-2) = 16$.

c) -2^4 It may be very tempting to say that the base is -2. However, there are no parentheses in this expression. Therefore, *2* is the base, and *4* is the exponent. To evaluate,

$$-2^4 = -1 \cdot 2^4 = -1 \cdot 2 \cdot 2 \cdot 2 \cdot 2$$
$$= -16$$

> **BE CAREFUL**
>
> Expressions of the form $(-a)^n$ and $-a^n$ are not always equal!

$$(-a)^n = \underbrace{(-a) \cdot (-a) \cdot (-a) \cdot \cdots \cdot (-a)}_{n \text{ factors of } -a}$$

$$-a^n = -1 \cdot \underbrace{a \cdot a \cdot a \cdot \cdots \cdot a}_{n \text{ factors of } a}$$

You Try 1

Identify the base and exponent in each expression and evaluate.

a) 5^3 b) -8^2 c) $\left(-\dfrac{2}{3}\right)^3$

2. Use the Product Rule for Exponents

Are there any rules we can use to *multiply* exponential expressions? For example, let's rewrite each product as a single power of the base using what we already know:

1) $2^3 \cdot 2^2 = \overbrace{2 \cdot 2 \cdot 2}^{\substack{3 \text{ factors} \\ \text{of } 2}} \cdot \overbrace{2 \cdot 2}^{\substack{2 \\ \text{factors} \\ \text{of } 2}}$
   $= \underbrace{2 \cdot 2 \cdot 2 \cdot 2 \cdot 2}_{5 \text{ factors of } 2}$
   $= 2^5$

2) $5^4 \cdot 5^3 = \overbrace{5 \cdot 5 \cdot 5 \cdot 5}^{\substack{4 \text{ factors} \\ \text{of } 5}} \cdot \overbrace{5 \cdot 5 \cdot 5}^{\substack{3 \\ \text{factors} \\ \text{of } 5}}$
   $= \underbrace{5 \cdot 5 \cdot 5 \cdot 5 \cdot 5 \cdot 5 \cdot 5}_{7 \text{ factors of } 5}$
   $= 5^7$

Let's summarize: $2^3 \cdot 2^2 = 2^5$, $5^4 \cdot 5^3 = 5^7$

Do you notice a pattern? *When you multiply expressions with the same base, keep the same base and add the exponents.* This is called the **product rule** for exponents.

> **Product Rule:** Let a be any real number and let m and n be positive integers. Then,
>
> $$a^m \cdot a^n = a^{m+n}$$

Example 2

Find each product.

a) $2^2 \cdot 2^4$ b) $x^9 \cdot x^6$ c) $5c^3 \cdot 7c^9$ d) $k^8 \cdot k \cdot k^{11}$

Solution

a) $2^2 \cdot 2^4 = 2^{2+4} = 2^6 = 64$ Since the bases are the same, add the exponents.

b) $x^9 \cdot x^6 = x^{9+6} = x^{15}$

c) $5c^3 \cdot 7c^9 = (5 \cdot 7)(c^3 \cdot c^9)$ Associative and commutative properties
   $= 35c^{12}$

d) $k^8 \cdot k \cdot k^{11} = k^{8+1+11} = k^{20}$

You Try 2

Find each product.

a) $3 \cdot 3^2$ b) $y^{10} \cdot y^4$ c) $-6m^5 \cdot 9m^{11}$ d) $h^4 \cdot h^6 \cdot h$

BE CAREFUL

Can the product rule be applied to $4^3 \cdot 5^2$? **No!** The bases are not the same, so we cannot simply add the exponents. To evaluate $4^3 \cdot 5^2$, we would evaluate $4^3 = 64$ and $5^2 = 25$, then multiply:

$$4^3 \cdot 5^2 = 64 \cdot 25 = 1600$$

3. Use the Power Rule $(a^m)^n = a^{mn}$

What does $(2^2)^3$ mean?

We can rewrite $(2^2)^3$ first as $2^2 \cdot 2^2 \cdot 2^2$. Then, using the product rule for exponents,

$$2^2 \cdot 2^2 \cdot 2^2 = 2^{2+2+2} \qquad \text{The bases are the same. Add the exponents.}$$
$$= 2^6$$
$$= 64$$

Notice that $(2^2)^3 = 2^{2+2+2}$. This is the same as $(2^2)^3 = 2^{2 \cdot 3}$. To simplify, the exponents can be multiplied. This leads us to the basic power rule for exponents.

> **Basic Power Rule:** Let a be any real number and let m and n be positive integers. Then,
>
> $$(a^m)^n = a^{mn}$$

Example 3

Simplify using the power rule.

a) $(3^8)^4$ b) $(n^3)^7$

Solution

a) $(3^8)^4 = 3^{8 \cdot 4} = 3^{32}$ b) $(n^3)^7 = n^{3 \cdot 7} = n^{21}$

You Try 3

Simplify using the power rule.   a) $(5^4)^3$ b) $(j^6)^5$

We can use another power rule to allow us to simplify an expression such as $(5c^2)^3$.

4. Use the Power Rule $(ab)^n = a^n b^n$

Power Rule for a Product: Let a and b be real numbers and let n be a positive integer.

$$(ab)^n = a^n b^n$$

 BE CAREFUL $(ab)^n = a^n b^n$ is different from $(a + b)^n$. $(a + b)^n \neq a^n + b^n$. We will study this in Chapter 6.

Example 4

Simplify each expression.

a) $(9y)^2$ b) $\left(\dfrac{1}{4}t\right)^3$ c) $(5c^2)^3$ d) $3(6ab)^2$

Solution

a) $(9y)^2 = 9^2 y^2 = 81y^2$

b) $\left(\dfrac{1}{4}t\right)^3 = \left(\dfrac{1}{4}\right)^3 \cdot t^3 = \dfrac{1}{64}t^3$

c) $(5c^2)^3 = 5^3 \cdot (c^2)^3 = 125c^{2 \cdot 3} = 125c^6$

d) $3(6ab)^2 = 3[6^2 \cdot (a)^2 \cdot (b)^2]$ The 3 is not in parentheses, therefore it will not be squared.

$\qquad\qquad = 3(36a^2 b^2)$
$\qquad\qquad = 108a^2 b^2$

 You Try 4

Simplify.

a) $(k^4)^7$ b) $(2k^{10}m^3)^6$ c) $(-r^2 s^8)^3$ d) $-4(3tu)^2$

5. Use the Power Rule $\left(\dfrac{a}{b}\right)^n = \dfrac{a^n}{b^n}$, Where $b \neq 0$

Power Rule for a Quotient: Let a and b be real numbers and let n be a positive integer. Then,

$$\left(\dfrac{a}{b}\right)^n = \dfrac{a^n}{b^n}, \text{ where } b \neq 0$$

Example 5

Simplify using the power rule for quotients.

a) $\left(\dfrac{3}{8}\right)^2$ b) $\left(\dfrac{5}{x}\right)^3$ c) $\left(\dfrac{t}{u}\right)^9$

Solution

a) $\left(\dfrac{3}{8}\right)^2 = \dfrac{3^2}{8^2} = \dfrac{9}{64}$

b) $\left(\dfrac{5}{x}\right)^3 = \dfrac{5^3}{x^3} = \dfrac{125}{x^3}$

c) $\left(\dfrac{t}{u}\right)^9 = \dfrac{t^9}{u^9}$

You Try 5

Simplify using the power rule for quotients.

a) $\left(\dfrac{5}{12}\right)^2$ b) $\left(\dfrac{2}{d}\right)^5$ c) $\left(\dfrac{u}{v}\right)^6$

The rules of exponents we have learned in this section can be summarized like this:

The Rules of Exponents

The **product rule** says that $a^m \cdot a^n = a^{m+n}$, where m and n are positive integers and a is a real number.

$$\text{Example: } p^4 \cdot p^{11} = p^{4+11} = p^{15}$$

There are three **power rules**. When a and b are real numbers and m and n are positive integers,

1) Basic Power Rule: $(a^m)^n = a^{mn}$

$$\text{Example: } (c^8)^3 = c^{8 \cdot 3} = c^{24}$$

2) Power Rule for a Product: $(ab)^n = a^n b^n$

$$\text{Example: } (3z)^4 = 3^4 \cdot z^4 = 81z^4$$

3) Power Rule for a Quotient: $\left(\dfrac{a}{b}\right)^n = \dfrac{a^n}{b^n}\ (b \neq 0)$

$$\text{Example: } \left(\dfrac{w}{2}\right)^4 = \dfrac{w^4}{2^4} = \dfrac{w^4}{16}$$

Answers to You Try Exercises

1) a) base: 5; exponent: 3; $5^3 = 125$ b) base: 8; exponent: 2; $-8^2 = -64$

c) base: $-\dfrac{2}{3}$; exponent: 3; $\left(-\dfrac{2}{3}\right)^3 = -\dfrac{8}{27}$ 2) a) 27   b) y^{14} c) $-54m^{16}$ d) h^{11}

3) a) 5^{12} b) j^{30} 4) a) k^{28} b) $64k^{60}m^{18}$ c) $-r^6s^{24}$ d) $-36t^2u^2$

5) a) $\dfrac{25}{144}$ b) $\dfrac{32}{d^5}$ c) $\dfrac{u^6}{v^6}$

2.2a Exercises

Boost your grade at mathzone.com!   MathZone   > Practice Problems   > NetTutor   > e-Professors   > Videos
                                                      > Self-Test

Objective 1

Rewrite each expression using exponents.

1) $9 \cdot 9 \cdot 9 \cdot 9$

2) $4 \cdot 4 \cdot 4 \cdot 4 \cdot 4 \cdot 4 \cdot 4$

3) $\left(\dfrac{1}{7}\right)\left(\dfrac{1}{7}\right)\left(\dfrac{1}{7}\right)\left(\dfrac{1}{7}\right)\left(\dfrac{1}{7}\right)$

4) $(0.8)(0.8)(0.8)$

5) $(-5)(-5)(-5)(-5)(-5)(-5)(-5)$

6) $(-c)(-c)(-c)(-c)(-c)$

7) $(-3y)(-3y)(-3y)(-3y)(-3y)(-3y)(-3y)(-3y)$

8) $\left(-\dfrac{5}{4}t\right)\left(-\dfrac{5}{4}t\right)\left(-\dfrac{5}{4}t\right)\left(-\dfrac{5}{4}t\right)$

Identify the base and the exponent in each.

9) 6^8

10) 9^4

11) $(0.05)^7$

12) $(0.3)^{10}$

13) $(-8)^5$

14) $(-7)^6$

15) $(9x)^8$

16) $(13k)^3$

17) $(-11a)^2$

18) $(-2w)^9$

19) $5p^4$

20) $-3m^5$

21) $-\dfrac{3}{8}y^2$

22) $\dfrac{5}{9}t^7$

 23) Evaluate $(3 + 4)^2$ and $3^2 + 4^2$. Are they equivalent? Why or why not?

24) Evaluate $(7 - 3)^2$ and $7^2 - 3^2$. Are they equivalent? Why or why not?

25) For any values of a and b, does $(a + b)^2 = a^2 + b^2$? Why or why not?

26) Does $-2^4 = (-2)^4$? Why or why not?

 27) Are $3t^4$ and $(3t)^4$ equivalent? Why or why not?

28) Is there any value of a for which $(-a)^2 = -a^2$? Support your answer with an example.

Evaluate.

29) 2^5

30) 9^2

31) $(11)^2$

32) 4^3

33) $(-2)^4$

34) $(-5)^3$

35) -3^4

36) -6^2

37) -2^3

38) -8^2

39) $\left(\dfrac{1}{5}\right)^3$

40) $\left(\dfrac{3}{2}\right)^4$

Objective 2

Evaluate the expression using the product rule, where applicable.

41) $2^2 \cdot 2^4$

42) $5^2 \cdot 5$

43) $3^2 \cdot 3^2$

44) $2^3 \cdot 2^3$

45) $5^2 \cdot 2^3$

46) $4^3 \cdot 3^2$

47) $\left(\dfrac{1}{2}\right)^4 \cdot \left(\dfrac{1}{2}\right)$

48) $\left(\dfrac{4}{3}\right) \cdot \left(\dfrac{4}{3}\right)^2$

Simplify the expression using the product rule. Leave your answer in exponential form.

49) $8^3 \cdot 8^9$

50) $6^4 \cdot 6^3$

51) $7^5 \cdot 7 \cdot 7^4$

52) $5^2 \cdot 5^3 \cdot 5^4$

53) $(-4)^2 \cdot (-4)^3 \cdot (-4)^2$

54) $(-3) \cdot (-3)^5 \cdot (-3)^2$

55) $a^2 \cdot a^3$

56) $t^5 \cdot t^4$

57) $k \cdot k^2 \cdot k^3$

58) $n^6 \cdot n^5 \cdot n^2$

59) $8y^3 \cdot y^2$

60) $10c^8 \cdot c^2 \cdot c$

61) $(9m^4)(6m^{11})$

62) $(-10p^8)(-3p)$

63) $(-6r)(7r^4)$

64) $(8h^5)(-5h^2)$

65) $(-7t^6)(t^3)(-4t^7)$

66) $(3k^2)(-4k^5)(2k^4)$

67) $\left(\frac{5}{3}x^2\right)(12x)(-2x^3)$

68) $\left(\frac{7}{10}y^9\right)(-2y^4)(3y^2)$

69) $\left(\frac{8}{21}b\right)(-6b^8)\left(-\frac{7}{2}b^6\right)$

70) $(12c^3)\left(\frac{14}{15}c^2\right)\left(\frac{5}{7}c^6\right)$

Objectives 3–5

Simplify the expression using one of the power rules.

71) $(x^4)^3$

72) $(w^5)^9$

73) $(t^6)^7$

74) $(a^2)^3$

75) $(2^3)^2$

76) $(3^2)^2$

77) $(-5^3)^2$

78) $(-4^5)^3$

79) $\left(\frac{1}{2}\right)^5$

80) $\left(\frac{3}{2}\right)^4$

81) $\left(\frac{4}{y}\right)^3$

82) $\left(\frac{n}{5}\right)^3$

83) $\left(\frac{d}{c}\right)^8$

84) $\left(\frac{t}{u}\right)^{10}$

85) $(5z)^3$

86) $(4u)^3$

87) $(-3p)^4$

88) $(2m)^5$

89) $(-4ab)^3$

90) $(-2cd)^4$

91) $6(xy)^3$

92) $-8(mn)^5$

93) $-9(tu)^4$

94) $2(ab)^6$

Objectives 2–5

95) Find the area and perimeter of each rectangle.

a) 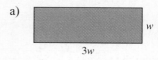 $3w$, w

b) 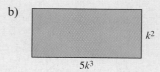 $5k^3$, k^2

96) Find the area.

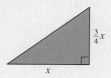 x, $\frac{5}{2}x$

97) Find the area.

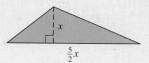 $\frac{3}{4}x$, x

98) The shape and dimensions of the Miller's family room is given below. They will have wall-to-wall carpeting installed, and the carpet they have chosen costs $2.50/ft^2$.

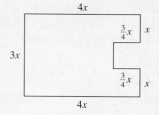 $4x$, $\frac{3}{4}x$, x, $3x$, $4x$

a) Write an expression for the amount of carpet they will need. (Include the correct units.)

b) Write an expression for the cost of carpeting the family room. (Include the correct units.)

Section 2.2b Combining the Rules of Exponents

Objective

1. Combine the Product Rule and Power Rules of Exponents

1. Combine the Product Rule and Power Rules of Exponents

Now that we have learned the product rule for exponents and the power rules for exponents, let's think about how to combine the rules.

If we were asked to evaluate $2^3 \cdot 3^2$, we would follow the order of operations. What would be the first step?

$$2^3 \cdot 3^2 = 8 \cdot 9 \qquad \text{Exponents}$$
$$= 72 \qquad \text{Multiply.}$$

When we combine the rules of exponents, we follow the order of operations.

Example 1

Simplify.

a) $(2c)^3(3c^8)^2$ 　　　 b) $2(5k^4m^3)^3$ 　　　 c) $\dfrac{(6t^5)^2}{(2u^4)^3}$

Solution

a) $(2c)^3(3c^8)^2$

The operations in this example are multiplication and exponents. Exponents precede multiplication in order of operations, so perform the exponential operations first.

$$(2c)^3(3c^8)^2 = (2^3c^3)(3^2)(c^8)^2 \qquad \text{Power rule}$$
$$= (8c^3)(9c^{16}) \qquad \text{Power rule and evaluate exponents}$$
$$= 72c^{19} \qquad \text{Product rule}$$

b) $2(5k^4m^3)^3$

Which operation should be performed first, multiplying $2 \cdot 5$ or simplifying $(5k^4m^3)^3$? In the order of operations, exponents are done before multiplication, so we will begin by simplifying $(5k^4m^3)^3$.

$$2(5k^4m^3)^3 = 2 \cdot (5)^3(k^4)^3(m^3)^3 \qquad \text{Order of operations and power rule}$$
$$= 2 \cdot 125k^{12}m^9 \qquad \text{Power rule}$$
$$= 250k^{12}m^9 \qquad \text{Multiply.}$$

c) $\dfrac{(6t^5)^2}{(2u^4)^3}$

Here, the operations are division and exponents. What comes first in the order of operations? **Exponents.**

$$\frac{(6t^5)^2}{(2u^4)^3} = \frac{36t^{10}}{8u^{12}} \qquad \text{Power rule}$$

$$= \frac{\overset{9}{\cancel{36}}t^{10}}{\underset{2}{\cancel{8}}u^{12}} \qquad \text{Divide out the common factor of 4.}$$

$$= \frac{9t^{10}}{2u^{12}}$$

BE CAREFUL

When simplifying the expression in Example 1c, $\dfrac{(6t^5)^2}{(2u^4)^3}$, it may be very tempting to reduce before applying the product rule, like this:

$$\frac{(\overset{3}{\cancel{6}}t^5)^2}{(\underset{1}{\cancel{2}}u^4)^3} \neq \frac{(3t^5)^2}{(u^4)^3} = \frac{9t^{10}}{u^{12}}$$

You can see, however, that because we did not follow the rules for the order of operations, we did not get the correct answer.

You Try 1

Simplify.

a) $-4(2a^9b^6)^4$ b) $(7x^{10}y)^2(-x^4y^5)^4$ c) $\dfrac{10(m^2n^3)^5}{(5p^4)^2}$ d) $\left(\dfrac{1}{6}w^7\right)^2(3w^{11})^3$

Answers to You Try Exercises

1) a) $-64a^{36}b^{24}$ b) $49x^{36}y^{22}$ c) $\dfrac{2m^{10}n^{15}}{5p^8}$ d) $\dfrac{3}{4}w^{47}$

2.2b Exercises

Boost your grade at mathzone.com! MathZone

> Practice Problems   > NetTutor   > e-Professors   > Videos
> Self-Test

Objective 1

1) When evaluating expressions involving exponents, always keep in mind the order of _____.

2) The first step in evaluating $(9-3)^2$ is _____.

Simplify.

3) $(k^9)^2(k^3)^2$

4) $(d^5)^3(d^2)^4$

5) $(5z^4)^2(2z^6)^3$

6) $(3r)^2(6r^8)^2$

7) $9pq(-p^{10}q^3)^5$

8) $-10a^4b(-ab^7)^4$

9) $(5+3)^2$

10) $(11-8)^2$

11) $(-4t^6u^2)^3(u^4)^5$

12) $(-m^2)^6(-2m^9)^4$

13) $8(6k^7l^2)^2$

14) $5(-7c^4d)^2$

15) $\left(\dfrac{3}{g^5}\right)^3\left(\dfrac{1}{6}\right)^2$

16) $\left(-\dfrac{2}{5}z^5\right)^3(10z)^2$

17) $\left(\dfrac{7}{8}n^2\right)^2(-4n^9)^2$

18) $\left(\dfrac{2}{3}d^8\right)^4\left(\dfrac{9}{2}d^3\right)^2$

19) $h^4(10h^3)^2(-3h^9)^2$

20) $-v^6(-2v^5)^5(-v^4)^3$

21) $3w^{11}(7w^2)^2(-w^6)^5$

22) $5z^3(-4z)^2(2z^3)^2$

23) $\dfrac{(12x^3)^2}{(10y^5)^2}$

24) $\dfrac{(-3a^4)^3}{(6b)^2}$

25) $\dfrac{(4d^9)^2}{(-2c^5)^6}$

26) $\dfrac{(-5m^7)^3}{(5n^{12})^2}$

27) $\dfrac{6(a^8b^3)^5}{(2c)^3}$

28) $\dfrac{(3x^8)^3}{15(y^2z^3)^4}$

29) $\dfrac{r^4(r^5)^7}{2t(11t^2)^2}$

30) $\dfrac{k^5(k^2)^3}{7m^{10}(2m^3)^2}$

31) $\left(\dfrac{4}{9}x^3y\right)^2\left(\dfrac{3}{2}x^6y^4\right)^3$

32) $(6s^8t^3)^2\left(-\dfrac{10}{3}st^4\right)^2$

33) $\left(-\dfrac{2}{5}c^9d^2\right)^3\left(\dfrac{5}{4}cd^6\right)^2$

34) $-\dfrac{11}{12}\left(\dfrac{3}{2}m^3n^{10}\right)^2$

35) $\left(\dfrac{5x^5y^2}{z^4}\right)^3$

36) $\left(-\dfrac{7a^4b}{8c^6}\right)^2$

37) $\left(-\dfrac{3t^4u^9}{2v^7}\right)^4$

38) $\left(\dfrac{2pr^8}{q^{11}}\right)^5$

39) $\left(\dfrac{12w^5}{4x^3y^6}\right)^2$

40) $\left(\dfrac{10b^3c^5}{15a}\right)^2$

41) The length of a side of a square is $5l^2$ units.

   a)   Write an expression for its perimeter.

   b)   Write an expression for its area.

42) The width of a rectangle is $2w$ units, and the length of the rectangle is $7w$ units.

   a)   Write an expression for its area.

   b)   Write an expression for its perimeter.

43) The length of a rectangle is x units, and the width of the rectangle is $\dfrac{3}{8}x$ units.

   a)   Write an expression for its area.

   b)   Write an expression for its perimeter.

44) The width of a rectangle is $4y^3$ units, and the length of the rectangle is $\dfrac{13}{2}y^3$ units.

   a)   Write an expression for its perimeter.

   b)   Write an expression for its area.

Section 2.3a Integer Exponents with Real Number Bases

Objectives

1. Use 0 as an Exponent

2. Evaluate Expressions Containing Real Number Bases and Negative Integer Exponents

Thus far, we have defined an exponential expression such as 2^3. The exponent of 3 indicates that $2^3 = 2 \cdot 2 \cdot 2$ (3 factors of 2) so that $2^3 = 2 \cdot 2 \cdot 2 = 8$.

   Is it possible to have an exponent of zero or a negative exponent? If so, what do they mean?

1. Use 0 as an Exponent

Definition   **Zero as an Exponent:** If $a \neq 0$, then $a^0 = 1$.

How can this be possible? Let's look at an example involving the product rule to help us understand why $a^0 = 1$.

   Let's evaluate $2^0 \cdot 2^3$. Using the product rule we get:

$$2^0 \cdot 2^3 = 2^{0+3} = 2^3 = 8$$

If we evaluate only 2^3, however, $2^3 = 8$. Therefore, if $2^0 \cdot 2^3 = 8$, then $2^0 = 1$. This is one way to understand that $a^0 = 1$.

Example 1

Evaluate.

   a) 5^0 b) -8^0 c) $(-7)^0$

Solution

   a) $5^0 = 1$

   b) $-8^0 = -1 \cdot 8^0 = -1 \cdot 1 = -1$

   c) $(-7)^0 = 1$

You Try 1

Evaluate.

a) 9^0 b) -2^0 c) $(-5)^0$

2. Evaluate Expressions Containing Real Number Bases and Negative Integer Exponents

So far we have worked with exponents that are zero or positive. What does a negative exponent mean?

Let's use the product rule to find $2^3 \cdot 2^{-3}$.

$$2^3 \cdot 2^{-3} = 2^{3+(-3)} = 2^0 = 1.$$

Remember that a number multiplied by its reciprocal is 1, and here we have that a quantity, 2^3, times another quantity, 2^{-3}, is 1. Therefore, 2^3 and 2^{-3} are reciprocals! This leads to the definition of a negative exponent.

Definition

Negative Exponent: If n is any integer and $a \neq 0$, then $a^{-n} = \left(\dfrac{1}{a}\right)^n = \dfrac{1}{a^n}$.

Therefore, to rewrite an expression of the form a^{-n} with a positive exponent, *take the reciprocal of the base and make the exponent positive.*

Example 2

Evaluate.

a) 2^{-3} b) $\left(\dfrac{3}{2}\right)^{-4}$ c) $\left(\dfrac{1}{5}\right)^{-3}$ d) $(-7)^{-2}$

Solution

a) 2^{-3}: The reciprocal of 2 is $\dfrac{1}{2}$, so

$$2^{-3} = \left(\dfrac{1}{2}\right)^3 = \dfrac{1^3}{2^3} = \dfrac{1}{8}$$

Above we found that $2^3 \cdot 2^{-3} = 1$ using the product rule, but now we can evaluate the product using the definition of a negative exponent.

$$2^3 \cdot 2^{-3} = 8 \cdot \left(\dfrac{1}{2}\right)^3 = 8 \cdot \dfrac{1}{8} = 1.$$

b) $\left(\dfrac{3}{2}\right)^{-4}$: The reciprocal of $\dfrac{3}{2}$ is $\dfrac{2}{3}$, so

$$\left(\frac{3}{2}\right)^{-4} = \left(\frac{2}{3}\right)^4 = \frac{2^4}{3^4} = \frac{16}{81}$$

 BE CAREFUL   Notice that a negative exponent does not make the answer negative!

c) $\left(\dfrac{1}{5}\right)^{-3}$: The reciprocal of $\dfrac{1}{5}$ is 5, so $\left(\dfrac{1}{5}\right)^{-3} = 5^3 = 125$.

d) $(-7)^{-2}$: The reciprocal of -7 is $-\dfrac{1}{7}$, so

$$(-7)^{-2} = \left(-\frac{1}{7}\right)^2 = \left(-1 \cdot \frac{1}{7}\right)^2 = (-1)^2 \left(\frac{1}{7}\right)^2 = 1 \cdot \frac{1^2}{7^2} = \frac{1}{49}.$$

 You Try 2

Evaluate.

a) $(10)^{-2}$ b) $\left(\dfrac{1}{4}\right)^{-2}$ c) $\left(\dfrac{2}{3}\right)^{-3}$ d) -5^{-3}

Answers to You Try Exercises

1) a) 1   b) -1 c) 1   2) a) $\dfrac{1}{100}$ b) 16   c) $\dfrac{27}{8}$ d) $-\dfrac{1}{125}$

2.3a Exercises

Objectives 1 and 2

1) True or False: Raising a positive base to a negative exponent will give a negative result. (Example: 2^{-4})

2) True or False: $9^0 = 0$.

3) True or False: The reciprocal of 8 is $\dfrac{1}{8}$.

4) True or False: $4^{-3} + 2^{-3} = 6^{-3}$.

Evaluate.

5) 6^0

6) $(-12)^0$

7) -4^0

8) -1^0

9) 0^4

10) $-(-7)^0$

11) $(5)^0 + (-5)^0$

12) $\left(\dfrac{4}{7}\right)^0 - \left(\dfrac{7}{4}\right)^0$

13) 6^{-2}

14) 5^{-2}

15) 2^{-4}

16) 11^{-2}

17) 5^{-3}

18) 2^{-5}

19) $\left(\dfrac{1}{8}\right)^{-2}$

20) $\left(\dfrac{1}{10}\right)^{-3}$

21) $\left(\dfrac{1}{2}\right)^{-5}$

22) $\left(\dfrac{1}{4}\right)^{-2}$

23) $\left(\dfrac{4}{3}\right)^{-3}$

24) $\left(\dfrac{2}{5}\right)^{-3}$

25) $\left(\dfrac{9}{7}\right)^{-2}$

26) $\left(\dfrac{10}{3}\right)^{-2}$

27) $\left(-\dfrac{1}{4}\right)^{-3}$

28) $\left(-\dfrac{1}{12}\right)^{-2}$

29) $\left(-\dfrac{3}{8}\right)^{-2}$

30) $\left(-\dfrac{5}{2}\right)^{-3}$

 31) -2^{-6}

32) -4^{-3}

33) -1^{-5}

34) -9^{-2}

35) $2^{-3} - 4^{-2}$

36) $5^{-2} + 2^{-2}$

37) $2^{-2} + 3^{-2}$

38) $4^{-1} - 6^{-2}$

39) $-9^{-2} + 3^{-3} + (-7)^0$

40) $6^0 - 9^{-1} + 4^0 + 3^{-2}$

Section 2.3b Integer Exponents with Variable Bases

Objectives

1. Evaluate an Expression Containing a Variable in the Base and a 0 Exponent

2. Rewrite an Exponential Expression Containing Variables in the Base with Only Positive Exponents

We will now work with bases which are variables.

1. Evaluate an Expression Containing a Variable in the Base and a 0 Exponent

Example 1

Evaluate. Assume the variable does not equal zero.

a) t^0 b) $(-k)^0$ c) $-(11p)^0$

Solution

a) $t^0 = 1$

b) $(-k)^0 = 1$

c) $-(11p)^0 = -1 \cdot (11p)^0$
$= -1 \cdot 1$
$= -1$

 You Try 1

Evaluate. Assume the variable does not equal zero.

a) p^0 b) $(10x)^0$ c) $-(7s)^0$

2. Rewrite an Exponential Expression Containing Variables in the Base with Only Positive Exponents

Next, let's apply the definition of a negative exponent to bases containing variables. As above, we will assume the variable does not equal zero since having zero in the denominator of a fraction will make the fraction undefined.

Recall that $2^{-4} = \left(\dfrac{1}{2}\right)^4 = \dfrac{1}{16}$. That is, to rewrite the expression with a positive exponent we take the reciprocal of the base.

What is the reciprocal of x? The reciprocal is $\dfrac{1}{x}$.

Example 2

Rewrite the expression with positive exponents. Assume the variable does not equal zero.

a) x^{-6} b) $\left(\dfrac{2}{n}\right)^{-6}$ c) $3a^{-2}$

Solution

a) $x^{-6} = \left(\dfrac{1}{x}\right)^6 = \dfrac{1^6}{x^6} = \dfrac{1}{x^6}$

b) $\left(\dfrac{2}{n}\right)^{-6} = \left(\dfrac{n}{2}\right)^6$ The reciprocal of $\dfrac{2}{n}$ is $\dfrac{n}{2}$.

$= \dfrac{n^6}{2^6} = \dfrac{n^6}{64}$

c) $3a^{-2} = 3 \cdot \left(\dfrac{1}{a}\right)^2$ Remember, the base is a, *not* $3a$, since there are no parentheses. Therefore, the exponent of -2 applies only to a.

$= 3 \cdot \dfrac{1}{a^2} = \dfrac{3}{a^2}$

You Try 2

Rewrite the expression with positive exponents. Assume the variable does not equal zero.

a) m^{-4} b) $\left(\dfrac{1}{z}\right)^{-7}$ c) $-2y^{-3}$

How could we rewrite $\dfrac{x^{-2}}{y^{-2}}$ with only positive exponents? One way would be to apply the power rule for exponents:

$$\dfrac{x^{-2}}{y^{-2}} = \left(\dfrac{x}{y}\right)^{-2} = \left(\dfrac{y}{x}\right)^2 = \dfrac{y^2}{x^2}$$

Let's do the same for $\dfrac{a^{-5}}{b^{-5}}$:

$$\dfrac{a^{-5}}{b^{-5}} = \left(\dfrac{a}{b}\right)^{-5} = \left(\dfrac{b}{a}\right)^5 = \dfrac{b^5}{a^5}$$

Notice that to rewrite the original expression with only positive exponents, the terms with the negative exponents "switch" their positions in the fraction. We can generalize this way:

Definition   If m and n are any integers and a and b are real numbers not equal to zero, then

$$\dfrac{a^{-m}}{b^{-n}} = \dfrac{b^n}{a^m}$$

Example 3

Rewrite the expression with positive exponents.

a) $\dfrac{c^{-8}}{d^{-3}}$ b) $\dfrac{5p^{-6}}{q^7}$ c) $t^{-2}u^{-1}$

Solution

a) $\dfrac{c^{-8}}{d^{-3}} = \dfrac{d^3}{c^8}$ To make the exponents positive, "switch" the positions of the terms in the fraction.

b) $\dfrac{5p^{-6}}{q^7}$

Since the exponent on q is positive, we do not change its position in the expression.

$$\frac{5p^{-6}}{q^7} = \frac{5}{p^6 q^7}$$

c) $t^{-2}u^{-1}$

This is the same as $\dfrac{t^{-2}u^{-1}}{1}$, so to rewrite this with positive exponents we move

$t^{-2}u^{-1}$ to the denominator to get $\dfrac{1}{t^2 u}$.

$$t^{-2}u^{-1} = \frac{1}{t^2 u}$$

 You Try 3

Rewrite the expression with positive exponents.

a) $\dfrac{n^{-6}}{y^{-2}}$ b) $\dfrac{z^{-9}}{3k^{-4}}$ c) $8x^{-5}y$

Answers to You Try Exercises

1) a) 1   b) 1   c) −1   2) a) $\dfrac{1}{m^4}$ b) z^7 c) $-\dfrac{2}{y^3}$ 3) a) $\dfrac{y^2}{n^6}$ b) $\dfrac{k^4}{3z^9}$ c) $\dfrac{8y}{x^5}$

2.3b Exercises

Boost your grade at mathzone.com! MathZone   > Practice Problems   > NetTutor > Self-Test   > e-Professors   > Videos

Objective 1

1) Identify the base in each expression.

   a) w^0 b) $-3n^{-5}$

   c) $(2p)^{-3}$ d) $4c^0$

2) True or False: $6^0 - 4^0 = (6 - 4)^0$

Evaluate. Assume the variable does not equal zero.

3) r^0 4) $(5m)^0$

5) $-2k^0$ 6) $-z^0$

7) $x^0 + (2x)^0$ 8) $\left(\dfrac{7}{8}\right)^0 - \left(\dfrac{3}{5}\right)^0$

Objective 2

Rewrite each expression with only positive exponents. Assume the variables do not equal zero.

9) y^{-4} 10) d^{-6}

11) p^{-1} 12) a^{-5}

video 13) $\dfrac{a^{-10}}{b^{-3}}$ 14) $\dfrac{h^{-2}}{k^{-1}}$

15) $\dfrac{y^{-8}}{x^{-5}}$ 16) $\dfrac{v^{-2}}{w^{-7}}$

17) $\dfrac{x^4}{10y^{-5}}$ 18) $\dfrac{7t^{-3}}{u^6}$

19) $9a^4b^{-3}$ 20) $\dfrac{1}{5}m^{-6}n^2$

21) $\dfrac{1}{c^{-5}d^{-8}}$ 22) $\dfrac{6r^{-3}s}{11t^2u^{-9}}$

video 23) $\dfrac{8a^6b^{-1}}{5c^{-10}d}$ 24) $\dfrac{17k^{-8}h^5}{20m^{-7}n^{-2}}$

25) $\dfrac{2z^4}{x^{-7}y^{-6}}$ 26) $\dfrac{1}{a^{-2}b^{-2}c^{-1}}$

video 27) $\left(\dfrac{a}{6}\right)^{-2}$ 28) $\left(\dfrac{3}{y}\right)^{-4}$

29) $\left(\dfrac{2n}{q}\right)^{-5}$ 30) $\left(\dfrac{w}{5v}\right)^{-3}$

video 31) $\left(\dfrac{12b}{cd}\right)^{-2}$ 32) $\left(\dfrac{2tu}{v}\right)^{-6}$

33) $-9k^{-2}$ 34) $3g^{-5}$

35) $3t^{-3}$ 36) $8h^{-4}$

37) $-m^{-9}$ 38) $-d^{-5}$

39) $\left(\dfrac{1}{z}\right)^{-10}$ 40) $\left(\dfrac{1}{k}\right)^{-6}$

41) $\left(\dfrac{1}{j}\right)^{-1}$ 42) $\left(\dfrac{1}{c}\right)^{-7}$

43) $5\left(\dfrac{1}{n}\right)^{-2}$ 44) $7\left(\dfrac{1}{t}\right)^{-8}$

video 45) $c\left(\dfrac{1}{d}\right)^{-3}$ 46) $x^2\left(\dfrac{1}{y}\right)^{-2}$

Section 2.4 The Quotient Rule

Objective

1. Use the Quotient Rule for Exponents

1. Use the Quotient Rule for Exponents

In this section, we will discuss how to simplify the quotient of two expressions containing exponents. Let's begin by simplifying $\dfrac{8^6}{8^4}$. One method for simplifying this expression is by writing the numerator and denominator without exponents:

$$\frac{8^6}{8^4} = \frac{\cancel{8} \cdot \cancel{8} \cdot \cancel{8} \cdot \cancel{8} \cdot 8 \cdot 8}{\cancel{8} \cdot \cancel{8} \cdot \cancel{8} \cdot \cancel{8}} \qquad \text{Divide out common factors.}$$

$$= 8 \cdot 8 = 8^2 = 64$$

Therefore,

$$\frac{8^6}{8^4} = 8^2 = 64$$

Do you notice a relationship between the exponents in the original expression and the exponent we get when we simplify?

$$\frac{8^6}{8^4} = 8^{6-4} = 8^2 = 64$$

That's right. We *subtracted* the exponents.

Quotient Rule for Exponents: If m and n are any integers and $a \neq 0$, then

$$\frac{a^m}{a^n} = a^{m-n}$$

Notice that the base in the numerator and denominator is a. *This means in order to apply the quotient rule, the bases must be the same. Subtract the exponent of the denominator from the exponent of the numerator.*

Example 1

Simplify.

a) $\dfrac{2^9}{2^3}$ b) $\dfrac{t^{10}}{t^4}$ c) $\dfrac{3}{3^{-2}}$ d) $\dfrac{n^5}{n^7}$ e) $\dfrac{3^2}{2^4}$

Solution

a) $\dfrac{2^9}{2^3} = 2^{9-3} = 2^6 = 64$ Since the bases are the same, subtract the exponents.

b) $\dfrac{t^{10}}{t^4} = t^{10-4} = t^6$ Since the bases are the same, subtract the exponents.

c) $\dfrac{3}{3^{-2}} = \dfrac{3^1}{3^{-2}} = 3^{1-(-2)}$ Since the bases are the same, subtract the exponents.

$\qquad\qquad = 3^3 = 27$ Be careful when subtracting the negative exponent!

d) $\dfrac{n^5}{n^7} = n^{5-7} = n^{-2}$ Same base, subtract the exponents.

$\qquad = \left(\dfrac{1}{n}\right)^2 = \dfrac{1}{n^2}$ Write with a positive exponent.

e) $\dfrac{3^2}{2^4} = \dfrac{9}{16}$ Since the bases are not the same, we cannot apply the quotient rule. Evaluate the numerator and denominator separately.

You Try 1

Simplify.

a) $\dfrac{5^7}{5^4}$ b) $\dfrac{c^4}{c^{-1}}$ c) $\dfrac{k^2}{k^{10}}$ d) $\dfrac{2^3}{2^7}$

We can apply the quotient rule to expressions containing more than one variable. Here are more examples:

Example 2

Simplify.

a) $\dfrac{x^8 y^7}{x^3 y^4}$ b) $\dfrac{12a^{-5}b^{10}}{8a^{-3}b^2}$

Solution

a) $\dfrac{x^8 y^7}{x^3 y^4} = x^{8-3} y^{7-4}$ Subtract the exponents.

$= x^5 y^3$

b) $\dfrac{12a^{-5}b^{10}}{8a^{-3}b^2}$ We will reduce $\dfrac{12}{8}$ in addition to applying the quotient rule.

$\dfrac{\overset{3}{\cancel{12}}a^{-5}b^{10}}{\underset{2}{\cancel{8}}a^{-3}b^2} = \dfrac{3}{2}a^{-5-(-3)}b^{10-2}$ Subtract the exponents.

$= \dfrac{3}{2}a^{-5+3}b^8$

$= \dfrac{3}{2}a^{-2}b^8$

$= \dfrac{3b^8}{2a^2}$

You Try 2

Simplify.

a) $\dfrac{r^4 s^{10}}{rs^3}$ b) $\dfrac{30m^6 n^{-8}}{42m^4 n^{-3}}$

Answers to You Try Exercises

1) a) 125   b) c^5 c) $\dfrac{1}{k^8}$ d) $\dfrac{1}{16}$ 2) a) $r^3 s^7$ b) $\dfrac{5m^2}{7n^5}$

2.4 Exercises

Boost your grade at mathzone.com! MathZone

> Practice Problems > NetTutor > e-Professors > Videos
> Self-Test

Objective 1

Simplify using the quotient rule.

1) $\dfrac{d^{12}}{d^5}$

2) $\dfrac{z^{10}}{z^4}$

3) $\dfrac{m^7}{m^5}$

4) $\dfrac{a^9}{a}$

5) $\dfrac{9t^{11}}{t^6}$

6) $\dfrac{4k^8}{k^5}$

7) $\dfrac{6^{15}}{6^{13}}$

8) $\dfrac{4^8}{4^5}$

9) $\dfrac{3^{11}}{3^7}$

10) $\dfrac{2^9}{2^4}$

11) $\dfrac{2^3}{2^7}$

12) $\dfrac{9^8}{9^{10}}$

 13) $\dfrac{5^6}{5^9}$

14) $\dfrac{8^3}{8^5}$

15) $\dfrac{10k^4}{k}$

16) $\dfrac{3b^9}{b^2}$

17) $\dfrac{20c^{11}}{30c^6}$

18) $\dfrac{48t^6}{56t^2}$

19) $\dfrac{z^2}{z^8}$

20) $\dfrac{n^5}{n^9}$

 21) $\dfrac{x^{-3}}{x^6}$

22) $\dfrac{w^{-14}}{w^{-3}}$

23) $\dfrac{r^{-5}}{r^{-3}}$

24) $\dfrac{y^8}{y^{15}}$

25) $\dfrac{a^{-1}}{a^9}$

26) $\dfrac{m^{-9}}{m^{-3}}$

27) $\dfrac{t^4}{t}$

28) $\dfrac{b^5}{b^{-1}}$

29) $\dfrac{15w^2}{w^{10}}$

30) $\dfrac{-8p^5}{p^{15}}$

video 31) $\dfrac{-6k}{k^4}$

32) $\dfrac{21h^3}{h^7}$

33) $\dfrac{a^4b^9}{ab^2}$

34) $\dfrac{p^5q^7}{p^2q^3}$

35) $\dfrac{5m^{-1}n^{-6}}{15m^{-5}n^2}$

36) $\dfrac{21tu^{-3}}{14t^7u^{-9}}$

37) $\dfrac{200x^8y^3}{20x^{10}y^{11}}$

38) $\dfrac{63a^{-2}b^2}{9a^7b^{10}}$

video 39) $\dfrac{6v^{-1}w}{54v^2w^{-5}}$

40) $\dfrac{3a^2b^{-5}}{15a^{-8}b^3}$

41) $\dfrac{3c^5d^{-2}}{8cd^{-3}}$

42) $\dfrac{9x^{-5}y^2}{4x^{-2}y^6}$

43) $\dfrac{(x+y)^6}{(x+y)^2}$

44) $\dfrac{(a+b)^9}{(a+b)^4}$

45) $\dfrac{(c+d)^{-5}}{(c+d)^{-11}}$

46) $\dfrac{(a+5b)^{-2}}{(a+5b)^{-3}}$

Mid-Chapter Summary: Putting the Rules Together

Objective

1. Combine the Rules of Exponents to Simplify Expressions

1. Combine the Rules of Exponents to Simplify Expressions

Now that we have learned many different rules for working with exponents, let's see how we can combine the rules to simplify expressions.

Example 1

Simplify using the rules of exponents. Assume all variables represent nonzero real numbers.

a) $(2t^{-6})^3(3t^2)^2$

b) $\left(\dfrac{7c^{10}d^7}{c^4d^2}\right)^2$

c) $\dfrac{w^{-3}\cdot w^4}{w^6}$

d) $\left(\dfrac{12a^{-2}b^9}{30ab^{-2}}\right)^{-3}$

Solution

a) $(2t^{-6})^3(3t^2)^2$ We must follow the order of operations. Therefore, do the exponents first.

$$\begin{aligned}
(2t^{-6})^3 \cdot (3t^2)^2 &= 2^3 t^{(-6)(3)} \cdot 3^2 t^{(2)(2)} & \text{Apply the power rule.}\\
&= 8t^{-18} \cdot 9t^4 & \text{Simplify.}\\
&= 72t^{-18+4} & \text{Multiply } 8 \cdot 9 \text{ and add the exponents.}\\
&= 72t^{-14} \\
&= \frac{72}{t^{14}} & \text{Write the answer using a positive exponent.}
\end{aligned}$$

b) $\left(\dfrac{7c^{10}d^7}{c^4 d^2}\right)^2$ How can we begin this problem? We can use the quotient rule to simplify the expression before squaring it.

$$\begin{aligned}
\left(\frac{7c^{10}d^7}{c^4 d^2}\right)^2 &= (7c^{10-4}d^{7-2})^2 & \text{Apply the quotient rule in the parentheses.}\\
&= (7c^6 d^5)^2 & \text{Simplify.}\\
&= 7^2 c^{(6)(2)} d^{(5)(2)} & \text{Apply the power rule.}\\
&= 49c^{12}d^{10}
\end{aligned}$$

c) $\dfrac{w^{-3} \cdot w^4}{w^6}$ Let's begin by simplifying the numerator:

$$\begin{aligned}
\frac{w^{-3} \cdot w^4}{w^6} &= \frac{w^{-3+4}}{w^6} & \text{Add the exponents in the numerator.}\\
&= \frac{w^1}{w^6}
\end{aligned}$$

Now, we can apply the quotient rule:

$$\begin{aligned}
&= w^{1-6} = w^{-5} & \text{Subtract the exponents.}\\
&= \frac{1}{w^5} & \text{Write the answer using a positive exponent.}
\end{aligned}$$

d) $\left(\dfrac{12a^{-2}b^9}{30ab^{-2}}\right)^{-3}$ Eliminate the negative exponent *outside* of the parentheses by taking the reciprocal of the base. Notice that we have *not* eliminated the negatives on the exponents *inside* the parentheses.

$$\left(\frac{12a^{-2}b^9}{30ab^{-2}}\right)^{-3} = \left(\frac{30ab^{-2}}{12a^{-2}b^9}\right)^3$$

We could apply the exponent of 3 to the quantity inside of the parentheses, but we could also reduce $\dfrac{30}{12}$ first and apply the quotient rule before cubing the quantity.

$$\begin{aligned}
&= \left(\frac{5}{2}a^{1-(-2)}b^{-2-9}\right)^3 & \text{Reduce } \frac{30}{12} \text{ and subtract the exponents.}\\
&= \left(\frac{5}{2}a^3 b^{-11}\right)^3
\end{aligned}$$

$$= \frac{125}{8}a^9 b^{-33} \qquad \text{Apply the power rule.}$$

$$= \frac{125a^9}{8b^{33}} \qquad \text{Write the answer using positive exponents.}$$

You Try I

Simplify using the rules of exponents.

a) $\left(\dfrac{m^{12}n^3}{m^4 n}\right)^4$ b) $(-p^{-5})^4 (6p^7)^2$ c) $\left(\dfrac{9x^4 y^{-5}}{54x^3 y}\right)^{-2}$

It is possible for variables to appear in exponents. The same rules apply.

Example 2

Simplify using the rules of exponents. Assume that the variables represent nonzero integers. Write your final answer so that the exponents have positive coefficients.

a) $c^{4x} \cdot c^{2x}$ b) $\dfrac{x^{5y}}{x^{9y}}$

Solution

a) $c^{4x} \cdot c^{2x} = c^{4x+2x} = c^{6x}$ The bases are the same, so apply the product rule. Add the exponents.

b) $\dfrac{x^{5y}}{x^{9y}} = x^{5y-9y}$ The bases are the same, so apply the quotient rule. Subtract the exponents.

$$= x^{-4y}$$

$$= \frac{1}{x^{4y}} \qquad \text{Write the answer with a positive exponent.}$$

You Try 2

Simplify using the rules of exponents. Assume that the variables represent nonzero integers. Write your final answer so that the exponents have positive coefficients.

a) $8^{2k} \cdot 8^k \cdot 8^{10k}$ b) $(w^3)^{-2p}$

Answers to You Try Exercises

1) a) $m^{32}n^8$ b) $\dfrac{36}{p^6}$ c) $\dfrac{36y^{12}}{x^2}$ 2) a) 8^{13k} b) $\dfrac{1}{w^{6p}}$

Mid-Chapter Summary Exercises

Boost your grade at mathzone.com!   MathZone

> Practice Problems
> NetTutor
> Self-Test
> e-Professors
> Videos

Objective 1

Use the rules of exponents to evaluate.

1) $\left(\dfrac{2}{3}\right)^4$

2) $(2^2)^3$

3) $\dfrac{6^9}{6^5 \cdot 6^4}$

4) $\dfrac{(-5)^4 \cdot (-5)^2}{(-5)^3}$

 5) $\left(\dfrac{10}{3}\right)^{-2}$

6) $\left(\dfrac{2}{5}\right)^{-3}$

7) $(9-4)^2$

8) $(3-7)^3$

9) 9^{-2}

10) 3^{-3}

11) $\dfrac{3^7}{3^{11}}$

12) $\dfrac{2^{19}}{2^{13}}$

13) $\left(-\dfrac{5}{3}\right)^7 \cdot \left(-\dfrac{5}{3}\right)^{-4}$

14) $\left(\dfrac{1}{2}\right)^{-4}$

15) $4^{-2} - 12^{-1}$

16) $3^{-2} + 2^{-3}$

Simplify. Assume all variables represent nonzero real numbers. The final answer should not contain negative exponents.

 17) $-10(-3g^4)^3$

18) $6(2d^3)^3$

19) $\dfrac{23t}{t^{11}}$

20) $\dfrac{r^{-6}}{r^{-2}}$

21) $\left(\dfrac{2xy^4}{3x^{-9}y^{-2}}\right)^3$

22) $\left(\dfrac{a^8b^3}{10a^5}\right)^3$

23) $\left(\dfrac{7r^3}{s^8}\right)^{-2}$

24) $\left(\dfrac{3n^{-6}}{m^2}\right)^{-4}$

25) $(-k^4)^3$

26) $(t^7)^8$

27) $(-2m^5n^2)^5$

28) $(13yz^6)^2$

29) $\left(-\dfrac{9}{4}z^5\right)\left(\dfrac{2}{3}z^{-1}\right)$

30) $(20w^2)\left(-\dfrac{3}{5}w^6\right)$

31) $\left(\dfrac{a^5}{b^4}\right)^{-6}$

32) $\dfrac{x^{-4}}{y^{10}}$

33) $(-ab^3c^5)^2\left(\dfrac{a^4}{bc}\right)^3$

34) $\dfrac{(4v^3)^2}{(6v^8)^2}$

35) $\left(\dfrac{48u^{-7}v^2}{36u^3v^{-5}}\right)^{-3}$

36) $\left(\dfrac{xy^5}{5x^{-2}y}\right)^{-3}$

37) $\left(\dfrac{-3t^4u}{t^2u^{-4}}\right)^4$

38) $\left(\dfrac{k^7m^7}{12k^{-1}m^6}\right)^2$

39) $(h^{-3})^7$

40) $(-n^4)^{-5}$

41) $\left(\dfrac{h}{2}\right)^5$

42) $17m^{-2}$

43) $-7c^8(-2c^2)^3$

44) $5p^3(4p^7)^2$

45) $(12a^7)^{-1}(6a)^2$

46) $(9c^2d)^{-1}$

47) $\left(\dfrac{9}{20}d^5\right)(2d^{-3})\left(\dfrac{4}{33}d^9\right)$

48) $\left(\dfrac{f^8 \cdot f^{-3}}{f^2 \cdot f^9}\right)^5$

49) $\left(\dfrac{56m^4n^8}{21m^4n^5}\right)^{-2}$

50) $\dfrac{(2x^4y)^{-2}}{(5xy^3)^2}$

Simplify. Assume that the variables represent nonzero integers. Write your final answer so that the exponents have positive coefficients.

51) $(p^{2n})^5$

52) $(3d^{4t})^2$

53) $y^m \cdot y^{10m}$

54) $t^{-6c} \cdot t^{9c}$

55) $x^{5a} \cdot x^{-8a}$

56) $b^{-2y} \cdot b^{-3y}$

57) $\dfrac{21c^{2x}}{35c^{8x}}$

58) $-\dfrac{5y^{-13a}}{8y^{-2a}}$

Section 2.5 Scientific Notation

Objectives

1. Understand How to Multiply a Number by a Power of Ten

2. Define and Identify Numbers in Scientific Notation

3. Rewrite a Number in Scientific Notation as a Number Without Exponents

4. Write a Number in Scientific Notation

5. Perform Operations with Numbers in Scientific Notation

The distance from the Earth to the Sun is approximately 150,000,000 km.

The gross domestic product of the United States in 2003 was $10,987,900,000,000. (Bureau of Economic Analysis)

A single rhinovirus (cause of the common cold) measures 0.00002 mm across.

Each of these is an example of a very large or very small number containing many zeros. Sometimes, performing operations with so many zeros can be difficult. This is why scientists and economists, for example, often work with such numbers in a different form called *scientific notation*.

Scientific notation is a short-hand method for writing very large and very small numbers. Writing numbers in scientific notation together with applying the rules of exponents can simplify calculations with very large and very small numbers.

1. Understand How to Multiply a Number by a Power of Ten

Before discussing scientific notation further, we need to understand some principles behind the notation. Let's look at multiplying numbers by positive powers of 10.

Example 1

Multiply.

a) 3.4×10^1 b) 0.0857×10^3 c) 97×10^2

Solution

a) $3.4 \times 10^1 = 3.4 \times 10 = 34$

b) $0.0857 \times 10^3 = 0.0857 \times 1000 = 85.7$

c) $97 \times 10^2 = 97 \times 100 = 9700$

Notice that when we multiply each of these numbers by a positive power of 10, the result is *larger* than the original number. In fact, the exponent determines how many places to the *right* the decimal point is moved.

$$3.40 \times 10^1 = 3.4 \times 10^1 = 34 \quad 0.0857 \times 10^3 = 85.7 \quad 97 \times 10^2 = 97.00 \times 10^2 = 9700$$

1 place to the right   3 places to right   2 places to right

You Try 1

Multiply by moving the decimal point the appropriate number of places.

a) 6.2×10^2 b) 5.31×10^5 c) 0.000122×10^4

What happens to a number when we multiply by a *negative* power of 10?

Example 2

Multiply.

a) 41×10^{-2} b) 367×10^{-4} c) 5.9×10^{-1}

Solution

a) $41 \times 10^{-2} = 41 \times \dfrac{1}{100} = \dfrac{41}{100} = 0.41$

b) $367 \times 10^{-4} = 367 \times \dfrac{1}{10,000} = \dfrac{367}{10,000} = 0.0367$

c) $5.9 \times 10^{-1} = 5.9 \times \dfrac{1}{10} = \dfrac{5.9}{10} = 0.59$

Is there a pattern? When we multiply each of these numbers by a negative power of 10, the result is *smaller* than the original number. The exponent determines how many places to the *left* the decimal point is moved:

$41 \times 10^{-2} = \underset{\text{2 places to the left}}{41.} \times 10^{-2} = 0.41$ $367 \times 10^{-4} = \underset{\text{4 places to the left}}{0367.} \times 10^{-4} = 0.0367$

$5.9 \times 10^{-1} = \underset{\text{1 place to the left}}{5.9} \times 10^{-1} = 0.59$

You Try 2

Multiply.

a) 83×10^{-2} b) 45×10^{-3}

It is important to understand the previous concepts to understand how to use scientific notation.

2. Define and Identify Numbers in Scientific Notation

Definition

A number is in **scientific notation** if it is written in the form $a \times 10^{n}$ where $1 \le |a| < 10$ and n is an integer.

$1 \le |a| < 10$ means that "a" is a number that has *one* nonzero digit to the left of the decimal point.

Here are some examples of numbers written in scientific notation: 3.82×10^{-5}, 1.2×10^{3}, 7×10^{-2}.

The following numbers are *not* in scientific notation:

$$51.94 \times 10^{4} \qquad\qquad 0.61 \times 10^{-3} \qquad\qquad 300 \times 10^{6}$$

↑ 2 digits to left of decimal point   ↑ Zero is to left of decimal point   ↑ 3 digits to left of decimal point

3. Rewrite a Number in Scientific Notation as a Number Without Exponents

We will continue our discussion by converting from scientific notation to a number without exponents.

Example 3

Rewrite without exponents.

a) 5.923×10^4 b) 7.4×10^{-3} c) 1.8875×10^3

Solution

a) $5.923 \times 10^4 \rightarrow 5.9230 = 59,230$ Remember, multiplying by a positive power of
     4 places to the right   10 will make the result *larger* than 5.923.

b) $7.4 \times 10^{-3} \rightarrow 007.4 = 0.0074$ Multiplying by a negative power of 10 will
     3 places to the left   make the result *smaller* than 7.4.

c) $1.8875 \times 10^3 \rightarrow 1.8875 = 1887.5$
     3 places to the right

You Try 3

Rewrite without exponents.

a) 3.05×10^4 b) 8.3×10^{-5} c) 6.91853×10^3

4. Write a Number in Scientific Notation

We will write the number 48,000 in scientific notation.

First, locate the decimal point in 48,000: 48,000.

Next, determine where the decimal point will be when the number is in scientific notation:

48,000.
∧
Decimal point
will be here.

Therefore, $48,000 = 4.8 \times 10^n$, where n is an integer. Will n be positive or negative? When we look at 4.8×10^n and understand that it must equal 48,000, we can see that 4.8 must be multiplied by a *positive* power of 10 to make it larger, 48,000. So, beginning with the original number

48000.
∧   ↖
Decimal point   Decimal point
will be here.   starts here.

We must move the decimal point four places.

48000.

Therefore, $48,000 = 4.8 \times 10^4$.

Example 4

Write each number in scientific notation.

Solution

a) The distance from the Earth to the Sun is approximately 150,000,000 km.

150,000,000.

Decimal point   Decimal point
will be here.   is here.

150,000,000.   Move decimal point eight places.

150,000,000 km = 1.5×10^8 km

b) A single rhinovirus measures 0.00002 mm across.

0.00002 mm

Decimal point
will be here.

0.00002 mm = 2×10^{-5} mm

Summary: How to Write a Number in Scientific Notation

1) Locate the decimal point in the original number.

2) Determine where the decimal point needs to be when converting to scientific notation. Remember, there will be *one* nonzero digit to the left of the decimal point.

3) Count how many places you must move the decimal point to take it from its original place to its position for scientific notation.

4) If the resulting number is *smaller* than the original number, you will multiply the result by a *positive* power of 10. Example: $350.9 = 3.509 \times 10^2$.

    If the resulting number is *larger* than the original number, you will multiply the result by a *negative* power of 10. Example: $0.0000068 = 6.8 \times 10^{-6}$.

 **You Try 4**

Write each number in scientific notation.

a) The gross domestic product of the United States in 2003 was approximately $10,987,900,000,000.

b) The diameter of a human hair is approximately 0.001 in.

5. Perform Operations with Numbers in Scientific Notation

We use the rules of exponents to perform operations with numbers in scientific notation.

Example 5

Perform the operations and simplify.

a) $(-2 \times 10^3)(4 \times 10^2)$ b) $\dfrac{9 \times 10^3}{2 \times 10^5}$

Solution

a) $(-2 \times 10^3)(4 \times 10^2) = (-2 \times 4)(10^3 \times 10^2)$ Commutative property
$= -8 \times 10^5$ Add the exponents.
$= -800,000$

b) $\dfrac{9 \times 10^3}{2 \times 10^5}$

$$\frac{9 \times 10^3}{2 \times 10^5} = \frac{9}{2} \times \frac{10^3}{10^5}$$
$$= 4.5 \times 10^{-2} = 0.045 \qquad \text{Subtract the exponents.}$$
$\uparrow$
Write $\dfrac{9}{2}$ in decimal form.

You Try 5

Perform the operations and simplify.

a) $(2.6 \times 10^2)(5 \times 10^4)$ b) $\dfrac{7.2 \times 10^{-9}}{6 \times 10^{-5}}$

Using Technology

A graphing calculator, like the TI-83 Plus or TI-84, can display numbers in scientific notation. First, the calculator must be set to scientific mode.

Press MODE to see the eight different options that allow you to change calculator settings, depending on what you need. The default settings are shown on the left. To change to scientific mode, use the arrow key, $>$, to highlight SCI, and press ENTER. This is shown on the right.

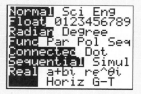

Scientific mode shows all your answers in scientific notation. If you enter the number 105,733 and press ENTER, the calculator will display 1.05733E5. This is how the calculator represents 1.05733×10^5. Enter the number 0.000759 and press ENTER to display the number in scientific notation, 7.59E-4, the calculator's equivalent of 7.59×10^{-4}.

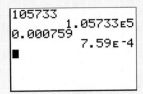

To enter a number that is already in scientific notation, use the EE key by pressing $\boxed{2^{nd}}$ $\boxed{,}$. For example, Avogadro's number is a very large number used in chemistry. To input Avogadro's number 6.022×10^{23} on a graphing calculator, input 6.022 $\boxed{2^{nd}}$ $\boxed{,}$ 23:

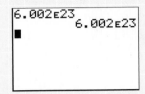

Predict what the calculator will display when you enter the following numbers, then verify your prediction on the calculator:

1)   5,230,000
2)   81,000,000
3)   0.00000164
4)   0.009522

What result would the calculator display when the following operations are entered?

5)   (3E4) $\times$ (2E3)
6)   (1E-6) $\times$ (5E2)
7)   (5.1E-9)/(3E-4)
8)   (8.4E2)/(4E5)

Answers to You Try Exercises

1) a) 620   b) 531,000   c) 1.22   2) a) 0.83   b) 0.045   3) a) 30,500   b) 0.000083

c) 6198.53   4) a) 1.09879×10^{13} dollars   b) 1.0×10^{-3} in.   5) a) 13,000,000   b) 0.00012

Answers to Technology Exercises

1) 5.23E6   2) 8.1E7   3) 1.64E-6   4) 9.522E-3   5) 6E7   6) 5E-4

7) 1.7E-5   8) 2.1E-3

2.5 Exercises

Boost your grade at mathzone.com! MathZone

> Practice Problems   > NetTutor   > e-Professors   > Videos
> Self-Test

Objectives 1 and 2

Determine if each number is in scientific notation.

1) 7.23×10^5
2) 24.0×10^{-3}
3) 0.16×10^{-4}
4) -2.8×10^4
5) -37×10^{-2}
6) 0.9×10^{-1}
7) -5×10^6
8) 7.5×2^{-10}

9) Explain, in your own words, how to determine if a number is expressed in scientific notation.

10) Explain, in your own words, how to write 4.1×10^{-3} without an exponent.

11) Explain, in your own words, how to write -7.26×10^4 without an exponent.

Objective 3

Write each number without an exponent.

12) 1.92×10^6
13) -6.8×10^{-5}
14) 2.03449×10^3
15) -5.26×10^4
16) -7×10^{-4}
17) 8×10^{-6}
18) -9.5×10^{-3}
19) 6.021967×10^5
20) 6×10^4
21) 3×10^6
22) -9.815×10^{-2}
23) -7.44×10^{-4}
24) 4.1×10^{-6}

Objective 4

Write each number in scientific notation.

25) 2110.5
26) 38.25
27) 0.000096
28) 0.00418
29) $-7,000,000$
30) 62,000
31) 3400
32) $-145,000$
33) 0.0008
34) -0.00000022
35) -0.076
36) 990
37) 6000
38) $-500,000$

Write each number in scientific notation.

39) The total weight of the Golden Gate Bridge is 380,800,000 kg. (www.goldengatebridge.com)

40) A typical hard drive may hold approximately 160,000,000,000 bytes of data.

41) The diameter of an atom is about 0.00000001 cm.

42) The oxygen-hydrogen bond length in a water molecule is 0.000000001 mm.

Objective 5

Perform the operation as indicated. Write the final answer without an exponent.

43) $\dfrac{6 \times 10^9}{2 \times 10^5}$
44) $(7 \times 10^2)(2 \times 10^4)$

45) $(2.3 \times 10^3)(3 \times 10^2)$
46) $\dfrac{8 \times 10^7}{4 \times 10^4}$

47) $\dfrac{8.4 \times 10^{12}}{-7 \times 10^9}$
48) $\dfrac{-4.5 \times 10^{-6}}{-1.5 \times 10^{-8}}$

49) $(-1.5 \times 10^{-8})(4 \times 10^6)$

50) $(-3 \times 10^{-2})(-2.6 \times 10^{-3})$

51) $\dfrac{-3 \times 10^5}{6 \times 10^8}$
52) $\dfrac{2 \times 10^1}{5 \times 10^4}$

53) $(9.75 \times 10^4) + (6.25 \times 10^4)$

54) $(4.7 \times 10^{-3}) + (8.8 \times 10^{-3})$

55) $(3.19 \times 10^{-5}) + (9.2 \times 10^{-5})$

56) $(2 \times 10^2) + (9.7 \times 10^2)$

For each problem, express each number in scientific notation, then solve the problem.

57) Humans shed about 1.44×10^7 particles of skin every day. How many particles would be shed in a year? (Assume 365 days in a year.)

58) Scientists send a lunar probe to land on the moon and send back data. How long will it take for pictures to reach the Earth if the distance between the Earth and the moon is 360,000 km and if the speed of light is $3 \cdot 10^5$ km/sec?

59) In Wisconsin in 2001, approximately 1,300,000 cows produced 2.21×10^{10} lb of milk. On average, how much milk did each cow produce? (www.nass.usda.gov)

60) The average snail can move 1.81×10^{-3} mi in 5 hrs. What is its rate of speed in miles per hour?

61) A photo printer delivers approximately 1.1×10^6 droplets of ink per square inch. How many droplets of ink would a 4 in. $\times$ 6 in. photo contain?

Chapter 2: Summary

Definition/Procedure	Example	Reference
Simplifying Expressions		2.1
Combining Like Terms **Like terms** contain the same variables with the same exponents.	Combine like terms and simplify. $4k^2 - 3k + 1 - 2(6k^2 - 5k + 7)$ $= 4k^2 - 3k + 1 - 12k^2 + 10k - 14$ $= -8k^2 + 7k - 13$	**p. 89**
Writing Mathematical Expressions	Write a mathematical expression for the following: *sixteen more than twice a number* Let x = the number. $\underbrace{\text{Sixteen}}_{16} \ \underbrace{\text{more than}}_{+} \ \underbrace{\text{twice a number}}_{2x}$ $16 + 2x$	**p. 91**
The Product Rule and Power Rules of Exponents		2.2a
Exponential Expression: $a^n = \underbrace{a \cdot a \cdot a \cdot \ldots \cdot a}_{n \text{ factors of } a}$ a is the **base**, n is the exponent	$5^4 = 5 \cdot 5 \cdot 5 \cdot 5$ 5 is the **base**, 4 is the exponent.	**p. 93**
Product Rule: $a^m \cdot a^n = a^{m+n}$	$x^8 \cdot x^2 = x^{10}$	**p. 94**
Basic Power Rule: $(a^m)^n = a^{mn}$	$(t^3)^5 = t^{15}$	**p. 95**
Power Rule for a Product: $(ab)^n = a^n b^n$	$(2c)^4 = 2^4 c^4 = 16c^4$	**p. 96**
Power Rule for a Quotient: $\left(\dfrac{a}{b}\right)^n = \dfrac{a^n}{b^n}$, where $b \neq 0$.	$\left(\dfrac{w}{5}\right)^3 = \dfrac{w^3}{5^3} = \dfrac{w^3}{125}$	**p. 96**
Combining the Rules of Exponents		2.2b
Remember to follow the order of operations.	Simplify. $(3y^4)^2(2y^9)^3$ $= 9y^8 \cdot 8y^{27}$ Exponents come before multiplication. $= 72y^{35}$ Use the product rule and multiply coefficients.	**p. 100**
Integer Exponents with Real Number Bases		2.3a
Zero Exponent: If $a \neq 0$, then $a^0 = 1$.	$(-9)^0 = 1$	**p. 102**

Definition/Procedure	Example	Reference
Negative Exponent: For $a \neq 0$, $a^{-n} = \left(\dfrac{1}{a}\right)^n = \dfrac{1}{a^n}$.	Evaluate. $\left(\dfrac{5}{2}\right)^{-3} = \left(\dfrac{2}{5}\right)^3 = \dfrac{2^3}{5^3} = \dfrac{8}{125}$	p. 103
Integer Exponents with Variable Bases		2.3b
	Rewrite p^{-10} with a positive exponent (assume $p \neq 0$). $p^{-10} = \left(\dfrac{1}{p}\right)^{10}$ Definition of negative exponent $\quad\quad = \dfrac{1}{p^{10}}$ Power rule	p. 105
If $a \neq 0$ and $b \neq 0$, then $\dfrac{a^{-m}}{b^{-n}} = \dfrac{b^n}{a^m}$.	Rewrite each expression with positive exponents. Assume the variables represent nonzero real numbers. a) $\dfrac{x^{-3}}{y^{-7}} = \dfrac{y^7}{x^3}$ b) $\dfrac{14m^{-6}}{n^{-1}} = \dfrac{14n}{m^6}$	p. 106
The Quotient Rule		2.4
Quotient Rule: If $a \neq 0$, then $\dfrac{a^m}{a^n} = a^{m-n}$.	Simplify. $\dfrac{4^9}{4^6} = 4^{9-6} = 4^3 = 64$	p. 109
Mid-Chapter Summary		
Putting the Rules Together	Simplify. $\left(\dfrac{a^4}{2a^7}\right)^{-5} = \left(\dfrac{2a^7}{a^4}\right)^5 = (2a^3)^5 = 32a^{15}$	p. 111
Scientific Notation		2.5
Scientific Notation A number is in **scientific notation** if it is written in the form $a \times 10^n$, where $1 \leq \lvert a \rvert < 10$ and n is an integer. That is, a is a number that has one nonzero digit to the left of the decimal point.	Write in scientific notation. a) $78{,}000 \rightarrow 78{,}000. \rightarrow 7.8 \times 10^4$ b) $0.00293 \rightarrow 0.00293 \rightarrow 2.93 \times 10^{-3}$	p. 116
Converting from Scientific Notation	Write without exponents. a) $5 \times 10^{-4} \rightarrow 0005. \rightarrow 0.0005$ b) $1.7 \times 10^6 = 1.700000 \rightarrow 1{,}700{,}000$	p. 117
Performing Operations	Multiply. $(4 \times 10^2)(2 \times 10^4)$ $\quad = (4 \times 2)(10^2 \times 10^4)$ $\quad = 8 \times 10^6$ $\quad = 8{,}000{,}000$	p. 119

Chapter 2: Review Exercises

(2.1) Combine like terms and simplify.

1) $15y^2 + 8y - 4 + 2y^2 - 11y + 1$

2) $7t + 10 - 3(2t + 3)$

3) $\frac{3}{2}(5n - 4) + \frac{1}{4}(n + 6)$

4) $1.4(a + 5) - (a + 2)$

Write a mathematical expression for each phrase, and combine like terms if possible. Let x represent the unknown quantity.

5) Ten less than a number

6) The sum of a number and twelve

7) The sum of a number and eight decreased by three

8) Six less than twice a number

(2.2a)

9) Write in exponential form

    a) $8 \cdot 8 \cdot 8 \cdot 8 \cdot 8 \cdot 8$ b) $(-7)(-7)(-7)(-7)$

10) Identify the base and the exponent.

    a) -6^5 b) $(4t)^3$

    c) $4t^3$ d) $-4t^3$

11) Use the rules of exponents to simplify.

    a) $2^3 \cdot 2^2$ b) $\left(\frac{1}{3}\right)^2 \cdot \left(\frac{1}{3}\right)$

    c) $(7^3)^4$ d) $(k^5)^6$

12) Use the rules of exponents to simplify.

    a) $(3^2)^2$ b) $8^3 \cdot 8^7$

    c) $(m^4)^9$ d) $p^9 \cdot p^7$

13) Simplify using the rules of exponents.

    a) $(5y)^3$ b) $(-7m^4)(2m^{12})$

    c) $\left(\frac{a}{b}\right)^6$ d) $6(xy)^2$

    e) $\left(\frac{10}{9}c^4\right)(2c)\left(\frac{15}{4}c^3\right)$

14) Simplify using the rules of exponents.

    a) $\left(\frac{x}{y}\right)^{10}$ b) $(-2z)^5$

    c) $(6t^7)\left(-\frac{5}{8}t^5\right)\left(\frac{2}{3}t^2\right)$

    d) $-3(ab)^4$ e) $(10j^6)(4j)$

(2.2b)

15) Simplify using the rules of exponents.

    a) $(z^5)^2(z^3)^4$ b) $-2(3c^5d^8)^2$

    c) $(9 - 4)^3$ d) $\dfrac{(10t^3)^2}{(2u^7)^3}$

16) Simplify using the rules of exponents.

    a) $\left(\dfrac{-20d^4c}{5b^3}\right)^3$ b) $(-2y^8z)^3(3yz^2)^2$

    c) $\dfrac{x^7 \cdot (x^2)^5}{(2y^3)^4}$ d) $(6 - 8)^2$

(2.3a)

17) Evaluate.

    a) 8^0 b) -3^0

    c) 9^{-1} d) $3^{-2} - 2^{-2}$

    e) $\left(\dfrac{4}{5}\right)^{-3}$

18) Evaluate.

    a) $(-12)^0$ b) $5^0 + 4^0$

    c) -6^{-2} d) 2^{-4}

    e) $\left(\dfrac{10}{3}\right)^{-2}$

(2.3b)

19) Rewrite the expression with positive exponents. Assume the variables do not equal zero.

    a) v^{-9} b) $\left(\dfrac{9}{c}\right)^{-2}$

    c) $\left(\dfrac{1}{y}\right)^{-8}$ d) $-7k^{-9}$

    e) $\dfrac{19z^{-4}}{a^{-1}}$ f) $20m^{-6}n^5$

    g) $\left(\dfrac{2j}{k}\right)^{-5}$

20) Rewrite the expression with positive exponents. Assume the variables do not equal zero.

    a) $\left(\dfrac{1}{x}\right)^{-5}$ b) $3p^{-4}$

    c) $a^{-8}b^{-3}$ d) $\dfrac{12k^{-3}r^5}{16mn^{-6}}$

e) $\dfrac{c^{-1}d^{-1}}{15}$

f) $\left(-\dfrac{m}{4n}\right)^{-3}$

g) $\dfrac{10b^4}{a^{-9}}$

(2.4)

21) Simplify using the rules of exponents. The final answer should not contain negative exponents. Assume the variables represent nonzero real numbers.

a) $\dfrac{3^8}{3^6}$

b) $\dfrac{r^{11}}{r^3}$

c) $\dfrac{48t^{-2}}{32t^3}$

d) $\dfrac{21xy^2}{35x^{-6}y^3}$

22) Simplify using the rules of exponents. The final answer should not contain negative exponents. Assume the variables represent nonzero real numbers.

a) $\dfrac{2^9}{2^{15}}$

b) $\dfrac{d^4}{d^{-10}}$

c) $\dfrac{m^{-5}n^3}{mn^8}$

d) $\dfrac{100a^8b^{-1}}{25a^7b^{-4}}$

23) Simplify by applying one or more of the rules of exponents. The final answer should not contain negative exponents. Assume the variables represent nonzero real numbers.

a) $(-3s^4t^5)^4$

b) $\dfrac{(2a^6)^5}{(4a^7)^2}$

c) $\left(\dfrac{z^4}{y^3}\right)^{-6}$

d) $(-x^3y)^5(6x^{-2}y^3)^2$

e) $\left(\dfrac{cd^{-4}}{c^8d^{-9}}\right)^5$

f) $\left(\dfrac{14m^5n^5}{7m^4n}\right)^3$

g) $\left(\dfrac{3k^{-1}t}{5k^{-7}t^4}\right)^{-3}$

h) $\left(\dfrac{40}{21}x^{10}\right)(3x^{-12})\left(\dfrac{49}{20}x^2\right)$

24) Simplify by applying one or more of the rules of exponents. The final answer should not contain negative exponents. Assume the variables represent nonzero real numbers.

a) $\left(\dfrac{4}{3}\right)^8\left(\dfrac{4}{3}\right)^{-2}\left(\dfrac{4}{3}\right)^{-3}$

b) $\left(\dfrac{k^{10}}{k^4}\right)^3$

c) $\left(\dfrac{x^{-4}y^{11}}{xy^2}\right)^{-2}$

d) $(-9z^5)^{-2}$

e) $\left(\dfrac{g^2\cdot g^{-1}}{g^{-7}}\right)^{-4}$

f) $(12p^{-3})\left(\dfrac{10}{3}p^5\right)\left(\dfrac{1}{4}p^2\right)^2$

g) $\left(\dfrac{30u^2v^{-3}}{40u^7v^{-7}}\right)^{-2}$

h) $-5(3h^4k^9)^2$

25) Simplify. Assume that the variables represent nonzero integers. Write your final answer so that the exponents have positive coefficients.

a) $y^{3k}\cdot y^{7k}$

b) $(x^{5p})^2$

c) $\dfrac{z^{12c}}{z^{5c}}$

d) $\dfrac{t^{6d}}{t^{11d}}$

Write each number without an exponent.

26) 9.38×10^5

27) -4.185×10^2

28) 9×10^3

29) 6.7×10^{-4}

30) 1.05×10^{-6}

31) 2×10^4

32) 8.8×10^{-2}

Write each number in scientific notation.

33) 0.0000575

34) $36,940$

35) $32,000,000$

36) 0.0000004

37) $178,000$

38) 66

39) 0.0009315

Write the number without exponents.

40) Golfer Tiger Woods earns over 7×10^7 dollars per year in product endorsements. (www.forbes.com)

Perform the operation as indicated. Write the final answer without an exponent.

41) $\dfrac{8 \times 10^6}{2 \times 10^{13}}$

42) $\dfrac{-1 \times 10^9}{5 \times 10^{12}}$

43) $(9 \times 10^{-8})(4 \times 10^7)$

44) $(5 \times 10^3)(3.8 \times 10^{-8})$

45) $\dfrac{-3 \times 10^{10}}{-4 \times 10^6}$

46) $(-4.2 \times 10^2)(3.1 \times 10^3)$

For each problem, write each of the numbers in scientific notation, then solve the problem. Write the answer without exponents.

47) Eight porcupines have a total of about 2.4×10^5 quills on their bodies. How many quills would one porcupine have?

48) In 2002, Nebraska had approximately 4.6×10^7 acres of farmland and about 50,000 farms. What was the average size of a Nebraska farm in 2002? (www.nass.usda.gov)

49) One molecule of water has a mass of 2.99×10^{-23} g. Find the mass of 100,000,000 molecules.

1) Combine like terms and simplify.

   a) $(-8k^2 + 3k - 5) + (2k^2 + k - 9)$

   b) $\frac{4}{3}(6c - 5) - \frac{1}{2}(4c + 3)$

2) Write a mathematical expression for "nine less than twice a number." Let x represent the number.

Evaluate.

3) 3^4

4) 8^0

5) 2^{-5}

6) $4^{-2} + 2^{-3}$

7) $\left(-\frac{3}{4}\right)^3$

8) $\left(\frac{10}{7}\right)^{-2}$

Simplify using the rules of exponents. Assume all variables represent nonzero real numbers. The final answer should not contain negative exponents.

9) $(5n^6)^3$

10) $(-3p^4)(10p^8)$

11) $\frac{m^{10}}{m^4}$

12) $\frac{a^9 b}{a^5 b^7}$

13) $\left(\frac{-12t^{-6}u^8}{4t^5u^{-1}}\right)^{-3}$

14) $(2y^{-4})^6\left(\frac{1}{2}y^5\right)^3$

15) Simplify $t^{10k} \cdot t^{3k}$. Assume that the variables represent nonzero integers.

16) Rewrite 7.283×10^5 without exponents.

17) Write 0.000165 in scientific notation.

18) Divide. Write the answer without exponents. $\frac{-7.5 \times 10^{12}}{1.5 \times 10^8}$

19) Write the number without an exponent: In 2002, the population of Texas was about 2.18×10^7. (U.S. Census Bureau)

20) An electron is a subatomic particle with a mass of 9.1×10^{-28} g. What is the mass of 2,000,000,000 electrons? Write the answer without exponents.

1) Write $\dfrac{90}{150}$ in lowest terms.

Perform the indicated operations. Write the answer in lowest terms.

2) $\dfrac{2}{15} + \dfrac{1}{10} + \dfrac{7}{20}$

3) $\dfrac{4}{15} \div \dfrac{20}{21}$

4) $-144 \div (-12)$

5) $-26 + 5 - 7$

6) -9^2

7) $(5 + 1)^2 - 2[17 + 5(10 - 14)]$

8) Glen Crest High School is building a new football field. The dimensions of a regulation-size field are $53\dfrac{1}{3}$ yd by 120 yd. (There are 10 yd of end zone on each end.) The sod for the field will cost $1.80/\text{yd}^2$.

   a) Find the perimeter of the field.

   b) How much will it cost to sod the field?

9) Evaluate $2p^2 - 11q$ when $p = 3$ and $q = -4$.

10) Rewrite $\dfrac{3}{4}(6m - 20n + 7)$ using the distributive property.

11) Combine like terms and simplify:
    $5(t^2 + 7t - 3) - 2(4t^2 - t + 5)$

12) Let x represent the unknown quantity, and write a mathematical expression for "thirteen less than half of a number."

Simplify using the rules of exponents. The answer should not contain negative exponents. Assume the variables represent nonzero real numbers.

13) $(4z^3)(-7z^5)$

14) $\dfrac{n^2}{n^9}$

15) $(-2a^{-6}b)^5$

16) Write 0.000729 in scientific notation.

17) Perform the indicated operation. Write the final answer without an exponent.

$$(6.2 \times 10^5)(9.4 \times 10^{-2})$$

CHAPTER 3

Linear Equations and Inequalities

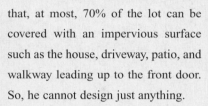

Algebra at Work: Landscape Architecture

A landscape architect must have excellent problem-solving skills.

Matthew is designing the driveway, patio, and walkway for this new home. The village has a building code that states that, at most, 70% of the lot can be covered with an impervious surface such as the house, driveway, patio, and walkway leading up to the front door. So, he cannot design just anything.

To begin, Matthew must determine the area of the land and find 70% of that number to determine how much land can be covered with these hard surfaces. He must subtract the area covered by the house to determine how much land he has left for the driveway, patio, and walkway. Using his design experience and problem-solving skills, he must come up with a plan for building the driveway, patio, and walkway that will not only please his client but will meet building codes as well.

In this chapter, we will learn different strategies for solving many different types of problems.

Section 3.1 Solving Linear Equations Part I

Objectives

1. Define a Linear Equation in One Variable
2. Use the Addition and Subtraction Properties of Equality
3. Use the Multiplication and Division Properties of Equality
4. Solve Equations of the Form $ax + b = c$
5. Combine Like Terms on One Side of the Equal Sign and Solve the Linear Equation

1. Define a Linear Equation in One Variable

What is an equation? It is a mathematical statement that two expressions are equal. $5 + 1 = 6$ is an equation.

> An equation contains an "=" sign and an expression does not.

$$5x + 4 = 7 \rightarrow \text{equation}$$
$$9y + 2y \rightarrow \text{expression}$$

We can **solve** equations, and we can **simplify** expressions.

In this section, we will begin our study of solving algebraic equations. Examples of algebraic equations include

$$a + 6 = 11, \qquad t^2 + 7t + 12 = 0, \qquad \sqrt{n - 4} = 16 - n$$

The first equation is an example of a linear equation; the second is a quadratic equation, and the third is a radical equation. In Sections 3.1 and 3.2 we will learn how to solve a linear equation. We will work with the other equations later in this book.

To **solve an equation** means to find the value or values of the variable that make the equation true.

$a + 6 = 11$:   The **solution** is $a = 5$ since we can substitute 5 for the variable and the equation is true:

$$a + 6 = 11$$
$$5 + 6 = 11 \quad \checkmark$$

> **Definition**
>
> A **linear equation in one variable** is an equation that can be written in the form
>
> $$ax + b = 0$$
>
> where a and b are real numbers and $a \neq 0$.

Notice that the exponent of the variable, x, is 1 in a linear equation. For this reason, these equations are also known as first-degree equations. Here are other examples of linear equations in one variable:

$$5y - 8 = 19, \qquad 2(c + 7) - 3 = 4c + 1, \qquad \frac{2}{3}k + \frac{1}{4} = k - 5$$

2. Use the Addition and Subtraction Properties of Equality

Begin with the true statement $8 = 8$. What happens if we add the same number, say 2, to each side? Is the statement still true?

$$8 = 8$$
$$8 + 2 = 8 + 2$$
$$10 = 10 \quad \text{True}$$

When we added 2 to each side of the equation, the statement was true. Will a statement remain true if we *subtract* the same number from each side? Let's begin with the true statement $5 = 5$ and subtract 9 from each side:

$$5 = 5$$
$$5 - 9 = 5 - 9$$
$$-4 = -4 \quad \text{True}$$

When we subtracted 9 from each side of the equation, the new statement was true.

$$8 = 8 \text{ and } 8 + 2 = 8 + 2 \text{ are } equivalent \ equations.$$
$$5 = 5 \text{ and } 5 - 9 = 5 - 9 \text{ are } equivalent \ equations \text{ as well.}$$

We can apply these principles to equations containing variables. This will help us solve equations.

Addition and Subtraction Properties of Equality

Let a, b, and c be expressions representing real numbers. Then,

1) If $a = b$, then $a + c = b + c$ Addition property of equality

2) If $a = b$, then $a - c = b - c$ Subtraction property of equality

---

Example 1

Solve and check each equation.

a) $x - 8 = 3$ b) $w + 9 = 2$ c) $-5 = m + 4$

Solution

Remember, to solve the equation means to find the value of the variable that makes the statement true. To do this, we want to get the variable by itself.

a) $x - 8 = 3$:   On the left side of the equal sign, the 8 is being **subtracted from** the x. Therefore, to get the x by itself, we perform the "opposite" operation—that is, we **add 8** to each side.

$$x - 8 = 3$$
$$x - 8 + 8 = 3 + 8 \qquad \text{Add 8.}$$
$$x = 11$$
$$\text{Check: } x - 8 = 3$$
$$11 - 8 = 3$$
$$3 = 3 \ \checkmark$$

The *solution set* is $\{11\}$. The **solution set** is the set of all solutions to an equation.

b) $w + 9 = 2$:   Here, **9** is being **added to** w. To get the w by itself, **subtract 9** from each side.

$$w + 9 = 2$$
$$w + 9 - 9 = 2 - 9 \qquad \text{Subtract 9.}$$
$$w = -7$$

Check: $w + 9 = 2$
$-7 + 9 = 2$
$2 = 2$ ✓

The solution set is $\{-7\}$.

c) $-5 = m + 4$

 BE CAREFUL   The variable does not always appear on the left-hand side of the equation. The 4 is being **added to** the m, so we will subtract **4** from each side.

$-5 = m + 4$
$-5 - 4 = m + 4 - 4$ Subtract 4.
$-9 = m$

Check: $-5 = m + 4$
$-5 = -9 + 4$
$-5 = -5$ ✓

The solution set is $\{-9\}$.

 You Try 1

Solve and check each equation.

a) $b - 8 = 9$ b) $-\dfrac{5}{2} = z + \dfrac{1}{2}$

3. Use the Multiplication and Division Properties of Equality

We have just seen that we can add a quantity to each side of an equation or subtract a quantity from each side of an equation, and we will obtain an equivalent equation. The same is true for multiplying and dividing.

Multiplication and Division Properties of Equality

Let a, b, and c be expressions representing real numbers. Then,

1)   If $a = b$, then $ac = bc$ Multiplication property of equality

2)   If $a = b$, then $\dfrac{a}{c} = \dfrac{b}{c}$ $(c \neq 0)$ Division property of equality

---

Example 2

Solve each equation.

a) $4k = -24$ b) $-m = 19$ c) $\dfrac{x}{3} = 5$ d) $\dfrac{3}{8}y = 12$

Solution

The goal is to get the variable on a side by itself.

a) $4k = -24$:   On the left-hand side of the equation, the k is being *multiplied* by 4. So, we will perform the "opposite" operation and *divide* each side by 4.

$$4k = -24$$
$$\frac{4k}{4} = \frac{-24}{4} \quad \text{Divide by 4.}$$
$$k = -6$$

Check: $4k = -24$
$$4(-6) = -24$$
$$-24 = -24 \ \checkmark$$

The solution set is $\{-6\}$.

b) $-m = 19$:   The negative sign in front of the m tells us that the coefficient of m is -1. Since m is being *multiplied* by -1, we will *divide* each side by -1.

$$-m = 19$$
$$\frac{-1m}{-1} = \frac{19}{-1} \quad \text{Divide by } -1.$$
$$m = -19$$

The check is left to the student.
The solution set is $\{-19\}$.

c) $\dfrac{x}{3} = 5$:   The x is being *divided* by 3. Therefore, we will *multiply* each side by 3.

$$\frac{x}{3} = 5$$
$$3 \cdot \frac{x}{3} = 3 \cdot 5 \quad \text{Multiply each side by 3.}$$
$$1x = 15 \quad\quad \text{Simplify.}$$
$$x = 15$$

The check is left to the student.
The solution set is $\{15\}$.

d) $\frac{3}{8}y = 12$:   On the left-hand side, the y is being *multiplied* by $\frac{3}{8}$. So, we could divide

each side by $\frac{3}{8}$. However, recall that dividing a quantity by a fraction is the

same as *multiplying by the reciprocal* of the fraction. Therefore, we will

multiply each side by the reciprocal of $\frac{3}{8}$.

$$\frac{3}{8}y = 12$$

$$\frac{8}{3} \cdot \frac{3}{8}y = \frac{8}{3} \cdot 12 \quad \text{The reciprocal of } \frac{3}{8} \text{ is } \frac{8}{3}.$$

$$1y = \frac{8}{\cancel{3}} \cdot \cancel{12}^{4}$$

$$y = 32$$

The check is left to the student.
The solution set is {32}.

You Try 2

Solve and check each equation.

a) $-8w = 6$ b) $\frac{5}{9}c = 20$

4. Solve Equations of the Form $ax + b = c$

So far we have not combined the properties of addition, subtraction, multiplication, and division to solve an equation. But that is exactly what we must do to solve equations like

$$3p + 7 = 31 \qquad \text{and} \qquad 4x + 9 - 6x + 2 = 17$$

Example 3

Solve each equation.

a) $3p + 7 = 31$ b) $\frac{4}{7}a - 5 = 1$ c) $-5.9 = 2.1y - 1.7$

Solution

a) $3p + 7 = 31$:   In this equation, there is a number, 7, being *added* to the term containing the variable, and the variable is being multiplied by a number, 3. *In general, we first eliminate the number being added to or subtracted from the variable.* Then, we eliminate the coefficient.

$$3p + 7 = 31$$

$$3p + 7 - 7 = 31 - 7 \qquad \text{Subtract 7 from each side.}$$

$$3p = 24 \qquad \text{Combine like terms.}$$

$$\frac{3p}{3} = \frac{24}{3} \qquad \text{Divide by 3.}$$

$$p = 8 \qquad \text{Simplify.}$$

Check: $3p + 7 = 31$

$$3(8) + 7 = 31$$

$$24 + 7 = 31$$

$$31 = 31 \quad \checkmark$$

The solution set is $\{8\}$.

b) $\frac{4}{7}a - 5 = 1$:

On the left-hand side, the a is being multiplied by $\frac{4}{7}$, and 5 is being subtracted from the a-term. To solve the equation, begin by eliminating the number being subtracted from the a-term.

$$\frac{4}{7}a - 5 = 1$$

$$\frac{4}{7}a - 5 + 5 = 1 + 5 \qquad \text{Add 5 to each side.}$$

$$\frac{4}{7}a = 6 \qquad \text{Combine like terms.}$$

$$\frac{7}{4} \cdot \frac{4}{7}a = \frac{7}{4} \cdot 6 \qquad \text{Multiply each side by the reciprocal of } \frac{4}{7}.$$

$$1a = \frac{7}{\overset{}{\underset{2}{4}}} \cdot \overset{3}{6} \qquad \text{Simplify.}$$

$$a = \frac{21}{2}$$

The check is left to the student.

The solution set is $\left\{ \dfrac{21}{2} \right\}$.

c) $-5.9 = 2.1y - 1.7$:

The variable is on the right-hand side of the equation. First, we will add 1.7 to each side, then we will divide by 2.1.

$$-5.9 = 2.1y - 1.7$$

$$-5.9 + 1.7 = 2.1y - 1.7 + 1.7 \qquad \text{Add 1.7 to each side.}$$

$$-4.2 = 2.1y \qquad \text{Combine like terms.}$$

$$\frac{-4.2}{2.1} = \frac{2.1y}{2.1} \qquad \text{Divide by 2.1.}$$

$$-2 = y \qquad \text{Simplify.}$$

Verify that -2 is the solution.
The solution set is $\{-2\}$.

You Try 3

Solve each equation.

a) $-2z + 9 = 15$ b) $20 = 13 + \dfrac{1}{6}n$ c) $-0.4t - 5.3 = 6.7$

5. Combine Like Terms on One Side of the Equal Sign and Solve the Linear Equation

Earlier, we were presented with the equation $4x + 9 - 6x + 2 = 17$. How do we solve such an equation? *We begin by combining like terms.* Then, we proceed as we did above.

Example 4

Solve each equation.

a) $4x + 9 - 6x + 2 = 17$ b) $2(1 - 3h) - 5(2h + 3) = -21$

c) $\dfrac{1}{2}(3b + 8) + \dfrac{3}{4} = -\dfrac{1}{2}$

Solution

a)
$$\begin{aligned}
4x + 9 - 6x + 2 &= 17 && \\
-2x + 11 &= 17 && \text{Combine like terms.} \\
-2x + 11 - 11 &= 17 - 11 && \text{Subtract 11 from each side.} \\
-2x &= 6 && \text{Combine like terms.} \\
\frac{-2x}{-2} &= \frac{6}{-2} && \text{Divide by } -2. \\
x &= -3 && \text{Simplify.}
\end{aligned}$$

$$\begin{aligned}
\text{Check: } 4x + 9 - 6x + 2 &= 17 \\
4(-3) + 9 - 6(-3) + 2 &= 17 \\
-12 + 9 + 18 + 2 &= 17 \\
17 &= 17 \ \checkmark
\end{aligned}$$

The solution set is $\{-3\}$.

b)
$$\begin{aligned}
2(1 - 3h) - 5(2h + 3) &= -21 && \\
2 - 6h - 10h - 15 &= -21 && \text{Distribute.} \\
-16h - 13 &= -21 && \text{Combine like terms.} \\
-16h - 13 + 13 &= -21 + 13 && \text{Add 13 to each side.} \\
-16h &= -8 && \text{Combine like terms.} \\
\frac{-16h}{-16} &= \frac{-8}{-16} && \text{Divide by } -16. \\
h &= \frac{1}{2} && \text{Simplify.}
\end{aligned}$$

The check is left to the student.

The solution set is $\left\{\dfrac{1}{2}\right\}$.

c) $\dfrac{1}{2}(3b + 8) + \dfrac{3}{4} = -\dfrac{1}{2}$

     $\dfrac{3}{2}b + 4 + \dfrac{3}{4} = -\dfrac{1}{2}$ Distribute.

     $\dfrac{3}{2}b + \dfrac{16}{4} + \dfrac{3}{4} = -\dfrac{1}{2}$ Get a common denominator for 4 and $\dfrac{3}{4}$.

                                   Rewrite 4 as $\dfrac{16}{4}$.

     $\dfrac{3}{2}b + \dfrac{19}{4} = -\dfrac{1}{2}$ Combine like terms.

$\dfrac{3}{2}b + \dfrac{19}{4} - \dfrac{19}{4} = -\dfrac{1}{2} - \dfrac{19}{4}$ Subtract $\dfrac{19}{4}$ from each side.

       $\dfrac{3}{2}b = -\dfrac{21}{4}$ Get a common denominator and subtract.

   $\dfrac{2}{3} \cdot \dfrac{3}{2}b = \dfrac{2}{3} \cdot \left(-\dfrac{21}{4}\right)$ Multiply both sides by the reciprocal of $\dfrac{3}{2}$.

     $b = \dfrac{\overset{1}{\cancel{2}}}{\underset{1}{\cancel{3}}} \cdot \left(-\dfrac{\overset{7}{\cancel{21}}}{\underset{2}{\cancel{4}}}\right)$ Perform the multiplication.

       $b = -\dfrac{7}{2}$

The check is left to the student.

The solution set is $\left\{-\dfrac{7}{2}\right\}$.

In Section 3.2 we will learn another way to solve an equation containing several fractions such as the equation above.

You Try 4

Solve.

a) $15 - 7u - 6 + 2u = -1$ b) $-3(4y - 3) + 4(y + 1) = 15$

Answers to You Try Exercises

1) a) $\{17\}$ b) $\{-3\}$ 2) a) $\left\{-\dfrac{3}{4}\right\}$ b) $\{36\}$ 3) a) $\{-3\}$ b) $\{42\}$ c) $\{-30\}$

4) a) $\{2\}$ b) $\left\{-\dfrac{1}{4}\right\}$

3.1 Exercises

Objective 1

Identify each as an expression or an equation.

1) $7t - 2 = 11$

2) $\dfrac{3}{4}k + 5(k - 6) = 2$

3) $8 - 10p + 4p + 5$

4) $9(2z - 7) + 3z$

5) Can we solve $3(c + 2) + 5(2c - 5)$? Why or why not?

6) Can we solve $3(c + 2) + 5(2c - 5) = -6$? Why or why not?

7) Which of the following are linear equations in one variable?

   a) $y^2 + 8y + 15 = 0$ b) $\dfrac{1}{2}w - 5(3w + 1) = 6$

   c) $8m - 7 + 2m + 1$ d) $0.3z + 0.2 = 1.5$

8) Which of the following are linear equations in one variable?

   a) $-7p = 0$

   b) $-2 = 5g - 4 + g + 10 + 3g - 1$

   c) $9x + 4y = 3$

   d) $10 - 6(4n - 1) + 7$

Determine if the given value is a solution to the equation.

9) $a - 4 = -9$; $a = 5$

10) $2d + 1 = 13$; $d = -6$

11) $-8p = 12$; $p = -\dfrac{3}{2}$

12) $15 = 21m$; $m = \dfrac{5}{7}$

13) $10 - 2(3y - 1) + y = -8$; $y = 4$

14) $-2w + 9 + w - 11 = -1$; $w = -3$

Objective 2

Solve and check each equation.

15) $r - 6 = 11$

16) $c + 2 = -5$

17) $b + 10 = 4$

18) $x - 3 = 9$

19) $-16 = k - 12$

20) $8 = t + 1$

21) $a + \dfrac{5}{8} = \dfrac{1}{2}$

22) $w - \dfrac{3}{4} = -\dfrac{1}{6}$

Objective 3

Solve and check each equation.

23) $3y = 30$

24) $2n = 8$

25) $-5z = 35$

26) $-8b = -24$

27) $-56 = -7v$

28) $-54 = 6m$

29) $\dfrac{a}{4} = 12$

30) $\dfrac{w}{5} = 4$

31) $-6 = \dfrac{k}{8}$

32) $30 = -\dfrac{x}{2}$

33) $\dfrac{2}{3}g = -10$

34) $\dfrac{7}{4}r = 42$

35) $-\dfrac{5}{3}d = -30$

36) $-\dfrac{1}{8}h = 3$

37) $\dfrac{11}{15} = \dfrac{1}{3}y$

38) $-\dfrac{5}{6} = -\dfrac{4}{9}x$

39) $0.5q = 6$

40) $0.3t = 3$

41) $-w = -7$

42) $-p = \dfrac{6}{7}$

43) $-12d = 0$

44) $4f = 0$

Objective 4

Solve and check each equation.

45) $3x - 7 = 17$

46) $2y - 5 = 3$

47) $7c + 4 = 18$

48) $5g + 19 = 4$

49) $8d - 15 = -15$

50) $-3r + 8 = 8$

51) $-11 = 5t - 9$

52) $4 = 7j - 8$

53) $10 = 3 - 7y$

54) $-6 = 9 - 3p$

55) $\dfrac{4}{9}w - 11 = 1$

56) $\dfrac{5}{3}a + 6 = 41$

57) $\dfrac{10}{7}m + 3 = 1$

58) $\dfrac{9}{10}x - 4 = 11$

59) $-\dfrac{1}{6}z + \dfrac{1}{2} = \dfrac{3}{4}$

60) $\dfrac{3}{5} - \dfrac{2}{3}k = 1$

61) $5 - 0.4p = 2.6$

62) $1.8 = 1.2n - 7.8$

Objective 5

Solve and check each equation.

63) $10v + 9 - 2v + 16 = 1$

64) $-8g - 7 + 6g + 1 = 20$

65) $5 - 3m + 9m + 10 - 7m = -4$

66) $t - 12 - 13t - 5 + 2t = -7$

 67) $5 = -12p + 7 + 4p - 12$

68) $12 = 9y + 11 - 3y - 7$

69) $2(5x + 3) - 3x + 4 = -11$

70) $6(2c - 1) + 3 - 7c = 42$

71) $-12 = 7(2a - 3) - (8a - 9)$

72) $20 = 5r - 3 + 2(9 - 3r)$

Section 3.2 Solving Linear Equations Part II

Objectives

1. Solve a Linear Equation in One Variable When Terms Containing the Variable Appear on Both Sides of the Equal Sign

2. Solve Equations Containing Fractions or Decimals

3. Solve Equations That Have No Solution or an Infinite Number of Solutions

4. Learn the Five Steps for Solving Applied Problems

5. Apply the Steps for Solving Applied Problems to Problems Involving Unknown Numbers

In Section 3.1, we learned how to solve equations such as

$$n - 3 = 18, \qquad 17 = \frac{4}{5}t + 9, \qquad 2(x + 5) - 9x = -11$$

Did you notice that all of the equations we solved in Section 3.1 contained variables on only one side of the equal sign? In this section, we will discuss how to solve equations containing variables on both sides of the equal sign.

1. Solve a Linear Equation in One Variable When Terms Containing the Variable Appear on Both Sides of the Equal Sign

To solve an equation such as $10y - 11 = 6y + 9$, we need to get the variables on one side of the equal sign and the constants on the other side. We will use the addition and subtraction properties of equality.

Example 1

Solve $10y - 11 = 6y + 9$.

Solution

Let's begin by getting the y-terms on the left side of the equal sign. To do this, we will subtract $6y$ from each side.

$$10y - 11 = 6y + 9$$
$$10y - 6y - 11 = 6y - 6y + 9 \qquad \text{Subtract } 6y.$$
$$4y - 11 = 9 \qquad \text{Combine like terms.}$$

The equation above looks like those we've solved previously. Next, add 11 to each side.

$$4y - 11 + 11 = 9 + 11 \qquad \text{Add 11.}$$
$$4y = 20$$
$$\frac{4y}{4} = \frac{20}{4} \qquad \text{Divide by 4.}$$
$$y = 5$$

Check: $10y - 11 = 6y + 9$
$$10(5) - 11 = 6(5) + 9$$
$$50 - 11 = 30 + 9$$
$$39 = 39 \quad \checkmark$$

The solution set is $\{5\}$.

We can summarize the steps used to solve a linear equation in one variable.

How to Solve a Linear Equation

1) If there are terms in parentheses, then distribute.

2) Combine like terms on each side of the equal sign.

3) Get the terms with the variables on one side of the equal sign and the constants on the other side using the addition and subtraction properties of equality.

4) Solve for the variable using the multiplication or division property of equality.

5) Check the solution in the original equation.

You Try 1

Solve $-3k + 4 = 8k - 15 - 6k - 11$.

There are some cases where we will not follow these steps exactly. We will see this with a couple of examples later in the section.

Example 2

Solve $8n + 5 - 3(2n + 9) = n + 2(2n + 7)$.

Solution

We will follow the steps to solve this equation.

1) Distribute.
$8n + 5 - 6n - 27 = n + 4n + 14$

2) Combine like terms.
$2n - 22 = 5n + 14$

3) Get the variables on one side of the equal sign and the constants on the other side.
Note: The variable can be on either side of the equal sign.

$$2n - 22 = 5n + 14$$
$$2n - 2n - 22 = 5n - 2n + 14$$ Get the variable on the right side of the $=$ sign by subtracting $2n$.

$$-22 = 3n + 14$$
$$-22 - 14 = 3n + 14 - 14$$ Get the constants on the left side of the $=$ sign by subtracting 14.

$$-36 = 3n$$

4) Divide both sides by 3 to solve for n.

$$\frac{-36}{3} = \frac{3n}{3}$$
$$-12 = n$$

5) Check $n = -12$ in the original equation.

$$8n + 5 - 3(2n + 9) = n + 2(2n + 7)$$
$$8(-12) + 5 - 3(2(-12) + 9) = -12 + 2(2(-12) + 7)$$
$$-96 + 5 - 3(-24 + 9) = -12 + 2(-24 + 7)$$
$$-91 - 3(-15) = -12 + 2(-17)$$
$$-91 + 45 = -12 + (-34)$$
$$-46 = -46 \ \checkmark$$

The solution set is $\{-12\}$.

You Try 2

Solve $5(3 - 2a) + 7a - 2 = 2(9 - 2a) + 15$.

2. Solve Equations Containing Fractions or Decimals

Some equations contain several fractions or decimals, which make them appear more difficult to solve. Here are two examples:

$$\frac{2}{9}x - \frac{1}{2} = \frac{1}{18}x + \frac{2}{3} \qquad \text{and} \qquad 0.05c + 0.4(c - 3) = -0.3$$

Before applying the steps for solving a linear equation, we can eliminate the fractions and decimals from the equations.

> To eliminate the fractions, determine the least common denominator for all of the fractions in the equation. Then, multiply both sides of the equation by the LCD.

Example 3

Solve $\dfrac{2}{9}x - \dfrac{1}{2} = \dfrac{1}{18}x + \dfrac{2}{3}$.

Solution

The least common denominator of all of the fractions in the equation is 18. Therefore, multiply both sides of the equation by 18.

$$18\left(\frac{2}{9}x - \frac{1}{2}\right) = 18\left(\frac{1}{18}x + \frac{2}{3}\right) \qquad \text{Multiply by 18 to eliminate the denominators.}$$

$$18 \cdot \frac{2}{9}x - 18 \cdot \frac{1}{2} = 18 \cdot \frac{1}{18}x + 18 \cdot \frac{2}{3} \qquad \text{Distribute.}$$

$$4x - 9 = x + 12$$
$$4x - x - 9 = x - x + 12 \qquad \text{Get the variable on the left side of the} = \text{sign by subtracting } x.$$
$$3x - 9 = 12$$
$$3x - 9 + 9 = 12 + 9 \qquad \text{Get the constants on the right side of the} = \text{sign by adding 9.}$$
$$3x = 21$$
$$\frac{3x}{3} = \frac{21}{3} \qquad \text{Divide by 3.}$$
$$x = 7 \qquad \text{Simplify.}$$

The check is left to the student.
The solution set is $\{7\}$.

You Try 3

Solve $\dfrac{1}{5}y + 1 = \dfrac{3}{10}y + \dfrac{1}{4}$.

Example 4

Solve $0.05c + 0.4(c - 3) = -0.3$.

Solution

We want to eliminate the decimals. The number containing a decimal place farthest to the right is 0.05. The 5 is in the hundredths place. Therefore, multiply both sides of the equation by 100 to eliminate all decimals in the equation.

$$100\,[0.05c + 0.4\,(c - 3)] = 100\,(-0.3)$$
$$100 \cdot (0.05c) + 100 \cdot [0.4\,(c - 3)] = 100\,(-0.3) \qquad \text{Distribute.}$$

Now we will distribute the 100 eliminate the decimals.

$$5c + \mathbf{40\,(c - 3)} = -30$$
$$5c + 40c - 120 = -30 \qquad \text{Distribute.}$$
$$45c - 120 = -30$$
$$45c - 120 + 120 = -30 + 120 \qquad \text{Get the constants on the right side of the} = \text{sign.}$$
$$45c = 90$$
$$\frac{45c}{45} = \frac{90}{45}$$
$$c = 2$$

The check is left to the student.
The solution set is $\{2\}$.

You Try 4

Solve $0.1d = 0.5 - 0.02(d - 5)$.

3. Solve Equations That Have No Solution or an Infinite Number of Solutions

Does every equation have a solution? Consider the next example:

Example 5

Solve $9a + 2 = 6a + 3(a - 5)$.

Solution

$$9a + 2 = 6a + 3(a - 5)$$
$$9a + 2 = 6a + 3a - 15 \qquad \text{Distribute.}$$
$$9a + 2 = 9a - 15 \qquad \text{Combine like terms.}$$
$$9a - 9a + 2 = 9a - 9a - 15 \qquad \text{Subtract } 9a.$$
$$2 = -15 \qquad \text{False}$$

Notice that the variable has "dropped out." Is $2 = -15$ a true statement? No! This means that the equation has *no solution*. We can say that the solution set is the **empty set**, denoted by $\varnothing$.

We have seen that a linear equation may have one solution or no solution. There is a third possibility—a linear equation may have an infinite number of solutions.

Example 6

Solve $p - 3p + 8 = 8 - 2p$.

Solution

$$p - 3p + 8 = 8 - 2p$$
$$-2p + 8 = 8 - 2p \qquad \text{Combine like terms.}$$
$$-2p + 2p + 8 = 8 - 2p + 2p \qquad \text{Add } 2p.$$
$$8 = 8 \qquad \text{True}$$

Here, the variable has "dropped out," and we are left with an equation, $8 = 8$, that is true. This means that any real number we substitute for p will make the original equation true. Therefore, this equation has an *infinite number of solutions*. The solution set is **{all real numbers}**.

Outcomes When Solving Linear Equations

There are three possible outcomes when solving a linear equation. The equation may have

1) **one solution.** Solution set: {a real number}. An equation that is true for some values and not for others is called a **conditional equation**.

   or

2) **no solution.** In this case, the variable will drop out, and there will be a false statement such as $2 = -15$. Solution set: $\varnothing$. An equation that has no solution is called a **contradiction**.

   or

3) **an infinite number of solutions.** In this case, the variable will drop out, and there will be a true statement such as $8 = 8$. Solution set: {all real numbers}. An equation that has all real numbers as its solution set is called an **identity**.

You Try 5

Solve.

a) $6 + 5x - 4 = 3x + 2 + 2x$ b) $3x - 4x + 9 = 5 - x$

4. Learn the Five Steps for Solving Applied Problems

Equations can be used to describe events that occur in the real world. Therefore, we need to learn how to translate information presented in English into an algebraic equation. We will begin slowly, then throughout the chapter we will work our way up to more challenging problems. Yes, it may be difficult at first, but with patience and persistence, you can do it!

How to Solve Applied Problems

While no single method will work for solving all applied problems, the following approach is suggested to help in the problem-solving process.

Steps for Solving Applied Problems

Step 1: Read the problem carefully. Then read it again. Draw a picture, if applicable.

Step 2: Identify what you are being asked to find. **Define the variable**; that is, assign a variable to represent an unknown quantity. Also,

- If there are other unknown quantities, define them in terms of the variable.
- Label the picture with the variable and other unknowns as well as with any given information in the problem.

Step 3: Translate from English to math. Some suggestions for doing so are

- Restate the problem in your own words.
- Read and think of the problem in "small parts."
- Make a chart to separate these "small parts" of the problem to help you translate to mathematical terms.
- Write an equation in English, then translate it to an algebraic equation.

Step 4: Solve the equation.

Step 5: Interpret the meaning of the solution as it relates to the problem. If there are other unknowns, find them. State the answer in a complete sentence. Be sure your answer makes sense in the context of the problem.

5. Apply the Steps for Solving Applied Problems to Problems Involving Unknown Numbers

Example 7

Write an equation and solve.
Six more than a number is nineteen. Find the number.

Solution

How should we begin?

Step 1: Read the problem carefully, then read it again.

Step 2: Identify what you are trying to find. *Define the variable*; that is, write down what the variable represents. Here, let x = a number.

Step 3: Translate the information that appears in English into an algebraic equation by rereading the problem slowly and "in parts."

In this problem, we will make a table to translate the statement from English to mathematical terms. Begin by thinking of the information in small parts and interpret their meanings.

In English	Meaning	In Math Terms
Six more than . . .	add 6	6 +
a number . . .	a number, the unknown	x
is . . .	equals	=
nineteen.	19	19

Therefore, "six more than a number is nineteen" becomes

$$6 + \qquad x \ = \ 19$$

Equation: $6 + x = 19$.

Step 4: Solve the equation.

$$6 + x = 19$$
$$x = 13$$

Step 5: State the answer in a complete sentence. Does the answer make sense?

$$6 + x = 19$$
$$6 + 13 = 19$$
$$19 = 19 \ \checkmark$$

The number is 13.

You Try 6

Write an equation and solve.
Five more than a number is twenty-one.

Sometimes, dealing with subtraction in a word problem can be confusing. So, let's look at an arithmetic problem first.

Example 8

What is four less than ten?

Solution

To solve this problem, do we subtract $10 - 4$ or $4 - 10$?

"Four less than ten" is written as $10 - 4$, and $10 - 4 = 6$. Six is four less than ten. The 4 is *subtracted from* the 10. Keep this problem in mind as you read the next example.

You Try 7

Write an equation and solve.
A number decreased by eight is twelve.

Example 9

Write an equation and solve.

Five less than twice a number is eleven. Find the number.

Solution

Step 1: Read the problem carefully, then reread it.

Step 2: Define the variable.

$$x = \text{a number}$$

Step 3: Translate from English to an algebraic equation. Remember, think about the statement in small parts.

We want to write an equation for "Five less than twice a number is eleven."

In English	Meaning	In Math Terms
Five less than . . .	subtract 5 *from* the quantity below	-5
twice a number . . .	two times a number, the unknown	$2x$
is . . .	equals	$=$
eleven.	11	11

So, "five less than twice a number is eleven" becomes

$$2x - 5 \qquad = \quad 11$$

Equation: $2x - 5 = 11$.

Step 4:   Solve the equation.

$$2x - 5 = 11$$
$$2x = 16$$
$$x = 8$$

Step 5:   The check is left for the student.
The number is 8.

You Try 8

Write an equation and solve.
The product of four and a number is thirty-six. Find the number.

Answers to You Try Exercises

1) $\{6\}$ 2) $\{20\}$ 3) $\left\{\dfrac{15}{2}\right\}$ 4) $\{5\}$ 5) a) $\{$all real numbers$\}$ b) $\varnothing$ 6) $x + 5 = 21; \{16\}$

7) $x - 8 = 12; \{20\}$ 8) $4x = 36; \{9\}$

3.2 Exercises

Boost your grade at mathzone.com!

> **Practice Problems**   > **NetTutor**   > **e-Professors**   > **Videos**
> **Self-Test**

Objectives 1 and 3

Solve each equation and check your answer.

1) $\dfrac{9}{2}c - 7 = 29$

2) $-6t + 1 = 16$

3) $2y + 7 = 5y - 2$

4) $8n - 21 = 3n - 1$

5) $6 - 7p = 2p + 33$

6) $z + 19 = 5 - z$

7) $-8x + 6 - 2x + 11 = 3 + 3x - 7x$

8) $10 - 13a + 2a - 16 = -5 + 7a + 11$

9) $4(2t + 5) - 7 = 5(t + 5)$

10) $3(2m + 10) = 6(m + 4) - 8m$

 11) $-9r + 4r - 11 + 2 = 3r + 7 - 8r + 9$

12) $3(4b - 7) + 8 = 6(2b + 5)$

13) $j - 15j + 8 = -3(4j - 3) - 2j - 1$

14) $n - 16 + 10n + 4 = 2(7n - 6) - 3n$

 15) $8(3t + 4) = 10t - 3 + 7(2t + 5)$

16) $2(9z - 1) + 7 = 10z - 14 + 8z + 2$

17) $8 - 7(2 - 3w) - 9w = 4(5w - 1) - 3w - 2$

18) $4m - (6m + 5) + 2 = 8m + 3(4 - 3m)$

19) $7y + 2(1 - 4y) = 8y - 5(y + 4)$

Objective 2

20) How can you eliminate the fractions from the equation

$$\dfrac{1}{6}x + \dfrac{5}{4} = \dfrac{1}{2}x - \dfrac{5}{12}?$$

Solve each equation by first clearing fractions or decimals.

21) $\dfrac{1}{6}x + \dfrac{5}{4} = \dfrac{1}{2}x - \dfrac{5}{12}$

22) $\dfrac{3}{4}n + \dfrac{1}{2} = \dfrac{1}{2}n + \dfrac{1}{4}$

23) $\dfrac{2}{3}d - 1 = \dfrac{1}{5}d + \dfrac{2}{5}$

24) $\dfrac{1}{5}c + \dfrac{2}{7} = 2 - \dfrac{1}{7}c$

25) $\dfrac{m}{3} + \dfrac{1}{2} = \dfrac{2m}{3} + 3$

26) $\dfrac{a}{8} - 1 = \dfrac{a}{3} - \dfrac{7}{12}$

 27) $\dfrac{1}{3} + \dfrac{1}{9}(k + 5) - \dfrac{k}{4} = 2$

28) $\dfrac{1}{2} = \dfrac{2}{9}(3x - 2) - \dfrac{x}{9} - \dfrac{x}{6}$

29) $0.05(t + 8) - 0.01t = 0.6$

30) $0.2(y - 3) + 0.05(y - 10) = -0.1$

31) $0.1x + 0.15(8 - x) = 0.125(8)$

32) $0.2(12) + 0.08z = 0.12(z + 12)$

33) $0.04s + 0.03(s + 200) = 27$

34) $0.06x + 0.1(x - 300) = 98$

Objectives 4 and 5

Solve using the five "Steps for Solving Applied Problems." See Examples 7–9.

35) Four more than a number is fifteen. Find the number.

36) Thirteen more than a number is eight. Find the number.

37) Seven less than a number is twenty-two. Find the number.

38) Nine less than a number is eleven. Find the number.

39) Twice a number is -16. Find the number.

40) The product of six and a number is fifty-four. Find the number.

41) Two-thirds of a number is ten. Find the number.

42) Three-fourths of a number is eighteen. Find the number.

43) Seven more than twice a number is thirty-five. Find the number.

44) Five more than twice a number is fifty-three. Find the number.

 45) Three times a number decreased by eight is forty. Find the number.

46) Twice a number decreased by seven is -13. Find the number.

47) Half of a number increased by ten is three. Find the number.

48) One-third of a number increased by four is one. Find the number.

49) Twice the sum of a number and five is sixteen. Find the number.

50) Twice the sum of a number and -8 is four. Find the number.

51) Three times a number is fifteen more than half the number. Find the number.

52) A number increased by fourteen is nine less than twice the number. Find the number.

53) A number decreased by six is five more than twice the number. Find the number.

54) A number divided by four is nine less than the number. Find the number.

Section 3.3 Applications of Linear Equations to General Problems, Consecutive Integers, and Fixed and Variable Cost

Objectives

1. Solve Problems Involving General Quantities

2. Solve Problems Involving Different Lengths

3. Solve Consecutive Integer Problems

4. Solve Problems Involving Fixed and Variable Costs

In the previous section you were introduced to solving applied problems. In this section we will build on these skills so that we can solve other types of problems.

1. Solve Problems Involving General Quantities

Example 1

Mrs. Ramirez has 26 students in her third-grade class. There are four more boys than girls. How many boys and girls are in her class?

Solution

Step 1: Read the problem carefully, twice.

Step 2: What are we asked to find?

We must find the number of boys and number of girls in the class.

Define the variable.

Look at the sentence, "There are four more boys than girls." The number of boys is expressed *in terms of* the number of girls. Therefore, let

$$x = \text{the number of girls}$$

Define the other unknown (the number of boys) in terms of x (the number of girls).

Since there are four *more* boys than girls,

The number of girls $+ 4 =$ the number of boys

$$x + 4 = \text{the number of boys}$$

We have defined the unknowns:

$$x = \text{the number of girls}$$
$$x + 4 = \text{the number of boys}$$

Step 3: Translate from English to an algebraic equation.

One suggestion given in the "Steps for Solving Applied Problems" on page 144 is to *restate the problem in your own words.*

One way to think of the situation in this problem is:

The number of girls plus the number of boys is 26.

Now, let's break down this statement into small parts:

In English	Meaning	In Math Terms
The number of girls . . .	number of girls	x
plus . . .	add	$+$
the number of boys . . .	number of boys	$x + 4$
is . . .	equals	$=$
26	26	26

"The number of girls plus the number of boys is 26" becomes

$\underbrace{\text{The number of girls}}_{x}$ $\underbrace{\text{plus}}_{+}$ $\underbrace{\text{the number of boys}}_{(x + 4)}$ $\underbrace{\text{is}}_{=}$ $\underbrace{26}_{26}$ becomes

Equation: $x + (x + 4) = 26$.

Step 4:   Solve the equation.

$$x + (x + 4) = 26$$
$$2x + 4 = 26$$
$$2x + 4 - 4 = 26 - 4 \qquad \text{Subtract 4.}$$
$$2x = 22$$
$$\frac{2x}{2} = \frac{22}{2} \qquad \text{Divide by 2.}$$
$$x = 11$$

Step 5:   Interpret the meaning of the solution as it relates to the problem.

$$x = \text{the number of girls}$$

There are 11 girls in the class.

Find any other unknown values we are asked to find.

$$x + 4 = \text{the number of boys}$$
$$11 + 4 = \text{the number of boys}$$

There are 15 boys in the class.

Does the answer make sense?

$$11 \text{ girls} + 15 \text{ boys} = 26 \text{ students}$$

Yes.

State the answer in a complete sentence.

There are 11 girls and 15 boys in this third-grade class.

You Try 1

The record low temperature for the month of January in Anchorage, Alaska, was (in °F) 84° less than January's record high temperature. The sum of the record low and record high is 16°. Find the lowest and highest temperatures ever recorded in Anchorage, Alaska, in January. (www.weather.com)

2. Solve Problems Involving Different Lengths

Example 2

A carpenter has a board that is 15 ft long. He needs to cut it into two pieces so that one piece is half as long as the other. What will be the length of each piece?

Solution

Step 1: Read the problem carefully, twice.

Step 2: We need to find the length of each piece when a 15-ft board is cut into two pieces. A picture will be very helpful in this problem.

Let x = the length of one piece

The other piece is half the length of the first. So,

$$\frac{1}{2}x = \text{the length of the second piece}$$

Step 3: Translate from English to an algebraic equation. Let's label the picture with the unknowns.

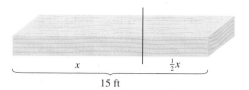

Look at the picture to develop an equation in English:

Length of one piece + Length of second piece = Total length of the board

$$x \qquad + \qquad \frac{1}{2}x \qquad = \qquad 15$$

Equation: $x + \frac{1}{2}x = 15$.

Step 4: Solve the equation.

$$x + \frac{1}{2}x = 15$$
$$\frac{3}{2}x = 15$$
$$\frac{2}{3} \cdot \frac{3}{2}x = \frac{2}{3} \cdot 15 \qquad \text{Multiply by the reciprocal of } \frac{3}{2}.$$
$$x = 10$$

Step 5: Interpret the meaning of the solution as it relates to the problem, and find the other unknown. One piece is 10 ft long. The length of the other piece is $\frac{1}{2}x$.

$$\frac{1}{2}(10) = 5$$

The second piece is 5 ft long.

This makes sense since 10 ft + 5 ft = 15 ft.

One board is 10 ft long, and the other board is 5 ft long.

You Try 2

A 24-ft chain must be cut into two pieces so that one piece is twice as long as the other. Find the length of each piece of chain.

3. Solve Consecutive Integer Problems

The idea of consecutive numbers comes up in many places. What *are* consecutive integers? "Consecutive" means one after the other, in order. In this section, we will look at consecutive integers, consecutive even integers, and consecutive odd integers.

Consecutive integers: An example of some consecutive integers is 5, 6, 7, 8. We will define these in terms of a variable. How do we do that?

Let $x =$ the first consecutive integer (In this case, that is 5.)

$$
\begin{array}{cccc}
5 & 6 & 7 & 8 \\
x & x + 1 & x + 2 & x + 3
\end{array}
$$

To get from 5 to 6, add 1 to 5. So, if $x = 5$, then $x + 1 = 6$.

To get from 5 to 7, add 2 to 5. So, if $x = 5$, then $x + 2 = 7$.

To get from 5 to 8, add 3 to 5. So, if $x = 5$, then $x + 3 = 8$.

Therefore, to define the unknowns for consecutive integers, let

$$
\begin{aligned}
x &= \text{the first integer} \\
x + 1 &= \text{the second integer} \\
x + 2 &= \text{the third integer} \\
x + 3 &= \text{the fourth integer}
\end{aligned}
$$

and so on.

Example 3

The sum of three consecutive integers is 126. Find the integers.

Solution

Step 1:   Read the problem carefully, twice.

Step 2:   Find three consecutive integers. Define the unknowns.

$$
\begin{aligned}
x &= \text{the first integer} \\
x + 1 &= \text{the second integer} \\
x + 2 &= \text{the third integer}
\end{aligned}
$$

Step 3:   Think of the information in small parts. Write an equation in English and then in terms of the variables.

We must *add* the three integers to get 126.

First integer + Second integer + Third integer = 126
$$x \quad + \quad (x + 1) \quad + \quad (x + 2) \quad = 126$$

Equation: $x + (x + 1) + (x + 2) = 126$

Step 4: Solve the equation.

$$x + (x + 1) + (x + 2) = 126$$

$3x + 3 = 126$	Combine like terms.
$3x = 123$	Subtract 3.
$x = 41$	Divide by 3.

Step 5: $x =$ the first integer $= 41$
$x + 1 =$ the second integer $= 42$
$x + 2 =$ the third integer $= 43$

If we add them, is the result 126?

$$41 + 42 + 43 = 126 \quad \checkmark \quad \text{Yes.}$$

The integers are 41, 42, and 43.

You Try 3

The sum of three consecutive integers is 177. Find the integers.

Next, let's look at consecutive even integers and consecutive odd integers. An example of each type of sequence of numbers is given here:

Consecutive even integers: $-10, -8, -6, -4$
Consecutive odd integers: $9, 11, 13, 15$

Consider the **even integers** first. Let $x =$ the first even integer.

-10	-8	-6	-4
x	$x + 2$	$x + 4$	$x + 6$

To get from -10 to -8, add 2 to -10. Therefore, if $x = -10$, $x + 2 = -8$.
To get from -10 to -6, add 4 to -10. If $x = -10$, $x + 4 = -6$.
To get from -10 to -4, add 6 to -10. So, if $x = -10$, $x + 6 = -4$.

Therefore, to define the unknowns for consecutive even integers, let

$$x = \text{the first even integer}$$
$$x + 2 = \text{the second even integer}$$
$$x + 4 = \text{the third even integer}$$
$$x + 6 = \text{the fourth even integer}$$

and so on.

Will the expressions for consecutive odd numbers be any different? No! When we count by consecutive odds, we are still counting by twos.

Returning to the sequence of consecutive odd integers 9, 11, 13, 15, if

$$x = \text{the first odd integer, then}$$

$$x + 2 = \text{the second odd integer}$$
$$x + 4 = \text{the third odd integer}$$
$$x + 6 = \text{the fourth odd integer}$$

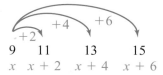

$$\begin{array}{cccc} 9 & 11 & 13 & 15 \\ x & x+2 & x+4 & x+6 \end{array}$$

Example 4

The sum of two consecutive odd integers is 47 more than three times the larger integer. Find the integers.

Solution

Step 1:   Read the problem slowly and carefully, twice.

Step 2:   Find two consecutive odd integers. Define the unknowns.

$$x = \text{the first odd integer}$$
$$x + 2 = \text{the second odd integer}$$

Step 3:   Write an equation. Think of the information in small parts.

In English	Meaning	In Math Terms
Sum of . . .	add together the expressions below	
Two consecutive odd integers . . .	the two consecutive odd integers are x and $x+2$	$x, x + 2$
is . . .	equals	$=$
47 more than . . .	47 added to . . .	$47 +$
Three times the larger integer	multiply the larger integer $(x + 2)$ by 3	$3(x + 2)$

Therefore, we get

The sum of two
consecutive odd integers   is   47 more than   three times the larger integer
$$x + (x + 2) \qquad = \qquad 47 + \qquad\qquad 3(x + 2)$$

Equation: $x + (x + 2) = 47 + 3(x + 2)$.

Step 4: Solve the equation.

$$x + (x + 2) = 47 + 3(x + 2)$$
$$2x + 2 = 47 + 3x + 6 \qquad \text{Distribute.}$$
$$2x + 2 = 53 + 3x \qquad \text{Combine like terms.}$$
$$-x + 2 = 53 \qquad \text{Subtract } 3x \text{ from each side.}$$
$$-x = 51 \qquad \text{Subtract 2 from each side.}$$
$$x = -51 \qquad \text{Divide by } -1.$$

Step 5:

$$x = \text{the first integer} = -51$$
$$x + 2 = \text{the second integer} = -49$$

The numbers are -51 and -49.

You Try 4

Twice the sum of three consecutive even integers is 18 more than five times the largest number. Find the integers.

4. Solve Problems Involving Fixed and Variable Costs

Many types of expenses are based on both a fixed cost and a variable cost. Renting a car is one such example.

Example 5

To rent a car for 1 day, Smooth-Ride Rental Car charges $19.95 plus $0.26 per mile to rent one of its economy cars. How much would it cost to rent the car for 1 day if you intend to drive 68 miles?

Solution

Driving for 1 day means there will be a **fixed cost** of $19.95 for the 1 day no matter how many miles you drive.

But, the cost of mileage is the **variable cost** because how much you pay *varies* depending on how many miles you drive.

$$\text{Cost of mileage} = (\text{Price per mile})(\text{Number of miles})$$
$$= (\$0.26)(68)$$
$$= \$17.68$$

Therefore,

$$\text{Total cost} = \text{Fixed cost} + \text{Cost of mileage}$$
$$= \$19.95 + (\$0.26)(68)$$
$$= \$19.95 + \$17.68$$
$$= \$37.63$$

It would cost $37.63 to rent the car.

You Try 5

To rent an SUV for 1 day, Smooth-Ride Rental Car charges $24.95 plus $0.30 per mile. How much would it cost to rent the SUV for 1 day if you intend to drive 33 miles?

Next, we will look at a problem that uses algebra.

Example 6

Tony's Truck Rental charges $55 per day plus $0.35 per mile to rent a 12-ft box truck. Jason rented one of these trucks for 2 days, and his bill came to $166. How many miles did Jason drive the truck?

Solution

Step 1: Read the problem carefully, twice.

Step 2: Find the number of miles Jason drove the truck. Define the unknown.

$$x = \text{number of miles Jason drove the truck}$$

Step 3: Translate from English to an algebraic equation. Use the idea in Example 5.

$$\text{Total cost} = \text{Fixed cost} + \text{Cost of mileage}$$

The *fixed cost* is how much it costs to rent the car for 2 days no matter how many miles are driven.

$$\overset{\text{2 days}\quad\text{Amount per day}}{\searrow\qquad\swarrow}$$
$$\text{Fixed cost for 2 days} = 2(\$55) = \$110$$

The cost of mileage varies depending on how many miles are driven:

$$\text{Cost of mileage} = \$0.35x$$
$$\overset{}{\nearrow\qquad\nwarrow}$$
$$\text{Cost per mile}\qquad\text{Number of miles}$$

The total cost is $166.

$$\text{Total cost} = \text{Fixed cost for 2 days} + \text{Cost of mileage}$$
$$\downarrow\qquad\qquad\downarrow\qquad\qquad\downarrow$$
$$166\quad=\qquad\quad 110\qquad +\qquad 0.35x$$

Equation: $166 = 110 + 0.35x$.

Step 4: Solve the equation.

$$166 = 110 + 0.35x$$
$$56 = 0.35x \qquad \text{Subtract 110.}$$
$$160 = x \qquad \text{Divide by 0.35.}$$

Step 5: Interpret the meaning of the solution as it relates to the problem.

Jason drove the truck 160 miles.
Does the answer make sense?

$$\text{Cost of renting the truck} = \$110 + \$0.35(160)$$
$$= \$110 + \$56$$
$$= \$166 \quad \text{Yes, it makes sense.}$$

Jason drove the truck 160 miles.

You Try 6

Tony's Truck Rental charges $46 per day plus $0.35 per mile to rent an 8-ft box truck. Jason rented one of these trucks for 3 days, and his bill came to $208. How many miles did Jason drive the truck?

Remember, apply the Steps for Solving Applied Problems on p. 144 when you are solving the problems in the exercises.

Answers to You Try Exercises

1) record low: −34°F, record high: 50°F   2) 8 ft and 16 ft   3) 58, 59, 60

4) 26, 28, 30   5) $34.85   6) 200 mi

3.3 Exercises

Boost your grade at mathzone.com! MathZone

> Practice Problems   > NetTutor   > e-Professors   > Videos
> Self-Test

1) On a baseball team, there are 5 more pitchers than catchers. If there are c catchers on the team, write an expression for the number of pitchers.

2) On Wednesday, the Snack Shack sold 23 more hamburgers than hot dogs. Write an expression for the number of hamburgers sold if h hot dogs were sold.

3) There were 31 fewer people on a flight from Chicago to New York than from Chicago to Los Angeles. Write an expression for the number of people on the flight to New York if there were p people on the flight to L.A.

4) The test average in Mr. Muscari's second-period class was 3.8 points lower than in his first-period class. If the average test score in the first-period class was a, write an expression for the test average in Mr. Muscari's second-period class.

5) A survey of adults aged 20–29 revealed that, of those who exercise regularly, three times as many men run as women. If w women run on a regular basis, write an expression for the number of male runners.

6) At Roundtree Elementary School, s students walk to school. One-third of that number ride their bikes.

Write an expression for the number of students who ride their bikes to class.

7) An electrician cuts a 14-ft wire into two pieces. If one is x ft long, how long is the other piece?

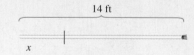

8) Jorie drives along a 142-mi stretch of highway from Oregon to California. If she drove m mi in Oregon, how far did she drive in California?

Objective 1

Solve using the five "Steps for Solving Applied Problems." See Example 1.

9) A 12-oz serving of Pepsi has 6.5 more teaspoons of sugar than a 12-oz serving of Gatorade. Together they contain 13.1 teaspoons of sugar. How much sugar is in each 12-oz drink? (www.dentalgentlecare.com)

10) Two of the smallest countries in the world are the Marshall Islands and Liechtenstein. The Marshall Islands covers 8 mi^2 more than Liechtenstein. Find the area of each country if together they encompass 132 mi^2. (www.infoplease.com)

11) In the 2004 Summer Olympics in Athens, Thailand won half as many medals as Greece. If they won a total of 24 medals, how many were won by each country? (www.olympics.org)

12) Latisha's golden retriever weighs twice as much as Janessa's border collie. Find the weight of each dog if they weigh 96 lb all together.

13) The Columbia River is 70 miles shorter than the Ohio River. Determine the length of each river if together they span 2550 miles. (ga.water.usgs.gov)

14) In the 2002 Alabama gubernatorial election, Don Siegelman had 3120 fewer votes than Bob Riley. If they received a total of 1,341,330 votes, how many people voted for each man? (www.sos.state.al.us)

Objective 2

Solve using the five "Steps for Solving Applied Problems." See Example 2.

15) A plumber has a 36-in. long pipe. He must cut it into two pieces so that one piece is 14 inches longer than the other. How long is each piece?

16) A builder has to install some decorative trim on the outside of a house. The piece she has is 75 in. long, but she needs to cut it so that one piece is 27 in. shorter than the other. Find the length of each piece.

17) Calida's mom found an 18-ft-long rope in the garage. She will cut it into two pieces so that one piece can be used for a long jump rope and the other for a short one. If the long rope is to be twice as long as the short one, find the length of each jump rope.

18) A 55-ft-long drainage pipe must be cut into two pieces before installation. One piece is two-thirds as long as the other. Find the length of each piece.

Objective 3

Solve using the five "Steps for Solving Applied Problems." See Examples 3 and 4.

19) The sum of three consecutive integers is 195. Find the integers.

20) The sum of two consecutive integers is 77. Find the integers.

21) Find two consecutive even integers such that twice the smaller is 10 more than the larger.

22) Find three consecutive odd integers such that four times the smallest is 56 more than the sum of the other two.

23) Find three consecutive odd integers such that their sum is five more than four times the largest integer.

24) The sum of two consecutive even integers is 52 less than three times the larger integer. Find the integers.

25) Two consecutive page numbers in a book add up to 345. Find the page numbers.

26) The addresses on the east side of Arthur Ave. are consecutive odd numbers. Two consecutive house numbers add up to 36. Find the addresses of these two houses.

Objective 4

Solve each problem. See Example 5.

27) Harlan's Rentals charges $25 plus $6 per hour to rent a machine to till the soil. What would it cost to rent the tiller for 4 hr?

28) Trusty Rental Cars charges $129.95 per week (7 days) plus $19.95 per day for additional days. What would it cost to rent a car for 10 days?

Solve using the five "Steps for Solving Applied Problems." See Example 6.

29) Industrial Rentals charges $95.00 per day plus $0.60 per mile to rent a dump truck. Charlie paid $133.40 for a 1-day rental. How far did he drive the dump truck?

30) Ciu Ling rented a cargo van to move her son to college. The rental company charges $42.00 per day plus $0.32 per mile. She used the van for 2 days, and her bill came to $158.88. How many miles did she drive the van?

31) A+ Rental Cars charges $39.00 per day plus $.15 per mile. If Luis spent $57.60 on a 1-day rental, how many miles did he drive?

32) Hannah rented a car for 1 day, and her bill amounted to $50.76. The rental company charges $42.00 per day and $0.12 per mile. How far did Hannah drive the car?

Objectives 1–4

Mixed Exercises

Write an equation and solve. Use the five "Steps for Solving Word Problems."

33) Irina was riding her bike when she got a flat tire. Then, she walked her bike 1 mi more than half the distance she rode it. How far did she ride her bike and how far did she walk if the total distance she traveled was 7 mi?

34) In 2002, Mia Hamm played in eight fewer U.S. National Team soccer matches than in 2003. During those 2 years she appeared in a total of 26 games. In how many games did she play in 2002 and in 2003? (www.ussoccer.com)

 35) One-sixth of the smallest of three consecutive even integers is three less than one-tenth the sum of the other even integers. Find the integers.

36) Five times the sum of two consecutive integers is two more than three times the larger integer. Find the integers.

37) In 2001, *Harry Potter* grossed $42 million more in theaters than *Shrek*. Together they earned $577.4 million.

How much did each movie earn in theaters in 2001? (www.boxofficereport.com)

38) Beach Net, an Internet cafe on the beach, charges customers a $2.00 sign-on fee plus $0.50 for each 15 min on the computer. For how long could a customer surf the net for $4.50?

39) Bonus Car Rental charges $39.95 per day plus $0.10 per mile to rent a midsize car. Jonas spent $54.75 on a 1-day rental. How far did he drive the car?

40) A 53-in. board is to be cut into three pieces so that one piece is 5 in. longer than the shortest piece, and the other piece is twice as long as the shortest piece. Find the length of each piece of board.

41) The three top-selling albums of 2002 were by Eminem, Nelly, and Avril Lavigne. Nelly sold 0.8 million more albums than Avril Lavigne while Eminem's album sold 3.5 million more than Avril's. Find the number of albums sold by each artist if together they sold 16.6 million copies. (www.andpop.com)

42) To produce its most popular skateboard, Top Dog Boards has a fixed cost in its factory of $5800 plus a cost of $6.20 for each skateboard built. The cost of production during the month of October was $11,008. How many skateboards were produced?

43) A 58-ft-long cable must be cut into three pieces. One piece will be 6 ft longer than the shortest piece, and the longest portion must be twice as long as the shortest. Find the length of the three pieces of cable.

44) A book is open to two pages numbered consecutively. The sum of the page numbers is 373. What are the page numbers?

Section 3.4 Applications of Linear Equations to Percent Increase/Decrease and Investment Problems

Objectives

1. Find the Sale Price of an Item
2. Solve Problems Involving Percent Change
3. Solve Problems Involving Simple Interest

Percents pop up everywhere. "Earn 4% simple interest on your savings account." "The unemployment rate decreased 1.2% this year." We've seen these statements in the newspaper, or we have heard about them on television. In this section, we will introduce applications involving percents.

Here's another type of percent problem we might see in a store: "Everything in the store is marked down 30%." Before tackling an algebraic percent problem, let's look at an arithmetic problem. Relating an algebra problem to an arithmetic problem can make it easier to solve an application that requires the use of algebra.

1. Find the Sale Price of an Item

Example 1

Jeans that normally sell for $28.00 are marked down 30%. What is the sale price?

Solution

Concentrate on the *procedure* used to obtain the answer. This is the same procedure we will use to solve algebra problems with percent increase and percent decrease.

$$\text{Sale price} = \text{Original price} - \text{Amount of discount}$$

How much is the discount? It is 30% of $28.00.
The amount of the discount is calculated by multiplying:

$$\text{Amount of discount} = (\text{Rate of discount})(\text{Original price})$$
$$\text{Amount of discount} = \quad (0.30) \quad \cdot \quad (\$28.00) \quad = \$8.40$$

$$
\begin{aligned}
\text{Sale price} &= \text{Original price} - \text{Amount of discount} \\
&= \quad \$28.00 \quad - \quad (0.30)(\$28.00) \\
&= \quad \$28.00 \quad - \quad \$8.40 \\
&= \quad \$19.60
\end{aligned}
$$

The sale price is $19.60.

You Try 1

A dress shirt that normally sells for $39.00 is marked down 25%. What is the sale price?

2. Solve Problems Involving Percent Change

Next, let's solve an algebra problem involving a markdown or percent decrease.

Example 2

The sale price of a Beyoncé CD is $14.80 after a 20% discount. What was the original price of the CD?

Solution

Step 1: Read the problem carefully, twice.

Step 2: Find the original price of the CD. Define the unknown.

$$x = \text{the original price of the CD}$$

Step 3: Translate from English to an algebraic equation.
Remember, one strategy is to think of the problem in another way. We can think of this problem as

$$\text{Sale price} = \text{Original price} - \text{Amount of discount}$$

The amount of the discount is calculated by multiplying:
(rate of discount)(original price). See Example 1.
So, the amount of the discount is $(0.20)(x)$.

$$
\begin{array}{ccccc}
\text{Sale price} & = & \text{Original price} & - & \text{Amount of discount} \\
14.80 & = & x & - & 0.20x
\end{array}
$$

Equation: $14.80 = x - 0.20x$.

Step 4: Solve the equation.

$$
\begin{aligned}
14.80 &= x - 0.20x \\
14.80 &= 0.80x \qquad \text{Combine like terms.} \\
18.5 &= x \qquad \text{Divide by 0.80.}
\end{aligned}
$$

Step 5: Interpret the meaning of the solution as it relates to the problem. The original price of the CD was $18.50.

Does the answer make sense?

$$
\begin{aligned}
(\$18.50)(0.20) &= \$3.70 \qquad \text{(Amount of 20\% discount)} \\
\$18.50 - \$3.70 &= \$14.80 \qquad \text{Yes.}
\end{aligned}
$$

The original price of the CD was $18.50.

 You Try 2

> A video game is on sale for $35.00 after a 30% discount. What was the original price of the video game?

3. Solve Problems Involving Simple Interest

When customers invest their money in bank accounts, their accounts earn interest. Why? Paying interest is a way for financial institutions to get people to deposit money.

There are different ways to calculate the amount of interest earned from an investment. In this section, we will discuss *simple interest*. **Simple interest** calculations are based on the initial amount of money deposited in an account. This initial amount of money is known as the **principal**.

The formula used to calculate simple interest is $I = PRT$, where

$$I = \text{interest earned (simple)}$$
$$P = \text{principal (initial amount invested)}$$
$$R = \text{annual interest rate (expressed as a decimal)}$$
$$T = \text{amount of time the money is invested (in years)}$$

We will begin with two arithmetic problems. The procedures used will help you understand more clearly how we arrive at the algebraic equation in Example 5.

Example 3

If \$500 is invested for 1 year in an account earning 6% simple interest, how much interest will be earned?

Solution

Use $I = PRT$ to find I, the interest earned.

$$P = \$500,\ R = 0.06,\ T = 1$$
$$I = PRT$$
$$I = (500)(0.06)(1)$$
$$I = 30$$

The interest earned will be \$30.

You Try 3

If \$3500 is invested for 2 years in an account earning 4.5% simple interest, how much interest will be earned?

Example 4

Tom invests \$2000 in an account earning 7% interest and \$9000 in an account earning 5% interest. After 1 year, how much interest will he have earned?

Solution

Tom will earn interest from two accounts. Therefore,

Total interest earned = Interest from 7% account + Interest from 5% account

Using $I = PRT$, we get

$$\text{Total interest earned} = (2000)(0.07)(1) + (9000)(0.05)(1)$$
$$\text{Total interest earned} = \qquad 140 \qquad + \qquad 450$$
$$= \$590$$

Tom will earn a total of \$590 in interest from the two accounts.

You Try 4

Donna invests $1500 in an account earning 4% interest and $4000 in an account earning 6.7% interest. After 1 year, how much interest will she have earned?

This idea of earning interest from different accounts is one we will use in Example 5.

Example 5

Last year, Neema Reddy had $10,000 to invest. She invested some of it in a savings account that paid 3% simple interest, and she invested the rest in a certificate of deposit that paid 5% simple interest. In 1 year, she earned a total of $380 in interest. How much did Neema invest in each account?

Solution

Step 1: Read the problem carefully, twice.

Step 2: Find the amount Neema invested in the 3% account and the amount she invested in the 5% account. Define the unknowns.

Let x = amount Neema invested in the 3% account.

How do we write an expression, in terms of x, for the amount invested in the 5% account?

Total invested   Amount invested in 3% account
$\downarrow$ $\downarrow$

10,000 $-$ x = Amount invested in the 5% account

We define the unknowns as

x = amount Neema invested in the 3% account
$10,000 - x$ = amount Neema invested in the 5% account

Step 3: Translate from English to an algebraic equation.

We will use the "English equation" we used in Example 4 since Neema is earning interest from two accounts.

Total interest earned = Interest from 3% account + Interest from 5% account

$$\underset{380}{} \quad = \quad \underset{x(0.03)(1)}{\overset{P \quad r \quad t}{}} \quad + \quad \underset{(10,000 - x)(0.05)(1)}{\overset{P \qquad r \quad t}{}}$$

The equation can be rewritten as

$$380 = 0.03x + 0.05(10,000 - x)$$

Step 4: Solve the equation.

Begin by multiplying both sides of the equation by 100 to eliminate the decimals.

$$100(380) = 100[0.03x + 0.05(10,000 - x)]$$
$$38,000 = 3x + 5(10,000 - x) \qquad \text{Multiply by 100.}$$

$$38,000 = 3x + 50,000 - 5x \quad \text{Distribute.}$$
$$38,000 = -2x + 50,000 \quad \text{Combine like terms.}$$
$$-12,000 = -2x \quad \text{Subtract 50,000.}$$
$$6000 = x \quad \text{Divide by } -2.$$

Step 5: Interpret the meaning of the solution as it relates to the problem.

Neema invested $6000 at 3% interest.

Find the other unknown. The amount invested at 5% is $10,000 - x$.

$$10,000 - x = 10,000 - 6000$$
$$= 4000$$

Neema invested $4000 at 5% interest.

Does the answer make sense?

Total interest earned = Interest from 3% account + Interest from 5% account

$$380 \quad = \quad 6000(0.03)(1) \quad + \quad 4000(0.05)(1)$$
$$= \quad 180 \quad + \quad 200$$
$$= 380$$

Yes, the answer makes sense.

Neema invested $6000 in the savings account earning 3% interest and $4000 in the certificate of deposit earning 5% interest.

You Try 5

Christine received an $8000 bonus from work. She invested part of it at 6% simple interest and the rest at 4% simple interest. Christine earned a total of $420 in interest after 1 year. How much did she deposit in each account?

Answers to You Try Exercises

1) $29.25   2) $50.00   3) $315   4) $328   5) $5000 at 6% and $3000 at 4%

3.4 Exercises

Boost your grade at mathzone.com! MathZone

> Practice Problems   > NetTutor   > Self-Test   > e-Professors   > Videos

Objective 1

Find the sale price of each item.

1) A cell phone that regularly sells for $75.00 is marked down 15%.

2) A baby stroller that retails for $69.00 is on sale at 20% off.

3) A sign reads, "Take 30% off the original price of all DVDs." The original price on the DVD you want to buy is $16.50.

4) The $120.00 basketball shoes James wants are now on sale at 15% off.

5) At the end of the summer, the bathing suit that sold for $29.00 is marked down 60%.

6) An advertisement states that a TV that regularly sells for $399.00 is being discounted 25%.

Objective 2

Solve using the five "Steps for Solving Applied Problems." See Example 2.

7) A digital camera is on sale for $119 after a 15% discount. What was the original price of the camera?

8) Marie paid $15.13 for a hardcover book that was marked down 15%. What was the original selling price of the book?

9) In February, a store discounted all of its calendars by 60%. If Ramon paid $4.38 for a calendar, what was its original price?

10) An appliance store advertises 20% off all of its refrigerators. Mr. Kotaris paid $399.20 for the fridge. Find its original price.

11) The sale price of a coffeemaker is $22.75. This is 30% off of the original price. What was the original price of the coffeemaker?

12) Katrina paid $20.40 for a backpack that was marked down 15%. Find the original retail price of the backpack.

13) One hundred forty countries participated in the 1984 Summer Olympics in Los Angeles. This was 75% more than the number of countries that took part in the Summer Olympics in Moscow 4 years earlier. How many countries participated in the 1980 Olympics in Moscow? (www.olympics.org)

14) In 2006 there were about 1224 acres of farmland in Custer County. This is 32% less than the number of acres of farmland in 2000. Calculate the number of acres of farmland in Custer County in 2000.

15) In 2004, there were 7569 Starbucks stores worldwide. This is approximately 1681% more stores than 10 years earlier. How many Starbucks stores were there in 1994? (Round to the nearest whole number.)
(www.starbucks.com)

16) Liu Fan's salary this year is 14% higher than it was 3 years ago. If he earns $37,050 this year, what did he earn 3 years ago?

Objective 3

Solve. See Examples 3 and 4.

17) Jenna invests $800 in an account for 1 year earning 4% simple interest. How much interest was earned from this account?

18) Last year, Mr. Jaworski deposited $11,000 in an account earning 7% simple interest for one year. How much interest was earned?

19) Sven Andersson invested $6500 in an account earning 6% simple interest. How much money will be in the account 1 year later?

20) If $3000 is deposited into an account for 1 year earning 5.5% simple interest, how much money will be in the account after 1 year?

21) Rachel Levin has a total of $5500 to invest for 1 year. She deposits $4000 into an account earning 6.5% annual simple interest and the rest into an account earning 8% annual simple interest. How much interest did Rachel earn?

22) Maurice plans to invest a total of $9000 for 1 year. In the account earning 5.2% simple interest he will deposit $6000, and in an account earning 7% simple interest he will deposit the rest. How much interest will Maurice earn?

Solve using the five "Steps for Solving Applied Problems." See Example 5.

23) Amir Sadat receives a $15,000 signing bonus on accepting his new job. He plans to invest some of it at 6% annual simple interest and the rest at 7% annual simple interest. If he will earn $960 in interest after 1 year, how much will Amir invest in each account?

24) Lisa Jenkins invested part of her $8000 inheritance in an account earning 5% simple interest and the rest in an account earning 4% simple interest. How much did

Lisa invest in each account if she earned $365 in total interest after 1 year?

25) Enrique's money earned $164 in interest after 1 year. He invested some of his money in an account earning 6% simple interest and $200 more than that amount in an account earning 5% simple interest. Find the amount Enrique invested in each account.

26) Saori Yamachi invested some money in an account earning 7.4% simple interest and twice that amount in an account earning 9% simple interest. She earned $1016 in interest after 1 year. How much did Saori invest in each account?

27) Last year, Clarissa invested a total of $7000 in two accounts earning simple interest. Some of it she invested at 9.5%, and the rest was invested at 7%. How much did she invest in each account if she earned a total of $560 in interest last year?

28) Ted has $3000 to invest. He deposits a portion of it in an account earning 5% simple interest and the rest at 6.5% simple interest. After 1 year he has earned $175.50 in interest. How much did Ted deposit in each account?

Objectives 2 and 3

Mixed Exercises

Solve using the five "Steps for Solving Applied Problems."

29) In her gift shop, Cheryl sells all stuffed animals for 60% more than what she paid her supplier. If one of these toys sells for $14.00 in her shop, what did it cost Cheryl?

30) Ivan has $8500 to invest. He will invest some of it in a long-term IRA paying 4% simple interest and the rest in a short-term CD earning 2.5% simple interest. After 1 year, Ivan's investments have earned $250 in interest. How much did Ivan invest in each account?

31) In Johnson County, 8330 people were collecting unemployment benefits in September 2006. This is 2% less than the number collecting the benefits in September 2005. How many people in Johnson County were getting unemployment benefits in September 2005?

32) Amari bought a new car for $13,640. This is 12% less than the car's sticker price. What was the sticker price of the car?

33) Tamara invests some money in three different accounts. She puts some of it in a CD earning 3% simple interest and twice as much in an IRA paying 4% simple interest. She also decides to invest $1000 more than what she's invested in the CD into a mutual fund earning 5% simple interest. Determine how much money Tamara invested in each account if she earned $290 in interest after 1 year.

34) Henry marks up the prices of his fishing poles by 50%. Determine what Henry paid his supplier for his best-selling fishing pole if Henry charges his customers $33.75.

35) Find the original price of a cell phone if it costs $63.20 after a 20% discount.

36) It is estimated that in 2003 the number of Internet users in Slovakia was 40% more than the number of users in Kenya. If Slovakia had 700,000 Internet users in 2003, how many people used the Internet in Kenya that year? (www.theodora.com)

37) Zoe's current salary is $40,144. This is 4% higher than last year's salary. What was Zoe's salary last year?

38) Sunil earns $375 in interest from 1-year investments. He invested some money in an account earning 6% simple interest, and he deposited $2000 more than that amount in an account paying 5% simple interest. How much did Sunil invest in each account?

39) On his 21st birthday, Jerry received $20,000 from a trust fund. He invested some of the money in an account earning 5% simple interest and put the rest in a high-risk investment paying 9% simple interest. After 1 year, he earned a total of $1560 in interest. How much did Jerry invest in each account?

40) In August 2007, there were 2600 entering freshmen at a state university. This is 4% higher than the number of freshmen entering the school in August of 2006. How many freshmen were enrolled in August 2006?

Section 3.5 Geometry Applications and Solving Formulas for a Specific Variable

Objectives

1. Solve Problems Using Formulas from Geometry
2. Solve Problems Involving Angle Measures
3. Solve Problems Involving Complementary and Supplementary Angles
4. Solve an Equation for a Specific Variable

In this section, we will make use of concepts and formulas from geometry (see Section 1.3) to solve applied problems. We will also learn how to solve a formula for a specific variable. We begin with geometry.

1. Solve Problems Using Formulas from Geometry

Example 1

The area of a rectangular room is 180 ft². Its width is 12 ft. What is the length of the room?

Solution

Step 1: Read the problem carefully, twice. Draw a picture.

12 ft

l   Area = 180 ft²

Step 2: Find the length of the room. Define the unknown.

$$l = \text{length of the room}$$

Step 3: Translate from English to an algebraic equation.

In geometry problems, our tactics will be slightly different because we are often dealing with known formulas. Let's identify the information given in the problem:

$$\text{Area of the rectangular room} = 180 \text{ ft}^2$$
$$\text{Width of the rectangular room} = 12 \text{ ft}$$

We are asked to find the length of the room. So, what formula involves the area, length, and width of a rectangle?

$$A = lw$$

where A = area, l = length, and w = width.

Next, substitute the known values into the formula.

$$A = lw$$
$$180 = l(12)$$

Step 4: Solve the equation.

We can rewrite the equation as

$$180 = 12l$$
$$\frac{180}{12} = \frac{12l}{12} \qquad \text{Divide by 12.}$$
$$15 = l$$

Step 5:   Interpret the meaning of the solution as it relates to the problem.
The length of the room is 15 ft.

> Remember to include the correct units!

Does the answer make sense?

$$\text{Area} = lw$$
$$180 = 15(12)$$
$$180 = 180 \qquad \text{Yes, the answer makes sense.}$$

You Try 1

The area of a rectangular ping-pong table is 45 ft². Its length is 9 ft. What is the width of the ping-pong table?

2. Solve Problems Involving Angle Measures

Example 2

Find the missing angle measures.

Solution

Step 1:   Read the problem carefully, twice.

Step 2:   Find the measures of the angles labeled as $x°$ and $(3x + 5)°$.

Step 3:   Translate from English to an algebraic equation.

To write an equation, we need to understand the relationship between the angles of a triangle. Recall from Section 1.3 that

The sum of the angle measures of a triangle is 180°.

We can write an equation in English:

Measure of one angle + Measure of second angle + Measure of third angle = 180
$$x \qquad + \qquad (3x + 5) \qquad + \qquad 83 \qquad = 180$$

Equation: $x + (3x + 5) + 83 = 180$.

Step 4:   Solve the equation.

$$x + (3x + 5) + 83 = 180$$
$$4x + 88 = 180 \qquad \text{Combine like terms.}$$
$$4x = 92 \qquad \text{Subtract 88.}$$
$$x = 23 \qquad \text{Divide by 4.}$$

Step 5:   Interpret the meaning of the solution as it relates to the problem.
The measure of one angle of the triangle is 23°.
Find the other unknown.

$$3x + 5 = \text{measure of the other unknown angle.}$$
$$3(23) + 5 = 69 + 5$$
$$= 74$$

The measure of the second unknown angle is 74°.
Does the answer make sense? Do the angle measures add to 180°?

$$23° + 74° + 83° = 180°$$

Yes, it makes sense.
The measure of one angle is 23°, and the measure of the other angle is 74°.

You Try 2

Find the missing angle measures.

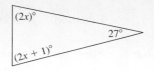

3. Solve Problems Involving Complementary and Supplementary Angles

Recall from Section 1.3 that

Two angles are **complementary** if the sum of their angles is 90°.
Two angles are **supplementary** if the sum of their angles is 180°.

If the measure of $\angle A$ is 52°, then

  a)   The measure of its complement is $90° - 52° = 38°$.
  b)   The measure of its supplement is $180° - 52° = 128°$.

Now, let's say an angle has a measure, $x°$. Using the same reasoning as above

  a)   The measure of its complement is $90 - x$.
  b)   The measure of its supplement is $180 - x$.

We will use these ideas to solve Example 3.

Example 3

Twice the complement of an angle is 18° less than the supplement of the angle. Find the measure of the angle.

Solution
Step 1:   Read the problem carefully, twice.
Step 2:   Find the measure of the angle. Define the unknowns.

$$x = \text{the measure of the angle}$$

Additionally, we need to write expressions for the complement and the supplement of the angle.

$$90 - x = \text{the measure of the complement}$$
$$180 - x = \text{the measure of the supplement}$$

Step 3:   Translate from English to an algebraic equation.
Think of this problem in "small parts."

In English	Meaning	In Math Terms
Twice the complement of an angle . . .	multiply the complement by 2	$2(90 - x)$
is . . .	equals	$=$
18° less than . . .	subtract 18 from the quantity below	-18
the supplement of the angle.	the supplement of the angle	$(180 - x)$

Twice the complement
of an angle   is   18° less than   the supplement
↓   ↓

Equation: $2(90 - x)$ $=$ $(180 - x)$ $-$ 18

Step 4:   Solve the equation.

$$2(90 - x) = (180 - x) - 18$$
$$180 - 2x = 162 - x \qquad \text{Distribute; combine like terms.}$$
$$180 - 2x + 2x = 162 - x + 2x \qquad \text{Add } 2x.$$
$$180 = 162 + x$$
$$18 = x \qquad \text{Subtract 162.}$$

Step 5:   Interpret the meaning of the solution as it relates to the problem.
The measure of the angle is 18°.

$$\text{Check:} \quad 2(90 - x) = (180 - x) - 18$$
$$2(90 - 18) = (180 - 18) - 18$$
$$2(72) = 162 - 18$$
$$144 = 144$$

The measure of the angle is 18°.

You Try 3

Ten times the complement of an angle is 9° less than the supplement of the angle. Find the measure of the angle.

4. Solve an Equation for a Specific Variable

The formula $P = 2l + 2w$ allows us to find the perimeter of a rectangle when we know its length (l) and width (w). But, what if we were solving problems where we repeatedly needed to find the value of w? Then, we could rewrite $P = 2l + 2w$ so that it is solved for w:

$$w = \frac{P - 2l}{2}$$

Doing this means that we have *solved the formula $P = 2l + 2w$ for the specific variable, w.*

Solving a formula for a specific variable may seem confusing at first because the formula contains more than one letter. Keep in mind that we will solve for a specific variable the same way we have been solving equations up to this point.

We'll start by solving $2x + 5 = 13$ step by step for x and then applying the same procedure to solving $mx + b = y$ for x.

Example 4

Solve $2x + 5 = 13$ and $mx + b = y$ for x.

Solution

Look at these equations carefully, and notice that they have the same form. Read the following steps in order.

Part 1: Solve $2x + 5 = 13$

Don't quickly run through the solution of this equation. *The emphasis here is on the steps used to solve the equation and why we use those steps!*

$$2\boxed{x} + 5 = 13$$

We are solving for x. We'll put a box around it. What is the first step? "Get rid of" what is being added to the $2x$, that is "get rid of" the 5 on the left. Subtract 5 from each side.

$$2\boxed{x} + 5 - 5 = 13 - 5$$

Combine like terms.

$$2\boxed{x} = 8$$

Part 2: Solve $mx + b = y$ for x.

Since we are solving for x, we'll put a box around it.

$$m\boxed{x} + b = y$$

The goal is to get the "x" on a side by itself. What do we do first? As in part 1, "get rid of" what is being added to the "mx" term, that is "get rid of" the "b" on the left. Since b is being added to "mx," we will subtract it from each side. (We are performing the same steps as in part 1!)

$$m\boxed{x} + b - b = y - b$$

Combine like terms.

$$m\boxed{x} = y - b$$

[We cannot combine the terms on the right, so the right remains $(y - b)$.]

Part 3:   We left off needing to solve $2\boxed{x} = 8$ for x. We need to eliminate the "2" on the left. Since x is being multiplied by 2, we will *divide* each side by 2.

$$\frac{2\boxed{x}}{2} = \frac{8}{2}$$

Simplify.

$$x = 4$$

Part 4:   Now, we have to solve $m\boxed{x} = y - b$ for x. We need to eliminate the "m" on the left. Since x is being multiplied by m, we will *divide* each side by m.

$$\frac{m\boxed{x}}{m} = \frac{y - b}{m}$$

These are the same steps used in part 3!

Simplify.
$$\frac{\cancel{m}\boxed{x}}{\cancel{m}} = \frac{y - b}{m}$$

$$x = \frac{y - b}{m} \quad \text{or} \quad x = \frac{y}{m} - \frac{b}{m}$$

To obtain the result $x = \dfrac{y}{m} - \dfrac{b}{m}$, we distributed the m in the denominator to each term in the numerator. Either form of the answer is correct.

When you are solving a formula for a specific variable, think about the steps you use to solve an equation in one variable.

You Try 4

Solve $an - pr = w$ for n.

Example 5

$R = \dfrac{\rho L}{A}$ is a formula used in physics. Solve this equation for L.

Solution

$$R = \frac{\rho \boxed{L}}{A} \qquad \text{Solve for } L. \text{ Put it in a box.}$$

$$AR = A \cdot \frac{\rho \boxed{L}}{A} \qquad \text{Begin by eliminating } A \text{ on the right.}$$

Since ρL is being divided by A, we will *multiply* by A on each side.

$$AR = \rho \boxed{L} \qquad \text{Next, eliminate } \rho.$$

$$\frac{AR}{\rho} = \frac{\rho \boxed{L}}{\rho} \qquad \text{Divide each side by } \rho.$$

$$\frac{AR}{\rho} = L$$

Example 6

$A = \dfrac{1}{2}h(b_1 + b_2)$ is the formula for the area of a trapezoid. Solve it for b_1.

Solution

$$A = \frac{1}{2}h(\boxed{b_1} + b_2) \qquad \text{There are two valid ways to solve this for } b_1. \text{ We'll look at both of them.}$$

Method 1

We will put b_1 in a box to remind us that this is what we must solve for. In method 1, we will start by eliminating the fraction. Multiply both sides by 2.

$$2A = 2 \cdot \frac{1}{2}h(\boxed{b_1} + b_2)$$

$$2A = h(\boxed{b_1} + b_2)$$

We are solving for b_1, which is in the parentheses. The quantity in parentheses is being multiplied by h, so we can *divide both sides by h* to eliminate it on the right.

$$\frac{2A}{h} = \frac{h(\boxed{b_1} + b_2)}{h}$$

$$\frac{2A}{h} = \boxed{b_1} + b_2$$

$$\frac{2A}{h} - b_2 = \boxed{b_1} + b_2 - b_2 \qquad \text{Subtract } b_2 \text{ from each side.}$$

$$\frac{2A}{h} - b_2 = b_1$$

Method 2

To solve $A = \frac{1}{2}h(b_1 + b_2)$ for b_1, we can begin by distributing $\frac{1}{2}h$ on the right.

We will put b_1 in a box to remind us that this is what we must solve for.

$$A = \frac{1}{2}h\,(b_1 + b_2)$$

$$A = \frac{1}{2}h\boxed{b_1} + \frac{1}{2}hb_2$$

$$2A = 2\left(\frac{1}{2}h\boxed{b_1} + \frac{1}{2}hb_2\right) \qquad \text{Multiply by 2 to eliminate the fractions.}$$

$$2A = h\boxed{b_1} + hb_2$$

$$2A - hb_2 = h\boxed{b_1} + hb_2 - hb_2 \qquad \text{Since } hb_2 \text{ is being added to } hb_1, \text{ subtract } hb_2 \text{ from each side.}$$

$$2A - hb_2 = h\boxed{b_1}$$

$$\frac{2A - hb_2}{h} = \frac{h\boxed{b_1}}{h} \qquad \text{Divide by } h.$$

$$\frac{2A - hb_2}{h} = b_1$$

Or, b_1 can be rewritten as

$$b_1 = \frac{2A}{h} - \frac{hb_2}{h}$$

$$b_1 = \frac{2A}{h} - b_2$$

So, $b_1 = \dfrac{2A - hb_2}{h}$ or $b_1 = \dfrac{2A}{h} - b_2.$

You Try 5

Solve for the indicated variable.

a) $t = \dfrac{qr}{s}$ for q b) $R = t(k - c)$ for c

Answers to You Try Exercises

1) 5 ft   2) 76°, 77°   3) 81°   4) $n = \dfrac{w + pr}{a}$ 5) a) $q = \dfrac{st}{r}$ b) $c = \dfrac{kt - R}{t}$ or $c = k - \dfrac{R}{t}$

3.5 Exercises   **Boost your grade**   MathZone   > **Practice**   > **NetTutor**   > **e-Professors**   > **Videos**
at mathzone.com!   Problems   > **Self-Test**

Objective 1

Use a formula from geometry to solve each problem. See Example 1.

1) The Torrence family has a rectangular, in-ground pool in their yard. It holds 1700 ft^3 of water. If it is 17 ft wide and 4 ft deep, what is the length of the pool?

2) A rectangular fish tank holds 2304 in^3 of water. Find the height of the tank if it is 24 in. long and 8 in. wide.

3) A computer printer is stocked with paper measuring 8.5 in. × 11 in. But, the printing covers only 48 in^2 of the page. If the length of the printed area is 8 in., what is the width?

4) The rectangular glass surface to produce copies on a copy machine has an area of 135 in^2. It is 9 in. wide. What is the longest length of paper that will fit on this copier?

5) The face of the clock on Big Ben in London has a radius of 11.5 ft. What is the area of this circular clock face? Use 3.14 for π. Round the answer to the nearest square foot. (www.bigben.freeservers.com)

 6) A lawn sprinkler sprays water in a circle of radius 6 ft. Find the area of grass watered by this sprinkler. Use 3.14 for π. Round the answer to the nearest square foot.

7) The "lane" on a basketball court is a rectangle of length 19 ft. The area of the lane is 228 ft^2. What is the width of the lane?

8) A trapezoidal plot of land has the dimensions pictured below. If the area is 9000 ft^2,
a) find the length of the missing side, x.

b) How much fencing would be needed to completely enclose the plot?

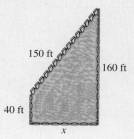

150 ft

160 ft

40 ft

x

9) A large can of tomato sauce is in the shape of a right circular cylinder. If its radius is 2 in. and its volume is 24π in^3, what is the height of the can?

10) A coffee can in the shape of a right circular cylinder has a volume of 50π in^3. Find the height of the can if its diameter is 5 in.

Objective 2

Find the missing angle measures. See Example 2.

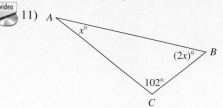 11) A

$x°$

$(2x)°$ B

102°

C

12)

A

59°

$(9x - 3)°$ $(4x + 7)°$

C   B

13)
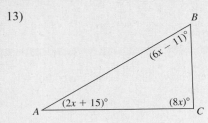

14)
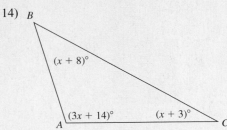

Find the measure of each indicated angle.

15)
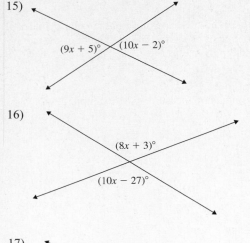

16)

17)
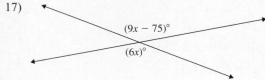

18)
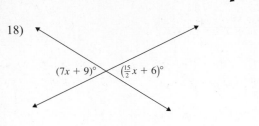

 19)
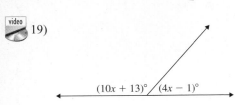

20)
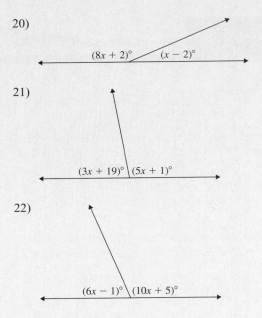

21)

22)

Objective 3

23) If x = the measure of an angle, write an expression for its supplement.

24) If x = the measure of an angle, write an expression for its complement.

Solve each problem. See Example 3.

25) Ten times the measure of an angle is 7° more than the measure of its supplement. Find the measure of the angle.

26) The measure of an angle is 12° more than twice its complement. Find the measure of the angle.

27) Four times the complement of an angle is 40° less than twice the angle's supplement. Find the angle, its complement, and its supplement.

28) Twice the supplement of an angle is 24° more than four times its complement. Find the angle, its complement, and its supplement.

29) The sum of an angle and three times its complement is 55° more than its supplement. Find the measure of the angle.

30) The sum of twice an angle and its supplement is 24° less than 5 times its complement. Find the measure of the angle.

31) The sum of 3 times an angle and twice its supplement is 400°. Find the angle.

32) The sum of twice an angle and its supplement is 253°. Find the angle.

Objective 4

Substitute the given values into the formula. Then, solve for the remaining variable.

33) $I = Prt$ (simple interest); if $I = 240$ when $P = 3000$ and $r = 0.04$, find t.

34) $I = Prt$ (simple interest); if $I = 156$ when $P = 650$ and $t = 3$, find r.

35) $V = lwh$ (volume of a rectangular box); if $V = 96$ when $l = 8$ and $h = 3$, find w.

36) $V = \dfrac{1}{3}Ah$ (volume of a pyramid); if $V = 60$ when $h = 9$, find A.

37) $P = 2l + 2w$ (perimeter of a rectangle); if $P = 50$ when $w = 7$, find l.

38) $P = s_1 + s_2 + s_3$ (perimeter of a triangle); if $P = 37$ when $s_1 = 17$ and $s_3 = 8$, find s_2.

39) $V = \dfrac{1}{3}\pi r^2 h$ (volume of a cone); if $V = 54\pi$ when $r = 9$, find h.

40) $V = \dfrac{1}{3}\pi r^2 h$ (volume of a cone); if $V = 32\pi$ when $r = 4$, find h.

41) $S = 2\pi r^2 + 2\pi rh$ (surface area of a right circular cylinder); if $S = 120\pi$ when $r = 5$, find h.

42) $S = 2\pi r^2 + 2\pi rh$ (surface area of a right circular cylinder); if $S = 66\pi$ when $r = 3$, find h.

43) $A = \dfrac{1}{2}h(b_1 + b_2)$ (area of a trapezoid); if $A = 790$ when $b_1 = 29$ and $b_2 = 50$, find h.

44) $A = \dfrac{1}{2}h(b_1 + b_2)$ (area of a trapezoid); if $A = 246.5$ when $h = 17$ and $b_2 = 16$, find b_1.

45) Solve for x.

   a) $x + 12 = 35$ b) $x + n = p$

   c) $x + q = v$

46) Solve for t.

   a) $t - 3 = 19$ b) $t - w = m$

   c) $t - v = j$

47) Solve for n.

   a) $5n = 30$ b) $yn = c$

   c) $wn = d$

48) Solve for y.

   a) $4y = 36$ b) $ay = x$

   c) $py = r$

49) Solve for c.

   a) $\dfrac{c}{3} = 7$ b) $\dfrac{c}{u} = r$

   c) $\dfrac{c}{x} = t$

50) Solve for m.

   a) $\dfrac{m}{8} = 2$ b) $\dfrac{m}{z} = p$

   c) $\dfrac{m}{q} = f$

 51) Solve for d.

   a) $8d - 7 = 17$ b) $kd - a = z$

52) Solve for g.

   a) $3g + 23 = 2$ b) $cg + k = \pi$

53) Solve for z.

   a) $6z + 19 = 4$ b) $yz + t = w$

54) Solve for p.

   a) $10p - 3 = 19$ b) $np - r = d$

Solve each formula for the indicated variable.

55) $F = ma$ for m (physics)

56) $C = 2\pi r$ for r (geometry)

57) $n = \dfrac{c}{v}$ for c (physics)

58) $f = \dfrac{R}{2}$ for R (physics)

59) $E = \sigma T^4$ for σ (meteorology)

60) $p = \rho gy$ for ρ (geology)

61) $V = \dfrac{1}{3}\pi r^2 h$ for h (geometry)

62) $U = \dfrac{1}{2}LI^2$ for L (physics)

63) $R = \dfrac{E}{I}$ for E (electronics)

64) $A = \dfrac{1}{2}bh$ for b (geometry)

65) $I = PRT$ for R (finance)

66) $I = PRT$ for P (finance)

67) $P = 2l + 2w$ for l (geometry)

68) $A = P + PRT$ for T (finance)

69) $H = \dfrac{D^2N}{2.5}$ for N (auto mechanics)

70) $V = \dfrac{AH}{3}$ for A (geometry)

71) $A = \dfrac{1}{2}h(b_1 + b_2)$ for b_2 (geometry)

72) $A = \pi(R^2 - r^2)$ for r^2 (geometry)

For exercises 73 and 74, refer to the following figure.

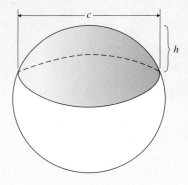

The surface area, S, of the spherical segment shown is given by $S = \dfrac{\pi}{4}(4h^2 + c^2)$, where h is the height of the segment and c is the diameter of the segment's base.

73) Solve the formula for h^2.

74) Solve the formula for c^2.

 75) The perimeter, P, of a rectangle is $P = 2l + 2w$, where l = length and w = width.

a) Solve $P = 2l + 2w$ for w.

b) Find the width of the rectangle with perimeter 28 cm and length 11 cm.

76) The area, A, of a triangle is $A = \dfrac{1}{2}bh$, where b = length of the base and h = height.

a) Solve $A = \dfrac{1}{2}bh$ for h.

b) Find the height of the triangle that has an area of 30 in^2 and a base of length 12 in.

77) $C = \dfrac{5}{9}(F - 32)$ is a formula that can be used to convert from degrees Fahrenheit, F, to degrees Celsius, C.

a) Solve this formula for F.

b) The average high temperature in Mexico City, Mexico, in April is 25°C. Use the result in part a) to find the equivalent temperature in degrees Fahrenheit. (www.bbc.co.uk)

78) The average low temperature in Stockholm, Sweden, in January is -5°C. Use the result in Exercise 77a) to find the equivalent temperature in degrees Fahrenheit. (www.bbc.co.uk)

Section 3.6 Applications of Linear Equations to Proportion, *d* = *rt*, and Mixture Problems

Objectives

1. Define Ratio and Proportion
2. Solve a Proportion
3. Solve Problems Involving a Proportion
4. Solve Problems Involving Money
5. Solve Mixture Problems
6. Solve Problems Involving Distance, Rate, and Time

1. Define Ratio and Proportion

A **ratio** is a quotient of two quantities. For example, if a survey revealed that 28 people said that their favorite candy was licorice, while 40 people said that their favorite candy was chocolate, then the ratio of people who prefer licorice *to* people who prefer chocolate is

$$\frac{\text{Number of licorice lovers}}{\text{Number of chocolate lovers}} = \frac{28}{40} \text{ or } \frac{7}{10}$$

A *ratio* is a way to compare two quantities. If two ratios are equivalent, like $\frac{28}{40}$ and $\frac{7}{10}$, we can set them equal to make a *proportion*. A **proportion** is a statement that two ratios are equal. $\frac{28}{40} = \frac{7}{10}$ is an example of a proportion. How can we be certain that a proportion is true?

2. Solve a Proportion

Example 1

Is the proportion $\frac{5}{9} = \frac{10}{18}$ true?

Solution

One way to decide is to find the **cross products**. If the cross products are equal, then the proportion is true:

$$\frac{5}{9} \diagdown\diagup \frac{10}{18}$$

Multiply.   Multiply.
$$5 \cdot 18 = 9 \cdot 10$$
$$90 = 90 \quad \text{True}$$

The cross products are equal, so $\frac{5}{9} = \frac{10}{18}$ is a true proportion.

If the cross products are not equal, then the proportion is not true.

You Try 1

Decide if the following proportions are true.

a) $\dfrac{7}{3} = \dfrac{35}{15}$ b) $\dfrac{3}{4} = \dfrac{7}{10}$

We can make a general statement about cross products:

> If $\dfrac{a}{b} = \dfrac{c}{d}$, then $ad = bc$ provided that $b \neq 0$ and $d \neq 0$.

This allows us to use cross products to solve equations.

Example 2

Solve each proportion.

a) $\dfrac{x}{6} = \dfrac{12}{9}$

b) $\dfrac{a + 4}{3} = \dfrac{a - 1}{4}$

Solution

a)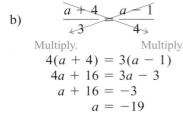

Multiply.   Multiply.
$$9x = 72$$
$$x = 8$$

The solution set is $\{8\}$.

b)
Multiply.   Multiply.
$$4(a + 4) = 3(a - 1)$$
$$4a + 16 = 3a - 3$$
$$a + 16 = -3$$
$$a = -19$$

The solution set is $\{-19\}$.

You Try 2

Solve each proportion.

a) $\dfrac{10}{14} = \dfrac{n}{21}$ b) $\dfrac{2k - 3}{5} = \dfrac{3k + 1}{10}$

3. Solve Problems Involving a Proportion

One application of proportions is in comparing prices.

Example 3

If 3 lb of potatoes cost $1.77, how much would 5 lb of potatoes cost?

Solution

Step 1:   Read the problem carefully, twice.

Step 2:   Find the cost of 5 lb of potatoes. Define the unknown.

$$x = \text{cost of 5 lb of potatoes}$$

Step 3:   Translate from English to an algebraic equation.
We will write a proportion.

$$
\begin{array}{r}
\text{3 lb of potatoes} \to \\
\text{Cost of 3 lb of potatoes} \to
\end{array}
\dfrac{3}{1.77} = \dfrac{5}{x}
\begin{array}{l}
\leftarrow \text{5 lb of potatoes} \\
\leftarrow \text{Cost of 5 lb of potatoes}
\end{array}
$$

Notice that if the number of pounds (3) is in the numerator on the left, the number of pounds (5) must be in the numerator on the right.

Equation: $\dfrac{3}{1.77} = \dfrac{5}{x}$

Step 4:   Solve the equation.

$$\frac{3}{1.77} = \frac{5}{x}$$
$$3x = (1.77)5 \qquad \text{Set the cross products equal.}$$
$$3x = 8.85$$
$$x = 2.95$$

Step 5:   Interpret the meaning of the solution as it relates to the problem.

The cost of 5 lb of potatoes is $2.95.

Does the answer make sense?
Check the numbers in the proportion.

$$\frac{3}{1.77} = \frac{5}{2.95}$$
$$3(2.95) = (1.77)5$$
$$8.85 = 8.85 \qquad \text{Yes, the answer makes sense.}$$

You Try 3

How much would a customer pay for 16 gal of unleaded gasoline if another customer paid $23.76 for 12 gal at the same gas station?

Another application of proportions is for solving similar triangles.

Similar Triangles

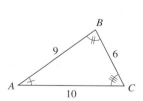   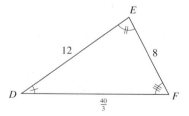

$$m\angle A = m\angle D, \quad m\angle B = m\angle E, \quad \text{and} \quad m\angle C = m\angle F$$

$\triangle ABC$ and $\triangle DEF$ are **similar triangles**. Two triangles are *similar* if they have the same shape. The corresponding angles have the same measure, and the corresponding sides are proportional.

Ratios of corresponding sides: $\dfrac{9}{12} = \dfrac{3}{4}$; $\dfrac{6}{8} = \dfrac{3}{4}$; $\dfrac{10}{\frac{40}{3}} = 10 \cdot \dfrac{3}{40} = \dfrac{3}{4}$.

The ratio of each of the corresponding sides is $\dfrac{3}{4}$.

We can set up and solve a proportion to find the length of an unknown side in two similar triangles.

Example 4

Given the following similar triangles, find *x*.

Solution

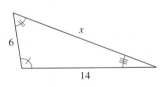

$\dfrac{6}{9} = \dfrac{x}{24}$ Set the ratios of two corresponding sides equal to each
                                          other. (Set up a proportion.)

$9x = 6 \cdot 24$ Solve the proportion.

$9x = 144$

$x = 16$

You Try 4

Given the following similar triangles, find *x*.

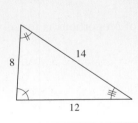
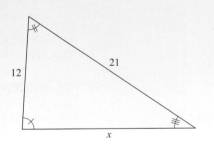

---

4. Solve Problems Involving Money

Many application problems involve thinking about the number of coins or bills and their values. Let's look at how arithmetic and algebra problems involving these ideas are related.

Example 5

Determine the amount of money you have in cents *and* in dollars if you have

a)   16 nickels   b)   7 quarters   c)   16 nickels and 7 quarters

d)   *n* nickels   e)   *q* quarters   f)   *n* nickels and *q* quarters

Solution

Concentrate on the *procedure* that is used to do the arithmetic so that we can apply the same idea to algebra.

a)   16 nickels: Amount of money,

in cents: 5 · (16) = 80¢   **in dollars:** 0.05 · (16) = $0.80

↑ Value of a nickel . ↑ Number of nickels = ↑ Value of 16 nickels   ↑ Value of a nickel . ↑ Number of nickels = ↑ Value of 16 nickels

b)   7 quarters: Amount of money,

in cents: 25 · (7) = 175¢   **in dollars:** 0.25 · (7) = $1.75

↑ Value of a quarter . ↑ Number of quarters = ↑ Value of 7 quarters   ↑ Value of a quarter . ↑ Number of quarters = ↑ Value of 7 quarters

c)   16 nickels and 7 quarters: Amount of money,

in cents: Value of 16 nickels + Value of 7 quarters = Value of 16 nickels and 7 quarters

from a & b: 5(16)   +   25(7)   =
              80   +   175   =   255¢

in dollars: Value of 16 nickels + Value of 7 quarters = Value of 16 nickels and 7 quarters

from a & b: 0.05(16)   +   0.25(7)   =
              0.80   +   1.75   = $2.55

d)   *n* nickels: Amount of money,

in cents: In part a, to find the value of 16 nickels, we multiplied

Value of a nickel → 5(16) ← Number of nickels

to get 80¢.

So, if there are *n* nickels, the value of *n* nickels is

Value of a nickel → 5n ← Number of nickels

n nickels have a value of 5*n* cents.

in dollars: To find the value of *n* nickels, in dollars, we will multiply.

Value of a nickel → 0.05n ← Number of nickels

n nickels have a value of 0.05*n* dollars.

e)   *q* quarters: Amount of money,

in cents: In part b, to find the value of 7 quarters, we multiplied

Value of a quarter → 25(7) ← Number of quarters

to get 175¢.

So, if there are q quarters, the value of q quarters is

Value of a quarter $\rightarrow 25q \leftarrow$ Number of quarters

q quarters have a value of $25q$ cents.

in dollars: To find the value of q quarters, in dollars, we will multiply.

Value of a quarter $\rightarrow 0.25q \leftarrow$ Number of quarters

q quarters have a value of $0.25q$ dollars.

f) n nickels and q quarters: Amount of money,

in cents: We will use the same idea that we used in part c:

Value of n nickels = Value of n nickels + Value of q quarters
and q quarters

$$= \qquad 5n \qquad + \qquad 25q \qquad \text{From (d) and (e)}$$

n nickels and q quarters have a value of $5n + 25q$ cents.

in dollars: Use the same procedure as above, except express the values in dollars.

Value of n nickels and q quarters $= 0.05n + 0.25q$

n nickels and q quarters have a value of $0.05n + 0.25q$ dollars.

You Try 5

Determine the amount of money you have in cents *and* in dollars if you have

a) 11 dimes   b) 20 pennies   c) 8 dimes and 46 pennies

d) d dimes   e) p pennies   f) d dimes and p pennies

Next, we'll apply this idea of the value of different denominations of money to an application problem.

Example 6

Jamaal has only dimes and quarters in his piggy bank. When he counts the change, he finds that he has $18.60 and that there are twice as many quarters as dimes. How many dimes and quarters are in his bank?

Solution

Step 1:   Read the problem carefully, twice.

Step 2:   Find the number of dimes and the number of quarters in the bank. Define the unknowns.

The number of quarters is defined *in terms of* the number of dimes, so we will let

$$d = \text{number of dimes}$$

Define the other unknown in terms of d.

$$2d = \text{number of quarters}$$

Step 3: Translate from English to an algebraic equation.

We will use the ideas presented in the previous example to first write an equation in English:

Value of the dimes + Value of the quarters = Total value of the coins in the bank
   0.10d   +   0.25(2d)   =   18.60
   ↑↑   ↑   ↑

Value of   Number   Value   Number
a dime   of dimes   of a   of quarters
(in   quarter
dollars)   (in dollars)

Note: We expressed all values in dollars because the total, $18.60, is in dollars.

Equation: $0.10d + 0.25(2d) = 18.60$.

Step 4: Solve the equation.

$$0.10d + 0.25(2d) = 18.60$$
$$100[0.10d + 0.25(2d)] = 100(18.60) \qquad \text{Multiply by 100 to eliminate the decimals.}$$
$$10d + 25(2d) = 1860$$
$$10d + 50d = 1860$$
$$60d = 1860$$
$$d = \frac{1860}{60} = 31$$

Step 5: Interpret the meaning of the solution as it relates to the problem.

There are 31 dimes in Jamaal's piggy bank.

Find the other unknown.

$$2d = 2(31) = 62$$

There are 62 quarters in the piggy bank.

Does the answer make sense?

Substitute the values into the equation $0.10d + 0.25(2d) = 18.60$

$$0.10(31) + 0.25(62) = 18.60$$
$$3.10 + 15.50 = 18.60$$

Yes, the answer makes sense.

Jamaal has 31 dimes and 62 quarters in the piggy bank.

 You Try 6

A collection of coins consists of pennies and nickels. There are 5 fewer nickels than there are pennies. If the coins are worth a total of $4.97, how many of each type of coin is in the collection?

5. Solve Mixture Problems

Mixture problems involve combining two or more substances to make a mixture of them. We will begin with an example from arithmetic then extend this concept to be used with algebra.

Example 7

The state of Illinois mixes ethanol (made from corn) in its gasoline to reduce pollution. If a customer purchases 15 gal of gasoline and it has a 10% ethanol content, how many gallons of ethanol are in the 15 gal of gasoline?

Solution

Think of the problem in English first:

$$\begin{array}{c} \text{Number of gallons} \\ \text{of ethanol in} \\ \text{the gasoline} \end{array} = \begin{pmatrix} \text{Concentration of ethanol} \\ \text{in the gasoline--change 10\%} \\ \text{to a decimal} \end{pmatrix} \begin{pmatrix} \text{Number of} \\ \text{gallons of gasoline} \end{pmatrix}$$

$$= (0.10) \cdot 15$$
$$= 1.5$$

There are 1.5 gal of ethanol in 15 gal of gasoline.

The idea above will be used to help us solve the next mixture problem.

Example 8

A chemist needs to make 30 liters (L) of a 6% acid solution. She will make it from some 4% acid solution and some 10% acid solution that is in the storeroom. How much of the 4% solution and 10% solution should she use?

Solution

Step 1: Read the problem carefully, twice.

Step 2: Find the number of liters of 4% acid solution and 10% acid solution she needs to make 30 L of a 6% acid solution. Define the unknowns.

We don't know how much of *either* solution is needed at this point in the problem. Begin by letting x represent one of the solutions we must find.

$$x = \text{number of liters of 4\% acid solution}$$

We need an expression for the number of liters of 10% solution. Since there is a total of 30 L in the final solution (6%), and x L of this is the 4% solution,

$$30 - x = \text{number of liters of 10\% acid solution}$$

Step 3: Translate from English to an algebraic equation.

Our equation will be based on the number of liters of *acid* in each solution. For example, the solution that is being made will be 30 L of a 6% acid solution. Therefore,

$$\begin{array}{c} \text{Number of liters} \\ \text{of acid in the 6\%} \\ \text{solution} \end{array} = (\text{Concentration}) \begin{pmatrix} \text{Number of} \\ \text{liters of 6\%} \\ \text{solution} \end{pmatrix}$$

$$= (0.06)(30)$$
$$= 1.8$$

The acid present in the final 6% acid solution must come from 4% and 10% solutions which were mixed. We get the following equation in English:

Number of liters of acid in the 4% solution	+	Number of liters of acid in the 10% solution	=	Number of liters of acid in the 6% solution

To help us translate this English equation to algebra, we will organize our information in a table:

		Concentration	Number of liters of solution	Number of liters of acid in the solution
Mix	4% solution	0.04	x	$0.04x$
these	10% solution	0.10	$30 - x$	$0.10(30 - x)$
to make →	6% solution	0.06	30	$0.06(30)$

Using the information in the table, we can rewrite the English equation in algebraic terms:

$$0.04x + 0.10(30 - x) = 0.06(30)$$

$$\uparrow \qquad\qquad \uparrow \qquad\qquad \uparrow$$

Liters of   Liters of acid   Liters of acid
acid in 4%   in 10% solution   in 6% solution
solution

Step 4: Solve the equation.

$$0.04x + 0.10(30 - x) = 0.06(30)$$
$$100[0.04x + 0.10(30 - x)] = 100[0.06(30)] \quad \text{Multiply by 100 to eliminate decimals.}$$
$$4x + 10(30 - x) = 6(30)$$
$$4x + 300 - 10x = 180 \qquad \text{Distribute.}$$
$$-6x + 300 = 180 \qquad \text{Combine like terms.}$$
$$-6x = -120 \qquad \text{Subtract 300 from each side.}$$
$$x = 20 \qquad \text{Divide by } -6.$$

Step 5: Interpret the meaning of the solution as it relates to the problem.

20 L of the 4% solution is needed.

Find the other unknown.

$$30 - x = 30 - 20 = 10 \text{ L of 10\% solution.}$$

Does the answer make sense?

Substitute into the equation relating the number of liters of acid:

$$20(0.04) + 10(0.10) = 0.06(30)$$
$$0.8 + 1 = 1.8$$
$$1.8 = 1.8 \qquad \text{Yes, it makes sense.}$$

20 L of a 4% acid solution and 10 L of a 10% acid solution are needed to make 30 L of a 6% acid solution.

You Try 7

How many milliliters (mL) of a 7% alcohol solution and how many milliliters of a 15% alcohol solution must be mixed to make 20 mL of a 9% alcohol solution?

6. Solve Problems Involving Distance, Rate, and Time

If you drive at 50 mph for 4 hr, how far will you drive? One way to get the answer is to use the formula

$$\text{Distance} = \text{Rate} \times \text{Time}$$
$$\text{or}$$
$$d = rt$$
$$d = (50 \text{ mph}) \cdot (4 \text{ hr})$$
$$\text{Distance traveled} = 200 \text{ mi}$$

Notice that the rate is in miles per *hour* and the time is in *hours*. The units must be consistent in this way. If the time in this problem had been expressed in minutes, it would have been necessary to convert minutes to hours.

The formula $d = rt$ will be used in Example 9.

Example 9

Alexandra and Jenny are taking a cross-country road-trip on their motorcycles. Jenny leaves a rest area first traveling at 60 mph. Alexandra leaves 30 min later, traveling on the same highway, at 70 mph. How long will it take Alexandra to catch Jenny?

Solution

Step 1: Read the problem carefully, twice.

Step 2: We must determine the amount of time it takes for Alexandra to catch Jenny. Define the unknowns.

The time Alexandra travels is in terms of how long Jenny travels. Therefore, let

$t = $ hours Jenny has been riding when Alexandra catches her

Alexandra leaves 30 minutes (1/2 hour) *after* Jenny, so the amount of time Alexandra actually travels is 1/2 hour *less than* Jenny.

$t - \dfrac{1}{2} = $ hours it takes Alexandra to catch Jenny

Let's draw a picture to help us visualize the situation described in the problem.

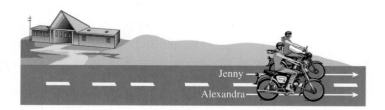

Step 3: Translate from English to an algebraic equation.

The picture helps us to see that our equation will result from the fact that

Jenny's distance = Alexandra's distance

Next, we will organize the information in a table.

First, fill in the girls' rates and times. Then, get the expressions for their distances using the fact that $d = rt$.

	d	$=$	r	t
Jenny	$60t$		60	t
Alexandra	$70(t - \frac{1}{2})$		70	$t - \frac{1}{2}$

To get the algebraic equation, use

Jenny's distance = Alexandra's distance

$$60t = 70\left(t - \frac{1}{2}\right)$$

Step 4: Solve the equation.

$$60t = 70\left(t - \frac{1}{2}\right)$$
$$60t = 70t - 35 \qquad \text{Distribute.}$$
$$-10t = -35 \qquad \text{Subtract } 70t.$$
$$t = \frac{-35}{-10} = 3\frac{1}{2} \text{ hr}$$

Step 5: Interpret the meaning of the solution as it relates to the problem.

We found that $t = 3\frac{1}{2}$ hr. Looking at the table, t is the amount of time Jenny has traveled. But, we are asked to determine how long it takes Alexandra to catch Jenny.

$$t - \frac{1}{2} = \text{Alexandra's time}$$
$$3\frac{1}{2} - \frac{1}{2} = 3 \text{ hr} = \text{Alexandra's time}$$

It takes Alexandra 3 hr to catch Jenny.

Check:
$$60t = 70\left(t - \frac{1}{2}\right)$$
$$60\left(3\frac{1}{2}\right) = 70\left(3\frac{1}{2} - \frac{1}{2}\right)$$
$$60\left(\frac{7}{2}\right) = 70(3)$$
$$210 = 210 \qquad \text{Yes, it makes sense.}$$

It takes Alexandra 3 hr to catch Jenny.

You Try 8

The Kansas towns of Topeka and Voda are 230 mi apart. Alberto left Topeka driving west on Interstate 70 at the same time Ramon left Voda driving east toward Topeka on I-70. Alberto and Ramon meet after 2 hr. If Alberto's speed was 5 mph faster than Ramon's speed, how fast was each of them driving?

Answers to You Try Exercises

1) a) true   b) false   2) a) $n = 15$ b) $k = 7$ 3) $31.68   4) 18   5) a) 110¢, $1.10
b) 20¢, $0.20   c) 126¢, $1.26   d) $10d$ cents, $0.10d$ dollars   e) $1p$ cents, $0.01p$ dollars
f) $10d + 1p$ cents, $0.10d + 0.01p$ dollars   6) 87 pennies, 82 nickels
7) 15 ml of 7%, 5 ml of 15%   8) Ramon: 55 mph, Alberto: 60 mph

---

3.6 Exercises

Boost your grade at mathzone.com!   MathZone

> Practice Problems   > NetTutor > Self-Test   > e-Professors   > Videos

Objective 1

Write as a ratio in lowest terms.

1) 15 tea drinkers to 25 coffee drinkers

2) 4 mi to 6 mi

3) 60 ft to 45 ft

4) 19 girls to 16 boys

 5) What is the difference between a ratio and a proportion?

 6) Is 0.45 equivalent to the ratio 9 to 20? Explain.

Determine if each proportion is true or false. See Example 1.

7) $\dfrac{2}{15} = \dfrac{8}{45}$

8) $\dfrac{50}{35} = \dfrac{10}{7}$

9) $\dfrac{42}{77} = \dfrac{6}{11}$

10) $\dfrac{20}{30} = \dfrac{4}{3}$

Objective 2

Solve each proportion. See Example 2.

11) $\dfrac{m}{9} = \dfrac{10}{45}$

12) $\dfrac{c}{3} = \dfrac{16}{12}$

13) $\dfrac{120}{50} = \dfrac{x}{4}$

14) $\dfrac{2}{27} = \dfrac{r}{36}$

15) $\dfrac{2a + 3}{8} = \dfrac{2}{16}$

16) $\dfrac{5d - 1}{2} = \dfrac{6d + 3}{3}$

17) $\dfrac{n - 4}{5} = \dfrac{5n - 2}{10}$

18) $\dfrac{2w - 1}{7} = \dfrac{3 - 6w}{4}$

Objective 3

For each problem, set up a proportion and solve. See Example 3.

19) David's hobby is archery. Usually, he can hit the bulls-eye 9 times out of 15 tries. If he shot 50 arrows, how many bulls-eyes could he expect to make?

20) Hector buys batteries for his Gameboy. He knows that 3 packs of batteries cost $4.26. How much should Hector expect to pay for 5 packs of batteries?

21) A 12-oz serving of Mountain Dew contains 55 mg of caffeine. How much caffeine is in an 18-oz serving of Mountain Dew? (www.nsda.org)

22) An 8-oz serving of Coca-Cola Classic contains 23 mg of caffeine. How much caffeine is in a 12-oz serving of Coke? (www.nsda.org)

23) The national divorce rate is about 4.8 divorces per 1000 people. How many divorces are expected in a town of 35,000 people if that town followed the national rate? (www.cobras.org)

24) If the exchange rate between the American dollar and the Norwegian krone is such that $2.00 = 13.60 krone, how many krone could be exchanged for $15.00? (moneycentral.msn.com)

Find the length of the indicated side, x, by setting up a proportion. See Example 4.

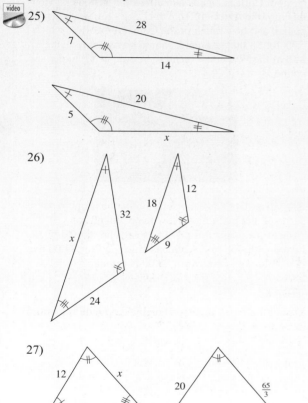

25)

26)

27)

28)

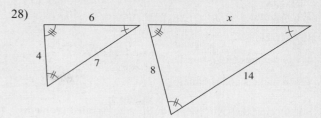

29)

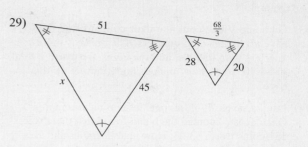

30)

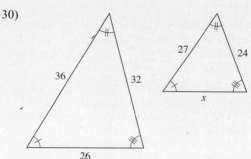

Objective 4

For Exercises 31–36, determine the amount of money a) in dollars and b) in cents given the following quantities.

31) 8 dimes

32) 32 nickels

33) 217 pennies

34) 12 quarters

35) 9 quarters and 7 dimes

36) 89 pennies and 14 nickels

For Exercises 37–42 write an expression which represents the amount of money in a) dollars and b) cents given the following quanities.

37) q quarters

38) p pennies

39) *d* dimes

40) *n* nickels

41) *p* pennies and *q* quarters

42) *n* nickels and *d* dimes

Solve using the five "Steps for Solving Applied Problems."
See Example 6.

43) Gino and Vince combine their coins to find they
have all nickels and quarters. They have 8 more
quarters than nickels, and the coins are worth a total
of $4.70. How many nickels and quarters do they
have?

44) Danika saves all of her pennies and nickels in a jar.
One day she counted them and found that there were
131 coins worth $3.43. How many pennies and how
many nickels were in the jar?

45) Kyung Soo has been saving her babysitting money.
She has $69.00 consisting of $5 bills and $1 bills. If
she has a total of 25 bills, how many $5 bills and how
many $1 bills does she have?

46) A bank employee is servicing the ATM after a busy
Friday night. She finds the machine contains only
$20 bills and $10 bills and that there are twice as
many $20 bills remaining as there are $10 bills. If
there is a total of $550.00 left in the machine, how
many of the bills are twenties, and how many are
tens?

 47) A movie theater charges $9.00 for adults and $7.00 for
children. The total revenue for a particular movie is
$475.00. Determine the number of each type of ticket
sold if the number of children's tickets sold was half
the number of adult tickets sold.

48) At the post office, Ronald buys 12 more 37¢ stamps
than 23¢ stamps. If he spends $9.24 on the stamps,
how many of each type did he buy?

Objective 5

Solve. See Example 7.

49) How many ounces of alcohol are in 40 oz of a 5%
alcohol solution?

50) How many milliliters of acid are in 30 mL of a 6%
acid solution?

51) Sixty milliliters of a 10% acid solution are mixed with
40 mL of a 4% acid solution. How much acid is in the
mixture?

52) Fifty ounces of an 8% alcohol solution are mixed with
20 ounces of a 6% alcohol solution. How much alcohol is in the mixture?

Solve using the five "Steps for Solving Applied Problems."
See Example 8.

 53) How many ounces of a 4% acid solution and how
many ounces of a 10% acid solution must be mixed to
make 24 oz of a 6% acid solution?

54) How many milliliters of a 15% alcohol solution must
be added to 80 mL of an 8% alcohol solution to make
a 12% alcohol solution?

55) How many liters of a 40% antifreeze solution must be
mixed with 5 L of a 70% antifreeze solution to make a
mixture that is 60% antifreeze?

56) How many milliliters of a 7% hydrogen peroxide solution and how many milliliters of a 1% hydrogen peroxide solution should be mixed to get 400 mL of a 3%
hydrogen peroxide solution?

57) Custom Coffees blends its coffees for customers. How
much of the Aztec coffee, which sells for $6.00 per
pound, and how much of the Cinnamon coffee, which
sells for $8.00 per pound, should be mixed to make
5 lb of the Winterfest blend to be sold at $7.20 per
pound?

58) All-Mixed-Up Nut Shop sells a mix consisting of
cashews and pistachios. How many pounds of
cashews, which sell for $7.00 per pound, should be
mixed with 4 lb of pistachios, which sell for $4.00 per
pound, to get a mix worth $5.00 per pound?

Objective 6

Solve using the five "Steps for Solving Applied Problems."
See Example 9.

59) A truck and a car leave the same intersection traveling
in the same direction. The truck is traveling at 35
mph, and the car is traveling at 45 mph. In how many
minutes will they be 6 mi apart?

60) At noon, a truck and a car leave the same intersection
traveling in the same direction. The truck is traveling
at 30 mph, and the car is traveling at 42 mph. At what
time will they be 9 mi apart?

61) Ajay is traveling north on a road while Rohan is traveling south on the same road. They pass by each other at 3 P.M., Ajay driving 30 mph and Rohan driving 40 mph. At what time will they be 105 miles apart?

62) When Lance and Jan pass each other on their bikes going in opposite directions, Lance is riding at 22 mph, and Jan is pedaling at 18 mph. If they continue at those speeds, after how long will they be 100 mi apart?

63) At noon, a cargo van crosses an intersection at 30 mph. At 12:30 P.M., a car crosses the same intersection traveling in the opposite direction. At 1 P.M., the van and car are 54 miles apart. How fast is the car traveling?

64) A freight train passes the Naperville train station at 9:00 A.M. going 30 mph. Ten minutes later a passenger train, headed in the same direction on an adjacent track, passes the same station at 45 mph. At what time will the passenger train catch the freight train?

Objectives 3–6

Mixed Exercises: Solve.

65) At the end of her shift, a cashier has a total of $6.30 in dimes and quarters. There are 7 more dimes than quarters. How many of each of these coins does she have?

66) An alloy that is 30% silver is mixed with 200 g of a 10% silver alloy. How much of the 30% alloy must be used to obtain an alloy that is 24% silver?

67) One-half cup of Ben and Jerry's Cherry Garcia Ice Cream has 260 calories. How many calories are in 3 cups of Cherry Garcia? (www.benjerry.com)

68) According to national statistics, 41 out of 200 Americans aged 28–39 are obese. In a group of 800 Americans in this age group, how many would be expected to be obese? (0-web10.epnet.com)

69) A jet flying at an altitude of 35,000 ft passes over a small plane flying at 10,000 ft headed in the same direction. The jet is flying twice as fast as the small plane, and thirty minutes later they are 100 mi apart. Find the speed of each plane.

70) Tickets for a high school play cost $3.00 each for children and $5.00 each for adults. The revenue from one performance was $663, and 145 tickets were sold. How many adult tickets and how many children's tickets were sold?

71) A pharmacist needs to make 20 cubic centimeters (cc) of a 0.05% steroid solution to treat allergic rhinitis. How much of a 0.08% solution and a 0.03% solution should she use?

72) Geri is riding her bike at 10 mph when Erin passes her going in the opposite direction at 14 mph. How long will it take before the distance between them is 6 mi?

73) How much pure acid must be added to 6 gal of a 4% acid solution to make a 20% acid solution?

74) If the exchange rate between the American dollar and the Japanese Yen is such that $4.00 = 442 yen, how many yen could be exchanged for $30.00? (moneycentral.msn.com)

Section 3.7 Solving Linear Inequalities in One Variable

Objectives

1. Graph a Linear Inequality on a Number Line and Use Set Notation and Interval Notation

2. Solve Linear Inequalities Using the Addition and Subtraction Properties of Inequality

3. Solve Linear Inequalities Using the Multiplication Property of Inequality

4. Solve Linear Inequalities Using a Combination of the Properties

5. Solve Compound Inequalities Containing Three Parts

6. Solve Applications Involving Linear Inequalities

Recall the inequality symbols

$<$ "is less than" $\leq$ "is less than or equal to"

$>$ "is greater than" $\geq$ "is greater than or equal to"

We will use the symbols to form *linear inequalities in one variable*.

While an equation states that two expressions are equal, an *inequality* states that two expressions are not necessarily equal. Inequalities look very similar to equations, and we solve inequalities very much like we solve equations. Here is a comparison of an equation and an inequality:

Equation	**Inequality**
$3x - 8 = 13$	$3x - 8 \leq 13$

> **Definition**   A **linear inequality in one variable** can be written in the form $ax + b < c$, $ax + b \leq c$, $ax + b > c$, or $ax + b \geq c$ where a, b, and c are real numbers and $a \neq 0$.

The solution to the linear equation $n + 9 = 6$ is $\{-3\}$. Often, however, the solution to a linear inequality is a set of numbers that can be represented in one of three ways:

1) On a graph

2) In set notation

3) In interval notation

In this section, we will learn how to solve linear inequalities in one variable and how to represent the solution in each of those three ways.

Graphing an Inequality and Using the Notations

1. Graph a Linear Inequality on a Number Line and Use Set Notation and Interval Notation

Example 1

Graph each inequality and express the solution in set notation and interval notation.

   a) $x \leq -1$ b) $t > 4$

Solution

a) $x \leq -1$:

Graphing $x \leq -1$ means that we are finding the solution set of $x \leq -1$. What value(s) of x will make the inequality true? The largest solution is -1. Then, any number *less than* -1 will make $x \leq -1$ true. We represent this **on the number line** as follows:

$$-4 \; -3 \; -2 \; -1 \;\; 0 \;\; 1 \;\; 2 \;\; 3 \;\; 4$$

The graph illustrates that the solution is the set of all numbers from and including -1 to $-\infty$ (the numbers get infinitely smaller without bound).

Notice that the dot on -1 is shaded. This tells us that -1 is included in the solution set. The shading to the left of -1 indicates that *any* real number in this region is a solution. Some examples would be $-1\frac{1}{2}$, -3, -5.2, etc. (The solutions are not just integers.)

We can express the solution set in **set notation** this way: $\{x \mid x \le -1\}$

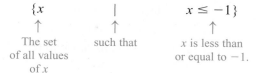

In **interval notation** we write

$$(-\infty, -1]$$

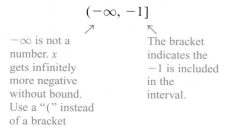

$-\infty$ is not a number. x gets infinitely more negative without bound. Use a "(" instead of a bracket

The bracket indicates the -1 is included in the interval.

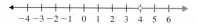

The variable does not appear anywhere in interval notation.

---

b) $t > 4$:

We will plot 4 as an *open circle* on the number line because the symbol is ">" and *not* "≥." The inequality $t > 4$ means that we must find the set of all numbers, t, *greater than* (but *not* equal to) 4. Shade to the right of 4.

$$\xleftarrow{\hspace{2em}}\overset{\displaystyle \quad\ \ \ \ \ \ \ \ \ \ \ \ \ \ \ \diamond}{\underset{-4\ -3\ -2\ -1\ \ 0\ \ 1\ \ 2\ \ 3\ \ 4\ \ 5\ \ 6}{\rule{0pt}{0pt}}}\xrightarrow{\hspace{2em}}$$

The graph illustrates that the solution is the set of all numbers from (but not including) 4 to ∞ (the numbers get infinitely bigger without bound).

We can express the solution set in *set notation* this way: $\{t \mid t > 4\}$

In *interval notation* we write

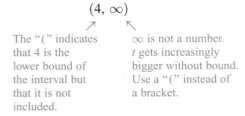

$$(4, \infty)$$

The "(" indicates that 4 is the lower bound of the interval but that it is not included.

∞ is not a number. t gets increasingly bigger without bound. Use a "(" instead of a bracket.

Hints for using interval notation:

1) The variable never appears in interval notation.
2) If a number is *included* in the solution set, it will get a bracket:
   $x \le -1 \to (-\infty, -1]$
3) If a number is *not included* in the solution set, it will get a parenthesis.

$$t > 4 \to (4, \infty)$$

4) $-\infty$ and ∞ *always* get parentheses.
5) The number of lesser value is always placed to the left. The number of greater value is placed to the right.
6) Even if we are not asked to graph the solution set, the graph may be helpful in writing the interval notation correctly.

You Try 1

Graph each inequality and express the solution in interval notation.

a) $k \ge -7$ b) $c < 5$

Solving Linear Inequalities

2. Solve Linear Inequalities Using the Addition and Subtraction Properties of Inequality

The addition and subtraction properties of equality help us to solve equations. The same properties hold for inequalities as well.

Addition and Subtraction Properties of Inequality

Let a, b, and c be real numbers. Then,

1) $a < b$ and $a + c < b + c$ are equivalent

and

2) $a < b$ and $a - c < b - c$ are equivalent.

The above properties hold for any of the inequality symbols.

Example 2

Solve $y - 8 \geq -5$. Graph the solution set and write the answer in interval and set notations.

Solution

$$y - 8 \geq -5$$
$$y - 8 + 8 \geq -5 + 8 \qquad \text{Add 8 to each side.}$$
$$y \geq 3$$

Interval notation: The solution set is $[3, \infty)$.

Set notation: $\{y | y \geq 3\}$.

You Try 2

Solve $k - 10 \geq -4$. Graph the solution set and write the answer in interval and set notations.

3. Solve Linear Inequalities Using the Multiplication Property of Inequality

While the addition and subtraction properties for solving equations and inequalities work the same way, this is not true for multiplication and division. Let's see why.

Begin with an inequality we know is true: $2 < 5$.

Multiply both sides by a *positive* number, say 3.

$$2 < 5 \qquad \qquad \text{True}$$
$$3(2) < 3(5) \qquad \text{Multiply by 3.}$$
$$6 < 15 \qquad \qquad \text{True}$$

Begin again with $2 < 5$. Multiply both sides by a *negative* number, say -3.

$$2 < 5 \qquad \qquad \text{True}$$
$$-3(2) < -3(5) \qquad \text{Multiply by } -3.$$
$$-6 < -15 \qquad \qquad \text{False}$$

To make $-6 < -15$ into a *true* statement, we must *reverse the direction of the inequality symbol.*

$$-6 > -15 \qquad \qquad \text{True}$$

If you begin with a true inequality and *divide* by a positive number or by a negative number, the results will be the same as above since division can be defined in terms of multiplication. This leads us to the multiplication property of inequality.

Multiplication Property of Inequality

Let a, b, and c be real numbers.

1)   If c is a *positive* number, then $a < b$ and $ac < bc$ are equivalent inequalities.

2)   If c is a *negative* number, then $a < b$ and $ac > bc$ are equivalent inequalities.

For the most part, the procedures used to solve linear inequalities are the same as those for solving linear equations **except** *when you multiply or divide an inequality by a negative number, you must reverse the direction of the inequality symbol.*

Example 3

Solve each inequality. Graph the solution set and write the answer in interval and set notations.

a) $-5w \leq 20$ b) $5w \leq -20$

Solution

a) $-5w \leq 20$

The first step, here, is to divide each side by -5. *Since we are dividing by a negative number, we must remember to reverse the direction of the inequality symbol.*

$$-5w \leq 20$$

$$\frac{-5w}{-5} \geq \frac{20}{-5} \qquad \text{Divide by } -5, \text{ so reverse the inequality symbol.}$$

$$w \geq -4$$

Interval notation: The solution set is $[-4, \infty)$.

Set notation: $\{w | w \geq -4\}$

b) $5w \leq -20$

Here, begin by dividing by 5. Since we are dividing by a *positive* number, the inequality symbol remains the same. (The negative number, -20, being divided by 5 has no effect on the inequality symbol.)

$$5w \leq -20$$

$$\frac{5w}{5} \leq \frac{-20}{5} \qquad \text{Divide by 5. Do } not \text{ reverse the inequality symbol.}$$

$$w \leq -4$$

Interval notation: The solution set is $(-\infty, -4]$.

Set notation: $\{w | w \leq -4\}$

You Try 3

Solve $-\dfrac{1}{4}m < 3$. Graph the solution set and write the answer in interval and set notations.

4. Solve Linear Inequalities Using a Combination of the Properties

Often it is necessary to combine the properties to solve an inequality.

Example 4

Solve $4(5 - 2d) + 11 < 2(d + 3)$. Graph the solution set and write the answer in interval and set notations.

Solution

$$4(5 - 2d) + 11 < 2(d + 3)$$

$$20 - 8d + 11 < 2d + 6 \qquad \text{Distribute.}$$

$$31 - 8d < 2d + 6 \qquad \text{Combine like terms.}$$

$$31 - 8d - 2d < 2d - 2d + 6 \qquad \text{Subtract } 2d \text{ from each side.}$$

$$31 - 10d < 6$$

$$31 - 31 - 10d < 6 - 31 \qquad \text{Subtract 31 from each side.}$$

$$-10d < -25$$

$$\frac{-10d}{-10} > \frac{-25}{-10} \qquad \begin{array}{l}\text{Divide both sides by } -10.\\ \text{Reverse the inequality symbol.}\end{array}$$

$$d > \frac{5}{2} \qquad \text{Simplify.}$$

To graph the inequality, think of $\dfrac{5}{2}$ as $2\dfrac{1}{2}$.

Interval notation: The solution set is $\left(\dfrac{5}{2}, \infty\right)$.

Set notation: $\left\{ d \mid d > \dfrac{5}{2} \right\}$

You Try 4

Solve $4(p + 2) + 1 > 2(3p + 10)$. Graph the solution set and write the answer in interval and set notations.

5. Solve Compound Inequalities Containing Three Parts

A **compound inequality** contains more than one inequality symbol. Some types of compound inequalities are

$$-5 < b + 4 < 1, \quad t \le \frac{1}{2} \text{ or } t \ge 3, \quad \text{and} \quad 2z + 9 < 5 \text{ and } z - 1 > 6$$

In this section, we will learn how to solve the first type of compound inequality. In Section 3.8 we will discuss the last two.

Consider the inequality $-2 \le x \le 3$. We can think of this in two ways:

1) x is *between* -2 and 3, and -2 and 3 are included in the interval.

or

2) We can break up $-2 \le x \le 3$ into the two inequalities $-2 \le x$ *and* $x \le 3$.

Either way we think about $-2 \le x \le 3$, the meaning is the same. On a number line, the inequality would be represented as

Notice that the **lower bound** of the interval on the number line is -2 (including -2), and the **upper bound** is 3 (including 3). Therefore, we can write the interval notation as

$$[-2, 3]$$

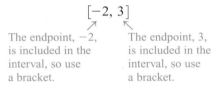

The endpoint, -2, is included in the interval, so use a bracket.

The endpoint, 3, is included in the interval, so use a bracket.

The set notation to represent $-2 \le x \le 3$ is $\{x | -2 \le x \le 3\}$.

Next, we will solve the inequality $-5 < b + 4 < 1$. To solve this type of compound inequality you must remember that *whatever operation you perform on one part of the inequality must be performed on all three parts.* All properties of inequalities apply.

| **Example 5** |

Solve $-5 < b + 4 < 1$. Graph the solution set, and write the answer in interval notation.

Solution

$$-5 < b + 4 < 1$$
$$-5 - 4 < b + 4 - 4 < 1 - 4 \qquad \text{To get the } b \text{ by itself subtract 4}$$
$$-9 < b < -3 \qquad\qquad\qquad \text{from each part of the inequality.}$$

Interval notation: The solution set is $(-9, -3)$.

Use parentheses here since -9 and -3 are not included in the solution set.

You Try 5

Solve $-2 \le 7k - 9 \le 19$. Graph the solution set, and write the answer in interval notation.

We can eliminate fractions in an inequality by multiplying by the LCD of all of the fractions.

Example 6

Solve $-\dfrac{7}{3} < \dfrac{1}{2}y - \dfrac{1}{3} \le \dfrac{1}{2}$. Graph the solution set, and write the answer in interval notation.

Solution

The LCD of the fractions is 6. Multiply by 6 to eliminate the fractions.

$$-\frac{7}{3} < \frac{1}{2}y - \frac{1}{3} \le \frac{1}{2}$$

$$6\left(-\frac{7}{3}\right) < 6\left(\frac{1}{2}y - \frac{1}{3}\right) \le 6\left(\frac{1}{2}\right)$$ Multiply all parts of the inequality by 6.

$$-14 < 3y - 2 \le 3$$

$$-14 + 2 < 3y - 2 + 2 \le 3 + 2$$ Add 2 to each part.

$$-12 < 3y \le 5$$ Combine like terms.

$$-\frac{12}{3} < \frac{3y}{3} \le \frac{5}{3}$$ Divide each part by 3.

$$-4 < y \le \frac{5}{3}$$ Simplify.

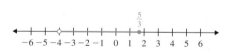

Interval notation: The solution set is $\left(-4, \dfrac{5}{3}\right]$.

You Try 6

Solve $-\dfrac{3}{4} < \dfrac{1}{3}z - \dfrac{3}{4} \le \dfrac{5}{4}$. Graph the solution set, and write the answer in interval notation.

Remember, if we multiply or divide an inequality by a negative number, we reverse the direction of the inequality symbol. When solving a compound inequality like these, reverse *both* symbols.

Example 7

Solve $11 < -3x + 2 < 17$. Graph the solution set, and write the answer in interval notation.

Solution

$$11 < -3x + 2 < 17$$

$$11 - 2 < -3x + 2 - 2 < 17 - 2 \qquad \text{Subtract 2 from each part.}$$
$$9 < -3x < 15$$
$$\frac{9}{-3} > \frac{-3x}{-3} > \frac{15}{-3} \qquad \text{When we divide by a negative number,}$$
$$\text{reverse the direction of the inequality symbol.}$$
$$-3 > x > -5 \qquad \text{Simplify.}$$

Think carefully about what $-3 > x > -5$ means. It means "x is less than -3 *and* x is greater than -5." This is especially important to understand when writing the correct interval notation.

The graph of the solution set is

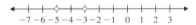

Even though we got $-3 > x > -5$ as our result, -5 is actually the lower bound of the solution set and -3 is the upper bound. The inequality $-3 > x > -5$ can also be written as $-5 < x < -3$.

Interval notation: The solution set is $(-5, -3)$.

        ↑ ↑

Lower bound on the left Upper bound on the right

You Try 7

Solve $4 < -2x - 4 < 10$. Graph the solution set, and write the answer in interval notation.

6. Solve Applications Involving Linear Inequalities

Certain phrases in applied problems indicate the use of inequality symbols:

at least:	$\geq$	no less than:	$\geq$
at most:	$\leq$	no more than:	$\leq$

There are others. Next, we will look at an example of a problem involving the use of an inequality symbol. We will use the same steps that were used to solve applications involving equations.

Example 8	

Joe Amici wants to have his son's birthday party at Kiddie Fun Factory. The cost of a party is $175 for the first 10 children plus $3.50 for each additional child. If Joe can spend at most $200, find the greatest number of children who can attend the party.

Solution

Step 1: Read the problem carefully, twice.

Step 2: We must determine the maximum number of kids who can attend the party. What will the variable represent? Define the unknown.

For the first 10 children the cost is $175. The change in cost comes when more than 10 kids attend. Each additional child, over the allowed 10, costs $3.50. So, if 1 more came, the extra cost would be $3.50(1) = $3.50. If 2 more attended, the additional cost would be $3.50(2) = $7.00, etc. Therefore,

$$x = \text{number of children } over \text{ the first 10 who attend the party}$$

Step 3: Translate from English to an algebraic inequality.

Let's put the situation into an equation in English:

Total cost of party = Cost of first 10 children + Cost of additional children

Total cost of party = $175 + 3.50x$

Also, Joe, can spend *at most* $200, so the cost must be ≤ 200.

Here is the inequality:

$$\text{Cost of party} \leq 200$$
$$175 + 3.50x \leq 200$$

Step 4: Solve the inequality.

$$175 + 3.50x \leq 200$$
$$3.50x \leq 25 \qquad \text{Subtract 175.}$$
$$x \leq 7.142\ldots \qquad \text{Divide by 3.50.}$$

Step 5: Interpret the meaning of the solution as it relates to the problem.

The result was $x \leq 7.142\ldots$ where x represents the number of children *over* the first 10 who can attend the party. Since it is not possible to have 7.142… children and $x \leq 7.142\ldots$, in order to stay within budget, at most 7 additional children can be invited.

Therefore, the greatest number of children who can attend the party is

$$
\begin{array}{ccccc}
\text{The first 10} & + & \text{Additional} & = & \text{Total} \\
\downarrow & & \downarrow & & \\
10 & + & 7 & = & 17
\end{array}
$$

At most, *17 children* can attend the party so that Joe Amici stays within the $200 budget. Does the answer make sense?

$$\begin{aligned} \text{Total cost of party} &= \$175 + \$3.50(7) \\ &= \$175 + \$24.50 \\ &= \$199.50 \end{aligned}$$

We can see that one more child (at a cost of $3.50) would put Mr. Amici over budget.

You Try 8

For $4.00 per month, Van can send or receive 200 text messages. Each additional message costs $0.05. If Van can spend at most $9.00 per month on text messages, find the greatest number he can send or receive each month.

Answers to You Try Exercises

1) a) $[-7, \infty)$

$-7\ -6\ -5\ -4\ -3\ -2\ -1\ \ 0\ \ 1\ \ 2\ \ 3$

b) $(-\infty, 5)$

$-4\ -3\ -2\ -1\ \ 0\ \ 1\ \ 2\ \ 3\ \ 4\ \ 5\ \ 6$

2) interval: $[6, \infty)$, set: $\{k|k \geq 6\}$

$0\ \ 1\ \ 2\ \ 3\ \ 4\ \ 5\ \ 6\ \ 7\ \ 8$

3) interval: $(-12, \infty)$, set: $\{m|m > -12\}$

$-12 \qquad\qquad\qquad\qquad 0$

4) interval: $\left(-\infty, -\dfrac{11}{2}\right)$, set: $\left\{p|p < -\dfrac{11}{2}\right\}$

$-\dfrac{11}{2}$

$-7\ -6\ -5\ -4\ -3\ -2\ -1\ \ 0\ \ 1\ \ 2$

5) $[1, 4]$

$0\ \ 1\ \ 2\ \ 3\ \ 4\ \ 5$

6) $(0, 6]$

$-6\ -5\ -4\ -3\ -2\ -1\ \ 0\ \ 1\ \ 2\ \ 3\ \ 4\ \ 5\ \ 6$

7) $(-7, -4)$

$-7\ -6\ -5\ -4\ -3\ -2\ -1\ \ 0\ \ 1\ \ 2\ \ 3$

8) 300

3.7 Exercises

Boost your grade at mathzone.com! MathZone

> Practice Problems   > NetTutor   > e-Professors   > Videos
> Self-Test

Objective 1

Graph the inequality. Express the solution in a) set notation and b) interval notation. See Examples 1–3.

1) $x \geq 3$

2) $t \geq -4$

3) $c < -1$

4) $r < \dfrac{5}{2}$

5) $w > -\dfrac{11}{3}$

6) $p \leq 2$

7) $1 \leq n \leq 4$

8) $-3 \leq g \leq 2$

9) $-2 < a < 1$

10) $-4 < d < 0$

 11) $\dfrac{1}{2} < z \leq 3$

12) $-2 \leq y < 3$

13) When do you use parentheses when writing a solution set in interval notation?

14) When do you use brackets when writing a solution set in interval notation?

Objectives 2 and 3

Solve each inequality. Graph the solution set and write the answer in a) set notation and b) interval notation. See Examples 2 and 3.

15) $n - 8 \leq -3$

16) $p + 6 \geq 4$

17) $y + 5 \geq 1$

18) $r - 9 \leq -5$

19) $3c > 12$

20) $8v > 24$

21) $15k < -55$

22) $16m < -28$

 23) $-4b \leq 32$

24) $-9a \geq 27$

25) $-14w > -42$

26) $-30t < -18$

27) $\dfrac{1}{3}x < -2$

28) $\dfrac{1}{5}z \geq -3$

29) $-\dfrac{2}{5}p \geq 4$

30) $-\dfrac{9}{4}y < -18$

Objective 4

Solve each inequality. Graph the solution set and write the answer in interval notation. See Example 4.

31) $8z + 19 > 11$

32) $5x - 2 \leq 18$

33) $12 - 7t \geq 15$

34) $-1 - 4p < 5$

35) $-23 - w < -20$

36) $16 - h \geq 9$

37) $7a + 4(5 - a) \leq 4 - 5a$

38) $6(7y + 4) - 10 > 2(10y + 13)$

39) $9c + 17 > 14c - 3$

40) $-11n + 6 \leq 16 - n$

 41) $\dfrac{8}{3}(2k + 1) > \dfrac{1}{6}k + \dfrac{8}{3}$

42) $\dfrac{1}{2}(c - 3) + \dfrac{3}{4}c \geq \dfrac{1}{2}(2c + 3) + \dfrac{3}{8}$

43) $0.04x + 0.12(10 - x) \geq 0.08(10)$

44) $0.09m + 0.05(8) \leq 0.07(m + 8)$

Objective 5

Solve each inequality. Graph the solution set and write the answer in interval notation. See Examples 5–7.

45) $-8 \leq a - 5 \leq -4$

46) $1 \leq t + 3 \leq 7$

47) $9 < 6n < 18$

48) $-10 < 2x < 7$

49) $-19 \leq 7p + 9 \leq 2$

50) $-5 \leq 3k - 11 \leq 4$

51) $-6 \leq 4c - 13 < -1$

52) $-11 < 6m + 1 \leq -3$

53) $2 < \dfrac{3}{4}u + 8 < 11$

54) $2 \leq \dfrac{5}{2}y - 3 \leq 7$

55) $-\dfrac{1}{2} \leq \dfrac{5d + 2}{6} \leq 0$

56) $2 < \dfrac{2b + 7}{3} < 5$

57) $-13 \leq 14 - 9h < 5$

58) $3 < 19 - 2j \leq 9$

59) $0 \leq 4 - 3w \leq 7$

60) $-6 < -5 - z < 0$

Objective 6

Write an inequality for each problem and solve. See Example 8.

61) Oscar makes a large purchase at Home Depot and plans to rent one of its trucks to take his supplies home. The most he wants to spend on the truck rental is $50.00. If Home Depot charges $19.00 for the first 75 min and $5.00 for each additional 15 min, for how long can Oscar keep the truck and remain within his budget? (www.homedepot.com)

62) Carson's Parking Garage charges $4.00 for the first 3 hr plus $1.50 for each additional half-hour. Ted has only $11.50 for parking. For how long can Ted park his car in this garage?

63) A taxi charges $2.00 plus $0.25 for every $\frac{1}{5}$ of a mile. How many miles can you go if you have $12.00?

64) A taxi charges $2.50 plus $0.20 for every $\frac{1}{4}$ of a mile. How many miles can you go if you have $12.50?

 65) Melinda's first two test grades in Psychology were 87 and 94. What does she need to make on the third test to maintain an average of at least 90?

66) Russell's first three test scores in Geography were 86, 72, and 81. What does he need to make on the fourth test to maintain an average of at least 80?

Section 3.8   Solving Compound Inequalities

Objectives

1. Find the Intersection and Union of Two Sets

2. Solve Compound Inequalities Containing the Word *And*

3. Solve Compound Inequalities Containing the Word *Or*

4. Solve Compound Inequalities That Have No Solution or That Have Solution Sets Consisting of All Real Numbers

Compound inequalities were introduced in Section 3.7. We learned how to solve a compound inequality like $-8 \leq 3x + 4 \leq 13$. But, we also saw examples of two different types of inequalities. They were

$$t \leq \frac{1}{2} \text{ or } t \geq 3 \quad \text{and} \quad 2z + 9 < 5 \text{ and } z - 1 > 6$$

In this section, we will discuss how to solve compound inequalities like these. But first, we must talk about set notation and operations.

1. Find the Intersection and Union of Two Sets

Example 1

Let $A = \{1, 2, 3, 4, 5, 6\}$ and $B = \{3, 5, 7, 9, 11\}$.

The **intersection** of sets A and B is the set of numbers which are elements of A **and** of B. The *intersection* of A and B is denoted by $A \cap B$.

$A \cap B = \{3, 5\}$ since 3 and 5 are found in both A and B.

The **union** of sets A and B is the set of numbers which are elements of A **or** of B. The *union* of A and B is denoted by $A \cup B$. The set $A \cup B$ consists of the elements in A or in B or in both.

$A \cup B = \{1, 2, 3, 4, 5, 6, 7, 9, 11\}$

> Although the elements 3 and 5 appear in both set A and in set B, we do not write them twice in the set A ∪ B.

So, the word "*and*" indicates *intersection*, while the word "*or*" indicates *union*. This same principle holds when solving compound inequalities involving "*and*" or "*or*."

You Try 1

Let $A = \{2, 4, 6, 8, 10\}$ and $B = \{1, 2, 5, 6, 9, 10\}$. Find $A \cap B$ and $A \cup B$.

2. Solve Compound Inequalities Containing the Word *And*

Example 2

Solve the compound inequality $c + 5 \geq 3$ and $8c \leq 32$. Graph the solution set, and write the answer in interval notation.

Solution

Step 1: Identify the inequality as "*and*" or "*or*" and understand what that means. These two inequalities are connected by "*and*." That means the solution set will consist of the values of c which make *both* inequalities true. The solution set will be the *intersection* of the solution sets of $c + 5 \geq 3$ and $8c \leq 32$.

Step 2: Solve each inequality separately.

$$c + 5 \geq 3 \quad \text{and} \quad 8c \leq 32$$
$$c \geq -2 \quad \text{and} \quad c \leq 4$$

Step 3: Graph the solution set to each inequality on its own number line even if the problem does not require you to graph the solution set. This will help you to visualize the solution set of the compound inequality.

$c \geq -2$:

$c \leq 4$:

Step 4: Look at the number lines and think about where the solution set for the compound inequality would be graphed.

Since this is an "*and*" inequality, the solution set of $c + 5 \geq 3$ and $8c \leq 32$ consists of the numbers that are solutions to *both* inequalities. We can visualize it this way: if we take the number line above representing $c \geq -2$ and place it on top of the number line representing $c \leq 4$, what shaded areas would overlap (intersect)?

$c \geq -2$ and $c \leq 4$: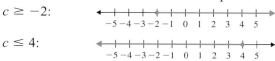

They intersect between -2 and 4, *including* those endpoints.

Step 5: Write the answer in interval notation.

The final number line illustrates that the solution to $c + 5 \geq 3$ and $8c \leq 32$ is $[-2, 4]$. The graph of the solution set is the final number line above.

Here are the steps to follow when solving a compound inequality.

Steps for Solving a Compound Inequality

1) Identify the inequality as "*and*" or "*or*" and understand what that means.

2) Solve each inequality separately.

3) Graph the solution set to each inequality on its own number line even if the problem does not explicitly tell you to graph the solution set. This will help you to visualize the solution to the compound inequality.

4) Use the separate number lines to graph the solution set of the compound inequality.

    a) If it is an "*and*" inequality, the solution set consists of the regions on the separate number lines that would *overlap* (intersect) if one number line was placed on top of the other.

    b) If it is an "*or*" inequality, the solution set consists of the *total* (union) of what would be shaded if you took the separate number lines and put one on top of the other.

5) Use the graph of the solution set to write the answer in interval notation.

You Try 2

Solve the compound inequality $y - 2 \leq 1$ and $7y > -28$. Graph the solution set, and write the answer in interval notation.

Example 3

Solve the compound inequality $7y + 2 > 37$ and $5 - \dfrac{1}{3}y < 6$. Write the solution set in interval notation.

Solution

Step 1: This is an "*and*" inequality. The solution set of this compound inequality will be the *intersection* of the solution sets of the separate inequalities

$$7y + 2 > 37 \text{ and } 5 - \frac{1}{3}y < 6.$$

Step 2: We must solve each inequality separately.

$$7y + 2 > 37 \quad \text{and} \quad 5 - \frac{1}{3}y < 6$$

$$7y > 35 \quad \text{and} \quad -\frac{1}{3}y < 1 \qquad \text{Multiply both sides by } -3.$$

$$y > 5 \quad \text{and} \quad y > -3 \qquad \begin{array}{l}\text{Reverse the direction of the}\\ \text{inequality symbol.}\end{array}$$

Step 3: Graph the solution sets separately so that it is easier to find their intersection.

$y > 5$:

$$-6\ -5\ -4\ -3\ -2\ -1\ \ 0\ \ 1\ \ 2\ \ 3\ \ 4\ \ 5\ \ 6$$

$y > -3$:

$$-6\ -5\ -4\ -3\ -2\ -1\ \ 0\ \ 1\ \ 2\ \ 3\ \ 4\ \ 5\ \ 6$$

Step 4: If we were to put the number lines above on top of each other, where do they intersect?

$y > 5$ and $y > -3$:

$$\underset{-6\ -5\ -4\ -3\ -2\ -1\ \ 0\ \ 1\ \ 2\ \ 3\ \ 4\ \ 5\ \ 6}{\xleftarrow{\hspace{2cm}}\text{—}\xrightarrow{\hspace{2cm}}}$$

Step 5: We can represent the shaded region above as $(5, \infty)$.

The solution to $7y + 2 > 37$ and $5 - \dfrac{1}{3}y < 6$ is $(5, \infty)$.

You Try 3

Solve each compound inequality and write the answer in interval notation.

a) $4x - 3 > 1$ and $x + 6 < 13$ b) $-\dfrac{4}{5}m > -8$ and $2m + 5 \leq 12$

3. Solve Compound Inequalities Containing the Word *Or*

Recall that the word "*or*" indicates the union of two sets.

Example 4

Solve the compound inequality $6p + 5 \leq -1$ or $p - 3 \geq 1$. Write the answer in interval notation.

Solution

Step 1: These two inequalities are joined by "*or*." Therefore, the solution set will consist of the values of p that are in the solution set of $6p + 5 \leq -1$ *or* that are in the solution set of $p - 3 \geq 1$ *or* that are in *both* solution sets.

Step 2: Solve each inequality separately.

$$
\begin{aligned}
6p + 5 &\leq -1 &\quad \text{or} \quad& p - 3 \geq 1 \\
6p &\leq -6 && \\
p &\leq -1 &\quad \text{or} \quad& p \geq 4
\end{aligned}
$$

Step 3: Graph the solution sets separately so that it is easier to find the *union* of the sets.

$p \leq -1$:

$$\underset{-5\ -4\ -3\ -2\ -1\ \ 0\ \ 1\ \ 2\ \ 3\ \ 4\ \ 5\ \ 6}{\xleftarrow{\hspace{3cm}}}$$

$p \geq 4$:

$$\underset{-5\ -4\ -3\ -2\ -1\ \ 0\ \ 1\ \ 2\ \ 3\ \ 4\ \ 5\ \ 6}{\xleftarrow{\hspace{3cm}}}$$

Step 4: The solution set of the compound inequality $6p + 5 \leq -1$ or $p - 3 \geq 1$ consists of the numbers which are solutions to the first inequality *or* the second inequality *or* both. We can visualize it this way: if we put the number lines on top of each other, the solution set of the compound inequality is the **total** (union) of what is shaded.

$p \leq -1$ or $p \geq 4$:
$$\underset{-6\ -5\ -4\ -3\ -2\ -1\ \ 0\ \ 1\ \ 2\ \ 3\ \ 4\ \ 5\ \ 6}{\xleftarrow{\hspace{3cm}}}$$

Step 5: The final number line illustrates that the solution set of $6p + 5 \leq -1$ or $p - 3 \geq 1$ is

$$(-\infty, -1] \cup [4, \infty)$$

↑

Use the *union* symbol for "or."

You Try 4

Solve $t + 8 \geq 14$ or $\dfrac{3}{2}t < 6$ and write the solution in interval notation.

4. Solve Compound Inequalities That Have No Solution or That Have Solution Sets Consisting of All Real Numbers

Example 5

Solve each compound inequality and write the answer in interval notation.

a) $k - 5 < -2$ or $4k + 9 > 6$ b) $\dfrac{1}{2}w \geq 3$ and $1 - w \geq 0$

Solution

a) $k - 5 < -2$ or $4k + 9 > 6$

Step 1: The solution to this "*or*" inequality is the *union* of the solution sets of $k - 5 < -2$ and $4k + 9 > 6$.

Step 2: Solve each inequality separately.

$$k - 5 < -2 \quad \text{or} \quad 4k + 9 > 6$$
$$4k > -3$$
$$k < 3 \quad \text{or} \quad k > -\frac{3}{4}$$

Step 3: $k < 3$:

$$\begin{array}{c}\xleftarrow{\hspace{1cm}}\!\!\!\!+\!\!+\!\!+\!\!+\!\!+\!\!+\!\!+\!\!\diamond\!\!+\!\!\rightarrow\\ {\scriptstyle -4\,-3\,-2\,-1\ \ 0\ \ 1\ \ 2\ \ 3\ \ 4}\end{array}$$

$k > -\dfrac{3}{4}$:

$$\begin{array}{c}\xleftarrow{\hspace{1cm}}\!\!+\!\!+\!\!+\!\!+\!\!\diamond\!\!\multimap\!\!+\!\!+\!\!+\!\!+\!\!\rightarrow\\ {\scriptstyle -4\,-3\,-2\,-1\ \ 0\ \ 1\ \ 2\ \ 3\ \ 4}\end{array}$$

Step 4: $k < 3$ or $k > -\dfrac{3}{4}$:

$$\begin{array}{c}\xleftarrow{\hspace{1cm}}\!\!+\!\!+\!\!+\!\!+\!\!+\!\!+\!\!+\!\!+\!\!+\!\!+\!\!\rightarrow\\ {\scriptstyle -4\,-3\,-2\,-1\ \ 0\ \ 1\ \ 2\ \ 3\ \ 4}\end{array}$$

If the number lines in step 3 were placed on top of each other, the *total* (union) of what would be shaded is the entire number line. This represents all real numbers.

Step 5: The solution set of $k - 5 < -2$ or $4k + 9 > 6$ is $(-\infty, \infty)$.

b) $\frac{1}{2}w \geq 3$ and $1 - w \geq 0$

Step 1: The solution to this "*and*" inequality is the *intersection* of the solution sets of $\frac{1}{2}w \geq 3$ and $1 - w \geq 0$.

Step 2: Solve each inequality separately.

$$\frac{1}{2}w \geq 3 \quad \text{and} \quad 1 - w \geq 0$$

$$\phantom{\frac{1}{2}w \geq 3 \quad \text{and} \quad} 1 \geq w \qquad \text{Add } w.$$

Multiply by 2. $\quad w \geq 6 \quad$ and $\quad w \leq 1 \qquad$ Rewrite $1 \geq w$ as $w \leq 1$.

Step 3: $w \geq 6$:

$$\begin{array}{c} \text{-1 0 1 2 3 4 5 6 7} \end{array}$$

$w \leq 1$:

$$\begin{array}{c} \text{-1 0 1 2 3 4 5 6 7} \end{array}$$

Step 4: $w \geq 6$ and $w \leq 1$:

$$\begin{array}{c} \text{-1 0 1 2 3 4 5 6 7} \end{array}$$

If the number lines in step 3 were placed on top of each other, the shaded regions would *not* intersect. Therefore, the solution set is the empty set, $\emptyset$.

Step 5: The solution set of $\frac{1}{2}w \geq 3$ and $1 - w \geq 0$ is $\emptyset$. (There is no solution.)

You Try 5

Solve the compound inequalities and write the solution in interval notation.

a) $-3w \leq w - 6$ and $5w < 4$ b) $9z - 8 \leq -8$ or $z + 7 \geq 2$

Answers to You Try Exercises

1) $A \cap B = \{2, 6, 10\}$, $A \cup B = \{1, 2, 4, 5, 6, 8, 9, 10\}$ 2)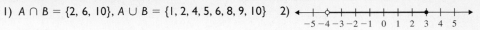
$$\begin{array}{c} \text{-5 -4 -3 -2 -1 0 1 2 3 4 5} \end{array}$$

$(-4, 3]$ 3) a) $(1, 7)$ b) $\left(-\infty, \dfrac{7}{2}\right]$ 4) $(-\infty, 4) \cup [6, \infty)$ 5) a) $\emptyset$ b) $(-\infty, \infty)$

3.8 Exercises

Boost your grade at mathzone.com! MathZone

> Practice Problems > NetTutor > Self-Test > e-Professors > Videos

Objective 1

1) Given sets A and B, explain how to find $A \cap B$.

2) Given sets X and Y, explain how to find $X \cup Y$.

Given sets $A = \{2, 4, 6, 8, 10\}$, $B = \{1, 3, 5\}$, $X = \{8, 10, 12, 14\}$, and $Y = \{5, 6, 7, 8, 9\}$ find

3) $X \cap Y$

4) $A \cap X$

5) $A \cup Y$

6) $B \cup Y$

7) $X \cap B$

8) $B \cap A$

9) $A \cup B$

10) $X \cup Y$

Each number line represents the solution set of an inequality. Graph the *intersection* of the solution sets and write the intersection in interval notation.

11) $x \geq -3$:

$x \leq 2$:

12) $n \leq 4$:

$n \geq 0$:

13) $t < 3$:

$t > -1$:

14) $y > -4$:

$y < -2$:

15) $c > 1$:

$c \geq 3$:

16) $p < 2$:

$p < -1$:

17) $z \leq 0$:

$z \geq 2$:

18) $g \geq -1$:

$g < -\dfrac{5}{2}$:

Objectives 2 and 4

Solve each compound inequality. Graph the solution set, and write the answer in interval notation.

19) $a \leq 5$ and $a \geq 2$

20) $k > -3$ and $k < 4$

21) $b - 7 > -9$ and $8b < 24$

22) $3x \leq 1$ and $x + 11 \geq 4$

 23) $5w + 9 \leq 29$ and $\dfrac{1}{3}w - 8 > -9$

24) $4y - 11 > -7$ and $\dfrac{3}{2}y + 5 \leq 14$

25) $2m + 15 \geq 19$ and $m + 6 < 5$

26) $d - 1 > 8$ and $3d - 12 < 4$

27) $r - 10 > -10$ and $3r - 1 > 8$

28) $2t - 3 \leq 6$ and $5t + 12 \leq 17$

29) $9 - n \leq 13$ and $n - 8 \leq -7$

30) $c + 5 \geq 6$ and $10 - 3c \geq -5$

Objective 1

Each number line represents the solution set of an inequality. Graph the *union* of the solution sets and write the union in interval notation.

31) $p < -1$:

$p > 5$:

32) $z < 2$:

$z > 6$:

33) $a \leq \dfrac{5}{3}$:

$a > 4$:

34) $v \leq -3$:

$v \geq \dfrac{11}{4}$:

35) $y > 1$:

$y > 3$:

36) $x \leq -6$:

$x \leq -2$:

37) $c < \dfrac{7}{2}$:

$c \geq -2$:

38) $q \leq 3$:

-4 -3 -2 -1 0 1 2 3 4

$q > -2.7$:

-4 -3 -2 -1 0 1 2 3 4

Objectives 3 and 4

Solve each compound inequality. Graph the solution set, and write the answer in interval notation.

39) $z < -1$ or $z > 3$

40) $x \leq -4$ or $x \geq 0$

 41) $6m \leq 21$ or $m - 5 > 1$

42) $a + 9 > 7$ or $8a \leq -44$

43) $3t + 4 > -11$ or $t + 19 > 17$

44) $5y + 8 \leq 13$ or $2y \leq -6$

45) $-2v - 5 \leq 1$ or $\dfrac{7}{3}v < -14$

46) $k - 11 < -4$ or $-\dfrac{2}{9}k \leq -2$

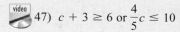 47) $c + 3 \geq 6$ or $\dfrac{4}{5}c \leq 10$

48) $\dfrac{8}{3}g \geq -12$ or $2g + 1 \leq 7$

49) $7 - 6n \geq 19$ or $n + 14 < 11$

50) $d - 4 > -7$ or $-6d \leq 2$

Definition/Procedure	Example	Reference
Solving Linear Equations Part I		3.1
The Addition and Subtraction Properties of Equality 1) If $a = b$, then $a + c = b + c$ 2) If $a = b$, then $a - c = b - c$	Solve. $a + 4 = 19$ $a + 4 - 4 = 19 - 4$ Subtract 4 from $a = 15$ each side. The solution set is $\{15\}$.	**p. 131**
The Multiplication and Division Properties of Equality 1) If $a = b$, then $ac = bc$ 2) If $a = b$, then $\dfrac{a}{c} = \dfrac{b}{c}$ $(c \neq 0)$	Solve. $\dfrac{3}{2}t = -30$ $\dfrac{2}{3} \cdot \dfrac{3}{2}t = -30 \cdot \dfrac{2}{3}$ Multiply each side by $\dfrac{2}{3}$. $t = -20$ The solution set is $\{-20\}$.	**p. 132**
Sometimes it is necessary to **combine like terms** as the first step in solving a linear equation.	Solve. $5y + 3 - 7y + 9 = 4$ $-2y + 12 = 4$ Combine like terms. $-2y + 12 - 12 = 4 - 12$ Subtract 12 from each side. $-2y = -8$ $\dfrac{-2y}{-2} = \dfrac{-8}{-2}$ Divide by -2. $y = 4$ The solution set is $\{4\}$.	**p. 136**
Solving Linear Equations Part II		3.2
How to Solve a Linear Equation 1) If applicable, clear the equation of fractions or decimals. 2) If there are terms in parentheses, distribute. 3) Use the properties of equality to get the variables on one side of the equation and the constants on the other side. Then, solve.	Solve. $4(c + 1) + 7 = 2c + 9$. $4c + 4 + 7 = 2c + 9$ Distribute. $4c + 11 = 2c + 9$ Combine like terms. $4c - 2c + 11 = 2c - 2c + 9$ Get variable terms on one side. $2c + 11 - 11 = 9 - 11$ Get constants on one side. $2c = -2$ $\dfrac{2c}{2} = \dfrac{-2}{2}$ Division property of equality. $c = -1$ The solution set is $\{-1\}$.	**p. 139**

Definition/Procedure	Example	Reference

Steps for Solving Applied Problems

1) Read and reread the problem. Draw a picture, if applicable.
2) Assign a variable to represent an unknown. Define other unknown quantities in terms of the variable.
3) Translate from English to math.
4) Solve the equation.
5) Interpret the meaning of the solution as it relates to the problem.

The sum of a number and fifteen is eight. Find the number.

1) Read the problem carefully, then read it again.
2) Define the unknown.
$$x = \text{the number}$$

3) "The sum of a number and fifteen is eight" means

The number plus fifteen equals eight.
$$x \quad + \quad 15 \quad = \quad 8$$
Equation: $x + 15 = 8$
4) Solve the equation.
$$x + 15 = 8$$
$$x + 15 - 15 = 8 - 15$$
$$x = -7$$
The number is -7.

p. 144

Applications of Linear Equations to General Problems, Consecutive Integers, and Fixed and Variable Cost

3.3

Apply the "**Steps for Solving Applied Problems**" to solve this application involving consecutive odd integers.

The sum of three consecutive odd integers is 87. Find the integers.

1) Read the problem carefully, twice.

2) Define the unknowns.
$$x = \text{the first odd integer}$$
$$x + 2 = \text{the second odd integer}$$
$$x + 4 = \text{the third odd integer}$$

3) "The sum of three consecutive odd integers is 87" means

First odd + Second odd + Third odd = 87
$$x \quad + \quad (x + 2) \quad + \quad (x + 4) = 87$$
Equation: $x + (x + 2) + (x + 4) = 87$

4) Solve $x + (x + 2) + (x + 4) = 87$.
$$3x + 6 = 87$$
$$3x + 6 - 6 = 87 - 6$$
$$3x = 81$$
$$\frac{3x}{3} = \frac{81}{3}$$
$$x = 27$$

5) Find the values of all of the unknowns.
$$x = 27, \; x + 2 = 29, \; x + 4 = 31$$
The numbers are 27, 29, and 31.

p. 152

Definition/Procedure	Example	Reference

Definition/Procedure	Example	Reference
Applications of Linear Equations to Proportions, $d = rt$, and Mixture Problems		3.6
A **proportion** is a statement that two ratios are equivalent.	If Malik can read 2 books in 3 weeks, how long will it take him to read 7 books? 1) Read the problem carefully, twice. 2) Define the unknown. x = number of weeks to read 7 books. 3) Set up a proportion. $$\frac{2 \text{ books}}{3 \text{ weeks}} = \frac{7 \text{ books}}{x \text{ weeks}}$$ Equation: $\dfrac{2}{3} = \dfrac{7}{x}$ 4) Solve $\dfrac{2}{3} = \dfrac{7}{x}$. $2x = 3(7)$ Set cross-products equal. $$\frac{2x}{2} = \frac{21}{2}$$ $$x = \frac{21}{2} = 10\frac{1}{2}$$ 5) It will take Malik $10\dfrac{1}{2}$ weeks to read 7 books.	p. 180
Solving Linear Inequalities in One Variable		3.7
We solve linear inequalities in very much the same way we solve linear equations *except when we multiply or divide by a negative number we must reverse the direction of the inequality symbol.* We can graph the solution set, write the solution in set notation, or write the solution in interval notation.	Solve $x - 5 \leq -3$. Graph the solution set and write the answer in both set notation and interval notation. $$x - 5 \leq -3$$ $$x - 5 + 5 \leq -3 + 5$$ $$x \leq 2$$ $\{x \mid x \leq 2\}$ Set notation $(-\infty, 2]$ Interval notation	p. 196

Definition/Procedure	Example	Reference
Solving Compound Inequalities		3.8
The solution set of a compound inequality joined by "**and**" will be the **intersection** of the solution sets of the individual inequalities.	Solve the compound inequality $5x - 2 \geq -17$ and $x + 8 \leq 9$. $5x - 2 \geq -17$ and $x + 8 \leq 9$ $5x \geq -15$ $x \geq -3$ and $x \leq 1$ $\xleftarrow{\hspace{0.5em}}\overset{}{\underset{-4\ -3\ -2\ -1\ \ 0\ \ 1\ \ 2\ \ 3\ \ 4}{\rule{8em}{0.4pt}}}\xrightarrow{\hspace{0.5em}}$ Solution in interval notation: $[-3, 1]$	**p. 207**
The solution set of a compound inequality joined by "**or**" will be the **union** of the solution sets of the individual inequalities.	Solve the compound inequality $x - 3 < -1$ or $7x > 42$. $x - 3 < -1$ or $7x > 42$ $x < 2$ or $x > 6$ $\xleftarrow{\hspace{0.5em}}\overset{}{\underset{0\ \ 1\ \ 2\ \ 3\ \ 4\ \ 5\ \ 6\ \ 7\ \ 8\ \ 9}{\rule{8em}{0.4pt}}}\xrightarrow{\hspace{0.5em}}$ Solution in interval notation: $(-\infty, 2) \cup (6, \infty)$	**p. 209**

(3.1) Determine if the given value is a solution to the equation.

1) $2n + 13 = 10; n = -\dfrac{3}{2}$

2) $5 + t = 3t - 1; t = 4$

Solve each equation.

3) $-9z = 30$

4) $p - 11 = -14$

5) $21 = k + 2$

6) $56 = \dfrac{8}{5}m$

7) $-\dfrac{4}{9}w = -\dfrac{10}{7}$

8) $-c = 4$

9) $21 = 0.6q$

10) $-12 = -0.4x$

11) $8b - 7 = 57$

12) $13 - 4y = 23$

13) $6 = 15 + \dfrac{9}{2}v$

14) $2.3a + 1.5 = 10.7$

15) $\dfrac{2}{7} - \dfrac{3}{4}k = -\dfrac{17}{14}$

16) $\dfrac{5}{9}t + \dfrac{1}{6} = -\dfrac{3}{2}$

17) $4p + 9 + 2(p - 12) = 15$

18) $5(2z + 3) - (11z - 4) = 3$

(3.2) Solve each equation.

19) $11x + 13 = 2x - 5$

20) $5(c + 3) - 2c = 4 + 3(2c + 1)$

21) $6 - 5(4d - 3) = 7(3 - 4d) + 8d$

22) $4k + 19 + 8k = 2(6k - 11)$

23) $0.05m + 0.11(6 - m) = 0.08(6)$

24) $1 - \dfrac{1}{6}(t + 5) = \dfrac{1}{2}$

Solve using the five "Steps for Solving Applied Problems."

25) Twelve less than a number is five. Find the number.

26) Two-thirds of a number is twenty-two. Find the number.

27) A number increased by nine is one less than twice the number. Find the number.

28) A number divided by four is the same as fifteen less than the number. Find the number.

(3.3)

29) Mr. Morrissey has 26 children in his kindergarten class, and c children attended pre-school. Write an expression for the number of children who did not attend pre-school.

30) In a parking lot, there were f foreign cars. Write an expression for the number of American cars in the lot if there were 14 more than the number of foreign cars.

Solve using the five "Steps for Solving Applied Problems."

31) In its first week, American Idol finalist Clay Aiken's debut CD sold 316,000 more copies than Idol winner Kelly Clarkson's debut CD. Together their debut CDs sold 910,000 copies. How many CDs did each of them sell during the first week? (www.top40.com)

32) The sum of three consecutive even integers is 258. Find the integers.

33) The road leading to the Sutter family's farmhouse is 500 ft long. Some of it is paved, and the rest is gravel. If the paved portion is three times as long as the gravel part of the road, determine the length of the gravel portion of the road.

34) A car wash charges its users $2.00 for the first 2 minutes plus $0.25 for each additional 30 seconds. If Jackie has $3.25, how long can she spend washing her car?

(3.4)

35) Before road construction began in front of her ice cream store, Imelda's average monthly revenue was about $8200. In the month since construction began, revenue has decreased by 18%. What was the revenue during the first month of construction?

36) In 2000, Americans consumed, per capita, 0.20 cups of gourmet coffee per day. In 2002, that number rose to 0.33 cups per day. Determine the percent increase in the number of cups of gourmet coffee consumed each day from 2000 to 2002. (www.fas.usda.gov)

37) The pet insurance market grew 342% from 1998 to 2002. The market was valued at $88 million in 2002. What was the value of the market in 1998? (Round to the tenths place.) (www.wattnet.com)

38) Jerome invested some money in an account earning 7% simple interest and $3000 more than that at 8% simple interest. After 1 yr, he earned $915 in interest. How much money did he invest in each account?

(3.5)

39) *Use a formula from geometry to solve:* The base of a triangle measures 9 cm. If the area of the triangle is 54 cm^2, find the height.

40) Find the missing angle measures.

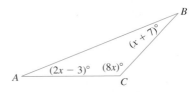

Find the measure of each indicated angle.

41)

42)

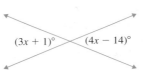

43) *Solve:* Three times the measure of an angle is 12° more than the measure of its supplement. Find the measure of the angle.

Solve for the indicated variable.

44) $y = mx + b$ for m

45) $pV = nRT$ for R

46) $C = \dfrac{1}{3}n(t + T)$ for t

(3.6) Solve each proportion.

47) $\dfrac{k}{12} = \dfrac{15}{9}$

48) $\dfrac{3a + 5}{6} = \dfrac{a - 1}{18}$

Set up a proportion and solve.

49) In 2002, 1 out of 5 females in eighth grade reported that they had used alcohol. In an eighth-grade class containing 85 girls, how many would be expected to have used alcohol? (www.cdc.gov)

50) Given these two similar triangles, find x.

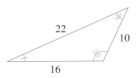

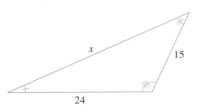

Solve.

51) A collection of coins contains 91 coins, all dimes and quarters. If the value of the coins is $14.05, determine the number of each type of coin in the collection.

52) To make a 6% acid solution, a chemist mixes some 4% acid solution with 12 L of 10% acid solution. How much of the 4% solution must be added to the 10% solution to make the 6% acid solution?

53) Peter leaves Mitchell's house traveling 30 mph. Mitchell leaves 15 min later, trying to catch up to Peter, going 40 mph. If they drive along the same route, how long will it take Mitchell to catch Peter?

(3.7) Solve each inequality. Graph the solution set and write the answer in interval notation.

54) $z + 6 \geq 14$

55) $-10y + 7 > 32$

56) $0.03c + 0.09(6 - c) > 0.06(6)$

57) $-15 < 4p - 7 \leq 5$

58) $-1 \leq \dfrac{5 - 3x}{2} \leq 0$

59) *Write an inequality and solve:* Renee's scores on her first three Chemistry tests were 95, 91, and 86. What does she need to make on her fourth test to maintain an average of at least 90?

(3.8)

60) $A = \{10, 20, 30, 40, 50\}$ $B = \{20, 25, 30, 35\}$
    a) Find $A \cup B$.
    b) Find $A \cap B$.

Solve each compound inequality. Graph the solution set and write the answer in interval notation.

61) $a + 6 \leq 9$ and $7a - 2 \geq 5$

62) $3r - 1 > 5$ or $-2r \geq 8$

63) $8 - y < 9$ or $\dfrac{1}{10}y > \dfrac{3}{5}$

64) $x + 12 \leq 9$ and $0.2x \geq 3$

Solve.

1) $-12q = 20$

2) $7p + 16 = 30$

3) $5 - 3(2c + 7) = 4c - 1 - 5c$

4) $\dfrac{5}{8}(3k + 1) - \dfrac{1}{4}(7k + 2) = 1$

5) $6(4t - 7) = 3(8t + 5)$

6) $\dfrac{3 + n}{5} = \dfrac{2n - 3}{8}$

Set up an equation and solve.

7) Nine less than twice a number is 33. Find the number.

8) The dosage for a certain medication is $\dfrac{1}{2}$ tsp for every 20 lb of body weight. What is the dosage for a child who weighs 90 lb?

9) Motor oil is available in three types: regular, synthetic, and synthetic blend, which is a mixture of regular oil with synthetic. Bob decides to make 5 qt of his own synthetic blend. How many quarts of regular oil costing $1.20 per quart and how many quarts of synthetic oil costing $3.40 per quart should he mix so that the blend is worth $1.86 per quart?

10) In 2004, Wisconsin had 9 drive-in theaters. This is 82% less than the number of drive-ins in 1967. Determine the number of drive-in theaters in Wisconsin in 1967. (www.drive-ins.com)

11) A contractor has 460 ft of fencing to enclose a rectangular construction site. If the length of the site is 160 ft, what is the width?

Solve for the indicated variable.

12) $R = \dfrac{kt}{5}$ for t

13) $S = 2\pi r^2 + 2\pi rh$ for h

Solve. Graph the solution set, and write the answer in interval notation.

14) $r + 7 \le 2$

15) $9 - 3x < 5x + 3$

16) $-1 < \dfrac{w - 5}{4} \le \dfrac{1}{2}$

17) Solve the compound inequality. Write the answer in interval notation. $y - 8 \le -5$ and $2y \ge 0$

18) *Write an inequality and solve:* Rawlings Builders will rent a forklift for $46.00 per day plus $9.00 per hour. If they have at most $100.00 allotted for a one-day rental, for how long can they keep the forklift and remain within budget?

Perform the operations and simplify.

1) $\dfrac{5}{12} - \dfrac{7}{9}$

2) $\dfrac{8}{15} \div 12$

3) $52 - 12 \div 4 + 3 \cdot 5$

Given the set of numbers
$\left\{ -13.7, \dfrac{19}{7}, 0, 8, \sqrt{17}, 0.\overline{61}, \sqrt{81}, -2 \right\}$ **identify**

4) the rational numbers.

5) the integers.

6) Which property is illustrated by
$9 + (4 + 1) = (9 + 4) + 1$?

Simplify. The answer should not contain any negative exponents.

7) $\dfrac{12k^{11}}{18k^4}$

8) $(-9y^5)^2(2y^8)^3$

9) $(-7m^8)\left(\dfrac{5}{21}m^{-6}\right)$

10) $\left(\dfrac{10a^{12}b^2}{5a^9b^{-3}}\right)^{-2}$

11) Write 279,000,000 in scientific notation.

12) Solve.
$$-31 = \dfrac{4}{7}z + 9$$

Solve. Graph the solution set and write the answer in interval notation.

13) $-14 < 6y + 10 < 3$

14) $8x \le -24$ or $4x - 5 \ge 6$

15) Solve for R: $A = P + PRT$

Write an equation and solve.

16) On Thursday, a pharmacist fills twice as many prescriptions with generic drugs as name-brand drugs. If he filled 72 prescriptions, how many were with generic drugs, and how many were with name-brand drugs?

17) Two friends start biking toward each other from opposite ends of a bike trail. One averages 8 mph, and the other averages 10 mph. If the distance between them when they begin is 9 mi, how long will it take them to meet?

CHAPTER 4

Linear Equations in Two Variables

Algebra at Work:
Landscape Architecture

We will take a final look at how mathematics is used in landscape architecture.

A landscape architect uses slope in many different ways. David explains that one important application of slope is in designing driveways after a new house has been built. Towns often have building codes that restrict the slope or steepness of a driveway. In this case, the rise of the land is the difference in height between the top and the bottom of the driveway. The run is the linear distance between those two points. By finding $\frac{rise}{run}$, a landscape architect knows how to design the driveway so that it meets the town's building code. This is especially important in cold weather climates, where if a driveway is too steep, a car will too easily slide into the street. If it doesn't meet the code, the driveway may have to be removed and rebuilt, or coils that radiate heat might have to be installed under the driveway to melt the snow in the wintertime. Either way, a mistake in calculating slope could cost the landscape company or the client a lot of extra money.

In Chapter 4, we will learn about slope, its meaning, and different ways to use it.

Section 4.1   Introduction to Linear Equations in Two Variables

Objectives

1. Define a Linear Equation in Two Variables
2. Decide Whether an Ordered Pair Is a Solution of a Given Equation
3. Complete Ordered Pairs for a Given Equation
4. Plot Ordered Pairs
5. Solve Applied Problems Involving Ordered Pairs

1. Define a Linear Equation in Two Variables

In Chapter 3, we learned how to solve equations such as $5y + 3 = 13$ and $2(t - 7) = 3t + 4$. They are examples of linear equations in one variable.

In this section, we will introduce linear equations in *two* variables. Let's begin with a definition.

> **Definition**   A **linear equation in two variables** can be written in the form $Ax + By = C$ where A, B, and C are real numbers and where both A and B do not equal zero.

Some examples of linear equations in two variables are

$$3x - 5y = 10 \qquad y = \frac{1}{2}x + 7 \qquad -9s + 2t = 4 \qquad x = -8$$

(We can write $x = -8$ as $x + 0y = -8$, therefore it is a linear equation in two variables.)

A solution to a linear equation in two variables is written as an ordered pair.

2. Decide Whether an Ordered Pair Is a Solution of a Given Equation

Example 1

Determine whether each ordered pair is a solution of $4x + 5y = 11$.

a) $(-1, 3)$ b) $\left(\dfrac{3}{2}, 5\right)$

Solution

a)   Solutions to the equation $4x + 5y = 11$ are written in the form (x, y) where (x, y) is called an *ordered pair*. Therefore, the ordered pair $(-1, 3)$ means that $x = -1$ and $y = 3$.

$$(-1, 3)$$
$$\nearrow \qquad \nwarrow$$
$$\text{x-coordinate} \qquad \text{y-coordinate}$$

To determine if $(-1, 3)$ is a solution of $4x + 5y = 11$, we substitute -1 for x and 3 for y. Remember to put these values in parentheses.

$$
\begin{aligned}
4x + 5y &= 11 \\
4(-1) + 5(3) &= 11 \qquad &&\text{Substitute $x = -1$ and $y = 3$.} \\
-4 + 15 &= 11 \qquad &&\text{Multiply.} \\
11 &= 11 \qquad &&\text{True}
\end{aligned}
$$

Since substituting $x = -1$ and $y = 3$ into the equation gives the true statement $11 = 11$, $(-1, 3)$ *is a solution* of $4x + 5y = 11$.

b)   The ordered pair $\left(\dfrac{3}{2}, 5\right)$ tells us that $x = \dfrac{3}{2}$ and $y = 5$.

$$4x + 5y = 11$$

$$4\left(\dfrac{3}{2}\right) + 5(5) = 11 \qquad \text{Substitute } \dfrac{3}{2} \text{ for } x \text{ and 5 for } y.$$

$$6 + 25 = 11 \qquad \text{Multiply.}$$

$$31 = 11 \qquad \text{False}$$

Since substituting $\left(\dfrac{3}{2}, 5\right)$ into the equation gives the false statement $31 = 11$, the ordered pair is *not* a solution to the equation.

You Try 1

Determine if each ordered pair is a solution of the equation $y = -\dfrac{3}{4}x + 5$.

a) $(12, -4)$ b) $(0, 7)$ c) $(-8, 11)$

3. Complete Ordered Pairs for a Given Equation

Often, we are given the value of one variable in an equation and we can find the value of the other variable which makes the equation true.

Example 2

Complete the ordered pair $(-3, \ )$ for $y = 2x + 10$.

Solution

To complete the ordered pair $(-3, \ )$, we must find the value of y from $y = 2x + 10$ when $x = -3$.

$$y = 2x + 10$$
$$y = 2(-3) + 10 \qquad \text{Substitute } -3 \text{ for } x.$$
$$y = -6 + 10$$
$$y = 4$$

When $x = -3$, $y = 4$.

The ordered pair is $(-3, 4)$.

You Try 2

Complete the ordered pair $(5, \ )$ for $y = 3x - 7$.

If we want to complete more than one ordered pair for a particular equation, we can organize the information in a **table of values**.

Complete the table of values for each equation and write the information as ordered pairs.

a) $-x + 3y = 8$

x	y
1	
	-4
	$\dfrac{2}{3}$

b) $y = 2$

x	y
7	
-5	
0	

Solution

a) $-x + 3y = 8$

x	y
1	
	-4
	$\dfrac{2}{3}$

i) In the first ordered pair, $x = 1$ and we must find y.

$$-x + 3y = 8$$
$$-(1) + 3y = 8 \qquad \text{Substitute 1 for } x.$$
$$-1 + 3y = 8$$
$$3y = 9 \qquad \text{Add 1 to each side.}$$
$$y = 3 \qquad \text{Divide by 3.}$$

The ordered pair is (1, 3).

ii) In the second ordered pair, $y = -4$ and we must find x.

$$-x + 3y = 8$$
$$-x + 3(-4) = 8 \qquad \text{Substitute } -4 \text{ for } y.$$
$$-x + (-12) = 8 \qquad \text{Multiply.}$$
$$-x = 20 \qquad \text{Add 12 to each side.}$$
$$x = -20 \qquad \text{Divide by } -1.$$

The ordered pair is $(-20, -4)$.

iii) In the third ordered pair, $y = \dfrac{2}{3}$ and we must find x.

$$-x + 3y = 8$$
$$-x + 3\left(\dfrac{2}{3}\right) = 8 \qquad \text{Substitute } \dfrac{2}{3} \text{ for } y.$$
$$-x + 2 = 8 \qquad \text{Multiply.}$$
$$-x = 6 \qquad \text{Subtract 2 from each side.}$$
$$x = -6 \qquad \text{Divide by } -1.$$

The ordered pair is $\left(-6, \dfrac{2}{3}\right)$.

As you complete each ordered pair, fill in the table of values. The completed table will look like this:

x	y
1	3
−20	−4
−6	$\dfrac{2}{3}$

The ordered pairs are $(1, 3)$, $(-20, -4)$, and $\left(-6, \dfrac{2}{3}\right)$.

b) $y = 2$

x	y
7	
−5	
0	

i) In the first ordered pair, $x = 7$ and we must find y. The equation $y = 2$ means that *no matter the value of x, y always equals 2*. Therefore, when $x = 7$, $y = 2$.

The ordered pair is $(7, 2)$.

Since $y = 2$ for every value of x, we can complete the table of values as follows:

x	y
7	2
−5	2
0	2

The ordered pairs are $(7, 2)$, $(-5, 2)$, and $(0, 2)$.

You Try 3

Complete the table of values for each equation and write the information as ordered pairs.

a) $x - 2y = 9$

x	y
5	
12	
	−7
	$\dfrac{5}{2}$

b) $x = -3$

x	y
	1
	3
	−8

4. Plot Ordered Pairs

When we completed the table of values for the last two equations, we were finding solutions to each linear equation in two variables.

How can we represent the solutions graphically?

To represent the solution to a linear equation in *one* variable we can use a number line:

$$\text{Solve} \quad x + 3 = 5$$
$$x = 2$$

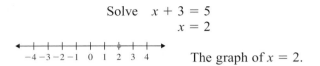

The graph of $x = 2$.

But we need a way to represent two values, an ordered pair, graphically. We will use the **Cartesian coordinate system**, or **rectangular coordinate system**, to graph the ordered pairs, (x, y).

In the Cartesian coordinate system, we have a horizontal number line, called the *x*-axis, and a vertical number line, called the *y*-axis.

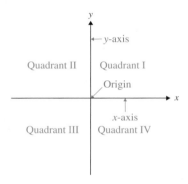

The *x*-axis and *y*-axis in the Cartesian coordinate system determine a flat surface called a **plane**. The axes divide this plane into four **quadrants**. (See figure above.) The point at which the *x*-axis and *y*-axis intersect is called the **origin**. The arrow at one end of the *x*-axis and one end of the *y*-axis indicates the positive direction on each axis.

Ordered pairs can be represented by **points** in the plane. Therefore, to graph the ordered pair $(4, 2)$ we *plot the point* $(4, 2)$. We will do this in Example 4.

Example 4

Plot the point $(4, 2)$.

Solution

Since $x = 4$, we say that the *x-coordinate* of the point is 4. Likewise, the *y-coordinate* is 2.

The *origin* has coordinates (0, 0). The coordinates of a point tell us how far from the origin to move in the *x*-direction and *y*-direction. So, the coordinates of the point (4, 2) tell us to do the following:

(4, 2)

First, from the origin, move 4 units to the right along the *x*-axis.

Then, from the current position, move 2 units up, parallel to the *y*-axis.

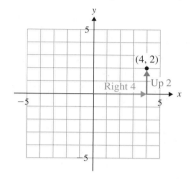

Example 5

Plot the points.

a) $(-2, 5)$ b) $(1, -4)$ c) $\left(\dfrac{5}{2}, 3\right)$

d) $(-5, -2)$ e) $(0, 1)$ f) $(-4, 0)$

Solution

The points are plotted on the following graph.

a) $(-2, 5)$ b) $(1, -4)$

<u>First</u>   <u>Then</u>   <u>First</u>   <u>Then</u>

From the origin, move left 2 units on the *x*-axis.

From the current position, move 5 units up, parallel to the *y*-axis.

From the origin, move right 1 unit on the *x*-axis.

From the current position, move 4 units down, parallel to the *y*-axis.

c) $\left(\dfrac{5}{2}, 3\right)$

Think of $\dfrac{5}{2}$ as $2\dfrac{1}{2}$. From the origin, move right $2\dfrac{1}{2}$ units, then up 3 units.

d) $(-5, -2)$ From the origin, move left 5 units, then down 2 units.

e) $(0, 1)$ The *x*-coordinate of 0 means that we don't move in the *x*-direction (horizontally). From the origin, move up 1 on the *y*-axis.

f) $(-4, 0)$ From the origin, move left 4 units. Since the y-coordinate is zero, we do not move in the y-direction (vertically).

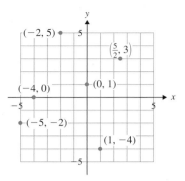

You Try 4

Plot the points.

a) $(3, 1)$ b) $(-2, 4)$ c) $(0, -5)$

d) $(2, 0)$ e) $(-4, -3)$ f) $\left(1, \dfrac{7}{2}\right)$

> The coordinate system should always be labeled to indicate how many units each mark represents.

We can graph sets of ordered pairs for a linear equation in two variables.

Example 6

Complete the table of values for $2x - y = 5$, then plot the points.

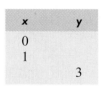

x	y
0	
1	
	3

Solution

i) In the first ordered pair, $x = 0$ and we must find y.

$$2x - y = 5$$
$$2(0) - y = 5 \quad \text{Substitute 0 for } x.$$
$$0 - y = 5$$
$$-y = 5$$
$$y = -5 \quad \text{Divide by } -1.$$

The ordered pair is $(0, -5)$.

ii) In the second ordered pair, $x = 1$ and we must find y.

$$2x - y = 5$$
$$2(1) - y = 5 \quad \text{Substitute 1 for } x.$$
$$2 - y = 5$$
$$-y = 3 \quad \text{Subtract 2 from each side.}$$
$$y = -3 \quad \text{Divide by } -1.$$

The ordered pair is $(1, -3)$.

iii) In the third ordered pair, $y = 3$ and we must find x.

$$2x - y = 5$$
$$2x - (3) = 5 \qquad \text{Substitute 3 for } y.$$
$$2x = 8 \qquad \text{Add 3 to each side.}$$
$$x = 4 \qquad \text{Divide by 2.}$$

The ordered pair is (4, 3).

Each of the points (0, −5), (1, −3), and (4, 3) satisfies the equation $2x - y = 5$.

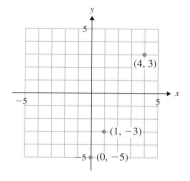

 You Try 5

Complete the table of values for $3x + y = 1$, then plot the points.

x	y
0	
−1	
	−5

5. Solve Applied Problems Involving Ordered Pairs

Next, we will look at an application of ordered pairs.

Example 7

The length of an 18-year-old female's hair is measured to be 250 millimeters (mm) (almost 10 in.). The length of her hair after x days can be approximated by

$$y = 0.30x + 250$$

where y is the length of her hair in millimeters.

a) Find the length of her hair (i) 10 days, (ii) 60 days, and (iii) 90 days after the initial measurement and write the results as ordered pairs.

b) Graph the ordered pairs.

c) How long would it take for her hair to reach a length of 274 mm (almost 11 in.)?

Solution

a)   The problem states that in the equation $y = 0.30x + 250,$

$$x = \text{number of days after the hair was measured}$$
$$y = \text{length of the hair (in millimeters)}$$

We must determine the length of her hair after 10 days, 60 days, and 90 days. We can organize the information in a table of values.

x	y
10	
60	
90	

i) $x = 10$: $y = 0.30x + 250$
              $y = 0.30(10) + 250$ Substitute 10 for x.
              $y = 3 + 250$ Multiply.
              $y = 253$

After 10 days, her hair is 253 mm long. We can write this as the ordered pair (10, 253).

ii) $x = 60$: $y = 0.30x + 250$
               $y = 0.30(60) + 250$ Substitute 60 for x.
               $y = 18 + 250$ Multiply.
               $y = 268$

After 60 days, her hair is 268 mm long. (Or, after about 2 months, her hair is about 10.5 in. long. It has grown about half of an inch.) We can write this as the ordered pair (60, 268).

iii) $x = 90$: $y = 0.30x + 250$
                $y = 0.30(90) + 250$ Substitute 90 for x.
                $y = 27 + 250$ Multiply.
                $y = 277$

After 90 days, her hair is 277 mm long. (Or, after about 3 months, her hair is about 10.9 in. long. It has grown about 0.9 in.) We can write this as the ordered pair (90, 277).

We can complete the table of values:

x	y
10	253
60	268
90	277

The ordered pairs are (10, 253), (60, 268), and (90, 277).

b)   Graph the ordered pairs.

The *x*-axis represents the number of days after the hair was measured. Since it does not make sense to talk about a negative number of days, we will not continue the *x*-axis in the negative direction.

The *y*-axis represents the length of the female's hair. Likewise, a negative number does not make sense in this situation, so we will not continue the *y*-axis in the negative direction.

The scales on the *x*-axis and *y*-axis are different. This is because the size of the numbers they represent are quite different.

Here are the ordered pairs we must graph: (10, 253), (60, 268), and (90, 277).

The *x*-values are 10, 60, and 90, so we will let each mark in the *x*-direction represent 10 units.

The *y*-values are 253, 268, and 277. While the numbers are rather large, they do not actually differ by much. We will begin labeling the *y*-axis at 250, but each mark in the *y*-direction will represent 3 units. Because there is a large jump in values from 0 to 250 on the *y*-axis, we indicate this with "÷" on the axis between the 0 and 250.

Notice, also, that we have labeled both axes. The ordered pairs are plotted on the following graph.

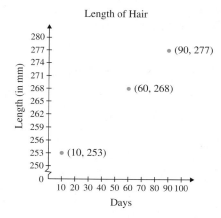

Length of Hair

c)   We must determine how many days it would take for the hair to grow to a length of 274 mm.

The length, 274 mm, is the *y*-value. We must find the value of *x* that corresponds to *y* = 274 since *x* represents the number of days.

The equation relating *x* and *y* is $y = 0.30x + 250$. We will substitute 274 for *y* and solve for *x*.

$$y = 0.30x + 250$$
$$274 = 0.30x + 250$$
$$24 = 0.30x$$
$$80 = x$$

It will take 80 days for her hair to grow to a length of 274 mm.

Answers to You Try Exercises

1) a) yes b) no c) yes   2) (5, 8)

3) a) $(5, -2)$, $(12, \frac{3}{2})$, $(-5, -7)$, $(14, \frac{5}{2})$ b) $(-3, 1)$, $(-3, 3)$, $(-3, -8)$

4) a) b) c) d) e) f)   5) $(0, 1)$, $(-1, 4)$, $(2, -5)$

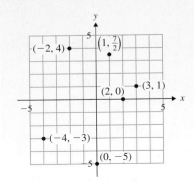

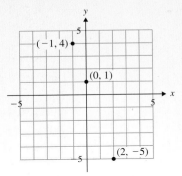

4.1 Exercises

Objectives 1 and 2

1) Explain the difference between a linear equation in one variable and a linear equation in two variables. Give an example of each.

2) True or False: $x^2 + 6y = -5$ is a linear equation in two variables.

Determine if each ordered pair is a solution of the given equation.

3) $7x + 2y = 4$; $(2, -5)$ 4) $3x - 5y = 1$; $(-2, 1)$

5) $-2x - y = 13$; $(-8, 3)$ 6) $-4x + 7y = -4$; $(8, 4)$

7) $y = 5x - 6$; $(3, 11)$ 8) $x = -y + 9$; $(4, -5)$

9) $y = -\dfrac{3}{2}x - 19$; $(-10, -4)$

10) $5y = \dfrac{2}{3}x + 2$; $(12, 2)$ 11) $x = 13$; $(5, 13)$

12) $y = -6$; $(-7, -6)$

Objective 3

Complete the ordered pair for each equation.

13) $y = -x + 6$; $(9,\ )$ 14) $y = 6x - 7$; $(2,\ )$

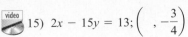 15) $2x - 15y = 13$; $\left(\ , -\dfrac{3}{4}\right)$

16) $-x + 12y = 7$; $\left(\ , \dfrac{3}{4}\right)$

17) $y = -4$; $(9,\ )$ 18) $x = 0$; $(\ , -14)$

Complete the table of values for each equation.

19) $y = -3x + 4$

x	y
0	
1	
2	
-1	

20) $y = 5x - 3$

x	y
0	
1	
-1	
-2	

21) $y = 12x + 10$

x	y
0	
1	
2	
	-2
	46

22) $y = -9x$

x	y
0	
5	
6	
	63
	-15

23) $3x - y = 8$

x	y
	0
0	
5	
	-7

24) $4x + 5y = -11$

x	y
0	
	0
1	
	1

25) $x = 5$

x	y
0	
4	
−1	
−8	

26) $y = -7$

x	y
	0
	−6
	5
	13

27) Explain, in words, how to complete the table of values for $x = -12$.

x	y
	0
	3
	−5

28) Explain, in words, how to complete the ordered pair (, −6) for $y = -x - 1$.

Objective 4

Name each point with an ordered pair, and identify the quadrant in which each point lies.

29)

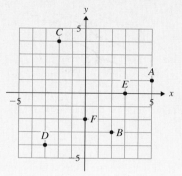

30)

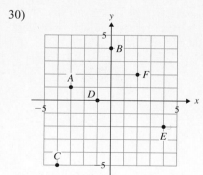

Graph each ordered pair and explain how you plotted the points.

31) (6, 2)

32) (3, 4)

33) (−1, 4)

34) (−2, −3)

Graph each ordered pair.

35) (−3, −5)

36) (−4, 5)

37) (2, −1)

38) (5, −2)

39) (−2, 0)

40) (0, 3)

41) (0, 5)

42) (−3, 0)

43) $\left(\dfrac{7}{2}, 1\right)$

44) $\left(-5, \dfrac{9}{2}\right)$

45) $\left(-3, \dfrac{5}{3}\right)$

46) $\left(4, -\dfrac{5}{8}\right)$

47) $\left(-\dfrac{11}{8}, -\dfrac{1}{2}\right)$

48) $\left(0, -\dfrac{17}{6}\right)$

Objectives 3 and 4

Complete the table of values for each equation, and plot the points.

 49) $y = -4x + 3$

x	y
0	
	0
2	
	7

50) $y = 2x - 5$

x	y
	0
	−1
0	
4	

51) $y = x$

x	y
	0
	4
−2	
−3	

52) $y = -x$

x	y
0	
−1	
	3
	−5

53) $3x - 2y = 6$

x	y
0	
	0
	3
−3	

54) $4x + 3y = 12$

x	y
0	
	0
1	
	8

55) $x = -6$

x	y
	0
	−5
	2
	4

56) $y - 2 = 0$

x	y
0	
1	
−4	
−2	

57) $y = -\frac{5}{4}x + 2$

58) $y = \frac{1}{2}x + 3$

x	y
0	
4	
8	
1	

x	y
0	
-2	
4	
-1	

59) For $y = \frac{2}{3}x - 7$,

a) find y when $x = 3$, $x = 6$, and $x = -3$. Write the results as ordered pairs.

b) find y when $x = 1$, $x = 5$, and $x = -2$. Write the results as ordered pairs.

c) why is it easier to find the y-values in part a) than in part b)?

60) What ordered pair is a solution to every linear equation of the form $y = mx$ where m is a real number?

Fill in the blank with *positive*, *negative*, or *zero*.

61) The x-coordinate of every point in quadrant I is

_____.

62) The y-coordinate of every point in quadrant I is

_____.

63) The y-coordinate of every point in quadrant II is

_____.

64) The x-coordinate of every point in quadrant III is

_____.

65) The y-coordinate of every point in quadrant IV is

_____.

66) The x-coordinate of every point in quadrant II is

_____.

67) The x-coordinate of every point on the y-axis is

_____.

68) The y-coordinate of every point on the x-axis is

_____

Objective 5

69) The number of drivers involved in fatal vehicle accidents in 2002 is given in the table.

Age	Number of Drivers
16	1300
17	1300
18	1600
19	1500

(Source: National Safety Council)

a) Write the information as ordered pairs (x, y), where x represents the age of the driver and y represents the number of drivers involved in fatal motor vehicle accidents.

b) Label a coordinate system, choose an appropriate scale, and graph the ordered pairs.

c) Explain the meaning of the ordered pair (18,1600) in the context of the problem.

70) The number of pounds of potato chips consumed per person is given in the table.

Year	Pounds of Chips
1998	14.8
1999	15.9
2000	16.0
2001	17.6
2002	17.0

(Source: U.S. Dept. of Agriculture)

a) Write the information as ordered pairs, (x, y), where x represents the year and y represents the number of pounds of potato chips consumed per person.

b) Label a coordinate system, choose an appropriate scale, and graph the ordered pairs.

c) Explain the meaning of the ordered pair (1999, 15.9) in the context of the problem.

71) Horton's Party Supplies rents a "moon jump" for $100 plus $20 per hour. This can be described by the equation

$$y = 20x + 100$$

where x represents the number of hours and y represents the cost.

a) Complete the table of values, and write the information as ordered pairs.

x	y
1	
3	
4	
6	

b) Label a coordinate system, choose an appropriate scale, and graph the ordered pairs.

c) Explain the meaning of the ordered pair (4, 180) in the context of the problem.

d) Look at the graph. Is there a pattern indicated by the points?

e) For how many hours could a customer rent the moon jump if she had $280?

72) Kelvin is driving from Los Angeles to Chicago to go to college. His distance from Los Angeles is given by the equation

$$y = 62x$$

where x represents the number of hours driven, and y represents his distance from Los Angeles.

a) Complete the table of values, and write the information as ordered pairs.

x	y
3	
8	
15	
20	

b) Label a coordinate system, choose an appropriate scale, and graph the ordered pairs.

c) Explain the meaning of the ordered pair (20,1240) in the context of the problem.

d) Look at the graph. Is there a pattern indicated by the points?

e) What does the *62* in $y = 62x$ represent?

f) How many hours of driving time will it take for Kelvin to get to Chicago if the distance between L.A. and Chicago is about 2034 miles? (Round to the nearest hour.)

Section 4.2 Graphing by Plotting Points and Finding Intercepts

Objectives

1. Graph a Linear Equation in Two Variables by Plotting Points

2. Graph a Linear Equation in Two Variables by Finding the Intercepts

3. Graph Linear Equations of the Forms $x = c$ and $y = d$

4. Model Data with a Linear Equation

In Section 4.1, we discussed completing a table of values for a linear equation in two variables. Shown next is a completed table of values for $4x - y = 5$.

x	y
2	3
0	−5
1	−1

Each of the ordered pairs (2, 3), (0, −5), and (1, −1) is a solution to the equation $4x - y = 5$. But, how many solutions does $4x - y = 5$ have? Like all linear equations in two variables, it has an infinite number of solutions, and the solutions are ordered pairs.

> Every linear equation in two variables has an infinite number of solutions, and the solutions are ordered pairs.

How can we represent these solutions? We can represent them with a graph, and that graph is a line.

> The graph of a linear equation in two variables, $Ax + By = C$, is a straight line. Each point on the line is a solution to the equation.

Let's go back to the last equation, add two more points to the table of values, and graph the line.

1. Graph a Linear Equation in Two Variables by Plotting Points

Example 1

Complete the table of values and graph $4x - y = 5$.

x	y
2	3
0	-5
1	-1
-1	
	0

Solution

When $x = -1$, we get

$$4x - y = 5$$
$$4(-1) - y = 5 \qquad \text{Substitute } -1 \text{ for } x.$$
$$-4 - y = 5$$
$$-y = 9$$
$$y = -9 \qquad \text{Solve for } y.$$

When $y = 0$, we get

$$4x - y = 5$$
$$4x - 0 = 5 \qquad \text{Substitute } 0 \text{ for } y.$$
$$4x = 5$$
$$x = \frac{5}{4} \qquad \text{Solve for } x.$$

The completed table of values is

x	y
2	3
0	-5
1	-1
-1	-9
$\frac{5}{4}$	0

This gives us the ordered pairs $(2, 3)$, $(0, -5)$, $(1, -1)$, $(-1, -9)$, and $\left(\dfrac{5}{4}, 0\right)$.

Each is a solution to the equation $4x - y = 5$.

Plot the points.

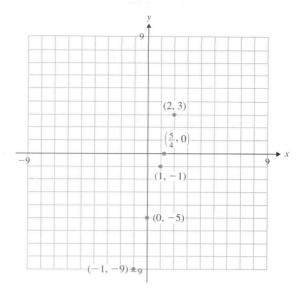

The points we have plotted lie on a straight line. We draw the line through these points to get the graph.

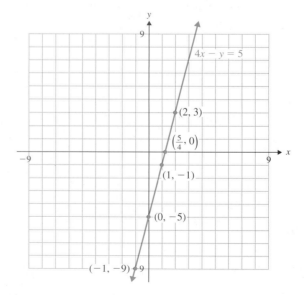

The line represents all solutions to the equation $4x - y = 5$. Every point on the line is a solution to the equation. The arrows on the ends of the line indicate that the line extends indefinitely in each direction. Although it is true that we need to find only two points to graph a line, it is best to plot at least three as a check.

You Try 1

Complete the table of values and graph $x - 2y = 3$.

x	y
1	−1
3	0
0	
	2
5	

Example 2

Graph $-x + 2y = 4$.

Solution

We will find three ordered pairs that satisfy the equation. Let's complete a tabl
values for $x = 0$, $y = 0$, and $x = 2$.

x	y
0	
	0
2	

$$x = 0: \quad -x + 2y = 4$$
$$-(0) + 2y = 4$$
$$2y = 4$$
$$y = 2$$

$$y = 0: \quad -x + 2y = 4 \qquad\qquad x = 2: \quad -x + 2y = 4$$
$$-x + 2(0) = 4 \qquad\qquad\qquad\qquad -(2) + 2y = 4$$
$$-x = 4 \qquad\qquad\qquad\qquad\qquad -2 + 2y = 4$$
$$x = -4 \qquad\qquad\qquad\qquad\qquad\quad 2y = 6$$
$$y = 3$$

We get the table of values

x	y
0	2
−4	0
2	3

Plot the points $(0, 2)$, $(-4, 0)$, and $(2, 3)$, and draw the line through them.

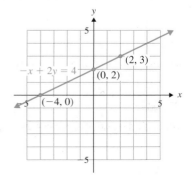

You Try 2

Graph each line.

a) $3x + 2y = 6$ b) $y = 4x - 8$

2. Graph a Linear Equation in Two Variables by Finding the Intercepts

In Example 2, the line crosses the x-axis at -4 and crosses the y-axis at 2. These points are called **intercepts**.

> The *x-intercept* of the graph of an equation is the point where the graph intersects the x-axis.
> The *y-intercept* of the graph of an equation is the point where the graph intersects the y-axis.

What is the y-coordinate of any point on the x-axis? It is zero. Likewise, the x-coordinate of any point on the y-axis is zero.

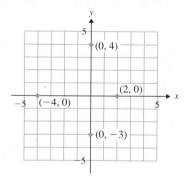

Therefore,

> To find the *x-intercept* of the graph of an equation, let y = 0 and solve for x.
> To find the *y-intercept* of the graph of an equation, let x = 0 and solve for y.

Finding intercepts is very helpful for graphing linear equations in two variables.

Example 3

Graph $y = -\dfrac{1}{2}x - 1$ by finding the intercepts and one other point.

Solution

We will begin by finding the intercepts.

x-intercept: Let $y = 0$, and solve for x.

$$y = -\frac{1}{2}x - 1$$

$$0 = -\frac{1}{2}x - 1$$

$$1 = -\frac{1}{2}x$$

$$-2 = x \qquad \text{Multiply both sides by } -2 \text{ to solve for } x.$$

The x-intercept is $(-2, 0)$.

y-intercept: Let $x = 0$, and solve for y.

$$y = -\frac{1}{2}x - 1$$

$$y = -\frac{1}{2}(0) - 1$$

$$y = 0 - 1$$
$$y = -1$$

The y-intercept is $(0, -1)$.

We must find another point. Let's look closely at the equation $y = -\frac{1}{2}x - 1$. The coefficient of x is $-\frac{1}{2}$. If we choose a value for x that is a multiple of 2 (the denominator of the fraction), then $-\frac{1}{2}x$ will not be a fraction.

Let $x = 2$.

$$y = -\frac{1}{2}x - 1$$

$$y = -\frac{1}{2}(2) - 1$$

$$y = -1 - 1$$
$$y = -2$$

The third point is $(2, -2)$.

Plot the points, and draw the line through them.

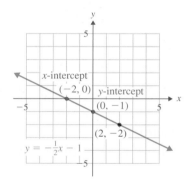

 **You Try 3**

Graph $y = 3x + 6$ by finding the intercepts and one other point.

3. Graph Linear Equations of the Forms $x = c$ and $y = d$

Remember from Section 4.1 that $x = c$ is a linear equation since it can be written in the form $x + 0y = c$. The same is true for $y = d$. It can be written as $0x + y = d$. Let's see how we can graph these equations.

Example 4

Graph $x = 3$.

Solution

The equation is $x = 3$. (This is the same as $x + 0y = 3$.) *$x = 3$ means that no matter the value of y, x always equals 3*. We can make a table of values where we choose any value for y, but x is always 3.

x	y
3	0
3	1
3	-2

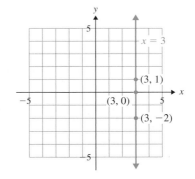

Plot the points, and draw a line through them. The graph of $x = 3$ is a *vertical line*. We can generalize the result as follows:

> If c is a constant, then the graph of $x = c$ is a *vertical line* going through the point $(c, 0)$.

 You Try 4

Graph $x = -4$.

Example 5

Graph $y = -2$.

Solution

The equation $y = -2$ is the same as $0x + y = -2$, therefore it is linear. *$y = -2$ means that no matter the value of x, y always equals -2.* Make a table of values where we choose any value for *x*, but *y* is always -2.

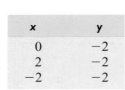

x	y
0	-2
2	-2
-2	-2

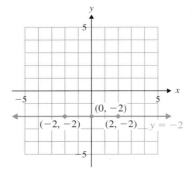

Plot the points, and draw a line through them. The graph of $y = -2$ is a *horizontal line.*

We can generalize the result as follows:

> If *d* is a constant, then the graph of $y = d$ is a *horizontal line* going through the point $(0, d)$.

You Try 5

Graph $y = 1$.

4. Model Data with a Linear Equation

Linear equations are often used to model (or describe mathematically) real-world data. We can use these equations to learn what has happened in the past or predict what will happen in the future.

Example 6

The average cost of college tuition at private, 4-year institutions can be modeled by

$$y = 464.2x + 9202$$

where *x* is the number of years after 1983 and *y* is the average tuition, in dollars.
(Source: The College Board)

a) Find the *y*-intercept of the graph of this equation and interpret its meaning.

b) Find the approximate cost of tuition in 1988, 1993, and 1998. Write the information as ordered pairs.

c) Graph $y = 464.2x + 9202$.

d) Use the graph to approximate the average cost of tuition in 2003. Is this the same result as when you use the equation to estimate the average cost?

Solution

a) To find the y-intercept, let $x = 0$.

$$y = 464.2(0) + 9202$$
$$y = 9202$$

The y-intercept is (0, 9202). What does this represent?

The problem states that x is the number of years *after* 1983. Therefore, $x = 0$ represents zero years after 1983, which is the year 1983.

The y-intercept (0, 9202) tells us that in 1983 the average cost of tuition at a private 4-year institution was $9202.

b) The approximate cost of tuition in

1988:   First, realize that $x \neq 1988$. x is the number of years *after* 1983. Since 1988 is 5 years after 1983, $x = 5$. Let $x = 5$ in $y = 464.2x + 9202$.

$$y = 464.2(5) + 9202$$
$$y = 2321 + 9202$$
$$y = 11{,}523$$

In 1988, the approximate cost of college tuition at these schools was $11,523. We can write this information as the ordered pair (5, 11,523).

1993:   Begin by finding x. 1993 is 10 years after 1983, so $x = 10$.

$$y = 464.2(10) + 9202$$
$$y = 4642 + 9202$$
$$y = 13{,}844$$

In 1993, the approximate cost of college tuition at private 4-year schools was $13,844.

The ordered pair (10, 13,844) can be written from this information.

1998:   1998 is 15 years after 1983, so $x = 15$.

$$y = 464.2(15) + 9202$$
$$y = 6963 + 9202$$
$$y = 16{,}165$$

The average cost of college tuition was approximately $16,165 in 1998. This can be represented by the ordered pair (15, 16,165).

c) We will plot the points (0, 9202), (5, 11,523), (10, 13,844), and (15, 16,165). Label the axes, and choose an appropriate scale for each.

The x-coordinates of the ordered pairs range from 0 to 15, so we will let each mark in the x-direction represent 3 units.

The y-coordinates of the ordered pairs range from 9202 to 16,165. We will let each mark in the y-direction represent 1000 units.

Average College Tuition at
Private 4-Year Schools

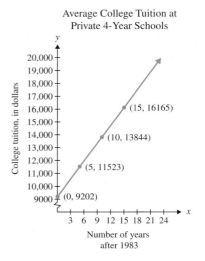

d) Using the graph to estimate the cost of tuition in 2003, we locate $x = 20$ on the
    x-axis since 2003 is 20 years after 1983. When $x = 20$ and we move straight up
    the graph to $y \approx 18{,}500$. Our approximation from the graph is \$18,500.
        If we use the equation and let $x = 20$, we get

$$y = 464.2x + 9202$$
$$y = 464.2(20) + 9202$$
$$y = 9284 + 9202$$
$$y = 18{,}486$$

From the equation we find that the cost of college tuition at private 4-year schools
was about \$18,486. The numbers are not exactly the same, but they are close.

Using Technology

A graphing calculator can be used to graph an equation and to verify information that we
find using algebra. We will graph the equation $y = -\dfrac{1}{2}x + 2$ and then find the intercepts
both algebraically and using the calculator.

First, we need to set the viewing window. This is the portion of the rectangular coordinate
system that will show on the screen. We will set the window so that the screen shows the
x-axis and y-axis from -5 to 5, with the scale set at 1.

Press WINDOW to set the window.

```
WINDOW
 Xmin=-5
 Xmax=5
 Xscl=1
 Ymin=-5
 Ymax=5
 Yscl=1
 Xres=1
```

Next, enter the equation into the calculator. To enter an equation, it must be solved for
y in terms of x. To enter $y = -\dfrac{1}{2}x + 2$, press Y= then enter $-\dfrac{1}{2}x + 2$, as shown next
on the left. Then press GRAPH to see the graph of the equation.

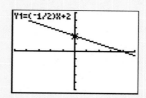

1. Find, algebraically, the *y*-intercept of the graph of $y = -\dfrac{1}{2}x + 2$. Is it consistent with the graph of the equation? *[(0, 2); yes, consistent]*

2. Find, algebraically, the *x*-intercept of the graph of $y = -\dfrac{1}{2}x + 2$. Is it consistent with the graph of the equation? *[(4, 0); yes, consistent]*

Now let's find the intercepts using the $\boxed{\text{ZOOM}}$ and $\boxed{\text{TRACE}}$ features on the graphing calculator.

Press $\boxed{\text{ZOOM}}$, then the down arrow to highlight 4:ZDecimal, and then press $\boxed{\text{ENTER}}$. The calculator will display the graph again, with a slightly different viewing window. Press $\boxed{\text{TRACE}}$ and use the arrow keys to move the cursor along the line. When the cursor is at the point where the graph crosses the *y*-axis, the screen displays (0, 2) as shown next on the left. Use the arrow key to move the cursor to the point where the line crosses the *x*-axis. The calculator displays (4, 0), as shown next on the right. This is consistent with the intercepts found in 1 and 2, using algebra.

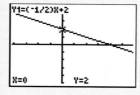
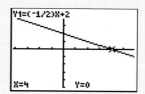

Use algebra to find the *x*- and *y*-intercepts of the graph of each equation. Then, use the graphing calculator to verify your results.

1. $y = 3x - 3$	2. $y = 2x + 4$	3. $5x - 2y = 10$
4. $3x - 2y = 6$	5. $3y - 4x = 3$	6. $2x + 5y = 5$

Answers to You Try Exercises

1) $\left(0, -\dfrac{3}{2}\right), (7, 2), (5, 1)$

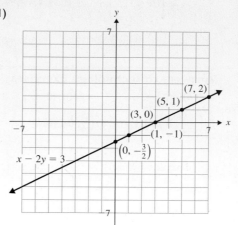

2) a)

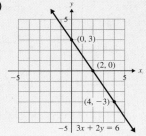

b)

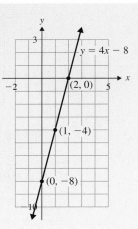

3) $(-2, 0), (0, 6), (-1, 3)$; Answers may vary.

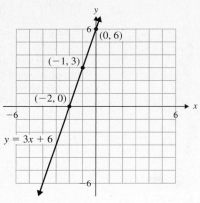

4)

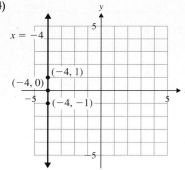

5)

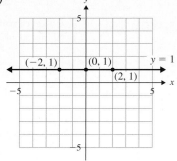

Answers to Technology Exercises

1. $(1, 0), (0, -3)$ 2. $(-2, 0), (0, 4)$ 3. $(2, 0), (0, -5)$ 4. $(2, 0), (0, -3)$

5. $\left(-\dfrac{3}{4}, 0\right), (0, 1)$ 6. $\left(\dfrac{5}{2}, 0\right), (0, 1)$

4.2 Exercises

Objective 1

1) Every linear equation in two variables has how many solutions?

2) The graph of a linear equation in two variables is a _____ .

Complete the table of values and graph each equation.

3) $y = 3x - 1$

x	y
0	
1	
2	
-1	

4) $y = -2x + 5$

x	y
0	
-1	
2	
3	

5) $y = -\frac{2}{3}x + 4$

x	y
0	
-3	
3	
6	

6) $y = \frac{5}{2}x + 6$

x	y
0	
2	
-2	
-4	

7) $-3x + 6y = 9$

x	y
0	
	0
	4
-1	

8) $4x = 1 - y$

x	y
	0
0	
$\frac{5}{2}$	
	5

9) $y + 4 = 0$

x	y
0	
-3	
-1	
2	

10) $x = -\frac{3}{2}$

x	y
	5
	0
	-1
	-2

Objectives 1–3

11) Explain how to find the y-intercept of the graph of an equation.

12) Explain how to find the x-intercept of the graph of an equation.

Graph each equation by finding the intercepts and at least one other point.

13) $y = -2x + 6$

14) $y = x - 3$

15) $3x - 4y = 12$

16) $5x + 2y = 10$

17) $x = \frac{1}{4}y - 1$

18) $x = -\frac{2}{3}y - 8$

19) $2x + 3y = -6$

20) $x - 4y = 6$

21) $y = -x$

22) $y = x$

23) $5y - 2x = 0$

24) $x + 3y = 0$

25) $x = 5$

26) $y = -1$

27) $y = 0$

28) $x = 0$

29) $y + 3 = 0$

30) $x - \frac{5}{2} = 0$

31) $x + 3y = 8$

32) $6x - y = 7$

Objective 4

Solve each application. See Example 6.

33) Concern about the Leaning Tower of Pisa in Italy led engineers to begin reinforcing the structure in the 1990s. The number of millimeters the tower was straightened can be described by

$$y = 1.5x$$

where x represents the number of days engineers worked on straightening the tower and y represents the number of millimeters (mm) the tower was moved toward vertical. (In the end, the tower will keep some of its famous lean.) (Source: Reuters-9/7/2000)

a) Make a table of values using $x = 0, 10, 20,$ and 60, and write the information as ordered pairs.

b) Explain the meaning of each ordered pair in the context of the problem.

c) Graph the equation using the information in a). Use an appropriate scale.

d) Engineers straightened the Leaning Tower of Pisa a total of about 450 mm. How long did this take?

34) The blood alcohol percentage of a 180-lb male can be modeled by

$$y = 0.02x$$

where x represents the number of drinks consumed (1 drink = 12 oz of beer, for example) and y represents the blood alcohol percentage. (Source: Taken from data from the U.S. Department of Health and Human Services)

a) Make a table of values using $x = 0, 1, 2,$ and 4, and write the information as ordered pairs.

b) Explain the meaning of each ordered pair in the context of the problem.

c) Graph the equation using the information in a). Use an appropriate scale.

d) If a 180-lb male had a blood alcohol percentage of 0.12, how many drinks did he have?

35) The relationship between altitude (in feet) and barometric pressure (in inches of mercury) can be modeled by

$$y = -0.001x + 29.86$$

for altitudes between sea level (0 ft) and 5000 ft. x represents the altitude and y represents the barometric pressure. (Source: From data taken from www.engineeringtoolbox.com)

Altitude and pressure

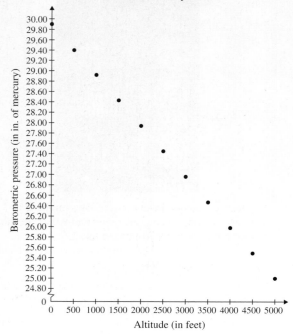

a) From the graph, estimate the pressure at the following altitudes: 0 ft (sea level), 1000 ft, 3500 ft, and 5000 ft.

b) Determine the barometric pressures at the same altitudes in a) using the equation. Are the numbers close?

c) Graph the line that models the data given on the original graph. Use an appropriate scale.

d) Can we use the equation $y = -0.001x + 29.86$ to determine the pressure at 10,000 ft? Why or why not?

36) The following graph shows the actual per-pupil spending on education in the state of Washington for the school years 1994–2003.

Per-pupil spending on education in the state of Washington

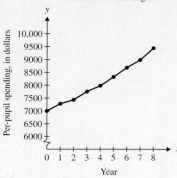

Algebraically, this can be modeled by

$$y = 298.3x + 6878.95$$

where x represents the number of years after the 1994–1995 school year. (So, $x = 0$ represents the 1994–1995 school year, $x = 1$ represents the 1995–1996 school year, etc.). y represents the per-pupil spending on education, in dollars. (www.effwa.org)

a) From the graph, estimate the amount spent per pupil during the following school years: 1994–1995, 1997–1998, and 2001–2002.

b) Determine the amount spent during the same school years as in a) using the equation. How do these numbers compare with the estimates from the graph?

Section 4.3 The Slope of a Line

Objectives

1. Understand the Concept of Slope
2. Find the Slope of a Line Given Two Points on the Line
3. Use Slope to Solve Applied Problems
4. Find the Slope of Horizontal and Vertical Lines
5. Use Slope and One Point on a Line to Graph the Line

1. Understand the Concept of Slope

In Section 4.2, we learned to graph lines by plotting points. You may have noticed that some lines are steeper than others. Their "slants" are different too.

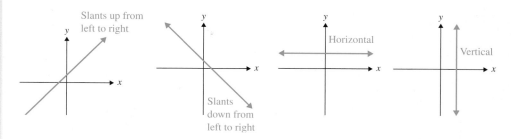

We can describe the steepness of a line with its *slope*.

> The **slope** of a line measures its steepness. It is the ratio of the vertical change in *y* to the horizontal change in *x*. Slope is denoted by *m*.

We can also think of slope as a rate of change. *Slope* is the rate of change between two points. More specifically, it describes the rate of change in *y* to the change in *x*.

For example, if a line has a slope of $\frac{3}{5}$, then the rate of change between two points on the line is a vertical change (change in *y*) of 3 units for every horizontal change (change in *x*) of 5 units. If a line has a slope of 4, or $\frac{4}{1}$, then the rate of change between two points on the line is a vertical change (change in *y*) of 4 units for every horizontal change (change in *x*) of 1 unit.

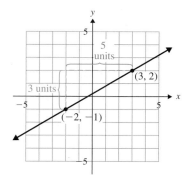

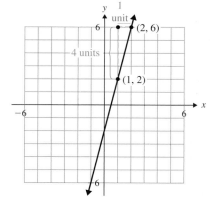

$$\text{Line with slope} = \frac{3}{5} \quad \begin{array}{l} \leftarrow \text{vertical change} \\ \leftarrow \text{horizontal change} \end{array} \qquad \text{Line with slope} = 4 \text{ or } \frac{4}{1} \quad \begin{array}{l} \leftarrow \text{vertical change} \\ \leftarrow \text{horizontal change} \end{array}$$

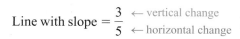

> As the magnitude of the slope gets larger the line gets steeper.

Here is an application of slope.

Example 1

A sign along a highway through the Rocky Mountains looks like this:

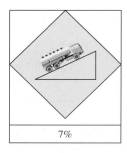

7%

What does this mean?

Solution

Percent means "out of 100." Therefore, we can write 7% as $\dfrac{7}{100}$. We can interpret $\dfrac{7}{100}$ as the ratio of the vertical change in the road to horizontal change in the road. $\left(\text{The slope of the road is } \dfrac{7}{100}.\right)$

$$m = \frac{7}{100} = \frac{\text{Vertical change}}{\text{Horizontal change}}$$

The highway rises 7 ft for every 100 horizontal feet.

You Try 1

The slope of a skateboard ramp is $\dfrac{7}{16}$ where the dimensions of the ramp are in inches. What does this mean?

2. Find the Slope of a Line Given Two Points on the Line

Here is line L. The points (x_1, y_1) and (x_2, y_2) are two points on line L. *We will find the ratio of the vertical change in y to the horizontal change in x between the points (x_1, y_1) and (x_2, y_2).*

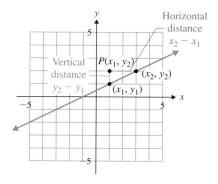

To get from (x_1, y_1) to (x_2, y_2), we move *vertically* to point P then *horizontally* to (x_2, y_2). The x-coordinate of point P is x_1 [the same as the point (x_1, y_1)], and the y-coordinate of P is y_2 [the same as (x_2, y_2)].

When we moved *vertically* from (x_1, y_1) to point $P(x_1, y_2)$, how far did we go? We moved a vertical distance $y_2 - y_1$.

The vertical change is $y_2 - y_1$ and is called the **rise**.

Then we moved *horizontally* from point $P(x_1, y_2)$ to (x_2, y_2). How far did we go? We moved a horizontal distance $x_2 - x_1$.

The horizontal change is $x_2 - x_1$ and is called the **run**.

We said that the slope of a line is the ratio of the vertical change (rise) to the horizontal change (run). Therefore,

The slope (m) of a line containing the points (x_1, y_1) and (x_2, y_2) is given by

$$m = \frac{\text{vertical change}}{\text{horizontal change}} = \frac{y_2 - y_1}{x_2 - x_1}$$

We can also think of slope as:

$$\frac{\text{rise}}{\text{run}} \quad \text{or} \quad \frac{\text{change in } y}{\text{change in } x}.$$

Let's look at some different ways to determine the slope of a line.

Example 2

Determine the slope of each line.

a)

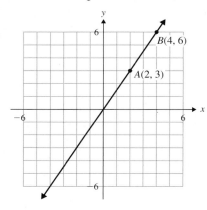

b)

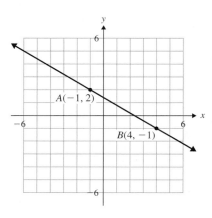

Solution

a)   We will find the slope in two ways.

  i) First, we will find the vertical change and the horizontal change by counting these changes as we go from A to B.

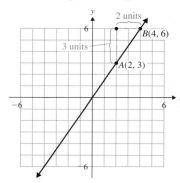

Vertical change (change in y) from A to B: 3 units

Horizontal change (change in x) from A to B: 2 units

$$\text{Slope} = \frac{\text{change in } y}{\text{change in } x} = \frac{3}{2} \quad \text{or} \quad m = \frac{3}{2}$$

  ii) We can also find the slope using the formula.

  Let $(x_1, y_1) = (2, 3)$ and $(x_2, y_2) = (4, 6)$.

$$m = \frac{y_2 - y_1}{x_2 - x_1} = \frac{6 - 3}{4 - 2} = \frac{3}{2}.$$

  You can see that we get the same result either way we find the slope.

b)   i) First, find the slope by counting the vertical change and horizontal change as we go from A to B.

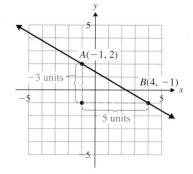

Vertical change (change in y) from A to B: -3 units

Horizontal change (change in x) from A to B: 5 units

$$\text{Slope} = \frac{\text{change in } y}{\text{change in } x} = \frac{-3}{5} = -\frac{3}{5}$$

$$\text{or} \quad m = -\frac{3}{5}$$

ii) We can also find the slope using the formula.

Let $(x_1, y_1) = (-1, 2)$ and $(x_2, y_2) = (4, -1)$.

$$m = \frac{y_2 - y_1}{x_2 - x_1} = \frac{-1 - 2}{4 - (-1)} = \frac{-3}{5} = -\frac{3}{5}.$$

Again, we obtain the same result using either method for finding the slope.

The slope of $-\dfrac{3}{5}$ can be thought of as $\dfrac{-3}{5}$, $\dfrac{3}{-5}$, or $-\dfrac{3}{5}$.

You Try 2

Determine the slope of each line by

a) counting the vertical change and horizontal change. b) using the formula.

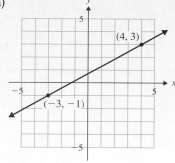

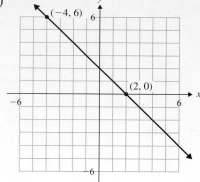

Notice that in Example 2a), the line has a positive slope and slants upward from left to right. As the value of x increases, the value of y increases as well. The line in 2b) has a negative slope and slants downward from left to right. Notice, in this case, that as the line goes from left to right, the value of x increases while the value of y decreases. We can summarize these results with the following general statements.

A line with a positive slope slants upward from left to right. As the value of x increases, the value of y increases as well.

A line with a negative slope slants downward from left to right. As the value of x increases, the value of y decreases.

3. Use Slope to Solve Applied Problems

| Example 3 |

The graph models the number of students at DeWitt High School from 2000 to 2006.

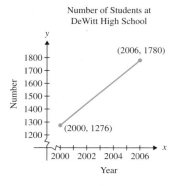

Number of Students at
DeWitt High School

a)   How many students attended the school in 2000? in 2006?

b)   What does the sign of the slope of the line segment mean in the context of the problem?

c)   Find the slope of the line segment, and explain what it means in the context of the problem.

Solution

a)   We can determine the number of students by reading the graph. In 2000, there were 1276 students, and in 2006 there were 1780 students.

b)   The positive slope tells us that from 2000 to 2006 the number of students was increasing.

c)   Let $(x_1, y_1) = (2000, 1276)$ and $(x_2, y_2) = (2006, 1780)$.

$$\text{Slope} = \frac{y_2 - y_1}{x_2 - x_1} = \frac{1780 - 1276}{2006 - 2000} = \frac{504}{6} = 84$$

The slope of the line is 84. Therefore, the number of students attending DeWitt High School between 2000 and 2006 increased by 84 per year.

4. Find the Slope of Horizontal and Vertical Lines

| Example 4 |

Find the slope of the line containing each pair of points.

a) $(-1, 2)$ and $(3, 2)$ b) $(-3, 4)$ and $(-3, -1)$

Solution

a)   Let $(x_1, y_1) = (-1, 2)$ and $(x_2, y_2) = (3, 2)$.

$$m = \frac{y_2 - y_1}{x_2 - x_1} = \frac{2 - 2}{3 - (-1)} = \frac{0}{4} = 0$$

If we plot the points, we see that they lie on a horizontal line. Each point on the line has a y-coordinate of 2, so $y_2 - y_1$ *always* equals zero. *The slope of every horizontal line is zero.*

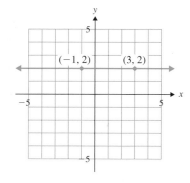

b)   Let $(x_1, y_1) = (-3, 4)$ and $(x_2, y_2) = (-3, -1)$.

$$m = \frac{y_2 - y_1}{x_2 - x_1} = \frac{-1 - 4}{-3 - (-3)} = \frac{-5}{0} \text{ undefined}$$

We say that the slope is undefined. Plotting these points gives us a vertical line. Each point on the line has an x-coordinate of -3, so $x_2 - x_1$ *always* equals zero.
   The slope of every vertical line is undefined.

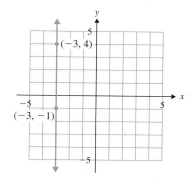

You Try 3

Find the slope of the line containing each pair of points.

a) $(5, 8)$ and $(-2, 8)$ b) $(4, 6)$ and $(4, 1)$

The slope of a horizontal line, $y = d$, is **zero**. The slope of a vertical line, $x = c$, is **undefined**. (c and d are constants.)

---

5. Use Slope and One Point on a Line to Graph the Line

We have seen how we can find the slope of a line given two points on the line. Now, we will see how we can use the slope and *one* point on the line to graph the line.

Example 5

Graph the line containing the point

a) $(2, -5)$ with a slope of $\dfrac{7}{2}$.   b) $(0, 4)$ with a slope of -3.

Solution

a)   The line contains the point $(2, -5)$ and has a slope of $\dfrac{7}{2}$.

   i) Begin by plotting the point.

   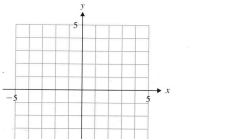

   ii) Analyze the slope. What does it mean?

   $$m = \frac{7}{2} = \frac{\text{change in } y}{\text{change in } x}$$

   To get from the point $(2, -5)$ to another point on the line, we will move up 7 units in the y-direction and right 2 units in the x-direction from $(2, -5)$.

iii) Start at the point $(2, -5)$ and use the slope to plot another point on the line.

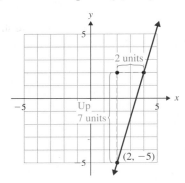

iv) Draw a line through the two points.

b) The line contains the point $(0, 4)$ with a slope of -3.

i) Begin by plotting the point $(0, 4)$.

ii) What does the slope, $m = -3$, mean? Since the slope describes $\dfrac{\text{change in } y}{\text{change in } x}$, we will think of the slope as a fraction. We can write -3 as $-\dfrac{3}{1}$ or $\dfrac{-3}{1}$ or $\dfrac{3}{-1}$.

Since we are using the slope to help us plot another point, we want to apply the negative sign to either the numerator or denominator. We will apply it to the numerator.

$$m = -3 = \frac{-3}{1} = \frac{\text{change in } y}{\text{change in } x}$$

To get from $(0, 4)$ to another point on the line, we will move *down* 3 units in the y-direction (since we have -3 in the numerator) and *right* 1 unit in the positive x-direction.

iii) Start at the point $(0, 4)$ and use the slope to plot another point on the line. This is the point $(1, 1)$.

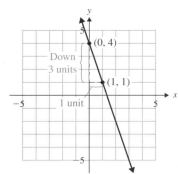

iv) Draw a line through the points.

You Try 4

Graph the line containing the point

a) $(1, 1)$ with a slope of $-\dfrac{2}{3}$.   b) $(0, -5)$ with a slope of 3.

c) $(-2, 2)$ with an undefined slope.

Using Technology

When we look at the graph of a linear equation, we should be able to estimate its slope. Use the equation $y = x$ as a guideline.

Step 1:   Graph the equation $y = x$.

We can make the graph a thick line (so we can tell it apart from the others) by moving the arrow all the way to the left and hitting enter:

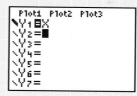   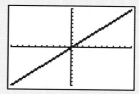

Step 2:   Keeping this equation, graph the equation $y = 2x$:

   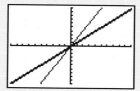

    a. Is the graph steeper or flatter than the graph of $y = x$?

    b. Make a guess as to whether $y = 3x$ will be steeper or flatter than $y = x$. Test your guess by graphing $y = 3x$.

Step 3:   Clear the equation $y = 2x$ and graph the equation $y = 0.5x$:

   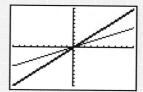

    a. Is the graph steeper or flatter than the graph of $y = x$?

    b. Make a guess as to whether $y = 0.65x$ will be steeper or flatter than $y = x$. Test your guess by graphing $y = 0.65x$.

Step 4:   Test similar situations, except with negative slopes: $y = -x$

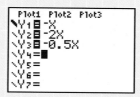   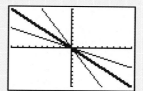

Did you notice that we have the same relationship, except in the opposite direction? That is, $y = 2x$ is steeper than $y = x$ in the positive direction, and $y = -2x$ is steeper than $y = -x$, but in the negative direction. And $y = 0.5x$ is flatter than $y = x$ in the positive direction, and $y = -0.5x$ is flatter than $y = -x$, but in the negative direction.

Answers to You Try Exercises

1) The ramp rises 7 in. for every 16 horizontal inches.   2) a) $m = \dfrac{4}{7}$ b) $m = -1$

3) a) $m = 0$ b) undefined

4) a) 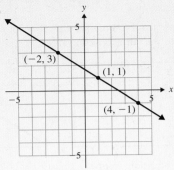   b)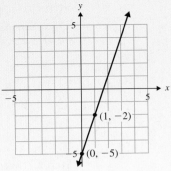

c)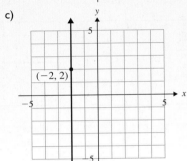

4.3 Exercises

Objective 1

1) Explain the meaning of slope.

2) Describe the slant of a line with a positive slope.

3) Describe the slant of a line with a negative slope.

4) The slope of a horizontal line is _____.

5) The slope of a vertical line is _____.

6) If a line contains the points (x_1, y_1) and (x_2, y_2), write the formula for the slope of the line.

Objectives 2 and 4

Determine the slope of each line by

    a) counting the vertical change and the horizontal change as you move from one point to the other on the line

    and

    b) using the slope formula. (See Example 2.)

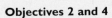

7)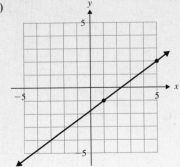

8)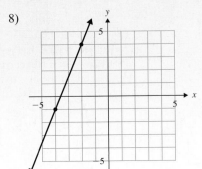

9)

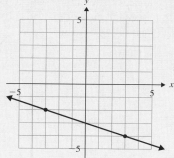

10)

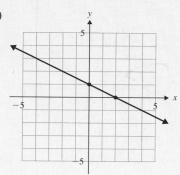

11)

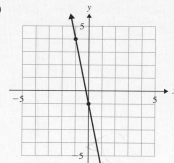

12)

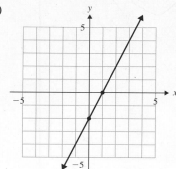

13)

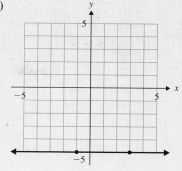

14)

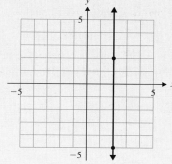

15) Graph a line with a positive slope and a negative *x*-intercept.

16) Graph a line with a negative slope and a positive *y*-intercept.

Use the slope formula to find the slope of the line containing each pair of points.

17) (3, 2) and (9, 5)

18) (4, 1) and (0, −5)

19) (−2, 8) and (2, 4)

20) (3, −2) and (−1, 6)

21) (9, 2) and (0, 4)

22) (−5, 1) and (2, −4)

23) (3, 5) and (−1, 5)

24) (−4, −4) and (−4, 10)

25) (3, 2) and (3, −1)

26) (0, −7) and (−2, −7)

27) $\left(\dfrac{3}{8}, -\dfrac{1}{3}\right)$ and $\left(\dfrac{1}{2}, \dfrac{1}{4}\right)$

28) $\left(\dfrac{3}{2}, \dfrac{7}{3}\right)$ and $\left(\dfrac{1}{3}, 4\right)$

29) (−1.7, −1.2) and (2.8, −10.2)

30) (4.8, −1.6) and (6, 1.4)

Objective 3

31) The slope of a roof is sometimes referred to as a *pitch*. A garage roof might have a *10-12 pitch*. The first number refers to the rise of the roof, and the second number refers to how far over you must go to attain that rise (the run).

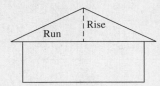

a) Find the slope of a roof with a 10-12 pitch.

b) Find the slope of a roof with an 8-12 pitch.

c) If a roof rises 8 in. in a 2-ft run, what is the slope of the roof? How do you write the slope in *x-12 pitch* form?

32) George will help his dad build a wheelchair ramp for his grandfather's house. Find the slope of the ramp.

9 in.

90 in.

33) Melissa purchased a new car in 2000. The graph shows the value of the car from 2000–2004.

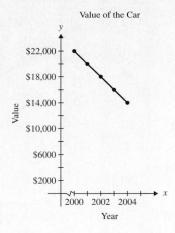

Value of the Car

a) What was the value of the car in 2000?

b) Without computing the slope, determine whether it is positive or negative.

c) What does the sign of the slope mean in the context of this problem?

d) Find the slope of the line segment, and explain what it means in the context of the problem.

34) The graph shows the approximate number of babies (in thousands) born to teenage girls from 1997–2001. (Source: U.S. Census Bureau)

Number of Births

a) Approximately how many babies were born to teen mothers in 1997? in 1998? in 2001?

b) Without computing the slope, determine whether it is positive or negative.

c) What does the sign of the slope mean in the context of the problem?

d) Find the slope of the line segment, and explain what it means in the context of the problem.

Objective 5

Graph the line containing the given point and with the given slope.

35) $(-3, -2); m = \dfrac{5}{2}$ 36) $(1, 3); m = \dfrac{1}{4}$

37) $(1, -4); m = \dfrac{1}{3}$ 38) $(-4, 2); m = \dfrac{2}{7}$

39) $(4, 5); m = -\dfrac{2}{3}$ 40) $(2, -1); m = -\dfrac{3}{2}$

41) $(-5, 1); m = 3$ 42) $(-3, -5); m = 2$

43) $(0, -3); m = -1$ 44) $(0, 4); m = -3$

45) $(6, 2); m = -4$ 46) $(2, 5); m = -1$

47) $(-2, -1); m = 0$ 48) $(-3, 3); m = 0$

49) $(4, 0);$ slope is undefined

50) $(1, 6);$ slope is undefined

51) $(0, 0); m = 1$

52) $(0, 0); m = -1$

Section 4.4 The Slope-Intercept Form of a Line

Objectives

1. Define the Slope-Intercept Form of a Line

2. Graph a Line Expressed in Slope-Intercept Form Using the Slope and y-Intercept

3. Rewrite an Equation in Slope-Intercept Form and Graph the Line

In Section 4.1, we learned that a linear equation in two variables can be written in the form $Ax + By = C$ (this is called **standard form**), where A, B, and C are real numbers and where both A and B do not equal zero. Equations of lines can take other forms, too, and we will look at one of those forms in this section.

1. Define the Slope-Intercept Form of a Line

We know that if (x_1, y_1) and (x_2, y_2) are points on a line, then the slope of the line is

$$m = \frac{y_2 - y_1}{x_2 - x_1}$$

Recall that to find the y-intercept of a line, we let $x = 0$ and solve for y. Let one of the points on a line be the y-intercept $(0, b)$, where b is a number. Let another point on the line be (x, y).

Since we will be using the slope formula

$$\text{let } (0, b) \text{ represent } (x_1, y_1) \text{ and}$$
$$\text{let } (x, y) \text{ represent } (x_2, y_2)$$

Begin with the slope formula and substitute the points $(0, b)$ and (x, y). Then, solve for y.

$$m = \frac{y_2 - y_1}{x_2 - x_1} = \frac{y - b}{x - 0} = \frac{y - b}{x}$$

Solve $m = \dfrac{y - b}{x}$ for y.

$$mx = \frac{y - b}{x} \cdot x \qquad \text{Multiply by } x \text{ to eliminate the fraction.}$$
$$mx = y - b$$
$$mx + b = y - b + b \qquad \text{Add } b \text{ to each side to solve for } y.$$
$$mx + b = y$$
$$\text{OR}$$
$$y = mx + b \qquad \text{Slope-intercept form}$$

Definition

The **slope-intercept form of a line** is $y = mx + b$, where m is the slope and $(0, b)$ is the y-intercept.

When an equation is in the form $y = mx + b$, we can quickly recognize the y-intercept and slope to graph the line.

2. Graph a Line Expressed in Slope-Intercept Form Using the Slope and *y*-Intercept

---

Example 1

Graph each equation.

a) $y = -\dfrac{4}{3}x + 2$ b) $y = 3x - 4$ c) $y = \dfrac{1}{2}x$

Solution

a) Graph $y = -\dfrac{4}{3}x + 2$.

The equation is in slope-intercept form. Identify the slope and *y*-intercept.

$$y = -\dfrac{4}{3}x + 2$$

Slope $= -\dfrac{4}{3}$,   *y*-intercept is $(0, 2)$.

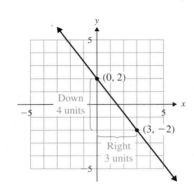

Graph the line by first plotting the *y*-intercept and then by using the slope to locate another point on the line.

b) Graph $y = 3x - 4$.
The equation is in slope-intercept form. Identify the slope and *y*-intercept.

$$y = 3x - 4$$
Slope $= 3$,   *y*-intercept is $(0, -4)$.

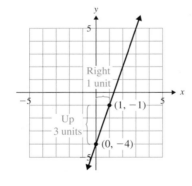

Plot the *y*-intercept first, then use the slope to locate another point on the line. Since the slope is 3, think of it as $\dfrac{3}{1}$. $\begin{array}{l}\leftarrow \text{change in } y \\ \leftarrow \text{change in } x\end{array}$

c) Graph $y = \dfrac{1}{2}x$.

The equation is in slope-intercept form. Identify the slope and *y*-intercept.

$y = \dfrac{1}{2}x$ is the same as $y = \dfrac{1}{2}x + 0$.

Slope $= \dfrac{1}{2}$,   *y*-intercept is $(0, 0)$.

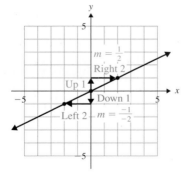

Plot the *y*-intercept, then use the slope to locate another point on the line.

$\dfrac{1}{2}$ is equivalent to $\dfrac{-1}{-2}$, so we can use $\dfrac{-1}{-2}$ as the slope to locate yet another point on the line.

 **You Try 1**

Graph each line using the slope and y-intercept.

a) $y = \dfrac{2}{5}x - 3$ b) $y = -4x + 5$ c) $y = -x$

3. Rewrite an Equation in Slope-Intercept Form and Graph the Line

A line is not always written in slope-intercept form. It may be written in *standard form* (like $5x + 3y = 12$) or in another form such as $2x = 2y + 10$. We can put equations like these into slope-intercept form by solving the equation for y.

Example 2

Put each equation into slope-intercept form, and graph.

a) $5x + 3y = 12$ b) $2x = 2y + 10$

Solution

a)   The slope-intercept form of a line is $y = mx + b$. We must solve the equation for y.

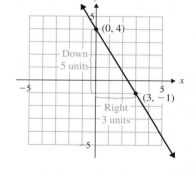

$$5x + 3y = 12$$
$$3y = -5x + 12 \qquad \text{Add } -5x \text{ to each side.}$$
$$y = -\frac{5}{3}x + 4 \qquad \text{Divide each side by 3.}$$

The slope-intercept form is $y = -\dfrac{5}{3}x + 4$.

$$\text{Slope} = -\frac{5}{3}, \quad y\text{-intercept is } (0, 4).$$

We can interpret the slope of $-\dfrac{5}{3}$ as either $\dfrac{-5}{3}$ or $\dfrac{5}{-3}$. Using either form will give us a point on the line. Here, we will use $\dfrac{-5}{3}$.

b)   We must solve $2x = 2y + 10$ for y.

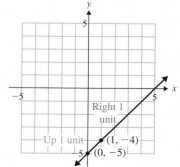

$$2x = 2y + 10$$
$$2x - 10 = 2y \qquad \text{Subtract 10 from each side.}$$
$$x - 5 = y \qquad \text{Divide each side by 2.}$$

The slope-intercept form is $y = x - 5$. We can also think of this as $y = 1x - 5$.

$$\text{slope} = 1, \quad y\text{-intercept is } (0, -5).$$

We will think of the slope as $\dfrac{1}{1}. \begin{array}{l} \leftarrow \text{change in } y \\ \leftarrow \text{change in } x \end{array}$ $\left(\text{We could also think of it as } \dfrac{-1}{-1}. \right)$

You Try 2

Put each equation into slope-intercept form, and graph.

a) $8x - 4y = 12$ b) $3x = 4 - 4y$

We have learned that we can use different methods for graphing lines. Given the equation of a line we can

1) make a table of values, plot the points, and draw the line through the points.

2) find the *x*-intercept by letting $y = 0$ and solving for *x*, and find the *y*-intercept by letting $x = 0$ and solving for *y*. Plot the points, then draw the line through the points.

3) put the equation into slope-intercept form, $y = mx + b$, identify the slope and *y*-intercept, then graph the line.

Answers to You Try Exercises

1) a)

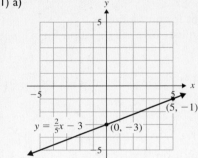

$y = \frac{2}{5}x - 3$ (0, −3)   (5, −1)

b)

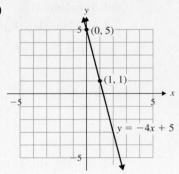

(0, 5)   (1, 1) $y = -4x + 5$

c)

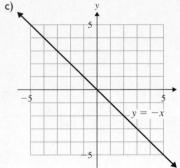

$y = -x$

2) a) $y = 2x - 3$

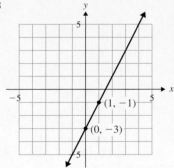

(1, −1)   (0, −3)

b) $y = -\dfrac{3}{4}x + 1$

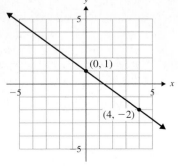

(0, 1)   (4, −2)

4.4 Exercises

Objectives 1 and 2

1) The slope-intercept form of a line is $y = mx + b$. What is the slope? What is the y-intercept?

2) How do you put an equation that is in standard form, $Ax + By = C$, into slope-intercept form?

Each of the following equations is in slope-intercept form. Identify the slope and the y-intercept, then graph each line using this information.

3) $y = \dfrac{2}{5}x - 6$ 4) $y = \dfrac{7}{4}x - 2$

5) $y = -\dfrac{5}{3}x + 4$ 6) $y = -\dfrac{1}{2}x + 5$

7) $y = \dfrac{3}{4}x + 1$ 8) $y = \dfrac{2}{3}x + 3$

9) $y = 4x - 2$ 10) $y = -3x - 1$

11) $y = -x + 5$ 12) $y = x$

13) $y = \dfrac{3}{2}x + \dfrac{1}{2}$ 14) $y = -\dfrac{3}{4}x - \dfrac{5}{2}$

15) $y = -2$ 16) $y = 4$

Objective 3

Put each equation into slope-intercept form, if possible, and graph.

17) $x + 3y = -6$ 18) $5x + 2y = 2$

19) $12x - 8y = 32$ 20) $y - x = 1$

21) $x + 9 = 2$ 22) $5 = x + 2$

23) $18 = 6y - 15x$ 24) $20x = 48 - 12y$

25) $y = 0$ 26) $y + 6 = 1$

27) Dave works in sales, and his income is a combination of salary and commission. He earns $34,000 per year plus 5% of his total sales. The equation $I = 0.05s + 34,000$ represents his total annual income, I, in dollars, when his sales total s dollars.

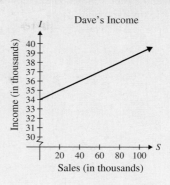

Dave's Income

a) What is the I-intercept? What does it mean in the context of this problem?

b) What is the slope? What does it mean in the context of the problem?

c) Use the graph to find Dave's income if his total sales are $80,000. Confirm your answer using the equation.

28) Li Mei gets paid hourly in her after-school job. Her income is given by $I = 7.50h$, where I represents her income in dollars and h represents the number of hours worked.

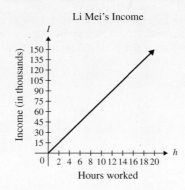

Li Mei's Income

a) What is the I-intercept? What does it mean in the context of the problem?

b) What is the slope? What does it mean in the context of the problem?

c) Use the graph to find how much Li Mei earns when she works 14 hr. Confirm your answer using the equation.

29) The per capita consumption of whole milk in the United States since 1945 can be modeled by $y = -0.59x + 40.53$ where x represents the number of years after 1945, and y represents the per capita consumption of whole milk in gallons.
(Source: U.S. Department of Agriculture)

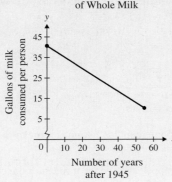

Per Capita Consumption
of Whole Milk

a) What is the y-intercept? What does it mean in the context of the problem?

b) What is the slope? What does it mean in the context of the problem?

c) Use the graph to estimate the per capita consumption of whole milk in the year 2000. Then, use the equation to determine this number.

30) On a certain day in 2004, the exchange rate between the American dollar and the Mexican peso was given by $p = 11.40d$, where d represents the number of dollars and p represents the number of pesos.

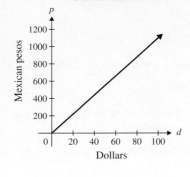

Dollars and Pesos

a) What is the p-intercept? What does it mean in the context of the problem?

b) What is the slope? What does it mean in the context of the problem?

c) Use the graph to estimate the value of 600 pesos in dollars. Then, use the equation to determine this number.

Write the slope-intercept form for the equation of a line with the given slope and y-intercept.

31) $m = -4$; y-int: $(0, 7)$

32) $m = 5$; y-int: $(0, 3)$

33) $m = \dfrac{8}{5}$; y-int: $(0, -6)$

34) $m = \dfrac{4}{9}$; y-int: $(0, -1)$

35) $m = \dfrac{1}{3}$; y-int: $(0, 5)$

36) $m = -\dfrac{1}{2}$; y-int: $(0, -3)$

37) $m = -1$; y-int: $(0, 0)$

38) $m = 1$; y-int: $(0, 4)$

39) $m = 0$; y-int: $(0, -2)$

40) $m = 0$; y-int: $(0, 0)$

Section 4.5 Writing an Equation of a Line

The focus of Chapter 4, thus far, has been on graphing lines given their equations. In this section, we will switch gears. Given information about a line, we will write an equation of that line.

Recall the forms of lines we have discussed so far.

1) **Standard Form:** $Ax + By = C$, where A, B, and C are real numbers and where both A and B do not equal zero.

We will now set an additional condition for when we write equations of lines in standard form:

A, B, and C must be integers and A must be positive.

2) **Slope-Intercept Form:** The slope-intercept form of a line is $y = mx + b$, where m is the slope, and the y-intercept is $(0, b)$.

1. Rewrite an Equation in Standard Form

In Section 4.4, we practiced writing equations of lines in slope-intercept form. Here we will discuss how to write a line in standard form.

Example 1

Rewrite each linear equation in standard form.

a) $3x + 8 = -2y$ b) $y = -\dfrac{3}{4}x + \dfrac{1}{6}$

Solution

a) To write $3x + 8 = -2y$ in standard form, the x and y terms must be on the same side of the equation.

$$3x + 8 = -2y$$
$$3x = -2y - 8 \qquad \text{Subtract 8 from each side.}$$
$$3x + 2y = -8 \qquad \text{Add } 2y \text{ to each side.}$$

$3x + 2y = -8$ is in standard form.

b) Since an equation $Ax + By = C$ is considered to be in standard form when A, B, and C are integers, the first step in writing $y = -\dfrac{3}{4}x + \dfrac{1}{6}$ in standard form is to eliminate the fractions.

$$y = -\dfrac{3}{4}x + \dfrac{1}{6}$$
$$12 \cdot y = 12\left(-\dfrac{3}{4}x + \dfrac{1}{6}\right) \qquad \text{Multiply both sides of the equation by 12.}$$
$$\qquad\qquad\qquad\qquad\qquad \text{This is the LCD of } -\dfrac{3}{4} \text{ and } \dfrac{1}{6}.$$
$$12y = -9x + 2$$
$$9x + 12y = 2 \qquad \text{Add } 9x \text{ to each side.}$$

The standard form is $9x + 12y = 2$.

You Try 1

Rewrite each equation in standard form.

a) $5x = 3 + 11y$ b) $y = \dfrac{1}{3}x - 7$

How to Write an Equation of a Line

"Write an equation of a line containing the points (6, 4) and (3, 2)" is a type of problem we will learn to solve in this section.

When we are asked to write an equation of a line, the information we are given about the line usually comes in one of three forms. We can be told

1) the slope and y-intercept of the line.

    or

2) the slope of the line and a point on the line.

    or

3) two points on a line.

The way we find an equation of the line depends on the type of information we are given. Let's begin with number 1.

2. Write an Equation of a Line Given Its Slope and y-Intercept

1) If we are given the slope and y-intercept of a line, use $y = mx + b$ and substitute those values into the equation.

Example 2

Find an equation of the line with slope $= -5$ and y-intercept (0, 9).

Solution

Since we are told the slope and y-intercept, use $y = mx + b$.

$$m = -5 \qquad \text{and} \qquad b = 9$$

Substitute these values into $y = mx + b$ to get $y = -5x + 9$.

You Try 2

Find an equation of the line with slope $= \dfrac{2}{3}$ and y-intercept (0, −6).

3. Use the Point-Slope Formula to Write an Equation of a Line Given Its Slope and a Point on the Line

2) If we are given the slope of the line and a point on the line, we can use the point-slope formula to find an equation of the line.

Point-Slope Formula If (x_1, y_1) is a point on line L and m is the slope of line L, then the equation of L is given by

$$y - y_1 = m(x - x_1)$$

This formula is called the **point-slope formula**.

The point-slope formula will help us write an equation of a line. We will not express our final answer in this form. We will write our answer in either slope-intercept form or in standard form.

Example 3

Find an equation of the line containing the point $(-4, 3)$ with slope $= \dfrac{1}{2}$. Express the answer in slope-intercept form.

Solution

First, ask yourself, *"What kind of information am I given?"* Since the problem tells us the slope of the line and a point on the line, we will use the point-slope formula.

Use $y - y_1 = m(x - x_1)$. Substitute $\dfrac{1}{2}$ for m. Substitute $(-4, 3)$ for (x_1, y_1).

(Notice we do *not* substitute anything for the x and y.)

$$y - y_1 = m(x - x_1)$$

$$y - 3 = \frac{1}{2}(x - (-4)) \qquad \text{Substitute } -4 \text{ for } x_1 \text{ and } 3 \text{ for } y_1.$$

$$y - 3 = \frac{1}{2}(x + 4)$$

$$y - 3 = \frac{1}{2}x + 2 \qquad \text{Distribute.}$$

Since we must express our answer in slope-intercept form, $y = mx + b$, we must solve the equation for y.

$$y = \frac{1}{2}x + 5 \qquad \text{Add 3 to each side.}$$

The equation is $y = \dfrac{1}{2}x + 5$.

You Try 3

Find an equation of the line containing the point $(5, 3)$ with slope $= 2$. Express the answer in slope-intercept form.

Example 4

A line has slope -3 and contains the point $(2, 1)$. Find the standard form for the equation of the line.

Solution

Although we are told to find the *standard form* for the equation of the line, we do not try to immediately "jump" to standard form.

First, ask yourself, *"What information am I given?"*

We are given the slope and a point on the line. Therefore, we will begin by using the point-slope formula. Our *last* step will be to put it in standard form.

Use $y - y_1 = m(x - x_1)$. Substitute -3 for m. Substitute $(2, 1)$ for (x_1, y_1).

$$y - y_1 = m(x - x_1)$$
$$y - 1 = -3(x - 2) \qquad \text{Substitute 2 for } x_1 \text{ and 1 for } y_1.$$
$$y - 1 = -3x + 6 \qquad \text{Distribute.}$$

Since we are asked to express the answer in standard form, we must get the x- and y-terms on the same side of the equation.

$$3x + y - 1 = 6 \qquad \text{Add } 3x \text{ to each side.}$$
$$3x + y = 7 \qquad \text{Add 1 to each side.}$$

The equation is $3x + y = 7$.

You Try 4

A line has slope -5 and contains the point $(-1, 3)$. Find the standard form for the equation of the line.

4. Use the Point-Slope Formula to Write an Equation of a Line Given Two Points on the Line

We are now ready to discuss how to write an equation of a line when we are given two points on a line.

3) **To write an equation of a line given two points on the line,**

   a) **use the points to find the slope of line**

      then

   b) **use the slope and *either one* of the points in the point-slope formula.**

We will solve the problem posed earlier in this section on p. 270.

Example 5

Write an equation of the line containing the points $(6, 4)$ and $(3, 2)$. Express the answer in slope-intercept form.

Solution

In this problem we are given two points on the line. Therefore, we will find the slope of the line, then use the point-slope formula.

Find the slope: $m = \dfrac{2-4}{3-6} = \dfrac{-2}{-3} = \dfrac{2}{3}$.

We will use the slope and *either one* of the points in the point-slope formula. (Each point will give the same result.) We will use $(6, 4)$.

$$\text{Point-slope formula: } y - y_1 = m(x - x_1)$$

Substitute $\dfrac{2}{3}$ for m. Substitute $(6, 4)$ for (x_1, y_1).

$$y - y_1 = m(x - x_1)$$

$$y - 4 = \frac{2}{3}(x - 6) \qquad \text{Substitute 6 for } x_1 \text{ and 4 for } y_1.$$

$$y - 4 = \frac{2}{3}x - 4 \qquad \text{Distribute.}$$

$$y = \frac{2}{3}x \qquad \text{Add 4 to each side to solve for } y.$$

The equation is $y = \dfrac{2}{3}x$.

 You Try 5

Find the standard form for the equation of the line containing the points $(-2, 5)$ and $(1, 1)$.

5. Write Equations of Horizontal and Vertical Lines

In Section 4.3, we discussed horizontal and vertical lines. We learned that the slope of a horizontal line is zero and that it has equation $y = d$, where d is a constant. The slope of a vertical line is undefined, and its equation is $x = c$, where c is a constant.

Definition

Equation of a Horizontal Line: The equation of a horizontal line containing the point (c, d) is $y = d$.

Equation of a Vertical Line: The equation of a vertical line containing the point (c, d) is $x = c$.

Example 6

Write an equation of the horizontal line containing the point $(5, -4)$.

Solution

The equation of a horizontal line has the form $y = d$, where d is the y-coordinate of the point. The equation of the line is $y = -4$.

 **You Try 6**

Write an equation of the horizontal line containing the point $(1, 7)$.

6. Write a Linear Equation to Model Real-World Data and to Solve an Applied Problem

As seen in previous sections of this chapter, equations of lines are used to describe many kinds of real-world situations. We will look at an example in which we must find the equation of a line given some data.

Example 7

Air Pollution

Since 1998, sulfur dioxide emissions in the United States have been decreasing by about 1080.5 thousand tons per year. In 2000, approximately 16,636 thousand tons of the pollutant were released into the air. (Source: *Statistical Abstract of the United States*)

a) Write a linear equation to model this data. Let x represent the number of years after 1998, and let y represent the amount of sulfur dioxide (in thousands of tons) released into the air.

b) How much sulfur dioxide was released into the air in 1998? in 2004?

Solution

a) Ask yourself, "What information is given in the problem?"

  i) ". . . emissions in the United States have been decreasing by about 1080.5 thousand tons per year" tells us the rate of change of emissions with respect to time. Therefore, this is the *slope*. It will be *negative* since emissions are decreasing.

$$m = -1080.5$$

  ii) "In 2000, approximately 16,636 thousand tons . . . were released into the air" gives us a point on the line.

  Let x = the number of years after 1998.

  The year 2000 corresponds to $x = 2$.

  y = amount of sulfur dioxide (in thousands of tons) released into the air.

  Then, 16,636 thousand tons corresponds to $y = 16{,}636$.

  A point on the line is **(2, 16,636)**.

  Now that we know the slope and a point on the line, we can write an equation of the line using the point-slope formula:

$$y - y_1 = m(x - x_1)$$

Substitute -1080.5 for m. Substitute $(2, 16{,}636)$ for (x_1, y_1).

$$y - y_1 = m(x - x_1)$$
$$y - 16{,}636 = -1080.5(x - 2) \qquad \text{Substitute 2 for } x_1 \text{ and } 16{,}636 \text{ for } y_1.$$
$$y - 16{,}636 = -1080.5x + 2161 \qquad \text{Distribute.}$$
$$y = -1080.5x + 18{,}797 \qquad \text{Add 16,636 to each side.}$$

The equation is $y = -1080.5x + 18{,}797$.

b) To determine the amount of sulfur dioxide released into the air in 1998, let $x = 0$ since $x =$ the number of years *after* 1998.

$$y = -1080.5(0) + 18{,}797 \qquad \text{Substitute } x = 0.$$
$$y = 18{,}797$$

In 1998, 18,797 thousand tons of sulfur dioxide were released. Notice, the equation is in slope-intercept form, $y = mx + b$, and our result is b. That is because to find the y-intercept, let $x = 0$.

To determine how much sulfur dioxide was released in 2004, let $x = 6$ since 2004 is 6 years after 1998.

$$y = -1080.5(6) + 18{,}797 \qquad \text{Substitute } x = 6.$$
$$y = -6483 + 18{,}797 \qquad \text{Multiply.}$$
$$y = 12{,}314$$

In 2004, 12,314 thousand tons of sulfur dioxide were released into the air.

Summary—Writing Equations of Lines

If you are given

1) **the slope and y-intercept of the line**, use $y = mx + b$ and substitute those values into the equation.

2) **the slope of the line and a point on the line**, use the point-slope formula:

$$y - y_1 = m(x - x_1)$$

Substitute the slope for m and the point you are given for (x_1, y_1). Write your answer in slope-intercept or standard form.

3) **two points on the line**, find the slope of the line and then use the slope and *either one* of the points in the point-slope formula. Write your answer in slope-intercept or standard form.

The equation of a **horizontal line** containing the point (c, d) is $\mathbf{y = d}$.

The equation of a **vertical line** containing the point (c, d) is $\mathbf{x = c}$.

Answers to You Try Exercises

1) a) $5x - 11y = 3$ b) $x - 3y = 21$ 2) $y = \dfrac{2}{3}x - 6$ 3) $y = 2x - 7$

4) $5x + y = -2$ 5) $4x + 3y = 7$ 6) $y = 7$

4.5 Exercises

Boost your grade at mathzone.com!   MathZone

> Practice Problems   > NetTutor   > e-Professors   > Videos
> Self-Test

Objective 1

Rewrite each equation in standard form.

1) $y = 8x + 3$ 2) $y = -5x - 7$

3) $x = -2y - 11$ 4) $x = y + 9$

5) $y = \dfrac{4}{5}x - 1$ 6) $y = \dfrac{7}{3}x + 2$

7) $y = -\dfrac{3}{2}x + \dfrac{1}{4}$ 8) $y = -\dfrac{1}{4}x - \dfrac{5}{6}$

Objective 2

9) Explain how to find an equation of a line when you are given the slope and y-intercept of the line.

Find an equation of the line with the given slope and y-intercept. Express your answer in the indicated form.

10) $m = 4$, y-int: $(0, -5)$; slope-intercept form

 11) $m = -7$, y-int: $(0, 2)$; slope-intercept form

12) $m = -2$, y-int: $(0, 8)$; standard form

13) $m = 1$, y-int: $(0, -3)$; standard form

14) $m = \dfrac{5}{2}$, y-int: $(0, -1)$; standard form

15) $m = -\dfrac{1}{3}$, y-int: $(0, -4)$; standard form

16) $m = -1$, y-int: $(0, 0)$; slope-intercept form

17) $m = 1$, y-int: $(0, 0)$; slope-intercept form

18) $m = \dfrac{4}{9}$, y-int: $\left(0, -\dfrac{1}{6}\right)$; slope-intercept form

Objective 3

19) a) If (x_1, y_1) is a point on a line with slope m, then the point-slope formula is _____.

b) Explain how to find an equation of a line when you are given the slope and a point on the line.

Find an equation of the line containing the given point with the given slope. Express your answer in the indicated form.

20) $(5, 8)$, $m = 3$; slope-intercept form

21) $(1, 6)$, $m = 5$; slope-intercept form

22) $(3, -2)$, $m = -2$; slope-intercept form

23) $(-9, 4)$, $m = -1$; slope-intercept form

24) $(-3, -7)$, $m = 1$; standard form

25) $(-2, -1)$, $m = 4$; standard form

26) $(-8, 6)$, $m = \dfrac{3}{4}$; standard form

27) $(12, 3)$, $m = -\dfrac{2}{3}$; standard form

28) $(2, 3)$, $m = -\dfrac{4}{5}$; slope-intercept form

29) $(-4, -5)$, $m = \dfrac{1}{6}$; slope-intercept form

30) $(-2, 0)$, $m = \dfrac{5}{8}$; standard form

31) $(6, 0)$, $m = -\dfrac{5}{9}$; standard form

32) $\left(\dfrac{1}{6}, -1\right)$, $m = 4$; slope-intercept form

Objective 4

33) Explain how to find an equation of a line when you are given two points on the line.

Find an equation of the line containing the two given points. Express your answer in the indicated form.

34) $(2, 5)$ and $(4, 1)$; slope-intercept form

35) $(3, 4)$ and $(7, 8)$; slope-intercept form

36) $(3, 2)$ and $(4, 5)$; slope-intercept form

 37) $(-2, 4)$ and $(1, 3)$; slope-intercept form

38) $(5, -2)$ and $(2, -4)$; slope-intercept form

39) $(-1, -5)$ and $(3, -2)$; standard form

40) $(-3, 0)$ and $(-5, 1)$; standard form

41) $(4, 1)$ and $(6, -3)$; standard form

42) $(2, 1)$ and $(4, 6)$; standard form

43) $(2.5, 4.2)$ and $(3.1, 7.2)$; slope-intercept form

44) $(-3.4, 5.8)$ and $(-1.8, 3.4)$; slope-intercept form

45) $(-6, 0)$ and $(2, -1)$; standard form

46) $(15, 9)$ and $(-5, -7)$; standard form

Objectives 2–5

Write the slope-intercept form of the equation of each line, if possible.

47)

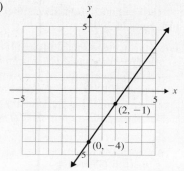

48)

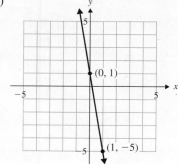

49)

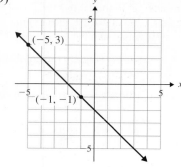

50)

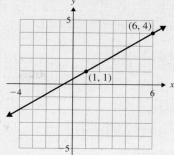

51)

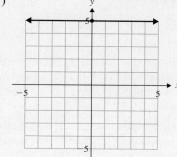

52)

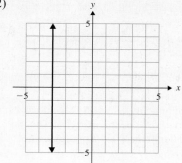

Mixed Exercises

Write the slope-intercept form of the equation of the line, if possible, given the following information.

53) $m = 4$ and contains $(-4, -1)$

54) contains $(3, 0)$ and $(7, -2)$

55) $m = \dfrac{8}{3}$ and y-intercept $(0, -9)$

56) $m = 3$ and contains $(4, 5)$

57) contains $(-1, -2)$ and $(-5, 1)$

58) y-intercept $(0, 6)$ and $m = 7$

59) vertical line containing $(3, 5)$

60) vertical line containing $\left(-\dfrac{3}{4}, 2\right)$

61) horizontal line containing $(0, -8)$

62) horizontal line containing $(7, 4)$

63) $m = -\dfrac{1}{2}$ and contains $(4, -5)$

64) $m = 2$ and y-intercept $(0, -11)$

65) contains $(0, 2)$ and $(6, 0)$

66) $m = -1$ and contains $(8, -8)$

67) $m = 1$ and y-intercept $(0, 0)$

68) contains $(-3, -1)$ and $(2, -3)$

Objective 6

Applications

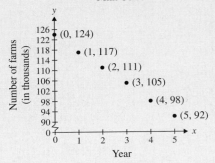

69) Since 1998 the population of Maine has been increasing by about 8700 people per year. In 2001, the population of Maine was about 1,284,000. (Source: *Statistical Abstract of the United States*)

   a) Write a linear equation to model this data. Let *x* represent the number of years after 1998, and let *y* represent the population of Maine.

   b) Explain the meaning of the slope in the context of the problem.

   c) According to the equation, how many people lived in Maine in 1998? in 2002?

   d) If the current trend continues, in what year would the population be 1,431,900?

70) Since 1997, the population of North Dakota has been decreasing by about 3290 people per year. The population was about 650,000 in 1997. (Source: *Statistical Abstract of the United States*)

   a) Write a linear equation to model this data. Let *x* represent the number of years after 1997, and let *y* represent the population of North Dakota.

   b) Explain the meaning of the slope in the context of the problem.

   c) According to the equation, how many people lived in North Dakota in 1999? in 2002?

   d) If the current trend holds, in what year would the population be 600,650?

71) The graph here shows the number of farms (in thousands) with milk cows between 1997 and 2002. *x* represents the number of years after 1997 so that $x = 0$ represents 1997, $x = 1$ represents 1998, and so on. Let *y* represent the number of farms (in thousands) with milk cows. (Source: USDA, National Agricultural Statistics Service)

Number of Farms With Milk Cows

a) Write a linear equation to model this data. Use the data points for 1997 and 2002.

b) Explain the meaning of the slope in the context of the problem.

c) If the current trend continues, find the number of farms with milk cows in 2004.

72) The graph here shows the average salary for a public school high school principal for several years beginning with 1990. *x* represents the number of years after 1990 so that $x = 0$ represents 1990, $x = 1$ represents 1991, and so on. Let *y* represent the average salary of a high school principal. (Source: *Statistical Abstract of the United States*)

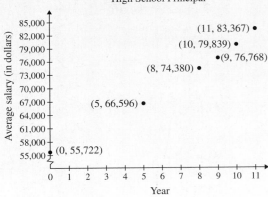

Average Salary for a Public High School Principal

a) Write a linear equation to model this data. Use the data points for 1990 and 2001. (Round the slope to the nearest whole number.)

b) Explain the meaning of the slope in the context of the problem.

c) Use the equation to estimate the average salary in 1993.

d) If the current trend continues, find the average salary in 2005.

73) The chart shows the number of hybrid vehicles registered in the United States from 2000 to 2003.
(Source: *American Demographics,* Sept 2004, Vol 26, Issue 7)

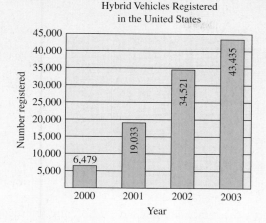

a)   Write a linear equation to model this data. Use the
data points for 2000 and 2003. Let x represent the
number of years after 2000, and let y represent the
number of hybrid vehicles registered in the United
States. Round the slope to the tenths place.

b)   Explain the meaning of the slope in the context of
the problem.

c)   Use the equation to determine the number of
vehicles registered in 2002. How does it compare
to the actual number on the chart?

d)   If the current trend continues, approximately how
many hybrid vehicles will be registered in 2010?

74) The chart shows the percentage of females in Belgium
(15–64 yr old) in the workforce between 1980 and 2000.
(Source: *Statistical Abstract of the United States*)

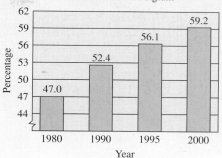

a)   Write a linear equation to model this data. Use
the data points for 1980 and 2000. Let x represent
the number of years after 1980, and let y represent the percent of females in Belgium in the
workforce.

b)   Explain the meaning of the slope in the context of
the problem.

c)   Use the equation to determine the percentage of
Belgian women in the workforce in 1990. How
does it compare to the actual number on the
chart?

d)   Do the same as part c) for the year 1995.

e)   In what year were 58% of Belgian women
working?

Section 4.6 Parallel and Perpendicular Lines

Objectives

1.   Use Slope to
Determine if Two
Lines Are Parallel

2.   Use Slope to
Determine if Two
Lines Are
Perpendicular

3.   Write an Equation
of a Line That Is
Parallel to a Given
Line

4.   Write an Equation
of a Line That Is
Perpendicular to a
Given Line

Recall that two lines in a plane are *parallel* if they do not intersect. If we are given the
equations of two lines, how can we determine if they are parallel?

Here are the equations of two lines, L_1 and L_2.

$$L_1: 2x - 3y = -3 \qquad L_2: y = \frac{2}{3}x - 5$$

We will graph each line.

L_1: Write $2x - 3y = -3$ in slope-intercept form.

$$-3y = -2x - 3 \qquad \text{Add } -2x \text{ to each side.}$$

$$y = \frac{-2}{-3}x - \frac{3}{-3} \qquad \text{Divide by } -3.$$

$$y = \frac{2}{3}x + 1 \qquad \text{Simplify.}$$

The slope-intercept form of L_1 is $y = \dfrac{2}{3}x + 1$, and L_2 is in slope-intercept form, $y = \dfrac{2}{3}x - 5$. Now, graph each line.

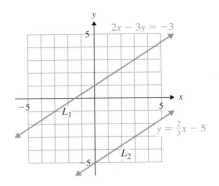

These lines are parallel. Their slopes are the same, but they have different y-intercepts. (If the y-intercepts were the same, they would be the same line.) This is how we determine if two (nonvertical) lines are parallel. They have the same slope, but different y-intercepts.

> Parallel lines have the same slope.

> If two lines are vertical they are parallel. However, their slopes are undefined.

1. Use Slope to Determine if Two Lines Are Parallel

Example 1

Determine whether each pair of lines is parallel.

a) $3x + 6y = 10$
$x + 2y = -12$

b) $y = -7x + 4$
$7x - y = 9$

Solution

a) To determine if the lines are parallel, we must find the slope of each line. If the slopes are the same, but the y-intercepts are different, the lines are parallel.

Write each equation in slope-intercept form.

$$
\begin{aligned}
3x + 6y &= 10 \\
6y &= -3x + 10 \\
y &= -\frac{3}{6}x + \frac{10}{6} \\
y &= -\frac{1}{2}x + \frac{5}{3} \\
m &= -\frac{1}{2}
\end{aligned}
\qquad\qquad
\begin{aligned}
x + 2y &= -12 \\
2y &= -x - 12 \\
y &= -\frac{x}{2} - \frac{12}{2} \\
y &= -\frac{1}{2}x - 6 \\
m &= -\frac{1}{2}
\end{aligned}
$$

Each line has a slope of $-\dfrac{1}{2}$. Their y-intercepts are different.

$3x + 6y = 10$ and $x + 2y = -12$ are parallel lines.

b) Again, we must find the slope of each line. $y = -7x + 4$ is already in slope-intercept form. Its slope is -7.

Write $7x - y = 9$ in slope-intercept form.

$$-y = -7x + 9 \qquad \text{Add } -7x \text{ to each side.}$$
$$y = \frac{-7}{-1}x + \frac{9}{-1} \qquad \text{Divide by } -1.$$
$$y = 7x - 9 \qquad \text{Simplify.}$$
$$m = 7$$

The slope of $y = -7x + 4$ is -7. The slope of $7x - y = 9$ is 7. The slopes are different, therefore the lines are not parallel.

2. Use Slope to Determine if Two Lines Are Perpendicular

The slopes of two lines can tell us about another relationship between the lines. The slopes can tell us if two lines are *perpendicular*.

Recall that two lines are **perpendicular** if they intersect at $90°$ angles.

Here are the graphs of two perpendicular lines and their equations. We will see how their slopes are related.

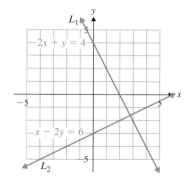

L_1 has equation $2x + y = 4$. L_2 has equation $x - 2y = 6$. Find the slopes of the lines by writing them in slope-intercept form.

$$L_1\colon 2x + y = 4$$

$$y = -2x + 4$$

$$m = -2$$

$$L_2\colon x - 2y = 6$$
$$-2y = -x + 6$$
$$y = \frac{-x}{-2} + \frac{6}{-2}$$
$$y = \frac{1}{2}x - 3$$
$$m = \frac{1}{2}$$

How are the slopes related? They are **negative reciprocals**. That is, if the slope of one line is a, then the slope of a line perpendicular to it is $-\dfrac{1}{a}$.

This is how we determine if two lines are perpendicular (where neither one is vertical).

Perpendicular lines have slopes that are negative reciprocals of each other.

Example 2

Determine whether each pair of lines is perpendicular.

a) $15x - 12y = -4$ b) $2x - 7y = 7$
    $4x - 5y = 10$ $21x + 6y = -2$

Solution

a) To determine if the lines are perpendicular, we must find the slope of each line. If the slopes are *negative reciprocals*, then the lines are perpendicular.

Write each equation in slope-intercept form.

$$15x - 12y = -4 \qquad\qquad 4x - 5y = 10$$
$$-12y = -15x - 4 \qquad\qquad -5y = -4x + 10$$
$$y = \frac{-15}{-12}x - \frac{4}{-12} \qquad\qquad y = \frac{-4}{-5}x + \frac{10}{-5}$$
$$y = \frac{5}{4}x + \frac{1}{3} \qquad\qquad y = \frac{4}{5}x - 2$$
$$m = \frac{5}{4} \qquad\qquad\qquad m = \frac{4}{5}$$

The slopes are reciprocals, but they are not *negative* reciprocals. Therefore, $15x - 12y = -4$ and $4x - 5y = 10$ are *not* perpendicular.

b) Begin by writing each equation in slope-intercept form so that we can find their slopes.

$$2x - 7y = 7 \qquad\qquad 21x + 6y = -2$$
$$-7y = -2x + 7 \qquad\qquad 6y = -21x - 2$$
$$y = \frac{-2}{-7}x + \frac{7}{-7} \qquad\qquad y = -\frac{21}{6}x - \frac{2}{6}$$
$$y = \frac{2}{7}x - 1 \qquad\qquad y = -\frac{7}{2}x - \frac{1}{3}$$
$$m = \frac{2}{7} \qquad\qquad\qquad m = -\frac{7}{2}$$

The slopes are negative reciprocals, therefore the lines are perpendicular.

 **You Try 1**

Determine if each pair of lines is parallel, perpendicular, or neither.

a) $x + 6y = 12$
   $y = 6x - 9$

b) $-4x + y = 1$
   $2x - 3y = 6$

c) $3x - 2y = -8$
   $15x - 10y = 2$

d) $x = 3$
   $y = 2$

Writing Equations of Parallel and Perpendicular Lines

We have learned that

1) Parallel lines have the same slope.

and

2) Perpendicular lines have slopes that are negative reciprocals of each other.

We will use this information to write the equation of a line that is parallel or perpendicular to a given line.

3. Write an Equation of a Line That Is Parallel to a Given Line

Example 3

A line contains the point (4, 2) and is parallel to the line $y = \dfrac{3}{2}x + 2$. Write the equation of the line in slope-intercept form.

Solution

Let's look at a graph to help us understand what is happening in this example.

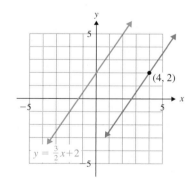

We must find the equation of the line in red. It is the line containing the point (4, 2) that is parallel to $y = \dfrac{3}{2}x + 2$.

The line $y = \dfrac{3}{2}x + 2$ has $m = \dfrac{3}{2}$. Therefore, the red line will have $m = \dfrac{3}{2}$ as well.

Now we know that the line for which we need to write an equation has $m = \dfrac{3}{2}$ and contains the point (4, 2). When we are given the slope and a point on the line, we use the point-slope formula to find its equation.

$$y - y_1 = m(x - x_1)$$

Substitute $\dfrac{3}{2}$ for m. Substitute (4, 2) for (x_1, y_1).

$$y - y_1 = m(x - x_1)$$
$$y - 2 = \frac{3}{2}(x - 4) \qquad \text{Substitute 4 for } x_1 \text{ and 2 for } y_1.$$
$$y - 2 = \frac{3}{2}x - 6 \qquad \text{Distribute.}$$
$$y = \frac{3}{2}x - 4 \qquad \text{Add 2 to each side.}$$

The equation is $y = \dfrac{3}{2}x - 4$.

You Try 2

A line contains the point (8, −5) and is parallel to the line $y = \dfrac{3}{4}x + \dfrac{2}{3}$. Write the equation of the line in slope-intercept form.

4. Write an Equation of a Line That Is Perpendicular to a Given Line

Example 4

Find the standard form for the equation of the line which contains the point (−5, 3) and which is perpendicular to $5x - 2y = 6$.

Solution

Begin by finding the slope of $5x - 2y = 6$ by putting it into slope-intercept form.

$$5x - 2y = 6$$
$$-2y = -5x + 6 \qquad \text{Add } -5x \text{ to each side.}$$
$$y = \frac{-5}{-2}x + \frac{6}{-2} \qquad \text{Divide by } -2.$$
$$y = \frac{5}{2}x - 3 \qquad \text{Simplify.}$$
$$m = \frac{5}{2}$$

Then, determine the slope of the line containing $(-5, 3)$ by finding the *negative reciprocal* of the slope of the given line.

$$m_{\text{perpendicular}} = -\frac{2}{5}$$

The line for which we need to write an equation has $m = -\frac{2}{5}$ and contains the point $(-5, 3)$. Use the point-slope formula to find an equation of the line.

$$y - y_1 = m(x - x_1)$$

Substitute $-\frac{2}{5}$ for m. Substitute $(-5, 3)$ for (x_1, y_1).

$$y - y_1 = m(x - x_1)$$
$$y - 3 = -\frac{2}{5}(x - (-5)) \qquad \text{Substitute } -5 \text{ for } x_1 \text{ and } 3 \text{ for } y_1.$$
$$y - 3 = -\frac{2}{5}(x + 5)$$
$$y - 3 = -\frac{2}{5}x - 2 \qquad \text{Distribute.}$$

Since we are asked to write the equation in standard form, eliminate the fraction by multiplying each side by 5.

$$5(y - 3) = 5\left(-\frac{2}{5}x - 2\right)$$
$$5y - 15 = -2x - 10 \qquad \text{Distribute.}$$
$$5y = -2x + 5 \qquad \text{Add 15 to each side.}$$
$$2x + 5y = 5 \qquad \text{Add } 2x \text{ to each side.}$$

The equation is $2x + 5y = 5$.

 You Try 3

Find the equation of the line perpendicular to $-3x + y = 2$ containing the point $(9, 4)$. Write the equation in standard form.

Using Technology

We can use a graphing calculator to explore what we have learned about perpendicular lines.

1. Graph the line $y = -2x + 4$. What is its slope?

2. Find the slope of the line perpendicular to the graph of $y = -2x + 4$.

3. Find the equation of the line perpendicular to $y = -2x + 4$ that passes through the point $(6, 0)$. Express the equation in slope-intercept form.

4. Graph both the original equation and the equation of the perpendicular line:

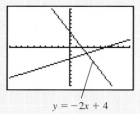

$y = -2x + 4$

5. Do the lines above appear to be perpendicular?

6. Hit ZOOM and choose 5:Zsquare:

$y = \frac{1}{2}x - 3$

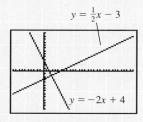

$y = -2x + 4$

7. Do the graphs look perpendicular now? Because the viewing window on a graphing calculator is a rectangle, *squaring* the window will give a more accurate picture of the graphs of the equations.

Answers to You Try Exercises

1) a) perpendicular   b) neither   c) parallel   d) perpendicular

2) $y = \frac{3}{4}x - 11$ 3) $x + 3y = 21$

Answers to Technology Exercises

1) -2 2) $\frac{1}{2}$ 3) $y = \frac{1}{2}x - 3$ 5) No because they do not meet at 90° angles.

7) Yes because they meet at 90° angles.

4.6 Exercises

Objectives 1 and 2

1) How do you know if two lines are perpendicular?

2) How do you know if two lines are parallel?

Determine if each pair of lines is parallel, perpendicular, or neither.

3) $y = -8x - 6$

$y = \frac{1}{8}x + 3$

4) $y = \frac{6}{11}x + 14$

$y = \frac{6}{11}x - 2$

 5) $y = \frac{2}{9}x + 4$

$4x - 18y = 9$

6) $y = -\frac{5}{4}x - \frac{1}{3}$

$-4x + 5y = 10$

7) $-3x + 2y = -10$

$3x - 4y = -2$

8) $x - 4y = -12$

$2x - 6y = 9$

9) $y = x$

$x + y = 7$

10) $-x + 5y = -35$

$y = 5x + 2$

11) $4y - 9x = 2$

$4x - 9y = 9$

12) $y - 2x = 3$

$x + 2y = 3$

13) $4x - 3y = 18$

$-8x + 6y = 5$

14) $x + y = 4$

$y = -x$

15) $x = 5$

$x = -2$

16) $y = 4$

$x = 3$

17) $x = 0$

$y = -3$

18) $y = -\frac{5}{2}$

$y = 0$

Lines L_1 and L_2 contain the given points. Determine if lines L_1 and L_2 are parallel, perpendicular, or neither.

19) L_1: (1, 2), (6, −13)

L_2: (−2, 5), (3, −10)

20) L_1: (3, −3), (1, 5)

L_2: (4, 3), (−12, −1)

 21) L_1: (−1, −7), (2, 8)

L_2: (10, 2), (0, 4)

22) L_1: (1, 7), (−2, −11)

L_2: (0, −8), (1, −5)

23) L_1: (5, −1), (7, 3)

L_2: (−6, 0), (4, 5)

24) L_1: (−3, 9), (4, 2)

L_2: (6, −8), (−10, 8)

25) L_1: (−3, 2), (0, 2)

L_2: (1, −1), (−2, −1)

26) L_1: (4, 1), (4, 3)

L_2: (1, 0), (1, −2)

27) L_1: (−5, 4), (−5, −1)

L_2: (−1, 2), (−3, 2)

28) L_1: (−2, −3), (1, −3)

L_2: (3, 3), (3, 5)

Objective 3

Write an equation of the line *parallel* to the given line and containing the given point. Write the answer in slope-intercept form or in standard form, as indicated.

 29) $y = 4x + 9$; (0, 2); slope-intercept form

30) $y = -3x - 1$; (0, 5); slope-intercept form

31) $y = \frac{1}{2}x - 5$; (4, 5); standard form

32) $y = 2x + 1$; (−2, −7); standard form

33) $4x + 3y = -6$; (−9, 4); standard form

34) $x + 4y = 32$; (−8, 5); standard form

35) $x + 5y = 10$; (15, 7); slope-intercept form

36) $18x - 3y = 9$; (2, −2); slope-intercept form

Objective 4

Write an equation of the line *perpendicular* to the given line and containing the given point. Write the answer in slope-intercept form or in standard form, as indicated.

37) $y = \frac{2}{3}x + 4$; (6, −3); slope-intercept form

38) $y = -\frac{4}{3}x + 2$; (8, 1); slope-intercept form

39) $y = -5x + 1$; (10, 0); standard form

40) $y = \frac{1}{4}x - 7$; (−2, 7); standard form

41) $x + y = 9$; (−5, −5); slope-intercept form

42) $y = x$; (10, −4); slope-intercept form

43) $24x - 15y = 10$; (16, −7); standard form

44) $2x + 5y = 11$; (4, 2); standard form

Objectives 3 and 4

Write the slope-intercept form (if possible) of the equation of the line meeting the given conditions.

45) perpendicular to $2x - 6y = -3$ containing (2, 2)

46) parallel to $6x + y = 4$ containing (−2, 0)

47) parallel to $y = 2x + 1$ containing (1, −3)

48) perpendicular to $y = -x - 8$ containing (4, 11)

49) parallel to $x = -4$ containing (−1, −5)

50) parallel to $y = 2$ containing (4, −3)

51) perpendicular to $y = 3$ containing (2, 1)

52) perpendicular to $x = 0$ containing $(5, 1)$

53) perpendicular to $21x - 6y = 2$ containing $(4, -1)$

54) parallel to $-3x + 4y = 8$ containing $(7, 4)$

55) parallel to $y = 0$ containing $\left(-3, -\dfrac{5}{2}\right)$

56) perpendicular to $y = \dfrac{3}{4}$ containing $(-2, 5)$

Section 4.7 Introduction to Functions

Objectives

1. Define and Identify Relation, Function, Domain, and Range

2. Use the Vertical Line Test to Determine Whether a Graph Represents a Function

3. Find the Domain and Range of a Relation from Its Graph

4. Given an Equation, Determine Whether y Is a Function of x

5. Given an Equation, Find the Domain of the Relation and Determine if y Is a Function of x

If you are driving on a highway at a constant speed of 60 miles per hour, the distance you travel depends on the amount of time spent driving.

Driving Time	Distance Traveled
1 hr	60 mi
2 hr	120 mi
2.5 hr	150 mi
3 hr	180 mi

We can express these relationships with the ordered pairs

$$(1, 60) \qquad (2, 120) \qquad (2.5, 150) \qquad (3, 180)$$

where the first coordinate represents the driving time (in hours), and the second coordinate represents the distance traveled (in miles).

We can also describe this relationship with the equation

$$y = 60x$$

where y is the distance traveled, in miles, and x is the number of hours spent driving.

The distance traveled *depends on* the amount of time spent driving. Therefore, the distance traveled is the **dependent variable**, and the driving time is the **independent variable**.

In terms of x and y, since the value of y *depends on* the value of x, y is the *dependent variable* and x is the *independent variable*.

1. Define and Identify Relation, Function, Domain, and Range

Relations and Functions

If we form a set of ordered pairs from the ones listed above

$$\{(1, 60), (2, 120), (2.5, 150), (3, 180)\}$$

we get a *relation*.

Definition   A **relation** is any set of ordered pairs.

> **Definition**
>
> The **domain** of a relation is the set of all values of the independent variable (the first coordinates in the set of ordered pairs). The **range** of a relation is the set of all values of the dependent variable (the second coordinates in the set of ordered pairs).

The domain of the last relation is $\{1, 2, 2.5, 3\}$.
The range of the relation is $\{60, 120, 150, 180\}$.

The relation $\{(1, 60), (2, 120), (2.5, 150), (3, 180)\}$ is also a *function* because every first coordinate corresponds to *exactly one* second coordinate. A function is a very important concept in mathematics.

> **Definition**
>
> A **function** is a special type of relation. If each element of the domain corresponds to *exactly one* element of the range, then the relation is a function.

Relations and functions can be represented in another way—as a *correspondence* or a *mapping* from one set to another.

In this representation, the domain is the set of all values in the first set, and the range is the set of all values in the second set. Our previously stated definition of a function still holds.

Example 1

Identify the domain and range of each relation, and determine whether each relation is a function.

a) $\{(2, 0), (3, 1), (6, 2), (6, -2)\}$

b) $\left\{(-2, -6), (0, -5), \left(1, -\dfrac{9}{2}\right), (4, -3), \left(5, -\dfrac{5}{2}\right)\right\}$

c)

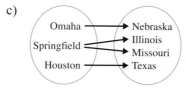

Solution

a)   The *domain* is the set of first coordinates, $\{2, 3, 6\}$. (We write the 6 in the set only once even though it appears in two ordered pairs.)

The *range* is the set of second coordinates, $\{0, 1, 2, -2\}$.

To determine whether or not $\{(2, 0), (3, 1), (6, 2), (6, -2)\}$ is a function ask yourself, *"Does every first coordinate correspond to exactly one second coordinate?"* No. In the ordered pairs $(6, 2)$ and $(6, -2)$, the same first coordinate, 6, corresponds to two different second coordinates, 2 and -2. Therefore, this relation is *not* a function.

b) $\left\{(-2, -6), (0, -5), \left(1, -\dfrac{9}{2}\right), (4, -3), \left(5, -\dfrac{5}{2}\right)\right\}$. The *domain* is $\{-2, 0, 1, 4, 5\}$.

The *range* is $\left\{-6, -5, -\dfrac{9}{2}, -3, -\dfrac{5}{2}\right\}$.

Ask yourself, "Does every first coordinate correspond to *exactly one* second coordinate?" *Yes.* This relation *is* a function.

c)

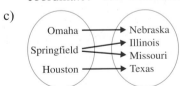

The *domain* is {Omaha, Springfield, Houston}.
The *range* is {Nebraska, Illinois, Missouri, Texas}.

One of the elements in the domain, Springfield, corresponds to *two* elements in the range, Illinois and Missouri. Therefore, this relation is *not* a function.

 You Try 1

Identify the domain and range of each relation, and determine whether each relation is a function.

a) $\{(-1, -3), (1, 1), (2, 3), (4, 7)\}$ b) $\{(-12, -6), (-12, 6), (-1, \sqrt{3}), (0, 0)\}$

c)

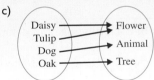

2. Use the Vertical Line Test to Determine Whether a Graph Represents a Function

We stated earlier that a relation is a function if each element of the domain corresponds to *exactly one* element of the range.

If the ordered pairs of a relation are such that the first coordinates represent x-values and the second coordinates represent y-values (the ordered pairs are in the form (x, y)), then we can think of the definition of a function in this way:

Definition   A relation is a **function** if each x-value corresponds to exactly one y-value.

What does a function look like when it is graphed?

Following are the graphs of the ordered pairs in the relations of Example 1a) and 1b).

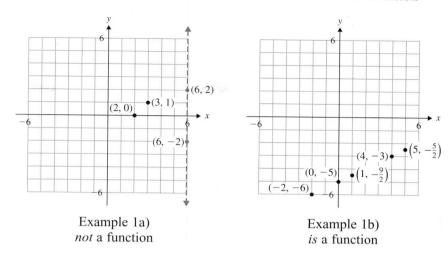

Example 1a)
not a function

Example 1b)
is a function

The relation in Example 1a) is *not* a function since the *x*-value of 6 corresponds to *two different y*-values, 2 and −2. If we draw a vertical line through the points on the graph of this relation, there is a vertical line that intersects the graph in more than one point—the line through (6, 2) and (6, −2).

The relation in Example 1b), however, *is* a function—each *x*-value corresponds to only one *y*-value. Anywhere we draw a vertical line through the points on the graph of this relation, the line intersects the graph in *exactly one point*.

This leads us to the **vertical line test** for a function.

The Vertical Line Test

If there is no vertical line that can be drawn through a graph so that it intersects the graph more than once, then the graph represents a function.

If a vertical line *can* be drawn through a graph so that it intersects the graph more than once, then the graph does *not* represent a function.

Example 2

Use the vertical line test to determine whether each graph, in blue, represents a function.

a)

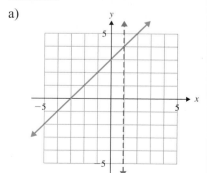

b)

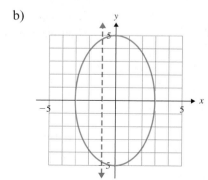

Solution

a)   Anywhere a vertical line is drawn through the graph, the line will intersect the graph only once. *This graph represents a function.*

b)   This graph fails the vertical line test because we can draw a vertical line through the graph that intersects it more than once. *This graph does* not *represent a function.*

We can identify the domain and range of a relation or function from its graph.

3. Find the Domain and Range of a Relation from Its Graph

Example 3

Identify the domain and range of each relation in Example 2.

Solution

a)

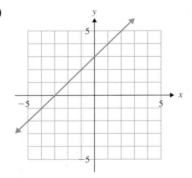

The arrows on the graph indicate that the graph continues without bound.

The domain of this relation (it is also a function) is the set of *x*-values on the graph. Since the graph continues indefinitely in the *x*-direction, the domain is the set of all real numbers. *The domain is* $(-\infty, \infty)$.

The range of this function is the set of *y*-values on the graph. Since the graph continues indefinitely in the *y*-direction, the range is the set of all real numbers. *The range is* $(-\infty, \infty)$.

b)

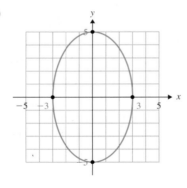

The set of *x*-values on the graph includes all real numbers from -3 to 3. *The domain is* $[-3, 3]$.

The set of *y*-values on the graph includes all real numbers from -5 to 5. *The range is* $[-5, 5]$.

You Try 2

Use the vertical line test to determine whether each relation is also a function. Then, identify the domain (D) and range (R).

a)

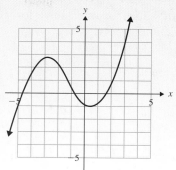

b)

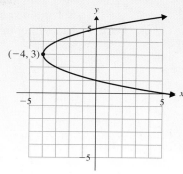

4. Given an Equation, Determine Whether *y* Is a Function of *x*

We can also represent relations and functions with equations. The example given at the beginning of the section illustrates this.

$y = 60x$ describes the distance traveled (y, in miles) after x hours of driving at 60 mph.

If $x = 2$, $y = 60(2) = 120$. If $x = 3$, $y = 60(3) = 180$, and so on. For *every* value of x that could be substituted into $y = 60x$ there is *exactly one* corresponding value of y. Therefore, $y = 60x$ is a function.

Furthermore, we can say that *y is a function of x*. In the function described by $y = 60x$, the value of y *depends on* the value of x. That is, x is the independent variable and y is the dependent variable.

> If a function describes the relationship between x and y so that x is the independent variable and y is the dependent variable, then we say that *y is a function of x*.

Example 4

Determine whether each relation describes y as a function of x.

a) $y = x + 2$ b) $y^2 = x$

Solution

a) To begin, substitute a couple of values for x and solve for y to get an idea of what is happening in this relation.

$x = 0$	$x = 3$	$x = -4$
$y = x + 2$	$y = x + 2$	$y = x + 2$
$y = 0 + 2$	$y = 3 + 2$	$y = -4 + 2$
$y = 2$	$y = 5$	$y = -2$

The ordered pairs $(0, 2)$, $(3, 5)$ and $(-4, -2)$ satisfy $y = x + 2$. Each of the values substituted for x has *one* corresponding y-value. Ask yourself, "For *any* value that could be substituted for x, how *many* corresponding values of y would there be?" In this case, when a value is substituted for x there will be *exactly* one corresponding value of y. Therefore, $y = x + 2$ *is* a function.

b) Substitute a couple of values for x and solve for y to get an idea of what is happening in this relation.

$x = 0$	$x = 4$	$x = 9$
$y^2 = x$	$y^2 = x$	$y^2 = x$
$y^2 = 0$	$y^2 = 4$	$y^2 = 9$
$y = 0$	$y = \pm 2$	$y = \pm 3$

The ordered pairs $(0, 0)$, $(4, 2)$, $(4, -2)$, $(9, 3)$, and $(9, -3)$ satisfy $y^2 = x$. Since $2^2 = 4$ and $(-2)^2 = 4$, $x = 4$ corresponds to two different y-values, 2 and -2. Likewise, $x = 9$ corresponds to the two different y-values of 3 and -3 since $3^2 = 9$ and $(-3)^2 = 9$. Finding one such example is enough to determine that $y^2 = x$ is *not* a function.

You Try 3

Determine whether each relation describes y as a function of x.

a) $y = 3x - 5$ b) $y^2 = x + 1$

5. Given an Equation, Find the Domain of the Relation and Determine if y Is a Function of x

We have seen how to determine the domain of a relation written as a set of ordered pairs, as a correspondence (or mapping), and as a graph. Next, we will discuss how to determine the domain of a relation written as an equation.

Sometimes, it is helpful to ask yourself, "Is there any number that *cannot* be substituted for x?"

Example 5

Determine the domain of each relation, and determine whether each relation describes y as a function of x.

a) $y = \dfrac{1}{x}$ b) $y = \dfrac{7}{x - 3}$ c) $y = -2x + 6$

Solution

a) To determine the domain of $y = \dfrac{1}{x}$ ask yourself, "Is there any number that *cannot* be substituted for x?" Yes. x *cannot equal zero because a fraction is undefined if its denominator equals zero.* Any other number can be substituted for x and $y = \dfrac{1}{x}$ will be defined.

The domain contains all real numbers *except* 0. We can write the domain in interval notation as $(-\infty, 0) \cup (0, \infty)$.

$y = \dfrac{1}{x}$ *is a function* since each value of x in the domain will have only one corresponding value of y.

b)   Ask yourself, "Is there any number that *cannot* be substituted for x in $y = \dfrac{7}{x - 3}$?" Look at the denominator. When will it equal 0? Set the denominator equal to 0 and solve for x to determine what value of x will make the denominator equal 0.

$$x - 3 = 0 \qquad \text{Set the denominator} = 0.$$
$$x = 3 \qquad \text{Solve.}$$

When $x = 3$, the denominator of $y = \dfrac{7}{x - 3}$ equals zero. The domain contains all real numbers *except* 3. Write the domain in interval notation as $(-\infty, 3) \cup (3, \infty)$. $y = \dfrac{7}{x - 3}$ *is a function*. For every value that can be substituted for x there is only one corresponding value of y.

c)   *Is there any number that cannot be substituted for x in* $y = -2x + 6$? *No.* Any real number can be substituted for x, and $y = -2x + 6$ will be defined.

The domain consists of all real numbers which can be written as $(-\infty, \infty)$. Every value substituted for x will have exactly one corresponding y-value. $y = -2x + 6$ *is a function.*

The domain of a relation that is written as an equation, where y is in terms of x is the set of all real numbers that can be substituted for the independent variable, x. When determining the domain of a relation, it can be helpful to keep these tips in mind.

1)   Ask yourself, "Is there any number that *cannot* be substituted for x?"

2)   If x is in the denominator of a fraction, determine what value of x will make the denominator equal 0 by setting the expression equal to zero. Solve for x. This x-value is *not* in the domain.

You Try 4

Determine the domain of each relation, and determine whether each relation describes y as a function of x.

a) $y = x - 9$ b) $y = -x^2 + 6$ c) $y = \dfrac{4}{x + 1}$

Answers to You Try Exercises

1) a) domain: $\{-1, 1, 2, 4\}$; range: $\{-3, 1, 3, 7\}$; yes   b) domain: $\{-12, -1, 0\}$; range: $\{-6, 6, \sqrt{3}, 0\}$; no
c) domain: {Daisy, Tulip, Dog, Oak}; range: {Flower, Animal, Tree}; yes   2) a) function; D: $(-\infty, \infty)$;
R: $(-\infty, \infty)$ b) not a function; D: $[-4, \infty)$; R: $(-\infty, \infty)$ 3) a) yes   b) no   4) a) $(-\infty, \infty)$;
function   b) $(-\infty, \infty)$; function   c) $(-\infty, -1) \cup (-1, \infty)$; function

4.7 Exercises

Boost your grade at mathzone.com!   MathZone

> Practice Problems   > NetTutor   > e-Professors   > Videos
> Self-Test

Objectives 1–3

1) a) What is a relation?

   b) What is a function?

   c) Give an example of a relation that is also a function.

2) Give an example of a relation that is *not* a function.

Identify the domain and range of each relation, and determine whether each relation is a function.

3) $\{(5, 13), (-2, 6), (1, 4), (-8, -3)\}$

4) $\{(0, -3), (1, -4), (1, -2), (16, -5), (16, -1)\}$

 5) $\{(9, -1), (25, -3), (1, 1), (9, 5), (25, 7)\}$

6) $\{(-6, 9), (-2, 1), (0, 0), (4, -11), (5, -13)\}$

7) $\{(-2, 9), (1, 6), (2, 9), (5, 30)\}$

8) $\left\{(-4, -2), \left(-3, -\frac{1}{2}\right), \left(-1, -\frac{1}{2}\right), (0, -2)\right\}$

9)
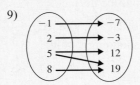

10)
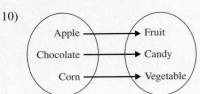

11)
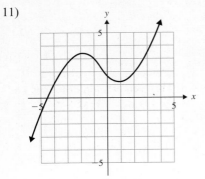

12)
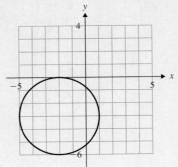

13)
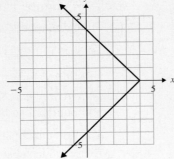

14)
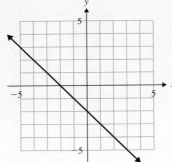

15)
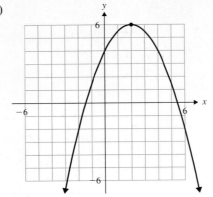

16)

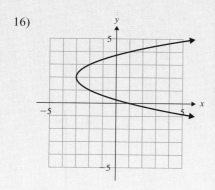

Objective 4

Determine whether each relation describes y as a function of x.

17) $y = x - 9$

18) $y = x + 4$

19) $y = 2x + 7$

20) $y = \dfrac{2}{3}x + 1$

 21) $x = y^4$

22) $x = y^2 - 3$

23) $y^2 = x - 4$

24) $y^2 = x + 9$

25) $y = \dfrac{10}{x + 2}$

26) $y = \dfrac{3}{2x + 5}$

Objective 5

Determine the domain of each relation, and determine whether each relation describes y as a function of x.

27) $y = x - 5$

28) $y = 2x + 1$

29) $y = x^3 + 2$

30) $y = -x^3 + 4$

31) $x = y^4$

32) $x = |y|$

33) $y = -\dfrac{8}{x}$

34) $y = \dfrac{5}{x}$

 35) $y = \dfrac{9}{x + 4}$

36) $y = \dfrac{2}{x - 7}$

37) $y = \dfrac{3}{x - 5}$

38) $y = \dfrac{1}{x + 10}$

39) $y = \dfrac{1}{x + 8}$

40) $y = \dfrac{12}{x - 11}$

41) $y = \dfrac{6}{5x - 3}$

42) $y = -\dfrac{4}{9x + 8}$

43) $y = \dfrac{15}{3x + 4}$

44) $y = \dfrac{5}{6x - 1}$

45) $y = \dfrac{1}{4x}$

46) $y = \dfrac{7}{6x}$

47) $y = \dfrac{8}{2x + 1}$

48) $y = -\dfrac{3}{8x - 5}$

49) $y = -\dfrac{5}{9 - 3x}$

50) $y = \dfrac{1}{-6 + 4x}$

Section 4.8 Function Notation and Linear Functions

Objectives

1. Understand Function Notation
2. Find Function Values for Real Number Values of the Variable
3. Evaluate a Function for a Given Variable or Expression
4. Define and Graph a Linear Function
5. Use a Linear Function to Solve an Applied Problem

1. Understand Function Notation

In Section 4.7 we said that if a function describes the relationship between x and y so that x is the independent variable and y is the dependent variable, then y *is a function of x.* That is, *the value of y depends on the value of x.*

A special notation is used to represent this relationship. It is called *function notation.*

> **Definition**
>
> $y = f(x)$ is called **function notation**, and it is read as, "y equals f of x."
> $y = f(x)$ means that y is a function of x (y depends on x).

If a relation is a function, then $f(x)$ can be used in place of y. *$f(x)$ is the same as y.*

For example, $y = x + 3$ is a function. We can also write $y = x + 3$ as $f(x) = x + 3$. *They mean the same thing.*

Example 1

a)   Evaluate $y = x + 3$ for $x = 2$.

b)   If $f(x) = x + 3$, find $f(2)$.

Solution

a)   To evaluate $y = x + 3$ for $x = 2$ means to substitute 2 for x and find the corresponding value of y.

$$y = x + 3$$
$$y = 2 + 3 \qquad \text{Substitute 2 for } x.$$
$$y = 5$$

When $x = 2$, $y = 5$. We can also say that the ordered pair $(2, 5)$ satisfies $y = x + 3$.

b)   To find $f(2)$ (read as "f of 2") means to find the value of the function when $x = 2$.

$$f(x) = x + 3$$
$$f(2) = 2 + 3 \qquad \text{Substitute 2 for } x.$$
$$f(2) = 5$$

We can also say that the ordered pair $(2, 5)$ satisfies $f(x) = x + 3$ where the ordered pair represents $(x, f(x))$.

Example 1 illustrates that evaluating $y = x + 3$ for $x = 2$ and finding $f(2)$ when $f(x) = x + 3$ is *exactly* the same thing. Remember, $f(x)$ is another name for y.

 You Try 1

a) Evaluate $y = -2x + 4$ for $x = 1$.   b) If $f(x) = -2x + 4$, find $f(1)$.

Different letters can be used to name functions. $g(x)$ is read as "g of x," $h(x)$ is read as "h of x," and so on. Also, the function notation does *not* indicate multiplication; $f(x)$ does *not* mean f times x.

 BE CAREFUL $f(x)$ does *not* mean f times x.

2. Find Function Values for Real Number Values of the Variable

Sometimes, we call evaluating a function for a certain value *finding a function value.*

Example 2

Let $f(x) = 6x - 5$ and $g(x) = x^2 - 8x + 3$. Find the following function values.

a) $f(3)$ b) $f(0)$ c) $g(-1)$

Solution

a)   "Find $f(3)$" means to find the value of the function when $x = 3$. Substitute 3 for x.

$$f(x) = 6x - 5$$
$$f(3) = 6(3) - 5 = 18 - 5 = 13$$
$$f(3) = 13$$

We can also say that the ordered pair $(3, 13)$ satisfies $f(x) = 6x - 5$.

b)   To find $f(0)$, substitute 0 for x in the function $f(x)$.

$$f(x) = 6x - 5$$
$$f(0) = 6(0) - 5 = 0 - 5 = -5$$
$$f(0) = -5$$

The ordered pair $(0, -5)$ satisfies $f(x) = 6x - 5$.

c)   To find $g(-1)$, substitute -1 for every x in the function $g(x)$.

$$g(x) = x^2 - 8x + 3$$
$$g(-1) = (-1)^2 - 8(-1) + 3 = 1 + 8 + 3 = 12$$
$$g(-1) = 12$$

The ordered pair $(-1, 12)$ satisfies $g(x) = x^2 - 8x + 3$.

 You Try 2

Let $f(x) = -4x + 1$ and $h(x) = 2x^2 + 3x - 7$. Find the following function values.

a) $f(5)$ b) $f(-2)$ c) $h(0)$ d) $h(3)$

3. Evaluate a Function for a Given Variable or Expression

Functions can be evaluated for variables or expressions.

Example 3

Let $h(x) = 5x + 3$. Find each of the following and simplify.

a) $h(c)$ b) $h(t - 4)$

Solution

a) Finding $h(c)$ (read as *h of c*) means to substitute c for x in the function h, and simplify the expression as much as possible.

$$h(x) = 5x + 3$$
$$h(c) = 5c + 3 \qquad \text{Substitute } c \text{ for } x.$$

b) Finding $h(t - 4)$ (read as *h of t minus 4*) means to substitute $t - 4$ for x in function h, and simplify the expression as much as possible. *Since $t - 4$ contains two terms, we must put it in parentheses.*

$$h(x) = 5x + 3$$
$$h(t - 4) = 5(t - 4) + 3 \qquad \text{Substitute } t - 4 \text{ for } x.$$
$$h(t - 4) = 5t - 20 + 3 \qquad \text{Distribute.}$$
$$h(t - 4) = 5t - 17 \qquad \text{Combine like terms.}$$

You Try 3

Let $f(x) = 2x - 7$. Find each of the following and simplify.

a) $f(k)$ b) $f(p + 3)$

4. Define and Graph a Linear Function

Earlier in this chapter, we learned that the graph of a linear equation in two variables, $Ax + By = C$, is a line. A linear equation is a function except when the line is vertical and has equation $x = c$. An alternative form for the equation of a line is *slope-intercept form*, $y = mx + b$. In this form, *y is expressed in terms of x*. We can replace y with $f(x)$ to rewrite the equation as a linear function.

Definition

A **linear function** has the form $f(x) = mx + b$, where m and b are real numbers, m is the *slope* of the line, and $(0, b)$ is the *y-intercept*.

Although m and b tell us the slope and y-intercept of the line, we can graph the function using any of the methods we discussed previously: making a table of values and plotting points, finding the x- and y-intercepts, or by using the slope and y-intercept.

Example 4

Graph $f(x) = -\dfrac{1}{3}x - 1$ using the slope and y-intercept.

Solution

$$f(x) = -\frac{1}{3}x - 1$$

$$m = -\frac{1}{3} \qquad y\text{-int: } (0, -1)$$

To graph this function, first plot the y-intercept, $(0, -1)$, then use the slope to locate another point on the line.

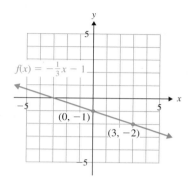

You Try 4

Graph $f(x) = \dfrac{3}{4}x - 2$ using the slope and y-intercept.

5. Use a Linear Function to Solve an Applied Problem

The independent variable of a function does not have to be x. When using functions to model real-life problems, we often choose a more "meaningful" letter to represent a quantity. For example, if the independent variable represents time, we may use the letter t instead of x. The same is true for naming the function.

No matter what letter is chosen for the independent variable, *the horizontal axis is used to represent the values of the independent variable, and the vertical axis represents the function values.*

Example 5

A compact disc is read at 44.1 kHz (kilohertz). This means that a CD player scans 44,100 samples of sound per second on a CD to produce the sound that we hear. The function

$$S(t) = 44.1t$$

tells us how many samples of sound, $S(t)$, in *thousands* of samples, are read after t seconds. (Source: www.mediatechnics.com)

a)   How many samples of sound are read after 20 sec?

b)   How many samples of sound are read after 1.5 min?

c)   How long would it take the CD player to scan 1,764,000 samples of sound?

d)   What is the smallest value t could equal in the context of this problem?

e)   Graph the function.

Solution

a)   To determine how much sound is read after 20 sec, let $t = 20$ and find $S(20)$.

$$S(t) = 44.1t$$
$$S(20) = 44.1(20) \qquad \text{Substitute 20 for } t.$$
$$S(20) = 882 \qquad\qquad \text{Multiply.}$$

$S(t)$ is in thousands, so the number of samples read is $882 \cdot 1000 = 882{,}000$ samples of sound.

b)   To determine how much sound is read after 1.5 min, do we let $t = 1.5$ and find $S(1.5)$? *No.* Recall that t is in *seconds*. Change 1.5 min to seconds before substituting for t. We must use the correct units in the function.

$$1.5 \text{ min} = 90 \text{ sec}$$

Let $t = 90$ and find $S(90)$.

$$S(t) = 44.1t$$
$$S(90) = 44.1(90)$$
$$S(90) = 3969$$

$S(t)$ is in thousands, so the number of samples read is $3969 \cdot 1000 = 3{,}969{,}000$ samples of sound.

c)   Since we are asked to determine how *long* it would take a CD player to scan 1,764,000 samples of sound, we will be solving for t. What do we substitute for $S(t)$? $S(t)$ is in *thousands*, so substitute $1{,}764{,}000 \div 1000 = 1764$ for $S(t)$. Find t when $S(t) = 1764$.

$$S(t) = 44.1t$$
$$1764 = 44.1t \qquad \text{Substitute 1764 for } S(t).$$
$$40 = t \qquad\qquad \text{Divide by 44.1.}$$

It will take 40 sec for the CD player to scan 1,764,000 samples of sound.

d)   Since t represents the number of seconds a CD has been playing, the smallest value that makes sense for t is 0.

e)   Since $S(t)$ is in thousands of samples, the information we obtained in parts a), b), and c) can be written as the ordered pairs (20, 882), (90, 3969), and (40, 1764). In addition, when $t = 0$ (from part d) we obtain $S(0) = 44.1(0) = 0$. (0, 0) is an additional ordered pair on the graph of the function.

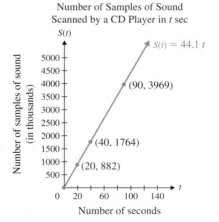

Number of Samples of Sound
Scanned by a CD Player in t sec

Answers to You Try Exercises

1) a) 2   b) 2   2) a) -19 b) 9   c) -7 d) 20   3) a) $f(k) = 2k - 7$ b) $f(p + 3) = 2p - 1$

4) $m = \dfrac{3}{4}$, y-int: $(0, -2)$

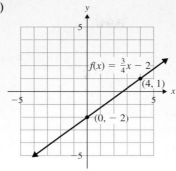

$f(x) = \dfrac{3}{4}x - 2$

$(4, 1)$

$(0, -2)$

4.8 Exercises

Boost your grade at mathzone.com!   MathZone

> Practice Problems
> NetTutor
> Self-Test
> e-Professors
> Videos

Objectives 1 and 2

1) Explain what it means when an equation is written in the form $y = f(x)$.

2) Does $y = f(x)$ mean "$y = f$ times x"? Explain.

3) a) Evaluate $y = 5x - 8$ for $x = 3$.

b) If $f(x) = 5x - 8$, find $f(3)$.

4) a) Evaluate $y = -3x - 2$ for $x = -4$.

b) If $f(x) = -3x - 2$, find $f(-4)$.

Let $f(x) = -4x + 7$ and $g(x) = x^2 + 9x - 2$. Find the following function values.

5) $f(5)$

6) $f(2)$

7) $f(0)$

8) $f\left(-\dfrac{3}{2}\right)$

9) $g(4)$

10) $g(1)$

11) $g(-1)$

12) $g(0)$

13) $g\left(-\dfrac{1}{2}\right)$

14) $g\left(\dfrac{1}{3}\right)$

For each function f in Problems 15–18, find $f(-1)$ and $f(4)$.

15) $f = \{(-3, 16), (-1, 10), (0, 7), (1, 4), (4, -5)\}$

16) $f = \left\{(-8, -1), \left(-1, \dfrac{5}{2}\right), (4, 5), (10, 8)\right\}$

video 17)

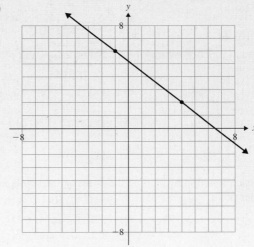

18)

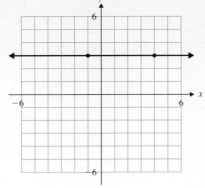

19) $f(x) = -3x - 2$. Find x so that $f(x) = 10$.

20) $f(x) = 5x + 4$. Find x so that $f(x) = 9$.

21) $g(x) = \dfrac{2}{3}x + 1$. Find x so that $g(x) = 5$.

22) $h(x) = -\dfrac{1}{2}x - 6$. Find x so that $h(x) = -2$.

Objective 3

23) $f(x) = -7x + 2$ and $g(x) = x^2 - 5x + 12$. Find each of the following and simplify.

   a) $f(c)$ b) $f(t)$

   c) $f(a + 4)$ d) $f(z - 9)$

   e) $g(k)$ f) $g(m)$

24) $f(x) = 5x + 6$ and $g(x) = x^2 - 3x - 11$. Find each of the following and simplify.

   a) $f(n)$ b) $f(p)$

   c) $f(w + 8)$ d) $f(r - 7)$

   e) $g(b)$ f) $g(s)$

Objective 4

Graph each function by making a table of values and plotting points.

25) $f(x) = x - 4$ 26) $f(x) = x + 2$

27) $f(x) = \dfrac{2}{3}x + 2$ 28) $g(x) = -\dfrac{3}{5}x + 2$

29) $h(x) = -3$ 30) $g(x) = 1$

Graph each function by finding the x- and y-intercepts and one other point.

31) $g(x) = 3x + 3$ 32) $k(x) = -2x + 6$

33) $f(x) = -\dfrac{1}{2}x + 2$ 34) $f(x) = \dfrac{1}{3}x + 1$

35) $h(x) = x$ 36) $f(x) = -x$

Graph each function using the slope and y-intercept.

37) $f(x) = -4x - 1$ 38) $f(x) = -x + 5$

39) $h(x) = \dfrac{3}{5}x - 2$ 40) $g(x) = -\dfrac{1}{4}x - 2$

41) $g(x) = 2x + \dfrac{1}{2}$ 42) $h(x) = 3x + 1$

43) $h(x) = -\dfrac{5}{2}x + 4$ 44) $g(x) = \dfrac{1}{2}x - \dfrac{3}{2}$

Graph each function

45) $s(t) = -\dfrac{1}{3}t - 2$ 46) $k(d) = d - 1$

47) $g(c) = -c + 4$ 48) $A(r) = -3r$

49) $C(h) = 1.5h$ 50) $N(t) = 3.5t + 1$

Objective 5

51) A truck on the highway travels at a constant speed of 54 mph. The distance, D (in miles), that the truck travels after t hr can be defined by the function

$$D(t) = 54t$$

   a) How far will the truck travel after 2 hr?

   b) How far will the truck travel after 4 hr?

   c) How long does it take the truck to travel 135 mi?

   d) Graph the function.

52) The velocity of an object, v (in feet per second), of an object during free-fall t sec after being dropped can be defined by the function

$$v(t) = 32t$$

   a) Find the velocity of an object 1 sec after being dropped.

   b) Find the velocity of an object 3 sec after being dropped.

   c) When will the object be traveling at 256 ft/sec?

   d) Graph the function.

53) Jenelle earns \$7.50 per hour at her part-time job. Her total earnings, E (in dollars), for working t hr can be defined by the function

$$E(t) = 7.50t$$

   a) Find $E(10)$, and explain what this means in the context of the problem.

   b) Find $E(15)$, and explain what this means in the context of the problem.

   c) Find t so that $E(t) = 210$, and explain what this means in the context of the problem.

54) If gasoline costs \$2.50 per gallon, then the cost, C (in dollars), of filling a gas tank with g gal of gas is defined by

$$C(g) = 2.50g$$

   a) Find $C(8)$, and explain what this means in the context of the problem.

b) Find $C(15)$, and explain what this means in the context of the problem.

c) Find g so that $C(g) = 30$, and explain what this means in the context of the problem.

55) A 16× DVD recorder can transfer 21.13 MB (megabytes) of data per second onto a recordable DVD. The function $D(t) = 21.13t$ describes how much data, D (in megabytes), is recorded on a DVD in t sec. (Source: www.osta.org)

a) How much data is recorded after 12 sec?

b) How much data is recorded after 1 min?

c) How long would it take to record 422.6 MB of data?

d) Graph the function.

56) The median hourly wage of an embalmer in Illinois in 2002 was $17.82. Seth's earnings, E (in dollars), for working t hr in a week can be defined by the function $E(t) = 17.82t$. (Source: www.igpa.uillinois.edu)

a) How much does Seth earn if he works 30 hr?

b) How much does Seth earn if he works 27 hr?

c) How many hours would Seth have to work to make $623.70?

d) If Seth can work at most 40 hr per week, what is the domain of this function?

e) Graph the function.

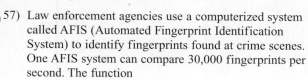

57) Law enforcement agencies use a computerized system called AFIS (Automated Fingerprint Identification System) to identify fingerprints found at crime scenes. One AFIS system can compare 30,000 fingerprints per second. The function

$$F(s) = 30s$$

describes how many fingerprints, $F(s)$ in thousands, are compared after s sec.

a) How many fingerprints can be compared in 2 sec?

b) How long would it take AFIS to search through 105,000 fingerprints?

58) Refer to the function in Problem 57 to answer the following questions.

a) How many fingerprints can be compared in 3 sec?

b) How long would it take AFIS to search through 45,000 fingerprints?

59) Refer to the function in Example 5 on p. 301 to determine the following.

a) Find $S(50)$, and explain what this means in the context of the problem.

b) Find $S(180)$, and explain what this means in the context of the problem.

c) Find t so that $S(t) = 2646$, and explain what this means in the context of the problem.

60) Refer to the function in Problem 55 to determine the following.

a) Find $D(10)$, and explain what this means in the context of the problem.

b) Find $D(120)$, and explain what this means in the context of the problem.

c) Find t so that $D(t) = 633.9$, and explain what this means in the context of the problem.

Definition/Procedure	Example	Reference
Introduction to Linear Equations in Two Variables		4.1
A linear equation in two variables can be written in the form $Ax + By = C$ where A, B, and C are real numbers and where both A and B do not equal zero. To determine if an ordered pair is a solution of an equation, substitute the values for the variables.	Is $(5, -3)$ a solution of $2x - 7y = 31$? Substitute 5 for x and -3 for y. $$2x - 7y = 31$$ $$2(5) - 7(-3) = 31$$ $$10 - (-21) = 31$$ $$10 + 21 = 31$$ $$31 = 31$$ Yes, $(5, -3)$ is a solution.	**p. 224**
Graphing by Plotting Points and Finding Intercepts		4.2
The graph of a linear equation in two variables, $Ax + By = C$, is a straight line. Each point on the line is a solution to the equation. We can graph the line by plotting the points and drawing the line through them.	Graph $y = \dfrac{1}{2}x - 4$ by plotting points. Make a table of values. Plot the points, and draw a line through them. | x	y
---	---
0	-4
2	-3
4	-2 | 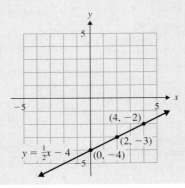	**p. 237**
The *x-intercept* of an equation is the point where the graph intersects the *x*-axis. To find the *x-intercept* of the graph of an equation, let $y = 0$ and solve for x. The *y-intercept* of an equation is the point where the graph intersects the *y*-axis. To find the *y-intercept* of the graph of an equation, let $x = 0$ and solve for y.	Graph $5x + 2y = 10$ by finding the intercepts and another point on the line. *x-intercept*: Let $y = 0$, and solve for x. $$5x + 2(0) = 10$$ $$5x = 10$$ $$x = 2$$ The *x-intercept* is $(2, 0)$.	**p. 241**

Definition/Procedure	Example	Reference

y-intercept: Let $x = 0$, and solve for y.

$$5x + 2y = 10$$
$$5(0) + 2y = 10$$
$$2y = 10$$
$$y = 5$$

The *y*-intercept is $(0, 5)$.
Another point on the line is $(4, -5)$.
Plot the points, and draw the line through them.

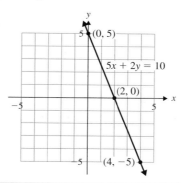

If c is a constant, then the graph of $x = c$ is a *vertical line* going through the point $(c, 0)$.

If d is a constant, then the graph of $y = d$ is a *horizontal line* going through the point $(0, d)$.

Graph $x = 2$.

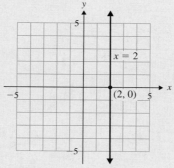

Graph $y = 3$.

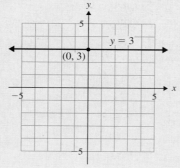

p. 243

Definition/Procedure	Example	Reference

The Slope of a Line

4.3

The *slope* of a line is the ratio of the vertical change in y to the horizontal change in x. Slope is denoted by m.

The slope of a line containing the points (x_1, y_1) and (x_2, y_2) is

$$m = \frac{y_2 - y_1}{x_2 - x_1}$$

The slope of a horizontal line is zero. The slope of a vertical line is undefined.

Find the slope of the line containing the points $(6, 9)$ and $(-2, 12)$.

$$m = \frac{y_2 - y_1}{x_2 - x_1}$$

$$= \frac{12 - 9}{-2 - 6} = \frac{3}{-8} = -\frac{3}{8}$$

The slope of the line is $-\frac{3}{8}$.

p. 251

If we know the slope of a line and a point on the line, we can graph the line.

Graph the line containing the point $(-5, -3)$ with a slope of $\frac{4}{7}$.

Start with the point $(-5, -3)$, and use the slope to plot another point on the line.

$$m = \frac{4}{7} = \frac{\text{change in } y}{\text{change in } x}$$

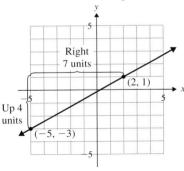

p. 257

The Slope-Intercept Form of a Line

4.4

The *slope-intercept form of a line* is $y = mx + b$ where m is the slope and $(0, b)$ is the y-intercept.

If a line is written in slope-intercept form, we can use the y-intercept and the slope to graph the line.

Write the equation in slope-intercept form and graph it.

$$6x + 4y = 16$$
$$4y = -6x + 16$$
$$y = -\frac{6}{4}x + \frac{16}{4}$$
$$y = -\frac{3}{2}x + 4 \quad \text{Slope-intercept form}$$

$$m = -\frac{3}{2}, y\text{-intercept } (0, 4)$$

p. 263

Definition/Procedure	Example	Reference

Plot $(0, 4)$, then use the slope to locate another point on the line. We will think of the slope as $m = \dfrac{-3}{2} = \dfrac{\text{change in } y}{\text{change in } x}$.

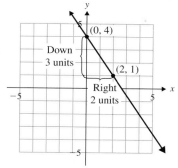

Writing an Equation of a Line

4.5

To write the equation of a line given its slope and y-intercept, *use $y = mx + b$ and substitute those values into the equation.*

Find an equation of the line with slope = 3 and y-intercept $(0, -8)$.

$y = mx + b$ Substitute 3 for
$y = 3x - 8$ m and -8 for b.

p. 270

If (x_1, y_1) is a point on a line and m is the slope of the line, then the equation of the line is given by $y - y_1 = m(x - x_1)$. This is the *point-slope formula.*

Given the slope of the line and a point on the line, we can use the point-slope formula to find an equation of the line.

Find an equation of the line containing the point $(1, -4)$ with slope = 2. Express the answer in standard form.

Use $y - y_1 = m(x - x_1)$.

Substitute 2 for m. Substitute $(1, -4)$ for (x_1, y_1).

$$y - (-4) = 2(x - 1)$$
$$y + 4 = 2x - 2$$
$$-2x + y = -6$$
$$2x - y = 6 \quad \text{Standard form}$$

p. 270

To write an equation of a line given two points on the line,

a) use the points to find the slope of the line

then

b) use the slope and *either one* of the points in the point-slope formula.

Find an equation of the line containing the points $(-2, 6)$ and $(4, 2)$. Express the answer in slope-intercept form.

$$m = \frac{2 - 6}{4 - (-2)} = \frac{-4}{6} = -\frac{2}{3}$$

We will use $m = -\dfrac{2}{3}$ and the point $(-2, 6)$ in the point-slope formula.

p. 272

Definition/Procedure	Example	Reference

$$y - y = m(x - x_1)$$

Substitute $-\dfrac{2}{3}$ for m. Substitute $(-2, 6)$ for (x_1, y_1).

$$y - 6 = -\frac{2}{3}(x - (-2)) \quad \text{Substitute.}$$

$$y - 6 = -\frac{2}{3}(x + 2)$$

$$y - 6 = -\frac{2}{3}x - \frac{4}{3} \qquad \text{Distribute.}$$

$$y = -\frac{2}{3}x + \frac{14}{3} \qquad \begin{array}{l}\text{Slope-intercept} \\ \text{form}\end{array}$$

Definition/Procedure	Example	Reference
The equation of a *horizontal line* containing the point (c, d) is $y = d$. The equation of a *vertical line* containing the point (c, d) is $x = c$.	The equation of a horizontal line containing the point $(7, -4)$ is $y = -4$. The equation of a vertical line containing the point $(9, 1)$ is $x = 9$.	**p. 273**
Parallel and Perpendicular Lines		4.6
Parallel lines have the same slope. *Perpendicular lines* have slopes that are negative reciprocals of each other.	Determine if the lines $5x + y = 3$ and $x - 5y = 20$ are parallel, perpendicular, or neither. Put each line into slope-intercept form to find their slopes. $\begin{aligned}5x + y &= 3 \\ y &= -5x + 3\end{aligned}$ $\qquad$ $\begin{aligned}x - 5y &= 20 \\ -5y &= -x + 20 \\ y &= \frac{1}{5}x - 4\end{aligned}$ $m = -5$ $\qquad\qquad$ $m = \dfrac{1}{5}$ The lines are *perpendicular* since their slopes are negative reciprocals of each other.	**p. 279**
To write an equation of the line parallel or perpendicular to a given line, we must find the slope of the given line first.	Write an equation of the line parallel to $2x - 3y = 21$ containing the point $(-6, -3)$. Express the answer in slope-intercept form. Find the slope of $2x - 3y = 21$. $\begin{aligned}2x - 3y &= 21 \\ -3y &= -2x + 21 \\ y &= \frac{2}{3}x - 7\end{aligned}$ $m = \dfrac{2}{3}$	**p. 283**

Definition/Procedure	Example	Reference

The slope of the parallel line is also $\frac{2}{3}$.

Since this line contains $(-6, -3)$, use the point-slope formula to write its equation.

$$y - y_1 = m(x - x_1)$$

$$y - (-3) = \frac{2}{3}(x - (-6)) \quad \text{Substitute values.}$$

$$y + 3 = \frac{2}{3}(x + 6)$$

$$y + 3 = \frac{2}{3}x + 4 \quad \text{Distribute.}$$

$$y = \frac{2}{3}x + 1 \quad \text{Slope-intercept form}$$

Introduction to Functions

4.7

A *relation* is any set of ordered pairs. A relation can also be represented as a correspondence or mapping from one set to another.

Relations:
a) $\{(-4, -12), (-1, -3), (3, 9), (5, 15)\}$
b)

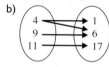

p. 288

The *domain* of a relation is the set of values of the independent variable (the first coordinates in the set of ordered pairs).

The *range* of a relation is the set of all values of the dependent variable (the second coordinates in the set of ordered pairs).

In a) above, the domain is $\{-4, -1, 3, 5\}$, and the range is $\{-12, -3, 9, 15\}$.

In b) above, the domain is $\{4, 9, 11\}$, and the range is $\{1, 6, 17\}$.

p. 289

A *function* is a relation in which each element of the domain corresponds to *exactly one* element of the range.

The relation above in a) *is* a function.

The relation above in b) *is not* a function.

p. 289

Definition/Procedure	Example	Reference

The Vertical Line Test

This graph represents a function. Anywhere a vertical line is drawn, it will intersect the graph only once.

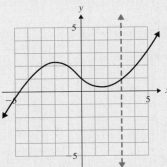

This is *not* the graph of a function. A vertical line can be drawn so that it intersects the graph more than once.

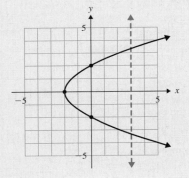

p. 290

The *domain* of a relation that is written as an equation, where *y* is in terms of *x*, is the set of all real numbers that can be substituted for the independent variable, *x*.

When determining the domain of a relation, it can be helpful to keep these tips in mind.

1) Ask yourself, "Is there any number that *cannot* be substituted for *x*?"

2) If *x* is in the denominator of a fraction, determine what value of *x* will make the denominator equal 0 by setting the denominator equal to zero. Solve for *x*. This *x*-value is *not* in the domain.

Determine the domain of $f(x) = \dfrac{9}{x + 8}$.

$x + 8 = 0$ Set the denominator $= 0$.
$\quad x = -8$ Solve.

When $x = -8$, the denominator of $f(x) = \dfrac{9}{x + 8}$ equals zero. The domain contains all real numbers *except* -8.

The domain of the function is $(-\infty, -8) \cup (-8, \infty)$.

p. 294

Definition/Procedure	Example	Reference
Function Notation and Linear Functions		4.8
If a function describes the relationship between x and y so that x is the independent variable and y is the dependent variable, then y is a function of x. $y = f(x)$ is called *function notation* and it is read as "y equals f of x." Finding a function value means evaluating the function for the given value of the variable.	If $f(x) = 9x - 4$, find $f(2)$. Substitute 2 for x and evaluate. $f(2) = 9(2) - 4 = 18 - 4 = 14$ $f(2) = 14$	**p. 297**
A *linear function* has the form $$f(x) = mx + b$$ where m and b are real numbers, m is the *slope* of the line, and $(0, b)$ is the *y-intercept*.	Graph $f(x) = -3x + 4$ using the slope and y-intercept. The slope is -3 and the y-intercept is $(0, 4)$. Plot the y-intercept and use the slope to locate another point on the line. 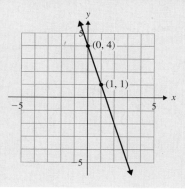	**p. 300**

(4.1) Determine if each ordered pair is a solution of the given equation.

1) $4x - y = 9$; $(1, -5)$

2) $3x + 2y = 20$; $(-4, 2)$

3) $y = \dfrac{5}{4}x + \dfrac{1}{2}$; $(2, 3)$

4) $x = 7$; $(7, -9)$

Complete the ordered pair for each equation.

5) $y = -6x + 10$; $(-3, \ )$

6) $y = \dfrac{2}{3}x + 5$; $(12, \ )$

7) $y = -8$; $(5, \ )$

8) $5x - 9y = 3$; $(\ , -2)$

Complete the table of values for each equation.

9) $y = x - 11$

x	y
0	
3	
-1	
-5	

10) $4x - 6y = 8$

x	y
	0
0	
3	
	-4

Plot the ordered pairs on the same coordinate system.

11) a) $(5, 2)$ b) $(-3, 0)$

c) $(-4, 3)$ d) $(6, -2)$

12) a) $(0, 1)$ b) $(-2, -5)$

c) $\left(\dfrac{5}{2}, 1\right)$ d) $\left(4, -\dfrac{1}{3}\right)$

13) The fine for an overdue book at the Hinsdale Public Library is given by

$$y = 0.10x$$

where x represents the number of days a book is overdue and y represents the amount of the fine, in dollars.

a) Complete the table of values, and write the information as ordered pairs.

x	y
1	
2	
7	
10	

b) Label a coordinate system, choose an appropriate scale, and graph the ordered pairs.

c) Explain the meaning of the ordered pair $(14, 1.40)$ in the context of the problem.

14) Fill in the blank with positive, negative, or zero.

a) The y-coordinate of every point in quadrant III is _____.

b) The x-coordinate of every point in quadrant IV is _____.

(4.2) Complete the table of values and graph each equation.

15) $y = -2x + 3$

x	y
0	
1	
2	
-2	

16) $3x + 2y = 4$

x	y
0	
	-2
	1
	4

Graph each equation by finding the intercepts and at least one other point.

17) $x - 2y = 6$

18) $5x + y = 10$

19) $y = -\dfrac{1}{6}x + 4$

20) $y = \dfrac{3}{4}x - 7$

21) $x = 5$

22) $y = -3$

(4.3) Determine the slope of each line.

23)

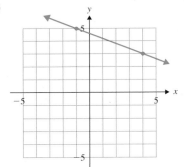

24)

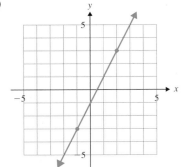

Use the slope formula to find the slope of the line containing each pair of points.

25) $(1, 7)$ and $(-4, 2)$

26) $(-2, -3)$ and $(3, -1)$

27) $(-2, 5)$ and $(3, -8)$

28) $(0, 4)$ and $(8, -2)$

29) $\left(\dfrac{3}{2}, -1\right)$ and $\left(-\dfrac{5}{2}, 7\right)$

30) $(2.5, 5.3)$ and $(-3.5, -1.9)$

31) $(9, 0)$ and $(9, 4)$

32) $(-7, 4)$ and $(1, 4)$

33) Paul purchased some shares of stock in 2002. The graph shows the value of one share of the stock from 2002–2006.

Value of One Share of Stock

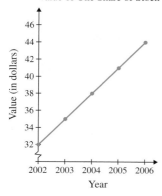

a) What was the value of one share of stock the year that Paul made his purchase?

b) Is the slope of the line segment positive or negative? What does the sign of the slope mean in the context of this problem?

c) Find the slope. What does it mean in the context of this problem?

Graph the line containing the given point and with the given slope.

34) $(-3, -2)$; $m = 4$

35) $(1, 5)$; $m = -3$

36) $(-2, 6)$; $m = -\dfrac{5}{2}$

37) $(-3, 2)$; slope undefined

38) $(5, 2)$; $m = 0$

(4.4) Identify the slope and y-intercept, then graph the line.

39) $y = x - 3$

40) $y = -2x + 7$

41) $y = -\dfrac{3}{4}x + 1$

42) $y = \dfrac{1}{4}x - 2$

43) $x - 3y = -6$

44) $2x - 7y = 35$

45) $x + y = 0$

46) $y + 3 = 4$

47) Personal consumption expenditures in the United States since 1998 can be modeled by $y = 371.5x + 5920.1$, where x represents the number of years after 1998, and y represents the personal consumption expenditure in billions of dollars. (Source: Bureau of Economic Analysis)

Personal Consumption Expenditures

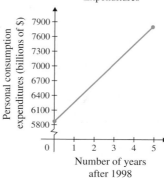

a) What is the y-intercept? What does it mean in the context of the problem?

b) Has the personal consumption expenditure been increasing or decreasing since 1998? By how much per year?

c) Use the graph to estimate the personal consumption expenditure in the year 2002. Then, use the equation to determine this number.

(4.5)

48) Write the point-slope formula for the equation of a line with slope m and which contains the point (x_1, y_1).

Write the *slope-intercept form* of the equation of the line, if possible, given the following information.

49) $m = 7$ and contains $(2, 5)$

50) $m = -8$ and y-intercept $(0, -1)$

51) $m = -\dfrac{4}{9}$ and y-intercept $(0, 2)$

52) contains $(-6, -5)$ and $(4, 10)$

53) contains $(3, -6)$ and $(-9, -2)$

54) $m = \dfrac{1}{2}$ and contains $(8, -3)$

55) horizontal line containing $(1, 9)$

56) vertical line containing $(4, 0)$

Write the *standard form* of the equation of the line given the following information.

57) contains $(-2, 2)$ and $(8, 7)$

58) $m = -1$ and contains $(4, -7)$

59) $m = -3$ and contains $\left(\dfrac{4}{3}, 1\right)$

60) contains $(15, -2)$ and $(-5, -10)$

61) $m = 6$ and y-intercept $(0, 0)$

62) $m = -\dfrac{5}{3}$ and y-intercept $(0, 2)$

63) contains $(1, 1)$ and $(-7, -5)$

64) $m = \dfrac{1}{6}$ and contains $(17, 2)$

65) The chart shows the number of wireless communication subscribers worldwide (in millions) from 2001–2004.
(Source: Dell'Oro Group and Standard and Poor's Industry Surveys)

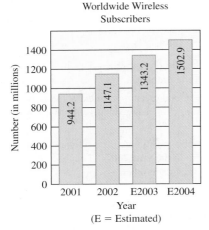

Worldwide Wireless Subscribers

a) Write a linear equation to model this data. Use the data points for 2001 and 2004. Let x represent the number of years after 2001, and let y represent the number of worldwide wireless subscribers, in millions. Round to the nearest tenth.

b) Explain the meaning of the slope in the context of the problem.

c) Use the equation to determine the number of subscribers in 2003. How does it compare to the value given on the chart?

(4.6) Determine if each pair of lines is parallel, perpendicular or neither.

66) $y = -\dfrac{1}{2}x + 7$

    $5x + 10y = 8$

67) $9x - 4y = -1$

    $-27x + 12y = 2$

68) $4x - 6y = -3$

    $-3x + 2y = -2$

69) $y = 6$

    $x = -2$

70) $x = 3$

    $x = 1$

71) $y = 6x - 7$

    $4x + y = 9$

72) $x + 2y = 22$

    $2x - y = 0$

Write an equation of the line *parallel* to the given line and containing the given point. Write the answer in slope-intercept form or in standard form, as indicated.

73) $y = 5x + 14$; $(-2, -4)$; slope-intercept form

74) $y = -3x + 1$; $(5, -19)$; slope-intercept form

75) $2x + y = 5$; $(1, -9)$; standard form

76) $x - 4y = 9$; $(5, 3)$; standard form

77) $5x - 3y = 7$; $(4, 8)$; slope-intercept form

78) $2x + 5y = -1$; $(12, -3)$; slope-intercept form

Write an equation of the line *perpendicular* to the given line and containing the given point. Write the answer in slope-intercept form or in standard form, as indicated.

79) $y = -\dfrac{1}{2}x + 9$; $(6, 5)$; slope-intercept form

80) $y = -x + 11$; $(-10, -8)$; slope-intercept form

81) $2x - 11y = 11$; $(2, -7)$; slope-intercept form

82) $3x + 5y = -45$; $(3, -1)$; slope-intercept form

83) $4x - y = -3$; $(-8, -3)$; standard form

84) $-2x + 3y = 15$; $(-5, 7)$; standard form

85) Write an equation of the line parallel to $x = 7$ containing $(2, 3)$.

86) Write an equation of the line parallel to $y = -5$ containing $(-4, 9)$.

87) Write an equation of the line perpendicular to $y = 6$ containing $(-1, -3)$.

88) Write an equation of the line perpendicular to $x = 7$ containing $(6, 0)$.

(4.7) Identify the domain and range of each relation, and determine whether each relation is a function.

89) $\{(-4, -9), (0, 3), (2, 9), (5, 18)\}$

90) $\{(-3, 1), (5, 3), (5, -3), (12, 4)\}$

91)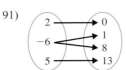

92)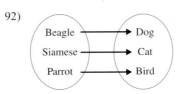

93)

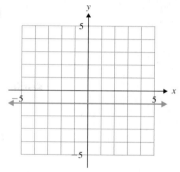

94)

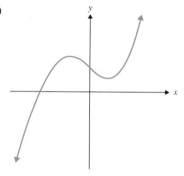

Determine the domain of each relation, and determine whether each relation describes y as a function of x.

95) $y = 4x - 7$

96) $y = \dfrac{8}{x + 3}$

97) $y = \dfrac{15}{x}$

98) $y^2 = x$

99) $y = x^2 - 6$

100) $y = \dfrac{5}{7x - 2}$

(4.8) For each function, f, find f(3) and f(−2).

101) $f = \{(-7, -2), (-2, -5), (1, -10), (3, -14)\}$

102)

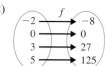

103)

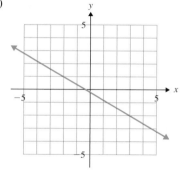

104)

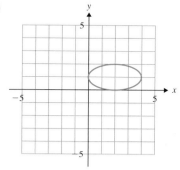

105) Let $f(x) = 5x - 12$, $g(x) = x^2 + 6x + 5$. Find each of the following and simplify.

   a) $f(4)$ b) $f(-3)$

   c) $g(3)$ d) $g(0)$

   e) $f(a)$ f) $g(t)$

   g) $f(k + 8)$ h) $f(c - 2)$

106) $h(x) = -3x + 7$. Find x so that $h(x) = 19$.

107) $g(x) = 4x - 9$. Find x so that $g(x) = 15$.

108) $f(x) = \dfrac{3}{2}x + 5$. Find x so that $f(x) = \dfrac{11}{2}$.

109) Graph $f(x) = -2x + 6$ by making a table of values and plotting points.

110) Graph $g(x) = \dfrac{3}{2}x + 3$ by finding the x- and y-intercepts and one other point.

Graph each function.

111) $h(c) = -\dfrac{5}{2}c + 4$

112) $D(t) = 3t$

113) A USB 2.0 device can transfer data at a rate of 480 MB/sec (megabytes/second). Let $f(t) = 480t$ represent the number of megabytes of data that can be transferred in t sec. (Source: www.usb.org)

   a) How many megabytes of a file can be transferred in 2 sec? in 6 sec?

   b) How long would it take to transfer a 1200 MB file?

114) A jet travels at a constant speed of 420 mph. The distance D (in miles) that the jet travels after t hr can be defined by the function

$$D(t) = 420t$$

   a) Find $D(2)$, and explain what this means in the context of the problem.

   b) Find t so that $D(t) = 2100$, and explain what this means in the context of the problem.

1) Is $(9, -13)$ a solution of $5x + 3y = 6$?

2) Complete the table of values and graph $y = -2x + 4$.

x	y
0	
3	
−1	
	0

3) Fill in the blanks with *positive* or *negative*. In quadrant II, the x-coordinate of every point is _____ and the y-coordinate is _____.

4) For $2x - 3y = 12$,

   a) find the x-intercept.

   b) find the y-intercept.

   c) find one other point on the line.

   d) graph the line.

5) Graph $x = -4$.

6) Find the slope of the line containing the points

   a) $(-8, -5)$ and $(4, -14)$

   b) $(9, 2)$ and $(3, 2)$

7) Graph the line containing the point $(-3, 6)$ with slope $= -\dfrac{2}{5}$.

8) Graph the line containing the point $(4, 1)$ with an undefined slope.

9) Put $6x - 2y = 8$ into slope-intercept form. Then, graph the line.

10) Write the slope-intercept form for the equation of the line with slope -4 and y-intercept $(0, 5)$.

11) Write the standard form for the equation of a line with slope $\dfrac{1}{2}$ containing the point $(3, 8)$.

12) Determine if $4x - 7y = -7$ and $14x + 8y = 3$ are parallel, perpendicular, or neither.

13) Find the slope-intercept form of the equation of the line

   a) perpendicular to $y = -3x + 11$ containing $(12, -5)$.

   b) parallel to $5x - 2y = 2$ containing $(8, 14)$.

14) Mr. Kumar owns a computer repair business. The graph below shows the annual profit since 2000. Let x represent the number of years after 2000, and let y represent the annual profit, in thousands.

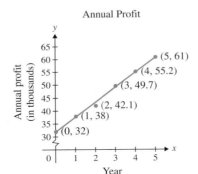

Annual Profit

   a) What was the profit in 2004?

   b) Write a linear equation (in slope-intercept form) to model this data. Use the data points for 2000 and 2005.

   c) What is the slope of the line? What does it mean in the context of the problem?

   d) What is the y-intercept? What does it mean in the context of the problem?

   e) If the profit continues to follow this trend, in what year can Mr. Kumar expect a profit of $90,000?

Identify the domain and range of each relation, and determine whether each relation is a function.

15) $\{(-2, -5), (1, -1), (3, 1), (8, 4)\}$

16)

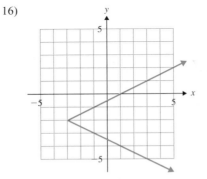

For each function, (a) determine the domain. (b) Is y a function of x?

17) $y = \dfrac{7}{3}x - 5$

18) $y = x + 8$

For each function, f, find $f(2)$.

19) $f = \{(-3, -8), (0, -5), (2, -3), (7, 2)\}$

20)

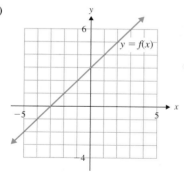

Let $f(x) = -4x + 2$ and $g(x) = x^2 - 3x + 7$. Find each of the following and simplify.

21) $f(6)$ 22) $g(2)$

23) $g(t)$

24) $f(h - 7)$

Graph the function.

25) $h(x) = -\dfrac{3}{4}x + 5$

26) A USB 1.1 device can transfer data at a rate of 12 MB/sec (megabytes/second). Let $f(t) = 12t$ represent the number of megabytes of data that can be transferred in t sec. (Source: www.usb.org)

   a) How many megabytes of a file can be transferred in 3 sec?

   b) How long would it take to transfer 132 MB?

Cumulative Review: Chapters 1–4

1) Write $\dfrac{252}{840}$ in lowest terms.

2) A rectangular picture frame measures 8 in. by 10.5 in. Find the perimeter of the frame.

Evaluate.

3) -2^6

4) $\dfrac{21}{40} \cdot \dfrac{25}{63}$

5) $3 - \dfrac{2}{5}$

6) Write an expression for "53 less than twice eleven" and simplify.

Simplify. The answer should not contain any negative exponents.

7) $(3t^4)(-7t^{10})$

8) $\left(\dfrac{54a^5}{72a^{11}}\right)^3$

9) Solve $12 - 5(2n + 9) = 3n + 2(n + 6)$

10) Solve. Write the solution in interval notation.
$19 - 4x \geq 25$

For 11–13, write an equation and solve.

11) One serving of Ben and Jerry's Chocolate Chip Cookie Dough Ice Cream has 10% fewer calories than one serving of their Chunky Monkey® Ice Cream. If one serving of Chocolate Chip Cookie Dough has 270 calories, how many calories are in one serving of Chunky Monkey? (www.benjerry.com)

12) Find the missing angle measures.

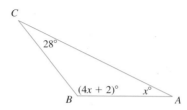

13) Lynette's age is 7 yr less than three times her daughter's age. If the sum of their ages is 57, how old is Lynette, and how old is her daughter?

14) Find the slope of the line containing the points $(-7, 8)$ and $(2, 17)$.

15) Graph $4x + y = 5$.

16) Write an equation of the line with slope $-\dfrac{5}{4}$ containing the point $(-8, 1)$. Express the answer in standard form.

17) Write an equation of the line perpendicular to $y = \dfrac{1}{3}x + 11$ containing the point $(4, -12)$. Express the answer in slope-intercept form.

18) Determine the domain of $y = \dfrac{3}{x + 7}$.

Let $f(x) = 8x + 3$. Find each of the following and simplify.

19) $f(-5)$

20) $f(a)$

21) $f(t + 2)$

22) Graph $f(x) = 2$.

CHAPTER 5

Solving Systems of Linear Equations

Algebra at Work: Custom Motorcycles

We will take another look at how algebra is used in a custom motorcycle shop.

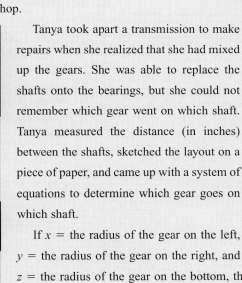

Tanya took apart a transmission to make repairs when she realized that she had mixed up the gears. She was able to replace the shafts onto the bearings, but she could not remember which gear went on which shaft. Tanya measured the distance (in inches) between the shafts, sketched the layout on a piece of paper, and came up with a system of equations to determine which gear goes on which shaft.

If x = the radius of the gear on the left, y = the radius of the gear on the right, and z = the radius of the gear on the bottom, then the system of equations Tanya must solve to determine where to put each gear is

$$x + y = 2.650$$
$$x + z = 2.275$$
$$y + z = 1.530$$

Solving this system, Tanya determines that $x = 1.698$ in., $y = 0.952$ in., and $z = 0.578$ in. Now she knows on which shaft to place each gear.

In this chapter, we will learn how to write and solve systems of two and three equations.

Section 5.1 Solving Systems of Linear Equations by Graphing

Objectives

1. Determine Whether a Given Ordered Pair Is a Solution of a System

2. Solve a Linear System by Graphing

3. Solve Linear Systems by Graphing When There Is No Solution or an Infinite Number of Solutions

4. Without Graphing, Determine Whether a System Has No Solution, One Solution, or an Infinite Number of Solutions

What is a system of linear equations? A **system of linear equations** consists of two or more linear equations with the same variables. In Sections 5.1–5.3, we will learn how to solve systems of two equations in two variables. Some examples of such systems are

$$2x + 5y = 5 \qquad\qquad y = \frac{1}{3}x - 8 \qquad\qquad -3x + y = 1$$
$$x + 4y = -1 \qquad\qquad 5x - 6y = 10 \qquad\qquad x = -2$$

In the third system, we see that $x = -2$ is written with only one variable. However, we can think of it as an equation in two variables by writing it as $x + 0y = -2$.

It is possible to solve systems of equations containing more than two variables. In Section 5.5, we will learn how to solve systems of linear equations in three variables.

1. Determine Whether a Given Ordered Pair Is a Solution of a System

We will begin our work with systems of equations by determining whether an ordered pair is a solution of the system.

Definition	A **solution of a system** of two equations in two variables is an ordered pair that is a solution of each equation in the system.

Example 1

Determine if $(-5, -1)$ is a solution of each system of equations.

a) $x - 3y = -2$
    $-4x + y = 19$

b) $2x - 9y = -1$
    $y = x + 8$

Solution

a) If $(-5, -1)$ is a solution of $\begin{array}{l} x - 3y = -2 \\ -4x + y = 19 \end{array}$ then when we substitute -5 for x and -1 for y, the ordered pair will make each equation true.

$x - 3y = -2$		$-4x + y = 19$	
$-5 - 3(-1) \stackrel{?}{=} -2$	Substitute.	$-4(-5) + (-1) \stackrel{?}{=} 19$	Substitute.
$-5 + 3 \stackrel{?}{=} -2$		$20 + (-1) \stackrel{?}{=} 19$	
$-2 = -2$	True	$19 = 19$	True

Since $(-5, -1)$ is a solution of each equation, it is a solution of the system.

b) We will substitute -5 for x and -1 for y to see if $(-5, -1)$ satisfies (is a solution of) each equation.

$2x - 9y = -1$		$y = x + 8$	
$2(-5) - 9(-1) \stackrel{?}{=} -1$	Substitute.	$-1 \stackrel{?}{=} (-5) + 8$	Substitute.
$-10 + 9 \stackrel{?}{=} -1$		$-1 = 3$	False
$-1 = -1$	True		

Although $(-5, -1)$ is a solution of the first equation, it does *not* satisfy $y = x + 8$. Therefore, $(-5, -1)$ is *not* a solution of the system.

You Try 1

Determine if $(6, 2)$ is a solution of each system.

a) $-x + 4y = 2$ b) $y = x - 4$
    $3x - 5y = 6$ $2x - 9y = -6$

We have learned how to determine if an ordered pair is a solution of a system. But, if we are given a system and no solution is given, how do we *find* the solution to the system of equations? In this chapter, we will discuss three methods for solving systems of equations.

We can solve a system of equations by

1) Graphing (this section)
2) Substitution (Section 5.2)
3) Elimination method (Section 5.3)

Let's begin with the graphing method.

2. Solve a Linear System by Graphing

To **solve a system of equations in two variables** means to find the ordered pair that satisfies each equation in the system.

Recall from Chapter 4 that the graph of $Ax + By = C$ is a line. The line represents the set of all points that satisfy $Ax + By = C$; that is, the line represents all solutions of the equation.

Let's look at Example 1a) again. We found that $(-5, -1)$ is a solution of the system $x - 3y = -2$.
                $-4x + y = 19$
How can we interpret this information from a graph?

Graph each line.

The graph of $x - 3y = -2$ represents all solutions of that equation. Likewise, all points on the line $-4x + y = 19$ are solutions of $-4x + y = 19$. Since $(-5, -1)$ is a point on each line, it is a solution of each equation.

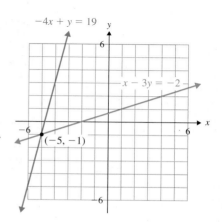

This is where the lines intersect. So the solution to the system $\begin{aligned} x - 3y &= -2 \\ -4x + y &= 19 \end{aligned}$ is the point of intersection $(-5, -1)$ of the graphs of each equation.

We can generalize this result for *any* lines by saying that if two lines intersect at a point, then that point is a solution of each equation. Therefore,

> When solving a system of equations by graphing, the point of intersection is the solution of the system. When a system has one solution, we say that the system is **consistent**.

Example 2

Solve the system by graphing.

$$y = x + 1$$
$$x + 2y = 8$$

Solution

To solve a system by graphing we must graph each line on the same axes. The point of intersection will be the solution of the system.

$$y = x + 1 \text{ is in slope-intercept form}$$
$$m = 1, \quad b = 1$$

We will use this information to graph this line.

We can graph $x + 2y = 8$ either by plotting points or by putting it into slope-intercept form. We will make a table of values and then plot the points.

$x + 2y = 8$

x	y
0	4
−2	5
2	3

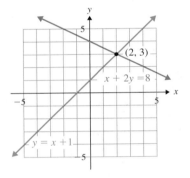

The point of intersection is (2, 3). Therefore, the solution to the system is (2, 3).

This is a consistent system.

> It is important that you use a straightedge to graph the lines. If the graph is not precise, it will be difficult to correctly locate the point of intersection. Furthermore, if the solution of a system contains numbers that are not integers, it may be impossible to accurately read the point of intersection. This is one reason why solving a system by graphing is not always the best way to find the solution. But it can be a useful method, and it is one that is used to solve problems not only in mathematics, but in areas such as business, economics, and chemistry as well.

You Try 2

Solve the system by graphing. $2x + 3y = 8$
$$y = -4x - 4$$

3. Solve Linear Systems by Graphing When There Is No Solution or an Infinite Number of Solutions

Do two lines *always* intersect? No! Then if we are trying to solve a system of two linear equations by graphing and the graphs do not intersect, what does this tell us about the solution to the system?

Example 3

Solve the system by graphing.

$$2y - x = 2$$
$$-x + 2y = -6$$

Solution

Graph each line on the same axes.

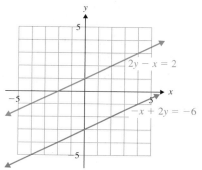

The lines are parallel; they will never intersect. Therefore, there is *no solution* to the system. We write the solution set as $\varnothing$.

> When solving a system of equations by graphing, if the lines are parallel, then the system has **no solution.** We write this as $\varnothing$. Furthermore, we say that a system that has no solution is an **inconsistent system.**

Example 4

Solve the system by graphing.

$$y = -\frac{3}{2}x + 2$$
$$6x + 4y = 8$$

Solution

Graph each line on the same axes.

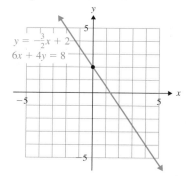

We can obtain the second equation by multiplying the first equation by 4 and rearranging the terms. So the second equation is a multiple of the first. This means that the graph of each equation is the same line. Therefore, each point on the line satisfies each equation. We say that there are an *infinite number of solutions* of the form $\left\{ (x, y) \,\middle|\, y = -\frac{3}{2}x + 2 \right\}$.

When solving a system of equations by graphing, if the graph of each equation is the same line, then the system has an **infinite number of solutions**. Furthermore, we say that the system is **dependent**.

We will summarize what we have learned so far about solving a system of linear equations by graphing:

Solving a System by Graphing

To solve a system by graphing, graph each line on the same axes.

1)   If the lines intersect at a single point, then the point of intersection is the solution of the system. We can say that the system is *consistent*. (See Figure 5.1a.)

2)   If the lines are parallel, then the system has *no solution*. We write the solution set as $\varnothing$. We say that the system is *inconsistent*. (See Figure 5.1b.)

3)   If the graphs are the same line, then the system has an *infinite number of solutions*. We say that the system is *dependent*. (See Figure 5.1c.)

Figure 5.1

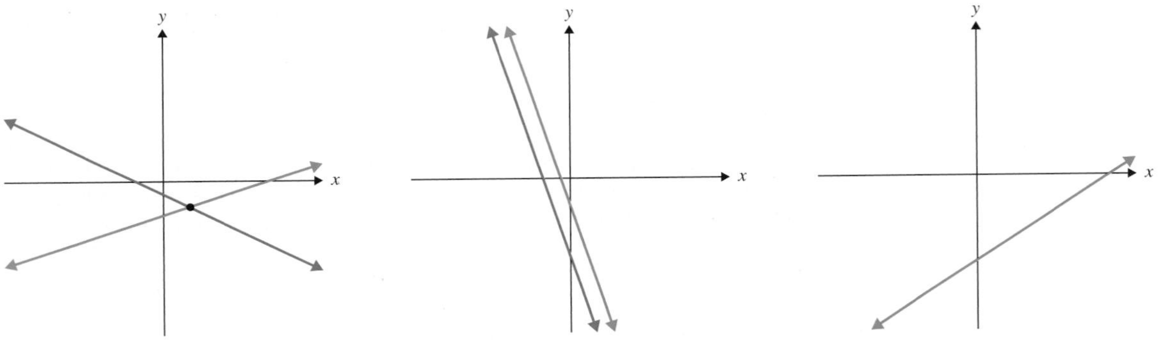

a)
One solution—the point
of intersection

b)
No solution

c)
Infinite number
of solutions

You Try 3

Solve each system by graphing.

a) $6x - 8y = 8$
      $3x - 4y = -12$

b) $x + 2y = 8$
      $-6y - 3x = -24$

---

4. Without Graphing, Determine Whether a System Has No Solution, One Solution, or an Infinite Number of Solutions

The graphs of lines can lead us to the solution of a system. But we can also learn something about the solution by looking at the equations of the lines *without* graphing them.

In Example 4 we saw that *one equation was a multiple of the other*. If we wrote each equation in slope-intercept form, we would get the same equation. This tells us that the system is *dependent,* and there are an *infinite number of solutions.*

Look at Figure 5.1b again. If two lines are parallel, what do we know about their slopes? *The slopes are the same.* So if we write each equation in slope-intercept form and the lines have the same slope but different y-intercepts, then the lines are parallel. Therefore the system has *no solution.*

We can see that the lines in Figure 1a have different slopes. Lines with different slopes will intersect at one point. Therefore, if the equations in a system have *different slopes*, the system will have *one solution.*

Example 5

Without graphing, determine whether each system has no solution, one solution, or an infinite number of solutions.

a) $y = \dfrac{2}{3}x + 5$

   $-4x + 3y = 6$

b) $8x - 12y = 4$

   $-6x + 9y = -3$

c) $10x + 4y = -9$

   $5x + 2y = 14$

Solution

a)   The first equation is written in slope-intercept form, so begin by also writing the second equation in slope-intercept form.

$$-4x + 3y = 6$$
$$3y = 4x + 6$$
$$y = \frac{4}{3}x + 2$$

The first equation is $y = \dfrac{2}{3}x + 5$, and the second can be written as $y = \dfrac{4}{3}x + 2$. The slopes are different, therefore this system has *one solution.*

b)   Write each equation in slope-intercept form.

$$8x - 12y = 4$$
$$-12y = -8x + 4$$
$$y = \frac{-8}{-12}x + \frac{4}{-12}$$
$$y = \frac{2}{3}x - \frac{1}{3}$$

$$-6x + 9y = -3$$
$$9y = 6x - 3$$
$$y = \frac{6}{9}x - \frac{3}{9}$$
$$y = \frac{2}{3}x - \frac{1}{3}$$

When we rewrite each equation in slope-intercept form, the equations are the same. Therefore, this system has an *infinite number of solutions.*

c)   Write each equation in slope-intercept form.

$$10x + 4y = -9$$
$$4y = -10x - 9$$
$$y = \frac{-10}{4}x - \frac{9}{4}$$
$$y = -\frac{5}{2}x - \frac{9}{4}$$

$$5x + 2y = 14$$
$$2y = -5x + 14$$
$$y = \frac{-5}{2}x + \frac{14}{2}$$
$$y = -\frac{5}{2}x + 7$$

When we rewrite each equation in slope-intercept form, we see that they have the same slope but different y-intercepts. If we graphed them, the lines would be parallel. Therefore, this system has *no solution.*

You Try 4

Without graphing, determine whether each system has no solution, one solution, or an infinite number of solutions.

a) $10x + y = 4$
   $20x + 2y = -7$

b) $y = -\dfrac{2}{3}x + \dfrac{1}{2}$
   $4x + 6y = 3$

Using Technology

In this section we have learned that the solution of a system of equations is the point at which their graphs intersect. We can solve a system by graphing using a graphing calculator. On the calculator we will solve the system that we solved in Example 2 on p. 324.

Solving the system by graphing:
$$y = x + 1$$
$$x + 2y = 8$$

Begin by entering each equation using the $\boxed{Y=}$ key. Before entering the second equation, we must solve for y.

$$y + 2y = 8$$
$$2y = -x + 8$$
$$y = -\frac{1}{2}x + 4$$

Enter the first equation next to $\backslash Y_1$ and enter $(-1/2)x + 4$ next to $\backslash Y_2$. Press $\boxed{\text{GRAPH}}$.

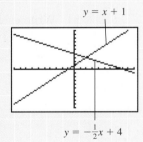

$y = x + 1$

$y = -\frac{1}{2}x + 4$

Since the lines intersect, the system has a solution. How can we find that solution?

Once you see from the graph that the lines intersect, press $\boxed{\text{2nd}}$ $\boxed{\text{TRACE}}$ and you will see this screen:

Use the arrow key to choose 5: INTERSECT, and press ENTER. Press ENTER three more times. The screen will have the solution displayed at the bottom, with the cursor at the point of intersection:

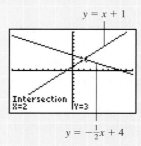

$$y = x + 1$$

$$y = -\frac{1}{2}x + 4$$

The solution to the system is $(2, 3)$. This is the same as the result we obtained in Example 2.

Use a graphing calculator to solve each system.

1. $y = -2x + 5$
   $y = x - 4$

2. $y = -3x - 4$
   $y = x$

3. $2x - y = 3$
   $y - 4x = -8$

4. $x + 2y = -1$
   $x + 4y = 2$

5. $3x + 2y = 4$
   $4x + 3y = 7$

6. $6x - 3y = 10$
   $-2x + y = 4$

Answers to You Try Exercises

1) a) no   b) yes

2) $(-2, 4)$

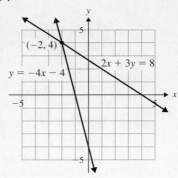

3) a) $\varnothing$

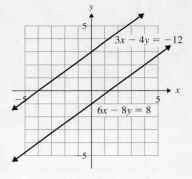

b) Infinite number of solutions of the form $\{(x, y) \mid x + 2y = 8\}$

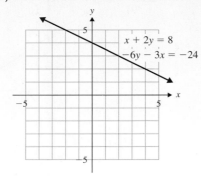

4) a) no solution   b) infinite number of solutions

Answers to Technology Exercises

1. $(3, -1)$ 2. $(-1, -1)$ 3. $(2.5, 2)$
4. $(-4, 1.5)$ 5. $(-2, 5)$ 6. no solution

5.1 Exercises

Boost your grade at mathzone.com!   MathZone

> Practice Problems   > NetTutor   > e-Professors   > Videos
> Self-Test

Objective 1

Determine if the ordered pair is a solution of the system of equations.

1) $2x - 3y = -15$
$-x + y = 4$
$(3, 7)$

2) $x + 2y = -2$
$5y - x = 9$
$(-4, 1)$

3) $3x + 2y = 4$
$4x - y = -3$
$(-2, 5)$

4) $-x + 7y = -6$
$2x - 3y = 9$
$(6, 0)$

5) $10x + 7y = -13$
$-6x - 5y = 11$
$\left(\dfrac{3}{2}, -4\right)$

6) $x = 8y + 3$
$12y = x - 6$
$\left(5, \dfrac{1}{4}\right)$

7) $y = 5x - 7$
$3x + 9 = y$
$(-1, -2)$

8) $x - 5y = 7$
$y = 2x + 13$
$(-8, -3)$

Objectives 2 and 3

9) If you are solving a system of equations by graphing, how do you know if the system has no solution?

10) If you are solving a system of equations by graphing, how do you know if the system has an infinite number of solutions?

Solve each system of equations by graphing. If the system is inconsistent or dependent, identify it as such.

11) $y = -\dfrac{2}{3}x + 3$
$y = x - 2$

12) $y = \dfrac{1}{2}x + 2$
$y = 2x - 1$

13) $y = 2x + 1$
$y = -3x - 4$

14) $y = \dfrac{3}{2}x - 4$
$y = -2x + 3$

15) $x + y = -1$
$x - 2y = 14$

16) $2x - 3y = 6$
$x + y = -7$

17) $x + 2y = 10$
$y = -\dfrac{1}{2}x + 1$

18) $x - 3y = 9$
$-x + 3y = 6$

19) $6x - 3y = 12$
$-2x + y = -4$

20) $2x + 2y = 8$
$x + y = 4$

21) $\dfrac{3}{4}x - y = 0$
$3x - 4y = 20$

22) $y = -x$
$4x + 4y = 2$

23) $x = 2y - 6$
$3x - 2y = 2$

24) $3x + y = 2$
$y = 2x - 3$

25) $y = -3x + 1$
$12x + 4y = 4$

26) $2x - y = 1$
$-2x + y = -3$

27) $y - 4x = 1$
$y = -3$

28) $5x + 2y = 6$
$-15x - 6y = -18$

29) $5x - y = -6$
$y = -x$

30) $y = -3x + 1$
$x = 1$

31) $y = \dfrac{3}{5}x - 6$
$-3x + 5y = 10$

32) $x - 4y = 1$
$x - y = 4$

Write a system of equations so that the given ordered pair is a solution of the system.

33) $(5, 1)$ 34) $(2, 4)$

35) $(-1, -4)$ 36) $(3, -2)$

37) $\left(-\dfrac{1}{2}, 3\right)$ 38) $\left(0, \dfrac{2}{3}\right)$

For Exercises 39–42, determine which ordered pair could not be a solution to the system of equations that is graphed. Explain why you chose that ordered pair.

39)

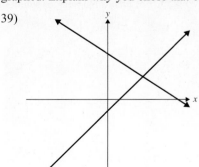

A. $(4, 2)$ C. $(7, 1)$

B. $\left(3, \dfrac{5}{2}\right)$ D. $(5, -3)$

40)

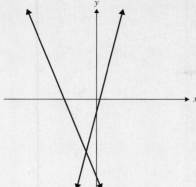

A. $(-3, -3)$ C. $(-1, -8)$

B. $(-2, 1)$ D. $\left(-\dfrac{3}{2}, -\dfrac{9}{2}\right)$

41)

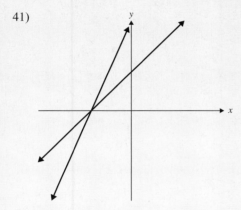

A. $(-6, 0)$ C. $(0, -5)$

B. $\left(-\dfrac{1}{2}, 0\right)$ D. $(-8.3, 0)$

42)

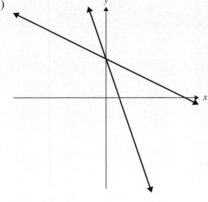

A. $(0, 4)$ C. $(0, 5.2)$

B. $\left(0, \dfrac{1}{3}\right)$ D. $(2, 0)$

Objective 4

43) How do you determine, *without graphing*, that a system of equations has exactly one solution?

44) How do you determine, *without graphing*, that a system of equations has no solution?

Without graphing, determine whether each system has no solution, one solution, or an infinite number of solutions.

45) $y = \dfrac{3}{2}x + \dfrac{7}{2}$
    $-9x + 6y = 21$

46) $y = 4x + 6$
    $8x - y = -7$

47) $5x - 2y = -11$
    $x + 6y = 18$

48) $5x - 8y = 24$
    $10x - 16y = -9$

 49) $x + y = 10$
    $-9x - 9y = 2$

50) $x - 4y = 2$
    $3x + y = 1$

51) $5x - y = -2$
    $x + 6y = 2$

52) $9x - 6y = 5$
    $y = \dfrac{3}{2}x - \dfrac{5}{6}$

53) $y = -2$
    $y = 3$

54) $y = x$
    $y - x = 4$

Objectives 2 and 3

55) The graph shows the number of people, seven years of age and older, who have participated more than once in snowboarding and ice/figure skating from 1997–2003. (Source: National Sporting Goods Association)

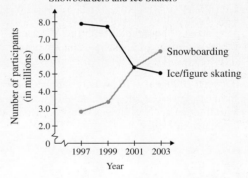

a) When were there more snowboarders than ice/figure skaters?

b) When did the number of snowboarders equal the number of skaters? How many people participated in each?

c) During which years did snowboarding see its greatest increase in participation?

d) During which years did skating see its greatest decrease in participation?

56) The graph shows the percentage of households in Delaware and Nevada with Internet access in various years from 1998 to 2003. (Source: www.census.gov)

Households with Internet Access

a) In 1998, which state had a greater percentage of households with Internet access? Approximately how many more were there?

b) In what year did both states have the same percentage of households with Internet access? What percentage of households had Internet access that year?

c) Between 2000 and 2001, which state had the greatest increase in the percentage of households with Internet access? How can this information be related to *slope*?

Solve by graphing. Given the functions $f(x)$ and $g(x)$, determine the value of x for which $f(x) = g(x)$.

57) $f(x) = x - 3, g(x) = -\dfrac{1}{4}x + 2$

58) $f(x) = x + 4, g(x) = -2x - 2$

59) $f(x) = 3x + 3, g(x) = x + 1$

60) $f(x) = -\dfrac{2}{3}x + 1, g(x) = x - 4$

61) $f(x) = 2x - 1, g(x) = -\dfrac{3}{2}x - 1$

62) $f(x) = x + 1, g(x) = 2x$

Solve each system using a graphing calculator.

63) $y = -2x + 4$
$y = x - 5$

64) $y = -x + 5$
$y = 3x + 1$

65) $-2x + y = 9$
$x - 3y = -2$

66) $4x + y = -6$
$x + 3y = 4$

67) $6x - 5y = 3.25$
$3x + y = -2.75$

68) $2x + 5y = 7.5$
$-3x + y = -11.25$

Section 5.2 Solving Systems of Linear Equations by Substitution

Objectives

1. Solve a Linear System by Substitution
2. Solve a Linear System Containing Fractions or Decimals
3. Solve an Inconsistent or a Dependent System by Substitution

In Section 5.1, we learned to solve a system of equations by graphing. Recall, however, that this is not always the *best* way to solve a system. If your graphs are not precise, you may read the solution incorrectly. And, if a solution consists of numbers which are not integers, like $\left(\dfrac{2}{3}, -\dfrac{1}{4}\right)$, it may not be possible to accurately identify the point of intersection of the graphs.

1. Solve a Linear System by Substitution

Another way to solve a system of equations is to use the **substitution method**. *This method is especially good when one of the variables has a coefficient of 1 or −1.*

Example 1

Solve the system using substitution.

$$2x + 3y = 14$$
$$y = 3x + 1$$

Solution

The second equation is already solved for y; it tells us that y *equals* $3x + 1$. Therefore, we can substitute $3x + 1$ for y in the first equation, then solve for x.

$$
\begin{array}{ll}
2x + 3y = 14 & \text{First equation} \\
2x + 3(3x + 1) = 14 & \text{Substitute.} \\
2x + 9x + 3 = 14 & \text{Distribute.} \\
11x + 3 = 14 & \\
11x = 11 & \\
x = 1 &
\end{array}
$$

We have found that $x = 1$, but we still need to find y. Substitute $x = 1$ in *either* equation, and solve for y. In this case, we will substitute $x = 1$ into the second equation since it is already solved for y.

$$
\begin{array}{ll}
y = 3x + 1 & \text{Second equation} \\
y = 3(1) + 1 & \text{Substitute.} \\
y = 3 + 1 & \\
y = 4 &
\end{array}
$$

Check $x = 1, y = 4$ in *both* equations.

$$
\begin{array}{ll}
2x + 3y = 14 & \\
2(1) + 3(4) \stackrel{?}{=} 14 & \text{Substitute.} \\
2 + 12 \stackrel{?}{=} 14 & \\
14 = 14 & \text{True}
\end{array}
\qquad
\begin{array}{ll}
y = 3x + 1 & \\
4 \stackrel{?}{=} 3(1) + 1 & \text{Substitute.} \\
4 \stackrel{?}{=} 3 + 1 & \\
4 = 4 & \text{True}
\end{array}
$$

We write the solution as an ordered pair. The solution to the system is $(1, 4)$.

If we solve the system in Example 1 by graphing, we can see that the lines intersect at $(1, 4)$ giving us the same solution we obtained using the substitution method.

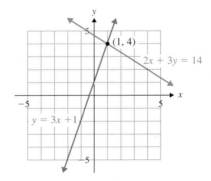

Let's summarize the steps we use to solve a system by the substitution method:

Solving a System by Substitution

1) Solve one of the equations for one of the variables. If possible, solve for a variable that has a coefficient of 1 or -1.

2) Substitute the expression in *step 1* into the *other* equation. The equation you obtain should contain only one variable.

3) Solve the equation in one variable from *step 2*.

4) Substitute the value found in *step 3* into either of the equations to obtain the value of the other variable.

5) Check the values in each of the original equations, and write the solution as an ordered pair.

Example 2

Solve the system by substitution.

$$x + 4y = 3 \qquad (1)$$
$$2x + 3y = -4 \qquad (2)$$

Solution

We will follow the steps listed above.

1) For which variable should we solve? The x in the first equation is the only variable with a coefficient of 1 or -1. Therefore, we will solve the first equation for x.

$$x + 4y = 3 \qquad (1) \qquad \text{First equation}$$
$$x = 3 - 4y \qquad\qquad \text{Subtract } 4y.$$

2) Substitute $3 - 4y$ for the x in equation (2).

$$2x + 3y = -4 \qquad (2) \qquad \text{Second equation}$$
$$2(3 - 4y) + 3y = -4 \qquad\qquad \text{Substitute.}$$

3) Solve the last equation for y.

$$2(3 - 4y) + 3y = -4$$
$$6 - 8y + 3y = -4 \qquad \text{Distribute.}$$
$$6 - 5y = -4$$
$$-5y = -10$$
$$y = 2$$

4) To determine the value of x we can substitute 2 for y in either equation. We will substitute it in equation (1).

$$x + 4y = 3 \qquad (1)$$
$$x + 4(2) = 3 \qquad \text{Substitute.}$$
$$x + 8 = 3$$
$$x = -5$$

5) The check is left to the reader. The solution of the system is $(-5, 2)$.

You Try 1

Solve the system by substitution.

$$10x + 3y = -4$$
$$8x + y = 1$$

If no variable in the system has a coefficient of 1 or -1, solve for any variable.

2. Solve a Linear System Containing Fractions or Decimals

If a system contains an equation with fractions, first multiply the equation by the least common denominator to eliminate the fractions. Likewise, if an equation in the system contains decimals, begin by multiplying the equation by the power of 10 that will eliminate the decimals.

Example 3

Solve the system by substitution.

$$\frac{2}{5}x - \frac{1}{3}y = 2 \qquad (1)$$
$$-\frac{1}{6}x + \frac{1}{2}y = \frac{4}{3} \qquad (2)$$

Solution

Before applying the steps for solving the system, eliminate the fractions in each equation.

$$\frac{2}{5}x - \frac{1}{3}y = 2$$
$$15\left(\frac{2}{5}x - \frac{1}{3}y\right) = 15 \cdot 2 \qquad \text{Multiply by the LCD: 15.}$$
$$6x - 5y = 30 \qquad (3) \quad \text{Distribute.}$$

$$-\frac{1}{6}x + \frac{1}{2}y = \frac{4}{3}$$
$$6\left(-\frac{1}{6}x + \frac{1}{2}y\right) = 6 \cdot \frac{4}{3} \qquad \text{Multiply by the LCD: 6.}$$
$$-x + 3y = 8 \qquad (4) \quad \text{Distribute.}$$

From the original equations we obtain an equivalent system of equations.

$$6x - 5y = 30 \qquad (3)$$
$$-x + 3y = 8 \qquad (4)$$

Now, we will work with equations (3) and (4).
Apply the steps:

1) The x in equation (4) has a coefficient of -1. Solve this equation for x.

$$-x + 3y = 8 \qquad (4)$$
$$-x = 8 - 3y \qquad \text{Subtract } 3y.$$
$$x = -8 + 3y \qquad \text{Divide by } -1.$$

2) Substitute $-8 + 3y$ for x in equation (3).

$$6x - 5y = 30 \qquad (3)$$
$$6(-8 + 3y) - 5y = 30 \qquad \text{Substitute.}$$

3) Solve the equation above for y.

$$6(-8 + 3y) - 5y = 30$$
$$-48 + 18y - 5y = 30 \qquad \text{Distribute.}$$
$$-48 + 13y = 30$$
$$13y = 78$$
$$y = 6 \qquad \text{Divide by } 13.$$

4) Find x by substituting 6 for y in either equation (3) or (4). Let's use equation (4) since it has smaller coefficients.

$$-x + 3y = 8 \qquad (4)$$
$$-x + 3(6) = 8 \qquad \text{Substitute.}$$
$$-x + 18 = 8$$
$$-x = -10$$
$$x = 10 \qquad \text{Divide by } -1.$$

5) Check $x = 10$ and $y = 6$ in the original equations. The solution of the system is $(10, 6)$.

You Try 2

Solve each system by substitution.

a) $\dfrac{2}{3}x - \dfrac{1}{9}y = -3$ b) $0.01x + 0.04y = 0.15$
$\qquad -\dfrac{5}{6}x + \dfrac{1}{2}y = \dfrac{1}{2}$ $\qquad\qquad 0.1x - 0.1y = 0.5$

3. Solve an Inconsistent or a Dependent System by Substitution

We saw in Section 5.1 that a system may be inconsistent (it has no solution) or a system may be dependent (it has an infinite number of solutions). If we are solving a system by graphing we know that a system has no solution if the lines are parallel, and a system has an infinite number of solutions if the graphs are the same line.

When we solve a system by substitution, how do we know if the system is inconsistent or dependent? Read Examples 4 and 5 to find out.

Example 4

Solve the system by substitution.

$$3x + y = 5 \qquad (1)$$
$$12x + 4y = -7 \qquad (2)$$

Solution

1) Solve equation (1) for y.

$$y = -3x + 5$$

2) Substitute $-3x + 5$ for y in equation (2).

$$12x + 4y = -7 \qquad (2)$$
$$12x + 4(-3x + 5) = -7 \qquad \text{Substitute.}$$

3) Solve the resulting equation for x.

$$12x + 4(-3x + 5) = -7$$
$$12x - 12x + 20 = -7 \qquad \text{Distribute.}$$
$$20 = -7 \qquad \text{False}$$

Since the variables drop out, and we get a false statement, there is no solution to the system. The system is inconsistent, so the solution set is $\varnothing$.

The graph of the equations in the system supports our work. The lines are parallel, therefore the system has no solution.

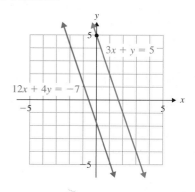

Example 5

Solve the system by substitution.

$$2x - 8y = 20 \qquad (1)$$
$$x = 4y + 10 \qquad (2)$$

Solution

1) Equation (2) is already solved for x.

2) Substitute $4y + 10$ for x in equation (1).

$$2x - 8y = 20 \qquad (1)$$
$$2(4y + 10) - 8y = 20 \qquad \text{Substitute.}$$

3) Solve the equation for y.

$$2(4y + 10) - 8y = 20$$
$$8y + 20 - 8y = 20 \qquad \text{Distribute.}$$
$$20 = 20 \qquad \text{True}$$

Since the variables drop out and we get a true statement, there are an infinite number of solutions to the system. The system is dependent, and we say that there are an infinite number of solutions of the form $\{(x, y) \mid 2x - 8y = 20\}$.

     The graph shows that the equations in the system are the same line, therefore the system has an infinite number of solutions.

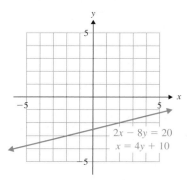

When you are solving a system of equations and the variables drop out,

1) If you get a *false statement*, like $3 = 5$, then the system has *no solution*.

2) If you get a *true statement*, like $-4 = -4$ then the system has an *infinite number of solutions*.

You Try 3

Solve each system by substitution.

a) $6x + y = -8$ b) $x - 3y = 5$
      $12x + 2y = -9$ $4x - 12y = 20$

Answers to You Try Exercises

1) $\left(\dfrac{1}{2}, -3\right)$ 2) a) $(-6, -9)$ b) $(7, 2)$ 3) a) $\varnothing$ b) infinite number of solutions of the form $\{(x, y) \mid x - 3y = 5\}$

5.2 Exercises

Boost your grade at mathzone.com!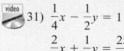

> Practice Problems   > NetTutor   > e-Professors   > Videos
> Self-Test

Objectives 1 and 3

1) If you were asked to solve this system by substitution, why would it be easiest to begin by solving for y in the second equation?

$$6x - 2y = -5$$
$$3x + y = 4$$

2) When solving a system by substitution, how do you know if the system has

   a) no solution?

   b) an infinite number of solutions?

Solve each system by substitution.

3) $y = 3x + 4$
   $-6x + y = -2$

4) $3x + y = 15$
   $y = 4x - 6$

5) $2x - y = 5$
   $x = y + 6$

6) $x = y - 8$
   $-3x - y = 12$

7) $2x + 5y = 8$
   $x - 6y = 4$

8) $x + 3y = -12$
   $3x + 4y = -6$

9) $2x + 30y = 9$
   $x = 6 - 15y$

10) $y = -3x - 20$
    $6x + 2y = 5$

11) $\quad 6x + y = -6$
    $-12x - 2y = 12$

12) $\quad x - 2y = 10$
    $3x - 6y = 30$

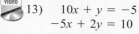
13) $\quad 10x + y = -5$
    $-5x + 2y = 10$

14) $2y - 7x = -14$
    $4x - y = 7$

15) $\quad x + 2y = 6$
    $x + 20y = -12$

16) $7x + 6y = 2$
    $x + 4y = 5$

17) $2x - 9y = -2$
    $6y - x = 0$

18) $5x - y = 8$
    $4y = 10 - x$

video
19) $9y - 18x = 5$
    $2x - y = 3$

20) $y = \dfrac{1}{3}x + 4$
    $3y - x = 12$

21) $2x - y = 6$
    $3y = -18 - x$

22) $y - 4x = -1$
    $8x + y = 2$

23) $2x - 5y = -4$
    $8x - 9y = 6$

24) $2x + 3y = 6$
    $5x + 2y = -7$

25) $9y - 2x = 22$
    $4x + 6y = -12$

26) $12x - 9y = -8$
    $-6x + 5y = 5$

27) $4y - 10x = -8$
    $15x - 6y = 12$

28) $28x + 7y = 1$
    $4x + y = -6$

Objective 2

29) If an equation in a system contains fractions, what should you do first to make the system easier to solve?

30) If an equation in a system contains decimals, what should you do first to make the system easier to solve?

Solve each system by substitution.

video
31) $\dfrac{1}{4}x - \dfrac{1}{2}y = 1$
    $\dfrac{2}{3}x + \dfrac{1}{6}y = \dfrac{25}{6}$

32) $\dfrac{2}{3}x + \dfrac{2}{3}y = 6$
    $\dfrac{3}{2}x - \dfrac{1}{4}y = \dfrac{13}{2}$

33) $\dfrac{x}{10} - \dfrac{y}{2} = \dfrac{13}{10}$
    $\dfrac{1}{3}x + \dfrac{5}{4}y = -\dfrac{3}{2}$

34) $\dfrac{4}{9}x - \dfrac{5}{3}y = -\dfrac{1}{9}$
    $\dfrac{x}{5} + \dfrac{y}{5} = -1$

35) $\dfrac{3}{4}x + \dfrac{5}{2}y = 5$
    $\dfrac{3}{2}x - \dfrac{1}{6}y = -\dfrac{1}{3}$

36) $\dfrac{1}{6}x + \dfrac{4}{3}y = \dfrac{13}{3}$
    $\dfrac{2}{5}x + \dfrac{3}{2}y = \dfrac{18}{5}$

37) $\dfrac{5}{3}x - \dfrac{4}{3}y = -\dfrac{4}{3}$
    $y = 2x + 4$

38) $\dfrac{3}{4}x + \dfrac{1}{2}y = 6$
    $x = 3y + 8$

video
39) $\quad 0.2x - 0.1y = 0.1$
    $0.01x + 0.04y = 0.23$

40) $0.01x + 0.10y = -0.11$
    $0.02x - 0.01y = 0.20$

41) $\quad 0.1x + 0.5y = 0.4$
    $-0.03x + 0.01y = 0.2$

42) $-0.02x + 0.05y = 0.4$
    $0.01x - 0.03y = -0.25$

43) $0.3x - 0.1y = 5$
    $0.15x + 0.1y = 4$

44) $0.2x - 0.1y = 1$
    $0.1x - 0.13y = 0.02$

Objective 1

Solve by substitution. Begin by combining like terms.

45) $5(2x - 3) + y - 6x = -24$
    $8y - 3(2y + 3) + x = -6$

46) $8 + 2(3x - 5) - 7x + 6y = 2$
    $9(y - 2) + 5x - 13y = -12$

47) $7x + 3(y - 2) = 7y + 6x - 1$
    $18 + 2(x - y) = 4(x + 2) - 5y$

48) $10(x + 3) - 7(y + 4) = 2(4x - 3y) + 3$
    $10 - 3(2x - 1) + 5y = 3y - 7x - 9$

49) $9y - 4(2y + 3) = -2(4x + 1)$
    $16 - 5(2x + 3) = 2(4 - y)$

50) $-(y + 3) = 5(2x + 1) - 7x$
    $x + 12 - 8(y + 2) = 6(2 - y)$

51) Noor needs to rent a car for one day while hers is being repaired. Rent-for-Less charges $0.40 per mile while Frugal Rentals charges $12 per day plus $0.30 per mile. Let x = the number of miles driven, and let y = the cost of the rental. The cost of renting a car from each company can be expressed with the following equations:

  Rent-for-Less: $y = 0.40x$
  Frugal Rentals: $y = 0.30x + 12$

a) How much would it cost Noor to rent a car from each company if she planned to drive 60 mi?

b) How much would it cost Noor to rent a car from each company if she planned to drive 160 mi?

c) Solve the system of equations using the substitution method, and explain the meaning of the solution.

d) Graph the system of equations, and explain when it is cheaper to rent a car from Rent-for-Less and when it is cheaper to rent a car from Frugal Rentals. When is the cost the same?

52) To rent a moving truck, Discount Van Lines charges $1.20 per mile while Comfort Ride Company charges $60 plus $1.00 per mile. Let x = the number of miles driven, and let y = the cost of the rental. The cost of renting a moving truck from each company can be expressed with the following equations:

  Discount Van Lines: $y = 1.20x$
  Comfort Ride: $y = 1.00x + 60$

a) How much would it cost to rent a truck from each company if the truck would be driven 100 mi?

b) How much would it cost to rent a truck from each company if the truck would be driven 400 mi?

c) Solve the system of equations using the substitution method, and explain the meaning of the solution.

d) Graph the system of equations, and explain when it is cheaper to rent a truck from Discount Van Lines and when it is cheaper to rent a truck from Comfort Ride. When is the cost the same?

Section 5.3 Solving Systems of Linear Equations by the Elimination Method

Objectives

1. Solve a Linear System Using the Elimination Method

2. Solve an Inconsistent or a Dependent System Using the Elimination Method

3. Use the Elimination Method Twice to Solve a Linear System

1. Solve a Linear System Using the Elimination Method

The next technique we will learn for solving a system of equations is the **elimination method**. (This is also called the **addition method**.) It is based on the addition property of equality which says that we can add the *same* quantity to each side of an equation and preserve the equality.

$$\text{If } a = b, \text{ then } a + c = b + c$$

We can extend this idea by saying that we can add *equal* quantities to each side of an equation and still preserve the equality.

$$\text{If } a = b \text{ and } c = d, \text{ then } a + c = b + d$$

The object of the elimination method is to add the equations (or multiples of one or both of the equations) so that one variable is eliminated. Then, we can solve for the remaining variable.

Example 1

Solve the system using the elimination method.

$$x + y = 9 \qquad (1)$$
$$x - y = -5 \qquad (2)$$

Solution

The left side of each equation is equal to the right side of each equation. Therefore, if we add the left sides together and add the right sides together, we can set them equal. We will add these equations vertically. The y-terms are eliminated, enabling us to solve for x.

$$
\begin{array}{ll}
\ x + y = 9 & (1) \\
+\ \ x - y = -5 & (2) \\
\hline
2x + 0y = 4 & \text{Add equations (1) and (2).} \\
2x = 4 & \text{Simplify.} \\
x = 2 & \text{Divide by 2.}
\end{array}
$$

We have determined that $x = 2$. But now we must find y. We can substitute $x = 2$ into either equation to find the value of y. Here, we will use equation (1).

$$
\begin{array}{ll}
x + y = 9 & (1) \\
2 + y = 9 & \text{Substitute 2 for } x. \\
y = 7 & \text{Subtract 2.}
\end{array}
$$

Check $x = 2$ and $y = 7$ in *both* equations.

$$
\begin{array}{ll}
x + y = 9 & \\
2 + 7 \overset{?}{=} 9 & \text{Substitute.} \\
9 = 9 & \text{True}
\end{array}
\qquad
\begin{array}{ll}
x - y = -5 & \\
2 - 7 \overset{?}{=} -5 & \text{Substitute.} \\
-5 = -5 & \text{True}
\end{array}
$$

The solution is (2, 7).

You Try 1

Solve the system using the elimination method.

$$x + 2y = -6$$
$$-x - 3y = 13$$

In Example 1, simply adding the equations eliminated a variable. But what do we do if we *cannot* eliminate a variable just by adding the equations together?

Example 2

Solve the system using the elimination method.

$$2x + 5y = 5 \quad (1)$$
$$x + 4y = 7 \quad (2)$$

Solution

Just adding these equations will *not* eliminate a variable. The multiplication property of equality tells us that multiplying both sides of an equation by the same quantity results in an equivalent equation. If we multiply equation (2) by -2, the coefficient of x will be -2.

$$-2(x + 4y) = -2(7) \qquad \text{-2 times (2)}$$
$$-2x - 8y = -14 \qquad \text{New, equivalent equation}$$

Original System		**Rewrite the System**
$2x + 5y = 5$	$\longrightarrow$	$2x + 5y = 5$
$x + 4y = 7$		$-2x - 8y = -14$

Add the equations in the rewritten system. The x is eliminated.

$$
\begin{array}{rl}
 2x + 5y = 5 & \\
+\ -2x - 8y = -14 & \\
\hline
0x - 3y = -9 & \text{Add equations.} \\
-3y = -9 & \text{Simplify.} \\
y = 3 & \text{Solve for } y.
\end{array}
$$

Substitute $y = 3$ into (1) or (2) to find x. We will use equation (2).

$$
\begin{array}{ll}
x + 4y = 7 & (2) \\
x + 4(3) = 7 & \text{Substitute 3 for } y. \\
x + 12 = 7 & \\
x = -5 &
\end{array}
$$

The solution is $(-5, 3)$. Check the solution in equations (1) and (2).

You Try 2

Solve the system using the elimination method.

$$8x - y = -5$$
$$-6x + 2y = 15$$

Here are the steps for solving a system of two linear equations using the elimination method.

Solving a System of Two Linear Equations by the Elimination Method

1) Write each equation in the form $Ax + By = C$.

2) Look at the equations carefully to determine which variable to eliminate. If necessary, multiply one or both of the equations by a number so that the coefficients of the variable to be eliminated are negatives of one another.

3) Add the equations, and solve for the remaining variable.

4) Substitute the value found in step 3 into either of the original equations to find the value of the other variable.

5) Check the solution in each of the original equations.

We will follow these steps to solve the next system.

Example 3

Solve the system using the elimination method.

$$3x = 7y + 5 \qquad (1)$$
$$2x - 3 = 5y \qquad (2)$$

Solution

1) **Write each equation in the form $Ax + By = C$.**

$$3x = 7y + 5 \qquad (1) \qquad\qquad\qquad 2x - 3 = 5y \qquad (2)$$
$$3x - 7y = 5 \quad \text{Subtract } 7y. \qquad\qquad 2x - 5y = 3 \quad \text{Subtract } 5y \text{ and add } 3.$$

When we rewrite the equations in the form $Ax + By = C$, we get

$$3x - 7y = 5 \qquad (3)$$
$$2x - 5y = 3 \qquad (4)$$

2) **Determine which variable to eliminate from equations (3) and (4).** Often, it is easier to eliminate the variable with the smaller coefficients. Therefore, *we will eliminate x.*

The least common multiple of 3 and 2 (the x-coefficients) is 6. Before we add the equations, one x-coefficient should be 6, and the other should be -6. Multiply equation (3) by 2 and equation (4) by -3.

Rewrite the System

$$2(3x - 7y) = 2(5) \qquad \text{2 times (3)} \qquad\qquad\qquad 6x - 14y = 10$$
$$-3(2x - 5y) = -3(3) \qquad \text{-3 times (4)} \qquad \longrightarrow \qquad -6x + 15y = -9$$

3) **Add the resulting equations to eliminate x. Solve for y.**

$$\begin{array}{r} 6x - 14y = 10 \\ + \ -6x + 15y = -9 \\ \hline y = 1 \end{array}$$

4) **Substitute $y = 1$ into equation (1) and solve for x.**

$$3x = 7y + 5 \qquad (1)$$
$$3x = 7(1) + 5 \qquad \text{Substitute.}$$
$$3x = 7 + 5$$
$$3x = 12$$
$$x = 4$$

5) **Check** to verify that $(4, 1)$ satisfies each of the original equations. The solution is $(4, 1)$.

You Try 3

Solve the system using the elimination method.

$$4x - 10 = -3y$$
$$5x = 4y - 3$$

2. Solve an Inconsistent or a Dependent System Using the Elimination Method

We have seen in Sections 5.1 and 5.2 that some systems have no solution, and some have an infinite number of solutions. How does the elimination method illustrate these results?

Example 4

Solve the system using the elimination method.

$$9y = 12x + 5 \qquad (1)$$
$$6y - 8x = -11 \qquad (2)$$

Solution

1) Write each equation in the form $Ax + By = C$.

$$9y = 12x + 5 \qquad \longrightarrow \qquad -12x + 9y = 5 \qquad (3)$$
$$6y - 8x = -11 \qquad\qquad -8x + 6y = -11 \qquad (4)$$

2) Determine which variable to eliminate from equations (3) and (4). Eliminate y. The least common multiple of 9 and 6 is 18. One y-coefficient must be 18, and the other must be -18.

Rewrite the System

$$-2(-12x + 9y) = -2(5) \qquad \longrightarrow \qquad 24x - 18y = -10$$
$$3(-8x + 6y) = 3(-11) \qquad\qquad -24x + 18y = -33$$

3) Add the equations.

$$24x - 18y = -10$$
$$+ \quad -24x + 18y = -33$$
$$\overline{\qquad\qquad 0 = -43 \qquad} \text{False}$$

The variables drop out, and we get a false statement. Therefore, the system is inconsistent, and the solution set is $\varnothing$.

You Try 4

Solve the system using the elimination method.

$$24x + 6y = -7$$
$$4y + 3 = -16x$$

Example 5

Solve the system using the elimination method.

$$12x - 18y = 9 \qquad (1)$$
$$y = \frac{2}{3}x - \frac{1}{2} \qquad (2)$$

Solution

1) Write equation (2) in the form $Ax + By = C$.

$$y = \frac{2}{3}x - \frac{1}{2} \qquad (2)$$
$$6y = 6\left(\frac{2}{3}x - \frac{1}{2}\right) \qquad \text{Multiply by 6 to eliminate fractions.}$$
$$6y = 4x - 3$$
$$-4x + 6y = -3 \qquad (3) \qquad \text{Rewrite as } Ax + By = C.$$

We can rewrite $y = \frac{2}{3}x - \frac{1}{2}$ as $-4x + 6y = -3$, equation (3).

2) Determine which variable to eliminate from equations (1) and (3).

$$12x - 18y = 9 \qquad (1)$$
$$-4x + 6y = -3 \qquad (3)$$

Eliminate x. Multiply equation (3) by 3.

$$12x - 18y = 9 \qquad (1)$$
$$-12x + 18y = -9 \qquad \text{3 times (3)}$$

3) Add the equations.

$$\begin{array}{r} 12x - 18y = 9 \\ + \ -12x + 18y = -9 \\ \hline 0 = 0 \qquad \text{True} \end{array}$$

The variables drop out, and we get a true statement. The system is dependent, so there are an infinite number of solutions of the form $\{(x, y)|12x - 18y = 9\}$.

You Try 5

Solve the system using the elimination method.

$$5y = 3x + 5$$
$$-6x + 10y = 10$$

3. Use the Elimination Method Twice to Solve a Linear System

Finally, there is a time when applying the elimination method *twice* is the best strategy. We illustrate this in the next example.

Example 6

Solve using the elimination method.

$$9x - 4y = 7 \qquad (1)$$
$$2x + 3y = -8 \qquad (2)$$

Solution

Each equation is written in the form $Ax + By = C$, so we begin with step 2.

2)   Determine which variable to eliminate from equations (1) and (2).

   Eliminate y.

Rewrite the System

$$3(9x - 4y) = 3(7)$$ $$27x - 12y = 21$$
$$4(2x + 3y) = 4(-8)$$ $\longrightarrow$ $$8x + 12y = -32$$

3)   Add the resulting equations to eliminate y. Solve for x.

$$
\begin{array}{r}
27x - 12y = 21 \\
+ \quad 8x + 12y = -32 \\
\hline
35x = -11
\end{array}
$$

$$x = -\frac{11}{35} \qquad \text{Solve for } x.$$

Normally, we would substitute $x = -\dfrac{11}{35}$ into equation (1) or equation (2) and solve for y. This time, however, working with a number like $-\dfrac{11}{35}$ would be difficult, so *we will use the elimination method a second time.*

   Go back to the original equations, (1) and (2), and use the elimination method again but eliminate the other variable, x. Then, solve for y.

Eliminate x from $$9x - 4y = 7 \qquad (1)$$
$$2x + 3y = -8 \qquad (2)$$

Rewrite the System

$$-2(9x - 4y) = -2(7)$$
$$9(2x + 3y) = 9(-8)$$

$\longrightarrow$

$$-18x + 8y = -14$$
$$18x + 27y = -72$$

Add the equations

$$\begin{array}{r} -18x + 8y = -14 \\ +\underline{18x + 27y = -72} \\ 35y = -86 \end{array}$$

$$y = -\frac{86}{35} \qquad \text{Solve for } y.$$

Check to verify that the solution is $\left(-\dfrac{11}{35}, -\dfrac{86}{35}\right)$.

 You Try 6

Solve using the elimination method.

$$5x + 7y = 20$$
$$2x - 3y = -4$$

Answers to You Try Exercises

1) $(8, -7)$ 2) $\left(\dfrac{1}{2}, 9\right)$ 3) $(1, 2)$ 4) $\varnothing$

5) Infinite number of solutions of the form $\{(x, y)|5y = 3x + 5\}$ 6) $\left(\dfrac{32}{29}, \dfrac{60}{29}\right)$

5.3 Exercises

Objectives 1 and 2

Solve each system using the elimination method.

1) $x - 3y = 1$
   $-x + y = -3$

2) $-2x - y = -1$
   $4x + y = -5$

3) $3x + 5y = -10$
   $7x - 5y = 10$

4) $4x - 3y = -5$
   $-4x + 5y = 11$

5) $7x + 6y = 3$
   $3x + 2y = -1$

6) $-8x + 5y = -16$
   $4x - 7y = 8$

7) $3x - y = 4$
   $-6x + 2y = -8$

8) $5x - 6y = -2$
   $10x - 12y = 7$

9) $3x + 2y = -9$
   $2x - 7y = 19$

10) $3x - 5y = -1$
    $-4x + 7y = 2$

11) $x = 12 - 4y$
    $2x - 7 = 9y$

12) $5x + 3y = -11$
    $y = 6x + 4$

13) $2x - 9 = 8y$
    $20y - 5x = 6$

14) $4x - 24y = -20$
    $x = 6y - 5$

15) $y = 6x - 10$
    $-4x + 5y = -11$

16) $8x = 6y - 1$
    $10y - 6 = -4x$

17) $3x + 4y = 9$
    $5x + 6y = 16$

18) $6x - 3y = -11$
    $9x - 2y = 1$

19) $-2x - 11 = 16y$
    $x = 3 - 8y$

20) $15x + 10y = 30$
    $4y = 12 - 6x$

21) $7x + 2y = 12$
    $24 - 14x = 4y$

22) $4 + 9y = -21x$
    $14x + 6y = -1$

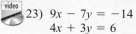 23) $9x - 7y = -14$
    $4x + 3y = 6$

24) $6x + 5y = 13$
    $5x + 3y = 5$

25) What is the first step in solving this system by the elimination method? *Do not solve.*

$$0.1x + 2y = -0.8$$
$$0.03x + 0.10y = 0.26$$

26) What is the first step in solving this system by the elimination method? *Do not solve.*

$$\frac{x}{4} + \frac{y}{2} = -1$$
$$\frac{3}{8}x + \frac{5}{3}y = -\frac{7}{12}$$

Solve each system using the elimination method.

27) $\dfrac{x}{4} + \dfrac{y}{2} = -1$

$\dfrac{3}{8}x + \dfrac{5}{3}y = -\dfrac{7}{12}$

28) $\dfrac{1}{2}x + \dfrac{2}{3}y = -\dfrac{29}{6}$

$-\dfrac{1}{3}x + y = -4$

29) $\dfrac{x}{2} - \dfrac{y}{5} = \dfrac{1}{10}$

$\dfrac{x}{3} + \dfrac{y}{4} = \dfrac{5}{6}$

30) $x + \dfrac{y}{4} = \dfrac{7}{2}$

$\dfrac{2}{5}x + \dfrac{1}{2}y = -1$

31) $x + \dfrac{3}{2}y = 13$

$-\dfrac{1}{8}x + \dfrac{1}{4}y = \dfrac{1}{8}$

32) $\dfrac{x}{12} - \dfrac{y}{6} = \dfrac{2}{3}$

$\dfrac{x}{4} + \dfrac{y}{3} = 2$

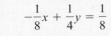

33) $0.1x + 2y = -0.8$
$0.03x + 0.10y = 0.26$

34) $0.6x - 0.1y = 0.5$
$0.10x - 0.03y = -0.01$

35) $0.02x + 0.07y = -0.24$
$0.05y - 0.04x = 0.10$

36) $0.8x - 0.3y = 0.5$
$0.07x + 0.05y = 0.12$

37) $2(y - 6) = 3y + 4(x - 5)$
$2(4x + 3) - 5 = 2(1 - y) + 5x$

38) $5(2x + 3) - 3x = 6(x + 2) + 4y$
$3(4y - 1) - x = 2(x - 4) + 7y$

39) $20 + 3(2y - 3) = 4(2y - 1) - 9x$
$5(3x - 4) + 8y = 3x + 7(y - 1)$

40) $6(3x + 4) - 8(x + 2) = 5 - 3y$
$6x - 2(5y + 2) = -7(2y - 1) - 4$

41) $6(x - 3) + x - 4y = 1 + 2(x - 9)$
$4(2y - 3) + 10x = 5(x + 1) - 4$

42) $12x - 4(2y + 3) = 5(x - y)$
$18 - 5(2x + 5) = -3(x + 4) - y + 1$

Objective 3

Solve each system using the elimination method twice. See Example 6.

43) $4x + 5y = -6$
$3x + 8y = 15$

44) $2x - 11y = 3$
$5x + 4y = -7$

45) $2x - 7y = -10$
$6x + 2y = 15$

46) $6x - 9y = -5$
$4x + 3y = 18$

47) $8x - 4y = -21$
$-5x + 6y = 12$

48) $3x + 7y = -16$
$8x - 5y = 20$

49) Given the following system of equations,

$$x + y = 8$$
$$x + y = c$$

find c so that the system has

a) an infinite number of solutions.

b) no solution.

50) Given the following system of equations,

$$x - y = 3$$
$$x - y = c$$

find c so that the system has

a) an infinite number of solutions.

b) no solution.

51) Given the following system of equations,

$$2x - 3y = 5$$
$$ax - 6y = 10$$

find a so that the system has

a) an infinite number of solutions.

b) exactly one solution.

52) Given the following system of equations,

$$-2x + by = 4$$
$$10x - 5y = -20$$

find b so that the system has

a) an infinite number of solutions.

b) exactly one solution.

53) Find b so that $(2, -1)$ is a solution to the system

$$3x - 4y = 10$$
$$-x + by = -7$$

54) Find a so that $(4, 3)$ is a solution to the system

$$ax + 5y = 3$$
$$2x - 3y = -1$$

Mid-Chapter Summary

Objectives (Summary)

1. Learn How to Recognize Which Method, Substitution or Elimination, Is Best for Solving a Given System of Linear Equations

1. Learn How to Recognize Which Method, Substitution or Elimination, Is Best for Solving a Given System of Linear Equations

We have learned three methods for solving systems of linear equations:

1) Graphing
2) Substitution
3) Elimination

How do we know which method is best for a particular system? We will answer this question by looking at a few examples, and then we will summarize our findings.

First, solving a system by graphing is the least desirable of the methods. The point of intersection can be difficult to read, especially if one of the numbers is a fraction. But, the graphing method is important in certain situations and is one you should know.

Example 1

Decide which method to use to solve each system, substitution or elimination, and explain why this method was chosen. Then, solve the system.

a) $-7x + 3y = 18$
    $x = 2y - 12$

b) $6x + y = -5$
    $-7x - 4y = 3$

c) $12x - 7y = -1$
    $8x + 3y = -16$

Solution

a) $-7x + 3y = 18$
    $x = 2y - 12$

The second equation in this system is solved for x, *and* there are no fractions in this equation. *Solve this system by substitution.*

$$-7x + 3y = 18 \qquad \text{First equation}$$
$$-7(2y - 12) + 3y = 18 \qquad \text{Substitute } 2y - 12 \text{ for } x.$$
$$-14y + 84 + 3y = 18 \qquad \text{Distribute.}$$
$$-11y + 84 = 18 \qquad \text{Combine like terms.}$$
$$-11y = -66 \qquad \text{Subtract 84.}$$
$$y = 6 \qquad \text{Solve for } y.$$

Substitute $y = 6$ into $x = 2y - 12$.

$$x = 2(6) - 12 \qquad \text{Substitute.}$$
$$x = 12 - 12 \qquad \text{Multiply.}$$
$$x = 0 \qquad \text{Solve for } x.$$

The solution is $(0, 6)$. The check is left to the student.

b) $6x + y = -5$
    $-7x - 4y = 3$

In the first equation, y has a coefficient of 1. Therefore, we could easily solve the equation for y and substitute the expression into the second equation.

(Remember, solving for one of the other variables would result in having fractions in the equation.) *Or*, since each equation is in the form $Ax + By = C$, the elimination method would work well too. *Either substitution or the elimination method would be a good choice to solve this system.* The student should choose whichever method he or she prefers. We will solve this system by substitution.

$$6x + y = -5 \qquad \text{First equation}$$
$$y = -6x - 5 \qquad \text{Solve the first equation for } y.$$

$$-7x - 4y = 3 \qquad \text{Second equation}$$
$$-7x - 4(-6x - 5) = 3 \qquad \text{Substitute } -6x - 5 \text{ for } y.$$
$$-7x + 24x + 20 = 3 \qquad \text{Distribute.}$$
$$17x + 20 = 3 \qquad \text{Combine like terms.}$$
$$17x = -17 \qquad \text{Subtract 20.}$$
$$x = -1 \qquad \text{Solve for } x.$$

Substitute $x = -1$ into $y = -6x - 5$.

$$y = -6(-1) - 5 \qquad \text{Substitute.}$$
$$y = 6 - 5 \qquad \text{Multiply.}$$
$$y = 1 \qquad \text{Solve for } x.$$

The check is left to the student. The solution is $(-1, 1)$.

c) $12x - 7y = -1$
    $8x + 3y = -16$

None of the variables has a coefficient of 1 or -1. Therefore, we do *not* want to solve for a variable and use substitution because we would have to work with fractions in the equation. *To solve a system like this, where none of the coefficients are 1 or -1, use the elimination method.*

Eliminate x.

Rewrite the System

$$-2(12x - 7y) = -2(-1) \qquad \longrightarrow \qquad -24x + 14y = 2$$
$$3(8x + 3y) = 3(-16) \qquad\qquad\qquad\quad + \underline{\quad 24x + 9y = -48\quad}$$
$$23y = -46$$
$$y = -2$$

Substitute $y = -2$ into $12x - 7y = -1$.

$$12x - 7(-2) = -1 \qquad \text{Substitute.}$$
$$12x + 14 = -1 \qquad \text{Multiply.}$$
$$12x = -15 \qquad \text{Subtract 14.}$$
$$x = -\frac{15}{12} = -\frac{5}{4} \qquad \text{Solve for } x.$$

Check $x = -\dfrac{5}{4}$, $y = -2$ in each of the original equations to verify that $\left(-\dfrac{5}{4}, -2\right)$ is the solution to the system.

Choosing Between Substitution and the Elimination Method to Solve a System

1) If at least one of the equations is solved for a variable and the equation contains no fractions, *use substitution*. See Example 1a).

$$-7x + 3y = 18$$
$$x = 2y - 12$$

2) If a variable has a coefficient of 1 or -1, you can solve for that variable and *use the substitution method*. See Example 1b).

$$6x + y = -5$$
$$-7x - 4y = 3$$

Or, leave each equation in the form Ax + By = C and solve using the elimination method. Either approach is good and is a matter of personal preference.

3) If none of the variables has a coefficient of 1 or -1, *use the elimination method*. See Example 1c).

$$12x - 7y = -1$$
$$8x + 3y = -16$$

Remember, if an equation contains fractions or decimals, begin by eliminating these. Then, decide which method to use following the guidelines listed here.

 ## You Try 1

Decide which method to use to solve each system, substitution or elimination, and explain why this method was chosen. Then, solve the system.

a) $6x - 7y = -9$
   $-5x + 2y = 19$

b) $2x + 5y = 3$
   $y = -3x + 11$

c) $x = -\dfrac{3}{4}y + \dfrac{3}{2}$
   $3x + 2y = 5$

d) $5x - 6y = 8$
   $x + 3y = 3$

Answers to You Try Exercises

1) a) $(-5, -3)$ b) $(4, -1)$ c) $(3, -2)$ d) $\left(2, \dfrac{1}{3}\right)$

Objective I

Decide which method to use to solve each system, substitution or elimination, and explain why this method was chosen. Then, solve the system.

1) $8x - 5y = 10$
   $2x - 3y = -8$

2) $y = 3x + 6$
   $9x + 4y = 10$

3) $x + 6y = -10$
   $3x - 8y = -4$

4) $3x + 7y = 12$
   $4x + 11y = 11$

5) $y - 4x = -11$
   $x = y + 8$

6) $4x - 5y = 4$
   $y = \dfrac{3}{4}x - \dfrac{1}{2}$

Solve each system by either the substitution or elimination method.

7) $9x - 2y = 8$
   $y = 2x + 1$

8) $-3x + 10y = 1$
   $2x + 5y = 11$

9) $8y - x = -11$
   $x + 10y = 2$

10) $x = 1 - 3y$
    $x + 2y = 4$

11) $10x + 4y = 7$
    $15x + 6y = -2$

12) $3x - 5y = 10$
    $-9x + 15y = -30$

13) $-2x - 7y = 5$
    $\dfrac{1}{3}x + \dfrac{3}{2}y = -\dfrac{1}{2}$

14) $\dfrac{1}{4}x = \dfrac{3}{2} - y$
    $3x - 4y = 14$

15) $y = -6$
    $5x + 2y = 3$

16) $7x - y = 5$
    $3x + 2y = -10$

17) $5y - 4x = 8$
    $10x + 3y = 11$

18) $4x + 5y = 13$
    $11y - 6x = -1$

19) $y = -6x + 5$
    $12x + 2y = 10$

20) $-28x + 7y = 3$
    $y = 4x - 9$

21) $0.01x + 0.02y = 0.28$
    $0.04x - 0.03y = 0.13$

22) $0.01x + 0.04y = 0.15$
    $0.02x - 0.01y = 0.03$

23) $6(2x - 3) = y + 4(x - 3)$
    $5(3x + 4) + 4y = 11 - 3y + 27x$

24) $y = 2x + 9$
    $2(5 - 2y) + 3x = 2(3x + 8) - 8y$

25) $y = \dfrac{5}{6}x + \dfrac{10}{3}$
    $\dfrac{1}{2}x + \dfrac{9}{8}y = -2$

26) $-\dfrac{7}{2}x - \dfrac{3}{2}y = \dfrac{19}{4}$
    $\dfrac{1}{3}x + \dfrac{3}{4}y = -\dfrac{1}{4}$

Solve each system by graphing.

27) $y = \dfrac{1}{2}x + 1$
    $x + y = 4$

28) $-x + 4y = 8$
    $y = 2x + 2$

29) $2x - y = -3$
    $y = -x - 3$

30) $y = -\dfrac{1}{3}x$
    $2x - 3y = -9$

31) $3x + 2y = 9$
    $x = 5$

32) $y = 1$
    $x - 2y = 2$

Section 5.4 Applications of Systems of Two Equations

Objectives

1. Learn the Steps for Solving an Applied Problem Using a System of Equations
2. Solve Problems Involving General Quantities
3. Solve Problems Involving Geometry
4. Solve Problems Involving Cost
5. Solve Mixture Problems
6. Solve Problems Involving Distance, Rate, and Time

1. Learn the Steps for Solving an Applied Problem Using a System of Equations

In Chapter 3 we introduced the five steps for solving applied problems. We defined unknown quantities in terms of *one* variable to write a linear equation to solve an applied problem. Sometimes, however, it is easier to use *two variables* to represent the unknown quantities. Then we must write a system of *two equations* which relate the variables and solve the system using one of the methods in this chapter. We will take the steps used in Chapter 3 to solve an equation in *one* variable and modify them so that we have a strategy for solving applied problems using a system of **two equations** in **two variables**.

Solving an Applied Problem Using a System of Equations

Step 1: **Read the problem carefully, twice.** Draw a picture if applicable.

Step 2: Identify what you are being asked to find. **Define the unknowns;** that is, **write down** what each variable represents. If applicable, label the picture with the variables.

Step 3: **Write a system of equations relating the variables.** It may be helpful to begin by writing the equations in words.

Step 4: **Solve the system.**

Step 5: **Interpret the meaning of the solution as it relates to the problem.** Be sure your answer makes sense.

---

2. Solve Problems Involving General Quantities

Example 1

Write a system of equations and solve.

In 2004 the prime-time TV shows of HBO received 60 more Emmy Award nominations than NBC. Together, their shows received a total of 188 nominations. How many Emmy nominations did HBO and NBC each receive in 2004? (Source: www.emmys.org)

Solution

Step 1: **Read the problem carefully, twice.**

Step 2: We must find the number of Emmy nominations received by HBO and by NBC.

Define the unknowns; assign a variable to represent each unknown quantity.

$$x = \text{number of nominations for HBO}$$
$$y = \text{number of nominations for NBC}$$

Step 3: **Write a system of equations relating the variables.**

We must write two equations. Let's think of the equations in English first. Then, we can translate them to algebraic equations.

i) *To get one equation*, use the information that says HBO and NBC received a total of 188 nominations. Write an equation in words, then translate it to an algebraic equation.

$$\underset{x}{\text{Number of HBO nominations}} + \underset{y}{\text{Number of NBC nominations}} = \underset{188}{\text{Total number of nominations}}$$

One equation is $x + y = 188$.

ii) *To get the second equation*, use the information that says that HBO received 60 more nominations than NBC.

The second equation is $x = 60 + y$.

The system of equations is $\begin{array}{l} x + y = 188 \\ x = 60 + y. \end{array}$

Step 4: **Solve the system.**

Use substitution.

$$x + y = 188$$
$$(60 + y) + y = 188 \qquad \text{Substitute.}$$
$$60 + 2y = 188$$
$$2y = 128$$
$$y = 64$$

Find x by substituting $y = 64$ into $x = 60 + y$.

$$x = 60 + 64$$
$$x = 124$$

The solution to the system is $(124, 64)$.

Step 5: **Interpret the meaning of the solution as it relates to the problem.**

HBO received 124 Emmy Award nominations, and NBC received 64.

Does the answer make sense? Yes. The total number of nominations for HBO and NBC was 188 and $124 + 64 = 188$. HBO received 60 more nominations than NBC and $60 + 64 = 124$.

The answer is correct as stated above.

You Try 1

Write a system of equations and solve.

The Turquoise Bay Motel has 51 rooms. Some of them have two double beds and half as many rooms have one queen-size bed. Determine how many rooms have two beds and how many rooms have one bed.

In Chapter 3 we solved problems involving geometry by defining the unknowns in terms of one variable. Now, we will see how we can use two variables and a system of equations to solve these types of applications.

3. Solve Problems Involving Geometry

Example 2

Write a system of equations and solve.

The Alvarez family purchased a rectangular lot on which they will build a new house. The lot is 70 ft longer than it is wide, and the perimeter is 460 ft. Find the length and width of the lot.

Solution

Step 1: **Read the problem carefully, twice. Draw a picture.**

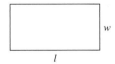

Step 2: We must find the length and width of the lot. **Define the unknowns; assign a variable to represent each unknown quantity.**

$$l = \text{length of the lot}$$
$$w = \text{width of the lot}$$

Label the picture with the variables.

Step 3: **Write a system of equations relating the variables.**

i) *To get one equation,* use the information that says the lot is 70 ft longer than it is wide. We can also think of this as telling us that the length is 70 ft more than the width.

Length	is	70 ft more than	width
↓	↓	↓	↓
l	$=$	$70+$	w

One equation is $l = 70 + w$.

ii) We are told that the perimeter of the lot is 460 feet. *Use this information to write the second equation.*

Recall that Perimeter = 2(length) + 2(width).

Therefore, $460 = 2l + 2w$

The second equation is $2l + 2w = 460$.

The system of equations is $\begin{array}{l} l = 70 + w \\ 2l + 2w = 460. \end{array}$

Step 4: **Solve the system.**

Use substitution.

$$2l + 2w = 460$$
$$2(70 + w) + 2w = 460 \qquad \text{Substitute.}$$
$$140 + 2w + 2w = 460 \qquad \text{Distribute.}$$
$$4w + 140 = 460$$
$$4w = 320$$
$$w = 80$$

Find l by substituting $w = 80$ into $l = 70 + w$.

$$l = 70 + 80$$
$$l = 150$$

The solution to the system is (150, 80). (The ordered pair is written as (l, w), in alphabetical order.)

Step 5: **Interpret the meaning of the solution as it relates to the problem.**
The length of the lot is 150 ft, and the width of the lot is 80 ft.

Check the solution to verify that the numbers make sense. The answer is correct as stated above.

You Try 2

Write a system of equations and solve.
   A rectangular mouse pad is 1.5 in. longer than it is wide. The perimeter of the mouse pad is 29 in. Find its dimensions.

Another common type of problem is one involving the cost of items along with the number of items. Solving these kinds of applications with a system of equations is generally a good strategy.

4. Solve Problems Involving Cost

Example 3

Write a system of equations and solve.
   In 2004, Usher and Alicia Keys each had concerts at the Allstate Arena near Chicago. Kayla and Levon sat in the same section for each performance. Kayla bought four tickets to see Usher and four to see Alicia Keys for $360. Levon spent $220.50 on two Usher tickets and three Alicia Keys tickets. Find the cost of a ticket to each concert. (Source: www.pollstaronline.com)

Solution

Step 1:   **Read the problem carefully, twice.**

Step 2:   We must find the cost of an Usher ticket and the cost of an Alicia Keys ticket. **Define the unknowns;** assign a variable to represent each unknown quantity.

$$x = \text{cost of one Usher ticket}$$
$$y = \text{cost of one Alicia Keys ticket}$$

Step 3:   **Write a system of equations relating the variables.**

     i)   *To get one equation*, use the information about Kayla's purchase.

$$\underset{\substack{\nearrow \quad \nwarrow \\ \text{Number} \quad \text{Cost of} \\ \text{of tickets} \quad \text{each ticket}}}{4x} \;+\; \underset{\substack{\nearrow \quad \nwarrow \\ \text{Number} \quad \text{Cost of} \\ \text{of tickets} \quad \text{each ticket}}}{4y} \;=\; 360.00$$

Cost of four Usher tickets + Cost of four Alicia Keys tickets = Total cost

One equation is $4x + 4y = 360.00$.

     ii)   *To get the second equation*, use the information about Levon's purchase.

$$\underset{\substack{\nearrow \quad \nwarrow \\ \text{Number} \quad \text{Cost of} \\ \text{of tickets} \quad \text{each ticket}}}{2x} \;+\; \underset{\substack{\nearrow \quad \nwarrow \\ \text{Number} \quad \text{Cost of} \\ \text{of tickets} \quad \text{each ticket}}}{3y} \;=\; 220.50$$

Cost of two Usher tickets + Cost of three Alicia Keys tickets = Total cost

The second equation is $2x + 3y = 220.50$.

The system of equations is $\begin{aligned} 4x + 4y &= 360.00 \\ 2x + 3y &= 220.50 \end{aligned}$.

Step 4:   **Solve the system.**

Use the elimination method. Multiply the second equation by -2 to eliminate x.

$$
\begin{aligned}
4x + 4y &= 360.00 \\
+\; -4x - 6y &= -441.00 \\
\hline
-2y &= {-81.00} \qquad \text{Add the equations.}\\
y &= 40.50
\end{aligned}
$$

Find x. We will substitute $y = 40.50$ into $4x + 4y = 360.00$

$$
\begin{aligned}
4x + 4(40.50) &= 360.00 \qquad \text{Substitute.}\\
4x + 162 &= 360.00 \\
4x &= 198.00 \\
x &= 49.50
\end{aligned}
$$

The solution to the system is $(49.50, 40.50)$.

Step 5: **Interpret the meaning of the solution as it relates to the problem.**

The cost of a ticket to see Usher in concert was $49.50, and the cost of a ticket to see Alicia Keys was $40.50.

Check the numbers to verify that they are correct.

You Try 3

Write a system of equations and solve.

At Julie's Jewelry Box, all necklaces sell for one price and all pairs of earrings sell for one price. Cailen buys three pairs of earrings and a necklace for $19.00, while Marcella buys one pair of earrings and two necklaces for $18.00. Find the cost of a pair of earrings and the cost of a necklace.

We have already seen how to solve mixture problems in terms of one variable and one equation. Now, we will discuss how to solve these same applications using two variables and a system of equations.

5. Solve Mixture Problems

Example 4

Write a system of equations and solve.

How many milliliters of an 8% hydrogen peroxide solution and how many milliliters of a 2% hydrogen peroxide solution must be mixed to get 300 ml of a 4% hydrogen peroxide solution?

Solution

Step 1: **Read the problem carefully, twice.**

Step 2: We must find the number of milliliters of the 8% hydrogen peroxide solution and the number of milliliters of 2% solution needed to make 300 ml of a 4% solution. **Define the unknowns;** assign a variable to represent each unknown quantity.

$$x = \text{number of milliliters of 8\% solution}$$
$$y = \text{number of milliliters of 2\% solution}$$

Step 3: **Write a system of equations relating the variables.**

Make a table to organize the information.

		Concentration	Number of Milliliters of the Solution	Number of Milliliters of Hydrogen Peroxide in the Solution
Mix these	8% solution	0.08	x	$0.08x$
	2% solution	0.02	y	$0.02y$
to make →	4% solution	0.04	300	0.04(300)

i) *To get one equation*, use the information in the *second column*. Since the 8% and 2% solutions are combined to make the 4% solution, we know that

Number of ml of 8% solution		Number of ml of 2% solution		Number of ml of 4% solution
↓		↓		↓
x	$+$	y	$=$	300

One equation is $x + y = 300$.

ii) *To get the second equation*, use the information in the *third column*. The number of milliliters of hydrogen peroxide in the final 4% must come from the hydrogen peroxide in the 8% and 2% solutions.

ml of hydrogen peroxide in 8% solution		ml of hydrogen peroxide in 2% solution		ml of hydrogen peroxide in 4% solution
↓		↓		↓
$0.08x$	$+$	$0.02y$	$=$	$0.04(300)$

The second equation is $0.08x + 0.02y = 0.04(300)$.

The system of equations is
$$\begin{aligned} x + y &= 300 \\ 0.08x + 0.02y &= 0.04(300). \end{aligned}$$

Step 4: **Solve the system.**

Multiply the second equation by 100 to eliminate the decimals. We get $8x + 2y = 4(300)$ or $8x + 2y = 1200$. The system we will work with, then, is

$$\begin{aligned} x + y &= 300 \\ 8x + 2y &= 1200 \end{aligned}$$

Use the elimination method. Multiply the first equation by -2 to eliminate y.

$$\begin{array}{rl} -2x - 2y = & -600 \\ + \quad 8x + 2y = & 1200 \\ \hline 6x = & 600 \\ x = & 100 \end{array} \qquad \text{Add the equations.}$$

Find y. Substitute $x = 100$ into $x + y = 300$.

$$\begin{aligned} 100 + y &= 300 \qquad \text{Substitute.} \\ y &= 200 \end{aligned}$$

The solution to the system is $(100, 200)$.

Step 5: **Interpret the meaning of the solution as it relates to the problem.**

100 ml of an 8% hydrogen peroxide solution and 200 ml of a 2% solution are needed to make 300 ml of a 4% hydrogen peroxide solution.

Check the answers in the equations to verify that they are correct.

You Try 4

Write a system of equations and solve.

How many milliliters of a 3% acid solution must be added to 60 ml of an 11% acid solution to make a 9% acid solution?

The next application we will look at involves distance, rate, and time.

6. Solve Problems Involving Distance, Rate, and Time

Example 5

Write a system of equations and solve.

Julia and Katherine start at the same point and begin biking in opposite directions. Julia rides 2 mph faster than Katherine. After 2 hr, the girls are 44 mi apart. How fast was each of them riding?

Solution

Step 1:   **Read the problem carefully, twice. Draw a picture.**

$$\text{Katherine} \longleftarrow \overset{\text{Start}}{\underset{\text{44 miles}}{\bullet}} \longrightarrow \text{Julia}$$

Step 2:   We must find the speed at which each girl was riding. **Define unknowns; assign a variable to represent each unknown quantity.**

$$x = \text{Julia's speed}$$
$$y = \text{Katherine's speed}$$

Step 3:   **Write a system of equations relating the variables.**

Make a table to organize the information. Recall that distance, rate, and time are related by the equation $d = rt$.

Fill in the information about their speeds (rates) x and y, and fill in 2 for the time since we know that they rode for 2 hr.

	d	$=$	r	t
Julia	$2x$		x	2
Katherine	$2y$		y	2

Since $d = rt$, Julia's distance $= x \cdot 2$ or $2x$ and Katherine's distance $= y \cdot 2$ or $2y$. Put these expressions in the table, and label the picture.

$$\text{Katherine} \longleftarrow \overset{2y \quad \text{Start} \quad 2x}{\underset{\text{44 miles}}{\bullet}} \longrightarrow \text{Julia}$$

i) *To get one equation*, look at the picture and think about the distance between the girls after 2 hr. We can write

Julia's distance	+	Katherine's distance	=	Distance between them
↓		↓		↓
$2x$	+	$2y$	=	44

One equation is $2x + 2y = 44$.

ii) *To get the second equation*, use the information that says Julia rides 2 mph faster than Katherine.

Julia's rate	is	2 mph more than	Katherine's rate
↓	↓	↓	↓
x	=	$2+$	y

The second equation is $x = 2 + y$.

The system of equations is $\begin{aligned} 2x + 2y &= 44 \\ x &= 2 + y. \end{aligned}$

Step 4: **Solve the system.**

Use substitution.

$$
\begin{aligned}
2x + 2y &= 44 \\
2(2 + y) + 2y &= 44 \qquad \text{Substitute } 2 + y \text{ for } x. \\
4 + 2y + 2y &= 44 \qquad \text{Distribute.} \\
4y + 4 &= 44 \\
4y &= 40 \\
y &= 10
\end{aligned}
$$

Find x by substituting $y = 10$ into $x = 2 + y$.

$$
\begin{aligned}
x &= 2 + 10 \\
x &= 12
\end{aligned}
$$

The solution to the system is (12, 10).

Step 5: **Interpret the meaning of the solution as it relates to the problem.**

Julia rides her bike at 12 mph, and Katherine rides hers at 10 mph.

Does the answer make sense? Julia's speed is 2 mph faster than Katherine's as stated in the problem. And, since each girl rode for 2 hr, the distance between them is

2(12)	+	2(10) = 24 + 20 = 44 mi
↓		↓
Julia's distance		Katherine's distance

Yes, the answer is correct as stated above.

 **You Try 5**

Write a system of equations and solve.

Kenny and Kyle start at the same point and begin biking in opposite directions. Kenny rides 1 mph faster than Kyle. After 3 hr, the boys are 51 mi apart. How fast was each of them riding?

Answers to You Try Exercises

1) 34 rooms have two beds, 17 rooms have one bed.   2) 6.5 in. by 8 in.   3) A pair of earrings costs $4.00 and a necklace costs $7.00.   4) 20 ml   5) Kyle: 8 mph; Kenny: 9 mph

5.4 Exercises

Boost your grade at mathzone.com!   MathZone

> Practice Problems   > NetTutor > Self-Test   > e-Professors   > Videos

Objective 2

Write a system of equations and solve.

1) The sum of two numbers is 36, and one number is two more than the other. Find the numbers.

2) One number is half another number. The sum of the two numbers is 108. Find the numbers.

3) In 2005, *The Aviator* was nominated for four more Academy Awards than the movie *Finding Neverland*. Together they received 18 nominations. How many Academy Award nominations did each movie receive? (Source: *Chicago Tribune*, 1/26/05)

4) Through 2005, the University of Southern California football team played in 10 more Rose Bowl games than the University of Michigan. All together, these teams have appeared in 48 Rose Bowls. How many appearances did each team make in the Rose Bowl? (Source: www.tournamentofroses.com)

5) There were a total of 2626 IHOP and Waffle House restaurants across the United States at the end of 2004. There were 314 fewer IHOPs than Waffle Houses. Determine the number of IHOP and Waffle House restaurants in the United States. (Source: www.ihop.com, www.ajc.com)

6) Through 2004, the University of Kentucky men's basketball team won four fewer NCAA championships than UCLA. The two teams won a total of 18 championship titles. How many championships did each team win? (Source: www.ncaasports.com)

7) The first NCAA Men's Basketball Championship game was played in 1939 in Evanston, Illinois. In 2004, 38,968 more people attended the final game in San Antonio than attended the first game in Evanston. The total number of people who attended these two games was 49,968. How many people saw the first NCAA championship in 1939, and how many were at the game in 2004? (Source: www.ncaasports.com)

8) Garth Brooks won two more Country Music Awards than Tim McGraw through 2004. They won a total of 20 awards. How many awards did each man win? (Source: www.cmaawards.com)

9) Annual per capita consumption of chicken in the United States in 2001 was 6.3 lb less than that of beef. Together, each American consumed, on average, 120.5 lb of beef and chicken in 2001. Find the amount of beef and the amount of chicken consumed, per person, in 2001. (Source: U.S. Department of Agriculture)

10) Mr. Chen has 27 students in his American History class. For their assignment on the Civil War, twice as many students chose to give a speech as chose to write a paper. How many students will be giving speeches, and how many will be writing papers?

Objective 3

11) The length of a rectangle is twice its width. Find the length and width of the rectangle if its perimeter is 78 in.

12) The width of a rectangle is 5 cm less than the length. If the perimeter is 46 cm, what are the dimensions of the rectangle?

13) Find the dimensions of a rectangular door that has a perimeter of 220 in. if the width is 50 in. less than the height of the door.

14) Yuki has a rectangular picture she wants to frame. Its perimeter is 42 in., and it is 4 in. longer than it is wide. Find the length and width of the picture.

15) An iPod Mini is rectangular in shape and has a perimeter of 28 cm. Its length is 4 cm more than its width. What are the dimensions of the iPod Mini?

16) Tiny Tots Day Care needs to put a new fence around their rectangular playground. They have determined they will need 220 ft of fencing. If the width of the playground is 30 ft less than the length, what are the playground's dimensions?

17) Find the measures of angles *x* and *y* if the measure of angle *x* is two-thirds the measure of angle *y* and if the angles are related according to the figure.

18) Find the measures of angles *x* and *y* if the measure of angle *y* is half the measure of angle *x* and if the angles are related according to the figure.

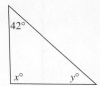

Objective 4

19) Jennifer and Carlos attended two concerts with their friends at the American Airlines Arena in Miami. Jennifer and her friends bought five tickets to see Marc Anthony and two tickets to the Santana concert for $563. Carlos' group purchased three Marc Anthony tickets and 6 for Santana for $657. Find the cost of a ticket to each concert. (Source: www.pollstaronline.com)

20) Both of the pop groups Train and Maroon 5 played at the House of Blues in North Myrtle Beach, SC in 2003. Three Maroon 5 tickets and two Train tickets would have cost $64, while four Maroon 5 tickets and four Train tickets would have cost $114. Find the cost of a ticket to each concert. (Source: www.pollstaronline.com)

21) Every Tuesday, Stacey goes to Panda Express to buy lunch for herself and her colleagues. One day she buys 3 two-item meals and 1 three-item meal for $21.96. The next Tuesday she spends $23.16 on 2 two-item meals and 2 three-item meals. What is the cost of a two-item meal and of a three-item meal? (Source: Panda Express menu)

22) Assume two families pay the average ticket price to attend a Green Bay Packers game in 2004. One family buys four tickets and a parking pass for $242.60. The other family buys six tickets and a parking pass for $351.40. Find the average ticket price and the cost of a parking pass to a Packers game in 2004. (Source: www.teammarketing.com)

23) On vacation, Wendell buys three key chains and five postcards for $10.00, and his sister buys two key chains and three postcards for $6.50. Find the cost of each souvenir.

24) At Sparkle Car Wash, two deluxe car washes and three regular car washes would cost $26.00. Four regular washes and one deluxe wash would cost $23.00. What is the cost of a deluxe car wash and of a regular wash?

25) Ella spends $7.50 on three cantaloupe and one watermelon at a farmers' market. Two cantaloupe and two watermelon would have cost $9.00. What is the price of a cantaloupe and the price of a watermelon?

26) One 12-oz serving of Coke and two 12-oz servings of Mountain Dew contain 31.3 tsp of sugar while three servings of Coke and one Mountain Dew contain 38.9 tsp of sugar. How much sugar is in a 12-oz serving of each drink? (Source: www.dentalgentlecare.com)

27) Carol orders six White Castle hamburgers and a small order of fries for $3.91, and Momar orders eight hamburgers and two small fries for $5.94. Find the cost of a hamburger and the cost of an order of french fries at White Castle. (Source: White Castle menu)

28) Six White Castle hamburgers and one small order of french fries contain 955 calories. Eight hamburgers and two orders of fries contain 1350. Determine how many calories are in a White Castle hamburger and in a small order of french fries. (Source: www.whitecastle.com)

Objective 5

29) How many ounces of a 9% alcohol solution and how many ounces of a 17% alcohol solution must be mixed to get 12 oz of a 15% alcohol solution?

30) How many milliliters of a 4% acid solution and how many milliliters of a 10% acid solution must be mixed to obtain 54 ml of a 6% acid solution?

31) How many pounds of peanuts that sell for $1.80 per pound should be mixed with cashews that sell for $4.50 per pound so that a 10-lb mixture is obtained that will sell for $2.61 per pound?

32) Raheem purchases 20 stamps. He buys some $0.37 stamps and some $0.23 stamps and spends $6.28. How many of each type of stamp did he buy?

33) Sally invested $4000 in two accounts, some of it at 3% simple interest, the rest in an account earning 5% simple interest. How much did she invest in each account if she earned $144 in interest after one year?

34) Diego inherited $20,000 and puts some of it into an account earning 4% simple interest and the rest in an account earning 7% simple interest. He earns a total of $1130 in interest after a year. How much did he deposit into each account?

35) Josh saves all of his quarters and dimes in a bank. When he opens it, he has 110 coins worth a total of $18.80. How many quarters and how many dimes does he have?

36) Mrs. Kowalski bought nine packages of batteries when they were on sale. The AA batteries cost $1.00 per package and the C batteries cost $1.50 per package. If she spent $11.50, how many packages of each type of battery did she buy?

37) How much pure acid and how many liters of a 10% acid solution should be mixed to get 12 L of a 40% acid solution?

38) How many ounces of pure orange juice and how many ounces of a citrus fruit drink containing 5% fruit juice should be mixed to get 76 oz of a fruit drink that is 25% fruit juice?

Objective 6

39) A car and a truck leave the same location, the car headed east and the truck headed west. The truck's speed is 10 mph less than the speed of the car. After 3 hr, the car and truck are 330 mi apart. Find the speed of each vehicle.

40) A passenger train and a freight train leave cities 400 mi apart and travel toward each other. The passenger train is traveling 20 mph faster than the freight train. Find the speed of each train if they pass each other after 5 hr.

41) Olivia can walk 8 mi in the same amount of time she can bike 22 mi. She bikes 7 mph faster than she walks. Find her walking and biking speeds.

42) A small, private plane can fly 400 mi in the same amount of time a jet can fly 1000 mi. If the jet's speed is 300 mph more than the speed of the small plane, find the speeds of both planes.

43) Nick and Scott leave opposite ends of a bike trail 13 mi apart and travel toward each other. Scott is traveling 2 mph slower than Nick. Find each of their speeds if they meet after 30 min.

44) Vashon can travel 120 mi by car in the same amount of time he can take the train 150 mi. If the train travels 10 mph faster than the car, find the speed of the car and the speed of the train.

Section 5.5 Systems of Linear Equations in Three Variables

Up to this point, we have discussed different ways to solve systems of linear equations in two variables. In this section we will learn how to solve a system of linear equations in *three* variables.

1. Understand the Concept of a System of Three Equations in Three Variables

> **Definition**
>
> A **linear equation in three variables** is an equation of the form $Ax + By + Cz = D$ where A, B, and C are not all zero and where A, B, C, and D are real numbers. Solutions to this type of an equation are **ordered triples** of the form (x, y, z).

An example of a linear equation in three variables is

$$2x - y + 3z = 12$$

Some solutions to the equation are

$(5, 1, 1)$	since	$2(5) - (1) + 3(1) = 12$
$(3, 0, 2)$	since	$2(3) - (0) + 3(2) = 12$
$(6, -3, -1)$	since	$2(6) - (-3) + 3(-1) = 12$

There are infinitely many solutions.

Ordered pairs, like $(2, 5)$, are plotted on a two-dimensional rectangular coordinate system. Ordered triples, however, like $(1, 2, 3)$ and $(3, 0, 2)$, are graphed on a three-dimensional coordinate system.

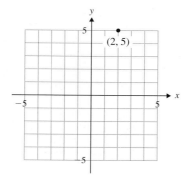

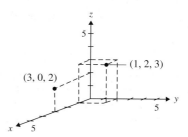

The graph of a linear equation in two variables is a line, but the graph of a linear equation in *three* variables is a *plane*.

A **solution to a system of linear equations in three variables** is an *ordered triple* that satisfies each equation in the system. Like systems of linear equations in two variables, systems of linear equations in *three* variables can have *one solution* (the system is consistent), *no solution* (the system is inconsistent), or *infinitely many solutions* (the system is dependent).

Here is an example of a system of linear equations in three variables:

$$x + 4y + 2z = 10$$
$$3x - y + z = 6$$
$$2x + 3y - z = -4$$

In Section 5.1 we solved systems of linear equations in *two* variables by graphing. Since the graph of an equation like $x + 4y + 2z = 10$ is a *plane*, however, solving a system like the one above by graphing would not be practical. But let's look at what happens, graphically, when a system of linear equations in three variables has one solution, no solution, or an infinite number of solutions.

One solution:

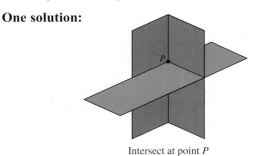

Intersect at point P

All three planes intersect at one point; this is the solution of the system.

No solution:

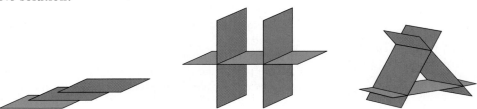

None of the planes may intersect or *two* of the planes may intersect, but if there is no solution to the system, *all three planes* do not have a common point of intersection.

Infinite number of solutions:

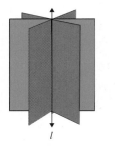 

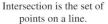

Intersection is the set of points on a line.

All points in common

The three planes may intersect so that they have a line in common—the infinite set of points on the line is the solution to the system. Or, the set of all points in a plane gives the infinite number of solutions.

2. Learn the Steps for Solving Systems of Linear Equations in Three Variables

We will discuss how to solve these systems by looking at two different cases.

Case 1: Every equation in the system contains three variables.

Case 2: At least one of the equations in the system contains only two variables.

We begin with Case 1.

Solving a System of Linear Equations in Three Variables

If each equation contains three variables, then to solve a system of linear equations in three variables, follow these steps.

1) **Label** the equations ①, ②, and ③.

2) **Choose a variable to eliminate. Eliminate** this variable from *two* sets of *two* equations using the elimination method. You will obtain two equations containing the same two variables. Label one of these new equations $\boxed{A}$ and the other $\boxed{B}$.

3) **Use the elimination method to eliminate a variable from equations** $\boxed{A}$ **and** $\boxed{B}$. You have now found the value of one variable.

4) **Find the value of another variable** by substituting the value found in 3) into either equation $\boxed{A}$ or $\boxed{B}$ and solving for the second variable.

5) **Find the value of the third variable** by substituting the values of the two variables found in 3) and 4) into equation ①, ②, or ③.

6) **Check** the solution in each of the original equations, and **write the solution as an ordered triple**.

---

3. *Case 1:* Solve a System of Three Equations in Three Variables When Each Equation Contains Three Variables

Example 1

Solve ① $x + 2y - 2z = 3$
② $2x + y + 3z = 1$
③ $x - 2y + z = -10$

Solution

1) **Label** the equations ①, ②, and ③.

2) **Choose a variable to eliminate:** *y.*

We will eliminate *y* from *two* sets of *two* equations.

a) *Equations* ① *and* ③. Add the equations to eliminate *y*. Label the resulting equation $\boxed{A}$.

$$
\begin{array}{rl}
① & x + 2y - 2z = 3 \\
③ \ + & x - 2y + \ z = -10 \\
\hline
\boxed{A} & 2x - \qquad z = -7
\end{array}
$$

b) *Equations* ② *and* ③. To eliminate y, multiply equation ② by 2 and add it to equation ③. Label the resulting equation $\boxed{B}$.

$$
\begin{array}{r}
2 \times ② \quad 4x + 2y + 6z = 2 \\
③ + \quad x - 2y + \ z = -10 \\
\hline
\boxed{B} \quad 5x + \qquad 7z = -8
\end{array}
$$

Equations $\boxed{A}$ and $\boxed{B}$ contain only *two* variables and they are the *same* variables, x and z.

3) **Use the elimination method to eliminate a variable from equations** $\boxed{A}$ **and** $\boxed{B}$.

$$
\begin{array}{r}
2x - \ z = -7 \ \boxed{A} \\
5x + 7z = -8 \ \boxed{B}
\end{array}
$$

We will eliminate z from $\boxed{A}$ and $\boxed{B}$. Multiply $\boxed{A}$ by 7 and add it to $\boxed{B}$.

$$
\begin{array}{r}
7 \times \boxed{A} \quad 14x - 7z = -49 \\
\boxed{B} + \ 5x + 7z = -8 \\
\hline
19x \qquad = -57 \\
\boxed{x = -3} \qquad \text{Divide by 19.}
\end{array}
$$

4) **Find the value of another variable** by substituting $x = -3$ into either equation $\boxed{A}$ or $\boxed{B}$.

We will substitute $x = -3$ into $\boxed{A}$ since it has smaller coefficients.

$$
\begin{array}{rl}
\boxed{A} \quad 2x - z = -7 & \\
2(-3) - z = -7 & \text{Substitute } -3 \text{ for } x. \\
-6 - z = -7 & \text{Multiply.} \\
-z = -1 & \text{Add 6.} \\
\boxed{z = 1} & \text{Divide by } -1.
\end{array}
$$

5) **Find the value of the third variable** by substituting $x = -3$ and $z = 1$ into either equation, ①, ②, or ③.

We will substitute $x = -3$ and $z = 1$ into ① to solve for y.

$$
\begin{array}{rl}
① \quad x + 2y - 2z = 3 & \\
-3 + 2y - 2(1) = 3 & \text{Substitute } -3 \text{ for } x \text{ and } 1 \text{ for } z. \\
-3 + 2y - 2 = 3 & \text{Multiply.} \\
2y - 5 = 3 & \text{Combine like terms.} \\
2y = 8 & \text{Add 5.} \\
\boxed{y = 4} & \text{Divide by 2.}
\end{array}
$$

6) **Check** the solution, $(-3, 4, 1)$ in each of the original equations, and **write the solution**.

$$① \quad x + 2y - 2z = 3$$
$$-3 + 2(4) - 2(1) \overset{?}{=} 3$$
$$-3 + 8 - 2 \overset{?}{=} 3$$
$$3 = 3 \qquad ✓ \text{ True}$$

$$② \quad 2x + y + 3z = 1$$
$$2(-3) + 4 + 3(1) \overset{?}{=} 1$$
$$-6 + 4 + 3 \overset{?}{=} 1$$
$$1 = 1 \qquad ✓ \text{ True}$$

$$③ \quad x - 2y + z = -10$$
$$-3 - 2(4) + 1 \overset{?}{=} -10$$
$$-3 - 8 + 1 \overset{?}{=} -10$$
$$-10 = -10 \qquad ✓ \text{ True}$$

The solution is $(-3, 4, 1)$.

You Try 1

Solve $\quad x + 2y + 3z = -11$
$\qquad\quad 3x - y + z = 0$
$\qquad -2x + 3y - z = 4$

4. Solve an Inconsistent or Dependent System of Three Equations in Three Variables

Example 2

Solve $① \quad -3x + 2y - z = 5$
$\qquad ② \qquad x + 4y + z = -4$
$\qquad ③ \quad 9x - 6y + 3z = -2$

Solution

Follow the steps.

1) *Label* the equations $①$, $②$, and $③$.

2) *Choose a variable to eliminate.* z will be the easiest. Eliminate z from *two* sets of *two* equations.

   a) *Equations $①$ and $②$.* Add the equations to eliminate z. Label the resulting equation $\boxed{A}$.

$$
\begin{array}{r}
① \quad -3x + 2y - z = 5 \\
② \; + \quad\underline{x + 4y + z = -4} \\
\boxed{A} \quad -2x + 6y \qquad\;\; = 1
\end{array}
$$

b) *Equations* ① *and* ③. To eliminate *z*, multiply equation ① by 3 and add it to equation ③. Label the resulting equation $\boxed{B}$.

$$① \ -3x + 2y - z = 5 \longrightarrow 3 \times ① \quad -9x + 6y - 3z = 15$$
$$③ \ + \quad 9x - 6y + 3z = -2$$
$$\boxed{B} \qquad 0 = 13 \quad \text{False}$$

Since the variables drop out and we get the false statement $0 = 13$, equations one and two have no ordered triple that satisfies each equation.
The system is inconsistent, so there is no solution. The solution set is $\varnothing$.

> When the variables drop out, and you get a false statement, there is *no solution* to the system. The system is inconsistent, so the solution set is $\varnothing$.

Example 3

Solve ① $\ -4x - 2y + 8z = -12$
② $\quad 2x + y - 4z = 6$
③ $\quad 6x + 3y - 12z = 18$

Solution

Follow the steps.

1) Label the equations ①, ②, and ③.

2) Choose a variable to eliminate. Eliminate *y* from *two* sets of *two* equations.

a) *Equations* ① *and* ②. To eliminate *y*, multiply equation ② by 2 and add it to equation ①. Label the resulting equation $\boxed{A}$.

$$2 \times ② \qquad 4x + 2y - 8z = \quad 12$$
$$① \ + \ -4x - 2y + 8z = -12$$
$$\boxed{A} \qquad\qquad 0 = 0 \quad \text{True}$$

The variables dropped out and we obtained the true statement $0 = 0$. This is because equation ① is a multiple of equation ②.

Notice, also, that equation ③ is a multiple of equation ②.
(Equation ③ = 3 × equation ②.)

The system is dependent. There are an infinite number of solutions of the form $\{(x, y, z) \mid 2x + y - 4z = 6\}$. The equations all have the same graph.

You Try 2

Solve each system of equations.

a) $\quad 8x + 20y - 4z = -16$ b) $\quad x + 4y - 3z = 2$
$\quad -6x - 15y + 3z = 12$ $\quad 2x - 5y + 2z = -8$
$\quad 2x + 5y - z = -4$ $\quad -3x - 12y + 9z = 7$

5. *Case 2:* Solve a System of Three Equations in Three Variables When at Least One Equation Is Missing a Variable

Example 4

Solve ① $5x - 2y = 6$
② $y + 2z = 1$
③ $3x - 4z = -8$

Solution

First, notice that while this *is* a system of three equations in three variables, none of the equations contains three variables. Furthermore, each equation is "missing" a different variable.

We will use many of the *ideas* outlined in the steps for solving a system of three equations, but after labeling our equations ①, ②, and ③, we will begin by using *substitution* rather than the elimination method.

1) Label the equations ①, ②, ③.

2) The goal of step 2 is to obtain two equations which contain the same two variables. We will modify this step from the way it was outlined on p. 367.

   In order to obtain *two* equations with the same *two* variables, we will use *substitution*.

   Look at equation ②. Since y is the only variable in the system with a coefficient of 1, we will solve equation ② for y.

$$② \quad y + 2z = 1$$
$$y = 1 - 2z \qquad \text{Subtract } 2z.$$

   Substitute $y = 1 - 2z$ into equation ① to obtain an equation containing the variables x and z. Simplify. Label the resulting equation $\boxed{A}$.

$$① \qquad\qquad 5x - 2y = 6$$
$$5x - 2(1 - 2z) = 6 \qquad \text{Substitute } 1 - 2z \text{ for } y.$$
$$5x - 2 + 4z = 6 \qquad \text{Distribute.}$$
$$\boxed{A} \qquad\qquad 5x + 4z = 8 \qquad \text{Add 2.}$$

3) The goal of step 3 is to solve for one of the variables. Equations $\boxed{A}$ and ③ contain only x and z.

   We will eliminate z from $\boxed{A}$ and ③. Add the two equations to eliminate z, then solve for x.

$$
\begin{array}{ll}
\boxed{A} & 5x + 4z = 8 \\
③ \ + & 3x - 4z = -8 \\
\hline
& 8x \qquad = 0 \\
& \boxed{x = 0} \qquad \text{Divide by 8.}
\end{array}
$$

4) Find the value of another variable by substituting $x = 0$ into either $\boxed{A}$, $\textcircled{1}$, or $\textcircled{3}$. We will substitute $x = 0$ into $\boxed{A}$.

$$\boxed{A} \quad 5x + 4z = 8$$
$$5(0) + 4z = 8 \qquad \text{Substitute 0 for } x.$$
$$0 + 4z = 8 \qquad \text{Multiply.}$$
$$4z = 8$$
$$\boxed{z = 2} \qquad \text{Divide by 4.}$$

5) Find the value of the third variable by substituting $x = 0$ into $\textcircled{1}$ or $z = 2$ into $\textcircled{2}$. We will substitute $x = 0$ into $\textcircled{1}$ to find y.

$$\textcircled{1} \quad 5x - 2y = 6$$
$$5(0) - 2y = 6 \qquad \text{Substitute 0 for } x.$$
$$0 - 2y = 6 \qquad \text{Multiply.}$$
$$-2y = 6$$
$$\boxed{y = -3} \qquad \text{Divide by } -2.$$

6) Check the solution $(0, -3, 2)$ in each of the original equations.

The check is left to the student.

The solution is $(0, -3, 2)$.

You Try 3

Solve $x + 2y = 8$
$\quad\quad\quad 2y + 3z = 1$
$\quad\quad\quad 3x - z = -3$

Summary of How to Solve a System of Three Linear Equations in Three Variables

1) *Case 1*: If every equation in the system contains three variables, then follow the steps on p. 367. See Example 1.

2) *Case 2*: If at least one of the equations in the systems contains only two variables, follow the steps on p. 367 but in a modified form. See Example 4.

6. Solve an Applied Problem Involving a System of Three Equations in Three Variables

To solve applications involving a system of three equations in three variables, we will use the process outlined in Section 5.4, *Solving an Applied Problem Using a System of Equations*, on p. 353.

Example 5

Write a system of equations and solve.

The top three gold-producing nations in 2002 were South Africa, the United States, and Australia. Together, these three countries produced 37% of the gold during that year. Australia's share was 2% less than that of the United States, while South Africa's percentage was 1.5 times Australia's percentage of world gold production. Determine what percentage of the world's gold supply was produced by each country in 2002. (Source: Market Share Reporter—2005, Vol. 1, "Mine Product" http://www.gold.org/value/market/supply-demand/min_production.html from World Gold Council)

Solution

Step 1: **Read the problem carefully, twice.**

Step 2: We must determine the percentage of the world's gold produced by South Africa, the United States, and Australia in 2002.

Define the unknowns; assign a variable to represent each unknown quantity.

x = percentage of world's gold supply produced by South Africa

y = percentage of world's gold supply produced by the United States

z = percentage of world's gold supply produced by Australia

Step 3: **Write a system of equations relating the variables.**

i) *To write one equation* we will use the information that says *together* the three countries produced 37% of the gold.

$$x + y + z = 37$$

ii) *To write a second equation* we will use the information that says Australia's share was 2% less than that of the United States.

$$z = y - 2$$

iii) *To write the third equation* we will use the statement that says South Africa's percentage was 1.5 times Australia's percentage.

$$x = 1.5z$$

The system is ① $x + y + z = 37$
② $z = y - 2$
③ $x = 1.5z$

Step 4: **Solve the system.**

Since two of the equations contain only two variables, we will solve the system following the steps in some modified form.

Label the equations ①, ②, and ③. Using substitution as our first step, we can rewrite the first equation in terms of a single variable, z, and solve for z.

Solve equation ② for y, and substitute $z + 2$ for y in equation ①.

$$② \quad z = y - 2$$
$$z + 2 = y \qquad \text{Solve for } y.$$

Using equation ③, substitute $1.5\,z$ for x in equation ①.

③ $x = 1.5z$
① $x + y + z = 37$
   $(1.5z) +$ $+ z = 37$ Substitute $1.5z$ for x and $z + 2$ for y.
              $3.5z + 2 = 37$ Combine like terms.
                 $3.5z = 35$ Subtract 2.
                  $\boxed{z = 10}$ Divide by 3.5.

To solve for x, we can substitute $z = 10$ into equation ③.

③ $x = 1.5z$
     $x = 1.5(10)$ Substitute 10 for z.
     $\boxed{x = 15}$ Multiply.

To solve for y, we can substitute $z = 10$ into equation 2.

② $z = y - 2$
     $10 = y - 2$ Substitute 10 for z.
     $\boxed{12 = y}$ Solve for y.

The solution of the system is $(15, 12, 10)$.

Step 5:   **Interpret the meaning of the solution as it relates to the problems.**

In 2002, South Africa produced 15% of the world's gold, the United States produced 12%, and Australia produced 10%.

The check is left to the student.

You Try 4

Write a system of equations and solve.

Amelia, Bella, and Carmen are sisters. Bella is 5 yr older than Carmen, and Amelia's age is 5 yr less than twice Carmen's age. The sum of their ages is 48. How old is each girl?

Answers to You Try Exercises

1) $(2, 1, -5)$ 2) a) infinite number of solutions of the form $\{(x, y, z) \mid 2x + 5y - z = -4\}$
b) $\varnothing$ 3) $(-2, 5, -3)$ 4) Amelia: 19; Bella: 17; Carmen: 12

5.5 Exercises

Boost your grade at mathzone.com! MathZone   > Practice Problems   > NetTutor   > e-Professors   > Videos
> Self-Test

Objective 1

Determine if the ordered triple is a solution of the system.

1) $4x + 3y - 7z = -6$
$x - 2y + 5z = -3$
$-x + y + 2z = 7$
$(-2, 3, 1)$

2) $3x + y + 2z = 2$
$-2x - y + z = 5$
$x + 2y - z = -11$
$(1, -5, 2)$

3) $-x + y - 2z = 2$
$3x - y + 5z = 4$
$2x + 3y - z = 7$
$(0, 6, 2)$

4) $6x - y + 4z = 4$
$-2x + y - z = 5$
$2x - 3y + z = 2$
$\left(-\dfrac{1}{2}, -3, 1\right)$

Objectives 2–4

Solve each system. See Examples 1–3.

5) $x + 3y + z = 3$
$4x - 2y + 3z = 7$
$-2x + y - z = -1$

6) $x - y + 2z = -7$
$-3x - 2y + z = -10$
$5x + 4y + 3z = 4$

7) $5x + 3y - z = -2$
$-2x + 3y + 2z = 3$
$x + 6y + z = -1$

8) $-2x - 2y + 3z = 2$
$3x + 3y - 5z = -3$
$-x + y - z = 9$

9) $3a + 5b - 3c = -4$
$a - 3b + c = 6$
$-4a + 6b + 2c = -6$

10) $a - 4b + 2c = -7$
$3a - 8b + c = 7$
$6a - 12b + 3c = 12$

11) $a - 5b + c = -4$
$3a + 2b - 4c = -3$
$6a + 4b - 8c = 9$

12) $-a + 2b - 12c = 8$
$-6a + 2b - 8c = -3$
$3a - b + 4c = 4$

13) $-15x - 3y + 9z = 3$
$5x + y - 3z = -1$
$10x + 2y - 6z = -2$

14) $-4x + 10y - 16z = -6$
$-6x + 15y - 24z = -9$
$2x - 5y + 8z = 3$

15) $-3a + 12b - 9c = -3$
$5a - 20b + 15c = 5$
$-a + 4b - 3c = -1$

16) $3x - 12y + 6z = 4$
$-x + 4y - 2z = 7$
$5x + 3y + z = -2$

Objective 5

Solve each system. See Example 4.

17) $5x - 2y + z = -5$
$x - y - 2z = 7$
$4y + 3z = 5$

18) $-x + z = 9$
$-2x + 4y - z = 4$
$7x + 2y + 3z = -1$

19) $a + 15b = 5$
$4a + 10b + c = -6$
$-2a - 5b - 2c = -3$

20) $2x - 6y - 3z = 4$
$-3y + 2z = -6$
$-x + 3y + z = -1$

21) $x + 2y + 3z = 4$
$-3x + y = -7$
$4y + 3z = -10$

22) $-3a + 5b + c = -4$
$a + 5b = 3$
$4a - 3c = -11$

23) $-5x + z = -3$
$4x - y = -1$
$3y - 7z = 1$

24) $a + b = 1$
$a - 5c = 2$
$b + 2c = -4$

25) $\begin{aligned} 4a + 2b &= -11 \\ -8a - 3c &= -7 \\ b + 2c &= 1 \end{aligned}$

26) $\begin{aligned} 3x + 4y &= -6 \\ -x + 3z &= 1 \\ 2y + 3z &= -1 \end{aligned}$

Objectives 2–5

Mixed Exercises

Solve each system.

27) $\begin{aligned} 6x + 3y - 3z &= -1 \\ 10x + 5y - 5z &= 4 \\ x - 3y + 4z &= 6 \end{aligned}$

28) $\begin{aligned} 2x + 3y - z &= 0 \\ x - 4y - 2z &= -5 \\ -4x + 5y + 3z &= -4 \end{aligned}$

29) $\begin{aligned} 7x + 8y - z &= 16 \\ -\frac{1}{2}x - 2y + \frac{3}{2}z &= 1 \\ \frac{4}{3}x + 4y - 3z &= -\frac{2}{3} \end{aligned}$

30) $\begin{aligned} 3a + b - 2c &= -3 \\ 9a + 3b - 6c &= -9 \\ -6a - 2b + 4c &= 6 \end{aligned}$

31) $\begin{aligned} 2a - 3b &= -4 \\ 3b - c &= 8 \\ -5a + 4c &= -4 \end{aligned}$

32) $\begin{aligned} 5x + y - 2z &= -2 \\ -\frac{1}{2}x - \frac{3}{4}y + 2z &= \frac{5}{4} \\ x - 6z &= 3 \end{aligned}$

33) $\begin{aligned} -4x + 6y + 3z &= 3 \\ -\frac{2}{3}x + y + \frac{1}{2}z &= \frac{1}{2} \\ 12x - 18y - 9z &= -9 \end{aligned}$

34) $\begin{aligned} x - \frac{5}{2}y + \frac{1}{2}z &= \frac{5}{4} \\ x + 3y - z &= 4 \\ -6x + 15y - 3z &= -1 \end{aligned}$

35) $\begin{aligned} a + b + 9c &= -3 \\ -5a - 2b + 3c &= 10 \\ 4a + 3b + 6c &= -15 \end{aligned}$

36) $\begin{aligned} 2x + 3y &= 2 \\ -3x + 4z &= 0 \\ y - 5z &= -17 \end{aligned}$

37) $\begin{aligned} 2x - y + 4z &= -1 \\ x + 3y + z &= -5 \\ -3x + 2y &= 7 \end{aligned}$

38) $\begin{aligned} a + 3b - 8c &= 2 \\ -2a - 5b + 4c &= -1 \\ 4a + b + 16c &= -4 \end{aligned}$

Objective 6

Write a system of equations and solve.

39) Moe buys two hot dogs, two orders of fries, and a large soda for $9.00. Larry buys two hot dogs, one order of fries, and two large sodas for $9.50, and Curly spends $11.00 on three hot dogs, two orders of fries, and a large soda. Find the price of a hot dog, an order of fries, and a large soda.

40) A movie theater charges $9.00 for an adult's ticket, $7.00 for a ticket for seniors 60 and over, and $6.00 for a child's ticket. For a particular movie, the theater sold a total of 290 tickets, which brought in $2400. The number of seniors' tickets sold was twice the number of children's tickets sold. Determine the number of adults', seniors', and children's tickets sold.

41) A Chocolate Chip Peanut Crunch Clif Bar contains 4 fewer grams of protein than a Chocolate Peanut Butter Balance Bar Plus. A Chocolate Peanut Butter Protein Plus PowerBar contains 9 more grams of protein than the Balance Bar Plus. All three bars contain a total of 50 g of protein. How many grams of protein are in each type of bar? (Source: www.clifbar.com, www.balance.com, www.powerbar.com)

42) A 1-tablespoon serving size of Hellman's Real Mayonnaise has twice as many calories as the same size serving of Hellman's Light Mayonnaise. One tablespoon of Miracle Whip has five fewer calories than 1 tablespoon of Hellman's Light Mayonnaise. If the three spreads have a total of 175 calories in one serving, determine the number of calories in one serving of each. (Source: Product labels)

43) The three NBA teams with the highest revenues in 2002–2003 were the New York Knicks, the Los Angeles Lakers, and the Chicago Bulls. Their revenues totalled $428 million. The Lakers took in $30 million more than the Bulls, and the Knicks took in $11 million more than the Lakers. Determine the revenue of each team during the 2002–2003 season. (Source: *Forbes,* Feb. 16, 2004, p. 66)

44) The best-selling paper towel brands in 2002 were Bounty, Brawny, and Scott. Together they accounted for 59% of the market. Bounty's market share was 25% more than Brawny's, and Scott's market share was 2% less than Brawny's. What percentage of the market did each brand hold in 2002? (Source: *USA Today,* Oct. 23, 2003, p. 3B from Information Resources, Inc.)

45) Ticket prices to a Cubs game at Wrigley Field vary depending on whether they are on a value date, a regular date, or a prime date. At the beginning of the 2005 season, Bill, Corrinne, and Jason bought tickets in the bleachers for several games. Bill spent $286 on four value dates, four regular dates, and three prime dates. Corrinne bought tickets for four value dates, three regular dates, and two prime dates for $220. Jason spent $167 on three value dates, three regular dates, and one prime date. How much did it cost to sit in the bleachers at Wrigley Field on a value date, regular date, and prime date in 2005? (Source: http://chicago.cubs.mlb.com)

46) To see the Boston Red Sox play at Fenway Park in 2005, two field box seats, three infield grandstand seats, and five bleacher seats cost $420. The cost of four field box seats, two infield grandstand seats, and three bleacher seats was $499. The total cost of buying one of each type of ticket was $153. What was the cost of each type of ticket during the 2005 season? (Source: http://boston.redsox.mlb.com)

47) The measure of the largest angle of a triangle is twice the middle angle. The smallest angle measures 28° less than the middle angle. Find the measures of the angles of the triangle. (Hint: Recall that the sum of the measures of the angles of a triangle is 180°.)

48) The measure of the smallest angle of a triangle is one-third the measure of the largest angle. The middle angle measures 30° less than the largest angle. Find the measures of the angles of the triangle. (Hint: Recall that the sum of the measures of the angles of a triangle is 180°.)

49) The smallest angle of a triangle measures 44° less than the largest angle. The sum of the two smaller angles is 20° more than the measure of the largest angle. Find the measures of the angles of the triangle.

50) The sum of the measures of the two smaller angles of a triangle is 40° less than the largest angle. The measure of the largest angle is twice the measure of the middle angle. Find the measures of the angles of the triangle.

51) The perimeter of a triangle is 29 cm. The longest side is 5 cm longer than the shortest side, and the sum of the two smaller sides is 5 cm more than the longest side. Find the lengths of the sides of the triangle.

52) The smallest side of a triangle is half the length of the longest side. The sum of the two smaller sides is 2 in. more than the longest side. Find the lengths of the sides if the perimeter is 58 in.

Definition/Procedure	Example	Reference
Solving Systems of Linear Equations by Graphing		5.1
A *system of linear equations* consists of two or more linear equations with the same variables. A *solution of a system* of two equations in two variables is an ordered pair that is a solution of each equation in the system.	Determine if $(6, 1)$ is a solution of the system $$x + 3y = 9$$ $$-2x + 7y = -5$$ $$x + 3y = 9$$ $$6 + 3(1) \stackrel{?}{=} 9 \qquad \text{Substitute.}$$ $$6 + 3 \stackrel{?}{=} 9$$ $$9 = 9 \qquad \text{True}$$ $$-2x + 7y = -5$$ $$-2(6) + 7(1) \stackrel{?}{=} -5 \qquad \text{Substitute.}$$ $$-12 + 7 \stackrel{?}{=} -5$$ $$-5 = -5 \qquad \text{True}$$ Since $(6, 1)$ is a solution of each equation in the system, *yes*, it is a solution of the system.	**p. 322**
To *solve a system by graphing*, graph each line on the same coordinate axes. a) If the lines intersect at a single point, then this point is the solution of the system. The system is *consistent*. b) If the lines are parallel, then the system has *no solution*. We write the solution set as ∅. The system is *inconsistent*. c) If the graphs are the same line, then the system has an *infinite number of solutions*. The system is *dependent*.	Solve by graphing. $\quad y = \dfrac{1}{4}x - 3$ $$3x + 4y = 4$$ 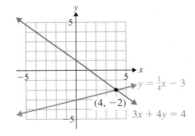 The solution of the system is $(4, -2)$. The system is consistent.	**p. 323**

Definition/Procedure	Example	Reference

| | 5.2

Steps for Solving a System by Substitution

1) Solve one of the equations for one of the variables. If possible, solve for a variable that has a coefficient of 1 or -1.

2) Substitute the expression in *step 1* into the *other* equation. The equation you obtain should contain only one variable.

3) Solve the equation in *step 2*.

4) Substitute the value found in *step 3* into one of the equations to obtain the value of the other variable.

5) Check the values in the original equations.

Solve by substitution. $2x - 7y = 2$
$\qquad\qquad\qquad\qquad x - 2y = -2$

1) Solve for x in the second equation since its coefficient is 1.

$$x = 2y - 2$$

2) Substitute $2y - 2$ for the x in the first equation.

$$2(2y - 2) - 7y = 2$$

3) Solve the equation in step 2 for y.

$$2(2y - 2) - 7y = 2$$
$$4y - 4 - 7y = 2 \quad \text{Distribute.}$$
$$-3y - 4 = 2 \quad \text{Combine like terms.}$$
$$-3y = 6 \quad \text{Add 4.}$$
$$y = -2 \quad \text{Divide by } -3.$$

4) Substitute $y = -2$ into the equation in step 1 to find x.

$$x = 2y - 2$$
$$x = 2(-2) - 2 \quad \text{Substitute } -2 \text{ for } y.$$
$$x = -4 - 2 \quad \text{Multiply.}$$
$$x = -6$$

5) The solution is $(-6, -2)$. Verify this by substituting $(-6, -2)$ into each of the original equations.

p. 333

If the variables drop out and a false equation is obtained, the system has *no solution. The system is inconsistent, and the solution set is $\varnothing$.*

Solve by substitution. $4x - 12y = 7$
$\qquad\qquad\qquad\qquad x = 3y - 1$

1) The second equation is solved for x.

2) Substitute $3y - 1$ for x in the first equation.

$$4(3y - 1) - 12y = 7$$

3) Solve the equation in step 2 for y.

$$4(3y - 1) - 12y = 7$$
$$12y - 4 - 12y = 7 \quad \text{Distribute.}$$
$$-4 = 7 \quad \text{False}$$

4) The system has no solution. The solution set is $\varnothing$.

p. 337

Definition/Procedure	Example	Reference
If the variables drop out and a true equation is obtained, the system has an *infinite number of solutions. The system is dependent.*	Solve by substitution. $\quad y = x + 4$ $\qquad\qquad\qquad\qquad 2x - 2y = -8$ 1) The first equation is solved for y. 2) Substitute $x + 4$ for y in the second equation. $$2x - 2(x + 4) = -8$$ 3) Solve the equation in step 2 for x. $$\begin{aligned} 2x - 2(x + 4) &= -8 \\ 2x - 2x - 8 &= -8 \quad \text{Distribute.} \\ -8 &= -8 \quad \text{True} \end{aligned}$$ 4) The system has an infinite number of solutions of the form $\{(x, y) \mid y = x + 4\}$.	**p. 338**

Solving Systems of Linear Equations by the Elimination Method

		5.3
Steps for Solving a System of Two Linear Equations by the Elimination Method 1) Write each equation in the form $Ax + By = C$. 2) Determine which variable to eliminate. If necessary, multiply one or both of the equations by a number so that the coefficients of the variable to be eliminated are negatives of one another. 3) Add the equations, and solve for the remaining variable. 4) Substitute the value found in step 3 into either of the original equations to find the value of the other variable. 5) Check the solution in each of the original equations.	Solve using the elimination method. $$\begin{aligned} 7x - 4y &= 1 \\ -4x + 3y &= 3 \end{aligned}$$ Eliminate y. Multiply the first equation by 3, and multiply the second equation by 4 to rewrite the system with equivalent equations. Rewrite the system $\begin{aligned} 3(7x - 4y) &= 3(1) \\ 4(-4x + 3y) &= 4(3) \end{aligned} \rightarrow \begin{aligned} 21x - 12y &= 3 \\ -16x + 12y &= 12 \end{aligned}$ Add the equations: $$\begin{aligned} 21x - 12y &= 3 \\ + \quad -16x + 12y &= 12 \\ \hline 5x &= 15 \\ x &= 3 \end{aligned}$$ Substitute $x = 3$ into either of the original equations and solve for y. $$\begin{aligned} 7x - 4y &= 1 \\ 7(3) - 4y &= 1 \\ 21 - 4y &= 1 \\ -4y &= -20 \\ y &= 5 \end{aligned}$$ The solution is $(3, 5)$. Verify this by substituting $(3, 5)$ into each of the original equations.	**p. 340**

Definition/Procedure	Example	Reference

Applications of Systems of Two Equations

5.4

Use the *five steps for solving applied problems* outlined in the section to solve an applied problem.

1) *Read the problem carefully, twice. Draw a picture, if applicable.*

2) Identify what you are being asked to find. *Define the unknown; write down what each variable represents.* If applicable, label the picture with the variables.

3) *Write a system of equations relating the variables.* It may be helpful to begin by writing an equation in words.

4) *Solve the system.*

5) *Interpret the meaning of the solution as it relates to the problem.*

Natalia spent $23.80 at an office supply store when she purchased boxes of pens and paper clips. The pens cost $3.50 per box, and the paper clips cost $0.70 per box. How many boxes of each did she buy if she purchased 10 items all together?

1) Read the problem carefully, twice.

2) Define the variables.

x = number of boxes of pens she bought
y = number of boxes of paper clips she bought

3) One equation involves the *cost* of the items:

Cost pens + Cost paper clips = Total cost
$3.50x$ + $0.70y$ = 23.80

The second equation involves the number of items:

Number of pens + Number of paper clips = Total number of items
x + y = 10

The system is

$$3.50x + 0.70y = 23.80$$
$$x + y = 10$$

4) Multiply by 10 to eliminate the decimals in the first equation, and then solve the system using substitution.

$10(3.50x + 0.70y) = 10(23.80)$ Eliminate decimals.

$$35x + 7y = 238$$

Solve the system $\begin{array}{l} 35x + 7y = 238 \\ x + y = 10 \end{array}$ to determine that the solution is (6, 4).

5) Natalia bought six boxes of pens and four boxes of paper clips. Verify the solution.

p. 353

Definition/Procedure	Example	Reference

<table>
<tr><td colspan="3">Systems of Linear Equations in Three Variables   5.5</td></tr>
</table>

Systems of Linear Equations in Three Variables

Reference: 5.5

A *linear equation in three variables* is an equation of the form $Ax + By + Cz = D$, where A, B, and C are not all zero and where A, B, C, and D are real numbers. Solutions to this type of an equation are *ordered triples* of the form (x, y, z).

Example:

$$5x + 3y + 9z = -2$$

One solution of this equation is $(-1, -2, 1)$ since substituting the values for $x, y,$ and z satisfies the equation.

$$5x + 3y + 9z = -2$$
$$5(-1) + 3(-2) + 9(1) \overset{?}{=} -2$$
$$-5 - 6 + 9 \overset{?}{=} -2$$
$$-2 = -2 \quad \text{True}$$

Reference: p. 365

Solving a System of Three Linear Equations in Three Variables

1) Label the equations ①, ②, and ③.

2) Choose a variable to eliminate. Eliminate this variable from *two sets* of *two* equations using the elimination method. You will obtain two equations containing the same two variables. Label one of these new equations $\boxed{A}$ and the other $\boxed{B}$.

3) Use the elimination method to eliminate a variable from equations $\boxed{A}$ and $\boxed{B}$. You have now found the value of one variable.

4) Find the value of another variable by substituting the value found in step 3 into either equation $\boxed{A}$ or $\boxed{B}$ and solving for the second variable.

5) Find the value of the third variable by substituting the values of the two variables found in steps 3 and 4 into equation ①, ②, or ③.

6) Check the solution in each of the original equations, and write the solution as an ordered triple.

Example:

Solve   ① $x + 2y + 3z = 5$
         ② $4x - 2y - z = -1$
         ③ $-3x + y + 4z = -12$

1) Label the equations ①, ②, and ③.

2) We will eliminate y from *two sets* of *two* equations.

a) *Equations* ① *and* ②. Add the equations to eliminate y. Label the resulting equation $\boxed{A}$.

   ① $x + 2y + 3z = 5$
   ② $+$ $4x - 2y - z = -1$
   $\boxed{A}$ $5x +   2z = 4$

b) *Equations* ② *and* ③. To eliminate y, multiply equation ③ by 2 and add it to equation ②. Label the resulting equation $\boxed{B}$.

 $2 \times$ ③ $-6x + 2y + 8z = -24$
   ② $+$ $4x - 2y - z = -1$
   $\boxed{B}$ $-2x +   7z = -25$

3) Eliminate x from $\boxed{A}$ and $\boxed{B}$. Multiply $\boxed{A}$ by 2 and $\boxed{B}$ by 5. Add the resulting equations.

 $2 \times \boxed{A}$ $10x + 4z = 8$
 $5 \times \boxed{B}$ $+$ $-10x + 35z = -125$
                   $39z = -117$
                    $\boxed{z = -3}$

Reference: p. 367

Definition/Procedure	Example	Reference

4) Substitute $z = -3$ into either $\boxed{A}$ or $\boxed{B}$. Substitute $z = -3$ into $\boxed{A}$ and solve for x.

$$\boxed{A} \qquad 5x + 2z = 4$$

$$5x + 2(-3) = 4 \qquad \text{Substitute } -3 \text{ for } z.$$

$$5x - 6 = 4 \qquad \text{Multiply.}$$

$$5x = 10 \qquad \text{Add 6.}$$

$$\boxed{x = 2} \qquad \text{Divide by 5.}$$

5) Substitute $x = 2$ and $z = -3$ into either equation ①, ②, or ③.

Substitute $x = 2$ and $z = -3$ into ① to solve for y.

$$① \qquad x + 2y + 3z = 5$$

$$2 + 2y + 3(-3) = 5 \qquad \text{Substitute.}$$

$$2 + 2y - 9 = 5 \qquad \text{Multiply.}$$

$$2y - 7 = 5 \qquad \text{Combine like terms.}$$

$$2y = 12 \qquad \text{Add 7.}$$

$$\boxed{y = 6} \qquad \text{Divide by 2.}$$

6) The solution is $(2, 6, -3)$.

The check is left to the student.

(5.1) Determine if the ordered pair is a solution of the system of equations.

1) $-2x + y = 3$
$3x - y = -17$
$(-4, -5)$

2) $5x - 2y = -6$
$-4x + 3y = 16$
$(2, 8)$

3) $3x + 4y = 0$
$9x + 2y = 5$
$\left(\dfrac{2}{3}, -\dfrac{1}{2}\right)$

Solve each system by graphing.

4) $y = \dfrac{1}{2}x - 2$
$y = 2x + 1$

5) $3x - y = 2$
$x + y = 2$

6) $-2x + 3y = 15$
$2x - 3y = -3$

7) $2x + 3y = 5$
$y = \dfrac{1}{2}x + 4$

8) $4x + y = -4$
$-2x - \dfrac{1}{2}y = 2$

Without graphing, determine whether each system has no solution, one solution, or an infinite number of solutions.

9) $x + 7y = -3$
$4x - 9y = 1$

10) $y = -\dfrac{2}{3}x - 3$
$4x + 6y = 5$

11) $5x - 4y = 2$
$y = \dfrac{5}{4}x - \dfrac{1}{2}$

12) $15x - 10y = 4$
$-9x + 6y = 1$

13) The graph shows the on-time departure percentages during the four quarters of 2003 at San Diego (Lindbergh) and Denver International Airports.

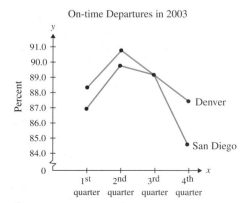

On-time Departures in 2003

(Source: www.census.gov)

a) When was Denver's on-time departure percentage better than San Diego's?

b) When were their percentages the same and, approximately, what percentage of flights left these airports on time?

(5.2) Solve each system by substitution.

14) $x + 8y = -2$
$2x + 11y = -9$

15) $y = \dfrac{5}{6}x - 2$
$6y - 5x = -12$

16) $-2x + y = -18$
$x + 7y = 9$

17) $6x - y = -3$
$15x + 2y = 15$

18) $\dfrac{5}{2}x + \dfrac{9}{2}y = \dfrac{1}{2}$
$\dfrac{1}{6}x + \dfrac{2}{3}y = -\dfrac{1}{3}$

19) $x + 8y = 2$
$x = 20 - 8y$

20) $2x - 3y = 3$
$3x + 4y = -21$

21) $2(2x - 3) = y + 3$
$8 + 5(y - 6) = 8(x + 3) - 9x - y$

22) $15 - y = y + 2(4x + 5)$
$2(x + 1) + 3 = 2(y + 4) + 10x$

(5.3) Solve each system using the elimination method.

23) $x - y = -8$
$3x + y = -12$

24) $-5x + 3y = 17$
$x + 2y = 20$

25) $4x - 5y = -16$
$-3x + 4y = 13$

26) $6x + 8y = 13$
$9x + 12y = -5$

27) $0.12x + 0.01y = 0.06$
$0.5x + 0.2y = -0.7$

28) $3(8 - y) = 5x + 3$
$x + 2(y - 3) = 2(4 - x)$

29) $\dfrac{3}{4}x - y = \dfrac{1}{2}$
$-\dfrac{x}{3} + \dfrac{y}{2} = -\dfrac{1}{6}$

30) $6x - 4y = 12$
$15x - 10y = 30$

Solve each system using the elimination method twice.

31) $2x + 7y = -8$
$3x - y = 13$

32) $x + 4y = 11$
$5x - 6y = 2$

(5.4) Write a system of equations and solve.

33) One day at the Village Veterinary Clinic, the doctors treated twice as many dogs as cats. If they treated a total of 51 cats and dogs, how many of each did the doctors see?

34) At Aurora High School, 183 sophomores study either Spanish or French. If there are 37 fewer French students than Spanish students, how many study each language?

35) At a Houston Texans football game in 2004, four hot dogs and two sodas cost $26.50, while three hot dogs and four sodas cost $28.00. Find the price of a hot dog and the price of a soda at a Texans game. (Source: www.teammarketing.com)

36) Two general admission tickets and two student tickets to a college performance of *Romeo and Juliet* cost $44.00. Four student tickets and five general admission tickets cost $101.00. Find the price of each type of ticket.

37) The perimeter of a rectangular computer monitor is 66 in. The length is 3 in. more than the width. What are the dimensions of the monitor?

38) Find the measures of angles x and y if the measure of x is twice the measure of y.

39) A store owner plans to make 10 pounds of a candy mix worth $1.92/lb. How many pounds of gummi bears worth $2.40/lb, and how many pounds of jelly beans worth $1.60/lb must be combined to make the candy mix?

40) How many milliliters of pure alcohol and how many milliliters of a 4% alcohol solution must be combined to make 480 milliliters of an 8% alcohol solution?

41) A car and a tour bus leave the same location and travel in opposite directions. The car's speed is 12 mph more than the speed of the bus. If they are 270 mi apart after $2\frac{1}{2}$ hr, how fast is each vehicle traveling?

42) Quinn can bike 64 mi in the same amount of time Clark can bike 56 mi. If Clark rides 2 mph slower than Quinn, find the speed of each rider.

(5.5) Determine if the ordered triple is a solution of the system.

43) $x - 6y + 4z = 13$
$5x + y + 7z = 8$
$2x + 3y - z = -5$
$(-3, -2, 1)$

44) $-4x + y + 2z = 1$
$x - 3y - 4z = 3$
$-x + 2y + z = -7$
$(0, -5, 3)$

Solve each system using one of the methods of section 5.5.

45) $2x - 5y - 2z = 3$
$x + 2y + z = 5$
$-3x - y + 2z = 0$

46) $x - 2y + 2z = 6$
$x + 4y - z = 0$
$5x + 3y + z = -3$

47) $5a - b + 2c = -6$
$-2a - 3b + 4c = -2$
$a + 6b - 2c = 10$

48) $2x + 3y - 15z = 5$
$-3x - y + 5z = 3$
$-x + 6y - 10z = 12$

49) $4x - 9y + 8z = 2$
$x + 3y = 5$
$6y + 10z = -1$

50) $-a + 5b - 2c = -3$
$3a + 2c = -3$
$2a + 10b = -2$

51) $x + 3y - z = 0$
$11x - 4y + 3z = 8$
$5x + 15y - 5z = 1$

52) $4x + 2y + z = 0$
$8x + 4y + 2z = 0$
$16x + 8y + 4z = 0$

53) $12a - 8b + 4c = 8$
$3a - 2b + c = 2$
$-6a + 4b - 2c = -4$

54) $3x - 12y - 6z = -8$
$x + y - z = 5$
$-4x + 16y + 8z = 10$

55) $5y + 2z = 6$
$-x + 2y = -1$
$4x - z = 1$

56) $2a - b = 4$
$3b + c = 8$
$-3a + 2c = -5$

57) $8x + z = 7$
$3y + 2z = -4$
$4x - y = 5$

58) $6y - z = -2$
$x + 3y = 1$
$-3x + 2z = 8$

Write a system of equations and solve.

59) One serving (8 fl oz) of Powerade has 20 mg more sodium than one serving of Propel. One serving of Gatorade Xtreme has twice as much sodium as the same serving size of Powerade. Together the three drinks have 200 mg of sodium. How much sodium is in one serving of each drink? (Source: Product labels)

60) In 2003, the top highway truck tire makers were Goodyear, Michelin, and Bridgestone. Together, they held 53% of the market. Goodyear's market share was 3% more than Bridgestone's, and Michelin's share was 1% less than Goodyear's. What percent of this tire market did each company hold in 2003? (Source: *Market Share Reporter*, Vol. I, 2005, p. 361: From: *Tire Business*, Feb. 2, 2004, p. 9)

61) In 2002, Verizon, Cingular, and T-Mobile had 64.3 million cellular phone subscribers. Verizon had 10.5 million more subscribers than Cingular, and T-Mobile had 11.9 million fewer subscribers than Cingular. Determine how many subscribers each company had in 2002. (Source: *Market Share Reporter*, Vol. II, 2005, p. 625: From: *New York Times*, June 25, 2003, p. C6, from The Yankee Group)

62) Since the introduction of compact discs (CDs), the sales of music on cassette tapes has plummeted. In 2000, 269.4 million fewer cassettes were shipped to stores than were shipped in 1994. The number sent to stores in 2003 was 58.8 million less than in 2000. In 1994, 2000, and 2003 the total number of cassette tapes shipped to stores was 438.6 million. How many cassettes were sent to be sold in 1994, 2000, and 2003? (Source: Recording Industry Association of America, 2003 Year-End Statistics)

63) A family of six people goes to an ice cream store every Sunday after dinner. One week, they order two ice cream cones, three shakes, and one sundae for $13.50. The next week they get three cones, one shake, and two sundaes for $13.00. The week after that they spend $11.50 on one shake, one sundae, and four ice cream cones. Find the price of an ice cream cone, a shake, and a sundae.

64) An outdoor music theater sells three types of seats—reserved, behind-the-stage, and lawn seats. Two reserved, three behind-the-stage, and four lawn seats cost $360. Four reserved, two behind-the-stage, and five lawn seats cost $470. One of each type of seat would total $130. Determine the cost of a reserved seat, a behind-the-stage seat, and a lawn seat.

65) The measure of the smallest angle of a triangle is one-third the measure of the middle angle. The measure of the largest angle is 70° more than the measure of the smallest angle. Find the measures of the angles of the triangle.

66) The perimeter of a triangle is 40 in. The longest side is twice the length of the shortest side, and the sum of the two smaller sides is four inches longer than the longest side. Find the lengths of the sides of the triangles.

1) Determine if $\left(\dfrac{3}{4}, -5\right)$ is a solution of the system.

$$8x + y = 1$$
$$-12x - 4y = 11$$

Solve each system by graphing.

2) $y = 2x - 3$
$3x + 2y = 8$

3) $x + y = 3$
$2x + 2y = -2$

Solve each system by substitution.

4) $5x + 9y = 3$
$x - 4y = -11$

5) $-9x + 12y = 21$
$y = \dfrac{3}{4}x + \dfrac{7}{4}$

Solve each system by the elimination method.

6) $4x - 3y = -14$
$x + 3y = 19$

7) $7x + 8y = 28$
$-5x + 6y = -20$

8) $-x + 4y + 3z = 6$
$3x - 2y + 6z = -18$
$x + y + 2z = -1$

Solve each system using any method.

9) $x - 8y = 1$
$-2x + 9y = -9$

10) $\dfrac{x}{6} + y = -\dfrac{1}{3}$
$-\dfrac{5}{8}x + \dfrac{3}{4}y = \dfrac{11}{4}$

11) $5(y + 4) - 9 = -3(x - 4) + 4y$
$13x - 2(3x + 2) = 3(1 - y)$

Write a system of equations and solve.

12) A 6-in. Turkey Breast Sandwich from Subway has 390 fewer calories than a Burger King Whopper. If the two sandwiches have a total of 950 calories, determine the number of calories in the Subway sandwich and the number of calories in a Whopper. (Source: www.subway.com, www.bk.com)

13) At a hardware store, three boxes of screws and two boxes of nails sell for $18 while one box of screws and four boxes of nails sell for $16. Find the price of a box of screws and the price of a box of nails.

14) The measure of the smallest angle of a triangle is 9° less than the measure of the middle angle. The largest angle is 30° more than the sum of the two smaller angles. Find the measures of the angles of the triangle.

Perform the operations and simplify.

1) $\dfrac{3}{10} - \dfrac{7}{15}$

2) $5\dfrac{5}{6} \div 1\dfrac{13}{15}$

3) $(5 - 8)^3 + 40 \div 10 - 6$

4) Find the area of the triangle.

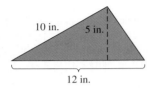

10 in.   5 in.

12 in.

5) Simplify $-8(3x^2 - x - 7)$

Simplify. The answer should not contain any negative exponents.

6) $(2y^5)^4$

7) $3c^2 \cdot 5c^{-8}$

8) $\dfrac{36a^{-2}b^5}{54ab^{13}}$

9) Write 0.00008319 in scientific notation.

10) Solve $11 - 3(2k - 1) = 2(6 - k)$.

11) Solve $0.04(3p - 2) - 0.02p = 0.1(p + 3)$.

12) Solve. Write the answer in interval notation.
$-47 \le 7t - 5 \le 6$

13) Write an equation and solve.

The number of plastic surgery procedures performed in the United States in 2003 was 293% more than the number performed in 1997. If approximately 8,253,000 cosmetic procedures were performed in 2003, how many took place in 1997? (Source: American Society for Aesthetic Plastic Surgery)

14) The area, A, of a trapezoid is $A = \dfrac{1}{2}h(b_1 + b_2)$

where $h = $ height of the trapezoid,
$b_1 = $ length of one base of the trapezoid, and
$b_2 = $ length of the second base of the trapezoid.

a) Solve the equation for h.

b) Find the height of the trapezoid that has an area of 39 cm^2 and bases of length 8 cm and 5 cm.

15) Graph $2x + 3y = 5$.

16) Find the x- and y-intercepts of the graph of $4x - 5y = 10$.

17) Write the slope-intercept form of the equation of the line containing $(-7, 4)$ and $(1, -3)$.

18) Determine whether the lines are parallel, perpendicular, or neither.

$$10x + 18y = 9$$
$$9x - 5y = 17$$

Solve each system of equations.

19) $9x + 7y = 7$
$3x + 4y = -11$

20) $3(2x - 1) - (y + 10) = 2(2x - 3) - 2y$
$3x + 13 = 4x - 5(y - 3)$

21) $\dfrac{5}{6}x - \dfrac{1}{2}y = \dfrac{2}{3}$
$-\dfrac{5}{4}x + \dfrac{3}{4}y = \dfrac{1}{2}$

Write a system of equations and solve.

22) Dhaval used twice as many 6-ft boards as 4-ft boards when he made a playhouse for his children. If he used a total of 48 boards, how many of each size did he use?

23) Through 2003, Aretha Franklin had won 10 more Grammy Awards than Whitney Houston, and Christina Aguilera had won half as many as Whitney Houston. All together these three singers had won 25 Grammy Awards. How many did each woman win? (Source: www.grammy.com)

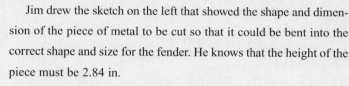

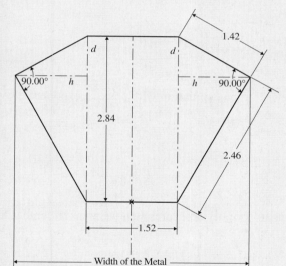

Polynomials

Algebra at Work: Custom Motorcycles

This is a final example of how algebra is used to build motorcycles in a custom chopper shop.

The support bracket for the fender of a custom motorcycle must be fabricated. To save money, Jim's boss told him to use a piece of scrap metal and not a new piece. So, he has to figure out how big of a piece of scrap metal he needs to be able to cut out the shape needed to make the fender.

Jim drew the sketch on the left that showed the shape and dimension of the piece of metal to be cut so that it could be bent into the correct shape and size for the fender. He knows that the height of the piece must be 2.84 in.

To determine the width of the piece of metal that he needs, Jim analyzes the sketch and writes the equation

$$[(1.42)^2 - d^2] + (2.84 - d)^2 = (2.46)^2$$

In order to solve this equation, he must know how to square the binomial $(2.84 - d)^2$, something we will learn in this chapter. When he solves the equation, Jim determines that $d \approx 0.71$ in. He uses this value of d to find that the width of the piece of metal that he must use to cut the correct shape for the fender is 3.98 in.

We will learn how to square binomials and perform other operations with polynomials in this chapter.

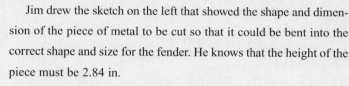

Section 6.1 Review of the Rules of Exponents

Objectives

1. Review the Rules of Exponents

In Chapter 2 we learned the rules of exponents. We will review them here to prepare us for the topics in the rest of this chapter—adding, subtracting, multiplying, and dividing polynomials.

1. Review the Rules of Exponents

Rules of Exponents

For real numbers a and b and integers m and n, the following rules apply:

Rule	Example
Product Rule: $a^m \cdot a^n = a^{m+n}$	$y^6 \cdot y^9 = y^{6+9} = y^{15}$
Basic Power Rule: $(a^m)^n = a^{mn}$	$(k^4)^7 = k^{28}$
Power Rule for a Product: $(ab)^n = a^n b^n$	$(9t)^2 = 9^2 t^2 = 81t^2$
Power Rule for a Quotient: $\left(\dfrac{a}{b}\right)^n = \dfrac{a^n}{b^n}$, where $b \neq 0$.	$\left(\dfrac{2}{r}\right)^5 = \dfrac{2^5}{r^5} = \dfrac{32}{r^5}$
Zero Exponent: If $a \neq 0$, then $a^0 = 1$.	$(7)^0 = 1$
Negative Exponent: For $a \neq 0$, $a^{-n} = \left(\dfrac{1}{a}\right)^n = \dfrac{1}{a^n}$.	$\left(\dfrac{3}{4}\right)^{-3} = \left(\dfrac{4}{3}\right)^3 = \dfrac{4^3}{3^3} = \dfrac{64}{27}$
If $a \neq 0$ and $b \neq 0$, then $\dfrac{a^{-m}}{b^{-n}} = \dfrac{b^n}{a^m}$.	Rewrite each expression with positive exponents. a) $\dfrac{x^{-6}}{y^{-3}} = \dfrac{y^3}{x^6}$ b) $\dfrac{8c^{-5}}{d^{-1}} = \dfrac{8d}{c^5}$
Quotient Rule: If $a \neq 0$, then $\dfrac{a^m}{a^n} = a^{m-n}$.	$\dfrac{2^9}{2^4} = 2^{9-4} = 2^5 = 32$

When we use the rules of exponents to simplify an expression, we must remember to use the order of operations.

Example 1

Simplify. Assume all variables represent nonzero real numbers. The answer should contain only positive exponents.

a) $(7k^{10})(-2k)$

b) $\dfrac{(-4)^5 \cdot (-4)^2}{(-4)^4}$

c) $\dfrac{10x^5 y^{-3}}{2x^2 y^5}$

d) $\left(\dfrac{c^2}{2d^4}\right)^{-5}$

e) $4(3p^5 q)^2$

Solution

a) $(7k^{10})(-2k) = -14k^{10+1}$ Multiply coefficients and add the exponents.

$\phantom{(7k^{10})(-2k)} = -14k^{11}$ Simplify.

b) $\dfrac{(-4)^5 \cdot (-4)^2}{(-4)^4} = \dfrac{(-4)^{5+2}}{(-4)^4} = \dfrac{(-4)^7}{(-4)^4}$ Product rule—the bases in the numerator are the same, so add the exponents.

$= (-4)^{7-4}$ Quotient rule

$= (-4)^3$

$= -64$ Evaluate.

c) $\dfrac{10x^5y^{-3}}{2x^2y^5} = 5x^{5-2}y^{-3-5}$ Divide coefficients and subtract the exponents.

$= 5x^3y^{-8}$ Simplify.

$= \dfrac{5x^3}{y^8}$ Write the answer with only positive exponents.

d) $\left(\dfrac{c^2}{2d^4}\right)^{-5} = \left(\dfrac{2d^4}{c^2}\right)^5$ Take the reciprocal of the base and make the exponent positive.

$= \dfrac{2^5 d^{20}}{c^{10}}$ Power rule

$= \dfrac{32d^{20}}{c^{10}}$ Simplify.

e)   In this expression, a quantity is raised to a power and that quantity is being multiplied by 4. Remember what the order of operations says: perform exponents before multiplication.

$4(3p^5q)^2 = 4(9p^{10}q^2)$ Apply the power rule *before* multiplying terms.

$= 36p^{10}q^2$ Multiply.

You Try 1

Simplify. Assume all variables represent nonzero real numbers. The answer should contain only positive exponents.

a) $(-6u^2)(-4u^3)$ b) $\dfrac{8^3 \cdot 8^4}{8^5}$ c) $\dfrac{8n^9}{12n^5}$

d) $(3y^{-9})^2(2y^7)$ e) $\left(\dfrac{3a^3b^{-4}}{2ab^6}\right)^{-4}$

Answers to You Try Exercises

1) a) $24u^5$ b) 64 c) $\dfrac{2n^4}{3}$ d) $\dfrac{18}{y^{11}}$ e) $\dfrac{16b^{40}}{81a^8}$

6.1 Exercises

Objective 1

Evaluate using the rules of exponents.

1) $2^2 \cdot 2^4$

2) $(-3)^2 \cdot (-3)$

3) $\dfrac{(-4)^8}{(-4)^5}$

4) $\dfrac{2^{10}}{2^6}$

5) 6^{-1}

6) $(12)^{-2}$

7) $\left(\dfrac{1}{9}\right)^{-2}$

8) $\left(-\dfrac{1}{5}\right)^{-3}$

9) $\left(\dfrac{3}{2}\right)^{-4}$

10) $\left(\dfrac{7}{9}\right)^{-2}$

11) $6^0 + \left(-\dfrac{1}{2}\right)^{-5}$

12) $\left(\dfrac{1}{4}\right)^{-2} + \left(\dfrac{1}{4}\right)^{0}$

13) $\dfrac{8^5}{8^7}$

14) $\dfrac{2^7}{2^{12}}$

Simplify. Assume all variables represent nonzero real numbers. The answer should not contain negative exponents.

15) $t^5 \cdot t^8$

16) $n^{10} \cdot n^6$

 17) $(-8c^4)(2c^5)$

18) $(3w^9)(-7w)$

19) $(z^6)^4$

20) $(y^3)^2$

21) $(5p^{10})^3$

22) $(-6m^4)^2$

23) $\left(-\dfrac{2}{3}a^7b\right)^3$

24) $\left(\dfrac{7}{10}r^2s^5\right)^2$

25) $\dfrac{f^{11}}{f^7}$

26) $\dfrac{u^9}{u^8}$

27) $\dfrac{35v^9}{5v^8}$

28) $\dfrac{36k^8}{12k^5}$

29) $\dfrac{9d^{10}}{54d^6}$

30) $\dfrac{7m^4}{56m^2}$

31) $\dfrac{x^3}{x^9}$

32) $\dfrac{v^2}{v^5}$

33) $\dfrac{m^2}{m^3}$

34) $\dfrac{t^3}{t^3}$

35) $\dfrac{45k^{-2}}{30k^2}$

36) $\dfrac{22n^{-9}}{55n^{-3}}$

 37) $5(2m^4n^7)^2$

38) $2(-3a^8b)^3$

39) $(6y^2)(2y^3)^2$

40) $(-c^4)(5c^9)^3$

41) $\left(\dfrac{7a^4}{b^{-1}}\right)^{-2}$

42) $\left(\dfrac{3t^{-3}}{2u}\right)^{-4}$

43) $\dfrac{a^{-12}b^7}{a^{-9}b^2}$

44) $\dfrac{mn^{-4}}{m^9n^7}$

45) $(xy^{-3})^{-5}$

46) $-(s^{-6}t^2)^{-4}$

47) $\left(\dfrac{a^2b}{4c^2}\right)^{-3}$

48) $\left(\dfrac{2s^3}{rt^4}\right)^{-5}$

 49) $\left(\dfrac{7h^{-1}k^9}{21h^{-5}k^5}\right)^{-2}$

50) $\left(\dfrac{24m^8n^{-3}}{16mn}\right)^{-3}$

51) $\left(\dfrac{15cd^{-4}}{5c^3d^{-10}}\right)^{-3}$

52) $\left(\dfrac{10x^{-5}y}{20x^5y^{-3}}\right)^{-2}$

Write expressions for the area and perimeter for each rectangle.

53)

2x
5x

54)

3y
y

 55)

$\frac{3}{4}p$
$\frac{1}{4}p$

56)

$\frac{5}{8}t$
$\frac{4}{3}t$

Simplify. Assume that the variables represent nonzero integers.

57) $k^{4a} \cdot k^{2a}$

58) $r^{9y} \cdot r^{y}$

59) $(g^{2x})^4$

60) $(t^{5c})^3$

61) $\dfrac{x^{7b}}{x^{4b}}$

62) $\dfrac{m^{10u}}{m^{3u}}$

Section 6.2 Addition and Subtraction of Polynomials

Objectives

1. Learn the Vocabulary Associated with Polynomials
2. Evaluate Polynomials
3. Add Polynomials in One Variable
4. Subtract Polynomials in One Variable
5. Add and Subtract Polynomials in More Than One Variable
6. Define and Evaluate a Polynomial Function

1. Learn the Vocabulary Associated with Polynomials

In Section 1.7 we defined an *algebraic expression* as a collection of numbers, variables, and grouping symbols connected by operation symbols such as $+$, $-$, $\times$, and $\div$.

An example of an algebraic expression is

$$4x^3 + 6x^2 - x + \frac{5}{2}$$

The *terms* of this algebraic expression are $4x^3$, $6x^2$, $-x$, and $\frac{5}{2}$. A *term* is a number or a variable or a product or quotient of numbers and variables.

Not only is $4x^3 + 6x^2 - x + \frac{5}{2}$ an expression, it is also a *polynomial*.

Definition   A **polynomial in x** is the sum of a finite number of terms of the form ax^n, where n is a whole number and a is a real number. (The exponents must be whole numbers.)

Let's look more closely at the polynomial $4x^3 + 6x^2 - x + \frac{5}{2}$.

1) The polynomial is written in **descending powers of x** since the powers of x decrease from left to right. Generally, we write polynomials in descending powers of the variable.

2) Recall that the term without a variable is called a **constant**. The constant is 5/2. The **degree of a term** equals the exponent on its variable. (If a term has more than one variable, the degree equals the *sum* of the exponents on the variables.) We will list each term, its coefficient, and its degree.

Term	Coefficient	Degree
$4x^3$	4	3
$6x^2$	6	2
$-x$	-1	1
$\dfrac{5}{2}$	$\dfrac{5}{2}$	$0 \left(\dfrac{5}{2} = \dfrac{5}{2}x^0 \right)$

3) The **degree of the polynomial** equals the highest degree of any nonzero term. The degree of $4x^3 + 6x^2 - x + \frac{5}{2}$ is 3. Or, we say that this is a **third-degree polynomial**.

Example 1

Decide if each expression *is* or *is not* a polynomial. If it is a polynomial, identify each term and the degree of each term. Then, find the degree of the polynomial.

a) $-7k^4 + 3.2k^3 - 8k^2 - 11$ b) $5n^2 - \dfrac{4}{3}n + 6 + \dfrac{9}{n^2}$

c) $x^3y^3 + 3x^2y^2 + 3xy + 1$ d) $12t^5$

Solution

a) $-7k^4 + 3.2k^3 - 8k^2 - 11$

This is a polynomial in k. Its terms have whole number exponents and real coefficients.

Term	Degree
$-7k^4$	4
$3.2k^3$	3
$-8k^2$	2
-11	0

The degree of this polynomial is 4.

b) $5n^2 - \dfrac{4}{3}n + 6 + \dfrac{9}{n^2}$

This is *not* a polynomial because one of its terms has a variable in the denominator.

c) $x^3y^3 + 3x^2y^2 + 3xy + 1$

This *is* a polynomial because the variables have whole number exponents and the coefficients are real numbers. Since this is a polynomial in two variables, we find the degree of each term by adding the exponents.

Term	Degree
x^3y^3	6
$3x^2y^2$	4
$3xy$	2
1	0

Add the exponents to get the degree.

The degree of this polynomial is 6.

d) $12t^5$

This *is* a polynomial even though it has only one term. The degree of the term is 5, and that is the degree of the polynomial as well.

You Try 1

Decide if each expression *is* or *is not* a polynomial. If it is a polynomial, identify each term and the degree of each term. Then, find the degree of the polynomial.

a) $g^3 + 8g^2 - \dfrac{5}{g}$ b) $6t^5 - \dfrac{4}{9}t^4 + t + 2$

c) $a^4b^3 - 9a^3b^3 - a^2b + 4a - 7$ d) $x + 5x^{1/2} + 6$

The polynomial in Example 1d) is $12t^5$ and has one term. We call $12t^5$ a *monomial*. A **monomial** is a polynomial that consists of one term ("mono" means one). Some other examples of monomials are

$$x^2, \qquad -9j^4, \qquad y, \qquad x^2y^2, \qquad \text{and} \qquad -2$$

A **binomial** is a polynomial that consists of exactly two terms ("bi" means two). Some examples are

$$n + 6, \qquad 5b^2 - 7, \qquad c^4 - d^4, \qquad \text{and} \qquad -12u^4v^3 + 8u^2v^2$$

A **trinomial** is a polynomial that consists of exactly three terms ("tri" means three). Here are some examples:

$$p^2 - 5p - 36, \qquad 3k^5 + 24k^2 - 6k, \qquad \text{and} \qquad 8r^4 + 10r^2s + 3s^2$$

It is important that you understand the meaning of these terms. We will use them throughout our study of algebra.

In Section 1.7 we saw that expressions have different values depending on the value of the variable(s). The same is true for polynomials.

2. Evaluate Polynomials

Example 2

Evaluate the trinomial $y^2 - 4y + 5$ when

a) $y = 3$ and   b) $y = -5$

Solution

a)   Substitute 3 for y in $y^2 - 4y + 5$. Remember to put 3 in parentheses.

$$y^2 - 4y + 5 = (3)^2 - 4(3) + 5 \qquad \text{Substitute.}$$
$$= 9 - 12 + 5$$
$$= -3 + 5 \qquad \text{Apply order of operations.}$$
$$= 2 \qquad \text{Add.}$$

b)   Substitute -5 for y in $y^2 - 4y + 5$. Put -5 in parentheses.

$$y^2 - 4y + 5 = (-5)^2 - 4(-5) + 5 \qquad \text{Substitute.}$$
$$= 25 + 20 + 5$$
$$= 50 \qquad \text{Add.}$$

You Try 2

Evaluate $n^2 - 4n - 2$ when

a) $n = 4$ b) $n = -3$

Adding and Subtracting Polynomials

Recall in Section 2.1 we said that **like terms** contain the same variables with the same exponents. We add or subtract like terms by adding or subtracting the coefficients and leaving the variable(s) and exponent(s) the same. We use the same idea for adding and subtracting polynomials.

3. Add Polynomials in One Variable

To add polynomials, add like terms.

Example 3

Add $11c^2 + 3c - 9$ and $2c^2 + 5c + 1$.

Solution

The addition problem can be set up horizontally or vertically. We will add these horizontally. Put the polynomials in parentheses since each contains more than one term. Use the associative and commutative properties to rewrite like terms together.

$$(11c^2 + 3c - 9) + (2c^2 + 5c + 1) = (11c^2 + 2c^2) + (3c + 5c) + (-9 + 1)$$
$$= 13c^2 + 8c - 8 \qquad \text{Combine like terms.}$$

Example 4

Add $(7y^3 - 10y^2 + y - 2) + (3y^3 + 4y^2 - 3)$.

Solution

We will add these vertically. Line up like terms in columns and add.

$$
\begin{array}{r}
7y^3 - 10y^2 + y - 2 \\
+ \quad 3y^3 + \; 4y^2 \quad\;\; - 3 \\
\hline
10y^3 - \; 6y^2 + y - 5
\end{array}
$$

You Try 3

Add $(t^3 + 2t^2 - 10t + 1) + (3t^3 - 9t^2 - t + 6)$.

4. Subtract Polynomials in One Variable

To subtract two polynomials such as $(6p^2 + 2p - 7) - (4p^2 - 9p + 3)$ we will be using the distributive property to clear the parentheses in the second polynomial.

Example 5

Subtract $(6p^2 + 2p - 7) - (4p^2 - 9p + 3)$.

Solution

$(6p^2 + 2p - 7) - (4p^2 - 9p + 3)$
$= (6p^2 + 2p - 7) - 1(4p^2 - 9p + 3)$
$= (6p^2 + 2p - 7) + (-1)(4p^2 - 9p + 3)$ Change -1 to $+(-1)$.
$= (6p^2 + 2p - 7) + (-4p^2 + 9p - 3)$ Distribute.
$= 2p^2 + 11p - 10$ Combine like terms.

In Example 5, when we apply the distributive property, the problem changes from subtracting two polynomials to adding two polynomials where the signs of the terms in the second polynomial have changed. This is one way we can think about subtracting two polynomials.

> To subtract two polynomials, change the sign of each term in the second polynomial. Then, add the polynomials.

Let's see how we apply this to subtracting polynomials both horizontally and vertically.

Example 6

Subtract $(-8k^3 - k^2 + 5k + 7) - (6k^3 - 3k^2 + k - 2)$.

a)   horizontally   b)   vertically

Solution

a) $(-8k^3 - k^2 + 5k + 7) - (6k^3 - 3k^2 + k - 2)$ Change the signs in the
     $= (-8k^3 - k^2 + 5k + 7) \oplus (-6k^3 + 3k^2 - k + 2)$ second polynomial and add.
     $= -14k^3 + 2k^2 + 4k + 9$ Combine like terms.

b)   To subtract vertically line up like terms in columns.

$$
\begin{array}{r}
-8k^3 - k^2 + 5k + 7 \\
- (6k^3 - 3k^2 + k - 2) \\
\end{array}
$$
Change the signs in the second polynomial and add the polynomials.

$$
\begin{array}{r}
-8k^3 - k^2 + 5k + 7 \\
\oplus (-6k^3 + 3k^2 - k + 2) \\
\hline
-14k^3 + 2k^2 + 4k + 9 \\
\end{array}
$$

You can see that adding and subtracting polynomials horizontally or vertically gives the same result.

You Try 4

Subtract $(9m^2 - 4m + 2) - (-m^2 + m - 6)$.

5. Add and Subtract Polynomials in More Than One Variable

Earlier in this section we introduced polynomials in more than one variable. To add and subtract these polynomials, remember that like terms contain the same variables with the same exponents.

Example 7

Perform the indicated operation.

a) $(x^2y^2 + 5x^2y - 8xy - 7) + (5x^2y^2 - 4x^2y - xy + 3)$

b) $(9cd - c + 3d + 1) - (6cd + 5c - 8)$

Solution

a) $(x^2y^2 + 5x^2y - 8xy - 7) + (5x^2y^2 - 4x^2y - xy + 3)$
   $= 6x^2y^2 + x^2y - 9xy - 4$ Combine like terms.

b) $(9cd - c + 3d + 1) - (6cd + 5c - 8) = (9cd - c + 3d + 1) + (-6cd - 5c + 8)$
   $= 3cd - 6c + 3d + 9$ Combine like terms.

You Try 5

Perform the indicated operation.

a) $(-10a^2b^2 + ab - 3b + 4) - (-9a^2b^2 - 6ab + 2b + 4)$

b) $(5.8t^3u^2 + 2.1tu - 7u) + (4.1t^3u^2 - 7.8tu - 1.6)$

6. Define and Evaluate a Polynomial Function

Look at the polynomial $2x^2 - 5x + 7$. If we substitute 3 for x, the *only* value of the expression is 10:

$$2(3)^2 - 5(3) + 7 = 2(9) - 15 + 7$$
$$= 18 - 15 + 7$$
$$= 10$$

It is true that polynomials have different values depending on what value is substituted for the variable. It is also true that for any value we substitute for x in a polynomial like $2x^2 - 5x + 7$ there will be *only one value* of the expression. Since each value substituted for the variable produces *only one value* of the expression, we can use function notation to represent a polynomial like $2x^2 - 5x + 7$.

$f(x) = 2x^2 - 5x + 7$ is a **polynomial function** since $2x^2 - 5x + 7$ is a polynomial.

Therefore, finding $f(3)$ when $f(x) = 2x^2 - 5x + 7$ is the same as evaluating $2x^2 - 5x + 7$ when $x = 3$.

Example 8

If $f(x) = x^3 - 4x^2 + 3x + 1$, find $f(-2)$.

Solution

Substitute -2 for x.

$$f(x) = x^3 - 4x^2 + 3x + 1$$
$$f(-2) = (-2)^3 - 4(-2)^2 + 3(-2) + 1$$
$$f(-2) = -8 - 4(4) - 6 + 1$$
$$f(-2) = -8 - 16 - 6 + 1$$
$$f(-2) = -29$$

You Try 6

If $g(t) = 2t^4 + t^3 - 7t^2 + 12$, find $g(-1)$.

Answers to You Try Exercises

1) a) not a polynomial   b) It is a polynomial of degree 5.   c) It is a polynomial of degree 7.

Term	Degree		Term	Degree
$6t^5$	5		a^4b^3	7
$-\dfrac{4}{9}t^4$	4		$-9a^3b^3$	6
t	1		$-a^2b$	3
2	0		$4a$	1
			-7	0

d) not a polynomial   2) a) -2 b) 19   3) $4t^3 - 7t^2 - 11t + 7$ 4) $10m^2 - 5m + 8$

5) a) $-a^2b^2 + 7ab - 5b$ b) $9.9t^3u^2 - 5.7tu - 7u - 1.6$ 6) 6

6.2 Exercises   **Boost your grade at mathzone.com!**   MathZone   > Practice Problems   > NetTutor   > Self-Test   > e-Professors   > Videos

Objective I

Is the given expression a polynomial? Why or why not?

1) $-5z^2 - 4z + 12$

2) $9t^3 + t^2 - t + \dfrac{3}{8}$

3) $g^3 + 3g^2 + 2g^{-1} - 5$

4) $6y^4$

5) $m^{2/3} + 4m^{1/3} + 4$

6) $8c - 5 + \dfrac{2}{c}$

Determine whether each is a monomial, a binomial, or a trinomial.

7) $3x - 7$

8) $-w^3$

9) $a^2b^2 + 10ab - 6$

10) $16r^2 + 9r$

11) 1

12) $v^4 + 7v^2 + 6$

13) How do you determine the degree of a polynomial in one variable?

14) Write a fourth-degree polynomial in one variable.

15) How do you determine the degree of a term in a polynomial in more than one variable?

16) Write a fifth-degree monomial in x and y.

For each polynomial, identify each term in the polynomial, the coefficient and degree of each term, and the degree of the polynomial.

17) $7y^3 + 10y^2 - y + 2$

18) $4d^2 + 12d - 9$

19) $-9r^3s^2 - r^2s^2 + \dfrac{1}{2}rs + 6s$

20) $8m^2n^2 + 0.5m^2n - mn + 3$

Objective 2
Evaluate each polynomial when a) $k = 2$ and b) $k = -3$.

21) $k^2 + 5k + 8$

22) $3k^3 - 10k - 11$

Evaluate each polynomial when $x = -4$ and $y = 3$.

 23) $2xy - 7x + 9$

24) $x^2y^2 + 2xy - x$

25) $-x^2y + 3xy + 10y - 1$

26) $\dfrac{1}{2}xy + x + 3y$

27) $\dfrac{2}{3}xy^2 - 3x + 4y$

28) $y^2 - x^2$

Objective 3
Add.

29) $(11w^2 + 2w - 13) + (-6w^2 + 5w + 7)$

30) $(4f^4 - 3f^2 + 8) + (2f^4 - f^2 + 1)$

31) $(-p + 16) + (-7p - 9)$

32) $(y^3 + 8y^2) + (y^3 - 11y^2)$

33) $\left(-7a^4 - \dfrac{3}{2}a + 1\right) + \left(2a^4 + 9a^3 - a^2 - \dfrac{3}{8}\right)$

34) $\left(2d^5 + \dfrac{1}{3}d^4 - 11\right) + (10d^5 - 2d^4 + 9d^2 + 4)$

35) $\left(\dfrac{11}{4}x^3 - \dfrac{5}{6}\right) + \left(\dfrac{3}{8}x^3 + \dfrac{11}{12}\right)$

36) $\left(\dfrac{3}{4}c + \dfrac{1}{8}\right) + \left(\dfrac{3}{2}c - \dfrac{5}{6}\right)$

37) $(6.8k^3 + 3.5k^2 - 10k - 3.3)$
    $+ (-4.2k^3 + 5.2k^2 + 2.7k - 1.1)$

38) $(0.6t^4 - 7.3t + 2.2) + (-1.8t^4 + 4.9t^3 + 8.1t + 7.1)$

Add.

39) $\begin{array}{r} 12x - 11 \\ + \ \underline{5x + 3} \end{array}$

40) $\begin{array}{r} -6n^3 + 1 \\ + \ \underline{4n^3 - 8} \end{array}$

 41) $\begin{array}{r} 9r^2 + 16r + 2 \\ + \ \underline{3r^2 - 10r + 9} \end{array}$

42) $\begin{array}{r} z^2 - 4z \\ + \ \underline{3z^2 + 9z + 4} \end{array}$

43) $\begin{array}{r} -2.6q^3 - q^2 + 6.9q - 1 \\ + \ \underline{4.1q^3 - 2.3q + 16} \end{array}$

44) $\begin{array}{r} 9a^4 + 5.3a^3 - 7a^2 - 1.2a + 6 \\ + \ \underline{-8a^4 - 2.8a^3 + 4a^2 - 3.9a + 5} \end{array}$

Objective 4
Subtract.

 45) $(8a^4 - 9a^2 + 17) - (15a^4 + 3a^2 + 3)$

46) $(16w^3 + 9w - 7) - (27w^3 - 3w - 4)$

47) $(j^2 + 18j + 2) - (-7j^2 + 6j + 2)$

48) $(-2m^2 + m + 5) - (3m^2 + m + 1)$

49) $(19s^5 - 11s^2) - (10s^5 + 3s^4 - 8s^2 - 2)$

50) $(h^5 + 7h^3 - 8h) - (-9h^5 + h^4 + 7h^3 - 8h - 6)$

51) $(-3b^4 - 5b^2 + b + 2) - (-2b^4 + 10b^3 - 5b^2 - 18)$

52) $(4t^3 - t^2 + 6) - (t^2 + 7t + 1)$

53) $\left(-\dfrac{5}{7}r^2 + \dfrac{4}{9}r + \dfrac{2}{3}\right) - \left(-\dfrac{5}{14}r^2 - \dfrac{5}{9}r + \dfrac{11}{6}\right)$

54) $\left(\dfrac{5}{6}y^3 + \dfrac{1}{2}y + 3\right) - \left(-\dfrac{1}{6}y^3 + y^2 - 3y\right)$

Subtract.

55) $\begin{array}{r} 17v + 3 \\ - \ \underline{2v + 9} \end{array}$

56) $\begin{array}{r} 10q - 7 \\ - \ \underline{4q + 8} \end{array}$

57) $\begin{array}{r} 2b^2 - 7b + 4 \\ - \ \underline{3b^2 + 5b - 3} \end{array}$

58) $\begin{array}{r} -3d^2 + 16d + 2 \\ - \ \underline{5d^2 + 7d - 3} \end{array}$

59) $\begin{array}{r} a^4 - 2a^3 + 6a^2 - 7a + 11 \\ - \ \underline{-2a^4 + 9a^3 - a^2 + 3} \end{array}$

60) $\begin{array}{r} 7y^4 + y^3 - 10y^2 + 6y - 2 \\ - \ \underline{-2y^4 + y^3 - 4y + 1} \end{array}$

61) Explain, in your own words, how to subtract two polynomials.

62) Do you prefer adding and subtracting polynomials vertically or horizontally? Why?

63) Will the sum of two trinomials always be a trinomial? Why or why not? Give an example.

64) Write a fourth-degree polynomial in x that does not contain a second-degree term.

Objectives 3 and 4

Perform the indicated operations.

65) $(-3b^4 + 4b^2 - 6) + (2b^4 - 18b^2 + 4)$
$+ (b^4 + 5b^2 - 2)$

66) $(-7m^2 - 14m + 56) + (3m^2 + 7m - 6)$
$+ (9m^2 - 10)$

67) $\left(n^3 - \dfrac{1}{2}n^2 - 4n + \dfrac{5}{8}\right) + \left(\dfrac{1}{4}n^3 - n^2 + 7n - \dfrac{3}{4}\right)$

68) $\left(\dfrac{2}{3}z^4 + z^3 - \dfrac{3}{2}z^2 + 1\right)$
$+ \left(z^4 - 2z^3 - \dfrac{1}{6}z^2 + 8z - 1\right)$

69) $(u^3 + 2u^2 + 1) - (4u^3 - 7u^2 + u + 9)$
$+ (8u^3 - 19u^2 + 2)$

70) $(21r^3 - 8r^2 + 3r + 2) + (-4r^2 + 5)$
$- (6r^3 - r^2 - 4r)$

71) $\left(\dfrac{3}{8}k^2 + k - \dfrac{1}{5}\right) - \left(2k^2 + k - \dfrac{7}{10}\right) + (k^2 - 9k)$

72) $\left(y + \dfrac{1}{4}\right) + \left(\dfrac{1}{2}y^2 - 3y + \dfrac{3}{4}\right) - \left(\dfrac{3}{4}y^2 - \dfrac{1}{2}y + 1\right)$

73) $(2t^3 - 8t^2 + t + 10)$
$- [(5t^3 + 3t^2 - t + 8) + (-6t^3 - 4t^2 + 3t + 5)]$

74) $(x^2 - 10x - 6)$
$- [(-8x^2 + 11x - 1) + (5x^2 - 9x - 3)]$

75) $(-12a^2 + 9) - (-9a^3 + 7a + 6) + (12a^2 - a + 10)$

76) $(5c + 7) - (c^2 + 4c - 2) - (-7c^3 - c + 4)$

Objective 5

Each of the polynomials below is a polynomial in two variables. Perform the indicated operation(s).

77) $(4a + 13b) - (a + 5b)$

78) $(-2g - 3h) + (6g + h)$

79) $\left(5m + \dfrac{5}{6}n + \dfrac{1}{2}\right) + \left(-6m + n - \dfrac{3}{4}\right)$

80) $\left(-2c - \dfrac{2}{3}d + 1\right) - \left(2c + \dfrac{1}{9}d - \dfrac{4}{7}\right)$

81) $(-12y^2z^2 + 5y^2z - 25yz^2 + 16)$
$+ (17y^2z^2 + 2y^2z - 15)$

82) $(-8u^2v^2 + 2uv + 3) - (-9u^2v^2 - 14uv + 18)$

83) $(8x^3y^2 - 7x^2y^2 + 7x^2y - 3) + (2x^3y^2 + x^2y - 1)$
$- (4x^2y^2 + 2x^2y + 8)$

84) $(r^3s^2 + r^2s^2 + 4) - (6r^3s^2 + 14r^2s^2 - 6)$
$+ (8r^3s^2 - 6r^2s^2 - 4)$

Objectives 3 and 4

Write an expression for each and perform the indicated operation(s).

85) Find the sum of $v^2 - 9$ and $4v^2 + 3v + 1$.

86) Add $11d - 12$ to $2d + 3$.

87) Subtract $g^2 - 7g + 16$ from $5g^2 + 3g + 6$.

88) Subtract $-9y^2 + 4y + 6$ from $2y^2 + y$.

89) Subtract the sum of $4n^2 + 1$ and $6n^2 - 10n + 3$ from $2n^2 + n + 4$.

90) Subtract $19x^3 + 4x - 12$ from the sum of $6x^3 + x^2 + x$ and $4x^3 - 3x - 8$.

Find the polynomial that represents the perimeter of each rectangle.

91)
$3x + 8$
$x - 1$

92)
$a^2 + 5a - 3$
$a^2 - 2a + 2$

93)
$3w^2 - 2w + 4$
$w - 7$

94)
$\dfrac{3}{4}t + 2$
$\dfrac{3}{4}t + 2$

Objective 6

95) If $f(x) = 5x^2 + 7x - 8$, find
a) $f(-3)$ b) $f(1)$

96) If $h(a) = -a^2 - 3a + 10$, find
a) $h(5)$ b) $h(-4)$

97) If $P(t) = t^3 - 3t^2 + 2t + 5$, find
a) $P(4)$ b) $P(0)$

98) If $G(c) = 3c^4 + c^2 - 9c - 4$, find
a) $G(0)$ b) $G(-1)$

99) If $H(z) = -4z + 9$, find z so that $H(z) = 11$.

100) If $f(x) = \dfrac{1}{3}x + 5$, find x so that $f(x) = 7$.

101) If $r(k) = \dfrac{2}{5}k - 3$, find k so that $r(k) = 13$.

102) If $Q(a) = 6a - 1$, find a so that $Q(a) = -9$.

Section 6.3 Multiplication of Polynomials

Objectives

1. Multiply a Monomial and a Polynomial

2. Multiply Two Polynomials

3. Multiply Two Binomials Using FOIL

4. Find the Product of More Than Two Polynomials

5. Find the Product of Binomials of the Form $(a + b)(a - b)$

6. Square a Binomial

7. Find Higher Powers of a Binomial

We have already learned that when multiplying two monomials, we multiply the coefficients and add the exponents of the same bases:

$$3x^4 \cdot 5x^2 = 15x^6 \qquad -2a^3b^2 \cdot 6ab^4 = -12a^4b^6$$

In this section we will discuss how to multiply other types of polynomials.

1. Multiply a Monomial and a Polynomial

When multiplying a monomial and a polynomial, we use the distributive property.

Example 1

Multiply $5n^2(2n^2 + 3n - 4)$.

Solution

$$5n^2(2n^2 + 3n - 4) = (5n^2)(2n^2) + (5n^2)(3n) + (5n^2)(-4) \qquad \text{Distribute.}$$
$$= 10n^4 + 15n^3 - 20n^2 \qquad \text{Multiply.}$$

 You Try 1

Multiply $6a^4(7a^3 - a^2 - 3a + 4)$.

2. Multiply Two Polynomials

To multiply two polynomials, multiply each term in the second polynomial by each term in the first polynomial. Then, combine like terms. The answer should be written in descending powers.

Example 2

Multiply.

a) $(2r - 3)(r^2 + 7r + 9)$ b) $(c^2 - 3c + 5)(2c^3 + c - 6)$

Solution

a) We will multiply each term in the second polynomial by the $2r$ in the first polynomial. Then, multiply each term in $(r^2 + 7r + 9)$ by the -3 in $(2r - 3)$. Then, add like terms.

$(2r - 3)(r^2 + 7r + 9)$
$= (2r)(r^2) + (2r)(7r) + (2r)(9) + (-3)(r^2) + (-3)(7r) + (-3)(9)$ Distribute.
$= 2r^3 + 14r^2 + 18r - 3r^2 - 21r - 27$ Multiply.
$= 2r^3 + 11r^2 - 3r - 27$ Combine like terms.

b) As in a), multiply each term in the second polynomial by each term in the first. Add like terms.

$$
\begin{aligned}
(c^2 &- 3c + 5)(2c^3 + c - 6) \\
&= (c^2)(2c^3) + (c^2)(c) + (c^2)(-6) + (-3c)(2c^3) + (-3c)(c) + (-3c)(-6) \\
&\quad + (5)(2c^3) + (5)(c) + (5)(-6) \qquad\qquad\text{Distribute.} \\
&= 2c^5 + c^3 - 6c^2 - 6c^4 - 3c^2 + 18c + 10c^3 + 5c - 30 \qquad\text{Multiply.} \\
&= 2c^5 - 6c^4 + 11c^3 - 9c^2 + 23c - 30 \qquad\qquad\text{Combine like terms.}
\end{aligned}
$$

Polynomials can be multiplied vertically as well. The process is very similar to the way we multiply whole numbers.

$$
\begin{array}{r}
458 \\
\times\ \ 32 \\
\hline
916 \\
13\ 74 \\
\hline
14{,}656
\end{array}
$$

Multiply 458 by 2.
Multiply 458 by 3.
Add.

In Example 3, we will find a product of polynomials by multiplying vertically.

Example 3

Multiply vertically. $(3n^2 + 4n - 1)(2n + 7)$

Solution

Set up the multiplication problem like you would for whole numbers:

$$
\begin{array}{r}
3n^2 +\ \ 4n - 1 \\
\times \qquad\quad 2n + 7 \\
\hline
21n^2 + 28n - 7 \\
6n^3 +\ \ 8n^2 -\ \ 2n \\
\hline
6n^3 + 29n^2 + 26n - 7
\end{array}
$$

Multiply each term in $3n^2 + 4n - 1$ by 7.
Multiply each term in $3n^2 + 4n - 1$ by $2n$.
Line up like terms in the same column. Add.

You Try 2

Multiply.

a) $(5x + 4)(7x^2 - 9x + 2)$ b) $\left(p^2 - \dfrac{3}{2}p - 6\right)(5p^2 + 8p - 4)$

3. Multiply Two Binomials Using FOIL

We can use the distributive property to multiply two binomials such as $(x + 8)(x - 3)$.

$$
\begin{aligned}
(x + 8)(x - 3) &= (x)(x) + (x)(-3) + (8)(x) + (8)(-3) \qquad\text{Distribute.} \\
&= x^2 - 3x + 8x - 24 \qquad\qquad\qquad\qquad\text{Multiply.} \\
&= x^2 + 5x - 24 \qquad\qquad\qquad\qquad\qquad\text{Add.}
\end{aligned}
$$

When using the distributive property, we must be sure that each term in the first binomial and each term in the second binomial get multiplied together.

We can use the distributive property in a special way to multiply two binomials by remembering the word *FOIL*. Using FOIL to multiply binomials guarantees that each term in one binomial has been multiplied by each term in the other binomial.

FOIL stands for

F	**O**	**I**	**L**
i	u	n	a
r	t	n	s
s	e	e	t
t	r	r	

The letters in the word FOIL tell us how to multiply the terms in the binomials. Then, after multiplying, add like terms.

Example 4

Use FOIL to multiply the binomials.

a) $(y + 4)(y + 7)$ b) $(3m - 5)(m - 2)$ c) $(x + 2y)(x - 6y)$

Solution

a) $(y + 4)(y + 7) = (y + 4)(y + 7)$

F: Multiply the **First** terms of the binomials: $(y)(y) = y^2$
O: Multiply the **Outer** terms of the binomials: $(y)(7) = 7y$
I: Multiply the **Inner** terms of the binomials: $(4)(y) = 4y$
L: Multiply the **Last** terms of the binomials: $(4)(7) = 28$

After multiplying, add like terms:

$$(y + 4)(y + 7) = y^2 + 7y + 4y + 28$$
$$= y^2 + 11y + 28$$

b) $(3m - 5)(m - 2) = (3m - 5)(m - 2)$

F: Multiply the **First** terms: $(3m)(m) = 3m^2$
O: Multiply the **Outer** terms: $(3m)(-2) = -6m$
I: Multiply the **Inner** terms: $(-5)(m) = -5m$
L: Multiply the **Last** terms: $(-5)(-2) = 10$

Add like terms:

$$(3m - 5)(m - 2) = 3m^2 - 6m - 5m + 10$$
$$= 3m^2 - 11m + 10$$

c) $(x + 2y)(x - 6y) = (x + 2y)(x - 6y)$

First ⌐ Last ⌐
Inner ⌐ Outer

F: Multiply the **First** terms: $(x)(x) = x^2$
O: Multiply the **Outer** terms: $(x)(-6y) = -6xy$
I: Multiply the **Inner** terms: $(2y)(x) = 2xy$
L: Multiply the **Last** terms: $(2y)(-6y) = -12y^2$

Add like terms:

$$(x + 2y)(x - 6y) = x^2 - 6xy + 2xy - 12y^2$$
$$= x^2 - 4xy - 12y^2$$

You Try 3

Use FOIL to multiply the binomials.

a) $(a + 9)(a + 2)$ b) $(4t + 3)(t - 5)$ c) $(c - 6d)(c - 3d)$

Did you notice that in each problem in Example 4 the outer and inner products were like terms, and therefore, could be added? This is often true, and we use this fact to combine the outer and inner products "in our heads."

Example 5

Multiply.

a) $(c + 5)(c + 3)$ b) $(2n - 1)(n - 6)$ c) $(3a - 2b)(a + b)$

Solution

a)   Try to multiply the binomials mentally using FOIL. Remember to add the products of the outer and inner terms.

F: c^2 L: 15
$(c + 5)(c + 3)$
I: $5c$
$+$ O: $3c$
─────
$8c$ Sum

Add the inner and outer terms to get the middle term of the product:

$$(c + 5)(c + 3) = c^2 + 8c + 15$$

b)
F: $2n^2$ L: 6

$(2n - 1)(n - 6)$

I: $-n$

+ O: $-12n$

$-13n$ Sum

Add the inner and outer terms to get the middle term of the product:

$$(2n - 1)(n - 6) = 2n^2 - 13n + 6$$

c)
F: $3a^2$ L: $-2b^2$

$(3a - 2b)(a + b)$

I: $-2ab$

+ O: $3ab$

ab Sum

Add the inner and outer terms to get the middle term of the product:

$$(3a - 2b)(a + b) = 3a^2 + ab - 2b^2$$

You Try 4

Multiply using FOIL.

a) $(x + 3)(x + 8)$ b) $(t - 7)(t - 6)$ c) $(3p - 2)(p + 4)$ d) $(x - 8y)(2x + 3y)$

While the polynomial multiplication problems we have seen so far are the most common types we encounter in algebra, there are other products we will see as well.

4. Find the Product of More Than Two Polynomials

Example 6

Multiply $5d^2(4d - 3)(2d - 1)$.

Solution

We can approach this problem a couple of ways.

Method I

Begin by multiplying the binomials, *then* multiply by the monomial.

$$5d^2(4d - 3)(2d - 1) = 5d^2(8d^2 - 4d - 6d + 3) \quad \text{Use FOIL to multiply the binomials.}$$
$$= 5d^2(8d^2 - 10d + 3) \quad \text{Combine like terms.}$$
$$= 40d^4 - 50d^3 + 15d^2 \quad \text{Distribute.}$$

Method 2
Begin by multiplying $5d^2$ by $(4d - 3)$, then multiply *that* product by $(2d - 1)$.

$$5d^2(4d - 3)(2d - 1) = (20d^3 - 15d^2)(2d - 1) \qquad \text{Multiply } 5d^2 \text{ by } (4d - 3).$$
$$= 40d^4 - 20d^3 - 30d^3 + 15d^2 \qquad \text{Use FOIL to multiply.}$$
$$= 40d^4 - 50d^3 + 15d^2 \qquad \text{Combine like terms.}$$

The result is the same. These may be multiplied by whichever method you prefer.

You Try 5

Multiply $-6x^3(x + 5)(3x - 4)$.

There are particular types of binomial products that come up often in algebra. We will look at those next.

5. Find the Product of Binomials of the Form $(a + b)(a - b)$

Let's find the product $(p + 5)(p - 5)$. Using FOIL we get

$$(p + 5)(p - 5) = p^2 - 5p + 5p - 25$$
$$= p^2 - 25$$

Notice that the "middle terms," the p-terms, drop out. In the result, $p^2 - 25$, the first term (p^2) is the square of p and the last term (25) is the square of 5. They are subtracted. The resulting binomial is a *difference of squares*. This pattern always holds when multiplying two binomials of the form $(a + b)(a - b)$.

$$(a + b)(a - b) = a^2 - b^2$$

Example 7

Multiply

a) $(r + 7)(r - 7)$ b) $(3 + y)(3 - y)$ c) $(2z - 5)(2z + 5)$

Solution

a) $(r + 7)(r - 7)$ is in the form $(a + b)(a - b)$ because $a = r$ and $b = 7$.

$$(r + 7)(r - 7) = r^2 - 7^2$$
$$= r^2 - 49$$

b) $(3 + y)(3 - y)$ is in the form $(a + b)(a - b)$ because $a = 3$ and $b = y$.

$$(3 + y)(3 - y) = 3^2 - y^2$$
$$= 9 - y^2$$

Be very careful on a problem like this. The answer is $9 - y^2$ *not* $y^2 - 9$; subtraction is not commutative.

c) $(2z - 5)(2z + 5)$ Since multiplication is commutative (the order in which we multiply does not affect the result), this is the same as $(2z + 5)(2z - 5)$, which is in the form $(a + b)(a - b)$ where $a = 2z$ and $b = 5$.

$$(2z + 5)(2z - 5) = (2z)^2 - 5^2 \qquad \text{Put } 2z \text{ in parentheses.}$$
$$= 4z^2 - 25$$

You Try 6

Find each product.

a) $(m + 9)(m - 9)$ b) $(4z + 3)(4z - 3)$ c) $(1 - w)(1 + w)$

6. Square a Binomial

Another type of special binomial product is a **binomial square** such as $(x + 6)^2$. $(x + 6)^2$ means $(x + 6)(x + 6)$. Therefore, we can use FOIL to multiply.

$$(x + 6)^2 = (x + 6)(x + 6) = x^2 + 6x + 6x + 36$$
$$= x^2 + 12x + 36$$

Let's square another binomial, $(y - 5)^2$.

$$(y - 5)^2 = (y - 5)(y - 5) = y^2 - 5y - 5y + 25$$
$$= y^2 - 10y + 25$$

In each case, notice that the outer and inner products are the same. When we add those terms, we see that the middle term of the result is *twice* the product of the terms in each binomial.

In the expansion of $(x + 6)^2$, $12x$ is $2(x)(6)$.

In the expansion of $(y - 5)^2$, $-10y$ is $2(y)(-5)$.

The *first* term in the result is the square of the *first* term in the binomial, and the *last* term in the result is the square of the *last* term in the binomial. We can express these relationships with these formulas:

$$(a + b)^2 = a^2 + 2ab + b^2$$
$$(a - b)^2 = a^2 - 2ab + b^2$$

We can think of the formulas in words as:

> To square a binomial, you square the first term, square the second term, then multiply 2 times the first term times the second term and add.

Finding the products $(a + b)^2 = a^2 + 2ab + b^2$ and $(a - b)^2 = a^2 - 2ab + b^2$ is also called *expanding* the binomial squares $(a + b)^2$ and $(a - b)^2$.

BE CAREFUL

$$(a + b)^2 \neq a^2 + b^2 \text{ and } (a - b)^2 \neq a^2 - b^2.$$

Example 8

Expand.

a) $(q + 4)^2$ 　　　 b) $(u - 10)^2$ 　　　 c) $(6n - 5)^2$

Solution

a) $(q + 4)^2 = q^2 \quad + \quad 2(q)(4) \quad + \quad 4^2$

　　　　　　 ↑ 　　　　　　 ↑ 　　　　 ↑

　　　Square the 　　Two times 　　Square the
　　　first term 　　first term 　　second term
　　　　　　　　 times second
　　　　　　　　 term

$= q^2 + 8q + 16$

Notice, $(q + 4)^2 \neq q^2 + 16$. Do not "distribute" the power of 2 to each term in the binomial!

b) $(u - 10)^2 = u^2 \quad - \quad 2(u)(10) \quad + \quad (10)^2$

　　　　　　　　 ↑ 　　　　　　 ↑ 　　　　 ↑

　　　　　Square the 　Two times 　　Square the
　　　　　first term 　first term 　　second term
　　　　　　　　　 times second
　　　　　　　　　 term

$= u^2 - 20u + 100$

c) $(6n - 5)^2 = (6n)^2 \quad - \quad 2(6n)(5) \quad + \quad (5)^2$

　　　　　　　　 ↑ 　　　　　　 ↑ 　　　　 ↑

　　　　　Square the 　Two times 　　Square the
　　　　　first term 　first term 　　second term
　　　　　　　　　 times second
　　　　　　　　　 term

$= 36n^2 - 60n + 25$

You Try 7

Expand.

a) $(t + 7)^2$ 　　 b) $(b - 12)^2$ 　　 c) $(3v + 4)^2$ 　　 d) $\left(\dfrac{3}{2}y - 1\right)^2$

7. Find Higher Powers of a Binomial

To find higher powers of binomials, we use techniques we have already discussed.

Example 9

Expand $(a + 2)^3$.

Solution

Just as $x^2 \cdot x = x^3$, $(a + 2)^2 \cdot (a + 2) = (a + 2)^3$. So we can think of $(a + 2)^3$ as $(a + 2)^2(a + 2)$.

$$(a + 2)^3 = (a + 2)^2(a + 2)$$
$$= (a^2 + 4a + 4)(a + 2) \qquad \text{Square the binomial.}$$
$$= a^3 + 2a^2 + 4a^2 + 8a + 4a + 8 \qquad \text{Multiply.}$$
$$= a^3 + 6a^2 + 12a + 8 \qquad \text{Combine like terms.}$$

You Try 8

Expand $(n - 3)^3$.

Answers to You Try Exercises

1) $42a^7 - 6a^6 - 18a^5 + 24a^4$ 2) a) $35x^3 - 17x^2 - 26x + 8$ b) $5p^4 + \frac{1}{2}p^3 - 46p^2 - 42p + 24$

3) a) $a^2 + 11a + 18$ b) $4t^2 - 17t - 15$ c) $c^2 - 9cd + 18d^2$

4) a) $x^2 + 11x + 24$ b) $t^2 - 13t + 42$ c) $3p^2 + 10p - 8$ d) $2x^2 - 13xy - 24y^2$

5) $-18x^5 - 66x^4 + 120x^3$ 6) a) $m^2 - 81$ b) $16z^2 - 9$ c) $1 - w^2$

7) a) $t^2 + 14t + 49$ b) $b^2 - 24b + 144$ c) $9v^2 + 24v + 16$ d) $\frac{9}{4}y^2 - 3y + 1$

8) $n^3 - 9n^2 + 27n - 27$

6.3 Exercises

Objective 1

1) Explain how to multiply two monomials.

2) Explain how to multiply a monomial by a trinomial.

Multiply.

3) $(7k^4)(2k^2)$

4) $(3z^7)(5z^3)$

5) $(-4t)(6t^8)$

6) $(8p^6)(-p^4)$

7) $\left(\frac{7}{10}d^9\right)\left(\frac{5}{2}d^2\right)$

8) $\left(-\frac{8}{9}c^5\right)\left(\frac{3}{10}c^7\right)$

Multiply.

9) $7y(4y - 9)$

10) $2a(11a + 5)$

11) $-4b(9b + 8)$

12) $-12m(11m - 4)$

 13) $6v^3(v^2 - 4v - 2)$

14) $3x^4(5x^3 + x - 7)$

15) $-3t^2(9t^3 - 6t^2 - 4t - 7)$

16) $-8u^5(9u^4 + 8u^3 + 12u - 1)$

17) $2x^3y(xy^2 + 8xy - 11y + 2)$

18) $5p^5q^2(-5p^2q + 12pq^2 - pq + 2q - 1)$

19) $-\frac{3}{4}t^4(20t^3 + 8t^2 - 5t)$

20) $\frac{2}{5}x^4(30x^2 - 15x + 7)$

Objective 2

Perform the indicated operations and simplify.

21) $2(10g^3 + 5g^2 + 4) - (2g^3 - 14g - 20)$

22) $6(7m^2 + 7m + 9) - 11(4m^2 + 7m + 1)$

23) $-(10r^3 - 14r + 27) + 3(3r^3 - 13r^2 - 15r + 6)$

24) $4(-w^3 - 5w^2 + 3w + 7)$
    $- (3w^3 - 7w^2 + 12w + 19)$

25) $7(a^3b^3 + 6a^3b^2 + 3)$
    $- 9(6a^3b^3 + 9a^3b^2 - 12a^2b + 8)$

26) $3(x^2y^2 + xy - 2) - 5(x^2y^2 + 6xy - 2)$
    $+ 2(x^2y^2 + 6xy - 2)$

27) $(q + 3)(5q^2 - 15q + 9)$

28) $(m + 9)(9m^2 + 4m - 7)$

29) $(p - 6)(2p^2 + 3p - 5)$

30) $(s - 5)(7s^2 - 3s - 11)$

31) $(5y^3 - y^2 + 8y + 1)(3y - 4)$

32) $(4n^3 + 4n^2 - 5n - 7)(7n - 6)$

 33) $\left(\frac{1}{2}k^2 + 3\right)(12k^2 + 5k - 10)$

34) $\left(\frac{2}{3}c^2 - 8\right)(6c^2 - 4c + 9)$

35) $(a^2 - a + 3)(a^2 + 4a - 2)$

36) $(r^2 + 2r + 5)(2r^2 - r - 3)$

37) $(3v^2 - v + 2)(-8v^3 + 6v^2 + 5)$

38) $(c^4 + 10c^2 - 7)(c^2 - 2c - 3)$

Multiply each horizontally and vertically. Which method do you prefer and why?

39) $(2x - 3)(4x^2 - 5x + 2)$

40) $(3n^2 + n - 4)(5n + 2)$

Objective 3

41) What do the letters in the word FOIL represent?

42) Can FOIL be used to expand $(x + 9)^2$? Explain your answer.

Use FOIL to multiply.

43) $(w + 8)(w + 7)$ 44) $(u + 4)(u + 6)$

45) $(k - 5)(k + 9)$ 46) $(n - 11)(n - 4)$

47) $(y - 6)(y - 1)$ 48) $(f + 5)(f - 9)$

49) $(4p + 5)(p - 3)$ 50) $(6t + 1)(t + 7)$

51) $(8n + 3)(3n + 4)$ 52) $(5b - 2)(8b + 5)$

53) $(6 - 5y)(2 - y)$ 54) $(3 - 2g)(5 - 4g)$

55) $(m + 7)(3 - m)$ 56) $(c + 5)(8 - c)$

 57) $(4a - 5b)(3a + 4b)$ 58) $(2x + 3y)(x - 6y)$

59) $(6p + 5q)(10p + 3q)$ 60) $(7m - 3n)(m - n)$

61) $\left(a + \dfrac{1}{4}\right)\left(a + \dfrac{4}{5}\right)$ 62) $\left(w + \dfrac{3}{2}\right)\left(w + \dfrac{4}{3}\right)$

Write an expression for a) the perimeter of each figure and b) the area of each figure.

63)

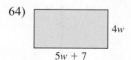
$y - 2$
$y + 6$

64)

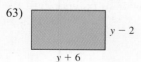
$4w$
$5w + 7$

65)

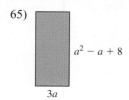
$a^2 - a + 8$
$3a$

66)

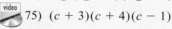
$2x^2 - 3$
$2x^2 - 3$

Objective 4

67) To find the product $2(n + 6)(n - 1)$, Raman begins by multiplying $2(n + 6)$ and then he multiplies that result by $(n - 1)$. Peggy begins by multiplying $(n + 6)(n - 1)$ and multiplies that result by 2. Who is right?

68) Find the product $(3a + 2)(a - 4)(a - 2)$
  a) by first multiplying $(3a + 2)(a - 4)$ and then multiplying that result by $(a - 2)$.
  b) by first multiplying $(a - 4)(a - 2)$ and then multiplying that result by $(3a + 2)$.
  c) What do you notice about the results?

Multiply. See Example 6.

69) $3(y + 4)(5y - 2)$

70) $8(x + 6)(2x + 1)$

71) $-18(3a - 1)(a + 2)$

72) $4(7 - 3z)(2z - 1)$

73) $-7r^2(r - 9)(r - 2)$

74) $12g^2(2g + 5)(-g + 1)$

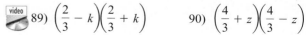 75) $(c + 3)(c + 4)(c - 1)$

76) $(x - 5)(x - 2)(x + 3)$

77) $(2p - 1)(p - 5)(p - 2)$

78) $(3r + 4)(r + 1)(r + 6)$

79) $10n\left(\dfrac{1}{2}n^2 + 3\right)(n^2 + 5)$

80) $12k\left(\dfrac{1}{4}k^2 - \dfrac{2}{3}\right)(k^2 + 1)$

Objectives 5 and 6

Find the following special products. See Examples 7 and 8.

81) $(x + 6)(x - 6)$ 82) $(z + 4)(z - 4)$

83) $(t - 3)(t + 3)$ 84) $(d - 10)(d + 10)$

85) $(2 - r)(2 + r)$ 86) $(9 + c)(9 - c)$

87) $\left(n + \dfrac{1}{2}\right)\left(n - \dfrac{1}{2}\right)$ 88) $\left(b - \dfrac{1}{5}\right)\left(b + \dfrac{1}{5}\right)$

89) $\left(\dfrac{2}{3} - k\right)\left(\dfrac{2}{3} + k\right)$ 90) $\left(\dfrac{4}{3} + z\right)\left(\dfrac{4}{3} - z\right)$

91) $(3m + 2)(3m - 2)$ 92) $(5y - 4)(5y + 4)$

93) $-(6a - b)(6a + b)$ 94) $-(2p + 7q)(2p - 7q)$

95) $(y + 8)^2$ 96) $(b + 6)^2$

97) $(t - 11)^2$ 98) $(g - 5)^2$

99) $(k - 2)^2$ 100) $(x - 8)^2$

101) $(4w + 1)^2$ 102) $(7n + 2)^2$

103) $(2d - 5)^2$ 104) $(3p - 5)^2$

105) Does $4(t + 3)^2 = (4t + 12)^2$? Why or why not?

106) Explain, in words, how to find the product $3(z - 4)^2$, then find the product.

Find the product.

107) $6(x + 1)^2$ 108) $2(k + 5)^2$

109) $2a(a + 3)^2$ 110) $-5c(c + 4)^2$

111) $-3(m - 1)^2$ 112) $4(y - 3)^2$

Objective 7

Expand.

113) $(r + 5)^3$ 114) $(w + 4)^3$

115) $(s - 2)^3$ 116) $(q - 1)^3$

117) $(c^2 - 9)^2$ 118) $(a^2 - 7)^2$

119) $\left(\dfrac{2}{3}n + 4\right)^2$ 120) $\left(\dfrac{3}{8}x + 2\right)^2$

121) $(y + 2)^4$ 122) $(b + 3)^4$

123) Does $(x + 5)^2 = x^2 + 25$? Why or why not?

124) Does $(y - 3)^3 = y^3 - 27$? Why or why not?

125) Express the volume of the cube as a polynomial.

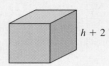

126) Express the area of the square as a polynomial.

127) Express the area of the shaded region as a polynomial.

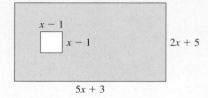

128) Express the area of the triangle as a polynomial.

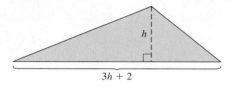

Section 6.4 Division of Polynomials

Objectives

1. Divide a Polynomial by a Monomial

2. Divide a Polynomial by a Polynomial

3. Divide a Polynomial by a Binomial Using Synthetic Division

The last operation with polynomials we need to discuss is division of polynomials. We will consider this in two parts:

1) Dividing a polynomial by a monomial

Examples: $\dfrac{12a^2 - a + 15}{3}$, $\dfrac{-48m^3 + 30m^2 - 8m}{8m^2}$

and

2) Dividing a polynomial by a polynomial

Examples: $\dfrac{n^2 + 14n + 48}{n + 6}$, $\dfrac{27z^3 - 1}{3z - 1}$

1. Divide a Polynomial by a Monomial

The procedure for dividing a polynomial by a monomial is based on the procedure for adding or subtracting fractions.

To add $\dfrac{4}{15} + \dfrac{7}{15}$, we do the following:

$$\frac{4}{15} + \frac{7}{15} = \frac{4 + 7}{15}$$ Add numerators, keep the common denominator.

$$= \frac{11}{15}$$

Reversing the process above we can write

$$\frac{11}{15} = \frac{4 + 7}{15}$$ Rewrite 11 as $4 + 7$.

$$= \frac{4}{15} + \frac{7}{15}$$ We can separate $\dfrac{4 + 7}{15}$ into two distinct fractions.

We can generalize this result and say that

$$\frac{a + b}{c} = \frac{a}{c} + \frac{b}{c} \quad (c \neq 0)$$

So, to divide a polynomial by a monomial, divide *each term* in the polynomial by the monomial and simplify.

Example 1

Divide.

a) $\dfrac{40a^2 - 25a + 10}{5}$ b) $\dfrac{9x^3 + 30x^2 + 3x}{3x}$

Solution

a) First, note that the polynomial is being divided by a *monomial.* That means we will divide each term in the numerator by 5.

$$\frac{40a^2 - 25a + 10}{5} = \frac{40a^2}{5} - \frac{25a}{5} + \frac{10}{5}$$
$$= 8a^2 - 5a + 2$$

The components of our division problem here are labeled the same way as when we divide with integers.

$$\begin{array}{c} \text{Dividend} \rightarrow \\ \text{Divisor} \rightarrow \end{array} \frac{40a^2 - 25a + 10}{5} = 8a^2 - 5a + 2 \leftarrow \text{Quotient}$$

We can check our answer by multiplying the quotient by the divisor. The answer should be the dividend.

$$\text{Check: } 5(8a^2 - 5a + 2) = 40a^2 - 25a + 10 \quad \checkmark$$

The quotient $8a^2 - 5a + 2$ is correct.

b) $\dfrac{9x^3 + 30x^2 + 3x}{3x} = \dfrac{9x^3}{3x} + \dfrac{30x^2}{3x} + \dfrac{3x}{3x}$ Divide each term in numerator by $3x$.

$\qquad\qquad\qquad\qquad = 3x^2 + 10x + 1$ Apply the quotient rule for exponents.

 BE CAREFUL Students will often just "cancel out" $\dfrac{3x}{3x}$ and get nothing. But $\dfrac{3x}{3x} = 1$ since a quantity divided by itself equals one.

$$\text{Check: } 3x(3x^2 + 10x + 1) = 9x^3 + 30x^2 + 3x \quad \checkmark$$

The quotient $3x^2 + 10x + 1$ is correct.

 **You Try 1**

Divide $\dfrac{24t^5 - 6t^4 - 54t^3}{6t^2}$.

Example 2

Divide $(15m + 45m^3 - 4 + 18m^2) \div (9m^2)$.

Solution

Although this example is written differently, it is the same as the previous examples. Notice, however, the terms in the numerator are not written in descending powers. Rewrite them in descending powers before dividing.

$$\dfrac{15m + 45m^3 - 4 + 18m^2}{9m^2} = \dfrac{45m^3 + 18m^2 + 15m - 4}{9m^2}$$

$$= \dfrac{45m^3}{9m^2} + \dfrac{18m^2}{9m^2} + \dfrac{15m}{9m^2} - \dfrac{4}{9m^2}$$

$$= 5m + 2 + \dfrac{5}{3m} - \dfrac{4}{9m^2} \quad \text{Apply quotient rule and simplify.}$$

The quotient is *not* a polynomial since m and m^2 appear in denominators. The quotient of polynomials is not necessarily a polynomial.

You Try 2

Divide $(6u^2 + 40 - 24u^3 - 8u) \div (8u^2)$.

2. Divide a Polynomial by a Polynomial

When dividing a polynomial by a polynomial containing two or more terms, we use **long division of polynomials**. This method is similar to long division of whole numbers. We will look at a long division problem here so that we can compare the procedure with polynomial long division.

Example 3

Divide 4593 by 8.

Solution

$$
\begin{array}{r}
5 \\
8\overline{)4593} \\
-40\!\downarrow \\
\hline
59
\end{array}
$$

1) How many times does 8 divide into 45 evenly?   5
2) Multiply $5 \times 8 = 40$.
3) Subtract $45 - 40 = 5$.
4) Bring down the 9.

Start the process again.

$$
\begin{array}{r}
57 \\
8\overline{)4593} \\
-40\!\downarrow \\
\hline
59 \\
-56 \\
\hline
33
\end{array}
$$

1) How many times does 8 divide into 59 evenly?   7
2) Multiply $7 \times 8 = 56$.
3) Subtract $59 - 56 = 3$.
4) Bring down the 3.

Do the procedure again.

$$
\begin{array}{r}
574 \\
8\overline{)4593} \\
-40 \\
\hline
59 \\
-56 \\
\hline
33 \\
-32 \\
\hline
1
\end{array}
$$

1) How many times does 8 divide into 33 evenly?   4
2) Multiply $4 \times 8 = 32$.
3) Subtract $33 - 32 = 1$.
4) There are no more numbers to bring down, so the remainder is 1.

Write the result.

$$4593 \div 8 = 574\frac{1}{8} \quad \begin{array}{l} \leftarrow \text{Remainder} \\ \leftarrow \text{Divisor} \end{array}$$

Check: $(8 \times 574) + 1 = 4592 + 1 = 4593$ ✓

Next we will divide two polynomials using a long division process similar to that of Example 3.

Example 4

Divide $\dfrac{3x^2 + 19x + 20}{x + 5}$.

Solution

First, notice that we are dividing by more than one term. That tells us to use long division of polynomials.

We will work with the x in $x + 5$ like we worked with the 8 in Example 3.

$$
\begin{array}{r}
3x \\
x + 5{\overline{\smash{\big)}\,3x^2 + 19x + 20}} \\
\underline{-(3x^2 + 15x)} \downarrow \\
4x + 20
\end{array}
$$

1) By what do we multiply x to get $3x^2$? $3x$
   Line up terms in the quotient according to exponents, so write $3x$ above $19x$.

2) Multiply $3x$ by $(x + 5)$. $3x(x + 5) = 3x^2 + 15x$

3) Subtract $(3x^2 + 19x) - (3x^2 + 15x) = 4x$.

4) Bring down the $+20$.

Start the process again. Remember, work with the x in $x + 5$ like we worked with the 8 in Example 3.

$$
\begin{array}{r}
3x + 4 \\
x + 5{\overline{\smash{\big)}\,3x^2 + 19x + 20}} \\
\underline{-(3x^2 + 15x)} \\
4x + 20 \\
\underline{-(4x + 20)} \\
0
\end{array}
$$

1) By what do we multiply x to get $4x$? 4
   Write $+4$ above $+20$.

2) Multiply 4 by $(x + 5)$. $4(x + 5) = 4x + 20$

3) Subtract $(4x + 20) - (4x + 20) = 0$.

4) There are no more terms. The remainder is 0.

Write the result.

$$\frac{3x^2 + 19x + 20}{x + 5} = 3x + 4$$

Check: $(x + 5)(3x + 4) = 3x^2 + 4x + 15x + 20 = 3x^2 + 19x + 20$ ✓

You Try 3

Divide.

a) $\dfrac{x^2 + 11x + 24}{x + 8}$ b) $\dfrac{2x^2 + 23x + 45}{x + 9}$

Next, we will look at a division problem with a remainder.

| Example 5 |

Divide $\dfrac{-11c + 16c^3 + 19 - 38c^2}{2c - 5}$.

Solution

When we write our long division problem, the polynomial in the numerator must be rewritten so that the exponents are in descending order. Then, perform the long division.

$$
\begin{array}{r}
8c^2 \\
2c - 5 \overline{)\,16c^3 - 38c^2 - 11c + 19} \\
-(16c^3 - 40c^2) \qquad \downarrow \\
\hline
2c^2 - 11c
\end{array}
$$

1) By what do we multiply $\underline{2c}$ to get $16c^3$? $8c^2$
2) Multiply $8c^2(2c - 5) = 16c^3 - 40c^2$.
3) Subtract.
$$(16c^3 - 38c^2) - (16c^3 - 40c^2)$$
$$= 16c^3 - 38c^2 - 16c^3 + 40c^2$$
$$= 2c^2$$
4) Bring down the $-11c$.

Repeat the process.

$$
\begin{array}{r}
8c^2 + \quad c \\
2c - 5 \overline{)\,16c^3 - 38c^2 - 11c + 19} \\
-(16c^3 - 40c^2) \qquad \downarrow \\
\hline
2c^2 - 11c \\
-(2c^2 - 5c) \\
\hline
-6c + 19
\end{array}
$$

1) By what do we multiply $\underline{2c}$ to get $2c^2$? c
2) Multiply $c(2c - 5) = 2c^2 - 5c$.
3) Subtract.
$$(2c^2 - 11c) - (2c^2 - 5c)$$
$$= 2c^2 - 11c - 2c^2 + 5c$$
$$= -6c$$
4) Bring down the $+19$.

Continue.

$$
\begin{array}{r}
8c^2 + \quad c - 3 \\
2c - 5 \overline{)\,16c^3 - 38c^2 - 11c + 19} \\
-(16c^3 - 40c^2) \\
\hline
2c^2 - 11c \\
-(2c^2 - 5c) \\
\hline
-6c + 19 \\
-(-6c + 15) \\
\hline
4
\end{array}
$$

1) By what do we multiply $\underline{2c}$ to get $-6c$? -3
2) Multiply $-3(2c - 5) = -6c + 15$.
3) Subtract.
$$(-6c + 19) - (-6c + 15)$$
$$= -6c + 19 + 6c - 15 = 4$$

We are done with the long division process. How do we know that? Since the degree of 4 (degree zero) is less than the degree of $2c - 5$ (degree one) we cannot divide anymore. *The remainder is 4.*

$$\frac{16c^3 - 38c^2 - 11c + 19}{2c - 5} = 8c^2 + c - 3 + \frac{4}{2c - 5}$$

Check: $(2c - 5)(8c^2 + c - 3) + 4 = 16c^3 + 2c^2 - 6c - 40c^2 - 5c + 15 + 4$
$$= 16c^2 - 38c^2 - 11c + 19 \quad \checkmark$$

You Try 4

Divide $-23t^2 - 2 + 20t^3 - 11t$ by $5t + 3$.

As we saw in Example 5, we must write our polynomials so that the exponents are in descending order. We have to watch out for something else as well—missing terms. If a polynomial is missing one or more terms, we put them into the polynomial with coefficients of zero.

Example 6

Divide $x^3 + 125$ by $x + 5$.

Solution

The degree of the polynomial $x^3 + 125$ is three, but it is missing the x^2-term and the x-term. We will insert these terms into the polynomial by giving them coefficients of zero.

$$x^3 + 125 = x^3 + 0x^2 + 0x + 25$$

Divide.
$$
\begin{array}{r}
x^2 - 5x + 25 \\
x + 5 \overline{)\, x^3 + 0x^2 + 0x + 125} \\
\underline{-(x^3 + 5x^2)} \\
-5x^2 + 0x \\
\underline{-(-5x^2 - 25x)} \\
25x + 125 \\
\underline{-(25x + 125)} \\
0
\end{array}
$$

$$(x^3 + 125) \div (x + 5) = x^2 - 5x + 25$$

Check: $(x + 5)(x^2 - 5x + 25) = x^3 - 5x^2 + 25x + 5x^2 - 25x + 125$
$$= x^3 + 125 \ \checkmark$$

You Try 5

Divide $\dfrac{2k^3 + 5k^2 + 91}{2k + 9}$.

3. Divide a Polynomial by a Binomial Using Synthetic Division

When we divide a polynomial by a binomial of the form $x - c$, another method called **synthetic division** can be used. Synthetic division uses only the numerical coefficients of the variables to find the quotient.

Consider the division problem $(3x^3 - 5x^2 - 6x + 13) \div (x - 2)$. On the left, we will illustrate the long division process as we have already presented it. On the right, we will show the process using only the coefficients of the variables.

$$
\begin{array}{r}
3x^2 \;\;+ x \;\; - 4 \\
x - 2\overline{)\; 3x^3 - 5x^2 - 6x + 13} \\
\underline{-(3x^3 - 6x^2)} \\
x^2 - 6x \\
\underline{-(x^2 - 2x)} \\
-4x + 13 \\
\underline{-(-4x + \;\; 8)} \\
5
\end{array}
\qquad
\begin{array}{r}
3 + 1 - \;\; 4 \\
1 - 2\overline{)\; 3 - 5 - 6 + 13} \\
\underline{-(3 - 6)} \\
1 - 6 \\
\underline{-(1 - 2)} \\
-4 + 13 \\
\underline{-(-4 + 8)} \\
5
\end{array}
$$

> As long as we keep the like terms lined up in the correct columns, the variables do not affect the numerical coefficients of the quotient. This process of using only the numerical coefficients to divide a polynomial by a binomial of the form $x - c$ is called *synthetic division*. Using synthetic division is often quicker than using the traditional long division process.

We will present the steps for performing synthetic division by looking at the above example again.

Example 7

Use synthetic division to divide $(3x^3 - 5x^2 - 6x + 13)$ by $(x - 2)$.

Solution

Remember, in order to be able to use synthetic division, the divisor must be in the form $x - c$. $x - 2$ is in the form $x - c$, and $c = 2$.

How to Perform Synthetic Division

Step 1: Write the dividend in descending powers of x. If a term of any degree is missing, insert the term into the polynomial with a coefficient of 0.

The dividend in the example is $3x^3 - 5x^2 - 6x + 13$. It is written in descending order, and no terms are missing.

Step 2: Write the value of c in an open box. Next to it, on the right, write the coefficients of the terms of the dividend. Skip a line and draw a horizontal line under the coefficients. Bring down the first coefficient.

In this example, $c = 2$.

$$
\underline{2|}\;\; 3 \quad -5 \quad -6 \quad 13
$$
$$
\overline{}
$$
$$
3
$$

Step 3: Multiply the number in the box by the coefficient under the horizontal line. (Here, that is $2 \cdot 3 = 6$.) Write the product under the next coefficient. (Write the 6 under the -5.) Then, *add* the numbers in the second column. (Here, we get 1.)

$$
\underline{2|}\;\; 3 \quad -5 \quad -6 \quad 13
$$
$$
 6 \qquad\qquad\qquad 2 \cdot 3 = 6;\; -5 + 6 = 1
$$
$$
\overline{}
$$
$$
3 \quad 1
$$

Step 4:   Multiply the number in the box by the number under the horizontal line in the second column. (Here, that is $2 \cdot 1 = 2$.) Write the product under the next coefficient. (Write the 2 under the -6.) Then, *add* the numbers in the third column. (Here, we get -4.)

$$\begin{array}{r|rrrr} 2) & 3 & -5 & -6 & 13 \\ & & 6 & 2 & \\ \hline & 3 & 1 & -4 & \end{array} \qquad 2 \cdot 1 = 2; \ -6 + 2 = -4$$

Step 5:   Repeat the procedure of step 4 with subsequent columns until there is a number in each column in the row under the horizontal line.

$$\begin{array}{r|rrrr} 2) & 3 & -5 & -6 & 13 \\ & & 6 & 2 & -8 \\ \hline & 3 & 1 & -4 & 5 \end{array} \qquad 2 \cdot (-4) = -8; \ 13 + (-8) = 5$$

The numbers in the last row represent the quotient and the remainder. The last number is the remainder. The numbers before it are the coefficients of the quotient. *The degree of the quotient is one less than the degree of the dividend.*

In our example, the dividend is a *third-degree* polynomial. Therefore, the quotient is a *second-degree* polynomial.

Since the 3 in the first row is the coefficient of x^3 in the dividend, the 3 in the last row is the coefficient of x^2 in the quotient, and so on.

$$\overset{\text{Dividend}}{3x^3 - 5x^2 - 6x + 13}$$

$$\begin{array}{r|rrrr} 2) & 3 & -5 & -6 & 13 \\ & & 6 & 2 & -8 \\ \hline & 3 & 1 & -4 & 5 \end{array} \rightarrow \text{Remainder}$$

$$\underset{\text{Quotient}}{3x^2 + 1x - 4}$$

$$(3x^3 - 5x^2 - 6x + 13) \div (x - 2) = 3x^2 + x - 4 + \frac{5}{x - 2}.$$

 You Try 6

Use synthetic division to divide $(2x^3 + x^2 - 16x - 7) \div (x - 3)$.

If the divisor is $x + 3$, then it can be written in the form $x - c$ as $x - (-3)$. So, $c = -3$.

BE CAREFUL

Synthetic division can be used only when dividing a polynomial by a binomial of the form $x - c$. If the divisor is not in the form $x - c$, use long division.

Synthetic division can be used to find $(4x^2 - 19x + 16) \div (x - 4)$ because $x - 4$ is in the form $x - c$.	Synthetic division *cannot* be used to find $(4x^2 - 19x + 16) \div (2x - 3)$ because $2x - 3$ is *not* in the form $x - c$. Use long division.

Summary of Polynomial Division

Remember, when asked to divide two polynomials, first identify which type of division problem it is.

1) To divide a *polynomial* by a *monomial,* divide *each term* in the polynomial by the monomial and simplify.

Monomial →
$$\frac{36r^3 + 60r^2 - 12r + 3}{12r} = \frac{36r^3}{12r} + \frac{60r^2}{12r} - \frac{12r}{12r} + \frac{3}{12r}$$
$$= 3r^2 + 5r - 1 + \frac{1}{4r}$$

3) To divide a *polynomial* by a *polynomial* containing two or more terms, use *long division.*

Binomial →
$$\frac{24x^3 + 14x^2 + 39x - 7}{6x - 1}$$

$$
\require{enclose}
\begin{array}{r}
4x^2 + 3x + 7 \\
6x - 1 \enclose{longdiv}{24x^3 + 14x^2 + 39x - 7} \\
\end{array}
$$

$$
\begin{array}{r}
-(24x^3 - 4x^2) \\
\hline
18x^2 + 39x \\
-(18x^2 - 3x) \\
\hline
42x - 7 \\
-(42x - 7) \\
\hline
0
\end{array}
$$

$$\frac{24x^3 + 14x^2 + 39x - 7}{6x - 1} = 4x^2 + 3x + 7$$

3) To divide a *polynomial* by a *binomial of the form $x - c$,* use long division or synthetic division.

Answers to You Try Exercises

1) $4t^3 - t^2 - 9t$ 2) $-3u + \dfrac{3}{4} - \dfrac{1}{u} + \dfrac{5}{u^2}$ 3) a) $x + 3$ b) $2x + 5$

4) $4t^2 - 7t + 2 - \dfrac{8}{5t + 3}$ 5) $k^2 - 2k + 9 + \dfrac{10}{2k + 9}$ 6) $2x^2 + 7x + 5 + \dfrac{8}{x - 3}$

6.4 Exercises

Label the dividend, divisor, and quotient of each division problem.

1) $\dfrac{12c^3 + 20c^2 - 4c}{4c} = 3c^2 + 5c - 1$

2) $2p + 3\overline{)10p^3 + p^2 - 25p - 6}$ $5p^2 - 7p - 2$

3) Explain, in your own words, how to divide a polynomial by a monomial.

4) When do you use long division to divide polynomials?

Objective I

Divide.

5) $\dfrac{4a^5 - 10a^4 + 6a^3}{2a^3}$

6) $\dfrac{28k^4 + 8k^3 - 40k^2}{4k^2}$

7) $\dfrac{18u^7 + 18u^5 + 45u^4 - 72u^2}{9u^2}$

8) $\dfrac{-15m^6 + 10m^5 + 20m^4 - 35m^3}{5m^3}$

9) $(35d^5 - 7d^2) \div (-7d^2)$

10) $(-32q^6 - 8q^3 + 4q^2) \div (-4q^2)$

11) $\dfrac{9w^5 + 42w^4 - 6w^3 + 3w^2}{6w^3}$

12) $\dfrac{-54j^5 + 30j^3 - 9j^2 + 15}{9j}$

13) $(10v^7 - 36v^5 - 22v^4 - 5v^2 + 1) \div (4v^4)$

14) $(60z^5 + 3z^4 - 10z) \div (5z^2)$

Divide.

15) $\dfrac{90a^4b^3 + 60a^3b^3 - 40a^3b^2 + 100a^2b^2}{10ab^2}$

16) $\dfrac{24x^6y^6 - 54x^5y^4 - x^3y^3 + 12x^3y^2}{6x^2y}$

17) $(9t^5u^4 - 63t^4u^4 - 108t^3u^4 + t^3u^2) \div (-9tu^2)$

18) $(-45c^8d^6 - 15c^6d^5 + 60c^3d^5 + 30c^3d^3) \div (-15c^3d^2)$

19) Irene divides $16t^3 - 36t^2 + 4t$ by $4t$ and gets a quotient of $4t^2 - 9t$. Is this correct? Why or why not?

20) Kinh divides $\dfrac{15x^2 + 12x}{12x}$ and gets a quotient of $15x^2$. What was his mistake? What is the correct answer?

Objective 2

Divide.

21) $\dfrac{g^2 + 9g + 20}{g + 5}$

22) $\dfrac{n^2 + 13n + 40}{n + 8}$

23) $\dfrac{p^2 + 8p + 12}{p + 2}$

24) $\dfrac{v^2 + 13v + 12}{v + 1}$

25) $\dfrac{k^2 + 4k - 45}{k + 9}$

26) $\dfrac{m^2 - 6m - 27}{m + 3}$

27) $\dfrac{h^2 + 5h - 24}{h - 3}$

28) $\dfrac{u^2 - 11u + 30}{u - 5}$

29) $\dfrac{4a^3 - 24a^2 + 29a + 15}{2a - 5}$

30) $\dfrac{28b^3 - 26b^2 + 41b - 15}{7b - 3}$

31) $(p + 45p^2 - 1 + 18p^3) \div (6p + 1)$

32) $(17z^2 - 10 - 12z^3 + 32z) \div (4z + 5)$

33) $(6t^2 - 7t + 4) \div (t - 5)$

34) $(7d^2 + 57d - 4) \div (d + 9)$

35) $\dfrac{61z + 12z^3 - 37 + 44z^2}{3z + 5}$

36) $\dfrac{23k^3 + 22k - 8 + 6k^4 + 44k^2}{6k - 1}$

37) $\dfrac{w^3 + 64}{w + 4}$ 38) $\dfrac{a^3 - 27}{a - 3}$

39) $(16r^3 + 58r^2 - 9) \div (8r - 3)$

40) $(50c^3 + 7c + 4) \div (5c + 2)$

Objective 3

41) Explain when synthetic division may be used to divide polynomials.

Can synthetic division be used to divide the polynomials in Exercises 42–44? Why or why not?

42) $(x^4 + 8x^3 - 10x + 3) \div (x - 6)$

43) $(2x^3 - 7x^2 - 2x + 10) \div (2x + 5)$

44) $\dfrac{x^3 - 15x^2 + 8x + 12}{x^2 - 2}$

Use synthetic division to divide the polynomials.

45) $(t^2 + 5t - 36) \div (t - 4)$

46) $(m^2 - 2m - 24) \div (m - 6)$

47) $\dfrac{5n^2 + 21n + 20}{n + 3}$ 48) $\dfrac{6k^2 + 4k - 19}{k + 2}$

49) $(2y^3 + 7y^2 - 10y + 21) \div (y + 5)$

50) $(4z^3 - 11z^2 - 6z + 10) \div (z - 3)$

51) $(4p - 3 - 10p^2 + 3p^3) \div (p - 3)$

52) $(10c^2 + 3c + 2c^3 - 20) \div (c + 4)$

53) $(2 + 5x^4 - 8x + 7x^3 - x^2) \div (x + 1)$

54) $(-4w^3 + w - 8 + w^4 + 7w^2) \div (w - 2)$

55) $\dfrac{r^3 - 3r^2 + 4}{r - 2}$ 56) $\dfrac{a^3 - 38a - 15}{a + 6}$

57) $\dfrac{m^4 - 81}{m - 3}$ 58) $\dfrac{h^4 - 1}{h + 1}$

59) $(2c^5 - 3c^4 - 11c) \div (c - 2)$

60) $(n^5 - 29n^2 - 2n) \div (n - 3)$

61) $(2x^3 + 7x^2 - 16x + 6) \div \left(x - \dfrac{1}{2}\right)$

62) $(3t^3 - 25t^2 + 14t - 2) \div \left(t - \dfrac{1}{3}\right)$

Objectives 1–3

Mixed Exercises

Divide.

63) $\dfrac{6x^4y^4 + 30x^4y^3 - x^2y^2 + 3xy}{6x^2y^2}$

64) $\dfrac{12v^2 - 23v + 14}{3v - 2}$

65) $\dfrac{-8g^4 + 49g^2 + 36 - 25g - 2g^3}{4g - 9}$

66) $(12c^2 + 6c - 30c^3 + 48c^4) \div (-6c)$

67) $\dfrac{6t^2 - 43t - 20}{t - 8}$

68) $\dfrac{-14u^3v^3 + 7u^2v^3 + 21uv + 56}{7u^2v}$

69) $(8n^3 - 125) \div (2n - 5)$

70) $(12a^4 - 19a^3 + 22a^2 - 9a - 20) \div (3a - 4)$

71) $(13x^2 - 7x^3 + 6 + 5x^4 - 14x) \div (x^2 + 2)$

72) $(18m^4 - 66m^3 + 39m^2 + 11m - 7) \div (6m^2 - 1)$

73) $\dfrac{-12a^3 + 9a^2 - 21a}{-3a}$ 74) $\dfrac{64r^3 + 27}{4r + 3}$

75) $\dfrac{10h^4 - 6h^3 - 49h^2 + 27h + 19}{2h^2 - 9}$

76) $\dfrac{16w^2 - 3 - 7w + 15w^4 - 5w^3}{5w^2 + 7}$

77) $\dfrac{6d^4 + 19d^3 - 8d^2 - 61d - 40}{2d^2 + 7d + 5}$

78) $\dfrac{8x^4 + 2x^3 - 13x^2 - 53x + 14}{2x^2 + 5x + 7}$

79) $\dfrac{9c^4 - 82c^3 - 41c^2 + 9c + 16}{c^2 - 10c + 4}$

80) $\dfrac{15n^4 - 16n^3 - 31n^2 + 50n - 22}{5n^2 - 7n + 2}$

81) $\dfrac{k^4 - 81}{k^2 + 9}$ 82) $\dfrac{b^4 - 16}{b^2 - 4}$

83) $\dfrac{49a^4 - 15a^2 - 14a^3 + 5a^6}{-7a^3}$

84) $\dfrac{9q^2 + 26q^4 + 8 - 6q - 4q^3}{2q^2}$

For each rectangle, find a polynomial that represents the missing side.

85)

$y - 6$

Find the length if the area is given by $4y^2 - 23y - 6$.

86)

$3x + 2$

Find the width if the area is given by $6x^2 + x - 2$.

87)

$9a^3$

Find the width if the area is given by
$18a^5 - 45a^4 + 9a^3$.

88)

$6w$

Find the length if the area is given by
$9w^3 + 6w^2 - 24w$.

89) Find the base of the triangle if the area is given by
$6h^3 + 3h^2 + h$.

90) Find the base of the triangle if the area is given by
$6n^3 - 2n^2 + 10n$.

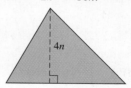

Definition/Procedure	Example	Reference

Review of the Rules of Exponents

For real numbers a and b and integers m and n, the following rules apply:

		6.1
Product Rule: $a^m \cdot a^n = a^{m+n}$	$p^4 \cdot p^6 = p^{4+6} = p^{10}$	p. 390
Power Rules: a) $(a^m)^n = a^{mn}$ b) $(ab)^n = a^n b^n$ c) $\left(\dfrac{a}{b}\right)^n = \dfrac{a^n}{b^n}$ $(b \neq 0)$	a) $(c^5)^3 = c^{5 \cdot 3} = c^{15}$ b) $(4z)^3 = 4^3 z^3 = 64z^3$ c) $\left(\dfrac{x}{3}\right)^4 = \dfrac{x^4}{3^4} = \dfrac{x^4}{81}$	p. 390
Zero Exponent: $a^0 = 1$ if $a \neq 0$	$5^0 = 1$	p. 390
Negative Exponent: a) $a^{-n} = \left(\dfrac{1}{a}\right)^n = \dfrac{1}{a^n}$ $(a \neq 0)$ b) $\dfrac{a^{-m}}{b^{-n}} = \dfrac{b^n}{a^m}$ $(a \neq 0, b \neq 0)$	a) $2^{-5} = \left(\dfrac{1}{2}\right)^5 = \dfrac{1}{2^5} = \dfrac{1}{32}$ b) $\dfrac{t^{-6}}{u^{-2}} = \dfrac{u^2}{t^6}$	p. 390
Quotient Rule: $\dfrac{a^m}{a^n} = a^{m-n}$ $(a \neq 0)$	$\dfrac{d^{12}}{d^4} = d^{12-4} = d^8$	p. 390

Addition and Subtraction of Polynomials — 6.2

Definition/Procedure	Example	Reference
A *polynomial in x* is the sum of a finite number of terms of the form ax^n where n is a whole number and a is a real number. The *degree of a term* equals the exponent on its variable. If a term has more than one variable, the degree equals the *sum* of the exponents on the variables. The *degree of the polynomial* equals the highest degree of any nonzero term.	Identify each term in the polynomial, the coefficient and degree of each term, and the degree of the polynomial. $5a^4b^2 - 16a^3b^2 - 4a^2b^3 + ab + 9b$ (see table below) The degree of the polynomial is 6.	p. 393

Term	Coefficient	Degree
$5a^4b^2$	5	6
$-16a^3b^2$	-16	5
$-4a^2b^3$	-4	5
ab	1	2
$9b$	9	1

Definition/Procedure	Example	Reference
To *add polynomials*, add like terms. Polynomials may be added horizontally or vertically.	Add the polynomials. $(6n^2 + 7n - 14) + (-2n^2 + 6n + 3)$ $= [6n^2 + (-2n^2)] + (7n + 6n) +$ $\quad (-14 + 3)$ $= 4n^2 + 13n - 11$	p. 396

Definition/Procedure	Example	Reference
To *subtract two polynomials*, change the sign of each term in the second polynomial. Then, add the polynomials.	Subtract. $(3h^3 - 7h^2 + 8h + 4) - (12h^3 - 8h^2 + 3h + 9)$ $= (3h^3 - 7h^2 + 8h + 4)$ $\oplus (-12h^3 + 8h^2 - 3h - 9)$ $= -9h^3 + h^2 + 5h - 5$	**p. 397**
$f(x) = 3x^2 + 8x - 4$ is an example of a *polynomial function* since $3x^2 + 8x - 4$ is a polynomial and since each real number that is substituted for x produces only one value for the expression. Finding $f(4)$ is the same as evaluating $3x^2 + 8x - 4$ when $x = 4$.	If $f(x) = 3x^2 + 8x - 4$, find $f(4)$. $f(4) = 3(4)^2 + 8(4) - 4$ $= 3(16) + 32 - 4$ $= 48 + 32 - 4$ $= 76$	**p. 398**

Multiplication of Polynomials — 6.3

Definition/Procedure	Example	Reference
When multiplying a *monomial* and a *polynomial*, use the distributive property.	Multiply. $4a^3(-3a^2 + 7a - 2)$ $= (4a^3)(-3a^2) + (4a^3)(7a) + (4a^3)(-2)$ $= -12a^5 + 28a^4 - 8a^3$	**p. 402**
To *multiply two polynomials*, multiply each term in the second polynomial by each term in the first polynomial. Then, combine like terms.	Multiply. $(2c + 5)(c^2 - 3c + 6)$ $= (2c)(c^2) + (2c)(-3c) + (2c)(6)$ $\quad + (5)(c^2) + (5)(-3c) + (5)(6)$ $= 2c^3 - 6c^2 + 12c + 5c^2 - 15c + 30$ $= 2c^3 - c^2 - 3c + 30$	**p. 402**
Multiplying Two Binomials We can use FOIL to multiply two binomials. **FOIL** stands for **F**irst **O**uter **I**nner **L**ast. Then, add like terms.	Use FOIL to multiply $(3k - 2)(k + 4)$. F:$3k^2$ L:-8 $(3k - 2)(k + 4)$ I:$-2k$ $+$ O:$12k$ $10k$ sum $(3k - 2)(k + 4) = 3k^2 + 10k - 8$	**p. 403**
Special Products a) $(a + b)(a - b) = a^2 - b^2$ b) $(a + b)^2 = a^2 + 2ab + b^2$ c) $(a - b)^2 = a^2 - 2ab + b^2$	a) Multiply: $(y + 6)(y - 6) = y^2 - 6^2$ $\qquad\qquad\qquad\qquad\quad = y^2 - 36$ b) Expand: $(t + 9)^2 = t^2 + 2(t)(9) + 9^2$ $\qquad\qquad\qquad\quad = t^2 + 18t + 81$ c) Expand $(4u - 3)^2$. $(4u - 3)^2 = (4u)^2 - 2(4u)(3) + 3^2$ $\qquad\qquad = 16u^2 - 24u + 9$	**p. 407**

Definition/Procedure	Example	Reference
Division of Polynomials		6.4
To *divide a polynomial by a monomial,* divide *each term* in the polynomial by the monomial and simplify.	Divide: $\dfrac{18r^4 + 2r^3 - 9r^2 + 6r - 10}{2r^2}$ $= \dfrac{18r^4}{2r^2} + \dfrac{2r^3}{2r^2} - \dfrac{9r^2}{2r^2} + \dfrac{6r}{2r^2} - \dfrac{10}{2r^2}$ $= 9r^2 + r - \dfrac{9}{2} + \dfrac{3}{r} - \dfrac{5}{r^2}$	**p. 413**
To *divide a polynomial by another polynomial* containing two or more terms, use *long division.*	Divide $\dfrac{12m^3 - 32m^2 - 17m + 25}{6m + 5}$. $$\begin{array}{r} 2m^2 - 7m + 3 \\ 6m+5\overline{)\,12m^3 - 32m^2 - 17m + 25} \\ \underline{-(12m^3 + 10m^2)} \quad\downarrow\qquad\downarrow \\ -42m^2 - 17m \\ \underline{-(-42m^2 - 35m)} \\ 18m + 25 \\ \underline{-(18m + 15)} \\ \text{Remainder} \leftarrow 10 \end{array}$$ $\dfrac{12m^3 - 32m^2 - 17m + 25}{6m + 5} =$ $\qquad 2m^2 - 7m + 3 + \dfrac{10}{6m + 5}$	**p. 415**
To *divide a polynomial by a binomial of the form $x - c$,* we can use either long division or *synthetic division.*	Use synthetic division to divide $(4x^3 - 17x^2 + 17x - 6) \div (x - 3)$. $$\begin{array}{r|rrrr} 3 & 4 & -17 & 17 & -6 \\ & & 12 & -15 & 6 \\ \hline & 4 & -5 & 2 & 0 \end{array} \rightarrow \text{Remainder}$$ $\underbrace{4x^2 - 5x + 2}_{\text{Quotient}}$ The degree of the quotient is one less than the degree of the dividend. $(4x^3 - 17x^2 + 17x - 6) \div (x - 3)$ $\qquad = 4x^2 - 5x + 2$	**p. 419**

(6.1) Evaluate using the rules of exponents.

1) $\dfrac{3^{10}}{3^6}$

2) 8^{-2}

3) $\left(\dfrac{5}{4}\right)^{-3}$

4) $-4^0 + 7^0$

Simplify. Assume all variables represent nonzero real numbers. The answer should not contain negative exponents.

5) $(z^6)^3$

6) $(4p^3)(-3p^7)$

7) $\dfrac{70r^9}{10r^4}$

8) $(-5c^4)^2$

9) $(-9t)(6t^6)$

10) $\dfrac{6m^{10}}{24m^6}$

11) $\dfrac{k^3}{k^{11}}$

12) $\dfrac{d^{-6}}{d^3}$

13) $(-2a^2b)^3(5a^{-12}b)$

14) $\dfrac{x^5y^{-3}}{x^8y^{-4}}$

15) $\left(\dfrac{3pq^{-10}}{2p^{-2}q^5}\right)^{-2}$

16) $(7c^{-8}d^2)(3c^{-2}d)^2$

17) $\dfrac{s^{-1}t^9}{st^{11}}$

18) $\left(\dfrac{2a^4b}{5a^9b^{-3}}\right)^{-3}$

Write expressions for the area and perimeter of each rectangle.

19)

$4x$

$2x$

20)

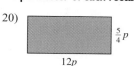

$\dfrac{5}{4}p$

$12p$

Simplify. Assume that the variables represent nonzero integers. Write the final answer so that the exponents have positive coefficients.

21) $x^{5t} \cdot x^{3t}$

22) $n^{4c} \cdot n^c$

23) $\dfrac{r^{9a}}{r^{3a}}$

24) $\dfrac{a^{11k}}{a^{8k}}$

25) $(y^{2p})^3$

26) $\dfrac{w^{-12a}}{w^{-3a}}$

(6.2) Identify each term in the polynomial, the coefficient and degree of each term, and the degree of the polynomial.

27) $4r^3 - 7r^2 + r + 5$

28) $x^3y + 6xy^2 - 8xy + 11y$

29) Evaluate $-x^2y^2 - 7xy + 2x + 5$ for $x = -3$ and $y = 2$.

30) Evaluate $p^3q^2 + 4p^2q^2 - pq - 3p + 8$ for $p = -1$ and $q = 3$.

31) If $h(x) = 4x^2 - x - 7$, find
   a) $h(-3)$
   b) $h(0)$

32) $f(t) = \dfrac{3}{4}t + 2$. Find t so that $f(t) = \dfrac{7}{2}$.

Add or subtract as indicated.

33) $(5t^2 + 11t - 4) - (7t^2 + t - 9)$

34) $(-3j^2 - j + 7) + (4j^2 - 8j + 1)$

35) $\begin{aligned} 5.8p^3 - 1.2p^2 + \quad\ p - 7.5 \\ + \ 2.1p^3 + 6.3p^2 + 3.8p + 3.9 \end{aligned}$

36) $\begin{aligned} -8.6n^3 + 10.9n^2 - 6.1n + 3.2 \\ - \ \ 2.7n^3 - \ \ 4.2n^2 + 2.3n + 9.5 \end{aligned}$

37) $\left(\dfrac{7}{4}k^2 + \dfrac{1}{6}k + 5\right) - \left(\dfrac{1}{2}k^2 + \dfrac{5}{6}k - 2\right)$

38) $\left(\dfrac{4}{9}w^2 - \dfrac{3}{8}w + \dfrac{2}{5}\right) + \left(\dfrac{2}{9}w^2 + \dfrac{5}{8}w - \dfrac{9}{20}\right)$

39) Subtract $3a^2b^2 - 10a^2b + ab + 6$ from $a^2b^2 + 7a^2b - 3ab + 11$.

40) Find the sum of $4c^3d^3 - 9c^2d^2 - c^2d + 9d + 1$ and $12c^3d^3 + 4c^2d^2 - 11cd - 9d - 7$.

41) Find the sum of $4m + 9n - 19$ and $-5m + 6n + 14$.

42) Subtract $-h^4 + 10j^4 - 6$ from $6h^4 - 2j^4 + 21$.

43) Subtract $4s^2 + 3s + 17$ from the sum of $4s - 11$ and $9s^2 - 19s + 2$.

44) Find the sum of $6xy + 4x - y - 10$ and $-4xy + 2y + 3$ and subtract it from $-6xy - 7x + y + 2$.

Find the polynomial that represents the perimeter of each rectangle.

45)

$d^2 + 3d + 5$

$d^2 - 5d + 2$

46)

$9m - 4$

$2m + 3$

(6.3) Multiply.

47) $4r(7r - 15)$

48) $-6m^3(9m^2 - 3m + 7)$

49) $(2w + 5)(-12w^3 + 6w^2 - 2w + 3)$

50) $\left(3t^2 - \dfrac{1}{4}\right)(-8t^2 + 3t - 20)$

51) $(y - 7)(y + 8)$ 52) $(f - 6)(f - 8)$

53) $(3n - 7)(2n - 9)$ 54) $(4p + 5)(2p + 1)$

55) $-(a - 11)(a + 12)$ 56) $-(4d + 3)(6d + 7)$

57) $7u^4v^2(-8u^2v + 7uv^2 + 12u - 3)$

58) $8fg^2(6f^3g^2 + 12f^2g^2 - fg + 2)$

59) $(3x - 8y)(2x + y)$ 60) $(9r + 2s)(r - s)$

61) $(ab + 5)(ab + 6)$ 62) $(7 - 3cd)(cd + 2)$

63) $(x^2 + 4x - 11)(12x^4 - 7x^2 + 9)$

64) $(2k^2 - 5k + 3)(k^2 + k - 6)$

65) $6c^3(4c - 5)(c - 2)$ 66) $-5(7w - 12)(w + 3)$

67) $(z + 4)(z + 1)(z + 5)$ 68) $(p + 3)(p - 6)(p + 2)$

69) $\left(\dfrac{3}{5}m + 2\right)\left(\dfrac{1}{3}m - 4\right)$ 70) $\left(\dfrac{2}{9}t - 5\right)\left(\dfrac{1}{10}t - 3\right)$

Expand.

71) $(b + 7)^2$ 72) $(x - 10)^2$

73) $(5q - 2)^2$ 74) $(7 - 3y)^2$

75) $(x - 2)^3$ 76) $(p + 10)^3$

Find the special products.

77) $(z + 9)(z - 9)$ 78) $(p - 12)(p + 12)$

79) $\left(\dfrac{1}{5}n - 2\right)\left(\dfrac{1}{5}n + 2\right)$ 80) $\left(\dfrac{9}{4} + \dfrac{3}{5}x\right)\left(\dfrac{9}{4} - \dfrac{3}{5}x\right)$

81) $\left(\dfrac{7}{8} - r^2\right)\left(\dfrac{7}{8} + r^2\right)$ 82) $\left(2a - \dfrac{1}{3}b\right)\left(2a + \dfrac{1}{3}b\right)$

83) $-2(3c - 4)^2$ 84) $6w(w + 3)^2$

85) Write an expression for the a) area and b) perimeter of the rectangle.

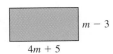

$m - 3$

$4m + 5$

86) Express the volume of the cube as a polynomial.

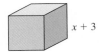

$x + 3$

(6.4) Divide.

87) $\dfrac{8t^5 - 14t^4 - 20t^3}{2t^3}$ 88) $\dfrac{16p^4 + 56p^3 - 32p^2 + 8p}{-8p}$

89) $\dfrac{c^2 + 8c - 20}{c - 2}$ 90) $\dfrac{y^2 - 15y + 56}{y - 8}$

91) $\dfrac{12r^3 - 13r^2 - 5r + 6}{3r + 2}$

92) $\dfrac{-66h^3 - 5h^2 + 45h - 4}{11h - 1}$

93) $\dfrac{30a^3 + 80a^2 - 15a + 20}{10a^2}$

94) $\dfrac{60d^4 - 40d^3 + 24d^2 + 6d}{12d^2}$

95) $(15x^4y^4 - 42x^3y^4 - 6x^2y + 10y) \div (-6x^2y)$

96) $(56a^6b^6 + 21a^4b^5 - 4a^3b^4 + a^2b - 7ab) \div (7a^3b^3)$

97) $(6q^2 + 2q - 35) \div (3q + 7)$

98) $(12r^2 - 16r + 11) \div (6r + 1)$

99) $\dfrac{23a - 7 + 15a^2}{5a - 4}$ 100) $\dfrac{-21 + 20x^2 - 37x}{4x - 9}$

101) $\dfrac{6m^4 + 2m^3 + 7m^2 + 5m - 20}{2m^2 + 5}$

102) $\dfrac{8t^4 + 32t^3 - 43t^2 - 44t + 48}{8t^2 - 11}$

103) $\dfrac{b^3 - 64}{b - 4}$ 104) $\dfrac{f^3 + 125}{f + 5}$

105) $\dfrac{-23 - 46w + 32w^3}{4w + 3}$ 106) $\dfrac{8k^2 - 8 + 15k^3}{3k - 2}$

107) $(7u^4 - 69u^3 + 15u^2 - 37u + 12) \div (u^2 - 10u + 3)$

108) $(6c^4 + 13c^3 - 21c^2 - 9c + 10) \div (2c^2 + 5c - 4)$

109) Find the base of the triangle if the area is given by $15y^2 + 12y$.

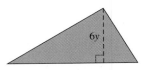

$6y$

110) Find the length of the rectangle if the area is given by $6x^3 - x^2 + 13x - 10$.

$3x - 2$

Evaluate.

1) $\left(\dfrac{5}{3}\right)^{-3}$

2) $\dfrac{2^9}{2^4}$

Simplify. Assume all variables represent nonzero real numbers. The answer should not contain negative exponents.

3) $(9d^4)(-3d^4)$

4) $(4t^5)^3$

5) $\dfrac{a^{12}b^{-5}}{a^8b^7}$

6) $\left(\dfrac{36xy^8}{54x^3y^{-1}}\right)^{-2}$

7) Given the polynomial $5p^3 - p^2 + 12p + 9$,

    a) What is the coefficient of p^2?

    b) What is the degree of the polynomial?

8) What is the degree of the polynomial
$4a^5b^3 + 13a^3b^3 - 6a^2b + 7ab - 1$?

9) Evaluate $-m^2 + 3n$ when $m = -5$ and $n = 8$.

Perform the indicated operation(s).

10) $6k^4(2k^2 - 7k + 5)$

11) $(10r^3s^2 + 7r^2s^2 - 11rs + 5)$
$+ (4r^3s^2 - 9r^2s^2 + 6rs + 3)$

12) Subtract $5j^2 - 2j + 9$ from $11j^2 - 10j + 3$.

13) $6(-n^3 + 4n - 2) - 3(2n^3 + 5n^2 + 8n - 1)$

14) $(c - 8)(c - 7)$

15) $(3y + 5)(2y + 1)$

16) $\left(u + \dfrac{3}{4}\right)\left(u - \dfrac{3}{4}\right)$

17) $(2a - 5b)(3a + b)$

18) $(3 - 8m)(2m^2 + 4m - 7)$

19) $3x(x + 4)^2$

Expand.

20) $(2z - 7)^2$

21) $(s - 4)^3$

Divide.

22) $\dfrac{r^2 + 10r + 21}{r + 7}$

23) $\dfrac{24t^5 - 60t^4 + 12t^3 - 8t^2}{12t^3}$

24) $(38v - 31 + 30v^3 - 51v^2) \div (5v - 6)$

25) $\dfrac{x^3 - 8}{x - 2}$

26) Write an expression for a) the area and b) the perimeter of the rectangle.

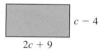

$c - 4$

$2c + 9$

27) Write an expression for the base of the triangle if the area is given by $12k^2 + 28k$.

$8k$

1) Given the set of numbers

$$\left\{\frac{6}{11}, -14, 2.7, \sqrt{19}, 43, 0.\overline{65}, 0, 8.21079314\ldots\right\}$$

list the

a) whole numbers

b) integers

c) rational numbers

2) Evaluate $-2^4 + 3 \cdot 8 \div (-2)$.

3) Divide $2\frac{6}{7} \div 1\frac{4}{21}$.

Simplify. Assume all variables represent nonzero real numbers. The answers should not contain negative exponents.

4) $-5(3w^4)^2$

5) $\left(\dfrac{2n^{-10}}{n^{-4}}\right)^3$

6) $p^{10k} \cdot p^{4k}$

Solve.

7) $-\dfrac{12}{5}c - 7 = 20$

8) $6(w + 4) + 2w = 1 + 8(w - 1)$

9) Solve the compound inequality and write the answer in interval notation.

$$3y + 16 < 4 \quad \text{or} \quad 8 - y \ge 7$$

10) *Write an equation in one variable and solve.* How many milliliters of a 12% alcohol solution and how many milliliters of a 4% alcohol solution must be mixed to obtain 60 ml of a 10% alcohol solution?

11) Find the x- and y-intercepts of $5x - 2y = 10$ and sketch a graph of the equation.

12) Graph $x = -3$.

13) Write an equation of the line containing the points $(-5, 8)$ and $(1, 2)$. Express the answer in standard form.

14) Solve this system using the elimination method.

$$-7x + 2y = 6$$
$$9x - y = 8$$

15) *Write a system of two equations in two variables and solve.* The length of a rectangle is 7 cm less than twice its width. The perimeter of the rectangle is 76 cm. What are the dimensions of the figure?

Perform the indicated operation(s).

16) $(4q^2 + 11q - 2) - 3(6q^2 - 5q + 4) + 2(-10q - 3)$

17) $(k - 9)(k + 6)$

18) $(4g - 9)(4g + 9)$

19) $\dfrac{8a^4b^4 - 20a^3b^2 + 56ab + 8b}{8a^3b^3}$

20) $\dfrac{17v^3 - 22v + 7v^4 + 24 - 47v^2}{7v - 4}$

21) $(2p^3 + 5p^2 - 11p + 9) \div (p + 4)$

22) $(3d^2 - 7)(4d^2 + 6d - 1)$

23) $(2c - 5)(c - 3)^2$

CHAPTER 7

Factoring Polynomials

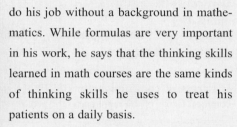

Algebra at Work: Ophthalmology

Mark is an ophthalmologist, a doctor specializing in the treatment of diseases of the eye. He says that he could not

do his job without a background in mathematics. While formulas are very important in his work, he says that the thinking skills learned in math courses are the same kinds of thinking skills he uses to treat his patients on a daily basis.

As a physician, Mark says that he must follow a very logical, analytical progression to form an accurate diagnosis and treatment plan. He examines a patient, performs tests, and then analyzes the results to form a diagnosis. Next, he must think of different ways to solve the problem and decide on the treatment plan that is best for that patient. He says that the skills he learned in his mathematics courses prepared him for the kind of problem

solving he must do every day to be an ophthalmologist. Factoring requires the kinds of skills Mark says are so important to him in his job—the ability to think through and solve a problem in an analytical and logical manner.

In this chapter, we will learn different techniques for factoring polynomials.

Section 7.1 The Greatest Common Factor and Factoring by Grouping

Objectives

1. Find the GCF of a Group of Monomials
2. Understand the Relationship Between Multiplying Polynomials and Factoring Polynomials
3. Factor Out the Greatest Common Monomial Factor
4. Factor Out the Greatest Common Binomial Factor
5. Factor by Grouping

In Section 1.1 we discussed writing a number as the product of factors:

$$12 \;=\; 3 \;\cdot\; 4$$

Product Factor Factor

To **factor** an integer is to write it as the product of two or more integers. Therefore, 12 can also be factored in other ways:

$$12 = 1 \cdot 12 \qquad 12 = 2 \cdot 6 \qquad 12 = -1 \cdot (-12)$$
$$12 = -2 \cdot (-6) \qquad 12 = -3 \cdot (-4) \qquad 12 = 2 \cdot 2 \cdot 3$$

The last **factorization**, $2 \cdot 2 \cdot 3$ or $2^2 \cdot 3$, is called the **prime factorization** of 12 since all of the factors are prime numbers. (See Section 1.1.) The factors of 12 are 1, 2, 3, 4, 6, 12, -1, -2, -3, -4, -6, and -12.

We can also write the factors as ± 1, ± 2, ± 3, ± 4, ± 6, and ± 12. (Read ± 1 as "plus or minus 1.")

In this chapter, we will learn how to factor polynomials, a skill that is used in many ways throughout algebra.

> The **greatest common factor (GCF)** of a group of two or more integers is the *largest* common factor of the numbers in the group.

For example, if we want to find the GCF of 12 and 20, we can list their positive factors.

$$12\text{: } 1, 2, 3, 4, 6, 12$$
$$20\text{: } 1, 2, 4, 5, 10, 20$$

The greatest common factor of 12 and 20 is 4. We can also use prime factors.

We begin our study of factoring polynomials by discussing how to find the greatest common factor of a group of two or more monomials. For this we will use prime factorization.

1. Find the GCF of a Group of Monomials

Example 1

Find the greatest common factor of x^5 and x^3.

Solution

We can begin by writing each monomial as the product of its prime factors. To find the GCF use each prime factor the *least* number of times it appears in any of the prime factorizations. Then, multiply.

We can write x^5 and x^3 as

$$x^5 = x \cdot x \cdot x \cdot x \cdot x$$
$$x^3 = x \cdot x \cdot x$$

In x^5, x appears as a factor *five times*.

In x^3, x appears as a factor *three times*.

The least number of times x appears as a factor is three. *There will be three factors of x in the GCF.*

$$\text{GCF} = x \cdot x \cdot x = x^3$$

In Example 1, notice that the power of 3 in the GCF is the smallest of the powers when comparing x^5 and x^3. This will always be true.

> The exponent on the variable in the GCF will be the *smallest* exponent appearing on the variable in the group of terms.

You Try 1

Find the greatest common factor of y^4 and y^7.

Example 2

Find the greatest common factor for each group of terms.

a) $30k^4, 10k^9, 50k^6$ b) $-12a^8b, 42a^5b^7$
c) $63c^5d^3, 18c^3, 27c^2d^2$

Solution

a) The GCF of the coefficients, 30, 10, and 50, is 10. The smallest exponent on k is 4, so k^4 will be part of the GCF.

The GCF of $30k^4, 10k^9,$ and $50k^6$ is $10k^4$.

b) The GCF of the coefficients, -12 and 42, is 6. The smallest exponent on a is 5, so a^5 will be part of the GCF. The smallest exponent on b is 1, so b will be part of the GCF.

The GCF of $-12a^8b$ and $42a^5b^7$ is $6a^5b$.

c) The GCF of the coefficients is 9. The smallest exponent on c is 2, so c^2 will be part of the GCF. There is no d in the term $18c^3$, so there will be no d in the GCF.

The GCF of $63c^5d^3, 18c^3,$ and $27c^2d^2$ is $9c^2$.

You Try 2

Find the greatest common factor for each group of terms.
a) $-16p^7, 8p^5, 40p^8$ b) $r^6s^5, 9r^8s^3, 12r^4s^4$

Factoring Out the Greatest Common Factor

Earlier we said that to **factor an integer** is to write it as the product of two or more integers.

To **factor a polynomial** is to write it as a product of two or more polynomials.

Throughout this chapter we will study different factoring techniques. We will begin by discussing how to factor out the greatest common factor.

2. Understand the Relationship Between Multiplying Polynomials and Factoring Polynomials

Factoring a polynomial is the opposite of multiplying polynomials. Let's see how these procedures are related.

Example 3

a) Multiply $2x(x + 5)$.   b) Factor out the GCF from $2x^2 + 10x$.

Solution

a) Use the distributive property to multiply.

$$2x(x + 5) = (2x)x + (2x)(5)$$
$$= 2x^2 + 10x$$

b) Use the distributive property to factor out the greatest common factor from $2x^2 + 10x$.
First, identify the GCF of $2x^2$ and $10x$.

$$\text{GCF} = 2x$$

Then, rewrite each term as a product of two factors with one factor being $2x$.

$$2x^2 = (2x)(x) \text{ and } 10x = (2x)(5)$$
$$2x^2 + 10x = (2x)(x) + (2x)(5)$$
$$= 2x(x + 5) \qquad \text{Distributive property}$$

When we factor $2x^2 + 10x$, we get $2x(x + 5)$. We can check our result by multiplying.

$$2x(x + 5) = 2x^2 + 10x \quad \checkmark$$

Steps for Factoring Out the Greatest Common Factor

1) Identify the GCF of all of the terms of the polynomial.

2) Rewrite each term as the product of the GCF and another factor.

3) Use the distributive property to factor out the GCF from the terms of the polynomial.

4) Check the answer by multiplying the factors. The result should be the original polynomial.

---

3. Factor Out the Greatest Common Monomial Factor

Example 4

Factor out the greatest common factor.

a) $12a^5 + 30a^4 + 6a^3$ b) $c^6 - 6c^2$

c) $4x^5y^3 + 12x^5y^2 - 28x^4y^2 - 4x^3y$

Solution

a)   Identify the GCF of all of the terms: GCF $= 6a^3$.

$$12a^5 + 30a^4 + 6a^3 = (6a^3)(2a^2) + (6a^3)(5a) + (6a^3)(1)$$ Rewrite each term using the GCF as one of the factors.

$$= 6a^3(2a^2 + 5a + 1)$$ Distributive property

Check: $6a^3(2a^2 + 5a + 1) = 12a^5 + 30a^4 + 6a^3$ ✓

b)   The GCF of all of the terms is c^2.

$$c^6 - 6c^2 = (c^2)(c^4) - (c^2)(6)$$ Rewrite each term using the GCF as one of the factors.
$$= c^2(c^4 - 6)$$ Distributive property

Check: $c^2(c^4 - 6) = c^6 - 6c^2$ ✓

c)   The GCF of all of the terms is $4x^3y$.

$$4x^5y^3 + 12x^5y^2 - 28x^4y^2 - 4x^3y$$
$$= (4x^3y)(x^2y^2) + (4x^3y)(3x^2y) - (4x^3y)(7xy) - (4x^3y)(1)$$ Rewrite each term using the GCF as one of the factors.

$$= 4x^3y(x^2y^2 + 3x^2y - 7xy - 1)$$ Distributive property

Check: $4x^3y(x^2y^2 + 3x^2y - 7xy - 1) = 4x^5y^3 + 12x^5y^2 - 28x^4y^2 - 4x^3y$ ✓

You Try 3

Factor out the greatest common factor.

a) $56k^4 - 24k^3 + 40k^2$ b) $3a^4b^4 - 12a^3b^4 + 18a^2b^4 - 3a^2b^3$

Sometimes we need to take out a negative factor.

Example 5

Factor out $-5d$ from $-10d^4 + 45d^3 - 15d^2 + 5d$.

Solution

$$-10d^4 + 45d^3 - 15d^2 + 5d$$
$$= (-5d)(2d^3) + (-5d)(-9d^2) + (-5d)(3d) + (-5d)(-1)$$ Rewrite each term using $-5d$ as one of the factors.

$$= -5d[2d^3 + (-9d^2) + 3d + (-1)]$$ Distributive property
$$= -5d(2d^3 - 9d^2 + 3d - 1)$$ Rewrite $+(-9d^2)$ as $-9d^2$ and $+(-1)$ as -1.

Check: $-5d(2d^3 - 9d^2 + 3d - 1) = -10d^4 + 45d^3 - 15d^2 + 5d$ ✓

When taking out a negative factor, be very careful with the signs!

You Try 4

Factor out $-p^2$ from $-p^5 - 7p^4 + 3p^3 + 11p^2$.

4. Factor Out the Greatest Common Binomial Factor

Until now, all of the GCFs have been monomials. Sometimes, however, the greatest common factor of terms is a *binomial*.

Recall that a *term* is a number or a variable or a product or quotient of numbers and variables.

So $x(y + 2)$ is a *term* because it is the *product* of x and $(y + 2)$.

We will use this idea to factor out a *binomial* as the greatest common factor of terms.

Example 6

Factor out the greatest common factor.

a) $x(y + 2) + 9(y + 2)$ b) $n^2(n + 6) - 3(n + 6)$

c) $r(s + 4) - (s + 4)$

Solution

a)   In the polynomial $x(y + 2) + 9(y + 2)$, $x(y + 2)$ is a term and $9(y + 2)$ is a

                Term   Term

term. What do these terms have in common? $y + 2$

The GCF of $x(y + 2)$ and $9(y + 2)$ is $(y + 2)$. Use the distributive property to factor out $y + 2$.

$$x(y + 2) + 9(y + 2) = (y + 2)(x + 9) \qquad \text{Distributive property}$$

Check: $(y + 2)(x + 9) = (y + 2)x + (y + 2)9 \qquad \text{Distribute.}$

The result $(y + 2)x + (y + 2)9$ is the same as $x(y + 2) + 9(y + 2)$ because multiplication is commutative. ✓

b)   Identify the terms of the polynomial.

$$n^2(n + 6) - 3(n + 6)$$

        Term   Term

The GCF is $n + 6$.

$$n^2(n + 6) - 3(n + 6) = (n + 6)(n^2 - 3) \qquad \text{Distributive property}$$

Check: $(n + 6)(n^2 - 3) = (n + 6)n^2 + (n + 6)(-3) \qquad \text{Distributive property}$
$$= n^2(n + 6) - 3(n + 6) \; ✓ \qquad \text{Commutative property}$$

c)   Let's begin by rewriting $r(s + 4) - (s + 4)$ as

$$r(s + 4) - 1(s + 4)$$

        Term   Term

The GCF is $s + 4$.

$$r(s + 4) - 1(s + 4) = (s + 4)(r - 1) \qquad \text{Distributive property}$$

BE CAREFUL

It is important to write -1 in front of $(s + 4)$. Otherwise, the following mistake is often made:

$$r(s + 4) - (s + 4) = (s + 4)r$$

The correct factor is $r - 1$ *not* r.

$$
\begin{aligned}
\text{Check: } (s + 4)(r - 1) &= (s + 4)r + (s + 4)(-1) \qquad \text{Distributive property} \\
&= r(s + 4) - 1(s + 4) \qquad\quad \text{Commutative property} \\
&= r(s + 4) - (s + 4) \quad \checkmark
\end{aligned}
$$

You Try 5

Factor out the GCF.

a) $t(u - 8) + 5(u - 8)$ b) $z(z^2 + 2) - 6(z^2 + 2)$ c) $2n(m + 7) - (m + 7)$

Taking out a binomial factor leads us to our next method of factoring—factoring by grouping.

5. Factor by Grouping

When we are asked to factor a polynomial containing four terms, we often try to **factor by grouping**.

Example 7

Factor by grouping.

a) $ab + 5a + 3b + 15$ b) $2pr - 5qr + 6p - 15q$
c) $x^3 + 6x^2 - 7x - 42$

Solution

a)   Begin by grouping two terms together so that each group has a common factor.

$$\underbrace{ab + 5a}_{\text{Group 1}} + \underbrace{3b + 15}_{\text{Group 2}}$$

From group 1 we can factor out a to get $a(b + 5)$
From group 2 we can factor out 3 to get $3(b + 5)$

$$\underbrace{ab + 5a}_{} + \underbrace{3b + 15}_{}$$

$$
\begin{aligned}
&= a(b + 5) + 3(b + 5) \qquad \text{Take out the common factor from each pair of terms.} \\
&= (b + 5)(a + 3) \qquad\quad\; \text{Factor out } (b + 5) \text{ using the distributive property.}
\end{aligned}
$$

Check: $(b + 5)(a + 3) = ab + 5a + 3b + 15$ ✓

b) Group two terms together so that each group has a common factor.

$$\underbrace{2pr - 5qr}_{\text{Group 1}} + \underbrace{6p - 15q}_{\text{Group 2}}$$

Factor out r from group 1 to get $r(2p - 5q)$

Factor out 3 from group 2 to get $3(2p - 5q)$

$$\underbrace{2pr - 5qr}_{\downarrow} + \underbrace{6p - 15q}_{\downarrow}$$

$= r(2p - 5q) + 3(2p - 5q)$ Take out the common factor from each pair of terms.
$= (2p - 5q)(r + 3)$ Factor out $(2p - 5q)$ using the distributive property.

Check: $(2p - 5q)(r + 3) = 2pr - 5qr + 6p - 15q$ ✓

c) Group two terms together so that each group has a common factor.

$$\underbrace{x^3 + 6x^2}_{\text{Group 1}} - \underbrace{7x - 42}_{\text{Group 2}}$$

Factor out x^2 from group 1 to get $x^2(x + 6)$

Factor out -7 from group 2 to get $-7(x + 6)$

We *must* factor out -7 *not* 7 from group 2 so that the binomial factors for groups 1 and 2 are the same! [If we had factored out 7, then the factorization of group 2 would have been $7(-x - 6)$.]

$$\underbrace{x^3 + 6x^2}_{\downarrow} - \underbrace{7x - 42}_{\downarrow}$$

$= x^2(x + 6) - 7(x + 6)$ Take out the common factor from each pair of terms.
$= (x + 6)(x^2 - 7)$ Factor out $(x + 6)$ using the distributive property.

Check: $(x + 6)(x^2 - 7) = x^3 + 6x^2 - 7x - 42$ ✓

You Try 6

Factor by grouping.

a) $2cd + 4d + 5c + 10$ b) $4k^2 - 36k + km - 9m$ c) $h^3 + 8h^2 - 5h - 40$

Often, we have to combine the two factoring techniques we have learned here. That is, we begin by factoring out the GCF and then we factor by grouping. Let's summarize how to factor a polynomial by grouping and then look at another example.

Steps for Factoring by Grouping

1)   Before trying to factor by grouping, look at each term in the polynomial and ask yourself, *"Can I factor out a GCF first?"* If so, factor out the GCF from all of the terms.

2)   Make two groups of two terms so that each group has a common factor.

3)   Take out the common factor in each group of terms.

4)   Factor out the common binomial factor using the distributive property.

5)   Check the answer by multiplying the factors.

Example 8

Factor completely. $4y^4 + 4y^3 - 20y^2 - 20y$

Solution

Notice that this polynomial has four terms. This is a clue for us to try factoring by grouping. *However,* look at the polynomial carefully and ask yourself, *"Can I factor out a GCF?"* Yes! *Therefore, the first step in factoring this polynomial is to factor out 4y.*

$$4y^4 + 4y^3 - 20y^2 - 20y = 4y(y^3 + y^2 - 5y - 5)$$ Factor out the GCF, $4y$.

The polynomial in parentheses has 4 terms. Try to factor it by grouping.

$$4y(\underbrace{y^3 + y^2}_{\text{Group 1}} \underbrace{- 5y - 5}_{\text{Group 2}})$$

$$= 4y[y^2(y + 1) - 5(y + 1)]$$ Take out the common factor in each group.

$$= 4y(y + 1)(y^2 - 5)$$ Factor out $(y + 1)$ using the distributive property.

Check: $4y(y + 1)(y^2 - 5) = 4y(y^3 + y^2 - 5y - 5)$

$$= 4y^4 + 4y^3 - 20y^2 - 20y \checkmark$$

You Try 7

Factor completely. $4ab + 14b + 8a + 28$

Remember, seeing a polynomial with four terms is a clue to try factoring by grouping. Not all polynomials will factor this way, however. We will learn other techniques later, and some polynomials must be factored using methods learned in later courses.

Answers to You Try Exercises

1) y^4 2) a) $8p^5$ b) r^4s^3 3) a) $8k^2(7k^2 - 3k + 5)$ b) $3a^2b^3(a^2b - 4ab + 6b - 1)$

4) $-p^2(p^3 + 7p^2 - 3p - 11)$ 5) a) $(u - 8)(t + 5)$ b) $(z^2 + 2)(z - 6)$ c) $(m + 7)(2n - 1)$

6) a) $(c + 2)(2d + 5)$ b) $(k - 9)(4k + m)$ c) $(h + 8)(h^2 - 5)$ 7) $2(b + 2)(2a + 7)$

7.1 Exercises

Objective 1

Find the greatest common factor of each group of terms.

1) $45m^3, 20m^2$

2) $18d^6, 21d^2$

3) $42k^5, 54k^7, 72k^9$

4) $25t^8, 55t, 30t^3$

5) $27x^4y, 45x^2y^3$

6) $24r^3s^6, 56r^2s^5$

7) $28u^2v^5, 20uv^3, -8uv^4$

8) $-6a^4b^3, 18a^2b^6, 12a^2b^4$

9) $21s^2t, 35s^2t^2, s^4t^2$

10) $p^4q^4, -p^3q^4, -p^3q$

11) $a(n - 7), 4(n - 7)$

12) $x^2(y + 9), z^2(y + 9)$

13) Explain how to find the GCF of a group of terms.

14) What does it mean to factor a polynomial?

Objectives 2–4

Factor out the greatest common factor. Be sure to check your answer.

15) $30s + 18$

16) $14a + 24$

17) $24z - 4$

18) $63f^2 - 49$

19) $3d^2 - 6d$

20) $20m - 5m^2$

21) $42y^2 + 35y^3$

22) $30b^3 - 5b$

23) $t^5 - t^4$

24) $r^9 + r^2$

25) $\frac{1}{2}c^2 + \frac{5}{2}c$

26) $\frac{1}{8}k^2 + \frac{7}{8}k$

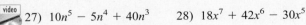

27) $10n^5 - 5n^4 + 40n^3$

28) $18x^7 + 42x^6 - 30x^5$

29) $2v^8 - 18v^7 - 24v^6 + 2v^5$

30) $12z^6 + 30z^5 - 15z^4 + 3z^3$

31) $8c^3 + 3d^2$

32) $m^5 - 5n^2$

33) $a^4b^2 + 4a^3b^3$

34) $20r^3s^3 - 14rs^4$

35) $50x^3y^3 - 70x^3y^2 + 40x^2y$

36) $21b^4d^3 + 15b^3d^3 - 27b^2d^2$

37) $m(n - 12) + 8(n - 12)$

38) $x(y + 5) + 3(y + 5)$

39) $p(8r - 3) - q(8r - 3)$

40) $a(9c + 4) - b(9c + 4)$

41) $y(z + 11) + (z + 11)$

42) $2u(v - 7) + (v - 7)$

43) $2k^2(3r + 4) - (3r + 4)$

44) $8p(3q + 5) - (3q + 5)$

45) Factor out -8 from $-64m - 40$.

46) Factor out -7 from $-14k + 21$.

47) Factor out $-5t^2$ from $-5t^3 + 10t^2$.

48) Factor out $-4v^3$ from $-4v^5 - 36v^3$.

49) Factor out $-a$ from $-3a^3 + 7a^2 - a$.

50) Factor out $-q$ from $-10q^3 - 4q^2 + q$.

51) Factor out -1 from $-b + 8$.

52) Factor out -1 from $-z - 6$.

Objective 5

Factor by grouping.

53) $kt + 3k + 8t + 24$

54) $uv + 5u + 10v + 50$

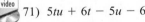

55) $fg - 7f + 4g - 28$

56) $cd - 5d + 8c - 40$

57) $2rs - 6r + 5s - 15$

58) $3jk - 7k + 6j - 14$

59) $3xy - 2y + 27x - 18$

60) $4ab + 32a + 3b + 24$

61) $8b^2 + 20bc + 2bc^2 + 5c^3$

62) $8u^2 - 16uv^2 + 3uv - 6v^3$

63) $4a^3 - 12ab + a^2b - 3b^2$

64) $5x^3 - 30x^2y^2 + xy - 6y^3$

65) $kt + 7t - 5k - 35$

66) $pq - 2q - 9p + 18$

67) $mn - 8m - 10n + 80$

68) $hk + 6k - 4h - 24$

69) $dg - d + g - 1$

70) $qr + 3q - r - 3$

71) $5tu + 6t - 5u - 6$

72) $4yz + 7z - 20y - 35$

73) $36g^4 + 3gh - 96g^3h - 8h^2$

74) $40j^3 + 72jk - 55j^2k - 99k^2$

75) Explain, in your own words, how to factor by grouping.

76) What should be the first step in factoring
$3xy + 6x + 15y + 30$?

Factor completely. You may need to begin by taking out the GCF first or by rearranging terms.

77) $2ab + 8a + 6b + 24$

78) $7pq + 28q + 14p + 56$

79) $8s^2t - 40st + 16s^2 - 80s$

80) $10hk^3 - 5hk^2 + 30k^3 - 15k^2$

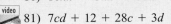

81) $7cd + 12 + 28c + 3d$

82) $9rs + 12 + 2s + 54r$

83) $42k^3 + 15d^2 - 18k^2d - 35kd$

84) $12x^3 + 2y^2 - 3x^2y - 8xy$

85) $9f^2j^2 + 45fj + 9fj^2 + 45f^2j$

86) $n^3m - 4n^2 + mn^2 - 4n^3$

87) $4x^4y - 14x^3 + 28x^4 - 2x^3y$

88) $12a^2c^2 - 20ac - 4ac^2 + 60a^2c$

Section 7.2 Factoring Trinomials of the Form $x^2 + bx + c$

Objectives

1. Practice Arithmetic Skills Needed to Factor a Trinomial of the Form $x^2 + bx + c$

2. Factor a Trinomial of the Form $x^2 + bx + c$

3. Factor Out the GCF and Then Factor a Trinomial of the Form $x^2 + bx + c$

4. Factor a Trinomial Containing Two Variables

One of the factoring problems encountered most often in algebra is the factoring of trinomials. In this section, we will discuss how to factor a trinomial of the form $x^2 + bx + c$; notice that the coefficient of the squared term is 1.

We will begin with a skill we need to be able to factor a trinomial of the form $x^2 + bx + c$.

1. Practice Arithmetic Skills Needed to Factor a Trinomial of the Form $x^2 + bx + c$

Example 1

Find two integers that

a) multiply to 10 and add up to 7.

b) multiply to 36 and add up to -13.

c) multiply to -27 and add up to 6.

Solution

a) If two numbers multiply to give us *positive* 10 and the numbers add up to *positive* 7, *then the two numbers will be positive.* (The product of two positive numbers is positive, and their sum is positive as well.)

First, list the pairs of *positive* integers which multiply to 10—the *factors* of 10. Then, find the *sum* of those factors.

Product of the Factors of 10	Sum of the Factors
$1 \cdot 10 = 10$	$1 + 10 = 11$
$2 \cdot 5 = 10$	$2 + 5 = 7$

2 and 5 multiply to 10 and add up to 7.

b) If two numbers multiply to *positive* 36 and the numbers add up to *negative* 13, *then the two numbers will be negative*. (The product of two negative numbers is positive, while the sum of two negative numbers is negative.)

First, list the pairs of negative integers which multiply to 36. Then, find the sum of those factors. You can stop making your list when you find the pair that works.

Product of the Factors of 36	Sum of the Factors
$-1 \cdot (-36) = 36$	$-1 + (-36) = -37$
$-2 \cdot (-18) = 36$	$-2 + (-18) = -20$
$-3 \cdot (-12) = 36$	$-3 + (-12) = -15$
$-4 \cdot (-9) = 36$	$-4 + (-9) = -13$

-4 and -9 multiply to 36 and add up to -13.

c) In order for two numbers to multiply to *negative* 27 and add up to *positive* 6, *one number must be positive and one number must be negative*. (The product of a positive number and a negative number is negative, while the sum of the numbers can be either positive *or* negative.)

First, list the pairs of integers which multiply to -27. Then, find the sum of those factors.

Product of the Factors of -27	Sum of the Factors
$-1 \cdot 27 = -27$	$-1 + 27 = 26$
$1 \cdot (-27) = -27$	$1 + (-27) = -26$
$-3 \cdot 9 = -27$	$-3 + 9 = 6$

The product of -3 and 9 is -27 and their sum is 6.

You Try 1

Find two integers whose

a) product is 42 and whose sum is 13.

b) product is -48 and whose sum is -2.

c) product is 60 and whose sum is -19.

You should try to get to the point where you can come up with the correct numbers *in your head* without making a list.

2. Factor a Trinomial of the Form $x^2 + bx + c$

In Section 7.1 we said that the process of factoring is the opposite of multiplying. Let's see how this will help us understand how to factor a trinomial of the form $x^2 + bx + c$.
   Multiply $(x + 4)(x + 7)$ using FOIL.

$$
\begin{aligned}
(x + 4)(x + 7) &= x^2 + 7x + 4x + 4 \cdot 7 \qquad \text{Multiply using FOIL.}\\
&= x^2 + (7 + 4)x + 28 \qquad \text{Use the distributive property and multiply } 4 \cdot 7.\\
&= x^2 + 11x + 28
\end{aligned}
$$

28 is the *product*
of 4 and 7.
↓

$$(x + 4)(x + 7) = x^2 + 11x + 28$$
↑
11 is the *sum*
of 4 and 7.

So, if we were asked to *factor* $x^2 + 11x + 28$, we need to think of two integers whose *product* is 28 and whose *sum* is 11. Those numbers are 4 and 7. The *factored form* of $x^2 + 11x + 28$ is $(x + 4)(x + 7)$.

Factoring a Polynomial of the Form $x^2 + bx + c$

If $x^2 + bx + c = (x + m)(x + n)$, then

1)   if b and c are positive, then both m and n must be positive.

2)   if c is positive and b is negative, then both m and n must be negative.

3)   if c is negative, then one integer, m, must be positive and the other integer, n, must be negative.

You can check the answer by multiplying the binomials. The result should be the original polynomial.

Example 2

Factor, if possible.
a) $x^2 + 14x + 40$ b) $y^2 - 8y + 15$
c) $k^2 + k - 56$ d) $t^2 - 4t - 12$
e) $z^2 - 14z + 49$ f) $r^2 + 7r + 9$

Solution

a) $x^2 + 14x + 40$

   We must find the two integers whose *product* is 40 and whose *sum* is 14. Both integers will be positive. Make a list of the pairs of integers whose product is 40 and find their sums.

Product of the Factors	Sum of the Factors
$1 \cdot 40 = 40$	$1 + 40 = 41$
$2 \cdot 20 = 40$	$2 + 20 = 22$
$4 \cdot 10 = 40$	$4 + 10 = 14$

The numbers are 4 and 10.

$$x^2 + 14x + 40 = (x + 4)(x + 10)$$

$Check$: $(x + 4)(x + 10) = x^2 + 10x + 4x + 40$
$$= x^2 + 14x + 40 \checkmark$$

> The order in which the factors are written does not matter. In this example,
> $(x + 4)(x + 10) = (x + 10)(x + 4)$.

b)　$y^2 - 8y + 15$

Find the two integers whose *product* is 15 and whose *sum* is -8. Since 15 is positive and the coefficient of y is a negative number, -8, both integers will be negative.

Product of the Factors	Sum of the Factors
$-1 \cdot (-15) = 15$	$-1 + (-15) = -16$
$-3 \cdot (-5) = 15$	$-3 + (-5) = -8$

The numbers are -3 and -5.

$$y^2 - 8y + 15 = (y - 3)(y - 5)$$

$Check$: $(y - 3)(y - 5) = y^2 - 5y - 3y + 15$
$$= y^2 - 8y + 15 \checkmark$$

c)　$k^2 + k - 56$

The coefficient of k is 1, so we can think of this trinomial as $k^2 + 1k - 56$.

Find the two integers whose *product* is -56 and whose *sum* is 1. Since the last term in the trinomial is negative, one of the integers must be positive and the other must be negative.

Try to find these integers mentally. The two numbers which first come to mind having a product of *positive* 56 are 7 and 8. We need a product of -56 so either the 7 is negative or the 8 is negative.

Product of the Factors	Sum of the Factors
$-7 \cdot 8 = -56$	$-7 + 8 = 1$

The numbers are -7 and 8.

$$k^2 + k - 56 = (k - 7)(k + 8)$$

Check: $(k - 7)(k + 8) = k^2 + 8k - 7k - 56$
$$= k^2 + k - 56 \ \checkmark$$

d) $t^2 - 4t - 12$

Find the two integers whose *product* is -12 and whose *sum* is -4. Since the last term in the trinomial is negative, one of the integers must be positive and the other must be negative.

Find the integers mentally. First, think about two integers whose product is *positive* 12: 1 and 12, 3 and 4, 6 and 2. One number must be positive and the other negative, however, to get our product of -12, and they must add up to -4.

Product of the Factors	Sum of the Factors
$6 \cdot (-2) = -12$	$6 + (-2) = 4$
$-6 \cdot 2 = -12$	$-6 + 2 = -4$

The numbers are -6 and 2.

$$t^2 - 4t - 12 = (t - 6)(t + 2)$$

The check is left to the student.

e) $z^2 - 14z + 49$

Since the *product*, 49, is positive and the *sum*, -14, is negative, both integers must be negative. The numbers that multiply to 49 and add to -14 are the same number, -7 and -7.

$$(-7) \cdot (-7) = 49 \text{ and } -7 + (-7) = -14$$
$$z^2 - 14z + 49 = (z - 7)(z - 7) \text{ or } (z - 7)^2$$

Either form of the factorization is correct.

f) $r^2 + 7r + 9$

Find the two integers whose *product* is 9 and whose *sum* is 7. We are looking for two positive numbers.

Factors of 9	Sum of the Factors
$1 \cdot 9 = 9$	$1 + 9 = 10$
$3 \cdot 3 = 9$	$3 + 3 = 6$

There are no such factors! Therefore, $r^2 + 7r + 9$ does not factor using the methods we have learned here. We say that it is **prime**. In later mathematics courses you may learn how to factor such a polynomial so that $r^2 + 7r + 9$ is not considered prime.

You Try 2

Factor, if possible.

a) $p^2 + 7p + 6$ b) $s^2 + 5s - 66$ c) $d^2 - 6d - 10$

d) $x^2 - 12x + 27$ e) $m^2 + 12m + 36$

3. Factor Out the GCF and Then Factor a Trinomial of the Form $x^2 + bx + c$

Sometimes it is necessary to factor out the GCF before applying this method for factoring trinomials.

> Therefore, from this point on, the *first* step in factoring *any* polynomial should be to ask yourself, *"Can I factor out a greatest common factor?"*
>   And since some polynomials can be factored more than once, after performing one factorization, ask yourself, *"Can I factor again?"* If so, factor again. If not, you know that the polynomial has been completely factored.

Example 3

Factor completely. $5y^3 - 15y^2 - 20y$

Solution

Ask yourself, *"Can I factor out a GCF?"* Yes. The GCF is $5y$.

$$5y^3 - 15y^2 - 20y = 5y(y^2 - 3y - 4)$$

Look at the trinomial and ask yourself, *"Can I factor again?"* Yes. The integers whose product is -4 and whose sum is -3 are -4 and 1. Therefore,

$$5y^3 - 15y^2 - 20y = 5y(y^2 - 3y - 4)$$
$$= 5y(y - 4)(y + 1)$$

We cannot factor again.

$$\text{Check: } 5y(y - 4)(y + 1) = 5y(y^2 + y - 4y - 4)$$
$$= 5y(y^2 - 3y - 4)$$
$$= 5y^3 - 15y^2 - 20y \; \checkmark$$

The completely factored form of $5y^3 - 15y^2 - 20y$ is $5y(y - 4)(y + 1)$.

You Try 3

Factor completely.

a) $6g^4 + 42g^3 + 60g^2$ b) $10r^2s - 40rs + 30s$

4. Factor a Trinomial Containing Two Variables

If a trinomial contains two variables and we cannot take out a GCF, the trinomial can still be factored according to the method outlined in this section.

Example 4

Factor completely. $a^2 + 9ab + 18b^2$

Solution

Ask yourself, *"Can I factor out a GCF?"* No. Notice that the first term is a^2. Let's rewrite the trinomial as

$$a^2 + 9ba + 18b^2$$

so that we can think of $9b$ as the coefficient of a. Find two expressions whose product is $18b^2$ and whose sum is $9b$. They are $3b$ and $6b$ since $3b \cdot 6b = 18b^2$ and $3b + 6b = 9b$

$$a^2 + 9ab + 18b^2 = (a + 3b)(a + 6b)$$

We cannot factor $(a + 3b)(a + 6b)$ any more, so this is the complete factorization. The check is left to the student.

You Try 4

Factor completely.

a) $x^2 + 15xy + 54y^2$ b) $3k^3 + 18ck^2 - 21c^2k$

Answers to You Try Exercises

1) a) 6, 7   b) −8, 6   c) −4, −15   2) a) $(p + 1)(p + 6)$ b) $(s + 11)(s - 6)$ c) prime
d) $(x - 9)(x - 3)$ e) $(m + 6)^2$ 3) a) $6g^2(g + 5)(g + 2)$ b) $10s(r - 1)(r - 3)$
4) a) $(x + 6y)(x + 9y)$ b) $3k(k + 7c)(k - c)$

7.2 Exercises

Objective 1

1) Find two integers whose

	PRODUCT IS	and whose SUM IS	ANSWER
a)	30	13	
b)	−28	3	
c)	−8	−7	
d)	56	−15	

2) Find two integers whose

	PRODUCT IS	and whose SUM IS	ANSWER
a)	45	−14	
b)	−66	5	
c)	12	13	
d)	−32	−14	

Objective 2

3) If $x^2 + bx + c$ factors to $(x + m)(x + n)$ and if c is positive and b is negative, what do you know about the signs of m and n?

4) If $x^2 + bx + c$ factors to $(x + m)(x + n)$ and if b and c are positive, what do you know about the signs of m and n?

5) When asked to factor a polynomial, what is the first question you should ask yourself?

6) What does it mean to say that a polynomial is prime?

7) After factoring a polynomial, what should you ask yourself to be sure that the polynomial is completely factored?

8) How do you check the factorization of a polynomial?

Complete the factorization.

9) $n^2 + 12n + 27 = (n + 9)(\quad)$

10) $p^2 + 11p + 24 = (p + 3)(\quad)$

11) $c^2 - 14c + 45 = (c - 5)(\quad)$

12) $t^2 - 5t + 4 = (t - 4)(\quad)$

13) $x^2 + x - 56 = (x - 7)(\quad)$

14) $r^2 - 4r - 21 = (r + 3)(\quad)$

Factor completely, if possible. Check your answer.

15) $g^2 + 8g + 12$ 16) $j^2 + 9j + 20$

17) $w^2 + 13w + 42$ 18) $t^2 + 15t + 36$

19) $c^2 - 13c + 36$ 20) $v^2 - 11v + 24$

21) $b^2 - 2b - 8$ 22) $s^2 + 3s - 28$

23) $u^2 + u - 132$ 24) $m^2 - m - 110$

25) $q^2 - 8q + 15$ 26) $z^2 - 10z + 24$

27) $y^2 + 9y + 10$ 28) $a^2 - 16a + 8$

29) $w^2 + 4w - 5$ 30) $z^2 - 11z - 12$

31) $p^2 - 20p + 100$ 32) $u^2 + 18u + 81$

33) $24 + 14d + d^2$ 34) $10 + 7k + k^2$

Objective 3

Factor completely by first taking out -1 and then by factoring the trinomial, if possible. Check your answer.

35) $-a^2 - 10a - 16$ 36) $-y^2 - 9y - 18$

37) $-h^2 + 2h + 15$ 38) $-j^2 - j + 56$

39) $-k^2 + 11k - 28$ 40) $-b^2 + 17b - 66$

41) $-x^2 - x + 90$ 42) $-c^2 + 14c + 15$

43) $-n^2 - 14n - 49$

44) $-z^2 + 4z - 4$

Objective 4

Factor completely. Check your answer.

45) $a^2 + 6ab + 5b^2$ 46) $v^2 + 7vw + 6w^2$

47) $m^2 + 4mn - 21n^2$ 48) $p^2 - 17pq + 72q^2$

49) $x^2 - 15xy + 36y^2$ 50) $r^2 - 9rs + 20s^2$

51) $f^2 - 10fg - 11g^2$ 52) $u^2 + 2uv - 48v^2$

53) $c^2 + 6cd - 55d^2$ 54) $w^2 + 17wx + 60x^2$

Objectives 2–4

Factor completely, if possible. Begin by asking yourself, *"Can I factor out a GCF?"*

55) $2r^2 + 8r + 6$ 56) $5b^2 - 30b + 40$

57) $4q^3 - 28q^2 + 48q$ 58) $6g^5 + 6g^4 - 12g^3$

59) $m^4n + 7m^3n^2 - 44m^2n^3$

60) $a^3b + 10a^2b^2 + 24ab^3$

61) $p^3q - 17p^2q^2 + 70pq^3$

62) $u^3v^2 - 2u^2v^3 - 15uv^4$

63) $18 - 11r + r^2$

64) $6 + 7t + t^2$

65) $7c^3d^2 - 7c^2d^2 - 14cd^2$

66) $2x^2y^4 + 18x^2y^3 - 72x^2y^2$

67) $s^2 + 4st + 5t^2$ 68) $p^2 - 7pq - 12q^2$

69) $2r^4 + 26r^3 + 84r^2$ 70) $10y^5 + 50y^4 + 60y^3$

71) $8x^4y^5 - 16x^3y^4 - 64x^2y^3$

72) $6b^3c^2 - 72b^2c^3 + 120bc^4$

73) $(a + b)k^2 + 7(a + b)k - 18(a + b)$

74) $(m - n)v^2 + 11(m - n)v + 30(m - n)$

75) $(x + y)t^2 - 4(x + y)t - 21(x + y)$

76) $(p - q)z^2 - 21(p - q)z + 110(p - q)$

77) Is $(2x + 8)(x + 5)$ the correct answer to the problem "Factor completely: $2x^2 + 18x + 40$"? Why or why not?

Section 7.3 Factoring Trinomials of the Form $ax^2 + bx + c$ $(a \neq 1)$

Objectives

1. Factor $ax^2 + bx + c$ $(a \neq 1)$ by Grouping

2. Factor $ax^2 + bx + c$ $(a \neq 1)$ by Trial and Error

If we are asked to factor each of these polynomials,

$$3x^2 + 18x + 24 \quad \text{and} \quad 3x^2 + 10x + 8$$

how do we begin? How do the polynomials differ?

The GCF of $3x^2 + 18x + 24$ is 3. To factor, begin by taking out the 3. We can factor using what we learned in Section 7.2.

$$3x^2 + 18x + 24 = 3(x^2 + 6x + 8) \qquad \text{Factor out 3.}$$
$$= 3(x + 4)(x + 2) \qquad \text{Factor the trinomial.}$$

In the second polynomial, $3x^2 + 10x + 8$, we *cannot* factor out the leading coefficient of 3 because the GCF is 1. In this section, we will discuss two methods for factoring a trinomial when the coefficient of the squared term is not 1.

1. Factor $ax^2 + bx + c(a \neq 1)$ by Grouping

To factor $x^2 + 6x + 8$, we think of two integers whose *product* is 8 and whose *sum* is 6. Those numbers are 4 and 2. So, $x^2 + 6x + 8 = (x + 4)(x + 2)$.

Sum is 10
↓

To factor $3x^2 + 10x + 8$, first find the product of 3 and 8. Then, find two integers

Product: $3 \cdot 8 = 24$

whose *product* is 24 and whose *sum* is 10. The numbers are 6 and 4. Rewrite the middle term, $10x$, as $6x + 4x$, then factor by grouping.

$$3x^2 + 10x + 8 = \underbrace{3x^2 + 6x}_{\text{Group}} + \underbrace{4x + 8}_{\text{Group}}$$
$$= 3x(x + 2) + 4(x + 2) \qquad \text{Take out the common factor from each group.}$$
$$= (x + 2)(3x + 4) \qquad \text{Factor out } (x + 2).$$

$$3x^2 + 10x + 8 = (x + 2)(3x + 4)$$

Check: $(x + 2)(3x + 4) = 3x^2 + 4x + 6x + 8$
$$= 3x^2 + 10x + 8 \ \checkmark$$

Example 1

Factor completely.

a) $4a^2 + 16a + 15$ b) $10p^2 - 13p + 4$ c) $5c^2 - 29cd - 6d^2$

Solution

a)   *Sum* is 16
↓

$$4a^2 + 16a + 15$$

Product: $4 \cdot 15 = 60$

Think of two integers whose *product* is 60 and whose *sum* is 16. *10 and 6*
Rewrite the middle term, 16a, as $10a + 6a$. Factor by grouping.

$$4a^2 + 16a + 15 = \underbrace{4a^2 + 10a}_{\text{Group}} + \underbrace{6a + 15}_{\text{Group}}$$

$$= 2a(2a + 5) + 3(2a + 5) \qquad \text{Take out the common factor from each group.}$$

$$= (2a + 5)(2a + 3) \qquad \text{Factor out } (2a + 5).$$

$$4a^2 + 16a + 15 = (2a + 5)(2a + 3)$$

Check by multiplying: $(2a + 5)(2a + 3) = 4a^2 + 16a + 15$ ✓

b)   *Sum is* -13
         ↓

$$10p^2 - 13p + 4$$

Product: $10 \cdot 4 = 40$

Think of two integers whose *product* is 40 and whose *sum* is -13. (Both numbers will be negative.) -8 *and* -5
Rewrite the middle term, $-13p$, as $-8p - 5p$. Factor by grouping.

$$10p^2 - 13p + 4 = \underbrace{10p^2 - 8p}_{\text{Group}} - \underbrace{5p + 4}_{\text{Group}}$$

$$= 2p(5p - 4) - 1(5p - 4) \qquad \text{Take out the common factor from each group.}$$

$$= (5p - 4)(2p - 1) \qquad \text{Factor out } (5p - 4).$$

$$10p^2 - 13p + 4 = (5p - 4)(2p - 1)$$

Check: $(5p - 4)(2p - 1) = 10p^2 - 13p + 4$ ✓

c)   *Sum is* -29
         ↓

$$5c^2 - 29cd - 6d^2$$

Product: $5 \cdot (-6) = -30$

The integers whose *product* is -30 and whose *sum* is -29 are -30 *and 1*.
Rewrite the middle term, $-29cd$, as $-30cd + cd$. Factor by grouping.

$$5c^2 - 29cd - 6d^2 = \underbrace{5c^2 - 30cd}_{\text{Group}} + \underbrace{cd - 6d^2}_{\text{Group}}$$

$$= 5c(c - 6d) + d(c - 6d) \qquad \text{Take out the common factor from each group.}$$

$$= (c - 6d)(5c + d) \qquad \text{Factor out } (c - 6d).$$

$$5c^2 - 29cd - 6d^2 = (c - 6d)(5c + d)$$

Check: $(c - 6d)(5c + d) = 5c^2 - 29cd - 6d^2$ ✓

You Try 1

Factor completely.

a) $2c^2 + 11c + 14$ b) $6n^2 - 23n - 4$ c) $8x^2 - 10xy + 3y^2$

Example 2

Factor completely.

$$24w^2 - 54w - 15$$

Solution

It is tempting to jump right in and multiply $24 \cdot (-15) = -360$ and try to think of two integers with a product of -360 and a sum of -54. However, before doing that, ask yourself, *"Can I factor out a GCF?"* Yes! We can factor out 3.

$$24w^2 - 54w - 15 = 3(\underbrace{8w^2 - 18w - 5}_{\text{Product: } 8 \cdot (-5) = -40})$$ Factor out 3.

Try to factor $8w^2 - 18w - 5$ by finding two integers whose *product* is -40 and whose *sum* is -18.

The numbers are -20 and 2.

$$= 3(\underbrace{8w^2 - 20w}_{\text{Group}} + \underbrace{2w - 5}_{\text{Group}})$$

$$= 3[4w(2w - 5) + 1(2w - 5)]$$ Take out common factor from each group.

$$= 3(2w - 5)(4w + 1)$$ Factor out $2w - 5$.

$$24w^2 - 54w - 15 = 3(2w - 5)(4w + 1)$$

Check by multiplying: $3(2w - 5)(4w + 1) = 3(8w^2 - 18w - 5)$
$$= 24w^2 - 54w - 15 \checkmark$$

You Try 2

Factor completely.

a) $20m^3 - 8m^2 + 4m$ b) $6z^2 + 20z + 16$

2. Factor $ax^2 + bx + c(a \neq 1)$ by Trial and Error

At the beginning of this section, we factored $3x^2 + 10x + 8$ by grouping. Now we will factor it by trial and error, which is just reversing the process of FOIL.

Example 3

Factor completely.

$$3x^2 + 10x + 8$$

Solution

Can we factor out a GCF? *No.* So, try to factor $3x^2 + 10x + 8$ as the product of two binomials. Notice that all terms are positive, so all factors will be positive.

Begin with the squared term, $3x^2$. What two expressions with integer coefficients do we multiply to get $3x^2$? *3x and x.* Put these in the binomials.

$$3x^2 + 10x + 8 = \underbrace{(3x}_{3x^2}\quad)(x\quad)$$

Next, look at the last term, 8. What are the pairs of integers that multiply to 8? *1 and 8, 4 and 2.*

Try these as the last terms of the binomials. The middle term, $10x$, comes from finding the sum of the products of the outer terms and inner terms.

$$3x^2 + 10x + 8 \overset{?}{=} (3x + 1)(x + 8)$$

These must both be $10x$. →
$$\begin{array}{r} 1x \\ + \ 24x \\ \hline 25x \end{array}$$

Since $25x \neq 10x$, $(3x + 1)(x + 8)$ is not the correct factorization of $3x^2 + 10x + 8$. Try switching the 1 and 8.

$$3x^2 + 10x + 8 \overset{?}{=} (3x + 8)(x + 1)$$

These must both be $10x$. →
$$\begin{array}{r} 8x \\ + \ 3x \\ \hline 11x \end{array}$$

$11x \neq 10x$, so $(3x + 8)(x + 1)$ is incorrect. Try 4 and 2.

$$3x^2 + 10x + 8 \overset{?}{=} (3x + 4)(x + 2)$$

These must both be $10x$. →
$$\begin{array}{r} 4x \\ + \ 6x \\ \hline 10x \end{array} \quad \text{Correct!}$$

$3x^2 + 10x + 8 = (3x + 4)(x + 2)$
Check by multiplying.

It may seem like this method is *nothing* but trial and error, but sometimes there are ways to narrow down the possibilities.

Example 4

Factor completely.

$$2r^2 - 13r + 20$$

Solution

Can we factor out a GCF? *No.* To get a product of $2r^2$, we will use $2r$ and r.

$$2r^2 - 13r + 20 = \underbrace{(2r \quad)(r \quad)}_{2r^2}$$

Since the last term is positive and the middle term is negative, we want pairs of *negative* integers that multiply to 20. The pairs are *−1 and −20, −2 and −10,* and *−4 and −5.*

Try these as the last terms of the binomials.

$$2r^2 - 13r + 20 \overset{?}{=} (2r - 1)(r - 20)$$

$$\begin{array}{c} -r \\ + \ (-40r) \\ \hline -41r \end{array} \quad \leftarrow \text{This must be } -13r. \text{ Incorrect}$$

Switch the -1 and -20.

$$2r^2 - 13r + 20 \overset{?}{=} (2r - 20)(r - 1)$$

Without even multiplying, we know this is incorrect. How? Notice that in the factor $(2r - 20)$ a 2 can be factored out to get $2(r - 10)$. But, at the very beginning we concluded that we could not take out a GCF from the original trinomial $2r^2 - 13r + 20$. Therefore, it will not be possible to take out a common factor from one of the binomial factors either.

> If you cannot take a factor out of the original trinomial, then you cannot take out a factor from one of the binomial factors either.

The next pair of numbers to try is -2 and -10.

$$2r^2 - 13r + 20 \overset{?}{=} (2r - 2)(r - 10)$$
$$\downarrow$$
$$2(r - 1) \quad \text{No. You can factor out 2.}$$

$$2r^2 - 13r + 20 \overset{?}{=} (2r - 10)(r - 2)$$
$$\downarrow$$
$$2(r - 5) \quad \text{No. You can factor out 2.}$$

Try -4 and -5. The -4 will not go in the binomial with $2r$ because then a 2 could be factored out: $2r - 4 = 2(r - 2)$.

$$2r^2 - 13r + 20 \stackrel{?}{=} (2r - 5)(r - 4)$$

$$\begin{array}{c} -5r \\ \underline{+\ (-8r)} \\ -13r \quad \text{Correct!} \end{array}$$

So, $2r^2 - 13r + 20 = (2r - 5)(r - 4)$. Check by multiplying.

We can use the reasoning in example 4 to eliminate some of the numbers to try in the binomials.

You Try 3

Factor completely.

a) $2m^2 + 11m + 12$ b) $3v^2 - 28v + 9$

Example 5

Factor completely.

a) $12d^2 + 46d - 8$ b) $-2h^2 + 9h + 56$

Solution

a) $12d^2 + 46d - 8$: Ask yourself, *"Can I take out a common factor?"* Yes!
GCF $= 2$.

$$12d^2 + 46d - 8 = 2(6d^2 + 23d - 4)$$

Now, try to factor $6d^2 + 23d - 4$. To get a product of $6d^2$, we can try either *6d and d* or *3d and 2d*. Let's start by trying *6d and d*.

$$6d^2 + 23d - 4 = (6d\quad)(d\quad)$$

List pairs of integers that multiply to -4: *4 and -1, -4 and 1, 2 and -2.*
    Try 4 and -1. Do not put 4 in the same binomial as $6d$ since then it would be possible to factor out 2. But, a 2 does not factor out of $6d^2 + 23d - 4$. Put the 4 in the same binomial as d.

$$6d^2 + 23d - 4 \stackrel{?}{=} (6d - 1)(d + 4)$$

$$\begin{array}{c} -d \\ \underline{+\ 24d} \\ 23d \quad \text{Correct} \end{array}$$

Don't forget that the very first step was to factor out a 2. Therefore,

$$12d^2 + 46d - 8 = 2(6d^2 + 23d - 4) = 2(6d - 1)(d + 4)$$

Check by multiplying.

b) $-2h^2 + 9h + 56$: Since the coefficient of the squared term is negative, begin by factoring out -1. (There is no other common factor except 1.)

$$-2h^2 + 9h + 56 = -1(2h^2 - 9h - 56)$$

Try to factor $2h^2 - 9h - 56$. To get a product of $2h^2$ we will use $2h$ and h in the binomials.

$$2h^2 - 9h - 56 = (2h \qquad)(h \qquad)$$

We need pairs of integers so that their product is -56. The ones that come to mind quickly involve 7 and 8 and 1 and 56: -7 and 8, 7 and -8, -1 and 56, 1 and -56. There are other pairs; if these do not work, we will list others.

Do *not* start with -1 and 56 or 1 and -56 because the middle term, $-9h$, is not very large. Using -1 and 56 or 1 and -56 would likely result in a larger middle term.

Try -7 and 8. Do not put 8 in the same binomial as $2h$ since then it would be possible to factor out 2.

$$2h^2 - 9h - 56 \overset{?}{=} (2h - 7)(h + 8)$$

$$\begin{array}{c} -7h \\ \underline{+ 16h} \\ 9h \end{array} \qquad \text{This must equal } -9h. \text{ Incorrect}$$

Only the sign of the sum is incorrect. *Change the signs in the binomials to get the correct sum.*

$$2h^2 - 9h - 56 \overset{?}{=} (2h + 7)(h - 8)$$

$$\begin{array}{c} 7h \\ \underline{+ (-16h)} \\ -9h \end{array} \qquad \text{Correct}$$

Remember that we factored out -1 to begin the problem.

$$-2h^2 + 9h + 56 = -1(2h^2 - 9h - 56) = -(2h + 7)(h - 8)$$

Check by multiplying.

You Try 4

Factor completely.

a) $15b^2 - 55b + 30$ b) $-4p^2 + 3p + 10$

We have seen two methods for factoring $ax^2 + bx + c$ $(a \neq 1)$: factoring by grouping and factoring by trial and error. In either case, remember to begin by taking out a common factor from all terms whenever possible.

Answers to You Try Exercises

1) a) $(2c + 7)(c + 2)$ b) $(6n + 1)(n - 4)$ c) $(4x - 3y)(2x - y)$

2) a) $4m(5m^2 - 2m + 1)$ b) $2(3z + 4)(z + 2)$

3) a) $(2m + 3)(m + 4)$ b) $(3v - 1)(v - 9)$

4) a) $5(3b - 2)(b - 3)$ b) $-(4p + 5)(p - 2)$

7.3 Exercises

Boost your grade at mathzone.com!   MathZone

> Practice Problems   > NetTutor   > e-Professors   > Videos
> Self-Test

Objective 1

1) Find two integers whose

	PRODUCT IS	**and whose SUM IS**	**ANSWER**
a)	-50	5	
b)	27	-28	
c)	12	8	
d)	-72	-6	

2) Find two integers whose

	PRODUCT IS	**and whose SUM IS**	**ANSWER**
a)	18	19	
b)	-132	1	
c)	-30	-13	
d)	63	-16	

Factor by grouping.

3) $2k^2 + 10k + 9k + 45$

4) $9m^2 + 54m + 2m + 12$

5) $7y^2 - 7y - 6y + 6$

6) $8c^2 - 8c + 11c - 11$

7) $8a^2 - 14ab + 12ab - 21b^2$

8) $10y^2 - 8yz - 15yz + 12z^2$

9) When asked to factor a polynomial, what is the first question you should ask yourself?

10) After factoring a polynomial, what should you ask yourself to be sure that the polynomial is factored completely?

11) How do we know that $(2x - 4)$ cannot be a factor of $2x^2 + 13x - 24$?

12) Find the polynomial which factors to $(3x - 8)(x + 2)$.

Complete the factorization.

13) $4a^2 + 17a + 18 = (4a + 9)(\quad)$

14) $5p^2 + 41p + 8 = (5p + 1)(\quad)$

15) $6k^2 - 5k - 21 = (3k - 7)(\quad)$

16) $12t^2 - 28t + 15 = (2t - 3)(\quad)$

17) $18x^2 - 17xy + 4y^2 = (2x - y)(\quad)$

18) $8c^2 + 14cd - 9d^2 = (4c + 9d)(\quad)$

Factor by grouping. See Example 1.

19) $2r^2 + 11r + 15$ 20) $3a^2 + 10a + 8$

21) $5p^2 - 21p + 4$ 22) $7j^2 - 30j + 8$

23) $11m^2 - 18m - 8$ 24) $5b^2 + 9b - 18$

25) $6v^2 + 11v - 7$ 26) $8x^2 - 14x + 3$

27) $10c^2 + 19c + 6$ 28) $15n^2 + 22n + 8$

29) $9x^2 - 13xy + 4y^2$ 30) $6a^2 + ab - 5b^2$

Objective 2

Factor by trial and error. See Examples 3 and 4.

31) $5w^2 + 11w + 6$ 32) $2g^2 + 13g + 18$

33) $3u^2 - 23u + 30$ 34) $7a^2 - 17a + 6$

35) $7k^2 + 15k - 18$ 36) $5z^2 - 18z - 35$

37) $8r^2 + 26r + 15$ 38) $6t^2 + 23t + 7$

39) $6v^2 - 19v + 14$ 40) $10m^2 + 47m - 15$

41) $21d^2 - 22d - 8$ 42) $12h^2 - 17h - 44$

43) $48v^2 + 64v + 5$

44) $18p^2 + 35p + 12$

45) $10a^2 - 13ab + 4b^2$

46) $8x^2 - 19xy + 6y^2$

Objectives 1 and 2

47) Factor $6t^2 + 5t - 4$ using each method. Do you get the same answer? Which method do you prefer? Why?

48) Factor $8m^2 + 69m - 27$ using each method. Do you get the same answer? Which method do you prefer? Why?

Factor completely.

49) $2y^2 - 19y + 24$

50) $3x^2 - 16x - 12$

51) $12c^3 + 15c^2 - 18c$

52) $4r^2 + 15r + 9$

53) $12t^2 - 28t - 5$

54) $30m^3 + 76m^2 + 14m$

55) $45h^2 + 57h + 18$

56) $21y^2 - 90y + 24$

57) $3b^2 - 7b + 5$

58) $5g^2 + g - 7$

59) $13t^2 + 17t - 18$

60) $42q^2 + 11q - 3$

61) $5c^2 + 23cd + 12d^2$

62) $7s^2 - 17st + 6t^2$

video
63) $2d^2 + 2d - 40$

64) $6z^2 + 42z + 72$

65) $8c^2d^3 + 4c^2d^2 - 60c^2d$

66) $30f^4g^2 + 23f^3g^2 + 3f^2g^2$

67) $36a^2 - 12a + 1$

68) $64x^2 - 112x + 49$

69) $3x^2(y + 6)^2 - 11x(y + 6)^2 - 20(y + 6)^2$

70) $2a^2(b - 5)^3 - a(b - 5)^3 - 21(b - 5)^3$

71) $9y^2(z - 10)^3 + 76y(z - 10)^3 + 32(z - 10)^3$

72) $14s^2(t - 13)^5 + 69s(t - 13)^5 + 27(t - 13)^5$

73) $8u^2(v + 8) - 38u(v + 8) - 33(v + 8)$

74) $12p^2(q - 1)^2 - 49p(q - 1)^2 + 49(q - 1)^2$

Factor completely by first taking out a negative common factor. See Example 5.

75) $-h^2 - 3h + 54$

76) $-r^2 + 11r - 28$

video
77) $-10z^2 + 19z - 6$

78) $-7a^2 + 18a - 8$

79) $-21v^2 + 54v + 27$

80) $-12c^2 - 26c - 10$

81) $-2j^3 - 32j^2 - 120j$

82) $-45p^3 + 18p^2 + 63p$

83) $-16y^2 - 34y + 15$

84) $-6x^3 - 54x^2 - 48x$

85) $-6c^3d + 27c^2d^2 - 12cd^3$

86) $-12s^4t^2 - 4s^3t^3 + 40s^2t^4$

Section 7.4 Factoring Binomials and Perfect Square Trinomials

Objectives

1. Factor a Perfect Square Trinomial

2. Factor the Difference of Two Squares

3. Factor the Sum and Difference of Two Cubes

1. Factor a Perfect Square Trinomial

Recall that we can square a binomial using the formulas

$$(a + b)^2 = a^2 + 2ab + b^2$$
$$(a - b)^2 = a^2 - 2ab + b^2$$

We can apply the first formula to find $(x + 5)^2$.

$$(x + 5)^2 = x^2 + 2x(5) + 5^2$$
$$= x^2 + 10x + 25$$

Since factoring a polynomial means writing the polynomial as a product of its factors, $x^2 + 10x + 25$ factors to $(x + 5)^2$.

$x^2 + 10x + 25$ is an example of a *perfect square trinomial*. A **perfect square trinomial** is a trinomial that results from squaring a binomial.

We can use the factoring method presented in Section 7.2 to factor a perfect square trinomial or we can learn to recognize the special pattern that appears in these trinomials.

Above we stated that $x^2 + 10x + 25$ factors to $(x + 5)^2$. How are the terms of the trinomial and binomial related?

Compare $x^2 + 10x + 25$ to $(x + 5)^2$.

x^2 is the square of x, the first term in the binomial.

25 is the square of 5, the last term in the binomial.

We get the term $10x$ by doing the following:

$$10x = 2 \quad \cdot \quad x \quad \cdot \quad 5$$

| Two times | First term in binomial | Last term in binomial |

This follows directly from how we found $(x + 5)^2$ using the formula.

Factoring a Perfect Square Trinomial

$$a^2 + 2ab + b^2 = (a + b)^2$$
$$a^2 - 2ab + b^2 = (a - b)^2$$

In order for a trinomial to be a perfect square, two of its terms must be perfect squares.

Example 1

Factor completely. $t^2 + 12t + 36$

Solution

We cannot take out a common factor, so let's see if this follows the pattern of a perfect square trinomial.

$$t^2 + 12t + 36$$

What do you square to get t^2? t $(t)^2$ $(6)^2$ What do you square to get 36? 6

Does the middle term equal $2 \cdot t \cdot 6$? *Yes.*

$$2 \cdot t \cdot 6 = 12t$$

Therefore, $t^2 + 12t + 36 = (t + 6)^2$.
Check by multiplying.

Example 2

Factor completely.

a) $n^2 - 14n + 49$ b) $4p^3 + 24p^2 + 36p$
c) $9k^2 + 30k + 25$ d) $4c^2 + 20c + 9$

Solution

a) We cannot take out a common factor. However, since the middle term is negative and the first and last terms are positive, the sign in the binomial will be a minus $(-)$ sign. Does this fit the pattern of a perfect square trinomial?

$$n^2 - 14n + 49$$

What do you square to get n^2? n $(n)^2$ $(7)^2$ What do you square to get 49? 7

Does the middle term equal $2 \cdot n \cdot 7$? *Yes.*

$$2 \cdot n \cdot 7 = 14n$$

Since there is a minus sign in front of $14n$, $n^2 - 14n + 49$ fits the pattern of $a^2 - 2ab + b^2 = (a - b)^2$ with $a = n$ and $b = 7$.

Therefore, $n^2 - 14n + 49 = (n - 7)^2$.
Check by multiplying.

b) From $4p^3 + 24p^2 + 36p$ we *can* begin by taking out the GCF of $4p$.

$$4p^3 + 24p^2 + 36p = 4p(p^2 + 6p + 9)$$

What do you square to get p^2? p $(p)^2$ $(3)^2$ What do you square to get 9? 3

Does the middle term equal $2 \cdot p \cdot 3$? *Yes.* $2 \cdot p \cdot 3 = 6p$.

$$4p^3 + 24p^2 + 36p = 4p(p^2 + 6p + 9)$$
$$= 4p(p + 3)^2$$

Therefore, $4p^3 + 24p^2 + 36p = 4p(p + 3)^2$. Check by multiplying.

c) We cannot take out a common factor. Since the first and last terms of $9k^2 + 30k + 25$ are perfect squares, let's see if this is a perfect square trinomial.

$$9k^2 + 30k + 25$$

What do you square to get $9k^2$? $3k$ $(3k)^2$ $(5)^2$ What do you square to get 25? 5

Does the middle term equal $2 \cdot 3k \cdot 5$? *Yes.*

$$2 \cdot 3k \cdot 5 = 30k$$

Therefore, $9k^2 + 30k + 25 = (3k + 5)^2$.
Check by multiplying.

d) We cannot take out a common factor. The first and last terms of $4c^2 + 20c + 9$ are perfect squares. Is this a perfect square trinomial?

$$4c^2 + 20c + 9$$

What do you square to get $4c^2$? $2c$ $(2c)^2$ $(3)^2$ What do you square to get 9? 3

Does the middle term equal $2 \cdot 2c \cdot 3$? *No.*

$$2 \cdot 2c \cdot 3 = 12c$$

This is *not* a perfect square trinomial. Applying a method from Section 7.3 we find that the trinomial does factor, however.

$$4c^2 + 20c + 9 = (2c + 9)(2c + 1)$$

Check by multiplying.

You Try 1

Factor completely.

a) $w^2 + 8w + 16$ b) $a^2 - 20a + 100$ c) $4d^2 - 36d + 81$

2. Factor the Difference of Two Squares

Another common type of factoring problem is a **difference of two squares**. Some examples of these types of binomials are

$$y^2 - 9 \qquad 25m^2 - 16n^2 \qquad 64 - t^2 \qquad h^4 - 16$$

Notice that in each binomial, the terms are being *subtracted*, and each term is a perfect square.

In Section 6.3, multiplication of polynomials, we saw that

$$(a + b)(a - b) = a^2 - b^2$$

If we reverse the procedure, we get the factorization of the difference of two squares.

Factoring the Difference of Two Squares
$$a^2 - b^2 = (a + b)(a - b)$$

Example 3

Factor completely.

a) $y^2 - 9$ b) $25m^2 - 16n^2$

Solution

a) First, notice that $y^2 - 9$ is the difference of two terms *and* those terms are perfect squares. We can use the formula $a^2 - b^2 = (a + b)(a - b)$.

Identify a and b.

$$y^2 - 9$$
$$\downarrow \qquad \downarrow$$

What do you square $(y)^2$ $(3)^2$ What do you square
to get y^2? y to get 9? 3

Then, $a = y$ and $b = 3$.

$$y^2 - 9 = (y + 3)(y - 3)$$

Check by multiplying: $(y + 3)(y - 3) = y^2 - 9$. ✔

b) Look carefully at $25m^2 - 16n^2$. Each term *is* a perfect square, and they are being subtracted.

Identify a and b.

$$25m^2 - 16n^2$$
$$\downarrow \qquad \downarrow$$

What do you square $(5m)^2$ $(4n)^2$ What do you square
to get $25m^2$? $5m$ to get $16n^2$? $4n$

Then, $a = 5m$ and $b = 4n$.

$$25m^2 - 16n^2 = (5m + 4n)(5m - 4n)$$

Check by multiplying: $(5m + 4n)(5m - 4n) = 25m^2 - 16n^2$. ✓

You Try 2

Factor completely.

a) $r^2 - 25$ b) $49p^2 - 121q^2$

Remember that sometimes we can factor out a GCF first. And, after factoring once, ask yourself, *"Can I factor again?"*

Example 4

Factor completely.

a) $128t - 2t^3$ b) $x^2 + 49$ c) $h^4 - 16$

Solution

a)   Ask yourself, *"Can I take out a common factor?"* Yes. Factor out $2t$.

$$128t - 2t^3 = 2t(64 - t^2)$$

Now ask yourself, *"Can I factor again?"* Yes. $64 - t^2$ is the difference of two squares. Identify a and b.

$$64 - t^2$$
$$\downarrow \quad \downarrow$$
$$(8)^2 \quad (t)^2$$

So, $a = 8$ and $b = t$.

$$64 - t^2 = (8 + t)(8 - t)$$

Therefore, $128t - 2t^3 = 2t(8 + t)(8 - t)$.
Check by multiplying.

BE CAREFUL

$(8 + t)(8 - t)$ is *not* the same as $(t + 8)(t - 8)$ because subtraction is not commutative.

While $8 + t = t + 8$, $8 - t$ *does not equal* $t - 8$. You must write the terms in the correct order.

Another way to see that they are not equivalent is to multiply $(t + 8)(t - 8)$.

$(t + 8)(t - 8) = t^2 - 64$. This is not the same as $64 - t^2$.

b) We cannot take out a common factor, but x^2 and 49 are perfect squares.

$$x^2 + 49$$
$$\downarrow \quad \downarrow$$
$$(x)^2 \ (7)^2$$

However, $x^2 + 49$ is the *sum* of two squares. *This binomial does not factor.*

$$x^2 + 49 \neq (x + 7)(x - 7) \text{ since } (x + 7)(x - 7) = x^2 - 49$$
$$x^2 + 49 \neq (x + 7)(x + 7) \text{ since } (x + 7)(x + 7) = x^2 + 14x + 49$$

If the sum of two squares does not contain a common factor, then it does not factor.

$x^2 + 49$ is *prime*.

c) The terms in $h^4 - 16$ have no common factors, but they are perfect squares. Identify a and b.

$$h^4 - 16$$
$$\downarrow \qquad \downarrow$$

What do you square to get h^4? h^2 $\qquad (h^2)^2 \ (4)^2 \qquad$ What do you square to get 16? 4

So, $a = h^2$ and $b = 4$.

$$h^4 - 16 = (h^2 + 4)(h^2 - 4)$$

Can we factor again?

$h^2 + 4$ is the *sum* of two squares. It will not factor.

$h^2 - 4$ is the difference of two squares, so it *will* factor.

$$h^2 - 4$$
$$\downarrow \quad \downarrow$$
$$(h)^2 \ (2)^2$$
$$a = h \text{ and } b = 2$$

$$h^2 - 4 = (h + 2)(h - 2)$$

Therefore,

$$h^4 - 16 = (h^2 + 4)(h^2 - 4)$$
$$= (h^2 + 4)(h + 2)(h - 2)$$

Check by multiplying.

You Try 3

Factor completely.

a) $12p^4 - 27p^2$ b) $100 - w^2$ c) $k^2 - \dfrac{25}{4}$ d) $a^2 + 36b^2$

3. Factor the Sum and Difference of Two Cubes

We can understand where we get the formulas for factoring the sum and difference of two cubes by looking at two products.

Multiply $(a + b)(a^2 - ab + b^2)$.

$$\begin{aligned}(a + b)(a^2 - ab + b^2) &= a(a^2 - ab + b^2) + b(a^2 - ab + b^2) &&\text{Distributive property}\\ &= a^3 - a^2b + ab^2 + a^2b - ab^2 + b^3 &&\text{Distribute.}\\ &= a^3 + b^3\end{aligned}$$

$$(a + b)(a^2 - ab + b^2) = a^3 + b^3$$

Multiply $(a - b)(a^2 + ab + b^2)$.

$$\begin{aligned}(a - b)(a^2 + ab + b^2) &= a(a^2 + ab + b^2) - b(a^2 + ab + b^2) &&\text{Distributive property}\\ &= a^3 + a^2b + ab^2 - a^2b - ab^2 - b^3 &&\text{Distribute.}\\ &= a^3 - b^3\end{aligned}$$

$$(a - b)(a^2 + ab + b^2) = a^3 - b^3$$

The formulas for factoring the sum and difference of two cubes, then, are as follows:

Factoring the Sum and Difference of Two Cubes
$$a^3 + b^3 = (a + b)(a^2 - ab + b^2)$$
$$a^3 - b^3 = (a - b)(a^2 + ab + b^2)$$

Notice that each factorization is the product of a binomial and a trinomial. To factor the sum and difference of two cubes

Step 1: Identify a and b.

Step 2: Place them in the binomial factor and write the trinomial based on a and b.

Step 3: Simplify.

We will see how this procedure works by analyzing the first formula.

To factor $a^3 + b^3$,

Step 1: Identify a and b.

$$a^3 + b^3$$

What do you cube to get a^3? a $(a)^3$ $(b)^3$ What do you cube to get b^3? b

Step 2: Write the binomial factor and then write the trinomial.

Square a.   Product of a and b Square b.

same sign

$$a^3 + b^3 = (a + b)[(a)^2 - (a)(b) + (b)^2]$$

opposite sign

Step 3: Simplify.

$$a^3 + b^3 = (a + b)(a^2 - ab + b^2)$$

Example 5

Factor completely.

a) $n^3 + 8$ b) $c^3 - 64$ c) $125r^3 + 27s^3$

Solution

a) Use steps 1–3 to factor.

Step 1: Identify a and b.

$$n^3 + 8$$

What do you cube to get n^3? n $(n)^3$ $(2)^3$ What do you cube to get 8? 2

So, $a = n$ and $b = 2$.

Step 2: Remember, $a^3 + b^3 = (a + b)(a^2 - ab + b^2)$.
Write the binomial factor, then write the trinomial.

Same sign   Square a.   Product of a and b Square b.

$$n^3 + 8 = (n + 2)[(n)^2 - (n)(2) + (2)^2]$$

Opposite sign

Step 3: Simplify.

$$n^3 + 8 = (n + 2)(n^2 - 2n + 4)$$

b) **Step 1:** Identify a and b.

$$c^3 - 64$$

What do you cube to get c^3? c $(c)^3$ $(4)^3$ What do you cube to get 64? 4

So, $a = c$ and $b = 4$.

Step 2: Write the binomial factor, then write the trinomial. Remember, $a^3 - b^3 = (a - b)(a^2 + ab + b^2)$.

Same sign   Square a.   Product of a and b Square b.

$$c^3 - 64 = (c - 4)[(c)^2 + (c)(4) + (4)^2]$$

Opposite sign

Step 3:   Simplify.

$$c^3 - 64 = (c - 4)(c^2 + 4c + 16)$$

c) $125r^3 + 27s^3$

Step 1:   Identify a and b.

$$125r^3 + 27s^3$$

$$\downarrow \qquad \downarrow$$

What do you cube
to get $125r^3$? $5r$ $(5r)^3$ $(3s)^3$ What do you cube
to get $27s^3$? $3s$

So, $a = 5r$ and $b = 3s$.

Step 2:   Write the binomial factor, then write the trinomial. Remember,
$a^3 + b^3 = (a + b)(a^2 - ab + b^2)$.

Square a.   Square b.
Product
of a and b
Same sign $\downarrow$ $\downarrow$

$$125r^3 + 27s^3 = (5r + 3s)[(5r)^2 - (5r)(3s) + (3s)^2]$$

Opposite
sign

Step 3:   Simplify.

$$125r^3 + 27s^3 = (5r + 3s)(25r^2 - 15rs + 9s^2)$$

You Try 4

Factor completely.

a) $r^3 + 1$ b) $p^3 - 1000$ c) $64x^3 - 125y^3$

Just as in the other factoring problems we've studied so far, the first step in factoring
any polynomial should be to ask ourselves, *"Can I factor out a GCF?"*

Example 6

Factor completely.

$$3d^3 - 81$$

Solution

"Can I factor out a GCF?" Yes. The GCF is 3.

$$3d^3 - 81 = 3(d^3 - 27)$$

Factor $d^3 - 27$. Use $a^3 - b^3 = (a - b)(a^2 + ab + b^2)$.

$$d^3 - 27 = (d - 3)[(d)^2 + (d)(3) + (3)^2]$$
$$\underset{\downarrow}{(d)^3} - \underset{\downarrow}{(3)^3} = (d - 3)(d^2 + 3d + 9)$$

$$3d^3 - 81 = 3(d^3 - 27)$$
$$= 3(d - 3)(d^2 + 3d + 9)$$

You Try 5

Factor completely.

a) $4t^3 + 4$ b) $72a^3 - 9b^6$

As always, the first thing you should do when factoring is ask yourself, *"Can I factor out a GCF?"* and the last thing you should do is ask yourself, *"Can I factor again?"* Now we will summarize the factoring methods discussed in this section.

Special Factoring Rules

Perfect square trinomials: $a^2 + 2ab + b^2 = (a + b)^2$
$\qquad\qquad\qquad\qquad\qquad a^2 - 2ab + b^2 = (a - b)^2$

Difference of two squares: $a^2 - b^2 = (a + b)(a - b)$

Sum of two cubes: $a^3 + b^3 = (a + b)(a^2 - ab + b^2)$

Difference of two cubes: $a^3 - b^3 = (a - b)(a^2 + ab + b^2)$

Answers to You Try Exercises

1) a) $(w + 4)^2$ b) $(a - 10)^2$ c) $(2d - 9)^2$ 2) a) $(r + 5)(r - 5)$ b) $(7p + 11q)(7p - 11q)$

3) a) $3p^2(2p + 3)(2p - 3)$ b) $(10 + w)(10 - w)$ c) $\left(k + \dfrac{5}{2}\right)\left(k - \dfrac{5}{2}\right)$ d) prime

4) a) $(r + 1)(r^2 - r + 1)$ b) $(p - 10)(p^2 + 10p + 100)$ c) $(4x - 5y)(16x^2 + 20xy + 25y^2)$
5) a) $4(t + 1)(t^2 - t + 1)$ b) $9(2a - b^2)(4a^2 + 2ab^2 + b^4)$

7.4 Exercises

Objectives 1 and 2

1) Find the following.

   a) 6^2

   c) 4^2

   e) 3^2

   b) 10^2

   d) 11^2

   f) 8^2

   g) 12^2

   i) $\left(\dfrac{3}{5}\right)^2$

   h) $\left(\dfrac{1}{2}\right)^2$

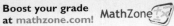 2) What is a perfect square trinomial?

3) Fill in the blank with a term that has a positive coefficient.

a) $(\_\_)^2 = n^4$ b) $(\_\_)^2 = 25t^2$

c) $(\_\_)^2 = 49k^2$ d) $(\_\_)^2 = 16p^4$

e) $(\_\_)^2 = \dfrac{1}{9}$ f) $(\_\_)^2 = \dfrac{25}{4}$

4) If x^n is a perfect square, then n is divisible by what number?

5) What perfect square trinomial factors to $(z + 9)^2$?

6) What perfect square trinomial factors to $(2b - 7)^2$?

7) Why isn't $9c^2 - 12c + 16$ a perfect square trinomial?

8) Why isn't $k^2 + 6k + 8$ a perfect square trinomial?

Objective 1

Factor completely.

9) $t^2 + 16t + 64$ 10) $x^2 + 12x + 36$

11) $g^2 - 18g + 81$ 12) $q^2 - 22q + 121$

13) $4y^2 + 12y + 9$ 14) $49r^2 + 14r + 1$

15) $9k^2 - 24k + 16$ 16) $16b^2 - 24b + 9$

17) $a^2 + \dfrac{2}{3}a + \dfrac{1}{9}$ 18) $m^2 + m + \dfrac{1}{4}$

19) $v^2 - 3v + \dfrac{9}{4}$ 20) $h^2 - \dfrac{4}{5}h + \dfrac{4}{25}$

21) $x^2 + 6xy + 9y^2$ 22) $9a^2 - 12ab + 4b^2$

23) $36t^2 - 60tu + 25u^2$ 24) $81k^2 + 18km + m^2$

25) $4f^2 + 24f + 36$ 26) $9j^2 - 18j + 9$

27) $2p^4 - 24p^3 + 72p^2$ 28) $5r^3 + 40r^2 + 80r$

29) $-18d^2 - 60d - 50$ 30) $-28z^2 + 28z - 7$

31) $12c^3 + 3c^2 + 27c$ 32) $100n^4 - 8n^3 + 64n^2$

Objective 2

33) What binomial factors to

a) $(x + 4)(x - 4)$?   b) $(4 + x)(4 - x)$?

34) What binomial factors to

a) $(y - 9)(y + 9)$?   b) $(9 - y)(9 + y)$?

Factor completely.

35) $x^2 - 9$ 36) $q^2 - 49$

37) $n^2 - 121$ 38) $d^2 - 81$

39) $m^2 + 64$ 40) $q^2 + 9$

41) $y^2 - \dfrac{1}{25}$ 42) $t^2 - \dfrac{1}{100}$

43) $c^2 - \dfrac{9}{16}$ 44) $m^2 - \dfrac{4}{25}$

45) $36 - h^2$ 46) $4 - b^2$

47) $169 - a^2$ 48) $121 - w^2$

49) $\dfrac{49}{64} - j^2$ 50) $\dfrac{144}{49} - r^2$

51) $100m^2 - 49$ 52) $36x^2 - 25$

53) $16p^2 - 81$ 54) $9a^2 - 1$

55) $4t^2 + 25$ 56) $64z^2 + 9$

57) $\dfrac{1}{4}k^2 - \dfrac{4}{9}$ 58) $\dfrac{1}{36}d^2 - \dfrac{4}{49}$

59) $b^4 - 64$ 60) $u^4 - 49$

61) $144m^2 - n^4$ 62) $64p^2 - 25q^4$

63) $r^4 - 1$ 64) $k^4 - 81$

65) $16h^4 - g^4$ 66) $b^4 - a^4$

67) $4a^2 - 100$ 68) $3p^2 - 48$

69) $2m^2 - 128$ 70) $6j^2 - 6$

71) $45r^4 - 5r^2$ 72) $32n^5 - 200n^3$

Objective 3

73) Find the following.

a) 4^3 b) 1^3

c) 10^3 d) 3^3

e) 5^3 f) 2^3

74) If x^n is a perfect cube, then n is divisible by what number?

75) Fill in the blank.

a) $(\_\_)^3 = y^3$ b) $(\_\_)^3 = 8c^3$

c) $(\_\_)^3 = 125r^3$ d) $(\_\_)^3 = x^6$

76) If x^n is a perfect square *and* a perfect cube, then n is divisible by what number?

Factor completely.

77) $d^3 + 1$ 78) $n^3 + 125$

79) $p^3 - 27$ 80) $g^3 - 8$

81) $k^3 + 64$ 82) $z^3 - 1000$

83) $27m^3 - 125$ 84) $64c^3 + 1$

85) $125y^3 - 8$

86) $27a^3 + 64$

87) $1000c^3 - d^3$

88) $125v^3 + w^3$

89) $8j^3 + 27k^3$

90) $125m^3 - 27n^3$

91) $64x^3 + 125y^3$

92) $27a^3 - 1000b^3$

93) $6c^3 + 48$

94) $9k^3 - 9$

95) $7v^3 - 7000w^3$

96) $216a^3 + 64b^3$

97) $h^6 - 64$

98) $p^6 - 1$

Extend the concepts of 7.1–7.4 to factor completely.

99) $(x + 5)^2 - (x - 2)^2$ 100) $(r - 6)^2 - (r + 1)^2$

101) $(2p + 3)^2 - (p + 4)^2$

102) $(3d - 2)^2 - (d - 5)^2$

103) $(t + 5)^3 + 8$ 104) $(c - 2)^3 + 27$

105) $(k - 9)^3 - 1$ 106) $(y + 3)^3 - 125$

Mid-Chapter Summary

Objectives

1. Learn Strategies for Factoring a Given Polynomial

1. Learn Strategies for Factoring a Given Polynomial

In this chapter, we have discussed several different types of factoring problems:

1) Factoring out a GCF (Section 7.1)
2) Factoring by grouping (Section 7.1)
3) Factoring a trinomial of the form $x^2 + bx + c$ (Section 7.2)
4) Factoring a trinomial of the form $ax^2 + bx + c$ (Section 7.3)
5) Factoring a perfect square trinomial (Section 7.4)
6) Factoring the difference of two squares (Section 7.4)
7) Factoring the sum and difference of two cubes (Section 7.4)

We have practiced the factoring methods separately in each section, but how do we know which factoring method to use given many different types of polynomials together? We will discuss some strategies in this section. First, recall the steps for factoring *any* polynomial:

To Factor a Polynomial

1) *Always* begin by asking yourself, *"Can I factor out a GCF?"* If so, factor it out.

2) Look at the expression to decide if it will factor further. Apply the appropriate method to factor. If there are

   a) *two terms*, see if it is a difference of two squares or the sum or difference of two cubes as in Section 7.4.

   b) *three terms*, see if it can be factored using the methods of Section 7.2 or Section 7.3 *or* determine if it is a perfect square trinomial (Section 7.4).

   c) *four terms*, see if it can be factored by grouping as in Section 7.1.

3) After factoring *always* look carefully at the result and ask yourself, *"Can I factor it again?"* If so, factor again.

Next, we will discuss how to decide which factoring method should be used to factor a particular polynomial.

Example 1

Factor completely.

a) $12a^2 - 27b^2$ b) $y^2 - y - 30$ c) $mn^2 - 4m + 5n^2 - 20$
d) $p^2 - 16p + 64$ e) $8x^2 + 26x + 20$ f) $27k^3 + 8$
g) $14c^3d + 63cd^2 - 7cd$

Solution

a) *"Can I factor out a GCF?"* is the first thing you should ask yourself. *Yes.* Factor out 3.

$$12a^2 - 27b^2 = 3(4a^2 - 9b^2)$$

Ask yourself, *"Can I factor again?"* Examine $4a^2 - 9b^2$. It has two terms that are being subtracted. (Now, think difference of squares or cubes.) Each term is a perfect square. $4a^2 - 9b^2$ is the difference of squares.

$$4a^2 - 9b^2 = (2a + 3b)(2a - 3b)$$
$$\quad\;\downarrow\qquad\downarrow$$
$$(2a)^2\;(3b)^2$$

$$12a^2 - 27b^2 = 3(4a^2 - 9b^2)$$
$$\qquad\qquad = 3(2a + 3b)(2a - 3b)$$

"Can I factor again?" No. It is completely factored.

b) *"Can I factor out a GCF?"* No. To factor $y^2 - y - 30$, think of two numbers whose *product* is -30 and *sum* is -1. The numbers are -6 and 5.

$$y^2 - y - 30 = (y - 6)(y + 5)$$

"Can I factor again?" No. It is completely factored.

c) Look at $mn^2 - 4m + 5n^2 - 20$. *"Can I factor out a GCF?"* No. Notice that this polynomial has *four terms*. When a polynomial has *four terms*, think about *factoring by grouping.*

$$\underbrace{mn^2 - 4m}\;\;\underbrace{+\;5n^2 - 20}$$
$$\qquad\downarrow\qquad\qquad\downarrow$$

$= m(n^2 - 4) + 5(n^2 - 4)$ Take out the common factor from each pair of terms.
$= (n^2 - 4)(m + 5)$ Factor out $(n^2 - 4)$ using the distributive property.

Examine $(n^2 - 4)(m + 5)$ and ask yourself, *"Can I factor again?"* Yes! $(n^2 - 4)$ is the difference of two squares. Factor again.

$$(n^2 - 4)(m + 5) = (n + 2)(n - 2)(m + 5)$$

"Can I factor again?" No. It is completely factored.

$$mn^2 - 4m + 5n^2 - 20 = (n + 2)(n - 2)(m + 5)$$

Seeing four terms is a clue to try factoring by grouping.

d) We cannot take out a GCF from $p^2 - 16p + 64$. It is a trinomial, and notice that the first and last terms are perfect squares. *Is this a perfect square trinomial?*

$$p^2 - 16p + 64$$
$$\downarrow \qquad\qquad \downarrow$$
$$(p)^2 \qquad (8)^2$$

Does the middle term equal $2 \cdot p \cdot (8)$? Yes.

$$2 \cdot p \cdot (8) = 16p$$

We will factor $p^2 - 16p + 64$ using $a^2 - 2ab + b^2 = (a - b)^2$ with $a = p$ and $b = 8$.
Then, $p^2 - 16p + 64 = (p - 8)^2$.
"Can I factor again?" No. It is completely factored.

e) It is tempting to jump right in and try to factor $8x^2 + 26x + 20$ as the product of two binomials, but ask yourself, *"Can I take out a GCF?"* Yes! Factor out 2.

$$8x^2 + 26x + 20 = 2(4x^2 + 13x + 10)$$

"Can I factor again?" Yes.

$$2(4x^2 + 13x + 10) = 2(4x + 5)(x + 2)$$

"Can I factor again?" No.

$$8x^2 + 26x + 20 = 2(4x + 5)(x + 2)$$

f) We cannot take out a GCF from $27k^3 + 8$. Notice that $27k^3 + 8$ has two terms, so think about squares and cubes. Neither term is a perfect square *and* the positive terms are being added, so this *cannot* be the difference of squares.

Is each term a perfect cube? *Yes!* $27k^3 + 8$ is the sum of two cubes. We will factor $27k^3 + 8$ using $a^3 + b^3 = (a + b)(a^2 - ab + b^2)$ with $a = 3k$ and $b = 2$.

$$27k^3 + 8 = (3k + 2)[(3k)^2 - (3k)(2) + (2)^2]$$
$$\downarrow \qquad \downarrow$$
$$(3k)^3 \quad (2)^3$$
$$= (3k + 2)(9k^2 - 6k + 4)$$

"Can I factor again?" No. It is completely factored.

g) Look at $14c^3d + 63cd^2 - 7cd$ and ask yourself, *"Can I factor out a GCF?"* *Yes!* The GCF is $7cd$.

$$14c^3d + 63cd^2 - 7cd = 7cd(2c^2 + 9d - 1)$$

"Can I factor again?" No. It is completely factored.

You Try 1

Factor completely.

a) $3p^2 + p - 10$ b) $2n^3 - n^2 + 12n - 6$ c) $4k^4 + 36k^3 + 32k^2$

d) $48 - 3y^4$ e) $8r^3 - 125$

Answers to You Try Exercises

1) a) $(3p - 5)(p + 2)$ b) $(n^2 + 6)(2n - 1)$ c) $4k^2(k + 8)(k + 1)$

d) $3(4 + y^2)(2 + y)(2 - y)$ e) $(2r - 5)(4r^2 + 10r + 25)$

Mid-Chapter Summary Exercises

Boost your grade at mathzone.com!   MathZone   > Practice Problems   > NetTutor   > e-Professors   > Videos
> Self-Test

Objective 1

Factor completely.

1) $m^2 + 16m + 60$

2) $h^2 - 36$

3) $uv + 6u + 9v + 54$

4) $2y^2 + 5y - 18$

5) $3k^2 - 14k + 8$

6) $n^2 - 14n + 49$

7) $16d^6 + 8d^5 + 72d^4$

8) $b^2 - 3bc - 4c^2$

9) $60w^3 + 70w^2 - 50w$

10) $7c^3 - 7$

11) $t^3 + 1000$

12) $pq - 6p + 4q - 24$

13) $49 - p^2$

14) $h^2 - 15h + 56$

15) $4x^2 + 4xy + y^2$

16) $27c - 18$

17) $3z^4 - 21z^3 - 24z^2$

18) $9a^2 + 6a - 8$

19) $4b^2 + 1$

20) $5abc - 15ac + 10bc - 30c$

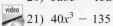

21) $40x^3 - 135$

22) $81z^2 + 36z + 4$

23) $c^2 - \dfrac{1}{4}$

24) $v^2 + 3v + 4$

25) $45s^2t + 4 - 36s^2 - 5t$

26) $12c^5d - 75cd^3$

27) $k^2 + 9km + 18m^2$

28) $64r^3 + 8$

29) $z^2 - 3z - 88$

30) $40f^4g^4 + 8f^3g^3 + 16fg^2$

31) $80y^2 - 40y + 5$

32) $4t^2 - t - 5$

33) $20c^2 + 26cd + 6d^2$

34) $x^2 - \dfrac{9}{49}$

35) $n^4 - 16m^4$

36) $k^2 - 21k + 108$

37) $2a^2 - 10a - 72$

38) $x^2y - 4y + 7x^2 - 28$

39) $r^2 - r + \dfrac{1}{4}$

40) $v^3 - 125$

41) $28gh + 16g - 63h - 36$

42) $-24x^3 + 30x^2 - 9x$

43) $8b^2 - 14b - 15$

44) $50u^2 + 60u + 18$

45) $55a^6b^3 + 35a^5b^3 - 10a^4b - 20a^2b$

46) $64 - u^2$

47) $2d^2 - 9d + 3$

48) $2v^4w + 14v^3w^2 + 12v^2w^3$

49) $9p^2 - 24pq + 16q^2$

50) $c^4 - 16$

51) $30y^2 + 37y - 7$

52) $g^2 + 49$

53) $80a^3 - 270b^3$

54) $26n^6 - 39n^4 + 13n^3$

55) $rt - r - t + 1$

56) $h^2 + 10h + 25$

57) $4g^2 - 4$

58) $25a^2 - 55ab + 24b^2$

59) $3c^2 - 24c + 48$

60) $9t^4 + 64u^2$

61) $144k^2 - 121$

62) $125p^3 - 64q^3$

63) $-48g^2 - 80g - 12$

64) $5d^2 + 60d + 55$

65) $q^3 + 1$

66) $9x^2 + 12x + 4$

67) $81u^4 - v^4$

68) $45v^2 + 9vw^2 + 30vw + 6w^3$

69) $11f^2 + 36f + 9$

70) $4y^3 - 4y^2 - 80y$

71) $2j^{11} - j^3$

72) $d^2 - \dfrac{169}{100}$

73) $w^2 - 2w - 48$

74) $16a^2 - 40a + 25$

75) $k^2 + 100$

76) $24y^3 + 375$

77) $m^2 + 4m + 4$

78) $r^2 - 15r + 54$

79) $9t^2 - 64$

80) $100c^4 - 36c^2$

Extend the concepts of 7.1–7.4 to factor completely.

 81) $(2z + 1)y^2 + 6(2z + 1)y - 55(2z + 1)$

82) $(a + b)c^2 - 5(a + b)c - 24(a + b)$

83) $(r - 4)^2 + 11(r - 4) + 28$

84) $(n + 3)^2 - 2(n + 3) - 35$

85) $(x + y)^2 - (2x - y)^2$ 86) $(3s - t)^2 - (2s + t)^2$

87) $n^2 + 12n + 36 - p^2$ 88) $a^2 + 2ab + b^2 - c^2$

Section 7.5 Solving Quadratic Equations by Factoring

Objectives

1. Solve a Quadratic Equation of the Form $ab = 0$ Where a and b Are Linear Factors

2. Solve $ax^2 + bx + c = 0$ by Factoring

3. Solve Other Quadratic Equations by Factoring

4. Solve Higher Degree Equations by Factoring

In Section 3.1 we began our study of linear equations in one variable. A *linear equation in one variable* is an equation that can be written in the form $ax + b = 0$, where a and b are real numbers and $a \neq 0$.

In this section, we will learn how to solve *quadratic equations*.

Definition	A **quadratic equation** can be written in the form $ax^2 + bx + c = 0$, where a, b, and c are real numbers and $a \neq 0$.

When a quadratic equation is written in the form $ax^2 + bx + c = 0$, we say that it is in **standard form**. But quadratic equations can be written in other forms too.

Some examples of quadratic equations are

$$x^2 + 12x + 27 = 0, \quad 2p(p - 5) = 0, \quad \text{and} \quad (c + 1)(c - 8) = 3.$$

Quadratic equations are also called *second-degree equations* because the highest power on the variable is 2.

There are many different ways to solve quadratic equations. In this section, we will learn how to solve them by factoring; other methods will be discussed later in this book.

Solving a quadratic equation by factoring is based on the *zero product rule*. The zero product rule states that if the product of two quantities is zero, then one or both of the quantities is zero.

For example, if $5y = 0$, then $y = 0$. If $p \cdot 4 = 0$, then $p = 0$. Or, if $ab = 0$, then either $a = 0$, $b = 0$, or *both* a and b equal zero.

Definition	**Zero product rule:** If $ab = 0$, then $a = 0$ or $b = 0$.

We will use this idea to solve quadratic equations by factoring.

1. Solve a Quadratic Equation of the Form $ab = 0$ Where a and b Are Linear Factors

Example 1

Solve $x(x + 8) = 0$.

Solution

Since the expression on the left is the product of two quantities x and $(x + 8)$, at least one of the quantities must equal zero for the *product* to equal zero.

$$x(x + 8) = 0$$

$$x = 0 \quad \text{or} \quad x + 8 = 0 \qquad \text{Set each factor equal to 0.}$$
$$x = -8 \qquad \text{Solve.}$$

Check in the original equation:

If $x = 0$:
$$0(0 + 8) \overset{?}{=} 0$$
$$0(8) = 0 \checkmark$$

If $x = -8$:
$$-8(-8 + 8) \overset{?}{=} 0$$
$$-8(0) = 0 \checkmark$$

The solution set is $\{-8, 0\}$.

 You Try 1

Solve $c(c - 9) = 0$.

2. Solve $ax^2 + bx + c = 0$ by Factoring

If the equation is set equal to 0 but the expression is not factored, begin by factoring the expression.

Example 2

Solve $m^2 - 6m - 40 = 0$.

Solution

$$m^2 - 6m - 40 = 0$$
$$(m - 10)(m + 4) = 0 \qquad \text{Factor.}$$

$$m - 10 = 0 \quad \text{or} \quad m + 4 = 0 \qquad \text{Set each factor equal to zero.}$$
$$m = 10 \quad \text{or} \quad m = -4 \qquad \text{Solve.}$$

Check in the original equation:

If $m = 10$:
$$(10)^2 - 6(10) - 40 \overset{?}{=} 0$$
$$100 - 60 - 40 = 0 \checkmark$$

If $m = -4$:
$$(-4)^2 - 6(-4) - 40 \overset{?}{=} 0$$
$$16 + 24 - 40 = 0 \checkmark$$

The solution set is $\{-4, 10\}$.

Here are the steps to use to solve a quadratic equation by factoring:

Solving a Quadratic Equation by Factoring

1) Write the equation in the form $ax^2 + bx + c = 0$ (standard form) so that all terms are on one side of the equal sign and zero is on the other side.

2) Factor the expression.

3) Set each factor equal to zero, and solve for the variable.

4) Check the answer(s).

You Try 2

Solve $h^2 + 9h + 18 = 0$.

---

3. Solve Other Quadratic Equations by Factoring

Example 3

Solve each equation by factoring.

a) $2r^2 + 3r = 20$ b) $6d^2 = -42d$ c) $k^2 = -12(k + 3)$

d) $2(x^2 + 5) + 5x = 6x(x - 1) + 16$ e) $(z - 8)(z - 4) = 5$

Solution

a) Begin by writing $2r^2 + 3r = 20$ in standard form, $ar^2 + br + c = 0$.

$$2r^2 + 3r - 20 = 0 \qquad \text{Standard form}$$
$$(2r - 5)(r + 4) = 0 \qquad \text{Factor.}$$

$$2r - 5 = 0 \quad \text{or} \quad r + 4 = 0 \qquad \text{Set each factor equal to zero.}$$
$$2r = 5$$
$$r = \frac{5}{2} \quad \text{or} \qquad r = -4 \qquad \text{Solve.}$$

Check in original equation:

If $r = \dfrac{5}{2}$:

$$2\left(\frac{5}{2}\right)^2 + 3\left(\frac{5}{2}\right) \stackrel{?}{=} 20$$

$$2\left(\frac{25}{4}\right) + \frac{15}{2} \stackrel{?}{=} 20$$

$$\frac{25}{2} + \frac{15}{2} \stackrel{?}{=} 20$$

$$\frac{40}{2} = 20 \quad \checkmark$$

If $r = -4$:

$$2(-4)^2 + 3(-4) \stackrel{?}{=} 20$$
$$2(16) - 12 \stackrel{?}{=} 20$$
$$32 - 12 = 20 \quad \checkmark$$

The solution set is is $\left\{-4, \dfrac{5}{2}\right\}$.

b)   Write $6d^2 = -42d$ in standard form.

$$6d^2 + 42d = 0 \qquad \text{Standard form}$$
$$6d(d + 7) = 0 \qquad \text{Factor.}$$

$$6d = 0 \quad \text{or} \quad d + 7 = 0 \qquad \text{Set each factor equal to zero.}$$
$$d = 0 \quad \text{or} \qquad d = -7 \qquad \text{Solve.}$$

The check is left to the student.

The solution set is $\{-7, 0\}$.

Since both terms in $6d^2 = -42d$ are divisible by 6, we could have started this problem by dividing by 6:

$$\frac{6d^2}{6} = \frac{-42d}{6} \qquad \text{Divide by 6.}$$
$$d^2 = -7d$$
$$d^2 + 7d = 0 \qquad \text{Write in standard form.}$$
$$d(d + 7) = 0 \qquad \text{Factor.}$$

$$d = 0 \quad \text{or} \quad d + 7 = 0 \qquad \text{Set each factor equal to zero.}$$
$$d = -7 \qquad \text{Solve.}$$

The solution set is $\{-7, 0\}$. We get the same result.

BE CAREFUL   We cannot divide by d even though each term contains a factor of d. Doing so would eliminate the solution of zero. *In general, we can divide an equation by a nonzero real number but we cannot divide an equation by a variable because we may eliminate a solution, and we may be dividing by zero.*

c)   To solve $k^2 = -12(k + 3)$, begin by writing the equation in standard form.

$$k^2 = -12k - 36 \qquad \text{Distribute.}$$
$$k^2 + 12k + 36 = 0 \qquad \text{Write in standard form.}$$
$$(k + 6)^2 = 0 \qquad \text{Factor.}$$

Since $(k + 6)^2 = 0$ means $(k + 6)(k + 6) = 0$, setting each factor equal to zero will result in the same value for k.

$$k + 6 = 0 \qquad \text{Set } k + 6 = 0.$$
$$k = -6 \qquad \text{Solve.}$$

The check is left to the student.

The solution set is $\{-6\}$.

d)   We will have to perform several steps to write the equation in standard form.

$$2(x^2 + 5) + 5x = 6x(x - 1) + 16$$
$$2x^2 + 10 + 5x = 6x^2 - 6x + 16 \qquad \text{Distribute.}$$

Move the terms on the left side of the equation to the right side so that the coefficient of x^2 is positive.

$$0 = 4x^2 - 11x + 6 \qquad \text{Write in standard form.}$$
$$0 = (4x - 3)(x - 2) \qquad \text{Factor.}$$

$$4x - 3 = 0 \quad \text{or} \quad x - 2 = 0 \qquad \text{Set each factor equal to zero.}$$
$$4x = 3$$
$$x = \frac{3}{4} \quad \text{or} \quad x = 2 \qquad \text{Solve.}$$

The check is left to the student.

The solution set is $\left\{\dfrac{3}{4}, 2\right\}$.

e)   It is tempting to solve $(z - 8)(z - 4) = 5$ like this:

$$(z - 8)(z - 4) = 5$$

$$z - 8 = 5 \quad \text{or} \quad z - 4 = 5$$

This is incorrect!

One side of the equation must equal zero in order to set each factor equal to zero. Begin by multiplying on the left.

$$(z - 8)(z - 4) = 5$$
$$z^2 - 12z + 32 = 5 \qquad \text{Multiply using FOIL.}$$
$$z^2 - 12z + 27 = 0 \qquad \text{Standard form.}$$
$$(z - 9)(z - 3) = 0 \qquad \text{Factor.}$$

$$z - 9 = 0 \quad \text{or} \quad z - 3 = 0 \qquad \text{Set each factor equal to zero.}$$
$$z = 9 \quad \text{or} \quad z = 3 \qquad \text{Solve.}$$

The check is left to the student.
The solution set is $\{3, 9\}$.

You Try 3

Solve.

a) $w^2 + 4w - 5 = 0$ b) $29b = 5(b^2 + 4)$ c) $(a + 6)(a + 4) = 3$
d) $t^2 = 8t$ e) $(2y + 1)^2 + 5 = y^2 + 2(y + 7)$

4. Solve Higher Degree Equations by Factoring

Sometimes, equations that are not quadratics can be solved by factoring as well.

Example 4

Solve $(2x - 1)(x^2 - 9x - 22) = 0$.

Solution

This is *not* a quadratic equation because if we multiplied the factors on the left we would get $2x^3 - 19x^2 - 35x + 22 = 0$. This is a *cubic* equation because the degree of the polynomial on the left is 3.

The original equation is the product of two factors so we can use the zero product rule to solve this equation.

$$(2x - 1)(x^2 - 9x - 22) = 0$$
$$(2x - 1)(x - 11)(x + 2) = 0 \qquad \text{Factor.}$$

$2x - 1 = 0 \quad$ or $\quad x - 11 = 0 \quad$ or $\quad x + 2 = 0 \qquad$ Set each factor equal to zero.
$\qquad 2x = 1$

$x = \dfrac{1}{2} \quad$ or $\qquad\qquad x = 11 \quad$ or $\qquad x = -2 \qquad$ Solve.

The check is left to the student.

The solution set is $\left\{ -2, \dfrac{1}{2}, 11 \right\}$.

You Try 4

Solve.

a) $(c + 10)(2c^2 + 5c - 7) = 0$ b) $r^4 = 25r^2$

In this section, it was possible to solve all of the equations by factoring. Below we show the relationship between solving a quadratic equation by factoring and solving it using a graphing calculator. In Chapter 11 we will learn other methods for solving quadratic equations.

Using Technology

In this section, we learned how to solve a quadratic equation by factoring. We can use a graphing calculator to solve a quadratic equation as well. Let's see how the two are related by using the equation $x^2 - x - 6 = 0$.

First, solve $x^2 - x - 6 = 0$ by factoring.

$$x^2 - x - 6 = 0$$
$$(x + 2)(x - 3) = 0$$

$x + 2 = 0 \quad$ or $\quad x - 3 = 0$
$\quad x = -2 \qquad\qquad x = 3$

The solution set is $\{-2, 3\}$.

Next, solve $x^2 - x - 6 = 0$ using a graphing calculator.

Recall from Chapter 4 that to find the x-intercepts of the graph of an equation we let $y = 0$ and solve the equation for x. If we let $y = x^2 - x - 6$, then solving $x^2 - x - 6 = 0$ is the same as finding the x-intercepts of the graph of $y = x^2 - x - 6$. x-intercepts are also called *zeroes* of the equation since they are the values of x that make $y = 0$.

Enter $y = x^2 - x - 6$ into the calculator, and press GRAPH. We obtain a graph called a parabola, and we can see that it has two x-intercepts. (We will study parabolas in more detail in Chapter 12.) If the scale for each tick mark on the graph is 1, then it appears that the x-intercepts are -2 and 3. To verify this, we can press 2nd TRACE and use the down arrow to highlight 2:zero and press ENTER. Move the cursor to the left of one of the intercepts and press ENTER, then move the cursor again, so that it is to the right of the same intercept and press ENTER. Press ENTER one more time, and the calculator will reveal the intercept and, therefore, the solution to the equation. Use the cursor to "surround" the other intercept and find its value.

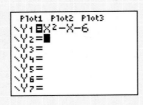   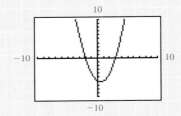

1. Solve $x^2 + 10x + 21 = 0$ by factoring.

2. Here is the graph of $y = x^2 + 10x + 21$. Each tick mark represents 1 unit. What are the zeroes of the equation?

3. Are the zeroes of $y = x^2 + 10x + 21$ the same as the values in the solution set of $x^2 + 10x + 21 = 0$?

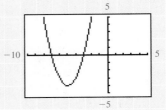

4. Solve $x^2 - 6x - 16 = 0$ by factoring.

5. Here is the graph of $y = x^2 - 6x - 16$. Each tick mark represents 2 units. What are the zeroes of the equation?

6. Are the zeroes of $y = x^2 - 6x - 16$ the same as the values in the solution set of $x^2 - 6x - 16 = 0$?

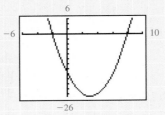

Answers to You Try Exercises

1) $\{0, 9\}$ 2) $\{-6, -3\}$ 3) a) $(-5, 1)$ b) $\left\{\frac{4}{5}, 5\right\}$ c) $\{-7, -3\}$ d) $\{0, 8\}$ e) $\left\{-2, \frac{4}{3}\right\}$

4) a) $\left\{-10, -\frac{7}{2}, 1\right\}$ b) $\{0, 5, -5\}$

Answers to Technology Exercises

1. $\{-7, -3\}$ 2. -7 and -3 3. yes   4. $\{-2, 8\}$ 5. -2 and 8 6. yes

7.5 Exercises

Boost your grade at mathzone.com!   MathZone

> Practice Problems   > NetTutor   > e-Professors   > Videos
> Self-Test

Objective I

1) Explain the zero product rule.

2) Can we solve $(y + 6)(y - 11) = 8$ by setting each factor equal to 8 like this: $y + 6 = 8$ or $y - 11 = 8$? Why or why not?

Solve each equation.

3) $(m + 9)(m - 8) = 0$ 4) $(a + 10)(a + 4) = 0$

5) $(q - 4)(q - 7) = 0$ 6) $(x - 5)(x + 2) = 0$

7) $(4z + 3)(z - 9) = 0$ 8) $(2n + 1)(n - 13) = 0$

9) $11s(s + 15) = 0$ 10) $-5r(r - 8) = 0$

11) $(6x - 5)^2 = 0$ 12) $(d + 7)^2 = 0$

13) $(4h + 7)(h + 3) = 0$

14) $(8p - 5)(3p - 11) = 0$

15) $\left(y + \dfrac{3}{2}\right)\left(y - \dfrac{1}{4}\right) = 0$

16) $\left(t - \dfrac{9}{8}\right)\left(t + \dfrac{5}{6}\right) = 0$

17) $q(q - 2.5) = 0$

18) $w(w + 0.8) = 0$

Objectives 2 and 3

Solve each equation.

19) $v^2 + 15v + 56 = 0$

20) $y^2 + 2y - 35 = 0$

21) $k^2 + 12k - 45 = 0$

22) $z^2 - 12z + 11 = 0$

23) $3y^2 - y - 10 = 0$

24) $4f^2 - 15f + 14 = 0$

25) $14w^2 + 8w = 0$

26) $10a^2 + 20a = 0$

27) $d^2 - 15d = -54$

28) $j^2 + 11j = -28$

29) $t^2 - 49 = 0$

30) $k^2 - 100 = 0$

31) $36 = 25n^2$

32) $16 = 169p^2$

33) $m^2 = 60 - 7m$

34) $g^2 + 20 = 12g$

35) $55w = -20w^2 - 30$

36) $4v = 14v^2 - 48$

37) $p^2 = 11p$ 38) $d^2 = d$

39) $45k + 27 = 18k^2$

40) $104r + 36 = 12r^2$

41) $b(b - 4) = 96$

42) $54 = w(15 - w)$

43) $-63 = 4j(j - 8)$

44) $g(3g + 11) = 70$

45) $10x(x + 1) - 6x = 9(x^2 + 5)$

46) $5r(3r + 7) = 2(4r^2 - 21)$

47) $3(h^2 - 4) = 5h(h - 1) - 9h$

48) $5(5 + u^2) + 10 = 3u(2u + 1) - u$

49) $\dfrac{1}{2}(m + 1)^2 = -\dfrac{3}{4}m(m + 5) - \dfrac{5}{2}$

50) $(2y - 3)^2 + y = (y - 5)^2 - 6$

51) $3t(t - 5) + 14 = 5 - t(t + 3)$

52) $\dfrac{1}{2}c(2 - c) - \dfrac{3}{2} = \dfrac{2}{5}c(c + 1) - \dfrac{7}{5}$

53) $33 = -m(14 + m)$

54) $-84 = s(s + 19)$

55) $(3w + 2)^2 - (w - 5)^2 = 0$

56) $(2j - 7)^2 - (j + 3)^2 = 0$

57) $(q + 3)^2 - (2q - 5)^2 = 0$

58) $(6n + 5)^2 - (3n + 4)^2 = 0$

Objective 4

The following equations are not quadratic but can be solved by factoring and applying the zero product rule. Solve each equation.

59) $8y(y + 4)(2y - 1) = 0$

60) $-13b(12b + 7)(b - 11) = 0$

61) $(9p - 2)(p^2 - 10p - 11) = 0$

62) $(4f + 5)(f^2 - 3f - 18) = 0$

63) $(2r - 5)(r^2 - 6r + 9) = 0$

64) $(3x - 1)(x^2 - 16x + 64) = 0$

65) $m^3 = 64m$

66) $r^3 = 81r$

67) $5w^2 + 36w = w^3$

68) $14a^2 - 49a = a^3$

69) $2g^3 = 120g - 14g^2$

70) $36z - 24z^2 = -3z^3$

71) $45h = 20h^3$

72) $64d^3 = 100d$

73) $2s^2(3s + 2) + 3s(3s + 2) - 35(3s + 2) = 0$

74) $10n^2(n - 8) + n(n - 8) - 2(n - 8) = 0$

75) $10a^2(4a + 3) + 2(4a + 3) = 9a(4a + 3)$

76) $12d^2(7d - 3) = 5d(7d - 3) + 2(7d - 3)$

Find the indicated values for the following polynomial functions.

77) $f(x) = x^2 + 10x + 21$. Find x so that $f(x) = 0$.

78) $h(t) = t^2 - 6t - 16$. Find t so that $h(t) = 0$.

79) $g(a) = 2a^2 - 13a + 24$. Find a so that $g(a) = 4$.

80) $Q(x) = 4x^2 - 4x + 9$. Find x so that $Q(x) = 8$.

81) $H(b) = b^2 + 3$. Find b so that $H(b) = 19$.

82) $f(z) = z^3 + 3z^2 - 54z + 5$. Find z so that $f(z) = 5$.

83) $h(k) = 5k^3 - 25k^2 + 20k$. Find k so that $h(k) = 0$.

84) $g(x) = 9x^2 - 10$. Find x so that $g(x) = -6$.

Section 7.6 Applications of Quadratic Equations

Objectives

1. Solve Problems Involving Geometry

2. Solve Problems Involving Consecutive Integers

3. Learn the Pythagorean Theorem

4. Solve an Applied Problem Using the Pythagorean Theorem

5. Solve an Applied Problem Using a Given Quadratic Equation

In Chapters 3 and 5 we explored applications of linear equations. In this section we will look at applications involving quadratic equations. Let's begin by restating the five Steps for Solving Applied Problems.

Steps for Solving Applied Problems

Step 1: Read and reread the problem. Draw a picture if applicable.

Step 2: Identify what you are being asked to find. *Define the variable*: assign a variable to represent an unknown quantity. Also,
- if there are other unknown quantities, define them in terms of the variable.
- label the picture with the variable and other unknowns as well as with any given information in the problem.

Step 3: Translate from English to math.

Step 4: Solve the equation.

Step 5: Interpret the meaning of the solution as it relates to the problem. Be sure the answer makes sense in the context of the problem.

1. Solve Problems Involving Geometry

Example 1

Solve. A rectangular vegetable garden is 7 ft longer than it is wide. What are the dimensions of the garden if it covers 60 ft^2?

Solution

Step 1: Read the problem carefully, twice. Draw a picture.

Step 2: We must find the length and width of the garden. We are told that the *area* is 60 ft^2. (The *square units* help us to identify this number as the area.) Since the length is defined in terms of the width, let

$$w = \text{width of the garden}$$
$$w + 7 = \text{length of the garden}$$

Label the picture.

Step 3: Translate from English to math. What is the formula for the area of a rectangle?

$$(\text{Length}) \cdot (\text{Width}) = \text{Area}$$
$$(w + 7) \cdot \quad w \quad = 60$$

The equation is $(w + 7)w = 60$.

Step 4: Solve the equation.

$$
\begin{aligned}
(w + 7)w &= 60 & \\
w^2 + 7w &= 60 & \text{Distribute.} \\
w^2 + 7w - 60 &= 0 & \text{Write in standard form.} \\
(w + 12)(w - 5) &= 0 & \text{Factor.} \\
w + 12 = 0 \quad \text{or} \quad w - 5 &= 0 & \text{Set each factor equal to zero.} \\
w = -12 \quad \text{or} \quad w &= 5 & \text{Solve.}
\end{aligned}
$$

Step 5: Interpret the meaning of the solution as it relates to the problem.

Since w represents the width of the garden, $w = -12$ *cannot* be an answer. But, $w = 5$ makes sense giving us that the width of the rectangle is 5 ft and the length is $5 + 7 = 12$ ft.

Check by multiplying: $(12 \text{ ft})(5 \text{ ft}) = 60 \text{ ft}^2$.

The length is 12 ft and the width is 5 ft.

You Try 1

Solve. The area of the surface of a desk is 8 ft². Find the dimensions of the desktop if the width is 2 ft less than the length.

2. Solve Problems Involving Consecutive Integers

In Chapter 3 we solved problems involving consecutive integers. Some applications involving consecutive integers lead to quadratic equations.

Example 2

Solve. Twice the sum of three consecutive odd integers is 9 less than the product of the smaller two. Find the integers.

Solution

Step 1:   Read the problem carefully, twice.

Step 2:   Find three consecutive odd integers.

$$x = \text{first odd integer}$$
$$x + 2 = \text{second odd integer}$$
$$x + 4 = \text{third odd integer}$$

Step 3:   Translate from English to math. Think of the information in small parts.

In English	Meaning	In Math Terms
Twice . . .	Two times	$2 \cdot$
the sum of three consecutive odd integers . . .	add three consecutive odd integers	$[x + (x + 2) + (x + 4)]$
is . . .	equals	$=$
9 less than . . .	subtract 9 from the quantity below	-9
the product of the smaller two	multiply the smaller two odd integers	$\uparrow$ $x(x + 2)$

The equation is $2[x + (x + 2) + (x + 4)] = x(x + 2) - 9$.

Step 4:   Solve the equation.

$$
\begin{aligned}
2[x + (x + 2) + (x + 4)] &= x(x + 2) - 9 \\
2(3x + 6) &= x(x + 2) - 9 && \text{Combine like terms.} \\
6x + 12 &= x^2 + 2x - 9 && \text{Distribute.} \\
0 &= x^2 - 4x - 21 && \text{Write in standard form.} \\
0 &= (x - 7)(x + 3) && \text{Factor.}
\end{aligned}
$$

$$x - 7 = 0 \quad \text{or} \quad x + 3 = 0 \qquad \text{Set each factor equal to 0.}$$
$$x = 7 \quad \text{or} \qquad x = -3 \qquad \text{Solve.}$$

Step 5: Interpret the meaning of the solution as it relates to the problem.

If $x = 7$, then $x + 2 = 7 + 2 = 9$ and $x + 4 = 7 + 4 = 11$.
If $x = -3$, then $x + 2 = -3 + 2 = -1$ and $x + 4 = -3 + 4 = 1$.
The odd integers are 7, 9, 11 or $-3, -1, 1$. Check the values in the equation.

 You Try 2

Solve. Find three consecutive even integers such that the product of the two smaller numbers is the same as twice the sum of the integers.

3. Learn the Pythagorean Theorem

A **right triangle** is a triangle that contains a 90° **(right)** angle. We label a right triangle as follows.

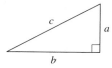

The side opposite the 90° angle is the longest side of the triangle and is called the **hypotenuse**. The other two sides are called the **legs**. The Pythagorean theorem states a relationship between the lengths of the sides of a right triangle. This is a very important relationship in mathematics and is one which is used in many different ways.

Pythagorean Theorem

Given a right triangle with legs of length a and b and hypotenuse of length c,

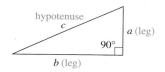

the Pythagorean Theorem states that $a^2 + b^2 = c^2$ (or $(\text{leg})^2 + (\text{leg})^2 = (\text{hypotenuse})^2$).

The Pythagorean theorem is true *only* for right triangles.

Example 3

Find the length of the missing side.

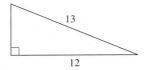

Solution

Since this is a right triangle, we can use the Pythagorean theorem to find the length of the side. Let a represent its length, and label the triangle.

The length of the hypotenuse is 13, so $c = 13$. a and 12 are legs. Let $b = 12$.

$$a^2 + b^2 = c^2 \qquad \text{Pythagorean theorem}$$
$$a^2 + (12)^2 = (13)^2 \qquad \text{Substitute values.}$$
$$a^2 + 144 = 169$$
$$a^2 - 25 = 0 \qquad \text{Write the equation in standard form.}$$
$$(a + 5)(a - 5) = 0 \qquad \text{Factor.}$$

$$a + 5 = 0 \quad \text{or} \quad a - 5 = 0 \qquad \text{Set each factor equal to 0.}$$
$$a = -5 \quad \text{or} \qquad a = 5 \qquad \text{Solve.}$$

$a = -5$ does not make sense as an answer because the length of a side of a triangle cannot be negative.

Therefore, $a = 5$.

Check: $5^2 + (12)^2 \overset{?}{=} (13)^2$
$$25 + 144 = 169 \quad \checkmark$$

 You Try 3

Find the length of the missing side.

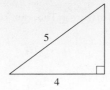

4. Solve an Applied Problem Using the Pythagorean Theorem

Example 4

Solve. An animal holding pen situated between two buildings will have walls as two of its sides and a fence on the longest side. The side with the fence is 20 ft longer than the shortest side, while the third side is 10 ft longer than the shortest side. Find the length of the fence.

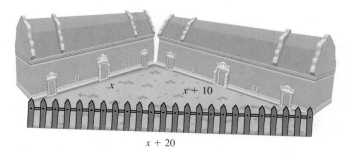

Solution

Step 1: Read the problem carefully, twice.

Step 2: Find the length of the fence.

$$x = \text{length of shortest side}$$
$$x + 10 = \text{length of the side along the other building}$$
$$x + 20 = \text{length of the fence}$$

Label the picture.

Step 3: Translate from English to math.

The pen is in the shape of a right triangle. The hypotenuse is $x + 20$ since it is the side across from the right angle. The legs are x and $x + 10$. From the Pythagorean theorem we get

$$a^2 + b^2 = c^2 \qquad \text{Pythagorean theorem}$$
$$x^2 + (x + 10)^2 = (x + 20)^2 \qquad \text{Substitute.}$$

Step 4: Solve the equation.

$$x^2 + (x + 10)^2 = (x + 20)^2$$
$$x^2 + x^2 + 20x + 100 = x^2 + 40x + 400 \qquad \text{Multiply using FOIL.}$$
$$2x^2 + 20x + 100 = x^2 + 40x + 400$$
$$x^2 - 20x - 300 = 0 \qquad \text{Write in standard form.}$$
$$(x - 30)(x + 10) = 0 \qquad \text{Factor.}$$

$$x - 30 = 0 \quad \text{or} \quad x + 10 = 0 \qquad \text{Set each factor equal to 0.}$$
$$x = 30 \quad \text{or} \qquad x = -10 \qquad \text{Solve.}$$

Step 5: Interpret the meaning of the solution as it relates to the problem.

x represents the length of the shortest side, so x cannot equal -10. Therefore, the length of the shortest side must be 30 ft.

The length of the fence $= x + 20$, so $30 + 20 = 50$ ft. The length of the fence is 50 ft.

Check: $x + 10 = 30 + 10 = 40$ ft $=$ length of the other leg.

$$a^2 + b^2 = c^2$$
$$(30)^2 + (40)^2 \overset{?}{=} (50)^2$$
$$900 + 1600 = 2500 \quad \checkmark$$

You Try 4

Solve. A wire is attached to the top of a pole. The wire is 4 ft longer than the pole, and the distance from the wire on the ground to the bottom of the pole is 4 ft less than the height of the pole. Find the length of the wire and the height of the pole.

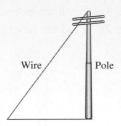

Next we will see how to use quadratic equations that model real-life situations.

5. Solve an Applied Problem Using a Given Quadratic Equation

Example 5

A Little League baseball player throws a ball upward. The height h of the ball (in feet) t sec after the ball is released is given by the quadratic equation

$$h = -16t^2 + 30t + 4$$

a) What is the initial height of the ball?

b) How long does it take the ball to reach a height of 18 ft?

c) How long does it take for the ball to hit the ground?

Solution

a) We are asked to find the height at which the ball is released. Since t represents the number of seconds after the ball is thrown, $t = 0$ at the time of release.

Let $t = 0$ and solve for h.

$$h = -16(0)^2 + 30(0) + 4 \qquad \text{Substitute 0 for } t.$$
$$= 0 + 0 + 4$$
$$= 4$$

The initial height of the ball is 4 ft.

b) We must find the *time* it takes for the ball to reach a height of 18 ft.

Find t when $h = 18$.

$$h = -16t^2 + 30t + 4$$
$$18 = -16t^2 + 30t + 4 \qquad \text{Substitute 18 for } h.$$
$$0 = -16t^2 + 30t - 14 \qquad \text{Write in standard form.}$$
$$0 = 8t^2 - 15t + 7 \qquad \text{Divide by } -2.$$
$$0 = (8t - 7)(t - 1) \qquad \text{Factor.}$$

$$8t - 7 = 0 \quad \text{or} \quad t - 1 = 0 \qquad \text{Set each factor equal to 0.}$$
$$8t = 7$$

$$t = \frac{7}{8} \quad \text{or} \qquad \quad t = 1 \qquad \text{Solve.}$$

How can two answers be possible? After $\dfrac{7}{8}$ sec the ball is 18 ft above the ground *on its way up,* and after 1 sec, the ball is 18 ft above the ground *on its way down.* The ball reaches a height of 18 ft after $\dfrac{7}{8}$ sec *and* after 1 sec.

c) We must determine the amount of time it takes for the ball to hit the ground. When the ball hits the ground, how high off of the ground is it? *It is 0 ft high.* Find t when $h = 0$.

$$h = -16t^2 + 30t + 4$$
$$0 = -16t^2 + 30t + 4 \qquad \text{Substitute 0 for } h.$$
$$0 = 8t^2 - 15t - 2 \qquad \text{Divide by } -2.$$
$$0 = (8t + 1)(t - 2) \qquad \text{Factor.}$$

$$8t + 1 = 0 \quad \text{or} \quad t - 2 = 0 \qquad \text{Set each factor equal to 0.}$$
$$8t = -1$$
$$t = -\dfrac{1}{8} \quad \text{or} \qquad t = 2 \qquad \text{Solve.}$$

Since t represents time, t cannot equal $-\dfrac{1}{8}$. We reject that as a solution.

Therefore, $t = 2$.

The ball will hit the ground after 2 sec.

Note: In Example 5, the equation can also be written using function notation $h(t) = -16t^2 + 30t + 4$ since the expression $-16t^2 + 30t + 4$ is a polynomial. Furthermore, $h(t) = -16t^2 + 30t + 4$ is a *quadratic function,* and we say that the height, h, is a function of the time, t. We will study quadratic functions in more detail in Chapter 12.

You Try 5

An object is thrown upward off of a building. The height h of the object (in feet) t sec after the object is released is given by the quadratic equation

$$h = -16t^2 + 36t + 36.$$

a) What is the initial height of the object?

b) How long does it take the object to reach a height of 44 ft?

c) How long does it take for the object to hit the ground?

Answers to You Try Exercises

1) width = 2 ft; length = 4 ft   2) 6, 8, 10 or $-2, 0, 2$ 3) 3

4) length of wire = 20 ft; height of pole = 16 ft

5) a) 36 ft   b) 0.25 sec and 2 sec   c) 3 sec

7.6 Exercises

Objective I

Find the length and width of each rectangle.

1) Area = 36 in^2

$x - 9$

x

2) Area = 40 cm^2

$x - 1$

$x + 2$

Find the base and height of each triangle.

3) Area = 12 cm^2

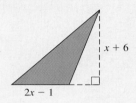

$x + 6$

$2x - 1$

4) Area = 42 in^2

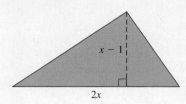

$x - 1$

$2x$

Find the base and height of each parallelogram.

5) Area = 18 in^2

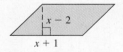

$x - 2$

$x + 1$

6) Area = 50 cm^2

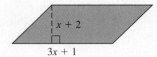

$x + 2$

$3x + 1$

7) The volume of the box is 240 in^3. Find its length and width.

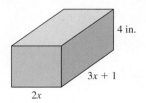

4 in.

$3x + 1$

$2x$

8) The volume of the box is 120 in^3. Find its width and height.

$x - 2$

8 in.

x

Write an equation and solve.

 video

9) A rectangular rug is 4 ft longer than it is wide. If its area is 45 ft^2, what is its length and width?

10) The surface of a rectangular bulletin board has an area of 300 in^2. Find its dimensions if it is 5 in. longer than it is wide.

11) Judy makes stained glass windows. She needs to cut a rectangular piece of glass with an area of 54 in^2 so that its width is 3 in. less than its length. Find the dimensions of the glass she must cut.

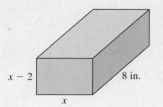

12) A rectangular painting is twice as long as it is wide. Find its dimensions if it has an area of 12.5 ft^2.

13) The volume of a rectangular storage box is 1440 in^3. It is 20 in. long, and it is half as tall as it is wide. Find the width and height of the box.

14) A rectangular aquarium is 15 in. high, and its length is 8 in. more than its width. Find the length and width if the volume of the aquarium is 3600 in^3.

15) The height of a triangle is 3 cm more than its base. Find the height and base if its area is 35 cm^2.

16) The area of a triangle is 16 cm^2. Find the height and base if its height is half the length of the base.

Objective 2

Write an equation and solve.

17) The product of two consecutive odd integers is 1 less than three times their sum. Find the integers.

18) The product of two consecutive integers is 19 more than their sum. Find the integers.

19) Find three consecutive even integers such that the sum of the smaller two is one-fourth the product of the second and third integers.

20) Find three consecutive integers such that the square of the smallest is 29 less than the product of the larger two.

21) Find three consecutive integers such that the square of the largest is 22 more than the product of the smaller two.

22) Find three consecutive odd integers such that the product of the smaller two is 15 more than four times the sum of the three integers.

Objective 3

 23) In your own words, explain the Pythagorean theorem.

24) Can the Pythagorean theorem be used to find *a* in this triangle? Why or why not?

Use the Pythagorean theorem to find the length of the missing side.

 25)

26)

27)

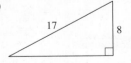

28)

29)

30)

Objective 4

Write an equation and solve.

31) The hypotenuse of a right triangle is 2 in. longer than the longer leg. The shorter leg measures 2 in. less than the longer leg. Find the measure of the longer leg of the triangle.

32) The longer leg of a right triangle is 7 cm more than the shorter leg. The length of the hypotenuse is 3 cm more than twice the length of the shorter leg. Find the length of the hypotenuse.

 33) A 13-ft ladder is leaning against a wall. The distance from the top of the ladder to the bottom of the wall is 7 ft more than the distance from the bottom of the ladder to the wall. Find the distance from the bottom of the ladder to the wall.

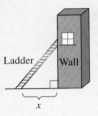

34) A wire is attached to the top of a pole. The pole is 2 ft shorter than the wire, and the distance from the wire on the ground to the bottom of the pole is 9 ft less than the length of the wire. Find the length of the wire and the height of the pole.

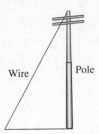

35) From a bike shop, Rana pedals due north while Yasmeen rides due west. When Yasmeen is 4 mi from the shop, the distance between her and Rana is two miles more than Rana's distance from the bike shop. Find the distance between Rana and Yasmeen.

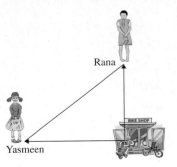

36) Henry and Allison leave home to go to work. Henry drives due west while his wife drives due south. At 8:30 am, Allison is 3 mi farther from home than Henry, and the distance between them is 6 mi more than Henry's distance from home. Find Henry's distance from his house.

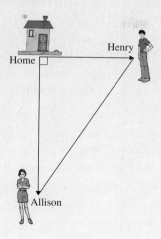

Objective 5

Solve.

37) A rock is dropped from a cliff and into the ocean. The height h (in feet) of the rock after t sec is given by $h = -16t^2 + 144$.

a)   What is the initial height of the rock?

b)   When is the rock 80 ft above the water?

c)   How long does it take the rock to hit the water?

38) An object is launched from a platform with an initial velocity of 32 ft/sec. The height h (in feet) of the object after t sec is given by $h = -16t^2 + 32t + 20$.

a)   What is the initial height of the object?

b)   When is the object 32 ft above the ground?

c)   How long does it take for the object to hit the ground?

39) Organizers of fireworks shows use quadratic and linear equations to help them design their programs. *Shells* contain the chemicals which produce the bursts we see in the sky. At a fireworks show the shells are shot from *mortars* and when the chemicals inside the shells ignite they explode, producing the brilliant bursts we see in the night sky. Shell size determines how high a shell will travel before exploding and how big its burst will be when it does explode. Large shell sizes go higher and produce larger bursts than small shell sizes. Pyrotechnicians take these factors into account when designing the shows so that they can determine the size of the safe zone for spectators and so that shows can be synchronized with music.

At a fireworks show, a 3-in. shell is shot from a mortar at an angle of 75°. The height, y (in feet), of the shell t sec after being shot from the mortar is given by the quadratic equation

$$y = -16t^2 + 144t$$

and the horizontal distance of the shell from the mortar, x (in feet), is given by the linear equation

$$x = 39t$$

(Source: http://library.thinkquest.org/15384/physics/physics.html)

a)   How high is the shell after 3 sec?

b)   What is the shell's horizontal distance from the mortar after 3 sec?

c)   The maximum height is reached when the shell explodes. How high is the shell when it bursts after 4.5 sec?

d)   What is the shell's horizontal distance from its launching point when it explodes? (Round to the nearest foot.)

When a 10-in. shell is shot from a mortar at an angle of 75°, the height, y (in feet), of the shell t sec after being shot from the mortar is given by

$$y = -16t^2 + 264t$$

and the horizontal distance of the shell from the mortar, x (in feet), is given by

$$x = 71t$$

e)   How high is the shell after 3 sec?

f)   Find the shell's horizontal distance from the mortar after 3 sec.

g)   The shell explodes after 8.25 sec. What is its height when it bursts?

h)   What is the shell's horizontal distance from its launching point when it explodes? (Round to the nearest foot.)

i) Compare your answers to a) and e). What is the difference in their heights after 3 sec?

j) Compare your answers to c) and g). What is the difference in the shells' heights when they burst?

k) Assuming that the technicians timed the firings of the 3-in. shell and the 10-in. shell so that they exploded at the same time, how far apart would their respective mortars need to be so that the 10-in. shell would burst directly above the 3-in. shell?

40) The senior class at Richmont High School is selling t-shirts to raise money for its prom. The equation $R(p) = -25p^2 + 600p$ describes the revenue, R, in dollars, as a function of the price, p, in dollars, of a t-shirt. That is, the revenue is a function of price.

a) Determine the revenue if the group sells each shirt for $10.

b) Determine the revenue if the group sells each shirt for $15.

c) If the senior class hopes to have a revenue of $3600, how much should it charge for each t-shirt?

41) A famous comedian will appear at a comedy club for one performance. The equation $R(p) = -5p^2 + 300p$ describes the relationship between the price of a ticket, p, in dollars, and the revenue, R, in dollars, from ticket sales. That is, the revenue is a function of price.

a) Determine the club's revenue from ticket sales if the price of a ticket is $40.

b) Determine the club's revenue from ticket sales if the price of a ticket is $25.

c) If the club is expecting its revenue from ticket sales to be $4500, how much should it charge for each ticket?

42) An object is launched upward from the ground with an initial velocity of 200 ft/sec. The height h (in feet) of the object after t sec is given by $h(t) = -16t^2 + 200t$.

a) Find the height of the object after 1 sec.

b) Find the height of the object after 4 sec.

c) When is the object to 400 ft above the ground?

d) How long does it take for the object to hit the ground?

Definition/Procedure	Example	Reference
The Greatest Common Factor and Factoring by Grouping		7.1

To *factor a polynomial* is to write it as a product of two or more polynomials: To factor out a greatest common factor (GCF), 1) Identify the GCF of all of the terms of the polynomial. 2) Rewrite each term as the product of the GCF and another factor. 3) Use the distributive property to factor out the GCF from the terms of the polynomial. 4) Check the answer by multiplying the factors.	Factor out the greatest common factor. $$16d^6 - 40d^5 + 72d^4$$ The GCF is $8d^4$. $16d^6 - 40d^5 + 72d^4$ $\quad = (8d^4)(2d^2) - (8d^4)(5d) + (8d^4)(9)$ $\quad = 8d^4(2d^2 - 5d + 9)$ *Check:* $8d^4(2d^2 - 5d + 9) = 16d^6 - 40d^5 + 72d^4$ ✓	**p. 434**
The first step in factoring any polynomial is to ask yourself, "*Can I factor out a GCF?*" The last step in factoring any polynomial is to ask yourself, "*Can I factor again?*" Try to *factor by grouping* when you are asked to factor a polynomial containing four terms. 1) Make two groups of two terms so that each group has a common factor. 2) Take out the common factor from each group of terms. 3) Factor out the common factor using the distributive property. 4) Check the answer by multiplying the factors.	Factor completely. $45tu + 27t + 20u + 12$ Since the four terms have a GCF of 1, we will not factor out a GCF. Begin by grouping two terms together so that each group has a common factor. $\underbrace{45tu + 27t}\ + \underbrace{20u + 12}$ $\qquad\downarrow\qquad\qquad\downarrow$ $= 9t(5u + 3) + 4(5u + 3)$ Take out the common factor. $= (5u + 3)(9t + 4)$ Factor out $(5u + 3)$. *Check:* $(5u + 3)(9t + 4) = 45tu + 27t + 20u + 12$ ✓	**p. 437**
Factoring Trinomials of the Form $x^2 + bx + c$		7.2
Factoring $x^2 + bx + c$ If $x^2 + bx + c = (x + m)(x + n)$, then 1) if b and c are positive, then both m and n must be positive. 2) if c is positive and b is negative, then both m and n must be negative. 3) if c is negative, then one integer, m, must be positive and the other integer, n, must be negative.	Factor completely. a) $y^2 + 7y + 12$ Think of two numbers whose *product* is 12 and whose *sum* is 7. *3 and 4*. Then, $$y^2 + 7y + 12 = (y + 3)(y + 4)$$ b) $2r^3 - 26r^2 + 60r$ Begin by factoring out the GCF of $2r$. $2r^3 - 26r^2 + 60r = 2r(r^2 - 13r + 30)$ $\qquad\qquad\qquad\qquad = 2r(r - 10)(r - 3)$	**p. 443**

Definition/Procedure	Example	Reference

Factoring Trinomials of the Form $ax^2 + bx + c$ ($a \neq 1$)   7.3

Factoring $ax^2 + bx + c$ by *grouping*

Factor completely. $5n^2 + 18n - 8$

$$\text{Sum is } 18$$
$$\downarrow$$
$$5n^2 + 18n - 8$$
$$\text{Product: } 5 \cdot (-8) = -40$$

Think of two integers whose *product* is -40 and whose *sum* is 18. *20 and -2.*

Factor by grouping.

$$5n^2 + 18n - 8 = \underbrace{5n^2 + 20n}_{\text{Group}} \underbrace{-2n - 8}_{\text{Group}} \quad \begin{array}{l}\text{Write}\\18n \text{ as}\\20n - 2n.\end{array}$$
$$= 5n(n + 4) - 2(n + 4)$$
$$= (n + 4)(5n - 2)$$

p. 449

Factoring $ax^2 + bx + c$ by *trial and error*

When approaching a problem in this way, we must keep in mind that we are reversing the FOIL process.

Factor completely. $4x^2 - 16x + 15$

$$4x^2 - 16x + 15 = (2x - 3)(2x - 5)$$

$$\begin{array}{c}+ -10x\\\hline -16x\end{array}$$

$$4x^2 - 16x + 15 = (2x - 3)(2x - 5)$$

p. 451

Factoring Binomials and Perfect Square Trinomials   7.4

A *perfect square trinomial* is a trinomial that results from squaring a binomial. Factoring a perfect square trinomial:

$$a^2 + 2ab + b^2 = (a + b)^2$$
$$a^2 - 2ab + b^2 = (a - b)^2$$

Factor completely.

a) $c^2 + 24c + 144 = (c + 12)^2$
   $a = c \quad b = 12$

b) $49p^2 - 56p + 16 = (7p - 4)^2$
   $a = 7p \quad b = 4$

p. 457

Factoring the *difference of two squares:*

$$a^2 - b^2 = (a + b)(a - b)$$

Factor completely.

$$d^2 - 16 = (d + 4)(d - 4)$$
$$\downarrow \quad \downarrow$$
$$(d)^2 \, (4)^2 \quad a = d, b = 4$$

p. 460

Factoring the *sum and difference of two cubes:*

$$a^3 + b^3 = (a + b)(a^2 - ab + b^2)$$
$$a^3 - b^3 = (a - b)(a^2 + ab + b^2)$$

Factor completely.

$$w^3 + 27 = (w + 3)((w)^2 - (w)(3) + (3)^2)$$
$$\downarrow \quad \downarrow$$
$$(w)^3 \, (3)^3 \quad a = w, b = 3$$
$$w^3 + 27 = (w + 3)(w^2 - 3w + 9)$$

p. 463

Definition/Procedure	Example	Reference

Solving Quadratic Equations by Factoring

7.5

Definition/Procedure	Example	Reference
A *quadratic equation* can be written in the form $ax^2 + bx + c = 0$, where $a, b,$ and c are real numbers and $a \neq 0$.	Some examples of quadratic equations are $$3x^2 - 5x + 9 = 0, \quad t^2 = 4t + 21,$$ $$2(p - 3)^2 = 8 - 7p$$	**p. 472**
To solve a quadratic equation by factoring, use the *zero product rule:* If $ab = 0$, then $a = 0$ or $b = 0$.	Solve $(y + 9)(y - 4) = 0$ $y + 9 = 0 \quad$ or $\quad y - 4 = 0 \quad$ Set each factor equal to zero. $y = -9 \quad$ or $\quad y = 4 \quad$ Solve. The solution set is $\{-9, 4\}$.	**p. 472**
Steps for Solving a Quadratic Equation by Factoring 1) Write the equation in the form $ax^2 + bx + c = 0$. 2) Factor the expression. 3) Set each factor equal to zero, and solve for the variable. 4) Check the answer(s).	Solve $4m^2 - 11 = m^2 + 2m - 3$. $$3m^2 - 2m - 8 = 0 \quad \text{Standard form}$$ $$(3m + 4)(m - 2) = 0 \quad \text{Factor.}$$ $3m + 4 = 0 \quad$ or $\quad m - 2 = 0$ $3m = -4$ $m = -\dfrac{4}{3} \quad$ or $\quad m = 2$ The solution set is $\left\{-\dfrac{4}{3}, 2\right\}$. Check the answers.	**p. 473**

Applications of Quadratic Equations

7.6

Definition/Procedure	Example	Reference
Pythagorean Theorem Given a right triangle with legs of length a and b and hypotenuse of length c, 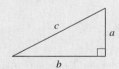 the Pythagorean theorem states that $$a^2 + b^2 = c^2$$	Find the length of side a. 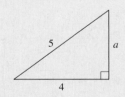 Let $b = 4$ and $c = 5$ in $a^2 + b^2 = c^2$. $$a^2 + (4)^2 = (5)^2$$ $$a^2 + 16 = 25$$ $$a^2 - 9 = 0$$ $$(a + 3)(a - 3) = 0$$ $a + 3 = 0 \quad$ or $\quad a - 3 = 0$ $a = -3 \quad$ or $\quad a = 3$ Reject -3 as a solution since the length of a side cannot be negative. Therefore, the length of side a is 3.	**p. 483**

(7.1) Find the greatest common factor of each group of terms.

1) $18, 27$

2) $56, 80, 24$

3) $33p^5q^3, 121p^4q^3, 44p^7q^4$

4) $42r^4s^3, 35r^2s^6, 49r^2s^4$

Factor out the greatest common factor.

5) $48y + 84$

6) $30a^4 - 9a$

7) $7n^5 - 21n^4 + 7n^3$

8) $72u^3v^3 - 42u^3v^2 - 24uv$

9) $a(b + 6) - 2(b + 6)$

10) $u(13w - 9) + v(13w - 9)$

Factor by grouping.

11) $mn + 2m + 5n + 10$

12) $jk + 7j - 5k - 35$

13) $5qr - 10q - 6r + 12$

14) $cd^2 + 5c - d^2 - 5$

15) Factor out $-4x$ from $-8x^3 - 12x^2 + 4x$.

16) Factor out -1 from $-r^2 + 6r - 2$.

(7.2) Factor completely.

17) $p^2 + 13p + 40$

18) $f^2 - 17f + 60$

19) $x^2 + xy - 20y^2$

20) $t^2 - 2tu - 63u^2$

21) $3c^2 - 24c + 36$

22) $4m^3n + 8m^2n^2 - 60mn^3$

(7.3) Factor completely.

23) $5y^2 + 11y + 6$

24) $3g^2 + g - 44$

25) $4m^2 - 16m + 15$

26) $6t^2 - 49t + 8$

27) $56a^3 + 4a^2 - 16a$

28) $18n^2 + 98n + 40$

29) $3s^2 + 11st - 4t^2$

30) $8f^2(g - 11)^3 - 6f(g - 11)^3 - 35(g - 11)^3$

(7.4) Factor completely.

31) $n^2 - 25$

32) $49a^2 - 4b^2$

33) $9t^2 + 16u^2$

34) $z^4 - 1$

35) $10q^2 - 810$

36) $48v - 27v^3$

37) $a^2 + 16a + 64$

38) $4x^2 - 20x + 25$

39) $h^3 + 8$

40) $q^3 - 1$

41) $27p^3 - 64q^3$

42) $16c^3 + 250d^3$

(7.1–7.4) Factor completely.

43) $7r^2 + 8r - 12$

44) $3y^2 + 60y + 300$

45) $\dfrac{9}{25} - x^2$

46) $81v^6 + 36v^5 - 9v^4$

47) $st - 5s - 8t + 40$

48) $n^2 - 11n + 30$

49) $w^5 - w^2$

50) $gh + 8g - 11h - 88$

51) $a^2 + 3a - 14$

52) $49k^2 - 144$

53) $(a - b)^2 - (a + b)^2$

54) $1000a^3 + 27b^3$

(7.5) Solve each equation.

55) $c(2c - 1) = 0$

56) $(4z + 7)^2 = 0$

57) $3x^2 + x = 2$

58) $f^2 - 1 = 0$

59) $n^2 = 12n + 45$

60) $10j^2 - 8 + 11j = 0$

61) $36 = 49d^2$

62) $-13w = w^2$

63) $8b + 64 = 2b^2$

64) $18 = a(9 - a)$

65) $y(5y - 9) = -4$

66) $(z + 2)^2 = -z(3z + 4) + 9$

67) $6a^3 - 3a^2 - 18a = 0$

68) $48 = 6r^2 + 12r$

69) $c(5c - 1) + 8 = 4(20 + c^2)$

70) $15t^3 + 40t = 70t^2$

71) $p^2(6p - 1) - 10p(6p - 1) + 21(6p - 1) = 0$

72) $k^2(4k - 3) - 3k(4k - 3) - 54(4k - 3) = 0$

(7.6)

73) Find the base and height if the area of the triangle is 15 in^2.

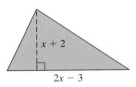

74) Find the length and width of the rectangle if its area is 28 cm^2.

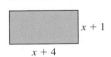

75) Find the height and length of the box if its volume is 96 in^3.

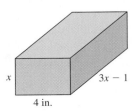

76) Find the length and width of the box if its volume is 360 in³.

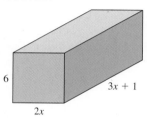

6

$3x + 1$

$2x$

Use the Pythagorean theorem to find the length of the missing side.

77)

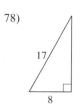

13

5

78)

17

8

Write an equation and solve.

79) A rectangular countertop has an area of 15 ft². If the width is 3.5 ft shorter than the length, what are the dimensions of the countertop?

80) Kelsey cuts a piece of fabric into a triangle to make a bandana for her dog. The base of the triangle is twice its height. Find the base and height if there is 144 in² of fabric.

81) The sum of three consecutive integers is one-third the square of the middle number. Find the integers.

82) Find two consecutive even integers such that their product is 6 more than 3 times their sum.

83) Seth builds a bike ramp in the shape of a right triangle. One leg is one inch shorter than the "ramp" while the other leg, the height of the ramp, is 8 in. shorter than the ramp. What is the height of the ramp?

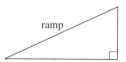

ramp

84) A car heads east from an intersection while a motorcycle travels south. After 20 min, the car is 2 mi farther from the intersection than the motorcycle. The distance between the two vehicles is 4 mi more than the motorcycle's distance from the intersection. What is the distance between the car and the motorcycle?

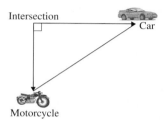

Intersection

Car

Motorcycle

85) An object is launched with an initial velocity of 95 ft/sec. The height h (in feet) of the object after t sec is given by $h = -16t^2 + 96t$.

a) From what height is the object launched?

b) When does the object reach a height of 128 ft?

c) How high is the object after 3 sec?

d) When does the object hit the ground?

1) What is the first thing you should do when you are asked to factor a polynomial?

Factor completely.

2) $n^2 - 11n + 30$

3) $16 - b^2$

4) $5a^2 - 13a - 6$

5) $56p^6q^6 - 77p^4q^4 + 7p^2q^3$

6) $y^3 - 8z^3$

7) $2d^3 + 14d^2 - 36d$

8) $r^2 + 25$

9) $9h^2 + 24h + 16$

10) $24xy - 36x + 22y - 33$

11) $s^2 - 3st - 28t^2$

12) $16s^4 - 81t^4$

13) $y^2(x + 3)^2 + 15y(x + 3)^2 + 56(x + 3)^2$

14) $12b^2 - 44b + 35$

15) $m^{12} + m^9$

Solve each equation.

16) $b^2 + 7b + 12 = 0$

17) $25k = k^3$

18) $144m^2 = 25$

19) $(c - 5)(c + 2) = 18$

20) $4q(q - 5) + 14 = 11(2 + q)$

21) $24y^2 + 80 = 88y$ 2

Write an equation and solve.

22) Find the width and height of the storage locker pictured below if its volume is 120 ft^3.

$x + 2$

3 ft

$\frac{1}{2}x$

23) Find three consecutive odd integers such that the sum of the three numbers is 60 less than the square of the largest integer.

24) Cory and Isaac leave an intersection with Cory jogging north and Isaac jogging west. When Isaac is 1 mi farther from the intersection than Cory, the distance between them is 2 mi more than Cory's distance from the intersection. How far is Cory from the intersection?

25) The length of a rectangular dog run is 4 ft more than twice its width. Find the dimensions of the run if it covers 96 ft^2.

26) An object is thrown upward with an initial velocity of 68 ft/sec. The height h (in feet) of the object t sec after it is thrown is given by

$$h = -16t^2 + 68t + 60$$

a) How long does it take for the object to reach a height of 120 ft?

b) What is the initial height of the object?

c) What is the height of the object after 2 sec?

d) How long does it take the object to hit the ground?

Perform the indicated operation(s) and simplify.

1) $\dfrac{3}{8} - \dfrac{5}{6} + \dfrac{7}{12}$

2) $-\dfrac{15}{32} \cdot \dfrac{12}{25}$

Simplify. The answer should not contain any negative exponents.

3) $\dfrac{54t^5u^2}{36tu^8}$

4) $(8k^6)(-3k^4)$

5) Write 4.813×10^5 without exponents.

6) Solve $\dfrac{1}{3}(n - 2) + \dfrac{1}{4} = \dfrac{5}{12} + \dfrac{1}{6}n$

7) Solve for R.
   $A = P + PRT$

8) *Write an equation and solve.*
   A Twix candy bar is half the length of a Toblerone candy bar. Together, they are 12 in. long. Find the length of each candy bar.

9) Graph $y = -\dfrac{3}{5}x + 7$.

10) Write the equation of the line perpendicular to $3x + y = 4$ containing the point $(-6, -1)$. Express the answer in slope-intercept form.

11) Use any method to solve this system of equations.

$$6(x + 2) + y = x - y - 2$$
$$5(2x - y + 1) = 2(x - y) - 5$$

Multiply and simplify.

12) $(6y + 5)(2y - 3)$

13) $(4p - 7)(2p^2 - 9p + 8)$

14) $(c + 8)^2$

15) Add $(4a^2b^2 - 17a^2b + 12ab - 11)$
    $+ (-a^2b^2 + 10a^2b - 5ab^2 + 7ab + 3)$.

Divide.

16) $\dfrac{12x^4 - 30x^3 - 14x^2 + 27x + 20}{2x - 5}$

17) $\dfrac{12r^3 + 4r^2 - 10r + 3}{4r^2}$

Factor completely, if possible.

18) $bc + 8b - 7c - 56$

19) $54q^2 - 144q + 42$

20) $y^2 + 1$

21) $t^4 - 81$

22) $x^3 - 125$

Solve.

23) $z^2 + 3z = 40$

24) $-12j(1 - 2j) = 16(5 + j)$

CHAPTER 8

Rational Expressions

Algebra at Work: Ophthalmology

At the beginning of Chapter 7 we saw how an ophthalmologist, a doctor specializing in diseases of the eye, uses mathematics every day to treat his patients. Here we will see another example of how math is used in this branch of medicine.

Some formulas in optics involve rational expressions. If Calvin determines that one of his patients needs glasses, he would use the following formula to figure out the proper prescription:

$$P = \frac{1}{f}$$

where f is the focal length, in meters, and P is the power of the lens, in diopters.

While computers now aid in these calculations, physicians believe that it is still important to double-check the calculations by hand.

In this chapter, we will learn how to perform operations with rational expressions and how to solve equations, like the one above, for a specific variable.

Section 8.1 Simplifying Rational Expressions

Objectives

1. Evaluate a Rational Expression for Given Values of the Variable

2. Find the Value(s) of the Variable That Make a Rational Expression Equal Zero and That Make a Rational Expression Undefined

3. Write a Rational Expression in Lowest Terms

4. Simplify a Rational Expression of the Form $\dfrac{a - b}{b - a}$

5. Write Equivalent Forms of a Rational Expression

6. Define a Rational Function and Find the Domain of a Rational Function

What Is a Rational Expression?

In Section 1.4 we defined a **rational number** as the quotient of two integers provided that the denominator does not equal zero. Some examples of rational numbers are

$$\frac{5}{12}, \qquad -\frac{3}{8}, \qquad 23 \left(\text{since } 23 = \frac{23}{1}\right)$$

We can define a rational expression in a similar way. A rational expression is a quotient of two polynomials provided that the denominator does not equal zero. We state the definition formally next.

Definition

A **rational expression** is an expression of the form $\dfrac{P}{Q}$, where P and Q are polynomials and where $Q \neq 0$.

Some examples of rational expressions are

$$\frac{9n^4}{2}, \qquad \frac{3x - 5}{x + 8}, \qquad \frac{6}{x^2 - 2c - 63}, \qquad -\frac{3a + 2b}{a^2 + b^2}$$

We can *evaluate* rational expressions for given values of the variable(s).

1. Evaluate a Rational Expression for Given Values of the Variable

Example 1

Evaluate $\dfrac{x^2 - 4}{x + 10}$ (if possible) for each value of x.

a) $x = 6$ b) $x = 2$ c) $x = -10$

Solution

a) Evaluate $\dfrac{x^2 - 4}{x + 10}$ when $x = 6$.

$$\frac{x^2 - 4}{x + 10} = \frac{(6)^2 - 4}{(6) + 10} \qquad \text{Substitute 6 for } x.$$
$$= \frac{36 - 4}{6 + 10}$$
$$= \frac{32}{16} = 2$$

b)   Evaluate $\dfrac{x^2 - 4}{x + 10}$ when $x = 2$.

$$\dfrac{x^2 - 4}{x + 10} = \dfrac{(2)^2 - 4}{(2) + 10} \qquad \text{Substitute 2 for } x.$$

$$= \dfrac{4 - 4}{2 + 10}$$

$$= \dfrac{0}{12} = 0$$

c)   Evaluate $\dfrac{x^2 - 4}{x + 10}$ when $x = -10$.

$$\dfrac{x^2 - 4}{x + 10} = \dfrac{(-10)^2 - 4}{(-10) + 10} \qquad \text{Substitute } -10 \text{ for } x.$$

$$= \dfrac{100 - 4}{-10 + 10}$$

$$= \dfrac{96}{0} \qquad \text{Undefined}$$

Remember, a fraction is **undefined** when its denominator equals zero. Therefore, we say that $\dfrac{x^2 - 4}{x + 10}$ is *undefined* when $x = -10$ since this value of x makes the denominator equal zero. So, x *cannot equal* 10 in this expression.

You Try 1

Evaluate $\dfrac{n^2 - 25}{n - 1}$ (if possible) for each value of n.

a) $n = 4$ b) $n = -2$ c) $n = 1$ d) $n = -5$

2. Find the Value(s) of the Variable That Make a Rational Expression Equal Zero and That Make a Rational Expression Undefined

Parts b) and c) in Example 1 remind us about two important aspects of fractions and rational expressions.

1)   A fraction (rational expression) equals zero when its *numerator* equals zero.

2)   A fraction (rational expression) is *undefined* when its denominator equals zero.

Finding the value(s) of the variable that make a rational expression equal zero and that make a rational expression undefined are just two of the skills we need to work with these expressions.

Example 2

For each rational expression, for what value(s) of the variable

i) does the expression equal zero?
ii) is the expression undefined?

a) $\dfrac{c + 6}{c - 2}$ b) $\dfrac{3k}{k^2 + 2k - 35}$ c) $\dfrac{4n^2 - 9}{7}$ d) $\dfrac{5}{2t + 1}$

Solution

a) i) $\dfrac{c + 6}{c - 2} = 0$ when its *numerator* equals zero. Set the numerator equal to zero, and solve for c.

$$c + 6 = 0$$
$$c = -6$$

$\dfrac{c + 6}{c - 2} = 0$ when $c = -6$.

ii) $\dfrac{c + 6}{c - 2}$ is *undefined* when its *denominator* equals zero. Set the denominator equal to zero, and solve for c.

$$c - 2 = 0$$
$$c = 2$$

$\dfrac{c + 6}{c - 2}$ is *undefined* when $c = 2$. This means that any real number *except* 2 can be substituted for c in this expression.

b) i) $\dfrac{3k}{k^2 + 2k - 35} = 0$ when its *numerator* equals zero. Set the numerator equal to zero, and solve for k.

$$3k = 0$$
$$k = \frac{0}{3} = 0$$

$\dfrac{3k}{k^2 + 2k - 35} = 0$ when $k = 0$.

ii) $\dfrac{3k}{k^2 + 2k - 35}$ is *undefined* when its *denominator* equals zero. Set the denominator equal to zero, and solve for k.

$$k^2 + 2k - 35 = 0$$ Factor.
$$(k + 7)(k - 5) = 0$$
$$k + 7 = 0 \quad \text{or} \quad k - 5 = 0$$ Set each factor equal to 0.
$$k = -7 \quad \text{or} \quad k = 5$$ Solve.

$\dfrac{3k}{k^2 + 2k - 35}$ is undefined when $k = -7$ or $k = 5$. All real numbers *except* -7 and 5 can be substituted for k in this expression.

c) i) To determine the values of n that make $\dfrac{4n^2 - 9}{7} = 0$, set $4n^2 - 9 = 0$ and solve.

$$4n^2 - 9 = 0$$
$$(2n + 3)(2n - 3) = 0 \qquad \text{Factor.}$$
$$2n + 3 = 0 \quad \text{or} \quad 2n - 3 = 0 \qquad \text{Set each factor equal to 0.}$$
$$2n = -3 \qquad\qquad 2n = 3$$
$$n = -\frac{3}{2} \quad \text{or} \quad n = \frac{3}{2} \qquad \text{Solve.}$$

$\dfrac{4n^2 - 9}{7} = 0$ when $n = -\dfrac{3}{2}$ or $n = \dfrac{3}{2}$.

ii) $\dfrac{4n^2 - 9}{7}$ is *undefined* when the denominator equals zero. However, the denominator is 7 and $7 \neq 0$. Therefore, there *is no value of n that makes* $\dfrac{4n^2 - 9}{7}$ undefined. *We say that* $\dfrac{4n^2 - 9}{7}$ *is defined for all real numbers.*

d) i) $\dfrac{5}{2t + 1} = 0$ when the numerator equals zero. The numerator is 5, and $5 \neq 0$. Therefore, $\dfrac{5}{2t + 1}$ will *never* equal zero.

ii) $\dfrac{5}{2t + 1}$ is *undefined* when $2t + 1 = 0$. Solve for t.

$$2t + 1 = 0$$
$$2t = -1$$
$$t = -\frac{1}{2}$$

$\dfrac{5}{2t + 1}$ is undefined when $t = -\dfrac{1}{2}$. So, $t \neq -\dfrac{1}{2}$ in the expression.

You Try 2

For each rational expression, determine which value(s) of the variable

i) make the expression equal zero.

ii) make the expression undefined.

a) $\dfrac{5x + 2}{x + 9}$ b) $\dfrac{2r^2 + 5r - 3}{6r}$ c) $\dfrac{2z^2 + 3}{8}$

All of the operations that can be performed with fractions can also be done with rational expressions. We begin our study of these operations with rational expressions by learning how to write a rational expression in lowest terms.

3. Write a Rational Expression in Lowest Terms

One way to think about writing a fraction such as $\dfrac{8}{12}$ in lowest terms is

$$\frac{8}{12} = \frac{2 \cdot 4}{3 \cdot 4} = \frac{2}{3} \cdot \frac{4}{4} = \frac{2}{3} \cdot 1 = \frac{2}{3}$$

This property that says $\dfrac{4}{4} = 1$ is what allows us to also approach reducing $\dfrac{8}{12}$ in the following way:

$$\frac{8}{12} = \frac{2 \cdot \cancel{4}}{3 \cdot \cancel{4}} = \frac{2}{3}$$

We can *factor* the numerator and denominator, then *divide* the numerator and denominator by the common factor, 4. This is the approach we use to write a rational expression in lowest terms.

Definition

Fundamental Property of Rational Expressions

If P, Q, and C are polynomials such that $Q \neq 0$ and $C \neq 0$, then

$$\frac{PC}{QC} = \frac{P}{Q}$$

This property mirrors the example above since

$$\frac{PC}{QC} = \frac{P}{Q} \cdot \frac{C}{C} = \frac{P}{Q} \cdot 1 = \frac{P}{Q}$$

Or, we can also think of the reducing procedure as dividing the numerator and denominator by the common factor, C.

$$\frac{P\cancel{C}}{Q\cancel{C}} = \frac{P}{Q}$$

Writing a Rational Expression in Lowest Terms

1) Completely **factor** the numerator and denominator.

2) **Divide** the numerator and denominator by the greatest common factor.

Example 3

Write each rational expression in lowest terms.

a) $\dfrac{20c^6}{5c^4}$ b) $\dfrac{4m + 12}{7m + 21}$ c) $\dfrac{3x^2 - 3}{x^2 + 9x + 8}$

Solution

a) $\dfrac{20c^6}{5c^4}$ is an expression we can simplify using the quotient rule presented in Chapter 2.

$$\frac{20c^6}{5c^4} = 4c^2$$

Divide 20 by 5 and use the quotient rule:
$\dfrac{c^6}{c^4} = c^{6-4} = c^2.$

b) $\dfrac{4m + 12}{7m + 21} = \dfrac{4\cancel{(m + 3)}}{7\cancel{(m + 3)}}$ Factor.

$$= \frac{4}{7}$$ Divide out the common factor, $m + 3$.

c) $\dfrac{3x^2 - 3}{x^2 + 9x + 8} = \dfrac{3(x^2 - 1)}{(x + 1)(x + 8)}$ Factor.

$$= \frac{3\cancel{(x + 1)}(x - 1)}{\cancel{(x + 1)}(x + 8)}$$ Factor completely.

$$= \frac{3(x - 1)}{x + 8}$$ Divide out the common factor, $x + 1$.

Notice that we divided by *factors* not *terms*.

$\dfrac{\cancel{x + 5}}{2\cancel{(x + 5)}} = \dfrac{1}{2}$

Divide by the *factor* $x + 5$.

$\dfrac{\cancel{x}}{x + 5} \neq \dfrac{1}{5}$

We cannot divide by x because the x in the denominator is a *term* in a sum.

You Try 3

Write each rational expression in lowest terms.

a) $\dfrac{6t - 48}{t^2 - 8t}$ b) $\dfrac{b - 2}{5b^2 - 6b - 8}$ c) $\dfrac{v^3 + 27}{4v^4 - 12v^3 + 36v^2}$

4. Simplify a Rational Expression of the Form $\dfrac{a - b}{b - a}$

Do you think that $\dfrac{x - 4}{4 - x}$ is in lowest terms? Let's look at it more closely to understand the answer.

$$\frac{x - 4}{4 - x} = \frac{x - 4}{-1(-4 + x)} \qquad \text{Factor } -1 \text{ out of the denominator.}$$

$$= \frac{1(x - 4)}{-1(x - 4)}$$

$$= -1$$

$$\frac{x - 4}{4 - x} = -1$$

We can generalize this result as

---

1) $b - a = -1(a - b)$

and

2) $\dfrac{a - b}{b - a} = -1$

The terms in the numerator and denominator in 2) differ only in sign. They divide out to -1.

Example 4

Write each rational expression in lowest terms.

a) $\dfrac{6 - d}{d - 6}$ b) $\dfrac{25z^2 - 9}{3 - 5z}$ c) $\dfrac{8u^3 - 16u^2 + 8u}{1 - u^2}$

Solution

a) $\dfrac{6 - d}{d - 6} = -1$ since $\dfrac{6 - d}{d - 6} = \dfrac{-1(d - 6)}{d - 6} = -1.$

b) $\dfrac{25z^2 - 9}{3 - 5z} = \dfrac{(5z + 3)\overset{-1}{\cancel{(5z - 3)}}}{\cancel{3 - 5z}}$ Factor.

$$= -1(5z + 3) \qquad \qquad \dfrac{5z - 3}{3 - 5z} = -1$$

$$= -5z - 3 \qquad \qquad \text{Distribute.}$$

c) $\dfrac{8u^3 - 16u^2 + 8u}{1 - u^2} = \dfrac{8u(u^2 - 2u + 1)}{(1 + u)(1 - u)}$ Factor.

$= \dfrac{8u(u - 1)^2}{(1 + u)(1 - u)}$ Factor completely.

$= \dfrac{8u\overset{-1}{\cancel{(u - 1)}}(u - 1)}{(1 + u)\cancel{(1 - u)}}$ $\dfrac{u - 1}{1 - u} = -1$

$= -\dfrac{8u(u - 1)}{1 + u}$ The negative sign can be written in front of the fraction.

You Try 4

Write each rational expression in lowest terms.

a) $\dfrac{x - y}{y - x}$ b) $\dfrac{15n - 5m}{2m - 6n}$ c) $\dfrac{12 - 3y^2}{y^2 - 10y + 16}$

5. Write Equivalent Forms of a Rational Expression

The answer to Example 4c) can be written in several different ways. You should be able to recognize equivalent forms of rational expressions because there isn't always just one way to write the correct answer.

Example 5

Write the answer to Example 4c), $-\dfrac{8u(u - 1)}{1 + u}$ in three different ways.

Solution

The negative sign in front of a fraction can also be applied to the numerator or to the denominator. $\left(\text{For example, } -\dfrac{4}{9} = \dfrac{-4}{9} = \dfrac{4}{-9}\right).$ Applying this concept to rational expressions can result in expressions that look quite different but that are, actually, equivalent.

i) Apply the negative sign to the denominator.

$$-\dfrac{8u(u - 1)}{1 + u} = \dfrac{8u(u - 1)}{-1(1 + u)}$$
$$= \dfrac{8u(u - 1)}{-1 - u} \qquad \text{Distribute.}$$

ii) Apply the negative sign to the numerator.

$$-\dfrac{8u(u - 1)}{1 + u} = \dfrac{-8u(u - 1)}{1 + u}$$

iii) Apply the negative sign to the numerator, but distribute the -1.

$$-\frac{8u(u-1)}{1+u} = \frac{(8u)(-1)(u-1)}{1+u}$$

$$= \frac{8u(-u+1)}{1+u} \qquad \text{Distribute.}$$

$$= \frac{8u(1-u)}{1+u} \qquad \text{Rewrite } -u+1 \text{ as } 1-u.$$

Therefore, $\dfrac{8u(u-1)}{-1-u}$, $\dfrac{-8u(u-1)}{1+u}$, and $\dfrac{8u(1-u)}{1+u}$ are *all* equivalent forms of $-\dfrac{8u(u-1)}{1+u}$.

Keep this idea of equivalent forms of rational expressions in mind when checking your answers against the answers in the back of the book. Sometimes students believe their answer is wrong because it "looks different" when, in fact, it is an *equivalent form* of the given answer!

You Try 5

Find three equivalent forms of $\dfrac{-(2-p)}{7p-9}$.

6. Define a Rational Function and Find the Domain of a Rational Function

We can combine what we have learned about rational expressions with what we have learned about functions in Section 4.7. $f(x) = \dfrac{x+3}{x-8}$ is an example of a **rational function** since $\dfrac{x+3}{x-8}$ is a rational expression and since each value that can be substituted for x will produce *only one* value for the expression.

Recall from Chapter 4 that the domain of a function $f(x)$ is the set of all real numbers that can be substituted for x. Since a rational expression is undefined when its denominator equals zero, we define the domain of a rational function as follows.

> The **domain of a rational function** consists of all real numbers except the value(s) of the variable that make(s) the denominator equal zero.

Therefore, to determine the domain of a rational function we set the denominator equal to zero and solve for the variable. Any value that makes the denominator equal to zero is *not* in the domain of the function.

To determine the domain of a rational function, sometimes it is helpful to ask yourself, "Is there any number that *cannot* be substituted for the variable?"

Example 6

Determine the domain of each rational function.

a) $f(x) = \dfrac{x + 3}{x - 8}$ b) $g(c) = \dfrac{6}{c^2 + 3c - 4}$

Solution

a) To determine the domain of $f(x) = \dfrac{x + 3}{x - 8}$ ask yourself, "Is there any number that *cannot* be substituted for x?" Yes, $f(x)$ is *undefined* when the denominator equals zero. Set the denominator equal to zero, and solve for x.

$$
\begin{array}{ll}
x - 8 = 0 & \text{Set the denominator} = 0. \\
x = 8 & \text{Solve.}
\end{array}
$$

When $x = 8$, the denominator of $f(x) = \dfrac{x + 3}{x - 8}$ equals zero. The domain contains all real numbers *except* 8. Write the domain in interval notation as $(-\infty, 8) \cup (8, \infty)$.

b) To determine the domain of $g(c) = \dfrac{6}{c^2 + 3c - 4}$, ask yourself, "Is there any number that *cannot* be substituted for c?   Yes, $g(c)$ is *undefined* when its *denominator* equals zero. Set the denominator equal to zero and solve for c.

$$
\begin{array}{ll}
c^2 + 3c - 4 = 0 & \text{Set the denominator} = 0. \\
(c + 4)(c - 1) = 0 & \text{Factor.} \\
c + 4 = 0 \quad \text{or} \quad c - 1 = 0 & \text{Set each factor equal to } 0. \\
c = -4 \quad \text{or} \quad c = 1 & \text{Solve.}
\end{array}
$$

When $c = -4$ or $c = 1$, the denominator of $g(c) = \dfrac{6}{c^2 + 3c - 4}$ equals zero. The domain contains all real numbers *except* -4 and 1. Write the domain in interval notation as $(-\infty, -4) \cup (-4, 1) \cup (1, \infty)$

You Try 6

Determine the domain of each rational function.

a) $h(t) = \dfrac{9}{t + 5}$ b) $f(x) = \dfrac{2x - 3}{x^2 - 8x + 12}$ c) $g(a) = \dfrac{a + 4}{10}$

Answers to You Try Exercises

1) a) -3 b) 7 c) undefined   d) 0 2) a) i) $-\dfrac{2}{5}$ ii) -9 b) i) $\dfrac{1}{2}$, -3 ii) 0

c) i) never equals zero   ii) defined for all real numbers

3) a) $\dfrac{6}{t}$ b) $\dfrac{1}{5b+4}$ c) $\dfrac{v+3}{4v^2}$ 4) a) -1 b) $-\dfrac{5}{2}$ c) $\dfrac{-3(y+2)}{y-8}$

5) Some possibilities are $\dfrac{p-2}{7p-9}$, $-\dfrac{2-p}{9-7p}$, $-\dfrac{2-p}{7p-9}$, $\dfrac{2-p}{-(7p-9)}$

6) a) $(-\infty, -5) \cup (-5, \infty)$ b) $(-\infty, 2) \cup (2, 6) \cup (6, \infty)$ c) $(-\infty, \infty)$

8.1 Exercises

Boost your grade at mathzone.com!   MathZone

> Practice Problems   > NetTutor   > e-Professors   > Videos
> Self-Test

Objective 1

1) When does a fraction or a rational expression equal 0?

2) When is a fraction or a rational expression undefined?

3) How do you determine the value of the variable for which a rational expression is undefined?

4) If $x^2 + 5$ is the numerator of a rational expression, can that expression equal zero? Give a reason.

Evaluate (if possible) for a) $x = 2$ and b) $x = -1$.

5) $\dfrac{7x+1}{3x-1}$

6) $\dfrac{2(3x^2+2)}{x^2+5x-1}$

Evaluate (if possible) for a) $z = -3$ and b) $z = 4$.

7) $\dfrac{(2z)^2}{z^2+5z+6}$

8) $\dfrac{2(z^2-16)}{z^2+11}$

9) $\dfrac{12+4z}{20-z^2}$

10) $\dfrac{5z-1}{z^2+z-20}$

Objective 2

Determine the value(s) of the variable for which

a)   the expression equals zero.

b)   the expression is undefined.

11) $\dfrac{x+2}{7x}$

12) $\dfrac{m}{m-10}$

13) $\dfrac{3r+1}{2r-9}$

14) $\dfrac{7s+11}{5s+11}$

15) $\dfrac{10c-c^2}{3c-4}$

16) $-\dfrac{z+5}{z^2-100}$

17) $\dfrac{9}{y}$

18) $\dfrac{5}{p-8}$

19) $-\dfrac{7a}{a^2+9a-36}$

20) $\dfrac{k-6}{k^2+8k+15}$

21) $\dfrac{r+10}{2r^2-5r-12}$

22) $\dfrac{2}{3d^2-7d+4}$

23) $\dfrac{t^2+8t+15}{4t}$

24) $\dfrac{2n-5}{9}$

25) $\dfrac{4y}{y^2+9}$

26) $\dfrac{x^2+49}{2}$

Objective 3

Write each rational expression in lowest terms.

27) $\dfrac{2y(y-11)}{3(y-11)}$

28) $\dfrac{30(f+2)}{6(f+2)(f-5)}$

29) $\dfrac{12d^5}{30d^8}$

30) $\dfrac{108g^4}{9g}$

31) $\dfrac{3c-12}{5c-20}$

32) $\dfrac{10d-5}{12d-6}$

33) $\dfrac{-18v-42}{15v+35}$

34) $\dfrac{-12h^2+15}{32h^2-40}$

35) $\dfrac{39q^2+26}{30q^2+20}$

36) $\dfrac{3u+15}{-7u-35}$

37) $\dfrac{b^2+b-56}{b+8}$

38) $\dfrac{g^2+9g+20}{g^2+2g-15}$

39) $\dfrac{r-4}{r^2-16}$

40) $\dfrac{t+2}{t^2-7t-18}$

41) $\dfrac{3k^2+28k+32}{k^2+10k+16}$

42) $\dfrac{3c^2-36c+96}{c-8}$

43) $\dfrac{p^2-16}{2p^2+7p-4}$

44) $\dfrac{6a^2-23a+20}{4a^2-25}$

45) $\dfrac{w^3+125}{5w^2-25w+125}$

46) $\dfrac{4m^3-4}{m^2+m+1}$

47) $\dfrac{8c + 24}{c^3 + 27}$

48) $\dfrac{2w^2 + 8w + 32}{w^3 - 64}$

49) $\dfrac{4m^2 - 20m + 4mn - 20n}{11m + 11n}$

50) $\dfrac{uv + 3u - 4v - 12}{v^2 - 9}$

51) $\dfrac{x^2 - y^2}{x^3 - y^3}$

52) $\dfrac{a^3 + b^3}{a^2 - b^2}$

53) Any rational expression of the form $\dfrac{a - b}{b - a}$ reduces to what expression?

54) Does $\dfrac{z + 9}{z - 9} = -1$?

Objective 4

Write each rational expression in lowest terms.

55) $\dfrac{12 - v}{v - 12}$

56) $\dfrac{q - 11}{11 - q}$

57) $\dfrac{k^2 - 49}{7 - k}$

58) $\dfrac{m - 10}{20 - 2m}$

59) $\dfrac{30 - 35x}{7x^2 + 8x - 12}$

60) $\dfrac{a^2 - 8a - 33}{11 - a}$

 61) $\dfrac{16 - 4b^2}{b - 2}$

62) $\dfrac{16 - 2w}{w^2 - 64}$

63) $\dfrac{8t^3 - 27}{9 - 4t^2}$

64) $\dfrac{r^3 - 3r^2 + 2r - 6}{21 - 7r}$

Objective 5

Find three equivalent forms of each rational expression.

65) $-\dfrac{b + 7}{b - 2}$

66) $-\dfrac{8y - 1}{2y + 5}$

 67) $-\dfrac{9 - 5t}{2t - 3}$

68) $\dfrac{-12m}{m^2 - 3}$

69) $\dfrac{w - 6}{-4w + 7}$

70) $\dfrac{-9d - 11}{10 - d}$

Reduce to lowest terms

   a) using long division.

   b) using the methods of this section.

71) $\dfrac{2x^2 + x - 28}{x + 4}$

72) $\dfrac{4k^2 - 11k + 6}{k - 2}$

73) $\dfrac{27t^3 - 8}{3t - 2}$

74) $\dfrac{8a^3 + 125}{2a + 5}$

Recall that the area of a rectangle is $A = lw$, where $w =$ width and $l =$ length. Solving for the width we get $w = \dfrac{A}{l}$ and solving for the length gives us $l = \dfrac{A}{w}$.

Find the missing side in each rectangle.

75) Area $= 5x^2 + 13x + 6$

                $x + 2$

Find the length.

76) Area $= 2y^2 - y - 15$

  $2y + 5$

Find the width.

77) Area $= c^3 - 2c^2 + 4c - 8$

  $c^2 + 4$

Find the width.

78) Area $= 2n^3 - 8n^2 + n - 4$

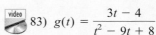
                $n - 4$

Find the length.

Objective 6

Determine the domain of each rational function.

79) $f(p) = \dfrac{1}{p - 7}$

80) $h(z) = \dfrac{z + 8}{z + 3}$

81) $k(r) = \dfrac{r}{5r + 2}$

82) $f(a) = \dfrac{6a}{7 - 2a}$

83) $g(t) = \dfrac{3t - 4}{t^2 - 9t + 8}$

84) $r(c) = \dfrac{c + 9}{c^2 - c - 42}$

85) $h(w) = \dfrac{w + 7}{w^2 - 81}$

86) $k(t) = \dfrac{t}{t^2 - 14t + 33}$

87) $A(c) = \dfrac{8}{c^2 + 6}$

88) $C(n) = \dfrac{3n + 1}{2}$

Section 8.2 Multiplying and Dividing Rational Expressions

Objectives

1. Multiply Fractions

2. Multiply Rational Expressions

3. Divide Rational Expressions

Multiplying Rational Expressions

We multiply rational expressions the same way we multiply rational numbers. Multiply numerators, multiply denominators, and simplify.

Multiplying Rational Expressions

If $\dfrac{P}{Q}$ and $\dfrac{R}{T}$ are rational expressions, then

$$\frac{P}{Q} \cdot \frac{R}{T} = \frac{PR}{QT}$$

To multiply two rational expressions, multiply their numerators, multiply their denominators, and simplify.

Let's begin by reviewing how we multiply two fractions.

1. Multiply Fractions

Example 1

Multiply $\dfrac{8}{15} \cdot \dfrac{5}{6}$.

Solution

We can multiply numerators, multiply denominators, then simplify by dividing out common factors *or* we can divide out the common factors before multiplying.

$$\frac{8}{15} \cdot \frac{5}{6} = \frac{2 \cdot 4}{3 \cdot \cancel{5}} \cdot \frac{\cancel{5}}{2 \cdot 3} \qquad \text{Factor and divide out common factors.}$$

$$= \frac{4}{3 \cdot 3} \qquad \text{Multiply.}$$

$$= \frac{4}{9} \qquad \text{Simplify.}$$

 You Try 1

Multiply $\dfrac{7}{32} \cdot \dfrac{20}{21}$.

Multiplying rational expressions works the same way.

2. Multiply Rational Expressions

> ### Multiplying Rational Expressions
> 1) Factor.
> 2) Reduce and multiply.
>
> *All products must be written in lowest terms.*

Example 2

Multiply.

a) $\dfrac{12a^4}{b^2} \cdot \dfrac{b^5}{6a^9}$

b) $\dfrac{8m + 48}{10m^6} \cdot \dfrac{m^2}{m^2 - 36}$

c) $\dfrac{3k^2 - 11k - 4}{k^2 + k - 20} \cdot \dfrac{k^2 + 10k + 25}{3k^2 + k}$

Solution

a) $\dfrac{12a^4}{b^2} \cdot \dfrac{b^5}{6a^9} = \dfrac{\cancel{12}a^4}{\cancel{b^2}} \cdot \dfrac{\cancel{b^2} \cdot b^3}{\cancel{6}a^4 \cdot a^5}$ Factor.

$\qquad = \dfrac{2b^3}{a^5}$ Reduce and multiply.

b) $\dfrac{8m + 48}{10m^6} \cdot \dfrac{m^2}{m^2 - 36} = \dfrac{\cancel{8}(m + 6)}{\cancel{10}m^2 \cdot m^4} \cdot \dfrac{\cancel{m^2}}{(m + 6)(m - 6)}$ Factor.

$\qquad\qquad = \dfrac{4}{5m^4(m - 6)}$ Reduce and multiply.

c) $\dfrac{3k^2 - 11k - 4}{k^2 + k - 20} \cdot \dfrac{k^2 + 10k + 25}{3k^2 + k} = \dfrac{(3k + 1)(k - 4)}{(k + 5)(k - 4)} \cdot \dfrac{(k + 5)^2}{k(3k + 1)}$ Factor.

$\qquad\qquad\qquad = \dfrac{k + 5}{k}$ Reduce and multiply.

You Try 2

Multiply.

a) $\dfrac{x^3}{14y^6} \cdot \dfrac{7y^2}{x^8}$

b) $\dfrac{r^2 - 25}{r^2 - 9r} \cdot \dfrac{r^2 + r - 90}{10 - 2r}$

3. Divide Rational Expressions

When we divide rational numbers we multiply by a reciprocal. For example,

$\dfrac{7}{4} \div \dfrac{3}{8} = \dfrac{7}{\cancel{4}} \cdot \dfrac{\cancel{8}^2}{3} = \dfrac{14}{3}$. We divide rational expressions the same way.

To divide rational expressions we multiply the first rational expression by the reciprocal of the second rational expression.

Dividing Rational Expressions

If $\dfrac{P}{Q}$ and $\dfrac{R}{T}$ are rational expressions with Q, R, and T not equal to zero, then

$$\frac{P}{Q} \div \frac{R}{T} = \frac{P}{Q} \cdot \frac{T}{R} = \frac{PT}{QR}$$

Multiply the first rational expression by the reciprocal of the second rational expression.

Example 3

Divide.

a) $\dfrac{36p^5}{q^4} \div \dfrac{4p^2}{q^{10}}$ b) $\dfrac{100a^3}{b^8} \div 20a^7$

c) $\dfrac{s^2 + 8s + 16}{s^2 - 11s + 24} \div \dfrac{s^2 + 5s + 4}{15 - 5s}$

Solution

a) $\dfrac{36p^5}{q^4} \div \dfrac{4p^2}{q^{10}} = \dfrac{\overset{9p^3}{\cancel{36p^5}}}{\cancel{q^4}} \cdot \dfrac{\overset{q^6}{\cancel{q^{10}}}}{\cancel{4p^2}}$ Multiply by the reciprocal.

$\qquad = 9p^3 q^6$ Reduce and multiply.

Notice that we used the *quotient rule* for exponents to reduce:

$$\frac{p^5}{p^2} = p^3, \quad \frac{q^{10}}{q^4} = q^6$$

b) $\dfrac{100a^3}{b^8} \div 20a^7 = \dfrac{\overset{5}{\cancel{100}}\,a^3}{b^8} \cdot \dfrac{1}{\underset{a^4}{\cancel{20a^7}}}$ Since $20a^7$ can be written as $\dfrac{20a^7}{1}$,

$\qquad = \dfrac{5}{a^4 b^8}$ its reciprocal is $\dfrac{1}{20a^7}$.

$\qquad\qquad\qquad\qquad\qquad$ Reduce and multiply.

Again, we used the *quotient rule* for exponents to simplify:

$$\frac{a^3}{a^7} = \frac{1}{a^4}$$

c) $\dfrac{s^2 + 8s + 16}{s^2 - 11s + 24} \div \dfrac{s^2 + 5s + 4}{15 - 5s}$

$\qquad = \dfrac{s^2 + 8s + 16}{s^2 - 11s + 24} \cdot \dfrac{15 - 5s}{s^2 + 5s + 4}$ Multiply by the reciprocal.

$\qquad = \dfrac{\overset{(s+4)}{\cancel{(s+4)^2}}}{(\cancel{s-3})(s-8)} \cdot \dfrac{\overset{-1}{5\cancel{(3-s)}}}{\cancel{(s+4)}(s+1)}$ Factor; $\dfrac{3-s}{s-3} = -1$.

$\qquad = \dfrac{-5(s+4)}{(s-8)(s+1)}$ Reduce and multiply.

You Try 3

Divide.

a) $\dfrac{r^6}{35t^9} \div \dfrac{r^4}{14t^3}$ b) $\dfrac{x^2 - 8x - 48}{7x^2 - 42x} \div \dfrac{x^2 - 16}{x^2 - 10x + 24}$

Remember that a fraction, itself, represents division. That is, $\dfrac{20}{4} = 20 \div 4 = 5$.

We can write division problems involving fractions and rational expressions in a similar way.

Example 4

Divide.

a) $\dfrac{\dfrac{7}{30}}{\dfrac{14}{25}}$ b) $\dfrac{\dfrac{2y+1}{3}}{\dfrac{4y^2-1}{6}}$

Solution

a) $\dfrac{\dfrac{7}{30}}{\dfrac{14}{25}}$ means $\dfrac{7}{30} \div \dfrac{14}{25}$. Then,

$$\dfrac{7}{30} \div \dfrac{14}{25} = \dfrac{7}{30} \cdot \dfrac{25}{14} \qquad \text{Multiply by the reciprocal.}$$

$$= \dfrac{\overset{1}{\cancel{7}}}{\underset{6}{\cancel{30}}} \cdot \dfrac{\overset{5}{\cancel{25}}}{\underset{2}{\cancel{14}}} \qquad \text{Divide 7 and 14 by 7. Divide 25 and 30 by 5.}$$

$$= \dfrac{5}{12} \qquad \text{Multiply.}$$

b) $\dfrac{\dfrac{2y+1}{3}}{\dfrac{4y^2-1}{6}}$ means $\dfrac{2y+1}{3} \div \dfrac{4y^2-1}{6}$. Then,

$$\dfrac{2y+1}{3} \div \dfrac{4y^2-1}{6} = \dfrac{2y+1}{3} \cdot \dfrac{6}{4y^2-1} \qquad \text{Multiply by the reciprocal.}$$

$$= \dfrac{\cancel{2y+1}}{\underset{1}{\cancel{3}}} \cdot \dfrac{\overset{2}{\cancel{6}}}{(\cancel{2y+1})(2y-1)} \qquad \text{Factor.}$$

$$= \dfrac{2}{2y-1} \qquad \text{Reduce and multiply.}$$

You Try 4

Divide.

a) $\dfrac{\dfrac{10}{27}}{\dfrac{20}{9}}$ b) $\dfrac{\dfrac{t^3 + 5t^2}{12}}{\dfrac{t + 5}{8}}$

Answers to You Try Exercises

1) $\dfrac{5}{24}$ 2) a) $\dfrac{1}{2x^5y^4}$ b) $-\dfrac{(r + 5)(r + 10)}{2r}$ 3) a) $\dfrac{2r^2}{5t^6}$ b) $\dfrac{x - 12}{7x}$ 4) a) $\dfrac{1}{6}$ b) $\dfrac{2t^2}{3}$

8.2 Exercises

Boost your grade at mathzone.com! MathZone > Practice Problems > NetTutor > e-Professors > Videos > Self-Test

Objective 1

Multiply.

1) $\dfrac{7}{8} \cdot \dfrac{1}{3}$

2) $\dfrac{2}{5} \cdot \dfrac{3}{11}$

3) $\dfrac{9}{14} \cdot \dfrac{7}{6}$

4) $\dfrac{4}{15} \cdot \dfrac{25}{36}$

Objective 2

Multiply.

5) $\dfrac{22}{15r^2} \cdot \dfrac{5r^4}{2}$

6) $\dfrac{27b^4}{4} \cdot \dfrac{3}{81b}$

7) $\dfrac{14u^5}{15v^2} \cdot \dfrac{20v^6}{7u^8}$

8) $\dfrac{15s^3}{21t^2} \cdot \dfrac{42t^4}{5s^{12}}$

9) $-\dfrac{12x}{24x^4} \cdot \dfrac{8x^9}{9}$

10) $\dfrac{42c}{9c^4} \cdot \dfrac{3}{35c^3}$

11) $\dfrac{a - 6}{10} \cdot \dfrac{2(a + 1)}{(a - 6)^2}$

12) $\dfrac{11(z + 3)}{6} \cdot \dfrac{8(z - 4)}{(z + 3)^2}$

 13) $\dfrac{5t^2}{(3t - 2)^2} \cdot \dfrac{3t - 2}{10t^3}$

14) $\dfrac{4u - 5}{9u^2} \cdot \dfrac{3u^6}{(4u - 5)^3}$

15) $\dfrac{8}{6p + 3} \cdot \dfrac{4p^2 - 1}{12}$

16) $\dfrac{n^2 + 7n + 12}{n + 3} \cdot \dfrac{4}{n + 4}$

17) $\dfrac{2v^2 + 15v + 18}{3v + 18} \cdot \dfrac{12v - 3}{8v + 12}$

18) $\dfrac{y^2 - 4y - 5}{3y^2 + y - 2} \cdot \dfrac{18y - 12}{4y^2}$

Objective 3

Divide.

19) $\dfrac{4}{5} \div \dfrac{8}{3}$

20) $\dfrac{16}{9} \div \dfrac{10}{3}$

21) $\dfrac{12}{7} \div 6$

22) $36 \div \dfrac{9}{4}$

Divide.

23) $\dfrac{42k^6}{25} \div \dfrac{12k^2}{35}$

24) $\dfrac{21}{8m^{10}} \div \dfrac{15}{4m^7}$

25) $\dfrac{c^2}{6b} \div \dfrac{c^8}{b}$

26) $-\dfrac{15g^3}{14h} \div \dfrac{40g}{7h^3}$

27) $\dfrac{30(x - 7)}{x + 8} \div \dfrac{18}{(x + 8)^2}$

28) $\dfrac{(k - 5)^2}{28} \div \dfrac{2(k - 5)}{21k^3}$

29) $\dfrac{2a - 1}{8a^3} \div \dfrac{(2a - 1)^2}{24a^5}$

30) $\dfrac{2p^4}{(p + 7)^2} \div \dfrac{12p^5}{p + 7}$

31) $\dfrac{18y - 45}{18} \div \dfrac{4y^2 - 25}{10}$

32) $\dfrac{q^2 + q - 56}{5} \div \dfrac{q - 7}{q}$

33) $\dfrac{j^2 - 25}{5j + 25} \div \dfrac{7j - 35}{5}$

34) $\dfrac{n^2 + 3n - 18}{5n^2 + 30n} \div \dfrac{4n - 12}{8n}$

35) $\dfrac{4c - 9}{2c^2 - 8c} \div \dfrac{12c - 27}{c^2 - 3c - 4}$

36) $\dfrac{p^2 + 15p}{p^2 - 6p - 16} \div \dfrac{p + 15}{p + 2}$

Divide.

37) $\dfrac{\frac{4}{15}}{\frac{8}{35}}$

38) $\dfrac{\frac{10}{21}}{\frac{25}{28}}$

39) $\dfrac{\frac{2}{9}}{\frac{8}{3}}$

40) $\dfrac{\frac{15}{4}}{\frac{5}{24}}$

41) $\dfrac{\frac{6s - 7}{4}}{\frac{6s - 7}{12}}$

42) $\dfrac{\frac{3d + 5}{18}}{\frac{3d + 5}{9}}$

43) $\dfrac{\frac{16r + 24}{r^3}}{\frac{12r + 18}{r}}$

44) $\dfrac{\frac{8m - 6}{m}}{\frac{40m - 30}{3m^2}}$

45) $\dfrac{\frac{4z - 20}{z^2}}{\frac{z^2 - 25}{z^6}}$

46) $\dfrac{\frac{a^2 - 49}{a^8}}{\frac{8a - 56}{2a^3}}$

47) $\dfrac{\frac{12}{9a^2 - 4}}{\frac{16a^2}{3a^2 + 2a}}$

48) $\dfrac{\frac{36x - 45}{6x^3}}{\frac{16x^2 - 25}{x^7}}$

Objectives 2 and 3

Multiply or divide as indicated.

49) $\dfrac{d^2 + d - 56}{d - 11} \cdot \dfrac{d^2 - 10d - 11}{4d + 32}$

50) $\dfrac{c^2 - 14c + 48}{3c + 15} \cdot \dfrac{c^2 + 10c + 25}{c^2 - 64}$

51) $\dfrac{b^2 - 4b + 4}{4b - 8} \div \dfrac{3b^2 - 4b - 4}{3b + 2}$

52) $\dfrac{2x - 3}{4x^2 - 9} \div \dfrac{3}{8x^2 + 10x - 3}$

53) $\dfrac{4n^2 - 1}{6n^5} \div \dfrac{2n^2 - 7n - 4}{10n^3}$

54) $\dfrac{14t^6}{t^2 - 4} \div \dfrac{7t^2}{3t^2 - 7t + 2}$

55) $\dfrac{a^2 - 4a}{6a + 54} \cdot \dfrac{a^2 + 13a + 36}{16 - a^2}$

56) $\dfrac{6h}{h^2 - 81} \cdot \dfrac{9h^2 - h^3}{8}$

57) $\dfrac{4t^2 + 12t + 36}{4t - 2} \cdot \dfrac{t^2 - 9}{t^3 - 27}$

58) $\dfrac{r^3 + 8}{r + 2} \cdot \dfrac{7}{4r^2 - 8r + 16}$

59) $\dfrac{64 - u^2}{40 - 5u} \div \dfrac{u^2 + 10u + 16}{2u + 3}$

60) $\dfrac{c^2 - 49}{c + 7} \div \dfrac{56 - 8c}{c + 3}$

61) $\dfrac{24x^3}{x^3 - 6x^2 + 5x - 30} \cdot \dfrac{3x^2 - 20x + 12}{18x^2 - 12x}$

62) $\dfrac{6a}{a^2 - 5a - 24} \cdot \dfrac{a^3 + 3a^2 + 2a + 6}{6a^2 + 12}$

63) $\dfrac{a^2 - b^2}{a^3 + b^3} \div \dfrac{9b - 9a}{8}$

64) $\dfrac{20}{y^2 - x^2} \div \dfrac{10x^2 + 10xy + 10y^2}{x^3 - y^3}$

65) $\dfrac{w^2 - 17w + 72}{6w} \div (w - 9)$

66) $\dfrac{3m^2 + 7m + 4}{8} \div (6m + 8)$

Perform the operations and simplify.

67) $\dfrac{a}{a^2 + 20a + 100} \div \left(\dfrac{a^2 - 7a - 18}{a^2 - 5a - 14} \cdot \dfrac{a^2 - 7a}{a^2 + a - 90} \right)$

68) $\dfrac{4j^2 - 37j + 9}{j^3} \div \left(\dfrac{3j + 2}{j^3 - j^2} \cdot \dfrac{j^2 - 10j + 9}{j} \right)$

69) $\dfrac{t^3 - 1}{t - 1} \div \left(\dfrac{5t + 1}{12t - 8} \cdot \dfrac{t^2 + t + 1}{t} \right)$

70) $\dfrac{x}{2x^2 - 8x + 32} \div \left(\dfrac{5x + 20}{x + 2} \cdot \dfrac{x^2 - 4}{x^3 + 64} \right)$

Section 8.3 Finding the Least Common Denominator

Objectives

1. Find the Least Common Denominator for a Group of Rational Expressions

2. For a Group of Fractions, Rewrite Each as an Equivalent Fraction With the LCD as Its Denominator

3. For a Group of Rational Expressions, Rewrite Each as an Equivalent Expression With the LCD as Its Denominator

1. Find the Least Common Denominator for a Group of Rational Expressions

Recall that to add or subtract fractions, they must have a common denominator. Similarly, rational expressions must have common denominators in order to be added or subtracted. In this section, we will discuss how to find the least common denominator (LCD) of rational expressions.

We begin by looking at the fractions $\frac{3}{8}$ and $\frac{5}{12}$. By inspection we can see that the LCD = 24. But, *why* is that true? Let's write each of the denominators, 8 and 12, as the product of their prime factors:

$$8 = 2 \cdot 2 \cdot 2 = 2^3$$
$$12 = 2 \cdot 2 \cdot 3 = 2^2 \cdot 3$$

The LCD will contain each factor the *greatest* number of times it appears in any single factorization.

> *2* appears as a factor *three* times in the factorization of 8 and it appears *twice* in the factorization of 12.
> *The LCD will contain 2^3.*
> *3* appears as a factor *one* time in the factorization of 12 but does not appear in the factorization of 8.
> *The LCD will contain 3.*

The LCD, then, is the product of the factors we have identified.

$$\text{LCD of } \frac{3}{8} \text{ and } \frac{5}{12} = 2^3 \cdot 3 = 8 \cdot 3 = 24$$

This is the same result as the one we obtained just by inspecting the two denominators.

The procedure we just illustrated is the one we use to find the least common denominator of rational expressions.

Finding the Least Common Denominator (LCD)

Step 1: Factor the denominators.

Step 2: The LCD will contain each unique factor the *greatest* number of times it appears in any single factorization.

Step 3: The LCD is the *product* of the factors identified in step 2.

| **Example 1** |

Find the LCD of each pair of rational expressions.

a) $\dfrac{7}{18}, \dfrac{11}{24}$ b) $\dfrac{11}{12a}, \dfrac{8}{15a}$ c) $\dfrac{4}{9t^3}, \dfrac{5}{6t^2}$

Solution

a) Follow the steps outlined for finding the least common denominator.

Step 1: Factor the denominators.

$$18 = 2 \cdot 3 \cdot 3 = 2 \cdot 3^2$$
$$24 = 2 \cdot 2 \cdot 2 \cdot 3 = 2^3 \cdot 3$$

Step 2: The LCD will contain each unique factor the *greatest* number of times it appears in any factorization. *The LCD will contain 2^3 and 3^2.*

Step 3: The LCD is the *product* of the factors in step 2.

$$LCD = 2^3 \cdot 3^2 = 8 \cdot 9 = 72$$

The least common denominator of $\dfrac{7}{18}$ and $\dfrac{11}{24}$ is 72.

b) Follow the steps to find the LCD of $\dfrac{11}{12a}$ and $\dfrac{8}{15a}$.

Step 1: Factor the denominators.

$$12a = 2 \cdot 2 \cdot 3 \cdot a = 2^2 \cdot 3 \cdot a$$
$$15a = 3 \cdot 5 \cdot a$$

Step 2: The LCD will contain each unique factor the *greatest* number of times it appears in any factorization. *It will contain 2^2, 3, 5, and a.*

Step 3: The LCD is the *product* of the factors in step 2.

$$LCD = 2^2 \cdot 3 \cdot 5 \cdot a = 60a.$$

The least common denominator of $\dfrac{11}{12a}$ and $\dfrac{8}{15a}$ is $60a$.

c) To find the LCD of $\dfrac{4}{9t^3}$ and $\dfrac{5}{6t^2}$,

Step 1: Factor the denominators.

$$9t^3 = 3 \cdot 3 \cdot t^3 = 3^2 \cdot t^3$$
$$6t^2 = 2 \cdot 3 \cdot t^2$$

Step 2: The LCD will contain each unique factor the *greatest* number of times it appears in any factorization. *It will contain 2, 3^2, and t^3.*

Step 3: The LCD is the *product* of the factors in step 2.

$$LCD = 2 \cdot 3^2 \cdot t^3 = 18t^3$$

The least common denominator of $\dfrac{4}{9t^3}$ and $\dfrac{5}{6t^2}$ is $18t^3$.

You Try 1

Find the LCD of each pair of rational expressions.

a) $\dfrac{4}{15}, \dfrac{9}{20}$ b) $\dfrac{8}{9k^3}, \dfrac{1}{12k^5}$

Example 2

Find the LCD of each group of rational expressions.

a) $\dfrac{6}{x}, \dfrac{2}{x+5}$ b) $\dfrac{10}{c-8}, \dfrac{4c}{c^2-5c-24}$

c) $\dfrac{9}{w^2+2w+1}, \dfrac{1}{2w^2+2w}$

Solution

a)   The denominators of $\dfrac{6}{x}$ and $\dfrac{2}{x+5}$ are already in simplest form. It is important
to recognize that *x and x + 5 are different factors.*

The LCD will be the product of x and $x + 5$.

The LCD of $\dfrac{6}{x}$ and $\dfrac{2}{x+5}$ is $x(x+5)$.

Usually, we leave the LCD in this form; we do not distribute.

b)   Follow the steps to find the LCD of $\dfrac{10}{c-8}$ and $\dfrac{4c}{c^2-5c-24}$.

Step 1:   Factor the denominators.

$$c - 8 \text{ cannot be factored}$$
$$c^2 - 5c - 24 = (c-8)(c+3)$$

Step 2:   The LCD will contain each unique factor the *greatest* number of times
it appears in any factorization. *It will contain c − 8 and c + 3.*

Step 3:   The LCD is the *product* of the factors identified in step 2.

$$\text{LCD} = (c-8)(c+3)$$

The LCD of $\dfrac{10}{c-8}$ and $\dfrac{4c}{c^2-5c-24}$ is $(c-8)(c+3)$.

c)   Find the LCD of $\dfrac{9}{w^2+2w+1}$ and $\dfrac{1}{2w^2+2w}$.

Step 1:   Factor the denominators.

$$w^2 + 2w + 1 = (w+1)^2$$
$$2w^2 + 2w = 2w(w+1)$$

Step 2:   The unique factors are 2, w, and $w + 1$ with *w + 1 appearing at most
twice. The factors we will use in the LCD are 2, w, and $(w+1)^2$.*

Step 3:   The LCD is the *product* of the factors identified in step 2.

$$LCD = 2w(w + 1)^2$$

The LCD of $\dfrac{9}{w^2 + 2w + 1}$ and $\dfrac{1}{2w^2 + 2w}$ is $2w(w + 1)^2$.

You Try 2

Find the LCD of each group of rational expressions.

a) $\dfrac{5}{k}, \dfrac{8k}{k + 2}$ b) $\dfrac{12}{p^2 - 7p}, \dfrac{6}{p - 7}$ c) $\dfrac{m}{m^2 - 25}, \dfrac{8}{m^2 + 10m + 25}$

At first glance it may appear that the least common denominator of $\dfrac{2}{y - 3}$ and $\dfrac{10}{3 - y}$ is $(y - 3)(3 - y)$. This is *not* the case. Recall from Section 8.1 that $a - b = -1(b - a)$.

We will use this idea to find the LCD of $\dfrac{2}{y - 3}$ and $\dfrac{10}{3 - y}$.

Example 3

Find the LCD of $\dfrac{2}{y - 3}$ and $\dfrac{10}{3 - y}$.

Solution

Since $3 - y = -(y - 3)$, we can rewrite $\dfrac{10}{3 - y}$ as

$$\frac{10}{3 - y} = \frac{10}{-(y - 3)} = -\frac{10}{y - 3}$$

Therefore, we can now think of our task as

Find the LCD of $\dfrac{2}{y - 3}$ and $-\dfrac{10}{y - 3}$.

The least common denominator is $y - 3$.

You Try 3

Find the LCD of $\dfrac{9}{r - 8}$ and $\dfrac{6}{8 - r}$.

2. For a Group of Fractions, Rewrite Each as an Equivalent Fraction With the LCD as Its Denominator

As we know from our previous work with fractions, after *determining* the least common denominator, we must *rewrite* those fractions as equivalent fractions with the LCD so that they can be added or subtracted.

Example 4

Identify the LCD of $\dfrac{7}{10}$ and $\dfrac{4}{15}$, and rewrite each as an equivalent fraction with the LCD as its denominator.

Solution

The LCD of $\dfrac{7}{10}$ and $\dfrac{4}{15}$ is 30. We must rewrite each fraction with a denominator of 30.

$\dfrac{7}{10}$:   By what number should we multiply 10 to get 30? 3

Multiply the numerator *and* denominator by 3 to obtain an equivalent fraction.

$$\frac{7}{10} \cdot \frac{3}{3} = \frac{21}{30}$$

$\dfrac{4}{15}$:   By what number should we multiply 15 to get 30? 2

Multiply the numerator *and* denominator by 2 to obtain an equivalent fraction.

$$\frac{4}{15} \cdot \frac{2}{2} = \frac{8}{30}$$

Therefore, the LCD of $\dfrac{7}{10}$ and $\dfrac{4}{15}$ is 30, and

$$\frac{7}{10} = \frac{21}{30} \quad \text{and} \quad \frac{4}{15} = \frac{8}{30}$$

You Try 4

Identify the LCD of $\dfrac{5}{8}$ and $\dfrac{1}{6}$, and rewrite each as an equivalent fraction with the LCD as its denominator.

3. For a Group of Rational Expressions, Rewrite Each as an Equivalent Expression With the LCD as Its Denominator

The procedure for rewriting rational expressions as equivalent expressions with the least common denominator is very similar to the process used in Example 4.

Writing Rational Expressions as Equivalent Expressions With the Least Common Denominator

Step 1: Identify and write down the LCD.

Step 2: Look at each rational expression (with its denominator in factored form) and compare its denominator with the LCD. Ask yourself, "What factors are missing?"

Step 3: Multiply the numerator and denominator by the "missing" factors to obtain an equivalent rational expression with the desired LCD. Multiply the terms in the numerator, but leave the denominator as the product of factors. (We will see why this is done in Section 8.4.)

Example 5

Identify the LCD of each pair of rational expressions, and rewrite each as an equivalent expression with the LCD as its denominator.

a) $\dfrac{7}{8n}, \dfrac{5}{6n^3}$

b) $\dfrac{t}{t-2}, \dfrac{4}{t+7}$

c) $\dfrac{3}{4a^2-24a}, \dfrac{9a}{a^2-11a+30}$

d) $\dfrac{r}{r-6}, \dfrac{2}{6-r}$

Solution

a) Follow the steps.

Step 1: Identify and write down the LCD of $\dfrac{7}{8n}$ and $\dfrac{5}{6n^3}$.

$$\text{LCD} = 24n^3$$

Step 2: Compare the denominators of $\dfrac{7}{8n}$ and $\dfrac{5}{6n^3}$ to the LCD and ask yourself, "What's missing?"

$\dfrac{7}{8n}$: 8n is "missing" the factors of 3 and n^2.

$\dfrac{5}{6n^3}$: $6n^3$ is "missing" the factor 4.

Step 3: Multiply the numerator and denominator by $3n^2$.

$$\frac{7}{8n} \cdot \frac{3n^2}{3n^2} = \frac{21n^2}{24n^3}$$

Multiply the numerator and denominator by 4.

$$\frac{5}{6n^3} \cdot \frac{4}{4} = \frac{20}{24n^3}$$

$$\frac{7}{8n} = \frac{21n^2}{24n^3} \qquad \text{and} \qquad \frac{5}{6n^3} = \frac{20}{24n^3}$$

b)   Follow the steps.

Step 1:   Identify and write down the LCD of $\dfrac{t}{t-2}$ and $\dfrac{4}{t+7}$.

$$LCD = (t-2)(t+7)$$

Step 2:   Compare the denominators of $\dfrac{t}{t-2}$ and $\dfrac{4}{t+7}$ to the LCD and ask yourself, "What's missing?"

$\dfrac{t}{t-2}$: *t − 2 is "missing" the*

factor t + 7.

$\dfrac{4}{t+7}$: *t + 7 is "missing"*

the factor t − 2.

Step 3:   Multiply the numerator and denominator by $t+7$.

$$\frac{t}{t-2} \cdot \frac{t+7}{t+7} = \frac{t(t+7)}{(t-2)(t+7)}$$
$$= \frac{t^2 + 7t}{(t-2)(t+7)}$$

Multiply the numerator and denominator by $t-2$.

$$\frac{4}{t+7} \cdot \frac{t-2}{t-2} = \frac{4(t-2)}{(t+7)(t-2)}$$
$$= \frac{4t-8}{(t-2)(t+7)}$$

Notice that we multiplied the factors in the numerator but left the denominator in factored form.

$$\frac{t}{t-2} = \frac{t^2 + 7t}{(t-2)(t+7)} \quad \text{and} \quad \frac{4}{t+7} = \frac{4t-8}{(t-2)(t+7)}$$

c)   Follow the steps.

Step 1:   Identify and write down the LCD of $\dfrac{3}{4a^2 - 24a}$ and $\dfrac{9a}{a^2 - 11a + 30}$.
First, we must factor the denominators.

$$\frac{3}{4a^2 - 24a} = \frac{3}{4a(a-6)}, \quad \frac{9a}{a^2 - 11a + 30} = \frac{9a}{(a-6)(a-5)}$$

We will work with the factored forms of the expressions.

$$LCD = 4a(a-6)(a-5)$$

Step 2:   Compare the denominators of $\dfrac{3}{4a(a-6)}$ and $\dfrac{9a}{(a-6)(a-5)}$ to the LCD and ask yourself, "What's missing?"

$\dfrac{3}{4a(a-6)}$: *4a(a − 6) is "missing"*

the factor a − 5.

$\dfrac{9a}{(a-6)(a-5)}$: *(a − 6)(a − 5) is*

"missing" 4a.

Step 3: Multiply the numerator and denominator by $a - 5$.

$$\frac{3}{4a(a - 6)} \cdot \frac{a - 5}{a - 5} = \frac{3(a - 5)}{4a(a - 6)(a - 5)}$$
$$= \frac{3a - 15}{4a(a - 6)(a - 5)}$$

Multiply the numerator and denominator by $4a$.

$$\frac{9a}{(a - 6)(a - 5)} \cdot \frac{4a}{4a} = \frac{36a^2}{4a(a - 6)(a - 5)}$$

$$\frac{3}{4a^2 - 24a} = \frac{3a - 15}{4a(a - 6)(a - 5)} \quad \text{and} \quad \frac{9a}{a^2 - 11a + 30} = \frac{36a^2}{4a(a - 6)(a - 5)}$$

d) Recall that $6 - r$ can be rewritten as $-(r - 6)$. So,

$$\frac{2}{6 - r} = \frac{2}{-(r - 6)} = -\frac{2}{r - 6}$$

Therefore, the LCD of $\dfrac{r}{r - 6}$ and $-\dfrac{2}{r - 6}$ is $r - 6$.

$\dfrac{r}{r - 6}$ already has the LCD, while $\dfrac{2}{6 - r} = -\dfrac{2}{r - 6}$.

You Try 5

Identify the least common denominator of each pair of rational expressions, and rewrite each as an equivalent expression with the LCD as its denominator.

a) $\dfrac{9}{7r^5}, \dfrac{4}{21r^2}$ b) $\dfrac{6}{y + 4}, \dfrac{8}{3y - 2}$ c) $\dfrac{d - 1}{d^2 + 2d}, \dfrac{3}{d^2 + 12d + 20}$

d) $\dfrac{k}{7 - k}, \dfrac{4}{k - 7}$

Answers to You Try Exercises

1) a) 60 b) $36k^5$ 2) a) $k(k + 2)$ b) $p(p - 7)$ c) $(m + 5)^2(m - 5)$ 3) $r - 8$

4) LCD = 24; $\dfrac{5}{8} = \dfrac{15}{24}, \dfrac{1}{6} = \dfrac{4}{24}$ 5) a) LCD = $21r^5$; $\dfrac{9}{7r^5} = \dfrac{27}{21r^5}, \dfrac{4}{21r^2} = \dfrac{4r^3}{21r^5}$

b) LCD = $(y + 4)(3y - 2)$; $\dfrac{6}{y + 4} = \dfrac{18y - 12}{(y + 4)(3y - 2)}, \dfrac{8}{3y - 2} = \dfrac{8y + 32}{(y + 4)(3y - 2)}$

c) LCD = $d(d + 2)(d + 10)$; $\dfrac{d - 1}{d^2 + 2d} = \dfrac{d^2 + 9d - 10}{d(d + 2)(d + 10)}, \dfrac{3}{d^2 + 12d + 20} = \dfrac{3d}{d(d + 2)(d + 10)}$

d) LCD = $k - 7$; $\dfrac{k}{7 - k} = -\dfrac{k}{k - 7}, \dfrac{4}{k - 7} = \dfrac{4}{k - 7}$

8.3 Exercises

Objective 1

Find the LCD of each group of rational expressions.

1) $\dfrac{5}{8}, \dfrac{9}{20}$

2) $\dfrac{11}{12}, \dfrac{5}{16}$

3) $\dfrac{17}{28}, \dfrac{1}{12}, \dfrac{8}{21}$

4) $\dfrac{23}{56}, \dfrac{9}{16}, \dfrac{13}{14}$

5) $\dfrac{6}{c^4}, \dfrac{7}{c^3}$

6) $\dfrac{4}{n^5}, \dfrac{1}{n^9}$

7) $\dfrac{8}{9p^3}, \dfrac{5}{12p^8}$

8) $\dfrac{4}{15r^6}, \dfrac{7}{6r^2}$

9) $\dfrac{10}{21w^5}, -\dfrac{6}{7w^7}$

10) $-\dfrac{2}{9z^4}, \dfrac{16}{45z^6}$

11) $-\dfrac{8}{3k^2}, \dfrac{5}{12k^5}$

12) $\dfrac{7}{10m}, \dfrac{9}{35m^4}$

13) $\dfrac{1}{8a^3b^3}, \dfrac{7}{12ab^4}$

14) $\dfrac{4}{27x^2y^2}, \dfrac{11}{3x^3y^2}$

15) $\dfrac{3}{n+4}, \dfrac{1}{2}$

16) $\dfrac{7}{9}, \dfrac{6}{z-8}$

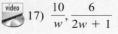

17) $\dfrac{10}{w}, \dfrac{6}{2w+1}$

18) $\dfrac{2}{y}, -\dfrac{9}{3y+5}$

19) $\dfrac{5}{6k+30}, \dfrac{11}{9k+45}$

20) $\dfrac{9}{4c-12}, \dfrac{8}{3c-9}$

21) $\dfrac{1}{12a^2-4a}, \dfrac{15}{6a^4-2a^3}$

22) $\dfrac{1}{10p^3-15p^2}, \dfrac{6}{2p^5-3p^4}$

23) $\dfrac{8}{r+7}, \dfrac{4}{r-2}$

24) $\dfrac{m}{m-6}, \dfrac{3}{m-5}$

25) $\dfrac{4x}{x^2-10x+16}, \dfrac{x}{x^2-5x-24}$

26) $\dfrac{12}{z^2+11z+18}, \dfrac{5z}{z^2-2z-8}$

27) $\dfrac{w}{w^2-3w-10}, \dfrac{5}{w^2-2w-15}, \dfrac{9w}{w^2+5w+6}$

28) $\dfrac{10t}{t^2-5t-14}, -\dfrac{8}{t^2-49}, \dfrac{t}{t^2+9t+14}$

29) $\dfrac{6}{b-4}, \dfrac{5}{4-b}$

30) $\dfrac{7}{a-6}, \dfrac{4}{6-a}$

31) $\dfrac{u}{v-u}, \dfrac{2}{u-v}$

32) $\dfrac{9}{y-x}, \dfrac{3y}{x-y}$

Objective 3

33) Explain, in your own words, how to rewrite $\dfrac{5}{x+8}$ as an equivalent rational expression with a denominator of $(x+8)(x-2)$.

34) Explain, in your own words, how to rewrite $\dfrac{6}{3-n}$ as an equivalent rational expression with a denominator of $n-3$.

Rewrite each rational expression with the indicated denominator.

35) $\dfrac{4}{3p^4} = \dfrac{}{6p^5}$

36) $\dfrac{2}{9k} = \dfrac{}{45k^3}$

37) $\dfrac{11}{4cd^2} = \dfrac{}{24c^3d^3}$

38) $\dfrac{7}{2t^3u} = \dfrac{}{6t^4u^5}$

39) $\dfrac{5}{m-9} = \dfrac{}{m(m-9)}$

40) $\dfrac{3}{2r+5} = \dfrac{}{r(2r+5)}$

41) $\dfrac{a}{8(3a+1)} = \dfrac{}{16a(3a+1)}$

42) $\dfrac{v}{5(v-6)} = \dfrac{}{20v^2(v-6)}$

43) $\dfrac{3b}{b+2} = \dfrac{}{(b+2)(b+5)}$

44) $\dfrac{10x}{x+9} = \dfrac{}{(x+9)(x-7)}$

45) $\dfrac{w+1}{4w-3} = \dfrac{}{(4w-3)(w-6)}$

46) $\dfrac{z-4}{3z-2} = \dfrac{}{(3z-2)(z+8)}$

47) $\dfrac{8}{5-n} = \dfrac{}{n-5}$

48) $\dfrac{6}{1-p} = \dfrac{}{p-1}$

49) $-\dfrac{a}{3a-2} = \dfrac{}{2-3a}$

50) $-\dfrac{4c}{5c-3} = \dfrac{}{3-5c}$

Identify the least common denominator of each group of rational expression, and rewrite each as an equivalent rational expression with the LCD as its denominator.

51) $\dfrac{3}{t}, \dfrac{8}{t^3}$

52) $\dfrac{10}{p^5}, \dfrac{7}{p^2}$

  53) $\dfrac{9}{8n^6}, \dfrac{2}{3n^2}$

54) $\dfrac{5}{6a}, \dfrac{7}{8a^5}$

55) $\dfrac{1}{x^3 y}, \dfrac{6}{5xy^5}$

56) $\dfrac{5}{6a^2 b^4}, \dfrac{5}{a^4 b}$

57) $\dfrac{t}{5t - 6}, \dfrac{10}{7}$

58) $\dfrac{r}{4}, \dfrac{3}{r - 6}$

59) $\dfrac{3}{c}, \dfrac{2}{c + 1}$

60) $\dfrac{8}{d}, \dfrac{2}{d - 4}$

61) $\dfrac{z}{z - 9}, \dfrac{4}{z}$

62) $\dfrac{m}{m + 3}, \dfrac{6}{m}$

63) $\dfrac{a}{24a + 36}, \dfrac{1}{18a + 27}$

64) $\dfrac{7}{12x - 4}, \dfrac{x}{18x - 6}$

65) $\dfrac{4}{h + 5}, \dfrac{7h}{h - 3}$

66) $\dfrac{1}{k - 10}, \dfrac{6k}{k + 4}$

67) $\dfrac{b}{3b - 2}, \dfrac{1}{b - 9}$

68) $\dfrac{8}{a + 9}, \dfrac{a}{2a + 7}$

video 69) $\dfrac{9y}{y^2 - y - 42}, \dfrac{3}{2y^2 + 12y}$

70) $\dfrac{4q}{3q^2 + 24q}, \dfrac{5}{q^2 + q - 56}$

71) $\dfrac{z}{z^2 - 10z + 25}, \dfrac{15z}{z^2 - 2z - 15}$

72) $\dfrac{c}{c^2 + 11c + 28}, \dfrac{6}{c^2 + 14c + 49}$

video 73) $\dfrac{11}{g - 3}, \dfrac{4}{9 - g^2}$

74) $\dfrac{3}{25 - n^2}, \dfrac{5}{n - 5}$

75) $\dfrac{10}{4k - 1}, \dfrac{k}{1 - 16k^2}$

76) $\dfrac{7}{2x - 3}, \dfrac{4x}{9 - 4x^2}$

77) $\dfrac{4}{w^2 - 4w}, \dfrac{6}{7w^2 - 28w}, \dfrac{11}{w^2 - 8w + 16}$

78) $\dfrac{8}{z^2 + 2z}, -\dfrac{2}{5z^2 + 10z}, \dfrac{5}{z^2 + 4z + 4}$

79) $-\dfrac{1}{a + 4}, \dfrac{a}{a^2 - 16}, \dfrac{3}{a^2 + 5a + 4}$

80) $\dfrac{t}{t^2 - 4t - 21}, \dfrac{2}{t + 3}, \dfrac{4}{t^2 - 49}$

Section 8.4 Adding and Subtracting Rational Expressions

Objectives

1. Add and Subtract Rational Expressions That Have a Common Denominator

2. Add and Subtract Rational Expressions With Different Denominators

3. Add and Subtract Rational Expressions With Denominators Containing Factors of the Form $a - b$ and $b - a$

We know that in order to add or subtract fractions, they must have a common denominator. The same is true for rational expressions.

1. Add and Subtract Rational Expressions That Have a Common Denominator

We will begin our discussion of adding and subtracting rational expressions by looking at fractions and rational expressions with common denominators.

Example 1

Add or subtract, as indicated.

a) $\dfrac{6}{7} - \dfrac{2}{7}$

b) $\dfrac{3a}{2a - 5} + \dfrac{4a + 1}{2a - 5}$

Solution

a)   Since $\dfrac{6}{7}$ and $\dfrac{2}{7}$ have the same denominator, subtract the terms in the numerator and keep the common denominator.

$$\frac{6}{7} - \frac{2}{7} = \frac{6 - 2}{7} \qquad \text{Subtract terms in the numerator.}$$

$$= \frac{4}{7} \qquad \text{Simplify.}$$

b)   Since $\dfrac{3a}{2a - 5}$ and $\dfrac{4a + 1}{2a - 5}$ have the same denominator, add the terms in the numerator and keep the common denominator.

$$\frac{3a}{2a - 5} + \frac{4a + 1}{2a - 5} = \frac{3a + (4a + 1)}{2a - 5} \qquad \text{Add terms in the numerator.}$$

$$= \frac{7a + 1}{2a - 5} \qquad \text{Combine like terms.}$$

We can generalize the procedure for adding and subtracting rational expressions that have a common denominator as follows.

Adding and Subtracting Rational Expressions

If $\dfrac{P}{Q}$ and $\dfrac{R}{Q}$ are rational expressions with $Q \neq 0$, then

1) $\dfrac{P}{Q} + \dfrac{R}{Q} = \dfrac{P + R}{Q}$

2) $\dfrac{P}{Q} - \dfrac{R}{Q} = \dfrac{P - R}{Q}$

 You Try 1

Add or subtract, as indicated.

a) $\dfrac{9}{11} - \dfrac{4}{11}$ b) $\dfrac{7c}{3c - 4} + \dfrac{2c + 5}{3c - 4}$

All answers to a sum or difference of rational expressions should be in lowest terms. Sometimes it is necessary to simplify our result to lowest terms by factoring the numerator and dividing the numerator and denominator by the greatest common factor.

Example 2

Add or subtract, as indicated.

a) $\dfrac{11}{12t} + \dfrac{7}{12t}$ b) $\dfrac{n^2 - 8}{n(n + 5)} - \dfrac{7 - 2n}{n(n + 5)}$

Solution

a) $\dfrac{11}{12t} + \dfrac{7}{12t} = \dfrac{11 + 7}{12t}$ Add terms in the numerator.

$= \dfrac{18}{12t}$ Combine terms.

$= \dfrac{3}{2t}$ Reduce to lowest terms.

b) $\dfrac{n^2 - 8}{n(n + 5)} - \dfrac{7 - 2n}{n(n + 5)} = \dfrac{(n^2 - 8) - (7 - 2n)}{n(n + 5)}$ Subtract terms in the numerator.

$= \dfrac{n^2 - 8 - 7 + 2n}{n(n + 5)}$ Distribute.

$= \dfrac{n^2 + 2n - 15}{n(n + 5)}$ Combine like terms.

$= \dfrac{\cancel{(n + 5)}(n - 3)}{n\cancel{(n + 5)}}$ Factor the numerator.

$= \dfrac{n - 3}{n}$ Reduce to lowest terms.

You Try 2

Add or subtract, as indicated.

a) $\dfrac{3}{20c^2} - \dfrac{9}{20c^2}$ b) $\dfrac{k^2 + 2k + 5}{(k + 4)(k - 1)} + \dfrac{5k + 7}{(k + 4)(k + 1)}$ c) $\dfrac{20d - 9}{4d(3d + 1)} - \dfrac{5d - 14}{4d(3d + 1)}$

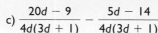

BE CAREFUL

After combining like terms in the numerator, ask yourself, "*Can I factor the numerator?*" If so, factor it. Sometimes, the expression can be reduced by dividing the numerator and denominator by the greatest common factor.

2. Add and Subtract Rational Expressions With Different Denominators

If we are asked to add or subtract rational expressions with different denominators, we must begin by rewriting each expression with the least common denominator. Then, add or subtract. Simplify the result.

Using the procedure studied in Section 8.3, here are the steps to follow to add or subtract rational expressions with different denominators.

Steps for Adding and Subtracting Rational Expressions With Different Denominators

1) Factor the denominators.

2) Write down the LCD.

3) Rewrite each rational expression as an equivalent rational expression with the LCD.

4) Add or subtract the numerators and keep the common denominator in factored form.

5) After combining like terms in the numerator ask yourself, *"Can I factor it?"* If so, factor.

6) Reduce the rational expression, if possible.

Example 3

Add or subtract, as indicated.

a) $\dfrac{m + 8}{3} + \dfrac{m - 1}{6}$ b) $\dfrac{3}{4x} - \dfrac{11}{10x^2}$ c) $\dfrac{4a - 6}{a^2 - 9} + \dfrac{a}{a + 3}$

Solution

a) The LCD is 6. $\dfrac{m - 1}{6}$ already has the LCD. Rewrite $\dfrac{m + 8}{3}$ with the LCD.

$$\frac{m + 8}{3} \cdot \frac{2}{2} = \frac{2(m + 8)}{6}$$

$$\frac{m + 8}{3} + \frac{m - 1}{6} = \frac{2(m + 8)}{6} + \frac{m - 1}{6} \qquad \text{Write each expression with the LCD.}$$

$$= \frac{2(m + 8) + (m - 1)}{6} \qquad \text{Add the numerators.}$$

$$= \frac{2m + 16 + m - 1}{6} \qquad \text{Distribute.}$$

$$= \frac{3m + 15}{6} \qquad \text{Combine like terms.}$$

Ask yourself, *"Can I factor the numerator?"* Yes.

$$= \frac{\cancel{3}(m + 5)}{\cancel{6}_{\,2}} \qquad \text{Factor.}$$

$$= \frac{m + 5}{2} \qquad \text{Reduce.}$$

b)   The LCD of $\dfrac{3}{4x}$ and $\dfrac{11}{10x^2}$ is $20x^2$. Rewrite each expression with the LCD.

$$\frac{3}{4x} \cdot \frac{5x}{5x} = \frac{15x}{20x^2}$$

$$\frac{11}{10x^2} \cdot \frac{2}{2} = \frac{22}{20x^2}$$

$$\frac{3}{4x} - \frac{11}{10x^2} = \frac{15x}{20x^2} - \frac{22}{20x^2} \qquad \text{Write each expression with the LCD.}$$

$$= \frac{15x - 22}{20x^2} \qquad \text{Subtract the numerators.}$$

"Can I factor the numerator?" No. The expression is in simplest form since the numerator and denominator have no common factors.

c)   Begin by factoring the denominator of $\dfrac{4a - 6}{a^2 - 9}$.

$$\frac{4a - 6}{a^2 - 9} = \frac{4a - 6}{(a + 3)(a - 3)}$$

The LCD of $\dfrac{4a - 6}{(a + 3)(a - 3)}$ and $\dfrac{a}{a + 3}$ is $(a + 3)(a - 3)$.

Rewrite $\dfrac{a}{a + 3}$ with the LCD.

$$\frac{a}{a + 3} \cdot \frac{a - 3}{a - 3} = \frac{a(a - 3)}{(a + 3)(a - 3)}$$

$$\frac{4a - 6}{a^2 - 9} + \frac{a}{a + 3} = \frac{4a - 6}{(a + 3)(a - 3)} + \frac{a}{a + 3} \qquad \begin{array}{l}\text{Factor the}\\ \text{denominator.}\end{array}$$

$$= \frac{4a - 6}{(a + 3)(a - 3)} + \frac{a(a - 3)}{(a + 3)(a - 3)} \qquad \begin{array}{l}\text{Write each expres-}\\ \text{sion with the LCD.}\end{array}$$

$$= \frac{4a - 6 + a(a - 3)}{(a + 3)(a - 3)} \qquad \text{Add the numerators.}$$

$$= \frac{4a - 6 + a^2 - 3a}{(a + 3)(a - 3)} \qquad \text{Distribute.}$$

$$= \frac{a^2 + a - 6}{(a + 3)(a - 3)} \qquad \text{Combine like terms.}$$

Ask yourself, *"Can I factor the numerator?"* Yes.

$$= \frac{\cancel{(a + 3)}(a - 2)}{\cancel{(a + 3)}(a - 3)} \qquad \text{Factor.}$$

$$= \frac{a - 2}{a - 3} \qquad \text{Reduce.}$$

 You Try 3

Add or subtract, as indicated.

a) $\dfrac{7}{12t^3} + \dfrac{4}{9t}$ b) $\dfrac{k - 3}{4} - \dfrac{k + 3}{6}$ c) $\dfrac{6}{r - 5} + \dfrac{r^2 - 17r}{r^2 - 25}$

Example 4

Subtract $\dfrac{4w}{w^2 + 9w + 14} - \dfrac{2w + 5}{w^2 + 3w - 28}$.

Solution

Factor the denominators, then write down the LCD.

$$\frac{4w}{w^2 + 9w + 14} = \frac{4w}{(w + 7)(w + 2)}, \qquad \frac{2w + 5}{w^2 + 3w - 28} = \frac{2w + 5}{(w + 7)(w - 4)}$$

Rewrite each expression with the LCD, $(w + 7)(w + 2)(w - 4)$.

$$\frac{4w}{(w + 7)(w + 2)} \cdot \frac{w - 4}{w - 4} = \frac{4w(w - 4)}{(w + 7)(w + 2)(w - 4)}$$

$$\frac{2w + 5}{(w + 7)(w - 4)} \cdot \frac{w + 2}{w + 2} = \frac{(2w + 5)(w + 2)}{(w + 7)(w + 2)(w - 4)}$$

$$\frac{4w}{w^2 + 9w + 14} - \frac{2w + 5}{w^2 + 3w - 28}$$

$$= \frac{4w}{(w + 7)(w + 2)} - \frac{2w + 5}{(w + 7)(w - 4)} \qquad \text{Factor denominators.}$$

$$= \frac{4w(w - 4)}{(w + 7)(w + 2)(w - 4)} - \frac{(2w + 5)(w + 2)}{(w + 7)(w + 2)(w - 4)} \qquad \text{Write each expression with the LCD.}$$

$$= \frac{4w(w - 4) - (2w + 5)(w + 2)}{(w + 7)(w + 2)(w - 4)} \qquad \text{Subtract the numerators.}$$

$$= \frac{4w^2 - 16w - (2w^2 + 9w + 10)}{(w + 7)(w + 2)(w - 4)} \qquad \text{Distribute. You must use parentheses.}$$

$$= \frac{4w^2 - 16w - 2w^2 - 9w - 10}{(w + 7)(w + 2)(w - 4)} \qquad \text{Distribute.}$$

$$= \frac{2w^2 - 25w - 10}{(w + 7)(w + 2)(w - 4)} \qquad \text{Combine like terms.}$$

Ask yourself, *"Can I factor the numerator?"* No. The expression is in simplest form since the numerator and denominator have no common factors.

BE CAREFUL

In Example 4, when you move from

$$\frac{4w(w - 4) - (2w + 5)(w + 2)}{(w + 7)(w + 2)(w - 4)} \quad \text{to} \quad \frac{4w^2 - 16w - (2w^2 + 9w + 10)}{(w + 7)(w + 2)(w - 4)}$$

you *must* use parentheses since the entire quantity $2w^2 + 9w + 10$ is being subtracted from $4w^2 - 16w$.

You Try 4

Subtract $\dfrac{3d}{d^2 + 13d + 40} - \dfrac{2d - 3}{d^2 + 7d - 8}$.

3. Add and Subtract Rational Expressions With Denominators Containing Factors of the Form $a - b$ and $b - a$

Example 5

Add or subtract, as indicated.

a) $\dfrac{s}{s - 3} - \dfrac{11}{3 - s}$

b) $\dfrac{3}{4 - h} + \dfrac{6}{h^2 - 16}$

Solution

a) Recall that $a - b = -(b - a)$. The least common denominator of $\dfrac{s}{s - 3}$ and $\dfrac{11}{3 - s}$ is $s - 3$ or $3 - s$. We will use LCD $= s - 3$.

$\dfrac{s}{s - 3}$ already has the LCD. Rewrite $\dfrac{11}{3 - s}$ with the LCD.

$$\frac{11}{3 - s} = \frac{11}{-(s - 3)} = -\frac{11}{s - 3}$$

$$\frac{s}{s - 3} - \frac{11}{3 - s} = \frac{s}{s - 3} - \left(-\frac{11}{s - 3}\right) \qquad \text{Write each expression with the LCD.}$$

$$= \frac{s}{s - 3} + \frac{11}{s - 3} \qquad \text{Distribute.}$$

$$= \frac{s + 11}{s - 3} \qquad \text{Add the numerators.}$$

b) Factor the denominator of $\dfrac{6}{h^2 - 16}$.

$$\frac{6}{h^2 - 16} = \frac{6}{(h + 4)(h - 4)}$$

Since $\dfrac{6}{(h + 4)(h - 4)}$ contains a factor of $h - 4$ and $\dfrac{3}{4 - h}$ contains $4 - h$ in its denominator, rewrite $\dfrac{3}{4 - h}$ with a denominator of $h - 4$.

$$\frac{3}{4 - h} = \frac{3}{-(h - 4)} = -\frac{3}{h - 4}$$

Now we must find the LCD of $\dfrac{6}{(h + 4)(h - 4)}$ and $-\dfrac{3}{h - 4}$.

$$LCD = (h + 4)(h - 4)$$

Rewrite $-\dfrac{3}{h - 4}$ with the LCD.

$$-\frac{3}{h - 4} \cdot \frac{h + 4}{h + 4} = -\frac{3(h + 4)}{(h + 4)(h - 4)} = \frac{-3(h + 4)}{(h + 4)(h - 4)}$$

$$
\begin{aligned}
\frac{3}{4 - h} + \frac{6}{h^2 - 16} &= -\frac{3}{h - 4} + \frac{6}{(h + 4)(h - 4)} \\
&= \frac{-3(h + 4)}{(h + 4)(h - 4)} + \frac{6}{(h + 4)(h - 4)} \qquad \text{Write each expression with the LCD.} \\
&= \frac{-3(h + 4) + 6}{(h + 4)(h - 4)} \qquad \text{Add the numerators.} \\
&= \frac{-3h - 12 + 6}{(h + 4)(h - 4)} \qquad \text{Distribute.} \\
&= \frac{-3h - 6}{(h + 4)(h - 4)} \qquad \text{Combine like terms.}
\end{aligned}
$$

Ask yourself, *"Can I factor the numerator?"* Yes.

$$= \frac{-3(h + 2)}{(h + 4)(h - 4)} \qquad \text{Factor.}$$

Although the numerator factors, the numerator and denominator do not contain any common factors. The result, $\dfrac{-3(h + 2)}{(h + 4)(h - 4)}$, is in simplest form.

You Try 5

Add or subtract, as indicated.

a) $\dfrac{5}{8 - y} + \dfrac{3}{y - 8}$ b) $\dfrac{1}{x^2 - 36} - \dfrac{x + 4}{6 - x}$

Answers to You Try Exercises

1) a) $\dfrac{5}{11}$ b) $\dfrac{9c + 5}{3c - 4}$ 2) a) $-\dfrac{3}{10c^2}$ b) $\dfrac{k + 3}{k - 1}$ c) $\dfrac{5}{4d}$ 3) a) $\dfrac{16t^2 + 21}{36t^3}$ b) $\dfrac{k - 15}{12}$ c) $\dfrac{r - 6}{r + 5}$

4) $\dfrac{d^2 - 10d + 15}{(d + 8)(d + 5)(d - 1)}$ 5) a) $-\dfrac{2}{y - 8}$ or $\dfrac{2}{8 - y}$ b) $\dfrac{(x + 5)^2}{(x + 6)(x - 6)}$

8.4 Exercises

Objective 1

Add or subtract, as indicated.

1) $\dfrac{8}{11} - \dfrac{3}{11}$

2) $\dfrac{4}{15} + \dfrac{7}{15}$

3) $\dfrac{7}{20} + \dfrac{9}{20}$

4) $\dfrac{11}{12} - \dfrac{5}{12}$

5) $\dfrac{8}{a} + \dfrac{2}{a}$

6) $\dfrac{3}{p} - \dfrac{9}{p}$

7) $\dfrac{10}{3k^2} - \dfrac{2}{3k^2}$

8) $\dfrac{6}{5c} + \dfrac{14}{5c}$

9) $\dfrac{n}{n+6} - \dfrac{9}{n+6}$

10) $\dfrac{5}{z-3} + \dfrac{2z}{z-3}$

 11) $\dfrac{8}{x+4} + \dfrac{2x}{x+4}$

12) $\dfrac{7m}{m+5} + \dfrac{35}{m+5}$

13) $\dfrac{7w-4}{w(3w-4)} - \dfrac{20-11w}{w(3w-4)}$

14) $\dfrac{10t+7}{t(2t+1)} - \dfrac{2t+3}{t(2t+1)}$

15) $\dfrac{2r+15}{(r-5)(r+2)} + \dfrac{r^2-10r}{(r-5)(r+2)}$

16) $\dfrac{d^2-12}{(d+4)(d+1)} + \dfrac{3-8d}{(d+4)(d+1)}$

Objective 2

17) For $\dfrac{8}{x-3}$ and $\dfrac{2}{x}$:

   a)   Find the LCD.

   b)   Explain, in your own words, how to rewrite each expression with the LCD.

   c)   Rewrite each expression with the LCD.

18) For $\dfrac{4}{9b^2}$ and $\dfrac{5}{6b^4}$:

   a)   Find the LCD.

   b)   Explain, in your own words, how to rewrite each expression with the LCD.

   c)   Rewrite each expression with the LCD.

Add or subtract as indicated.

19) $\dfrac{5}{8} + \dfrac{1}{6}$

20) $\dfrac{9}{10} - \dfrac{5}{6}$

21) $\dfrac{5x}{12} - \dfrac{4x}{15}$

22) $\dfrac{9t}{5} + \dfrac{3}{4}$

23) $\dfrac{5}{8u} - \dfrac{2}{3u^2}$

24) $\dfrac{11}{4h^2} + \dfrac{2}{3h}$

25) $\dfrac{3}{2a} + \dfrac{6}{7a^2}$

26) $\dfrac{3}{2f^2} - \dfrac{7}{f}$

27) $\dfrac{2}{k} + \dfrac{9}{k+10}$

28) $\dfrac{3}{y+1} + \dfrac{11}{y}$

 29) $\dfrac{15}{d-8} - \dfrac{4}{d}$

30) $\dfrac{8}{r-7} - \dfrac{4}{r}$

31) $\dfrac{1}{z+6} + \dfrac{4}{z+2}$

32) $\dfrac{6}{c-5} + \dfrac{5}{c+3}$

33) $\dfrac{x}{2x+1} - \dfrac{3}{x+5}$

34) $\dfrac{m}{3m+4} - \dfrac{2}{m-9}$

35) $\dfrac{t}{t+7} + \dfrac{11t-21}{t^2-49}$

36) $\dfrac{-3u-5}{u^2-1} + \dfrac{u}{u+1}$

37) $\dfrac{b}{b^2-16} + \dfrac{10}{b^2-5b-36}$

38) $\dfrac{7g}{g^2-9g-10} + \dfrac{4}{g^2-100}$

39) $\dfrac{3c}{c^2+4c-12} - \dfrac{2c-5}{c^2+2c-24}$

40) $\dfrac{4a}{a^2-5a-24} - \dfrac{2a+3}{a^2-10a+16}$

41) $\dfrac{4m}{m^2+m-6} - \dfrac{7}{m^2+4m-12}$

42) $\dfrac{3x}{x^2-x-56} - \dfrac{6}{x^2-10x+16}$

 43) $\dfrac{4b+1}{3b-12} + \dfrac{5b}{b^2-b-12}$

44) $\dfrac{k+9}{2k-24} + \dfrac{4k}{k^2-15k+36}$

Objective 3

45) Is $(x - 7)(7 - x)$ the LCD for $\dfrac{5}{x - 7} + \dfrac{2}{7 - x}$?
Why or why not?

46) What is the LCD of $\dfrac{n}{5 - 3n} - \dfrac{8}{3n - 5}$?

Add or subtract, as indicated.

47) $\dfrac{9}{z - 6} + \dfrac{2}{6 - z}$

48) $\dfrac{15}{q - 8} + \dfrac{9}{8 - q}$

49) $\dfrac{5}{c - d} - \dfrac{3}{d - c}$

50) $\dfrac{7}{f - 5} - \dfrac{20}{5 - f}$

51) $\dfrac{10}{m - 3} + \dfrac{m + 11}{3 - m}$

52) $\dfrac{9}{x - 7} + \dfrac{x - 3}{7 - x}$

53) $\dfrac{5}{2 - n} + \dfrac{n + 3}{n - 2}$

54) $\dfrac{6}{4 - a} + \dfrac{a + 2}{a - 4}$

55) $\dfrac{2c}{12b - 7c} - \dfrac{13}{7c - 12b}$

56) $\dfrac{2}{4u - 3v} - \dfrac{6u}{3v - 4u}$

57) $\dfrac{5}{8 - t} + \dfrac{10}{t^2 - 64}$

58) $\dfrac{8}{r^2 - 9} + \dfrac{2}{3 - r}$

 59) $\dfrac{a}{4a^2 - 9} - \dfrac{4}{3 - 2a}$

60) $\dfrac{2y}{9y^2 - 25} - \dfrac{2}{5 - 3y}$

Objectives 2 and 3

Perform the indicated operations.

61) $\dfrac{2}{j^2 + 8j} + \dfrac{2j}{j + 8} - \dfrac{1}{3j}$

62) $\dfrac{4}{w^2 - 3w} + \dfrac{9}{w} - \dfrac{10w}{w - 3}$

63) $\dfrac{2k + 7}{k^2 - 4k} + \dfrac{9k}{2k^2 - 15k + 28} + \dfrac{15}{2k^2 - 7k}$

64) $\dfrac{3b - 1}{b^2 + 8b} + \dfrac{b}{3b^2 + 25b + 8} + \dfrac{2}{3b^2 + b}$

 65) $\dfrac{c}{c^2 - 8c + 16} - \dfrac{5}{c^2 - c - 12}$

66) $\dfrac{n}{n^2 + 11n + 30} - \dfrac{3}{n^2 + 10n + 25}$

67) $\dfrac{1}{x + y} + \dfrac{x}{x^2 - y^2} - \dfrac{4}{2x - 2y}$

68) $\dfrac{8}{3a + 3b} + \dfrac{3}{a - b} - \dfrac{3a}{a^2 - b^2}$

69) $\dfrac{n + 5}{4n^2 + 7n - 2} - \dfrac{n - 4}{3n^2 + 7n + 2}$

70) $\dfrac{3v - 4}{6v^2 - v - 5} - \dfrac{v - 2}{3v^2 + v - 4}$

71) $\dfrac{y + 6}{y^2 - 4y} + \dfrac{y}{2y^2 - 13y + 20} - \dfrac{1}{2y^2 - 5y}$

72) $\dfrac{g - 5}{5g^2 - 30g} + \dfrac{g}{2g^2 - 17g + 30} - \dfrac{6}{2g^2 - 5g}$

For each rectangle, find a rational expression in simplest form to represent its a) area and b) perimeter.

73) $\dfrac{4}{x - 3}$; $\dfrac{x + 1}{2}$

74) $\dfrac{x - 4}{6}$; $\dfrac{10}{x + 1}$

75) $\dfrac{1}{w^2 - 4}$; $\dfrac{w}{w + 2}$

76) $\dfrac{2}{t^2 + 9t + 20}$; $\dfrac{t}{t + 5}$

Mid-Chapter Summary

Objective

1. Review the Concepts Presented in Sections 8.1–8.4

1. Review the Concepts Presented in Sections 8.1–8.4

In Section 8.1, we defined a rational expression, and we evaluated expressions. We also discussed how to write a rational expression in lowest terms.

Example 1

Write in lowest terms: $\dfrac{4x^2 - 28x}{x^2 - 15x + 56}$.

Solution

$$\frac{4x^2 - 28x}{x^2 - 15x + 56} = \frac{4x(x - 7)}{(x - 7)(x - 8)} \qquad \text{Factor.}$$

$$= \frac{4x}{x - 8} \qquad \text{Divide by } x - 7.$$

Recall that a rational expression *equals zero* when its *numerator equals zero*. A rational expression is *undefined* when its *denominator equals zero*.

Example 2

Determine the value(s) of a for which $\dfrac{a + 6}{a^2 - 81}$

a)   equals zero.   b)   is undefined.

Solution

a) $\dfrac{a + 6}{a^2 - 81}$ equals zero when its numerator equals zero.

Let $a + 6 = 0$, and solve for a.

$$a + 6 = 0$$
$$a = -6$$

$\dfrac{a + 6}{a^2 - 81}$ equals zero when $a = -6$.

b) $\dfrac{a + 6}{a^2 - 81}$ is undefined when its denominator equals zero.

Let $a^2 - 81 = 0$, and solve for a.

$$a^2 - 81 = 0$$
$$(a + 9)(a - 9) = 0 \qquad \text{Factor.}$$
$$a + 9 = 0 \quad \text{or} \quad a - 9 = 0 \qquad \text{Set each factor equal to zero.}$$
$$a = -9 \quad \text{or} \qquad a = 9 \qquad \text{Solve.}$$

$\dfrac{a + 6}{a^2 - 81}$ is undefined when $a = 9$ or $a = -9$. So, $a \neq 9$ and $a \neq -9$ in the expression.

In Sections 8.2–8.4 we learned how to multiply, divide, add, and subtract rational expressions. Now we will practice these operations together so that we will learn to recognize the techniques needed to perform these operations.

Example 3

Divide $\dfrac{y^2 - 6y - 27}{25y^2 - 49} \div \dfrac{y^2 - 9y}{14 - 10y}$.

Solution

Do we need a common denominator to divide? *No.* A common denominator is needed to add or subtract but not to multiply or divide.

To divide, multiply the first rational expression by the reciprocal of the second expression, then factor, reduce, and multiply.

$$\frac{y^2 - 6y - 27}{25y^2 - 49} \div \frac{y^2 - 9y}{14 - 10y} = \frac{y^2 - 6y - 27}{25y^2 - 49} \cdot \frac{14 - 10y}{y^2 - 9y} \qquad \text{Multiply by the reciprocal.}$$

$$= \frac{\cancel{(y - 9)}(y + 3)}{(5y + 7)\cancel{(5y - 7)}} \cdot \frac{\overset{-1}{\cancel{2(7 - 5y)}}}{y\cancel{(y - 9)}} \qquad \text{Factor.}$$

$$= -\frac{2(y + 3)}{y(5y + 7)} \qquad \text{Reduce and multiply.}$$

Recall that $\dfrac{7 - 5y}{5y - 7} = -1$.

Example 4

Add $\dfrac{n}{n - 6} + \dfrac{2}{n + 1}$.

Solution

To add or subtract rational expressions we need a common denominator. We do not need to factor these denominators, so we are ready to identify the LCD.

$$\text{LCD} = (n - 6)(n + 1)$$

Rewrite each expression with the LCD.

$$\frac{n}{n - 6} \cdot \frac{n + 1}{n + 1} = \frac{n(n + 1)}{(n - 6)(n + 1)}, \qquad \frac{2}{n + 1} \cdot \frac{n - 6}{n - 6} = \frac{2(n - 6)}{(n - 6)(n + 1)}$$

$$\frac{n}{n - 6} + \frac{2}{n + 1} = \frac{n(n + 1)}{(n - 6)(n + 1)} + \frac{2(n - 6)}{(n - 6)(n + 1)} \qquad \text{Write each expression with the LCD.}$$

$$= \frac{n(n + 1) + 2(n - 6)}{(n - 6)(n + 1)} \qquad \text{Add the numerators.}$$

$$= \frac{n^2 + n + 2n - 12}{(n - 6)(n + 1)} \qquad \text{Distribute.}$$

$$= \frac{n^2 + 3n - 12}{(n - 6)(n + 1)} \qquad \text{Combine like terms.}$$

Although this numerator will not factor, remember that sometimes it *is* possible to factor the numerator and simplify the result.

a) Write in lowest terms: $\dfrac{2m^2 - 7m + 5}{1 - m^2}$.

b) Subtract $\dfrac{a}{a+3} - \dfrac{2}{a}$.   c) Multiply $\dfrac{x^3 - 8}{9} \cdot \dfrac{3x+6}{x^2-4}$.

d) Determine the values of k for which $\dfrac{3k-4}{k^2+5k}$ i) equals zero.   ii) is undefined.

Answers to You Try Exercises

1) a) $\dfrac{5-2m}{m+1}$ b) $\dfrac{a^2 - 2a - 6}{a(a+3)}$ c) $\dfrac{x^2 + 2x + 4}{3}$ d) i) $\dfrac{4}{3}$ ii) $-5, 0$

**Mid-Chapter
Summary Exercises**

**Boost your grade
at mathzone.com!**   MathZone   > **Practice
Problems**   > **NetTutor**
> **Self-Test**   > **e-Professors**   > **Videos**

Objective 1

Evaluate, if possible, for a) $x = -4$ and b) $x = 3$.

1) $\dfrac{x}{2x - 6}$

2) $\dfrac{x+4}{3x+5}$

3) $-\dfrac{x^2}{x^2 - 10}$

4) $\dfrac{5x - 3}{x^2 + 6x + 8}$

Determine the value(s) of the variable for which

a)   the expression is undefined.

b)   the expression equals zero.

5) $-\dfrac{4w}{w^2 - 9}$

6) $\dfrac{m+8}{2m^2 - 3m - 14}$

7) $\dfrac{3 - 5b}{b^2 + 2b - 8}$

8) $\dfrac{13k - 11}{49 - k^2}$

9) $\dfrac{t - 6}{t^2 + 1}$

10) $\dfrac{8}{5r}$

Write each rational expression in lowest terms.

11) $\dfrac{36n^9}{27n^{12}}$

12) $\dfrac{8w^{12}}{16w^7}$

13) $\dfrac{2j + 5}{2j^2 - 3j - 20}$

14) $\dfrac{m^2 + 5m - 24}{2m^2 + 2m - 24}$

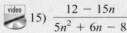

15) $\dfrac{12 - 15n}{5n^2 + 6n - 8}$

16) $\dfrac{-x - y}{xy + y^2 + 3x + 3y}$

Perform the operations and simplify.

17) $\dfrac{5}{f + 8} - \dfrac{2}{f}$

18) $\dfrac{2c^2 - 4c - 6}{c + 1} \div \dfrac{3c - 9}{7}$

19) $\dfrac{9a^3}{10b} \cdot \dfrac{40b^2}{81a}$

20) $\dfrac{4j}{j^2 - 81} + \dfrac{3}{j^2 - 3j - 54}$

21) $\dfrac{3}{q^2 - q - 20} + \dfrac{8q}{q^2 + 11q + 28}$

22) $\dfrac{12y^3}{4z} \cdot \dfrac{8z^4}{72y^6}$

23) $\dfrac{16 - m^2}{m + 4} \div \dfrac{8m - 32}{m + 7}$

24) $\dfrac{12p}{4p^2 + 11p + 6} - \dfrac{5}{p^2 - 4p - 12}$

25) $\dfrac{13}{r - 8} + \dfrac{4}{8 - r}$

26) $\dfrac{n - 4}{4n - 40} \cdot \dfrac{100 - n^2}{n + 10}$

27) $\dfrac{\dfrac{10d}{d + 11}}{\dfrac{5d^4}{2d + 22}}$

28) $\dfrac{xy - 4x + 3y - 12}{y^2 - 16} \div \dfrac{x^3 + 27}{6}$

29) $\dfrac{a^2 - 4}{a^3 + 8} \cdot \dfrac{5a^2 - 10a + 20}{3a - 6}$

30) $\dfrac{7}{d + 9} + \dfrac{6}{d^2}$

31) $\dfrac{13}{5z} - \dfrac{1}{3z}$

32) $\dfrac{\dfrac{9k^2 - 1}{14k}}{\dfrac{3k + 1}{21k^3}}$

33) $\dfrac{12a^4}{10a - 20} \div \dfrac{3a}{a^3 - 2a^2 + 5a - 10}$

34) $\dfrac{2w}{25 - w^2} + \dfrac{w - 3}{w^2 - 12w + 35}$

35) $\dfrac{10}{x - 8} + \dfrac{4}{x + 3}$

36) $\dfrac{1}{4y} + \dfrac{7}{6y^2}$

37) $\dfrac{10q}{8p - 10q} + \dfrac{8p}{10q - 8p}$

38) $\dfrac{m}{7m - 4n} - \dfrac{20n}{4n - 7m}$

39) $\dfrac{6u + 1}{3u^2 - 2u} - \dfrac{u}{3u^2 + u - 2} + \dfrac{10}{u^2 + u}$

40) $\dfrac{2p + 3}{p^2 + 7p} - \dfrac{4p}{p^2 - p - 56} + \dfrac{5}{p^2 - 8p}$

41) $\dfrac{\dfrac{6v - 30}{4}}{\dfrac{v - 5}{2}}$

42) $\dfrac{\dfrac{3c^3}{8c + 24}}{\dfrac{6c}{c + 3}}$

43) $\dfrac{x}{2x^2 - 7x - 4} - \dfrac{x + 3}{4x^2 + 4x + 1}$

44) $\dfrac{f - 12}{f - 7} - \dfrac{5}{7 - f}$

45) $\left(\dfrac{2c}{c + 8} + \dfrac{4}{c - 2} \right) \div \dfrac{6}{5c + 40}$

46) $\left(\dfrac{3n}{3n - 1} - \dfrac{4}{n + 4} \right) \cdot \dfrac{9n^2 - 1}{21n^2 + 28}$

47) $\dfrac{3}{w^2 - w} + \dfrac{4}{5w} - \dfrac{3}{w - 1}$

48) $\dfrac{4}{k^2 + 4k} - \dfrac{3}{2k} + \dfrac{1}{k + 4}$

For each rectangle, find a rational expression in simplest form to represent its a) area and b) perimeter.

49) $\dfrac{x}{2}$

$\dfrac{x - 3}{4}$

50) $\dfrac{8}{z + 1}$

$\dfrac{z}{z + 5}$

Section 8.5 Simplifying Complex Fractions

Objectives

1. Simplify a Complex Fraction That Has One Term in the Numerator and One Term in the Denominator

2. Simplify a Complex Fraction That Has More Than One Term in the Numerator and/or Denominator by Rewriting It as a Division Problem

3. Simplify a Complex Fraction That Has More Than One Term in the Numerator and/or Denominator by Multiplying by the LCD

In algebra we sometimes encounter fractions that contain fractions in their numerators, denominators, or both. Such fractions are called *complex fractions*. Some examples of complex fractions are

$$\frac{\dfrac{5}{8}}{\dfrac{3}{4}}, \qquad \frac{\dfrac{1}{2}+\dfrac{1}{3}}{3-\dfrac{5}{4}}, \qquad \frac{\dfrac{2}{x^2 y}}{\dfrac{1}{y}-\dfrac{y}{x}}, \qquad \frac{\dfrac{7k-28}{3}}{\dfrac{k-4}{k}}$$

> **Definition** A **complex fraction** is a rational expression that contains one or more fractions in its numerator, its denominator, or both.

A complex fraction is not considered to be an expression in simplest form. In this section, we will learn how to simplify complex fractions to lowest terms.

We begin by looking at two different types of complex fractions.

1) Complex fractions with *one term* in the numerator and *one term* in the denominator.
2) Complex fractions that have *more than one term* in their numerators, their denominators, or both.

1. Simplify a Complex Fraction That Has One Term in the Numerator and One Term in the Denominator

We studied these expressions in Section 8.2 when we learned how to divide fractions. We will look at another example.

Example 1

Simplify $\dfrac{\dfrac{7k-28}{3}}{\dfrac{k-4}{k}}$.

Solution

There is one term in the numerator: $\dfrac{7k-28}{3}$

There is one term in the denominator: $\dfrac{k-4}{k}$

To simplify, rewrite as a division problem then carry out the division.

$$\frac{\dfrac{7k-28}{3}}{\dfrac{k-4}{k}} = \frac{7k-28}{3} \div \frac{k-4}{k}$$ Rewrite the complex fraction as a division problem.

$$= \frac{7k-28}{3} \cdot \frac{k}{k-4}$$ Change division to multiplication by the reciprocal of $\dfrac{k-4}{k}$.

$$= \frac{7\cancel{(k-4)}}{3} \cdot \frac{k}{\cancel{k-4}}$$ Factor and divide numerator and denominator by $k-4$ to simplify.

$$= \frac{7k}{3}$$ Multiply.

 You Try 1

Simplify $\dfrac{\dfrac{6n}{n^2-81}}{\dfrac{n^2}{2n+18}}$.

To simplify a complex fraction containing one term in the numerator and one term in the denominator:

1)   Rewrite the complex fraction as a division problem.

2)   Perform the division by multiplying the first fraction by the reciprocal of the second.

(We are multiplying the numerator of the complex fraction by the reciprocal of the denominator.)

---

2. Simplify a Complex Fraction That Has More Than One Term in the Numerator and/or Denominator by Rewriting It as a Division Problem

When a complex fraction has more than one term in the numerator and/or the denominator, we can use one of two methods to simplify.

Method 1

1)   Combine the terms in the numerator and combine the terms in the denominator so that each contains only one fraction.

2)   Rewrite as a division problem.

3)   Perform the division by multiplying the first fraction by the reciprocal of the second.

Example 2

Simplify.

a) $\dfrac{\dfrac{1}{2} + \dfrac{1}{3}}{3 - \dfrac{5}{4}}$ b) $\dfrac{\dfrac{2}{x^2 y}}{\dfrac{1}{y} - \dfrac{y}{x}}$

Solution

a) The numerator is $\dfrac{1}{2} + \dfrac{1}{3}$, and it contains two terms.

The denominator is $3 - \dfrac{5}{4}$, and it contains two terms.

We will add the terms in the numerator and subtract the terms in the denominator so that the numerator and denominator will each contain one fraction.

$$\dfrac{\dfrac{1}{2} + \dfrac{1}{3}}{3 - \dfrac{5}{4}} = \dfrac{\dfrac{3}{6} + \dfrac{2}{6}}{\dfrac{12}{4} - \dfrac{5}{4}}$$ Rewrite the fractions in the numerator with the LCD of 6.
Rewrite the terms in the denominator with the LCD of 4.

$$= \dfrac{\dfrac{5}{6}}{\dfrac{7}{4}}$$ Add the fractions in the numerator.
Subtract the fractions in the denominator.

$$= \dfrac{5}{6} \div \dfrac{7}{4}$$ Rewrite as a division problem.

$$= \dfrac{5}{\cancel{6}_{3}} \cdot \dfrac{\cancel{4}^{2}}{7}$$ Change division to multiplication by the reciprocal of $\dfrac{7}{4}$.

$$= \dfrac{10}{21}$$ Multiply and simplify.

b) For $\dfrac{\dfrac{2}{x^2 y}}{\dfrac{1}{y} - \dfrac{y}{x}}$:

The numerator is $\dfrac{2}{x^2 y}$, and it contains one term.

The denominator is $\dfrac{1}{y} - \dfrac{y}{x}$, and it contains two terms.

We will subtract the terms in the denominator so that it, like the numerator, will contain only one term.

$$\dfrac{\dfrac{2}{x^2 y}}{\dfrac{1}{y} - \dfrac{y}{x}} = \dfrac{\dfrac{2}{x^2 y}}{\dfrac{x}{xy} - \dfrac{y^2}{xy}}$$ Rewrite the terms in the denominator with the LCD of xy.

$$= \dfrac{\dfrac{2}{x^2 y}}{\dfrac{x - y^2}{xy}}$$ Subtract the rational expressions in the denominator.

$$= \dfrac{2}{x^2 y} \div \dfrac{x - y^2}{xy}$$ Rewrite as a division problem.

$$= \dfrac{2}{\underset{x}{\cancel{x^2 y}}} \cdot \dfrac{\cancel{xy}}{x - y^2}$$ Change division to multiplication by the reciprocal of $\dfrac{x - y^2}{xy}$.

$$= \dfrac{2}{x(x - y^2)}$$ Multiply and simplify.

You Try 2

Simplify.

a) $\dfrac{\dfrac{9}{8} - \dfrac{3}{4}}{\dfrac{2}{3} + \dfrac{1}{4}}$

b) $\dfrac{\dfrac{5}{a} + \dfrac{3}{ab}}{\dfrac{1}{ab} + 2}$

3. Simplify a Complex Fraction That Has More Than One Term in the Numerator and/or Denominator by Multiplying by the LCD

Another method we can use to simplify complex fractions involves multiplying the numerator and denominator of the complex fraction by the LCD of *all* of the fractions in the expression.

Method 2

Step 1: Identify and write down the LCD of *all* of the fractions in the complex fraction.

Step 2: Multiply the numerator and denominator of the complex fraction by the LCD.

Step 3: Simplify.

We will simplify the complex fractions we simplified in Example 2 using this other method.

Example 3

Simplify using method 2.

a) $\dfrac{\dfrac{1}{2} + \dfrac{1}{3}}{3 - \dfrac{5}{4}}$

b) $\dfrac{\dfrac{2}{x^2 y}}{\dfrac{1}{y} - \dfrac{y}{x}}$

Solution

a) **Step 1:** Look at all of the fractions in the complex fraction. They are $\dfrac{1}{2}, \dfrac{1}{3}$, and $\dfrac{5}{4}$. Write down their LCD. LCD = 12.

Step 2: Multiply the numerator and denominator of the complex fraction by the LCD, 12. (Note: We are multiplying the fraction by $\dfrac{12}{12}$, which equals 1.)

$$\dfrac{12\left(\dfrac{1}{2} + \dfrac{1}{3}\right)}{12\left(3 - \dfrac{5}{4}\right)}$$

Step 3: Simplify.

$$\dfrac{12\left(\dfrac{1}{2} + \dfrac{1}{3}\right)}{12\left(3 - \dfrac{5}{4}\right)} = \dfrac{12 \cdot \dfrac{1}{2} + 12 \cdot \dfrac{1}{3}}{12 \cdot 3 - 12 \cdot \dfrac{5}{4}} \qquad \text{Distribute.}$$

$$= \dfrac{6 + 4}{36 - 15} \qquad \text{Multiply.}$$

$$= \dfrac{10}{21} \qquad \begin{array}{l}\text{Add terms in the numerator, subtract} \\ \text{terms in the denominator.}\end{array}$$

This is the same result we obtained in Example 2 using Method 1.

> In the denominator we multiplied the 3 by 12 even though 3 is not a fraction. Remember, *all* terms, not just the fractions, must be multiplied by the LCD.

b) **Step 1:** Look at all of the fractions in the complex fraction. They are $\dfrac{2}{x^2 y}, \dfrac{1}{y}$, and $\dfrac{y}{x}$. Write down their LCD. LCD = $x^2 y$.

Step 2:   Multiply the numerator and denominator of the complex fraction by the LCD, x^2y.

$$\dfrac{x^2y\left(\dfrac{2}{x^2y}\right)}{x^2y\left(\dfrac{1}{y}-\dfrac{y}{x}\right)}$$

We are multiplying the fraction by $\dfrac{x^2y}{x^2y}$, which equals 1.

Step 3:   Simplify.

$$\dfrac{x^2y\left(\dfrac{2}{x^2y}\right)}{x^2y\left(\dfrac{1}{y}-\dfrac{y}{x}\right)}=\dfrac{x^2y\cdot\dfrac{2}{x^2y}}{x^2y\cdot\dfrac{1}{y}-x^2y\cdot\dfrac{y}{x}}\qquad\text{Distribute.}$$

$$=\dfrac{2}{x^2-xy^2}\qquad\text{Multiply.}$$

$$=\dfrac{2}{x(x-y^2)}\qquad\text{Factor.}$$

If the numerator and denominator factor, factor them. Sometimes, you can divide by a common factor to simplify.

Notice that the result is the same as what was obtained in Example 2 using method 1.

 You Try 3

Simplify using method 2.

a) $\dfrac{\dfrac{9}{8}-\dfrac{3}{4}}{\dfrac{2}{3}+\dfrac{1}{4}}$ b) $\dfrac{\dfrac{5}{a}+\dfrac{3}{ab}}{\dfrac{1}{ab}+2}$

You should be familiar with both methods for simplifying complex fractions containing two terms in the numerator or denominator. After a lot of practice, you will be able to decide which method works best for a particular problem.

Answers to You Try Exercises

1) $\dfrac{12}{n(n-9)}$ 2) a) $\dfrac{9}{22}$ b) $\dfrac{5b+3}{2ab+1}$ 3) a) $\dfrac{9}{22}$ b) $\dfrac{5b+3}{2ab+1}$

8.5 Exercises

> **Practice Problems** > **NetTutor** > **Self-Test** > **e-Professors** > **Videos**

1) Explain, in your own words, two ways to simplify $\dfrac{\frac{2}{9}}{\frac{5}{18}}$.

 Then, simplify it both ways. Which method do you prefer and why?

2) Explain, in your own words, two ways to simplify $\dfrac{\frac{3}{2} - \frac{1}{5}}{\frac{1}{10} + \frac{3}{5}}$. Then, simplify it both ways. Which method do you prefer and why?

Objective 1

Simplify completely.

3) $\dfrac{\frac{7}{10}}{\frac{5}{4}}$

4) $\dfrac{\frac{3}{8}}{\frac{4}{3}}$

5) $\dfrac{\frac{a^2}{b}}{\frac{a}{b^3}}$

6) $\dfrac{\frac{u^5}{v^2}}{\frac{u^2}{v}}$

7) $\dfrac{\frac{s^3}{t^3}}{\frac{s^4}{t}}$

8) $\dfrac{\frac{x^4}{y}}{\frac{x^2}{y^3}}$

9) $\dfrac{\frac{14m^5n^4}{9}}{\frac{35mn^6}{3}}$

10) $\dfrac{\frac{11b^4c^2}{4}}{\frac{55bc}{8}}$

11) $\dfrac{\frac{t-6}{5}}{\frac{t-6}{t}}$

12) $\dfrac{\frac{m-3}{m}}{\frac{m-3}{16}}$

13) $\dfrac{\frac{8}{y^2-64}}{\frac{6}{y+8}}$

14) $\dfrac{\frac{g^2-36}{15}}{\frac{g-6}{45}}$

15) $\dfrac{\frac{25w-35}{w^5}}{\frac{30w-42}{w}}$

16) $\dfrac{\frac{d^3}{16d-24}}{\frac{d}{40d-60}}$

17) $\dfrac{\frac{2x}{x+7}}{\frac{2}{x^2+4x-21}}$

18) $\dfrac{\frac{c^2-7c-8}{6c}}{\frac{c-8}{c}}$

Objectives 2 and 3

Simplify using method 1 then by using method 2. Think about which method you prefer and why.

19) $\dfrac{\frac{1}{4} + \frac{3}{2}}{\frac{2}{3} + \frac{1}{2}}$

20) $\dfrac{\frac{7}{9} - \frac{1}{3}}{2 + \frac{1}{9}}$

21) $\dfrac{\frac{7}{c} + \frac{2}{d}}{1 - \frac{5}{c}}$

22) $\dfrac{\frac{r}{s} - 2}{\frac{1}{s} + \frac{3}{r}}$

23) $\dfrac{\frac{5}{z-2} - \frac{1}{z+1}}{\frac{1}{z-2} + \frac{4}{z+1}}$

24) $\dfrac{\frac{6}{w+4} + \frac{4}{w-1}}{\frac{5}{w-1} + \frac{3}{w+4}}$

Simplify using either method 1 or method 2.

25) $\dfrac{9 + \frac{5}{y}}{\frac{9y+5}{8}}$

26) $\dfrac{4 - \frac{12}{m}}{\frac{4m-12}{9}}$

27) $\dfrac{x - \frac{7}{x}}{x - \frac{11}{x}}$

28) $\dfrac{\frac{4}{c} - c}{3 + \frac{8}{c}}$

29) $\dfrac{\frac{4}{3} + \frac{2}{5}}{\frac{1}{6} - \frac{2}{3}}$

30) $\dfrac{\frac{1}{4} - \frac{5}{6}}{\frac{3}{8} + \frac{1}{3}}$

31) $\dfrac{\dfrac{2}{a} - \dfrac{2}{b}}{\dfrac{1}{a^2} - \dfrac{1}{b^2}}$

32) $\dfrac{\dfrac{4}{x} - \dfrac{4}{y}}{\dfrac{3}{x^2} - \dfrac{3}{y^2}}$

41) $\dfrac{\dfrac{6}{x+3} - \dfrac{4}{x-1}}{\dfrac{2}{x-1} + \dfrac{1}{x+2}}$

42) $\dfrac{\dfrac{c^2}{d} + \dfrac{2}{c^2 d}}{\dfrac{d}{c} - \dfrac{c}{d}}$

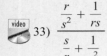

33) $\dfrac{\dfrac{r}{s^2} + \dfrac{1}{rs}}{\dfrac{s}{r} + \dfrac{1}{r^2}}$

34) $\dfrac{\dfrac{n}{m^3} + \dfrac{m}{n}}{\dfrac{3}{n} - \dfrac{m}{n^4}}$

43) $\dfrac{\dfrac{r^2 - 6}{20}}{r - \dfrac{6}{r}}$

44) $\dfrac{\dfrac{1}{6}}{\dfrac{7}{8}}$

35) $\dfrac{1 - \dfrac{4}{t+5}}{\dfrac{4}{t^2 - 25} + \dfrac{t}{t-5}}$

36) $\dfrac{1 + \dfrac{4}{t-3}}{\dfrac{t}{t-3} + \dfrac{2}{t^2 - 9}}$

45) $\dfrac{\dfrac{a-4}{12}}{\dfrac{a-4}{a}}$

46) $\dfrac{\dfrac{8}{w} - w}{1 + \dfrac{6}{w}}$

Objectives 1–3

Mixed Exercises

Simplify completely.

47) $\dfrac{\dfrac{5}{6}}{\dfrac{9}{15}}$

48) $\dfrac{\dfrac{5}{h+2} + \dfrac{7}{2h-3}}{\dfrac{1}{h-3} + \dfrac{3}{2h-3}}$

37) $\dfrac{b + \dfrac{1}{b}}{b - \dfrac{3}{b}}$

38) $\dfrac{\dfrac{z+6}{4}}{\dfrac{z+6}{z}}$

49) $\dfrac{\dfrac{5}{2n+1} + 1}{\dfrac{1}{n+3} + \dfrac{2}{2n+1}}$

50) $\dfrac{\dfrac{y^4}{z^3}}{\dfrac{y^6}{z^4}}$

39) $\dfrac{\dfrac{m}{n^2}}{\dfrac{m^4}{n}}$

40) $\dfrac{\dfrac{z^2 + 1}{5}}{z + \dfrac{1}{z}}$

Section 8.6 Solving Rational Equations

Objectives

1. Understand the Difference Between Adding and Subtracting Rational Expressions and Solving a Rational Equation

2. Solve Rational Equations

3. Solve a Proportion

4. Solve an Equation for a Specific Variable

A **rational equation** is an equation that contains a rational expression. Some examples of rational equations are

$$\frac{1}{3}x + \frac{3}{4} = \frac{5}{6}x - 2, \qquad \frac{5}{c-3} - \frac{c}{c+1} = 4, \qquad \frac{2t}{t^2 + 6t + 8} + \frac{7}{t+2} = \frac{4}{t+4}$$

1. Understand the Difference Between Adding and Subtracting Rational Expressions and Solving a Rational Equation

In Chapter 3 we solved rational equations like the first one above, and we learned how to add and subtract rational expressions in Section 8.4. Let's summarize the difference between the two because this is often a point of confusion for students.

> 1) *The sum or difference of rational expressions does not contain an = sign.* To add or subtract, rewrite each expression with the LCD, and *keep the denominator* while performing the operations.
>
> 2) *An equation contains an = sign.* To solve an equation containing rational expressions, *multiply* the equation by the LCD of all fractions to *eliminate* the denominators, then solve.

Example 1

Determine whether each is an equation or is a sum or difference of expressions. Then, solve the equation or find the sum or difference.

a) $\dfrac{k-2}{5} - \dfrac{k}{2} = \dfrac{7}{5}$ 　　　 b) $\dfrac{k-2}{5} - \dfrac{k}{2}$

Solution

a) This is an *equation* because it contains an $=$ sign. We will *solve* for k using the method we learned in Chapter 3: Eliminate the denominators by multiplying by the LCD of all of the expressions.   LCD $= 10$.

$$10\left(\dfrac{k-2}{5} - \dfrac{k}{2}\right) = 10 \cdot \dfrac{7}{5} \qquad \text{Multiply by LCD of 10 to eliminate the denominators.}$$
$$2(k-2) - 5k = 14 \qquad \text{Distribute and eliminate denominator.}$$
$$2k - 4 - 5k = 14$$
$$-3k - 4 = 14$$
$$-3k = 18$$
$$k = -6$$

Check:
$$\dfrac{-6-2}{5} - \dfrac{(-6)}{2} \overset{?}{=} \dfrac{7}{5} \qquad \text{Substitute } -6 \text{ for } k \text{ in the original equation.}$$
$$-\dfrac{8}{5} + 3 \overset{?}{=} \dfrac{7}{5}$$
$$-\dfrac{8}{5} + \dfrac{15}{5} \overset{?}{=} \dfrac{7}{5}$$
$$\dfrac{7}{5} = \dfrac{7}{5} \quad ✓$$

The solution is $\{-6\}$.

b) $\dfrac{k-2}{5} - \dfrac{k}{2}$ is *not* an equation to be solved because it does *not* contain an $=$ sign.

It is a difference of rational expressions. Rewrite each expression with the LCD, then subtract, *keeping the denominators* while performing the operations.

LCD $= 10$

$$\dfrac{(k-2)}{5} \cdot \dfrac{2}{2} = \dfrac{2(k-2)}{10}, \qquad \dfrac{k}{2} \cdot \dfrac{5}{5} = \dfrac{5k}{10}$$

$$\dfrac{k-2}{5} - \dfrac{k}{2} = \dfrac{2(k-2)}{10} - \dfrac{5k}{10} \qquad \text{Rewrite each expression with a denominator of 10.}$$
$$= \dfrac{2(k-2) - 5k}{10} \qquad \text{Subtract the numerators.}$$
$$= \dfrac{2k - 4 - 5k}{10} \qquad \text{Distribute.}$$
$$= \dfrac{-3k - 4}{10} \qquad \text{Combine like terms.}$$

You Try 1

Determine whether each is an equation or is a sum or difference of expressions. Then solve the equation or find the sum or difference.

a) $\dfrac{n}{9} + \dfrac{n-11}{6}$ b) $\dfrac{n}{9} + \dfrac{n-11}{6} = -1$

Here are the steps we use to solve a rational equation.

How to Solve a Rational Equation

1)   If possible, factor all denominators.

2)   Write down the LCD of all of the expressions.

3)   Multiply both sides of the equation by the LCD to *eliminate* the denominators.

4)   Solve the equation.

5)   Check the solution(s) in the original equation. If a proposed solution makes a denominator equal 0, then it is rejected as a solution.

Let's look at more examples of how to solve equations containing rational expressions.

2. Solve Rational Equations

Example 2

Solve $\dfrac{a}{3} + \dfrac{4}{a} = \dfrac{13}{3}$.

Solution

Since this is an equation, we will eliminate the denominators by multiplying the equation by the LCD of all of the expressions.

LCD $= 3a$

$$3a\left(\dfrac{a}{3} + \dfrac{4}{a}\right) = 3a\left(\dfrac{13}{3}\right) \qquad \text{Multiply both sides of the equation by the LCD, } 3a.$$

$$3a\left(\dfrac{a}{3}\right) + 3a\left(\dfrac{4}{a}\right) = 3a\left(\dfrac{13}{3}\right) \qquad \text{Distribute and divide out common factors.}$$

$$a^2 + 12 = 13a$$

$$a^2 - 13a + 12 = 0 \qquad \text{Subtract } 13a.$$

$$(a - 12)(a - 1) = 0 \qquad \text{Factor.}$$

$$a - 12 = 0 \quad \text{or} \quad a - 1 = 0$$

$$a = 12 \quad \text{or} \quad a = 1$$

Check:

$a = 12$

$$\frac{a}{3} + \frac{4}{a} \stackrel{?}{=} \frac{13}{3}$$

$$\frac{12}{3} + \frac{4}{12} \stackrel{?}{=} \frac{13}{3}$$

$$\frac{12}{3} + \frac{1}{3} = \frac{13}{3} \quad \checkmark$$

$a = 1$

$$\frac{a}{3} + \frac{4}{a} \stackrel{?}{=} \frac{13}{3}$$

$$\frac{1}{3} + \frac{4}{1} \stackrel{?}{=} \frac{13}{3}$$

$$\frac{1}{3} + \frac{12}{3} = \frac{13}{3} \quad \checkmark$$

The solution set is $\{1, 12\}$.

You Try 2

Solve $\dfrac{c}{2} + 1 = \dfrac{24}{c}$.

It is *very* important to check the proposed solution. Sometimes, what appears to be a solution actually is not.

Example 3

Solve $3 - \dfrac{3}{x + 3} = \dfrac{x}{x + 3}$.

Solution

Since this is an equation, we will eliminate the denominators by multiplying the equation by the LCD of all of the expressions.

$LCD = x + 3$

$$(x + 3)\left(3 - \frac{3}{x + 3}\right) = (x + 3)\left(\frac{x}{x + 3}\right)$$ Multiply both sides of the equation by the LCD, $x + 3$.

$$(x + 3)3 - \cancel{(x + 3)} \cdot \frac{3}{\cancel{x + 3}} = \cancel{(x + 3)}\left(\frac{x}{\cancel{x + 3}}\right)$$ Distribute and divide out common factors.

$$3x + 9 - 3 = x$$ Multiply.

$$3x + 6 = x$$

$$6 = -2x$$ Subtract $3x$.

$$-3 = x$$ Divide by -2.

Check: $3 - \dfrac{3}{(-3) + 3} \stackrel{?}{=} \dfrac{-3}{(-3) + 3}$ Substitute -3 for x in the original equation.

$$3 - \frac{3}{0} = \frac{-3}{0}$$

Since $x = -3$ makes the denominator equal zero, -3 cannot be a solution to the equation. Therefore, this equation has no solution. The solution set is $\varnothing$.

BE CAREFUL *Always* check what *appears* to be the solution or solutions to an equation containing rational expressions. If one of these values makes a denominator zero, then it *cannot* be a solution to the equation.

You Try 3

Solve each equation.

a) $\dfrac{7}{s + 1} + \dfrac{2s}{s + 1} = 3$ b) $\dfrac{3m}{m - 4} - 1 = \dfrac{12}{m - 4}$

Example 4

Solve $\dfrac{1}{3} - \dfrac{1}{t + 2} = \dfrac{t + 14}{3t^2 - 12}$.

Solution

This is an equation. Eliminate the denominators by multiplying by the LCD. Begin by factoring the denominator of $\dfrac{t + 14}{3t^2 - 12}$.

$$\frac{1}{3} - \frac{1}{t + 2} = \frac{t + 14}{3(t + 2)(t - 2)} \qquad \text{Factor the denominator.}$$

$$\text{LCD} = 3(t + 2)(t - 2) \qquad \text{Write down the LCD of all of the expressions.}$$

$$3(t + 2)(t - 2)\left(\frac{1}{3} - \frac{1}{t + 2}\right) = 3(t + 2)(t - 2)\left(\frac{t + 14}{3(t + 2)(t - 2)}\right) \qquad \begin{array}{l}\text{Multiply both sides}\\ \text{of the equation by}\\ \text{the LCD.}\end{array}$$

$$3(t + 2)(t - 2)\left(\frac{1}{3}\right) - 3(t + 2)(t - 2)\left(\frac{1}{t + 2}\right) = 3(t + 2)(t - 2)\left(\frac{t + 14}{3(t + 2)(t - 2)}\right)$$

Distribute and divide out common factors.

$$(t + 2)(t - 2) - 3(t - 2) = t + 14 \qquad \text{Multiply.}$$
$$t^2 - 4 - 3t + 6 = t + 14 \qquad \text{Distribute.}$$
$$t^2 - 3t + 2 = t + 14 \qquad \text{Combine like terms.}$$
$$t^2 - 4t - 12 = 0 \qquad \text{Subtract } t \text{ and subtract 14.}$$
$$(t - 6)(t + 2) = 0 \qquad \text{Factor.}$$
$$t - 6 = 0 \quad \text{or} \quad t + 2 = 0 \qquad \text{Set each factor equal to zero.}$$
$$t = 6 \quad \text{or} \qquad t = -2 \qquad \text{Solve.}$$

Look at the factored form of the equation. If $t = 6$, no denominator will equal zero. If $t = -2$, however, two of the denominators will equal zero. Therefore, we must reject $t = -2$ as a solution. Check only $t = 6$.

$$\text{Check: } \frac{1}{3} - \frac{1}{6 + 2} \stackrel{?}{=} \frac{6 + 14}{3(6)^2 - 12} \qquad \text{Substitute } t = 6 \text{ into the original equation.}$$

$$\frac{1}{3} - \frac{1}{8} \stackrel{?}{=} \frac{20}{108 - 12} \qquad \text{Simplify.}$$

$$\frac{1}{3} - \frac{1}{8} \stackrel{?}{=} \frac{20}{96} \qquad \text{Simplify.}$$

$$\frac{8}{24} - \frac{3}{24} \stackrel{?}{=} \frac{5}{24} \qquad \text{Get a common denominator and reduce } \frac{20}{96}.$$

$$\frac{5}{24} = \frac{5}{24} \ \checkmark \qquad \text{Subtract.}$$

The solution set is $\{6\}$.

The previous problem is a good example of why it is necessary to check all "solutions" to equations containing rational expressions.

You Try 4

Solve $\dfrac{w}{w + 6} - \dfrac{3}{w + 2} = \dfrac{6 - 3w}{w^2 + 8w + 12}$.

3. Solve a Proportion

Example 5

Solve $\dfrac{20}{d + 6} = \dfrac{8}{d}$.

Solution

This equation is a *proportion*. A **proportion** is a statement that two ratios are equal. We can solve this proportion as we have solved the other equations in this section, by multiplying both sides of the equation by the LCD. Or, recall from Section 3.6 that *we can solve a proportion by setting the cross products equal to each other*.

$$\frac{20}{d + 6} = \frac{8}{d}$$

Multiply.　　Multiply.

$$20d = 8(d + 6) \qquad \text{Set the cross product equal to each other.}$$
$$20d = 8d + 48 \qquad \text{Distribute.}$$
$$12d = 48 \qquad \text{Subtract } 8d.$$
$$d = 4 \qquad \text{Solve.}$$

$$Check: \frac{20}{4+6} \stackrel{?}{=} \frac{8}{4} \qquad \text{Substitute } d = 4 \text{ into the original equation.}$$

$$\frac{20}{10} \stackrel{?}{=} 2$$

$$2 = 2 \; \checkmark$$

The solution is {4}.

You Try 5

Solve $\dfrac{9}{y} = \dfrac{5}{y-2}$.

4. Solve an Equation for a Specific Variable

In Section 3.5, we learned how to solve an equation for a specific variable. For example, to solve $2l + 2w = P$ for w, we do the following:

$$2l + 2\boxed{w} = P \qquad \text{Put a box around } w, \text{ the variable for which we are solving.}$$

$$2\boxed{w} = P - 2l \qquad \text{Subtract } 2l.$$

$$w = \frac{P - 2l}{2} \qquad \text{Divide by 2.}$$

Next we discuss how to solve for a specific variable in a rational expression.

Example 6

Solve $m = \dfrac{x}{a - A}$ for a.

Solution

Note that the equation contains a lowercase a and an uppercase A. These represent different quantities, so students should be sure to write them correctly. Put a in a box.

Since a is in the denominator of the rational expression, multiply both sides of the equation by $a - A$ to eliminate the denominator.

$$m = \frac{x}{\boxed{a} - A} \qquad \text{Put } a \text{ in a box.}$$

$$(\boxed{a} - A)m = (\boxed{a} - A)\left(\frac{x}{\boxed{a} - A}\right) \qquad \begin{array}{l}\text{Multiply both sides by } a - A \text{ to}\\ \text{eliminate the denominator.}\end{array}$$

$$\boxed{a}\,m - Am = x \qquad \text{Distribute.}$$

$$\boxed{a}\,m = x + Am \qquad \text{Add } Am.$$

$$a = \frac{x + Am}{m} \qquad \text{Divide by } m.$$

You Try 6

Solve $y = \dfrac{c}{r + d}$ for d.

Example 7

Solve $\dfrac{1}{a} + \dfrac{1}{b} = \dfrac{1}{c}$ for b.

Solution

Put the b in a box. The LCD of all of the fractions is abc. Multiply both sides of the equation by abc.

$$\dfrac{1}{a} + \dfrac{1}{\boxed{b}} = \dfrac{1}{c} \qquad \text{Put } b \text{ in a box.}$$

$$a\boxed{b}c\left(\dfrac{1}{a} + \dfrac{1}{\boxed{b}}\right) = a\boxed{b}c\left(\dfrac{1}{c}\right) \qquad \text{Multiply both sides by } abc \text{ to eliminate the denominator.}$$

$$\not{a}\boxed{b}c \cdot \dfrac{1}{\not{a}} + a\boxed{\not{b}}c \cdot \dfrac{1}{\boxed{\not{b}}} = a\boxed{b}\not{c}\left(\dfrac{1}{\not{c}}\right) \qquad \text{Distribute.}$$

$$\boxed{b}c + ac = a\boxed{b} \qquad \text{Divide out common factors.}$$

Since we are solving for b and there are terms containing b on each side of the equation, we must get bc and ab on one side of the equation and ac on the other side.

$$ac = a\boxed{b} - \boxed{b}c \qquad \text{Subtract } bc \text{ from each side.}$$

To isolate b, we will *factor* b out of each term on the right-hand side of the equation.

$$ac = \boxed{b}(a - c) \qquad \text{Factor out } b.$$

$$\dfrac{ac}{a - c} = b \qquad \text{Divide by } a - c.$$

You Try 7

Solve $\dfrac{1}{a} + \dfrac{1}{b} = \dfrac{1}{c}$ for a.

Answers to You Try Exercises

1) a) sum; $\dfrac{5n - 33}{9}$ b) equation; $n = 3$ 2) $\{-8, 6\}$ 3) a) $\{4\}$ b) $\varnothing$

4) $\{4\}$ 5) $\left\{\dfrac{9}{2}\right\}$ 6) $d = \dfrac{c - yr}{y}$ 7) $a = \dfrac{bc}{b - c}$

8.6 Exercises

Objectives 1 and 2

1) When solving an equation containing rational expressions, do you keep the LCD throughout the problem or do you eliminate the denominators?

2) When adding or subtracting two rational expressions, do you keep the LCD throughout the problem or do you eliminate the denominators?

Determine whether each is an equation or is a sum or difference of expressions. Then, solve the equation or find the sum or difference.

3) $\dfrac{m}{8} + \dfrac{m-7}{4}$

4) $\dfrac{2r+9}{3} - \dfrac{r}{9}$

5) $\dfrac{2f-19}{20} = \dfrac{f}{4} + \dfrac{2}{5}$

6) $\dfrac{2h}{5} + \dfrac{2}{3} = \dfrac{h+3}{5}$

7) $\dfrac{z}{z-6} - \dfrac{4}{z}$

8) $\dfrac{2}{a^2} + \dfrac{1}{a+9}$

9) $1 + \dfrac{4}{c+2} = \dfrac{9}{c+2}$

10) $\dfrac{10}{b-8} - 4 = \dfrac{2}{b-8}$

Values that make the denominators equal zero cannot be solutions of an equation. Find *all* of the values that make the denominators zero and that, therefore, cannot be solutions of each equation. Do *not* solve the equation.

11) $\dfrac{t}{t+10} - \dfrac{5}{t} = 3$

12) $\dfrac{k+2}{k-3} + 1 = \dfrac{7}{k}$

13) $\dfrac{2}{d^2-81} + \dfrac{8}{d} = \dfrac{6}{d+9}$

14) $\dfrac{4}{p+2} - \dfrac{5}{p} = \dfrac{p}{p^2-4}$

15) $\dfrac{v+7}{v^2-13v+36} - \dfrac{5}{3v-12} = \dfrac{v}{v-9}$

16) $\dfrac{3h}{h^2+3h-40} + \dfrac{1}{h+8} = \dfrac{h-1}{4h-20}$

Objectives 2 and 3

Solve each equation.

17) $\dfrac{y}{3} - \dfrac{1}{2} = \dfrac{1}{6}$

18) $\dfrac{a}{2} + \dfrac{3}{10} = \dfrac{4}{5}$

19) $\dfrac{1}{2}h + h = -3$

20) $\dfrac{1}{6}j - j = -10$

21) $\dfrac{7u+12}{15} = \dfrac{2u}{5} - \dfrac{3}{5}$

22) $\dfrac{8m-7}{24} = \dfrac{m}{6} - \dfrac{7}{8}$

23) $\dfrac{4}{3t+2} = \dfrac{2}{2t-1}$

24) $\dfrac{9}{4x+1} = \dfrac{3}{x+2}$

25) $\dfrac{w}{3} = \dfrac{2w-5}{12}$

26) $\dfrac{r-2}{2} = \dfrac{3r-5}{4}$

27) $\dfrac{12}{a} - 2 = \dfrac{6}{a}$

28) $\dfrac{16}{z} + 7 = -\dfrac{12}{z}$

29) $\dfrac{n}{n+2} + 3 = \dfrac{8}{n+2}$

30) $\dfrac{4q}{q+1} - 2 = \dfrac{3}{q+1}$

 31) $\dfrac{2}{s+6} + 4 = \dfrac{2}{s+6}$

32) $\dfrac{u}{u-4} + 3 = \dfrac{4}{u-4}$

33) $\dfrac{c}{c-7} - 4 = \dfrac{10}{c-7}$

34) $\dfrac{2b}{b+9} - 5 = \dfrac{3}{b+9}$

35) $\dfrac{32}{g} + 10 = -\dfrac{8}{g}$

36) $\dfrac{9}{r} - 1 = \dfrac{6}{r}$

Solve each equation.

37) $\dfrac{1}{m-1} + \dfrac{2}{m+3} = \dfrac{4}{m+3}$

38) $\dfrac{4}{c+2} + \dfrac{2}{c-6} = \dfrac{5}{c+2}$

39) $\dfrac{4}{w-8} - \dfrac{10}{w+8} = \dfrac{40}{w^2-64}$

40) $\dfrac{4}{p-5} + \dfrac{7}{p+5} = \dfrac{18}{p^2-25}$

41) $\dfrac{3}{a+3} + \dfrac{14}{a^2-4a-21} = \dfrac{5}{a-7}$

42) $\dfrac{5}{k+4} - \dfrac{3}{k+2} = \dfrac{8}{k^2+6k+8}$

43) $\dfrac{9}{t+4} + \dfrac{8}{t^2-16} = \dfrac{1}{t-4}$

44) $\dfrac{12}{g^2-9} + \dfrac{2}{g+3} = \dfrac{7}{g-3}$

45) $\dfrac{4}{x^2+2x-15} = \dfrac{8}{x-3} + \dfrac{2}{x+5}$

46) $\dfrac{4}{p-2} - \dfrac{9}{p^2-8p+12} = \dfrac{9}{p-6}$

47) $\dfrac{k^2}{3} = \dfrac{k^2+2k}{4}$

48) $\dfrac{x^2}{2} = \dfrac{x^2-5x}{3}$

49) $\dfrac{5}{m^2-25} = \dfrac{4}{m^2+5m}$

50) $\dfrac{3}{t^2} = \dfrac{6}{t^2+5t}$

51) $\dfrac{10v}{3v-12} - \dfrac{v+6}{v-4} = \dfrac{v}{3}$

52) $\dfrac{b-2}{2b-12} - \dfrac{b+2}{b-6} = \dfrac{b}{2}$

53) $\dfrac{w}{5} = \dfrac{w-3}{w+1} + \dfrac{12}{5w+5}$

54) $\dfrac{3y-2}{y+2} = \dfrac{y}{4} + \dfrac{1}{4y+8}$

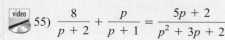 55) $\dfrac{8}{p+2} + \dfrac{p}{p+1} = \dfrac{5p+2}{p^2+3p+2}$

56) $\dfrac{6}{x+1} + \dfrac{x}{x-3} = \dfrac{3x+14}{x^2-2x-3}$

57) $\dfrac{11}{c+9} = \dfrac{c}{c-4} - \dfrac{36-8c}{c^2+5c-36}$

58) $\dfrac{3}{f+2} = \dfrac{f}{f+6} - \dfrac{2}{f^2+8f+12}$

59) $\dfrac{8}{3g^2-7g-6} + \dfrac{4}{g-3} = \dfrac{8}{3g+2}$

60) $\dfrac{1}{r-1} + \dfrac{2}{5r-3} = \dfrac{37}{5r^2-8r+3}$

61) $\dfrac{h}{h^2+2h-8} + \dfrac{4}{h^2+8h-20} = \dfrac{4}{h^2+14h+40}$

62) $\dfrac{b}{b^2+b-6} + \dfrac{3}{b^2+9b+18} = \dfrac{8}{b^2+4b-12}$

63) $\dfrac{u}{8} = \dfrac{2}{10-u}$

64) $-\dfrac{a}{4} = \dfrac{3}{a+7}$

65) $\dfrac{5}{r+4} - \dfrac{2}{r} = -1$

66) $\dfrac{6}{c-5} - \dfrac{2}{c} = 1$

67) $\dfrac{q}{q^2+4q-32} + \dfrac{2}{q^2-14q+40} = \dfrac{6}{q^2-2q-80}$

68) $\dfrac{r}{r^2+8r+15} - \dfrac{2}{r^2+r-6} = \dfrac{2}{r^2+3r-10}$

Objective 4

Solve for the indicated variable.

69) $V = \dfrac{nRT}{P}$ for P

70) $W = \dfrac{CA}{m}$ for m.

71) $y = \dfrac{kx}{z}$ for z

72) $a = \dfrac{rt}{2b}$ for b

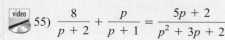 73) $B = \dfrac{t+u}{3x}$ for x

74) $Q = \dfrac{n-k}{5r}$ for r

75) $z = \dfrac{a}{b+c}$ for b

76) $d = \dfrac{t}{l-n}$ for n

77) $A = \dfrac{4r}{q-t}$ for t

78) $h = \dfrac{3A}{r+s}$ for s

79) $w = \dfrac{na}{kc+b}$ for c

80) $r = \dfrac{kx}{y-az}$ for y

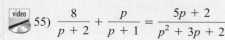 81) $\dfrac{1}{t} = \dfrac{1}{r} - \dfrac{1}{s}$ for r

82) $\dfrac{1}{R_1} + \dfrac{1}{R_2} = \dfrac{1}{R_3}$ for R_2

83) $\dfrac{2}{A} + \dfrac{1}{C} = \dfrac{3}{B}$ for C

84) $\dfrac{5}{x} = \dfrac{1}{y} - \dfrac{4}{z}$ for z

Section 8.7 Applications of Equations Containing Rational Expressions

Objectives

1. Solve Problems Involving Proportions
2. Solve Problems Involving Distance, Rate, and Time
3. Solve Problems Involving Work

We have studied applications of linear and quadratic equations. Now we turn our attention to applications involving equations with rational expressions. We will continue to use the Steps for Solving Applied Problems outlined in Section 3.2.

1. Solve Problems Involving Proportions

We first solved application problems involving proportions in Section 3.6. We begin this section with a problem involving a proportion.

Example 1

Write an equation and solve.

One morning at a coffee shop, the ratio of the number of customers who ordered regular coffee to the number who ordered decaffeinated coffee was 4 to 1. If the number of people who ordered regular coffee was 126 more than the number who ordered decaf, how many people ordered each type of coffee?

Solution

Step 1: Read the problem carefully, twice.

Step 2: Find the number of customers who ordered regular coffee and the number who ordered decaffeinated coffee.

Define the unknowns:

$$x = \text{number who ordered decaffeinated coffee}$$
$$x + 126 = \text{number who ordered regular coffee}$$

Step 3: Translate from English to an algebraic equation.

We will write a proportion.

Number who ordered regular coffee $\rightarrow \dfrac{4}{1} = \dfrac{x + 126}{x} \leftarrow$ Number who ordered regular coffee
Number who ordered decaffeinated coffee $\rightarrow$ $\leftarrow$ Number who ordered decaffeinated coffee

Equation: $\dfrac{4}{1} = \dfrac{x + 126}{x}$.

Step 4: Solve the equation.

$$\dfrac{4}{1} = \dfrac{x + 126}{x}$$

Multiply.   Multiply.

$4x = 1(x + 126)$ Set the cross products equal to each other.
$4x = x + 126$
$3x = 126$ Subtract x.
$x = 42$ Divide by 3.

Step 5: Find the other unknown and interpret the meaning of the solution in the context of the problem.

$$x = 42$$
$$x + 126 = 168$$

42 customers ordered decaffeinated coffee, and 168 customers ordered regular coffee.

This checks since $4(42) = 168$ giving us the 4 to 1 ratio.

You Try 1

Write an equation and solve.

During one week at a bookstore, the ratio of the number of romance novels sold to the number of travel books sold was 5 to 3. Determine the number of each type of book sold if customers bought 106 more romance novels than travel books.

2. Solve Problems Involving Distance, Rate, and Time

In Section 3.6 we solved problems involving distance (d), rate (r), and time (t). The basic formula is $d = rt$. We can solve this formula for r and then for t to obtain

$$r = \frac{d}{t} \quad \text{and} \quad t = \frac{d}{r}$$

In this section, we will encounter problems involving boats going with and against a current, and planes going with and against the wind. Both scenarios use the same idea.

Say a boat's speed is 18 mph in still water. If that same boat had a 4 mph current pushing *against* it, how fast would it be traveling? (The current will cause the boat to slow down.) A boat traveling *against* the current is said to be traveling *upstream*.

$$\text{Speed } against \text{ the current} = 18 \text{ mph} - 4 \text{ mph}$$
$$= 14 \text{ mph}$$

$$\frac{\text{Speed } against}{\text{the current}} = \frac{\text{Speed in}}{\text{still water}} - \frac{\text{Speed of}}{\text{the current}}$$

If the speed of the boat in still water is 18 mph and a 4 mph current is *pushing* the boat, how fast would the boat be traveling *with* the current? (The current will cause the boat to travel faster.) A boat traveling *with* the current is said to be traveling *downstream*.

$$\text{Speed } with \text{ the current} = 18 \text{ mph} + 4 \text{ mph}$$
$$= 22 \text{ mph}$$

$$\frac{\text{Speed } with}{\text{the current}} = \frac{\text{Speed in}}{\text{still water}} + \frac{\text{Speed of}}{\text{the current}}$$

We will use these ideas in Example 2.

Example 2

Write an equation and solve.

A boat can travel 8 mi downstream in the same amount of time it can travel 6 mi upstream. If the speed of the current is 2 mph, what is the speed of the boat in still water?

Solution

Step 1: Read the problem carefully, twice. Let us clarify that "8 mi downstream" means *8 mi with the current,* and "6 mi upstream" means *6 mi against the current.*

Step 2: Find the speed of the boat in still water. Define the unknowns.

$$x = \text{speed of the boat in still water}$$
$$x + 2 = \text{speed of the boat } with \text{ the current (downstream)}$$
$$x - 2 = \text{speed of the boat } against \text{ the current (upstream)}$$

Step 3:   Translate from English to an algebraic equation.

We will make a table to organize the information.

	d	r	t
Downstream	8	$x + 2$	
Upstream	6	$x - 2$	

Before we fill in the last column, we need to understand how we will write an equation. Our equation will come from the fact that the "boat can travel 8 mi downstream *in the same amount of time* it can travel 6 mi upstream." Therefore, we can write an equation in English as

$$\text{Time for boat to go } 8 \text{ mi downstream} = \text{Time for boat to go } 6 \text{ mi upstream}$$

In the last column of the table, we want to write expressions for the time in terms of x.

Since $d = rt$, we can solve for t and obtain

$$t = \frac{d}{r}$$

Substituting the information in the table we get

$$\text{Downstream: } t = \frac{d}{r} \rightarrow t = \frac{8}{x + 2}$$

$$\text{Upstream: } \quad t = \frac{d}{r} \rightarrow t = \frac{6}{x - 2}$$

Put these values into the table.

	d	r	t
Downstream	8	$x + 2$	$\dfrac{8}{x + 2}$
Upstream	6	$x - 2$	$\dfrac{6}{x - 2}$

Looking at the table, we are ready to write the equation.

$$\text{Time for boat to go } 8 \text{ mi downstream} = \text{Time for boat go } 6 \text{ mi upstream}$$

Equation: $$\frac{8}{x + 2} = \frac{6}{x - 2}$$

Step 4:   Solve the equation.

The equation is a proportion, so we can solve it by setting the cross products equal to each other.

$$\frac{8}{x+2} \times \frac{6}{x-2}$$

Multiply   Multiply

$8(x - 2) = 6(x + 2)$ Set the cross products equal to each other.

$8x - 16 = 6x + 12$ Distribute.

$2x = 28$ Subtract $6x$ and add 16.

$x = 14$ Divide by 2.

Step 5:   Interpret the meaning of the solution as it relates to the problem.

The speed of the boat in still water is 14 mph.

We can check the answer by checking to see that the time spent traveling downstream is the same as the time spent traveling upstream.

Downstream: The *speed* of the boat traveling 8 mi downstream is $14 + 2 = 16$ mph.

$$\text{The } time \text{ spent traveling downstream is } t = \frac{d}{r} = \frac{8}{16} = \frac{1}{2} \text{ hr.}$$

Upstream: The *speed* of the boat traveling 6 mi upstream is $14 - 2 = 12$ mph.

$$\text{The } time \text{ spent traveling upstream is } t = \frac{d}{r} = \frac{6}{12} = \frac{1}{2} \text{ hr.}$$

So, time upstream = time downstream.   ✓

 You Try 2

Write an equation and solve.

It takes a boat the same amount of time to travel 10 mi upstream as it does to travel 15 mi downstream. Find the speed of the boat in still water if the speed of the current is 4 mph.

3. Solve Problems Involving Work

Suppose it takes Brian 5 hr to paint his bedroom. What is the *rate* at which he does the job?

$$\text{rate} = \frac{1 \text{ room}}{5 \text{ hr}} = \frac{1}{5} \text{ room/hr}$$

Brian works at a rate of $\frac{1}{5}$ of a room per hour.

In general, then, we can say that if it takes t units of time to do a job, then the *rate* at which the job is done is $\frac{1}{t}$ job per unit of time.

This idea of *rate* is what we use to determine how long it can take for 2 or more people or things to do a job.

Let's assume, again, that Brian can paint his room in 5 hr. At this rate, how much of the job can he do in 2 hr?

$$\begin{aligned}
\text{Fractional part} \atop \text{of the job done} &= \text{Rate of} \atop \text{work} \cdot \text{Amount of} \atop \text{time worked} \\
&= \frac{1}{5} \cdot 2 \\
&= \frac{2}{5}
\end{aligned}$$

$\frac{2}{5}$ of the room can be painted in 2 hr.

The basic equation used to solve work problems is:

$$\text{Fractional part of a job} \atop \text{done by one person or thing} + \text{Fractional part of a job} \atop \text{done by another person or thing} = 1 \text{ (whole job)}$$

Example 3

Write an equation and solve.

If Brian can paint his bedroom in 5 hr, but his brother, Doug, could paint the room on his own in 4 hr, how long would it take for the two of them to paint the room together?

Solution

Step 1: Read the problem carefully, twice.

Step 2: Determine how long it will take for Brian and Doug to paint the room together.

$$t = \text{number of hours to paint the room together}$$

Step 3: Translate from English to an algebraic equation.

Let's write down their rates:

$$\text{Brian's rate} = \frac{1}{5} \text{ room/hr (since the job takes him 5 hr)}$$

$$\text{Doug's rate} = \frac{1}{4} \text{ room/hr (since the job takes him 4 hr)}$$

It takes them t hours to paint the room together. Recall that

$$\text{Fractional part} \atop \text{of job done} = \text{Rate of} \atop \text{work} \cdot \text{Amount of} \atop \text{time worked}$$

$$\text{Brian's fractional part} = \frac{1}{5} \cdot t = \frac{1}{5}t$$

$$\text{Doug's fractional part} = \frac{1}{4} \cdot t = \frac{1}{4}t$$

The equation we can write comes from

$$\underset{\text{job done by Brian}}{\text{Fractional part of the}} + \underset{\text{job done by Doug}}{\text{Fractional part of the}} = 1 \text{ whole job}$$

$$\frac{1}{5}t \qquad + \qquad \frac{1}{4}t \qquad = \qquad 1$$

Equation: $\dfrac{1}{5}t + \dfrac{1}{4}t = 1$.

Step 4: Solve the equation.

$$20\left(\frac{1}{5}t + \frac{1}{4}t\right) = 20(1) \qquad \text{Multiply by the LCD of 20 to eliminate the fractions.}$$

$$20\left(\frac{1}{5}t\right) + 20\left(\frac{1}{4}t\right) = 20(1) \qquad \text{Distribute.}$$

$$4t + 5t = 20 \qquad \text{Multiply.}$$

$$9t = 20 \qquad \text{Combine like terms.}$$

$$t = \frac{20}{9} \qquad \text{Divide by 9.}$$

Step 5: Interpret the meaning of the solution as it relates to the problem.

Brian and Doug could paint the room together in $\dfrac{20}{9}$ hr or $2\dfrac{2}{9}$ hr.

Check: $\underset{\text{done by Brian}}{\text{Part of job}} = \dfrac{1}{5}t = \dfrac{1}{\cancel{5}_{1}} \cdot \dfrac{\cancel{20}^{4}}{9} = \dfrac{4}{9}$

$\underset{\text{done by Doug}}{\text{Part of job}} = \dfrac{1}{4}t = \dfrac{1}{\cancel{4}_{1}} \cdot \dfrac{\cancel{20}^{5}}{9} = \dfrac{5}{9}$

$$\frac{4}{9} + \frac{5}{9} = \frac{9}{9} = 1$$

The answer checks since together they did 1 whole job.

You Try 3

Write an equation and solve.

Krutesh can mow a lawn in 2 hr while it takes Stefan 3 hr to mow the same lawn. How long would it take for them to mow the lawn if they worked together?

Answers to You Try Exercises

1) 159 travel books, 265 romance novels   2) 20 mph   3) $\dfrac{6}{5}$ hr or $1\dfrac{1}{5}$ hr

Objective 1

Solve the following proportions.

1) $\dfrac{12}{7} = \dfrac{60}{x}$

2) $\dfrac{30}{18} = \dfrac{45}{y}$

3) $\dfrac{6}{13} = \dfrac{x}{x + 56}$

4) $\dfrac{15}{8} = \dfrac{x}{x - 63}$

Write an equation for each and solve. See Example 1.

5) At a motocross race, the ratio of male spectators to female spectators was 10 to 3. If there were 370 male spectators, how many females were in the crowd?

6) The ratio of students in a history lecture who took notes in pen to those who took notes in pencil was 8 to 3. If 72 students took notes in pen, how many took notes in pencil?

7) In a gluten-free flour mixture, the ratio of potato-starch flour to tapioca flour is 2 to 1. If a mixture contains 3 more cups of potato-starch flour than tapioca flour how much of each type of flour is in the mixture?

8) Rosa Cruz won an election over her opponent by a ratio of 6 to 5. If her opponent received 372 fewer votes than she did, how many votes did each candidate receive?

9) The ancient Greeks believed that the rectangle most pleasing to the eye, the golden rectangle, had sides in which the ratio of its length to its width was approximately 8 to 5. They erected many buildings using this golden ratio, including the Parthenon. The marble floor of a museum foyer is to be designed as a golden rectangle. If its width is to be 18 ft less than its length, find the length and width of the foyer.

10) To obtain a particular color, a painter mixed two colors in a ratio of 7 parts blue to 3 parts yellow. If he used 8 fewer gallons of yellow than blue, how many gallons of blue paint did he use?

11) Ms. Hiramoto has invested her money so that the ratio of the amount in bonds to the amount in stocks is 3 to 2.

If she has $4000 more invested in bonds than in stocks, how much does she have invested in each?

12) At a wildlife refuge, the ratio of deer to rabbits is 4 to 9. Determine the number of deer and rabbits at the refuge if there are 40 more rabbits than deer.

13) In a small town, the ratio of households with pets to those without pets is 5 to 4. If 271 more households have pets than do not, how many households have pets?

14) An industrial cleaning solution calls for 5 parts water to 2 parts concentrated cleaner. If a worker uses 15 more quarts of water than concentrated cleaner to make a solution,

a) how much concentrated cleaner did she use?

b) how much water did she use?

c) how much solution did she make?

Objective 2

15) If the speed of a boat in still water is 10 mph,

a) what is its speed going *against* a 3 mph current?

b) What is its speed *with* a 3 mph current?

16) If an airplane travels at a constant rate of 300 mph,

a) what is its speed going *into* a 25 mph wind?

b) what is its speed going *with* a 25 mph wind?

17) If an airplane travels at a constant rate of x mph,

a) what is its speed going *with* a 30 mph wind?

b) what is its speed going *against* a 30 mph wind?

18) If the speed of a boat in still water is 13 mph,

a) what is its speed going *against* a current with a rate of x mph?

b) what is its speed going *with* a current with a rate of x mph?

Write an equation for each and solve. See Example 2.

19) A current flows at 5 mph. A boat can travel 20 mi downstream in the same amount of time it can go 12 mi upstream. What is the speed of the boat in still water?

20) With a current flowing at 4 mph, a boat can travel 32 mi with the current in the same amount of time it can go 24 mi against the current. Find the speed of the boat in still water.

21) A boat travels at 16 mph in still water. It takes the same amount of time for the boat to travel 15 mi downstream as to go 9 mi upstream. Find the speed of the current.

22) A boat can travel 12 mi downstream in the time it can go 6 mi upstream. If the speed of the boat in still water is 9 mph, what is the speed of the current?

23) An airplane flying at constant speed can fly 350 mi with the wind in the same amount of time it can fly 300 mi against the wind. What is the speed of the plane if the wind blows at 20 mph?

24) When the wind is blowing at 25 mph, a plane flying at a constant speed can travel 500 mi with the wind in the same amount of time it can fly 400 mi against the wind. Find the speed of the plane.

25) In still water the speed of a boat is 10 mph. Against the current it can travel 4 mi in the same amount of time it can travel 6 mi with the current. What is the speed of the current?

26) Flying at a constant speed, a plane can travel 800 mi with the wind in the same amount of time it can fly 650 mi against the wind. If the wind blows at 30 mph, what is the speed of the plane?

Objective 3

27) Toby can finish a computer programming job in 4 hr. What is the rate at which he does the job?

28) It takes Crystal 3 hr to paint her backyard fence. What is the rate at which she works?

29) Eloise can fertilize her lawn in t hr. What is the rate at which she does this job?

30) It takes Manu twice as long to clean a pool as it takes Anders. If it takes Anders t hr to clean the pool, at what rate does Manu do the job?

Write an equation for each and solve. See Example 3.

31) It takes Arlene 2 hr to trim the bushes at a city park while the same job takes Andre 3 hr. How long would it take for them to do the job together?

32) A hot water faucet can fill a sink in 8 min while it takes the cold water faucet only 6 min. How long would it take to fill the sink if both faucets were on?

33) Jermaine and Sue must put together notebooks for each person attending a conference. Working alone it would take Jermaine 5 hr while it would take Sue 8 hr. How long would it take for them to assemble the notebooks together?

34) The Williams family has two printers on which they can print out their vacation pictures. The larger printer can print all of the pictures in 3 hr, while it would take 5 hr on the smaller printer. How long would it take to print the vacation pictures using both printers?

35) A faucet can fill a tub in 12 min. The leaky drain can empty the tub in 30 min. If the faucet is on and the drain is leaking, how long would it take to fill the tub?

36) It takes Deepak 50 min to shovel snow from his sidewalk and driveway. When he works with his brother, Kamal, it takes only 30 min. How long would it take Kamal to do the shoveling himself?

37) Fatima and Antonio must cut out shapes for an art project at a day-care center. Fatima can do the job twice as fast as Antonio. Together, it takes 2 hr to cut out all of the shapes. How long would it take Fatima to cut out the shapes herself?

38) It takes Burt three times as long as Phong to set up a new alarm system. Together they can set it up in 90 min. How long would it take Phong to set up the alarm system by himself?

39) Working together it takes 2 hr for a new worker and an experienced worker to paint a billboard. If the new employee worked alone, it would take him 6 hr. How long would it take the experienced worker to paint the billboard by himself?

40) Audrey can address party invitations in 40 min, while it would take her mom 1 hr. How long would it take for them to address the invitations together?

Definition/Procedure	Example	Reference

Simplifying Rational Expressions — **8.1**

A *rational expression* is an expression of the form $\frac{P}{Q}$, where P and Q are polynomials and where $Q \neq 0$.

We can *evaluate* rational expressions.

Evaluate $\frac{4a - 9}{a + 2}$ for $a = 3$.

$$\frac{4(3) - 9}{3 + 2} = \frac{12 - 9}{5} = \frac{3}{5}$$

p. 502

How to Determine When a Rational Expression Equals Zero and When it is Undefined

1) To determine what value of the variable makes the expression equal zero, set the numerator equal to zero and solve for the variable.

2) To determine what value of the variable makes the expression undefined, set the denominator equal to zero and solve for the variable.

For what value(s) of x is $\frac{x - 9}{x + 8}$

a) equal to zero? b) undefined?

a) $\frac{x - 9}{x + 8} = 0$ when $x - 9 = 0$.

$$x - 9 = 0$$
$$x = 9$$

When $x = 9$, the expression equals zero.

b) $\frac{x - 9}{x + 8}$ is undefined when its denominator equals zero. Solve $x + 8 = 0$.

$$x + 8 = 0$$
$$x = -8$$

When $x = -8$, the expression is undefined.

p. 503

To Write an Expression in Lowest Terms

1) Completely *factor* the numerator and denominator.

2) *Divide* the numerator and denominator by the greatest common factor.

Simplify $\frac{2r^2 - 11r + 15}{4r^2 - 36}$.

$$\frac{2r^2 - 11r + 15}{4r^2 - 36} = \frac{(2r - 5)(r - 3)}{4(r + 3)(r - 3)}$$
$$= \frac{2r - 5}{4(r + 3)}$$

p. 506

Simplifying $\frac{a - b}{b - a}$.

A rational expression of the form $\frac{a - b}{b - a}$ will simplify to -1.

Simplify $\frac{4 - w}{w^2 - 16}$.

$$\frac{4 - w}{w^2 - 16} = \frac{\overset{-1}{4 - w}}{(w + 4)(w - 4)}$$
$$= -\frac{1}{w + 4}$$

p. 508

Definition/Procedure	Example	Reference
Rational Functions The *domain* of a rational function consists of all real numbers except the value(s) of the variable which make the denominator equal zero.	Determine the domain of the rational function $$f(x) = \frac{x-9}{x+2}.$$ $x + 2 = 0$ Set the denominator $= 0$. $x = -2$ Solve. When $x = -2$, the denominator of $f(x) = \dfrac{x-9}{x+2}$ equals zero. The domain contains all real numbers *except* -2. Write the domain in interval notation as $(-\infty, -2) \cup (-2, \infty)$.	**p. 510**
Multiplying and Dividing Rational Expressions		**8.2**
Multiplying Rational Expressions 1) Factor numerators and denominators. 2) Reduce and multiply.	Multiply $\dfrac{15v^3}{v^2 + 8v + 12} \cdot \dfrac{2v + 12}{5v}$. $\dfrac{15v^3}{v^2 + 8v + 12} \cdot \dfrac{2v + 12}{5v}$ $\qquad = \dfrac{\overset{3}{\cancel{15}}v^2 \cdot \cancel{v}}{(v+2)\cancel{(v+6)}} \cdot \dfrac{2\cancel{(v+6)}}{\cancel{5v}}$ $\qquad = \dfrac{6v^2}{v+2}$	**p. 515**
Dividing Rational Expressions To divide rational expressions, multiply the first expression by the reciprocal of the second.	Divide $\dfrac{3x^2 + 4x}{x+1} \div \dfrac{9x^2 - 16}{21x - 28}$. $\dfrac{3x^2 + 4x}{x+1} \div \dfrac{9x^2 - 16}{21x - 28}$ $\qquad = \dfrac{3x^2 + 4x}{x+1} \cdot \dfrac{21x - 28}{9x^2 - 16}$ $\qquad = \dfrac{x\cancel{(3x+4)}}{x+1} \cdot \dfrac{7\cancel{(3x-4)}}{\cancel{(3x+4)}\cancel{(3x-4)}}$ $\qquad = \dfrac{7x}{x+1}$	**p. 516**

Definition/Procedure	Example	Reference

Finding the Least Common Denominator

8.3

To Find the Least Common Denominator (LCD)

Step 1: Factor the denominators.

Step 2: The LCD will contain each unique factor the greatest number of times it appears in any single factorization.

Step 3: The LCD is the *product* of the factors identified in step 2.

Find the LCD of $\dfrac{5a}{a^2 + 7a}$ and $\dfrac{2}{a^2 + 14a + 49}$.

Step 1: $a^2 + 7a = a(a + 7)$
$a^2 + 14a + 49 = (a + 7)^2$

Step 2: The factors we will use in the LCD are a and $(a + 7)^2$.

Step 3: LCD $= a(a + 7)^2$

p. 520

Adding and Subtracting Rational Expressions

8.4

Adding and Subtracting Rational Expressions

1) Factor the denominators.

2) Write down the LCD.

3) Rewrite each rational expression as an equivalent expression with the LCD.

4) Add or subtract the numerators and keep the common denominator in factored form.

5) After combining like terms in the numerator ask yourself, "Can I factor it?" If so, factor.

6) Reduce the rational expression, if possible.

Add $\dfrac{y}{y + 5} + \dfrac{4y - 30}{y^2 - 25}$.

1) Factor the denominator of $\dfrac{4y - 30}{y^2 - 25}$.

$$\dfrac{4y - 30}{y^2 - 25} = \dfrac{4y - 30}{(y + 5)(y - 5)}$$

2) The LCD is $(y + 5)(y - 5)$.

3) Rewrite $\dfrac{y}{y + 5}$ with the LCD.

$$\dfrac{y}{y + 5} \cdot \dfrac{y - 5}{y - 5} = \dfrac{y(y - 5)}{(y + 5)(y - 5)}$$

4) $\dfrac{y}{y + 5} + \dfrac{4y - 30}{y^2 - 25}$

$$= \dfrac{y(y - 5)}{(y + 5)(y - 5)} + \dfrac{4y - 30}{(y + 5)(y - 5)}$$

$$= \dfrac{y(y - 5) + 4y - 30}{(y + 5)(y - 5)}$$

$$= \dfrac{y^2 - 5y + 4y - 30}{(y + 5)(y - 5)}$$

$$= \dfrac{y^2 - y - 30}{(y + 5)(y - 5)}$$

5) $\qquad = \dfrac{\cancel{(y + 5)}(y - 6)}{\cancel{(y + 5)}(y - 5)}$ Factor.

6) $\qquad = \dfrac{y - 6}{y - 5}$ Reduce.

p. 532

Definition/Procedure	Example	Reference

A *complex fraction* is a rational expression that contains one or more fractions in its numerator, its denominator, or both.

Some examples of complex fractions are

$$\frac{\dfrac{9}{10}}{\dfrac{3}{2}}, \quad \frac{\dfrac{b+5}{2}}{\dfrac{4b+20}{7}}, \quad \frac{\dfrac{1}{x}-\dfrac{1}{y}}{1-\dfrac{x}{y}}$$

p. 543

To simplify a complex fraction containing one term in the numerator and one term in the denominator:

1) Rewrite the complex fraction as a division problem.

2) Perform the division by multiplying the first fraction by the reciprocal of the second.

Simplify $\dfrac{\dfrac{b+5}{2}}{\dfrac{4b+20}{7}}$.

$$\frac{\dfrac{b+5}{2}}{\dfrac{4b+20}{7}} = \frac{b+5}{2} \div \frac{4b+20}{7}$$

$$= \frac{b+5}{2} \cdot \frac{7}{4(b+5)}$$

$$= \frac{\cancel{b+5}}{2} \cdot \frac{7}{4\cancel{(b+5)}}$$

$$= \frac{7}{8}$$

p. 543

To simplify complex fractions containing more than one term in the numerator and/or the denominator,

Method 1

1) Combine the terms in the numerator and combine the terms in the denominator so that each contains only one fraction.

2) Rewrite as a division problem.

3) Perform the division.

Method 1

Simplify $\dfrac{\dfrac{1}{x}-\dfrac{1}{y}}{1-\dfrac{x}{y}}$.

1) $\dfrac{\dfrac{1}{x}-\dfrac{1}{y}}{1-\dfrac{x}{y}} = \dfrac{\dfrac{y}{xy}-\dfrac{x}{xy}}{\dfrac{y}{y}-\dfrac{x}{y}} = \dfrac{\dfrac{y-x}{xy}}{\dfrac{y-x}{y}}$

$$= \frac{y-x}{xy} \div \frac{y-x}{y}$$

$$= \frac{\cancel{y-x}}{xy} \cdot \frac{\cancel{y}}{\cancel{y-x}} = \frac{1}{x}$$

p. 544

Method 2

Step 1: Write down the LCD of *all* of the fractions in the complex fraction.

Step 2: Multiply the numerator and denominator of the complex fraction by the LCD.

Step 3: Simplify.

Method 2

Simplify $\dfrac{\dfrac{1}{x}-\dfrac{1}{y}}{1-\dfrac{x}{y}}$.

Step 1: LCD = xy

Definition/Procedure	Example	Reference

Step 2:
$$\frac{\dfrac{1}{x} - \dfrac{1}{y}}{1 - \dfrac{x}{y}} = \frac{xy\left(\dfrac{1}{x} - \dfrac{1}{y}\right)}{xy\left(1 - \dfrac{x}{y}\right)}$$

Step 3:
$$= \frac{xy \cdot \dfrac{1}{x} - xy \cdot \dfrac{1}{y}}{xy \cdot 1 - xy \cdot \dfrac{x}{y}} \qquad \text{Distribute.}$$

$$= \frac{y - x}{xy - x^2} \qquad \text{Simplify.}$$

$$= \frac{\cancel{y - x}}{x(\cancel{y - x})}$$

$$= \frac{1}{x}$$

Solving Rational Equations   8.6

An equation contains an = sign.

To solve a rational equation, *multiply* the equation by the LCD to *eliminate* the denominators, then solve.

Always check the answer to be sure it does not make a denominator equal zero.

Solve $\dfrac{n}{n + 4} + 1 = \dfrac{20}{n + 4}$.   **p. 552**

This is an equation because it contains an = sign. We must eliminate the denominator. Identify the LCD of all of the expressions in the equation.

$$LCD = (n + 4)$$

Multiply both sides of the equation by $(n + 4)$.

$$(n + 4)\left(\frac{n}{n + 4} + 1\right) = (n + 4)\left(\frac{20}{n + 4}\right)$$

$$\cancel{(n + 4)} \cdot \frac{n}{\cancel{n + 4}} + (n + 4) \cdot 1 = \cancel{(n + 4)} \cdot \frac{20}{\cancel{n + 4}}$$

$$n + n + 4 = 20$$
$$2n + 4 = 20$$
$$2n = 16$$
$$n = 8$$

The check is left to the student.

Solve an Equation for a Specific Variable

Solve $x = \dfrac{2a}{n + m}$ for n.   **p. 556**

Since we are solving for n, put it in a box.

$$x = \frac{2a}{\boxed{n} + m}$$

$$(\boxed{n} + m)x = \cancel{(\boxed{n} + m)} \cdot \frac{2a}{\cancel{\boxed{n} + m}}$$

$$(\boxed{n} + m)x = 2a$$

$$\boxed{n}x + mx = 2a$$

$$\boxed{n}x = 2a - mx$$

$$n = \frac{2a - mx}{x}$$

Definition/Procedure	Example	Reference

Applications of Equations Containing Rational Expressions

8.7

Use the Steps for Solving Word Problems outlined in Section 3.2.

Write an equation and solve.

Dimos can put up the backyard pool in 6 hr, but it takes his father only 4 hr to put up the pool. How long would it take the two of them to put up the pool together?

Step 1: Read the problem carefully, twice.

Step 2: t = number of hours to put up the pool together.

Step 3: Translate from English to an algebraic equation.

$$\text{Dimos'} \atop \text{rate} = \frac{1}{6} \text{ pool/hr} \qquad \text{Father's} \atop \text{rate} = \frac{1}{4} \text{ pool/hr}$$

Fractional part = rate · time

$$\text{Dimos' part} = \frac{1}{6} \cdot t = \frac{1}{6}t$$

$$\text{Father's part} = \frac{1}{4} \cdot t = \frac{1}{4}t$$

$$\text{Fractional} \atop \text{job by Dimos} + \text{Fractional} \atop \text{job by his father} = 1 \text{ whole job}$$

$$\frac{1}{6}t \quad + \quad \frac{1}{4}t \quad = \quad 1$$

Equation: $\frac{1}{6}t + \frac{1}{4}t = 1$.

Step 4: Solve the equation.

$$12\left(\frac{1}{6}t + \frac{1}{4}t\right) = 12(1) \qquad \text{Multiply by 12, the LCD.}$$

$$12 \cdot \frac{1}{6}t + 12 \cdot \frac{1}{4}t = 12(1) \qquad \text{Distribute.}$$

$$2t + 3t = 12 \qquad \text{Multiply.}$$

$$5t = 12$$

$$t = \frac{12}{5}$$

Step 5: Interpret the meaning of the solution as it relates to the problem.

Dimos and his father could put up the pool together in $\frac{12}{5}$ hr or $2\frac{2}{5}$ hr.

The check is left to the student.

p. 559

(8.1) Evaluate, if possible, for a) $n = 4$ and b) $n = -3$.

1) $\dfrac{5n - 2}{n^2 - 9}$

2) $\dfrac{n^2 - 3n - 4}{2n + 1}$

Determine the value(s) of the variable for which
a) the expression equals zero.
b) the expression is undefined.

3) $\dfrac{k + 5}{k - 1}$

4) $\dfrac{s}{4s + 9}$

5) $\dfrac{2c^2 - 11c - 6}{c^2 - 3c}$

6) $\dfrac{12}{4t^2 - 1}$

7) $\dfrac{14 - 7d}{d^2 + 9}$

8) $\dfrac{3m^2 - 13m - 10}{m^2 + 36}$

Write each rational expression in lowest terms.

9) $\dfrac{63a^2}{9a^{11}}$

10) $\dfrac{88k^9}{8k^2}$

11) $\dfrac{15c - 55}{33c - 121}$

12) $\dfrac{r^2 - 15r + 54}{3r^2 - 18r}$

13) $\dfrac{2z - 7}{6z^2 - 19z - 7}$

14) $\dfrac{4t^4 - 32t}{(t - 2)(t^2 + 2t + 4)}$

15) $\dfrac{10 - x}{x^2 - 100}$

16) $\dfrac{y^2 + 9y - yz - 9z}{yz - 12y - z^2 + 12z}$

Find three equivalent forms of each rational expression.

17) $-\dfrac{u - 6}{u + 2}$

18) $-\dfrac{3n + 2}{5 - 4n}$

Find the missing side in each rectangle.

19) Area $= 2l^2 - 5l - 3$

$l - 3$

Find the length.

20) Area $= 3b^2 + 17b + 20$

$3b + 5$

Find the width.

Determine the domain of each rational function.

21) $g(x) = \dfrac{4}{x - 5}$

22) $f(c) = \dfrac{c + 2}{3c + 8}$

23) $h(a) = \dfrac{9a}{a^2 - 2a - 24}$

24) $k(t) = \dfrac{6t - 1}{t^2 + 7}$

(8.2) Perform the operations and simplify.

25) $\dfrac{10}{9} \cdot \dfrac{6}{25}$

26) $\dfrac{27}{56} \div \dfrac{45}{64}$

27) $\dfrac{16k^4}{3m^2} \div \dfrac{4k^2}{27m}$

28) $\dfrac{t + 4}{6} \cdot \dfrac{3(t - 2)}{(t + 4)^2}$

29) $\dfrac{6w - 1}{6w^2 + 5w - 1} \cdot \dfrac{3w + 3}{12w}$

30) $\dfrac{3x^2 + 14x + 16}{15x + 40} \div \dfrac{11x + 22}{x - 5}$

31) $\dfrac{25 - a^2}{4a^2 + 12a} \div \dfrac{a^3 - 125}{a^2 + 3a}$

32) $\dfrac{3p^5}{20q^2} \cdot \dfrac{4q^3}{21p^7}$

Divide.

33) $\dfrac{\dfrac{8}{9}}{\dfrac{12}{5}}$

34) $\dfrac{\dfrac{4s + 9}{10}}{\dfrac{4s + 9}{2}}$

35) $\dfrac{\dfrac{16m - 8}{m^2}}{\dfrac{12m - 6}{m^3}}$

36) $\dfrac{\dfrac{2r + 6}{r^4}}{\dfrac{r^2 - 9}{5r}}$

(8.3) Find the LCD of each group of fractions.

37) $\dfrac{5}{6}, \dfrac{4}{9}$

38) $\dfrac{1}{10}, \dfrac{8}{15}, \dfrac{5}{6}$

39) $\dfrac{2}{k^2}, \dfrac{9}{k}$

40) $\dfrac{4}{9x^2y}, \dfrac{13}{4xy^4}$

41) $\dfrac{3}{m}, \dfrac{4}{m + 5}$

42) $\dfrac{r}{4r - 8}, \dfrac{5}{6r - 12}$

43) $\dfrac{8}{2d^2 - d}, \dfrac{11}{6d - 3}$

44) $\dfrac{1}{3x + 7}, \dfrac{6x}{x - 9}$

45) $\dfrac{9}{2 - b}, \dfrac{4b}{b - 2}$

46) $\dfrac{w}{w - 7}, \dfrac{6}{7 - w}$

47) $\dfrac{5c - 1}{c^2 + 10c + 24}, \dfrac{9c}{c^2 - 3c - 28}$

48) $\dfrac{5f}{f^2 - 9f - 22}, \dfrac{-12f}{2f^2 - 19f + 33}$

49) $\dfrac{6}{x^2 + 8x}, \dfrac{1}{3x^2 + 24x}, \dfrac{13}{x^2 + 16x + 64}$

50) $\dfrac{3}{a^2 - 13a + 40}, \dfrac{a + 12}{a^2 - 7a - 8}, \dfrac{1}{a^2 - 4a - 5}$

51) $\dfrac{8c}{c^2 - d^2}, \dfrac{d}{d - c}$

52) $\dfrac{15}{x - y}, \dfrac{x}{y^2 - x^2}$

Rewrite each rational expression with the indicated denominator.

53) $\dfrac{6}{5r} = \dfrac{}{20r^3}$

54) $\dfrac{1}{6xy^2} = \dfrac{}{42x^4y^3}$

55) $\dfrac{8}{3z + 4} = \dfrac{}{z(3z + 4)}$

56) $\dfrac{5k}{k - 8} = \dfrac{}{(k - 6)(k - 8)}$

57) $\dfrac{t - 3}{2t + 1} = \dfrac{}{(2t + 1)(t + 5)}$

58) $\dfrac{n}{4 - n} = \dfrac{}{n - 4}$

Identify the LCD of each group of fractions, and rewrite each as an equivalent fraction with the LCD as its denominator.

59) $\dfrac{4}{9z^3}, \dfrac{7}{12z^5}$

60) $\dfrac{3}{8a^3b}, \dfrac{6}{5ab^5}$

61) $\dfrac{8}{p + 7}, \dfrac{2}{p}$

62) $\dfrac{9c}{c^2 + 6c - 16}, \dfrac{4}{c^2 - 4c + 4}$

63) $\dfrac{1}{g - 10}, \dfrac{3}{10 - g}$

64) $\dfrac{7}{2r^2 - 12r}, \dfrac{3r}{36 - r^2}, \dfrac{r - 5}{2r^2 + 12r}$

(8.4) Add or subtract, as indicated.

65) $\dfrac{5}{18} - \dfrac{7}{12}$

66) $\dfrac{3}{8c} + \dfrac{7}{8c}$

67) $\dfrac{4m}{m - 3} - \dfrac{5}{m - 3}$

68) $\dfrac{2}{5z^2} + \dfrac{9}{10z}$

69) $\dfrac{8}{t + 4} + \dfrac{3}{t}$

70) $\dfrac{n}{2n - 5} - \dfrac{4}{n}$

71) $\dfrac{5}{y - 2} - \dfrac{6}{y + 3}$

72) $\dfrac{8d - 3}{d^2 - 3d - 28} + \dfrac{2d}{5d - 35}$

73) $\dfrac{10p + 3}{4p + 4} - \dfrac{8}{p^2 - 6p - 7}$

74) $\dfrac{k - 3}{k^2 + 14k + 49} - \dfrac{2}{k^2 + 7k}$

75) $\dfrac{2}{m - 11} + \dfrac{19}{11 - m}$

76) $\dfrac{t + 10}{t - 20} - \dfrac{8}{20 - t}$

77) $\dfrac{1}{8 - r} + \dfrac{16}{r^2 - 64}$

78) $\dfrac{x^2}{x^2 - y^2} + \dfrac{x}{y - x}$

79) $\dfrac{8}{w^2 + 7w} + \dfrac{3w}{w + 7} + \dfrac{2}{5w}$

80) $\dfrac{3}{g^2 - 7g} + \dfrac{2g}{5g - 35} - \dfrac{6}{5g}$

81) $\dfrac{d + 4}{d^2 + 2d} + \dfrac{d}{5d^2 + 7d - 6} - \dfrac{10}{5d^2 - 3d}$

82) $\dfrac{a}{9a^2 - 4} + \dfrac{a + 1}{6a^2 - 4a} - \dfrac{1}{6a + 4}$

For each rectangle, find a rational expression in simplest form to represent its a) area and b) perimeter.

83)

$\dfrac{x}{8}$

$\dfrac{12}{x - 4}$

84)

$\dfrac{2}{x^2}$

$\dfrac{x}{x + 1}$

(8.5) Simplify completely.

85) $\dfrac{\dfrac{2}{5}}{\dfrac{7}{15}}$

86) $\dfrac{\dfrac{f}{g}}{\dfrac{f^2}{g}}$

87) $\dfrac{p + \dfrac{6}{p}}{\dfrac{8}{p} + p}$

88) $\dfrac{\dfrac{a}{b} - \dfrac{2a}{b^2}}{\dfrac{4}{ab} - \dfrac{a}{b}}$

89) $\dfrac{\dfrac{n}{6n + 48}}{\dfrac{n^2}{4n + 32}}$

90) $\dfrac{\dfrac{2}{3} - \dfrac{4}{5}}{\dfrac{1}{6} + \dfrac{1}{2}}$

91) $\dfrac{1 - \dfrac{1}{y - 9}}{\dfrac{2}{y + 3} + 1}$

92) $\dfrac{\dfrac{4q}{7q + 63}}{\dfrac{q^2}{8q + 72}}$

93) $\dfrac{\dfrac{c}{c + 2} + \dfrac{1}{c^2 - 4}}{1 - \dfrac{3}{c + 2}}$

94) $\dfrac{1 + \dfrac{b}{a - b}}{\dfrac{b}{a^2 - b^2} + \dfrac{1}{a + b}}$

(8.6) Solve each equation.

95) $\dfrac{5w}{6} - \dfrac{1}{2} = -\dfrac{1}{6}$

96) $\dfrac{4a + 3}{10} = \dfrac{a}{5} + \dfrac{3}{2}$

97) $\dfrac{4}{y - 6} = \dfrac{12}{y + 2}$

98) $\dfrac{m}{6} = \dfrac{4}{m + 5}$

99) $\dfrac{r}{r + 6} + 3 = \dfrac{10}{r + 6}$

100) $\dfrac{3}{x - 5} + \dfrac{2}{2x + 1} = \dfrac{1}{2x^2 - 9x - 5}$

101) $\dfrac{16}{9t - 27} + \dfrac{2t - 4}{t - 3} = \dfrac{t}{9}$

102) $\dfrac{5}{j + 4} + \dfrac{j}{j - 2} = \dfrac{2j^2 - 2j}{j^2 + 2j - 8}$

103) $\dfrac{3}{b + 2} = \dfrac{16}{b^2 - 4} - \dfrac{4}{b - 2}$

104) $\dfrac{3k}{k + 9} = \dfrac{3}{k + 1}$

105) $\dfrac{c}{c^2 + 3c - 28} - \dfrac{5}{c^2 + 15c + 56} = \dfrac{5}{c^2 + 4c - 32}$

106) $\dfrac{a}{a^2 - 1} + \dfrac{4}{a^2 + 9a + 8} = \dfrac{8}{a^2 + 7a - 8}$

Solve for the indicated variable.

107) $A = \dfrac{2p}{c}$ for c

108) $R = \dfrac{s + T}{D}$ for D

109) $n = \dfrac{t}{a + b}$ for a

110) $w = \dfrac{N}{c - ak}$ for k

111) $\dfrac{1}{r} = \dfrac{1}{s} + \dfrac{1}{t}$ for s

112) $\dfrac{1}{R_1} + \dfrac{1}{R_2} = \dfrac{1}{R_3}$ for R_1

(8.7) Write an equation and solve.

113) The ratio of saturated fat to total fat in a Starbucks tall Caramel Frappuccino is 2 to 3. If there are 4 more grams of total fat in the drink than there are grams of saturated fat, how much total fat is in a Caramel Frappuccino? (*Source:* Starbucks brochure)

114) A boat can travel 9 mi downstream in the same amount of time it can travel 6 mi upstream. If the speed of the boat in still water is 10 mph, find the speed of the current.

115) When the wind is blowing at 40 mph, a plane flying at a constant speed can travel 800 mi with the wind in the same amount of time it can fly 600 mi against the wind. Find the speed of the plane.

116) Wayne can clean the carpets in his house in 4 hr, but it would take his son, Garth, 6 hr to clean them on his own. How long would it take both of them to clean the carpets together?

1) Evaluate, if possible, for $t = -5$.

$$\frac{3t + 5}{t^2 + 25}$$

Determine the value(s) of the variable for which

    a) the expression is undefined.

    b) the expression equals zero.

2) $\dfrac{4m - 3}{m + 7}$

3) $\dfrac{k^2 + 4}{k^2 + 2k - 48}$

Write each rational expression in lowest terms.

4) $\dfrac{54w^3}{24w^8}$

5) $\dfrac{7v^2 + 55v - 8}{v^2 - 64}$

6) Write three equivalent forms of $\dfrac{9 - h}{2h - 3}$.

7) Identify the LCD of $\dfrac{3}{b}$ and $\dfrac{b}{b + 8}$.

Perform the operations and simplify.

8) $\dfrac{7}{12z} + \dfrac{5}{12z}$

9) $-\dfrac{21m^4}{n} \div \dfrac{12m^8}{n^3}$

10) $\dfrac{6v}{7} - \dfrac{2v}{21}$

11) $\dfrac{r}{2r + 1} + \dfrac{3}{r + 5}$

12) $\dfrac{a^3 - 8}{6a - 66} \cdot \dfrac{a^2 - 9a - 22}{4 - a^2}$

13) $\dfrac{c - 3}{c - 15} + \dfrac{c + 8}{15 - c}$

14) $\dfrac{x}{x^2 - 49} - \dfrac{3}{x^2 - 2x - 63}$

Simplify completely.

15) $\dfrac{1 + \dfrac{2}{d - 3}}{\dfrac{-2d}{d - 3} - d}$

16) $\dfrac{\dfrac{15}{7}}{\dfrac{20}{21}}$

Solve each equation.

17) $\dfrac{7t}{12} + \dfrac{t - 4}{6} = \dfrac{7}{3}$

18) $\dfrac{30}{x^2 - 9} = \dfrac{5}{x - 3} - \dfrac{2}{x + 3}$

19) $\dfrac{5}{n^2 + 10n + 24} + \dfrac{5}{n^2 + 3n - 18} = \dfrac{n}{n^2 + n - 12}$

20) Solve for p.

$$\frac{1}{p} + \frac{1}{q} = \frac{1}{r}$$

Write an equation for each and solve.

21) Leticia can assemble a swing set in 3 hr while it takes Betty 5 hr. How long would it take for them to assemble the swing set together?

22) A river flows at 3 mph. If a boat can travel 16 mi downstream in the same amount of time it can go 10 mi upstream, find the speed of the boat in still water.

Find the area and perimeter of each figure.

1)

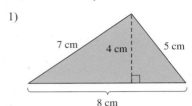

2)

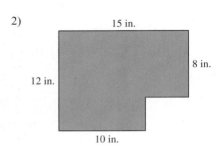

Simplify. The answer should not contain negative exponents.

3) $(3y^2)^4$

4) $(2k^3)^{-5}$

Solve each inequality. Write the answer in interval notation.

5) $-9 \le 4c + 5 \le 9$

6) $\frac{1}{3}t - 7 > 2$ or $4 - t > 6$

7) Find the x- and y-intercepts of $3x + 5y = 10$, and graph the equation.

8) Find the slope of the line containing the points $(-7, 4)$ and $(3, -1)$.

9) Solve using the elimination method.

$$10x + 3y = -3$$
$$6x + 4y = 7$$

10) *Write a system of two equations in two variables and solve.*
Four cans of cat food and three cans of dog food cost $3.85, while two cans of dog food and one can of cat food cost $1.90. Find the cost of a can of dog food and the cost of a can of cat food.

Multiply and simplify.

11) $(m - 7)^2$

12) $(3g - 8)(2g^2 + 7g - 9)$

Divide.

13) $\dfrac{36c^4 + 56c^3 - 8c^2 + 6c}{8c^2}$

14) $\dfrac{3x^3 + 14x^2 - 26x - 16}{x + 6}$

Factor completely.

15) $8h^3 + 125$

16) $3p^2 + 2p - 8$

17) $ab + 3b - 2a - 6$

18) Solve $s(s - 10) = 5s - 54$.

19) For what value(s) of z is the expression $\dfrac{9z + 1}{z^2 - 4z}$

a) undefined?

b) equal to zero?

20) Write in lowest terms. $\dfrac{4b^2 + 32b + 48}{b^2 - b - 42}$

Perform the operations and simplify.

21) $\dfrac{9 - d^2}{d^3 + 3d^2} \cdot \dfrac{14d^5}{2d^2 - 5d - 3}$

22) $\dfrac{4}{c - 5} + \dfrac{c + 1}{2c}$

23) Simplify $\dfrac{\dfrac{2}{x - 9} - 1}{1 + \dfrac{5}{x - 9}}$.

Solve.

24) $\dfrac{3w}{2w - 1} - 6 = \dfrac{4}{2w - 1}$

25) $\dfrac{10}{m + 7} + \dfrac{m}{m - 10} = -\dfrac{40}{m^2 - 3m - 70}$

CHAPTER 9

Absolute Value Equations and Inequalities and More on Solving Systems

Algebra at Work: Finance

When a client comes to a financial planner for help investing money, the adviser may recommend many different types of places to invest the money. Stocks and bonds are just two of the investments the financial planner could suggest.

Theresa has a client who has six different stocks and four different bonds in her investment portfolio. The closing prices of the stocks change each day as do the interest rates the bonds yield. Theresa keeps track of these closing figures each day and then organizes this information weekly so that she knows how much her client's portfolio is worth at any given time.

To make it easier to organize this information, Theresa puts all of the information into matrices (the plural of matrix). Each week she creates a matrix containing information about the closing price of each stock each day of the week. She creates another matrix that contains information about the interest rates each of the bonds yield each day of the week.

When there is a lot of information to organize, financial planners often turn to matrices to help them keep track of all of their information.

In this chapter, we will learn about using augmented matrices to solve systems of equations.

Section 9.1 Solving Absolute Value Equations

Objectives

1. Understand the Meaning of an Absolute Value Equation
2. Solve an Equation of the Form $|ax + b| = k$ for $k > 0$
3. Solve an Equation of the Form $|ax + b| + t = c$
4. Solve an Equation of the Form $|ax + b| = k$ for $k \leq 0$
5. Solve an Equation of the Form $|ax + b| = |cx + d|$

In Section 1.4 we learned that the absolute value of a number describes its *distance from zero*.

$$|5| = 5 \quad \text{and} \quad |-5| = 5$$

5 units from zero   5 units from zero

$-7\ -6\ -5\ -4\ -3\ -2\ -1\ \ \ 0\ \ \ 1\ \ \ 2\ \ \ 3\ \ \ 4\ \ \ 5\ \ \ 6\ \ \ 7$

We use this idea of *distance from zero* to solve absolute value equations and inequalities.

1. Understand the Meaning of an Absolute Value Equation

Example 1

Solve $|x| = 3$.

Solution

Since the equation contains an absolute value, **solve $|x| = 3$** means *"Find the number or numbers whose distance from zero is 3."*

3 units from zero   3 units from zero

$-6\ -5\ -4\ -3\ -2\ -1\ \ \ 0\ \ \ 1\ \ \ 2\ \ \ 3\ \ \ 4\ \ \ 5\ \ \ 6$

Those numbers are 3 and -3. Each of them is 3 units from zero.

The solution set is $\{-3, 3\}$.

Check: $|3| = 3, |-3| = 3$ ✓

You Try 1

Solve $|y| = 8$.

2. Solve an Equation of the Form $|ax + b| = k$ for $k > 0$

Example 2

Solve $|m + 1| = 5$.

Solution

Solving this absolute value equation means, *"Find the number or numbers that can be substituted for m so that the quantity m + 1 is 5 units from 0."*

$m + 1$ will be 5 units from zero if $m + 1 = 5$ or if $m + 1 = -5$, since both 5 and -5 are 5 units from zero. Therefore, we can solve the equation this way:

$$|m + 1| = 5$$

$$m + 1 = 5 \quad \text{or} \quad m + 1 = -5 \qquad \text{Set the quantity inside the absolute}$$
$$\qquad\qquad\qquad\qquad\qquad\qquad\qquad\qquad\qquad \text{value equal to 5 and } -5.$$
$$m = 4 \quad \text{or} \qquad\quad m = -6 \qquad \text{Solve.}$$

$$\text{Check: } m = 4: |4 + 1| \overset{?}{=} 5 \qquad\qquad m = -6: \quad |-6 + 1| \overset{?}{=} 5$$
$$|5| = 5 \;\checkmark \qquad\qquad\qquad\qquad\qquad |-5| = 5 \;\checkmark$$

The solution set is $\{-6, 4\}$.

Solving an Absolute Value Equation

If P represents an expression and k is a positive real number, then to solve $|P| = k$ we rewrite the absolute value equation as the *compound equation*

$$P = k \quad \text{or} \quad P = -k$$

and solve for the variable.

P can represent expressions like x, $3a + 2$, $\frac{1}{4}t - 9$, and so on.

You Try 2

Solve $|c - 4| = 3$.

Example 3

Solve $|5r - 3| = 13$.

Solution

Solving this equation means, "*Find the number or numbers that can be substituted for r so that the quantity $5r - 3$ is 13 units from zero.*"

$$|5r - 3| = 13$$

$$5r - 3 = 13 \quad \text{or} \quad 5r - 3 = -13 \qquad \text{Set the quantity inside the absolute}$$
$$5r = 16 \qquad\qquad 5r = -10 \qquad\quad \text{value equal to 13 and } -13.$$
$$r = \frac{16}{5} \qquad \text{or} \qquad\qquad r = -2 \qquad\quad \text{Solve.}$$

$$|5r - 3| = 13 \qquad\qquad |5r - 3| = 13$$

Check: $r = \dfrac{16}{5}$: $\left|5\left(\dfrac{16}{5}\right) - 3\right| \stackrel{?}{=} 13$ $\qquad$ $r = -2$: $|5(-2) - 3| \stackrel{?}{=} 13$

$$|16 - 3| \stackrel{?}{=} 13 \qquad\qquad |-10 - 3| \stackrel{?}{=} 13$$

$$|13| = 13 \;\; \checkmark \qquad\qquad |-13| = 13 \;\; \checkmark$$

The solution set is $\left\{-2, \dfrac{16}{5}\right\}$.

 You Try 3

Solve $|2k + 1| = 9$.

3. Solve an Equation of the Form $|ax + b| + t = c$

Example 4

Solve $\left|\dfrac{3}{2}t + 7\right| + 5 = 6$.

Solution

Before we rewrite this equation as a compound equation, we must *isolate* the absolute value (get the absolute value on a side by itself).

$$\left|\dfrac{3}{2}t + 7\right| + 5 = 6$$

$$\left|\dfrac{3}{2}t + 7\right| = 1 \qquad \text{Subtract 5 to isolate the absolute value.}$$

$$\dfrac{3}{2}t + 7 = 1 \qquad \text{or} \qquad \dfrac{3}{2}t + 7 = -1 \qquad \text{Set the quantity inside the absolute value equal to 1 and } -1.$$

$$\dfrac{3}{2}t = -6 \qquad\qquad \dfrac{3}{2}t = -8 \qquad \text{Subtract 7.}$$

$$\dfrac{2}{3} \cdot \dfrac{3}{2}t = \dfrac{2}{3} \cdot (-6) \qquad \dfrac{2}{3} \cdot \dfrac{3}{2}t = \dfrac{2}{3} \cdot (-8) \qquad \text{Multiply by } \dfrac{2}{3} \text{ to solve for } t.$$

$$t = -4 \qquad \text{or} \qquad t = -\dfrac{16}{3} \qquad \text{Solve.}$$

The check is left to the student.

The solution set is $\left\{-\dfrac{16}{3}, -4\right\}$.

You Try 4

Solve $\left|\frac{1}{4}n - 3\right| + 2 = 5$.

4. Solve an Equation of the Form $|ax + b| = k$ for $k \leq 0$

Example 5

Solve $|4y - 11| = -9$.

Solution

This equation says that the absolute value of the quantity $4y - 11$ equals *negative* 9. Can an absolute value be negative? No! This equation has *no solution*.

The solution set is $\varnothing$.

You Try 5

Solve $|d + 3| = -5$.

5. Solve an Equation of the Form $|ax + b| = |cx + d|$

Another type of absolute value equation involves two absolute values.

> If P and Q are expressions, then to solve $|P| = |Q|$, we rewrite the absolute value equation as the *compound equation*.
>
> $$P = Q \quad \text{or} \quad P = -Q$$
>
> and solve for the variable.

Example 6

Solve $|2w - 3| = |w + 9|$.

Solution

This equation is true when the quantities inside the absolute values are the *same* or when they are *negatives* of each other.

$$|2w - 3| = |w + 9|$$

The quantities are the same   **or**   the quantities are negatives of each other

$2w - 3 = w + 9$	$2w - 3 = -(w + 9)$
$w = 12$	$2w - 3 = -w - 9$
	$3w = -6$
	$w = -2$

Check: $w = 12$: $|2(12) - 3| \overset{?}{=} |12 + 9|$ $w = -2$: $|2(-2) - 3| \overset{?}{=} |-2 + 9|$
$|24 - 3| \overset{?}{=} |21|$ $|-4 - 3| \overset{?}{=} |7|$
$|21| = 21$ ✓ $|-7| = 7$ ✓

The solution set is $\{-2, 12\}$.

 BE CAREFUL   In Example 6 and other examples like it, you *must* put parentheses around the expression with the negative as in $-(w + 9)$.

 You Try 6

Solve $|c + 7| = |3c - 1|$.

 Using Technology

We can use a graphing calculator to solve an equation by entering one side of the equation as Y_1 and the other side as Y_2. Then graph the equations. Remember that absolute value equations like the ones found in this section can have 0, 1, or 2 solutions. *The x-coordinates of their points of intersection are the solutions to the equation.*

We will solve $|3x - 1| = 5$ algebraically and by using a graphing calculator, and then compare the results.

First, use algebra to solve $|3x - 1| = 5$. You should get $\left\{ -\dfrac{4}{3}, 2 \right\}$.

Next, use a graphing calculator to solve $|3x - 1| = 5$.
We will enter $|3x - 1|$ as Y_1 and 5 as Y_2. To enter $Y_1 = |3x - 1|$,

1.   Press the $\boxed{Y=}$ key, so that the cursor is to the right of $\backslash Y_1 =$.
2.   Press $\boxed{\text{MATH}}$ and then press the right arrow, to highlight **NUM**. Also highlighted is 1:abs (which stands for *absolute value*).
3.   Press $\boxed{\text{ENTER}}$ and you are now back on the $\backslash Y_1 =$ screen. Enter $3x - 1$ with a closing parentheses so that you have now entered $Y_1 = \text{abs}(3x - 1)$.
4.   Press the down arrow to enter $\backslash Y_2 = 5$.
5.   Press $\boxed{\text{GRAPH}}$.

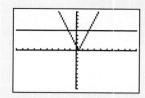

The graphs intersect at two points because there are two solutions to this equation. *Remember that the solutions to the equation are the x-coordinates of the points of intersection.*

To find these x-coordinates we will use the INTERSECT feature introduced in Chapter 5.

To find the left-hand intersection point, press $\boxed{\text{2nd}}$ $\boxed{\text{TRACE}}$ and select 5:intersect. Press $\boxed{\text{ENTER}}$. Move the cursor close to the point on the left and press $\boxed{\text{ENTER}}$ three times. You get the result in the screen below on the left.

To find the right-hand intersect point, press $\boxed{\text{2nd}}$ $\boxed{\text{TRACE}}$, select 5:intersect, and press $\boxed{\text{ENTER}}$. Move the cursor close to the point, and press $\boxed{\text{ENTER}}$ three times. You will see the screen that is below on the right.

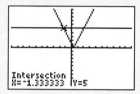
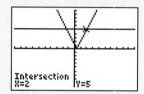

The screen on the left shows $x = -1.333333$. This is the calculator's approximation of $x = -1.\overline{3}$, the decimal equivalent of $x = -\dfrac{4}{3}$, one of the solutions found using algebra.

The screen on the right shows $x = 2$ as a solution, the same solution we obtained algebraically.

The calculator gives us a solution set of $\{-1.333333, 2\}$, while the solution set found using algebra is $\left\{-\dfrac{4}{3}, 2\right\}$.

Solve each equation algebraically, then verify your answer using a graphing calculator.

1. $|x - 1| = 2$
2. $|x + 4| = 6$
3. $|2x + 3| = 3$

4. $|4x - 5| = 1$
5. $|3x + 7| - 6 = -8$
6. $|6 - x| + 3 = 3$

Answers to You Try Exercises

1) $\{-8, 8\}$ 2) $\{1, 7\}$ 3) $\{-5, 4\}$ 4) $\{0, 24\}$

5) $\varnothing$ 6) $\left\{-\dfrac{3}{2}, 4\right\}$

Answers to Technology Exercises

1. $\{-1, 3\}$ 2. $\{-10, 2\}$ 3. $\{-3, 0\}$
4. $\{1, 1.5\}$ 5. $\varnothing$ 6. $\{6\}$

9.1 Exercises

Objective 1

1) In your own words, explain the *meaning* of the absolute value of a number.

2) Does $|x| = -8$ have a solution? Why or why not?

Objective 2

Solve.

3) $|q| = 6$

4) $|z| = 7$

5) $|q - 5| = 3$

6) $|a + 2| = 13$

7) $|4t - 5| = 7$

8) $|9x - 8| = 10$

9) $|12c + 5| = 1$

10) $|10 + 7g| = 8$

11) $|1 - 8m| = 9$

12) $|4 - 5k| = 11$

13) $\left|\dfrac{2}{3}b + 3\right| = 13$

14) $\left|\dfrac{3}{4}h + 8\right| = 7$

15) $\left|4 - \dfrac{3}{5}d\right| = 6$

16) $\left|4 - \dfrac{1}{6}w\right| = 1$

17) $\left|\dfrac{3}{4}y - 2\right| = \dfrac{3}{5}$

18) $\left|\dfrac{3}{2}r + 5\right| = \dfrac{3}{4}$

Objectives 3 and 4

Solve.

19) $|m - 5| = -3$

20) $|2k + 7| = -15$

21) $|10p + 2| = 0$

22) $|4c - 11| = 0$

23) $|z - 6| + 4 = 20$

24) $|q + 3| - 1 = 14$

25) $|2a + 5| + 8 = 13$

26) $|6t - 11| + 5 = 10$

27) $|w + 14| = 0$

28) $|5h + 7| = -5$

29) $|8n + 11| = -1$

30) $|4p - 3| = 0$

31) $|5b + 3| + 6 = 19$

32) $1 = |7 - 8x| - 4$

33) $|3m - 1| + 5 = 2$

34) $\left|\dfrac{5}{4}k + 2\right| + 9 = 7$

Objective 5

Solve the following equations containing two absolute values.

35) $|s + 9| = |2s + 5|$

36) $|j - 8| = |4j - 7|$

37) $|3z + 2| = |6 - 5z|$

38) $|1 - 2a| = |10a + 3|$

39) $\left|\dfrac{3}{2}x - 1\right| = |x|$

40) $|y| = \left|\dfrac{4}{7}y + 12\right|$

41) $|7c + 10| = |5c + 2|$

42) $|4 - 11r| = |5r + 3|$

43) $\left|\dfrac{1}{4}t - \dfrac{5}{2}\right| = \left|5 - \dfrac{1}{2}t\right|$

44) $\left|k + \dfrac{1}{6}\right| = \left|\dfrac{2}{3}k + \dfrac{1}{2}\right|$

Objective 1

45) Write an absolute value equation that means *x is 9 units from zero*.

46) Write an absolute value equation that means *y is 6 units from zero*.

47) Write an absolute value equation that has a solution set of $\left\{-\dfrac{1}{2}, \dfrac{1}{2}\right\}$.

48) Write an absolute value equation that has a solution set of $\{-1.4, 1.4\}$.

Section 9.2 Solving Absolute Value Inequalities

Objectives

1. Solve Absolute Value Inequalities Containing $<$ or $\leq$

2. Solve Absolute Value Inequalities Containing $>$ or $\geq$

3. Solve Special Cases of Absolute Value Inequalities

4. Solve an Applied Problem Using an Absolute Value Inequality

In Section 9.1 we learned how to solve absolute value equations. In this section, we will learn how to solve **absolute value inequalities**. Some examples of absolute value inequalities are

$$|t| < 6, \qquad |n + 2| \leq 5, \qquad |3k - 1| > 11, \qquad \left|5 - \frac{1}{2}y\right| \geq 3$$

1. Solve Absolute Value Inequalities Containing $<$ or $\leq$

What does it mean to solve $|x| \leq 3$? It means to find the set of all real numbers whose distance from zero is *3 units or less.*

3 is 3 units from 0.

-3 is 3 units from 0.

Any number *between* 3 and -3 is less than 3 units from zero. For example, if $x = 1$, $|1| \leq 3$. If $x = -2, |-2| \leq 3$. We can represent the solution set on a number line as

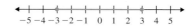

We can write the solution set in interval notation as $[-3, 3]$.

> Let P be an expression and let k be a positive real number. To solve $|P| \leq k$, solve the three-part inequality $-k \leq P \leq k$. ($<$ may be substituted for $\leq$.)

Example 1

Solve $|t| < 6$. Graph the solution set and write the answer in interval notation.

Solution

We must find the set of all real numbers whose distance from zero is less than 6. We can do this by solving the three-part inequality

$$-6 < t < 6.$$

We can represent this on a number line as

We can write the solution set in interval notation as $(-6, 6)$. Any number between -6 and 6 will satisfy the inequality.

 **You Try 1**

Solve. Graph the solution set and write the answer in interval notation.

$$|u| < 9$$

Example 2

Solve $|n + 2| \leq 5$. Graph the solution set and write the answer in interval notation.

Solution

We must find the set of all real numbers, n, so that $n + 2$ is less than or equal to 5 units from zero. To solve $|n + 2| \leq 5$, we must solve the three-part inequality

$$-5 \leq n + 2 \leq 5$$
$$-7 \leq n \leq 3 \qquad \text{Subtract 2.}$$

The number line representation is

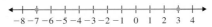

In interval notation, the solution set is $[-7, 3]$. Any number between -7 and 3 will satisfy the inequality.

 You Try 2

Solve. Graph the solution set and write the answer in interval notation.

$$|6k + 5| \leq 13$$

Example 3

Solve $|4 - 5p| < 16$. Graph the solution set and write the answer in interval notation.

Solution

Solve the three-part inequality.

$$-16 < 4 - 5p < 16$$
$$-20 < -5p < 12 \qquad \text{Subtract 4.}$$
$$4 > p > -\frac{12}{5} \qquad \text{Divide by } -5 \text{ and change the direction of the inequality symbols.}$$

This inequality means *p is less than 4 and greater than* $-\frac{12}{5}$. We can rewrite it as

$$-\frac{12}{5} < p < 4$$

The number line representation of the solution set is

In interval notation, we write $\left(-\frac{12}{5}, 4\right)$.

You Try 3

Solve. Graph the solution set and write the answer in interval notation.

$$|9 - 2w| \leq 3$$

2. Solve Absolute Value Inequalities Containing $>$ or $\geq$

To solve $|x| \geq 4$ means to find the set of all real numbers whose distance from zero is *4 units or more*.

4 is 4 units from 0.

-4 is 4 units from 0.

Any number greater than 4 *or* less than -4 is more than 4 units from zero.

For example, if $x = 6$, $|6| \geq 4$. If $x = -5$, then $|-5| = 5$ and $5 \geq 4$. We can represent the solution set to $|x| \geq 4$ as

The solution set consists of two separate regions, so we can write a compound inequality using *or*.

$x \leq -4$ or $x \geq 4$

In interval notation, we write $(-\infty, -4] \cup [4, \infty)$.

> Let P be an expression and let k be a positive, real number. To solve $|P| \geq k$ ($>$ may be substituted for $\geq$), solve the compound inequality $P \geq k$ or $P \leq -k$.

Example 4

Solve $|r| > 2$. Graph the solution set and write the answer in interval notation.

Solution

We must find the set of all real numbers whose distance from zero is greater than 2. The solution is the compound inequality

$r > 2$ or $r < -2$

On the number line, we can represent the solution set as

‹—+—+—+—◇—+—+—+—◇—+—+—›
$-5 -4 -3 -2 -1 \; 0 \; 1 \; 2 \; 3 \; 4 \; 5$

In interval notation, we write $(-\infty, -2) \cup (2, \infty)$. Any number in the shaded region will satisfy the inequality. For example, to the right of 2, if $r = 3$, then $|3| > 2$. To the left of -2, if $r = -4$, then $|-4| > 2$.

You Try 4

Solve. Graph the solution set and write the answer in interval notation.

$$|d| \geq 5$$

Example 5

Solve the inequality. Graph the solution set and write the answer in interval notation.

$$|3k - 1| > 11$$

Solution

To solve $|3k - 1| > 11$ means to find the set of all real numbers, k, so that $3k - 1$ is more than 11 units from zero on the number line. We will solve the compound inequality.

$3k - 1 > 11$	or	$3k - 1 < -11$
↑		↑

$3k - 1$ is more than 11 units away from zero to the *right* of zero. $3k - 1$ is more than 11 units away from zero to the *left* of zero.

$3k - 1 > 11$	or	$3k - 1 < -11$	
$3k > 12$	or	$3k < -10$	Add 1.
$k > 4$	or	$k < -\dfrac{10}{3}$	Divide by 3.

On the number line, we get

From the number line, we can write the interval notation $\left(-\infty, -\dfrac{10}{3}\right) \cup (4, \infty)$.

Any number in the shaded region will satisfy the inequality.

You Try 5

Solve. Graph the solution set and write the answer in interval notation.
a) $|8q + 9| \geq 7$ b) $|3 - w| > 2$

As we did when solving absolute value equations, we do not rewrite the inequality without the absolute value until we have isolated the absolute value.

Example 6

Solve $|c + 6| + 10 > 12$. Graph the solution set and write the answer in interval notation.

Solution

Begin by getting the absolute value on a side by itself.

$$|c + 6| + 10 > 12$$
$$|c + 6| > 2 \qquad \text{Subtract 10.}$$

$$c + 6 > 2 \quad \text{or} \quad c + 6 < -2 \qquad \text{Rewrite as a compound inequality.}$$
$$c > -4 \quad \text{or} \qquad c < -8 \qquad \text{Subtract 6.}$$

The graph of the solution set is

$$\xleftarrow{\quad} \!\!\!\! {-10\ -9\ -8\ -7\ -6\ -5\ -4\ -3\ -2\ -1\ \ 0\ \ 1\ \ 2} \!\!\!\! \xrightarrow{\quad}$$

The interval notation is $(-\infty, -8) \cup (-4, \infty)$.

You Try 6

Solve $|k + 8| - 5 \geq 9$. Graph the solution set and write the answer in interval notation.

Example 7 illustrates why it is important to understand what the absolute value inequality means before trying to solve it.

3. Solve Special Cases of Absolute Value Inequalities

Example 7

Solve each inequality.

a) $|z + 3| < -6$ b) $|2s - 1| \geq 0$ c) $|4d + 7| + 9 \leq 9$

Solution

a) Look carefully at this inequality, $|z + 3| < -6$. It says that the absolute value of a quantity, $z + 3$, is *less than* a negative number. Since the absolute value of a quantity is always zero or positive, this inequality has *no solution*.

The solution set is $\varnothing$.

b) $|2s - 1| \geq 0$ says that the absolute value of a quantity, $2s - 1$, is greater than or equal to zero. An absolute value is *always* greater than or equal to zero, so *any* value of s will make the inequality true.

The solution set consists of all real numbers, which we can write in interval notation as $(-\infty, \infty)$.

c)   Begin by isolating the absolute value.

$$|4d + 7| + 9 \le 9$$
$$|4d + 7| \le 0 \qquad \text{Subtract 9.}$$

The absolute value of a quantity can *never be less than zero* but it *can equal zero*. To solve this, we must solve $4d + 7 = 0$.

$$4d + 7 = 0$$
$$4d = -7$$
$$d = -\frac{7}{4}$$

The solution set is $\left\{ -\dfrac{7}{4} \right\}$.

You Try 7

Solve each inequality.

a) $|p + 4| \ge 0$ b) $|5n - 7| < -2$ c) $|6y - 1| + 3 \le 3$

4. Solve an Applied Problem Using an Absolute Value Inequality

Example 8

On an assembly line, a machine is supposed to fill a can with 19 oz of soup. However, the possibility for error is ± 0.25 oz. Let x represent the range of values for the amount of soup in the can. Write an absolute value inequality to represent the range for the number of ounces of soup in the can, then solve the inequality and explain the meaning of the answer.

Solution

If the *actual* amount of soup in the can is x and there is supposed to be 19 oz in the can, then the error in the amount of soup in the can is $|x - 19|$. If the possible error is ± 0.25 oz, then we can write the inequality

$$|x - 19| \le 0.25$$

Solve. $$-0.25 \le x - 19 \le 0.25$$
$$18.75 \le x \le 19.25$$

The actual amount of soup in the can is between 18.75 and 19.25 oz.

Answers to You Try Exercises

1) $(-9, 9)$

2) $\left[-3, \dfrac{4}{3}\right]$

3) $[3, 6]$

4) $(-\infty, -5] \cup [5, \infty)$

5) a) $(-\infty, -2] \cup \left[-\dfrac{1}{4}, \infty\right)$

 b) $(-\infty, 1) \cup (5, \infty)$

6) $(-\infty, -22] \cup [6, \infty)$

7) a) $(-\infty, \infty)$ b) $\varnothing$ c) $\left\{\dfrac{1}{6}\right\}$

9.2 Exercises

Boost your grade at mathzone.com!   MathZone

> Practice Problems   > NetTutor   > e-Professors   > Videos
> Self-Test

Graph each inequality on a number line and represent the sets of numbers using interval notation.

1) $-1 \le p \le 5$

2) $7 < t < 11$

3) $y < 2 \text{ or } y > 9$

4) $a \le -8 \text{ or } a \ge \dfrac{1}{2}$

5) $n \le -\dfrac{9}{2} \text{ or } n \ge \dfrac{3}{5}$

6) $-\dfrac{1}{4} \le q \le \dfrac{11}{4}$

7) $4 < b < \dfrac{17}{3}$

8) $x < -12 \text{ or } x > -9$

Objectives 1 and 3

Solve each inequality. Graph the solution set and write the answer in interval notation.

9) $|m| \le 7$

10) $|c| < 1$

11) $|3k| < 12$

12) $\left|\dfrac{5}{4}z\right| \le 30$

13) $|w - 2| < 4$

14) $|k - 6| \le 2$

15) $|3r + 10| \le 4$

16) $|4a + 1| \le 12$

17) $|7 - 6p| \le 3$

18) $|17 - 9d| < 8$

19) $|5q + 11| < 0$

20) $|6t + 16| < 0$

21) $|2x + 7| \le -12$

22) $|8m - 15| \le -5$

23) $|8c - 3| + 15 < 20$

24) $|2v + 5| + 3 < 14$

25) $\left|\dfrac{3}{2}h + 6\right| - 2 \le 10$

26) $7 + \left|\dfrac{8}{3}u - 9\right| < 12$

Objectives 2 and 3

Solve each inequality. Graph the solution set and write the answer in interval notation.

27) $|t| \ge 7$

28) $|p| > 3$

29) $|5a| > 2$

30) $|2c| \ge 11$

31) $|d + 10| \ge 4$

32) $|q - 7| > 12$

33) $|4v - 3| \ge 9$

34) $|6a + 19| > 11$

35) $|17 - 6x| > 5$

36) $|1 - 4g| \ge 10$

37) $|8k + 5| \ge 0$

38) $|5b - 6| \ge 0$

39) $|z - 3| \ge -5$

40) $|3r + 10| > -11$

41) $|2m - 1| + 4 > 5$

42) $|w + 6| - 4 \ge 2$

43) $-3 + \left|\dfrac{5}{6}n + \dfrac{1}{2}\right| \ge 1$

44) $\left|\dfrac{3}{2}y - \dfrac{5}{4}\right| + 9 \ge 11$

45) Explain why $|3t - 7| < 0$ has no solution.

46) Explain why $|4l + 9| \le -10$ has no solution.

47) Explain why the solution to $|2x + 1| \ge -3$ is $(-\infty, \infty)$.

48) Explain why the solution to $|7y - 3| \ge 0$ is $(-\infty, \infty)$.

Objectives 1–3

The following exercises contain absolute value equations, linear inequalities, and both types of absolute value inequalities. Solve each. Write the solution set for equations in set notation and use interval notation for inequalities.

49) $|2v + 9| > 3$

50) $\left|\dfrac{5}{3}a + 2\right| = 8$

51) $3 = |4t + 5|$

52) $|4k + 9| \le 5$

53) $9 \le |7 - 8q|$

54) $|2p - 5| - 12 = 11$

55) $2(x - 8) + 10 < 4x$

56) $\dfrac{1}{2}n + 11 < 8$

57) $|8 - r| \le 5$

58) $|d + 6| > 7$

59) $|6y + 5| \le -9$

60) $8 \le |5v + 2|$

61) $\left|\dfrac{4}{3}x + 1\right| = \left|\dfrac{5}{3}x + 8\right|$

62) $|7z - 8| \le 0$

63) $|3m - 8| - 11 > -3$

64) $|6c - 1| = -14$

65) $|4 - 9t| + 2 = 1$

66) $|5b - 11| - 18 < -10$

 67) $-\dfrac{3}{5} \ge \dfrac{5}{2}a - \dfrac{1}{2}$

68) $4 + 3(2r - 5) > 9 - 4r$

69) $|6k + 17| > -4$

70) $|5 - w| \ge 3$

71) $5 \ge |c + 8| - 2$

72) $0 \le |4a + 1|$

73) $|5h - 8| > 7$

74) $\left|\dfrac{2}{3}y - 1\right| = \left|\dfrac{3}{2}y + 4\right|$

75) $\left|\dfrac{1}{2}d - 4\right| + 7 = 13$

76) $|9q - 8| < 0$

77) $|5j + 3| + 1 \le 9$

78) $|7 - x| \le 12$

Objective 4

 79) A gallon of milk should contain 128 oz. The possible error in this measurement, however, is ± 0.75 oz. Let a represent the range of values for the amount of milk in the container. Write an absolute value inequality to represent the range for the number of ounces of milk in the container, then solve the inequality and explain the meaning of the answer.

80) Dawn buys a 27-oz box of cereal. The possible error in this amount, however, is ± 0.5 oz. Let c represent the range of values for the amount of cereal in the box. Write an absolute value inequality to represent the range for the number of ounces of cereal in the box, then solve the inequality and explain the meaning of the answer.

81) Emmanuel spent $38 on a birthday gift for his son. He plans on spending within $5 of that amount on his daughter's birthday gift. Let b represent the range of values for the amount he will spend on his daughter's gift. Write an absolute value inequality to represent the range for the amount of money Emmanuel will spend on his daughter's birthday gift, then solve the inequality and explain the meaning of the answer.

82) An employee at a home-improvement store is cutting a window shade for a customer. The customer wants the shade to be 32 in. wide. If the machine's possible error in cutting the shade is $\pm\dfrac{1}{16}$ in., write an absolute value inequality to represent the range for the width of the window shade, and solve the inequality. Explain the meaning of the answer. Let w represent the range of values for the width of the shade.

Section 9.3 Linear Inequalities in Two Variables

Objectives

1. Define a Linear Inequality in Two Variables and Identify Points in the Solution Set

2. Graph a Linear Inequality in Two Variables

3. Graph the Solution Set of a Compound Linear Inequality in Two Variables

4. Solve an Applied Problem Using a System of Linear Inequalities in Two Variables

In Chapter 3, we first learned how to solve linear inequalities in *one variable* such as $2x - 3 \geq 5$. Let's review how to solve it.

$$2x - 3 \geq 5$$
$$2x \geq 8 \qquad \text{Add 3.}$$
$$x \geq 4 \qquad \text{Divide by 2.}$$

We can graph the solution set on a number line:

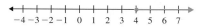

And we can write the solution set in interval notation: $[4, \infty)$.

In this section, we will learn how to graph the solution set of linear inequalities in *two variables*.

1. Define a Linear Inequality in Two Variables and Identify Points in the Solution Set

> **Definition**
>
> A **linear equality in two variables** is an inequality that can be written in the form $Ax + By \geq C$ or $Ax + By \leq C$ where A, B, and C are real numbers and where A and B are not both zero. ($>$ and $<$ may be substituted for $\geq$ and $\leq$.)

Here are some examples of linear inequalities in two variables.

$$5x - 3y \geq 6, \qquad y < \frac{1}{4}x + 3, \qquad x \leq 2, \qquad y > -4$$

> $x \leq 2$ can be considered a linear inequality in two variables because it can be written as $x + 0y \leq 2$. Likewise $y > -4$ can be written as $0x + y > -4$.

Earlier we found that the solution set of $2x - 3 \geq 5$, a linear inequality in *one* variable, is $[4, \infty)$. This means that any real number greater than or equal to 4 will make the inequality true. [For example, if $x = 6$, then $2(6) - 3 \geq 9$.] We graphed the solution set on a number line.

The solutions to linear inequalities in *two* variables, such as $x + y \geq 3$, are *ordered pairs* of the form (x, y) that make the inequality true. We graph a linear inequality in two variables on a rectangular coordinate system.

Example I

Shown here is the graph of $x + y \geq 3$. Find three points that solve $x + y \geq 3$, and find three points that are not in the solution set.

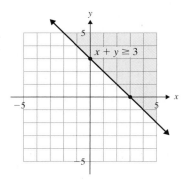

Solution

The solution set of $x + y \geq 3$ consists of all points either on the line or in the shaded region. *Any* point on the line or in the shaded region will make $x + y \geq 3$ true.

Solutions of $x + y \geq 3$	Check by Substituting into $x + y \geq 3$
(5, 2)	$5 + 2 \geq 3$ ✓
(1, 4)	$1 + 4 \geq 3$ ✓
(3, 0) (on the line)	$3 + 0 \geq 3$ ✓

(5, 2), (−2, 8), and (3, 0) are just some points that satisfy $x + y \geq 3$. There are infinitely many solutions.

Not in the Solution Set of $x + y \geq 3$	Verify by Substituting into $x + y \geq 3$
(0, 0)	$0 + 0 \geq 3$ False
(−4, 1)	$-4 + 1 \geq 3$ False
(2, −3)	$2 + (-3) \geq 3$ False

(0, 0), (−4, 1), and (2, −3) are just three points that do not satisfy $x + y \geq 3$. There are infinitely many such points.

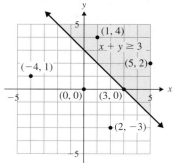

Points in the shaded region and on the line are in the solution set.

The points in the unshaded region are *not* in the solution set.

If the inequality in Example 1 had been $x + y > 3$, then the line would have been drawn as a *dotted line* and all points on the line would *not* be part of the solution set.

You Try 1

Shown here is the graph of $5x - 3y \geq -15$. Find three points that solve $5x - 3y \geq -15$, and find three points that are not in the solution set.

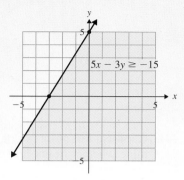

2. Graph a Linear Inequality in Two Variables

As you saw in the graph in Example 1, the line divides the *plane* into two regions or **half planes**. The line $x + y = 3$ is the **boundary** between the two half planes. We can use this boundary and two different methods to graph a linear inequality in two variables. The first method we will discuss is the **test point** method.

**Graphing a Linear Inequality in Two Variables
Using the Test Point Method**

1) **Graph the boundary line.** If the inequality contains $\geq$ or $\leq$, make it a *solid line*. If the inequality contains $>$ or $<$, make it a *dotted line*.

2) **Choose a test point not on the line, and shade the appropriate region.** Substitute the test point into the inequality.

    a) If it *makes the inequality true,* shade the side of the line *containing* the test point. All points in the shaded region are part of the solution set.

    b) If the test point *does not satisfy the inequality,* shade the *other* side of the line. All points in the shaded region are part of the solution set.

If $(0, 0)$ is not on the line, it is an easy point to test in the inequality.

Example 2

Graph $3x + 4y \leq -8$.

Solution

1) Graph the boundary line $3x + 4y = -8$ as a solid line.

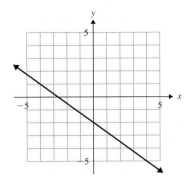

2) Choose a test point not on the line and substitute it into the inequality to determine whether or not it makes the inequality true.

Test Point	Substitute into $3x + 4y \leq -8$
$(0, 0)$	$3(0) + 4(0) \leq -8$
	$0 + 0 \leq -8$
	$0 \leq -8$ False

Since the test point $(0, 0)$ does *not* satisfy the inequality we will shade the side of the line that does *not* contain the point $(0, 0)$.

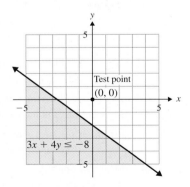

All points on the line and in the shaded region satisfy the inequality $3x + 4y \leq -8$.

Example 3

Graph $-x + 2y > -4$.

Solution

1) Since the inequality symbol is $>$ and not $\geq$, graph the boundary line $-x + 2y = -4$ as a dotted line. (This means that the points *on* the line are not part of the solution set.)

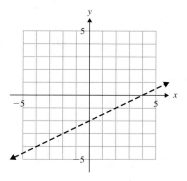

2) Choose a test point not on the line and substitute it into the inequality to determine whether or not it makes the inequality true.

Test Point	Substitute into $-x + 2y > -4$
$(0, 0)$	$-(0) + 2(0) > -4$
	$0 + 0 > -4$
	$0 > -4$ True

Since the test point $(0, 0)$ satisfies the inequality, shade the side of the line containing the point $(0, 0)$.

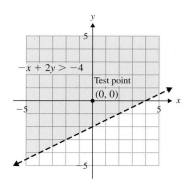

All points in the shaded region satisfy the inequality $-x + 2y > -4$.

You Try 2

Graph each inequality.

a) $2x + y \leq 4$ b) $x + 4y > 12$

There is another method we can use to graph linear inequalities in two variables. It involves writing the boundary line in *slope-intercept form* and working with the inequality in the form $y \geq mx + b$ ($y > mx + b$) or $y \leq mx + b$ ($y < mx + b$).

Using Slope-Intercept Form to Graph a Linear Inequality in Two Variables

1) *Write the inequality in the form* $y \geq mx + b$ ($y > mx + b$) *or* $y \leq mx + b$ ($y < mx + b$).

2) *Graph the boundary line,* $y = mx + b$.

    a) *If the inequality contains* $\geq$ *or* $\leq$, *make it a* solid line.

    b) *If the inequality contains* $>$ *or* $<$, *make it a* dotted line.

3) *Shade the appropriate side of the line.*

    a) *If the inequality is in the form* $y \geq mx + b$ *or* $y > mx + b$, *shade* above *the line.*

    b) *If the inequality is in the form* $y \leq mx + b$ *or* $y < mx + b$, *shade* below *the line.*

Example 4

Graph each inequality using the slope-intercept method.

a) $y < -\dfrac{1}{3}x + 5$　　　　b) $2x - y \leq -2$

Solution

a) 1) $y < -\dfrac{1}{3}x + 5$ is already in the correct form.

     2) Graph the boundary line $y = -\dfrac{1}{3}x + 5$ as a *dotted line*.

     3) Since $y < -\dfrac{1}{3}x + 5$ has a *less than* symbol, shade *below* the line.

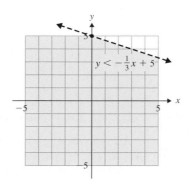

All points in the shaded region satisfy $y < -\dfrac{1}{3}x + 5$. We can choose a point in the shaded region as a check. $(0, 0)$ is in the shaded region. Substitute this point into $y < -\dfrac{1}{3}x + 5$.

$$0 < -\frac{1}{3}(0) + 5$$
$$0 < 0 + 5$$
$$0 < 5 \qquad \text{True}$$

b) 1) Solve $2x - y \leq -2$ for y.

$$2x - y \leq -2$$
$$-y \leq -2x - 2 \qquad \text{Subtract } 2x.$$
$$y \geq 2x + 2 \qquad \text{Divide by } -1 \text{ and change the direction of the inequality symbol.}$$

2) Graph $y = 2x + 2$ as a *solid line*.

3) Since $y \geq 2x + 2$ has a *greater than or equal to* symbol, shade *above* the line.

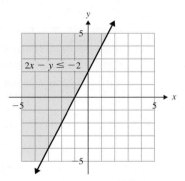

All points on the line and in the shaded region satisfy $2x - y \leq -2$.

 You Try 3

Graph each inequality using the slope-intercept method.

a) $y \geq -\frac{3}{4}x - 6$ b) $5x - 2y > 4$

3. Graph the Solution Set of a Compound Linear Inequality in Two Variables

In Section 3.8, we solved compound linear inequalities like

1) $x + 2 \geq 5$ and $4x \leq 24$

and

2) $x < -2$ or $3x + 1 > 4$

and we graphed their solution sets on a number line. The solution set and graph for $x + 2 \geq 5$ and $4x \leq 24$ are

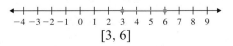

$$[3, 6]$$

Recall that the solution set of a compound inequality containing *and* is the *intersection* of the solution sets of the inequalities.

The solution set and graph for $x < -2$ or $3x + 1 > 4$ are

$$(-\infty, -2) \cup (1, \infty)$$

Recall that the solution set of a compound inequality containing *or* is the *union* of the solution sets of the inequalities.

Graphing compound linear inequalities in *two* variables works in a very similar way.

Steps for Graphing Compound Linear Inequalities in Two Variables

1) Graph each inequality separately on the same axes. Shade lightly.

2) If the inequality contains *and,* the solution set is the *intersection* of the shaded regions. Heavily shade this region.

3) If the inequality contains *or,* the solution set is the *union* (total) of the shaded regions. Heavily shade this region.

Example 5

Graph $x \leq 2$ and $2x + 3y \geq 3$.

Solution

To graph $x \leq 2$, graph the boundary line $x = 2$ as a solid line. The *x*-values are *less than* 2 to the *left* of 2, so shade to the left of the line $x = 2$.

Graph $2x + 3y \geq 3$. The region shaded blue in the third graph is the *intersection* of the shaded regions and the solution set of the compound inequality.

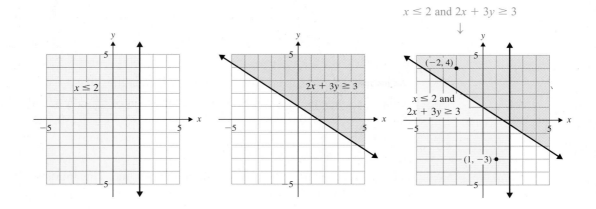

Any point inside the blue area will satisfy *both* inequalities. For example, the point $(-2, 4)$ is in this region.

Test Point	Verify		
$(-2, 4)$	$x \le 2$:	$-2 \le 2$ ✓	True
	$2x + 3y \ge 3$: $2(-2) + 3(4) \ge 3$		
	$-4 + 12 \ge 3$		
	$8 \ge 3$ ✓		True

> We show this test point on the graph.

Any point outside this region will not satisfy *both* inequalities and is *not* part of the solution set. One such point is $(1, -3)$.

Test Point	Verify		
$(1, -3)$	$x \le 2$:	$1 \le 2$ ✓	True
	$2x + 3y \ge 3$: $2(1) + 3(-3) \ge 3$		
	$2 - 9 \ge 3$		
	$-7 \ge 3$ ✓		False

Although we show three separate graphs in this example, it is customary to graph everything on the same axes, shading lightly at first, then to heavily shade the region that is the graph of the compound inequality.

You Try 4

Graph the compound inequality $y \le 3x - 1$ and $y + 2x \le 4$.

Example 6

Graph $y \le \dfrac{1}{2}x$ or $2x + y \ge 2$.

Solution

Graph each inequality separately. The solution set of the compound inequality will be the *union* (total) of the shaded regions.

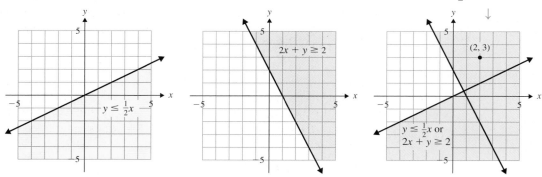

Any point in the shaded region of the third graph will be a solution to the compound inequality $y \le \frac{1}{2}x$ or $2x + y \ge 2$. This means the point must satisfy $y \le \frac{1}{2}x$ *or* $2x + y \ge 2$ *or* both. One point in the shaded region is $(2, 3)$.

Test Point	Verify	
$(2, 3)$	$y \le \frac{1}{2}x$:	$3 \le \frac{1}{2}(2)$
		$3 \le 1$ False
	$2x + y \ge 2$:	$2(2) + 3 \ge 2$
		$4 + 3 \ge 2$
		$7 \ge 2$ ✓ True

Although $(2, 3)$ does not satisfy $y \le \frac{1}{2}x$, it *does* satisfy $2x + y \ge 2$. Therefore, it *is* a solution of the compound inequality.

It is left to the reader to choose a point in the region that is *not* shaded to verify that it does not satisfy either inequality.

You Try 5

Graph the compound inequality $x \ge -4$ or $x - 3y \le -3$.

4. Solve an Applied Problem Using a System of Linear Inequalities in Two Variables

A practical application of linear inequalities in two variables is a process called **linear programming**. Companies use linear programming to determine the best way to use their machinery, employees, and other resources.

A linear programming problem may consist of several inequalities called **constraints**. The constraints form a system of inequalities. Constraints describe the conditions that the variables must meet. The graph of the solution set to this system of inequalities is called the **feasible region**—the ordered pairs which are the possible solutions to the system of inequalities.

Example 7

A manufacturer has to reduce the number of hours its factory will run. Instead of laying off two of its employees, it is decided Harvey and Amy will share the available hours. During a particular week, the company wants them to work at most 40 hours between them.

   Let x = the number of hours Harvey works

   y = the number of hours Amy works

a)   Write a system of linear inequalities to describe the constraints on the number of hours available to work.

b)   Graph the feasible region (solution set of the system), which describes the possible number of hours each person can work.

c)   Find three points in the feasible region and discuss their meanings.

d)   Find one point outside of the feasible region and discuss its meaning.

Solution

a)   Since x and y represent the number of hours worked, x and y cannot be negative. We can write $x \geq 0$ and $y \geq 0$.

   Together they can work at most 40 hours. We can write $x + y \leq 40$.

   The system of inequalities to describe the constraints on the number of hours available is

$$x \geq 0$$
$$y \geq 0$$
$$x + y \leq 40$$

   We can also think of this as $x \geq 0$ *and* $y \geq 0$ *and* $x + y \leq 40$. Therefore, the solution to this system is the *intersection* of the solution sets of each inequality.

b)   The graphs of $x \geq 0$ *and* $y \geq 0$ give us the set of points in the first quadrant since x and y are both positive here.

   Graph $x + y \leq 40$. This will be the region *below and including* the line $x + y = 40$ in quadrant I.

   The feasible region is graphed next.

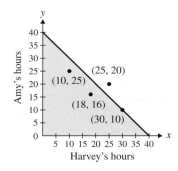

c) Three points in the feasible region are (10, 25), (18, 16), and (30, 10).

(10, 25) represents Harvey working 10 hours and Amy working 25 hours. It satisfies all three inequalities.

$$
\begin{array}{lll}
x \geq 0: & 10 \geq 0 \quad \checkmark & \text{True} \\
y \geq 0: & 25 \geq 0 \quad \checkmark & \text{True} \\
x + y \leq 40: & 10 + 25 \leq 40 \quad \checkmark & \text{True}
\end{array}
$$

(18, 16) represents Harvey working 18 hours and Amy working 16 hours. It satisfies all three inequalities.

$$
\begin{array}{lll}
x \geq 0: & 18 \geq 0 \quad \checkmark & \text{True} \\
y \geq 0: & 16 \geq 0 \quad \checkmark & \text{True} \\
x + y \leq 40: & 18 + 16 \leq 40 \quad \checkmark & \text{True}
\end{array}
$$

(30, 10) represents Harvey working 30 hours and Amy working 10 hours. It satisfies all three inequalities. This point is on the line $x + y = 40$, so notice that these hours total exactly 40.

$$
\begin{array}{lll}
x \geq 0: & 30 \geq 0 \quad \checkmark & \text{True} \\
y \geq 0: & 10 \geq 0 \quad \checkmark & \text{True} \\
x + y \leq 40: & 30 + 10 \leq 40 \quad \checkmark & \text{True}
\end{array}
$$

d) (25, 20) is outside the feasible region. It represents Harvey working 25 hours and Amy working 20 hours. This is not possible since it does not satisfy the inequality $x + y \leq 40$.

$$
\begin{array}{lll}
x \geq 0: & 25 \geq 0 \quad \checkmark & \text{True} \\
y \geq 0: & 20 \geq 0 \quad \checkmark & \text{True} \\
x + y \leq 40: & 25 + 20 \leq 40 & \text{False}
\end{array}
$$

Answers to You Try Exercises

1) Answers may vary.   2) a)   b)

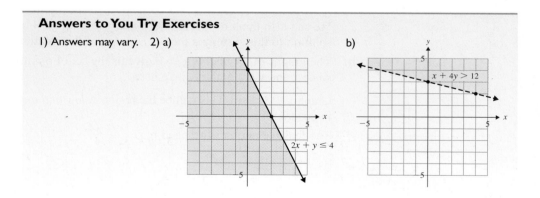

3) a)

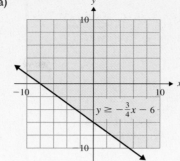

$$y \geq -\frac{3}{4}x - 6$$

b)

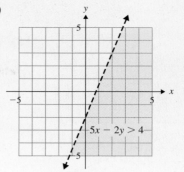

$$5x - 2y > 4$$

4)

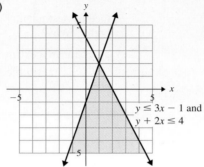

$$y \leq 3x - 1 \text{ and}$$
$$y + 2x \leq 4$$

5)

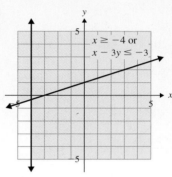

$$x \geq -4 \text{ or}$$
$$x - 3y \leq -3$$

9.3 Exercises

Boost your grade at mathzone.com!   MathZone

> Practice Problems
> NetTutor
> Self-Test
> e-Professors
> Videos

Objective 1

The graphs of linear inequalities are given next. For each, find three points that satisfy the inequality and three that are not in the solution set.

1) $x - 4y \geq 4$

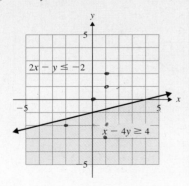

$$2x - y \leq -2$$

$$x - 4y \geq 4$$

2) $2x + 3y \leq 18$

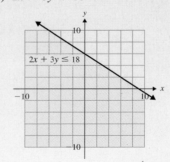

$$2x + 3y \leq 18$$

3) $y < 3x + 4$ (x, y)

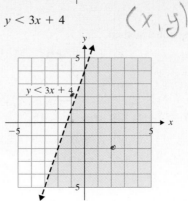

$$y < 3x + 4$$

4) $y > -\dfrac{2}{5}x - 3$

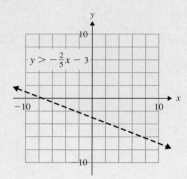

5) $x + y < 0$

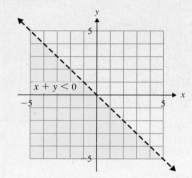

6) $y - 2x \geq 0$

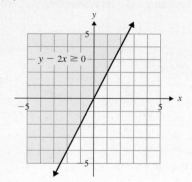

Objective 2

Graph using the test point method.

7) $2x + y \geq 6$ 8) $4x + y \leq 3$

9) $y < x + 2$ 10) $y > \dfrac{1}{2}x - 1$

11) $2x - 7y \leq 14$ 12) $4x + 3y < 15$

13) $y < x$ 14) $y \geq 3x$

15) $y \geq -5$ 16) $x < 1$

Use the slope-intercept form to graph each inequality.

17) $y \leq 4x - 3$ 18) $y \geq \dfrac{5}{2}x - 8$

19) $y > \dfrac{2}{5}x - 4$ 20) $y < \dfrac{1}{4}x + 1$

21) $6x + y > 3$ 22) $2x + y > -5$

 23) $9x - 3y \leq -21$ 24) $3x + 5y < -20$

25) $x > 2y$ 26) $x - y \leq 0$

27) To graph an inequality like $y \geq \dfrac{1}{3}x + 2$, which method, test point or slope-intercept, would you prefer? Why?

28) To graph an inequality like $7x + 2y < 10$, which method, test point or slope-intercept, would you prefer? Why?

Graph using either the test point or slope-intercept method.

29) $y > -\dfrac{3}{4}x + 1$ 30) $y \leq \dfrac{1}{3}x - 6$

31) $5x + 2y < -8$ 32) $4x + y < 7$

33) $9x - 3y \leq 21$ 34) $5x - 3y \geq -9$

35) $x > 2$ 36) $y \leq 4$

37) $3x - 4y > 12$ 38) $6x - y \leq 2$

Objective 3

The graphs of compound linear inequalities in two variables are given next. For each, find three points that are in the solution set and three that are not.

39) $y \geq \dfrac{4}{5}x + 2$ and $y < 5$

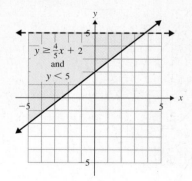

40) $x > 4$ and $y \le -\dfrac{2}{3}x + 2$

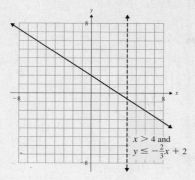

$x > 4$ and
$y \le -\frac{2}{3}x + 2$

41) $-x + 4y \ge 16$ or $2x + 3y \ge 15$

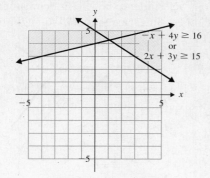

$-x + 4y \ge 16$
or
$2x + 3y \ge 15$

42) $5x + 2y \le -6$ or $2x + 5y \ge 10$

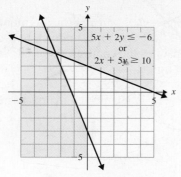

$5x + 2y \le -6$
or
$2x + 5y \ge 10$

43) Is $(3, 5)$ in the solution set of the compound inequality $x - y \ge -6$ and $2x + y < 7$? Why or why not?

44) Is $(3, 5)$ in the solution set of the compound inequality $x - y \ge -6$ or $2x + y < 7$? Why or why not?

Graph each compound inequality.

 45) $x \le 4$ and $y \ge -\dfrac{3}{2}x + 3$

46) $y \le \dfrac{1}{4}x + 2$ and $y \ge -1$

47) $y < x + 4$ and $y \ge -3$

48) $x < 3$ and $y > \dfrac{2}{3}x - 1$

49) $2x - 3y < -9$ and $x + 6y < 12$

50) $5x - 3y > 9$ and $2x + 3y \le 12$

51) $y \le -x - 1$ or $x \ge 6$

52) $y \le 2$ or $y \le \dfrac{4}{5}x + 2$

53) $y \le 4$ or $4y - 3x \ge -8$

54) $x + 3y \ge 3$ or $x \ge -2$

55) $y > -\dfrac{2}{3}x + 1$ or $-2x + 5y \le 0$

56) $y > x - 4$ or $3x + 2y \ge 12$

57) $x \ge 5$ and $y \le -3$ 58) $x \le 6$ and $y \ge 1$

59) $y < 4$ or $x \ge -3$ 60) $x \ge 2$ or $y \ge -6$

61) $2x + 5y < 15$ or $y \le \dfrac{3}{4}x - 1$

62) $y - 2x \le 1$ and $y \ge -\dfrac{1}{5}x - 2$

63) $y \ge \dfrac{2}{3}x - 4$ and $4x + y \le 3$

64) $y < 5x + 2$ or $x + 4y < 12$

Objective 4

65) During the school year, Tazia earns money by babysitting and tutoring. She can work at most 15 hr per week.

Let x = number of hours Tazia babysits
    y = number of hours Tazia tutors

a) Write a system of linear inequalities to describe the constraints on the number of hours Tazia can work per week.

b) Graph the feasible region that describes how her hours can be distributed between babysitting and tutoring.

c) Find three points in the feasible region and discuss their meanings.

d) Find one point outside the feasible region and discuss its meaning.

66) A machine in a factory can be calibrated to fill either large or small bags of potato chips. The machine will run at most 12 hr per day.

Let x = number of hours the machine fills large bags
y = number of hours the machine fills small bags

a) Write a system of linear inequalities to describe the constraints on the number of hours the machine fills the bags each day.

b) Graph the feasible region that describes how the hours can be distributed between filling the large and small bags of chips.

c) Find three points in the feasible region and discuss their meanings.

d) Find one point outside the feasible region and discuss its meaning.

Section 9.4 Solving Systems of Equations Using Matrices

Objectives

1. Learn the Vocabulary Associated with Gaussian Elimination

2. Solve a System of Linear Equations Using Gaussian Elimination

In Chapter 5 we learned how to solve systems of linear equations by graphing, substitution, and the elimination method. In this section, we will learn how to use **row operations** and **Gaussian elimination** to solve systems of linear equations. We begin by defining some terms.

1. Learn the Vocabulary Associated with Gaussian Elimination

A **matrix** is a rectangular array of numbers. (The plural of *matrix* is *matrices*.) Each number in the matrix is an **element** of the matrix. An example of a matrix is

$$
\begin{array}{ccc}
\text{Column} & \text{Column} & \text{Column} \\
1 & 2 & 3 \\
\downarrow & \downarrow & \downarrow
\end{array}
$$

$$
\begin{array}{l}
\text{Row 1} \rightarrow \\
\text{Row 2} \rightarrow
\end{array}
\begin{bmatrix}
3 & -1 & 4 \\
0 & 2 & -5
\end{bmatrix}
$$

We can represent a system of equations in an *augmented matrix*. An **augmented matrix** does not contain the variables and has a vertical line to distinguish between different parts of the equation. This vertical line separates the coefficients of the variables from the constants on the other side of the = sign.

For example, we can represent the system $\begin{aligned} 5x + 4y &= 1 \\ x - 3y &= 6 \end{aligned}$ in augmented matrix form as

$$
\begin{bmatrix}
5 & 4 & | & 1 \\
1 & -3 & | & 6
\end{bmatrix}
\begin{array}{l}
\leftarrow \text{This row represents the first equation.} \\
\leftarrow \text{This row represents the second equation.}
\end{array}
$$

We can also think of the matrix in terms of the columns.

$$
\begin{array}{cc}
\text{Coefficients} & \\
\text{of } x & \text{Constants} \\
\downarrow & \downarrow
\end{array}
$$

$$
\begin{bmatrix}
5 & 4 & | & 1 \\
1 & -3 & | & 6
\end{bmatrix}
$$

$$
\begin{array}{c}
\uparrow \\
\text{Coefficients} \\
\text{of } y
\end{array}
$$

Gaussian elimination is the process of using row operations on augmented matrices to solve a system of linear equations. It is a variation of the elimination method and can be very efficient. Computers often use augmented matrices and row operations to solve systems.

The goal of Gaussian elimination is to obtain a matrix of the form $\begin{bmatrix} 1 & a & b \\ 0 & 1 & c \end{bmatrix}$ when

solving a system of two equations or to obtain a matrix of the form $\begin{bmatrix} 1 & a & b & d \\ 0 & 1 & c & e \\ 0 & 0 & 1 & f \end{bmatrix}$

when solving a system of three equations, where *a, b, c, d, e,* and *f* are real numbers. Notice that there are 1's along the **diagonal** of the matrix and zeros below the diagonal. These matrices are in *row echelon form*. A matrix is in **row echelon form** when it has 1's along the diagonal and zeros below the diagonal. We get matrices in row echelon form by performing row operations.

2. Solve a System of Linear Equations Using Gaussian Elimination

The row operations we can perform on augmented matrices are similar to the operations we use to solve a system of equations using the elimination method.

Matrix Row Operations

Performing the following row operations on a matrix produces an equivalent matrix.

1) Interchanging two rows

2) Multiplying every element in a row by a nonzero real number

3) Replacing a row by the sum of it and the multiple of another row

---

Let's use these operations to solve a system using Gaussian elimination. Notice the similarities between this method and the elimination method we learned in Chapter 5.

Example 1

Solve using Gaussian elimination.

$$x + 5y = -1$$
$$2x - y = 9$$

Solution

Begin by writing the system in an augmented matrix.

$$\begin{bmatrix} 1 & 5 & -1 \\ 2 & -1 & 9 \end{bmatrix}$$

We will use the 1 to make the element below it a zero. Use a row operation.

$$\text{Use this} \searrow \begin{bmatrix} \text{①} & 5 & | & -1 \\ \text{②} & -1 & | & 9 \end{bmatrix} \nearrow \text{to make} \atop \text{this zero.}$$

If we multiply the 1 by -2 (to get -2) and add it to the 2, we get zero. We must do this operation to the entire row. Denote this as $-2R_1 + R_2 \rightarrow R_2$. (Read as, "$-2$ times row 1 plus row 2 makes the new row 2.") So, we are obtaining a new row 2.

$$\text{Use this} \searrow \begin{bmatrix} \text{①} & 5 & | & -1 \\ \text{②} & -1 & | & 9 \end{bmatrix} \begin{array}{c} -2R_1 + R_2 \rightarrow R_2 \end{array} \begin{bmatrix} 1 & 5 & | & -1 \\ -2(1)+2 & -2(5)+(-1) & | & -2(-1)+9 \end{bmatrix}$$
$$\text{to make} \nearrow \atop \text{this zero.}$$

$$= \begin{bmatrix} 1 & 5 & | & -1 \\ 0 & -11 & | & 11 \end{bmatrix}$$

Multiply each element of row 1 by -2 and add it to the corresponding element of row 2.

We are not *making* a new row 1, so it stays the same.

Next get a 1 on the diagonal in row 2.

This column is in the correct form.
$\downarrow$

$$\begin{bmatrix} 1 & 5 & | & -1 \\ 0 & \boxed{-11} & | & 11 \end{bmatrix} \begin{array}{c} -\frac{1}{11}R_2 \rightarrow R_2 \end{array} \begin{bmatrix} 1 & 5 & | & -1 \\ 0 & 1 & | & -1 \end{bmatrix}$$

Multiply each element of row 2 by $-\dfrac{1}{11}$ to get a 1 on the diagonal.

$\uparrow$
Make this 1.

We have obtained the final matrix because there are 1's on the diagonal and a zero below. The matrix is in row echelon form. From this matrix, write a system of equations. The last row gives us the value of y.

$$\begin{bmatrix} 1 & 5 & | & -1 \\ 0 & 1 & | & -1 \end{bmatrix} \qquad \begin{array}{c} 1x + 5y = -1 \\ 0x + 1y = -1 \end{array} \quad \text{or} \quad \begin{array}{c} x + 5y = -1 \\ y = -1 \end{array}$$

Substitute $y = -1$ into $x + 5y = -1$.

$$\begin{array}{ll} x + 5(-1) = -1 & \text{Substitute } -1 \text{ for } y. \\ x - 5 = -1 & \text{Multiply.} \\ x = 4 & \text{Add 5.} \end{array}$$

The solution is $(4, -1)$. Check by substituting $(4, -1)$ into both equations of the original system.

Here are the steps for using Gaussian elimination to solve a system of equations. Remember that our goal is to obtain a matrix with 1's along the diagonal and zeros below—row echelon form.

How to Solve a System of Equations Using Gaussian Elimination

Step 1: Write the system in an augmented matrix.

Step 2: Get the first element of the first column to be 1.

Step 3: The goal of step 3 is to obtain zeros below the 1 in the first column. Use row operations and the 1 in step 2 to make all entries below it zero.

Step 4: Turn your attention to the second column. Get the second element of the second column to be a 1. (We are getting 1's along the diagonal.)

Step 5: The goal of step 5 is to get zeros below the 1 in the second column. Use row operations and the 1 in step 4 (in the second column) to make all entries below it zero.

Step 6: Continue this procedure until the matrix is in row echelon form—1's along the diagonal and zeros below.

Step 7: Write a system of equations from the matrix that is in row echelon form.

Step 8: Solve the system. The last equation in the system will give you the value of one of the variables. Find the values of all of the variables using substitution.

Step 9: Check the solution in each equation of the original system.

 **You Try 1**

Solve the system using Gaussian elimination. $x - y = -1$
$$-3x + 5y = 9$$

Next we will solve a system of three equations using Gaussian elimination.

Example 2

Solve using Gaussian elimination.

$$2x + y - z = -3$$
$$x + 2y - 3z = 1$$
$$-x - y + 2z = 2$$

Solution

Step 1:   Write the system in an augmented matrix.

$$\begin{bmatrix} 2 & 1 & -1 & | & -3 \\ 1 & 2 & -3 & | & 1 \\ -1 & -1 & 2 & | & 2 \end{bmatrix}$$

Step 2: *Get the first element of the first column to be 1.*

We *could* multiply row 1 by $\dfrac{1}{2}$ to make the first entry 1, but this would make the rest of the entries in the first row fractions. Instead, recall that we can interchange two rows. If we interchange row 1 and row 2 (denote this as $R_1 \leftrightarrow R_2$), the first entry in the first column will be 1.

Original matrix $\qquad\qquad\qquad$ Matrix 2

$$\begin{bmatrix} 2 & 1 & -1 & -3 \\ 1 & 2 & -3 & 1 \\ -1 & -1 & 2 & 2 \end{bmatrix} \quad R_1 \leftrightarrow R_2 \quad \begin{bmatrix} 1 & 2 & -3 & 1 \\ 2 & 1 & -1 & -3 \\ -1 & -1 & 2 & 2 \end{bmatrix} \quad \text{Interchange row 1 and row 2.}$$

Step 3: *Obtain zeros below the 1 in the first column of matrix 2.*

Use this → Matrix 2

to make each →
of these zero. ↳

$$\begin{bmatrix} ① & 2 & -3 & 1 \\ ② & 1 & -1 & -3 \\ \boxed{-1} & -1 & 2 & 2 \end{bmatrix}$$

To obtain a zero in place of the 2, multiply the 1 by -2 (to get -2) and add it to the 2. Perform that same operation on the entire row to obtain the new row 2: $-2R_1 + R_2 \rightarrow R_2$.

Use this ↘ Matrix 2 $\qquad\qquad\qquad\qquad$ Matrix 3

to make each →
of these zero. ↳

$$\begin{bmatrix} ① & 2 & -3 & 1 \\ ② & 1 & -1 & -3 \\ \boxed{-1} & -1 & 2 & 2 \end{bmatrix} \quad -2R_1 + R_2 \rightarrow R_2 \quad \begin{bmatrix} 1 & 2 & -3 & 1 \\ 0 & -3 & 5 & -5 \\ -1 & -1 & 2 & 2 \end{bmatrix} \quad \begin{array}{l}\text{Multiply row 1 by} \\ -2 \text{ and add the} \\ \text{result to row 2.}\end{array}$$

To obtain a zero in place of the -1, add the 1 and -1. Perform that same operation on the entire row to obtain a new row 3: $R_1 + R_3 \rightarrow R_3$.

Use this ↘ Matrix 3 $\qquad\qquad\qquad\qquad$ Matrix 4

to make
this zero. ↘

$$\begin{bmatrix} ① & 2 & -3 & 1 \\ 0 & -3 & 5 & -5 \\ \boxed{-1} & -1 & 2 & 2 \end{bmatrix} \quad R_1 + R_3 \rightarrow R_3 \quad \begin{bmatrix} 1 & 2 & -3 & 1 \\ 0 & -3 & 5 & -5 \\ 0 & 1 & -1 & 3 \end{bmatrix} \quad \begin{array}{l}\text{Add rows} \\ 1 \text{ and 3.}\end{array}$$

We have completed step 3 because each entry below the 1 in the first column of matrix 4 is zero.

Step 4: Turn your attention to column 2. *The second element in the second column needs to be 1.* (We are getting 1's along the diagonal.)

We *could* multiply row 2 by $-\dfrac{1}{3}$ to obtain the 1, but the rest of the entries would be fractions. Instead, if we interchange row 2 and row 3, we will get 1 on the diagonal *and keep the zeros in the first column.* (Sometimes fractions are unavoidable and you must work with them.)

Matrix 4 $\qquad\qquad\qquad\qquad$ Matrix 5

$$\begin{bmatrix} 1 & 2 & -3 & 1 \\ 0 & -3 & 5 & -5 \\ 0 & 1 & -1 & 3 \end{bmatrix} \quad R_2 \leftrightarrow R_3 \quad \begin{bmatrix} 1 & 2 & -3 & 1 \\ 0 & 1 & -1 & 3 \\ 0 & -3 & 5 & -5 \end{bmatrix} \quad \begin{array}{l}\text{Interchange row 2 and row 3.}\end{array}$$

Step 5: *Obtain a zero below the 1 in the second column of matrix 5.*

Matrix 5

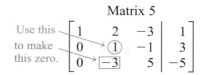

To obtain a zero in place of -3, multiply the 1 above it by 3 (to get 3) and add it to -3. Perform that same operation on the entire row to obtain the new row 3: $3R_2 + R_3 \rightarrow R_3$

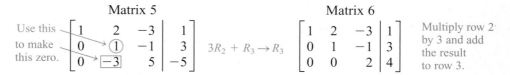

We have completed step 5 because the only entry below the 1 in the second column on the diagonal in matrix 6 is a zero.

Step 6: *Continue the procedure until the matrix is in row echelon form*—until it has 1's along the diagonal and zeros below.

Turn your attention to column 3. *The last element in the third column needs to be a 1.* (This is the last 1 we need along the diagonal.)

Multiply row 3 by $\dfrac{1}{2}$ to obtain the 1.

Matrix 6

$$\begin{bmatrix} 1 & 2 & -3 & | & 1 \\ 0 & 1 & -1 & | & 3 \\ 0 & 0 & 2 & | & 4 \end{bmatrix} \quad \tfrac{1}{2}R_3 \rightarrow R_3$$

Matrix 7

$$\begin{bmatrix} 1 & 2 & -3 & | & 1 \\ 0 & 1 & -1 & | & 3 \\ 0 & 0 & 1 & | & 2 \end{bmatrix}$$

Multiply row 2 by $\dfrac{1}{2}$ to obtain the last 1 along the diagonal.

We are done performing row operations because matrix 7 is in row echelon form.

Step 7: *Write a system of equations* from the matrix that is in row echelon form.

Matrix 7

$$\begin{bmatrix} 1 & 2 & -3 & | & 1 \\ 0 & 1 & -1 & | & 3 \\ 0 & 0 & 1 & | & 2 \end{bmatrix} \qquad \begin{aligned} 1x + 2y - 3z &= 1 \\ 0x + 1y - 1z &= 3 \\ 0x + 0y + 1z &= 2 \end{aligned} \quad \text{or} \quad \begin{aligned} x + 2y - 3z &= 1 \\ y - z &= 3 \\ z &= 2 \end{aligned}$$

Step 8: *Solve the system.* The last row of the matrix tells us that $z = 2$. Substitute $z = 2$ into the equation above it ($y - z = 3$) to get the value of y.

$$\begin{aligned} y - z &= 3 \\ y - 2 &= 3 \qquad \text{Substitute 2 for } z. \\ y &= 5 \qquad \text{Add 2 to each side.} \end{aligned}$$

Substitute $y = 5$ and $z = 2$ into $x + 2y - 3z = 1$ to solve for x.

$$
\begin{aligned}
x + 2y - 3z &= 1 \\
x + 2(5) - 3(2) &= 1 \qquad \text{Substitute values.} \\
x + 10 - 6 &= 1 \qquad \text{Multiply.} \\
x + 4 &= 1 \qquad \text{Subtract.} \\
x &= -3 \qquad \text{Subtract 4 from each side.}
\end{aligned}
$$

The solution of the system is $(-3, 5, 2)$.

Step 9: *Check the solution* in each equation of the original system. The check is left to the student.

This procedure may seem long and complicated at first, but as you practice and become more comfortable with the steps, you will see that it is actually quite efficient.

You Try 2

Solve the system using Gaussian elimination.
$$
\begin{aligned}
x + 3y - 2z &= 10 \\
3x + 2y + z &= 9 \\
-x + 4y - z &= -1
\end{aligned}
$$

If we are performing Gaussian elimination and obtain a matrix that produces a false equation like this

$$
\left[\begin{array}{ccc|c}
1 & 4 & 7 & 5 \\
0 & 1 & -6 & 9 \\
0 & 0 & 0 & 8
\end{array}\right] \qquad 0x + 0y + 0z = 8 \quad \text{False}
$$

then the system has *no solution*.

If, however, we obtain a matrix that produces a row of zeros as shown below,

$$
\left[\begin{array}{ccc|c}
1 & 3 & -4 & 2 \\
0 & 1 & 5 & -1 \\
0 & 0 & 0 & 0
\end{array}\right] \qquad 0x + 0y + 0z = 0 \quad \text{True}
$$

then the system has an *infinite number of solutions*.

Answers to You Try Exercises

1) $(2, 3)$ 2) $(4, 0, -3)$

9.4 Exercises

Objective I

Write each system in an augmented matrix.

1) $x - 7y = 15$
   $4x + 3y = -1$

2) $x + 6y = 4$
   $-5x + y = -3$

 3) $x + 6y - z = -2$
   $3x + y + 4z = 7$
   $-x - 2y + 3z = 8$

4) $x + 2y - 7z = 3$
   $3x - 5y = -1$
   $-x + 2z = -4$

Write a system of linear equations in x and y represented by each augmented matrix.

5) $\begin{bmatrix} 3 & 10 & -4 \\ 1 & -2 & 5 \end{bmatrix}$

6) $\begin{bmatrix} 1 & -1 & 6 \\ -4 & 7 & 2 \end{bmatrix}$

7) $\begin{bmatrix} 1 & -6 & 8 \\ 0 & 1 & -2 \end{bmatrix}$

8) $\begin{bmatrix} 1 & 2 & 11 \\ 0 & 1 & 3 \end{bmatrix}$

Write a system of linear equations in x, y, and z represented by each augmented matrix.

9) $\begin{bmatrix} 1 & -3 & 2 & 7 \\ 4 & -1 & 3 & 0 \\ -2 & 2 & -3 & -9 \end{bmatrix}$

10) $\begin{bmatrix} 1 & 4 & -3 & -5 \\ -1 & 2 & 5 & 8 \\ 6 & -2 & -1 & 3 \end{bmatrix}$

11) $\begin{bmatrix} 1 & 5 & 2 & 14 \\ 0 & 1 & -8 & 2 \\ 0 & 0 & 1 & -3 \end{bmatrix}$

12) $\begin{bmatrix} 1 & 4 & -7 & -11 \\ 0 & 1 & 3 & -1 \\ 0 & 0 & 1 & 6 \end{bmatrix}$

Objective 2

Solve each system using Gaussian elimination.

 13) $x + 4y = -1$
   $3x + 5y = 4$

14) $x - 3y = 1$
   $-3x + 7y = 3$

15) $x - 3y = 9$
   $-6x + 5y = 11$

16) $x + 4y = -6$
   $2x + 5y = 0$

 17) $4x - 3y = 6$
   $x + y = -2$

18) $-4x + 5y = -3$
   $x - 8y = -6$

 19) $x + y - z = -5$
   $4x + 5y - 2z = 0$
   $8x - 3y + 2z = -4$

20) $x - 2y + 2z = 3$
   $2x - 3y + z = 13$
   $-4x - 5y - 6z = 8$

21) $x - 3y + 2z = -1$
   $3x - 8y + 4z = 6$
   $-2x - 3y - 6z = 1$

22) $x - 2y + z = -2$
   $2x - 3y + z = 3$
   $3x - 6y + 2z = 1$

23) $-4x - 3y + z = 5$
   $x + y - z = -7$
   $6x + 4y + z = 12$

24) $6x - 9y - 2z = 7$
   $-3x + 4y + z = -4$
   $x - y - z = 1$

25) $x - 3y + z = -4$
   $4x + 5y - z = 0$
   $2x - 6y + 2z = 1$

26) $x - y + 3z = 1$
   $5x - 5y + 15z = 5$
   $-4x + 4y - 12z = -4$

Definition/Procedure	Example	Reference
Solving Absolute Value Equations		9.1
If P represents an expression and k is a positive, real number, then to solve $\lvert P \rvert = k$ we rewrite the absolute value equation as the *compound equation* $P = k$ or $P = -k$ and solve for the variable.	Solve $\lvert 4a + 10 \rvert = 18$. $$\lvert 4a + 10 \rvert = 18$$ $$\swarrow \quad \searrow$$ $4a + 10 = 18$ or $4a + 10 = -18$ $\qquad 4a = 8 \qquad\qquad\quad 4a = -28$ $\qquad\quad a = 2$ or $\qquad\quad a = -7$ Check the solutions in the original equation. The solution set is $\{-7, 2\}$.	**p. 580**
Solving Absolute Value Inequalities		9.2
Inequalities Containing $<$ or $\leq$ Let P be an expression and let k be a positive, real number. To solve $\lvert P \rvert \leq k$, solve the three-part inequality $$-k \leq P \leq k$$ ($<$ may be substituted for $\leq$.)	Solve $\lvert x - 3 \rvert \leq 2$. Graph the solution set and write the answer in interval notation. $$-2 \leq x - 3 \leq 2$$ $$1 \leq x \leq 5$$ In interval notation, we write $[1, 5]$.	**p. 587**
Inequalities Containing $>$ or $\geq$ Let P be an expression and let k be a positive, real number. To solve $\lvert P \rvert \geq k$, ($>$ may be substituted for $\geq$) solve the compound inequality $$P \geq k \text{ or } P \leq -k$$	Solve $\lvert 2n - 5 \rvert > 1$. Graph the solution set and write the answer in interval notation. $2n - 5 > 1$ or $2n - 5 < -1$ Solve. $\quad 2n > 6$ or $\qquad 2n < 4$ Add 5. $\quad\, n > 3$ or $\qquad\quad n < 2$ Divide by 2. In interval notation, we write $(-\infty, 2) \cup (3, \infty)$.	**p. 589**
Linear Inequalities in Two Variables		9.3
A *linear inequality in two variables* is an inequality that can be written in the form $Ax + By \geq C$ or $Ax + By \leq C$, where A, B, and C are real numbers and where A and B are not both zero. ($>$ and $<$ may be substituted for $\geq$ and $\leq$.)	Some examples of linear inequalities in two variables are $$x + 3y \leq 2, \quad y > -\frac{2}{3}x + 5, \quad y \geq -1, \quad x < 4$$	**p. 595**

Definition/Procedure	Example	Reference

Graphing a Linear Inequality in Two Variables Using the Test Point Method

1) Graph the boundary line.
   a) If the inequality contains $\geq$ or $\leq$, make it a solid line.
   b) If the inequality contains $>$ or $<$, make it a dotted line.
2) Choose a test point not on the line, and shade the appropriate region. Substitute the test point into the inequality.
   a) If it *makes the inequality true*, shade the side of the line *containing* the test point. All points in the shaded region are part of the solution set.
   b) If the test point *does not satisfy the inequality*, shade the *other* side of the line. All points in the shaded region are part of the solution set.

Note that if $(0, 0)$ is not on the line, it is an easy point to test in the inequality.

Graph using the test point method.

$$2x + y > -3$$

1) Graph the boundary line as a *dotted* line.
2) Choose a test point not on the line and substitute it into the inequality to determine whether or not it makes the inequality true.

Test Point	Substitute into $2x + y > -3$
$(0, 0)$	$2(0) + (0) > -3$
	$0 + 0 > -3$
	$0 > -3$ True

Since the test point satisfies the inequality, shade the side of the line containing $(0, 0)$.

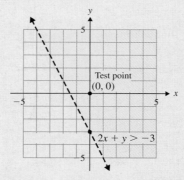

All points in the shaded region satisfy $2x + y > -3$.

p. 597

Using Slope-Intercept Form to Graph a Linear Inequality in Two Variables

1) Write the inequality in the form
   $y \geq mx + b$ ($y > mx + b$) or $y \leq mx + b$
   ($y < mx + b$).

2) Graph the boundary line $y = mx + b$.
   a) If the inequality contains $\geq$ or $\leq$, make it a solid line.
   b) If the inequality contains $>$ or $<$, make it a dotted line.

Graph using the slope-intercept method.

$$-x + 3y \leq 6$$

1) Solve $-x + 3y \leq 6$ for y.

$$3y \leq x + 6$$

$$y \leq \frac{1}{3}x + 2$$

2) Graph $y = \frac{1}{3}x + 2$ as a *solid line*.

p. 600

Definition/Procedure	Example	Reference

3) *Shade the appropriate side of the line.*
   a) If the inequality is in the form $y \geq mx + b$ or $y > mx + b$, shade *above* the line.

   b) If the inequality is in the form $y \leq mx + b$ or $y < mx + b$, shade *below* the line.

3) Since $y \leq \dfrac{1}{3}x + 2$ has a *less than or equal to* symbol, shade *below* the line.

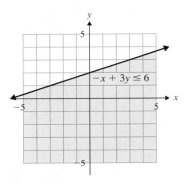

All points on the line and in the shaded region satisfy $-x + 3y \leq 6$.

Graphing Compound Linear Inequalities in Two Variables

1) Graph each inequality separately on the same axes. Shade lightly.

2) If the inequality contains *and*, the solution set is the *intersection* of the shaded regions. (Heavily shade this region.)

3) If the inequality contains *or*, the solution set is the *union* (total) of the shaded regions. Heavily shade this region.

Graph the compound inequality $y \geq -4x + 3$ and $y \geq 1$.

1) Graph each inequality separately on the same axes, shading lightly.

2) Since the inequality contains *and*, the solution set is the *intersection* of the shaded regions.

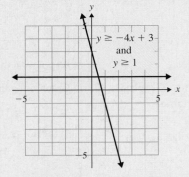

$y \geq -4x + 3$
and
$y \geq 1$

Any point in the shaded area will satisfy *both* inequalities.

p. 602

Solving Systems of Equations Using Matrices

9.4

An *augmented matrix* contains a vertical line to separate different parts of the matrix.

An example of an augmented matrix is

$$\left[\begin{array}{cc|c} 1 & 4 & -9 \\ 2 & -3 & 8 \end{array}\right].$$

p. 610

Definition/Procedure	Example	Reference

Matrix Row Operations

Performing the following row operations on a matrix produces an equivalent matrix.

1) Interchanging two rows

2) Multiplying every element in a row by a nonzero real number

3) Replacing a row by the sum of it and the multiple of another row

Gaussian elimination is the process of performing row operations on a matrix to put it into *row echelon form*.

A matrix is in row echelon form when it has 1's along the diagonal and 0's below.

$$\begin{bmatrix} 1 & a & | & b \\ 0 & 1 & | & c \end{bmatrix} \qquad \begin{bmatrix} 1 & a & b & | & d \\ 0 & 1 & c & | & e \\ 0 & 0 & 1 & | & f \end{bmatrix}$$

Solve using Gaussian elimination.

$$\begin{aligned} x - y &= 5 \\ 2x + 7y &= 1 \end{aligned}$$

Write the system in an augmented matrix. Then, perform row operations to get it into row echelon form.

$$\begin{bmatrix} 1 & -1 & | & 5 \\ 2 & 7 & | & 1 \end{bmatrix} \xrightarrow{-2R_1 + R_2 \to R_2} \begin{bmatrix} 1 & -1 & | & 5 \\ 0 & 9 & | & -9 \end{bmatrix}$$

$$\begin{bmatrix} 1 & -1 & | & 5 \\ 0 & 9 & | & -9 \end{bmatrix} \xrightarrow{\frac{1}{9}R_2 \to R_2} \begin{bmatrix} 1 & -1 & | & 5 \\ 0 & 1 & | & -1 \end{bmatrix}$$

The matrix is in row echelon form since it has 1's on the diagonal and a zero below.

Write a system of equations from the matrix that is in row echelon form.

$$\begin{bmatrix} 1 & -1 & | & 5 \\ 0 & 1 & | & -1 \end{bmatrix} \quad \begin{aligned} 1x - 1y &= 5 \\ 0x + 1y &= -1 \end{aligned} \quad \text{or} \quad \begin{aligned} x - y &= 5 \\ y &= -1 \end{aligned}$$

Solving the system we obtain the solution $(4, -1)$.

p. 611

(9.1) Solve.

1) $|m| = 9$

2) $\left|\dfrac{1}{2}c\right| = 5$

3) $|7t + 3| = 4$

4) $|4 - 3y| = 12$

5) $|8p + 11| - 7 = -3$

6) $|5k + 3| - 8 = 4$

7) $\left|4 - \dfrac{5}{3}x\right| = \dfrac{1}{3}$

8) $\left|\dfrac{2}{3}w + 6\right| = \dfrac{5}{2}$

9) $|7r - 6| = |8r + 2|$

10) $|3z - 4| = |5z - 6|$

11) $|2a - 5| = -10$

12) $|h + 6| - 12 = -20$

13) $|9d + 4| = 0$

14) $|6q - 7| = 0$

15) Write an absolute value equation which means *a is 4 units from zero.*

16) Write an absolute value equation which means *t is 7 units from zero.*

(9.2) Solve each inequality. Graph the solution set and write the answer in interval notation.

17) $|c| \le 3$

18) $|w + 1| < 11$

19) $|4t| > 8$

20) $|2v - 7| \ge 15$

21) $|12r + 5| \ge 7$

22) $|3k - 11| < 4$

23) $|4 - a| < 9$

24) $|2 - 5q| > 6$

25) $|4c + 9| - 8 \le -2$

26) $|3m + 5| + 2 \ge 7$

27) $|5y + 12| - 15 \ge -8$

28) $3 + |z - 6| \le 13$

29) $|k + 5| > -3$

30) $|4q - 9| < 0$

31) $|12s + 1| \le 0$

32) A radar gun indicated that a pitcher threw a 93 mph fastball. The radar gun's possible error in measuring the speed of a pitch is ± 1 mph. Write an absolute value inequality to represent the range for the speed of the pitch, and solve the inequality. Explain the meaning of the answer. Let *s* represent the range of values for the speed of the pitch.

(9.3) Graph each linear inequality in two variables.

33) $y \le -2x + 7$

34) $y \ge -\dfrac{3}{2}x + 2$

35) $y > -\dfrac{1}{3}x - 4$

36) $y < \dfrac{3}{4}x - 5$

37) $-3x + 4y > 12$

38) $5x - 2y \ge 8$

39) $4x - y > -5$

40) $y < x$

41) $x \ge 4$

42) $y \le 3$

Graph each compound inequality.

43) $y \ge \dfrac{3}{4}x - 4$ and $y \le -5$

44) $y \le -\dfrac{1}{3}x - 2$ and $x \le 4$

45) $y \le -\dfrac{1}{2}x + 7$ and $x \le 1$

46) $y \ge -\dfrac{2}{3}x - 4$ or $x < 1$

47) $y < \dfrac{5}{4}x - 5$ or $y < -3$

48) $4x - y < -1$ or $y > \dfrac{1}{2}x + 5$

49) $2x + y \le 3$ or $6x + y > 4$

50) $2x + 5y \le 10$ and $y \ge \dfrac{1}{3}x + 4$

51) $4x + 2y \ge -6$ and $y \le 2$

52) $3x - 4y < 20$ or $y < -2$

(9.4) Solve each system using Gaussian elimination.

53) $\begin{aligned} x - y &= -11 \\ 2x + 9y &= 0 \end{aligned}$

54) $\begin{aligned} x - 8y &= -13 \\ 4x + 9y &= -11 \end{aligned}$

55) $\begin{aligned} 5x + 3y &= 5 \\ -x + 8y &= -1 \end{aligned}$

56) $\begin{aligned} 3x + 5y &= 5 \\ -4x - 9y &= 5 \end{aligned}$

57) $\begin{aligned} x - 3y - 3z &= -7 \\ 2x - 5y - 3z &= 2 \\ -3x + 5y + 4z &= -1 \end{aligned}$

58) $\begin{aligned} x - 3y + 5z &= 3 \\ 2x - 5y + 6z &= -3 \\ 3x + 2y + 2z &= 3 \end{aligned}$

Solve.

1) $|4y - 9| = 11$

2) $|d + 6| - 3 = 7$

3) $|3k + 5| = |k - 11|$

4) $\left|\dfrac{1}{2}n - 1\right| = -8$

5) Write an absolute value equation that means *x is 8 units from zero.*

Solve each inequality. Graph the solution set and write the answer in interval notation.

6) $|c| > 4$

7) $|2z - 7| \leq 9$

8) $|4m + 9| - 8 \geq 5$

9) A scale in a doctor's office has a possible error of ± 0.75 lb. If Thanh's weight is measured as 168 lb, write an absolute value inequality to represent the range for his weight, and solve the inequality. Let *w* represent the range of values for Thanh's weight. Explain the meaning of the answer.

Graph each inequality.

10) $y \geq 3x + 1$

11) $2x - 5y > 10$

Graph the compound inequality.

12) $-2x + 3y \geq -12$ and $x \leq 3$

13) $y < -x$ or $2x - y > 1$

Solve using Gaussian elimination.

14) $x + 5y = -4$
    $3x + 2y = 14$

15) $-3x + 5y + 8z = 0$
    $x - 3y + 4z = 8$
    $2x - 4y - 3z = 3$

Perform the operations and simplify.

1) $5 \times 6 - 36 \div 3^2$

2) $\dfrac{5}{12} - \dfrac{7}{8}$

Evaluate.

3) 3^4

4) 2^5

5) $\left(\dfrac{1}{8}\right)^2$

6) 4^{-3}

7) Write 0.00000914 in scientific notation.

8) Solve $8 - 3(2y - 5) = 4y + 1$

9) Solve $3 - \dfrac{2}{7}n \geq 9$. Write the answer in interval notation.

10) *Write an equation and solve.*

How many ounces of a 9% alcohol solution must be added to 8 oz of a 3% alcohol solution to obtain a 5% alcohol solution?

Find the slope of the line containing each pair of points.

11) $(-6, 4)$ and $(4, 2)$

12) $(5, 3)$ and $(5, 9)$

13) Write the slope-intercept form of the line containing $(7, 2)$ with slope $\dfrac{1}{3}$.

14) Solve by graphing.

$$2x + y = -1$$
$$y = 3x - 6$$

Multiply and simplify.

15) $-4p^2(3p^2 - 7p - 1)$

16) $(2k + 5)(2k - 5)$

17) $(t + 8)^2$

18) Divide $\dfrac{6c^3 + 7c^2 - 38c + 24}{3c - 4}$.

Factor completely.

19) $9m^2 - 121$

20) $z^2 - 14z + 48$

Solve.

21) $a^2 + 6a + 9 = 0$

22) $2(x^2 - 4) = -(7x + 4)$

23) Subtract $\dfrac{1}{r^2 - 25} - \dfrac{r + 3}{2r + 10}$.

24) Multiply and simplify $\dfrac{w^2 - 3w - 54}{w^3 - 8w^2} \cdot \dfrac{w}{w + 6}$.

25) Solve $\left|\dfrac{1}{4}q - 7\right| - 8 = -5$.

26) Solve. Graph the solution set and write the answer in interval notation.

$$|9v + 4| > 14$$

27) Graph the compound inequality

$$3x + 4y > 16 \text{ or } y < \dfrac{1}{5}x + 1$$

28) Solve the system.

$$4x + 7y \qquad = -3$$
$$2x - \qquad 3z = 7$$
$$3y - z = 7$$

Radicals and Rational Exponents

Algebra at Work: Forensics

Forensic scientists use mathematics in many ways to help them analyze evidence and solve crimes. To help him reconstruct an accident scene, Keith can use this formula containing a radical to estimate the minimum speed of a vehicle when the accident occurred:

$$S = \sqrt{30fd}$$

where f = the drag factor, based on the type of road surface

d = the length of the skid, in feet

S = the speed of the vehicle in miles per hour

Keith is investigating an accident in a residential neighborhood where the speed limit is 25 mph. The car involved in the accident left skid marks 60 ft long. Tests showed that the drag factor of the asphalt road was 0.80. Was the driver speeding at the time of the accident?

Substitute the values into the equation and evaluate it to determine the minimum speed of the vehicle at the time of the accident:

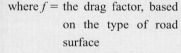

$$S = \sqrt{30fd}$$
$$S = \sqrt{30(0.80)(60)}$$
$$S = \sqrt{1440} \approx 38 \text{ mph}$$

The driver was going at least 38 mph when the accident occurred. This is well over the speed limit of 25 mph.

We will learn how to simplify radicals in this chapter as well as how to work with equations like the one given here.

Section 10.1 Finding Roots

Objectives

1. Find the Square Root of a Rational Number

2. Approximate the Square Root of a Whole Number

3. Find the Higher Roots of Rational Numbers

In Section 1.2 we introduced the idea of exponents as representing repeated multiplication. For example,

$$3^2 \text{ means } 3 \cdot 3, \text{ so } 3^2 = 9.$$
$$2^4 \text{ means } 2 \cdot 2 \cdot 2 \cdot 2, \text{ so } 2^4 = 16.$$

In this chapter we will study the opposite procedure, finding **roots** of numbers.

1. Find the Square Root of a Rational Number

 **Example 1**

Find all square roots of 25.

Solution

To find a *square* root of 25 ask yourself, "What number do I *square* to get 25?" Or, "What number multiplied by itself equals 25?" One number is 5 since $5^2 = 25$. Another number is -5 since $(-5)^2 = 25$.

     5 is a square root of 25.

   -5 is a square root of 25.

 You Try 1

Find all square roots of 64.

The $\sqrt{}$ symbol represents the *positive* square root of a number. For example,

$$\sqrt{25} = 5$$

 $\sqrt{25} = 5$ but $\sqrt{25} \neq -5$. The $\sqrt{}$ symbol represents *only* the positive square root.

To find the negative square root of a number we must put a $-$ in front of the $\sqrt{}$. For example,

$$-\sqrt{25} = -5$$

Next we will define some terms associated with the $\sqrt{}$ symbol.

$\sqrt{}$ is the **square root symbol** or the **radical sign**. The number under the radical sign is the **radicand**.

$$\text{Radical sign} \rightarrow \sqrt{25}$$
$$\uparrow$$
$$\text{Radicand}$$

The entire expression, $\sqrt{25}$, is called a **radical**.

Example 2

Find each square root.

a) $\sqrt{100}$ b) $-\sqrt{16}$ c) $\sqrt{\dfrac{4}{25}}$ d) $-\sqrt{\dfrac{81}{49}}$

Solution

a) $\sqrt{100} = 10$ since $(10)^2 = 100$.

b) $-\sqrt{16}$ means $-1 \cdot \sqrt{16}$. Therefore,

$$-\sqrt{16} = -1 \cdot \sqrt{16} = -1 \cdot (4) = -4$$

c) Since $\sqrt{4} = 2$ and $\sqrt{25} = 5$, $\sqrt{\dfrac{4}{25}} = \dfrac{2}{5}$.

d) $-\sqrt{\dfrac{81}{49}}$ means $-1 \cdot \sqrt{\dfrac{81}{49}}$. Therefore,

$$-\sqrt{\dfrac{81}{49}} = -1 \cdot \sqrt{\dfrac{81}{49}} = -1 \cdot \left(\dfrac{9}{7}\right) = -\dfrac{9}{7}$$

You Try 2

Find each square root.

a) $-\sqrt{144}$ b) $\sqrt{\dfrac{25}{36}}$ c) $-\sqrt{\dfrac{1}{64}}$

Example 3

Find $\sqrt{-9}$.

Solution

Recall that to find $\sqrt{-9}$ you can ask yourself, "What number do I *square* to get -9?" or "What number multiplied by itself equals -9?" *There is no such real number* since $3^2 = 9$ and $(-3)^2 = 9$. Therefore, $\sqrt{-9}$ is not a real number.

You Try 3

Find $\sqrt{-36}$.

Let's review what we know about a square root and the radicand and add a third fact.

1) If the radicand is a perfect square, the *square* root is a *rational* number.

Example: $\sqrt{16} = 4$ 16 is a perfect square.

$$\sqrt{\frac{100}{49}} = \frac{10}{7}$$ $\frac{100}{49}$ is a perfect square.

2) If the radicand is a negative number, the square root is *not* a real number.

Example: $\sqrt{-25}$ is *not* a real number.

3) If the radicand is positive and *not* a perfect square, then the square root is an *irrational* number.

Example: $\sqrt{13}$ is irrational.   13 is not a perfect square.

The square root of such a number is a real number that is a nonrepeating, nonterminating decimal. It is important to be able to approximate such square roots because sometimes it is necessary to estimate their places on a number line or on a Cartesian coordinate system when graphing.

For the purposes of graphing, approximating a radical to the nearest tenth is sufficient. A calculator with a $\sqrt{}$ key will give a better approximation of the radical.

2. Approximate the Square Root of a Whole Number

Example 4

Approximate $\sqrt{13}$ to the nearest tenth and plot it on a number line.

Solution

What is the largest perfect square that is *less than* 13?   **9**

What is the smallest perfect square that is *greater than* 13?   **16**

Since 13 is between 9 and 16 ($9 < 13 < 16$), it is true that $\sqrt{13}$ is between $\sqrt{9}$ and $\sqrt{16}$.

$$(\sqrt{9} < \sqrt{13} < \sqrt{16})$$
$$\sqrt{9} = 3$$
$$\sqrt{13} = ?$$
$$\sqrt{16} = 4$$

$\sqrt{13}$ must be between 3 and 4. Numerically, 13 is closer to 16 than it is to 9. So, $\sqrt{13}$ will be closer to $\sqrt{16}$ than to $\sqrt{9}$. Check to see if 3.6 is a good approximation of $\sqrt{13}$. ($\approx$ means approximately equal to.)

$$\text{If } \sqrt{13} \approx 3.6, \text{ then } (3.6)^2 \approx 13$$
$$(3.6)^2 = (3.6) \cdot (3.6) = 12.96$$

Is 3.7 a better approximation of $\sqrt{13}$?

$$\text{If } \sqrt{13} \approx 3.7, \text{ then } (3.7)^2 \approx 13$$
$$(3.7)^2 = (3.7) \cdot (3.7) = 13.69$$

3.6 is a better approximation of $\sqrt{13}$.

$$\sqrt{13} \approx 3.6$$

A calculator evaluates $\sqrt{13}$ as 3.6055513. Remember that this is only an approximation.

You Try 4

Approximate $\sqrt{29}$ to the nearest tenth and plot it on a number line.

3. Find the Higher Roots of Rational Numbers

We saw in Example 2a) that $\sqrt{100} = 10$ since $(10)^2 = 100$. Finding a $\sqrt{}$ is the *opposite* of squaring a number. Similarly, we can find higher roots of numbers like $\sqrt[3]{a}$ (read as "the cube root of a"), $\sqrt[4]{a}$ (read as "the fourth root of a"), $\sqrt[5]{a}$ (the fifth root of a), etc.

Example 5

Find each root.
a) $\sqrt[3]{125}$ b) $\sqrt[4]{81}$ c) $\sqrt[5]{32}$

Solution

a) To find $\sqrt[3]{125}$ (read as "the cube root of 125") ask yourself, "What number do I *cube* to get 125?" That number is 5.

$$\sqrt[3]{125} = 5 \text{ since } 5^3 = 125$$

Finding the cube root of a number is the *opposite* of cubing a number.

b) To find $\sqrt[4]{81}$ (read as "the fourth root of 81") ask yourself, "What number do I raise to the *fourth power* to get 81?" That number is 3.

$$\sqrt[4]{81} = 3 \text{ since } 3^4 = 81$$

Finding the fourth root of a number is the *opposite* of raising a number to the fourth power.

c)   To find $\sqrt[5]{32}$ (read as "the fifth root of 32") ask yourself, "What number do I raise to the *fifth power* to get 32?" That number is 2.

$$\sqrt[5]{32} = 2 \text{ since } 2^5 = 32$$

Finding the fifth root of a number is the *opposite* of raising a number to the fifth power.

You Try 5

Find each root.

a) $\sqrt[4]{16}$ b) $\sqrt[3]{27}$

We can use a general notation for writing roots of numbers.

> The $\sqrt[n]{a}$ is read as "the *nth* root of *a*." If $\sqrt[n]{a} = b$, then $b^n = a$.
>
> *n* is the **index** of the radical.

> When finding square roots we do not write $\sqrt[2]{a}$. The square root of *a* is written as $\sqrt{a}$, and the index is understood to be 2.

In Section 1.2 we first presented the powers of numbers that students are expected to know. ($2^2 = 4$, $2^3 = 8$, etc.) Use of these powers was first necessary in the study of the rules of exponents in Chapter 2. Knowing these powers is necessary for finding roots as well, so the student can refer to p. 21 to review this list of powers.

While it is true that the square root of a negative number is not a real number, sometimes it *is* possible to find the *higher* root of a negative number.

Example 6

Find each root, if possible.

a) $\sqrt[3]{-64}$ b) $\sqrt[5]{-32}$ c) $-\sqrt[4]{16}$ d) $\sqrt[4]{-16}$

Solution

a)   To find $\sqrt[3]{-64}$ ask yourself, "What number do I *cube* to get -64?" That number is -4.

$$\sqrt[3]{-64} = -4 \text{ since } (-4)^3 = -64$$

b) To find $\sqrt[5]{-32}$ ask yourself, "What number do I raise to the *fifth power* to get -32?" That number is -2.

$$\sqrt[5]{-32} = -2 \text{ since } (-2)^5 = -32$$

c) $-\sqrt[4]{16}$ means $-1 \cdot \sqrt[4]{16}$. Therefore,

$$-\sqrt[4]{16} = -1 \cdot \sqrt[4]{16} = -1 \cdot 2 = -2$$

d) To find $\sqrt[4]{-16}$ ask yourself, "What number do I raise to the *fourth power* to get -16?" *There is no such real number* since $2^4 = 16$ and $(-2)^4 = 16$.

$$\sqrt[4]{-16} \text{ is not a real number.}$$

We can summarize what we have seen in example 6 as follows:

1) The *odd root* of a negative number is a negative number.

2) The *even root* of a negative number is not a real number.

You Try 6

Find each root, if possible.

a) $\sqrt[6]{-64}$ b) $\sqrt[3]{-125}$ c) $-\sqrt[4]{81}$

Answers to You Try Exercises

1) 8 and -8 2) a) -12 b) $\dfrac{5}{6}$ c) $-\dfrac{1}{8}$ 3) not a real number

4) 5.4 $\begin{array}{c} \sqrt{29} \\ \leftarrow\!\!+\!\!+\!\!+\!\!+\!\!+\!\!+\!\!+\!\!\bullet\!\!+\!\!\rightarrow \\ \ \ 0\ \ 1\ \ 2\ \ 3\ \ 4\ \ 5\ \ 6 \end{array}$ 5) a) 2 b) 3

6) a) not a real number   b) -5 c) -3

10.1 Exercises

Boost your grade at mathzone.com!   MathZone

> Practice Problems   > NetTutor > Self-Test   > e-Professors   > Videos

Objectives 1 and 3

Decide if each statement is true or false. If it is false, explain why.

1) $\sqrt{121} = 11$ and -11 2) $\sqrt{81} = 9$

3) The cube root of a negative number is a negative number.

4) The square root of a negative number is a negative number.

5) The even root of a negative number is a negative number.

6) The odd root of a negative number is a negative number.

Objective 1

Find all square roots of each number.

7) 49   8) 144

9) 1   10) 81

11) 400   12) 900

13) 2500   14) 4900

15) $\dfrac{4}{9}$ 16) $\dfrac{36}{25}$

17) $\dfrac{1}{81}$ 18) $\dfrac{1}{16}$

Find each square root, if possible.

19) $\sqrt{49}$ 20) $\sqrt{144}$

21) $\sqrt{1}$ 22) $\sqrt{81}$

23) $\sqrt{169}$ 24) $\sqrt{36}$

25) $\sqrt{-4}$ 26) $\sqrt{-100}$

27) $\sqrt{\dfrac{81}{25}}$ 28) $\sqrt{\dfrac{16}{169}}$

29) $\sqrt{\dfrac{49}{64}}$ 30) $\sqrt{\dfrac{121}{4}}$

 31) $-\sqrt{36}$ 32) $-\sqrt{64}$

33) $-\sqrt{\dfrac{1}{121}}$ 34) $-\sqrt{\dfrac{1}{100}}$

Objective 2

Approximate each square root to the nearest tenth and plot it on a number line.

 35) $\sqrt{11}$ 36) $\sqrt{2}$

37) $\sqrt{46}$ 38) $\sqrt{22}$

39) $\sqrt{17}$ 40) $\sqrt{69}$

41) $\sqrt{5}$ 42) $\sqrt{35}$

43) $\sqrt{61}$ 44) $\sqrt{8}$

Objectives 1 and 3

45) Explain how to find $\sqrt[3]{64}$.

46) Explain how to find $\sqrt[4]{16}$.

47) Does $\sqrt[4]{-81} = -3$? Why or why not?

48) Does $\sqrt[3]{-8} = -2$? Why or why not?

Find each root, if possible.

49) $\sqrt[3]{8}$ 50) $\sqrt[3]{1}$

51) $\sqrt[3]{125}$ 52) $\sqrt[3]{27}$

53) $\sqrt[3]{-1}$ 54) $\sqrt[3]{-8}$

55) $\sqrt[4]{81}$ 56) $\sqrt[4]{16}$

57) $\sqrt[4]{-1}$ 58) $\sqrt[4]{-81}$

59) $-\sqrt[4]{16}$ 60) $-\sqrt[4]{1}$

 61) $\sqrt[5]{-32}$ 62) $-\sqrt[6]{64}$

63) $-\sqrt[3]{-27}$ 64) $-\sqrt[3]{-1000}$

65) $\sqrt[6]{-64}$ 66) $\sqrt[4]{-16}$

67) $\sqrt[3]{\dfrac{8}{125}}$ 68) $\sqrt[4]{\dfrac{81}{16}}$

 69) $\sqrt{60 - 11}$ 70) $\sqrt{100 + 21}$

71) $\sqrt[3]{100 + 25}$ 72) $\sqrt[3]{9 - 36}$

73) $\sqrt{1 - 9}$ 74) $\sqrt{25 - 36}$

75) $\sqrt{5^2 + 12^2}$ 76) $\sqrt{3^2 + 4^2}$

Section 10.2 Rational Exponents

Objectives

1. Define and Evaluate Expressions of the Form $a^{1/n}$

2. Define and Evaluate Expressions of the Form $a^{m/n}$

3. Define and Evaluate Expressions of the Form $a^{-m/n}$

4. Combine the Rules of Exponents to Simplify Expressions

5. Convert a Radical Expression to Exponential Form and Simplify

1. Define and Evaluate Expressions of the Form $a^{1/n}$

In this section, we will explain the relationship between radicals and rational exponents (fractional exponents). Sometimes, converting between these two forms makes it easier to simplify expressions.

If a is a nonnegative number and n is a positive integer greater than 1, then

$$a^{1/n} = \sqrt[n]{a}$$

(The denominator of the fractional exponent is the index of the radical.)

Example 1

Write in radical form and evaluate.

a) $8^{1/3}$ b) $49^{1/2}$ c) $81^{1/4}$

Solution

a) The denominator of the fractional exponent is the index of the radical. Therefore,

$$8^{1/3} = \sqrt[3]{8} = 2$$

b) The denominator in the exponent of $49^{1/2}$ is 2, so the index on the radical is 2, meaning *square* root.

$$49^{1/2} = \sqrt{49} = 7$$

c) $81^{1/4} = \sqrt[4]{81} = 3$

 **You Try 1**

Write in radical form and evaluate.

a) $16^{1/4}$ b) $121^{1/2}$

2. Define and Evaluate Expressions of the Form $a^{m/n}$

We can add another relationship between rational exponents and radicals.

> If a is a nonnegative number and m and n are integers such that $n > 1$,
>
> $$a^{m/n} = (a^{1/n})^m = (\sqrt[n]{a})^m$$
>
> (The denominator of the fractional exponent is the index of the radical, and the numerator is the power to which we raise the radical expression.)

Alternatively, we can think of $a^{m/n}$ in this way:

$$a^{m/n} = (a^m)^{1/n} = \sqrt[n]{a^m}$$

Example 2

Write in radical form and evaluate.

a) $25^{3/2}$ b) $-64^{2/3}$

Solution

a)   According to the definition, the denominator of the fractional exponent is the index of the radical, and the numerator is the power to which we raise the radical expression.

$$
\begin{aligned}
25^{3/2} &= (25^{1/2})^3 && \text{Use the definition to rewrite the exponent.} \\
&= (\sqrt{25})^3 && \text{Rewrite as a radical.} \\
&= 5^3 && \sqrt{25} = 5 \\
&= 125
\end{aligned}
$$

b)   To evaluate $-64^{2/3}$, *first* evaluate $64^{2/3}$, *then* take the negative of that result.

$$
\begin{aligned}
-64^{2/3} = -(64^{2/3}) &= -(64^{1/3})^2 && \text{Use the definition to rewrite the exponent.} \\
&= -(\sqrt[3]{64})^2 && \text{Rewrite as a radical.} \\
&= -(4)^2 && \sqrt[3]{64} = 4 \\
&= -16
\end{aligned}
$$

You Try 2

Write in radical form and evaluate.

a) $32^{2/5}$ b) $-100^{3/2}$

3. Define and Evaluate Expressions of the Form $a^{-m/n}$

Recall the definition of a negative exponent from Section 2.3a.

> If n is any integer and $a \neq 0$, then
>
> $$a^{-n} = \left(\frac{1}{a}\right)^n = \frac{1}{a^n}.$$

That is, to rewrite the expression with a *positive* exponent, take the reciprocal of the base. For example,

$$2^{-4} = \left(\frac{1}{2}\right)^4 = \frac{1}{16}$$

We can extend this to rational exponents.

If a is a positive number and m and n are integers such that $n > 1$,

$$a^{-m/n} = \left(\frac{1}{a}\right)^{m/n} = \frac{1}{a^{m/n}}.$$

(To rewrite the expression with a *positive* exponent, take the reciprocal of the base.)

Example 3

Rewrite with a positive exponent and evaluate.

a) $36^{-1/2}$ b) $32^{-2/5}$ c) $\left(\dfrac{125}{64}\right)^{-2/3}$

Solution

a) To write $36^{-1/2}$ with a positive exponent, take the reciprocal of the base.

$$36^{-1/2} = \left(\frac{1}{36}\right)^{1/2}$$ The reciprocal of 36 is $\frac{1}{36}$.

$$= \sqrt{\frac{1}{36}}$$ The denominator of the fractional exponent is the index of the radical.

$$= \frac{1}{6}$$

b)

$$32^{-2/5} = \left(\frac{1}{32}\right)^{2/5}$$ The reciprocal of 32 is $\frac{1}{32}$.

$$= \left(\sqrt[5]{\frac{1}{32}}\right)^2$$ The denominator of the fractional exponent is the index of the radical.

$$= \left(\frac{1}{2}\right)^2$$ $\sqrt[5]{\frac{1}{32}} = \frac{1}{2}$

$$= \frac{1}{4}$$

c)

$$\left(\frac{125}{64}\right)^{-2/3} = \left(\frac{64}{125}\right)^{2/3}$$ The reciprocal of $\frac{125}{64}$ is $\frac{64}{125}$.

$$= \left(\sqrt[3]{\frac{64}{125}}\right)^2$$ The denominator of the fractional exponent is the index of the radical.

$$= \left(\frac{4}{5}\right)^2$$ $\sqrt[3]{\frac{64}{125}} = \frac{4}{5}$

$$= \frac{16}{25}$$

BE CAREFUL The negative exponent does not make the expression negative!

 You Try 3

Rewrite with a positive exponent and evaluate.

a) $144^{-1/2}$ b) $16^{-3/4}$ c) $\left(\dfrac{8}{27}\right)^{-2/3}$

We can combine the rules presented in this section with the rules of exponents we learned in Chapter 2 to simplify expressions containing numbers or variables.

4. Combine the Rules of Exponents to Simplify Expressions

Example 4

Simplify completely. The answer should contain only positive exponents.

a) $(6^{1/5})^2$ b) $25^{3/4} \cdot 25^{-1/4}$ c) $\dfrac{8^{2/9}}{8^{11/9}}$

Solution

a) $(6^{1/5})^2 = 6^{2/5}$ Multiply exponents.

b) $25^{3/4} \cdot 25^{-1/4} = 25^{\frac{3}{4}+\left(-\frac{1}{4}\right)}$ Add exponents.

$\qquad\qquad = 25^{2/4}$ Add $\dfrac{3}{4}+\left(-\dfrac{1}{4}\right)$.

$\qquad\qquad = 25^{1/2}$ Reduce $\dfrac{2}{4}$.

$\qquad\qquad = 5$ Evaluate.

c) $\dfrac{8^{2/9}}{8^{11/9}} = 8^{\frac{2}{9}-\frac{11}{9}}$ Subtract exponents.

$\qquad = 8^{-9/9}$ Subtract $\dfrac{2}{9}-\dfrac{11}{9}$.

$\qquad = 8^{-1}$ Reduce $-\dfrac{9}{9}$.

$\qquad = \left(\dfrac{1}{8}\right)^1$ Rewrite with a positive exponent.

$\qquad = \dfrac{1}{8}$

 You Try 4

Simplify completely. The answer should contain only positive exponents.

a) $49^{3/8} \cdot 49^{1/8}$ b) $(16^{1/12})^3$ c) $\dfrac{7^{2/5}}{7^{4/5}}$

Example 5

Simplify completely. Assume the variables represent positive real numbers. The answer should contain only positive exponents.

a) $r^{1/8} \cdot r^{3/8}$

b) $\left(\dfrac{x^{2/3}}{y^{1/4}}\right)^6$

c) $\dfrac{n^{-5/6} \cdot n^{1/3}}{n^{-1/6}}$

Solution

a) $r^{1/8} \cdot r^{3/8} = r^{\frac{1}{8}+\frac{3}{8}}$ Add exponents.

$\qquad\qquad\quad = r^{4/8}$

$\qquad\qquad\quad = r^{1/2}$ Reduce $\dfrac{4}{8}$.

b) $\left(\dfrac{x^{2/3}}{y^{1/4}}\right)^6 = \dfrac{x^{\frac{2}{3}\cdot 6}}{y^{\frac{1}{4}\cdot 6}}$ Multiply exponents.

$\qquad\qquad\quad = \dfrac{x^4}{y^{3/2}}$ Multiply and reduce.

c) $\dfrac{n^{-5/6} \cdot n^{1/3}}{n^{-1/6}} = \dfrac{n^{-\frac{5}{6}+\frac{1}{3}}}{n^{-1/6}}$ Add exponents.

$\qquad\qquad\quad = \dfrac{n^{-\frac{5}{6}+\frac{2}{6}}}{n^{-1/6}}$ Get a common denominator.

$\qquad\qquad\quad = \dfrac{n^{-3/6}}{n^{-1/6}}$ Add exponents.

$\qquad\qquad\quad = n^{-\frac{3}{6}-(-\frac{1}{6})}$ Subtract exponents.

$\qquad\qquad\quad = n^{-2/6}$ $-\dfrac{3}{6}-\left(-\dfrac{1}{6}\right) = -\dfrac{3}{6}+\dfrac{1}{6} = -\dfrac{2}{6}$

$\qquad\qquad\quad = n^{-1/3}$ Reduce.

$\qquad\qquad\quad = \dfrac{1}{n^{1/3}}$ Rewrite with a positive exponent.

You Try 5

Simplify completely. Assume the variables represent positive real numbers. The answer should contain only positive exponents.

a) $(a^3 b^{1/5})^{10}$

b) $\dfrac{t^{3/10}}{t^{7/10}}$

c) $\dfrac{s^{3/4}}{s^{1/2} \cdot s^{-5/4}}$

5. Convert a Radical Expression to Exponential Form and Simplify

Some radicals can be simplified by first putting them into rational exponent form and then converting them back to radicals.

Example 6

Rewrite each radical in exponential form, then simplify. Write the answer in simplest (or radical) form. Assume the variable represents a nonnegative real number.

a) $\sqrt[8]{9^4}$

b) $\sqrt[6]{s^4}$

c) $(\sqrt{7})^2$

d) $\sqrt[3]{5^3}$

Solution

a) Since the index of the radical is the denominator of the rational exponent and the power is the numerator, we can write

$$\sqrt[8]{9^4} = 9^{4/8} \qquad \text{Write with a rational exponent.}$$
$$= 9^{1/2} \qquad \text{Reduce } \frac{4}{8}.$$
$$= 3 \qquad \text{Evaluate.}$$

b)
$$\sqrt[6]{s^4} = s^{4/6} \qquad \text{Write with a rational exponent.}$$
$$= s^{2/3} \qquad \text{Reduce } \frac{4}{6}.$$
$$= \sqrt[3]{s^2} \qquad \text{Write in radical form.}$$

$\sqrt[6]{s^4}$ is not in simplest form because the 4 and the 6 contain a common factor of 2. $\sqrt[3]{s^2}$ *is* in simplest form because 2 and 3 do not have any common factors besides 1.

c)
$$(\sqrt{7})^2 = (7^{1/2})^2 \qquad \text{Write with a rational exponent.}$$
$$= 7^{\frac{1}{2} \cdot 2} \qquad \text{Multiply exponents.}$$
$$= 7^1$$
$$= 7$$

d)
$$(\sqrt[3]{5^3}) = (5^3)^{1/3} \qquad \text{Write with a rational exponent.}$$
$$= 5^{3 \cdot \frac{1}{3}} \qquad \text{Multiply exponents.}$$
$$= 5^1$$
$$= 5$$

You Try 6

Rewrite each radical in exponential form, then simplify. Write the answer in simplest (or radical) form. Assume the variable represents a nonnegative real number.

a) $\sqrt[6]{125^2}$ b) $\sqrt[10]{p^4}$ c) $(\sqrt{6})^2$ d) $\sqrt[3]{9^3}$

Example 6 c) and d) illustrate an important and useful property of radicals.

> If a is a nonnegative real number and n is a positive integer greater than 1, then
>
> $$(\sqrt[n]{a})^n = a \text{ and } \sqrt[n]{a^n} = a.$$

We can show, in general, why this is true by writing the expressions with rational exponents.

$$(\sqrt[n]{a})^n = a^{n/n} = a^1 = a$$
$$\sqrt[n]{a^n} = a^{n/n} = a^1 = a$$

> $\sqrt[n]{a^n} = |a|$ if a is negative and n is even.

Example 7

Simplify. Assume the variable represents a nonnegative real number.

a) $(\sqrt[3]{4})^3$ b) $\sqrt[4]{5^4}$ c) $(\sqrt{x})^2$

Solution

a) Since the index of the radical is the same as the exponent, $(\sqrt[3]{4})^3 = 4$.
b) $\sqrt[4]{5^4} = 5$
c) $(\sqrt{x})^2 = x$

 You Try 7

Simplify. Assume the variable represents a nonnegative real number.

a) $(\sqrt{10})^2$ b) $\sqrt[3]{7^3}$ c) $\sqrt{t^2}$

 Using Technology

Now that we have discussed the relationship between rational exponents and radicals, let's learn how to evaluate these expressions on a graphing calculator.

Example I   Evaluate $9^{1/2}$.

Using the properties in this chapter, we get

$$9^{1/2} = \sqrt{9} = 3$$

We will look at two ways to correctly enter the expression into the calculator.

1.   The $\boxed{\wedge}$ key indicates that an exponent follows. To enter and evaluate $9^{1/2}$, press this sequence of keys: $\boxed{9}\boxed{\wedge}\boxed{(}\boxed{1}\boxed{\div}\boxed{2}\boxed{)}\boxed{\text{ENTER}}$. The result is 3.

   You *must* use parentheses! If you forget the parentheses and enter $\boxed{9}\boxed{\wedge}\boxed{1}\boxed{\div}\boxed{2}$, the result will be 4.5 because the calculator is following the order of operations.

2.   To enter $9^{1/2}$ as $\sqrt{9}$, press this sequence of keys: $\boxed{2^{nd}}\boxed{x^2}\boxed{9}\boxed{)}\boxed{\text{ENTER}}$. Again, the result is 3. Notice that the calculator inserted the left parenthesis automatically. You need to close the parentheses so that the calculator knows where the square root ends.

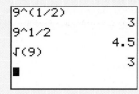

Example 2   Evaluate $8^{1/3}$.

Using the properties in this chapter, we get

$$8^{1/3} = \sqrt[3]{8} = 2$$

1.   To evaluate $8^{1/3}$ on the calculator using the $\boxed{\wedge}$ key, press $\boxed{8}\boxed{\wedge}\boxed{(}\boxed{1}\boxed{\div}\boxed{3}\boxed{)}\boxed{\text{ENTER}}$. The result is 2.

2. Roots higher than square roots can be found by pressing the ⌈MATH⌉ key. To enter $\sqrt[3]{8}$, press ⌈MATH⌉ and use the down arrow to highlight 4: $\sqrt[3]{}$ (. Press ⌈ENTER⌉. Then press ⌈8⌉ ⌈)⌉ ⌈ENTER⌉, and you will see that the result is 2.

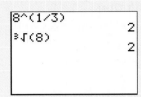

Evaluate each expression using the properties learned in this chapter, then verify your answer using the graphing calculator.

1. $49^{1/2}$
2. $144^{1/2}$
3. $81^{1/4}$
4. $32^{1/5}$

5. $125^{1/3}$
6. $64^{1/3}$
7. $16^{3/2}$
8. $27^{4/3}$

Answers to You Try Exercises

1) a) 2 b) 11   2) a) 4 b) -1000 3) a) $\dfrac{1}{12}$ b) $\dfrac{1}{8}$ c) $\dfrac{9}{4}$ 4) a) 7 b) 2 c) $\dfrac{1}{7^{2/5}}$

5) a) $a^{30}b^2$ b) $\dfrac{1}{t^{2/5}}$ c) $s^{3/2}$ 6) a) 5 b) $\sqrt[5]{p^2}$ c) 6 d) 9   7) a) 10 b) 7 c) t

Answers to Technology Exercises

1. 7   2. 12   3. 3   4. 2   5. 5   6. 4   7. 64   8. 81

Objective 1

1) Explain how to write $25^{1/2}$ in radical form.

2) Explain how to write $1^{1/3}$ in radical form.

Write in radical form and evaluate.

3) $9^{1/2}$

4) $64^{1/2}$

5) $1000^{1/3}$

6) $27^{1/3}$

7) $32^{1/5}$

8) $81^{1/4}$

9) $-125^{1/3}$

10) $-64^{1/6}$

11) $\left(\dfrac{4}{121}\right)^{1/2}$

12) $\left(\dfrac{4}{9}\right)^{1/2}$

 13) $\left(\dfrac{125}{64}\right)^{1/3}$

14) $\left(\dfrac{16}{81}\right)^{1/4}$

15) $-\left(\dfrac{36}{169}\right)^{1/2}$

16) $-\left(\dfrac{100}{9}\right)^{1/2}$

Objective 2

17) Explain how to write $16^{3/4}$ in radical form.

18) Explain how to write $100^{3/2}$ in radical form.

Write in radical form and evaluate.

19) $8^{4/3}$

20) $81^{3/4}$

21) $125^{2/3}$

22) $32^{3/5}$

23) $64^{5/6}$

24) $1000^{2/3}$

25) $-27^{4/3}$

26) $-36^{3/2}$

27) $\left(\dfrac{16}{81}\right)^{3/4}$

28) $\left(\dfrac{64}{125}\right)^{2/3}$

29) $-\left(\dfrac{1000}{27}\right)^{2/3}$

30) $-\left(\dfrac{8}{27}\right)^{4/3}$

Objective 3

Decide whether each statement is true or false. Explain your answer.

31) $81^{-1/2} = -9$

32) $\left(\dfrac{1}{100}\right)^{-3/2} = \left(\dfrac{1}{100}\right)^{2/3}$

Rewrite with a positive exponent and evaluate.

33) $49^{-1/2}$

34) $100^{-1/2}$

35) $1000^{-1/3}$

36) $27^{-1/3}$

37) $\left(\dfrac{1}{81}\right)^{-1/4}$

38) $\left(\dfrac{1}{32}\right)^{-1/5}$

39) $-\left(\dfrac{1}{64}\right)^{-1/3}$

40) $-\left(\dfrac{1}{125}\right)^{-1/3}$

41) $64^{-5/6}$

42) $81^{-3/4}$

43) $125^{-2/3}$

44) $64^{-2/3}$

45) $\left(\dfrac{25}{4}\right)^{-3/2}$

46) $\left(\dfrac{9}{100}\right)^{-3/2}$

47) $\left(\dfrac{64}{125}\right)^{-2/3}$

48) $\left(\dfrac{81}{16}\right)^{-3/4}$

Objective 4

Simplify completely. The answer should contain only positive exponents.

49) $2^{2/3} \cdot 2^{7/3}$

50) $5^{3/4} \cdot 5^{5/4}$

51) $(9^{1/4})^2$

52) $(7^{2/3})^3$

53) $8^{7/5} \cdot 8^{-3/5}$

54) $6^{-4/3} \cdot 6^{5/3}$

55) $\dfrac{4^{10/3}}{4^{4/3}}$

56) $\dfrac{2^{23/4}}{2^{3/4}}$

57) $\dfrac{5^{3/2}}{5^{9/2}}$

58) $\dfrac{32^{3/5}}{32^{7/5}}$

59) $\dfrac{6^{-1}}{6^{1/2} \cdot 6^{-5/2}}$

60) $\dfrac{10^{-5/2} \cdot 10^{3/2}}{10^{-4}}$

61) $\dfrac{7^{4/9} \cdot 7^{1/9}}{7^{2/9}}$

62) $\dfrac{4^{2/5}}{4^{6/5} \cdot 4^{3/5}}$

Simplify completely. Assume the variables represent positive real numbers. The answer should contain only positive exponents.

63) $k^{7/4} \cdot k^{3/4}$

64) $z^{1/6} \cdot z^{5/6}$

65) $j^{-3/5} \cdot j^{3/10}$

66) $h^{1/6} \cdot h^{-3/4}$

67) $(-9v^{5/8})(8v^{3/4})$

68) $(-3x^{-1/3})(8x^{4/9})$

69) $\dfrac{a^{5/9}}{a^{4/9}}$

70) $\dfrac{x^{1/6}}{x^{5/6}}$

71) $\dfrac{20c^{-2/3}}{72c^{5/6}}$

72) $\dfrac{48w^{3/10}}{10w^{2/5}}$

73) $(q^{4/5})^{10}$

74) $(t^{3/8})^{16}$

75) $(x^{-2/9})^3$

76) $(n^{-2/7})^3$

77) $(z^{1/5})^{2/3}$

78) $(r^{4/3})^{5/2}$

79) $(81u^{8/3}v^4)^{3/4}$

80) $(64x^6y^{12/5})^{5/6}$

81) $(32r^{1/3}s^{4/9})^{3/5}$

82) $(125a^9b^{1/4})^{2/3}$

83) $\left(\dfrac{f^{6/7}}{27g^{-5/3}}\right)^{1/3}$

84) $\left(\dfrac{16c^{-8}}{b^{-11/3}}\right)^{3/4}$

85) $\left(\dfrac{x^{-5/3}}{w^{3/2}}\right)^{-6}$

86) $\left(\dfrac{t^{-3/2}}{u^{1/4}}\right)^{-4}$

87) $\dfrac{y^{1/2} \cdot y^{-1/3}}{y^{5/6}}$

88) $\dfrac{t^5}{t^{1/2} \cdot t^{3/4}}$

89) $\left(\dfrac{a^4b^3}{32a^{-2}b^4}\right)^{2/5}$

90) $\left(\dfrac{16c^{-8}d^3}{c^4d^5}\right)^{3/2}$

Objective 5

Rewrite each radical in exponential form, then simplify. Write the answer in simplest (or radical) form. Assume all variables represent nonnegative real numbers.

91) $\sqrt[6]{49^3}$

92) $\sqrt[9]{8^3}$

93) $\sqrt[6]{1000^2}$

94) $\sqrt[4]{81^2}$

95) $(\sqrt{5})^2$

96) $(\sqrt{14})^2$

97) $\sqrt{3^2}$

98) $\sqrt{7^2}$

99) $(\sqrt[3]{12})^3$

100) $(\sqrt[3]{10})^3$

101) $(\sqrt[4]{15})^4$

102) $(\sqrt[4]{9})^4$

103) $\sqrt[3]{x^{12}}$

104) $\sqrt[4]{t^8}$

105) $\sqrt[6]{k^2}$

106) $\sqrt[9]{w^6}$

107) $\sqrt[4]{z^2}$

108) $\sqrt[8]{m^4}$

109) $\sqrt{d^4}$

110) $\sqrt{s^6}$

Section 10.3 Simplifying Expressions Containing Square Roots

Objectives

1. Multiply Square Roots
2. Simplify the Square Root of a Whole Number
3. Use the Quotient Rule for Square Roots
4. Simplify Square Root Expressions Containing Variables with Even Exponents
5. Simplify Square Root Expressions Containing Variables with Odd Exponents
6. Multiply, Divide, and Simplify Expressions Containing Square Roots

In this section, we will introduce rules for finding the product and quotient of square roots as well as for simplifying expressions containing square roots.

1. Multiply Square Roots

We begin with an example that we can evaluate using the order of operations:

Evaluate $\sqrt{4} \cdot \sqrt{9}$.

$$\sqrt{4} \cdot \sqrt{9} = 2 \cdot 3$$
$$= 6$$

Since both roots are *square* roots, however, we can also evaluate the product this way:

$$\sqrt{4} \cdot \sqrt{9} = \sqrt{4 \cdot 9}$$
$$= \sqrt{36}$$
$$= 6$$

We obtain the same result. This leads us to the product rule for multiplying expressions containing square roots.

Definition	**Product Rule for Square Roots**
	Let a and b be nonnegative real numbers. Then, $$\sqrt{a} \cdot \sqrt{b} = \sqrt{a \cdot b}$$

Example 1

Multiply. Assume the variable represents a nonnegative real number.

a) $\sqrt{5} \cdot \sqrt{2}$ b) $\sqrt{3} \cdot \sqrt{x}$

Solution

a) $\sqrt{5} \cdot \sqrt{2} = \sqrt{5 \cdot 2} = \sqrt{10}$
b) $\sqrt{3} \cdot \sqrt{x} = \sqrt{3 \cdot x} = \sqrt{3x}$

> **BE CAREFUL**
> We can multiply radicals this way *only if* the indices are the same. We will see later how to multiply radicals with different indices such as $\sqrt{5} \cdot \sqrt[3]{t}$.

 You Try 1

Multiply. Assume the variable represents a nonnegative real number.

a) $\sqrt{6} \cdot \sqrt{5}$ b) $\sqrt{10} \cdot \sqrt{r}$

2. Simplify the Square Root of a Whole Number

Knowing how to simplify radicals is very important in the study of algebra. We begin by discussing how to simplify expressions containing square roots.

How do we know when a square root is simplified?

When Is a Square Root Simplified?

An expression containing a square root is simplified when all of the following conditions are met:

1) The radicand does not contain any factors (other than 1) that are perfect squares.

2) The radicand does not contain any fractions.

3) There are no radicals in the denominator of a fraction.

Note: Condition 1) implies that the radical cannot contain variables with exponents greater than or equal to 2, the index of the square root.

We will discuss higher roots in Section 10.4.

To simplify expressions containing square roots we reverse the process of multiplying. That is, we use the product rule that says

$$\sqrt{a \cdot b} = \sqrt{a} \cdot \sqrt{b}$$

where a or b are perfect squares.

Example 2

Simplify completely.

a) $\sqrt{18}$ b) $\sqrt{500}$ c) $\sqrt{21}$ d) $\sqrt{48}$

Solution

a) The radical $\sqrt{18}$ is not in simplest form since 18 contains a factor (other than 1) that is a perfect square. Think of two numbers that multiply to 18 so that at least one of the numbers is a perfect square.

$$18 = 9 \cdot 2$$

(While it is true that $18 = 6 \cdot 3$, neither 6 nor 3 is a perfect square.) Rewrite $\sqrt{18}$:

$$\begin{aligned} \sqrt{18} &= \sqrt{9 \cdot 2} && \text{9 is a perfect square.} \\ &= \sqrt{9} \cdot \sqrt{2} && \text{Product rule} \\ &= 3\sqrt{2} && \sqrt{9} = 3 \end{aligned}$$

$3\sqrt{2}$ is completely simplified because 2 does not have any factors that are perfect squares.

b) Does 500 have a factor that is a perfect square? Yes! $500 = 100 \cdot 5$. To simplify $\sqrt{500}$, rewrite it as

$$\begin{aligned} \sqrt{500} &= \sqrt{100 \cdot 5} && \text{100 is a perfect square.} \\ &= \sqrt{100} \cdot \sqrt{5} && \text{Product rule} \\ &= 10\sqrt{5} && \sqrt{100} = 10 \end{aligned}$$

$10\sqrt{5}$ is completely simplified because 5 does not have any factors that are perfect squares.

c) To simplify $\sqrt{21}$, think of two numbers that multiply to 21 so that at least one of the numbers is a perfect square.

$$21 = 3 \cdot 7 \qquad \text{Neither 3 nor 7 is a perfect square.}$$
$$21 = 1 \cdot 21 \qquad \text{While 1 is a perfect square, this will not help us simplify } \sqrt{21}.$$

$\sqrt{21}$ is in simplest form.

d) There are two good ways to simplify $\sqrt{48}$. We will look at both of them.

i) Two numbers that multiply to 48 are 16 and 3 with 16 being a perfect square. We can write

$$\sqrt{48} = \sqrt{16 \cdot 3} \qquad \text{16 is a perfect square.}$$
$$= \sqrt{16} \cdot \sqrt{3} \qquad \text{Product rule}$$
$$= 4\sqrt{3} \qquad \sqrt{16} = 4$$

$4\sqrt{3}$ is completely simplified because 3 does not have any factors that are perfect squares.

ii) We can also think of 48 as $4 \cdot 12$ since 4 is a perfect square. We can write

$$\sqrt{48} = \sqrt{4 \cdot 12} \qquad \text{4 is a perfect square.}$$
$$= \sqrt{4} \cdot \sqrt{12} \qquad \text{Product rule}$$
$$= 2\sqrt{12} \qquad \sqrt{4} = 2$$

Therefore, $\sqrt{48} = 2\sqrt{12}$. Is $\sqrt{12}$ in simplest form? *No, because 12 = 4 · 3 and 4 is a perfect square.* We must continue to simplify.

$$\sqrt{48} = 2\sqrt{12}$$
$$= 2\sqrt{4 \cdot 3} \qquad \text{4 is a perfect square.}$$
$$= 2\sqrt{4} \cdot \sqrt{3} \qquad \text{Product rule}$$
$$= 2 \cdot 2 \cdot \sqrt{3} \qquad \sqrt{4} = 2$$
$$= 4\sqrt{3} \qquad \text{Multiply 2 · 2.}$$

$4\sqrt{3}$ is completely simplified because 3 does not have any factors that are perfect squares.

You can see in Example 2d) that using either $\sqrt{48} = \sqrt{16 \cdot 3}$ or $\sqrt{48} = \sqrt{4 \cdot 12}$ leads us to the same result. Furthermore, this example illustrates that a radical is not always *completely* simplified after just one iteration of the simplification process. It is necessary to always examine the radical to determine whether or not it can be simplified more.

After simplifying a radical, look at the result and ask yourself, "*Is the radical in simplest form?*" If it is not, simplify again. Asking yourself this question will help you to be sure that the radical *is* completely simplified.

You Try 2

Simplify completely.

a) $\sqrt{28}$ b) $\sqrt{75}$ c) $\sqrt{72}$

3. Use the Quotient Rule for Square Roots

We can simplify a quotient like $\dfrac{\sqrt{36}}{\sqrt{9}}$ in one of two ways:

$$\frac{\sqrt{36}}{\sqrt{9}} = \frac{6}{3} = 2 \quad \text{or} \quad \frac{\sqrt{36}}{\sqrt{9}} = \sqrt{\frac{36}{9}}$$
$$= \sqrt{4}$$
$$= 2$$

Either way we obtain the same result. This leads us to the quotient rule for dividing expressions containing square roots.

> **Definition**
>
> **Quotient Rule for Square Roots**
> Let a and b be nonnegative real numbers such that $b \neq 0$. Then,
>
> $$\sqrt{\frac{a}{b}} = \frac{\sqrt{a}}{\sqrt{b}}$$

Example 3

Simplify completely.

a) $\sqrt{\dfrac{9}{49}}$ b) $\sqrt{\dfrac{200}{2}}$ c) $\dfrac{\sqrt{72}}{\sqrt{6}}$ d) $\sqrt{\dfrac{5}{81}}$

Solution

a) Since 9 and 49 are each perfect squares, simplify $\sqrt{\dfrac{9}{49}}$ by finding the square root of each separately.

$$\sqrt{\frac{9}{49}} = \frac{\sqrt{9}}{\sqrt{49}} \qquad \text{Quotient rule}$$
$$= \frac{3}{7} \qquad \sqrt{9} = 3 \text{ and } \sqrt{49} = 7$$

b) Neither 200 nor 2 is a perfect square, but if we simplify the fraction $\dfrac{200}{2}$ we get 100, which *is* a perfect square.

$$\sqrt{\frac{200}{2}} = \sqrt{100} \qquad \text{Simplify } \frac{200}{2}.$$
$$= 10$$

c)   We can simplify $\dfrac{\sqrt{72}}{\sqrt{6}}$ using two different methods.

  i)   Begin by applying the quotient rule to obtain a fraction under *one* radical and simplify the fraction.

$$\dfrac{\sqrt{72}}{\sqrt{6}} = \sqrt{\dfrac{72}{6}} \qquad \text{Quotient rule}$$
$$= \sqrt{12} \qquad \text{Simplify } \dfrac{72}{6}.$$
$$= \sqrt{4 \cdot 3} \qquad \text{4 is a perfect square.}$$
$$= \sqrt{4} \cdot \sqrt{3} \qquad \text{Product rule}$$
$$= 2\sqrt{3} \qquad \sqrt{4} = 2$$

  ii)   We can apply the product rule to rewrite $\sqrt{72}$ then simplify the fraction.

$$\dfrac{\sqrt{72}}{\sqrt{6}} = \dfrac{\sqrt{6} \cdot \sqrt{12}}{\sqrt{6}} \qquad \text{Product rule}$$
$$= \dfrac{\overset{1}{\cancel{\sqrt{6}}} \cdot \sqrt{12}}{\cancel{\sqrt{6}}} \qquad \text{Divide out the common factor.}$$
$$= \sqrt{12} \qquad \text{Simplify.}$$
$$= \sqrt{4 \cdot 3} \qquad \text{4 is a perfect square.}$$
$$= \sqrt{4} \cdot \sqrt{3} \qquad \text{Product rule}$$
$$= 2\sqrt{3} \qquad \sqrt{4} = 2$$

Either method will produce the same result.

d)   The fraction $\dfrac{5}{81}$ does not reduce and 81 *is* a perfect square. Begin by applying the quotient rule.

$$\sqrt{\dfrac{5}{81}} = \dfrac{\sqrt{5}}{\sqrt{81}} \qquad \text{Quotient rule}$$
$$= \dfrac{\sqrt{5}}{9} \qquad \sqrt{81} = 9$$

You Try 3

Simplify completely.

a) $\sqrt{\dfrac{100}{169}}$ b) $\sqrt{\dfrac{27}{3}}$ c) $\dfrac{\sqrt{250}}{\sqrt{5}}$ d) $\sqrt{\dfrac{11}{36}}$

4. Simplify Square Root Expressions Containing Variables with Even Exponents

Recall that one condition that an expression containing a square root must meet to be simplified is that the radicand cannot contain any factors (other than 1) that are perfect squares.

For a radicand containing variables this means that the radical is simplified if the power on the variable is less than 2. For example, $\sqrt{r^6}$ is not in simplified form. If r represents a nonnegative real number, then we can use rational exponents to simplify $\sqrt{r^6}$.

$$\sqrt{r^6} = (r^6)^{1/2} = r^{6 \cdot \frac{1}{2}} = r^{6/2} = r^3$$

Multiplying $6 \cdot \dfrac{1}{2}$ *is the same as* dividing 6 by 2. We can generalize this with the following statement:

If a is a nonnegative real number and m is an integer, then

$$\sqrt{a^m} = a^{m/2}$$

We can combine this property with the product and quotient rules to simplify radical expressions.

Example 4

Simplify completely. Assume all variables represent positive real numbers.

a) $\sqrt{z^2}$ b) $\sqrt{49t^2}$ c) $\sqrt{18b^{14}}$ d) $\sqrt{\dfrac{32}{n^{20}}}$

Solution

a) $\sqrt{z^2} = z^{2/2} = z^1 = z$

b) $\sqrt{49t^2} = \sqrt{49} \cdot \sqrt{t^2}$ Product rule
$\phantom{\sqrt{49t^2}} = 7 \cdot t^{2/2}$ Evaluate.
$\phantom{\sqrt{49t^2}} = 7t$ Simplify.

c) $\sqrt{18b^{14}} = \sqrt{18} \cdot \sqrt{b^{14}}$ Product rule
$\phantom{\sqrt{18b^{14}}} = \sqrt{9} \cdot \sqrt{2} \cdot b^{14/2}$ 9 is a perfect square.
$\phantom{\sqrt{18b^{14}}} = 3\sqrt{2} \cdot b^7$ Simplify.
$\phantom{\sqrt{18b^{14}}} = 3b^7\sqrt{2}$ Rewrite using the commutative property.

d) $\sqrt{\dfrac{32}{n^{20}}} = \dfrac{\sqrt{32}}{\sqrt{n^{20}}}$ Quotient rule

$\phantom{\sqrt{\dfrac{32}{n^{20}}}} = \dfrac{\sqrt{16} \cdot \sqrt{2}}{n^{20/2}}$ 16 is a perfect square.

$\phantom{\sqrt{\dfrac{32}{n^{20}}}} = \dfrac{4\sqrt{2}}{n^{10}}$ Simplify.

You Try 4

Simplify completely. Assume all variables represent positive real numbers.

a) $\sqrt{y^{10}}$ b) $\sqrt{144p^{16}}$ c) $\sqrt{\dfrac{45}{w^4}}$

5. Simplify Square Root Expressions Containing Variables with Odd Exponents

How do we simplify an expression containing a square root if the power under the square root does not divide evenly by 2? We can use the product rule for radicals and fractional exponents to help us understand how to simplify such expressions.

Example 5

Simplify completely. Assume all variables represent nonnegative real numbers.

a) $\sqrt{x^7}$ b) $\sqrt{c^{11}}$ c) $\sqrt{p^{17}}$

Solution

a) To simplify $\sqrt{x^7}$, write x^7 as the product of two factors so that the exponent of one of the factors is the *largest* number less than 7 that is divisible by 2 (the index of the radical).

$$\begin{aligned} \sqrt{x^7} &= \sqrt{x^6 \cdot x^1} & \text{6 is the largest number less than 7 that is divisible by 2.} \\ &= \sqrt{x^6} \cdot \sqrt{x} & \text{Product rule} \\ &= x^{6/2} \cdot \sqrt{x} & \text{Use a fractional exponent to simplify.} \\ &= x^3 \sqrt{x} & 6 \div 2 = 3 \end{aligned}$$

b) To simplify $\sqrt{c^{11}}$, write c^{11} as the product of two factors so that the exponent of one of the factors is the *largest* number less than 11 that is divisible by 2 (the index of the radical).

$$\begin{aligned} \sqrt{c^{11}} &= \sqrt{c^{10} \cdot c^1} & \text{10 is the largest number less than 11 that is divisible by 2.} \\ &= \sqrt{c^{10}} \cdot \sqrt{c} & \text{Product rule} \\ &= c^{10/2} \cdot \sqrt{c} & \text{Use a fractional exponent to simplify.} \\ &= c^5 \sqrt{c} & 10 \div 2 = 5 \end{aligned}$$

c) To simplify $\sqrt{p^{17}}$, write p^{17} as the product of two factors so that the exponent of one of the factors is the *largest* number less than 17 that is divisible by 2 (the index of the radical).

$$\begin{aligned} \sqrt{p^{17}} &= \sqrt{p^{16} \cdot p^1} & \text{16 is the largest number less than 17 that is divisible by 2.} \\ &= \sqrt{p^{16}} \cdot \sqrt{p} & \text{Product rule} \\ &= p^{16/2} \cdot \sqrt{p} & \text{Use a fractional exponent to simplify.} \\ &= p^8 \sqrt{p} & 16 \div 2 = 8 \end{aligned}$$

You Try 5

Simplify completely. Assume all variables represent nonnegative real numbers.

a) $\sqrt{m^5}$ b) $\sqrt{z^{19}}$

We used the product rule to simplify each radical above. During the simplification, however, we always divided an exponent by 2. This idea of division gives us another way to simplify radical expressions. Once again let's look at the radicals and their simplified forms in Example 5 to see how we can simplify radical expressions using division.

$$\sqrt{x^7} = x^3\sqrt{x^1} = x^3\sqrt{x} \qquad \sqrt{c^{11}} = c^5\sqrt{c^1} = c^5\sqrt{c} \qquad \sqrt{p^{17}} = p^8\sqrt{p^1} = p^8\sqrt{p}$$

Index of radical → 2) 7 3 → Quotient
-6
1 → Remainder

Index of radical → 2) 11 5 → Quotient
-10
1 → Remainder

Index of radical → 2) 17 8 → Quotient
-16
1 → Remainder

To simplify a radical expression containing variables:

1) Divide the original exponent in the radicand by the index of the radical.

2) The exponent on the variable *outside* of the radical will be the *quotient* of the division problem.

3) The exponent on the variable *inside* of the radical will be the *remainder* of the division problem.

Example 6

Simplify completely. Assume all variables represent nonnegative real numbers.

a) $\sqrt{t^9}$ b) $\sqrt{16b^5}$ c) $\sqrt{45y^{21}}$

Solution

a) To simplify $\sqrt{t^9}$, divide: 2) 9   4 → Quotient
-8
1 → Remainder

$$\sqrt{t^9} = t^4\sqrt{t^1} = t^4\sqrt{t}$$

b) $\sqrt{16b^5} = \sqrt{16} \cdot \sqrt{b^5}$ Product rule
$\quad = 4 \cdot b^2\sqrt{b^1}$ $5 \div 2$ gives a quotient of 2 and a remainder of 1.
$\quad = 4b^2\sqrt{b}$

c) $\sqrt{45y^{21}} = \sqrt{45} \cdot \sqrt{y^{21}}$ Product rule
$\quad = \sqrt{9} \cdot \sqrt{5} \cdot y^{10}\sqrt{y^1}$
$\qquad\qquad\uparrow\qquad\quad\uparrow$
$\qquad$ Product rule $21 \div 2$ gives a quotient of 10 and a remainder of 1.

$\quad = 3\sqrt{5} \cdot y^{10}\sqrt{y}$ $\sqrt{9} = 3$
$\quad = 3y^{10} \cdot \sqrt{5} \cdot \sqrt{y}$ Use the commutative property to rewrite the expression.
$\quad = 3y^{10}\sqrt{5y}$ Use the product rule to write the expression with one radical.

You Try 6

Simplify completely. Assume all variables represent nonnegative real numbers.

a) $\sqrt{m^{13}}$ b) $\sqrt{100v^7}$ c) $\sqrt{32a^3}$

If a radical contains more than one variable, apply the product or quotient rule.

Example 7

Simplify completely. Assume all variables represent positive real numbers.

a) $\sqrt{8a^{15}b^3}$ b) $\sqrt{\dfrac{5r^{27}}{s^8}}$

Solution

a) $\sqrt{8a^{15}b^3} = \sqrt{8} \cdot \sqrt{a^{15}} \cdot \sqrt{b^3}$

$= \sqrt{4} \cdot \sqrt{2} \cdot a^7\sqrt{a^1} \cdot b^1\sqrt{b^1}$

Product rule   15 ÷ 2 gives a quotient   3 ÷ 2 gives a quotient
of 7 and a remainder of 1.   of 1 and a remainder of 1.

$= 2\sqrt{2} \cdot a^7\sqrt{a} \cdot b\sqrt{b}$ $\sqrt{4} = 2$

$= 2a^7b \cdot \sqrt{a} \cdot \sqrt{b}$ Use the commutative property to rewrite
the expression.

$= 2a^7b\sqrt{ab}$ Use the product rule to write the expression
with one radical.

b) $\sqrt{\dfrac{5r^{27}}{s^8}} = \dfrac{\sqrt{5r^{27}}}{\sqrt{s^8}}$ Quotient rule

$= \dfrac{\sqrt{5} \cdot \sqrt{r^{27}}}{s^4}$ → Product rule
→ 8 ÷ 2 = 4

$= \dfrac{\sqrt{5} \cdot r^{13}\sqrt{r^1}}{s^4}$ 27 ÷ 2 gives a quotient of 13 and a
remainder of 1.

$= \dfrac{r^{13} \cdot \sqrt{5} \cdot \sqrt{r}}{s^4}$ Use the commutative property to
rewrite the expression.

$= \dfrac{r^{13}\sqrt{5r}}{s^4}$ Use the product rule to write the
expression with one radical.

 You Try 7

Simplify completely. Assume all variables represent positive real numbers.

a) $\sqrt{c^5d^{12}}$ b) $\sqrt{27x^{10}y^9}$ c) $\sqrt{\dfrac{40u^{13}}{v^{20}}}$

6. Multiply, Divide, and Simplify Expressions Containing Square Roots

Earlier in this section we discussed multiplying and dividing radicals such as

$$\sqrt{5} \cdot \sqrt{2} \quad \text{and} \quad \sqrt{\dfrac{200}{2}}$$

Next we will look at some examples of multiplying and dividing radical expressions that also contain variables. Remember to always look at the result and ask yourself, "*Is the radical in simplest form?*" If it is not, simplify completely.

| Example 8 |

Perform the indicated operation and simplify completely. Assume all variables represent positive real numbers.

a) $\sqrt{6t} \cdot \sqrt{3t}$ b) $\sqrt{2a^3b} \cdot \sqrt{8a^2b^5}$ c) $\dfrac{\sqrt{20x^5}}{\sqrt{5x}}$

Solution

a) $\sqrt{6t} \cdot \sqrt{3t} = \sqrt{6t \cdot 3t}$ Product rule

$\phantom{\sqrt{6t} \cdot \sqrt{3t}} = \sqrt{18t^2}$

$\phantom{\sqrt{6t} \cdot \sqrt{3t}} = \sqrt{18} \cdot \sqrt{t^2}$ Product rule

$\phantom{\sqrt{6t} \cdot \sqrt{3t}} = \sqrt{9 \cdot 2} \cdot t$ 9 is a perfect square; $\sqrt{t^2} = t$.

$\phantom{\sqrt{6t} \cdot \sqrt{3t}} = \sqrt{9} \cdot \sqrt{2} \cdot t$ Product rule

$\phantom{\sqrt{6t} \cdot \sqrt{3t}} = 3\sqrt{2} \cdot t$ $\sqrt{9} = 3$

$\phantom{\sqrt{6t} \cdot \sqrt{3t}} = 3t\sqrt{2}$ Use the commutative property to rewrite the expression.

b) $\sqrt{2a^3b} \cdot \sqrt{8a^2b^5}$

There are two good methods for multiplying these radicals.

i) Multiply the radicands to obtain one radical.

$\sqrt{2a^3b} \cdot \sqrt{8a^2b^5} = \sqrt{2a^3b \cdot 8a^2b^5}$ Product rule

$\phantom{\sqrt{2a^3b} \cdot \sqrt{8a^2b^5}} = \sqrt{16a^5b^6}$ Multiply.

Is the radical in simplest form? *No.*

$\phantom{\sqrt{2a^3b} \cdot} = \sqrt{16} \cdot \sqrt{a^5} \cdot \sqrt{b^6}$ Product rule

$\phantom{\sqrt{2a^3b} \cdot} = 4 \cdot a^2\sqrt{a} \cdot b^3$ Evaluate.

$\phantom{\sqrt{2a^3b} \cdot} = 4a^2b^3\sqrt{a}$ Commutative property

ii) Simplify each radical, then multiply.

$\sqrt{2a^3b} = \sqrt{2} \cdot \sqrt{a^3} \cdot \sqrt{b}$ $\sqrt{8a^2b^5} = \sqrt{8} \cdot \sqrt{a^2} \cdot \sqrt{b^5}$

$\phantom{\sqrt{2a^3b}} = \sqrt{2} \cdot a\sqrt{a} \cdot \sqrt{b}$ $\phantom{\sqrt{8a^2b^5}} = 2\sqrt{2} \cdot a \cdot b^2\sqrt{b}$

$\phantom{\sqrt{2a^3b}} = a\sqrt{2ab}$ $\phantom{\sqrt{8a^2b^5}} = 2ab^2\sqrt{2b}$

Then, $\sqrt{2a^3b} \cdot \sqrt{8a^2b^5} = a\sqrt{2ab} \cdot 2ab^2\sqrt{2b}$

$\phantom{\sqrt{2a^3b} \cdot \sqrt{8a^2b^5}} = a \cdot 2ab^2 \cdot \sqrt{2ab} \cdot \sqrt{2b}$ Commutative property

$\phantom{\sqrt{2a^3b} \cdot \sqrt{8a^2b^5}} = 2a^2b^2\sqrt{4ab^2}$ Multiply.

$\phantom{\sqrt{2a^3b} \cdot \sqrt{8a^2b^5}} = 2a^2b^2 \cdot \sqrt{4} \cdot \sqrt{a} \cdot \sqrt{b^2}$ Product rule

$\phantom{\sqrt{2a^3b} \cdot \sqrt{8a^2b^5}} = 2a^2b^2 \cdot 2 \cdot \sqrt{a} \cdot b$ Evaluate.

$\phantom{\sqrt{2a^3b} \cdot \sqrt{8a^2b^5}} = 2a^2b^2 \cdot 2 \cdot b \cdot \sqrt{a}$ Commutative property

$\phantom{\sqrt{2a^3b} \cdot \sqrt{8a^2b^5}} = 4a^2b^3\sqrt{a}$ Multiply.

Both methods give the same result.

c) $\dfrac{\sqrt{20x^5}}{\sqrt{5x}}$ can be simplified two different ways.

   i)   Begin by using the quotient rule to obtain one radical and simplify the radicand.

$$
\begin{aligned}
\frac{\sqrt{20x^5}}{\sqrt{5x}} &= \sqrt{\frac{20x^5}{5x}} && \text{Quotient rule} \\
&= \sqrt{4x^4} && \text{Simplify.} \\
&= \sqrt{4} \cdot \sqrt{x^4} && \text{Product rule} \\
&= 2x^2 && \text{Simplify.}
\end{aligned}
$$

   ii)   Begin by simplifying the radical in the numerator.

$$
\begin{aligned}
\frac{\sqrt{20x^5}}{\sqrt{5x}} &= \frac{\sqrt{20} \cdot \sqrt{x^5}}{\sqrt{5x}} && \text{Product rule} \\
&= \frac{\sqrt{4} \cdot \sqrt{5} \cdot x^2\sqrt{x}}{\sqrt{5x}} && \text{Product rule; simplify } \sqrt{x^5}. \\
&= \frac{2\sqrt{5} \cdot x^2\sqrt{x}}{\sqrt{5x}} && \sqrt{4} = 2 \\
&= \frac{2x^2\sqrt{5x}}{\sqrt{5x}} && \text{Product rule} \\
&= 2x^2 && \text{Divide out the common factor.}
\end{aligned}
$$

Both methods give the same result. In this case, the second method was longer. Sometimes, however, this method *can* be more efficient.

You Try 8

Perform the indicated operation and simplify completely. Assume all variables represent positive real numbers.

a) $\sqrt{2n^3} \cdot \sqrt{6n}$ b) $\sqrt{15cd^5} \cdot \sqrt{3c^2d}$ c) $\dfrac{\sqrt{128k^9}}{\sqrt{2k}}$

Answers to You Try Exercises

1) a) $\sqrt{30}$ b) $\sqrt{10r}$ 2) a) $2\sqrt{7}$ b) $5\sqrt{3}$ c) $6\sqrt{2}$ 3) a) $\dfrac{10}{13}$ b) 3   c) $5\sqrt{2}$ d) $\dfrac{\sqrt{11}}{6}$

4) a) y^5 b) $12p^8$ c) $\dfrac{3\sqrt{5}}{w^2}$ 5) a) $m^2\sqrt{m}$ b) $z^9\sqrt{z}$ 6) a) $m^6\sqrt{m}$ b) $10v^3\sqrt{v}$ c) $4a\sqrt{2a}$

7) a) $c^2d^6\sqrt{c}$ b) $3x^5y^4\sqrt{3y}$ c) $\dfrac{2u^6\sqrt{10u}}{v^{10}}$ 8) a) $2n^2\sqrt{3}$ b) $3cd^3\sqrt{5c}$ c) $8k^4$

10.3 Exercises

Boost your grade at mathzone.com! MathZone > **Practice Problems** > **NetTutor** > **Self-Test** > **e-Professors** > **Videos**

Objective 1

Unless otherwise stated, assume all variables represent nonnegative real numbers.

Multiply and simplify.

1) $\sqrt{3} \cdot \sqrt{7}$
2) $\sqrt{11} \cdot \sqrt{5}$
3) $\sqrt{10} \cdot \sqrt{3}$
4) $\sqrt{7} \cdot \sqrt{2}$
5) $\sqrt{6} \cdot \sqrt{y}$
6) $\sqrt{5} \cdot \sqrt{p}$

Objective 2

Label each statement as true or false. Give a reason for your answer.

7) $\sqrt{20}$ is in simplest form.
8) $\sqrt{35}$ is in simplest form.
9) $\sqrt{42}$ is in simplest form.
10) $\sqrt{63}$ is in simplest form.

Objective 2

Simplify completely. If the radical is already simplified, then say so.

11) $\sqrt{20}$
12) $\sqrt{12}$
13) $\sqrt{54}$
14) $\sqrt{63}$
15) $\sqrt{33}$
16) $\sqrt{15}$
17) $\sqrt{8}$
18) $\sqrt{24}$
19) $\sqrt{80}$
20) $\sqrt{108}$
21) $\sqrt{98}$
22) $\sqrt{96}$
23) $\sqrt{38}$
24) $\sqrt{46}$
25) $\sqrt{400}$
26) $\sqrt{900}$

Objective 3

Simplify completely.

27) $\sqrt{\dfrac{144}{25}}$
28) $\sqrt{\dfrac{16}{81}}$
29) $\dfrac{\sqrt{4}}{\sqrt{49}}$
30) $\dfrac{\sqrt{64}}{\sqrt{121}}$
31) $\sqrt{\dfrac{8}{2}}$
32) $\sqrt{\dfrac{75}{3}}$
33) $\dfrac{\sqrt{54}}{\sqrt{6}}$
34) $\dfrac{\sqrt{48}}{\sqrt{3}}$

35) $\sqrt{\dfrac{60}{5}}$
36) $\sqrt{\dfrac{40}{5}}$
37) $\dfrac{\sqrt{120}}{\sqrt{6}}$
38) $\dfrac{\sqrt{54}}{\sqrt{3}}$
39) $\dfrac{\sqrt{30}}{\sqrt{2}}$
40) $\dfrac{\sqrt{35}}{\sqrt{5}}$
41) $\sqrt{\dfrac{6}{49}}$
42) $\sqrt{\dfrac{2}{81}}$
43) $\sqrt{\dfrac{45}{16}}$
44) $\sqrt{\dfrac{60}{49}}$

Objective 4

Simplify completely. Assume all variables represent positive real numbers.

45) $\sqrt{x^8}$
46) $\sqrt{q^6}$
47) $\sqrt{m^{10}}$
48) $\sqrt{a^2}$
49) $\sqrt{w^{14}}$
50) $\sqrt{t^{16}}$
51) $\sqrt{100c^2}$
52) $\sqrt{9z^8}$
53) $\sqrt{64k^6}$
54) $\sqrt{25p^{20}}$
55) $\sqrt{28r^4}$
56) $\sqrt{27z^{12}}$
57) $\sqrt{300q^{22}}$
58) $\sqrt{50n^4}$
59) $\sqrt{\dfrac{81}{c^6}}$
60) $\sqrt{\dfrac{h^2}{169}}$
61) $\dfrac{\sqrt{40}}{\sqrt{t^8}}$
62) $\dfrac{\sqrt{18}}{\sqrt{m^{30}}}$
63) $\sqrt{\dfrac{75}{y^{12}}}$
64) $\sqrt{\dfrac{44}{w^2}}$

Objective 5

Simplify completely. Assume all variables represent positive real numbers.

65) $\sqrt{a^5}$
66) $\sqrt{c^7}$
67) $\sqrt{g^{13}}$
68) $\sqrt{k^{15}}$
69) $\sqrt{b^{25}}$
70) $\sqrt{h^{31}}$
71) $\sqrt{72x^3}$
72) $\sqrt{100a^5}$
73) $\sqrt{13q^7}$
74) $\sqrt{20c^9}$
75) $\sqrt{75t^{11}}$
76) $\sqrt{45p^{17}}$
77) $\sqrt{c^8d^2}$
78) $\sqrt{r^4s^{12}}$

79) $\sqrt{a^4 b^3}$

80) $\sqrt{x^2 y^9}$

81) $\sqrt{u^5 v^7}$

82) $\sqrt{f^3 g^9}$

83) $\sqrt{36 m^9 n^4}$

84) $\sqrt{4 t^6 u^5}$

85) $\sqrt{44 x^{12} y^5}$

86) $\sqrt{63 c^7 d^4}$

87) $\sqrt{32 t^5 u^7}$

88) $\sqrt{125 k^3 l^9}$

89) $\sqrt{\dfrac{a^7}{81 b^6}}$

90) $\sqrt{\dfrac{x^5}{49 y^6}}$

91) $\sqrt{\dfrac{3 r^9}{s^2}}$

92) $\sqrt{\dfrac{17 h^{11}}{k^8}}$

97) $\sqrt{w} \cdot \sqrt{w^5}$

98) $\sqrt{d^3} \cdot \sqrt{d^{11}}$

99) $\sqrt{n^3} \cdot \sqrt{n^4}$

100) $\sqrt{a^{10}} \cdot \sqrt{a^3}$

101) $\sqrt{2k} \cdot \sqrt{8 k^5}$

102) $\sqrt{5 z^9} \cdot \sqrt{5 z^3}$

103) $\sqrt{6 x^4 y^3} \cdot \sqrt{2 x^5 y^2}$

104) $\sqrt{5 a^6 b^5} \cdot \sqrt{10 a b^4}$

105) $\sqrt{8 c^9 d^2} \cdot \sqrt{5 c d^7}$

106) $\sqrt{6 t^3 u^3} \cdot \sqrt{3 t^7 u^4}$

video 107) $\dfrac{\sqrt{18 k^{11}}}{\sqrt{2 k^3}}$

108) $\dfrac{\sqrt{48 m^{15}}}{\sqrt{3 m^9}}$

109) $\dfrac{\sqrt{120 h^8}}{\sqrt{3 h^2}}$

110) $\dfrac{\sqrt{72 c^{10}}}{\sqrt{6 c^2}}$

111) $\dfrac{\sqrt{50 a^{16}}}{\sqrt{5 a^7}}$

112) $\dfrac{\sqrt{21 z^{18}}}{\sqrt{3 z^{13}}}$

Objective 6

Perform the indicated operation and simplify. Assume all variables represent positive real numbers.

93) $\sqrt{5} \cdot \sqrt{10}$

94) $\sqrt{8} \cdot \sqrt{6}$

95) $\sqrt{21} \cdot \sqrt{3}$

96) $\sqrt{2} \cdot \sqrt{14}$

Section 10.4 Simplifying Expressions Containing Higher Roots

Objectives

1. Multiply Higher Roots

2. Simplify Higher Roots of Integers

3. Use the Quotient Rule for Higher Roots

4. Simplify Radical Expressions Containing Variables

5. Multiply and Divide Radicals with Different Indices

In Section 10.1 we first discussed finding higher roots like $\sqrt[4]{16} = 2$ and $\sqrt[3]{-27} = -3$. In this section, we will extend what we learned about multiplying, dividing, and simplifying *square* roots to doing the same with higher roots.

1. Multiply Higher Roots

> **Definition**
>
> **Product Rule for Higher Roots**
> If a and b are real numbers such that the roots exist, then
> $$\sqrt[n]{a} \cdot \sqrt[n]{b} = \sqrt[n]{a \cdot b}$$

This rule enables us to multiply and simplify radicals with any index in å way that is similar to multiplying and simplifying square roots.

Example 1

Multiply. Assume the variable represents a nonnegative real number.

a) $\sqrt[3]{2} \cdot \sqrt[3]{7}$ b) $\sqrt[4]{t} \cdot \sqrt[4]{10}$

Solution

a) $\sqrt[3]{2} \cdot \sqrt[3]{7} = \sqrt[3]{2 \cdot 7} = \sqrt[3]{14}$

b) $\sqrt[4]{t} \cdot \sqrt[4]{10} = \sqrt[4]{t \cdot 10} = \sqrt[4]{10t}$

You Try 1

Multiply. Assume the variable represents a nonnegative real number.

a) $\sqrt[4]{6} \cdot \sqrt[4]{5}$ b) $\sqrt[5]{8} \cdot \sqrt[5]{k^2}$

BE CAREFUL

Remember that we can apply the product rule in this way *only* if the indices of the radicals are the same. Later in this section we will discuss how to multiply radicals with different indices.

2. Simplify Higher Roots of Integers

In Section 10.3 we said that one condition that must be met for a *square root* to be simplified is that its radicand does not contain any *perfect squares*. Likewise, to be considered in simplest form, the radicand of an expression containing a **cube root** cannot contain any **perfect cubes,** the radicand of an expression containing a **fourth root** cannot contain any **perfect fourth powers,** and so on.

Next we list the conditions that determine when a radical with *any* index is in simplest form.

When Is a Radical Simplified?

Let P be an expression and let n be an integer greater than 1. Then $\sqrt[n]{P}$ is completely simplified when all of the following conditions are met:

1)   The radicand does not contain any factors (other than 1) which are perfect nth powers.

2)   The radicand does not contain any fractions.

3)   There are no radicals in the denominator of a fraction.

> Condition 1) implies that the radical cannot contain variables with exponents greater than or equal to n, the index of the radical.

To simplify radicals with any index, we reverse the process of multiplying radicals where a or b is an nth power.

$$\sqrt[n]{a \cdot b} = \sqrt[n]{a} \cdot \sqrt[n]{b}$$

Remember, to be certain that a radical is simplified completely always look at the radical carefully and ask yourself, "*Is the radical in simplest form?*"

| Example 2 |

Simplify completely.

a) $\sqrt[3]{250}$ b) $\sqrt[4]{48}$

Solution

a)   We will look at two methods for simplifying $\sqrt[3]{250}$.

  i)   Since we must simplify the *cube* root of 250, think of two numbers that multiply to 250 so that at least one of the numbers is a *perfect cube*.

$$250 = 125 \cdot 2$$

Rewrite $\sqrt[3]{250}$:

$$\begin{aligned}\sqrt[3]{250} &= \sqrt[3]{125 \cdot 2} && \text{125 is a perfect cube.}\\ &= \sqrt[3]{125} \cdot \sqrt[3]{2} && \text{Product rule}\\ &= 5\sqrt[3]{2} && \sqrt[3]{125} = 5\end{aligned}$$

Is $5\sqrt[3]{2}$ in simplest form? Yes, because 2 does not have any factors that are perfect cubes.

  ii)   Begin by using a factor tree to find the prime factorization of 250.

$$250 = 2 \cdot 5^3$$

Rewrite $\sqrt[3]{250}$ using the product rule.

$$\begin{aligned}\sqrt[3]{250} &= \sqrt[3]{2 \cdot 5^3} && 2 \cdot 5^3 \text{ is the prime factorization of 250.}\\ &= \sqrt[3]{2} \cdot \sqrt[3]{5^3} && \text{Product rule}\\ &= \sqrt[3]{2} \cdot 5 && \sqrt[3]{5^3} = 5\\ &= 5\sqrt[3]{2} && \text{Commutative property}\end{aligned}$$

We obtain the same result using either method.

b)   We will use two methods for simplifying $\sqrt[4]{48}$.

  i)   Since we must simplify the *fourth* root of 48, think of two numbers that multiply to 48 so that at least one of the numbers is a *perfect fourth power*.

$$48 = 16 \cdot 3$$

Rewrite $\sqrt[4]{48}$:

$$\begin{aligned}\sqrt[4]{48} &= \sqrt[4]{16 \cdot 3} && \text{16 is a perfect fourth power.}\\ &= \sqrt[4]{16} \cdot \sqrt[4]{3} && \text{Product rule}\\ &= 2\sqrt[4]{3} && \sqrt[4]{16} = 2\end{aligned}$$

Is $2\sqrt[4]{3}$ in simplest form? Yes, because 3 does not have any factors that are perfect fourth powers.

  ii)   Begin by using a factor tree to find the prime factorization of 48.

$$48 = 2^4 \cdot 3$$

Rewrite $\sqrt[4]{48}$ using the product rule.

$$\begin{aligned}\sqrt[4]{48} &= \sqrt[4]{2^4 \cdot 3} && 2^4 \cdot 3 \text{ is the prime factorization of 48.}\\ &= \sqrt[4]{2^4} \cdot \sqrt[4]{3} && \text{Product rule}\\ &= 2\sqrt[4]{3} && \sqrt[4]{2^4} = 2\end{aligned}$$

Once again, both methods give us the same result.

You Try 2

Simplify completely.

a) $\sqrt[3]{40}$ b) $\sqrt[5]{63}$

3. Use the Quotient Rule for Higher Roots

Definition

Quotient Rule for Higher Roots

If a and b are real numbers such that the roots exist and $b \neq 0$, then

$$\sqrt[n]{\frac{a}{b}} = \frac{\sqrt[n]{a}}{\sqrt[n]{b}}$$

We apply the quotient rule when working with nth-roots the same way we apply it when working with square roots.

Example 3

Simplify completely.

a) $\sqrt[3]{-\dfrac{81}{3}}$ b) $\dfrac{\sqrt[3]{96}}{\sqrt[3]{2}}$

Solution

a)   We can think of $-\dfrac{81}{3}$ as $\dfrac{-81}{3}$ or $\dfrac{81}{-3}$. Let's think of it as $\dfrac{-81}{3}$.

Neither -81 nor 3 is a perfect cube. But if we simplify $\dfrac{-81}{3}$ we get -27, which *is* a perfect cube.

$$\sqrt[3]{-\frac{81}{3}} = \sqrt[3]{-27} \qquad \text{Simplify } -\frac{81}{3}.$$
$$= -3$$

b)   Let's begin by applying the quotient rule to obtain a fraction under *one* radical and simplify the fraction.

$$\frac{\sqrt[3]{96}}{\sqrt[3]{2}} = \sqrt[3]{\frac{96}{2}} \qquad \text{Quotient rule}$$
$$= \sqrt[3]{48} \qquad \text{Simplify } \frac{96}{2}.$$
$$= \sqrt[3]{8 \cdot 6} \qquad \text{8 is a perfect cube.}$$
$$= \sqrt[3]{8} \cdot \sqrt[3]{6} \qquad \text{Product rule}$$
$$= 2\sqrt[3]{6} \qquad \sqrt[3]{8} = 2$$

Is $2\sqrt[3]{6}$ in simplest form? Yes, because 6 does not have any factors that are perfect cubes.

658 Chapter 10 Radicals and Rational Exponents

You Try 3

Simplify completely.

a) $\sqrt[4]{\dfrac{1}{81}}$ b) $\dfrac{\sqrt[3]{162}}{\sqrt[3]{3}}$

4. Simplify Radical Expressions Containing Variables

In Section 10.2 we discussed the relationship between radical notation and fractional exponents. Recall that

> If a is a nonnegative number and m and n are integers such that $n > 1$, then
>
> $$\sqrt[n]{a^m} = a^{m/n}.$$
>
> That is, the index of the radical becomes the denominator of the fractional exponent, and the power in the radicand becomes the numerator of the fractional exponent.

This is the principle we use to simplify radicals with indices greater than 2.

Example 4

Simplify completely. Assume all variables represent positive real numbers.

a) $\sqrt[3]{y^{15}}$ b) $\sqrt[4]{16t^{24}u^8}$ c) $\sqrt[5]{\dfrac{c^{10}}{d^{30}}}$

Solution

a) $\sqrt[3]{y^{15}} = y^{15/3} = y^5$

b) $\sqrt[4]{16t^{24}u^8} = \sqrt[4]{16} \cdot \sqrt[4]{t^{24}} \cdot \sqrt[4]{u^8}$ Product rule
$= 2 \cdot t^{24/4} \cdot u^{8/4}$ Write with rational exponents.
$= 2t^6u^2$ Simplify exponents.

c) $\sqrt[5]{\dfrac{c^{10}}{d^{30}}} = \dfrac{\sqrt[5]{c^{10}}}{\sqrt[5]{d^{30}}}$ Quotient rule

$= \dfrac{c^{10/5}}{d^{30/5}}$ Write with rational exponents.

$= \dfrac{c^2}{d^6}$ Simplify exponents.

You Try 4

Simplify completely. Assume all variables represent positive real numbers.

a) $\sqrt[3]{a^3b^{21}}$ b) $\sqrt[4]{\dfrac{m^{12}}{16n^{20}}}$

To simplify a radical expression if the power in the radicand does not divide evenly by the index, we use the same methods we used in Section 10.3 for simplifying similar expressions with square roots. We can use the product rule or we can use the idea of quotient and remainder in a division problem.

Example 5

Simplify $\sqrt[4]{x^{23}}$ completely in two ways: 1) use the product rule and 2) divide the exponent by the index and use the quotient and remainder. Assume the variable represents a nonnegative real number.

Solution

1) Using the product rule:

   To simplify $\sqrt[4]{x^{23}}$, write x^{23} as the product of two factors so that the exponent of one of the factors is the *largest* number less than 23 that is divisible by 4 (the index).

$$\sqrt[4]{x^{23}} = \sqrt[4]{x^{20} \cdot x^3}$$ 20 is the largest number less than 23 that is divisible by 4.
$$= \sqrt[4]{x^{20}} \cdot \sqrt[4]{x}$$ Product rule
$$= x^{20/4} \cdot \sqrt[4]{x}$$ Use a fractional exponent to simplify.
$$= x^5 \sqrt[4]{x}$$ $20 \div 4 = 5$

2) Using the quotient and remainder:

   To simplify $\sqrt[4]{x^{23}}$, divide $4\overline{)\,23}$

   $$\begin{array}{r} 5 \leftarrow \text{Quotient} \\ 4\overline{)\,23} \\ -20 \\ \hline 3 \leftarrow \text{Remainder} \end{array}$$

   Recall from our work with square roots in Section 10.3 that

   i) the exponent on the variable *outside* of the radical will be the *quotient* of the division problem.

       and

   ii) the exponent on the variable *inside* of the radical will be the *remainder* of the division problem.

$$\sqrt[4]{x^{23}} = x^5 \sqrt[4]{x^3}$$

Is $x^5\sqrt[4]{x^3}$ in simplest form? Yes, because the exponent inside of the radical is less than the index.

 **You Try 5**

Simplify $\sqrt[5]{r^{32}}$ completely using both methods shown in Example 5. Assume r represents a nonnegative real number.

We can apply the product and quotient rules together with the methods above to simplify certain radical expressions.

Example 6

Completely simplify $\sqrt[3]{56a^{16}b^8}$. Assume the variables represent nonnegative real numbers.

Solution

$$\sqrt[3]{56a^{16}b^8} = \sqrt[3]{56} \cdot \sqrt[3]{a^{16}} \cdot \sqrt[3]{b^8} \qquad \text{Product rule}$$
$$= \sqrt[3]{8} \cdot \sqrt[3]{7} \cdot a^5\sqrt[3]{a^1} \cdot b^2\sqrt[3]{b^2}$$

| Product rule | $16 \div 3$ gives a quotient of 5 and a remainder of 1. | $8 \div 3$ gives a quotient of 2 and a remainder of 2. |

$$= 2\sqrt[3]{7} \cdot a^5\sqrt[3]{a} \cdot b^2\sqrt[3]{b^2} \qquad \text{Simplify } \sqrt[3]{8}.$$
$$= 2a^5b^2 \cdot \sqrt[3]{7} \cdot \sqrt[3]{a} \cdot \sqrt[3]{b^2} \qquad \text{Use the commutative property to rewrite the expression.}$$

$$= 2a^5b^2\sqrt[3]{7ab^2} \qquad \text{Product rule}$$

You Try 6

Simplify completely. Assume the variables represent positive real numbers.

a) $\sqrt[4]{48x^{15}y^{22}}$ b) $\sqrt[3]{\dfrac{r^{19}}{27s^{12}}}$

5. Multiply and Divide Radicals with Different Indices

The product and quotient rules for radicals apply only when the radicals have the same indices. To multiply or divide radicals with different indices, we first change the radical expressions to rational exponent form.

Example 7

Multiply the expressions, and write the answer in simplest radical form. Assume the variables represent positive real numbers.

$$\sqrt[3]{x^2} \cdot \sqrt{x}$$

Solution

The indices of $\sqrt[3]{x^2}$ and $\sqrt{x}$ are not the same, so we *cannot* use the product rule right now. Rewrite each radical as a fractional exponent, use the product rule for *exponents*, then convert the answer back to radical form.

$$\sqrt[3]{x^2} \cdot \sqrt{x} = x^{2/3} \cdot x^{1/2} \qquad \text{Change radicals to fractional exponents.}$$
$$= x^{4/6} \cdot x^{3/6} \qquad \text{Get a common denominator to add exponents.}$$
$$= x^{\frac{4}{6} + \frac{3}{6}} \qquad \text{Add exponents.}$$
$$= x^{7/6}$$
$$= \sqrt[6]{x^7} \qquad \text{Rewrite in radical form.}$$
$$= x\sqrt[6]{x} \qquad \text{Simplify.}$$

You Try 7

Perform the indicated operation, and write the answer in simplest radical form. Assume the variables represent positive real numbers.

a) $\sqrt[4]{y} \cdot \sqrt[6]{y}$ b) $\dfrac{\sqrt[3]{c^2}}{\sqrt{c}}$

Answers to You Try Exercises

1) a) $\sqrt[4]{30}$ b) $\sqrt[5]{8k^2}$ 2) a) $2\sqrt[3]{5}$ b) simplified   3) a) $\dfrac{1}{3}$ b) $3\sqrt[3]{2}$ 4) a) ab^7 b) $\dfrac{m^3}{2n^5}$

5) $r^6\sqrt[5]{r^2}$ 6) a) $2x^3y^5\sqrt[4]{3x^3y^2}$ b) $\dfrac{r^6\sqrt[3]{r}}{3s^4}$ 7) a) $\sqrt[12]{y^5}$ b) $\sqrt[6]{c}$

10.4 Exercises

Boost your grade at mathzone.com!

> Practice Problems
> NetTutor
> Self-Test
> e-Professors
> Videos

1) In your own words, explain the product rule for radicals.

2) In your own words, explain the quotient rule for radicals.

3) How do you know that a radical expression containing a cube root is completely simplified?

4) How do you know that a radical expression containing a fourth root is completely simplified?

Objective 1

Multiply. Assume the variable represents a nonnegative real number.

5) $\sqrt[3]{5} \cdot \sqrt[3]{4}$

6) $\sqrt[5]{6} \cdot \sqrt[5]{2}$

7) $\sqrt[5]{9} \cdot \sqrt[5]{m^2}$

8) $\sqrt[4]{11} \cdot \sqrt[4]{h^3}$

9) $\sqrt[3]{a^2} \cdot \sqrt[3]{b}$

10) $\sqrt[5]{t^2} \cdot \sqrt[5]{u^4}$

Objectives 2 and 3

Simplify completely.

11) $\sqrt[3]{24}$

12) $\sqrt[3]{48}$

13) $\sqrt[4]{64}$

14) $\sqrt[4]{32}$

15) $\sqrt[3]{54}$

16) $\sqrt[3]{88}$

17) $\sqrt[3]{2000}$

18) $\sqrt[3]{108}$

19) $\sqrt[5]{64}$

20) $\sqrt[4]{162}$

21) $\sqrt[4]{\dfrac{1}{16}}$

22) $\sqrt[3]{\dfrac{1}{125}}$

23) $\sqrt[3]{-\dfrac{54}{2}}$

24) $\sqrt[4]{\dfrac{48}{3}}$

25) $\dfrac{\sqrt[3]{48}}{\sqrt[3]{2}}$

26) $\dfrac{\sqrt[3]{500}}{\sqrt[3]{2}}$

27) $\dfrac{\sqrt[4]{240}}{\sqrt[4]{3}}$

28) $\dfrac{\sqrt[3]{8000}}{\sqrt[3]{4}}$

Objective 4

Simplify completely. Assume all variables represent positive real numbers.

29) $\sqrt[3]{d^6}$

30) $\sqrt[3]{g^9}$

31) $\sqrt[3]{n^{20}}$

32) $\sqrt[4]{t^{36}}$

33) $\sqrt[5]{x^5y^{15}}$

34) $\sqrt[6]{a^{12}b^6}$

35) $\sqrt[3]{w^{14}}$

36) $\sqrt[3]{b^{19}}$

37) $\sqrt[4]{y^9}$

38) $\sqrt[4]{m^7}$

39) $\sqrt[3]{d^5}$

40) $\sqrt[3]{c^{29}}$

41) $\sqrt[3]{u^{10}v^{15}}$

42) $\sqrt[3]{x^9y^{16}}$

43) $\sqrt[3]{b^{16}c^5}$

44) $\sqrt[4]{r^{15}s^9}$

45) $\sqrt[4]{m^3n^{18}}$

46) $\sqrt[3]{a^{11}b}$

47) $\sqrt[3]{24x^{10}y^{12}}$

48) $\sqrt[3]{54y^{10}z^{24}}$

49) $\sqrt[3]{250w^4x^{16}}$

50) $\sqrt[3]{72t^{17}u^7}$

51) $\sqrt[4]{\dfrac{m^8}{81}}$

52) $\sqrt[4]{\dfrac{16}{x^{12}}}$

53) $\sqrt[5]{\dfrac{32a^{23}}{b^{15}}}$

54) $\sqrt[3]{\dfrac{h^{17}}{125k^{21}}}$

75) $\sqrt[3]{\dfrac{h^{14}}{h^2}}$

76) $\sqrt[3]{\dfrac{a^{20}}{a^{14}}}$

55) $\sqrt[4]{\dfrac{t^9}{81s^{24}}}$

56) $\sqrt[5]{\dfrac{32c^9}{d^{20}}}$

77) $\sqrt[3]{\dfrac{c^{11}}{c^4}}$

78) $\sqrt[3]{\dfrac{z^{16}}{z^5}}$

57) $\sqrt[3]{\dfrac{u^{28}}{v^3}}$

58) $\sqrt[4]{\dfrac{m^{13}}{n^8}}$

79) $\sqrt[4]{\dfrac{162d^{21}}{2d^2}}$

80) $\sqrt[4]{\dfrac{48t^{11}}{3t^6}}$

Perform the indicated operation and simplify. Assume the variables represent positive real numbers.

59) $\sqrt[3]{6} \cdot \sqrt[3]{4}$

60) $\sqrt[3]{4} \cdot \sqrt[3]{10}$

61) $\sqrt[3]{9} \cdot \sqrt[3]{12}$

62) $\sqrt[3]{9} \cdot \sqrt[3]{6}$

63) $\sqrt[3]{20} \cdot \sqrt[3]{4}$

64) $\sqrt[3]{28} \cdot \sqrt[3]{2}$

65) $\sqrt[3]{m^4} \cdot \sqrt[3]{m^5}$

66) $\sqrt[3]{t^5} \cdot \sqrt[3]{t}$

67) $\sqrt[4]{k^7} \cdot \sqrt[4]{k^9}$

68) $\sqrt[4]{a^9} \cdot \sqrt[4]{a^{11}}$

69) $\sqrt[3]{r^7} \cdot \sqrt[3]{r^4}$

70) $\sqrt[3]{y^2} \cdot \sqrt[3]{y^{17}}$

71) $\sqrt[5]{p^{14}} \cdot \sqrt[5]{p^9}$

72) $\sqrt[5]{c^{17}} \cdot \sqrt[5]{c^9}$

73) $\sqrt[3]{9z^{11}} \cdot \sqrt[3]{3z^8}$

74) $\sqrt[3]{2h^4} \cdot \sqrt[3]{4h^{16}}$

Objective 5

The following radical expressions do not have the same indices. Perform the indicated operation, and write the answer in simplest radical form. Assume the variables represent positive real numbers.

81) $\sqrt{p} \cdot \sqrt[3]{p}$

82) $\sqrt[3]{y^2} \cdot \sqrt[4]{y}$

83) $\sqrt[4]{n^3} \cdot \sqrt{n}$

84) $\sqrt[5]{k^4} \cdot \sqrt{k}$

85) $\sqrt[5]{c^3} \cdot \sqrt[3]{c^2}$

86) $\sqrt[3]{a^2} \cdot \sqrt[5]{a^2}$

87) $\dfrac{\sqrt{w}}{\sqrt[4]{w}}$

88) $\dfrac{\sqrt[4]{m^3}}{\sqrt{m}}$

89) $\dfrac{\sqrt[5]{t^4}}{\sqrt[3]{t^2}}$

90) $\dfrac{\sqrt[4]{h^3}}{\sqrt[3]{h^2}}$

Section 10.5 Adding and Subtracting Radicals

Objectives

1. Add and Subtract Radical Expressions

2. Simplify Radical Expressions Containing Integers, Then Add or Subtract Them

3. Simplify Radical Expressions Containing Variables, Then Add or Subtract Them

Just as we can add and subtract like terms such as

$$4x + 6x = 10x$$

we can add and subtract *like radicals* such as

$$4\sqrt{3} + 6\sqrt{3}$$

Like radicals have the same index and the same radicand.

Some examples of like radicals are

$$4\sqrt{3} \text{ and } 6\sqrt{3}, \qquad -\sqrt[3]{5} \text{ and } 8\sqrt[3]{5}, \qquad \sqrt{x} \text{ and } 7\sqrt{x}, \qquad 2\sqrt[3]{a^2b} \text{ and } \sqrt[3]{a^2b}$$

1. Add and Subtract Radical Expressions

In order to add or subtract radicals, they must be *like* radicals.

We add and subtract like radicals in the same way we add and subtract like terms—add or subtract the "coefficients" of the radicals and multiply that result by the radical. Recall that we are using the distributive property when we are combining like terms in this way.

Example 1

Add.

a) $4x + 6x$ b) $4\sqrt{3} + 6\sqrt{3}$

Solution

a) First notice that $4x$ and $6x$ are like terms. Therefore, they can be added.

$$4x + 6x = (4 + 6)x \qquad \text{Distributive property}$$
$$= 10x \qquad \text{Simplify.}$$

Or, we can say that by just adding the coefficients, $4x + 6x = 10x$.

b) Before attempting to add $4\sqrt{3}$ and $6\sqrt{3}$, we must be certain that they are like radicals. Since they *are* like, they can be added.

$$4\sqrt{3} + 6\sqrt{3} = (4 + 6)\sqrt{3} \qquad \text{Distributive property}$$
$$= 10\sqrt{3} \qquad \text{Simplify.}$$

Or, we can say that by just adding the coefficients of $\sqrt{3}$, we get $4\sqrt{3} + 6\sqrt{3} = 10\sqrt{3}$.

You Try 1

Add.

a) $9c + 7c$ b) $9\sqrt{10} + 7\sqrt{10}$

Example 2

Perform the operations and simplify. Assume all variables represent nonnegative real numbers.

a) $8 + \sqrt{5} - 12 + 3\sqrt{5}$ b) $6\sqrt{x} + 11\sqrt[3]{x} + 2\sqrt{x} - 6\sqrt[3]{x}$

Solution

a) Begin by writing like terms together.

$$8 + \sqrt{5} - 12 + 3\sqrt{5} = 8 - 12 + \sqrt{5} + 3\sqrt{5} \qquad \text{Commutative property}$$
$$= -4 + (1 + 3)\sqrt{5} \qquad \text{Subtract; distributive property}$$
$$= -4 + 4\sqrt{5} \qquad \text{Add.}$$

Is $-4 + 4\sqrt{5}$ in simplest form? *Yes*. The terms are not like so they cannot be combined further, *and* $\sqrt{5}$ is in simplest form.

b) Begin by noticing that there are *two* different types of radicals: $\sqrt{x}$ and $\sqrt[3]{x}$. Write the like radicals together.

$$6\sqrt{x} + 11\sqrt[3]{x} + 2\sqrt{x} - 6\sqrt[3]{x} = 6\sqrt{x} + 2\sqrt{x} + 11\sqrt[3]{x} - 6\sqrt[3]{x}$$
$$= (6 + 2)\sqrt{x} + (11 - 6)\sqrt[3]{x}$$
$$= 8\sqrt{x} + 5\sqrt[3]{x}$$

Is $8\sqrt{x} + 5\sqrt[3]{x}$ in simplest form? *Yes*. The radicals are not like (they have different indices) so they cannot be combined further. Also, each radical, $\sqrt{x}$ and $\sqrt[3]{x}$, is in simplest form.

You Try 2

Perform the operations and simplify. Assume all variables represent nonnegative real numbers.

a) $9 + \sqrt[3]{4} - 2 + 10\sqrt[3]{4}$ b) $8\sqrt[3]{2n} - 3\sqrt{2n} + 5\sqrt{2n} + 5\sqrt[3]{2n}$

2. Simplify Radical Expressions Containing Integers, Then Add or Subtract Them

Sometimes it looks like two radicals cannot be added or subtracted. But if the radicals can be *simplified* and they turn out to be *like* radicals, then we can add or subtract them.

Steps for Adding and Subtracting Radicals

1) Write each radical expression in simplest form.

2) Combine like radicals.

Example 3

Perform the operations and simplify.

a) $\sqrt{8} + 5\sqrt{2}$ b) $6\sqrt{18} + 3\sqrt{50} - \sqrt{45}$ c) $-7\sqrt[3]{40} + \sqrt[3]{5}$

Solution

a) $\sqrt{8}$ and $5\sqrt{2}$ are not like radicals. Can either radical be simplified? *Yes.* We can simplify $\sqrt{8}$.

$$
\begin{aligned}
\sqrt{8} + 5\sqrt{2} &= \sqrt{4 \cdot 2} + 5\sqrt{2} && \text{4 is a perfect square.}\\
&= \sqrt{4} \cdot \sqrt{2} + 5\sqrt{2} && \text{Product rule}\\
&= 2\sqrt{2} + 5\sqrt{2} && \sqrt{4} = 2\\
&= 7\sqrt{2} && \text{Add like radicals.}
\end{aligned}
$$

b) $6\sqrt{18}$, $3\sqrt{50}$, and $\sqrt{45}$ are not like radicals. In this case, *each* radical should be simplified to determine if they can be combined.

$$
\begin{aligned}
6\sqrt{18} + 3\sqrt{50} - \sqrt{45} &= 6\sqrt{9 \cdot 2} + 3\sqrt{25 \cdot 2} - \sqrt{9 \cdot 5} && \text{Factor.}\\
&= 6\sqrt{9} \cdot \sqrt{2} + 3\sqrt{25} \cdot \sqrt{2} - \sqrt{9} \cdot \sqrt{5} && \text{Product rule}\\
&= 6 \cdot 3 \cdot \sqrt{2} + 3 \cdot 5 \cdot \sqrt{2} - 3\sqrt{5} && \text{Simplify radicals.}\\
&= 18\sqrt{2} + 15\sqrt{2} - 3\sqrt{5} && \text{Multiply.}\\
&= 33\sqrt{2} - 3\sqrt{5} && \text{Add like radicals.}
\end{aligned}
$$

$33\sqrt{2} - 3\sqrt{5}$ is in simplest form since they aren't like expressions.

c) $
\begin{aligned}
-7\sqrt[3]{40} + \sqrt[3]{5} &= -7\sqrt[3]{8 \cdot 5} + \sqrt[3]{5} && \text{8 is a perfect cube.}\\
&= -7\sqrt[3]{8} \cdot \sqrt[3]{5} + \sqrt[3]{5} && \text{Product rule}\\
&= -7 \cdot 2 \cdot \sqrt[3]{5} + \sqrt[3]{5} && \sqrt[3]{8} = 2\\
&= -14\sqrt[3]{5} + \sqrt[3]{5} && \text{Multiply.}\\
&= -13\sqrt[3]{5} && \text{Add like radicals.}
\end{aligned}
$

You Try 3

Perform the operations and simplify.

a) $7\sqrt{3} - \sqrt{12}$ b) $2\sqrt{63} - 11\sqrt{28} + 2\sqrt{21}$ c) $\sqrt[3]{54} + 5\sqrt[3]{16}$

3. Simplify Radical Expressions Containing Variables, Then Add or Subtract Them

We follow the same steps to add and subtract radicals containing variables.

Example 4

Perform the operations and simplify. Assume all variables represent nonnegative real numbers.

a) $10\sqrt{8t} - 9\sqrt{2t}$ b) $\sqrt[3]{xy^6} + \sqrt[3]{x^7}$

Solution

a) $\sqrt{2t}$ is simplified, but $\sqrt{8t}$ is not. We must simplify $\sqrt{8t}$.

$$\begin{aligned}
\sqrt{8t} &= \sqrt{8} \cdot \sqrt{t} && \text{Product rule} \\
&= \sqrt{4} \cdot \sqrt{2} \cdot \sqrt{t} && \text{4 is a perfect square.} \\
&= 2\sqrt{2} \cdot \sqrt{t} && \sqrt{4} = 2 \\
&= 2\sqrt{2t} && \text{Product rule}
\end{aligned}$$

Substitute $2\sqrt{2t}$ for $\sqrt{8t}$ in the original expression.

$$\begin{aligned}
10\sqrt{8t} - 9\sqrt{2t} &= 10(2\sqrt{2t}) - 9\sqrt{2t} && \text{Substitute } 2\sqrt{2t} \text{ for } \sqrt{8t}. \\
&= 20\sqrt{2t} - 9\sqrt{2t} && \text{Multiply.} \\
&= 11\sqrt{2t} && \text{Subtract.}
\end{aligned}$$

b) Each radical in the expression $\sqrt[3]{xy^6} + \sqrt[3]{x^7}$ must be simplified.

$$\sqrt[3]{xy^6} = y^2\sqrt[3]{x} \qquad \sqrt[3]{x^7} = x^2\sqrt[3]{x^1}$$

$6 \div 3 = 2$ | $7 \div 3$ gives a quotient of 2 and a remainder of 1.

Then, substitute the simplified radicals in the original expression.

$$\begin{aligned}
\sqrt[3]{xy^6} + \sqrt[3]{x^7} &= y^2\sqrt[3]{x} + x^2\sqrt[3]{x} && \text{Substitute} \\
&= \sqrt[3]{x}(y^2 + x^2) && \text{Factor out } \sqrt[3]{x} \text{ from each term.}
\end{aligned}$$

In this problem, we cannot *add* $y^2\sqrt[3]{x} + x^2\sqrt[3]{x}$ like we added radicals in previous examples, but we *can* factor out $\sqrt[3]{x}$.

$\sqrt[3]{x}(y^2 + x^2)$ is the completely simplified form of the sum.

You Try 4

Perform the operations and simplify. Assume all variables represent nonnegative real numbers.

a) $2\sqrt{6k} + 4\sqrt{54k}$ b) $\sqrt[4]{mn^{11}} + \sqrt[4]{81mn^3}$

Answers to You Try Exercises

1) a) $16c$ b) $16\sqrt{10}$ 2) a) $7 + 11\sqrt[3]{4}$ b) $13\sqrt[3]{2n} + 2\sqrt{2n}$

3) a) $5\sqrt{3}$ b) $-16\sqrt{7} + 2\sqrt{21}$ c) $13\sqrt[3]{2}$ 4) a) $14\sqrt{6k}$ b) $\sqrt[4]{mn^3}\,(n^2 + 3)$

10.5 Exercises

Boost your grade at mathzone.com!   MathZone

> Practice Problems   > NetTutor > Self-Test   > e-Professors   > Videos

Objectives 1–3

1) How do you know if two radicals are *like* radicals?

2) What are the steps for adding or subtracting radicals?

Perform the operation and simplify. Assume all variables represent nonnegative real numbers.

3) $5\sqrt{2} + 9\sqrt{2}$

4) $11\sqrt{7} + 7\sqrt{7}$

5) $4\sqrt{3} - 9\sqrt{3}$

6) $16\sqrt{2} - 20\sqrt{2}$

7) $7\sqrt[3]{4} + 8\sqrt[3]{4}$

8) $10\sqrt[3]{5} - 2\sqrt[3]{5}$

9) $6 - \sqrt{13} + 5 - 2\sqrt{13}$

10) $-8 + 3\sqrt{6} - 4\sqrt{6} + 9$

11) $15\sqrt[3]{z^2} - 20\sqrt[3]{z^2}$

12) $7\sqrt[3]{p} - 4\sqrt[3]{p}$

13) $2\sqrt[3]{n^2} + 9\sqrt[5]{n^2} - 11\sqrt[3]{n^2} + \sqrt[5]{n^2}$

14) $5\sqrt[4]{s} - 3\sqrt[3]{s} + 2\sqrt[3]{s} + 4\sqrt[4]{s}$

15) $\sqrt{5c} - 8\sqrt{6c} + \sqrt{5c} + 6\sqrt{6c}$

16) $10\sqrt{2m} + 6\sqrt{3m} - \sqrt{2m} + 8\sqrt{3m}$

17) $6\sqrt{3} - \sqrt{12}$

18) $\sqrt{45} + 4\sqrt{5}$

19) $\sqrt{48} + \sqrt{3}$

20) $\sqrt{44} - 8\sqrt{11}$

21) $\sqrt{20} + 4\sqrt{45}$

22) $\sqrt{28} - 2\sqrt{63}$

23) $3\sqrt{98} + 4\sqrt{50}$

24) $3\sqrt{50} - 4\sqrt{8}$

25) $\sqrt{32} - 3\sqrt{8}$

26) $3\sqrt{24} + \sqrt{96}$

27) $\sqrt{12} + \sqrt{75} - \sqrt{3}$

28) $\sqrt{2} - \sqrt{98} + \sqrt{50}$

29) $\sqrt{20} - 2\sqrt{45} - \sqrt{80}$

30) $\sqrt{96} + \sqrt{24} - 5\sqrt{54}$

31) $8\sqrt[3]{9} + \sqrt[3]{72}$

32) $5\sqrt[3]{88} + 2\sqrt[3]{11}$

33) $2\sqrt[3]{81} - 14\sqrt[3]{3}$

34) $6\sqrt[3]{3} - 3\sqrt[3]{81}$

35) $\sqrt[3]{6} - \sqrt[3]{48}$

36) $11\sqrt[3]{16} + 7\sqrt[3]{2}$

37) $6q\sqrt{q} + 7\sqrt{q^3}$

38) $11\sqrt{m^3} + 8m\sqrt{m}$

39) $4d^2\sqrt{d} - 24\sqrt{d^5}$

40) $16k^4\sqrt{k} - 13\sqrt{k^9}$

41) $9\sqrt{n^5} - 4n\sqrt{n^3}$

42) $8w\sqrt{w^5} + 4\sqrt{w^7}$

43) $9t^3\sqrt[3]{t} - 5\sqrt[3]{t^{10}}$

44) $8r^4\sqrt[3]{r} - 16\sqrt[3]{r^{13}}$

45) $5a\sqrt[4]{a^7} + \sqrt[4]{a^{11}}$

46) $-3\sqrt[4]{c^{11}} + 6c^2\sqrt[4]{c^3}$

47) $11\sqrt{5z} + 2\sqrt{20z}$

48) $8\sqrt{3r} - 5\sqrt{12r}$

49) $2\sqrt{8p} - 6\sqrt{2p}$

50) $4\sqrt{63t} + 6\sqrt{7t}$

51) $7\sqrt[3]{81a^5} + 4a\sqrt[3]{3a^2}$

52) $3\sqrt[3]{40x} - 12\sqrt[3]{5x}$

53) $4c^2\sqrt[3]{108c} - 15\sqrt[3]{32c^7}$

54) $16\sqrt[3]{128h^2} + 4\sqrt[3]{16h^2}$

55) $\sqrt{xy^3} + 3y\sqrt{xy}$

56) $5a\sqrt{ab} + 2\sqrt{a^3b}$

57) $6c^2\sqrt{8d^3} - 9d\sqrt{2c^4d}$

58) $11v\sqrt{5u^3} - 2u\sqrt{45uv^2}$

59) $3\sqrt{75m^3n} + m\sqrt{12mn}$

60) $y\sqrt{54xy} - 6\sqrt{24xy^3}$

61) $18a^5\sqrt[3]{7a^2b} + 2a^3\sqrt[3]{7a^8b}$

62) $8p^2q\sqrt[3]{11pq^2} + 3p^2\sqrt[3]{88pq^5}$

63) $15cd\sqrt[4]{9cd} - \sqrt[4]{9c^5d^5}$

64) $7yz^2\sqrt[4]{11y^4z} + 3z\sqrt[4]{11y^8z^5}$

65) $\sqrt{m^5} + \sqrt{mn^2}$

66) $\sqrt{z^3} - \sqrt{y^6z}$

67) $\sqrt[3]{a^9b} - \sqrt[3]{b^7}$

68) $\sqrt[3]{c^8} + \sqrt[3]{c^2d^3}$

69) $\sqrt[3]{u^2v^6} + \sqrt[3]{u^2}$

70) $\sqrt[3]{s} - \sqrt[3]{r^6s^4}$

Section 10.6 Combining Multiplication, Addition, and Subtraction of Radicals

Objectives

1. Multiply a Binomial Containing Radical Expressions by a Monomial
2. Multiply Two Binomials Containing Radical Expressions Using FOIL
3. Square a Binomial Containing Radical Expressions
4. Multiply Two Binomials of the Form $(a + b)(a - b)$ Containing Radicals

In Section 10.3 we learned to multiply radicals like $\sqrt{6} \cdot \sqrt{2}$. In this section, we will learn how to simplify expressions that combine multiplication, addition, and subtraction of radicals.

1. Multiply a Binomial Containing Radical Expressions by a Monomial

Example 1

Multiply and simplify. Assume all variables represent nonnegative real numbers.

a) $4(\sqrt{5} - \sqrt{20})$ b) $\sqrt{2}(\sqrt{10} + \sqrt{15})$ c) $\sqrt{x}(\sqrt{x} + \sqrt{32y})$

Solution

a) Since $\sqrt{20}$ can be simplified, we will do that first.

$$\begin{aligned} \sqrt{20} &= \sqrt{4 \cdot 5} && \text{4 is a perfect square.} \\ &= \sqrt{4} \cdot \sqrt{5} && \text{Product rule} \\ &= 2\sqrt{5} && \sqrt{4} = 2 \end{aligned}$$

Substitute $2\sqrt{5}$ for $\sqrt{20}$ in the original expression.

$$\begin{aligned} 4(\sqrt{5} - \sqrt{20}) &= 4(\sqrt{5} - 2\sqrt{5}) && \text{Substitute } 2\sqrt{5} \text{ for } \sqrt{20}. \\ &= 4(-\sqrt{5}) && \text{Subtract.} \\ &= -4\sqrt{5} && \text{Multiply.} \end{aligned}$$

b) Neither $\sqrt{10}$ nor $\sqrt{15}$ can be simplified. Begin by applying the distributive property.

$$\begin{aligned} \sqrt{2}(\sqrt{10} + \sqrt{15}) &= \sqrt{2} \cdot \sqrt{10} + \sqrt{2} \cdot \sqrt{15} && \text{Distribute.} \\ &= \sqrt{20} + \sqrt{30} && \text{Product rule} \end{aligned}$$

Is $\sqrt{20} + \sqrt{30}$ in simplest form? *No.* $\sqrt{20}$ can be simplified.

$$\begin{aligned} &= \sqrt{4 \cdot 5} + \sqrt{30} && \text{4 is a perfect square.} \\ &= \sqrt{4} \cdot \sqrt{5} + \sqrt{30} && \text{Product rule} \\ &= 2\sqrt{5} + \sqrt{30} && \sqrt{4} = 2 \end{aligned}$$

$2\sqrt{5} + \sqrt{30}$ is in simplest form. They are not like radicals, so they cannot be combined.

c) Since $\sqrt{32y}$ can be simplified, we will do that first.

$$\begin{aligned} \sqrt{32y} &= \sqrt{32} \cdot \sqrt{y} && \text{Product rule} \\ &= \sqrt{16 \cdot 2} \cdot \sqrt{y} && \text{16 is a perfect square.} \\ &= \sqrt{16} \cdot \sqrt{2} \cdot \sqrt{y} && \text{Product rule} \\ &= 4\sqrt{2y} && \sqrt{16} = 4; \text{ multiply } \sqrt{2} \cdot \sqrt{y}. \end{aligned}$$

Substitute $4\sqrt{2y}$ for $\sqrt{32y}$ in the original expression.

$$\sqrt{x}(\sqrt{x} + \sqrt{32y}) = \sqrt{x}(\sqrt{x} + 4\sqrt{2y}) \qquad \text{Substitute } 4\sqrt{2y} \text{ for } \sqrt{32y}.$$
$$= \sqrt{x} \cdot \sqrt{x} + \sqrt{x} \cdot 4\sqrt{2y} \qquad \text{Distribute.}$$
$$= x + 4\sqrt{2xy} \qquad \text{Multiply.}$$

You Try 1

Multiply and simplify. Assume all variables represent nonnegative real numbers.

a) $6(\sqrt{75} + 2\sqrt{3})$ b) $\sqrt{3}(\sqrt{3} + \sqrt{21})$ c) $\sqrt{c}(\sqrt{c^3} - \sqrt{100d})$

2. Multiply Two Binomials Containing Radical Expressions Using FOIL

In Chapter 6, we first multiplied binomials using **FOIL** (**F**irst **O**uter **I**nner **L**ast).

$$(2x + 3)(x + 4) = \underset{\text{F}}{2x \cdot x} + \underset{\text{O}}{2x \cdot 4} + \underset{\text{I}}{3 \cdot x} + \underset{\text{L}}{3 \cdot 4}$$
$$= 2x^2 + 8x + 3x + 12$$
$$= 2x^2 + 11x + 12$$

We can multiply binomials containing radicals the same way.

Example 2

Multiply and simplify. Assume all variables represent nonnegative real numbers.

a) $(2 + \sqrt{5})(4 + \sqrt{5})$ b) $(2\sqrt{3} + \sqrt{2})(\sqrt{3} - 5\sqrt{2})$
c) $(\sqrt{r} + \sqrt{3s})(\sqrt{r} + 8\sqrt{3s})$

Solution

a)   Since we must multiply two binomials, we will use FOIL.

$$(2 + \sqrt{5})(4 + \sqrt{5}) = \underset{\text{F}}{2 \cdot 4} + \underset{\text{O}}{2 \cdot \sqrt{5}} + \underset{\text{I}}{4 \cdot \sqrt{5}} + \underset{\text{L}}{\sqrt{5} \cdot \sqrt{5}}$$
$$= 8 + 2\sqrt{5} + 4\sqrt{5} + 5 \qquad \text{Multiply.}$$
$$= 13 + 6\sqrt{5} \qquad \text{Combine like terms.}$$

b) $(2\sqrt{3} + \sqrt{2})(\sqrt{3} - 5\sqrt{2})$

$$= \underset{\text{F}}{2\sqrt{3} \cdot \sqrt{3}} + \underset{\text{O}}{2\sqrt{3} \cdot (-5\sqrt{2})} + \underset{\text{I}}{\sqrt{2} \cdot \sqrt{3}} + \underset{\text{L}}{\sqrt{2} \cdot (-5\sqrt{2})}$$
$$= 2 \cdot 3 + (-10\sqrt{6}) + \sqrt{6} + (-5 \cdot 2) \qquad \text{Multiply.}$$
$$= 6 - 10\sqrt{6} + \sqrt{6} - 10 \qquad \text{Multiply.}$$
$$= -4 - 9\sqrt{6} \qquad \text{Combine like terms.}$$

c) $(\sqrt{r} + \sqrt{3s})(\sqrt{r} + 8\sqrt{3s})$

$$= \sqrt{r} \cdot \sqrt{r} + \sqrt{r} \cdot 8\sqrt{3s} + \sqrt{3s} \cdot \sqrt{r} + \sqrt{3s} \cdot 8\sqrt{3s}$$

$$ \quad\quad \text{F} \quad\quad\quad \text{O} \quad\quad\quad \text{I} \quad\quad\quad \text{L}$$

$$= \quad r \quad + \quad 8\sqrt{3rs} \quad + \quad \sqrt{3rs} \quad + \quad 8 \cdot 3s \quad\quad\quad \text{Multiply.}$$

$$= \quad r \quad + \quad 8\sqrt{3rs} \quad + \quad \sqrt{3rs} \quad + \quad 24s \quad\quad\quad \text{Multiply.}$$

$$= r + 9\sqrt{3rs} + 24s \quad\quad\quad\quad\quad\quad\quad\quad\quad\quad \text{Combine like terms.}$$

You Try 2

Multiply and simplify. Assume all variables represent nonnegative real numbers.

a) $(6 - \sqrt{7})(5 + \sqrt{7})$ b) $(\sqrt{2} + 4\sqrt{5})(3\sqrt{2} + \sqrt{5})$

c) $(\sqrt{6p} - \sqrt{2q})(\sqrt{6p} - 3\sqrt{2q})$

3. Square a Binomial Containing Radical Expressions

Recall, again, from Chapter 6, that we can use FOIL to square a binomial or we can use these special formulas:

$$(a + b)^2 = a^2 + 2ab + b^2$$
$$(a - b)^2 = a^2 - 2ab + b^2$$

For example,

$$(k + 7)^2 = (k)^2 + 2(k)(7) + (7)^2$$
$$= k^2 + 14k + 49$$

and

$$(2p - 5)^2 = (2p)^2 - 2(2p)(5) + (5)^2$$
$$= 4p^2 - 20p + 25$$

To square a binomial containing radicals, we can either use FOIL or we can use the formulas above. Understanding how to use the formulas to square a binomial will make it easier to solve radical equations in Section 10.7.

Example 3

Multiply and simplify. Assume all variables represent nonnegative real numbers.

a) $(\sqrt{10} + 3)^2$ b) $(2\sqrt{5} - 6)^2$ c) $(\sqrt{x} + \sqrt{7})^2$

Solution

a)   Use $(a + b)^2 = a^2 + 2ab + b^2$.

$$(\sqrt{10} + 3)^2 = (\sqrt{10})^2 + 2(\sqrt{10})(3) + (3)^2 \quad\quad \text{Substitute } \sqrt{10} \text{ for } a \text{ and 3 for } b.$$
$$= 10 + 6\sqrt{10} + 9 \quad\quad\quad\quad\quad\quad\quad \text{Multiply.}$$
$$= 19 + 6\sqrt{10} \quad\quad\quad\quad\quad\quad\quad\quad\quad \text{Combine like terms.}$$

b) Use $(a - b)^2 = a^2 - 2ab + b^2$.

$$
\begin{aligned}
(2\sqrt{5} - 6)^2 &= (2\sqrt{5})^2 - 2(2\sqrt{5})(6) + (6)^2 &&\text{Substitute } 2\sqrt{5} \text{ for } a \text{ and } 6 \text{ for } b. \\
&= (4 \cdot 5) - (4\sqrt{5})(6) + 36 &&\text{Multiply.} \\
&= 20 - 24\sqrt{5} + 36 &&\text{Multiply.} \\
&= 56 - 24\sqrt{5} &&\text{Combine like terms.}
\end{aligned}
$$

c) Use $(a + b)^2 = a^2 + 2ab + b^2$.

$$
\begin{aligned}
(\sqrt{x} + \sqrt{7})^2 &= (\sqrt{x})^2 + 2(\sqrt{x})(\sqrt{7}) + (\sqrt{7})^2 &&\text{Substitute } \sqrt{x} \text{ for } a \text{ and } \sqrt{7} \text{ for } b. \\
&= x + 2\sqrt{7x} + 7 &&\text{Square; product rule}
\end{aligned}
$$

You Try 3

Multiply and simplify. Assume all variables represent nonnegative real numbers.

a) $(\sqrt{6} + 5)^2$ b) $(3\sqrt{2} - 4)^2$ c) $(\sqrt{w} + \sqrt{11})^2$

4. Multiply Two Binomials of the Form $(a + b)(a - b)$ Containing Radicals

We will review one last rule from Chapter 6 on multiplying binomials. We will use this in Section 10.7 when we divide radicals.

$$(a + b)(a - b) = a^2 - b^2$$

For example,

$$
\begin{aligned}
(t + 8)(t - 8) &= (t)^2 - (8)^2 \\
&= t^2 - 64
\end{aligned}
$$

The same rule applies when we multiply binomials containing radicals.

Example 4

Multiply and simplify. Assume all variables represent nonnegative real numbers.

a) $(2 + \sqrt{5})(2 - \sqrt{5})$ b) $(\sqrt{x} + \sqrt{y})(\sqrt{x} - \sqrt{y})$

Solution

a) Use $(a + b)(a - b) = a^2 - b^2$.

$$
\begin{aligned}
(2 + \sqrt{5})(2 - \sqrt{5}) &= (2)^2 - (\sqrt{5})^2 &&\text{Substitute 2 for } a \text{ and } \sqrt{5} \text{ for } b. \\
&= 4 - 5 &&\text{Square each term.} \\
&= -1 &&\text{Subtract.}
\end{aligned}
$$

b) Use $(a + b)(a - b) = a^2 - b^2$.

$$
\begin{aligned}
(\sqrt{x} + \sqrt{y})(\sqrt{x} - \sqrt{y}) &= (\sqrt{x})^2 - (\sqrt{y})^2 &&\text{Substitute } \sqrt{x} \text{ for } a \text{ and } \sqrt{y} \text{ for } b. \\
&= x - y &&\text{Square each term.}
\end{aligned}
$$

In each case, when we multiply expressions containing square roots of the form $(a + b)(a - b)$ the radicals are eliminated. *This will always be true.*

You Try 4

Multiply and simplify. Assume all variables represent nonnegative real numbers.

a) $(4 + \sqrt{10})(4 - \sqrt{10})$ b) $(\sqrt{5h} + \sqrt{k})(\sqrt{5h} - \sqrt{k})$

Answers to You Try Exercises

1) a) $42\sqrt{3}$ b) $3 + 3\sqrt{7}$ c) $c^2 - 10\sqrt{cd}$
2) a) $23 + \sqrt{7}$ b) $26 + 13\sqrt{10}$ c) $6p - 8\sqrt{3pq} + 6q$
3) a) $31 + 10\sqrt{6}$ b) $34 - 24\sqrt{2}$ c) $w + 2\sqrt{11w} + 11$
4) a) 6 b) $5h - k$

10.6 Exercises

Objective 1

Multiply and simplify. Assume all variables represent nonnegative real numbers.

1) $3(x + 5)$

2) $8(k + 3)$

3) $7(\sqrt{6} + 2)$

4) $5(4 - \sqrt{7})$

5) $\sqrt{10}(\sqrt{3} - 1)$

6) $\sqrt{2}(9 + \sqrt{11})$

 7) $-6(\sqrt{32} + \sqrt{2})$

8) $10(\sqrt{12} - \sqrt{3})$

9) $4(\sqrt{45} - \sqrt{20})$

10) $-3(\sqrt{18} + \sqrt{50})$

11) $\sqrt{5}(\sqrt{24} - \sqrt{54})$

12) $\sqrt{2}(\sqrt{20} + \sqrt{45})$

13) $\sqrt{3}(4 + \sqrt{6})$

14) $\sqrt{5}(\sqrt{15} + \sqrt{2})$

15) $\sqrt{7}(\sqrt{24} + \sqrt{5})$

16) $\sqrt{18}(\sqrt{8} - 8)$

17) $\sqrt{t}(\sqrt{t} - \sqrt{81u})$

18) $\sqrt{s}(\sqrt{12r} + \sqrt{7s})$

19) $\sqrt{ab}(\sqrt{5a} + \sqrt{27b})$

20) $\sqrt{2xy}(\sqrt{2y} - y\sqrt{x})$

Objectives 2–4

21) How are the problems *Multiply (x + 8)(x + 3)* and *Multiply (3 + √2)(1 + √2)* similar? What method can be used to multiply each of them?

22) How are the problems *Multiply (y − 5)²* and *Multiply (√7 − 2)²* similar? What method can be used to multiply each of them?

23) What formula can be used to multiply $(5 + \sqrt{6})(5 - \sqrt{6})$?

24) What happens to the radical terms whenever we multiply $(a + b)(a - b)$ where the binomials contain square roots?

Multiply and simplify. Assume all variables represent nonnegative real numbers.

25) $(p + 7)(p + 6)$

26) $(z - 8)(z + 2)$

 27) $(6 + \sqrt{7})(2 + \sqrt{7})$

28) $(3 + \sqrt{5})(1 + \sqrt{5})$

29) $(\sqrt{2} + 8)(\sqrt{2} - 3)$

30) $(\sqrt{6} - 7)(\sqrt{6} + 2)$

31) $(\sqrt{5} - 4\sqrt{3})(2\sqrt{5} - \sqrt{3})$

32) $(5\sqrt{2} - \sqrt{3})(2\sqrt{3} - \sqrt{2})$

33) $(3\sqrt{6} - 2\sqrt{2})(\sqrt{2} + 5\sqrt{6})$

34) $(2\sqrt{10} + 3\sqrt{2})(\sqrt{10} - 2\sqrt{2})$

35) $(5 + 2\sqrt{3})(\sqrt{7} + \sqrt{2})$

36) $(\sqrt{5} + 4)(\sqrt{3} - 6\sqrt{2})$

37) $(\sqrt{x} + \sqrt{2y})(\sqrt{x} + 5\sqrt{2y})$

38) $(\sqrt{5a} + \sqrt{b})(3\sqrt{5a} + \sqrt{b})$

39) $(\sqrt{6p} - 2\sqrt{q})(8\sqrt{q} + 5\sqrt{6p})$

40) $(4\sqrt{3r} + \sqrt{s})(3\sqrt{s} - 2\sqrt{3r})$

41) $(5y - 4)^2$

42) $(3k + 2)^2$

43) $(\sqrt{3} + 1)^2$

44) $(2 + \sqrt{5})^2$

45) $(\sqrt{11} - \sqrt{5})^2$

46) $(\sqrt{3} + \sqrt{13})^2$

47) $(2\sqrt{3} + \sqrt{10})^2$

48) $(3\sqrt{2} - \sqrt{3})^2$

49) $(\sqrt{2} - 4\sqrt{6})^2$

50) $(\sqrt{5} + 3\sqrt{10})^2$

51) $(\sqrt{h} + \sqrt{7})^2$

52) $(\sqrt{m} + \sqrt{3})^2$

53) $(\sqrt{x} - \sqrt{y})^2$

54) $(\sqrt{b} - \sqrt{a})^2$

55) $(c + 9)(c - 9)$

56) $(g - 7)(g + 7)$

57) $(\sqrt{2} + 3)(\sqrt{2} - 3)$

58) $(\sqrt{3} + 1)(\sqrt{3} - 1)$

59) $(6 - \sqrt{5})(6 + \sqrt{5})$

60) $(4 - \sqrt{7})(4 + \sqrt{7})$

61) $(4\sqrt{3} + \sqrt{2})(4\sqrt{3} - \sqrt{2})$

62) $(2\sqrt{2} - 2\sqrt{7})(2\sqrt{2} + 2\sqrt{7})$

63) $(\sqrt{11} + 5\sqrt{3})(\sqrt{11} - 5\sqrt{3})$

64) $(\sqrt{15} + 5\sqrt{2})(\sqrt{15} - 5\sqrt{2})$

65) $(\sqrt{c} + \sqrt{d})(\sqrt{c} - \sqrt{d})$

66) $(\sqrt{2y} + \sqrt{z})(\sqrt{2y} - \sqrt{z})$

67) $(5 - \sqrt{t})(5 + \sqrt{t})$

68) $(6 - \sqrt{q})(6 + \sqrt{q})$

69) $(8\sqrt{f} - \sqrt{g})(8\sqrt{f} + \sqrt{g})$

70) $(\sqrt{a} + 3\sqrt{4b})(\sqrt{a} - 3\sqrt{4b})$

Extension

Multiply and simplify.

71) $(\sqrt[3]{2} - 3)(\sqrt[3]{2} + 3)$

72) $(1 + \sqrt[3]{6})(1 - \sqrt[3]{6})$

73) $(1 + 2\sqrt[3]{5})(1 - 2\sqrt[3]{5} + 4\sqrt[3]{25})$

74) $(3 + \sqrt[3]{2})(9 - 3\sqrt[3]{2} + \sqrt[3]{4})$

75) $[(\sqrt{3} + \sqrt{6}) + \sqrt{2}][(\sqrt{3} + \sqrt{6}) - \sqrt{2}]$

76) $[(\sqrt{5} - \sqrt{2}) + \sqrt{2}][(\sqrt{5} - \sqrt{2}) - \sqrt{2}]$

Section 10.7 Dividing Radicals

Objectives

1. Rationalize a Denominator Containing One Square Root Term

2. Rationalize a Denominator Containing One Term That Is a Higher Root

3. Rationalize a Denominator Containing Two Terms

4. Simplify a Radical Expression by Dividing Out Common Factors from the Numerator and Denominator

It is generally agreed in mathematics that a radical expression is *not* in simplest form if its denominator contains a radical.

$$\frac{1}{\sqrt{3}} \text{ is not simplified} \qquad \frac{\sqrt{3}}{3} \text{ is simplified}$$

This comes from the days before calculators when it was much easier to perform calculations if there were no radicals in the denominator of an expression.

Later we will show that $\dfrac{1}{\sqrt{3}} = \dfrac{\sqrt{3}}{3}$. The process of eliminating radicals from the denominator of an expression is called **rationalizing the denominator**. We will look at two types of rationalizing problems.

1) Rationalizing a denominator containing one term.
2) Rationalizing a denominator containing two terms.

The way we rationalize a denominator is based on the fact that if we multiply the numerator and denominator of a fraction by the same quantity we are actually multiplying the fraction by 1. Therefore, the fractions may look different, but they are equivalent.

$$\frac{2}{3} \cdot \frac{4}{4} = \frac{8}{12} \qquad \frac{2}{3} \text{ and } \frac{8}{12} \text{ are equivalent}$$

We use the same idea to rationalize the denominator of a radical expression.

1. Rationalize a Denominator Containing One Square Root Term

The goal of rationalizing is to eliminate the radical from the denominator. With regard to square roots, recall that $\sqrt{a} \cdot \sqrt{a} = \sqrt{a^2} = a$ for $a \geq 0$. For example,

$$\sqrt{2} \cdot \sqrt{2} = \sqrt{2^2} = 2, \quad \sqrt{19} \cdot \sqrt{19} = \sqrt{(19)^2} = 19, \quad \sqrt{t} \cdot \sqrt{t} = \sqrt{t^2} = t \ (t \geq 0)$$

We will use this property to rationalize the denominators of the following expressions.

Example 1

Rationalize the denominator of each expression.

a) $\dfrac{1}{\sqrt{3}}$ b) $\dfrac{36}{\sqrt{18}}$ c) $\dfrac{5\sqrt{3}}{\sqrt{2}}$

Solution

a) To eliminate the square root from the denominator of $\dfrac{1}{\sqrt{3}}$, ask yourself, "By what do I multiply $\sqrt{3}$ to get a *perfect square* under the square root?" The answer is $\sqrt{3}$ since $\sqrt{3} \cdot \sqrt{3} = \sqrt{3^2} = \sqrt{9} = 3$. Multiply by $\sqrt{3}$ in the numerator *and* denominator. (We are actually multiplying $\dfrac{1}{\sqrt{3}}$ by 1.)

$$\frac{1}{\sqrt{3}} = \frac{1}{\sqrt{3}} \cdot \frac{\sqrt{3}}{\sqrt{3}} \qquad \text{Rationalize the denominator.}$$

$$= \frac{\sqrt{3}}{\sqrt{3^2}} \qquad \text{Multiply.}$$

$$= \frac{\sqrt{3}}{\sqrt{9}}$$

$$= \frac{\sqrt{3}}{3} \qquad \sqrt{3^2} = \sqrt{9} = 3$$

 BE CAREFUL

$\dfrac{\sqrt{3}}{3}$ is in simplest form. We cannot reduce terms inside and outside of the radical.

$$\text{Wrong:} \quad \frac{\sqrt{3}}{3} = \frac{\sqrt{3}^1}{3_1} = \sqrt{1} = 1$$

b) Notice that we can simplify the denominator of $\dfrac{36}{\sqrt{18}}$. We will do that first.

$$\sqrt{18} = \sqrt{9} \cdot \sqrt{2} = 3\sqrt{2}$$

$$\frac{36}{\sqrt{18}} = \frac{36}{3\sqrt{2}} \qquad \text{Substitute } 3\sqrt{2} \text{ for } \sqrt{18}.$$

$$= \frac{12}{\sqrt{2}} \qquad \text{Simplify } \frac{36}{3}.$$

To eliminate the square root from the denominator of $\dfrac{12}{\sqrt{2}}$, ask yourself, "By what do I multiply $\sqrt{2}$ to get a *perfect square* under the square root?" The answer is $\sqrt{2}$.

$$= \frac{12}{\sqrt{2}} \cdot \frac{\sqrt{2}}{\sqrt{2}} \qquad \text{Rationalize the denominator.}$$

$$= \frac{12\sqrt{2}}{\sqrt{2^2}}$$

$$= \frac{12\sqrt{2}}{2} \qquad \sqrt{2^2} = 2$$

$$= 6\sqrt{2} \qquad \text{Simplify } \frac{12}{2}.$$

c) To rationalize $\dfrac{5\sqrt{3}}{\sqrt{2}}$, multiply the numerator and denominator by $\sqrt{2}$.

$$\frac{5\sqrt{3}}{\sqrt{2}} = \frac{5\sqrt{3}}{\sqrt{2}} \cdot \frac{\sqrt{2}}{\sqrt{2}}$$

$$= \frac{5\sqrt{6}}{2}$$

You Try 1

Rationalize the denominator of each expression.

a) $\dfrac{1}{\sqrt{7}}$ b) $\dfrac{15}{\sqrt{27}}$ c) $\dfrac{9\sqrt{6}}{\sqrt{5}}$

Sometimes we will apply the quotient or product rule before rationalizing.

Example 2

Simplify completely.

a) $\sqrt{\dfrac{3}{24}}$ b) $\sqrt{\dfrac{5}{14}} \cdot \sqrt{\dfrac{7}{3}}$

Solution

a) Begin by simplifying the fraction $\dfrac{3}{24}$ under the radical.

$$\sqrt{\frac{3}{24}} = \sqrt{\frac{1}{8}} \qquad \text{Simplify.}$$

$$= \frac{\sqrt{1}}{\sqrt{8}} \qquad \text{Quotient rule}$$

$$= \frac{1}{\sqrt{4} \cdot \sqrt{2}} \qquad \text{Product rule}$$

$$= \frac{1}{2\sqrt{2}} \qquad \sqrt{4} = 2$$

$$= \frac{1}{2\sqrt{2}} \cdot \frac{\sqrt{2}}{\sqrt{2}} \qquad \text{Rationalize the denominator.}$$

$$= \frac{\sqrt{2}}{2 \cdot 2} \qquad \sqrt{2} \cdot \sqrt{2} = 2$$

$$= \frac{\sqrt{2}}{4} \qquad \text{Multiply.}$$

b)   Begin by using the product rule to multiply the radicands.

$$\sqrt{\frac{5}{14}} \cdot \sqrt{\frac{7}{3}} = \sqrt{\frac{5}{14} \cdot \frac{7}{3}} \qquad \text{Product rule}$$

Multiply the fractions under the radical.

$$= \sqrt{\frac{5}{\overset{}{\underset{2}{14}}} \cdot \frac{\overset{1}{7}}{3}}$$

$$= \sqrt{\frac{5}{6}} \qquad \text{Multiply.}$$

$$= \frac{\sqrt{5}}{\sqrt{6}} \qquad \text{Quotient rule}$$

$$= \frac{\sqrt{5}}{\sqrt{6}} \cdot \frac{\sqrt{6}}{\sqrt{6}} \qquad \text{Rationalize the denominator.}$$

$$= \frac{\sqrt{30}}{6} \qquad \text{Multiply.}$$

You Try 2

Simplify completely.

a) $\sqrt{\dfrac{10}{35}}$ b) $\sqrt{\dfrac{21}{10}} \cdot \sqrt{\dfrac{2}{7}}$

We work with radical expressions containing variables the same way.

Example 3

Simplify completely. Assume all variables represent positive real numbers.

a) $\dfrac{2}{\sqrt{x}}$ b) $\sqrt{\dfrac{12m^3}{7n}}$ c) $\sqrt{\dfrac{6cd^2}{cd^3}}$

Solution

a)   Ask yourself, "By what do I multiply $\sqrt{x}$ to get a *perfect square* under the square root?" The perfect square we want to get is $\sqrt{x^2}$.

$$\sqrt{x} \cdot \sqrt{?} = \sqrt{x^2} = x$$
$$\sqrt{x} \cdot \sqrt{x} = \sqrt{x^2} = x$$

$$\frac{2}{\sqrt{x}} = \frac{2}{\sqrt{x}} \cdot \frac{\sqrt{x}}{\sqrt{x}} \qquad \text{Rationalize the denominator.}$$

$$= \frac{2\sqrt{x}}{\sqrt{x^2}} \qquad \text{Multiply.}$$

$$= \frac{2\sqrt{x}}{x} \qquad \sqrt{x^2} = x$$

b)   Before rationalizing, apply the quotient rule, and simplify the numerator.

$$\sqrt{\frac{12m^3}{7n}} = \frac{\sqrt{12m^3}}{\sqrt{7n}} \qquad \text{Quotient rule}$$

$$= \frac{2m\sqrt{3m}}{\sqrt{7n}} \qquad \text{Simplify } \sqrt{12m^3}.$$

Rationalize the denominator. "By what do I multiply $\sqrt{7n}$ to get a *perfect square* under the square root?" The perfect square we want to get is $\sqrt{7^2 n^2}$ or $\sqrt{49n^2}$.

$$\sqrt{7n} \cdot \sqrt{?} = \sqrt{7^2 n^2} = 7n$$
$$\sqrt{7n} \cdot \sqrt{7n} = \sqrt{7^2 n^2} = 7n$$

$$\sqrt{\frac{12m^3}{7n}} = \frac{2m\sqrt{3m}}{\sqrt{7n}}$$

$$= \frac{2m\sqrt{3m}}{\sqrt{7n}} \cdot \frac{\sqrt{7n}}{\sqrt{7n}} \qquad \text{Rationalize the denominator.}$$

$$= \frac{2m\sqrt{21mn}}{7n} \qquad \text{Multiply.}$$

c)   Before rationalizing $\sqrt{\dfrac{6cd^2}{cd^3}}$, simplify the radicand.

$$\sqrt{\frac{6cd^2}{cd^3}} = \sqrt{\frac{6}{d}} \qquad \text{Quotient rule for exponents}$$

$$= \frac{\sqrt{6}}{\sqrt{d}} \qquad \text{Quotient rule for radicals}$$

$$= \frac{\sqrt{6}}{\sqrt{d}} \cdot \frac{\sqrt{d}}{\sqrt{d}} \qquad \text{Rationalize the denominator.}$$

$$= \frac{\sqrt{6d}}{d} \qquad \text{Multiply.}$$

 You Try 3

Simplify completely. Assume all variables represent positive real numbers.

a) $\dfrac{5}{\sqrt{p}}$ b) $\sqrt{\dfrac{18k^5}{10m}}$ c) $\sqrt{\dfrac{20r^3 s}{s^2}}$

2. Rationalize a Denominator Containing One Term That Is a Higher Root

Many students assume that to rationalize denominators like we have up until this point, all you have to do is multiply the numerator and denominator of the expression by the denominator as in

$$\frac{4}{\sqrt{3}} = \frac{4}{\sqrt{3}} \cdot \frac{\sqrt{3}}{\sqrt{3}} = \frac{4\sqrt{3}}{3}$$

We will see, however, why this reasoning is incorrect.

To rationalize an expression like $\dfrac{4}{\sqrt{3}}$ we asked ourselves, "By what do I multiply $\sqrt{3}$ to get a *perfect square* under the *square root?*"

To rationalize an expression like $\dfrac{5}{\sqrt[3]{2}}$ we must ask ourselves, "By what do I multiply $\sqrt[3]{2}$ to get a *perfect cube* under the *cube root?*" The perfect cube we want is 2^3 (since we began with 2) so that $\sqrt[3]{2} \cdot \sqrt[3]{2^2} = \sqrt[3]{2^3} = 2$.

We will practice some fill-in-the-blank problems to eliminate radicals before we move on to rationalizing.

Example 4

Fill in the blank.

a) $\sqrt[3]{5} \cdot \sqrt[3]{?} = \sqrt[3]{5^3} = 5$ b) $\sqrt[3]{3} \cdot \sqrt[3]{?} = \sqrt[3]{3^3} = 3$

c) $\sqrt[3]{x^2} \cdot \sqrt[3]{?} = \sqrt[3]{x^3} = x$ d) $\sqrt[5]{8} \cdot \sqrt[5]{?} = \sqrt[5]{2^5} = 2$

e) $\sqrt[4]{27} \cdot \sqrt[4]{?} = \sqrt[4]{3^4} = 3$

Solution

a) Ask yourself, "By what do I multiply $\sqrt[3]{5}$ to get $\sqrt[3]{5^3}$?" The answer is $\sqrt[3]{5^2}$.

$$\sqrt[3]{5} \cdot \sqrt[3]{?} = \sqrt[3]{5^3} = 5$$
$$\sqrt[3]{5} \cdot \sqrt[3]{5^2} = \sqrt[3]{5^3} = 5$$

b) "By what do I multiply $\sqrt[3]{3}$ to get $\sqrt[3]{3^3}$?" $\sqrt[3]{3^2}$

$$\sqrt[3]{3} \cdot \sqrt[3]{?} = \sqrt[3]{3^3} = 3$$
$$\sqrt[3]{3} \cdot \sqrt[3]{3^2} = \sqrt[3]{3^3} = 3$$

c) "By what do I multiply $\sqrt[3]{x^2}$ to get $\sqrt[3]{x^3}$?" $\sqrt[3]{x}$

$$\sqrt[3]{x^2} \cdot \sqrt[3]{?} = \sqrt[3]{x^3} = x$$
$$\sqrt[3]{x^2} \cdot \sqrt[3]{x} = \sqrt[3]{x^3} = x$$

d) In this example, $\sqrt[5]{8} \cdot \sqrt[5]{?} = \sqrt[5]{2^5} = 2$, why are we trying to obtain $\sqrt[5]{2^5}$ instead of $\sqrt[5]{8^5}$? Because in the first radical, $\sqrt[5]{8}$, 8 *is a power of 2*. Before attempting to fill in the blank, rewrite 8 as 2^3.

$$\sqrt[5]{8} \cdot \sqrt[5]{?} = \sqrt[5]{2^5} = 2$$
$$\sqrt[5]{2^3} \cdot \sqrt[5]{?} = \sqrt[5]{2^5} = 2$$
$$\sqrt[5]{2^3} \cdot \sqrt[5]{2^2} = \sqrt[5]{2^5} = 2$$

e) Fill in the blank for this problem using the same approach as the problem above. Since 27 is a power of 3, rewrite $\sqrt[4]{27}$ as $\sqrt[4]{3^3}$.

$$\sqrt[4]{27} \cdot \sqrt[4]{?} = \sqrt[4]{3^4} = 3$$
$$\sqrt[4]{3^3} \cdot \sqrt[4]{?} = \sqrt[4]{3^4} = 3$$
$$\sqrt[4]{3^3} \cdot \sqrt[4]{3} = \sqrt[4]{3^4} = 3$$

You Try 4

Fill in the blank.

a) $\sqrt[3]{2} \cdot \sqrt[3]{?} = \sqrt[3]{2^3} = 2$ b) $\sqrt[5]{t^2} \cdot \sqrt[5]{?} = \sqrt[5]{t^5} = t$ c) $\sqrt[4]{125} \cdot \sqrt[4]{?} = \sqrt[4]{5^4} = 5$

We will use the technique presented in Example 4 to rationalize denominators with indices higher than 2.

Example 5

Rationalize the denominator. Assume the variable represents a positive real number.

a) $\dfrac{7}{\sqrt[3]{3}}$ b) $\sqrt[5]{\dfrac{3}{4}}$ c) $\dfrac{7}{\sqrt[4]{n}}$

Solution

a) To rationalize the denominator of $\dfrac{7}{\sqrt[3]{3}}$, *first* identify what we want to get as the denominator *after* multiplying. **We want to obtain $\sqrt[3]{3^3}$ since $\sqrt[3]{3^3} = 3$.**

$$\frac{7}{\sqrt[3]{3}} \cdot \frac{}{\underset{\uparrow}{}} = \frac{}{\sqrt[3]{3^3}} \qquad \leftarrow \text{This is what we want to get.}$$

What is needed here?

Ask yourself, "By what do I multiply $\sqrt[3]{3}$ to get $\sqrt[3]{3^3}$?" $\sqrt[3]{3^2}$

$$\frac{7}{\sqrt[3]{3}} \cdot \frac{\sqrt[3]{3^2}}{\sqrt[3]{3^2}} = \frac{7\sqrt[3]{3^2}}{\sqrt[3]{3^3}} \qquad \text{Multiply.}$$
$$= \frac{7\sqrt[3]{9}}{3} \qquad \text{Simplify.}$$

b) Use the quotient rule for radicals to rewrite $\sqrt[5]{\dfrac{3}{4}}$ as $\dfrac{\sqrt[5]{3}}{\sqrt[5]{4}}$. Then, write 4 as 2^2 to get

$$\frac{\sqrt[5]{3}}{\sqrt[5]{4}} = \frac{\sqrt[5]{3}}{\sqrt[5]{2^2}}$$

Working with $\dfrac{\sqrt[5]{3}}{\sqrt[5]{2^2}}$, identify what we want to get as the denominator *after* multiplying. **We want to obtain $\sqrt[5]{2^5}$ since $\sqrt[5]{2^5} = 2$.**

$$\frac{\sqrt[5]{3}}{\sqrt[5]{2^2}} \cdot \frac{}{\underset{\uparrow}{}} = \frac{}{\sqrt[5]{2^5}} \qquad \leftarrow \text{This is what we want to get.}$$

What is needed here?

"By what do I multiply $\sqrt[5]{2^2}$ to get $\sqrt[5]{2^5}$?" $\sqrt[5]{2^3}$

$$\frac{\sqrt[5]{3}}{\sqrt[5]{2^2}} \cdot \frac{\sqrt[5]{2^3}}{\sqrt[5]{2^3}} = \frac{\sqrt[5]{3} \cdot \sqrt[5]{2^3}}{\sqrt[5]{2^5}} \qquad \text{Multiply.}$$

$$= \frac{\sqrt[5]{3} \cdot \sqrt[5]{8}}{2}$$

$$= \frac{\sqrt[5]{24}}{2} \qquad \text{Multiply.}$$

c) To rationalize the denominator of $\dfrac{7}{\sqrt[4]{n}}$, *first* identify what we want to get as the denominator *after* multiplying. We want to obtain $\sqrt[4]{n^4}$ since $\sqrt[4]{n^4} = n$.

$$\frac{7}{\sqrt[4]{n}} \cdot \frac{}{} = \frac{}{\sqrt[4]{n^4}} \qquad \leftarrow \text{This is what we want to get.}$$

$$\underset{\text{What is needed here?}}{\uparrow}$$

Ask yourself, "By what do I multiply $\sqrt[4]{n}$ to get $\sqrt[4]{n^4}$?" $\sqrt[4]{n^3}$

$$\frac{7}{\sqrt[4]{n}} \cdot \frac{\sqrt[4]{n^3}}{\sqrt[4]{n^3}} = \frac{7\sqrt[4]{n^3}}{\sqrt[4]{n^4}} \qquad \text{Multiply.}$$

$$= \frac{7\sqrt[4]{n^3}}{n} \qquad \text{Simplify.}$$

You Try 5

Rationalize the denominator. Assume the variable represents a positive number.

a) $\dfrac{4}{\sqrt[3]{7}}$ b) $\sqrt[4]{\dfrac{2}{27}}$ c) $\sqrt[5]{\dfrac{8}{w^3}}$

3. Rationalize a Denominator Containing Two Terms

To rationalize the denominator of an expression like $\dfrac{1}{5 + \sqrt{3}}$, we multiply the numerator and the denominator of the expression by the *conjugate* of $5 + \sqrt{3}$.

Definition   The **conjugate** of a binomial is the binomial obtained by changing the sign between the two terms.

Expression	Conjugate
$5 + \sqrt{3}$	$5 - \sqrt{3}$
$\sqrt{7} - 2\sqrt{5}$	$\sqrt{7} + 2\sqrt{5}$
$\sqrt{a} + \sqrt{b}$	$\sqrt{a} - \sqrt{b}$

In Section 10.6 we applied the formula

$$(a + b)(a - b) = a^2 - b^2$$

to multiply binomials containing square roots. Recall that the terms containing the square roots were eliminated.

Example 6

Multiply $8 - \sqrt{6}$ by its conjugate.

Solution

The conjugate of $8 - \sqrt{6}$ is $8 + \sqrt{6}$. We will first multiply using FOIL to show *why* the radical drops out, then we will multiply using the formula

$$(a + b)(a - b) = a^2 - b^2$$

i) Use FOIL to multiply.

$$(8 - \sqrt{6})(8 + \sqrt{6}) = 8 \cdot 8 + 8 \cdot \sqrt{6} - 8 \cdot \sqrt{6} - \sqrt{6} \cdot \sqrt{6}$$
$$\qquad\qquad\qquad\quad\; F \qquad\quad O \qquad\quad I \qquad\quad L$$
$$= 64 - 6$$
$$= 58$$

ii) Use $(a + b)(a - b) = a^2 - b^2$.

$$(8 - \sqrt{6})(8 + \sqrt{6}) = (8)^2 - (\sqrt{6})^2 \qquad \text{Substitute 8 for } a \text{ and } \sqrt{6} \text{ for } b.$$
$$= 64 - 6$$
$$= 58$$

Each method gives the same result.

You Try 6

Multiply $2 + \sqrt{11}$ by its conjugate.

> To rationalize the denominator of an expression in which the denominator contains two terms, multiply the numerator and denominator of the expression by the conjugate of the denominator.

Example 7

Rationalize the denominator and simplify completely. Assume the variables represent positive real numbers.

a) $\dfrac{3}{5 + \sqrt{3}}$

b) $\dfrac{\sqrt{a} + b}{\sqrt{b} - a}$

Solution

a) First, notice that the denominator of $\dfrac{3}{5 + \sqrt{3}}$ has two terms. To rationalize the denominator we must multiply the numerator and denominator by $5 - \sqrt{3}$, the conjugate of the denominator.

$$\frac{3}{5 + \sqrt{3}} \cdot \frac{5 - \sqrt{3}}{5 - \sqrt{3}} \qquad \text{Multiply by the conjugate.}$$

$$= \frac{3(5 - \sqrt{3})}{(5)^2 - (\sqrt{3})^2} \qquad (a + b)(a - b) = a^2 - b^2$$

$$= \frac{15 - 3\sqrt{3}}{25 - 3} \qquad \text{Simplify.}$$

$$= \frac{15 - 3\sqrt{3}}{22} \qquad \text{Subtract.}$$

b) The conjugate of $\sqrt{b} - a$ is $\sqrt{b} + a$.

$$\frac{\sqrt{a} + b}{\sqrt{b} - a} \cdot \frac{\sqrt{b} + a}{\sqrt{b} + a} \qquad \text{Multiply by the conjugate.}$$

In the numerator we must multiply $(\sqrt{a} + b)(\sqrt{b} + a)$. Since these are binomials, we will use FOIL.

$$\frac{\sqrt{a} + b}{\sqrt{b} - a} \cdot \frac{\sqrt{b} + a}{\sqrt{b} + a} = \frac{\sqrt{ab} + a\sqrt{a} + b\sqrt{b} + ab}{(\sqrt{b})^2 - (a)^2} \qquad \begin{array}{l} \text{FOIL} \\ (a + b)(a - b) = a^2 - b^2 \end{array}$$

$$= \frac{\sqrt{ab} + a\sqrt{a} + b\sqrt{b} + ab}{b - a^2} \qquad \text{Square the terms.}$$

 You Try 7

Rationalize the denominator and simplify completely. Assume the variables represent positive real numbers.

a) $\dfrac{1}{\sqrt{7} - 2}$ b) $\dfrac{c + \sqrt{d}}{c - \sqrt{d}}$

4. Simplify a Radical Expression by Dividing Out Common Factors from the Numerator and Denominator

Sometimes it is necessary to simplify a radical expression by dividing out common factors from the numerator and denominator. This is a skill we will need in Chapter 11 when we are solving quadratic equations, so we will look at an example here.

Example 8

Simplify completely: $\dfrac{4\sqrt{5} + 12}{4}$.

Solution

It is tempting to do one of the following:

$$\frac{\cancel{4}\sqrt{5} + 12}{\cancel{4}} = \sqrt{5} + 12$$

or

$$\frac{4\sqrt{5} + \overset{3}{\cancel{12}}}{\cancel{4}} = 4\sqrt{5} + 3$$

Each is incorrect because $4\sqrt{5}$ is a *term* in a sum and 12 is a *term* in a sum.

The correct way to simplify $\dfrac{4\sqrt{5} + 12}{4}$ is to begin by factoring out a 4 in the numerator and *then* divide the numerator and denominator by any common factors.

$$\frac{4\sqrt{5} + 12}{4} = \frac{4(\sqrt{5} + 3)}{4} \qquad \text{Factor out 4 from the numerator.}$$

$$= \frac{\overset{1}{\cancel{4}}(\sqrt{5} + 3)}{\underset{1}{\cancel{4}}} \qquad \text{Divide by 4.}$$

$$= \sqrt{5} + 3 \qquad \text{Simplify.}$$

We can divide numerator and denominator by 4 in $\dfrac{4(\sqrt{5} + 3)}{4}$ because the 4 in the numerator is part of a *product* not a sum or difference.

You Try 8

Simplify completely.

a) $\dfrac{5\sqrt{7} - 40}{5}$ b) $\dfrac{20 + 6\sqrt{2}}{4}$

Answers to You Try Exercises

1) a) $\dfrac{\sqrt{7}}{7}$ b) $\dfrac{5\sqrt{3}}{3}$ c) $\dfrac{9\sqrt{30}}{5}$ 2) a) $\dfrac{\sqrt{14}}{7}$ b) $\dfrac{\sqrt{15}}{5}$ 3) a) $\dfrac{5\sqrt{p}}{p}$ b) $\dfrac{3k^2\sqrt{5km}}{5m}$

c) $\dfrac{2r\sqrt{5rs}}{s}$ 4) a) 2^2 or 4 b) t^3 c) 5 5) a) $\dfrac{4\sqrt[3]{49}}{7}$ b) $\dfrac{\sqrt[4]{6}}{3}$ c) $\dfrac{\sqrt[5]{8w^2}}{w}$ 6) -7

7) a) $\dfrac{\sqrt{7}+2}{3}$ b) $\dfrac{c^2+2c\sqrt{d}+d}{c^2-d}$ 8) a) $\sqrt{7}-8$ b) $\dfrac{10+3\sqrt{2}}{2}$

10.7 Exercises

Objective 1

1) What does it mean to rationalize the denominator of a radical expression?

2) In your own words, explain how to rationalize the denominator of an expression containing one term in the denominator.

Rationalize the denominator of each expression.

3) $\dfrac{1}{\sqrt{5}}$

4) $\dfrac{1}{\sqrt{6}}$

5) $\dfrac{3}{\sqrt{2}}$

6) $\dfrac{5}{\sqrt{3}}$

7) $\dfrac{9}{\sqrt{6}}$

8) $\dfrac{25}{\sqrt{10}}$

 9) $-\dfrac{20}{\sqrt{8}}$

10) $-\dfrac{18}{\sqrt{45}}$

11) $\dfrac{\sqrt{3}}{\sqrt{28}}$

12) $\dfrac{\sqrt{8}}{\sqrt{27}}$

13) $\sqrt{\dfrac{20}{60}}$

14) $\sqrt{\dfrac{12}{80}}$

15) $\sqrt{\dfrac{18}{26}}$

16) $\sqrt{\dfrac{42}{35}}$

17) $\dfrac{\sqrt{56}}{\sqrt{48}}$

18) $\dfrac{\sqrt{66}}{\sqrt{12}}$

Multiply and simplify.

 19) $\sqrt{\dfrac{10}{7}}\cdot\sqrt{\dfrac{7}{3}}$

20) $\sqrt{\dfrac{11}{5}}\cdot\sqrt{\dfrac{5}{2}}$

21) $\sqrt{\dfrac{6}{5}}\cdot\sqrt{\dfrac{1}{8}}$

22) $\sqrt{\dfrac{1}{6}}\cdot\sqrt{\dfrac{4}{5}}$

23) $\sqrt{\dfrac{6}{7}}\cdot\sqrt{\dfrac{7}{3}}$

24) $\sqrt{\dfrac{11}{10}}\cdot\sqrt{\dfrac{8}{11}}$

Simplify completely. Assume all variables represent positive real numbers.

25) $\dfrac{8}{\sqrt{y}}$

26) $\dfrac{4}{\sqrt{w}}$

27) $\dfrac{\sqrt{5}}{\sqrt{t}}$

28) $\dfrac{\sqrt{2}}{\sqrt{m}}$

29) $\sqrt{\dfrac{10f^3}{g}}$

30) $\sqrt{\dfrac{12s^6}{r}}$

video 31) $\sqrt{\dfrac{64v^7}{5w}}$

32) $\sqrt{\dfrac{81c^5}{2d}}$

33) $\sqrt{\dfrac{a^3b^3}{3ab^4}}$

34) $\sqrt{\dfrac{m^2n^5}{7m^3n}}$

35) $-\dfrac{\sqrt{75}}{\sqrt{b^3}}$

36) $-\dfrac{\sqrt{24}}{\sqrt{v^3}}$

37) $\dfrac{\sqrt{13}}{\sqrt{j^5}}$

38) $\dfrac{\sqrt{22}}{\sqrt{w^7}}$

Objective 2

Fill in the blank. Assume all variables represent positive real numbers.

39) $\sqrt[3]{2}\cdot\sqrt[3]{?}=\sqrt[3]{2^3}=2$

40) $\sqrt[3]{5}\cdot\sqrt[3]{?}=\sqrt[3]{5^3}=5$

41) $\sqrt[3]{9}\cdot\sqrt[3]{?}=\sqrt[3]{3^3}=3$

42) $\sqrt[3]{4}\cdot\sqrt[3]{?}=\sqrt[3]{2^3}=2$

43) $\sqrt[3]{c}\cdot\sqrt[3]{?}=\sqrt[3]{c^3}=c$

44) $\sqrt[3]{p}\cdot\sqrt[3]{?}=\sqrt[3]{p^3}=p$

45) $\sqrt[5]{4}\cdot\sqrt[5]{?}=\sqrt[5]{2^5}=2$

46) $\sqrt[5]{16}\cdot\sqrt[5]{?}=\sqrt[5]{2^5}=2$

47) $\sqrt[4]{m^3}\cdot\sqrt[4]{?}=\sqrt[4]{m^4}=m$

48) $\sqrt[4]{k}\cdot\sqrt[4]{?}=\sqrt[4]{k^4}=k$

Rationalize the denominator of each expression. Assume all variables represent positive real numbers.

 49) $\dfrac{4}{\sqrt[3]{3}}$

50) $\dfrac{26}{\sqrt[3]{5}}$

51) $\dfrac{12}{\sqrt[3]{2}}$

52) $\dfrac{21}{\sqrt[3]{3}}$

53) $\dfrac{9}{\sqrt[3]{25}}$

54) $\dfrac{6}{\sqrt[3]{4}}$

55) $\sqrt[4]{\dfrac{5}{9}}$

56) $\sqrt[4]{\dfrac{2}{25}}$

57) $\sqrt[5]{\dfrac{3}{8}}$

58) $\sqrt[5]{\dfrac{7}{4}}$

59) $\sqrt[4]{\dfrac{2}{9}}$

60) $\sqrt[4]{\dfrac{10}{27}}$

61) $\dfrac{10}{\sqrt[3]{z}}$

62) $\dfrac{6}{\sqrt[3]{u}}$

63) $\sqrt[3]{\dfrac{3}{n^2}}$

64) $\sqrt[3]{\dfrac{5}{x^2}}$

65) $\dfrac{\sqrt[3]{7}}{\sqrt[3]{2k^2}}$

66) $\dfrac{\sqrt[3]{2}}{\sqrt[3]{25t}}$

67) $\dfrac{9}{\sqrt[5]{a^3}}$

68) $\dfrac{8}{\sqrt[5]{h^2}}$

69) $\sqrt[4]{\dfrac{c}{d^3}}$

70) $\sqrt[4]{\dfrac{x^2}{y}}$

 71) $\sqrt[4]{\dfrac{5}{2m}}$

72) $\sqrt[4]{\dfrac{2}{3t^2}}$

Rationalize the denominator and simplify completely. Assume the variables represent positive real numbers.

81) $\dfrac{3}{2 + \sqrt{3}}$

82) $\dfrac{8}{6 - \sqrt{5}}$

83) $\dfrac{10}{9 - \sqrt{2}}$

84) $\dfrac{5}{4 + \sqrt{6}}$

85) $\dfrac{\sqrt{8}}{\sqrt{3} + \sqrt{2}}$

86) $\dfrac{\sqrt{32}}{\sqrt{5} - \sqrt{7}}$

87) $\dfrac{\sqrt{3} - \sqrt{5}}{\sqrt{10} - \sqrt{3}}$

88) $\dfrac{\sqrt{3} + \sqrt{6}}{\sqrt{2} + \sqrt{5}}$

 89) $\dfrac{\sqrt{m}}{\sqrt{m} + \sqrt{n}}$

90) $\dfrac{\sqrt{u}}{\sqrt{u} - \sqrt{v}}$

91) $\dfrac{b - 25}{\sqrt{b} - 5}$

92) $\dfrac{d - 9}{\sqrt{d} + 3}$

93) $\dfrac{\sqrt{x} + \sqrt{y}}{\sqrt{x} - \sqrt{y}}$

94) $\dfrac{\sqrt{f} - \sqrt{g}}{\sqrt{f} + \sqrt{g}}$

Objective 4
Simplify completely.

95) $\dfrac{5 + 10\sqrt{3}}{5}$

96) $\dfrac{18 - 6\sqrt{7}}{6}$

97) $\dfrac{30 - 18\sqrt{5}}{4}$

98) $\dfrac{36 + 20\sqrt{2}}{12}$

 99) $\dfrac{\sqrt{45} + 6}{9}$

100) $\dfrac{\sqrt{48} + 28}{4}$

101) $\dfrac{-10 - \sqrt{50}}{5}$

102) $\dfrac{-35 + \sqrt{200}}{15}$

Objective 3

73) How do you find the conjugate of a binomial?

74) When you multiply a binomial containing a square root by its conjugate, what happens to the radical?

Find the conjugate of each binomial. Then, multiply the binomial by its conjugate.

75) $(5 + \sqrt{2})$

76) $(\sqrt{5} - 4)$

77) $(\sqrt{2} + \sqrt{6})$

78) $(\sqrt{3} - \sqrt{10})$

79) $(\sqrt{t} - 8)$

80) $(\sqrt{p} + 5)$

Section 10.8 Solving Radical Equations

In this section, we will discuss how to solve *radical equations*.

An equation containing a variable in the radicand is a **radical equation**. Some examples of radical equations are

$$\sqrt{c} = 3 \qquad \sqrt[3]{n} = 2 \qquad \sqrt{2x + 1} + 1 = x \qquad \sqrt{5w + 6} - \sqrt{4w + 1} = 1$$

1. Understand the Steps Used to Solve a Radical Equation

Before we present the steps for solving radical equations, we need to answer two important questions which may arise from those steps.

1) How do we eliminate the radical?

2) Why is it *absolutely necessary* to check the proposed solution(s) in the original equation?

1) How do we eliminate the radical?

To understand how to eliminate a radical, we will revisit the relationship between roots and powers. When the value of the radical is a real number,

$$(\sqrt{x})^2 = x \text{ since } (\sqrt{x})^2 = (x^{1/2})^2 = x$$
$$(\sqrt[3]{x})^3 = x \text{ since } (\sqrt[3]{x})^3 = (x^{1/3})^3 = x$$
$$(\sqrt[4]{x})^4 = x \text{ since } (\sqrt[4]{x})^4 = (x^{1/4})^4 = x$$

To eliminate a radical from expressions like those above, raise the expression to the power equal to the index.

We must also keep in mind when solving an equation that whatever operation is done to one side of the equation must be done to the other side as well.

2) Why is it absolutely necessary to check the proposed solution(s) in the original equation?

There may be *extraneous solutions*. In the process of solving an equation we obtain new equations. *An extraneous solution is a value that satisfies one of the new equations but does not satisfy the original equation.* Extraneous solutions occur frequently when solving radical equations, so we *must* check all "solutions" in the original equation and discard those that are extraneous.

Steps for Solving Radical Equations

Step 1: Get a radical on a side by itself.

Step 2: Eliminate a radical by raising both sides of the equation to the power equal to the index of the radical.

Step 3: Combine like terms on each side of the equation.

Step 4: If the equation still contains a radical, repeat steps 1–3.

Step 5: Solve the equation.

Step 6: Check the proposed solutions *in the original equation* and discard extraneous solutions.

2. Solve an Equation Containing One Radical Expression

> **Example 1** —————————————————————————————

Solve.

a) $\sqrt{c} = 3$ b) $\sqrt[3]{n} = 2$ c) $\sqrt{t + 5} + 6 = 0$

Solution

a) **Step 1:** The radical *is* on a side by itself.

$$\sqrt{c} = 3$$

Step 2: *Square* both sides to eliminate the *square root*.

$$(\sqrt{c})^2 = 3^2 \qquad \text{Square both sides.}$$
$$c = 9$$

Steps 3 and 4 do not apply because there are no like terms to combine and no radicals remain.

Step 5: The solution obtained above is $c = 9$.

Step 6: Check $c = 9$ in the *original* equation.

$$\sqrt{c} = 3$$
$$\sqrt{9} \overset{?}{=} 3 \quad \checkmark$$

The solution set is $\{9\}$.

b) **Step 1:** The radical *is* on a side by itself.

$$\sqrt[3]{n} = 2$$

Step 2: *Cube* both sides to eliminate the *cube root*.

$$(\sqrt[3]{n})^3 = 2^3 \qquad \text{Cube both sides.}$$
$$n = 8$$

Steps 3 and 4 do not apply because there are no like terms to combine and no radicals remain.

Step 5: The solution obtained is $n = 8$.

Step 6: Check $n = 8$ in the *original* equation.

$$\sqrt[3]{n} = 2$$
$$\sqrt[3]{8} \overset{?}{=} 2 \quad \checkmark$$

The solution set is $\{8\}$.

c) **Step 1:** Get the radical on a side by itself.

$$\sqrt{t + 5} + 6 = 0$$
$$\sqrt{t + 5} = -6 \qquad \text{Subtract 6 from each side.}$$

Step 2: *Square* both sides to eliminate the square root.

$$(\sqrt{t + 5})^2 = (-6)^2 \qquad \text{Square both sides.}$$
$$t + 5 = 36 \qquad \text{The square root is eliminated.}$$

Steps 3 and 4 do not apply because there are no like terms to combine on each side and no radicals remain.

Step 5: Solve the equation.

$$t + 5 = 36$$
$$t = 31 \qquad \text{Subtract 5 from each side.}$$

Step 6: Check $t = 31$ in the *original* equation.

$$\sqrt{t + 5} + 6 = 0$$
$$\sqrt{31 + 5} + 6 \overset{?}{=} 0$$
$$\sqrt{36} + 6 \overset{?}{=} 0$$
$$6 + 6 = 0 \quad \text{False}$$

$t = 31$ does not satisfy the *original* equation. It is an extraneous solution. There is no real solution to this equation.

The solution is $\varnothing$.

You Try 1

Solve.

a) $\sqrt{a} = 9$ b) $\sqrt[3]{y + 3} = 4$ c) $\sqrt{m - 7} + 3 = 0$

Example 2

Solve $\sqrt{2x + 1} + 1 = x$.

Solution

Step 1: Get the radical on a side by itself.

$$\sqrt{2x + 1} + 1 = x$$
$$\sqrt{2x + 1} = x - 1 \qquad \text{Subtract 1 from each side.}$$

Step 2: *Square* both sides to eliminate the *square root*.

$$\left(\sqrt{2x + 1}\right)^2 = (x - 1)^2 \qquad \text{Square both sides.}$$

Use FOIL or the formula $(a - b)^2 = a^2 - 2ab + b^2$ to square the right side!

$$2x + 1 = x^2 - 2x + 1$$

Steps 3 and 4 do not apply because there are no like terms to combine on each side and no radicals remain.

Step 5: Solve the equation.

$$2x + 1 = x^2 - 2x + 1$$
$$0 = x^2 - 4x \qquad \text{Subtract } 2x \text{ and subtract 1.}$$
$$0 = x(x - 4) \qquad \text{Factor.}$$

$$x = 0 \quad \text{or} \quad x - 4 = 0 \qquad \text{Set each factor equal to zero.}$$
$$x = 0 \quad \text{or} \qquad x = 4 \qquad \text{Solve.}$$

Step 6:   Check $x = 0$ and $x = 4$ in the *original* equation.

$x = 0$:

$$\sqrt{2x + 1} + 1 = x$$
$$\sqrt{2(0) + 1} + 1 \overset{?}{=} 0$$
$$\sqrt{1} + 1 \overset{?}{=} 0$$
$$2 \overset{?}{=} 0 \quad \text{False}$$

$x = 4$:

$$\sqrt{2x + 1} + 1 = x$$
$$\sqrt{2(4) + 1} + 1 \overset{?}{=} 4$$
$$\sqrt{8 + 1} + 1 \overset{?}{=} 4$$
$$\sqrt{9} + 1 \overset{?}{=} 4$$
$$3 + 1 = 4 \quad \text{True}$$

$x = 4$ *is* a solution but $x = 0$ is *not* because $x = 0$ does not satisfy the original equation.

The solution set is $\{4\}$.

You Try 2

Solve.

a) $\sqrt{3p + 10} - 4 = p$ b) $\sqrt{4h - 3} - h = -2$

3. Solve an Equation Containing Two Radicals That Does Not Require Squaring a Binomial

Now we turn our attention to solving equations containing two radicals. We use the same steps presented earlier, but in some cases it will be necessary to repeat steps 1–3 if a radical remains after performing these steps once.

Example 3

Solve $\sqrt[3]{7a + 1} - 2\sqrt[3]{a - 1} = 0$.

Solution

Step 1:   Get a radical on a side by itself.

$$\sqrt[3]{7a + 1} = 2\sqrt[3]{a - 1} \qquad \text{Add } 2\sqrt[3]{a - 1}.$$

Step 2:   *Cube* both sides to eliminate the *cube roots*.

$$\left(\sqrt[3]{7a + 1}\right)^3 = \left(2\sqrt[3]{a - 1}\right)^3 \qquad \text{Cube both sides.}$$
$$7a + 1 = 8(a - 1) \qquad 2^3 = 8$$

Steps 3 and 4 do not apply because there are no like terms to combine on each side and no radicals remain.

Step 5:   Solve the equation.

$$7a + 1 = 8(a - 1)$$
$$7a + 1 = 8a - 8 \qquad \text{Distribute.}$$
$$1 = a - 8 \qquad \text{Subtract } 7a.$$
$$9 = a \qquad \text{Add } 8.$$

Step 6:   Verify that $a = 9$ satisfies the original equation.

The solution set is $\{9\}$.

You Try 3

Solve $3\sqrt[3]{r-4} - \sqrt[3]{5r+2} = 0$.

4. Square a Binomial Containing a Radical Expression

Recall that to square a binomial we can either use FOIL or one of the formulas

$$(a+b)^2 = a^2 + 2ab + b^2$$
$$(a-b)^2 = a^2 - 2ab + b^2$$

Example 4

Square the binomial and simplify $\left(3 - \sqrt{m+2}\right)^2$.

Solution

Use the formula $(a-b)^2 = a^2 - 2ab + b^2$.

$$\left(3 - \sqrt{m+2}\right)^2 = (3)^2 - 2(3)\left(\sqrt{m+2}\right) + \left(\sqrt{m+2}\right)^2 \qquad \text{Substitute 3 for } a \\ \text{and } \sqrt{m+2} \text{ for } b.$$

$$= 9 - 6\sqrt{m+2} + (m+2)$$
$$= m + 11 - 6\sqrt{m+2} \qquad \text{Combine like terms.}$$

You Try 4

Square each binomial and simplify.

a) $(\sqrt{z} - 4)^2$ b) $(5 + \sqrt{3d-1})^2$

5. Solve an Equation Containing Two Radicals That Requires Squaring a Binomial

Example 5

Solve each equation.

a) $\sqrt{x} - \sqrt{x-7} = 1$ b) $\sqrt{5w+6} - \sqrt{4w+1} = 1$

Solution

a) **Step 1:** Get a radical on a side by itself.

$$\sqrt{x} - \sqrt{x-7} = 1$$
$$\sqrt{x} = 1 + \sqrt{x-7} \qquad \text{Add } \sqrt{x-7} \text{ to each side.}$$

Step 2: Square both sides of the equation to eliminate a radical.

$$\left(\sqrt{x}\right)^2 = \left(1 + \sqrt{x - 7}\right)^2 \qquad \text{Square both sides.}$$
$$x = (1)^2 + 2(1)\left(\sqrt{x - 7}\right) + \left(\sqrt{x - 7}\right)^2 \qquad \text{Use the formula } (a + b)^2 = a^2 + 2ab + b^2.$$
$$x = 1 + 2\sqrt{x - 7} + x - 7$$

Step 3: Combine like terms on the right side.

$$x = x - 6 + 2\sqrt{x - 7} \qquad \text{Combine like terms.}$$

Step 4: The equation still contains a radical, so repeat steps 1–3.

Step 1: Get the radical on a side by itself.

$$x = x - 6 + 2\sqrt{x - 7}$$
$$0 = -6 + 2\sqrt{x - 7} \qquad \text{Subtract } x \text{ from each side.}$$
$$6 = 2\sqrt{x - 7} \qquad \text{Add 6 to each side.}$$
$$3 = \sqrt{x - 7} \qquad \text{Divide both sides by 2 to make the numbers smaller to work with.}$$

Step 2: Square both sides to eliminate the square root.

$$3^2 = \left(\sqrt{x - 7}\right)^2$$
$$9 = x - 7 \qquad \text{Square both sides.}$$

Steps 3 and 4 don't apply.

Step 5: Solve the equation.

$$9 = x - 7$$
$$16 = x \qquad \text{Add 7 to each side.}$$

Step 6: Verify that $x = 16$ satisfies the original equation.

The solution set is $\{16\}$.

b) **Step 1:** Get a radical on a side by itself.

$$\sqrt{5w + 6} - \sqrt{4w + 1} = 1$$
$$\sqrt{5w + 6} = 1 + \sqrt{4w + 1} \qquad \text{Add } \sqrt{4w + 1} \text{ to each side.}$$

Step 2: Square both sides of the equation to eliminate a radical.

$$\left(\sqrt{5w + 6}\right)^2 = \left(1 + \sqrt{4w + 1}\right)^2 \qquad \text{Square both sides.}$$
$$5w + 6 = (1)^2 + 2(1)\left(\sqrt{4w + 1}\right) + \left(\sqrt{4w + 1}\right)^2 \qquad \text{Use the formula } (a + b)^2 = a^2 + 2ab + b^2.$$
$$5w + 6 = 1 + 2\sqrt{4w + 1} + 4w + 1$$

Step 3: Combine like terms on the right side.

$$5w + 6 = 4w + 2 + 2\sqrt{4w + 1} \qquad \text{Combine like terms.}$$

Step 4: The equation still contains a radical, so repeat steps 1–3.

Step 1: Get the radical on a side by itself.

$$5w + 6 = 4w + 2 + 2\sqrt{4w + 1}$$
$$w + 6 = 2 + 2\sqrt{4w + 1} \qquad \text{Subtract } 4w.$$
$$w + 4 = 2\sqrt{4w + 1} \qquad \text{Subtract } 2.$$

We do not need to eliminate the 2 from in front of the radical before squaring both sides. The radical must not be a part of a *sum* or *difference* when we square.

Step 2: Square both sides of the equation to eliminate the radical.

$$(w + 4)^2 = \left(2\sqrt{4w + 1}\right)^2 \qquad \text{Square both sides.}$$
$$w^2 + 8w + 16 = 4(4w + 1)$$

On the left, we squared the binomial using the formula $(a + b)^2 = a^2 + 2ab + b^2$. On the right, don't forget to square the 2 in front of the square root.

Steps 3 and 4 do not apply.

Step 5: Solve the equation.

$$w^2 + 8w + 16 = 4(4w + 1)$$
$$w^2 + 8w + 16 = 16w + 4 \qquad \text{Distribute.}$$
$$w^2 - 8w + 12 = 0 \qquad \text{Get all terms on the left.}$$
$$(w - 6)(w - 2) = 0 \qquad \text{Factor.}$$

$$w - 6 = 0 \quad \text{or} \quad w - 2 = 0 \qquad \text{Set each factor equal to zero.}$$
$$w = 6 \quad \text{or} \qquad w = 2 \qquad \text{Solve.}$$

Step 6: Verify that $w = 6$ and $w = 2$ each satisfy the original equation.

The solution set is $\{2, 6\}$.

Watch out for two common mistakes that students make when solving an equation like the one in Example 5b.

1) Do not square both sides before getting a radical on a side by itself.

   This is incorrect:

   $$\left(\sqrt{5w + 6} - \sqrt{4w + 1}\right)^2 = 1^2$$
   $$5w + 6 - (4w + 1) = 1$$

2) The *second* time we perform step 2, watch out for this common error:

   This is incorrect:

   $$(w + 4)^2 = \left(2\sqrt{4w + 1}\right)^2$$
   $$w^2 + 16 = 2(4w + 1)$$

On the left we must multiply using FOIL or the formula $(a + b)^2 = a^2 + 2ab + b^2$ and on the right we must remember to square the 2.

You Try 5

Solve each equation.

a) $\sqrt{2y + 1} - \sqrt{y} = 1$ b) $\sqrt{3t + 4} + \sqrt{t + 2} = 2$

Using Technology

In Chapter 9 we learned how to solve an absolute value equation using a graphing calculator. Now we will see how the algebraic solution of an equation containing a radical is related to graphs.

In Example 2 of this section we solved $\sqrt{2x + 1} + 1 = x$. The proposed solutions we obtained were 0 and 4, but when we checked these values in the original equation, only 4 satisfied the equation.

On a graphing calculator we can enter the left side of the equation as Y_1 and the right side as Y_2. Press $\boxed{\text{GRAPH}}$ to see the graphs of each equation on the same screen.

   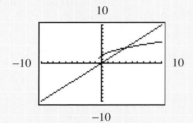

Notice that the graphs intersect at only one point: $(4, 4)$. Recall that the *solution to the equation is the x-coordinate of the point of intersection.* We obtain the same result that was obtained by solving the equation algebraically. The solution set is $\{4\}$.

Use a graphing calculator to verify the results of Examples 1, 3, and 5. Solve each equation.

1. $\sqrt{c} = 3$ (Example 1a)
2. $\sqrt[3]{n} = 2$ (Example 1b)
3. $\sqrt{t + 5} + 6 = 0$ (Example 1c)
4. $\sqrt[3]{7a + 1} - 2\sqrt[3]{a - 1} = 0$ (Example 3)
5. $\sqrt{x} - \sqrt{x - 7} = 1$ (Example 5a)
6. $\sqrt{5w + 6} - \sqrt{4w + 1} = 1$ (Example 5b)

Answers to You Try Exercises

1) a) $\{81\}$ b) $\{61\}$ c) $\varnothing$ 2) a) $\{-3, -2\}$ b) $\{7\}$ 3) $\{5\}$

4) a) $z - 8\sqrt{z} + 16$ b) $3d + 24 + 10\sqrt{3d - 1}$ 5) a) $\{0, 4\}$ b) $\{-1\}$

10.8 Exercises

Objectives 1–3

1) Why is it necessary to check the proposed solutions to a radical equation in the original equation?

2) How do you eliminate the radical from an equation like $\sqrt[3]{x} = 2$?

Solve.

3) $\sqrt{q} = 7$

4) $\sqrt{z} = 10$

5) $\sqrt{w} - \dfrac{2}{3} = 0$

6) $\sqrt{r} - \dfrac{3}{5} = 0$

7) $\sqrt{a} + 5 = 3$

8) $\sqrt{k} + 8 = 2$

9) $\sqrt[3]{y} = 5$

10) $\sqrt[3]{c} = 3$

11) $\sqrt[3]{m} = -4$

12) $\sqrt[3]{t} = -2$

13) $\sqrt{b - 11} - 3 = 0$

14) $\sqrt{d + 3} - 5 = 0$

15) $\sqrt{4g - 1} + 7 = 1$

16) $\sqrt{3l + 4} + 10 = 6$

17) $\sqrt{3f + 2} + 9 = 11$

18) $\sqrt{5u - 4} + 12 = 17$

19) $\sqrt[3]{2x - 5} + 3 = 1$

20) $\sqrt[3]{4a + 1} + 7 = 4$

21) $\sqrt{2c + 3} = \sqrt{5c}$

22) $\sqrt{5w - 2} = \sqrt{4w + 2}$

23) $\sqrt[3]{6j - 2} = \sqrt[3]{j - 7}$

24) $\sqrt[3]{m + 3} = \sqrt[3]{2m - 11}$

25) $5\sqrt{1 - 5h} = 4\sqrt{1 - 8h}$

26) $3\sqrt{6a - 2} = 4\sqrt{3a + 3}$

27) $3\sqrt{3x + 6} = 2\sqrt{9x - 9}$

28) $5\sqrt{q + 11} = 2\sqrt{8q + 25}$

29) Multiply $(x + 3)^2$.

30) Multiply $(2y - 5)^2$.

Solve.

31) $m = \sqrt{m^2 - 3m + 6}$

32) $b = \sqrt{b^2 + 4b - 24}$

33) $p + 6 = \sqrt{12 + p}$

34) $c - 7 = \sqrt{2c + 1}$

35) $\sqrt{r^2 - 8r - 19} = r - 9$

36) $\sqrt{x^2 + x + 4} = x + 8$

37) $6 + \sqrt{c^2 + 3c - 9} = c$

38) $-4 + \sqrt{z^2 + 5z - 8} = z$

39) $w - \sqrt{10w + 6} = -3$

40) $3 - \sqrt{8t + 9} = -t$

41) $3v = 8 + \sqrt{3v + 4}$

42) $4k = 3 + \sqrt{10k + 5}$

Objective 4

Use $(a + b)^2 = a^2 + 2ab + b^2$ or $(a - b)^2 = a^2 - 2ab + b^2$ to multiply each of the following binomials.

43) $(\sqrt{x} + 5)^2$

44) $(\sqrt{y} - 8)^2$

45) $(9 - \sqrt{a + 4})^2$

46) $(4 + \sqrt{p + 5})^2$

47) $(2\sqrt{3n - 1} + 7)^2$

48) $(5 - 3\sqrt{2k - 3})^2$

Objective 5

Solve.

49) $\sqrt{2y - 1} = 2 + \sqrt{y - 4}$

50) $\sqrt{3n + 4} = \sqrt{2n + 1} + 1$

51) $1 + \sqrt{3s - 2} = \sqrt{2s + 5}$

52) $\sqrt{4p + 12} - 1 = \sqrt{6p - 11}$

53) $\sqrt{3k + 1} - \sqrt{k - 1} = 2$

54) $\sqrt{4z - 3} - \sqrt{5z + 1} = -1$

55) $\sqrt{3x + 4} - 5 = \sqrt{3x - 11}$

56) $\sqrt{4c - 7} = \sqrt{4c + 1} - 4$

57) $\sqrt{3v + 3} - \sqrt{v - 2} = 3$

58) $\sqrt{2y + 1} - \sqrt{y} = 1$

59) $\sqrt{5a + 19} - \sqrt{a + 12} = 1$

60) $\sqrt{2u + 3} - \sqrt{5u + 1} = -1$

Extension

Solve.

61) $\sqrt{13 + \sqrt{r}} = \sqrt{r} + 7$

62) $\sqrt{m - 1} = \sqrt{m} - \sqrt{m - 4}$

63) $\sqrt{y + \sqrt{y + 5}} = \sqrt{y + 2}$

64) $\sqrt{2d - \sqrt{d + 6}} = \sqrt{d + 6}$

Objectives 1–3

Solve for the indicated variable.

65) $v = \sqrt{\dfrac{2E}{m}}$ for E

66) $V = \sqrt{\dfrac{300VP}{m}}$ for P

67) $c = \sqrt{a^2 + b^2}$ for b^2

68) $r = \sqrt{\dfrac{A}{\pi}}$ for A

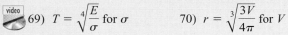

69) $T = \sqrt[4]{\dfrac{E}{\sigma}}$ for σ

70) $r = \sqrt[3]{\dfrac{3V}{4\pi}}$ for V

71) The speed of sound is proportional to the square root of the air temperature in still air. The speed of sound is given by the formula

$$V_S = 20\sqrt{T + 273}$$

where V_S is the speed of sound in meters/second and T is the temperature of the air in °Celsius.

a) What is the speed of sound when the temperature is $-17°C$ (about 1°F)?

b) What is the speed of sound when the temperature is 16°C (about 61°F)?

c) What happens to the speed of sound as the temperature increases?

d) Solve the equation for T.

72) If the area of a square is A and each side has length l, then the length of a side is given by

$$l = \sqrt{A}$$

A square rug has an area of 25 ft².

a) Find the dimensions of the rug.

b) Solve the equation for A.

73) Let V represent the volume of a cylinder, h represent its height, and r represent its radius. V, h, and r are related according to the formula

$$r = \sqrt{\dfrac{V}{\pi h}}$$

a) A cylindrical soup can has a volume of 28π in³. It is 7 in. high. What is the radius of the can?

b) Solve the equation for V.

74) For shallow water waves, the wave velocity is given by

$$c = \sqrt{gH}$$

where g is the acceleration due to gravity (32 ft/sec²) and H is the depth of the water (in feet).

a) Find the velocity of a wave in 8 ft of water.

b) Solve the equation for H.

Definition/Procedure	Example	Reference
Finding Roots		10.1
If the radicand is a perfect square, then the square root is a *rational* number.	$\sqrt{49} = 7$ since $7^2 = 49$.	**p. 628**
If the radicand is a negative number, then the square root is *not* a real number.	$\sqrt{-36}$ is not a real number.	**p. 628**
If the radicand is positive and not a perfect square, then the square root is an *irrational* number.	$\sqrt{7}$ is irrational because 7 is not a perfect square.	**p. 628**
The $\sqrt[n]{a}$ is read as "the *n*th root of *a*." If $\sqrt[n]{a} = b$, then $b^n = a$. *n* is the *index* of the radical.	$\sqrt[5]{32} = 2$ since $2^5 = 32$.	**p. 630**
The *odd root* of a negative number is a negative number.	$\sqrt[3]{-125} = -5$ since $(-5)^3 = 125$.	**p. 631**
The *even root* of a negative number is not a real number.	$\sqrt[4]{-16}$ is not a real number.	**p. 631**
Rational Exponents		10.2
If *a* is a nonnegative number and *n* is a positive integer greater than 1, then $a^{1/n} = \sqrt[n]{a}$.	$8^{1/3} = \sqrt[3]{8} = 2$	**p. 633**
If *a* is a nonnegative number and *m* and *n* are integers such that $n > 1$, then $a^{m/n} = (a^{1/n})^m = (\sqrt[n]{a})^m$.	$16^{3/4} = (\sqrt[4]{16})^3 = 2^3 = 8$	**p. 634**
If *a* is positive number and *m* and *n* are integers such that $n > 1$, then $a^{-m/n} = \left(\dfrac{1}{a}\right)^{m/n} = \dfrac{1}{a^{m/n}}$.	$25^{-3/2} = \left(\dfrac{1}{25}\right)^{3/2} = \left(\sqrt{\dfrac{1}{25}}\right)^3 = \left(\dfrac{1}{5}\right)^3 = \dfrac{1}{125}$ **BE CAREFUL** The negative exponent does not make the expression negative.	**p. 635**
If *a* is a nonnegative real number and *n* is a positive integer greater than 1, then $(\sqrt[n]{a})^n = a$ and $\sqrt[n]{a^n} = a$.	$(\sqrt{19})^2 = 19$ $\sqrt[4]{t^4} = t$ (provided *t* represents a nonnegative real number)	**p. 638**
Simplifying Expressions Containing Square Roots		10.3
Product Rule for Square Roots Let *a* and *b* be nonnegative real numbers. Then, $\sqrt{a} \cdot \sqrt{b} = \sqrt{a \cdot b}$.	$\sqrt{5} \cdot \sqrt{7} = \sqrt{5 \cdot 7} = \sqrt{35}$	**p. 642**

Definition/Procedure	Example	Reference

An expression containing a square root is simplified when all of the following conditions are met:

1) The radicand does not contain any factors (other than 1) which are perfect squares.
2) The radicand does not contain any fractions.
3) There are no radicals in the denominator of a fraction.

To *simplify square roots,* reverse the process of multiplying radicals, where a or b is a perfect square.

$$\sqrt{a \cdot b} = \sqrt{a} \cdot \sqrt{b}$$

After simplifying a radical, look at the result and ask yourself, "*Is the radical in simplest form?*" If it is not, simplify again.

Simplify $\sqrt{24}$.

$$\begin{aligned} \sqrt{24} &= \sqrt{4 \cdot 6} &&\text{4 is a perfect square.} \\ &= \sqrt{4} \cdot \sqrt{6} &&\text{Product rule} \\ &= 2\sqrt{6} &&\sqrt{4} = 2 \end{aligned}$$

p. 643

Quotient Rule for Square Roots

Let a and b be nonnegative real numbers such that $b \neq 0$. Then, $\sqrt{\dfrac{a}{b}} = \dfrac{\sqrt{a}}{\sqrt{b}}$.

$$\begin{aligned} \sqrt{\frac{72}{25}} &= \frac{\sqrt{72}}{\sqrt{25}} &&\text{Quotient rule} \\ &= \frac{\sqrt{36} \cdot \sqrt{2}}{5} &&\text{Product rule; } \sqrt{25} = 5 \\ &= \frac{6\sqrt{2}}{5} &&\sqrt{36} = 6 \end{aligned}$$

p. 645

If a is a nonnegative real number and m is an integer, then $\sqrt{a^m} = a^{m/2}$.

$\sqrt{k^{18}} = k^{18/2} = k^9$

(provided k represents a nonnegative real number)

p. 647

Two Approaches to Simplifying Radical Expressions Containing Variables

Let a represent a nonnegative real number. To simplify $\sqrt[n]{a^n}$ where n is odd and positive,

i) Method 1:

Write a^n as the product of two factors so that the exponent of one of the factors is the *largest* number less than n that is divisible by 2 (the index of the radical).

ii) Method 2:

1) Divide the exponent in the radicand by the index of the radical.
2) The exponent on the variable *outside* of the radical will be the *quotient* of the division problem.
3) The exponent on the variable *inside* of the radical will be the *remainder* of the division problem.

i) Simplify $\sqrt{x^9}$.

$$\begin{aligned} \sqrt{x^9} &= \sqrt{x^8 \cdot x^1} \\ &= \sqrt{x^8} \cdot \sqrt{x} &&\text{Product rule} \\ &= x^{8/2}\sqrt{x} \\ &= x^4\sqrt{x} &&8 \div 2 = 4 \end{aligned}$$

8 is the largest number less than 9 that is divisible by 2.

ii) Simplify $\sqrt{p^{15}}$.

$$\begin{aligned} \sqrt{p^{15}} &= p^7\sqrt{p^1} \\ &= p^7\sqrt{p} \end{aligned}$$

$15 \div 2$ gives a quotient of 7 and a remainder of 1.

p. 648

Definition/Procedure	Example	Reference
Simplifying Expressions Containing Higher Roots		10.4
Product Rule for Higher Roots If a and b are real numbers such that the roots exist, then $\sqrt[n]{a} \cdot \sqrt[n]{b} = \sqrt[n]{a \cdot b}$.	$\sqrt[3]{3} \cdot \sqrt[3]{5} = \sqrt[3]{15}$	**p. 654**
Let P be an expression and let n be a positive integer greater than 1. Then $\sqrt[n]{P}$ is *completely simplified* when all of the following conditions are met: 1) The radicand does not contain any factors (other than 1) that are perfect nth powers. 2) The radicand does not contain any fractions. 3) There are no radicals in the denominator of a fraction. To *simplify radicals with any index*, reverse the process of multiplying radicals, where a or b is an nth power. $$\sqrt[n]{a \cdot b} = \sqrt[n]{a} \cdot \sqrt[n]{b}$$	Simplify $\sqrt[3]{40}$. **Method 1:** Think of two numbers that multiply to 40 so that one of them is a *perfect cube*. $\qquad 40 = 8 \cdot 5 \qquad$ 8 is a perfect cube. Then, $\sqrt[3]{40} = \sqrt[3]{8 \cdot 5}$ $\qquad\qquad = \sqrt[3]{8} \cdot \sqrt[3]{5} \qquad$ Product rule $\qquad\qquad = 2\sqrt[3]{5} \qquad\quad \sqrt[3]{8} = 2$ **Method 2:** Begin by using a factor tree to find the prime factorization of 40. $\qquad\quad 40 = 2^3 \cdot 5$ $\sqrt[3]{40} = \sqrt[3]{2^3 \cdot 5}$ $\qquad = \sqrt[3]{2^3} \cdot \sqrt[3]{5} \qquad$ Product rule $\qquad = 2\sqrt[3]{5} \qquad\quad \sqrt[3]{2^3} = 2$	**p. 655**
Quotient Rule for Higher Roots If a and b are real numbers such that the roots exist and $b \neq 0$, then $\sqrt[n]{\dfrac{a}{b}} = \dfrac{\sqrt[n]{a}}{\sqrt[n]{b}}$.	$\sqrt[4]{\dfrac{32}{81}} = \dfrac{\sqrt[4]{32}}{\sqrt[4]{81}} = \dfrac{\sqrt[4]{16} \cdot \sqrt[4]{2}}{3} = \dfrac{2\sqrt[4]{2}}{3}$	**p. 657**
Simplifying Higher Roots with Variables in the Radicand If a is a nonnegative number and m and n are integers such that $n > 1$, then $\sqrt[n]{a^m} = a^{m/n}$.	Simplify $\sqrt[4]{a^{12}}$. $\sqrt[4]{a^{12}} = a^{12/4} = a^3$	**p. 658**
If the exponent does not divide evenly by the index, we can use two methods for simplifying the radical expression. If a is a nonnegative number and m and n are integers such that $n > 1$, then **i) Method 1:** Use the product rule. To simplify $\sqrt[n]{a^m}$, write a^m as the product of two factors so that the exponent of one of the factors is the *largest* number less than m that is divisible by n (the index). **ii) Method 2:** Use the quotient and remainder (presented in Section 10.3).	i) Simplify $\sqrt[5]{c^{17}}$. $\sqrt[5]{c^{17}} = \sqrt[5]{c^{15} \cdot c^2} \qquad$ 15 is the largest number less than 17 that is divisible by 5. $\qquad = \sqrt[5]{c^{15}} \cdot \sqrt[5]{c^2} \qquad$ Product rule $\qquad = c^{15/5} \cdot \sqrt[5]{c^2}$ $\qquad = c^3 \sqrt[5]{c^2} \qquad$ $15 \div 5 = 3$ ii) Simplify $\sqrt[4]{m^{11}}$ $\sqrt[4]{m^{11}} = m^2 \sqrt[4]{m^3} \qquad$ $11 \div 4$ gives a quotient of 2 and a remainder of 3.	**p. 659**

Definition/Procedure	Example	Reference
Adding and Subtracting Radicals		10.5

Like radicals have the same index and the same radicand.
 	
    In order to add or subtract radicals, they must be like radicals.
 	
Steps for Adding and Subtracting Radicals
 	
1) Write each radical expression in simplest form.
2) Combine like radicals.
	Perform the operations and simplify.
	
	a) $5\sqrt{2} + 9\sqrt{7} - 3\sqrt{2} + 4\sqrt{7}$
	   $= 2\sqrt{2} + 13\sqrt{7}$
	b) $\sqrt{72} + \sqrt{18} - \sqrt{45}$
	   $= \sqrt{36} \cdot \sqrt{2} + \sqrt{9} \cdot \sqrt{2} - \sqrt{9} \cdot \sqrt{5}$
	   $= 6\sqrt{2} + 3\sqrt{2} - 3\sqrt{5}$
	   $= 9\sqrt{2} - 3\sqrt{5}$
		p. 662

Combining Multiplication, Addition, and Subtraction of Radicals		10.6

Multiply expressions containing radicals using the same techniques that are used for multiplying polynomials.	Multiply and simplify. Assume all variables represent nonnegative real numbers.
	
	a) $\sqrt{m}(\sqrt{2m} + \sqrt{n})$
	   $= \sqrt{m} \cdot \sqrt{2m} + \sqrt{m} \cdot \sqrt{n}$
	   $= \sqrt{2m^2} + \sqrt{mn}$
	   $= m\sqrt{2} + \sqrt{mn}$
	b) $(\sqrt{k} + \sqrt{6})(\sqrt{k} - \sqrt{2})$
	
	Since we are multiplying two binomials, multiply using FOIL.
	
	$(\sqrt{k} + \sqrt{6})(\sqrt{k} - \sqrt{2})$
	$= \underset{F}{\sqrt{k} \cdot \sqrt{k}} - \underset{O}{\sqrt{2} \cdot \sqrt{k}} + \underset{I}{\sqrt{6} \cdot \sqrt{k}} - \underset{L}{\sqrt{6} \cdot \sqrt{2}}$
	$= k^2 - \sqrt{2k} + \sqrt{6k} - \sqrt{12}$ Product rule
	$= k^2 - \sqrt{2k} + \sqrt{6k} - 2\sqrt{3}$ $\sqrt{12} = 2\sqrt{3}$
		p. 667

To square a binomial we can either use FOIL or one of the special formulas from Chapter 6:
 	
   $(a + b)^2 = a^2 + 2ab + b^2$
   $(a - b)^2 = a^2 - 2ab + b^2$
	$(\sqrt{7} + 5)^2 = (\sqrt{7})^2 + 2(\sqrt{7})(5) + (5)^2$
	           $= 7 + 10\sqrt{7} + 25$
	           $= 32 + 10\sqrt{7}$
		p. 669

To multiply binomials of the form $(a + b)(a - b)$ use the formula $(a + b)(a - b) = a^2 - b^2$.	$(3 + \sqrt{10})(3 - \sqrt{10}) = (3)^2 - (\sqrt{10})^2$
	                  $= 9 - 10$
	                  $= -1$
		p. 670

Dividing Radicals		10.7
The process of eliminating radicals from the denominator of an expression is called *rationalizing the denominator*. First, we give examples of rationalizing denominators containing one term.	Rationalize the denominator of each expression. a) $\dfrac{9}{\sqrt{2}} = \dfrac{9}{\sqrt{2}} \cdot \dfrac{\sqrt{2}}{\sqrt{2}} = \dfrac{9\sqrt{2}}{2}$ b) $\dfrac{5}{\sqrt[3]{2}} = \dfrac{5}{\sqrt[3]{2}} \cdot \dfrac{\sqrt[3]{2^2}}{\sqrt[3]{2^2}} = \dfrac{5\sqrt[3]{2^2}}{\sqrt[3]{2^3}} = \dfrac{5\sqrt[3]{4}}{2}$	**p. 672**

Definition/Procedure	Example	Reference
The *conjugate* of a binomial is the binomial obtained by changing the sign between the two terms.	$\sqrt{11} - 4$ conjugate: $\sqrt{11} + 4$ $-8 + \sqrt{5}$ conjugate: $-8 - \sqrt{5}$	**p. 679**

Rationalizing a Denominator with Two Terms

To rationalize the denominator of an expression containing two terms, like $\dfrac{4}{\sqrt{2} - 3}$, multiply the numerator and denominator of the expression by the conjugate of $\sqrt{2} - 3$.

Rationalize the denominator of $\dfrac{4}{\sqrt{2} - 3}$.

$$\frac{4}{\sqrt{2} - 3} = \frac{4}{\sqrt{2} - 3} \cdot \frac{\sqrt{2} + 3}{\sqrt{2} + 3} \quad \text{Multiply by the conjugate.}$$

$$= \frac{4(\sqrt{2} + 3)}{(\sqrt{2})^2 - (3)^2} \quad (a + b)(a - b) = a^2 - b^2$$

$$= \frac{4(\sqrt{2} + 3)}{2 - 9} \quad \text{Square the terms.}$$

$$= \frac{4(\sqrt{2} + 3)}{-7} \quad \text{Subtract.}$$

$$= -\frac{4\sqrt{2} + 12}{7} \quad \text{Distribute.}$$

p. 679

Solving Radical Equations

10.8

Steps for Solving Radical Equations

Step 1: Get a radical on a side by itself.

Step 2: Eliminate a radical by raising both sides of the equation to the power equal to the index of the radical.

Step 3: Combine like terms on each side of the equation.

Step 4: If the equation still contains a radical, repeat steps 1–3.

Step 5: Solve the equation.

Step 6: Check the proposed solutions *in the original equation* and discard extraneous solutions.

Solve $t = 2 + \sqrt{2t - 1}$.

$$t - 2 = \sqrt{2t - 1} \quad \text{Get the radical by itself.}$$
$$(t - 2)^2 = (\sqrt{2t - 1})^2 \quad \text{Square both sides.}$$
$$t^2 - 4t + 4 = 2t - 1$$
$$t^2 - 6t + 5 = 0 \quad \text{Get all terms on the same side.}$$
$$(t - 5)(t - 1) = 0 \quad \text{Factor.}$$
$$t - 5 = 0 \quad \text{or} \quad t - 1 = 0$$
$$t = 5 \quad \text{or} \quad t = 1$$

Check $t = 5$ and $t = 1$ in the *original* equation.

$t = 5$: $t = 2 + \sqrt{2t - 1}$ | $t = 1$: $t = 2 + \sqrt{2t - 1}$
$5 \stackrel{?}{=} 2 + \sqrt{2(5) - 1}$ | $1 \stackrel{?}{=} 2 + \sqrt{2(1) - 1}$
$5 \stackrel{?}{=} 2 + \sqrt{9}$ | $1 \stackrel{?}{=} 2 + 1$
$5 = 2 + 3$ | $1 = 3$
True | False

$t = 5$ *is* a solution, but $t = 1$ is *not* because $t = 1$ does not satisfy the original equation.

The solution set is $\{5\}$.

p. 685

(10.1) Find each root, if possible.

1) $\sqrt{25}$

2) $\sqrt{-16}$

3) $-\sqrt{81}$

4) $\sqrt{\dfrac{169}{4}}$

5) $\sqrt[3]{64}$

6) $\sqrt[5]{32}$

7) $\sqrt[3]{-1}$

8) $-\sqrt[4]{81}$

9) $\sqrt[6]{-64}$

10) $\sqrt{9-16}$

Approximate each square root to the nearest tenth and plot it on a number line.

11) $\sqrt{34}$

12) $\sqrt{52}$

(10.2)

13) Explain how to write $8^{2/3}$ in radical form.

14) Explain how to eliminate the negative from the exponent in an expression like $9^{-1/2}$.

Evaluate.

15) $36^{1/2}$

16) $32^{1/5}$

17) $\left(\dfrac{27}{125}\right)^{1/3}$

18) $\left(\dfrac{16}{49}\right)^{1/2}$

19) $-16^{1/4}$

20) $-1000^{1/3}$

21) $125^{2/3}$

22) $32^{3/5}$

23) $\left(\dfrac{64}{27}\right)^{2/3}$

24) $\left(\dfrac{125}{64}\right)^{2/3}$

25) $81^{-1/2}$

26) $\left(\dfrac{1}{27}\right)^{-1/3}$

27) $81^{-3/4}$

28) $1000^{-2/3}$

29) $\left(\dfrac{27}{1000}\right)^{-2/3}$

30) $\left(\dfrac{25}{16}\right)^{-3/2}$

Simplify completely. Assume the variables represent positive real numbers. The answer should contain only positive exponents.

31) $3^{6/7} \cdot 3^{8/7}$

32) $(169^4)^{1/8}$

33) $(8^{1/5})^{10}$

34) $\dfrac{8^2}{8^{11/3}}$

35) $\dfrac{7^2}{7^{5/3} \cdot 7^{1/3}}$

36) $(2k^{-5/6})(3k^{1/2})$

37) $(64a^4b^{12})^{5/6}$

38) $\left(\dfrac{t^4u^3}{7t^7u^5}\right)^{-2}$

39) $\left(\dfrac{81c^{-5}d^9}{16c^{-1}d^2}\right)^{-1/4}$

Rewrite each radical in exponential form, then simplify. Write the answer in simplest (or radical) form. Assume all variables represent nonnegative real numbers.

40) $\sqrt[4]{36^2}$

41) $\sqrt[12]{27^4}$

42) $(\sqrt{17})^2$

43) $\sqrt[3]{7^3}$

44) $\sqrt[5]{t^{20}}$

45) $\sqrt[4]{k^{28}}$

46) $\sqrt{x^{10}}$

47) $\sqrt{w^6}$

(10.3) Simplify completely. Assume all variables represent positive real numbers.

48) $\sqrt{28}$

49) $\sqrt{1000}$

50) $\dfrac{\sqrt{63}}{\sqrt{7}}$

51) $\sqrt{\dfrac{18}{49}}$

52) $\dfrac{\sqrt{48}}{\sqrt{121}}$

53) $\sqrt{k^{12}}$

54) $\sqrt{\dfrac{40}{m^4}}$

55) $\sqrt{x^9}$

56) $\sqrt{y^5}$

57) $\sqrt{45t^2}$

58) $\sqrt{80n^{21}}$

59) $\sqrt{72x^7y^{13}}$

60) $\sqrt{\dfrac{m^{11}}{36n^2}}$

Perform the indicated operation and simplify. Assume all variables represent positive real numbers.

61) $\sqrt{5} \cdot \sqrt{3}$

62) $\sqrt{6} \cdot \sqrt{15}$

63) $\sqrt{2} \cdot \sqrt{12}$

64) $\sqrt{b^7} \cdot \sqrt{b^3}$

65) $\sqrt{11x^5} \cdot \sqrt{11x^8}$

66) $\sqrt{5a^2b} \cdot \sqrt{15a^6b^4}$

67) $\dfrac{\sqrt{200k^{21}}}{\sqrt{2k^5}}$

68) $\dfrac{\sqrt{63c^{17}}}{\sqrt{7c^9}}$

(10.4) Simplify completely. Assume all variables represent positive real numbers.

69) $\sqrt[3]{16}$

70) $\sqrt[3]{250}$

71) $\sqrt[4]{48}$

72) $\sqrt[3]{\dfrac{81}{3}}$

73) $\sqrt[4]{z^{24}}$

74) $\sqrt[5]{p^{40}}$

75) $\sqrt[3]{a^{20}}$

76) $\sqrt[5]{x^{14}y^7}$

77) $\sqrt[3]{16z^{15}}$

78) $\sqrt[3]{80m^{17}n^{10}}$

79) $\sqrt[4]{\dfrac{h^{12}}{81}}$

80) $\sqrt[5]{\dfrac{c^{22}}{32d^{10}}}$

Perform the indicated operation and simplify. Assume the variables represent positive real numbers.

81) $\sqrt[3]{3} \cdot \sqrt[3]{7}$

82) $\sqrt[3]{25} \cdot \sqrt[3]{10}$

83) $\sqrt[4]{4t^7} \cdot \sqrt[4]{8t^{10}}$

84) $\sqrt[5]{\dfrac{x^{21}}{x^{16}}}$

85) $\sqrt[3]{n} \cdot \sqrt{n}$

86) $\dfrac{\sqrt[4]{a^3}}{\sqrt[3]{a}}$

(10.5) Perform the operations and simplify. Assume all variables represent nonnegative real numbers.

87) $8\sqrt{5} + 3\sqrt{5}$

88) $\sqrt{125} + \sqrt{80}$

89) $\sqrt{80} - \sqrt{48} + \sqrt{20}$

90) $9\sqrt[3]{72} - 8\sqrt[3]{9}$

91) $3p\sqrt{p} - 7\sqrt{p^3}$

92) $9n\sqrt{n} - 4\sqrt{n^3}$

93) $10d^2\sqrt{8d} - 32d\sqrt{2d^3}$

(10.6) Multiply and simplify. Assume all variables represent nonnegative real numbers.

94) $\sqrt{6}(\sqrt{7} - \sqrt{6})$

95) $3\sqrt{k}(\sqrt{20k} + \sqrt{2})$

96) $(5 - \sqrt{3})(2 + \sqrt{3})$

97) $(\sqrt{2r} + 5\sqrt{s})(3\sqrt{s} + 4\sqrt{2r})$

98) $(2\sqrt{5} - 4)^2$

99) $(1 + \sqrt{y+1})^2$

100) $(\sqrt{6} - \sqrt{5})(\sqrt{6} + \sqrt{5})$

(10.7) Rationalize the denominator of each expression. Assume all variables represent positive real numbers.

101) $\dfrac{14}{\sqrt{3}}$

102) $\dfrac{20}{\sqrt{6}}$

103) $\dfrac{\sqrt{18k}}{\sqrt{n}}$

104) $\dfrac{\sqrt{45}}{\sqrt{m^5}}$

105) $\dfrac{7}{\sqrt[3]{2}}$

106) $-\dfrac{15}{\sqrt[3]{9}}$

107) $\dfrac{\sqrt[3]{x^2}}{\sqrt[3]{y}}$

108) $\sqrt[4]{\dfrac{3}{4k^2}}$

109) $\dfrac{2}{3 + \sqrt{3}}$

110) $\dfrac{z - 4}{\sqrt{z} + 2}$

Simplify completely.

111) $\dfrac{8 - 24\sqrt{2}}{8}$

112) $\dfrac{-\sqrt{48} - 6}{10}$

(10.8) Solve.

113) $\sqrt{x + 8} = 3$

114) $10 - \sqrt{3r - 5} = 2$

115) $\sqrt{3j + 4} = -\sqrt{4j - 1}$

116) $\sqrt[3]{6d - 14} = -2$

117) $a = \sqrt{a + 8} - 6$

118) $1 + \sqrt{6m + 7} = 2m$

119) $\sqrt{4a + 1} - \sqrt{a - 2} = 3$

120) $\sqrt{6x + 9} - \sqrt{2x + 1} = 4$

121) Solve for V: $r = \sqrt{\dfrac{3V}{\pi h}}$

122) The velocity of a wave in shallow water is given by $c = \sqrt{gH}$, where g is the acceleration due to gravity (32 ft/sec^2) and H is the depth of the water (in feet). Find the velocity of a wave in 10 ft of water.

Find each root, if possible.

1) $\sqrt{144}$

2) $\sqrt[3]{125}$

3) $\sqrt[3]{-27}$

4) $\sqrt{-16}$

Evaluate.

5) $16^{1/4}$

6) $27^{4/3}$

7) $(49)^{-1/2}$

8) $\left(\dfrac{8}{125}\right)^{-2/3}$

Simplify completely. Assume the variables represent positive numbers. The answer should contain only positive exponents.

9) $m^{3/8} \cdot m^{1/4}$

10) $\dfrac{35a^{1/6}}{14a^{5/6}}$

11) $(2x^{3/10}y^{-2/5})^{-5}$

Simplify completely.

12) $\sqrt{75}$

13) $\sqrt[3]{48}$

14) $\sqrt{\dfrac{24}{2}}$

Simplify completely. Assume all variables represent nonnegative real numbers.

15) $\sqrt{y^6}$

16) $\sqrt[4]{p^{24}}$

17) $\sqrt{t^9}$

18) $\sqrt{63m^5n^8}$

19) $\sqrt[3]{c^{23}}$

20) $\sqrt[3]{\dfrac{a^{14}b^7}{27}}$

Perform the operations and simplify. Assume all variables represent positive real numbers.

21) $\sqrt{3} \cdot \sqrt{12}$

22) $\sqrt[3]{z^4} \cdot \sqrt[3]{z^6}$

23) $\dfrac{\sqrt{120w^{15}}}{\sqrt{2w^4}}$

24) $9\sqrt{7} - 3\sqrt{7}$

25) $\sqrt{12} - \sqrt{108} + \sqrt{18}$

26) $2h^3\sqrt[4]{h} - 16\sqrt[4]{h^{13}}$

Multiply and simplify. Assume all variables represent nonnegative real numbers.

27) $\sqrt{6}(\sqrt{2} - 5)$

28) $(3 - 2\sqrt{5})(\sqrt{2} + 1)$

29) $(\sqrt{7} + \sqrt{3})(\sqrt{7} - \sqrt{3})$

30) $(\sqrt{2p + 1} + 2)^2$

31) $2\sqrt{t}(\sqrt{t} - \sqrt{3u})$

Rationalize the denominator of each expression. Assume the variable represents a positive real number.

32) $\dfrac{2}{\sqrt{5}}$

33) $\dfrac{8}{\sqrt{7} + 3}$

34) $\dfrac{\sqrt{6}}{\sqrt{a}}$

35) $\dfrac{5}{\sqrt[3]{9}}$

36) Simplify completely. $\dfrac{2 - \sqrt{48}}{2}$

Solve.

37) $\sqrt{5h + 4} = 3$

38) $z = \sqrt{1 - 4z} - 5$

39) $\sqrt{3k + 1} - \sqrt{2k - 1} = 1$

40) In the formula $r = \sqrt{\dfrac{V}{\pi h}}$, V represents the volume of a cylinder, h represents the height of cylinder, and r represents the radius.

a) A cylindrical container has a volume of 72π in³. It is 8 in. high. What is the radius of the container?

b) Solve the formula for V.

1) Combine like terms.

   $4x - 3y + 9 - \dfrac{2}{3}x + y - 1$

2) Write in scientific notation.

   8,723,000

3) Solve $3(2c - 1) + 7 = 9c + 5(c + 2)$.

4) Graph $3x + 2y = 12$.

5) Write the equation of the line containing the points $(5, 3)$ and $(1, -2)$. Write the equation in slope-intercept form.

6) Solve by substitution.

   $2x + 7y = -12$
   $x - 4y = -6$

7) Multiply.

   $(5p^2 - 2)(3p^2 - 4p - 1)$

8) Divide.

   $\dfrac{8n^3 - 1}{2n - 1}$

Factor completely.

9) $4w^2 + 5w - 6$

10) $8 - 18t^2$

11) Solve $3(k^2 + 20) - 4k = 2k^2 + 11k + 6$.

12) *Write an equation and solve.* The width of a rectangle is 5 in. less than its length. The area is 84 in^2. Find the dimensions of the rectangle.

Perform the operations and simplify.

13) $\dfrac{5a^2 + 3}{a^2 + 4a} - \dfrac{3a - 2}{a + 4}$

14) $\dfrac{10m^2}{9n} \cdot \dfrac{6n^2}{35m^5}$

15) Solve $\dfrac{3}{r^2 + 8r + 15} - \dfrac{4}{r + 3} = 1$.

16) Solve $|6g + 1| \geq 11$. Write the answer in interval notation.

17) Solve using Gaussian elimination.

   $x + 3y + \ z = 3$
   $2x - \ y - 5z = -1$
   $-x + 2y + 3z = 0$

18) Simplify. Assume all variables represent nonnegative real numbers.

   a) $\sqrt{500}$ b) $\sqrt[3]{56}$

   c) $\sqrt{p^{10}q^7}$ d) $\sqrt[4]{32a^{15}}$

19) Evaluate.

   a) $81^{1/2}$ b) $8^{4/3}$

   c) $(27)^{-1/3}$

20) Multiply and simplify $2\sqrt{3}(5 - \sqrt{3})$.

21) Rationalize the denominator. Assume the variables represent positive real numbers.

   a) $\sqrt{\dfrac{20}{50}}$ b) $\dfrac{6}{\sqrt[3]{2}}$

   c) $\dfrac{x}{\sqrt[3]{y^2}}$ d) $\dfrac{\sqrt{a} - 2}{1 - \sqrt{a}}$

22) Solve.

   a) $\sqrt{2b - 1} + 7 = 6$

   b) $\sqrt{3z + 10} = 2 - \sqrt{z + 4}$

CHAPTER 11

Quadratic Equations

Algebra at Work: Ophthalmology

We have already seen two applications of mathematics to ophthalmology, and here we have a third. An ophthalmologist can use a quadratic equation to convert between a prescription for glasses and a prescription for contact lenses.

After having reexamined her patient for contact lens use, Sarah can use the following quadratic equation to double-check the prescription for the contact lenses based on the prescription her patient currently has for her glasses.

$$D_c = s(D_g)^2 + D_g$$

where D_g = power of the glasses, in diopters

s = distance of the glasses to the eye, in meters

D_c = power of the contact lenses, in diopters

For example, if the power of a patient's eyeglasses is +9.00 diopters and the glasses rest 1 cm or 0.01 m from the eye, the power the patient would need in her contact lenses would be

$$D_c = 0.01(9)^2 + 9$$
$$D_c = 0.01(81) + 9$$
$$D_c = 0.81 + 9$$
$$D_c = 9.81 \text{ diopters}$$

An eyeglass power of +9.00 diopters would convert to a contact lens power of +9.81 diopters. In this chapter, we will learn different ways to solve quadratic equations.

Section 11.1 Review of Solving Equations by Factoring

Objective

1. Review How to Solve a Quadratic Equation by Factoring

We defined a quadratic equation in Chapter 7. Let's restate the definition:

> A quadratic equation can be written in the form $ax^2 + bx + c = 0$, where $a, b,$ and c are real numbers and $a \neq 0$.

In Section 7.5 we learned how to solve quadratic equations by factoring. We will not be able to solve all quadratic equations by factoring, however. Therefore, we need to learn other methods for solving quadratic equations. In this chapter, we will discuss the following four methods for solving quadratic equations.

Four Methods for Solving Quadratic Equations

1) Factoring

2) Square root property

3) Completing the square

4) Quadratic formula

---

1. Review How to Solve a Quadratic Equation by Factoring

We begin by reviewing how to solve an equation by factoring.

Solving a Quadratic Equation by Factoring

1) Write the equation in the form $ax^2 + bx + c = 0$ so that all terms are on one side of the equal sign and zero is on the other side.

2) Factor the expression.

3) Set each factor equal to zero, and solve for the variable.

4) Check the answer(s).

Example 1

Solve by factoring.

a) $8t^2 + 3 = -14t$ b) $(a - 9)(a + 7) = -15$ c) $3x^3 + 10x^2 = 8x$

Solution

a) Begin by writing $8t^2 + 3 = -14t$ in standard form.

$$8t^2 + 14t + 3 = 0 \qquad \text{Standard form}$$
$$(4t + 1)(2t + 3) = 0 \qquad \text{Factor.}$$

$$4t + 1 = 0 \quad \text{or} \quad 2t + 3 = 0 \qquad \text{Set each factor equal to zero.}$$
$$4t = -1 \qquad\qquad 2t = -3$$
$$t = -\frac{1}{4} \quad \text{or} \qquad t = -\frac{3}{2} \qquad \text{Solve.}$$

The check is left to the student.

The solution set is $\left\{-\dfrac{3}{2}, -\dfrac{1}{4}\right\}$.

b) You may want to solve $(a - 9)(a + 7) = -15$ like this:

$$(a - 9)(a + 7) = -15$$

$$a - 9 = -15 \quad \text{or} \quad a + 7 = -15$$
$$a = -6 \quad \text{or} \quad a = -22$$

This is incorrect!
One side of the equation must equal *zero* and the other side factored to be able to apply the zero product rule and set each factor equal to zero.
   Begin by multiplying the binomials using FOIL.

$$(a - 9)(a + 7) = -15$$
$$a^2 - 2a - 63 = -15 \qquad \text{Multiply using FOIL.}$$
$$a^2 - 2a - 48 = 0 \qquad \text{Write in standard form.}$$
$$(a + 6)(a - 8) = 0 \qquad \text{Factor.}$$

$$a + 6 = 0 \quad \text{or} \quad a - 8 = 0 \qquad \text{Set each factor equal to zero.}$$
$$a = -6 \qquad\qquad a = 8 \qquad \text{Solve.}$$

The check is left to the student.
The solution set is $\{-6, 8\}$.

c) Although this is a cubic equation and not quadratic, we *can* solve it by factoring.

$$3x^3 + 10x^2 = 8x$$
$$3x^3 + 10x^2 - 8x = 0 \qquad \text{Get zero on one side of the equal sign.}$$
$$x(3x^2 + 10x - 8) = 0 \qquad \text{Factor out } x.$$
$$x(3x - 2)(x + 4) = 0 \qquad \text{Factor } 3x^2 + 10x - 8.$$

$$x = 0 \quad \text{or} \quad 3x - 2 = 0 \quad \text{or} \quad x + 4 = 0 \qquad \text{Set each factor equal to zero.}$$
$$3x = 2$$

$$x = 0 \quad \text{or} \quad x = \dfrac{2}{3} \quad \text{or} \quad x = -4 \qquad \text{Solve.}$$

The check is left to the student.

The solution set is $\left\{-4, 0, \dfrac{2}{3}\right\}$.

You Try 1

Solve by factoring.

a) $c^2 - 12 = c$ b) $p(7p + 18) + 8 = 0$

c) $(k - 7)(k - 5) = -1$ d) $2x^3 + 30x = 16x^2$

Answers to You Try Exercises

1) a) $\{-3, 4\}$ b) $\left\{-2, -\dfrac{4}{7}\right\}$ c) $\{6\}$ d) $\{0, 3, 5\}$

11.1 Exercises

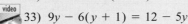

Objective 1

Solve each equation.

1) $(t + 7)(t - 6) = 0$ 2) $3z(2z - 9) = 0$

3) $u^2 + 15u + 44 = 0$ 4) $n^2 + 10n - 24 = 0$

 5) $x^2 = x + 56$ 6) $c^2 + 3c = 54$

7) $1 - 100w^2 = 0$

8) $9j^2 = 49$

9) $5m^2 + 8 = 22m$

10) $19a + 20 = -3a^2$

11) $23d = -10 - 6d^2$

12) $8h^2 + 12 = 35h$

13) $2r = 7r^2$

14) $5n^2 = -6n$

Identify each equation as linear or quadratic.

15) $9m^2 - 2m + 1 = 0$ 16) $17 = 3z - z^2$

17) $13 - 4x = 19$

18) $10 - 2(3d + 1) = 5d + 19$

19) $y(2y - 5) = 3y + 1$

20) $3(4y - 3) = y(y + 1)$

21) $-4(b + 7) + 5b = 2b + 9$

22) $6 + 2k(k - 1) = 5k$

In this section, there is a mix of linear and quadratic equations as well as equations of higher degree. Solve each equation.

23) $13c = 2c^2 + 6$ 24) $12x - 1 = 2x + 9$

25) $2p(p + 4) = p^2 + 5p + 10$

26) $z^2 - 20 = 22 - z$

27) $5(3n - 2) - 11n = 2n - 1$

28) $5a^2 = 45a$ 29) $3t^3 + 5t = -8t^2$

30) $6(2k - 3) + 10 = 3(2k - 5)$

31) $2(r + 5) = 10 - 4r^2$

32) $3d - 4 = d(d + 8)$

33) $9y - 6(y + 1) = 12 - 5y$

34) $3m(2m + 5) - 8 = 2m(3m + 5) + 2$

35) $\dfrac{1}{16}w^2 + \dfrac{1}{8}w = \dfrac{1}{2}$ 36) $6h = 4h^3 + 5h^2$

37) $12n + 3 = -12n^2$ 38) $u = u^2$

39) $3b^2 - b - 7 = 4b(2b + 3) - 1$

40) $\dfrac{1}{2}q^2 + \dfrac{3}{4} = \dfrac{5}{4}q$

41) $t^3 + 7t^2 - 4t - 28 = 0$

42) $5m^3 + 2m^2 - 5m - 2 = 0$

Write an equation and solve.

43) The length of a rectangle is 5 in. more than its width. Find the dimensions of the rectangle if its area is 14 in².

44) The width of a rectangle is 3 cm shorter than its length. If the area is 70 cm², what are the dimensions of the rectangle?

45) The length of a rectangle is 1 cm less than twice its width. The area is 45 cm². What are the dimensions of the rectangle?

46) A rectangle has an area of 32 in². Its length is 4 in. less than three times its width. Find the length and width.

Find the base and height of each triangle.

47)

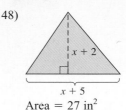

$x + 1$

$x + 6$

Area = 18 in²

48)

$x + 2$

$x + 5$

Area = 27 in²

49)

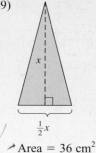

$\frac{1}{2}x$

Area = 36 cm²

50)

2x

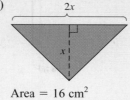

Area = 16 cm²

53)

2x

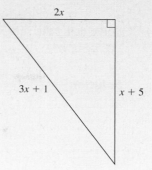

3x + 1   x + 5

Find the lengths of the sides of the following right triangles.

51)

x   x + 1

x − 7

52)

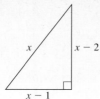

x   x − 2

x − 1

54)

x + 1

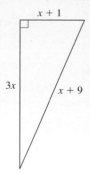

3x   x + 9

Section 11.2 Solving Quadratic Equations Using the Square Root Property

Objectives

1. Solve an Equation of the Form $x^2 = k$

2. Solve an Equation of the Form $(ax + b)^2 = k$

3. Use the Distance Formula

The next method we will discuss for solving quadratic equations is the square root property.

1. Solve an Equation of the Form $x^2 = k$

Definition	**The Square Root Property**
	Let k be a constant. If $x^2 = k$, then $x = \sqrt{k}$ or $x = -\sqrt{k}$.
	(The solution is often written as $x = \pm\sqrt{k}$, read as "x equals plus or minus the square root of k.")

We can use the square root property to solve an equation containing a squared quantity and a constant.

Example 1

Solve using the square root property.

a) $x^2 = 9$ b) $t^2 - 20 = 0$ c) $3a^2 - 20 = 4$ d) $y^2 + 29 = 4$

Solution

a)
$$x^2 = 9$$

$$x = \sqrt{9} \quad \text{or} \quad x = -\sqrt{9} \qquad \text{Square root property}$$
$$x = 3 \quad \text{or} \quad x = -3$$

Check:

$$x = 3: \qquad x^2 = 9 \qquad\qquad x = -3: \qquad x^2 = 9$$
$$(3)^2 \overset{?}{=} 9 \qquad\qquad\qquad (-3)^2 \overset{?}{=} 9$$
$$9 = 9 \; \checkmark \qquad\qquad\qquad\qquad 9 = 9 \; \checkmark$$

The solution set is $\{-3, 3\}$.
An equivalent way to solve $x^2 = 9$ is to write it as

$$x^2 = 9$$
$$x = \pm\sqrt{9} \qquad \text{Square root property}$$
$$x = \pm 3$$

We will use this approach when solving using the square root property.
The solution set is $\{-3, 3\}$.

b) To solve $t^2 - 20 = 0$, begin by getting t^2 on a side by itself.

$$t^2 - 20 = 0$$
$$t^2 = 20 \qquad \text{Add 20 to each side.}$$
$$t = \pm\sqrt{20} \qquad \text{Square root property}$$
$$t = \pm\sqrt{4} \cdot \sqrt{5} \qquad \text{Product rule for radicals}$$
$$t = \pm 2\sqrt{5} \qquad \sqrt{4} = 2$$

Check:

$$t = 2\sqrt{5}: \qquad t^2 - 20 = 0 \qquad\qquad t = -2\sqrt{5}: \qquad t^2 - 20 = 0$$
$$(2\sqrt{5})^2 - 20 \overset{?}{=} 0 \qquad\qquad\qquad (-2\sqrt{5})^2 - 20 \overset{?}{=} 0$$
$$(4 \cdot 5) - 20 \overset{?}{=} 0 \qquad\qquad\qquad (4 \cdot 5) - 20 \overset{?}{=} 0$$
$$20 - 20 = 0 \; \checkmark \qquad\qquad\qquad 20 - 20 = 0 \; \checkmark$$

The solution set is $\{-2\sqrt{5}, 2\sqrt{5}\}$.

c) To solve $3a^2 - 20 = 4$, begin by getting $3a^2$ on a side by itself.

$$3a^2 - 20 = 4$$
$$3a^2 = 24 \qquad \text{Add 20 to each side.}$$
$$a^2 = 8 \qquad \text{Divide by 3.}$$
$$a = \pm\sqrt{8} \qquad \text{Square root property}$$
$$a = \pm\sqrt{4} \cdot \sqrt{2} \qquad \text{Product rule for radicals}$$
$$a = \pm 2\sqrt{2} \qquad \sqrt{4} = 2$$

The check is left to the student.
The solution set is $\{-2\sqrt{2}, 2\sqrt{2}\}$.

d)
$$y^2 + 29 = 4$$
$$y^2 = -25 \qquad \text{Subtract 29 from each side.}$$
$$y = \pm\sqrt{-25} \qquad \text{Square root property}$$

Since $\sqrt{-25}$ is not a real number, there is no real number solution to $y^2 + 29 = 4$.

You Try 1

Solve using the square root property.

a) $p^2 = 100$ b) $w^2 - 32 = 0$ c) $5c^2 + 3 = 23$ d) $z^2 + 13 = 7$

Can we solve $(w - 4)^2 = 25$ using the square root property? Yes. The equation has a *squared quantity* and a *constant*.

2. Solve an Equation of the Form $(ax + b)^2 = k$

Example 2

Solve $x^2 = 25$ and $(w - 4)^2 = 25$ using the square root property.

Solution

While the equation $(w - 4)^2 = 25$ has a *binomial* that is being squared, the two equations are actually in the same form.

$$x^2 = 25 \qquad\qquad (w - 4)^2 = 25$$
$$\uparrow \quad\ \ \uparrow \qquad\qquad\qquad \uparrow \qquad\quad \uparrow$$
$$x \text{ squared} = \text{constant} \qquad (w - 4) \text{ squared} = \text{constant}$$

Solve $x^2 = 25$:

$$x^2 = 25$$
$$x = \pm\sqrt{25} \qquad \text{Square root property}$$
$$x = \pm 5$$

The solution set is $\{-5, 5\}$.
We solve $(w - 4)^2 = 25$ in the same way with some additional steps.

$$(w - 4)^2 = 25$$
$$w - 4 = \pm\sqrt{25} \qquad \text{Square root property}$$
$$w - 4 = \pm 5$$

This means $w - 4 = 5$ or $w - 4 = -5$. Solve both equations.

$$w - 4 = 5 \quad \text{or} \quad w - 4 = -5$$
$$w = 9 \quad \text{or} \qquad\quad w = -1 \qquad \text{Add 4 to each side.}$$

Check:

$$w = 9: \quad (w - 4)^2 = 25$$
$$(9 - 4)^2 \overset{?}{=} 25$$
$$5^2 \overset{?}{=} 25$$
$$25 = 25 \ \checkmark$$

$$w = -1: \quad (w - 4)^2 = 25$$
$$(-1 - 4)^2 \overset{?}{=} 25$$
$$(-5)^2 \overset{?}{=} 25$$
$$25 = 25 \ \checkmark$$

The solution set is $\{-1, 9\}$.

You Try 2

Solve $(c + 6)^2 = 81$ using the square root property.

Example 3

Solve.

a) $(3t + 4)^2 = 9$ b) $(2m - 5)^2 = 12$

Solution

a) $(3t + 4)^2 = 9$

$$3t + 4 = \pm\sqrt{9} \qquad \text{Square root property}$$
$$3t + 4 = \pm 3$$

This means $3t + 4 = 3$ or $3t + 4 = -3$. Solve both equations.

$$3t + 4 = 3 \qquad \text{or} \qquad 3t + 4 = -3$$
$$3t = -1 \qquad\qquad 3t = -7 \qquad \text{Subtract 4 from each side.}$$
$$t = -\frac{1}{3} \quad \text{or} \qquad t = -\frac{7}{3} \qquad \text{Divide by 3.}$$

The solution set is $\left\{ -\dfrac{7}{3}, -\dfrac{1}{3} \right\}$.

b) $(2m - 5)^2 = 12$

$$2m - 5 = \pm\sqrt{12} \qquad\quad \text{Square root property}$$
$$2m - 5 = \pm 2\sqrt{3} \qquad\quad \text{Simplify } \sqrt{12}.$$
$$2m = 5 \pm 2\sqrt{3} \qquad\quad \text{Add 5 to each side.}$$
$$m = \frac{5 \pm 2\sqrt{3}}{2} \qquad\quad \text{Divide by 2.}$$

One solution is $\dfrac{5 + 2\sqrt{3}}{2}$, and the other is $\dfrac{5 - 2\sqrt{3}}{2}$.

The solution set is $\left\{ \dfrac{5 - 2\sqrt{3}}{2}, \dfrac{5 + 2\sqrt{3}}{2} \right\}$. This can also be written as $\left\{ \dfrac{5 \pm 2\sqrt{3}}{2} \right\}$.

You Try 3

Solve.

a) $(7q + 1)^2 = 36$ b) $(5a - 3)^2 = 24$

3. Use the Distance Formula

In mathematics, we sometimes need to find the distance between two points in a plane. The **distance formula** enables us to do that. We can use the Pythagorean theorem to develop the distance formula.

Let us say that our goal is to find the distance between the points with coordinates (x_1, y_1) and (x_2, y_2) as pictured here. [We also include the point (x_2, y_1) in our drawing so that we get a right triangle.]

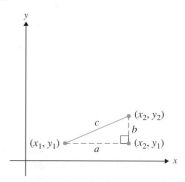

We will label the legs a and b and the hypotenuse c. Recall that a and b represent the *lengths* of the legs and c represents the *length* of the hypotenuse. Our goal is to find the *distance* between (x_1, y_1) and (x_2, y_2), *which is the same as* finding the length of c.

How long is side a? $x_2 - x_1$

How long is side b? $y_2 - y_1$

The Pythagorean theorem states that $a^2 + b^2 = c^2$. Substitute $(x_2 - x_1)$ for a and $(y_2 - y_1)$ for b, then solve for c.

$$a^2 + b^2 = c^2 \qquad \text{Pythagorean theorem}$$
$$(x_2 - x_1)^2 + (y_2 - y_1)^2 = c^2 \qquad \text{Substitute values.}$$
$$\pm\sqrt{(x_2 - x_1)^2 + (y_2 - y_1)^2} = c \qquad \text{Solve for } c.$$

The distance between the two points (x_1, y_1) and (x_2, y_2) above is $c = \sqrt{(x_2 - x_1)^2 + (y_2 - y_1)^2}$. We only want the positive square root since c is a length.

Since we are using this formula to find the *distance* between two points, we usually use the letter d instead of c.

Definition

The Distance Formula
The distance between two points with coordinates (x_1, y_1) and (x_2, y_2) is given by

$$d = \sqrt{(x_2 - x_1)^2 + (y_2 - y_1)^2}.$$

Example 4

Find the distance between the points $(-4, 1)$ and $(2, 5)$.

Solution

Begin by labeling the points: $\overset{x_1, y_1}{(-4, 1)} \; \overset{x_2, y_2}{(2, 5)}$

Substitute the values into the distance formula.

$$d = \sqrt{(x_2 - x_1)^2 + (y_2 - y_1)^2}$$
$$= \sqrt{[2 - (-4)]^2 + (5 - 1)^2} \qquad \text{Substitute values.}$$
$$= \sqrt{(2 + 4)^2 + (4)^2}$$
$$= \sqrt{(6)^2 + (4)^2}$$
$$= \sqrt{36 + 16}$$
$$= \sqrt{52}$$
$$= 2\sqrt{13} \qquad\qquad\qquad \text{Simplify.}$$

 You Try 4

Find the distance between the points $(1, 2)$ and $(7, -3)$.

Answers to You Try Exercises

1) a) $\{-10, 10\}$ b) $\{-4\sqrt{2}, 4\sqrt{2}\}$ c) $\{-2, 2\}$ d) no real number solution   2) $\{-15, 3\}$

3) a) $\left\{-1, \dfrac{5}{7}\right\}$ b) $\left\{\dfrac{3 - 2\sqrt{6}}{5}, \dfrac{3 + 2\sqrt{6}}{5}\right\}$ 4) $\sqrt{61}$

11.2 Exercises

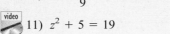

Boost your grade at mathzone.com! MathZone

> Practice Problems   > NetTutor   > e-Professors   > Videos
> Self-Test

Objective 1

1) What are two methods that can be used to solve $y^2 - 16 = 0$? Solve the equation using both methods.

2) If k is a negative number and $x^2 = k$, what can you conclude about the solution to the equation?

Solve using the square root property.

3) $b^2 = 36$

4) $h^2 = 64$

5) $t^2 = -25$

6) $s^2 = -1$

7) $r^2 - 27 = 0$

8) $a^2 - 30 = 0$

9) $n^2 = \dfrac{4}{9}$

10) $v^2 = \dfrac{121}{16}$

11) $z^2 + 5 = 19$

12) $q^2 - 3 = 15$

13) $2d^2 + 5 = 55$

14) $4m^2 + 1 = 37$

15) $4p^2 + 9 = 89$

16) $3j^2 - 11 = 70$

17) $1 = 7 - 6h^2$

18) $145 = 2w^2 - 55$

19) $2 = 11 + 9x^2$

20) $7 = 19 + 6k^2$

Objective 2

Solve using the square root property.

21) $(r + 10)^2 = 4$

22) $(x - 5)^2 = 81$

23) $(q - 7)^2 = 1$

24) $(c + 12)^2 = 25$

25) $(a + 1)^2 = 22$

26) $(t - 8)^2 = 26$

27) $(p + 4)^2 - 18 = 0$

28) $(d + 2)^2 - 7 = 13$

29) $(5y - 2)^2 + 6 = 22$

30) $29 = 4 + (3m + 1)^2$

31) $20 = (2w + 1)^2$

32) $(5b - 6)^2 = 11$

33) $8 = (3q - 10)^2 - 6$

34) $22 = (6x + 11)^2 + 4$

35) $(p + 3)^2 + 4 = 2$

36) $(4d - 19)^2 + 3 = 1$

37) $(10v + 7)^2 - 9 = 15$

38) $(3m + 8)^2 + 16 = 61$

39) $\left(\dfrac{3}{4}n - 8\right)^2 = 4$

40) $\left(\dfrac{2}{3}j + 10\right)^2 = 16$

41) $14 = 10 + (5c + 4)^2$

42) $-6 = 3 - (2q - 9)^2$

Objective 1

Use the Pythagorean theorem and the square root property to find the length of the missing side.

43)

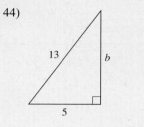

44)

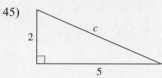

45)

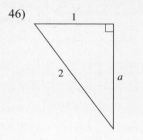

46)

Write an equation and solve. (Hint: Draw a picture.)

47) The width of a rectangle is 4 in., and its diagonal is $2\sqrt{13}$ in. long. What is the length of the rectangle?

48) The length of a rectangle is $3\sqrt{5}$ cm, and its width is 2 cm. How long is the rectangle's diagonal?

49) Find the length of the diagonal of the rectangle if it has a width of 5 cm and a length of $4\sqrt{2}$ cm.

50) The width of a rectangle is 3 in., and its diagonal is $\sqrt{73}$ in. long. Find the length of the rectangle.

Write an equation and solve.

 51) A 13-ft ladder is leaning against a wall so that the base of the ladder is 5 ft away from the wall. How high on the wall does the ladder reach?

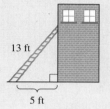

52) Salma is flying a kite. It is 30 ft from her horizontally, and it is 40 ft above her hand. How long is the kite string?

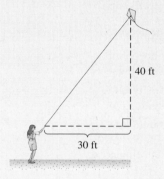

Objective 3

Find the distance between the given points.

53) $(7, -1)$ and $(3, 2)$ 54) $(3, 10)$ and $(12, 6)$

55) $(-5, -6)$ and $(-2, -8)$

56) $(5, -2)$ and $(-3, 4)$ 57) $(0, 13)$ and $(0, 7)$

58) $(-8, 3)$ and $(2, 1)$ 59) $(-4, 11)$ and $(2, 6)$

60) $(0, 3)$ and $(3, -1)$ 61) $(3, -3)$ and $(5, -7)$

62) $(-5, -6)$ and $(-1, 2)$

Section 11.3   Complex Numbers

Objectives

1. Define a Complex Number
2. Find the Square Root of a Negative Number
3. Multiply and Divide Square Roots Containing Negative Numbers
4. Add and Subtract Complex Numbers
5. Multiply Complex Numbers
6. Multiply a Complex Number by Its Conjugate
7. Divide Complex Numbers
8. Use the Square Root Property to Solve an Equation of the Form $(ax + b)^2 = k$, Where $k < 0$

1. Define a Complex Number

In Section 11.2, we saw that not all quadratic equations have real number solutions. For example, there is no real number solution to $x^2 = -1$ since the square of a real number will not be negative.

$x^2 = -1$ *does* have a solution under another set of numbers called *complex numbers.*

Before we define a complex number, we must define the number *i*. *i* is called an *imaginary number*.

Definition	The **imaginary number** *i* is defined as

$$i = \sqrt{-1}$$

Therefore, squaring both sides gives us

$$i^2 = -1$$

$i = \sqrt{-1}$ and $i^2 = -1$ are two *very* important facts to remember. We will be using them often!

Definition	A **complex number** is a number of the form $a + bi$, where a and b are real numbers. a is called the **real part** and b is called the **imaginary part.**

The following table lists some examples of complex numbers and their real and imaginary parts.

Complex Number	Real Part	Imaginary Part
$5 + 2i$	5	2
$\dfrac{1}{3} - 7i$	$\dfrac{1}{3}$	-7
$8i$	0	8
4	4	0

$8i$ can be written in the form $a + bi$ as $0 + 8i$. Likewise, besides being a real number, 4 is a complex number since it can be written as $4 + 0i$.

Since all real numbers, a, can be written in the form $a + 0i$, all real numbers are also complex numbers.

> The set of real numbers is a subset of the set of complex numbers.

Since we defined i as $i = \sqrt{-1}$, we can now evaluate square roots of negative numbers.

2. Find the Square Root of a Negative Number

Example 1

Simplify.

a) $\sqrt{-9}$ b) $\sqrt{-7}$ c) $\sqrt{-12}$

Solution

a) $\sqrt{-9} = \sqrt{-1 \cdot 9} = \sqrt{-1} \cdot \sqrt{9} = i \cdot 3 = 3i$
b) $\sqrt{-7} = \sqrt{-1 \cdot 7} = \sqrt{-1} \cdot \sqrt{7} = i\sqrt{7}$
c) $\sqrt{-12} = \sqrt{-1 \cdot 12} = \sqrt{-1} \cdot \sqrt{12} = i\sqrt{4}\sqrt{3} = i \cdot 2\sqrt{3} = 2i\sqrt{3}$

> In Example 1b) we wrote $i\sqrt{7}$ instead of $\sqrt{7}i$, and in Example 1c) we wrote $2i\sqrt{3}$ instead of $2\sqrt{3}i$. That is because we want to avoid the confusion of thinking that the i is under the radical. It is good practice to write the i *before* the radical.

You Try 1

Simplify.

a) $\sqrt{-36}$ b) $\sqrt{-13}$ c) $\sqrt{-20}$

3. Multiply and Divide Square Roots Containing Negative Numbers

When multiplying or dividing radicals with negative radicands, write each radical in terms of i first. Remember, also, that since $i = \sqrt{-1}$ it follows that $i^2 = -1$. We must keep this in mind when simplifying expressions.

> Whenever an i^2 appears in an expression, replace it with -1.

Example 2

Multiply and simplify. $\sqrt{-8} \cdot \sqrt{-2}$

Solution

$$\begin{aligned}
\sqrt{-8} \cdot \sqrt{-2} &= i\sqrt{8} \cdot i\sqrt{2} && \text{Write each radical in terms of } i \text{ before multiplying.} \\
&= i^2\sqrt{16} && \text{Multiply.} \\
&= (-1)(4) && \text{Replace } i^2 \text{ with } -1. \\
&= -4
\end{aligned}$$

You Try 2

Perform the operation and simplify.

a) $\sqrt{-6} \cdot \sqrt{-3}$ b) $\dfrac{\sqrt{-72}}{\sqrt{-2}}$

4. Add and Subtract Complex Numbers

Just as we can add, subtract, multiply, and divide real numbers, we can perform all of these operations with complex numbers.

Adding and Subtracting Complex Numbers

1) To add complex numbers, add the real parts and add the imaginary parts.

2) To subtract complex numbers, apply the distributive property and combine the real parts and combine the imaginary parts.

Example 3

Add or subtract.

a) $(8 + 3i) + (4 + 2i)$ b) $(7 + i) - (3 - 4i)$

Solution

$$\begin{aligned}
\text{a)} \quad (8 + 3i) + (4 + 2i) &= (8 + 4) + (3 + 2)i && \text{Add real parts; add imaginary parts.} \\
&= 12 + 5i
\end{aligned}$$

$$\begin{aligned}
\text{b)} \quad (7 + i) - (3 - 4i) &= 7 + i - 3 + 4i && \text{Distributive property} \\
&= (7 - 3) + (1 + 4)i && \text{Add real parts; add imaginary parts.} \\
&= 4 + 5i
\end{aligned}$$

You Try 3

Add or subtract.

a) $(-10 + 6i) + (1 + 8i)$ b) $(2 - 5i) - (-1 + 6i)$

5. Multiply Complex Numbers

We multiply complex numbers just like we would multiply polynomials. There may be an additional step, however. Remember to replace i^2 with -1.

Example 4

Multiply and simplify.

a) $5(-2 + 3i)$　　　　b) $(8 + 3i)(-1 + 4i)$　　　　c) $(6 + 2i)(6 - 2i)$

Solution

a) $5(-2 + 3i) = -10 + 15i$　　　Distributive property

b) Look carefully at $(8 + 3i)(-1 + 4i)$. Each complex number is a *binomial* [similar to, say, $(x + 3)(x + 4)$]. How can we multiply two binomials? We can use FOIL.

$$(8 + 3i)(-1 + 4i) = (8)(-1) + (8)(4i) + (3i)(-1) + (3i)(4i)$$
$$\quad \text{F} \qquad\quad \text{O} \qquad\quad \text{I} \qquad\quad \text{L}$$
$$= \quad -8 \quad + \quad 32i \quad - \quad 3i \quad + \quad 12i^2$$
$$= -8 + 29i + 12(-1) \qquad\qquad \text{Replace } i^2 \text{ with } -1.$$
$$= -8 + 29i - 12$$
$$= -20 + 29i$$

c) $(6 + 2i)$ and $(6 - 2i)$ are each binomials, so we will multiply them using FOIL.

$$(6 + 2i)(6 - 2i) = (6)(6) + (6)(-2i) + (2i)(6) + (2i)(-2i)$$
$$\quad \text{F} \qquad\quad \text{O} \qquad\quad \text{I} \qquad\quad \text{L}$$
$$= \quad 36 \quad - \quad 12i \quad + \quad 12i \quad - \quad 4i^2$$
$$= 36 - 4(-1) \qquad\qquad\qquad \text{Replace } i^2 \text{ with } -1.$$
$$= 36 + 4$$
$$= 40$$

You Try 4

Multiply and simplify.

a) $-3(6 - 7i)$　　　b) $(5 - i)(4 + 8i)$　　　c) $(-2 - 9i)(-2 + 9i)$

---

6. Multiply a Complex Number by Its Conjugate

In Chapter 10, we learned about conjugates of radical expressions. For example, the conjugate of $3 + \sqrt{5}$ is $3 - \sqrt{5}$.

The complex numbers in Example 4c, $6 + 2i$ and $6 - 2i$, are **complex conjugates**.

Definition　　The **conjugate** of $a + bi$ is $a - bi$.

We found that $(6 + 2i)(6 - 2i) = 40$ — the result is a real number. The product of a complex number and its conjugate is *always* a real number, as illustrated next.

$$(a + bi)(a - bi) = (a)(a) + (a)(-bi) + (bi)(a) + (bi)(-bi)$$
$$ \underset{F}{} \quad \underset{O}{} \quad \underset{I}{} \quad \underset{L}{}$$
$$= a^2 \ - \ abi \ + \ abi \ - \ b^2 i^2$$
$$= a^2 - b^2(-1) \qquad \text{Replace } i^2 \text{ with } -1.$$
$$= a^2 + b^2$$

We can summarize these facts about complex numbers and their conjugates as follows:

Complex Conjugates

1)   The conjugate of $a + bi$ is $a - bi$.

2)   The product of $a + bi$ and $a - bi$ is a real number.

3)   We can find the product $(a + bi)(a - bi)$ by using FOIL or by using $(a + bi)(a - bi) = a^2 + b^2$.

Example 5

Multiply $-3 + 4i$ by its conjugate using the formula $(a + bi)(a - bi) = a^2 + b^2$.

Solution

The conjugate of $-3 + 4i$ is $-3 - 4i$.

$$(-3 + 4i)(-3 - 4i) = (-3)^2 + (4)^2 \qquad a = -3, b = 4$$
$$= 9 + 16$$
$$= 25$$

 You Try 5

Multiply $2 - 9i$ by its conjugate using the formula $(a + bi)(a - bi) = a^2 + b^2$.

7. Divide Complex Numbers

To rationalize the denominator of a radical expression like $\dfrac{2}{3 + \sqrt{5}}$, we multiply the numerator and denominator by $3 - \sqrt{5}$, the conjugate of the denominator. We divide complex numbers in the same way.

Dividing Complex Numbers

To divide complex numbers, multiply the numerator and denominator by the *conjugate of the denominator*. Write the result in the form $a + bi$.

Example 6

Divide. Write the result in the form $a + bi$.

a) $\dfrac{3}{4 - 5i}$

b) $\dfrac{6 - 2i}{-7 + i}$

Solution

a) $\dfrac{3}{4 - 5i} = \dfrac{3}{(4 - 5i)} \cdot \dfrac{(4 + 5i)}{(4 + 5i)}$ Multiply the numerator and denominator by the conjugate of the denominator.

$= \dfrac{12 + 15i}{16 + 20i - 20i - 25i^2}$ Multiply numerators.
Multiply the denominators using FOIL.

$= \dfrac{12 + 15i}{16 - 25(-1)}$ Replace i^2 with -1.

$= \dfrac{12 + 15i}{16 + 25}$

$= \dfrac{12 + 15i}{41}$

$= \dfrac{12}{41} + \dfrac{15}{41}i$ Write the result in the form $a + bi$.

Recall that we can find the product $(4 - 5i)(4 + 5i)$ using FOIL *or* by using the formula $(a + bi)(a - bi) = a^2 + b^2$.

b) $\dfrac{6 - 2i}{-7 + i} = \dfrac{(6 - 2i)}{(-7 + i)} \cdot \dfrac{(-7 - i)}{(-7 - i)}$ Multiply the numerator and denominator by the conjugate of the denominator.

$= \dfrac{-42 - 6i + 14i + 2i^2}{49 + 7i - 7i - i^2}$ Multiply using FOIL.

$= \dfrac{-42 + 8i + 2(-1)}{49 - (-1)}$ Replace i^2 with -1.

$= \dfrac{-42 + 8i - 2}{49 + 1} = \dfrac{-44 + 8i}{50} = -\dfrac{44}{50} + \dfrac{8}{50}i = -\dfrac{22}{25} + \dfrac{4}{25}i$

You Try 6

Divide. Write the result in the form $a + bi$.

a) $\dfrac{6}{-2 + i}$

b) $\dfrac{5 + 3i}{-6 - 4i}$

Now that we have defined i as $\sqrt{-1}$, we can find complex solutions to quadratic equations.

8. Use the Square Root Property to Solve an Equation of the Form $(ax + b)^2 = k$, Where $k < 0$

Example 7

Solve.

a) $x^2 = -64$ b) $(z + 8)^2 + 11 = 7$ c) $(6k - 5)^2 + 20 = 0$

Solution

a) $x^2 = -64$
$x = \pm\sqrt{-64}$ Square root property
$x = \pm 8i$

Check:
$x = 8i$: $x^2 = -64$ $x = -8i$: $x^2 = -64$
$(8i)^2 \stackrel{?}{=} -64$ $(-8i)^2 \stackrel{?}{=} -64$
$64i^2 \stackrel{?}{=} -64$ $64i^2 \stackrel{?}{=} -64$
$64(-1) \stackrel{?}{=} -64$ $64(-1) \stackrel{?}{=} -64$
$-64 = -64$ ✓ $-64 = -64$ ✓

The solution set is $\{-8i, 8i\}$.

b) $(z + 8)^2 + 11 = 7$
$(z + 8)^2 = -4$ Subtract 11 from each side.
$z + 8 = \pm\sqrt{-4}$ Square root property
$z + 8 = \pm 2i$
$z = -8 \pm 2i$ Subtract 8 from each side.

The check is left to the student.
The solution set is $\{-8 - 2i, -8 + 2i\}$.

c) $(6k - 5)^2 + 20 = 0$
$(6k - 5)^2 = -20$ Subtract 20 from each side.
$6k - 5 = \pm\sqrt{-20}$ Square root property
$6k - 5 = \pm 2i\sqrt{5}$ Simplify $\sqrt{-20}$.
$6k = 5 \pm 2i\sqrt{5}$ Add 5 to each side.
$k = \dfrac{5 \pm 2i\sqrt{5}}{6}$ Divide by 6.

The check is left to the student.

The solution set is $\left\{\dfrac{5 - 2i\sqrt{5}}{6}, \dfrac{5 + 2i\sqrt{5}}{6}\right\}$.

Did you notice that in each example a complex number *and* its conjugate were the solutions to the equation? This will always be true provided that the variables in the equation have real number coefficients.

> If $a + bi$ is a solution of a quadratic equation having only real coefficients, then $a - bi$ is also a solution.

You Try 7

Solve.

a) $m^2 = -144$ b) $(c - 7)^2 + 100 = 0$ c) $(2y + 3)^2 - 5 = -23$

Answers to You Try Exercises

1) a) $6i$ b) $i\sqrt{13}$ c) $2i\sqrt{5}$ 2) a) $-3\sqrt{2}$ b) 6   3) a) $(-9 + 14i)$ b) $3 - 11i$

4) a) $-18 + 21i$ b) $28 + 36i$ c) 40   5) 85   6) a) $-\dfrac{12}{5} - \dfrac{6}{5}i$ b) $-\dfrac{21}{26} + \dfrac{1}{26}i$

7) a) $\{-12i, 12i\}$ b) $\{7 + 10i, 7 - 10i\}$ c) $\left\{ \dfrac{-3 - 3i\sqrt{2}}{2}, \dfrac{-3 + 3i\sqrt{2}}{2} \right\}$

11.3 Exercises

Objectives 1 and 2

Determine if each statement is true or false.

1) Every complex number is a real number.

2) Every real number is a complex number.

3) Since $i = \sqrt{-1}$, it follows that $i^2 = -1$.

4) In the complex number $-6 + 5i$, -6 is the real part and $5i$ is the imaginary part.

Simplify.

5) $\sqrt{-81}$

6) $\sqrt{-16}$

7) $\sqrt{-25}$

8) $\sqrt{-169}$

9) $\sqrt{-6}$

10) $\sqrt{-30}$

 11) $\sqrt{-27}$

12) $\sqrt{-75}$

13) $\sqrt{-60}$

14) $\sqrt{-28}$

Objective 3

Find the error in each of the following exercises, then find the correct answer.

15) $\sqrt{-5} \cdot \sqrt{-10} = \sqrt{-5 \cdot (-10)}$
$\qquad = \sqrt{50}$
$\qquad = \sqrt{25} \cdot \sqrt{2}$
$\qquad = 5\sqrt{2}$

16) $(\sqrt{-7})^2 = \sqrt{(-7)^2}$
$\qquad = \sqrt{49}$
$\qquad = 7$

Perform the indicated operation and simplify.

17) $\sqrt{-1} \cdot \sqrt{-5}$

18) $\sqrt{-5} \cdot \sqrt{-15}$

19) $\sqrt{-12} \cdot \sqrt{-3}$

20) $\sqrt{-20} \cdot \sqrt{-5}$

21) $\dfrac{\sqrt{-60}}{\sqrt{-15}}$

22) $\dfrac{\sqrt{-2}}{\sqrt{-128}}$

23) $(\sqrt{-13})^2$

24) $(\sqrt{-1})^2$

Objective 4

25) Explain how to add complex numbers.

26) How is multiplying $(1 + 3i)(2 - 7i)$ similar to multiplying $(x + 3)(2x - 7)$?

27) When i^2 appears in an expression, it should be replaced with what?

28) Explain how to divide complex numbers.

Perform the indicated operations.

29) $(-4 + 9i) + (7 + 2i)$

30) $(6 + i) + (8 - 5i)$

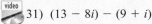 31) $(13 - 8i) - (9 + i)$

32) $(-12 + 3i) - (-7 - 6i)$

33) $\left(-\dfrac{3}{4} - \dfrac{1}{6}i \right) - \left(-\dfrac{1}{2} + \dfrac{2}{3}i \right)$

34) $\left(\dfrac{1}{2} + \dfrac{7}{9}i \right) - \left(\dfrac{7}{8} - \dfrac{1}{6}i \right)$

35) $16i - (3 + 10i) + (3 + i)$

36) $(-6 - 5i) + (2 + 6i) - (-4 + i)$

Objective 5

Multiply and simplify.

37) $3(8 - 5i)$

38) $-6(8 - i)$

39) $\frac{2}{3}(-9 + 2i)$

40) $\frac{1}{2}(18 + 7i)$

41) $6i(5 + 6i)$

42) $-4i(6 + 11i)$

43) $(2 + 5i)(1 + 6i)$

44) $(2 + i)(10 + 5i)$

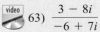 45) $(-1 + 3i)(4 - 6i)$

46) $(-4 - 9i)(3 - i)$

47) $(5 - 3i)(9 - 3i)$

48) $(3 - 4i)(6 + 7i)$

49) $\left(\frac{3}{4} + \frac{3}{4}i\right)\left(\frac{2}{5} + \frac{1}{5}i\right)$

50) $\left(\frac{1}{3} - \frac{4}{3}i\right)\left(\frac{3}{4} + \frac{2}{3}i\right)$

Objective 6

Identify the conjugate of each complex number, then multiply the number and its conjugate.

51) $11 + 4i$

52) $-1 - 2i$

53) $-3 - 7i$

54) $4 + 9i$

55) $-6 + 4i$

56) $6 - 5i$

Objective 7

Divide. Write the result in the form $a + bi$.

57) $\dfrac{4}{2 - 3i}$

58) $\dfrac{-10}{8 - 9i}$

59) $\dfrac{8i}{4 + i}$

60) $\dfrac{i}{6 - 5i}$

61) $\dfrac{2i}{-3 + 7i}$

62) $\dfrac{9i}{-4 + 10i}$

 63) $\dfrac{3 - 8i}{-6 + 7i}$

64) $\dfrac{-5 + 2i}{4 - i}$

65) $\dfrac{2 + 3i}{5 - 6i}$

66) $\dfrac{1 + 6i}{5 + 2i}$

67) $\dfrac{9}{i}$

68) $\dfrac{16 + 3i}{-i}$

Objective 8

Solve.

69) $q^2 = -4$

70) $w^2 = -121$

71) $z^2 + 3 = 0$

72) $h^2 + 14 = -23$

73) $5f^2 + 39 = -21$

74) $2y^2 + 56 = 0$

75) $63 = 7x^2$

76) $4 = 9r^2$

77) $(c + 3)^2 - 4 = -29$

78) $(u - 15)^2 - 4 = -8$

79) $1 = 15 + (k - 2)^2$

80) $2 = 14 + (g + 4)^2$

81) $-3 = (m - 9)^2 + 5$

82) $-8 = (t + 10)^2 - 40$

83) $36 + (4p - 5)^2 = 6$

84) $(3k - 1)^2 + 20 = 4$

85) $(6g + 11)^2 + 50 = 1$

86) $9 = 38 + (9s - 4)^2$

87) $n^3 + 49n = 0$

88) $d^3 + 25d = 0$

89) $q^3 = -54q$

90) $-125a = a^3$

Section 11.4 Solving Quadratic Equations by Completing the Square

Objectives

1. Complete the Square for an Expression of the Form $ax^2 + bx$, Where b Is an Even Integer

2. Complete the Square for an Expression of the Form $ax^2 + bx$, Where b Is an Odd Integer or Where b Is a Fraction

3. Solve an Equation of the Form $ax^2 + bx + c = 0$ by Completing the Square

The third method we will learn for solving a quadratic equation is **completing the square**. We need to review an idea first presented in Section 7.4.

Completing the Square

A **perfect square trinomial** is a trinomial whose factored form is the square of a binomial. Some examples of perfect square trinomials are

Perfect Square Trinomials	Factored Form
$x^2 + 10x + 25$	$(x + 5)^2$
$d^2 - 8d + 16$	$(d - 4)^2$

In the trinomial $x^2 + 10x + 25$, x^2 is called the *quadratic term*, $10x$ is called the *linear term*, and 25 is called the *constant*.

1. Complete the Square for an Expression of the Form $ax^2 + bx$, Where b Is an Even Integer

In a perfect square trinomial where the coefficient of the quadratic term is 1, the constant term is related to the coefficient of the linear term in the following way: *If you find half of the linear coefficient and square the result, you will get the constant term.*

$x^2 + 10x + 25$: The constant, 25, is obtained by

1)   finding half of the coefficient of x

$$\frac{1}{2}(10) = 5$$

2)   then squaring the result.

$$5^2 = 25 \text{ (the constant)}$$

$d^2 - 8d + 16$: The constant, 16, is obtained by

1)   finding half of the coefficient of d

$$\frac{1}{2}(-8) = -4$$

2)   then squaring the result.

$$(-4)^2 = 16 \text{ (the constant)}$$

We can generalize this procedure so that we can find the constant needed to obtain the perfect square trinomial for any quadratic expression of the form $x^2 + bx$. Finding this perfect square trinomial is called *completing the square* because the trinomial will factor to the square of a binomial.

To find the constant needed to complete the square for $x^2 + bx$:

Step 1: Find half of the coefficient of x: $\left(\dfrac{1}{2}b\right)$.

Step 2: Square the result: $\left(\dfrac{1}{2}b\right)^2$

Step 3: Then add it to $x^2 + bx$ to get $x^2 + bx + \left(\dfrac{1}{2}b\right)^2$.

 BE CAREFUL   The coefficient of the squared term *must* be 1 before you complete the square!

Example 1

Complete the square for each expression to obtain a perfect square trinomial. Then, factor.

a) $y^2 + 6y$ b) $t^2 - 14t$

Solution

a) Find the constant needed to complete the square for $y^2 + 6y$.

    Step 1: Find half of the coefficient of y:

$$\frac{1}{2}(6) = 3$$

    Step 2: Square the result:

$$3^2 = 9$$

    Step 3: Add 9 to $y^2 + 6y$:

$$y^2 + 6y + 9$$

    The perfect square trinomial is $y^2 + 6y + 9$.
    The factored form is $(y + 3)^2$.

b) Find the constant needed to complete the square for $t^2 - 14t$.

    Step 1: Find half of the coefficient of t:

$$\frac{1}{2}(-14) = -7$$

    Step 2: Square the result:

$$(-7)^2 = 49$$

Step 3: Add 49 to $t^2 - 14t$:

$$t^2 - 14t + 49$$

The perfect square trinomial is $t^2 - 14t + 49$.
The factored form is $(t - 7)^2$.

You Try 1

Complete the square for each expression to obtain a perfect square trinomial. Then, factor.
a) $w^2 + 2w$ b) $z^2 - 16z$

At the beginning of the section and in Example 1 we saw the following perfect square trinomials and their factored forms. We will look at the relationship between the constant in the factored form and the coefficient of the linear term.

Perfect Square Trinomial		Factored Form
$x^2 + 10x + 25$	5 is $\frac{1}{2}(10)$.	$(x + 5)^2$
$d^2 - 8d + 16$	-4 is $\frac{1}{2}(-8)$.	$(d - 4)^2$
$y^2 + 6y + 9$	3 is $\frac{1}{2}(6)$.	$(d + 3)^2$
$t^2 - 14t + 49$	-7 is $\frac{1}{2}(-14)$.	$(t - 7)^2$

This pattern will always hold true and can be helpful in factoring some perfect square trinomials.

2. Complete the Square for an Expression of the Form $ax^2 + bx$, Where b Is an Odd Integer or Where b Is a Fraction

Example 2

Complete the square for each expression to obtain a perfect square trinomial. Then, factor.

a) $n^2 + 5n$ b) $k^2 - \dfrac{2}{3}k$

Solution

a) Find the constant needed to complete the square for $n^2 + 5n$.

Step 1: Find half of the coefficient of n:

$$\frac{1}{2}(5) = \frac{5}{2}$$

Step 2:   Square the result:

$$\left(\frac{5}{2}\right)^2 = \frac{25}{4}$$

Step 3:   Add $\frac{25}{4}$ to $n^2 + 5n$.

The perfect square trinomial is $n^2 + 5n + \frac{25}{4}$.

The factored form is $\left(n + \frac{5}{2}\right)^2$

$\frac{5}{2}$ is $\frac{1}{2}(5)$, the coefficient of n.

Check: $\left(n + \frac{5}{2}\right)^2 = n^2 + 2n\left(\frac{5}{2}\right) + \left(\frac{5}{2}\right)^2 = n^2 + 5n + \frac{25}{4}$

b)   Find the constant needed to complete the square for $k^2 - \frac{2}{3}k$.

Step 1:   Find half of the coefficient of k:

$$\frac{1}{2}\left(-\frac{2}{3}\right) = -\frac{1}{3}$$

Step 2:   Square the result:

$$\left(-\frac{1}{3}\right)^2 = \frac{1}{9}$$

Step 3:   Add $\frac{1}{9}$ to $k^2 - \frac{2}{3}k$:

$$k^2 - \frac{2}{3}k + \frac{1}{9}$$

The perfect square trinomial is $k^2 - \frac{2}{3}k + \frac{1}{9}$.

The factored form is $\left(k - \frac{1}{3}\right)^2$

$-\frac{1}{3}$ is $\frac{1}{2}\left(-\frac{2}{3}\right)$, the coefficient of k.

Check: $\left(k - \frac{1}{3}\right)^2 = k^2 + 2k\left(-\frac{1}{3}\right) + \left(-\frac{1}{3}\right)^2 = k^2 - \frac{2}{3}k + \frac{1}{9}$.

You Try 2

Complete the square for each expression to obtain a perfect square trinomial. Then, factor.

a) $p^2 - 3p$ b) $m^2 + \frac{3}{4}m$

3. Solve an Equation of the Form $ax^2 + bx + c = 0$ by Completing the Square

Any quadratic equation of the form $ax^2 + bx + c = 0$ $(a \neq 0)$ can be written in the form $(x - h)^2 = k$ by completing the square. Once an equation is in this form, we can use the square root property to solve for the variable.

Steps for Solving a Quadratic Equation ($ax^2 + bx + c = 0$) by Completing the Square

Step 1: **The coefficient of the squared term must be 1.** If it is not 1, divide both sides of the equation by a to obtain a leading coefficient of 1.

Step 2: **Get the variables on one side of the equal sign and the constant on the other side.**

Step 3: **Complete the square.** Find half of the linear coefficient, then square the result. Add that quantity to *both* sides of the equation.

Step 4: **Factor.**

Step 5: **Solve using the square root property.**

Example 3

Solve by completing the square.

a) $x^2 + 6x + 8 = 0$ b) $k^2 - 2k + 17 = 0$ c) $12h + 4h^2 = 6$

Solution

a) $x^2 + 6x + 8 = 0$

   Step 1:   The coefficient of x^2 is already 1.

   Step 2:   Get the variables on one side of the equal sign and the constant on the other side.

$$x^2 + 6x = -8$$

   Step 3:   Complete the square.

$$\frac{1}{2}(6) = 3$$
$$3^2 = 9$$

   Add 9 to both sides of the equation.

$$x^2 + 6x + 9 = -8 + 9$$
$$x^2 + 6x + 9 = 1$$

   Step 4:   Factor.

$$(x + 3)^2 = 1$$

Step 5: Solve using the square root property.

$$(x + 3)^2 = 1$$
$$x + 3 = \pm\sqrt{1}$$
$$x + 3 = \pm 1$$

$$x + 3 = 1 \quad \text{or} \quad x + 3 = -1$$
$$x = -2 \quad \text{or} \quad x = -4$$

The check is left to the student.
The solution set is $\{-4, -2\}$.

Notice that we would have obtained the same result if we had solved the equation by factoring.

$$x^2 + 6x + 8 = 0$$
$$(x + 4)(x + 2) = 0$$

$$x + 4 = 0 \quad \text{or} \quad x + 2 = 0$$
$$x = -4 \quad \text{or} \quad x = -2$$

---

b) $k^2 - 2k + 17 = 0$

Step 1: The coefficient of k^2 is already 1.

Step 2: Get the variables on one side of the equal sign and the constant on the other side.

$$k^2 - 2k = -17$$

Step 3: Complete the square.

$$\frac{1}{2}(-2) = -1$$
$$(-1)^2 = 1$$

Add 1 to both sides of the equation.

$$k^2 - 2k + 1 = -17 + 1$$
$$k^2 - 2k + 1 = -16$$

Step 4: Factor.

$$(k - 1)^2 = -16$$

Step 5: Solve using the square root property.

$$(k - 1)^2 = -16$$
$$k - 1 = \pm\sqrt{-16}$$
$$k - 1 = \pm 4i$$
$$k = 1 \pm 4i$$

The check is left to the student.
The solution set is $\{1 - 4i, 1 + 4i\}$.

c) $12h + 4h^2 = 6$

Step 1: Since the coefficient of h^2 is *not* 1, divide the whole equation by 4.

$$\frac{12h}{4} + \frac{4h^2}{4} = \frac{6}{4}$$

$$3h + h^2 = \frac{3}{2}$$

Step 2: The constant is on a side by itself. Rewrite the left side of the equation.

$$h^2 + 3h = \frac{3}{2}$$

Step 3: Complete the square.

$$\frac{1}{2}(3) = \frac{3}{2}$$

$$\left(\frac{3}{2}\right)^2 = \frac{9}{4}$$

Add $\frac{9}{4}$ to both sides of the equation.

$$h^2 + 3h + \frac{9}{4} = \frac{3}{2} + \frac{9}{4}$$

$$h^2 + 3h + \frac{9}{4} = \frac{6}{4} + \frac{9}{4} \qquad \text{Get a common denominator.}$$

$$h^2 + 3h + \frac{9}{4} = \frac{15}{4}$$

Step 4: Factor.

$$\left(h + \frac{3}{2}\right)^2 = \frac{15}{4}$$

$$\uparrow$$

$$\frac{3}{2} \text{ is } \frac{1}{2}(3), \text{ the coefficient of } h.$$

Step 5: Solve using the square root property.

$$\left(h + \frac{3}{2}\right)^2 = \frac{15}{4}$$

$$h + \frac{3}{2} = \pm\sqrt{\frac{15}{4}}$$

$$h + \frac{3}{2} = \pm\frac{\sqrt{15}}{2}$$

$$h = -\frac{3}{2} \pm \frac{\sqrt{15}}{2}$$

The check is left to the student.

The solution set is $\left\{ -\frac{3}{2} - \frac{\sqrt{15}}{2}, -\frac{3}{2} + \frac{\sqrt{15}}{2} \right\}$.

You Try 3

Solve by completing the square.

a) $q^2 + 10q - 24 = 0$ b) $r^2 - 8r - 1 = 0$ c) $2m^2 + 16 = 10m$

Answers to You Try Exercises

1) a) $w^2 + 2w + 1; (w + 1)^2$ b) $z^2 - 16z + 64; (z - 8)^2$ 2) a) $p^2 - 3p + \dfrac{9}{4}; \left(p - \dfrac{3}{2}\right)^2$

b) $m^2 + \dfrac{3}{4}m + \dfrac{9}{64}; \left(m + \dfrac{3}{8}\right)^2$ 3) a) $\{-12, 2\}$ b) $\{4 - \sqrt{17}, 4 + \sqrt{17}\}$

c) $\left\{\dfrac{5}{2} - \dfrac{\sqrt{7}}{2}i, \dfrac{5}{2} + \dfrac{\sqrt{7}}{2}i\right\}$

11.4 Exercises

Objectives 1 and 2

1) What is a perfect square trinomial? Give an example.

2) In $2x^2 - 8x + 10$, what is the

  a) quadratic term?

  b) linear term?

  c) constant?

3) Can you complete the square on $3y^2 + 15y$ as it is given? Why or why not?

4) What is the first thing you should do if you want to solve $2p^2 - 7p = 8$ by completing the square?

Complete the square for each expression to obtain a perfect square trinomial. Then, factor.

5) $a^2 + 12a$ 6) $g^2 + 4g$

7) $c^2 - 18c$ 8) $k^2 - 16k$

9) $r^2 + 3r$ 10) $z^2 - 7z$

11) $b^2 - 9b$ 12) $t^2 + 5t$

13) $x^2 + \dfrac{1}{3}x$ 14) $y^2 - \dfrac{3}{5}y$

Objective 3

15) What are the steps used to solve a quadratic equation by completing the square?

16) Can $x^3 + 10x - 3 = 0$ be solved by completing the square? Give a reason for your answer.

Solve by completing the square.

17) $x^2 + 6x + 8 = 0$ 18) $t^2 + 12t - 13 = 0$

19) $k^2 - 8k + 15 = 0$ 20) $v^2 - 6v - 27 = 0$

21) $s^2 + 10 = -10s$ 22) $u^2 - 9 = 2u$

23) $t^2 = 2t - 9$ 24) $p^2 = -10p - 26$

25) $v^2 + 4v + 8 = 0$ 26) $c^2 + 4c = 14$

27) $d^2 + 3 = 12d$ 28) $a^2 + 19 = 8a$

29) $m^2 + 3m - 40 = 0$

30) $p^2 + 5p + 4 = 0$

31) $x^2 - 7x + 12 = 0$

32) $d^2 + d - 72 = 0$

33) $r^2 - r = 3$

34) $y^2 - 3y = 7$

35) $c^2 + 5c + 7 = 0$

36) $b^2 + 12 = 7b$

[video] 37) $3k^2 - 6k + 12 = 0$

38) $4f^2 + 16f + 48 = 0$

39) $4r^2 + 24r = 8$

40) $3h^2 + 12h = 15$

41) $10d = 2d^2 + 12$

42) $54x - 6x^2 = 72$

43) $2n^2 + 8 = 5n$

44) $2t^2 + 3t + 4 = 0$

45) $4a^2 - 7a + 3 = 0$

46) $n + 2 = 3n^2$

47) $(y + 5)(y - 3) = 5$

48) $(b - 4)(b + 10) = -17$

49) $(2m + 1)(m - 3) = -7$

50) $(3c + 4)(c + 2) = 3$

Solve each problem by writing an equation and solving it by completing the square.

51) The length of a rectangle is 8 in. more than its width. Find the dimensions of the rectangle if it has an area of 153 in^2.

52) The area of a rectangle is 253 cm^2. Its width is 12 cm less than its length. What are the dimensions of the rectangle?

Section 11.5 Solving Equations Using the Quadratic Formula

Objectives

1. Derive the Quadratic Formula

2. Solve a Quadratic Equation Using the Quadratic Formula

3. Determine the Number and Type of Solutions to a Quadratic Equation Using the Discriminant

4. Solve an Applied Problem Using the Quadratic Formula

1. Derive the Quadratic Formula

In Section 11.4 we saw that any quadratic equation of the form $ax^2 + bx + c = 0$ $(a \neq 0)$ can be solved by completing the square. Therefore, we can solve equations like $x^2 - 8x + 5 = 0$ and $2x^2 + 3x - 1 = 0$ using this method.

We can develop *another* method for solving quadratic equations if we can complete the square on the general quadratic equation $ax^2 + bx + c = 0$ $(a \neq 0)$. We will derive the *quadratic formula*.

The steps we use to complete the square on $ax^2 + bx + c = 0$ are *exactly* the same steps we use to solve an equation like $2x^2 + 3x - 1 = 0$. We will do these steps side by side so that you can more easily understand how we are solving $ax^2 + bx + c = 0$ by completing the square.

Solve by Completing the Square

$$2x^2 + 3x - 1 = 0 \qquad\qquad ax^2 + bx + c = 0$$

Step 1: **The coefficient of the squared term must be 1.**

$$2x^2 + 3x - 1 = 0 \qquad\qquad ax^2 + bx + c = 0$$

$$\frac{2x^2}{2} + \frac{3x}{2} - \frac{1}{2} = \frac{0}{2} \quad \text{Divide by 2.} \qquad \frac{ax^2}{a} + \frac{bx}{a} + \frac{c}{a} = \frac{0}{a} \quad \text{Divide by } a.$$

$$x^2 + \frac{3}{2}x - \frac{1}{2} = 0 \quad \text{Simplify.} \qquad x^2 + \frac{b}{a}x + \frac{c}{a} = 0 \quad \text{Simplify.}$$

Step 2: **Get the constant on the other side of the equal sign.**

$$x^2 + \frac{3}{2}x = \frac{1}{2} \quad \text{Add } \frac{1}{2}. \qquad\qquad x^2 + \frac{b}{a}x = -\frac{c}{a} \quad \text{Subtract } \frac{c}{a}.$$

Step 3: **Complete the square.**

$$\frac{1}{2}\left(\frac{3}{2}\right) = \frac{3}{4} \qquad \frac{1}{2} \text{ of } x\text{-coefficient}$$

$$\left(\frac{3}{4}\right)^2 = \frac{9}{16} \qquad \text{Square the result.}$$

Add $\dfrac{9}{16}$ to both sides of the equation.

$$\frac{1}{2}\left(\frac{b}{a}\right) = \frac{b}{2a} \qquad \frac{1}{2} \text{ of } x\text{-coefficient}$$

$$\left(\frac{b}{2a}\right)^2 = \frac{b^2}{4a^2} \qquad \text{Square the result.}$$

Add $\dfrac{b^2}{4a^2}$ to both sides of the equation.

$$x^2 + \frac{3}{2}x + \frac{9}{16} = \frac{1}{2} + \frac{9}{16}$$

$$x^2 + \frac{3}{2}x + \frac{9}{16} = \frac{8}{16} + \frac{9}{16} \qquad \text{Get a common denominator.}$$

$$x^2 + \frac{3}{2}x + \frac{9}{16} = \frac{17}{16} \qquad \text{Add.}$$

$$x^2 + \frac{b}{a}x + \frac{b^2}{4a^2} = -\frac{c}{a} + \frac{b^2}{4a^2}$$

$$x^2 + \frac{b}{a}x + \frac{b^2}{4a^2} = -\frac{4ac}{4a^2} + \frac{b^2}{4a^2} \qquad \text{Get a common denominator.}$$

$$x^2 + \frac{b}{a}x + \frac{b^2}{4a^2} = \frac{b^2 - 4ac}{4a^2} \qquad \text{Add.}$$

Step 4: **Factor.**

$$\left(x + \frac{3}{4}\right)^2 = \frac{17}{16}$$
$$\uparrow$$
$$\frac{3}{4} \text{ is } \frac{1}{2}\left(\frac{3}{2}\right), \text{ the coefficient of } x.$$

$$\left(x + \frac{b}{2a}\right)^2 = \frac{b^2 - 4ac}{4a^2}$$
$$\uparrow$$
$$\frac{b}{2a} \text{ is } \frac{1}{2}\left(\frac{b}{a}\right), \text{ the coefficient of } x.$$

Step 5: **Solve using the square root property.**

$$\left(x + \frac{3}{4}\right)^2 = \frac{17}{16}$$

$$x + \frac{3}{4} = \pm\sqrt{\frac{17}{16}}$$

$$x + \frac{3}{4} = \frac{\pm\sqrt{17}}{4} \qquad \sqrt{16} = 4$$

$$x = -\frac{3}{4} \pm \frac{\sqrt{17}}{4} \qquad \text{Subtract } \frac{3}{4}.$$

$$x = \frac{-3 \pm \sqrt{17}}{4} \qquad \begin{array}{l}\text{Same denomi-}\\ \text{nators, add}\\ \text{numerators.}\end{array}$$

$$\left(x + \frac{b}{2a}\right)^2 = \frac{b^2 - 4ac}{4a^2}$$

$$x + \frac{b}{2a} = \pm\sqrt{\frac{b^2 - 4ac}{4a^2}}$$

$$x + \frac{b}{2a} = \frac{\pm\sqrt{b^2 - 4ac}}{2a} \qquad \sqrt{4a^2} = 2a$$

$$x = -\frac{b}{2a} \pm \frac{\sqrt{b^2 - 4ac}}{2a} \qquad \text{Subtract } \frac{b}{2a}.$$

$$x = \frac{-b \pm \sqrt{b^2 - 4ac}}{2a} \qquad \begin{array}{l}\text{Same denomi-}\\ \text{nators, add}\\ \text{numerators.}\end{array}$$

The result on the right is called the *quadratic formula.*

The Quadratic Formula

The solutions of any quadratic equation of the form $ax^2 + bx + c = 0$ $(a \neq 0)$ are

$$x = \frac{-b \pm \sqrt{b^2 - 4ac}}{2a}$$

---

1) The equation to be solved *must* be written in the form $ax^2 + bx + c = 0$ so that a, b, and c can be identified correctly.

2) $x = \dfrac{-b \pm \sqrt{b^2 - 4ac}}{2a}$ represents the two solutions $x = \dfrac{-b + \sqrt{b^2 - 4ac}}{2a}$ and

   $x = \dfrac{-b - \sqrt{b^2 - 4ac}}{2a}$.

3) Notice that the fraction bar continues under $-b$ and does not end at the radical.

$$x = \frac{-b \pm \sqrt{b^2 - 4ac}}{2a} \qquad\qquad x = -b \pm \frac{\sqrt{b^2 - 4ac}}{2a}$$

   Correct   Incorrect

4) The quadratic formula is a *very* important result and one that we will use often. *It should be memorized!*

2. Solve a Quadratic Equation Using the Quadratic Formula

Example 1

Solve using the quadratic formula.

a) $2x^2 + 3x - 1 = 0$ b) $k^2 = 10k - 29$

Solution

a) Is $2x^2 + 3x - 1 = 0$ in the form $ax^2 + bx + c = 0$? Yes. Identify the values of a, b, and c, and substitute them into the quadratic formula.

$$a = 2 \qquad b = 3 \qquad c = -1$$

$$
\begin{aligned}
x &= \frac{-b \pm \sqrt{b^2 - 4ac}}{2a} && \text{Quadratic formula} \\[2mm]
&= \frac{-(3) \pm \sqrt{(3)^2 - 4(2)(-1)}}{2(2)} && \text{Substitute } a = 2, b = 3, \text{ and } c = -1. \\[2mm]
&= \frac{-3 \pm \sqrt{9 - (-8)}}{4} && \text{Perform the operations.} \\[2mm]
&= \frac{-3 \pm \sqrt{17}}{4} && 9 - (-8) = 9 + 8 = 17
\end{aligned}
$$

The solution set is $\left\{ \dfrac{-3 - \sqrt{17}}{4}, \dfrac{-3 + \sqrt{17}}{4} \right\}$.

This is the same result we obtained when we solved this equation by completing the square at the beginning of the section.

b)   Is $k^2 = 10k - 29$ in the form $ax^2 + bx + c = 0$? *No.* Begin by writing the equation in the correct form.

$$k^2 = 10k - 29$$
$$k^2 - 10k + 29 = 0 \qquad \text{Subtract } 10k \text{ and add } 29 \text{ to both sides.}$$

Identify a, b, and c.

$$a = 1 \qquad b = -10 \qquad c = 29$$

$$
\begin{aligned}
k &= \frac{-b \pm \sqrt{b^2 - 4ac}}{2a} && \text{Quadratic formula} \\[2mm]
&= \frac{-(-10) \pm \sqrt{(-10)^2 - 4(1)(29)}}{2(1)} && \text{Substitute } a = 1, b = -10, \text{ and } c = 29. \\[2mm]
&= \frac{10 \pm \sqrt{100 - 116}}{2} && \text{Perform the operations.} \\[2mm]
&= \frac{10 \pm \sqrt{-16}}{2} && 100 - 116 = -16 \\[2mm]
&= \frac{10 \pm 4i}{2} && \sqrt{-16} = 4i \\[2mm]
&= \frac{10}{2} \pm \frac{4}{2}i && \text{Write in the form } a + bi. \\[2mm]
&= 5 \pm 2i && \text{Reduce.}
\end{aligned}
$$

The solution set is $\{5 - 2i, 5 + 2i\}$.

You Try 1

Solve using the quadratic formula.

a) $n^2 + 9n + 18 = 0$ b) $5t^2 + t - 2 = 0$

Equations in various forms may be solved using the quadratic formula.

Example 2

Solve using the quadratic formula.

a) $(3p - 1)(3p + 4) = 3p - 5$ b) $\dfrac{1}{2}w^2 + \dfrac{2}{3}w - \dfrac{1}{3} = 0$

Solution

a)   Is $(3p - 1)(3p + 4) = 3p - 5$ in the form $ax^2 + bx + c = 0$? *No.* Before we can apply the quadratic formula, we must write it in that form.

$$
\begin{aligned}
(3p - 1)(3p + 4) &= 3p - 5 \\
9p^2 + 9p - 4 &= 3p - 5 && \text{Multiply using FOIL.} \\
9p^2 + 6p + 1 &= 0 && \text{Subtract } 3p \text{ and add 5 to both sides.}
\end{aligned}
$$

The equation is in the correct form. Identify a, b, and c.

$$a = 9 \qquad b = 6 \qquad c = 1$$

$$p = \frac{-b \pm \sqrt{b^2 - 4ac}}{2a} \qquad \text{Quadratic formula}$$

$$= \frac{-(6) \pm \sqrt{(6)^2 - 4(9)(1)}}{2(9)} \qquad \text{Substitute } a = 9, b = 6, \text{ and } c = 1.$$

$$= \frac{-6 \pm \sqrt{36 - 36}}{18} \qquad \text{Perform the operations.}$$

$$= \frac{-6 \pm \sqrt{0}}{18}$$

$$= \frac{-6 \pm 0}{18} \qquad \sqrt{0} = 0$$

$$= \frac{-6}{18} = -\frac{1}{3}$$

The solution set is $\left\{ -\dfrac{1}{3} \right\}$.

b) Is $\dfrac{1}{2}w^2 + \dfrac{2}{3}w - \dfrac{1}{3} = 0$ in the form $ax^2 + bx + c = 0$? Yes. However, working with fractions in the quadratic formula would be difficult. *Eliminate the fractions by multiplying the equation by 6, the least common denominator of the fractions.*

$$6\left(\frac{1}{2}w^2 + \frac{2}{3}w - \frac{1}{3}\right) = 6 \cdot 0 \qquad \text{Multiply by 6 to eliminate the fractions.}$$

$$3w^2 + 4w - 2 = 0$$

Identify a, b, and c from this form of the equation.

$$a = 3 \qquad b = 4 \qquad c = -2$$

$$w = \frac{-b \pm \sqrt{b^2 - 4ac}}{2a} \qquad \text{Quadratic formula}$$

$$= \frac{-(4) \pm \sqrt{(4)^2 - 4(3)(-2)}}{2(3)} \qquad \text{Substitute } a = 3, b = 4, \text{ and } c = -2.$$

$$= \frac{-4 \pm \sqrt{16 - (-24)}}{6} \qquad \text{Perform the operations.}$$

$$= \frac{-4 \pm \sqrt{40}}{6} \qquad 16 - (-24) = 16 + 24 = 40$$

$$= \frac{-4 \pm 2\sqrt{10}}{6} \qquad \sqrt{40} = 2\sqrt{10}$$

$$= \frac{2(-2 \pm \sqrt{10})}{6} \qquad \text{Factor out 2 in the numerator.}$$

$$= \frac{-2 \pm \sqrt{10}}{3} \qquad \begin{array}{l}\text{Divide numerator and denominator by 2 to} \\ \text{simplify.}\end{array}$$

The solution set is $\left\{ \dfrac{-2 - \sqrt{10}}{3}, \dfrac{-2 + \sqrt{10}}{3} \right\}$.

You Try 2

Solve using the quadratic formula.

a) $3 - 2z = -2z^2$ b) $(d + 6)(d - 2) = -10$ c) $\dfrac{5}{4}r^2 + \dfrac{1}{5} = r$

3. Determine the Number and Type of Solutions to a Quadratic Equation Using the Discriminant

We can find the solutions of any quadratic equation of the form $ax^2 + bx + c = 0$ $(a \neq 0)$ using the quadratic formula.

$$x = \frac{-b \pm \sqrt{b^2 - 4ac}}{2a}$$

The expression under the radical, $b^2 - 4ac$, is called the **discriminant**. The discriminant tells us what kind of solution a quadratic equation has.

1) If $b^2 - 4ac$ is *positive and the square of an integer*, the equation has *two rational solutions*.

2) If $b^2 - 4ac$ is *positive but not a perfect square*, the equation has *two irrational solutions*.

3) If $b^2 - 4ac$ is *negative*, the equation has *two complex solutions of the form a + bi and a − bi* where $b \neq 0$.

4) If $b^2 - 4ac = 0$, the equation has *one rational solution*.

Example 3

Find the value of the discriminant. Then, determine the number and type of solutions of each equation.

a) $z^2 + 6z - 4 = 0$ b) $5h^2 = 6h - 2$

Solution

a) Is $z^2 + 6z - 4 = 0$ in the form $ax^2 + bx + c = 0$? *Yes.* Identify a, b, and c.

$$a = 1 \qquad b = 6 \qquad c = -4$$

$$\text{Discriminant} = b^2 - 4ac = (6)^2 - 4(1)(-4)$$
$$= 36 + 16$$
$$= 52$$

Since 52 is positive but *not* a perfect square, the equation will have *two irrational solutions*. ($\sqrt{52} = 2\sqrt{13}$ will appear in the solution, and $2\sqrt{13}$ is irrational.)

b)   Is $5h^2 = 6h - 2$ in the form $ax^2 + bx + c = 0$? *No.* Rewrite the equation in that form, and identify a, b, and c.

$$5h^2 - 6h + 2 = 0$$

$$a = 5 \qquad b = -6 \qquad c = 2$$

$$\text{Discriminant} = b^2 - 4ac = (-6)^2 - 4(5)(2)$$
$$= 36 - 40$$
$$= -4$$

Since the discriminant $= -4$, the equation will have *two complex solutions of the form $a + bi$ and $a - bi$, where $b \neq 0$.*

BE CAREFUL   The discriminant is $b^2 - 4ac$ *not* $\sqrt{b^2 - 4ac}$.

You Try 3

Find the value of the discriminant. Then, determine the number and type of solutions of each equation.

a) $2x^2 + x + 5 = 0$ b) $m^2 + 5m = 24$ c) $-3v^2 = 4v - 1$

d) $4r(2r - 3) = -1 - 6r - r^2$

4. Solve an Applied Problem Using the Quadratic Formula

Example 4

A ball is thrown upward from a height of 20 ft. The height h of the ball (in feet) t sec after the ball is released is given by

$$h = -16t^2 + 16t + 20.$$

a)   How long does it take the ball to reach a height of 8 ft?

b)   How long does it take the ball to hit the ground?

Solution

a)   Find the *time* it takes for the ball to reach a height of 8 ft.

Find t when $h = 8$.

$$
\begin{aligned}
h &= -16t^2 + 16t + 20 \\
8 &= -16t^2 + 16t + 20 && \text{Substitute 8 for } h. \\
0 &= -16t^2 + 16t + 12 && \text{Write in standard form.} \\
0 &= 4t^2 - 4t - 3 && \text{Divide by } -4.
\end{aligned}
$$

Solve using the quadratic formula.

$$a = 4 \quad b = -4 \quad c = -3$$

$$t = \frac{-b \pm \sqrt{b^2 - 4ac}}{2a}$$ 　Quadratic formula

$$= \frac{-(-4) \pm \sqrt{(-4)^2 - 4(4)(-3)}}{2(4)}$$ 　Substitute $a = 4$, $b = -4$, and $c = -3$.

$$= \frac{4 \pm \sqrt{16 + 48}}{8}$$ 　Perform the operations.

$$= \frac{4 \pm \sqrt{64}}{8}$$

$$= \frac{4 \pm 8}{8}$$ 　$\sqrt{64} = 8$

The equation has two rational solutions:

$$t = \frac{4 + 8}{8} \qquad \text{or} \qquad t = \frac{4 - 8}{8}$$

$$t = \frac{12}{8} = \frac{3}{2} \qquad \text{or} \qquad t = \frac{-4}{8} = -\frac{1}{2}$$

Since t represents time, t cannot equal $-\frac{1}{2}$. We reject that as a solution.

Therefore, $t = \frac{3}{2}$ sec or 1.5 sec.

The ball will be 8 ft above the ground after 1.5 sec.

b)   When the ball hits the ground, it is 0 ft above the ground.

Find t when $h = 0$.

$$h = -16t^2 + 16t + 20$$
$$0 = -16t^2 + 16t + 20$$ 　Substitute 0 for h.
$$0 = 4t^2 - 4t - 5$$ 　Divide by -4.

Solve using the quadratic formula.

$$a = 4 \quad b = -4 \quad c = -5$$

$$t = \frac{-(-4) \pm \sqrt{(-4)^2 - 4(4)(-5)}}{2(4)}$$ 　Substitute $a = 4$, $b = -4$, and $c = -5$.

$$= \frac{4 \pm \sqrt{16 + 80}}{8}$$ 　Perform the operations.

$$= \frac{4 \pm \sqrt{96}}{8}$$

$$= \frac{4 \pm 4\sqrt{6}}{8}$$ 　$\sqrt{96} = \sqrt{16} \cdot \sqrt{6} = 4\sqrt{6}$

$$= \frac{4(1 \pm \sqrt{6})}{8}$$ 　Factor out 4 in the numerator.

$$t = \frac{1 \pm \sqrt{6}}{2}$$ Divide numerator and denominator by 4 to simplify.

The equation has two irrational solutions:

$$t = \frac{1 + \sqrt{6}}{2} \quad \text{or} \quad t = \frac{1 - \sqrt{6}}{2}$$

Since $\sqrt{6} \approx 2.4$, we get

$$t \approx \frac{1 + 2.4}{2} \quad \text{or} \quad t \approx \frac{1 - 2.4}{2}$$

$$t \approx \frac{3.4}{2} = 1.7 \quad \text{or} \quad t \approx -0.7$$

Since t represents time, t cannot equal $\dfrac{1 - \sqrt{6}}{2}$. We reject this as a solution.

Therefore, $t = \dfrac{1 + \sqrt{6}}{2}$ sec or $t \approx 1.7$ sec.

The ball will hit the ground after about 1.7 sec.

You Try 4

An object is thrown upward from a height of 12 ft. The height h of the object (in feet) t sec after the object is thrown is given by

$$h = -16t^2 + 56t + 12$$

a) How long does it take the object to reach a height of 36 ft?

b) How long does it take the object to hit the ground?

Answers to You Try Exercises

1) a) $\{-6, -3\}$ b) $\left\{ \dfrac{-1 - \sqrt{41}}{10}, \dfrac{-1 + \sqrt{41}}{10} \right\}$ 2) a) $\left\{ \dfrac{1}{2} - \dfrac{\sqrt{5}}{2}i, \dfrac{1}{2} + \dfrac{\sqrt{5}}{2}i \right\}$

b) $\{-2 - \sqrt{6}, -2 + \sqrt{6}\}$ c) $\left\{ \dfrac{2}{5} \right\}$ 3) a) -39; two complex solutions

b) 121; two rational solutions   c) 28; two irrational solutions   d) 0; one rational solution

4) a) It takes $\dfrac{1}{2}$ sec to reach 36 ft on its way up and 3 sec to reach 36 ft on its way down.

b) $\dfrac{7 + \sqrt{61}}{4}$ sec or approximately 3.7 sec.

11.5 Exercises

Objectives 2 and 3

Find the error in each, and correct the mistake.

1) The solution to $ax^2 + bx + c = 0$ $(a \neq 0)$ can be found using the quadratic formula.

$$x = -b \pm \frac{\sqrt{b^2 - 4ac}}{2a}$$

2) In order to solve $5n^2 - 3n = 1$ using the quadratic formula, a student substitutes a, b, and c into the formula in this way: $a = 5 \quad b = -3 \quad c = 1$

$$n = \frac{-(-3) \pm \sqrt{(-3)^2 - 4(5)(1)}}{2(5)}$$

3) $\dfrac{-2 \pm 6\sqrt{11}}{2} = -1 \pm 6\sqrt{11}$

4) The discriminant of $3z^2 - 4z + 1 = 0$ is

$$\sqrt{b^2 - 4ac} = \sqrt{(-4)^2 - 4(3)(1)}$$
$$= \sqrt{16 - 12}$$
$$= \sqrt{4}$$
$$= 2.$$

Objective 2

Solve using the quadratic formula.

5) $x^2 + 4x + 3 = 0$

6) $v^2 - 8v + 7 = 0$

7) $3t^2 + t - 10 = 0$

8) $6q^2 + 11q + 3 = 0$

 9) $k^2 + 2 = 5k$

10) $n^2 = 5 - 3n$

11) $y^2 = 8y - 25$

12) $-4x + 5 = -x^2$

13) $3 - 2w = -5w^2$

14) $2d^2 = -4 - 5d$

15) $r^2 + 7r = 0$

16) $p^2 - 10p = 0$

17) $3v(v + 3) = 7v + 4$

18) $2k(k - 3) = -3$

19) $(2c - 5)(c - 5) = -3$

20) $-11 = (3z - 1)(z - 5)$

21) $\dfrac{1}{6}u^2 + \dfrac{4}{3}u = \dfrac{5}{2}$

22) $\dfrac{5}{2}r^2 + 3r + 2 = 0$

 23) $m^2 + \dfrac{4}{3}m + \dfrac{5}{9} = 0$

24) $\dfrac{1}{6}h + \dfrac{1}{2} = \dfrac{3}{4}h^2$

25) $2(p + 10) = (p + 10)(p - 2)$

26) $(t - 8)(t - 3) = 3(3 - t)$

27) $4g^2 + 9 = 0$

28) $25q^2 - 1 = 0$

29) $x(x + 6) = -34$

30) $c(c - 4) = -22$

 31) $(2s + 3)(s - 1) = s^2 - s + 6$

32) $(3m + 1)(m - 2) = (2m - 3)(m + 2)$

33) $3(3 - 4y) = -4y^2$

34) $5a(5a + 2) = -1$

35) $-\dfrac{1}{6} = \dfrac{2}{3}p^2 + \dfrac{1}{2}p$

36) $\dfrac{1}{2}n = \dfrac{3}{4}n^2 + 2$

37) $4q^2 + 6 = 20q$

38) $4w^2 = 6w + 16$

Objective 3

Find the value of the discriminant. Then, determine the number and type of solutions of each equation. *Do not solve.*

39) $10d^2 - 9d + 3 = 0$

40) $3j^2 + 8j + 2 = 0$

41) $4y^2 + 49 = -28y$

42) $3q = 1 + 5q^2$

43) $-5 = u(u + 6)$

44) $g^2 + 4 = 4g$

 45) $2w^2 - 4w - 5 = 0$

46) $3 + 2p^2 - 7p = 0$

Find the value of a, b, or c so that each equation has only one rational solution.

47) $z^2 + bz + 16 = 0$

48) $k^2 + bk + 49 = 0$

49) $4y^2 - 12y + c = 0$

50) $25t^2 - 20t + c = 0$

51) $ap^2 + 12p + 9 = 0$

52) $ax^2 - 6x + 1 = 0$

Objective 4

Write an equation and solve.

53) One leg of a right triangle is 1 in. more than twice the other leg. The hypotenuse is $\sqrt{29}$ in. long. Find the lengths of the legs.

54) The hypotenuse of a right triangle is $\sqrt{34}$ in. long. The length of one leg is 1 in. less than twice the other leg. Find the lengths of the legs.

Solve.

 55) An object is thrown upward from a height of 24 ft. The height h of the object (in feet) t sec after the object is released is given by $h = -16t^2 + 24t + 24$.

  a) How long does it take the object to reach a height of 8 ft?

  b) How long does it take the object to hit the ground?

56) A ball is thrown upward from a height of 6 ft. The height h of the ball (in feet) t sec after the ball is released is given by $h = -16t^2 + 44t + 6$

  a) How long does it take the ball to reach a height of 16 ft?

  b) How long does it take the object to hit the ground?

Mid-Chapter Summary

We have learned four methods for solving quadratic equations.

Methods for Solving Quadratic Equations

1) Factoring
2) Square root property
3) Completing the square
4) Quadratic formula

While it is true that the quadratic formula can be used to solve *every* quadratic equation of the form $ax^2 + bx + c = 0$ $(a \neq 0)$, it is not always the most *efficient* method. In this section we will discuss how to decide which method to use to solve a quadratic equation.

1. Learn How to Decide Which Method to Use to Solve a Given Quadratic Equation

Example 1

Solve.

a) $p^2 - 6p = 16$ b) $m^2 - 8m + 13 = 0$

c) $3t^2 + 8t + 7 = 0$ d) $(2z - 7)^2 - 6 = 0$

Solution

a) Write $p^2 - 6p = 16$ in standard form.

$$p^2 - 6p - 16 = 0$$

Does $p^2 - 6p - 16$ factor? Yes. *Solve by factoring.*

$$(p - 8)(p + 2) = 0$$

$$p - 8 = 0 \quad \text{or} \quad p + 2 = 0 \qquad \text{Set each factor equal to 0.}$$
$$p = 8 \quad \text{or} \qquad p = -2 \qquad \text{Solve.}$$

The solution set is $\{-2, 8\}$.

b) To solve $m^2 - 8m + 13 = 0$ ask yourself, "Can I factor $m^2 - 8m + 13$?" No, it does not factor. We could solve this using the quadratic formula, but *completing the square* is also a good method for solving this equation. Why?

Completing the square is a desirable method for solving a quadratic equation when the coefficient of the squared term is 1 or −1 and when the coefficient of the linear term is even.
   We will solve $m^2 - 8m + 13 = 0$ by completing the square.

Step 1: The coefficient of m^2 is 1.

Step 2: Get the variables on one side of the equal sign and the constant on the other side.

$$m^2 - 8m = -13$$

Step 3: Complete the square:

$$\frac{1}{2}(-8) = -4$$
$$(-4)^2 = 16$$

Add 16 to both sides of the equation.

$$m^2 - 8m + 16 = -13 + 16$$
$$m^2 - 8m + 16 = 3$$

Step 4: Factor:

$$(m - 4)^2 = 3$$

Step 5: Solve using the square root property:

$$(m - 4)^2 = 3$$
$$m - 4 = \pm\sqrt{3}$$
$$m = 4 \pm \sqrt{3}$$

The solution set is $\{4 - \sqrt{3}, 4 + \sqrt{3}\}$.

Completing the square works well when the coefficient of the squared term is 1 or -1 and when the coefficient of the linear term is *even* because when we complete the square in step 3, we will not obtain a fraction. (Half of an even number is an integer.)

---

c) Ask yourself, "Can I solve $3t^2 + 8t + 7 = 0$ by factoring?" No, $3t^2 + 8t + 7$ does not factor. Completing the square would not be a very efficient way to solve the equation because the coefficient of t^2 is 3, and dividing the equation by 3 would give us $t^2 + \dfrac{8}{3}t + \dfrac{7}{3} = 0$.

We will solve $3t^2 + 8t + 7 = 0$ using the quadratic formula.

Identify a, b, and c.

$$a = 3 \qquad b = 8 \qquad c = 7$$

$$t = \frac{-b \pm \sqrt{b^2 - 4ac}}{2a} \qquad \text{Quadratic formula}$$

$$= \frac{-(8) \pm \sqrt{(8)^2 - 4(3)(7)}}{2(3)} \qquad \text{Substitute } a = 3, b = 8, \text{ and } c = 7.$$

$$= \frac{-8 \pm \sqrt{64 - 84}}{6} \qquad \text{Perform the operations.}$$

$$= \frac{-8 \pm \sqrt{-20}}{6}$$

$$= \frac{-8 \pm 2i\sqrt{5}}{6} \qquad \sqrt{-20} = i\sqrt{4}\sqrt{5} = 2i\sqrt{5}$$

$$= \frac{2(-4 \pm i\sqrt{5})}{6} \qquad \text{Factor out 2 in the numerator.}$$

$$= \frac{-4 \pm i\sqrt{5}}{3} \qquad \begin{array}{l}\text{Divide numerator and}\\\text{denominator by 2 to simplify.}\end{array}$$

$$= -\frac{4}{3} \pm \frac{\sqrt{5}}{3}i \qquad \text{Write in the form } a + bi.$$

The solution set is $\left\{-\dfrac{4}{3} - \dfrac{\sqrt{5}}{3}i, -\dfrac{4}{3} + \dfrac{\sqrt{5}}{3}i\right\}$.

d) What method should we use to solve $(2z - 7)^2 - 6 = 0$?

We *could* square the binomial, combine like terms, then solve, possibly, by factoring or using the quadratic formula. However, this would be very inefficient. The equation contains a squared quantity and a constant.
We will solve $(2z - 7)^2 - 6 = 0$ using the square root property.

$$(2z - 7)^2 - 6 = 0$$
$$(2z - 7)^2 = 6 \qquad \text{Add 6 to each side.}$$
$$2z - 7 = \pm\sqrt{6} \qquad \text{Square root property}$$
$$2z = 7 \pm \sqrt{6} \qquad \text{Add 7 to each side.}$$
$$z = \frac{7 \pm \sqrt{6}}{2} \qquad \text{Divide by 2.}$$

The solution set is $\left\{ \dfrac{7 - \sqrt{6}}{2}, \dfrac{7 + \sqrt{6}}{2} \right\}$.

 You Try 1

Solve.

a) $2k^2 + 3 = 9k$ b) $2r^2 + 3r - 2 = 0$ c) $(n - 8)^2 + 9 = 0$ d) $y^2 + 4y = -10$

Answers to You Try Exercises

1) a) $\left\{ \dfrac{9 - \sqrt{57}}{4}, \dfrac{9 + \sqrt{57}}{4} \right\}$ b) $\left\{ -2, \dfrac{1}{2} \right\}$ c) $\{8 - 3i, 8 + 3i\}$ d) $\{-2 - i\sqrt{6}, -2 + i\sqrt{6}\}$

 Mid-Chapter Summary Exercises

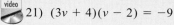 **Boost your grade at mathzone.com!** MathZone > Practice Problems > NetTutor > Self-Test > e-Professors > Videos

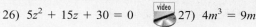

Objective 1

Keep in mind the four methods we have learned for solving quadratic equations: *factoring, the square root property, completing the square, and the quadratic formula.* Solve the equations using one of these methods.

1) $z^2 - 50 = 0$

3) $a(a + 1) = 20$

5) $u^2 + 7u + 9 = 0$

7) $2k(2k + 7) = 3(k + 1)$

9) $m^2 + 14m + 60 = 0$

11) $10 + (3b - 1)^2 = 4$

13) $1 = \dfrac{x^2}{12} - \dfrac{x}{3}$

15) $r^2 - 4r = 3$

2) $j^2 - 6j = 8$

4) $2x^2 + 6 = 3x$

6) $3p^2 - p - 4 = 0$

8) $2 = (w + 3)^2 + 8$

10) $\dfrac{1}{2}y^2 = \dfrac{3}{4} - \dfrac{1}{2}y$

12) $c^2 + 8c + 25 = 0$

14) $100 = 4d^2$

16) $2t^3 + 108t = -30t^2$

17) $p(p + 8) = 3(p^2 + 2) + p$

18) $h^2 = h$

19) $\dfrac{10}{z} = 1 + \dfrac{21}{z^2}$

20) $2s(2s + 3) = 4s + 5$

21) $(3v + 4)(v - 2) = -9$

22) $34 = 6y - y^2$

23) $(c - 5)^2 + 16 = 0$

24) $(2b + 1)(b + 5) = -7$

25) $3g = g^2$

26) $5z^2 + 15z + 30 = 0$

27) $4m^3 = 9m$

28) $\dfrac{9}{2a^2} = \dfrac{1}{6} + \dfrac{1}{a}$

29) $\dfrac{1}{3}q^2 + \dfrac{5}{6}q + \dfrac{4}{3} = 0$

30) $-3 = (12d + 5)^2 + 6$

Section 11.6 Equations in Quadratic Form

Objectives

1. Solve Quadratic Equations That Result from Equations Containing Rational or Radical Expressions

2. Solve an Equation in Quadratic Form by Factoring

3. Solve an Equation in Quadratic Form Using Substitution and Then Factoring

4. Solve an Equation in Quadratic Form Using Substitution and the Quadratic Formula

5. Use Substitution for a Binomial to Solve a Quadratic Equation

In Chapters 8 and 10, we solved some equations that were *not* quadratic but could be rewritten in the form of a quadratic equation, $ax^2 + bx + c = 0$. Two such examples are:

$$\frac{10}{x} - \frac{7}{x + 1} = \frac{2}{3} \quad \text{and} \quad r + \sqrt{r} = 12$$

Rational equation (Ch. 8)   Radical equation (Ch. 10)

We will review how to solve each type of equation.

1. Solve Quadratic Equations That Result from Equations Containing Rational or Radical Expressions

Example 1

Solve $\dfrac{10}{x} - \dfrac{7}{x + 1} = \dfrac{2}{3}$.

Solution

To solve an equation containing rational expressions, *multiply the equation by the LCD of all of the fractions to eliminate the denominators,* then solve.

$$\text{LCD} = 3x(x + 1)$$

$$3x(x + 1)\left(\frac{10}{x} - \frac{7}{x + 1}\right) = 3x(x + 1)\left(\frac{2}{3}\right)$$

Multiply both sides of the equation by the LCD of the fractions.

$$3\cancel{x}(x + 1) \cdot \frac{10}{\cancel{x}} - 3x\cancel{(x + 1)} \cdot \frac{7}{\cancel{x + 1}} = \cancel{3}x(x + 1) \cdot \left(\frac{2}{\cancel{3}}\right)$$

Distribute and divide out common factors.

$$30(x + 1) - 3x(7) = 2x(x + 1)$$

Distribute.

$$30x + 30 - 21x = 2x^2 + 2x$$

Combine like terms.

$$9x + 30 = 2x^2 + 2x$$

$$0 = 2x^2 - 7x - 30$$

Write in the form $ax^2 + bx + c = 0$.

$$0 = (2x + 5)(x - 6)$$

Factor.

$$2x + 5 = 0 \quad \text{or} \quad x - 6 = 0$$

Set each factor equal to zero.

$$2x = -5$$

$$x = -\frac{5}{2} \quad \text{or} \quad x = 6$$

Solve.

Recall that you *must* check the proposed solutions in the original equation to be certain they do not make a denominator equal zero.

The solution set is $\left\{-\dfrac{5}{2}, 6\right\}$.

You Try 1

Solve $\dfrac{1}{m} = \dfrac{1}{2} + \dfrac{m}{m+4}$.

Example 2

Solve $r + \sqrt{r} = 12$.

Solution

The first step in solving a radical equation is getting a radical on a side by itself.

$$r + \sqrt{r} = 12$$
$$\sqrt{r} = 12 - r \qquad\qquad \text{Subtract } r \text{ from each side.}$$
$$(\sqrt{r})^2 = (12 - r)^2 \qquad\qquad \text{Square both sides.}$$
$$r = 144 - 24r + r^2$$
$$0 = r^2 - 25r + 144 \qquad\qquad \text{Write in the form } ax^2 + bx + c = 0.$$
$$0 = (r - 16)(r - 9) \qquad\qquad \text{Factor.}$$

$$r - 16 = 0 \quad \text{or} \quad r - 9 = 0 \qquad \text{Set each factor equal to zero.}$$
$$r = 16 \quad \text{or} \quad r = 9 \qquad \text{Solve.}$$

Recall that you *must* check the proposed solutions in the original equation.

Check r = 16:
$$r + \sqrt{r} = 12$$
$$16 + \sqrt{16} \stackrel{?}{=} 12$$
$$16 + 4 = 12 \qquad \text{False}$$

Check r = 9:
$$r + \sqrt{r} = 12$$
$$9 + \sqrt{9} \stackrel{?}{=} 12$$
$$9 + 3 = 12 \qquad \text{True}$$

16 is an extraneous solution.
The solution set is $\{9\}$.

You Try 2

Solve $y + 3\sqrt{y} = 10$.

Equations in Quadratic Form

2. Solve an Equation in Quadratic Form by Factoring

Some equations that are not quadratic can be solved using the same methods that can be used to solve quadratic equations. These are called **equations in quadratic form**. Some examples of equations in quadratic form are:

$$x^4 - 10x^2 + 9 = 0, \qquad t^{2/3} + t^{1/3} - 6 = 0, \qquad 2n^4 - 5n^2 = -1$$

Let's compare the equations above to *quadratic equations* to understand why they are said to be in quadratic form.

<table>
<tr><td colspan="3" align="center">**COMPARE**</td></tr>
<tr><td>**An Equation in Quadratic Form**</td><td align="center">**to**</td><td>**A Quadratic Equation**</td></tr>
</table>

An Equation in Quadratic Form	A Quadratic Equation
This exponent is *twice* this exponent.	This exponent is *twice* this exponent.
$x^4 - 10x^2 + 9 = 0$	$x^2 - 10x^1 + 9 = 0$
This exponent is *twice* this exponent.	This exponent is *twice* this exponent.
$t^{2/3} + t^{1/3} - 6 = 0$	$t^2 + t^1 - 6 = 0$
This exponent is *twice* this exponent.	This exponent is *twice* this exponent.
$2n^4 - 5n^2 = -1$	$2n^2 - 5n^1 = -1$

This pattern enables us to work with equations in quadratic form like we can work with quadratic equations.

Example 3

Solve.

a) $x^4 - 10x^2 + 9 = 0$ b) $t^{2/3} + t^{1/3} - 6 = 0$

Solution

a) Compare $x^4 - 10x^2 + 9 = 0$ to $x^2 - 10x + 9 = 0$.

We can factor $x^2 - 10x + 9$: $(x - 9)(x - 1)$

Confirm by multiplying $(x - 9)(x - 1)$ using FOIL:

$$(x - 9)(x - 1) = x^2 - 1x - 9x + 9 = x^2 - 10x + 9$$

We can factor $x^4 - 10x^2 + 9$ in a similar way since the exponent of the first term (4) is *twice* the exponent of the second term (2).

$$x^4 - 10x^2 + 9 = (x^2 - 9)(x^2 - 1)$$

Confirm by multiplying using FOIL:

$$(x^2 - 9)(x^2 - 1) = x^4 - 1x^2 - 9x^2 + 9 = x^4 - 10x^2 + 9$$

We can solve $x^4 - 10x^2 + 9 = 0$ by factoring.

$$x^4 - 10x^2 + 9 = 0$$
$$(x^2 - 9)(x^2 - 1) = 0 \qquad \text{Factor.}$$
$$x^2 - 9 = 0 \quad \text{or} \quad x^2 - 1 = 0 \qquad \text{Set each factor equal to 0.}$$
$$x^2 = 9 \qquad\qquad x^2 = 1 \qquad \text{Square root property}$$
$$x = \pm 3 \qquad\qquad x = \pm 1$$

The check is left to the student.
The solution set is $\{-3, -1, 1, 3\}$.

b)   Compare $t^{2/3} + t^{1/3} - 6 = 0$ to $t^2 + t - 6 = 0$.

$t^2 + t - 6$ factors to $(t + 3)(t - 2)$.

Confirm by multiplying $(t + 3)(t - 2)$ using FOIL:

$$(t + 3)(t - 2) = t^2 - 2t + 3t - 6 = t^2 + t - 6$$

We can factor $t^{2/3} + t^{1/3} - 6$ in a similar way since the exponent of the first term (2/3) is *twice* the exponent of the second term (1/3).

$$t^{2/3} + t^{1/3} - 6 = (t^{1/3} + 3)(t^{1/3} - 2)$$

Confirm by multiplying $(t^{1/3} + 3)(t^{1/3} - 2)$ using FOIL:

$$(t^{1/3} + 3)(t^{1/3} - 2) = t^{2/3} - 2t^{1/3} + 3t^{1/3} - 6 = t^{2/3} + t^{1/3} - 6.$$

We can solve $t^{2/3} + t^{1/3} - 6 = 0$ by factoring.

$$t^{2/3} + t^{1/3} - 6 = 0$$
$$(t^{1/3} + 3)(t^{1/3} - 2) = 0 \qquad \text{Factor.}$$

$t^{1/3} + 3 = 0$	or $\quad t^{1/3} - 2 = 0$	Set each factor equal to 0.
$t^{1/3} = -3$	$t^{1/3} = 2$	Isolate the constant.
$\sqrt[3]{t} = -3$	$\sqrt[3]{t} = 2$	$t^{1/3} = \sqrt[3]{t}$
$(\sqrt[3]{t})^3 = (-3)^3$	$(\sqrt[3]{t})^3 = 2^3$	Cube both sides.
$t = -27 \quad$ or	$t = 8$	Solve.

The check is left to the student.
The solution set is $\{-27, 8\}$.

You Try 3

Solve.

a) $r^4 - 13r^2 + 36 = 0$ b) $c^{2/3} + 4c^{1/3} - 5 = 0$

3. Solve an Equation in Quadratic Form Using Substitution and Then Factoring

The equations in Example 3 can also be solved using a method called **substitution**. We will illustrate the method in Example 4.

Example 4

Solve $x^4 - 10x^2 + 9 = 0$ using substitution.

Solution

$$x^4 - 10x^2 + 9 = 0$$
$$\downarrow$$
$$x^4 = (x^2)^2$$

To rewrite $x^4 - 10x^2 + 9 = 0$ in quadratic form, let $u = x^2$.

$$\text{If } u = x^2, \text{ then}$$
$$u^2 = x^4.$$

$$x^4 - 10x^2 + 9 = 0$$
$$u^2 - 10u + 9 = 0 \qquad \text{Substitute } u^2 \text{ for } x^4 \text{ and } u \text{ for } x^2.$$
$$(u - 9)(u - 1) = 0 \qquad \text{Solve by factoring.}$$

$$u - 9 = 0 \quad \text{or} \quad u - 1 = 0 \qquad \text{Set each factor equal to 0.}$$
$$u = 9 \quad \text{or} \qquad u = 1 \qquad \text{Solve for } u.$$

$u = 9$ and $u = 1$ are *not* the solutions to $x^4 - 10x^2 + 9 = 0$. We must solve for x. Above we let $u = x^2$. *To solve for x, substitute 9 for u and solve for x and then substitute 1 for u and solve for x.*

Substitute 9 for u.
Square root property

$$u = x^2 \qquad u = x^2$$
$$9 = x^2 \qquad 1 = x^2$$
$$\pm 3 = x \qquad \pm 1 = x$$

Substitute 1 for u.
Square root property

The solution set is $\{-3, -1, 1, 3\}$. This is the same as the result we obtained in Example 3a).

You Try 4

Solve by substitution.

a) $r^4 - 13r^2 + 36 = 0$ b) $c^{2/3} + 4c^{1/3} - 5 = 0$

4. Solve an Equation in Quadratic Form Using Substitution and the Quadratic Formula

If an equation cannot be solved by factoring, we can use the quadratic formula.

Example 5

Solve $2n^4 - 5n^2 = -1$.

Solution

Write the equation in standard form: $2n^4 - 5n^2 + 1 = 0$.
Can we solve the equation by factoring? *No.*

We will solve $2n^4 - 5n^2 + 1 = 0$ using the quadratic formula.
Begin with substitution.

$$\text{If } u = n^2, \text{ then}$$
$$u^2 = n^4.$$

$$2n^4 - 5n^2 + 1 = 0$$
$$2u^2 - 5u + 1 = 0$$

Solve $2u^2 - 5u + 1 = 0$ for u using the quadratic formula.

$$u = \frac{5 \pm \sqrt{(-5)^2 - 4(2)(1)}}{2(2)} \qquad a = 2, b = -5, c = 1$$

$$u = \frac{5 \pm \sqrt{25 - 8}}{4} = \frac{5 \pm \sqrt{17}}{4}$$

$u = \dfrac{5 \pm \sqrt{17}}{4}$ does not solve the *original* equation. We must solve for x using the

fact that $u = x^2$. Since $u = \dfrac{5 \pm \sqrt{17}}{4}$ means $u = \dfrac{5 + \sqrt{17}}{4}$ or $u = \dfrac{5 - \sqrt{17}}{4}$, we get

$$u = x^2 \qquad\qquad\qquad\qquad u = x^2$$

$$\frac{5 + \sqrt{17}}{4} = x^2 \qquad\qquad\qquad \frac{5 - \sqrt{17}}{4} = x^2$$

$$\pm\sqrt{\frac{5 + \sqrt{17}}{4}} = x \qquad\qquad \pm\sqrt{\frac{5 - \sqrt{17}}{4}} = x \qquad \text{Square root property}$$

$$\frac{\pm\sqrt{5 + \sqrt{17}}}{2} = x \qquad\qquad \frac{\pm\sqrt{5 - \sqrt{17}}}{2} = x \qquad \sqrt{4} = 2$$

The solution set is $\left\{ \dfrac{\sqrt{5 + \sqrt{17}}}{2}, -\dfrac{\sqrt{5 + \sqrt{17}}}{2}, \dfrac{\sqrt{5 - \sqrt{17}}}{2}, -\dfrac{\sqrt{5 - \sqrt{17}}}{2} \right\}$.

You Try 5

Solve $2k^4 + 3 = 9k^2$.

5. Use Substitution for a Binomial to Solve a Quadratic Equation

We can use substitution to solve an equation like the one in Example 6.

Example 6

Solve $2(3a + 1)^2 - 7(3a + 1) - 4 = 0$.

Solution

The binomial $3a + 1$ appears as a *squared quantity* and as a *linear quantity*. Begin by using substitution.

$$\text{Let } u = 3a + 1. \quad \text{Then,}$$
$$u^2 = (3a + 1)^2.$$

Substitute.

$$2(3a + 1)^2 - 7(3a + 1) - 4 = 0$$
$$2u^2 \qquad\quad - 7u \qquad\quad - 4 = 0$$

Does $2u^2 - 7u - 4 = 0$ factor? *Yes.* Solve by factoring.

$$(2u + 1)(u - 4) = 0 \qquad \text{Factor } 2u^2 - 7u - 4 = 0.$$

$$2u + 1 = 0 \quad \text{or} \quad u - 4 = 0 \qquad \text{Set each factor equal to 0.}$$

$$u = -\frac{1}{2} \quad \text{or} \quad u = 4 \qquad \text{Solve for } u.$$

Solve for a using $u = 3a + 1$.

When $u = -\dfrac{1}{2}$:

$$u = 3a + 1$$
$$-\frac{1}{2} = 3a + 1$$

Subtract 1.　$-\dfrac{3}{2} = 3a$

Multiply by $\dfrac{1}{3}$.　$-\dfrac{1}{2} = a$

When $u = 4$:

$$u = 3a + 1$$
$$4 = 3a + 1$$

$3 = 3a$　Subtract 1.

$1 = a$　Divide by 3.

The solution set is $\left\{ -\dfrac{1}{2}, 1 \right\}$.

You Try 6

Solve $3(2p - 1)^2 - 11(2p - 1) + 10 = 0.$

BE CAREFUL　Don't forget to solve for the variable in the *original* equation.

Answers to You Try Exercises

1) $\left\{ -2, \dfrac{4}{3} \right\}$　　2) $\{4\}$　　3) a) $\{-3, -2, 2, 3\}$　b) $\{-125, 1\}$　　4) a) $\{-3, -2, 2, 3\}$

b) $\{-125, 1\}$　　5) $\left\{ \dfrac{\sqrt{9 + \sqrt{57}}}{2}, -\dfrac{\sqrt{9 + \sqrt{57}}}{2}, \dfrac{\sqrt{9 - \sqrt{57}}}{2}, -\dfrac{\sqrt{9 - \sqrt{57}}}{2} \right\}$　　6) $\left\{ \dfrac{4}{3}, \dfrac{3}{2} \right\}$

11.6 Exercises

Boost your grade at mathzone.com! MathZone

> Practice Problems   > NetTutor   > e-Professors   > Videos
> Self-Test

Objective 1

Solve.

1) $t - \dfrac{48}{t} = 8$

2) $z + 11 = -\dfrac{24}{z}$

3) $\dfrac{2}{x} + \dfrac{6}{x-2} = -\dfrac{5}{2}$

4) $\dfrac{3}{y} - \dfrac{6}{y-1} = \dfrac{1}{2}$

5) $1 = \dfrac{2}{c} + \dfrac{1}{c-5}$

6) $\dfrac{2}{g} = 1 + \dfrac{g}{g+5}$

7) $\dfrac{3}{2v+2} + \dfrac{1}{v} = \dfrac{3}{2}$

8) $\dfrac{1}{b+3} + \dfrac{1}{b} = \dfrac{1}{3}$

9) $\dfrac{9}{n^2} = 5 + \dfrac{4}{n}$

10) $3 - \dfrac{16}{a^2} = \dfrac{8}{a}$

11) $\dfrac{5}{6r} = 1 - \dfrac{r}{6r-6}$

12) $\dfrac{7}{4} - \dfrac{x}{4x+4} = \dfrac{1}{x}$

Solve.

13) $g = \sqrt{g+20}$

14) $c = \sqrt{7c-6}$

15) $a = \sqrt{\dfrac{14a-8}{5}}$

16) $k = \sqrt{\dfrac{6-11k}{2}}$

17) $p - \sqrt{p} = 6$

18) $v + \sqrt{v} = 2$

19) $x = 5\sqrt{x} - 4$

20) $10 = m - 3\sqrt{m}$

21) $2 + \sqrt{2y-1} = y$

22) $1 - \sqrt{5t+1} = -t$

23) $2 = \sqrt{6k+4} - k$

24) $\sqrt{10-3q} - 6 = q$

Objectives 2–4

Determine whether or not each is an equation in quadratic form. Do *not* solve.

25) $n^4 - 12n^2 + 32 = 0$

26) $p^6 + 8p^3 - 9 = 0$

27) $2t^6 + 3t^3 - 5 = 0$

28) $a^4 - 4a - 3 = 0$

29) $c^{2/3} - 4c - 6 = 0$

30) $3z^{2/3} + 2z^{1/3} + 1 = 0$

31) $m + 9m^{1/2} = 4$

32) $2x^{1/2} - 5x^{1/4} = 2$

33) $5k^4 + 6k - 7 = 0$

34) $r^{-2} = 10 - 4r^{-1}$

Solve.

35) $x^4 - 10x^2 + 9 = 0$

36) $d^4 - 29d^2 + 100 = 0$

37) $p^4 - 11p^2 + 28 = 0$

38) $k^4 - 9k^2 + 8 = 0$

39) $a^4 + 12a^2 = -35$

40) $c^4 + 9c^2 = -18$

41) $b^{2/3} + 3b^{1/3} + 2 = 0$

42) $z^{2/3} + z^{1/3} - 12 = 0$

43) $t^{2/3} - 6t^{1/3} = 40$

44) $p^{2/3} - p^{1/3} = 6$

45) $2n^{2/3} = 7n^{1/3} + 15$

46) $10k^{1/3} + 8 = -3k^{2/3}$

47) $v - 8v^{1/2} + 12 = 0$

48) $j - 6j^{1/2} + 5 = 0$

49) $4h^{1/2} + 21 = h$

50) $s + 12 = -7s^{1/2}$

51) $2a - 5a^{1/2} - 12 = 0$

52) $2w = 9w^{1/2} + 18$

53) $9n^4 = -15n^2 - 4$

54) $4h^4 + 19h^2 + 12 = 0$

55) $z^4 - 2z^2 = 15$

56) $a^4 + 2a^2 = 24$

57) $w^4 - 6w^2 + 2 = 0$

58) $p^4 - 8p^2 + 3 = 0$

59) $2m^4 + 1 = 7m^2$

60) $8x^4 + 2 = 9x^2$

61) $t^{-2} - 4t^{-1} - 12 = 0$

62) $d^{-2} + d^{-1} - 6 = 0$

63) $4 = 13y^{-1} - 3y^{-2}$

64) $14h^{-1} + 3 = 5h^{-2}$

Objective 5

Solve.

65) $(x-2)^2 + 11(x-2) + 24 = 0$

66) $(r+1)^2 - 3(r+1) - 10 = 0$

67) $2(3q+4)^2 - 13(3q+4) + 20 = 0$

68) $4(2b-3)^2 - 9(2b-3) - 9 = 0$

69) $(5a-3)^2 + 6(5a-3) = -5$

70) $(3z-2)^2 - 8(3z-2) = 20$

71) $3(k+8)^2 + 5(k+8) = 12$

72) $5(t+9)^2 + 37(t+9) + 14 = 0$

73) $1 - \dfrac{8}{2w+1} = -\dfrac{16}{(2w+1)^2}$

74) $1 - \dfrac{8}{4p+3} = -\dfrac{12}{(4p+3)^2}$

75) $1 + \dfrac{2}{h-3} = \dfrac{1}{(h-3)^2}$

76) $\dfrac{2}{(c+6)^2} + \dfrac{2}{(c+6)} = 1$

Section 11.7 Formulas and Applications

Objectives

1. Solve a Formula for a Particular Variable
2. Solve an Applied Problem Involving Volume
3. Solve an Applied Problem Involving Area
4. Solve an Applied Problem Modeled by a Given Quadratic Equation

Sometimes, solving a formula for a variable involves using one of the techniques we've learned for solving a quadratic equation or for solving an equation containing a radical.

1. Solve a Formula for a Particular Variable

Example 1

Solve $v = \sqrt{\dfrac{300VP}{m}}$ for m.

Solution

Put a box around the m. The goal is to get m on a side by itself.

$$v = \sqrt{\dfrac{300VP}{\boxed{m}}}$$

$$v^2 = \dfrac{300VP}{\boxed{m}} \qquad \text{Square both sides.}$$

Since we are solving for m and it is in the denominator, multiply both sides by m to eliminate the denominator.

$$\boxed{m}\, v^2 = 300VP \qquad \text{Multiply both sides by } m.$$

$$m = \dfrac{300VP}{v^2} \qquad \text{Divide both sides by } v^2.$$

 You Try 1

Solve $v = \sqrt{\dfrac{2E}{m}}$ for m.

It may be necessary to use the quadratic formula to solve a formula for a variable.

Compare the following two equations. Each equation is *quadratic in x* because each is written in the form $ax^2 + bx + c = 0$.

$$8x^2 + 3x - 2 = 0 \qquad \text{and} \qquad 8x^2 + tx - z = 0$$
$$a = 8 \quad b = 3 \quad c = -2 \qquad\qquad a = 8 \quad b = t \quad c = -z$$

To solve the equations for x, we can use the quadratic formula.

Example 2

Solve for x.

a) $8x^2 + 3x - 2 = 0$ b) $8x^2 + tx - z = 0$

Solution

a) $8x^2 + 3x - 2$ does not factor, so we will solve using the quadratic formula.

$$8x^2 + 3x - 2 = 0$$
$$a = 8 \quad b = 3$$

$$x = \frac{-3 \pm \sqrt{(3)^2 - 4(8)}}{2(8)} \qquad\qquad x = \frac{-b \pm \sqrt{b^2 - 4ac}}{2a}$$

$$= \frac{-3 \pm \sqrt{9 + 64}}{16} \qquad\qquad \text{Perform the operations.}$$

$$= \frac{-3 \pm \sqrt{73}}{16}$$

The solution set is $\left\{ \dfrac{-3 - \sqrt{73}}{16}, \dfrac{-3 + \sqrt{73}}{16} \right\}$.

b) Solve $8x^2 + tx - z = 0$ for x using the quadratic formula.

$$a = 8 \qquad b = t$$

$$x = \frac{-t \pm \sqrt{t^2 - 4(8)}}{2(8)} \qquad\qquad x = \frac{-b \pm \sqrt{b^2 - 4ac}}{2a}$$

$$= \frac{-t \pm \sqrt{t^2 + 32z}}{16} \qquad\qquad \text{Perform the operations.}$$

The solution set is $\left\{ \dfrac{-t - \sqrt{t^2 + 32z}}{16}, \dfrac{-t + \sqrt{t^2 + 32z}}{16} \right\}$.

 You Try 2

Solve for n.

a) $3n^2 + 5n - 1 = 0$ b) $3n^2 + pn - r = 0$

2. Solve an Applied Problem Involving Volume

Example 3

A rectangular piece of cardboard is 5 in. longer than it is wide. A square piece that measures 2 in. on each side is cut from each corner, then the sides are turned up to make an uncovered box with volume 252 in³. Find the length and width of the original piece of cardboard.

Solution

Step 1: Read the problem carefully, twice.

Step 2: Draw a picture. Define the unknowns. Label the picture.

$$x = \text{width of cardboard}$$
$$x + 5 = \text{length of the cardboard}$$

The volume of a box is (length)(width)(height). We will use the formula (length)(width)(height) = 252.

Original Cardboard   **Box**

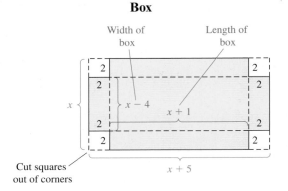

Figure 1

Figure 2

Figure 1 shows the original piece of cardboard with the sides labeled. Figure 2 illustrates how to label the box when the squares are cut out of the corners. When the sides are folded along the dotted lines, we must label the length, width, and height of the box.

$$\text{Length of box} = \begin{matrix}\text{Length of original}\\\text{cardboard}\end{matrix} - \begin{matrix}\text{Length of}\\\text{side cut out}\\\text{on the left}\end{matrix} - \begin{matrix}\text{Length of}\\\text{side cut out}\\\text{on the right}\end{matrix}$$
$$= \quad x + 5 \quad - \quad 2 \quad - \quad 2$$
$$= \quad x + 1$$

$$\text{Width of box} = \begin{matrix}\text{Width of original}\\\text{cardboard}\end{matrix} - \begin{matrix}\text{Length of}\\\text{side cut out}\\\text{on top}\end{matrix} - \begin{matrix}\text{Length of}\\\text{side cut out}\\\text{on bottom}\end{matrix}$$
$$= \quad x \quad - \quad 2 \quad - \quad 2$$
$$= \quad x - 4$$

$$\text{Height of box} = \text{Length of side cut out}$$
$$= \quad 2$$

We define the length, width, and height of the box as

$$\text{Length of box} = x + 1$$
$$\text{Width of box} = x - 4$$
$$\text{Height of box} = 2$$

Step 3: Translate from English to math to write an equation.

$$\text{Volume of box} = (\text{length})(\text{width})(\text{height})$$
$$252 = (x + 1)(x - 4)(2)$$

Step 4: Solve the equation.

$252 = (x + 1)(x - 4)(2)$	
$126 = (x + 1)(x - 4)$	Divide both sides by 2.
$126 = x^2 - 3x - 4$	Multiply.
$0 = x^2 - 3x - 130$	Write in standard form.
$0 = (x + 10)(x - 13)$	Factor.
$x + 10 = 0 \quad\text{or}\quad x - 13 = 0$	Set each factor equal to zero.
$x = -10 \quad\text{or}\quad x = 13$	Solve.

Step 5: Interpret the meaning of the solution as it relates to the problem.

x represents the width of the original piece of cardboard, so x cannot equal -10. Therefore, the width of the original piece of cardboard is 13 in.

The length of the cardboard is $x + 5$, so $13 + 5 = 18$ in.

Width of cardboard $= 13$ in.   Length of cardboard $= 18$ in.

Check:

$$\text{Width of box} = 13 - 4 = 9 \text{ in.}$$
$$\text{Length of box} = 18 - 4 = 14 \text{ in.}$$
$$\text{Height of box} = 2 \text{ in.}$$
$$\text{Volume of box} = 9(14)(2) = 252 \text{ in}^3.$$

You Try 3

The width of a rectangular piece of cardboard is 2 in. less than its length. A square piece that measures 3 in. on each side is cut from each corner, then the sides are turned up to make a box with volume 504 in³. Find the length and width of the original piece of cardboard.

3. Solve an Applied Problem Involving Area

Example 4

A rectangular pond is 20 ft long and 12 ft wide. The pond is bordered by a strip of grass of uniform (the same) width. The area of the grass is 320 ft². How wide is the border of grass around the pond?

Solution

Step 1: Read the problem carefully, twice.

Step 2: Draw a picture. Define the unknowns. Label the picture.

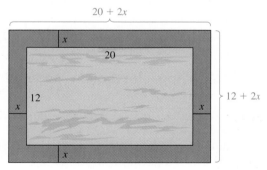

$x = $ width of the strip of grass
$20 + 2x = $ length of pond plus two strips of grass
$12 + 2x = $ width of pond plus two strips of grass

Step 3: Translate from English to math to write an equation.

We are told that the area of the grass border is 320 ft². We can also calculate the area of the pond since we know its length and width. The pond plus grass border forms a large rectangle of length $20 + 2x$ and width $12 + 2x$. The equation will come from the following relationship:

$$\begin{array}{ccc} \text{Area of pond} & \text{Area of} & \text{Area of} \\ \text{plus grass} & - \quad \text{pond} & = \quad \text{grass border} \\ (20 + 2x)(12 + 2x) & - \quad 20(12) = & 320 \end{array}$$

Step 4: Solve the equation.

$$\begin{aligned} (20 + 2x)(12 + 2x) - 20(12) &= 320 \\ 240 + 64x + 4x^2 - 240 &= 320 && \text{Multiply.} \\ 4x^2 + 64x &= 320 && \text{Combine like terms.} \\ x^2 + 16x &= 80 && \text{Divide by 4.} \\ x^2 + 16x - 80 &= 0 && \text{Write in standard form.} \\ (x + 20)(x - 4) &= 0 && \text{Factor.} \\ x + 20 = 0 \quad \text{or} \quad x - 4 &= 0 && \text{Set each factor equal to 0.} \\ x = -20 \quad \text{or} \quad x &= 4 && \text{Solve.} \end{aligned}$$

Step 5: Interpret the meaning of the solution as it relates to the problem.

x represents the width of the strip of grass, so x cannot equal -20.

The width of the strip of grass is 4 ft.

Check: Substitute $x = 4$ into the equation written in step 3.

$$\begin{aligned} (20 + 2(4))(12 + 2(4)) - 20(12) &\stackrel{?}{=} 320 \\ (28)(20) - 240 &\stackrel{?}{=} 320 \\ 560 - 240 &\stackrel{?}{=} 320 \\ 320 &= 320 \quad \checkmark \end{aligned}$$

You Try 4

A rectangular pond is 6 ft wide and 10 ft long and is surrounded by a concrete border of uniform width. The area of the border is 80 ft^2. Find the width of the border.

4. Solve an Applied Problem Modeled by a Given Quadratic Equation

Example 5

The total tourism-related output in the United States from 2000 to 2004 can be modeled by

$$y = 16.4x^2 - 50.6x + 896$$

where x is the number of years since 2000 and y is the total tourism-related output in billions of dollars. (Source: www.bea.gov)

a) According to the model, how much money was generated in 2002 due to tourism-related output?

b) In what year was the total tourism-related output about $955 billion?

Solution

a) Since x is the number of years *after* 2000, the year 2002 corresponds to $x = 2$.

$$y = 16.4x^2 - 50.6x + 896$$
$$y = 16.4(2)^2 - 50.6(2) + 896 \qquad \text{Substitute 2 for } x.$$
$$y = 860.4$$

The total tourism-related output in 2002 was approximately $860.4 billion.

b) y represents the total tourism-related output (in billions), so substitute 955 for y and solve for x.

$$y = 16.4x^2 - 50.6x + 896$$
$$955 = 16.4x^2 - 50.6x + 896 \qquad \text{Substitute 955 for } y.$$
$$0 = 16.4x^2 - 50.6x - 59 \qquad \text{Write in standard form.}$$

Use the quadratic formula to solve for x.

$$a = 16.4 \quad b = -50.6$$
$$x = \frac{50.6 \pm \sqrt{(-50.6)^2 - 4(16.4)}}{2(16.4)} \qquad \begin{array}{l}\text{Substitute the values into}\\ \text{the quadratic formula.}\end{array}$$
$$x \approx 3.99 \approx 4 \text{ or } x \approx -0.90$$

The negative value of x does not make sense in the context of the problem. Use $x \approx 4$.

The total tourism-related output was about $955 billion in 2004.

Answers to You Try Exercises

1) $m = \dfrac{2E}{v^2}$ 2) a) $\left\{ \dfrac{-5 - \sqrt{37}}{6}, \dfrac{-5 + \sqrt{37}}{6} \right\}$ b) $\left\{ \dfrac{-p - \sqrt{p^2 + 12r}}{6}, \dfrac{-p + \sqrt{p^2 + 12r}}{6} \right\}$

3) length $= 20$ in., width $= 18$ in. 4) 2 ft

11.7 Exercises

Boost your grade at mathzone.com! MathZone > Practice Problems > NetTutor > Self-Test > e-Professors > Videos

Objective 1

Solve for the indicated variable.

1) $A = \pi r^2$ for r

2) $V = \dfrac{1}{3}\pi r^2 h$ for r

3) $a = \dfrac{v^2}{r}$ for v

4) $K = \dfrac{1}{2}Iw^2$ for w

 5) $E = \dfrac{I}{d^2}$ for d

6) $L = \dfrac{2U}{I^2}$ for I

7) $F = \dfrac{kq_1 q_2}{r^2}$ for r

8) $E = \dfrac{kq}{r^2}$ for r

9) $d = \sqrt{\dfrac{4A}{\pi}}$ for A

10) $d = \sqrt{\dfrac{12V}{\pi h}}$ for V

11) $T_p = 2\pi\sqrt{\dfrac{l}{g}}$ for l

12) $V = \sqrt{\dfrac{3RT}{M}}$ for T

 13) $T_p = 2\pi\sqrt{\dfrac{l}{g}}$ for g

14) $V = \sqrt{\dfrac{3RT}{M}}$ for M

15) Compare the equations $3x^2 - 5x + 4 = 0$ and $rx^2 + 5x + s = 0$.

    a) How are the equations alike?

    b) How can both equations be solved for x?

16) What method could be used to solve
$2t^2 + 7t + 1 = 0$ and $kt^2 + mt + n = 0$ for t? Why?

Solve for the indicated variable.

17) $rx^2 - 5x + s = 0$ for x

18) $cx^2 + dx - 3 = 0$ for x

19) $pz^2 + rz - q = 0$ for z

20) $hr^2 - kr + j = 0$ for r

21) $da^2 - ha = k$ for a

22) $kt^2 + mt = -n$ for t

23) $s = \dfrac{1}{2}gt^2 + vt$ for t

24) $s = 2\pi rh + \pi r^2$ for r

Objectives 2 and 3

Write an equation and solve.

25) The length of a rectangular piece of sheet metal is 3 in. longer than its width. A square piece that measures 1 in. on each side is cut from each corner, then the sides are turned up to make a box with volume 70 in^3. Find the length and width of the original piece of sheet metal.

26) The width of a rectangular piece of cardboard is 8 in. less than its length. A square piece that measures 2 in. on each side is cut from each corner, then the sides are turned up to make a box with volume 480 in^3. Find the length and width of the original piece of cardboard.

27) A rectangular swimming pool is 60 ft wide and 80 ft long. A nonskid surface of uniform width is to be installed around the pool. If there is 576 ft^2 of the nonskid material, how wide can the strip of the nonskid surface be?

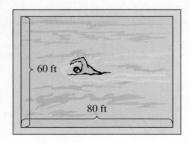

60 ft

80 ft

28) A picture measures 10 in. by 12 in. Emilio will get it framed with a border around it so that the total area of the picture plus the frame of uniform width is 168 in^2. How wide is the border?

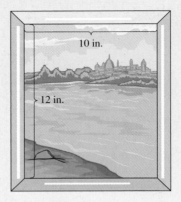

10 in.

12 in.

29) The height of a triangular sail is 1 ft less than twice the base of the sail. Find its height and the length of its base if the area of the sail is 60 ft^2.

30) Chandra cuts fabric into isosceles triangles for a quilt. The height of each triangle is 1 in. less than the length of the base. The area of each triangle is 15 in^2. Find the height and base of each triangle.

31) Valerie makes a bike ramp in the shape of a right triangle. The base of the ramp is 4 in. more than twice its height, and the length of the incline is 4 in. less than three times its height. How high is the ramp?

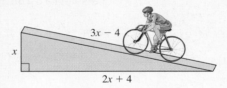

$3x - 4$

x

$2x + 4$

32) The width of a widescreen TV is 10 in. less than its length. The diagonal of the rectangular screen is 10 in. more than the length. Find the length and width of the screen.

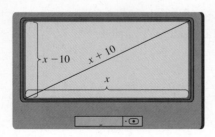

$x - 10$ $x + 10$

x

Objective 4

Solve.

33) An object is propelled upward from a height of 4 ft. The height h of the object (in feet) t sec after the object is released is given by

$$h = -16t^2 + 60t + 4$$

a) How long does it take the object to reach a height of 40 ft?

b) How long does it take the object to hit the ground?

34) An object is launched from the ground. The height h of the object (in feet) t sec after the object is released is given by

$$h = -16t^2 + 64t$$

When will the object be 48 ft in the air?

35) Attendance at Broadway plays from 1996–2000 can be modeled by

$$y = -0.25x^2 + 1.5x + 9.5$$

where x represents the number of years after 1996 and y represents the number of people who attended a Broadway play (in millions).
(Source: *Statistical Abstracts of the United States*)

a) Approximately how many people saw a Broadway play in 1996?

b) In what year did approximately 11.75 million people see a Broadway play?

36) The illuminance E (measure of the light emitted, in lux) of a light source is given by

$$E = \frac{I}{d^2}.$$

where I is the luminous intensity (measured in candela) and d is the distance, in meters, from the light source. The luminous intensity of a lamp is 2700 candela at a distance of 3 m from the lamp. Find the illuminance, E, in lux.

37) A sandwich shop has determined that the demand for its turkey sandwich is $\dfrac{65}{P}$ per day, where P is the price of the sandwich in dollars. The daily supply is given by $10P + 3$. Find the price at which the demand for the sandwich equals the supply.

38) A hardware store determined that the demand for shovels one winter was $\dfrac{2800}{P}$, where P is the price of the shovel in dollars. The supply was given by $12P + 32$. Find the price at which demand for the shovels equals the supply.

Definition/Procedure	Example	Reference
Review of Solving Quadratic Equations by Factoring		11.1
Steps for Solving a Quadratic Equation by Factoring 1) Write the equation in the form $ax^2 + bx + c = 0$. 2) Factor the expression. 3) Set each factor equal to zero, and solve for the variable. 4) Check the answer(s).	Solve $5z^2 - 7z = 6$. $$5z^2 - 7z - 6 = 0$$ $$(5z + 3)(z - 2) = 0$$ $$5z + 3 = 0 \quad \text{or} \quad z - 2 = 0$$ $$5z = -3$$ $$z = -\frac{3}{5} \quad \text{or} \quad z = 2$$ The check is left to the student. The solution set is $\left\{ -\dfrac{3}{5}, 2 \right\}$.	**p. 706**
Solving Quadratic Equations Using the Square Root Property		11.2
The Square Root Property Let k be a constant. If $x^2 = k$, then $x = \sqrt{k}$ or $x = -\sqrt{k}$.	Solve $6p^2 = 54$. $$p^2 = 9 \qquad \text{Divide by 6.}$$ $$p = \pm\sqrt{9} \qquad \text{Square root property}$$ $$p = \pm 3 \qquad \sqrt{9} = 3$$ The solution set is $\{-3, 3\}$.	**p. 709**
The Pythagorean Theorem If the lengths of the legs of a right triangle are a and b and the hypotenuse has length c, then $a^2 + b^2 = c^2$. 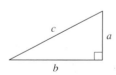	Find the length of the missing side. 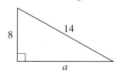 Use the Pythagorean theorem, $a^2 + b^2 = c^2$. $$b = 8, c = 14 \qquad \text{Find } a.$$ $$a^2 + (8)^2 = (14)^2 \qquad \text{Substitute values.}$$ $$a^2 + 64 = 196$$ $$a^2 = 132 \qquad \text{Subtract 64 from each side.}$$ $$a = \pm\sqrt{132} \qquad \text{Square root property}$$ $$a = \pm 2\sqrt{33} \qquad \sqrt{132} = \sqrt{4}\sqrt{33}$$ $$= 2\sqrt{33}$$ Discard $-2\sqrt{33}$ as a solution since the length of a side of a triangle cannot be negative. $$a = 2\sqrt{33}$$	**p. 713**

Definition/Procedure	Example	Reference

The Distance Formula

The distance between two points with coordinates (x_1, y_1) and (x_2, y_2) is given by $d = \sqrt{(x_2 - x_1)^2 + (y_2 - y_1)^2}$.

Find the distance between the points $(6, -2)$ and $(0, 2)$.

Label the points: $\overset{x_1}{(6,} \overset{y_1}{-2)} \overset{x_2}{(0,} \overset{y_2}{2)}$

Substitute the values into the distance formula.

$$d = \sqrt{(0 - 6)^2 + (2 - (-2))^2}$$
$$= \sqrt{(-6)^2 + (4)^2}$$
$$= \sqrt{36 + 16}$$
$$= \sqrt{52}$$
$$= 2\sqrt{13}$$

p. 713

Complex Numbers

11.3

Definition of i:

$$i = \sqrt{-1}$$

Therefore,

$$i^2 = -1$$

A *complex number* is a number of the form $a + bi$ where a and b are real numbers. a is called the *real part* and b is called the *imaginary part*. The set of real numbers is a subset of the set of complex numbers.

Examples of complex numbers:

$$-2 + 7i$$
$$5 \quad \text{(since it can be written } 5 + 0i)$$
$$4i \quad \text{(since it can be written } 0 + 4i)$$

p. 716

Simplifying Complex Numbers

Simplify $\sqrt{-25}$.

$$\sqrt{-25} = \sqrt{-1} \cdot \sqrt{25}$$
$$= i \cdot 5$$
$$= 5i$$

p. 717

When multiplying or dividing radicals with negative radicands, write each radical in terms of i first.

Multiply $\sqrt{-12} \cdot \sqrt{-3}$.

$$\sqrt{-12} \cdot \sqrt{-3} = i\sqrt{12} \cdot i\sqrt{3} = i^2\sqrt{36}$$
$$= -1 \cdot 6 = -6$$

p. 717

Adding and Subtracting Complex Numbers

To add and subtract complex numbers, combine the real parts and combine the imaginary parts.

Subtract $(10 + 7i) - (-2 + 4i)$.

$$(10 + 7i) - (-2 + 4i) = 10 + 7i + 2 - 4i$$
$$= 12 + 3i$$

p. 718

Multiply complex numbers like we multiply polynomials. Remember to replace i^2 with -1.

Multiply and simplify.

a) $4(9 + 5i) = 36 + 20i$

b) $(-3 + i)(2 - 7i) = -6 + 21i + 2i - 7i^2$

$$\qquad\qquad\qquad\quad \text{F} \quad \text{O} \quad \text{I} \quad \text{L}$$
$$= -6 + 23i - 7(-1)$$
$$= -6 + 23i + 7$$
$$= 1 + 23i$$

p. 719

Definition/Procedure	Example	Reference
Complex Conjugates 1) The conjugate of $a + bi$ is $a - bi$. 2) The product of $a + bi$ and $a - bi$ is a real number. 3) Find the product $(a + bi)(a - bi)$ using FOIL or recall that $(a + bi)(a - bi) = a^2 + b^2$.	Multiply $-5 - 3i$ by its conjugate. The conjugate of $-5 - 3i$ is $-5 + 3i$. i) Multiply using FOIL. $$(-5 - 3i)(-5 + 3i) = 25 - 15i + 15i - 9i^2$$ $$\text{F} \quad \text{O} \quad \text{I} \quad \text{L}$$ $$= 25 - 9(-1)$$ $$= 25 + 9$$ $$= 34$$ ii) Use $(a + bi)(a - bi) = a^2 + b^2$. $$(-5 - 3i)(-5 + 3i) = (-5)^2 + (3)^2$$ $$= 25 + 9$$ $$= 34$$	**p. 719**
Dividing Complex Numbers To divide complex numbers, multiply the numerator and denominator by the *conjugate of the denominator*. Write the result in the form $a + bi$.	Divide $\dfrac{6i}{2 + 5i}$. Write the result in the form $a + bi$. $$\frac{6i}{2 + 5i} = \frac{6i}{2 + 5i} \cdot \frac{(2 - 5i)}{(2 - 5i)}$$ $$= \frac{12i - 30i^2}{2^2 + 5^2}$$ $$= \frac{12i - 30(-1)}{29}$$ $$= \frac{30}{29} + \frac{12}{29}i$$	**p. 720**
We can find complex solutions to quadratic equations. If $a + bi$ is a solution of a quadratic equation having only real coefficients, then $a - bi$ is also a solution.	Solve $(h - 7)^2 + 20 = 4$. $(h - 7)^2 = -16$ Subtract 20 from each side. $h - 7 = \pm\sqrt{-16}$ Square root property $h - 7 = \pm 4i$ $\sqrt{-16} = 4i$ $h = 7 \pm 4i$ Add 7 to each side. The solution set is $\{7 - 4i, 7 + 4i\}$.	**p. 722**
Solving Quadratic Equations by Completing the Square		11.4
A *perfect square trinomial* is a trinomial whose factored form is the square of a binomial.	**Perfect Square Trinomial** **Factored Form** $y^2 + 8y + 16$ $(y + 4)^2$ $9t^2 - 30t + 25$ $(3t - 5)^2$	**p. 725**
To find the constant needed to complete the square for $x^2 + bx$, **Step 1:** Find half of the coefficient of x: $\left(\dfrac{1}{2}b\right)$ **Step 2:** Square the result: $\left(\dfrac{1}{2}b\right)^2$. **Step 3:** Add it to $x^2 + bx$: $x^2 + bx + \left(\dfrac{1}{2}b\right)^2$.	Complete the square for $x^2 + 12x$ to obtain a perfect square trinomial. Then, factor. **Step 1:** Find half of the coefficient of x: $\dfrac{1}{2}(12) = 6$ **Step 2:** Square the result: $6^2 = 36$ **Step 3:** Add 36 to $x^2 + 12x$: $x^2 + 12x + 36$ The perfect square trinomial is $x^2 + 12x + 36$. The factored form is $(x + 6)^2$.	**p. 725**

Definition/Procedure	Example	Reference

Solving a Quadratic Equation ($ax^2 + bx + c = 0$) by Completing the Square

Step 1: *The coefficient of the squared term must be 1. If it is not 1, divide both sides of the equation by a to obtain a leading coefficient of 1.*

Step 2: *Get the variables on one side of the equal sign and the constant on the other side.*

Step 3: *Complete the square. Find half of the linear coefficient, then square the result. Add that quantity to both sides of the equation.*

Step 4: *Factor.*

Step 5: *Solve using the square root property.*

Solve $x^2 + 6x + 7 = 0$ by completing the square.

$$x^2 + 6x + 7 = 0 \quad \text{The coefficient of } x^2 \text{ is 1.}$$
$$x^2 + 6x = -7 \quad \text{Get the constant on the other side of the equal sign.}$$

Complete the square.

$$\frac{1}{2}(6) = 3$$
$$(3)^2 = 9$$

Add 9 to both sides of the equation.

$$x^2 + 6x + 9 = -7 + 9$$
$$(x + 3)^2 = 2 \qquad \text{Factor.}$$
$$x + 3 = \pm\sqrt{2} \qquad \text{Square root property}$$
$$x = -3 \pm \sqrt{2}$$

The solution set is $\{-3 - \sqrt{2}, -3 + \sqrt{2}\}$.

p. 729

Solving Equations Using the Quadratic Formula

11.5

The Quadratic Formula

The solutions of any quadratic equation of the form $ax^2 + bx + c = 0 \ (a \neq 0)$ are

$$x = \frac{-b \pm \sqrt{b^2 - 4ac}}{2a}$$

Solve $2x^2 - 5x - 2 = 0$ using the quadratic formula.

$$a = 2 \quad b = -5 \quad c = -2$$

Substitute the values into the quadratic formula.

$$x = \frac{-(-5) \pm \sqrt{(-5)^2 - 4(2)(-2)}}{2(2)}$$
$$x = \frac{5 \pm \sqrt{25 + 16}}{4} = \frac{5 \pm \sqrt{41}}{4}$$

The solution set is $\left\{\dfrac{5 - \sqrt{41}}{4}, \dfrac{5 + \sqrt{41}}{4}\right\}$.

p. 735

The *discriminant* is the expression under the radical in the quadratic formula, $b^2 - 4ac$.

1) If $b^2 - 4ac$ is *positive and the square of an integer*, the equation has *two rational solutions*.

2) If $b^2 - 4ac$ is *positive but not a perfect square*, the equation has *two irrational solutions*.

3) If $b^2 - 4ac$ is *negative*, the equation has *two complex solutions of the form $a + bi$ and $a - bi$, where $b \neq 0$.*

4) If $b^2 - 4ac = 0$, the equation has *one rational solution.*

Find the value of the discriminant for $3m^2 + 4m + 5 = 0$ and determine the number and type of solutions of the equation.

$$a = 3 \quad b = 4 \quad c = 5$$
$$b^2 - 4ac = (4)^2 - 4(3)(5) = 16 - 60 = -44$$

Discriminant $= -44$. The equation has two complex solutions of the form $a + bi$ and $a - bi$.

p. 738

Definition/Procedure	Example	Reference

Equations in Quadratic Form   11.6

Some equations that are not quadratic can be solved using the same methods that can be used to solve quadratic equations. These are called *equations in quadratic form*.	Solve. $r^4 + 2r^2 - 24 = 0$ $(r^2 - 4)(r^2 + 6) = 0$ Factor. $r^2 - 4 = 0$ or $r^2 + 6 = 0$ $r^2 = 4$ $r^2 = -6$ $r = \pm\sqrt{4}$ $r = \pm\sqrt{-6}$ $r = \pm 2$ $r = \pm i\sqrt{6}$ The solution set is $\{-i\sqrt{6}, i\sqrt{6}, -2, 2\}$.	**p. 747**

Formulas and Applications   11.7

Solve a formula for a particular variable.	Solve for s: $g = \dfrac{10}{s^2}$ $s^2 g = 10$ Multiply both sides by s^2. $s^2 = \dfrac{10}{g}$ Divide both sides by g. $s = \pm\sqrt{\dfrac{10}{g}}$ Square root property $s = \dfrac{\pm\sqrt{10}}{\sqrt{g}} \cdot \dfrac{\sqrt{g}}{\sqrt{g}}$ Rationalize the denominator. $s = \dfrac{\pm\sqrt{10g}}{g}$	**p. 754**
Solving application problems using a given quadratic equation.	A woman dives off of a cliff 49 m above the ocean. Her height, h, in meters, above the water is given by $$h = -9.8t^2 + 49$$ where t is the time, in seconds, after she leaves the cliff. When will she hit the water? Let $h = 0$ and solve for t. $h = -9.8t^2 + 49$ $0 = -9.8t^2 + 49$ Substitute 0 for h. $9.8t^2 = 49$ Add $9.8t^2$ to each side. $t^2 = 5$ Divide by 9.8. $t = \pm\sqrt{5}$ Square root property Since t represents time, we discard $-\sqrt{5}$. Use $t = \sqrt{5}$. She will hit the water after $\sqrt{5}$ sec or about 2.2 sec.	**p. 758**

(11.1) Solve by factoring.

1) $a^2 - 3a - 54 = 0$

2) $2t^2 + 9t + 10 = 0$

3) $\frac{2}{3}c^2 = \frac{2}{3}c + \frac{1}{2}$

4) $4k = 12k^2$

5) $x^3 + 3x^2 - 16x - 48 = 0$

6) $3p - 16 = p(p - 7)$

Write an equation and solve.

7) A rectangle has an area of 96 cm^2. Its width is 4 cm less than its length. Find the length and width.

8) Find the base and height of the triangle if its area is 30 in^2.

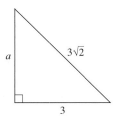

(11.2) Solve using the square root property.

9) $d^2 = 144$

10) $m^2 = 75$

11) $v^2 + 4 = 0$

12) $2c^2 - 11 = 25$

13) $(b - 3)^2 = 49$

14) $(6y + 7)^2 - 15 = 0$

15) $27k^2 - 30 = 0$

16) $(j - 14)^2 + 5 = 0$

17) Find the length of the missing side.

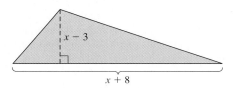

18) A rectangle has a length of $5\sqrt{2}$ in. and a width of 4 in. How long is its diagonal?

Find the distance between the given points.

19) $(2, 3)$ and $(7, 5)$

20) $(-2, 5)$ and $(-1, -3)$

21) $(3, -1)$ and $(0, 3)$

22) $(-5, 8)$ and $(2, 3)$

(11.3) Simplify.

23) $\sqrt{-49}$

24) $\sqrt{-8}$

25) $\sqrt{-2} \cdot \sqrt{-8}$

26) $\sqrt{-6} \cdot \sqrt{-3}$

Perform the indicated operations.

27) $(2 + i) + (10 - 4i)$

28) $(4 + 3i) - (11 - 4i)$

29) $\left(\frac{4}{5} - \frac{1}{3}i\right) - \left(\frac{1}{2} + i\right)$

30) $\left(-\frac{3}{8} - 2i\right) + \left(\frac{5}{8} + \frac{3}{2}i\right) - \left(\frac{1}{4} - \frac{1}{2}i\right)$

Multiply and simplify.

31) $5(-6 + 7i)$

32) $-8i(4 + 3i)$

33) $3i(-7 + 12i)$

34) $(3 - 4i)(2 + i)$

35) $(4 - 6i)(3 - 6i)$

36) $\left(\frac{1}{5} - \frac{2}{3}i\right)\left(\frac{3}{2} - \frac{2}{3}i\right)$

Identify the conjugate of each complex number, then multiply the number and its conjugate.

37) $2 - 7i$

38) $-2 + 3i$

Divide. Write the result in the form $a + bi$.

39) $\dfrac{6}{2 + 5i}$

40) $\dfrac{-12}{4 - 3i}$

41) $\dfrac{8}{i}$

42) $\dfrac{4i}{1 - 3i}$

43) $\dfrac{9 - 4i}{6 - i}$

44) $\dfrac{5 - i}{-2 + 6i}$

(11.4) Complete the square for each expression to obtain a perfect square trinomial. Then, factor.

45) $r^2 + 10r$

46) $z^2 - 12z$

47) $c^2 - 5c$

48) $x^2 + x$

49) $a^2 + \frac{2}{3}a$

50) $d^2 - \frac{5}{2}d$

Solve by completing the square.

51) $p^2 - 6p - 16 = 0$

52) $w^2 - 2w - 35 = 0$

53) $n^2 + 10n = 6$

54) $t^2 + 9 = -4t$

55) $f^2 + 3f + 1 = 0$

56) $j^2 - 7j = 4$

57) $-3q^2 + 7q = 12$

58) $6v^2 - 15v + 3 = 0$

(11.5) Solve using the quadratic formula.

59) $m^2 + 4m - 12 = 0$

60) $3y^2 = 10y - 8$

61) $10g - 5 = 2g^2$

62) $20 = 4x - 5x^2$

63) $\frac{1}{6}t^2 - \frac{1}{3}t + \frac{2}{3} = 0$ 64) $(s-3)(s-5) = 9$

65) $(6r+1)(r-4) = -2(12r+1)$

66) $z^2 - \frac{3}{2}z + \frac{13}{16} = 0$

Find the value of the discriminant. Then, determine the number and type of solutions of each equation. Do not solve.

67) $3n^2 - 2n - 5 = 0$ 68) $5x^2 + 5x + \frac{5}{4} = 0$

69) $t^2 = -3(t+2)$ 70) $3 - 7y = -y^2$

71) Find the value of b so that $4k^2 + bk + 9 = 0$ has only one rational solution.

72) A ball is thrown upward from a height of 4 ft. The height h of the ball (in feet) t sec after the ball is released is given by $h = -16t^2 + 52t + 4$.

   a) How long does it take the ball to reach a height of 16 ft?

   b) How long does it take the ball to hit the ground?

(11.6) Solve.

73) $z + 2 = \frac{15}{z}$ 74) $\frac{5k}{k+1} = 3k - 4$

75) $\frac{10}{m} = 3 + \frac{8}{m^2}$ 76) $f = \sqrt{7f - 12}$

77) $x - 4\sqrt{x} = 5$ 78) $n^4 - 17n^2 + 16 = 0$

79) $b^4 + 5b^2 - 14 = 0$ 80) $q^{2/3} + 2q^{1/3} - 3 = 0$

81) $y + 2 = 3y^{1/2}$ 82) $2r^4 = 7r^2 - 2$

83) $2(v+2)^2 + (v+2) - 3 = 0$

84) $(2k-5)^2 - 5(2k-5) - 6 = 0$

(11.7) Solve for the indicated variable.

85) $F = \frac{mv^2}{r}$ for v 86) $U = \frac{1}{2}kx^2$ for x

87) $r = \sqrt{\frac{A}{\pi}}$ for A 88) $r = \sqrt{\frac{V}{\pi l}}$ for V

89) $kn^2 - ln - m = 0$ for n

90) $2p^2 + t = rp$ for p

Write an equation and solve.

91) Ayesha is making a pillow sham by sewing a border onto an old pillow case. The rectangular pillow case measures 18 in. by 27 in. When she sews a border of uniform width around the pillowcase, the total area of the surface of the pillow sham will be 792 in². How wide is the border?

92) The width of a rectangular piece of cardboard is 4 in. less than its length. A square piece that measures 2 in. on each side is cut from each corner, then the sides are turned up to make a box with volume 280 in³. Find the length and width of the original piece of cardboard.

93) A flower shop determined that the demand for its tulip bouquet is $\frac{240}{P}$ per week, where P is the price of the bouquet in dollars. The weekly supply is given by $4P - 2$. Find the price at which demand for the tulips equals the supply.

94) U.S. sales of a certain brand of wine can be modeled by

$$y = -0.20x^2 + 4.0x + 8.4$$

for the years 1990–2005. x is the number of years after 1990 and y is the number of bottles sold, in millions.

   a) How many bottles were sold in 1990?

   b) How many bottles were sold in 2003?

   c) In what year did sales reach 28.4 million bottles?

1) Find the distance between the points $(-6, 2)$ and $(4, 3)$.

Simplify.

2) $\sqrt{-64}$

3) $\sqrt{-18}$

4) $\sqrt{-5} \cdot \sqrt{-10}$

5) Subtract $(-3 + 7i) - (-8 + 4i)$

Multiply and simplify.

6) $7i(9 - 7i)$

7) $\left(\dfrac{1}{2} + 3i\right)\left(\dfrac{2}{3} - i\right)$

8) $(9 - 2i)(9 + 2i)$

Divide. Write the result in the form $a + bi$.

9) $\dfrac{8}{6 + i}$

10) $\dfrac{1 - 2i}{8 - 3i}$

Solve by completing the square.

11) $b^2 + 4b - 7 = 0$

12) $2x^2 - 6x + 14 = 0$

13) Solve $x^2 - 8x + 17 = 0$ using the quadratic formula.

Solve using any method.

14) $(c + 5)^2 + 8 = 2$

15) $3q^2 + 2q = 8$

16) $(4n + 1)^2 + 9(4n + 1) + 18 = 0$

17) $(2t - 3)(t - 2) = 2$

18) $p^4 + p^2 - 72 = 0$

19) $\dfrac{3}{10x} = \dfrac{x}{x - 1} - \dfrac{4}{5}$

20) Find the value of the discriminant. Then, determine the number and type of solutions of the equation. *Do not solve.*

$$5z^2 - 6z - 1 = 0$$

21) Find the length of the missing side.

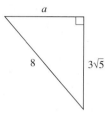

22) A ball is projected upward from the top of a 200 ft tall building. The height h of the ball (in feet) t sec after the ball is released is given by $h = -16t^2 + 24t + 200$.

a) When will the ball be 40 ft above the ground?

b) When will the ball hit the ground?

23) Solve $r = \sqrt{\dfrac{3V}{\pi h}}$ for V.

24) Solve $rt^2 - st = 6$ for t.

Write an equation and solve.

25) A rectangular piece of sheet metal is 6 in. longer than it is wide. A square piece that measures 3 in. on each side is cut from each corner, then the sides are turned up to make a box with volume 273 in³. Find the length and width of the original piece of sheet metal.

Perform the operations and simplify.

1) $\dfrac{4}{15} + \dfrac{1}{6} - \dfrac{3}{5}$

2) $24 - 8 \div 2 - |3 - 10|$

3) Find the area and perimeter of this figure.

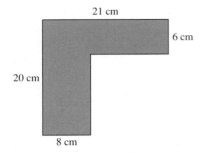

21 cm

6 cm

20 cm

8 cm

Simplify. The final answer should contain only positive exponents.

4) $(-2d^5)^3$

5) $(5x^4y^{-10})(3xy^3)^2$

6) $\left(\dfrac{40a^{-3}b}{8a^{-8}b^4}\right)^{-3}$

7) *Write an equation and solve.*
In December 2007, an electronics store sold 108 digital cameras. This is a 20% increase over their sales in December 2006. How many digital cameras did they sell in December 2006?

8) Solve $y = mx + b$ for m.

9) Find the x- and y-intercepts of $2x - 5y = 8$ and graph.

10) Identify the slope, y-intercept, and graph $y = -\dfrac{3}{4}x + 1$.

11) *Write a system of two equations in two variables and solve.*
Two bags of chips and three cans of soda cost $3.85 while one bag of chips and two cans of soda cost $2.30. Find the cost of a bag of chips and a can of soda.

12) Subtract
$(4x^2y^2 - 11x^2y + xy + 2) - (x^2y^2 - 6x^2y + 3xy^2 + 10xy - 6)$.

13) Multiply and simplify $3(r - 5)^2$.

Factor completely.

14) $4p^3 + 14p^2 - 8p$

15) $a^3 + 125$

16) Add $\dfrac{z - 8}{z + 4} + \dfrac{3}{z}$.

17) Simplify $\dfrac{2 + \dfrac{6}{c}}{\dfrac{2}{c^2} - \dfrac{8}{c}}$

18) Solve $|4k - 3| = 9$.

19) Solve this system: $\begin{aligned} 4x - 2y + z &= -7 \\ -3x + y - 2z &= 5 \\ 2x + 3y + 5z &= 4 \end{aligned}$

Simplify. Assume all variables represent nonnegative real numbers.

20) $\sqrt{75}$

21) $\sqrt[3]{40}$

22) $\sqrt{63x^7y^4}$

23) Simplify $64^{2/3}$.

24) Rationalize the denominator: $\dfrac{5}{2 + \sqrt{3}}$.

25) Multiply and simplify $(10 + 3i)(1 - 8i)$.

Solve.

26) $3k^2 - 4 = 20$

27) $\dfrac{3}{5}x^2 + \dfrac{1}{5} = \dfrac{1}{5}x$

28) $1 - \dfrac{1}{3h - 2} = \dfrac{20}{(3h - 2)^2}$

29) $p^2 + 6p = 27$

30) Solve $r = \sqrt{\dfrac{V}{\pi h}}$ for V.

Functions and Their Graphs

Algebra at Work: Forensics

When forensic scientists are called to a crime scene where a skeleton has been found, they look at many different features to work with police to piece together a profile of the victim. The hips can reveal whether the person was male or female.

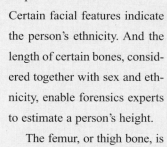

Certain facial features indicate the person's ethnicity. And the length of certain bones, considered together with sex and ethnicity, enable forensics experts to estimate a person's height.

The femur, or thigh bone, is the largest bone in the human body and is one that is commonly used to estimate a person's height.

Raul arrives at a crime scene to help identify skeletal remains. After taking measurements on the skull and hips, he determines that the victim was a white female. To estimate her height, Raul can use the linear function

$$H(f) = 2.47f + 54.10$$

where $f =$ the length of the femur, in centimeters

$H(f) =$ the height of the victim, in centimeters

The length of the femur is 44 cm. Substituting 44 for f in the function above, Raul estimates the height of the victim:

$$H(44) = 2.47(44) + 54.10$$
$$H(44) = 162.78 \text{ cm}$$

The measurements that Raul took on the skeletal remains indicate that the victim was a white female and that she was about 162.78 cm or 64 in. tall.

Section 12.1 Relations and Functions

Objectives

1. Define and Identify Relation, Function, Domain, and Range

2. Given an Equation, Find the Domain of the Relation and Determine if *y* Is a Function of *x*

3. Determine the Domain of and Graph a Linear Function

4. Define a Polynomial Function and a Quadratic Function

5. Evaluate Linear and Quadratic Functions for a Given Variable or Expression

6. Define and Determine the Domain of a Rational Function

7. Define and Determine the Domain of a Square Root Function

8. Summarize Strategies for Determining the Domain of a Given Function

1. Define and Identify Relation, Function, Domain, and Range

We first studied functions in Chapter 4. Now, we will review some of the concepts first presented in Sections 4.7 and 4.8, and we will extend what we have learned to include new functions. Recall the following definitions.

> **Definition**
>
> A **relation** is any set of ordered pairs. The **domain** of a relation is the set of all values of the independent variable (the first coordinates in the set of ordered pairs). The **range** of a relation is the set of all values of the dependent variable (the second coordinates in the set of ordered pairs).

> **Definition**
>
> A **function** is a special type of relation. If each element of the domain corresponds to *exactly one* element of the range, then the relation is a function.

Example 1

Identify the domain and range of each relation, and determine whether each relation is a function.

a) $\{(-2, -7), (0, -1), (1, 2), (5, 14)\}$

b)
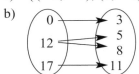

c)
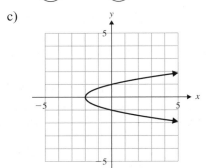

Solution

a)   The *domain* is the set of first coordinates, $\{-2, 0, 1, 5\}$.
The *range* is the set of second coordinates, $\{-7, -1, 2, 14\}$.

Ask yourself, "Does every first coordinate correspond to *exactly one* second coordinate?" *Yes.* This relation *is* a function.

b)   The *domain* is $\{0, 12, 17\}$.
The *range* is $\{3, 5, 8, 11\}$.

One of the elements in the domain, 12, corresponds to *two* elements in the range, 5 and 8. Therefore, this relation is *not* a function.

c)   The domain is $[-2, \infty)$. The range is $(-\infty, \infty)$.

To determine if this graph represents a function, recall that we can use the *vertical line test.*

The **vertical line test** says that if there is no vertical line that can be drawn through a graph so that it intersects the graph more than once, then the graph represents a function.

This graph fails the vertical line test because we can draw a vertical line through the graph that intersects it more than once. *This graph does* not *represent a function.*

You Try 1

Identify the domain and range of each relation, and determine whether each relation is a function.

a) $\{(-8, 1), (-5, 2), (-5, -2), (7, -4)\}$ b)

c)

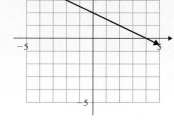

Next, we will look at relations and functions written as equations.

2. Given an Equation, Find the Domain of the Relation and Determine if *y* Is a Function of *x*

If a relation is written as an equation so that y is in terms of x, then the *domain* is the set of all real numbers that can be substituted for the independent variable, x. The resulting set of real numbers that are obtained for y, the dependent variable, is the *range*.

To determine the domain, sometimes it is helpful to ask yourself, *"Is there any number that cannot be substituted for x?"*

Example 2

Determine whether each relation describes y as a function of x, and determine the domain of the relation.

a) $y = -3x + 4$ b) $y^2 = x$

Solution

a)   *Every* value substituted for x will have *exactly one* corresponding value of y. For example, if we substitute 5 for x, the only value of y is -11. Therefore, $y = -3x + 4$ *is a function.*

To determine the domain, ask yourself, *"Is there any number that cannot be substituted for x in $y = -3x + 4$?"* No. Any real number can be substituted for x, and $y = -3x + 4$ will be defined.

The domain consists of all real numbers. This can be written as $(-\infty, \infty)$.

b)   If we substitute a number such as 4 for x and solve for y, we get

$$y^2 = 4$$
$$y = \pm\sqrt{4}$$
$$y = \pm 2$$

The ordered pairs $(4, 2)$ and $(4, -2)$ satisfy $y^2 = x$. Since $x = 4$ corresponds to two *different* y-values, $y^2 = x$ is *not* a function.

To determine the domain of this relation, ask yourself, *"Is there any number that cannot be substituted for x in $y^2 = x$?"* In this case, let's first look at y. Since y is squared, any real number substituted for y will produce a number that is greater than or equal to zero. Therefore, in the equation, x will equal a number that is greater than or equal to zero.

The domain is $[0, \infty)$.

You Try 2

Determine whether each relation describes y as a function of x, and determine the domain of the relation.

a) $y = x^2 + 5$ b) $y = x^3$

Recall that if a function describes the relationship between x and y so that x is the independent variable and y is the dependent variable, then *y is a function of x.* That is, *the value of y depends on the value of x.*

Recall also that we can use *function notation* to represent this relationship.

Definition

$y = f(x)$ is called **function notation**, and it is read as, "y equals f of x."
$y = f(x)$ means that y is a function of x (that is, y depends on x).

If a relation is a function, then $f(x)$ can be used in place of y. **$f(x)$ is the same as y.**

In Example 2, we concluded that $y = -3x + 4$ is a function. Using function notation, we can write $y = -3x + 4$ as $f(x) = -3x + 4$. *They mean the same thing.*

Types of Functions

3. Determine the Domain of and Graph a Linear Function

In addition to being a function, $y = -3x + 4$ is a linear equation in two variables. Because it is also a function, we call it a *linear function*. We first studied linear functions in Chapter 4, and we restate its definition here.

Definition

A **linear function** has the form $f(x) = mx + b$, where m and b are real numbers, m is the *slope* of the line, and $(0, b)$ is the *y-intercept*.

Let's look at some functions and learn how to determine their domains.

Example 3

$$f(x) = \frac{1}{2}x - 3$$

a)   What is the domain of f?   b)   Graph the function.

Solution

a)   The domain is the set of all real numbers that can be substituted for x. Ask yourself, "*Is there any number that cannot be substituted for x in $f(x) = \frac{1}{2}x - 3$?*"

   No. Any real number can be substituted for x, and $f(x) = \frac{1}{2}x - 3$ will be defined. The domain consists of all real numbers. This can be written as $(-\infty, \infty)$.

b)   The y-intercept is $(0, -3)$, and the slope of the line is $\frac{1}{2}$. Use this information to graph the line.

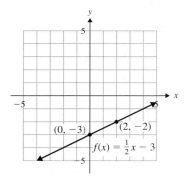

You Try 3

$g(x) = -x + 2$

a) What is the domain of g?   b) Graph the function.

4. Define a Polynomial Function and a Quadratic Function

Another function often used in mathematics is a **polynomial function**.

   The expression $x^3 + 2x^2 - 9x + 5$ is a polynomial. For each real number that is substituted for x, there will be *only one value* of the expression. For example, if we substitute 2 for x, the *only value* of the expression is 3. (Try this yourself to verify the result!) Since each value substituted for the variable produces *only one value* of the expression, we can use function notation to represent a polynomial like $x^3 + 2x^2 - 9x + 5$.

$f(x) = x^3 + 2x^2 - 9x + 5$ is a **polynomial function** since $x^3 + 2x^2 - 9x + 5$ is a polynomial.

The domain of a polynomial function is all real numbers, $(-\infty, \infty)$.

In Chapter 4, we learned how to find a function value. For example, if $f(x) = x^3 + 2x^2 - 9x + 5$, then to find $f(-3)$ means to substitute -3 for x and evaluate.

$$\begin{aligned} f(x) &= x^3 + 2x^2 - 9x + 5 \\ f(-3) &= (-3)^3 + 2(-3)^2 - 9(-3) + 5 \qquad \text{Substitute } -3 \text{ for } x. \\ &= -27 + 18 + 27 + 5 \qquad \text{Evaluate.} \\ &= 23 \end{aligned}$$

We can also say that the ordered pair $(-3, 23)$ satisfies $f(x) = x^3 + 2x^2 - 9x + 5$.

A quadratic function is a special type of polynomial function.

Definition

A polynomial function of the form $f(x) = ax^2 + bx + c$, where a, b, and c are real numbers and $a \neq 0$ is a **quadratic function**.

An example of a *quadratic function* is $f(x) = 3x^2 - x - 8$. (Notice that this is similar to a quadratic equation, an equation of the form $ax^2 + bx + c = 0$.)

5. Evaluate Linear and Quadratic Functions for a Given Variable or Expression

Above we found $f(-3)$ for $f(x) = x^3 + 2x^2 - 9x + 5$. We can also evaluate functions for variables and expressions.

Example 4

Let $g(x) = 4x - 10$ and $h(x) = x^2 + 2x - 11$. Find each of the following and simplify.

a) $g(k)$ b) $g(n + 3)$ c) $h(p)$ d) $h(w - 4)$

Solution

a) Finding $g(k)$ (read as *g of k*) means to substitute k for x in the function g, and simplify the expression as much as possible.

$$\begin{aligned} g(x) &= 4x - 10 \\ g(k) &= 4k - 10 \qquad \text{Substitute } k \text{ for } x. \end{aligned}$$

b)   Finding $g(n + 3)$ (read as *g of n plus 3*) means to substitute $n + 3$ for x in the function g, and simplify the expression as much as possible. *Since $n + 3$ contains two terms, we must put it in parentheses.*

$$g(x) = 4x - 10$$
$$g(n + 3) = 4(n + 3) - 10 \qquad \text{Substitute } n + 3 \text{ for } x.$$
$$g(n + 3) = 4n + 12 - 10 \qquad \text{Distribute.}$$
$$g(n + 3) = 4n + 2 \qquad \text{Combine like terms.}$$

c)   To find $h(p)$, substitute p for x in function h.

$$h(x) = x^2 + 2x - 11$$
$$h(p) = p^2 + 2p - 11 \qquad \text{Substitute } p \text{ for } x.$$

d)   To find $h(w - 4)$, substitute $w - 4$ for x in function h, and simplify the expression as much as possible. *When we substitute, we must put $w - 4$ in parentheses since $w - 4$ consists of more than one term.*

$$h(x) = x^2 + 2x - 11$$
$$h(w - 4) = (w - 4)^2 + 2(w - 4) - 11 \qquad \text{Substitute } w - 4 \text{ for } x.$$
$$h(w - 4) = w^2 - 8w + 16 + 2w - 8 - 11 \qquad \text{Multiply.}$$
$$h(w - 4) = w^2 - 6w - 3 \qquad \text{Combine like terms.}$$

You Try 4

Let $f(x) = -9x + 2$ and $k(x) = x^2 + 5x + 8$. Find each of the following and simplify.

a) $f(c)$ b) $f(r - 5)$ c) $k(z)$ d) $k(m + 2)$

6. Define and Determine the Domain of a Rational Function

Some functions contain rational expressions. Like a polynomial, each real number that can be substituted for the variable in a rational expression will produce *only one* value for the expression. Therefore, we can use function notation to represent a rational expression.

$$f(x) = \frac{4x - 3}{x + 5} \text{ is an example of a } \textbf{rational function}.$$

Since a fraction is undefined when its denominator equals zero, it follows that a rational expression is undefined when its denominator equals zero. For example, in the function $f(x) = \dfrac{4x - 3}{x + 5}$, x *cannot* equal -5 because then we would get zero in the denominator. This is important to keep in mind when we are trying to find the domain of a rational function.

Definition

The **domain of a rational function** consists of all real numbers *except* the value(s) of the variable that make the denominator equal zero.

Therefore, to determine the domain of a rational function we set the denominator equal to zero and solve for the variable. Any values that make the denominator equal to zero are *not* in the domain of the function.

Remember, to determine the domain of a rational function, sometimes, it is helpful to ask yourself, *"Is there any number that cannot be substituted for the variable?"*

Example 5

Determine the domain of each rational function.

a) $f(x) = \dfrac{5x}{x - 7}$
b) $g(t) = \dfrac{t + 4}{2t + 1}$
c) $r(x) = \dfrac{2x + 9}{6}$

Solution

a) To determine the domain of $f(x) = \dfrac{5x}{x - 7}$ ask yourself, *"Is there any number that cannot be substituted for x?"* Yes, $f(x)$ is *undefined* when the denominator equals zero. Set the denominator equal to zero, and solve for x.

$$x - 7 = 0 \qquad \text{Set the denominator} = 0.$$
$$x = 7 \qquad \text{Solve.}$$

When $x = 7$, the denominator of $f(x) = \dfrac{5x}{x - 7}$ equals zero. The domain contains all real numbers *except* 7. Write the domain in interval notation as $(-\infty, 7) \cup (7, \infty)$.

b) Ask yourself, *"Is there any number that cannot be substituted for t in* $g(t) = \dfrac{t + 4}{2t + 1}$?" Look at the denominator. When will it equal 0? Set the denominator equal to 0 and solve for t to determine what value of t will make the denominator equal 0.

$$2t + 1 = 0 \qquad \text{Set the denominator} = 0.$$
$$2t = -1$$
$$t = -\frac{1}{2} \qquad \text{Solve.}$$

When $t = -\dfrac{1}{2}$, the denominator of $g(t) = \dfrac{t + 4}{2t + 1}$ equals zero. The domain contains all real numbers *except* $-\dfrac{1}{2}$. Write the domain in interval notation as

$$\left(-\infty, -\frac{1}{2}\right) \cup \left(-\frac{1}{2}, \infty\right).$$

c) *Is there any number that cannot be substituted for x in* $r(x) = \dfrac{2x + 9}{6}$? The denominator is a constant, 6, and it can never equal zero. Therefore, any real number can be substituted for x and the function will be defined.

The domain consists of all real numbers, which can be written as $(-\infty, \infty)$.

You Try 5

Determine the domain of each rational function.

a) $f(x) = \dfrac{9}{x}$ b) $k(c) = \dfrac{c}{3c + 4}$ c) $g(n) = \dfrac{n - 2}{n^2 - 9}$

7. Define and Determine the Domain of a Square Root Function

When real numbers are substituted for the variable in radical expressions like $\sqrt{x}$ and $\sqrt{4r + 1}$, each value that is substituted will produce *only one* value for the expression. Function notation can be used to represent radical expressions too.

$$f(x) = \sqrt{x} \text{ and } g(r) = \sqrt{4r + 1} \text{ are examples of } \textbf{square root functions}.$$

When we determine the domain of a square root function, we are determining all *real* numbers that may be substituted for the variable so that the range contains only *real* numbers. (Complex numbers are *not* included.) All values substituted for the independent variable that produce complex numbers as function values are *not* in the domain of the function. This means that there may be *many* values that are excluded from the domain of a square root function if it is to be real valued.

Example 6

Determine the domain of each square root function.

a) $f(x) = \sqrt{x}$ b) $g(r) = \sqrt{4r + 1}$

Solution

a)   Ask yourself, *"Is there any number that cannot be substituted for x in $f(x) = \sqrt{x}$?"* Yes. There are many values that cannot be substituted for x. Since we are only considering real numbers in the domain and range, x cannot be a negative number because then $f(x)$ would be imaginary. For example, if $x = -4$, then $f(-4) = \sqrt{-4} = 2i$. Therefore, x must be greater than or equal to 0 in order to produce a real number value for $f(x)$. The domain consists of $x \geq 0$ or $[0, \infty)$.

b)   In part a) we saw that in order for $f(x)$ to be a real number, the *quantity under the radical (radicand)* must be 0 or positive. In $g(r) = \sqrt{4r + 1}$, the *radicand* is $4r + 1$. In order for $g(r)$ to be defined, $4r + 1$ must be 0 or positive. Mathematically we write this as $4r + 1 \geq 0$. To determine the domain of $g(r) = \sqrt{4r + 1}$, solve the inequality $4r + 1 \geq 0$.

$$4r + 1 \geq 0 \qquad \text{The value of the radicand must be } \geq 0.$$
$$4r \geq -1$$
$$r \geq -\frac{1}{4} \qquad \text{Solve.}$$

Any value of r that satisfies $r \geq -\dfrac{1}{4}$ will make the radicand $4r + 1$ greater than or equal to zero. Write the domain as $\left[-\dfrac{1}{4}, \infty \right)$.

> To determine the domain of a square root function, set up an inequality so that the radicand ≥ 0. Solve for the variable. These are the real numbers in the domain of the function.

You Try 6

Determine the domain of each square root function.

a) $h(x) = \sqrt{x - 9}$ b) $k(t) = \sqrt{7t + 2}$

8. Summarize Strategies for Determining the Domain of a Given Function

Let's summarize what we have learned about determining the domain of a function.

Determining the Domain of a Function

The **domain of a function in x** is the set of all real numbers that can be substituted for the independent variable, x. When determining the domain of a function, it can be helpful to keep these tips in mind.

1) Ask yourself, "Is there any number that *cannot* be substituted for x?"

2) The domain of a **linear function** is all real numbers, $(-\infty, \infty)$.

3) The domain of a **polynomial function** is all real numbers, $(-\infty, \infty)$.

4) To find the domain of a **rational function**, determine what value(s) of x will make the denominator equal 0 by setting the expression in the denominator equal to zero. Solve for x. These x-value(s) are *not* in the domain.

5) To determine the domain of a **square root function**, set up an inequality so that the radicand ≥ 0. Solve for x. These are the values of x in the domain.

Another function we encounter often in algebra is an absolute value function like $f(x) = |x|$. We will study absolute value functions in Section 12.2.

Answers to You Try Exercises

1) a) domain: $\{-8, -5, 7\}$; range: $\{-4, -2, 1, 2\}$; no   b) domain: {Chicago, Mexico City, Montreal}; range: {USA, Mexico, Canada}; yes   c) domain: $(-\infty, \infty)$; range: $(-\infty, \infty)$; yes
2) a) is a function; domain: $(-\infty, \infty)$ b) is a function; domain: $(-\infty, \infty)$
3) a) $(-\infty, \infty)$ b)

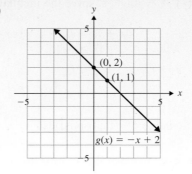

4) a) $f(c) = -9c + 2$ b) $f(r - 5) = -9r + 47$ c) $k(z) = z^2 + 5z + 8$
d) $k(m + 2) = m^2 + 9m + 22$

5) a) $(-\infty, 0) \cup (0, \infty)$ b) $\left(-\infty, -\dfrac{4}{3}\right) \cup \left(-\dfrac{4}{3}, \infty\right)$ c) $(-\infty, -3) \cup (-3, 3) \cup (3, \infty)$

6) a) $[9, \infty)$ b) $\left[-\dfrac{2}{7}, \infty\right)$

12.1 Exercises

Objective 1

1) What is a function?

2) What is the domain of a relation?

Identify the domain and range of each relation, and determine whether each relation is a function.

3) $\{(5, 0), (6, 1), (14, 3), (14, -3)\}$

4) $\{(-5, 7), (-4, 5), (0, -3), (0.5, -4), (3, -9)\}$

5)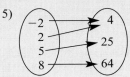

6)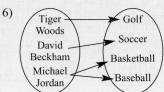

7)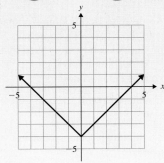

8)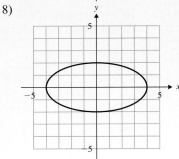

Objective 2

Determine whether each relation describes y as a function of x.

9) $y = 5x + 17$

10) $y = 4x^2 - 10x + 3$

11) $y = \dfrac{x}{x + 6}$

12) $y^2 = x + 2$

13) $y^2 = x - 8$

14) $y = \sqrt{x + 3}$

15) $x = |y|$

16) $y = |x|$

Answer true or false. If the answer is false, explain why.

17) $f(x)$ is read as "f times x."

18) $f(x) = -4x + 1$ is an example of a linear function.

Objective 3

Graph each function.

19) $f(x) = x - 5$

20) $g(x) = -x + 3$

21) $h(a) = -2a + 1$

22) $r(t) = 3t - 2$

23) $g(x) = -\dfrac{3}{2}x - 1$

24) $f(x) = \dfrac{1}{4}x + 2$

25) $k(c) = c$

26) $h(x) = -3$

Objectives 4 and 5

Let $f(x) = 3x - 7$ and $g(x) = x^2 - 4x - 9$. Find each of the following and simplify.

27) $f(6)$

28) $f(0)$

29) $g(3)$

30) $g(-2)$

31) $f(a)$

32) $f(z)$

33) $g(d)$

34) $g(r)$

35) $f(c + 4)$

36) $f(w + 9)$

37) $g(t + 2)$

38) $g(a + 3)$

39) $g(h - 1)$

40) $g(p - 5)$

Let $f(x) = -5x + 2$ and $g(x) = x^2 + 7x + 2$. Find each of the following and simplify.

41) $f(4)$

42) $f(7)$

43) $g(-6)$

44) $g(3)$

45) $f(-3k)$

46) $f(9a)$

47) $g(5t)$

48) $g(-8n)$

49) $f(b + 1)$

50) $f(t - 6)$

51) $g(r + 4)$

52) $g(a - 9)$

53) $f(x) = 4x + 3$. Find x so that $f(x) = 23$.

54) $g(x) = -9x + 1$. Find x so that $g(x) = -17$.

55) $h(x) = -2x - 5$. Find x so that $h(x) = 0$.

56) $k(x) = 8x - 6$. Find x so that $k(x) = 0$.

57) $p(x) = x^2 - 6x - 16$. Find x so that $p(x) = 0$.

58) $g(x) = 2x^2 - 5x - 9$. Find x so that $g(x) = -6$.

Objectives 2–8

59) What is the domain of a linear function?

60) What is the domain of a polynomial function?

61) How do you find the domain of a rational function?

62) How do you find the domain of a square root function?

Determine the domain of each function.

63) $f(x) = x + 10$

64) $h(x) = -8x - 2$

65) $p(a) = 8a^2 + 4a - 9$

66) $r(t) = t^3 - 7t^2 + t + 4$

67) $f(x) = \dfrac{6}{x + 8}$

68) $k(x) = \dfrac{2x}{x - 9}$

69) $h(x) = \dfrac{10}{x}$

70) $Q(r) = \dfrac{7}{2r}$

71) $g(c) = \dfrac{3c}{2c - 1}$

72) $f(x) = \dfrac{4x + 3}{5x + 2}$

73) $R(t) = -\dfrac{t - 4}{7t + 3}$

74) $k(n) = \dfrac{8}{1 - 3n}$

75) $h(x) = \dfrac{9x + 2}{4}$

76) $p(c) = \dfrac{c - 2}{7}$

77) $k(x) = \dfrac{1}{x^2 + 11x + 24}$

78) $f(t) = \dfrac{5}{t^2 - 7t + 6}$

79) $r(c) = \dfrac{c + 3}{c^2 - 5c - 36}$

80) $g(a) = \dfrac{4}{2a^2 + 3a}$

81) $f(x) = \sqrt{x}$

82) $r(t) = -\sqrt{t}$

83) $h(n) = \sqrt{n + 2}$

84) $g(c) = \sqrt{c + 10}$

85) $p(a) = \sqrt{a - 8}$

86) $f(a) = \sqrt{a - 1}$

87) $k(x) = \sqrt{2x - 5}$

88) $r(k) = \sqrt{3k + 7}$

89) $g(t) = \sqrt{-t}$

90) $h(x) = \sqrt{3 - x}$

91) $r(a) = \sqrt{9 - a}$

92) $g(c) = \sqrt{8 - 5c}$

93) $f(x) = |x|$

94) $k(t) = |-t|$

95) If a certain carpet costs $22 per square yard, then the cost C, in dollars, of y yards of carpet is given by the function $C(y) = 22y$.

a) Find the cost of 20 yd^2 of carpet.

b) Find the cost of 56 yd^2 of carpet.

c) If a customer spent $770 on carpet, how many square yards of carpet did he buy?

d) Graph the function.

96) A freight train travels at a constant speed of 32 mph. The distance D, in miles, that the train travels after t hr is given by the function $D(t) = 32t$.

a) How far will the train travel after 3 hr?

b) How far will the train travel after 8 hr?

c) How long does it take for the train to travel 208 mi?

d) Graph the function.

97) For labor only, the Arctic Air-Conditioning Company charges $40 to come to the customer's home plus $50 per hour. These labor charges can be described by the function $L(h) = 50h + 40$, where h is the time, in hours, and L is the cost of labor, in dollars.

a) Find $L(1)$ and explain what this means in the context of the problem.

b) Find $L(1.5)$ and explain what this means in the context of the problem.

c) Find h so that $L(h) = 165$, and explain what this means in the context of the problem.

98) For labor only, a plumber charges $30 for a repair visit plus $60 per hour. These labor charges can be described by the function $L(h) = 60h + 30$, where h is the time, in hours, and L is the cost of labor, in dollars.

a) Find $L(2)$ and explain what this means in the context of the problem.

b) Find $L(1)$ and explain what this means in the context of the problem.

c) Find h so that $L(h) = 210$, and explain what this means in the context of the problem.

99) The area, A, of a circle is a function of its radius, r.

a) Write an equation using function notation to describe this relationship between A and r.

b) If the radius is given in centimeters, find $A(3)$ and explain what this means in the context of the problem.

c) If the radius is given in inches, find $A(5)$ and explain what this means in the context of the problem.

d) What is the radius of a circle with an area of 64π in^2?

100) The perimeter, P, of a square is a function of the length of its side, s.

a) Write an equation using function notation to describe this relationship between P and s.

b) If the length of a side is given in feet, find $P(2)$ and explain what this means in the context of the problem.

c) If the length of a side is given in centimeters, find $P(11)$ and explain what this means in the context of the problem.

d) What is the length of each side of a square that has a perimeter of 18 inches?

Section 12.2 Graphs of Functions and Transformations

Objectives

1. Illustrate Vertical Shifts with Absolute Value Functions
2. Illustrate Horizontal Shifts with Quadratic Functions
3. Illustrate Reflecting a Graph About the x-Axis with Square Root Functions
4. Graph a Function Using a Combination of the Transformations
5. Graph a Piecewise Function
6. Define the Greatest Integer Function, $f(x) = [\![x]\!]$, and Find Function Values
7. Graph Greatest Integer Functions
8. Represent an Applied Problem with the Graph of a Greatest Integer Function

Some functions and their graphs appear often when studying algebra. We will look at the basic graphs of

1. the absolute value function, $f(x) = |x|$.
2. the quadratic function, $f(x) = x^2$.
3. the square root function, $f(x) = \sqrt{x}$.

It is possible to obtain the graph of any function by plotting points. But we will also see how we can graph other, similar functions by transforming the graphs of the functions above.

First, we will graph two absolute value functions. We will begin by plotting points so that we can observe the pattern that develops.

1. Illustrate Vertical Shifts with Absolute Value Functions

Example 1

Graph $f(x) = |x|$ and $g(x) = |x| + 2$ on the same axes. Identify the domain and range.

Solution

$f(x) = |x|$		$g(x) = |x| + 2$	
x	$f(x)$	x	$g(x)$
0	0	0	2
1	1	1	3
2	2	2	4
-1	1	-1	3
-2	2	-2	4

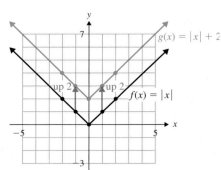

The domain of $f(x) = |x|$ is $(-\infty, \infty)$. The range is $[0, \infty)$.

The domain of $g(x) = |x| + 2$ is $(-\infty, \infty)$. The range is $[2, \infty)$.

Absolute value functions like these have V-shaped graphs. We can see from the tables of values that although the x-values are the same in each table, the corresponding y-values in the table for $g(x)$ are *2 more than* the y-values in the first table.

$$f(x) = |x| \qquad \begin{aligned} g(x) &= |x| + 2 \\ g(x) &= f(x) + 2 \qquad \text{Substitute } f(x) \text{ for } |x|. \end{aligned}$$

The y-coordinates of the ordered pairs of $g(x)$ will be *2 more than* the y-coordinates of the ordered pairs of $f(x)$ when the ordered pairs of f and g have the same x-coordinates. This means that *the graph of g will be the same shape as the graph of f but g will be shifted up 2 units.*

Definition   Given the graph of any function $f(x)$, if $g(x) = f(x) + k$, where k is a constant, the graph of $g(x)$ will be the same shape as the graph of $f(x)$, but g will be **shifted vertically** k units.

In Example 1, $k = 2$.

$$f(x) = |x| \quad \text{and} \quad g(x) = |x| + 2$$
$$\text{or} \qquad g(x) = f(x) + 2$$

The graph of g is the same shape as the graph of f, but the graph of g is shifted *up* 2 units. We say that we can graph $g(x) = |x| + 2$ by *transforming* the graph of $f(x) = |x|$. This vertical shifting works not only for absolute value functions but for any function.

You Try 1

Graph $g(x) = |x| - 1$.

---

2. Illustrate Horizontal Shifts with Quadratic Functions

In the previous section we said that a quadratic fuction can be written in the form $f(x) = ax^2 + bx + c$ where a, b, and c are real numbers and $a \neq 0$. Here we begin our discussion of graphing quadratic functions.

The graph of a quadratic function is called a **parabola**. Let's look at the simplest form of a quadratic function, $f(x) = x^2$, and a variation of it. In section 12.3 we will discuss graphing quadratic functions in much greater detail.

Example 2

Graph $f(x) = x^2$ and $g(x) = (x + 3)^2$ on the same axes. Identify the domain and range.

Solution

$f(x) = x^2$	
x	**f(x)**
0	0
1	1
2	4
-1	1
-2	4

$g(x) = (x + 3)^2$	
x	**g(x)**
-3	0
-2	1
-1	4
-4	1
-5	4

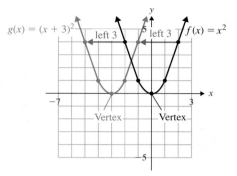

Notice that the graphs of $f(x)$ and $g(x)$ open upward. The lowest point on a parabola that opens upward or the highest point on a parabola that opens downward is called the **vertex**. The vertex of the graph of $f(x)$ is $(0, 0)$, and the vertex of the graph of $g(x)$ is $(-3, 0)$. When graphing a quadratic function by plotting points, it is important to locate the vertex. *The x-coordinate of the vertex is the value of x that makes the expression that is being squared equal to zero.*

The domain of $f(x) = x^2$ is $(-\infty, \infty)$. The range is $[0, \infty)$.

The domain of $g(x) = (x + 3)^2$ is $(-\infty, \infty)$. The range is $[0, \infty)$.

We can see from the tables of values that although the y-values are the same in each table, the corresponding x-values in the table for $g(x)$ are *3 less than* the x-values in the first table.

The x-coordinates of the ordered pairs of $g(x)$ will be *3 less than* the x-coordinates of the ordered pairs of $f(x)$ when the ordered pairs of f and g have the same y-coordinates. This means that *the graph of g will be the same shape as the graph of f but the graph of g will be shifted left 3 units.*

Definition

Given the graph of any function $f(x)$, if $g(x) = f(x - h)$, where h is a constant, then the graph of $g(x)$ will be the same shape as the graph of $f(x)$ but the graph of g will be **shifted horizontally** h units.

In Example 2, $h = -3$.

$$f(x) = x^2 \quad \text{and} \quad g(x) = (x + 3)^2$$
$$\text{or} \quad g(x) = f(x - (-3))$$

The graph of g is the same shape as the graph of f but the graph of g is shifted -3 units horizontally or 3 units to the *left*. This horizontal shifting works for any function, not just quadratic functions.

 You Try 2

Graph $g(x) = (x + 4)^2$.

BE CAREFUL It is important to distinguish between the graph of an absolute value function and the graph of a quadratic function. The absolute value functions we will study have V-shaped graphs. The graph of a quadratic function is *not* shaped like a V. It is a parabola.

The next type of transformation we will discuss is reflecting the graph of a function about the x-axis.

3. Illustrate Reflecting a Graph About the x-Axis with Square Root Functions

Example 3

Graph $f(x) = \sqrt{x}$ and $g(x) = -\sqrt{x}$ on the same axes. Identify the domain and range.

Solution

The domain of each function is $[0, \infty)$. From the graphs we can see that the range of $f(x) = \sqrt{x}$ is $[0, \infty)$, while the range of $g(x) = -\sqrt{x}$ is $(-\infty, 0]$.

$f(x) = \sqrt{x}$		$g(x) = -\sqrt{x}$	
x	$f(x)$	x	$g(x)$
0	0	0	0
1	1	1	-1
4	2	4	-2
9	3	9	-3

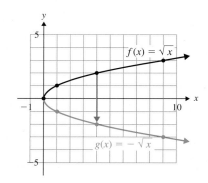

The tables of values show us that although the x-values are the same in each table, the corresponding y-values in the table for $g(x)$ are the *negatives* of the y-values in the first table.

We say that *the graph of g is the reflection of the graph of f about the x-axis.* (g is the mirror image of f.)

Definition

Reflection about the x-axis: Given the graph of any function $f(x)$, if $g(x) = -f(x)$ then the graph of $g(x)$ will be the **reflection of the graph of f about the x-axis**. That is, obtain the graph of g by keeping the x-coordinate of each point on the graph of f the same but take the negative of the y-coordinate.

In Example 5,

$$f(x) = \sqrt{x} \quad \text{and} \quad g(x) = -\sqrt{x}$$
$$\text{or} \quad g(x) = -f(x)$$

The graph of g is the mirror image of the graph of f with respect to the x-axis. This is true for any functions where $g(x) = -f(x)$.

You Try 3

Graph $g(x) = -x^2$.

We can combine the techniques used in the transformation of the graphs of functions to help us graph more complicated functions.

4. Graph a Function Using a Combination of the Transformations

Example 4

Graph $h(x) = |x + 2| - 3$.

Solution

The graph of $h(x)$ will be the same shape as the graph of $f(x) = |x|$. So, let's see what the constants in $h(x)$ tell us about transforming the graph of $f(x) = |x|$.

$$h(x) = |x + 2| - 3$$

Shift $f(x)$ left 2.   Shift $f(x)$ down 3.

Sketch the graph of $f(x) = |x|$, including some key points, then *move every point on the graph of f left 2 and down 3 to obtain the graph of h.*

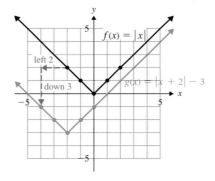

You Try 4

Graph $g(x) = -x^2 + 5$.

Graphs of Other Functions

5. Graph a Piecewise Function

Definition   A **piecewise function** is a single function defined by two or more different rules.

Example 5

Graph the piecewise function

$$f(x) = \begin{cases} 2x - 4, & x \geq 3 \\ -x + 2, & x < 3 \end{cases}$$

Solution

This is a piecewise function because $f(x)$ is defined by two different rules. *The rule we use to find $f(x)$ depends on what value is substituted for x.*

Graph $f(x)$ by making two separate tables of values, one for each rule.

When $x \geq 3$, use the rule

$$f(x) = 2x - 4$$

The first x-value we will put in the table of values is 3 because it is the smallest number (lower bound) of the domain of $f(x) = 2x - 4$. *The other values we choose for x must be greater than 3 because this is when we use the rule $f(x) = 2x - 4$.* **This part of the graph will not extend to the left of (3, 2).**

$$f(x) = 2x - 4$$
$$(x \geq 3)$$

x	f(x)
3	2
4	4
5	6
6	8

When $x < 3$, use the rule

$$f(x) = -x + 2$$

The first x-value we will put in the table of values is 3 because it is the upper bound of the domain. *Notice that 3 is not included in the domain (the inequality is <, **not** ≤) so that the point $(3, f(3))$ will be represented as an open circle on the graph. The other values we choose for x must be less than 3 because this is when we use the rule $f(x) = -x + 2$.* **This part of the graph will not extend to the right of (3, −1).**

$$f(x) = -x + 2$$
$$(x < 3)$$

x	f(x)	
3	−1	(3, −1) will be
2	0	an open circle.
1	1	
0	2	

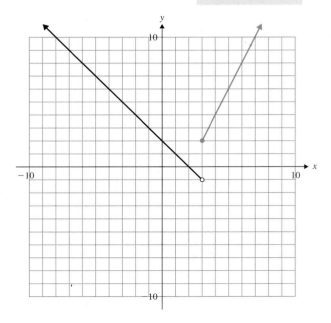

You Try 5

Graph.

$$f(x) = \begin{cases} -2x + 3, & x \le -2 \\ \dfrac{3}{2}x - 1, & x > -2 \end{cases}$$

6. Define the Greatest Integer Function, $f(x) = [\![x]\!]$, and Find Function Values

Another function that has many practical applications is the greatest integer function. Before we look at a graph and an application, we need to understand what the greatest integer function means.

Definition

The **greatest integer function**

$$f(x) = [\![x]\!]$$

represents the largest integer less than or equal to x.

Example 6

Let $f(x) = [\![x]\!]$. Find the following function values.

a) $f\left(9\dfrac{1}{2}\right)$ b) $f(6)$ c) $f(-2.3)$

Solution

a) $f\left(9\dfrac{1}{2}\right) = \left[\!\left[9\dfrac{1}{2}\right]\!\right]$, which means to find the largest integer that is *less than or equal to* $9\dfrac{1}{2}$. That number is 9.

$$f\left(9\dfrac{1}{2}\right) = \left[\!\left[9\dfrac{1}{2}\right]\!\right] = 9$$

b) $f(x) = [\![6]\!] = 6$ since the largest integer *less than or equal to* 6 is 6.

c) $f(-2.3) = [\![-2.3]\!]$

To help us understand how to find this function value we will locate -2.3 on a number line.

The largest integer *less than or equal to* -2.3 is -3.
$f(-2.3) = [\![-2.3]\!] = -3$

You Try 6

Let $f(x) = [\![x]\!]$. Find the following function values.

a) $f(5.1)$ b) $f(0)$ c) $f\left(-5\dfrac{1}{4}\right)$

7. Graph Greatest Integer Functions

Example 7

Graph $f(x) = [\![x]\!]$.

Solution

To understand what produces the pattern in the graph of this function, we begin by *closely* examining what occurs between $x = 0$ and $x = 1$ (when $0 \le x \le 1$).

$$f(x) = [\![x]\!]$$

x	f(x)	
0	0	
$\frac{1}{4}$	0	For all values of *x greater than or equal to 0* and
$\frac{1}{2}$	0	*less than 1*, the function value, $f(x)$, equals zero.
$\frac{3}{4}$	0	
⋮	0	
1	1	→ When $x = 1$ the function value changes to 1.

The graph will have an open circle at $(1, 0)$ because for $x < 1, f(x) = 0$. That means that x can get *very close to* 1 and the function value will be zero, but $f(1) \ne 0$.

   This pattern will continue so that for the x-values in the interval $[1, 2)$, the function values are 1. The graph will have an open circle at $(2, 1)$.

   For the x-values in the interval $[2, 3)$, the function values are 2. The graph will have an open circle at $(3, 2)$.

   Continuing in this way we get the graph below.

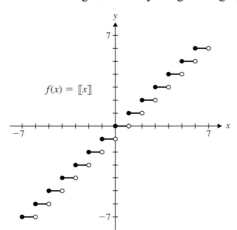

The domain of the function is $(-\infty, \infty)$. The range is the set of all integers $\{\ldots, -3, -2, -1, 0, 1, 2, 3, \ldots\}$.

Because of the appearance of the graph, $f(x) = [\![x]\!]$ is also called a **step function**.

You Try 7

Graph $f(x) = [\![x]\!] - 3$.

8. Represent an Applied Problem with the Graph of a Greatest Integer Function

Example 8

To mail a letter within the United States in 2005, the U.S. Postal Service charged $0.39 for the first ounce and $0.24 for each additional ounce or fraction of an ounce. Let $C(x)$ represent the cost of mailing a letter within the United States, and let x represent the weight of the letter, in ounces. Graph $C(x)$ for any letter weighing up to (and including) 5 ounces. (Source: www.usps.com)

Solution

If a letter weighs between 0 and 1 ounce ($0 < x \le 1$) the cost, $C(x)$, is $0.39.

If a letter weighs more than 1 oz but less than or equal to 2 oz ($1 < x \le 2$) the cost, $C(x)$, is $0.39 + $0.24 = $0.63.

The pattern will continue, and we get the graph below.

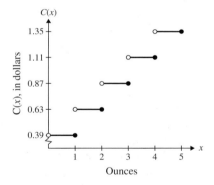

You Try 8

To mail a package within the United States at *book rate* in 2005, the U.S. Postal Service charged $1.59 for the first pound and $0.48 for each additional pound or fraction of a pound. Let $C(x)$ represent the cost of mailing a package at book rate and let x represent the weight of the package, in pounds. Graph $C(x)$ for any package weighing up to (and including) 5 pounds. (Source: www.usps.com)

Using Technology

The graphing calculator screens show the graphs of quadratic functions. Match each equation to its graph. Then, identify the domain and range of each function.

1. $f(x) = (x - 2)^2$

2. $f(x) = |x| - 2$

3. $f(x) = -x^2$

4. $f(x) = \sqrt{x + 2}$

5. $f(x) = -|x| + 2$

6. $f(x) = (x + 2)^2 - 3$

a)

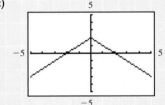

b)

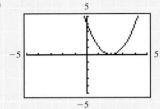

c)

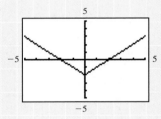

d)

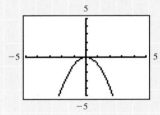

e)

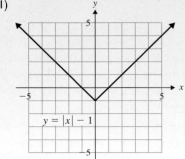

f)

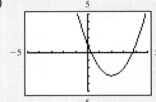

Answers to You Try Exercises

1)

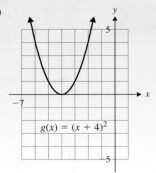

$y = |x| - 1$

2)

$g(x) = (x + 4)^2$

3)

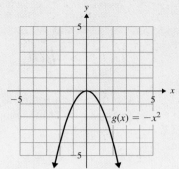

$g(x) = -x^2$

4)

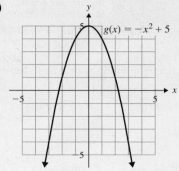

$g(x) = -x^2 + 5$

5)

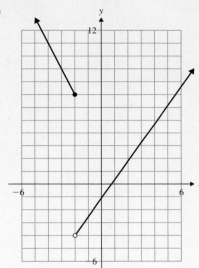

6) a) 5   b) 0   c) −6

7)

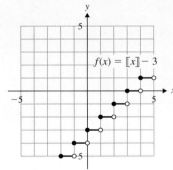

$f(x) = [\![x]\!] - 3$

8)

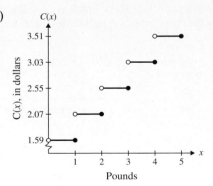

Answers to Technology Exercises

1. b); domain $(-\infty, \infty)$; range $[0, \infty)$

2. e); domain $(-\infty, \infty)$; range $[-2, \infty)$

3. f); domain $(-\infty, \infty)$; range $(-\infty, 0]$

4. a); domain $[-2, \infty)$; range $[0, \infty)$

5. c); domain $(-\infty, \infty)$; range $(-\infty, 2]$

6. d); domain $(-\infty, \infty)$; range $[-3, \infty)$

12.2 Exercises

Boost your grade at mathzone.com!   MathZone   > Practice Problems   > NetTutor   > Self-Test   > e-Professors   > Videos

Objectives 1–4

Graph each function by plotting points, and identify the domain and range.

1) $f(x) = |x| + 3$

2) $g(x) = |x - 2|$

 3) $k(x) = \dfrac{1}{2}|x|$

4) $g(x) = 2|x|$

5) $g(x) = x^2 - 4$

6) $h(x) = (x - 2)^2$

7) $f(x) = -x^2 - 1$

8) $f(x) = (x - 2)^2 - 5$

9) $f(x) = \sqrt{x + 3}$

10) $g(x) = \sqrt{x} + 2$

11) $f(x) = 2\sqrt{x}$

12) $h(x) = -\dfrac{1}{2}\sqrt{x}$

Given the following pairs of functions, explain how the graph of $g(x)$ can be obtained from the graph of $f(x)$ using the transformation techniques discussed in this section.

13) $f(x) = |x|, g(x) = |x| - 2$

14) $f(x) = |x|, g(x) = |x| + 1$

15) $f(x) = x^2, g(x) = (x + 2)^2$

16) $f(x) = x^2, g(x) = (x - 3)^2$

17) $f(x) = x^2, g(x) = -x^2$

18) $f(x) = \sqrt{x}, g(x) = -\sqrt{x}$

Sketch the graph of $f(x)$. Then, graph $g(x)$ on the same axes using the transformation techniques discussed in this section.

19) $f(x) = |x|$
    $g(x) = |x| - 2$

20) $f(x) = |x|$
    $g(x) = |x| + 1$

21) $f(x) = |x|$
    $g(x) = |x| + 3$

22) $f(x) = |x|$
    $g(x) = |x| - 4$

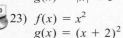 23) $f(x) = x^2$
    $g(x) = (x + 2)^2$

24) $f(x) = x^2$
    $g(x) = (x - 3)^2$

25) $f(x) = x^2$
    $g(x) = (x - 4)^2$

26) $f(x) = x^2$
    $g(x) = (x + 1)^2$

27) $f(x) = x^2$
    $g(x) = -x^2$

28) $f(x) = \sqrt{x}$
    $g(x) = -\sqrt{x}$

29) $f(x) = \sqrt{x + 1}$
    $g(x) = -\sqrt{x + 1}$

30) $f(x) = \sqrt{x - 2}$
    $g(x) = -\sqrt{x - 2}$

31) $f(x) = |x - 3|$
    $g(x) = -|x - 3|$

32) $f(x) = |x + 4|$
    $g(x) = -|x + 4|$

Use the transformation techniques discussed in this section to graph each of the following functions.

33) $f(x) = |x| - 5$

34) $f(x) = \sqrt{x} + 3$

35) $y = \sqrt{x - 4}$

36) $y = (x - 2)^2$

 37) $g(x) = |x + 2| + 3$

38) $h(x) = |x + 1| - 5$

39) $y = (x - 3)^2 + 1$

40) $f(x) = (x + 2)^2 - 3$

41) $f(x) = \sqrt{x + 4} - 2$

42) $y = \sqrt{x - 3} + 2$

43) $h(x) = -x^2 + 6$

44) $y = -(x - 1)^2$

45) $g(x) = -|x - 1| + 3$

46) $h(x) = -|x + 3| - 2$

47) $f(x) = -\sqrt{x + 5}$

48) $y = -\sqrt{x + 2}$

Match each function to its graph.

49) $f(x) = x^2 - 3, g(x) = (x - 3)^2,$
    $h(x) = -(x + 3)^2, k(x) = -x^2 + 3$

a)

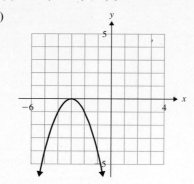

b)

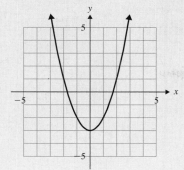

b)

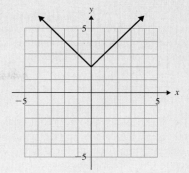

c)

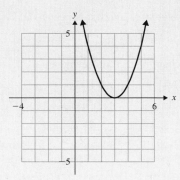

c)

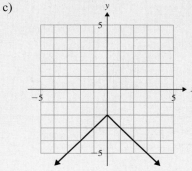

d)

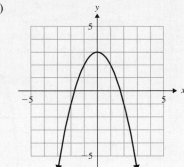

d)

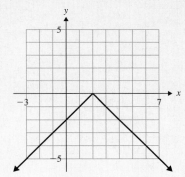

50) $f(x) = -|x - 2|$, $g(x) = |x + 2|$,
   $h(x) = -|x| - 2$, $k(x) = |x| + 2$

a)

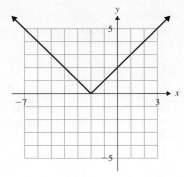

If the following transformations are performed on the graph of $f(x)$ to obtain the graph of $g(x)$, write the equation of $g(x)$.

51) $f(x) = \sqrt{x}$ is shifted 5 units to the left.

52) $f(x) = \sqrt{x}$ is shifted down 6 units.

53) $f(x) = |x|$ is shifted left 2 units and down 1 unit.

54) $f(x) = |x|$ is shifted right 1 unit and up 4 units. $x - \ell + 4$

55) $f(x) = x^2$ is shifted left 3 units and up $\dfrac{1}{2}$ unit.

56) $f(x) = x^2$ is shifted right 5 units and down 1.5 units.

57) $f(x) = x^2$ is reflected about the x-axis.

58) $f(x) = |x|$ is reflected about the x-axis.

59) Graph $f(x) = x^3$ by plotting points. (Hint: Make a table of values and choose 0, positive, and negative numbers for x.) Then, use the transformation techniques discussed in this section to graph each of the following functions.

a) $g(x) = (x + 2)^3$ b) $h(x) = x^3 - 3$

c) $k(x) = -x^3$ d) $r(x) = (x - 1)^3 - 2$

60) Graph $f(x) = \sqrt[3]{x}$ by plotting points. (Hint: Make a table of values and choose 0, positive, and negative numbers for x.) Then, use the transformation techniques discussed in this section to graph each of the following functions.

a) $g(x) = \sqrt[3]{x} + 4$ b) $h(x) = -\sqrt[3]{x}$

c) $k(x) = \sqrt[3]{x - 2}$ d) $r(x) = -\sqrt[3]{x} - 3$

Objective 5

Graph the following piecewise functions.

61) $f(x) = \begin{cases} -x - 3, & x \le -1 \\ 2x + 2, & x > -1 \end{cases}$

62) $g(x) = \begin{cases} x - 1, & x \ge 2 \\ -3x + 3, & x < 2 \end{cases}$

63) $h(x) = \begin{cases} -x + 5, & x \ge 3 \\ \dfrac{1}{2}x + 1, & x < 3 \end{cases}$

64) $f(x) = \begin{cases} 2x + 13, & x \le -4 \\ -\dfrac{1}{2}x + 1, & x > -4 \end{cases}$

65) $g(x) = \begin{cases} -\dfrac{3}{2}x - 3, & x < 0 \\ 1, & x \ge 0 \end{cases}$

66) $h(x) = \begin{cases} -\dfrac{2}{3}x - \dfrac{7}{3}, & x \ge -1 \\ 2, & x < -1 \end{cases}$

67) $k(x) = \begin{cases} x + 1, & x \ge -2 \\ 2x + 8, & x < -2 \end{cases}$

68) $g(x) = \begin{cases} x, & x \le 0 \\ 2x + 3, & x > 0 \end{cases}$

69) $f(x) = \begin{cases} 2x - 4, & x > 1 \\ -\dfrac{1}{3}x - \dfrac{5}{3}, & x \le 1 \end{cases}$

70) $k(x) = \begin{cases} \dfrac{1}{2}x + \dfrac{5}{2}, & x < 3 \\ -x + 7, & x \ge 3 \end{cases}$

Objective 6

Let $f(x) = [\![x]\!]$. Find the following function values.

71) $f\left(3\dfrac{1}{4}\right)$

72) $f\left(10\dfrac{3}{8}\right)$

73) $f(9.2)$

74) $f(7.8)$

75) $f(8)$

76) $f\left(\dfrac{4}{5}\right)$

77) $f\left(-6\dfrac{2}{5}\right)$

78) $f\left(-1\dfrac{3}{4}\right)$

79) $f(-8.1)$

80) $f(-3.6)$

Objective 7

Graph the following greatest integer functions.

81) $f(x) = [\![x]\!] + 1$

82) $g(x) = [\![x]\!] - 2$

83) $h(x) = [\![x]\!] - 4$

84) $k(x) = [\![x]\!] + 3$

85) $g(x) = [\![x + 2]\!]$

86) $h(x) = [\![x - 1]\!]$

87) $k(x) = \left[\!\left[\dfrac{1}{2}x\right]\!\right]$

88) $f(x) = [\![2x]\!]$

Objective 8

89) To ship small packages within the United States a shipping company charges \$3.75 for the first pound and \$1.10 for each additional pound or fraction of a pound. Let $C(x)$ represent the cost of shipping a package, and let x represent the weight of the package. Graph $C(x)$ for any package weighing up to (and including) 6 lb.

90) To deliver small packages overnight, an express delivery service charges \$15.40 for the first pound and \$4.50 for each additional pound or fraction of a pound. Let $C(x)$ represent the cost of shipping a package overnight, and let x represent the weight of the package. Graph $C(x)$ for any package weighing up to (and including) 6 lb.

91) Visitors to downtown Hinsdale must pay the parking meters to park their cars. The cost of parking is 5¢ for the first 12 min and 5¢ for each additional 12 min or fraction of this time. Let $P(t)$ represent the cost of parking, and let t represent the number of minutes the car is parked at the meter. Graph $P(t)$ for parking a car for up to (and including) 1 hr.

92) To consult with an attorney costs \$35 for every 10 min or fraction of this time. Let $C(t)$ represent the cost of meeting an attorney, and let t represent the length of the meeting, in minutes. Graph $C(t)$ for meeting with the attorney for up to (and including) 1 hr.

Section 12.3 Quadratic Functions and Their Graphs

Objectives

1. Graph a Quadratic Function by Shifting the Graph of $f(x) = x^2$

2. Graph $f(x) = a(x - h)^2 + k$ Using the Vertex, Axis of Symmetry, and Other Characteristics

3. Graph $f(x) = ax^2 + bx + c$ by Completing the Square to Rewrite It in the Form $f(x) = a(x - h)^2 + k$

4. Graph $f(x) = ax^2 + bx + c$ by Identifying the Vertex Using $\left(-\dfrac{b}{2a}, f\left(-\dfrac{b}{2a}\right)\right)$

In this section, we will study quadratic functions in more detail and see how the rules we learned in Section 12.2 apply specifically to these functions. We restate the definition of a quadratic function here.

> **Definition**   A **quadratic function** is a function that can be written in the form
>
> $$f(x) = ax^2 + bx + c$$
>
> where a, b, and c are real numbers and $a \neq 0$. An example is $f(x) = x^2 + 6x + 10$. The graph of a quadratic function is called a **parabola**. The lowest point on a parabola that opens upward or the highest point on a parabola that opens downward is called the **vertex**.

Quadratic functions can be written in other forms as well. One common form is $f(x) = a(x - h)^2 + k$. An example is $f(x) = 2(x - 3)^2 + 1$.

We will study the form $f(x) = a(x - h)^2 + k$ first since graphing parabolas from this form comes directly from the transformation techniques we learned earlier.

1. Graph a Quadratic Function by Shifting the Graph of $f(x) = x^2$

Example 1

Graph $g(x) = (x - 2)^2 - 1$.

Solution

If we compare $g(x)$ to $f(x) = x^2$, what do the constants that have been added to $g(x)$ tell us about transforming the graph of $f(x)$?

$$g(x) = (x - 2)^2 - 1$$

Shift $f(x)$ right 2.   Shift $f(x)$ down 1.

Sketch the graph of $f(x) = x^2$, then move every point on the graph of f right 2 and down 1 to obtain the graph of $g(x)$. This moves the vertex from $(0, 0)$ to $(2, -1)$.

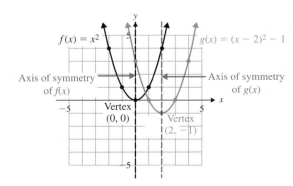

Every parabola has symmetry. If we were to fold the paper along the y-axis, one half of the graph of $f(x) = x^2$ would fall exactly on the other half. The y-axis (the line $x = 0$) is the **axis of symmetry** of $f(x) = x^2$.

Look at the graph of $g(x) = (x - 2)^2 - 1$. This parabola is symmetric with respect to the vertical line $(x = 2)$ through its vertex $(2, -1)$. If we were to fold the paper along the line $x = 2$, half of the graph of $g(x)$ would fall exactly on the other half. The line $x = 2$ is the *axis of symmetry* of $g(x) = (x - 2)^2 - 1$.

2. Graph $f(x) = a(x - h)^2 + k$ Using the Vertex, Axis of Symmetry, and Other Characteristics

When a quadratic function is in the form $f(x) = a(x - h)^2 + k$, we can read the vertex directly from the equation. Furthermore, the value of a tells us if the parabola opens upward or downward and whether the graph is narrower, wider, or the same width as $y = x^2$.

Graphing a Quadratic Function of the Form $f(x) = a(x - h)^2 + k$

1) The vertex of the parabola is (h, k).

2) The axis of symmetry is the vertical line with equation $x = h$.

3) If a is positive, the parabola opens upward.

   If a is negative, the parabola opens downward.

4) If $|a| < 1$, then the graph of $f(x) = a(x - h)^2 + k$ is *wider* than the graph of $y = x^2$.

   If $|a| > 1$, then the graph of $f(x) = a(x - h)^2 + k$ is *narrower* than the graph of $y = x^2$.

   If $a = 1$ or $a = -1$, the graph is the *same* width as $y = x^2$.

Example 2

Graph $y = \dfrac{1}{2}(x + 3)^2 - 2$. Also find the x- and y-intercepts.

Solution

Here is the information we can get from the equation.

1) $h = -3$ and $k = -2$. The vertex is $(-3, -2)$.

2) The axis of symmetry is $x = -3$.

3) $a = +\dfrac{1}{2}$. Since a is positive, the parabola opens upward.

4) Since $a = \dfrac{1}{2}$ and $\dfrac{1}{2} < 1$, the graph of $y = \dfrac{1}{2}(x + 3)^2 - 2$ is *wider* than the graph of $y = x^2$.

To graph the function, start by putting the vertex on the axes. Then, choose a couple of values of x to the left or right of the vertex to plot more points. Use the axis of symmetry to find more points on the graph of $y = \frac{1}{2}(x + 3)^2 - 2$.

x	y
−1	0
1	6

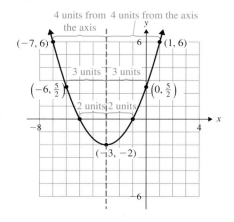

We can read the x-intercepts from the graph: $(-5, 0)$ and $(-1, 0)$. To find the y-intercept, let $x = 0$ and solve for y.

$$y = \frac{1}{2}(x + 3)^2 - 2$$

$$y = \frac{1}{2}(0 + 3)^2 - 2$$

$$y = \frac{1}{2}(9) - 2 = \frac{9}{2} - 2 = \frac{9}{2} - \frac{4}{2} = \frac{5}{2}$$

The y-intercept is $\left(0, \frac{5}{2}\right)$.

You Try 1

Graph $y = 2(x - 1)^2 - 2$. Also find the x- and y-intercepts.

Example 3

Graph $f(x) = -(x - 1)^2 + 5$. Find the x- and y-intercepts.

Solution

Here is the information we can get from the equation.

1) $h = 1$ and $k = 5$. The vertex is $(1, 5)$.
2) The axis of symmetry is $x = 1$.
3) $a = -1$. Since a is negative, the parabola opens downward.
4) Since $a = -1$, the graph of $f(x) = -(x - 1)^2 + 5$ is the *same* width as $y = x^2$.

Put the vertex on the axes. Choose a couple of x-values to the left or right of the vertex to plot more points. Use the axis of symmetry to find more points on the graph of $f(x) = -(x - 1)^2 + 5$.

x	y
2	4
3	1

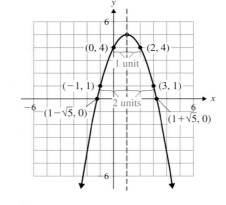

We can read the y-intercept from the graph: $(0, 4)$.

To find the x-intercepts, let $f(x) = 0$ and solve for x.

$$f(x) = -(x - 1)^2 + 5$$
$$0 = -(x - 1)^2 + 5 \qquad \text{Substitute 0 for } f(x).$$
$$-5 = -(x - 1)^2 \qquad \text{Subtract 5.}$$
$$5 = (x - 1)^2 \qquad \text{Divide by } -1.$$
$$\pm\sqrt{5} = x - 1 \qquad \text{Square root property}$$
$$1 \pm \sqrt{5} = x \qquad \text{Add 1.}$$

The x-intercepts are $(1 + \sqrt{5}, 0)$ and $(1 - \sqrt{5}, 0)$.

You Try 2

Graph $f(x) = -(x + 3)^2 + 2$. Find the x- and y-intercepts.

Graphing Parabolas from the Form $f(x) = ax^2 + bx + c$

When a quadratic function is written in the form $f(x) = ax^2 + bx + c$, there are two methods we can use to graph the function.

Method 1: Rewrite $f(x) = ax^2 + bx + c$ in the form $f(x) = a(x - h)^2 + k$ by *completing the square*.

Method 2: Use the formula $x = -\dfrac{b}{2a}$ to find the x-coordinate of the vertex. Then, the vertex has

coordinates $\left(-\dfrac{b}{2a}, f\left(-\dfrac{b}{2a} \right) \right)$.

We will begin with Method 1. We will modify the steps we used in Section 11.4 to solve quadratic equations by completing the square.

3. Graph $f(x) = ax^2 + bx + c$ by Completing the Square to Rewrite It in the Form $f(x) = a(x - h)^2 + k$

> ### Rewriting $f(x) = ax^2 + bx + c$ in the Form $f(x) = a(x - h)^2 + k$ by Completing the Square
>
> **Step 1:** **The coefficient of the square term must be 1.** If it is not 1, multiply or divide both sides of the equation (*including f(x)*) by the appropriate value to obtain a leading coefficient of 1.
>
> **Step 2:** **Separate the constant from the terms containing the variables by grouping the variable terms with parentheses.**
>
> **Step 3:** **Complete the square for the quantity in the parentheses.** Find half of the linear coefficient, then square the result. *Add* that quantity inside the parentheses and *subtract* the quantity from the constant. (Adding and subtracting the same number on the same side of an equation is like adding 0 to the equation.)
>
> **Step 4:** **Factor the expression inside the parentheses.**
>
> **Step 5:** **Solve for $f(x)$.**

Example 4

Graph each function. Begin by completing the square to rewrite each function in the form $f(x) = a(x - h)^2 + k$. Include the intercepts.

a) $f(x) = x^2 + 6x + 10$

b) $g(x) = -\dfrac{1}{2}x^2 + 4x - 6$

Solution

a) ***Step 1:*** The coefficient of x^2 is 1.

  Step 2: Separate the constant from the variable terms using parentheses.

$$f(x) = (x^2 + 6x) + 10$$

  Step 3: Complete the square for the quantity in the parentheses.

$$\frac{1}{2}(6) = 3$$
$$3^2 = 9$$

  Add 9 inside the parentheses and subtract 9 from the 10. This is like adding 0 to the equation.

$$f(x) = (x^2 + 6x + 9) + 10 - 9$$
$$f(x) = (x^2 + 6x + 9) + 1$$

  Step 4: Factor the expression inside the parentheses.

$$f(x) = (x + 3)^2 + 1$$

  Step 5: The equation *is* solved for $f(x)$.

From the equation $f(x) = (x + 3)^2 + 1$ we can see that

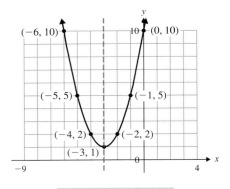

  i) The vertex is $(-3, 1)$.

  ii) The axis of symmetry is $x = -3$.

  iii) $a = 1$ so the parabola opens upward.

  iv) Since $a = 1$, the graph is the same width as $y = x^2$.

Find some other points on the parabola. Use the axis of symmetry.

To find the x-intercepts, let $f(x) = 0$ and solve for x. Use *either* form of the equation. We will use $f(x) = (x + 3)^2 + 1$.

x	$f(x)$
-2	2
-1	5

$$\begin{aligned}
0 &= (x + 3)^2 + 1 & &\text{Let } f(x) = 0. \\
-1 &= (x + 3)^2 & &\text{Subtract 1.} \\
\pm\sqrt{-1} &= x + 3 & &\text{Square root property} \\
-3 \pm i &= x & &\sqrt{-1} = i; \text{ subtract 3.}
\end{aligned}$$

Since the solutions to $f(x) = 0$ are *not* real numbers, *there are no x-intercepts.* To find the y-intercepts, let $x = 0$ and solve for $f(0)$.

$$\begin{aligned}
f(x) &= (x + 3)^2 + 1 \\
f(0) &= (0 + 3)^2 + 1 \\
f(0) &= 9 + 1 = 10
\end{aligned}$$

The y-intercept is $(0, 10)$.

b) ***Step 1:*** The coefficient of x^2 is $-\dfrac{1}{2}$. Multiply both sides of the equation [including the $g(x)$] by -2 so that the coefficient of x^2 will be 1.

$$\begin{aligned}
g(x) &= -\frac{1}{2}x^2 + 4x - 6 \\
-2g(x) &= -2\left(-\frac{1}{2}x^2 + 4x - 6\right) & &\text{Multiply by } -2. \\
-2g(x) &= x^2 - 8x + 12 & &\text{Distribute.}
\end{aligned}$$

Step 2: Separate the constant from the variable terms using parentheses.

$$-2g(x) = (x^2 - 8x) + 12$$

Step 3: Complete the square for the quantity in parentheses.

$$\begin{aligned}
\frac{1}{2}(-8) &= -4 \\
(-4)^2 &= 16
\end{aligned}$$

Add 16 inside the parentheses and subtract 16 from the 12.

$$-2g(x) = (x^2 - 8x + 16) + 12 - 16$$
$$-2g(x) = (x^2 - 8x + 16) - 4$$

Step 4:   Factor the expression inside the parentheses.

$$-2g(x) = (x - 4)^2 - 4$$

Step 5:   Solve the equation for $g(x)$ by dividing by -2.

$$\frac{-2g(x)}{-2} = \frac{(x - 4)^2}{-2} - \frac{4}{-2}$$

$$g(x) = -\frac{1}{2}(x - 4)^2 + 2$$

From $g(x) = -\frac{1}{2}(x - 4)^2 + 2$ we can see that

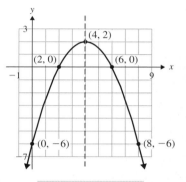

i)   The vertex is (4, 2).

ii)   The axis of symmetry is $x = 4$.

iii) $a = -\frac{1}{2}$ [the same as in the form

$g(x) = -\frac{1}{2}x^2 + 4x - 6$] so the parabola

opens downward.

iv)   Since $a = -\frac{1}{2}$, the graph of $g(x)$ will be

wider than $y = x^2$.

Find some other points on the parabola.
Use the axis of symmetry.

x	g(x)
6	0
8	-6

Using the axis of symmetry we can see that the x-intercepts are (6, 0) and (2, 0) and that the y-intercept is (0, -6).

You Try 3

Graph each function. Begin by completing the square to rewrite each function in the form $f(x) = a(x - h)^2 + k$. Include the intercepts.

a) $f(x) = x^2 + 4x + 3$ b) $g(x) = -2x^2 + 12x - 8$

4. Graph $f(x) = ax^2 + bx + c$ by Identifying the Vertex Using $\left(-\dfrac{b}{2a}, f\left(-\dfrac{b}{2a}\right)\right)$

We can also graph quadratic functions of the form $f(x) = ax^2 + bx + c$ by using $h = -\dfrac{b}{2a}$ to find the x-coordinate of the vertex. The formula comes from completing the square on $f(x) = ax^2 + bx + c$.

Although there is a formula for k, it is only necessary to remember the formula for h. The y-coordinate of the vertex, then, is $k = f\left(-\dfrac{b}{2a}\right)$. The axis of symmetry is $x = h$.

Example 5

Graph $f(x) = x^2 - 6x + 3$ using the vertex formula. Include the intercepts.

Solution

$a = 1$, $b = -6$, $c = 3$. Since $a = +1$, the graph opens upward. The x-coordinate (h) of the vertex is

$$h = -\frac{b}{2a}$$

$$h = -\frac{(-6)}{2(1)} = \frac{6}{2} = 3$$

$h = 3$. Then, the y-coordinate (k) of the vertex is $k = f(3)$.

$$f(x) = x^2 - 6x + 3$$
$$f(3) = 3^2 - 6(3) + 3$$
$$= 9 - 18 + 3$$
$$= -6$$

The vertex is $(3, -6)$. The axis of symmetry is $x = 3$.

Find more points on the graph of $f(x) = x^2 - 6x + 3$, then use the axis of symmetry to find other points on the parabola.

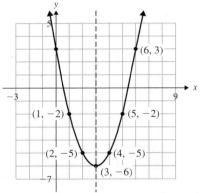

x	f(x)
4	-5
5	-2
6	3

To find the x-intercepts, let $f(x) = 0$ and solve for x.

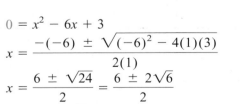

$$0 = x^2 - 6x + 3$$
$$x = \frac{-(-6) \pm \sqrt{(-6)^2 - 4(1)(3)}}{2(1)} \qquad \text{Solve using the quadratic formula.}$$
$$x = \frac{6 \pm \sqrt{24}}{2} = \frac{6 \pm 2\sqrt{6}}{2} \qquad \text{Simplify.}$$
$$x = 3 \pm \sqrt{6}$$

The x-intercepts are $(3 + \sqrt{6}, 0)$ and $(3 - \sqrt{6}, 0)$.

We can see from the graph that the *y*-intercept is (0, 3).

You Try 4

Graph $f(x) = -x^2 - 8x - 13$ using the vertex formula. Include the intercepts.

Using Technology

In Section 7.5 we said that the solutions of the equation $x^2 - x - 6 = 0$ are the *x*-intercepts of the graph of $y = x^2 - x - 6$. The *x*-intercepts are also called the *zeroes* of the equation since they are the values of *x* that make $y = 0$.

Use the zeroes of each function and its transformation from $y = x^2$ to match each equation with its graph:

1. $f(x) = x^2 - 2x + 2$ 4. $f(x) = x^2 - 4$
2. $f(x) = x^2 - 4x - 5$ 5. $f(x) = x^2 - 6x + 9$
3. $f(x) = -(x + 1)^2 + 4$ 6. $f(x) = -x^2 - 8x - 19$

a)
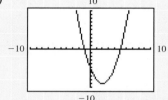

b)
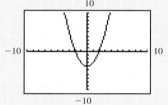

c)
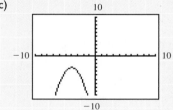

d)
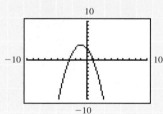

e)
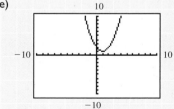

f)
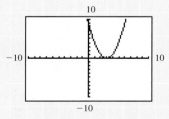

Answers to You Try Exercises

1) *x*-int: $(0, 0)$, $(2, 0)$; *y*-int: $(0, 0)$

2) *x*-int: $(-3 + \sqrt{2}, 0)$, $(-3 - \sqrt{2}, 0)$; *y*-int: $(0, -7)$

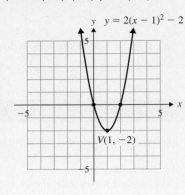

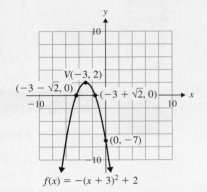

3) a) *x*-ints: $(-3, 0)$, $(-1, 0)$; *y*-int: $(0, 3)$

b) *x*-ints: $(3 + \sqrt{5}, 0)$, $(3 - \sqrt{5}, 0)$; *y*-int: $(0, -8)$

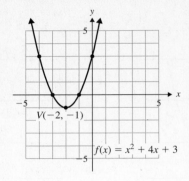

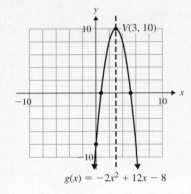

4) *x*-ints: $(-4 + \sqrt{3}, 0)$, $(-4 - \sqrt{3}, 0)$; *y*-int: $(0, -13)$

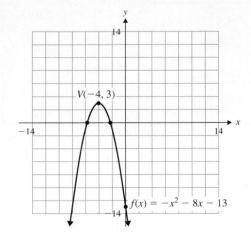

Answers to Technology Exercises

1. e)   2. a)   3. d)   4. b)   5. f)   6. c)

12.3 Exercises

Objectives 1 and 2

Given a quadratic function of the form $f(x) = a(x - h)^2 + k$, answer the following.

1) What is the vertex?

2) What is the equation of the axis of symmetry?

3) How do you know if the parabola opens upward?

4) How do you know if the parabola opens downward?

5) How do you know if the parabola is narrower than the graph of $y = x^2$?

6) How do you know if the parabola is wider than the graph of $y = x^2$?

For each quadratic function, identify the vertex, axis of symmetry, and x- and y-intercepts. Then, graph the function.

7) $f(x) = (x + 1)^2 - 4$ 8) $g(x) = (x - 3)^2 - 1$

9) $g(x) = (x - 2)^2_, + 3$ 10) $h(x) = (x + 2)^2 + 7$

11) $y = (x - 4)^2 - 2$ 12) $y = (x + 1)^2 - 5$

13) $f(x) = -(x + 3)^2 + 6$

14) $g(x) = -(x - 3)^2 + 2$

15) $y = -(x + 1)^2 - 5$

16) $f(x) = -(x - 2)^2 - 4$

17) $f(x) = 2(x - 1)^2 - 8$

18) $y = 2(x + 1)^2 - 2$

19) $h(x) = \overset{-1}{\frac{1}{2}}(x + 4)^2$

20) $g(x) = \frac{1}{4}x^2 - 1$

21) $y = -x^2 + 5$

22) $h(x) = -(x - 3)^2$

23) $f(x) = -\frac{1}{3}(x + 4)^2 + 3$

24) $y = -\frac{1}{2}(x - 4)^2 + 2$

25) $g(x) = 3(x + 2)^2 + 5$

26) $f(x) = 2(x - 3)^2 + 3$

Objective 3

Rewrite each function in the form $f(x) = a(x - h)^2 + k$ by completing the square. Then, graph the function. Include the intercepts.

27) $f(x) = x^2 - 2x - 3$ 28) $g(x) = x^2 + 6x + 8$

29) $y = x^2 + 6x + 7$ 30) $h(x) = x^2 - 4x + 1$

31) $g(x) = x^2 + 4x$ 32) $y = x^2 - 8x + 18$

33) $h(x) = -x^2 - 4x + 5$

34) $f(x) = -x^2 - 2x + 3$

35) $y = -x^2 + 6x - 10$

36) $g(x) = -x^2 - 4x - 6$

37) $f(x) = 2x^2 - 8x + 4$

38) $y = 2x^2 - 8x + 2$

39) $g(x) = -\frac{1}{3}x^2 - 2x - 9$

40) $h(x) = -\frac{1}{2}x^2 - 3x - \frac{19}{2}$

41) $y = x^2 - 3x + 2$

42) $f(x) = x^2 + 5x + \frac{21}{4}$

Objective 4

Graph each function using the vertex formula. Include the intercepts.

43) $y = x^2 + 2x - 3$ 44) $g(x) = x^2 - 6x + 8$

45) $f(x) = -x^2 - 8x - 13$

46) $y = -x^2 + 2x + 2$

47) $g(x) = 2x^2 - 4x + 4$

48) $f(x) = -4x^2 - 8x - 6$

49) $y = -3x^2 + 6x + 1$ 50) $h(x) = 2x^2 - 12x + 9$

51) $f(x) = \frac{1}{2}x^2 - 4x + 5$ 52) $y = \frac{1}{2}x^2 + 2x - 3$

53) $h(x) = -\frac{1}{3}x^2 - 2x - 5$

54) $g(x) = \frac{1}{5}x^2 - 2x + 8$

Section 12.4 Applications of Quadratic Functions and Graphing Other Parabolas

Objectives

1. Find the Maximum or Minimum Value of a Quadratic Function of the Form $f(x) = ax^2 + bx + c$

2. Given a Quadratic Function, Solve an Applied Problem Involving a Maximum or Minimum Value

3. Write a Quadratic Function to Model an Applied Problem Involving a Maximum or Minimum Value, and Solve the Problem

4. Graph Parabolas Given in the Form $x = a(y - k)^2 + h$

5. Rewrite $x = ay^2 + by + c$ as $x = a(y - k)^2 + h$ by Completing the Square

6. Find the Vertex of the Graph of $x = ay^2 + by + c$ Using $y = -\dfrac{b}{2a}$, and Graph the Equation

1. Find the Maximum or Minimum Value of a Quadratic Function of the Form $f(x) = ax^2 + bx + c$

From our work with quadratic functions we have seen that the vertex is either the lowest point or the highest point on the graph depending on whether the parabola opens upward or downward.

If the parabola opens upward, the vertex is the *lowest* point on the parabola.

If the parabola opens downward, the vertex is the *highest* point on the parabola.

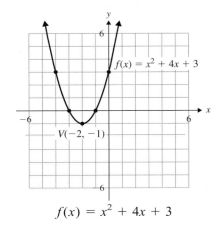

$$f(x) = x^2 + 4x + 3$$

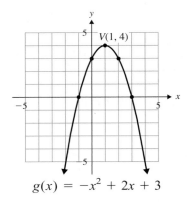

$$g(x) = -x^2 + 2x + 3$$

The y-coordinate of the vertex, -1, is the *smallest* y-value the function will have. We say that **-1 is the minimum value of the function**. $f(x)$ has no maximum because the graph continues upward indefinitely—the y-values get larger without bound.

The y-coordinate of the vertex, 4, is the *largest* y-value the function will have. We say that **4 is the maximum value of the function**. $g(x)$ has no minimum because the graph continues downward indefinitely—the y-values get smaller without bound.

Maximum and Minimum Values of a Quadratic Function

Let $f(x) = ax^2 + bx + c$.

1) If a is *positive*, the graph of $f(x)$ opens upward, so the vertex is the lowest point on the parabola. The y-coordinate of the vertex is the *minimum* value of the function $f(x)$.

2) If a is *negative*, the graph of $f(x)$ opens downward, so the vertex is the highest point on the parabola. The y-coordinate of the vertex is the *maximum* value of the function $f(x)$.

We can use this information about the vertex to help us solve problems.

Example 1

Let $f(x) = -x^2 + 4x + 2$.

a) Does the function attain a minimum or maximum value at its vertex?
b) Find the vertex of the graph of $f(x)$.
c) What is the minimum or maximum value of the function?
d) Graph the function to verify parts a)−c).

Solution

a) Since $a = -1$, the graph of $f(x)$ will open downward. Therefore, the vertex will be the *highest* point on the parabola. The function will attain its *maximum* value at the vertex.

b) Use $x = -\dfrac{b}{2a}$ to find the x-coordinate of the vertex. For $f(x) = -x^2 + 4x + 2$,

$$a = -1 \quad b = 4 \quad c = 2$$
$$x = -\frac{b}{2a} = -\frac{(4)}{2(-1)} = 2$$

The y-coordinate of the vertex is $f(2)$.

$$f(2) = -(2)^2 + 4(2) + 2$$
$$= -4 + 8 + 2$$
$$= 6$$

The vertex is $(2, 6)$.

c) $f(x)$ has no minimum value. The *maximum* value of the function is 6, the y-coordinate of the vertex. (The largest y-value of the function is 6.)
    We say that the maximum value of the function is 6 and that it occurs at $x = 2$ (the x-coordinate of the vertex).

d) From the graph of $f(x)$, we can see that our conclusions in parts a)−c) make sense.

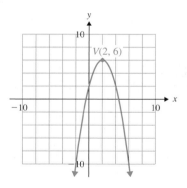

You Try 1

Let $f(x) = x^2 + 6x + 7$. Repeat parts a)−d) from Example 1.

2. Given a Quadratic Function, Solve an Applied Problem Involving a Maximum or Minimum Value

Example 2

A ball is thrown upward from a height of 24 ft. The height h of the ball (in feet) t sec after the ball is released is given by

$$h(t) = -16t^2 + 16t + 24.$$

a)   How long does it take the ball to reach its maximum height?

b)   What is the maximum height attained by the ball?

Solution

a)   Begin by understanding what the function $h(t)$ tells us: $a = -16$, so the graph of h would open downward. Therefore, the vertex is the highest point on the parabola. The maximum value of the function occurs at the vertex. The ordered pairs that satisfy $h(t)$ are of the form $(t, h(t))$.

   To determine how long it takes the ball to reach its maximum height we must find the t-coordinate of the vertex.

$$t = -\frac{b}{2a}$$

$$t = -\frac{16}{2(-16)} = \frac{1}{2}$$

   The ball will reach its maximum height after $\frac{1}{2}$ sec.

b)   The maximum height the ball reaches is the y-coordinate [or $h(t)$-coordinate] of the vertex. Since the ball attains its maximum height when $t = \frac{1}{2}$, find $h\left(\frac{1}{2}\right)$.

$$h\left(\frac{1}{2}\right) = -16\left(\frac{1}{2}\right)^2 + 16\left(\frac{1}{2}\right) + 24$$

$$= -16\left(\frac{1}{4}\right) + 8 + 24$$

$$= -4 + 32$$

$$= 28$$

   The ball reaches a maximum height of 28 ft.

 **You Try 2**

An object is propelled upward from a height of 10 ft. The height h of the object (in feet) t sec after the ball is released is given by

$$h(t) = -16t^2 + 32t + 10$$

a) How long does it take the object to reach its maximum height?

b) What is the maximum height attained by the object?

3. Write a Quadratic Function to Model an Applied Problem Involving a Maximum or Minimum Value, and Solve the Problem

| **Example 3** |

Ayesha plans to put a fence around her rectangular garden. If she has 32 ft of fencing, what is the maximum area she can enclose?

Solution

Begin by drawing a picture.

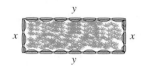

Let $x =$ the width of the garden

Let $y =$ the length of the garden

Label the picture.

We will write two equations for a problem like this:

1) *The maximize or minimize equation;* this equation describes what we are trying to maximize or minimize.

2) *The constraint equation;* this equation describes the restrictions on the variables or the conditions the variables must meet.

Here is how we will get the equations.

1) We will write a *maximize* equation because we are trying to find the *maximum area* of the garden.

$$\text{Let } A = \text{area of the garden}$$

The area of the rectangle above is xy. Our equation is

$$\text{Maximize: } A = xy$$

2) To write the *constraint* equation, think about the restriction put on the variables. We cannot choose *any* two numbers for x and y. Since Ayesha has 32 ft of fencing, the distance around the garden is 32 ft. This is the perimeter of the rectangular garden. The perimeter of the rectangle drawn above is $2x + 2y$, and it must equal 32 ft.

The constraint equation is

$$\text{Constraint: } 2x + 2y = 32$$

Set up this maximization problem as

$$\text{Maximize: } A = xy$$
$$\text{Constraint: } 2x + 2y = 32$$

Solve the constraint for a variable, and then substitute the expression into the maximize equation.

$$
\begin{aligned}
2x + 2y &= 32 \\
2y &= 32 - 2x \\
y &= 16 - x \qquad \text{Solve the constraint for } y.
\end{aligned}
$$

Substitute $y = 16 - x$ into $A = xy$.

$$A = x(16 - x)$$
$$A = 16x - x^2 \qquad \text{Distribute.}$$
$$A = -x^2 + 16x \qquad \text{Write in descending powers.}$$

Look carefully at $A = -x^2 + 16x$. This is a quadratic function! Its graph is a parabola that opens downward (since $a = -1$). At the vertex, the function attains its maximum. The ordered pairs that satisfy this function are of the form $(x, A(x))$, where x represents the width and $A(x)$ represents the area of the rectangular garden. *The second coordinate of the vertex is the maximum area we are looking for.*

$$A = -x^2 + 16x$$

Use $x = -\dfrac{b}{2a}$ with $a = -1$ and $b = 16$ to find the x-coordinate of the vertex (the width of the rectangle that produces the maximum area).

$$x = -\frac{16}{2(-1)} = 8$$

Substitute $x = 8$ into $A = -x^2 + 16x$ to find the maximum area.

$$A = -(8)^2 + 16(8)$$
$$A = -64 + 128$$
$$A = 64$$

The graph of $A = -x^2 + 16x$ is a parabola that opens downward with vertex $(8, 64)$.

The maximum area of the garden is 64 ft^2, and this will occur when the width of the garden is 8 ft. (The length will be 8 ft as well.)

Steps for Solving a Max/Min Problem Like Example 3

1) Draw a picture, if applicable.

2) Define the unknowns. Label the picture.

3) Write the max/min equation.

4) Write the constraint equation.

5) Solve the constraint for a variable. Substitute the expression into the max/min equation to obtain a quadratic function.

6) Find the vertex of the parabola using the vertex formula, $x = -\dfrac{b}{2a}$.

7) Answer the question being asked.

You Try 3

Find the maximum area of a rectangle that has a perimeter of 28 in.

Parabolas That Are Not Functions

4. Graph Parabolas Given in the Form $x = a(y - k)^2 + h$

Not all parabolas are functions. Parabolas can open in the x-direction as illustrated below. Clearly, these fail the vertical line test for functions.

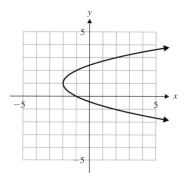
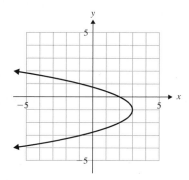

Parabolas which open in the y-direction, or vertically, result from the functions

$$y = a(x - h)^2 + k \qquad \text{or} \qquad y = ax^2 + bx + c.$$

If we interchange the x and y, we obtain the equations

$$x = a(y - k)^2 + h \qquad \text{or} \qquad x = ay^2 + by + c.$$

The graphs of these equations are parabolas which open in the x-direction, or horizontally.

Graphing an Equation of the Form $x = a(y - k)^2 + h$

1) The vertex of the parabola is (h, k). (Notice, however, that h and k have changed their positions when compared to a quadratic function.)

2) The axis of symmetry is the horizontal line $y = k$.

3) If a is positive, the graph opens to the right.
   If a is negative, the graph opens to the left.

Example 4

Graph each equation. Find the x- and y-intercepts.

a) $x = (y + 2)^2 - 1$ b) $x = -2(y - 2)^2 + 4$

Solution

a) 1) $h = -1$ and $k = -2$. The vertex is $(-1, -2)$.

2) The axis of symmetry is $y = -2$.

3) $a = +1$, so the parabola opens to the right. It is the same width as $y = x^2$.

To find the x-intercept, let $y = 0$ and solve for x.

$$x = (y + 2)^2 - 1$$
$$x = (0 + 2)^2 - 1$$
$$x = 4 - 1 = 3$$

The x-intercept is $(3, 0)$.

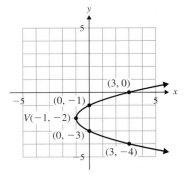

Find the y-intercepts by substituting 0 for x and solving for y.

$$x = (y + 2)^2 - 1$$
$$0 = (y + 2)^2 - 1 \qquad \text{Substitute 0 for } x.$$
$$1 = (y + 2)^2 \qquad \text{Add 1.}$$
$$\pm 1 = y + 2 \qquad \text{Square root property}$$

$$1 = y + 2 \qquad \text{or} \qquad -1 = y + 2$$
$$-1 = y \qquad\qquad\qquad -3 = y \qquad \text{Solve.}$$

The y-intercepts are $(0, -3)$ and $(0, -1)$.
Use the axis of symmetry to locate the point $(3, -4)$ on the graph.

b) $x = -2(y - 2)^2 + 4$

1) $h = 4$ and $k = 2$. The vertex is $(4, 2)$.

2) The axis of symmetry is $y = 2$.

3) $a = -2$, so the parabola opens to the left. It is narrower than $y = x^2$.

To find the x-intercept, let $y = 0$
and solve for x.

$$x = -2(y - 2)^2 + 4$$
$$x = -2(0 - 2)^2 + 4$$
$$x = -2(4) + 4 = -4$$

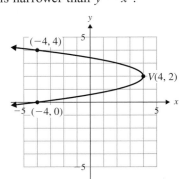

The x-intercept is $(-4, 0)$.

Find the y-intercepts by substituting 0 for x and solving for y.

$$x = -2(y-2)^2 + 4$$
$$0 = -2(y-2)^2 + 4 \qquad \text{Substitute 0 for } x.$$
$$-4 = -2(y-2)^2 \qquad \text{Subtract 4.}$$
$$2 = (y-2)^2 \qquad \text{Divide by } -2.$$
$$\pm\sqrt{2} = y - 2 \qquad \text{Square root property}$$
$$2 \pm \sqrt{2} = y \qquad \text{Add 2.}$$

The y-intercepts are $(0, 2 - \sqrt{2})$ and $(0, 2 + \sqrt{2})$.
Use the axis of symmetry to locate the point $(-4, 4)$ on the graph.

You Try 4

Graph $x = -(y + 1)^2 - 3$. Find the x- and y-intercepts.

Graphing Parabolas from the Form $x = ay^2 + by + c$

We can use two methods to graph $x = ay^2 + by + c$.

Method 1: Rewrite $x = ay^2 + by + c$ in the form $x = a(y - k)^2 + h$ by *completing the square*.

Method 2: Use the formula $y = -\dfrac{b}{2a}$ to find the *y-coordinate* of the vertex. Find the *x-coordinate* by substituting the *y*-value into the equation $x = ay^2 + by + c$.

---

5. Rewrite $x = ay^2 + by + c$ as $x = a(y - k)^2 + h$ by Completing the Square

Example 5

Rewrite $x = 2y^2 - 4y + 8$ in the form $x = a(y - k)^2 + h$ by completing the square.

Solution

To complete the square follow the same procedure used for quadratic functions. (This is outlined on p. 801 in Section 12.3.)

Step 1: Divide the equation by 2 so that the coefficient of y^2 is 1.

$$\frac{x}{2} = y^2 - 2y + 4$$

Step 2: Separate the constant from the variable terms using parentheses.

$$\frac{x}{2} = (y^2 - 2y) + 4$$

Step 3: Complete the square for the quantity in parentheses. Add 1 *inside* the parentheses and *subtract* 1 from the 4.

$$\frac{x}{2} = (y^2 - 2y + 1) + 4 - 1$$

$$\frac{x}{2} = (y^2 - 2y + 1) + 3$$

Step 4: Factor the expression inside the parentheses.

$$\frac{x}{2} = (y - 1)^2 + 3$$

Step 5: Solve the equation for x by multiplying by 2.

$$2\left(\frac{x}{2}\right) = 2[(y - 1)^2 + 3]$$
$$x = 2(y - 1)^2 + 6$$

You Try 5

Rewrite $x = -y^2 - 6y - 1$ in the form $x = a(y - k)^2 + h$ by completing the square.

6. Find the Vertex of the Graph of $x = ay^2 + by + c$ Using $y = -\dfrac{b}{2a}$, and Graph the Equation

Example 6

Graph $x = y^2 - 2y + 5$. Find the vertex using the vertex formula. Find the x- and y-intercepts.

Solution

Since this equation is solved for x and is quadratic in y, it opens in the x-direction. $a = 1$, so it opens to the right. Use the vertex formula to find the *y-coordinate* of the vertex.

$$y = -\frac{b}{2a}$$
$$y = -\frac{-2}{2(1)} = 1 \qquad a = 1, b = -2$$

Substitute $y = 1$ into $x = y^2 - 2y + 5$ to find the x-coordinate of the vertex.

$$x = (1)^2 - 2(1) + 5$$
$$x = 1 - 2 + 5 = 4$$

The vertex is (4, 1). Since the vertex is (4, 1) and the parabola opens to the right, the graph has *no y-intercepts.*

To find the *x*-intercept, let $y = 0$ and solve for *x*.

$$x = y^2 - 2y + 5$$
$$x = 0^2 - 2(0) + 5$$
$$x = 5$$

The x-intercept is (5, 0).

Find another point on the parabola by choosing a value for *y* that is close to the *y*-coordinate of the vertex. Let $y = -1$. Find *x*.

$$x = (-1)^2 - 2(-1) + 5$$
$$x = 1 + 2 + 5 = 8$$

Another point on the parabola is (8, −1). Use the axis of symmetry to locate the additional points (5, 2) and (8, 3).

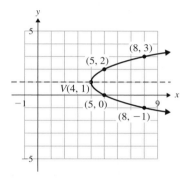

You Try 6

Graph $x = y^2 + 6y + 3$. Find the vertex using the vertex formula. Find the *x*- and *y*-intercepts.

Using Technology

To graph a parabola that is a function, just enter the equation and press GRAPH.

Example 1: Graph $f(x) = -x^2 + 2$

Enter $Y_1 = -x^2 + 2$ to graph the function on a calculator.

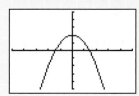

To graph an equation on a calculator, it must be entered so that y is a function of x. Since a parabola that opens horizontally is not a function, we must solve for y in terms of x so that the equation is represented by two different functions.

Example 2: Graph $x = y^2 - 4$ on a calculator.

Solve for y.

$$x = y^2 - 4$$
$$x + 4 = y^2$$
$$\pm\sqrt{x + 4} = y$$

Now the equation $x = y^2 - 4$ is rewritten so that y is in terms of x. In the graphing calculator, enter $y = \sqrt{x + 4}$ as Y_1. This represents the top half of the parabola since the y-values are positive above the x-axis. Enter $y = -\sqrt{x + 4}$ as Y_2. This represents the bottom half of the parabola since the y-values are negative below the x-axis. Set an appropriate window and press $\boxed{\text{GRAPH}}$.

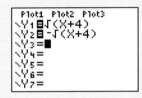
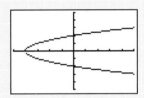

Graph each parabola on a graphing calculator. Where appropriate, rewrite the equation for y in terms of x. These problems come from the homework exercises so that the graphs can be found in the Answers to Selected Exercises appendix.

1. $f(x) = x^2 + 6x + 9$; Exercise 3
2. $x = y^2 + 2$; Exercise 27
3. $x = \dfrac{1}{4}(y + 2)^2$; Exercise 33

4. $f(x) = -\dfrac{1}{2}x^2 + 4x - 6$; Exercise 5
5. $x = -(y - 4)^2 + 5$; Exercise 29
6. $x = y^2 - 4y + 5$; Exercise 35

Answers to You Try Exercises

1) a) minimum value   b) vertex $(-3, -2)$
c) The minimum value of the function is -2.

2) a) I sec   b) 26 ft   3) 49 in^2
4) $V(-3, -1)$; x-int: $(-4, 0)$; y-int: none

d)

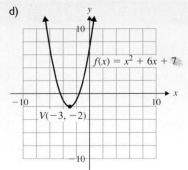

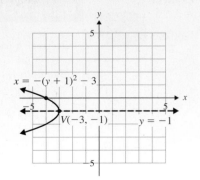

5) $x = -(y + 3)^2 + 8$

6) $V(-6, -3)$; x-int: $(3, 0)$; y-int: $(0, -3 - \sqrt{6})$, $(0, -3 + \sqrt{6})$

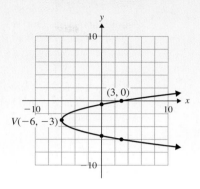

$(3, 0)$

$V(-6, -3)$

Answers to Technology Exercises

1. The equation can be entered as it is.

2. $Y_1 = \sqrt{x - 2}$; $Y_2 = -\sqrt{x - 2}$

3. $Y_1 = -2 + \sqrt{4x}$; $Y_2 = -2 - \sqrt{4x}$

4. The equation can be entered as it is.

5. $Y_1 = 4 + \sqrt{5 - x}$; $Y_2 = 4 - \sqrt{5 - x}$

6. $Y_1 = 2 + \sqrt{x - 1}$; $Y_2 = 2 - \sqrt{x - 1}$

12.4 Exercises

Objective 1

1) Let $f(x) = ax^2 + bx + c$. How do you know if the function has a maximum or minimum value at the vertex?

2) Is there a maximum value of the function $y = 2x^2 + 12x + 11$? Explain your answer.

For Problems 3–6, answer parts a)–d) for each function, $f(x)$.

  a) Does the function attain a minimum or maximum value at its vertex?

  b) Find the vertex of the graph of $f(x)$.

  c) What is the minimum or maximum value of the function?

  d) Graph the function to verify parts a)–c).

3) $f(x) = x^2 + 6x + 9$ 4) $f(x) = -x^2 + 2x + 4$

 5) $f(x) = -\dfrac{1}{2}x^2 + 4x - 6$

6) $f(x) = 2x^2 + 4x$

Objectives 2 and 3

Solve.

7) An object is fired upward from the ground so that its height h (in feet) t sec after being fired is given by

$$h(t) = -16t^2 + 320t$$

  a) How long does it take the object to reach its maximum height?

  b) What is the maximum height attained by the object?

  c) How long does it take the object to hit the ground?

8) An object is thrown upward from a height of 64 ft so that its height h (in feet) t sec after being thrown is given by

$$h(t) = -16t^2 + 48t + 64$$

  a) How long does it take the object to reach its maximum height?

b) What is the maximum height attained by the object?

c) How long does it take the object to hit the ground?

9) The number of guests staying at the Cozy Inn from January to December 2007 can be approximated by

$$N(x) = -10x^2 + 120x + 120$$

where x represents the number of months after January 2007 ($x = 0$ represents January, $x = 1$ represents February, etc.), and $N(x)$ represents the number of guests who stayed at the inn. During which month did the inn have the greatest number of guests? How many people stayed at the inn during that month?

10) The average number of traffic tickets issued in a city on any given day Sunday−Saturday can be approximated by

$$T(x) = -7x^2 + 70x + 43$$

where x represents the number of days after Sunday ($x = 0$ represents Sunday, $x = 1$ represents Monday, etc.), and $T(x)$ represents the number of traffic tickets issued. On which day are the most tickets written? How many tickets are issued on that day?

11) The number of babies born to teenage mothers from 1989 to 2002 can be approximated by

$$N(t) = -0.721t^2 + 2.75t + 528$$

where t represents the number of years after 1989 and $N(t)$ represents the number of babies born (in thousands). According to this model, in what year was the number of babies born to teen mothers the greatest? How many babies were born that year? (Source: U.S. Census Bureau)

12) The number of violent crimes in the United States from 1985 to 1999 can be modeled by

$$C(x) = -49.2x^2 + 636x + 12{,}468$$

where x represents the number of years after 1985 and $C(x)$ represents the number of violent crimes (in thousands). During what year did the greatest number of violent crimes occur, and how many were there? (Source: U.S. Census Bureau)

13) Every winter Rich makes a rectangular ice rink in his backyard. He has 100 ft of material to use as the border. What is the maximum area of the ice rink?

14) Find the dimensions of the rectangular garden of greatest area that can be enclosed with 40 ft of fencing.

15) The Soo family wants to fence in a rectangular area to hold their dogs. One side of the pen will be their barn. Find the dimensions of the pen of greatest area that can be enclosed with 48 ft of fencing.

16) A farmer wants to enclose a rectangular area with 120 ft of fencing. One side is a river and will not require a fence. What is the maximum area that can be enclosed?

17) Find two integers whose sum is 18 and whose product is a maximum.

18) Find two integers whose sum is 26 and whose product is a maximum.

19) Find two integers whose difference is 12 and whose product is a minimum.

20) Find two integers whose difference is 30 and whose product is a minimum.

Objective 4

Given a quadratic equation of the form $x = a(y - k)^2 + h$, answer the following.

21) What is the vertex?

22) What is the equation of the axis of symmetry?

23) If a is negative, which way does the parabola open?

24) If a is positive, which way does the parabola open?

For each equation, identify the vertex, axis of symmetry, and x- and y-intercepts. Then, graph the equation.

25) $x = (y - 1)^2 - 4$ 26) $x = (y + 3)^2 - 1$

27) $x = y^2 + 2$ 28) $x = (y - 4)^2$

29) $x = -(y - 4)^2 + 5$ 30) $x = -(y + 1)^2 - 7$

31) $x = -2(y - 2)^2 - 9$ 32) $x = -\dfrac{1}{2}(y - 4)^2 + 7$

33) $x = \dfrac{1}{4}(y + 2)^2$ 34) $x = 2y^2 + 3$

Objective 5

Rewrite each equation in the form $x = a(y - k)^2 + h$ by completing the square and graph it.

35) $x = y^2 - 4y + 5$ 36) $x = y^2 + 4y - 6$

37) $x = -y^2 + 6y + 6$ 38) $x = -y^2 - 2y - 5$

39) $x = \dfrac{1}{3}y^2 + \dfrac{8}{3}y - \dfrac{5}{3}$ 40) $x = 2y^2 - 4y + 5$

41) $x = -4y^2 - 8y - 10$ 42) $x = \dfrac{1}{2}y^2 + 4y - 1$

Objective 6

Graph each equation using the vertex formula. Find the x- and y-intercepts.

43) $x = y^2 - 4y + 3$ 44) $x = -y^2 + 4y$

45) $x = -y^2 + 2y + 2$ 46) $x = y^2 + 6y - 4$

47) $x = -2y^2 + 4y - 6$ 48) $x = 3y^2 + 6y - 1$

49) $x = 4y^2 - 16y + 13$ 50) $x = 2y^2 + 4y + 8$

51) $x = \dfrac{1}{4}y^2 - \dfrac{1}{2}y + \dfrac{25}{4}$ 52) $x = -\dfrac{3}{4}y^2 + \dfrac{3}{2}y - \dfrac{11}{4}$

Section 12.5 The Algebra of Functions

Objectives

1. Add, Subtract, Multiply, and Divide Functions
2. Solve an Applied Problem Using Operations on Functions
3. Find the Composition of Functions
4. Solve an Applied Problem Using the Composition of Functions

1. Add, Subtract, Multiply, and Divide Functions

We have learned that we can add, subtract, multiply, and divide polynomials. These same operations can be performed with functions.

Operations Involving Functions

Given the functions $f(x)$ and $g(x)$, the sum, difference, product, and quotient of f and g are defined by

1) $(f + g)(x) = f(x) + g(x)$

2) $(f - g)(x) = f(x) - g(x)$

3) $(fg)(x) = f(x) \cdot g(x)$

4) $\left(\dfrac{f}{g}\right)(x) = \dfrac{f(x)}{g(x)}$, where $g(x) \neq 0$

The domain of $f + g$, $f - g$, fg, and $\dfrac{f}{g}$ is the *intersection* of the domains of $f(x)$ and $g(x)$.

Example 1

Let $f(x) = x^2 - 2x + 7$ and $g(x) = 4x - 3$. Find each of the following.

a) $(f + g)(x)$ b) $(f - g)(x)$ and $(f - g)(-1)$ c) $(fg)(x)$

d) $\left(\dfrac{f}{g}\right)(x)$

Solution

a) $(f + g)(x) = f(x) + g(x)$

$\qquad = (x^2 - 2x + 7) + (4x - 3)$ Substitute the functions.

$\qquad = x^2 + 2x + 4$ Combine like terms.

b) $(f - g)(x) = f(x) - g(x)$

$\qquad = (x^2 - 2x + 7) - (4x - 3)$ Substitute the functions.

$\qquad = x^2 - 2x + 7 - 4x + 3$ Distribute.

$\qquad = x^2 - 6x + 10$ Combine like terms.

Use the result above to find $(f - g)(-1)$.

$\qquad (f - g)(x) = x^2 - 6x + 10$

$\qquad (f - g)(-1) = (-1)^2 - 6(-1) + 10$ Substitute -1 for x.

$\qquad\qquad\qquad = 1 + 6 + 10$

$\qquad\qquad\qquad = 17$

We can also find $(f - g)(-1)$ using the rule this way:

$$
\begin{aligned}
(f - g)(-1) &= f(-1) - g(-1) \\
&= ((-1)^2 - 2(-1) + 7) - (4(-1) - 3) \qquad \text{Substitute } -1 \text{ for } x \text{ in} \\
&= (1 + 2 + 7) - (-4 - 3) \qquad\qquad\qquad f(x) \text{ and } g(x). \\
&= 10 - (-7) \\
&= 17
\end{aligned}
$$

c) $\quad(fg)(x) = f(x) \cdot g(x)$

$$
\begin{aligned}
&= (x^2 - 2x + 7)(4x - 3) \qquad\qquad \text{Substitute the functions.} \\
&= 4x^3 - 3x^2 - 8x^2 + 6x + 28x - 21 \qquad \text{Multiply.} \\
&= 4x^3 - 11x^2 + 34x - 21 \qquad\qquad \text{Combine like terms.}
\end{aligned}
$$

d) $\quad \left(\dfrac{f}{g}\right)(x) = \dfrac{f(x)}{g(x)}, \text{ where } g(x) \neq 0$

$$
= \frac{x^2 - 2x + 7}{4x - 3}, \text{ where } x \neq \frac{3}{4} \qquad \text{Substitute the functions.}
$$

You Try 1

Let $f(x) = 3x^2 - 8$ and $g(x) = 2x + 1$. Find

a) $(f + g)(x)$ and $(f + g)(-2)$ b) $(f - g)(x)$ c) $(fg)(x)$ d) $\left(\dfrac{f}{g}\right)(3)$

2. Solve an Applied Problem Using Operations on Functions

Example 2

A publisher sells paperback romance novels to a large bookstore chain for $4.00 per book. Therefore, the publisher's revenue, in dollars, is defined by the function

$$
R(x) = 4x
$$

where x is the number of books sold to the retailer. The publisher's cost, in dollars, to produce x books is

$$
C(x) = 2.5x + 1200
$$

In business, profit is defined as revenue − cost. In terms of functions this is written as $P(x) = R(x) - C(x)$, where $P(x)$ is the profit function.

a) Find the profit function, $P(x)$, that describes the publisher's profit from the sale of x books.

b) If the publisher sells 10,000 books to this chain of bookstores, what is the publisher's profit?

Solution

a) $\quad P(x) = R(x) - C(x)$

$$
\begin{aligned}
&= 4x - (2.5x + 1200) \qquad \text{Substitute the functions.} \\
&= 1.5x - 1200 \\
P(x) &= 1.5x - 1200
\end{aligned}
$$

b) Find $P(10,000)$.

$$P(10,000) = 1.5(10,000) - 1200$$
$$= 15,000 - 1200$$
$$= 13,800$$

The publisher's profit is $13,800.

You Try 2

A candy company sells its Valentine's Day candy to a grocery store retailer for $6.00 per box. The candy company's revenue, in dollars, is defined by $R(x) = 6x$, where x is the number of boxes sold to the retailer. The company's cost, in dollars, to produce x boxes of candy is $C(x) = 4x + 900$.

a) Find the profit function, $P(x)$, that defines the company's profit from the sale of x boxes of candy.

b) Find the candy company's profit from the sale of 2000 boxes of candy.

3. Find the Composition of Functions

Another operation that can be performed with functions is called the **composition of functions**.

Definition

Given the functions $f(x)$ and $g(x)$, the **composition function** $f \circ g$ (read "f of g") is defined as

$$(f \circ g)(x) = f(g(x))$$

where $g(x)$ is in the domain of f.

Example 3

Let $f(x) = 3x + 5$ and $g(x) = x - 2$. Find $(f \circ g)(x)$.

Solution

$$
\begin{aligned}
(f \circ g)(x) &= f(g(x)) & \\
&= f(x - 2) & \text{Substitute } x - 2 \text{ for } g(x). \\
&= 3(x - 2) + 5 & \text{Substitute } x - 2 \text{ for } x \text{ in } f(x). \\
&= 3x - 6 + 5 & \text{Distribute.} \\
&= 3x - 1 &
\end{aligned}
$$

The composition of functions can also be explained this way. Finding $(f \circ g)(x) = f(g(x))$ in Example 3 meant that the function $g(x)$ was substituted into the function $f(x)$. $g(x)$ was the innermost function. Therefore, *first* the function g performs an operation on x. The result is $g(x)$. Then, the function f performs an operation on $g(x)$. The result is $f(g(x))$.

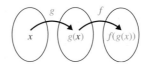

You Try 3

Let $f(x) = -6x + 2$ and $g(x) = 5x + 1$. Find $(f \circ g)(x)$.

Example 4

Let $f(x) = x + 8$ and $g(x) = 2x - 5$. Find

a) $g(3)$ b) $(f \circ g)(3)$ c) $(f \circ g)(x)$

Solution

a) $g(x) = 2x - 5$
    $g(3) = 2(3) - 5 = 1$

b) $(f \circ g)(3) = f(g(3))$
    $\qquad\qquad = f(1)$ $g(3) = 1$ from a)
    $\qquad\qquad = 1 + 8$ Substitute 1 for x in $f(x) = x + 8$.
    $\qquad\qquad = 9$

c) $(f \circ g)(x) = f(g(x))$
    $\qquad\qquad = f(2x - 5)$ Substitute $2x - 5$ for $g(x)$.
    $\qquad\qquad = (2x - 5) + 8$ Substitute $2x - 5$ for x in $f(x)$.
    $\qquad\qquad = 2x + 3$

We can also find $(f \circ g)(3)$ by substituting 3 into the expression for $(f \circ g)(x)$.

$$(f \circ g)(x) = 2x + 3$$
$$(f \circ g)(3) = 2(3) + 3 = 9$$

Notice that this is the same as the result we obtained in b).

You Try 4

Let $f(x) = x - 10$ and $g(x) = 3x + 4$. Find

a) $g(-1)$ b) $(f \circ g)(-1)$ c) $(f \circ g)(x)$

Example 5

Let $f(x) = 4x - 1$, $g(x) = x^2$, and $h(x) = x^2 + 5x - 2$. Find

a) $(f \circ g)(x)$ b) $(g \circ f)(x)$ c) $(h \circ f)(x)$

Solution

a) $(f \circ g)(x) = f(g(x))$
    $\qquad\qquad = f(x^2)$ Substitute x^2 for $g(x)$.
    $\qquad\qquad = 4(x^2) - 1$ Substitute x^2 for x in $f(x)$.
    $\qquad\qquad = 4x^2 - 1$

b) $(g \circ f)(x) = g(f(x))$
$$= g(4x - 1) \qquad \text{Substitute } 4x - 1 \text{ for } f(x).$$
$$= (4x - 1)^2 \qquad \text{Substitute } 4x - 1 \text{ for } x \text{ in } g(x).$$
$$= 16x^2 - 8x + 1 \qquad \text{Expand the binomial.}$$

> In general, $(f \circ g)(x) \neq (g \circ f)(x).$

c) $(h \circ f)(x) = h(f(x))$
$$= h(4x - 1) \qquad \text{Substitute } 4x - 1 \text{ for } f(x).$$
$$= (4x - 1)^2 + 5(4x - 1) - 2 \qquad \text{Substitute } 4x - 1 \text{ for } x \text{ in } h(x).$$
$$= 16x^2 - 8x + 1 + 20x - 5 - 2 \qquad \text{Distribute.}$$
$$= 16x^2 + 12x - 6 \qquad \text{Combine like terms.}$$

You Try 5

Let $f(x) = x^2 + 6$, $g(x) = 2x - 3$, and $h(x) = x^2 - 4x + 9$. Find

a) $(g \circ f)(x)$ b) $(f \circ g)(x)$ c) $(h \circ g)(x)$

4. Solve an Applied Problem Using the Composition of Functions

Example 6

The area, A, of a square expressed in terms of its perimeter, P, is defined by the function

$$A(P) = \frac{1}{16}P^2$$

The perimeter of a square, which has a side of length x, is defined by the function

$$P(x) = 4x$$

a) Find $(A \circ P)(x)$ and explain what it represents.

b) Find $(A \circ P)(3)$ and explain what it represents.

Solution

a) $(A \circ P)(x) = A(P(x))$
$$= A(4x) \qquad \text{Substitute } 4x \text{ for } P(x).$$
$$= \frac{1}{16}(4x)^2 \qquad \text{Substitute } 4x \text{ for } P \text{ in } A(P) = \frac{1}{16}P^2.$$
$$= \frac{1}{16}(16x^2)$$
$$= x^2$$

$(A \circ P)(x) = x^2$. This is the formula for the area of a square in terms of the length of a side, x.

b)   To find $(A \circ P)(3)$, use the result obtained in a).

$$(A \circ P)(x) = x^2$$
$$(A \circ P)(3) = 3^2 = 9$$

A square that has a side of length 3 units has an area of 9 square units.

You Try 6

Let $f(x) = 100x$ represent the number of centimeters in x meters. Let $g(y) = 1000y$ represent the number of meters in y kilometers.

a)   Find $(f \circ g)(y)$ and explain what it represents.

b)   Find $(f \circ g)(4)$ and explain what it represents.

Answers to You Try Exercises

1) a) $3x^2 + 2x - 7; 1$ b) $3x^2 - 2x - 9$ c) $6x^3 + 3x^2 - 16x - 8$ d) $\dfrac{19}{7}$

2) a) $P(x) = 2x - 900$ b) 3100 3) $(f \circ g)(x) = -30x - 4$ 4) a) 1 b) -9 c) $3x - 6$

5) a) $2x^2 + 9$ b) $4x^2 - 12x + 15$ c) $4x^2 - 20x + 30$ 6) a) $(f \circ g)(y) = 100,000y$; this tells us the number of centimeters in y km.   b) $(f \circ g)(4) = 400,000$; there are $400,000$ cm in 4 km.

12.5 Exercises

Boost your grade at **mathzone.com!**   MathZone   > Practice Problems   > NetTutor   > Self-Test   > e-Professors   > Videos

Objective 1

For each pair of functions, find a) $(f + g)(x)$, b) $(f + g)(5)$, c) $(f - g)(x)$, and d) $(f - g)(2)$.

1) $f(x) = -3x + 1, g(x) = 2x - 11$

2) $f(x) = 5x - 9, g(x) = x + 4$

3) $f(x) = 4x^2 - 7x - 1, g(x) = x^2 + 3x - 6$

4) $f(x) = -2x^2 + x + 8, g(x) = 3x^2 - 4x - 6$

For each pair of functions, find a) $(fg)(x)$ and b) $(fg)(-3)$.

5) $f(x) = x, g(x) = -x + 5$

6) $f(x) = -2x, g(x) = 3x + 1$

7) $f(x) = 2x + 3, g(x) = 3x + 1$

8) $f(x) = 4x + 7, g(x) = x - 5$

For each pair of functions, find a) $\left(\dfrac{f}{g}\right)(x)$ and b) $\left(\dfrac{f}{g}\right)(-2)$.

Identify any values that are not in the domain of $\left(\dfrac{f}{g}\right)(x)$.

9) $f(x) = 6x + 9, g(x) = x + 4$

10) $f(x) = 3x - 8, g(x) = x - 1$

11) $f(x) = x^2 - 5x - 24, g(x) = x - 8$

12) $f(x) = x^2 - 15x + 54, g(x) = x - 9$

13) $f(x) = 3x^2 + 14x + 8, g(x) = 3x + 2$

14) $f(x) = 2x^2 + x - 15, g(x) = 2x - 5$

Objective 2

15) A manufacturer's revenue, $R(x)$ in dollars, from the sale of x calculators is given by $R(x) = 12x$. The company's cost, $C(x)$ in dollars, to produce x calculators is $C(x) = 8x + 2000$.

a)   Find the profit function, $P(x)$, that defines the manufacturer's profit from the sale of x calculators.

b)   What is the profit from the sale of 1500 calculators?

16) $R(x) = 80x$ is the revenue function for the sale of x bicycles, in dollars. The cost to manufacture x bikes, in dollars, is $C(x) = 60x + 7000$.

a)   Find the profit function, $P(x)$, that describes the manufacturer's profit from the sale of x bicycles.

b)   What is the profit from the sale of 500 bicycles?

17) $R(x) = 18x$ is the revenue function for the sale of x toasters, in dollars. The cost to manufacture x toasters, in dollars, is $C(x) = 15x + 2400$.

a)   Find the profit function, $P(x)$, that describes the profit from the sale of x toasters.

b)   What is the profit from the sale of 800 toasters?

18) A company's revenue, $R(x)$ in dollars, from the sale of x dog houses is given by $R(x) = 60x$. The company's cost, $C(x)$ in dollars, to produce x dog houses is $C(x) = 45x + 6000$.

a)   Find the profit function, $P(x)$, that describes the company's profit from the sale of x dog houses.

b)   What is the profit from the sale of 300 dog houses?

Objective 3

19) Let $f(x) = 5x - 4$ and $g(x) = x + 7$. Find

a) $(f \circ g)(x)$ b) $(g \circ f)(x)$

c) $(f \circ g)(3)$

20) Let $f(x) = x - 10$ and $g(x) = 4x + 3$. Find

a) $(f \circ g)(x)$ b) $(g \circ f)(x)$

c) $(f \circ g)(-6)$

21) Let $h(x) = -2x + 9$ and $k(x) = 3x - 1$. Find

a) $(k \circ h)(x)$ b) $(h \circ k)(x)$

c) $(k \circ h)(-1)$

22) Let $r(x) = 6x + 2$ and $v(x) = -7x - 5$. Find

a) $(v \circ r)(x)$ b) $(r \circ v)(x)$

c) $(r \circ v)(2)$

23) Let $g(x) = x^2 - 6x + 11$ and $h(x) = x - 4$. Find

a) $(h \circ g)(x)$ b) $(g \circ h)(x)$

c) $(g \circ h)(4)$

24) Let $f(x) = x^2 + 7x - 9$ and $g(x) = x + 2$. Find

a) $(g \circ f)(x)$ b) $(f \circ g)(x)$

c) $(g \circ f)(3)$

25) Let $f(x) = 2x^2 + 3x - 10$ and $g(x) = 3x - 5$. Find

a) $(f \circ g)(x)$ b) $(g \circ f)(x)$

c) $(f \circ g)(1)$

26) Let $h(x) = 3x^2 - 8x + 2$ and $k(x) = 2x - 3$. Find

a) $(h \circ k)(x)$ b) $(k \circ h)(x)$

c) $(k \circ h)(0)$

27) Let $m(x) = x + 8$ and $n(x) = -x^2 + 3x - 8$. Find

a) $(n \circ m)(x)$ b) $(m \circ n)(x)$

c) $(m \circ n)(0)$

28) Let $f(x) = -x^2 + 10x + 4$ and $g(x) = x + 1$. Find

a) $(g \circ f)(x)$ b) $(f \circ g)(x)$

c) $(f \circ g)(-2)$

Objective 4

29) Oil spilled from a ship off the coast of Alaska with the oil spreading out in a circle across the surface of the water. The radius of the oil spill is given by $r(t) = 4t$ where t is the number of minutes after the leak began and $r(t)$ is in feet. The area of the spill is given by $A(r) = \pi r^2$ where r represents the radius of the oil slick. Find each of the following and explain their meanings.

a) $r(5)$ b) $A(20)$ c) $A(r(t))$ d) $A(r(5))$

30) The radius of a circle is half its diameter. We can express this with the function $r(d) = \dfrac{1}{2}d$, where d is the diameter of a circle and r is the radius. The area of a circle in terms of its radius is $A(r) = \pi r^2$. Find each of the following and explain their meanings.

a) $r(6)$ b) $A(3)$ c) $A(r(d))$ d) $A(r(6))$

Section 12.6 Variation

Objectives

1. Understand the Meaning of Direct Variation

2. For a Given Problem, Write an Equation Involving Direct Variation and Solve

3. Understand the Meaning of Inverse Variation

4. For a Given Problem, Write an Equation Involving Inverse Variation and Solve

5. For a Given Problem, Write an Equation Involving Joint Variation and Solve

6. For a Given Problem, Write an Equation Involving Combined Variation and Solve

Direct Variation

1. Understand the Meaning of Direct Variation

In Section 4.7 we discussed the following situation:

If you are driving on a highway at a constant speed of 60 mph, the distance you travel depends on the amount of time spent driving.

Let y = the distance traveled, in miles, and let x = the number of hours spent driving. An equation relating x and y is

$$y = 60x$$

and y is a function of x.

We can make a table of values relating x and y.

x	y
1	60
1.5	90
2	120
3	180

As the value of x increases, the value of y also increases. (The more hours you drive, the farther you will go.) Likewise, as the value of x decreases, the value of y also decreases. (The fewer hours you drive, the shorter the distance you will travel.)

We can say that the distance traveled, y, is *directly proportional to* the time spent traveling, x. Or y *varies directly as* x.

Definition

Direct Variation: *y* varies directly as *x* (or *y* is directly proportional to *x*) means

$$y = kx$$

where k is a nonzero real number. **k is called the constant of variation.**

If two quantities vary directly and $k > 0$, then as one quantity increases the other increases as well. And, as one quantity decreases, the other decreases.

In our example of driving distance, $y = 60x$, *60 is the constant of variation.*

2. For a Given Problem, Write an Equation Involving Direct Variation and Solve

Given information about how variables are related, we can write an equation and solve a variation problem.

Example 1

Suppose y varies directly as x. If $y = 18$ when $x = 3$,

a) Find the constant of variation, k.

b) Write a variation equation relating x and y using the value of k found in a).

c) Find y when $x = 11$.

Solution

a) To find the constant of variation, write a *general* variation equation relating x and y. *y varies directly as x means*

$$y = kx$$

We are told that $y = 18$ when $x = 3$. Substitute these values into the equation and solve for k.

$$y = kx$$
$$18 = k(3) \qquad \text{Substitute 3 for } x \text{ and 18 for } y.$$
$$6 = k \qquad \text{Divide by 3.}$$

b) The *specific* variation equation is the equation obtained when we substitute 6 for k in $y = kx$.

$$y = 6x$$

c) To find y when $x = 11$, substitute 11 for x in $y = 6x$ and evaluate.

$$y = 6x$$
$$= 6(11) \qquad \text{Substitute 11 for } x.$$
$$= 66 \qquad \text{Multiply.}$$

Steps for Solving a Variation Problem

Step 1:   Write the *general* variation equation.

Step 2:   Find k by substituting the known values into the equation and solving for k.

Step 3:   Write the *specific* variation equation by substituting the value of k into the *general* variation equation.

Step 4:   Use the specific variation equation to solve the problem.

You Try 1

Suppose y varies directly as x. If $y = 40$ when $x = 5$,

a) Find the constant of variation.

b) Write the specific variation equation relating x and y.

c) Find y when $x = 3$.

Example 2

Suppose p varies directly as the square of z. If $p = 12$ when $z = 2$, find p when $z = 5$.

Solution

Step 1:   Write the *general* variation equation.
p varies directly as the *square* of z means

$$p = kz^2.$$

Step 2:   Find k using the known values: $p = 12$ when $z = 2$.

$$p = kz^2$$
$$12 = k(2)^2 \qquad \text{Substitute 2 for } z \text{ and 12 for } p.$$
$$12 = k(4)$$
$$3 = k$$

Step 3:   Substitute $k = 3$ into $p = kz^2$ to get the *specific* variation equation.

$$p = 3z^2$$

Step 4:   We are asked to find p when $z = 5$. Substitute $z = 5$ into $p = 3z^2$ to get p.

$$p = 3z^2$$
$$= 3(5)^2 \qquad \text{Substitute 5 for } z.$$
$$= 3(25)$$
$$= 75$$

You Try 2

Suppose w varies directly as the cube of n. If $w = 135$ when $n = 3$, find w when $n = 2$.

Application

Example 3

A theater's nightly revenue varies directly as the number of tickets sold. If the revenue from the sale of 80 tickets is $3360, find the revenue from the sale of 95 tickets.

Solution

Let n = the number of tickets sold
   R = revenue

We will follow the four steps for solving a variation problem.

Step 1:   Write the *general* variation equation.

$$R = kn$$

Step 2: Find k using the known values: $R = 3360$ when $n = 80$.

$$R = kn$$
$$3360 = k(80) \qquad \text{Substitute 80 for } n \text{ and 3360 for } R.$$
$$42 = k \qquad \text{Divide by 80.}$$

Step 3: Substitute $k = 42$ into $R = 42n$ to get the *specific* variation equation.

$$R = 42n$$

Step 4: We must find the revenue from the sale of 95 tickets. Substitute $n = 95$ into $R = 42n$ to find R.

$$R = 42n$$
$$R = 42(95)$$
$$R = 3990$$

The revenue is $3990.

You Try 3

The cost to carpet a room varies directly as the area of the room. If it costs $525.00 to carpet a room of area 210 ft², how much would it cost to carpet a room of area 288 ft²?

Inverse Variation

3. Understand the Meaning of Inverse Variation

If two quantities vary *inversely* (are *inversely* proportional) then as one value increases, the other decreases. Likewise, as one value decreases, the other increases.

Definition

Inverse Variation: *y* varies inversely as *x* (or *y* is inversely proportional to *x*) means

$$y = \frac{k}{x}$$

where *k* is a nonzero real number. ***k* is the constant of variation.**

A good example of inverse variation is the relationship between the time, t, it takes to travel a given distance, d, as a function of the rate (or speed), r. We can define this relationship as $t = \dfrac{d}{r}$. As the rate, r, increases, the time, t, that it takes to travel d mi decreases. Likewise, as r decreases, the time, t, that it takes to travel d mi increases. Therefore, t varies *inversely* as r.

4. For a Given Problem, Write an Equation Involving Inverse Variation and Solve

Example 4

Suppose q varies inversely as h. If $q = 4$ when $h = 15$, find q when $h = 10$.

Solution

Step 1: Write the *general* variation equation.

$$q = \frac{k}{h}$$

Step 2: Find k using the known values: $q = 4$ when $h = 15$.

$$q = \frac{k}{h}$$
$$4 = \frac{k}{15} \qquad \text{Substitute 15 for } h \text{ and 4 for } q.$$
$$60 = k \qquad \text{Multiply by 15.}$$

Step 3: Substitute $k = 60$ into $q = \dfrac{k}{h}$ to get the *specific* variation equation.

$$q = \frac{60}{h}$$

Step 4: Substitute 10 for h in $q = \dfrac{60}{h}$ to find q.

$$q = \frac{60}{10}$$
$$= 6$$

 You Try 4

Suppose m varies inversely as the square of v. If $m = 1.5$ when $v = 4$, find m when $v = 2$.

Application

Example 5

The intensity of light (in lumens) varies inversely as the square of the distance from the source. If the intensity of the light is 40 lumens 5 ft from the source, what is the intensity of the light 4 ft from the source?

Solution

Let d = distance from the source (in feet)

  I = intensity of the light (in lumens)

Step 1: Write the *general* variation equation.

$$I = \frac{k}{d^2}$$

Step 2: Find k using the known values: $I = 40$ when $d = 5$.

$$I = \frac{k}{d^2}$$

$$40 = \frac{k}{(5)^2} \qquad \text{Substitute 5 for } d \text{ and 40 for } I.$$

$$40 = \frac{k}{25}$$

$$1000 = k \qquad \text{Multiply by 25.}$$

Step 3: Substitute $k = 1000$ into $I = \dfrac{k}{d^2}$ to get the *specific* variation equation.

$$I = \frac{1000}{d^2}$$

Step 4: Find the intensity, I, of the light 4 ft from the source. Substitute $d = 4$ into $I = \dfrac{1000}{d^2}$ to find I.

$$I = \frac{1000}{(4)^2}$$

$$= \frac{1000}{16}$$

$$= 62.5$$

The intensity of the light is 62.5 lumens.

You Try 5

If the voltage in an electrical circuit is held constant (stays the same), then the current in the circuit varies inversely as the resistance. If the current is 40 amps when the resistance is 3 ohms, find the current when the resistance is 8 ohms.

Joint Variation

5. For a Given Problem, Write an Equation Involving Joint Variation and Solve

If a variable varies directly as the *product* of two or more other variables, the first variable *varies jointly* as the other variables.

Definition	**Joint Variation:** *y* varies jointly as **x** and **z** means
	$$y = kxz$$
	where *k* is a nonzero real number.

Example 6

For a given amount invested in a bank account (called the principal), the interest earned varies jointly as the interest rate (expressed as a decimal) and the time the principal is in the account. If Graham earns $80 in interest when he invests his money for 1 yr at 4%, how much interest would the same principal earn if he invested it at 5% for 2 yr?

Solution

Let $r =$ interest rate (as a decimal)

$t =$ the number of years the principal is invested

$I =$ interest earned

Step 1:   Write the *general* variation equation.

$$I = krt$$

Step 2:   Find k using the known values: $I = 80$ when $t = 1$ and $r = 0.04$.

$$I = krt$$
$$80 = k(0.04)(1) \qquad \text{Substitute the values into } I = krt.$$
$$80 = 0.04k$$
$$2000 = k \qquad \text{Divide by 0.04.}$$

(The amount he invested, the principal, is $2000.)

Step 3:   Substitute $k = 2000$ into $I = krt$ to get the *specific* variation equation.

$$I = 2000rt$$

Step 4:   Find the interest Graham would earn if he invested $2000 at 5% interest for 2 yr. Let $r = 0.05$ and $t = 2$. Solve for I.

$$I = 2000(0.05)(2) \qquad \text{Substitute 0.05 for } r \text{ and 2 for } t.$$
$$= 200 \qquad \text{Multiply.}$$

Graham would earn $200.

You Try 6

The volume of a box of constant height varies jointly as its length and width. A box with a volume of 9 ft^3 has a length of 3 ft and a width of 2 ft. Find the volume of a box with the same height, if its length is 4 ft and its width is 3 ft.

Combined Variation

6. For a Given Problem, Write an Equation Involving Combined Variation and Solve

A combined variation problem involves both direct and inverse variation.

Example 7

Suppose y varies directly as the square root of x and inversely as z. If $y = 12$ when $x = 36$ and $z = 5$, find y when $x = 81$ and $z = 15$.

Solution

Step 1: Write the *general* variation equation.

$$y = \frac{k\sqrt{x}}{z} \qquad \begin{array}{l} \leftarrow y \text{ varies directly as the square root of } x. \\ \leftarrow y \text{ varies inversely as } z. \end{array}$$

Step 2: Find k using the known values: $y = 12$ when $x = 36$ and $z = 5$.

$$12 = \frac{k\sqrt{36}}{5} \qquad \text{Substitute the values.}$$
$$60 = 6k \qquad \text{Multiply by 5; } \sqrt{36} = 6.$$
$$10 = k \qquad \text{Divide by 6.}$$

Step 3: Substitute $k = 10$ into $y = \frac{k\sqrt{x}}{z}$ to get the specific variation equation.

$$y = \frac{10\sqrt{x}}{z}$$

Step 4: Find y when $x = 81$ and $z = 15$.

$$y = \frac{10\sqrt{81}}{15} \qquad \text{Substitute 81 for } x \text{ and 15 for } z.$$
$$y = \frac{10 \cdot 9}{15} = \frac{90}{15} = 6$$

You Try 7

Suppose a varies directly as b and inversely as the square of c. If $a = 28$ when $b = 12$ and $c = 3$, find a when $b = 36$ and $c = 4$.

Answers to You Try Exercises

1) a) 8   b) $y = 8x$ c) 24   2) 40   3) \$720.00   4) 6   5) 15 amps
6) 18 ft^3 7) 47.25

12.6 Exercises   **Boost your grade at mathzone.com!**   MathZone

> **Practice Problems**   > **NetTutor**   > **e-Professors**   > **Videos**
> **Self-Test**

Objectives 1, 3, 5 and 6

1) If z varies directly as y, then as y increases, the value of z _____.

2) If a varies inversely as b, then as b increases, the value of a _____.

Decide whether each equation represents direct, inverse, joint, or combined variation.

3) $y = 6x$

4) $c = 4ab$

5) $f = \dfrac{15}{t}$

6) $z = 3\sqrt{x}$

7) $p = \dfrac{8q^2}{r}$

8) $w = \dfrac{11}{v^2}$

Objectives 2 and 4–6

Write a general variation equation using k as the constant of variation.

9) M varies directly as n.

10) q varies directly as r.

11) h varies inversely as j.

12) R varies inversely as B.

13) T varies inversely as the square of c.

14) b varies directly as the cube of w.

15) s varies jointly as r and t.

16) C varies jointly as A and D.

17) Q varies directly as the square root of z and inversely as m.

18) r varies directly as d and inversely as the square of L.

 19) Suppose z varies directly as x. If $z = 63$ when $x = 7$,

   a) find the constant of variation.

   b) write the specific variation equation relating z and x.

   c) find z when $x = 6$.

20) Suppose A varies directly as D. If $A = 12$ when $D = 3$,

   a) find the constant of variation.

   b) write the specific variation equation relating A and D.

   c) find A when $D = 11$.

21) Suppose N varies inversely as y. If $N = 4$ when $y = 12$,

   a) find the constant of variation.

   b) write the specific variation equation relating N and y.

   c) find N when $y = 3$.

22) Suppose j varies inversely as m. If $j = 7$ when $m = 9$,

   a) find the constant of variation.

   b) write the specific variation equation relating j and m.

   c) find j when $m = 21$.

 23) Suppose Q varies directly as the square of r and inversely as w. If $Q = 25$ when $r = 10$ and $w = 20$,

   a) find the constant of variation.

   b) write the specific variation equation relating Q, r, and w.

   c) find Q when $r = 6$ and $w = 4$.

24) Suppose y varies jointly as a and the square root of b. If $y = 42$ when $a = 3$ and $b = 49$,

   a) find the constant of variation.

   b) write the specific variation equation relating y, a, and b.

   c) find y when $a = 4$ and $b = 9$.

Solve.

25) If B varies directly as R, and $B = 35$ when $R = 5$, find B when $R = 8$.

26) If q varies directly as p, and $q = 10$ when $p = 4$, find q when $p = 10$.

27) If L varies inversely as the square of h, and $L = 8$ when $h = 3$, find L when $h = 2$.

28) If w varies inversely as d, and $w = 3$ when $d = 10$, find w when $d = 5$.

 29) If y varies jointly as x and z, and $y = 60$ when $x = 4$ and $z = 3$, find y when $x = 7$ and $z = 2$.

30) If R varies directly as P and inversely as the square of Q, and $R = 5$ when $P = 10$ and $Q = 4$, find R when $P = 18$ and $Q = 3$.

Solve each problem by writing a variation equation.

31) Kosta is paid hourly at his job. His weekly earnings vary directly as the number of hours worked. If Kosta earned $437.50 when he worked 35 hr, how much would he earn if he worked 40 hr?

32) The cost of manufacturing a certain brand of spiral notebook is inversely proportional to the number produced. When 16,000 notebooks are produced, the cost per notebook is $0.80. What is the cost of each notebook when 12,000 are produced?

33) If distance is held constant, the time it takes to travel that distance is inversely proportional to the speed at which one travels. If it takes 14 hr to travel the given distance at 60 mph, how long would it take to travel the same distance at 70 mph?

34) The surface area of a cube varies directly as the square of the length of one of its sides. A cube has a surface area of 54 in^2 when the length of each side is 3 in. What is the surface area of a cube with a side of length 6 in.?

35) The power in an electrical system varies jointly as the current and the square of the resistance. If the power is 100 watts when the current is 4 amps and the resistance is 5 ohms, what is the power when the current is 5 amps and the resistance is 6 ohms?

36) The force exerted on an object varies jointly as the mass and acceleration of the object. If a 20-Newton force is exerted on an object of mass 10 kg and an acceleration of 2 m/sec^2, how much force is exerted on a 50 kg object with an acceleration of 8 m/sec^2?

37) The kinetic energy of an object varies jointly as its mass and the square of its speed. When a roller coaster car with a mass of 1000 kg is traveling at 15 m/sec, its kinetic energy is 112,500 J (joules). What is the kinetic energy of the same car when it travels at 18 m/sec?

38) The volume of a cylinder varies jointly as its height and the square of its radius. The volume of a cylindrical can is 108π cm^3 when its radius is 3 cm and it is 12 cm high. Find the volume of a cylindrical can with a radius of 4 cm and a height of 3 cm.

39) The frequency of a vibrating string varies inversely as its length. If a 5-ft-long piano string vibrates at 100 cycles/sec, what is the frequency of a piano string that is 2.5 ft long?

40) The amount of pollution produced varies directly as the population. If a city of 500,000 people produces 800,000 tons of pollutants, how many tons of pollutants would be produced by a city of 1,000,000 people?

41) The resistance of a wire varies directly as its length and inversely as its cross-sectional area. A wire of length 40 cm and cross-sectional area 0.05 cm^2 has a resistance of 2 ohms. Find the resistance of 60 cm of the same type of wire.

42) When a rectangular beam is positioned horizontally, the maximum weight that it can support varies jointly as its width and the square of its thickness and inversely as its length. A beam is $\frac{3}{4}$ ft wide, $\frac{1}{3}$ ft thick, and 8 ft long, and it can support 17.5 tons. How much weight can a similar beam support if it is 1 ft wide, $\frac{1}{2}$ ft thick and 12 ft long?

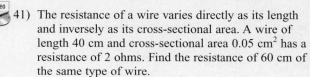

43) Hooke's law states that the force required to stretch a spring is proportional to the distance that the spring is stretched from its original length. A force of 200 lb is required to stretch a spring 5 in. from its natural length. How much force is needed to stretch the spring 8 in. beyond its natural length?

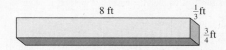

5 in.

8 in.

44) The weight of an object on Earth varies inversely as the square of its distance from the center of the earth. If an object weighs 300 lb on the surface of the earth (4000 mi from the center), what is the weight of the object if it is 800 mi above the earth? (Round to the nearest pound.)

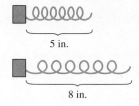

Definition/Procedure	Example	Reference
Relations and Functions		**12.1**
A *relation* is any set of ordered pairs. A relation can also be represented as a correspondence or mapping from one set to another or as an equation.	a) $\{(-6, 2), (-3, 1), (0, 0), (9, -3)\}$ b) $y^2 = x$	**p. 772**
The *domain* of a relation is the set of values of the independent variable (the first coordinates in the set of ordered pairs). The *range* of a relation is the set of all values of the dependent variable (the second coordinates in the set of ordered pairs). If a relation is written as an equation so that y is in terms of x, then the *domain* is the set of all real numbers that can be substituted for the independent variable, x. The resulting set of real numbers that are obtained for y, the dependent variable, is the *range*.	In a) above, the domain is $\{-6, -3, 0, 9\}$, and the range is $\{-3, 0, 1, 2\}$. In b) above, the domain is $[0, \infty)$, and the range is $(-\infty, \infty)$.	**p. 772**
A *function* is a relation in which each element of the domain corresponds to *exactly one* element of the range.	The relation in a) *is* a function. The relation in b) *is not* a function since there are elements of the domain that correspond to more than one element in the range. For example, if $x = 4$, then $y = 2$ or $y = -2$.	**p. 772**
$y = f(x)$ is called *function notation*, and it is read as, "*y* equals *f* of *x*."	Let $f(x) = 7x + 2$. Find $f(-2)$. $\begin{aligned} f(-2) &= 7(-2) + 2 & \text{Substitute } -2 \text{ for } x. \\ &= -14 + 2 & \text{Multiply.} \\ &= -12 & \text{Add.} \end{aligned}$	**p. 774**
A *linear function* has the form $f(x) = mx + b$, where m and b are real numbers, m is the *slope* of the line, and $(0, b)$ is the *y-intercept*.	Given $f(x) = -3x + 1$, determine the domain of f and graph the function. To determine the domain, ask yourself, "*Is there any number that cannot be substituted for x* in $f(x) = -3x + 1$?" *No.* Any number can be substituted for x and the function will be defined. Therefore, the domain consists of all real numbers. This is written as $(-\infty, \infty)$. This is a linear function, so its graph is a line.	**p. 775**

Definition/Procedure	Example	Reference
Polynomial Functions The domain of a polynomial function is all real numbers, $(-\infty, \infty)$, since any real number can be substituted for the variable.	$f(x) = 2x^3 - 9x^2 + 8x + 4$ is an example of a *polynomial function* since $2x^3 - 9x^2 + 8x + 4$ is a polynomial.	**p. 775**
Rational Functions The *domain of a rational function* consists of all real numbers *except* the value(s) of the variable that make the denominator equal zero.	$g(x) = \dfrac{7x}{x - 2}$ is an example of a *rational function* since $\dfrac{7x}{x - 2}$ is a rational expression. To determine the domain, ask yourself, *"Is there any number that cannot be substituted for x in* $g(x) = \dfrac{7x}{x - 2}$?" Yes. We cannot substitute 2 for x because the denominator would equal zero so the function would be undefined. The domain of $g(x)$ is $(-\infty, 2) \cup (2, \infty)$.	**p. 777**
Square Root Functions When we determine the domain of a square root function, we are determining all *real* numbers that may be substituted for the variable so that the range contains only *real* numbers. Therefore, a square root function is defined only when its radicand is nonnegative. This means that there may be *many* values that are excluded from the domain of a square root function.	$h(x) = \sqrt{x + 8}$ is an example of a *square root function*. Determine the domain of $h(x) = \sqrt{x + 8}$. Solve the inequality $x + 8 \geq 0$. $x + 8 \geq 0$ The value of the radicand must be ≥ 0. $\quad x \geq -8$ Solve. The domain is $[-8, \infty)$.	**p. 779**
Graphs of Functions and Transformations		**12.2**
Vertical Shifts Given the graph of $f(x)$, if $g(x) = f(x) + k$, where k is a constant, then the graph of $g(x)$ will be the same shape as the graph of $f(x)$, but the graph of g will be shifted *vertically* k units.	The graph of $g(x) = |x| + 2$ will be the same shape as the graph of $f(x) = |x|$, but $g(x)$ is shifted *up* 2 units. Functions f and g are *absolute value functions*.	**p. 783**

Definition/Procedure	Example	Reference

Horizontal Shifts

Given the graph of $f(x)$, if $g(x) = f(x - h)$, where h is a constant, the graph of $g(x)$ will be the same shape as the graph of $f(x)$, but the graph of g will be shifted *horizontally* h units.

The graph of $g(x) = (x + 3)^2$ will be the same shape as the graph of $f(x) = x^2$, but g is shifted *left* 3 units.

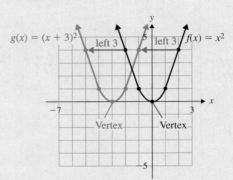

Functions f and g are *quadratic functions,* and their graphs are called *parabolas.*

p. 784

Reflection About the *x*-Axis

Given the graph of $f(x)$, if $g(x) = -f(x)$, then the graph of $g(x)$ will be the *reflection of the graph of f about the x-axis.* That is, obtain the graph of g by keeping the *x*-coordinate of each point on the graph of f the same but take the negative of the *y*-coordinate.

Let $f(x) = \sqrt{x}$ and $g(x) = -\sqrt{x}$. The graph of $g(x)$ will be the reflection of $f(x)$ about the *x*-axis.

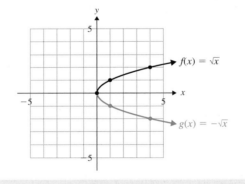

p. 786

A *piecewise function* is a single function defined by two or more different rules.

$$f(x) = \begin{cases} x + 3, & x > -2 \\ -\dfrac{1}{2}x + 2, & x \le -2 \end{cases}$$

p. 787

The *greatest integer function,* $f(x) = [\![x]\!]$, represents the largest integer less than or equal to x.

$[\![8.3]\!] = 8$

$\left[\!\left[-4\dfrac{3}{8} \right]\!\right] = -5$

p. 789

Definition/Procedure	Example	Reference

Quadratic Functions and Their Graphs

12.3

A *quadratic function* is a function that can be written in the form $f(x) = ax^2 + bx + c$, where a, b, and c are real numbers and $a \neq 0$. The graph of a quadratic function is called a *parabola*.

$f(x) = 5x^2 + 7x - 9$

p. 797

A quadratic function can also be written in the form $f(x) = a(x - h)^2 + k$.

From this form we can learn a great deal of information.

1) The vertex of the parabola is (h, k).
2) The axis of symmetry is the vertical line with equation $x = h$.
3) If a is positive, the parabola opens upward. If a is negative, the parabola opens downward.
4) If $|a| < 1$, then the graph of $f(x) = a(x - h)^2 + k$ is *wider* than the graph of $y = x^2$.

If $|a| > 1$, then the graph of $f(x) = a(x - h)^2 + k$ is *narrower* than the graph of $y = x^2$.

Graph $f(x) = -(x + 3)^2 + 4$.

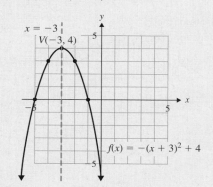

Vertex: $(-3, 4)$

Axis of symmetry: $x = -3$

$a = -1$, so graph opens downward.

p. 798

When a quadratic function is written in the form $f(x) = ax^2 + bx + c$, there are two methods we can use to graph the function.

Method 1: Rewrite $f(x) = ax^2 + bx + c$ in the form $f(x) = a(x - h)^2 + k$ by *completing the square*.

Method 2: Use the formula $h = -\dfrac{b}{2a}$ to find the x-coordinate of the vertex. Then, the vertex has coordinates
$$\left(-\frac{b}{2a}, f\left(-\frac{b}{2a}\right)\right).$$

Graph $f(x) = x^2 + 4x + 5$.

Method 1: Complete the square

$f(x) = x^2 + 4x + 5$
$f(x) = (x^2 + 4x + 2^2) + 5 - 2^2$
$f(x) = (x^2 + 4x + 4) + 5 - 4$
$f(x) = (x + 2)^2 + 1$

The vertex of the parabola is $(-2, 1)$.

Method 2: Use the formula $h = -\dfrac{b}{2a}$.

$h = -\dfrac{4}{2(1)} = -2$. Then, $f(-2) = 1$.

The vertex of the parabola is $(-2, 1)$.

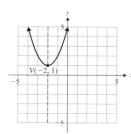

p. 800

Definition/Procedure	Example	Reference

12.4

Applications of Quadratic Functions and Graphing Other Parabolas

Maximum and Minimum Values of a Quadratic Function

Let $f(x) = ax^2 + bx + c$.

1) If a is *positive*, the y-coordinate of the vertex is the *minimum* value of the function $f(x)$.

2) If a is *negative*, the y-coordinate of the vertex is the *maximum* value of the function $f(x)$.

Find the minimum value of the function $f(x) = 2x^2 + 12x + 7$.

$a = 2$. Since a is positive, the function attains its minimum value at the vertex.

x-coordinate of vertex: $h = -\dfrac{b}{2a} = -\dfrac{12}{2(2)} = -3$

y-coordinate of vertex:
$$f(-3) = 2(-3)^2 + 12(-3) + 7$$
$$= 18 - 36 + 7 = -11$$

The minimum value of the function is -11.

p. 808

The graph of the quadratic equation $x = ay^2 + by + c$ is a parabola that opens in the x-direction, or horizontally.

The quadratic equation $x = ay^2 + by + c$ can also be written in the form $x = a(y - k)^2 + h$. When it is written in this form we can find the following.

1) The vertex of the parabola is (h, k).

2) The axis of symmetry is the horizontal line $y = k$.

3) If a is positive, the graph opens to the right.

If a is negative, the graph opens to the left.

Graph $x = \dfrac{1}{2}(y + 4)^2 - 2$.

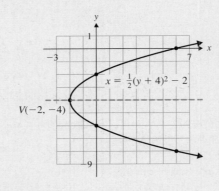

$x = \frac{1}{2}(y+4)^2 - 2$

$V(-2, -4)$

Vertex: $(-2, -4)$

Axis of symmetry: $y = -4$

$a = \dfrac{1}{2}$, so the graph opens to the right.

p. 813

The Algebra of Functions

12.5

Operations Involving Functions

Given the functions $f(x)$ and $g(x)$, we can find their sum, difference, product, and quotient.

Let $f(x) = 5x - 1$ and $g(x) = x + 4$.

$$(f + g)(x) = f(x) + g(x)$$
$$= (5x - 1) + (x + 4)$$
$$= 6x + 3$$

p. 821

Definition/Procedure	Example	Reference
Composition of Functions Given the functions $f(x)$ and $g(x)$, the *composition function* $f \circ g$ (read "*f* of *g*") is defined as $$(f \circ g)(x) = f(g(x))$$ where $g(x)$ is in the domain of f.	$f(x) = 4x - 10$ and $g(x) = -3x + 2$. $$\begin{aligned}(f \circ g)(x) &= f(g(x))\\ &= f(-3x + 2)\\ &= 4(-3x + 2) - 10 \quad \text{Substitute } -3x+2 \\ &= -12x + 8 - 10 \quad \text{for } x \text{ in } f(x).\\ &= -12x - 2\end{aligned}$$	**p. 823**
Variation		**12.6**
Direct Variation *y varies directly as x* (or *y is directly proportional to x*) means $$y = kx$$ where k is a nonzero real number. k is called the *constant of variation*.	The circumference, C, of a circle is given by $C = 2\pi r$. C varies directly as r where $k = 2\pi$.	**p. 828**
Inverse Variation *y varies inversely as x* (or *y is inversely proportional to x*) means $$y = \frac{k}{x}$$ where k is a nonzero real number.	The time, t (in hours), it takes to drive 600 mi is inversely proportional to the rate, r, at which you drive. $$t = \frac{600}{r}$$ The constant of variation, k, equals 600.	**p. 831**
Joint Variation *y varies jointly as x and z* means $y = kxz$ where k is a nonzero real number.	For a given amount, called the principal, deposited in a bank account, the interest earned, I, varies jointly as the interest rate, r, and the time, t, the principal is in the account. $$I = 1000rt$$ The constant of variation, k, equals 1000. This is the amount of money deposited in the account, the principal.	**p. 833**

Definition/Procedure	Example	Reference

Steps for Solving a Variation Problem

Step 1: Write the *general* variation equation.

Step 2: Find k by substituting the known values into the equation and solving for k.

Step 3: Write the *specific* variation equation by substituting the value of k into the *general* variation equation.

Step 4: Substitute the given values into the *specific* equation to find the required value.

The cost of manufacturing a certain soccer ball is inversely proportional to the number produced. When 15,000 are made, the cost per ball is $4.00. What is the cost to manufacture each soccer ball when 25,000 are produced?

Let n = number of soccer balls produced
C = cost of producing each ball

Step 1: *General* variation equation: $C = \dfrac{k}{n}$

Step 2: Find k using $C = 4$ when $n = 15,000$.

$$4 = \frac{k}{15,000}$$
$$60,000 = k$$

Step 3: *Specific* variation equation: $C = \dfrac{60,000}{n}$

Step 4: Find the cost, C, per ball when $n = 25,000$.

$$C = \frac{60,000}{25,000} \qquad \text{Substitute 25,000 for } n.$$
$$= 2.4$$

The cost per ball is $2.40.

p. 829

(12.1) Identify the domain and range of each relation, and determine whether each relation is a function.

1) $\{(-7, -4), (-5, -1), (2, 3), (2, 5), (4, 9)\}$

2)

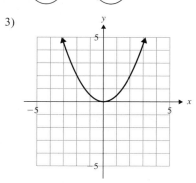

3)

4) $y = 6x - 4$

Determine whether each relation describes y as a function of x.

5) $y = -\dfrac{5}{2}x + 3$

6) $y = \dfrac{x + 4}{2x - 3}$

7) $x = |y|$

8) $y = |x|$

9) $y = -\sqrt{3x - 8}$

10) $y^2 = x - 1$

11) $y = x^2 - 10x + 7$

12) $y = \sqrt{-x}$

Graph each function.

13) $f(x) = 2x - 5$

14) $h(x) = -\dfrac{1}{3}x + 4$

15) $g(t) = t$

16) $r(a) = 2$

17) Let $f(x) = -8x + 3$ and $g(x) = x^2 + 7x - 12$. Find each of the following and simplify.

   a) $f(5)$

   b) $f(-4)$

   c) $g(-2)$

   d) $g(3)$

   e) $f(c)$

   f) $g(r)$

   g) $f(p - 3)$

   h) $g(t + 4)$

18) $h(x) = 3x + 10$. Find x so that $h(x) = -8$.

19) $k(x) = -\dfrac{2}{3}x + 8$. Find x so that $k(x) = 0$.

20) $g(x) = x^2 + 3x - 28$. Find x so that $g(x) = 0$.

21) $p(x) = x^2 - 8x + 15$. Find x so that $p(x) = 3$.

22) How do you find the domain of a

   a)   square root function?

   b)   rational function?

Determine the domain of each function.

23) $f(x) = \dfrac{9x}{x - 5}$

24) $P(a) = 3a^4 - 7a^3 + 12a^2 + a - 8$

25) $C(n) = -4n - 9$

26) $g(x) = \sqrt{x}$

27) $h(t) = \sqrt{5t - 7}$

28) $R(c) = \dfrac{c + 8}{2c + 3}$

29) $g(x) = \dfrac{6}{x}$

30) $f(p) = \sqrt{3 - 4p}$

31) $k(c) = c^2 + 11c + 2$

32) $Q(x) = -\dfrac{4x + 1}{10}$

33) $h(a) = \dfrac{a + 6}{a^2 - 7a - 8}$

34) $s(r) = \sqrt{r + 12}$

35) To rent a car for one day, a company charges its customers a flat fee of \$26 plus \$0.20 per mile. This can be described by the function $C(m) = 0.20m + 26$, where m is the number of miles driven and C is the cost of renting a car, in dollars.

   a)   What is the cost of renting a car that is driven 30 mi?

   b)   What is the cost of renting a car that is driven 100 mi?

   c)   If a customer paid \$56 to rent a car, how many miles did she drive?

   d)   If a customer paid \$42 to rent a car, how many miles did he drive?

36) The area, A, of a square is a function of the length of its side, s.

   a)   Write an equation using function notation to describe this relationship between A and s.

   b)   If the length of a side is given in inches, find $A(4)$ and explain what this means in the context of the problem.

   c)   If the length of a side is given in feet, find $A(1)$ and explain what this means in the context of the problem.

   d)   What is the length of each side of a square that has an area of 49 cm^2?

(12.2) Graph each function, and identify the domain and range.

37) $f(x) = \sqrt{x}$

38) $g(x) = |x|$

39) $h(x) = (x + 4)^2$

40) $f(x) = \sqrt{x} - 3$

41) $k(x) = -|x| + 5$

42) $h(x) = |x - 1| + 2$

43) $g(x) = \sqrt{x - 2} - 1$

44) $f(x) = \frac{1}{2}\sqrt{x}$

Graph each piecewise function.

45) $f(x) = \begin{cases} -\dfrac{1}{2}x - 2, & x \le 2 \\ x - 3, & x > 2 \end{cases}$

46) $g(x) = \begin{cases} 1, & x < -3 \\ x + 4, & x \ge -3 \end{cases}$

Let $f(x) = [\![x]\!]$. Find the following function values.

47) $f\left(7\dfrac{2}{3}\right)$

48) $f(2.1)$

49) $f\left(-8\dfrac{1}{2}\right)$

50) $f(-5.8)$

51) $f\left(\dfrac{3}{8}\right)$

Graph each greatest integer function.

52) $f(x) = [\![x]\!]$

53) $g(x) = \left[\!\!\left[\dfrac{1}{2}x\right]\!\!\right]$

54) To mail a letter from the United States to Mexico in 2005 cost $0.60 for the first ounce, $0.85 over one ounce but less than or equal to 2 oz, then $0.40 for each additional ounce or fraction of an ounce. Let $C(x)$ represent the cost of mailing a letter from the U.S. to Mexico, and let x represent the weight of the letter, in ounces. Graph $C(x)$ for any letter weighing up to (and including) 5 oz. (www.usps.com)

If the following transformations are performed on the graph of $f(x)$ to obtain the graph of $g(x)$, write the equation of $g(x)$.

55) $f(x) = |x|$ is shifted right 5 units.

56) $f(x) = \sqrt{x}$ is shifted left 2 units and up 1 unit.

(12.3 and 12.4)

57) Given a quadratic function in the form $f(x) = a(x - h)^2 + k$, answer the following.

   a)   What is the vertex?

   b)   What is the equation of the axis of symmetry?

   c)   What does the sign of a tell us about the graph of f?

58) What are two ways to find the vertex of the graph of $f(x) = ax^2 + bx + c$?

59) Given a quadratic equation of the form $x = a(y - k)^2 + h$, answer the following.

   a)   What is the vertex?

   b)   What is the equation of the axis of symmetry?

   c)   What does the sign of a tell us about the graph of the equation?

60) What are two ways to find the vertex of the graph of $x = ay^2 + by + c$?

For each quadratic equation, identify the vertex, axis of symmetry, and x- and y-intercepts. Then, graph the equation.

61) $f(x) = (x + 2)^2 - 1$

62) $g(x) = -\dfrac{1}{2}(x - 3)^2 - 2$

63) $x = -y^2 - 1$

64) $y = 2x^2$

65) $x = -(y - 3)^2 + 11$

66) $x = (y + 1)^2 - 5$

Rewrite each equation in the form $f(x) = a(x - h)^2 + k$ or $x = a(y - k)^2 + h$ by completing the square. Then, graph the function. Include the intercepts.

67) $x = y^2 + 8y + 7$

68) $f(x) = -2x^2 - 8x + 2$

69) $y = \dfrac{1}{2}x^2 - 4x + 9$

70) $x = -y^2 + 4y - 4$

Graph each equation using the vertex formula. Include the intercepts.

71) $f(x) = x^2 - 2x - 4$

72) $x = 3y^2 - 12y$

73) $x = -\dfrac{1}{2}y^2 - 3y - \dfrac{5}{2}$

74) $y = -x^2 - 6x - 10$

Solve.

75) An object is thrown upward from a height of 240 ft so that its height h (in feet) t sec after being thrown is given by

$$h(t) = -16t^2 + 32t + 240$$

   a)   How long does it take the object to reach its maximum height?

   b)   What is the maximum height attained by the object?

   c)   How long does it take the object to hit the ground?

76) A restaurant wants to add outdoor seating to its inside service. It has 56 ft of fencing to enclose a rectangular, outdoor café. Find the dimensions of the outdoor café of maximum area if the building will serve as one side of the café.

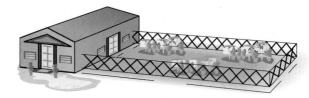

(12.5) Let $f(x) = 5x + 2$, $g(x) = -x + 4$, $h(x) = 3x^2 - 7$, and $k(x) = x^2 - 7x - 8$. Find each of the following.

77) $(f + g)(x)$

78) $(h - k)(x)$

79) $(g - h)(2)$

80) $(f + k)(-3)$

81) $(fg)(x)$

82) $(gk)(1)$

For each pair of functions, find a) $\left(\dfrac{f}{g}\right)(x)$ and b) $\left(\dfrac{f}{g}\right)(3)$.

Identify any values that are not in the domain of $\left(\dfrac{f}{g}\right)(x)$.

83) $f(x) = 6x - 5$, $g(x) = x + 4$

84) $f(x) = 3x^2 - 5x + 2$, $g(x) = 3x - 2$

85) $R(x) = 20x$ is the revenue function for the sale of x children's soccer uniforms, in dollars. The cost to produce x soccer uniforms, in dollars, is

$$C(x) = 14x + 400$$

a) Find the profit function, $P(x)$, that describes the profit from the sale of x uniforms.

b) What is the profit from the sale of 200 uniforms?

86) Let $f(x) = x + 6$ and $g(x) = 2x - 9$. Find

a) $(g \circ f)(x)$

b) $(f \circ g)(x)$

c) $(f \circ g)(5)$

87) Let $g(x) = 4x - 3$ and $h(x) = x^2 + 5$. Find

a) $(g \circ h)(x)$

b) $(h \circ g)(x)$

c) $(h \circ g)(-2)$

88) Let $h(x) = 2x - 1$ and $k(x) = x^2 + 5x - 4$. Find

a) $(k \circ h)(x)$

b) $(h \circ k)(x)$

c) $(h \circ k)(-3)$

89) Antoine's gross weekly pay, G, in terms of the number of hours, h, he worked is given by $G(h) = 12h$. His net weekly pay, N, in terms of his gross pay is given by $N(G) = 0.8G$.

a) Find $(N \circ G)(h)$ and explain what it represents.

b) Find $(N \circ G)(30)$ and explain what it represents.

c) What is his net pay if he works 40 hr in 1 week?

(12.6)

90) Suppose c varies directly as m. If $c = 56$ when $m = 8$, find c when $m = 3$.

91) Suppose A varies jointly as t and r. If $A = 15$ when $t = \dfrac{1}{2}$ and $r = 5$, find A when $t = 3$ and $r = 4$.

92) Suppose p varies directly as n and inversely as the square of d. If $p = 42$ when $n = 7$ and $d = 2$, find p when $n = 12$ and $d = 3$.

Solve each problem by writing a variation equation.

93) The weight of a ball varies directly as the cube of its radius. If a ball with a radius of 2 in. weighs 0.96 lb, how much would a ball made out of the same material weigh if it had a radius of 3 in.?

94) If the temperature remains the same, the volume of a gas is inversely proportional to the pressure. If the volume of a gas is 10 L (liters) at a pressure of 1.25 atm (atmospheres), what is the volume of the gas at 2 atm?

1) What is a function?

2) Given the relation $\{(-8, -1), (2, 3), (5, 3), (7, 10)\}$,

   a)   Determine the domain.

   b)   Determine the range.

   c)   Is this a function?

3) For $y = \sqrt{3x + 7}$,

   a)   Determine the domain.

   b)   Is y a function of x?

Determine the domain of each function.

4) $f(x) = \dfrac{4}{9}x - 2$

5) $g(t) = \dfrac{t + 10}{7t - 8}$

Let $f(x) = 4x + 3$ and $g(x) = x^2 - 6x + 10$. Find each of the following and simplify.

6) $g(4)$

7) $f(c)$

8) $f(n - 7)$

9) $g(k + 5)$

10) Let $h(x) = -2x + 6$. Find x so that $h(x) = 9$.

11) A garden supply store charges $50 per cubic yard plus a $60 delivery fee to deliver cedar mulch. This can be described by the function $C(m) = 50m + 60$ where m is the amount of cedar mulch delivered, in cubic yards, and C is the cost, in dollars.

   a)   Find $C(3)$ and explain what it means in the context of the problem.

   b)   If a customer paid $360 to have cedar mulch delivered to his home, how much did he order?

Graph each function and identify the domain and range.

12) $f(x) = |x| - 4$

13) $g(x) = \sqrt{x + 3}$

14) $h(x) = -\dfrac{1}{3}x - 1$

15) Graph $f(x) = \begin{cases} x + 3, & x > -1 \\ -2x - 5, & x \le -1 \end{cases}$

Graph each equation. Identify the vertex, axis of symmetry, and intercepts.

16) $f(x) = -(x + 2)^2 + 4$

17) $x = y^2 - 3$

18) $x = 3y^2 - 6y + 5$

19) $g(x) = x^2 - 6x + 8$

20) A rock is thrown upward from a cliff so that it falls into the ocean below. The height h (in feet) of the rock t sec after being thrown is given by

$$h(t) = -16t^2 + 64t + 80$$

   a)   What is the maximum height attained by the rock?

   b)   How long does it take the rock to hit the water?

Let $f(x) = 2x + 7$ and $g(x) = x^2 + 5x - 3$. Find each of the following.

21) $(g - f)(x)$

22) $(f + g)(-1)$

23) $(f \circ g)(x)$

24) $(g \circ f)(x)$

25) Suppose n varies jointly as r and the square of s. If $n = 72$ when $r = 2$ and $s = 3$, find n when $r = 3$ and $s = 5$.

26) The loudness of a sound is inversely proportional to the square of the distance between the source of the sound and the listener. If the sound level measures 112.5 decibels (dB) 4 ft from a speaker, how loud is the sound 10 ft from the speaker?

What property is illustrated by each statement? Choose from commutative, associative, distributive, inverse, and identity.

1) $4 \cdot \dfrac{1}{4} = 1$

2) $16 + 5 = 5 + 16$

Evaluate.

3) $\left(\dfrac{1}{2}\right)^5$

4) 5^{-3}

5) 10^0

6) Solve $6(2y + 1) - 4y = 5(y + 2)$.

7) Solve the compound inequality

$$x + 8 \le 6 \text{ or } 1 - 2x \le -5$$

Graph the solution set and write the answer in interval notation.

8) Write the equation of the line parallel to $4x + 3y = 15$ containing the point $(-5, 6)$. Express it in standard form.

9) Solve this system using any method.

$$x - \dfrac{1}{4}y = \dfrac{5}{2}$$
$$\dfrac{1}{2}x + \dfrac{1}{3}y = \dfrac{13}{6}$$

10) Multiply and simplify $(p - 8)(p + 7)$.

11) Divide $\dfrac{12r - 40r^2 + 6r^3 + 4}{4r^2}$.

Factor completely.

12) $k^2 - 15k + 54$

13) $100 - 9m^2$

14) Solve $(a + 4)(a + 1) = 18$.

15) Divide $\dfrac{c - 8}{2c^2 - 5c - 12} \div \dfrac{3c - 24}{c^2 - 16}$.

16) Solve $\dfrac{4x + 2}{x + 5} = 10$.

17) Solve $|7y + 6| \le -8$.

18) Graph the compound inequality

$$y \le -\dfrac{1}{2}x + 4 \text{ and } 2x - y \le 2$$

Simplify. Assume all variables represent nonnegative real numbers.

19) $\sqrt{60}$

20) $\sqrt[4]{16}$

21) $\sqrt{18c^6d^{11}}$

22) $(100)^{-3/2}$

23) Add $\sqrt{12} + \sqrt{3} + \sqrt{48}$.

24) Solve $x = -6 + \sqrt{x + 8}$.

25) Divide $\dfrac{4 - 2i}{2 + 3i}$.

Solve.

26) $(3n - 4)^2 + 9 = 0$

27) $4(y^2 + 2y) = 5$

28) $q - 8q^{1/2} + 7 = 0$

29) Let $g(x) = x + 1$ and $h(x) = x^2 + 4x + 3$.

   a) Find $g(7)$.

   b) Find $\left(\dfrac{h}{g}\right)(x)$. Identify any values that are not in the domain of $\left(\dfrac{h}{g}\right)(x)$.

   c) Find x so that $g(x) = 5$.

   d) Find $(g \circ h)(x)$.

30) Graph $f(x) = -x^2 + 4$ and identify the domain and range.

31) Graph $x = -y^2 - 2y - 3$.

32) Suppose y varies inversely as the square of x. If $y = 12$ when $x = 2$, find y when $x = 4$.

CHAPTER 13

Inverse, Exponential, and Logarithmic Functions

Algebra at Work: Finances

Financial planners use mathematical formulas involving exponential functions every day to help clients invest their money. Interest on investments can be paid in various ways—monthly, weekly, continuously, etc.

Isabella has a client who wants to invest $5000 for 4 yr. He would like to know if it is better to invest it at 4.8% compounded continuously or at 5% compounded monthly.

To determine how much the investment will be worth after 4 yr if it is put in the account with interest compounded continuously, Isabella uses the formula $A = Pe^{rt}$ ($e \approx 2.71828$). To determine how much money her client will have after 4 yr if he puts his $5000 into the account earning 5% interest compounded monthly, Isabella uses the formula $A = P\left(1 + \dfrac{r}{n}\right)^{nt}$.

Compounded Continuously

$$A = Pe^{rt}$$

$$A = 5000e^{(0.048)(4)}$$

$$A = \$6058.35$$

Compounded Monthly

$$A = P\left(1 + \frac{r}{n}\right)^{nt}$$

$$A = (5000)\left(1 + \frac{0.05}{12}\right)^{(12)(4)}$$

$$A = \$6104.48$$

If her client invests $5000 in the account paying 4.8% interest compounded continuously, his investment will grow to $6058.35 after 4 yr. If he puts the money in the account paying 5% compounded monthly, he will have $6104.48. Isabella advises him to invest his money in the 5% account.

We will work with these and other exponential functions in this chapter.

Section 13.1 Inverse Functions

Objectives

1. Decide if a Function Is One-to-One
2. Use the Horizontal Line Test to Determine if a Function Is One-to-One
3. Find the Inverse of a One-to-One Function
4. Given the Graph of $f(x)$, Graph $f^{-1}(x)$
5. Given the Equations of $f(x)$ and $f^{-1}(x)$, Show That $(f^{-1} \circ f)(x) = x$ and $(f \circ f^{-1})(x) = x$

In this chapter, we will study inverse functions and two very useful types of functions in mathematics: exponential and logarithmic functions. But, before we can begin our study of exponential functions (Section 13.2) and logarithmic functions (Section 13.3), we must learn about one-to-one and inverse functions. This is because exponential and logarithmic functions are related in a special way: they are *inverses* of one another.

One-to-One Functions

1. Decide if a Function Is One-to-One

Recall from Sections 4.7 and 12.1 that a relation is a *function* if each x-value corresponds to exactly one y-value. Let's look at two functions, f and g.

$$f = \{(1, -3), (2, -1), (4, 3), (7, 9)\} \quad g = \{(0, 3), (1, 4), (-1, 4), (2, 7)\}$$

In functions f and g, each x-value corresponds to exactly one y-value. That is why they are functions. In function f, each *y-value also corresponds to exactly one x-value*. Therefore, f is a *one-to-one function*. In function g, however, each y-value does *not* correspond to exactly one x-value. (The y-value of 4 corresponds to $x = 1$ and $x = -1$.) Therefore, g is *not* a one-to-one function.

Definition	In order for a function to be a **one-to-one function**, each x-value corresponds to exactly one y-value, and each y-value corresponds to exactly one x-value.

Alternatively we can say that a function is one-to-one if each value in its domain corresponds to exactly one value in its range *and* if each value in its range corresponds to exactly one value in its domain.

Example 1

Determine whether each function is one-to-one.

a) $f = \{-1, 9), (1, -3), (2, -6), (4, -6)\}$

b) $g = \{(-3, 13), (-1, 5), (5, -19), (8, -31)\}$

c)

State	Number of Representatives in U.S. House of Representatives (2005)
Alaska	1
California	52
Connecticut	5
Delaware	1
Ohio	18

d)

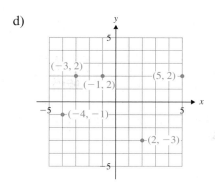

Solution

a) *f* is *not* a one-to-one function since the *y*-value −6 corresponds to two different *x*-values: (2, −6) and (4, −6).

b) *g is* a one-to-one function since each *y*-value corresponds to exactly one *x*-value.

c) The information in the table does *not* represent a one-to-one function since the value 1 in the range corresponds to two different values in the domain, Alaska and Delaware.

d) The graph does *not* represent a one-to-one function since three points have the same *y*-value: (−3, 2), (−1, 2), and (5, 2).

 You Try 1

Determine whether each function is one-to-one.

a) $f = \{(-2, -13), (0, -7), (4, 5), (5, 8)\}$ b) $g = \{(-4, 2), (-1, 1), (0, 2), (3, 5)\}$

c)

Element	Atomic Mass (in amu)
Hydrogen	1.00794
Lithium	6.941
Sulfur	32.066
Lead	207.2

d)

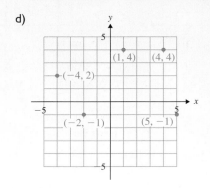

2. Use the Horizontal Line Test to Determine if a Function Is One-to-One

Just as we can use the vertical line test to determine if a graph represents a function, we can use the *horizontal line test* to determine if a function is one-to-one.

Definition	
	Horizontal Line Test: If every horizontal line that could be drawn through a function would intersect the graph at most once, then the function is one-to-one.

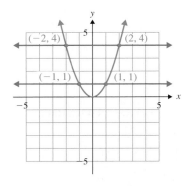

Look at the graph of the given function. We can see that if a horizontal line intersects the graph more than once, then one *y*-value corresponds to more than one *x*-value. This means that the function is not one-to-one. For example, the *y*-value of 1 corresponds to $x = 1$ and $x = -1$.

Example 2	

Determine whether each graph represents a one-to-one function.

a)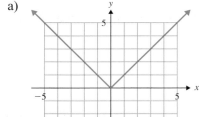

b)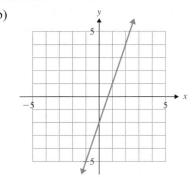

Solution

a) *Not* one-to-one. It is possible to draw a horizontal line through the graph so that it intersects the graph more than once.

b) *Is* one-to-one. Every horizontal line that could be drawn through the graph would intersect the graph at most once.

You Try 2

Determine whether each graph represents a one-to-one function.

a)

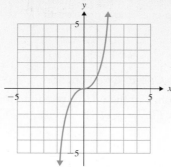

b)

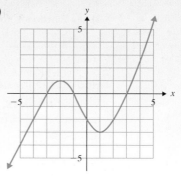

Inverse Functions

3. Find the Inverse of a One-to-One Function

We began this section with the discussion of one-to-one functions because they lead to other special functions—inverse functions. *A one-to-one function has an inverse function.*

To find the inverse of a one-to-one function, we interchange the coordinates of the ordered pairs.

Example 3

Find the inverse function of $f = \{(4, 2), (9, 3), (36, 6)\}$.

Solution

To find the inverse of f, switch the x- and y-coordinates of each ordered pair. The inverse of f is $\{(2, 4), (3, 9), (6, 36)\}$.

You Try 3

Find the inverse function of $f = \{(-5, -1), (-3, 2), (0, 7), (4, 13)\}$.

We use special notation to represent the inverse of a function. If f is a one-to-one function, then f^{-1} (read "f inverse") represents the inverse of f. For Example 3, we can write the inverse as $f^{-1} = \{(2, 4), (3, 9), (6, 36)\}$.

Definition

Inverse Function: Let f be a one-to-one function. The **inverse** of f, denoted by f^{-1}, is a one-to-one function that contains the set of all ordered pairs (y, x), where (x, y) belongs to f.

1) f^{-1} is read "f inverse" *not* "f to the negative one."

2) f^{-1} does *not* mean $\dfrac{1}{f}$.

3) If a function is not one-to-one, it does not have an inverse.

Finding the Inverse of a One-to-One Function

We said that if (x, y) belongs to the one-to-one function $f(x)$, then (y, x) belongs to its inverse, $f^{-1}(x)$ (read as *f inverse of x*). We use this idea to find the equation for the inverse of $f(x)$.

How to Find an Equation of the Inverse of $y = f(x)$

Step 1: Replace $f(x)$ with y.

Step 2: Interchange x and y.

Step 3: Solve for y.

Step 4: Replace y with the inverse notation, $f^{-1}(x)$.

Example 4

Find an equation of the inverse of $f(x) = 3x + 4$.

Solution

$$f(x) = 3x + 4$$
$$y = 3x + 4 \qquad \text{Replace } f(x) \text{ with } y.$$
$$x = 3y + 4 \qquad \text{Interchange } x \text{ and } y.$$

Solve for y.

$$x - 4 = 3y \qquad \text{Subtract 4.}$$
$$\frac{x - 4}{3} = y \qquad \text{Divide by 3.}$$
$$\frac{1}{3}x - \frac{4}{3} = y \qquad \text{Simplify.}$$

$$f^{-1}(x) = \frac{1}{3}x - \frac{4}{3} \qquad \text{Replace } y \text{ with } f^{-1}(x).$$

You Try 4

Find an equation of the inverse of $f(x) = -5x + 10$.

In Example 5, we will look more closely at the relationship between a function and its inverse.

Example 5

Find the equation of the inverse of $f(x) = 2x - 4$. Then, graph $f(x)$ and $f^{-1}(x)$ on the same axes.

Solution

$$f(x) = 2x - 4$$
$$y = 2x - 4 \qquad \text{Replace } f(x) \text{ with } y.$$
$$x = 2y - 4 \qquad \text{Interchange } x \text{ and } y.$$

Solve for y.

$$x + 4 = 2y \qquad \text{Add 4.}$$

$$\frac{x + 4}{2} = y \qquad \text{Divide by 2.}$$

$$\frac{1}{2}x + 2 = y \qquad \text{Simplify.}$$

$$f^{-1}(x) = \frac{1}{2}x + 2 \qquad \text{Replace } y \text{ with } f^{-1}(x).$$

We will graph $f(x)$ and $f^{-1}(x)$ by making a table of values for each. Then we can see another relationship between the two functions.

$f(x) = 2x - 4$		$f^{-1}(x) = \frac{1}{2}x + 2$	
x	$y = f(x)$	x	$y = f^{-1}(x)$
0	−4	−4	0
1	−2	−2	1
2	0	0	2
5	6	6	5

Notice that the x- and y-coordinates have switched when we compare the tables of values. Graph $f(x)$ and $f^{-1}(x)$.

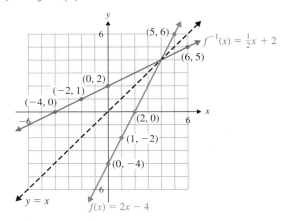

You Try 5

Find the equation of the inverse of $f(x) = -3x + 1$. Then, graph $f(x)$ and $f^{-1}(x)$ on the same axes.

4. Given the Graph of $f(x)$, Graph $f^{-1}(x)$

Look again at the tables in Example 5. The x-values for $f(x)$ become the y-values of $f^{-1}(x)$, and the y-values of $f(x)$ become the x-values of $f^{-1}(x)$. This is true not only for the values in the tables but for *all* values of x and y. That is, for all ordered pairs (x, y) that belong to $f(x)$, (y, x) belongs to $f^{-1}(x)$. Another way to say this is

> The domain of f becomes the range of f^{-1}, and the range of f becomes the domain of f^{-1}.

Let's turn our attention to the graph in Example 5. The graphs of $f(x)$ and $f^{-1}(x)$ are mirror images of one another with respect to the line $y = x$. We say that *the graphs of* $f(x)$ *and* $f^{-1}(x)$ *are symmetric with respect to the line* $y = x$. This is true for every function $f(x)$ and its inverse, $f^{-1}(x)$.

> The graphs of $f(x)$ and $f^{-1}(x)$ are symmetric with respect to the line $y = x$.

Example 6

Given the graph of $f(x)$, graph $f^{-1}(x)$.

Solution

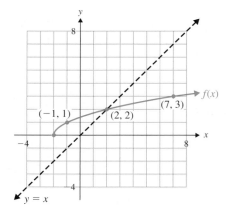

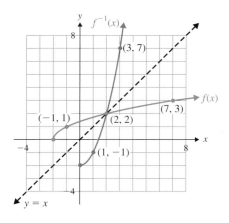

Some points on the graph of $f(x)$ are $(-2, 0)$, $(-1, 1)$, $(2, 2)$, and $(7, 3)$. We can obtain points on the graph of $f^{-1}(x)$ by interchanging the x- and y-values.

Some points on the graph of $f^{-1}(x)$ are $(0, -2)$, $(1, -1)$, $(2, 2)$, and $(3, 7)$. Plot these points to get the graph of $f^{-1}(x)$. Notice that the graphs are symmetric with respect to the line $y = x$.

You Try 6

Given the graph of $f(x)$, graph $f^{-1}(x)$.

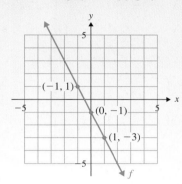

5. Given the Equations of $f(x)$ and $f^{-1}(x)$, Show That $(f^{-1} \circ f)(x) = x$ and $(f \circ f^{-1})(x) = x$

Going back to the tables in Example 5, we see from the first table that $f(0) = -4$ and from the second table that $f^{-1}(-4) = 0$. The first table also shows that $f(1) = -2$ while the second table shows that $f^{-1}(-2) = 1$. That is, putting x into the function f produces $f(x)$. And putting $f(x)$ into $f^{-1}(x)$ produces x.

$$
\begin{array}{ccc}
f(0) = -4 & \text{and} & f^{-1}(-4) = 0 \\
f(1) = -2 & \text{and} & f^{-1}(-2) = 1 \\
\uparrow \quad \uparrow & & \uparrow \qquad \uparrow \\
x \quad f(x) & & f^{-1}(f(x)) = x
\end{array}
$$

This leads us to another fact about functions and their inverses.

Let f be a one-to-one function. Then, f^{-1} is the inverse of f such that $(f^{-1} \circ f)(x) = x$ and $(f \circ f^{-1})(x) = x$.

Example 7

If $f(x) = 4x + 3$, show that $f^{-1}(x) = \dfrac{1}{4}x - \dfrac{3}{4}$.

Solution

Show that $(f^{-1} \circ f)(x) = x$ and $(f \circ f^{-1})(x) = x$.

$$
\begin{aligned}
(f^{-1} \circ f)(x) &= f^{-1}(f(x)) \\
&= f^{-1}(4x + 3) && \text{Substitute } 4x + 3 \text{ for } f(x). \\
&= \frac{1}{4}(4x + 3) - \frac{3}{4} && \text{Evaluate.} \\
&= x + \frac{3}{4} - \frac{3}{4} && \text{Distribute.} \\
&= x
\end{aligned}
$$

$$
\begin{aligned}
(f \circ f^{-1})(x) &= f(f^{-1}(x)) \\
&= f\!\left(\frac{1}{4}x - \frac{3}{4}\right) && \text{Substitute } \frac{1}{4}x - \frac{3}{4} \text{ for } f^{-1}(x). \\
&= 4\!\left(\frac{1}{4}x - \frac{3}{4}\right) + 3 && \text{Evaluate.} \\
&= x - 3 + 3 && \text{Distribute.} \\
&= x
\end{aligned}
$$

You Try 7

If $f(x) = -6x + 2$, show that $f^{-1}(x) = -\dfrac{1}{6}x + \dfrac{1}{3}$.

Using Technology

A graphing calculator can list tables of values on one screen for more than one equation. The graphing calculator screen below is the table of values generated when the equation of one line is entered as Y_1 and the equation of another line is entered as Y_2.

X	Y1	Y2
0	4	-2
2	8	-1
4	12	0
6	16	1
8	20	2
10	24	3
12	28	4

X=0

We read the table as follows:

- The points $(0, 4), (2, 8), (4, 12), (6, 16), (8, 20), (10, 24)$, and $(12, 28)$ are points on the line entered as Y_1.
- The points $(0, -2), (2, -1)$ $(4, 0), (6, 1), (8, 2), (10, 3)$, and $(12, 4)$ are points on the line entered as Y_2.

Equations Y_1 and Y_2 are linear functions, and they are inverses.

1. Looking at the table of values, what evidence is there that the functions Y_1 and Y_2 are inverses of each other?
2. Find the equations of the lines Y_1 and Y_2.
3. Graph Y_1 and Y_2. Is there evidence from their graphs that they are inverses?
4. Using the methods of this chapter, show that Y_1 and Y_2 are inverses.

Answers to You Try Exercises

1) a) yes   b) no   c) yes   d) no   2) a) yes   b) no

3) $\{(-1, -5), (2, -3), (7, 0), (13, 4)\}$

4) $f^{-1}(x) = -\dfrac{1}{5}x + 2$

5) $f^{-1}(x) = -\dfrac{1}{3}x + \dfrac{1}{3}$ 6)

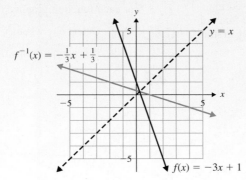

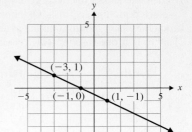

7) Show that $(f^{-1} \circ f)(x) = x$ and $(f \circ f^{-1})(x) = x$

Answers to Technology Exercises

1. If Y_1 and Y_2 are inverses, then if (x, y) is a point on Y_1, (y, x) is a point on Y_2. We see this is true with $(0, 4)$ on Y_1 and $(4, 0)$ on Y_2, with $(2, 8)$ on Y_1 and $(8, 2)$ on Y_2, and with $(4, 12)$ on Y_1 and $(12, 4)$ on Y_2.
2. $Y_1 = 2x + 4$, $Y_2 = 0.5x - 2$
3. Yes. They appear to be symmetric with respect to the line $y = x$.
4. Let $f(x) = Y_1$ and $f^{-1}(x) = Y_2$. We can show that $(f \circ f^{-1})(x) = x$ and $(f^{-1} \circ f)(x) = x$.

13.1 Exercises

Boost your grade at mathzone.com! MathZone

> Practice Problems   > NetTutor   > e-Professors   > Videos
> Self-Test

Objective 1

Determine whether each function is one-to-one. If it is one-to-one, find its inverse.

1) $f = \{(-4, 3), (-2, -3), (2, -3), (6, 13)\}$

2) $g = \{(0, -7), (1, -6), (4, -5), (25, -2)\}$

3) $h = \{(-5, -16), (-1, -4), (3, 8)\}$

4) $f = \{(-6, 3), (-1, 8), (4, 3)\}$

5) $g = \{(2, 1), (5, 2), (7, 14), (10, 19)\}$

6) $h = \{(-1, 4), (0, -2), (5, 1), (9, 4)\}$

Determine whether each function is one-to-one.

7) The table shows the average temperature during selected months in Tulsa, Oklahoma. The function matches each month with the average temperature, in °F. Is it one-to-one? (www.noaa.gov)

Month	Average Temp. (°F)
Jan.	36.4
Apr.	60.8
July	83.5
Oct.	62.6

8) The table shows some NCAA conferences and the number of schools in the conference. The function matches each conference with the number of schools it contains. Is it one-to-one?

Conference	Number of Member Schools
ACC	12
Big 10	11
Big 12	12
MVC	10
Pac10	10

Objectives 1, 2, and 4

9) Do all functions have inverses? Explain your answer.

10) What test can be used to determine whether the graph of a function has an inverse?

Determine whether each statement is true or false. If it is false, rewrite the statement so that it is true.

11) $f^{-1}(x)$ is read as "f to the negative one of x."

12) If f^{-1} is the inverse of f, then $(f^{-1} \circ f)(x) = x$ and $(f \circ f^{-1})(x) = x$.

13) The domain of f is the range of f^{-1}.

14) If f is one-to-one and $(5, 9)$ is on the graph of f, then $(-5, -9)$ is on the graph of f^{-1}.

15) The graphs of $f(x)$ and $f^{-1}(x)$ are symmetric with respect to the x-axis.

16) Let $f(x)$ be one-to-one. If $f(7) = 2$, then $f^{-1}(2) = 7$.

For each function graphed here, answer the following.

a) Determine whether it is one-to-one.

b) If it is one-to-one, graph its inverse.

17)

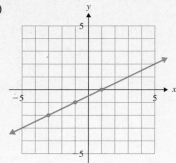

18)

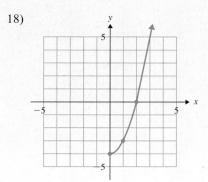

19)

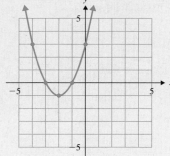

20)

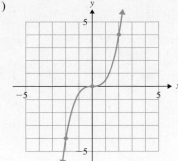

21)

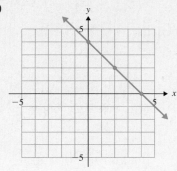

22)

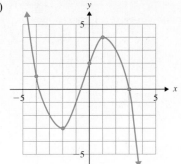

Objective 3

Find the inverse of each one-to-one function. Then, graph the function and its inverse on the same axes.

23) $g(x) = x - 6$

24) $h(x) = x + 3$

25) $f(x) = -2x + 5$

26) $g(x) = 4x - 9$

27) $g(x) = \dfrac{1}{2}x$

28) $h(x) = -\dfrac{1}{3}x$

29) $f(x) = x^3$

30) $g(x) = \sqrt[3]{x} + 4$

Find the inverse of each one-to-one function.

31) $f(x) = 2x - 6$

32) $g(x) = -4x + 8$

33) $h(x) = -\dfrac{3}{2}x + 4$

34) $f(x) = \dfrac{2}{5}x + 1$

35) $g(x) = \sqrt[3]{x + 2}$

36) $h(x) = \sqrt[3]{x - 7}$

37) $f(x) = \sqrt{x}, x \geq 0$

38) $g(x) = \sqrt{x + 3}, x \geq -3$

Objective 5

Given the one-to-one function $f(x)$, find the function values *without* finding the equation of $f^{-1}(x)$. Find the value in a) before b).

39) $f(x) = 5x - 2$

    a) $f(1)$ b) $f^{-1}(3)$

40) $f(x) = 3x + 7$

    a) $f(-4)$ b) $f^{-1}(-5)$

41) $f(x) = -\dfrac{1}{3}x + 5$

    a) $f(9)$ b) $f^{-1}(2)$

42) $f(x) = \dfrac{1}{2}x - 1$

    a) $f(6)$ b) $f^{-1}(2)$

43) $f(x) = -x + 3$

    a) $f(-7)$ b) $f^{-1}(10)$

44) $f(x) = -\dfrac{5}{4}x + 2$

    a) $f(8)$ b) $f^{-1}(-8)$

45) $f(x) = 2^x$

    a) $f(3)$ b) $f^{-1}(8)$

46) $f(x) = 3^x$

    a) $f(-2)$ b) $f^{-1}\left(\dfrac{1}{9}\right)$

47) If $f(x) = x + 9$, show that $f^{-1}(x) = x - 9$.

48) If $f(x) = x - 12$, show that $f^{-1}(x) = x + 12$.

video 49) If $f(x) = -6x + 4$, show that $f^{-1}(x) = -\dfrac{1}{6}x + \dfrac{2}{3}$.

50) If $f(x) = -\dfrac{1}{7}x + \dfrac{2}{7}$, show that $f^{-1}(x) = -7x + 2$.

51) If $f(x) = \dfrac{3}{2}x - 9$, show that $f^{-1}(x) = \dfrac{2}{3}x + 6$.

52) If $f(x) = -\dfrac{5}{8}x + 10$, show that
$$f^{-1}(x) = -\dfrac{8}{5}x + 16.$$

53) If $f(x) = \sqrt[3]{x - 10}$, show that $f^{-1}(x) = x^3 + 10$.

54) If $f(x) = x^3 - 1$, show that $f^{-1}(x) = \sqrt[3]{x + 1}$.

Section 13.2 Exponential Functions

Objectives

1. Define an Exponential Function
2. Graph an Exponential Function of the Form $f(x) = a^x$
3. Graph an Exponential Function of the Form $f(x) = a^{x+c}$
4. Define the Irrational Number e and Graph $f(x) = e^x$
5. Solve an Exponential Equation by Expressing Both Sides of the Equation with the Same Base
6. Solve an Applied Problem Using a Given Exponential Function

In Chapter 12 we studied the following types of functions:

Linear functions like $f(x) = 2x + 5$
Quadratic functions like $g(x) = x^2 - 6x + 8$
Absolute value functions like $h(x) = |x|$
Square root functions like $k(x) = \sqrt{x - 3}$

1. Define an Exponential Function

In this section, we will learn about *exponential functions*.

Definition   An **exponential function** is a function of the form

$$f(x) = a^x$$

where $a > 0$, $a \neq 1$, and x is a real number.

---

1) We stipulate that a is a positive number ($a > 0$) because if a were a negative number, some expressions would not be real numbers.

   Example: If $a = -2$ and $x = \dfrac{1}{2}$, we get

   $$f(x) = (-2)^{1/2} = \sqrt{-2} \text{ (not real)}$$

   Therefore, a *must* be a positive number.

2) We add the condition that $a \neq 1$ because if $a = 1$, the function would be linear, not exponential.

   Example: If $a = 1$, then $f(x) = 1^x$. This is equivalent to $f(x) = 1$, which is a linear function.

2. Graph an Exponential Function of the Form $f(x) = a^x$

We can graph exponential functions by plotting points. *It is important to choose many values for the variable so that we obtain positive numbers, negative numbers, and zero in the exponent.*

Example 1

Graph $f(x) = 2^x$ and $g(x) = 3^x$ on the same axes.

Solution

Make a table of values for each function. Be sure to choose values for x that will give us *positive numbers, negative numbers, and zero* in the exponent.

$f(x) = 2^x$			$g(x) = 3^x$	
x	$f(x)$		x	$g(x)$
0	1		0	1
1	2		1	3
2	4		2	9
3	8		3	27
-1	$\dfrac{1}{2}$		-1	$\dfrac{1}{3}$
-2	$\dfrac{1}{4}$		-2	$\dfrac{1}{9}$

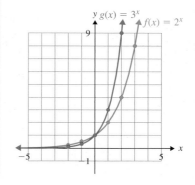

Plot each set of points and connect them with a smooth curve. Note that the larger the value of a, the more rapidly the y-values increase. Additionally, as x increases, the value of y also increases. Here are some other interesting facts to note about the graphs of these functions.

1) Each graph passes the vertical line test so the graphs *do* represent functions.
2) Each graph passes the horizontal line test, so the functions are one-to-one.
3) The y-intercept of each function is $(0, 1)$.
4) The domain of each function is $(-\infty, \infty)$, and the range is $(0, \infty)$.

You Try 1

Graph $f(x) = 4^x$.

Example 2

Graph $f(x) = \left(\dfrac{1}{2}\right)^x$.

Solution

Make a table of values and be sure to choose values for x that will give us *positive numbers, negative numbers, and zero* in the exponent.

$f(x) = \left(\dfrac{1}{2}\right)^x$	
x	**f(x)**
0	1
1	$\dfrac{1}{2}$
2	$\dfrac{1}{4}$
−1	2
−2	4
−3	8

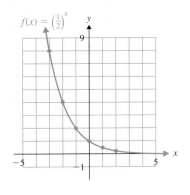

Like the graphs of $f(x) = 2^x$ and $g(x) = 3^x$ in Example 1, the graph of $f(x) = \left(\dfrac{1}{2}\right)^x$ passes both the vertical and horizontal line tests making it a one-to-one function. The *y*-intercept is (0, 1). The domain is $(-\infty, \infty)$, and the range is $(0, \infty)$.

In the case of $f(x) = \left(\dfrac{1}{2}\right)^x$, however, as the value of *x* increases, the value of *y decreases*. This is because $0 < a < 1$.

You Try 2

Graph $g(x) = \left(\dfrac{1}{3}\right)^x$.

We can summarize what we have learned so far about exponential functions:

Characteristics of $f(x) = a^x$, where $a > 0$ and $a \neq 1$

1) If $f(x) = a^x$ where $a > 1$, the value of *y* increases as the value of *x* increases.

2) If $f(x) = a^x$, where $0 < a < 1$, the value of *y* decreases as the value of *x* increases.

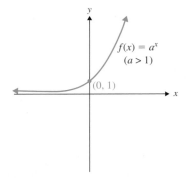

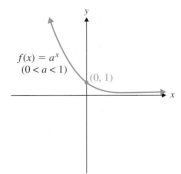

3) The function is one-to-one.

4) The *y*-intercept is (0, 1).

5) The domain is $(-\infty, \infty)$, and the range is $(0, \infty)$.

3. Graph an Exponential Function of the Form $f(x) = a^{x+c}$

Next we will graph an exponential function with an expression other than x as its exponent.

Example 3

Graph $f(x) = 3^{x-2}$.

Solution

Remember, for the table of values we want to choose values of x that will give us positive numbers, negative numbers, and zero *in the exponent*. First we will determine what value of x will make the exponent equal zero.

$$x - 2 = 0$$
$$x = 2$$

If $x = 2$, the exponent equals zero. Choose a couple of numbers *greater than* 2 and a couple that are *less than* 2 to get positive and negative numbers in the exponent.

$$f(x) = 3^{x-2}$$

	x	$x - 2$	$f(x) = 3^{x-2}$	Plot
Values greater than 2	2	0	$3^0 = 1$	$(2, 1)$
	3	1	$3^1 = 3$	$(3, 3)$
	4	2	$3^2 = 9$	$(4, 9)$
Values less than 2	1	-1	$3^{-1} = \dfrac{1}{3}$	$\left(1, \dfrac{1}{3}\right)$
	0	-2	$3^{-2} = \dfrac{1}{9}$	$\left(0, \dfrac{1}{9}\right)$

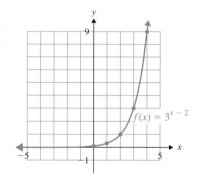

Note that the y-intercept is not $(0, 1)$ because the exponent is $x - 2$ *not* x, as in $f(x) = a^x$. The graph of $f(x) = 3^{x-2}$ is the same shape as the graph of $g(x) = 3^x$ except that the graph of f is shifted 2 units to the right. This is because $f(x) = g(x - 2)$.

You Try 3

Graph $f(x) = 2^{x+4}$.

4. Define the Irrational Number e and Graph $f(x) = e^x$

Next we will introduce a special exponential function, one with a base of e.

Like the number π, e is an irrational number that has many uses in mathematics. In the 1700s the work of Swiss mathematician Leonhard Euler led him to the approximation of e.

| Definition | **Approximation of e** |

$$e \approx 2.718281828459045235$$

One of the questions Euler set out to answer was, what happens to the value of $\left(1 + \dfrac{1}{n}\right)^n$ as n gets larger and larger? He found that as n gets larger, $\left(1 + \dfrac{1}{n}\right)^n$ gets closer to a fixed number. This number is e. Euler approximated e to the 18 decimal places above, and the letter e was chosen to represent this number in his honor. It should be noted that there are other ways to generate e. Finding the value that $\left(1 + \dfrac{1}{n}\right)^n$ approaches as n gets larger and larger is just one way. Also, since e is irrational, it is a nonterminating, nonrepeating decimal.

Example 4

Graph $f(x) = e^x$.

Solution

A calculator is needed to generate a table of values. We will use either the $\boxed{e^x}$ key or the two keys $\boxed{\text{INV}}$ (or $\boxed{\text{2nd}}$) and $\boxed{\ln x}$ to find powers of e. (Calculators will approximate powers of e to a few decimal places.)

For example, if a calculator has an $\boxed{e^x}$ key, find e^2 by pressing the following keys:

$$\boxed{2}\,\boxed{e^x} \quad \text{or} \quad \boxed{e^x}\,\boxed{2}\,\boxed{\text{ENTER}}$$

To four decimal places, $e^2 \approx 7.3891$.

If a calculator has an $\boxed{\ln x}$ key with e^x written above it, find e^2 by pressing the following keys:

$$\boxed{2}\,\boxed{\text{INV}}\,\boxed{\ln x} \quad \text{or} \quad \boxed{\text{INV}}\,\boxed{\ln x}\,\boxed{2}\,\boxed{\text{ENTER}}$$

The same approximation for e^2 is obtained.

Remember to choose positive numbers, negative numbers, and zero for x when making the table of values. We will approximate the values of e^x to four decimal places.

$f(x) = e^x$	
x	**f(x)**
0	1
1	2.7183
2	7.3891
3	20.0855
-1	0.3679
-2	0.1353

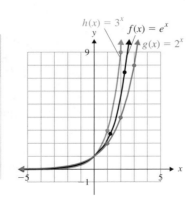

Notice that the graph of $f(x) = e^x$ is between the graphs of $g(x) = 2^x$ and $h(x) = 3^x$. This is because $2 < e < 3$, so e^x grows more quickly than 2^x, but e^x grows more slowly than 3^x.

We will study e^x and its special properties in more detail later in the chapter.

Solving Exponential Equations

An **exponential equation** is an equation that has a variable in the exponent. Some examples of exponential equations are

$$2^x = 8, \qquad 3^{a-5} = \frac{1}{9}, \qquad e^t = 14, \qquad 5^{2y-1} = 6^{y+4}$$

In this section, we will learn how to solve exponential equations like the first two examples. We can solve those equations by getting the same base.

5. Solve an Exponential Equation by Expressing Both Sides of the Equation with the Same Base

We know that the exponential function $f(x) = a^x$ ($a > 0$, $a \neq 1$) is one-to-one. This leads to the following property that enables us to solve many exponential equations.

$$\text{If } a^x = a^y, \text{ then } x = y. \ (a > 0, a \neq 1)$$

This property says that if two sides of an equation have the same base, set the exponents equal and solve for the unknown variable.

Solving an Exponential Equation

Step 1: **If possible, express each side of the equation with the same base.** If it is *not* possible to get the same base, a different method must be used. (This is presented in Section 13.6.)

Step 2: **Use the rules of exponents to simplify the exponents.**

Step 3: **Set the exponents equal and solve for the variable.**

Example 5

Solve each equation.

a) $2^x = 8$ b) $49^{c+3} = 7^{3c}$ c) $9^{6n} = 27^{n-4}$ d) $3^{a-5} = \frac{1}{9}$

Solution

a) ***Step 1:*** Express each side of the equation with the same base.

$$2^x = 8$$
$$2^x = 2^3 \qquad \text{Rewrite 8 with a base of 2: } 8 = 2^3.$$

Step 2: The exponents are simplified.

Step 3: Since the bases are the same, set the exponents equal and solve.

$$x = 3$$

The solution set is $\{3\}$.

b) **Step 1:** Express each side of the equation with the same base.

$$49^{c+3} = 7^{3c}$$
$$(7^2)^{c+3} = 7^{3c} \qquad \text{Both sides are powers of 7; } 49 = 7^2.$$

Step 2: Use the rules of exponents to simplify the exponents.

$$7^{2(c+3)} = 7^{3c} \qquad \text{Power rule for exponents}$$
$$7^{2c+6} = 7^{3c} \qquad \text{Distribute.}$$

Step 3: Since the bases are the same, set the exponents equal and solve.

$$2c + 6 = 3c \qquad \text{Set the exponents equal.}$$
$$6 = c \qquad \text{Subtract } 2c.$$

The solution set is $\{6\}$.

c) **Step 1:** Express each side of the equation with the same base. *9 and 27 are each powers of 3.*

$$9^{6n} = 27^{n-4}$$
$$(3^2)^{6n} = (3^3)^{n-4} \qquad 9 = 3^2; \, 27 = 3^3$$

Step 2: Use the rules of exponents to simplify the exponents.

$$3^{2(6n)} = 3^{3(n-4)} \qquad \text{Power rule for exponents}$$
$$3^{12n} = 3^{3n-12} \qquad \text{Multiply.}$$

Step 3: Since the bases are the same, set the exponents equal and solve.

$$12n = 3n - 12 \qquad \text{Set the exponents equal.}$$
$$9n = -12 \qquad \text{Subtract } 3n.$$
$$n = -\frac{12}{9} = -\frac{4}{3} \qquad \text{Divide by 9; simplify.}$$

The solution set is $\left\{-\dfrac{4}{3}\right\}$.

d) **Step 1:** Express each side of the equation $3^{a-5} = \dfrac{1}{9}$ with the same base. $\dfrac{1}{9}$ *can be expressed with a base of 3:* $\dfrac{1}{9} = \left(\dfrac{1}{3}\right)^2 = 3^{-2}$

$$3^{a-5} = \frac{1}{9}$$
$$3^{a-5} = 3^{-2} \qquad \text{Rewrite } \frac{1}{9} \text{ with a base of 3.}$$

Step 2:   The exponents are simplified.

Step 3:   Set the exponents equal and solve.

$$a - 5 = -2 \qquad \text{Set the exponents equal.}$$
$$a = 3 \qquad \text{Add 5.}$$

The solution set is $\{3\}$.

You Try 4

Solve each equation.

a) $(12)^x = 144$ b) $6^{t-5} = 36^{t+4}$ c) $32^{2w} = 8^{4w-1}$ d) $8^k = \dfrac{1}{64}$

6. Solve an Applied Problem Using a Given Exponential Function

Example 6

The value of a car depreciates (decreases) over time. The value, $V(t)$, in dollars, of a sedan t yr after it is purchased is given by

$$V(t) = 18{,}200(0.794)^t$$

a) What was the purchase price of the car?
b) What will the car be worth 5 yr after purchase?

Solution

a) To find the purchase price of the car, let $t = 0$.
    Evaluate $V(0)$.

$$\begin{aligned} V(0) &= 18{,}200(0.794)^0 \\ &= 18{,}200(1) \\ &= 18{,}200 \end{aligned}$$

The purchase price of the car was \$18,200.

b) To find the value of the car after 5 yr, let $t = 5$. Use a calculator to find $V(5)$.

$$\begin{aligned} V(5) &= 18{,}200(0.794)^5 \\ &= 5743.46 \end{aligned}$$

The car will be worth about \$5743.46.

You Try 5

The value, $V(t)$, in dollars, of a pickup truck t yr after it is purchased is given by

$$V(t) = 23{,}500(0.785)^t$$

a) What was the purchase price of the pickup?

b) What will the pickup truck be worth 4 yr after purchase?

Answers to You Try Exercises

1)

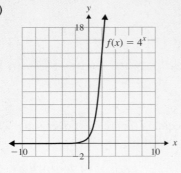

2) $g(x) = \left(\frac{1}{3}\right)^x$

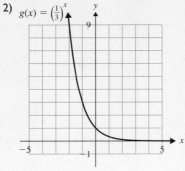

3)

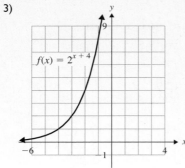

4) a) $\{2\}$ b) $\{-13\}$ c) $\left\{\dfrac{3}{2}\right\}$ d) $\{-2\}$ 5) a) \$23,500   b) \$8923.73

13.2 Exercises

Boost your grade at mathzone.com!   MathZone

> Practice Problems   > NetTutor   > Self-Test   > e-Professors   > Videos

Objectives 1 and 2

1) When making a table of values to graph an exponential function, what kind of values should be chosen for the variable?

 video

2) What is the y-intercept of the graph of $f(x) = a^x$ where $a > 0$ and $a \neq 1$?

Graph each exponential function.

3) $f(x) = 5^x$

4) $g(x) = 4^x$

5) $y = 2^x$

6) $f(x) = 3^x$

7) $h(x) = \left(\frac{1}{3}\right)^x$

8) $y = \left(\frac{1}{4}\right)^x$

For an exponential function of the form $f(x) = a^x$ ($a > 0$, $a \neq 1$), answer the following.

9) What is the domain?

10) What is the range?

Objective 3

Graph each exponential function.

11) $g(x) = 2^{x+1}$

12) $y = 3^{x+2}$

13) $f(x) = 3^{x-4}$

14) $h(x) = 2^{x-3}$

15) $y = 4^{x+3}$

16) $g(x) = 4^{x-1}$

17) $f(x) = 2^{2x}$

18) $h(x) = 3^{\frac{1}{2}x}$

19) $y = 2^x + 1$

20) $f(x) = 2^x - 3$

21) $g(x) = 3^x - 2$

22) $h(x) = 3^x + 1$

23) $y = -2^x$

24) $f(x) = -\left(\dfrac{1}{3}\right)^x$

Objective 4

25) What is the approximate value of e to four decimal places?

26) Is e a rational or an irrational number? Explain your answer.

For Exercises 27–30 match each exponential function with its graph.

A.

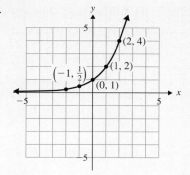

B.

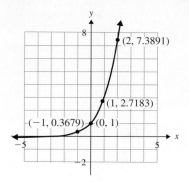

C.

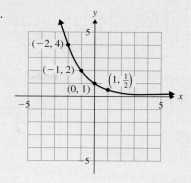

D.

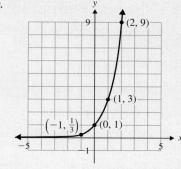

27) $f(x) = e^x$

28) $g(x) = 2^x$

29) $h(x) = 3^x$

30) $k(x) = \left(\dfrac{1}{2}\right)^x$

Objective 5

Solve each exponential equation.

31) $9^x = 81$

32) $4^y = 16$

33) $5^{4d} = 125$

34) $4^{3a} = 64$

35) $16^{m-2} = 2^{3m}$

36) $3^{5t} = 9^{t+4}$

37) $7^{2k-6} = 49^{3k+1}$

38) $(1000)^{2p-3} = 10^{4p+1}$

39) $32^{3c} = 8^{c+4}$

40) $(125)^{2x-9} = 25^{x-3}$

41) $100^{5z-1} = (1000)^{2z+7}$

42) $32^{y+1} = 64^{y+2}$

43) $81^{3n+9} = 27^{2n+6}$

44) $27^{5v} = 9^{v+4}$

45) $6^x = \dfrac{1}{36}$

46) $11^t = \dfrac{1}{121}$

47) $2^a = \dfrac{1}{8}$

48) $3^z = \dfrac{1}{81}$

49) $9^r = \dfrac{1}{27}$

50) $16^c = \dfrac{1}{8}$

51) $\left(\dfrac{3}{4}\right)^{5k} = \left(\dfrac{27}{64}\right)^{k+1}$

52) $\left(\dfrac{3}{2}\right)^{y+4} = \left(\dfrac{81}{16}\right)^{y-2}$

53) $\left(\dfrac{5}{6}\right)^{3x+7} = \left(\dfrac{36}{25}\right)^{2x}$ 54) $\left(\dfrac{7}{2}\right)^{5w} = \left(\dfrac{4}{49}\right)^{4w+3}$

Objective 6

 Solve each application.

55) The value of a car depreciates (decreases) over time. The value, $V(t)$, in dollars, of an SUV t yr after it is purchased is given by

$$V(t) = 32{,}700(0.812)^t$$

a) What was the purchase price of the SUV?

b) What will the SUV be worth 3 yr after purchase?

56) The value, $V(t)$, in dollars, of a compact car t yr after it is purchased is given by

$$V(t) = 10{,}150(0.784)^t$$

a) What was the purchase price of the car?

b) What will the car be worth 5 yr after purchase?

57) The value, $V(t)$, in dollars, of a minivan t yr after it is purchased is given by

$$V(t) = 16{,}800(0.803)^t$$

a) What was the purchase price of the minivan?

b) What will the minivan be worth 6 yr after purchase?

58) The value, $V(t)$, in dollars, of a sports car t yr after it is purchased is given by

$$V(t) = 48{,}600(0.820)^t$$

a) What was the purchase price of the sports car?

b) What will the sports car be worth 4 yr after purchase?

59) From 1995–2005, the value of homes in a suburb increased by 3% per year. The value, $V(t)$, in dollars, of a particular house t yr after 1995 is given by

$$V(t) = 185{,}200(1.03)^t$$

a) How much was the house worth in 1995?

b) How much was the house worth in 2002?

60) From 1995–2005, the value of condominiums in a big city high rise building increased by 2% per year. The value, $V(t)$, in dollars, of a particular condo t yr after 1995 is given by

$$V(t) = 420{,}000(1.02)^t$$

a) How much was the condominium worth in 1995?

b) How much was the condominium worth in 2005?

 An *annuity* is an account into which money is deposited every year. The amount of money, A in dollars, in the account after t yr of depositing c dollars at the beginning of every year earning an interest rate r (as a decimal) is

$$A = c\left(\frac{(1 + r)^t - 1}{r}\right)(1 + r)$$

Use the formula for Exercises 61–64.

61) After Fernando's daughter is born, he decides to begin saving for her college education. He will deposit $2000 every year in an annuity for 18 yr at a rate of 9%. How much will be in the account after 18 yr?

62) To save for retirement, Susan plans to deposit $6000 per year in an annuity for 30 yr at a rate of 8.5%. How much will be in the account after 30 yr?

63) Patrice will deposit $4000 every year in an annuity for 10 yr at a rate of 7%. How much will be in the account after 10 yr?

64) Haeshin will deposit $3000 every year in an annuity for 15 yr at a rate of 8%. How much will be in the account after 15 yr?

 65) After taking a certain antibiotic, the amount of amoxicillin $A(t)$, in milligrams, remaining in the patient's system t hr after taking 1000 mg of amoxicillin is

$$A(t) = 1000e^{-0.5332t}$$

How much amoxicillin is in the patient's system 6 hr after taking the medication?

 66) Some cockroaches can reproduce according to the formula

$$y = 2(1.65)^t$$

where y is the number of cockroaches resulting from the mating of two cockroaches and their offspring t months after the first two cockroaches mate.

   If Morris finds two cockroaches in his kitchen (assuming one is male and one is female) how large can the cockroach population become after 12 months?

Section 13.3 Logarithmic Functions

1. Define a Logarithm

In Section 13.2 we graphed $f(x) = 2^x$ by making a table of values and plotting the points. The graph passes the horizontal line test, making the function one-to-one. Recall that if (x, y) is on the graph of a function, then (y, x) is on the graph of its inverse. We can graph the inverse of $f(x) = 2^x, f^{-1}(x)$, by switching the x- and y-coordinates in the table of values and plotting the points.

$f(x) = 2^x$	
x	$y = f(x)$
0	1
1	2
2	4
3	8
-1	$\dfrac{1}{2}$
-2	$\dfrac{1}{4}$

$f^{-1}(x)$	
x	$y = f^{-1}(x)$
1	0
2	1
4	2
8	3
$\dfrac{1}{2}$	-1
$\dfrac{1}{4}$	-2

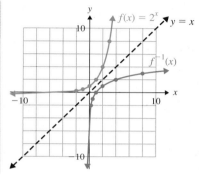

Above is the graph of $f(x) = 2^x$ and its inverse. Notice that, like the graphs of all functions and their inverses, they are symmetric with respect to the line $y = x$.

What is the equation of $f^{-1}(x)$? We will use the procedure outlined in Section 13.1 to find the equation of $f^{-1}(x)$.

If $f(x) = 2^x$, then find the equation of $f^{-1}(x)$ as follows.

Step 1: Replace $f(x)$ with y.

$$y = 2^x$$

Step 2: Interchange x and y.

$$x = 2^y$$

Step 3: Solve for y.

How do we solve $x = 2^y$ for y? To answer this question we must introduce another concept called *logarithms*.

Definition

Definition of Logarithm: If $a > 0$, $a \neq 1$, and $x > 0$, then for every real number y,

$$y = \log_a x \text{ means } x = a^y$$

> It is very important to note that the base of the logarithm must be positive and not equal to 1, and that x must be positive as well.

The word **log** is an abbreviation for **logarithm**. We read $\log_a x$ as "log of x to the base a" or "log to the base a of x." *This definition of a logarithm should be memorized!*

The relationship between the logarithmic form of an equation ($y = \log_a x$) and the exponential form of an equation ($x = a^y$) is one that has many uses. Notice the relationship between the two forms.

Logarithmic Form	Exponential Form
Value of the logarithm ↓ $y = \log_a x$ ↑ Base	Exponent ↓ $x = a^y$ ↑ Base

From the above you can see that *a logarithm is an exponent.* $\log_a x$ is the power to which we raise a to get x.

2. Rewrite an Equation in Logarithmic Form as an Equation in Exponential Form

Much of our work with logarithms involves converting between logarithmic and exponential notation. After working with logs and exponential form, we will come back to the question of how to solve $x = 2^y$ for y.

Example 1

Write in exponential form.

a) $\log_6 36 = 2$ b) $\log_4 \dfrac{1}{64} = -3$ c) $\log_7 1 = 0$

Solution

a) $\log_6 36 = 2$ means that 2 is the power to which we raise 6 to get 36. The exponential form is $6^2 = 36$.

$$\log_6 36 = 2 \text{ means } 6^2 = 36.$$

b) $\log_4 \dfrac{1}{64} = -3$ means $4^{-3} = \dfrac{1}{64}$.

c) $\log_7 1 = 0$ means $7^0 = 1$.

You Try 1

Write in exponential form.

a) $\log_3 81 = 4$ b) $\log_5 \dfrac{1}{25} = -2$ c) $\log_{64} 8 = \dfrac{1}{2}$ d) $\log_{13} 13 = 1$

3. Rewrite an Equation in Exponential Form as an Equation in Logarithmic Form

Example 2

Write in logarithmic form.

a) $10^4 = 10{,}000$ b) $9^{-2} = \dfrac{1}{81}$ c) $8^1 = 8$

d) $\sqrt{25} = 5$

Solution

a) $10^4 = 10{,}000$ means $\log_{10} 10{,}000 = 4$.

b) $9^{-2} = \dfrac{1}{81}$ means $\log_9 \dfrac{1}{81} = -2$.

c) $8^1 = 8$ means $\log_8 8 = 1$.

d) To write $\sqrt{25} = 5$ in logarithmic form, rewrite $\sqrt{25}$ as $25^{1/2}$.

$$\sqrt{25} = 5 \text{ is the same as } 25^{1/2} = 5$$

$$25^{1/2} = 5 \text{ means } \log_{25} 5 = \frac{1}{2}.$$

When working with logarithms, we will often change radical notation to the equivalent fractional exponent. This is because a logarithm *is* an exponent.

You Try 2

Write in logarithmic form.

a) $7^2 = 49$ b) $5^{-4} = \dfrac{1}{625}$ c) $19^0 = 1$ d) $\sqrt{144} = 12$

Solving Logarithmic Equations

4. Solve a Logarithmic Equation of the Form $\log_a b = c$

A **logarithmic equation** is an equation in which at least one term contains a logarithm. In this section, we will learn how to solve a logarithmic equation of the form $\log_a b = c$. We will learn how to solve other types of logarithmic equations in Sections 13.5 and 13.6.

To solve a logarithmic equation of the form

$$\log_a b = c$$

write the equation in exponential form ($a^c = b$) and solve for the variable.

Example 3

Solve each logarithmic equation.

a) $\log_{10} r = 3$ b) $\log_3(7a + 18) = 4$ c) $\log_w 25 = 2$

d) $\log_2 16 = c$ e) $\log_{36} \sqrt[4]{6} = x$

Solution

a) To solve $\log_{10} r = 3$, write the equation in exponential form and solve for r.

$$\log_{10} r = 3 \quad \text{means} \quad 10^3 = r$$
$$1000 = r$$

The solution set is $\{1000\}$.

b) To solve $\log_3(7a + 18) = 4$, write the equation in exponential form and solve for a.

$$\log_3(7a + 18) = 4 \quad \text{means} \quad 3^4 = 7a + 18$$
$$81 = 7a + 18$$
$$63 = 7a \qquad \text{Subtract 18.}$$
$$9 = a \qquad \text{Divide by 9.}$$

The solution set is $\{9\}$.

c) Write $\log_w 25 = 2$ in exponential form and solve for w.

$$\log_w 25 = 2 \quad \text{means} \quad w^2 = 25$$
$$w = \pm 5 \qquad \text{Square root property}$$

Although we get $w = 5$ or $w = -5$ when we solve $w^2 = 25$, recall that the base of a logarithm must be a positive number. Therefore, $w = -5$ is *not* a solution. The solution set is $\{5\}$.

d) Write $\log_2 16 = c$ in exponential form and solve for c.

$$\log_2 16 = c \quad \text{means} \quad 2^c = 16$$
$$c = 4$$

The solution set is $\{4\}$.

e) $\log_{36} \sqrt[4]{6} = x$ means $36^x = \sqrt[4]{6}$

$$(6^2)^x = 6^{1/4} \qquad \text{Express each side with the same base;}$$
$$\text{rewrite the radical as a fractional exponent.}$$
$$6^{2x} = 6^{1/4} \qquad \text{Power rule for exponents}$$
$$2x = \frac{1}{4} \qquad \text{Set the exponents equal.}$$
$$x = \frac{1}{8} \qquad \text{Divide by 2.}$$

The solution set is $\left\{ \dfrac{1}{8} \right\}$.

You Try 3

Solve each logarithmic equation.

a) $\log_2 y = 5$ b) $\log_5(3p + 11) = 3$ c) $\log_x 169 = 2$ d) $\log_6 36 = n$

e) $\log_{64} \sqrt[5]{8} = k$

5. Evaluate a Logarithm

Often when working with logarithms we are asked to *evaluate* them or to find the value of a log.

Example 4

Evaluate.

a) $\log_3 9$ b) $\log_2 8$ c) $\log_{10} \dfrac{1}{10}$ d) $\log_{25} 5$

Solution

a)   To evaluate (or find the value of) $\log_3 9$ means to find the power to which we raise 3 to get 9. That power is **2**.

$$\log_3 9 = 2 \quad \text{since} \quad 3^2 = 9$$

b)   To evaluate $\log_2 8$ means to find the power to which we raise 2 to get 8. That power is **3**.

$$\log_2 8 = 3 \quad \text{since} \quad 2^3 = 8$$

c)   To evaluate $\log_{10} \dfrac{1}{10}$ means to find the power to which we raise 10 to get $\dfrac{1}{10}$. That power is **−1**.

If you don't see that this is the answer, set the expression $\log_{10} \dfrac{1}{10}$ equal to x, write the equation in exponential form, and solve for x, as in Example 3.

$$\log_{10} \frac{1}{10} = x \quad \text{means} \quad 10^x = \frac{1}{10}$$

$$10^x = 10^{-1} \qquad \frac{1}{10} = 10^{-1}$$
$$x = -1$$

Then, $\log_{10} \dfrac{1}{10} = -1$.

d)   To evaluate $\log_{25} 5$ means to find the power to which we raise 25 to get 5. That power is $\dfrac{1}{2}$.

Once again, we can also find the value of $\log_{25} 5$ by setting it equal to x, writing the equation in exponential form, and solving for x.

$$\log_{25} 5 = x \quad \text{means} \quad 25^x = 5$$
$$(5^2)^x = 5 \qquad \text{Express each side with the same base.}$$
$$5^{2x} = 5^1 \qquad \text{Power rule; } 5 = 5^1$$
$$2x = 1 \qquad \text{Set the exponents equal.}$$
$$x = \frac{1}{2} \qquad \text{Divide by 2.}$$

Therefore, $\log_{25} 5 = \dfrac{1}{2}$.

You Try 4

Evaluate.

a) $\log_{10} 100$ b) $\log_3 81$ c) $\log_8 \dfrac{1}{8}$ d) $\log_{144} 12$

6. Evaluate Common Logarithms, and Solve Equations of the Form log b = c

Logarithms have many applications not only in mathematics but also in other areas such as chemistry, biology, engineering, and economics.

Since our number system is a base 10 system, logarithms to the base 10 are very widely used and are called **common logarithms** or **common logs**. A base 10 log has a special notation—$\log_{10} x$ is written as log x. *When a log is written in this way, the base is assumed to be 10.*

$$\log x \text{ means } \log_{10} x$$

We must keep this in mind when evaluating logarithms and when solving logarithmic equations.

Example 5

Evaluate log 100.

Solution

log 100 is equivalent to $\log_{10} 100$. To evaluate log 100 means to find the power to which we raise 10 to get 100. That power is **2**.

$$\log 100 = 2$$

You Try 5

Evaluate log 1000.

Example 6

Solve $\log(3x - 8) = 1$.

Solution

$\log(3x - 8) = 1$ is equivalent to $\log_{10}(3x - 8) = 1$. Write the equation in exponential form and solve for x.

$$\log(3x - 8) = 1 \text{ means } 10^1 = 3x - 8$$
$$10 = 3x - 8$$
$$18 = 3x \qquad \text{Add 8.}$$
$$6 = x \qquad \text{Divide by 3.}$$

The solution set is $\{6\}$.

You Try 6

Solve $\log(12q + 16) = 2$

We will study common logs in more depth in section 13.5.

7. Use the Properties $\log_a a = 1$ and $\log_a 1 = 0$

There are a couple of properties of logarithms which can simplify our work with them.
If a is any real number, then $a^1 = a$. Furthermore, if $a \neq 0$, then $a^0 = 1$. Write $a^1 = a$ and $a^0 = 1$ in logarithmic form to obtain these two properties of logarithms:

Properties of Logarithms

If $a > 0$ and $a \neq 1$,

1) $\log_a a = 1$

2) $\log_a 1 = 0$

Example 7

Use the properties of logarithms to evaluate each.

a) $\log_{12} 1$ b) $\log_3 3$ c) $\log 10$ d) $\log_{\sqrt{5}} 1$

Solution

a) By property 2, $\log_{12} 1 = 0$.

b) By property 1, $\log_3 3 = 1$.

c) The base of $\log 10$ is 10. Therefore, $\log 10 = \log_{10} 10$. By property 1, $\log 10 = 1$.

d) By property 2, $\log_{\sqrt{5}} 1 = 0$.

You Try 7

Use the properties of logarithms to evaluate each.

a) $\log_{16} 16$ b) $\log_{1/3} 1$ c) $\log_{\sqrt{11}} \sqrt{11}$

8. Define and Graph a Logarithmic Function

Next we define a logarithmic function.

Definition

For $a > 0$, $a \neq 1$, and $x > 0$, $f(x) = \log_a x$ is the **logarithmic function with base a**.

$f(x) = \log_a x$ can also be written as $y = \log_a x$. Changing $y = \log_a x$ to exponential form, we get

$$a^y = x$$

Remembering that a is a positive number not equal to 1, it follows that

1) any real number may be substituted for y. Therefore, **the range of $y = \log_a x$ is $(-\infty, \infty)$.**

2) x must be a positive number. So, **the domain of $y = \log_a x$ is $(0, \infty)$.**

Let's return to the problem of finding the equation of the inverse of $f(x) = 2^x$ that was first introduced on p. 875.

Example 8

Find the equation of the inverse of $f(x) = 2^x$.

Solution

Step 1: Replace $f(x)$ with y. $y = 2^x$

Step 2: Interchange x and y. $x = 2^y$

Step 3: Solve for y.

To solve $x = 2^y$ for y, write the equation in logarithmic form.

$$x = 2^y \quad \text{means} \quad y = \log_2 x$$

Step 4: Replace y with $f^{-1}(x)$.

$$f^{-1}(x) = \log_2 x$$

The inverse of the exponential function $f(x) = 2^x$ is $f^{-1}(x) = \log_2 x$.

You Try 8

Find the equation of the inverse of $f(x) = 6^x$.

The inverse of the exponential function $f(x) = a^x$ (where $a > 0$, $a \neq 1$, and x is any real number) is $f^{-1}(x) = \log_a x$. Furthermore,

1) The domain of $f(x)$ is the range of $f^{-1}(x)$.
2) The range of $f(x)$ is the domain of $f^{-1}(x)$.

Their graphs are symmetric with respect to $y = x$.

To graph a logarithmic function, write it in exponential form first. Then, make a table of values, plot the points, and draw the curve through the points.

Example 9

Graph $f(x) = \log_2 x$.

Solution

Substitute y for $f(x)$ and write the equation in exponential form.

$$y = \log_2 x \quad \text{means} \quad 2^y = x$$

To make a table of values, it will be easier to *choose values for y* and compute the corresponding values of x. Remember to choose values of y that will give us positive numbers, negative numbers, and zero in the exponent.

$2^y = x$	
x	**y**
1	0
2	1
4	2
8	3
$\dfrac{1}{2}$	-1
$\dfrac{1}{4}$	-2

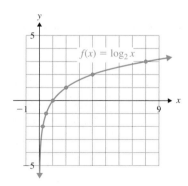

From the graph we can see that the domain of f is $(0, \infty)$, and the range of f is $(-\infty, \infty)$.

You Try 9

Graph $f(x) = \log_4 x$.

Example 10

Graph $f(x) = \log_{1/3} x$.

Solution

Substitute y for $f(x)$ and write the equation in exponential form.

$$y = \log_{1/3} x \quad \text{means} \quad \left(\frac{1}{3}\right)^y = x$$

For the table of values, *choose values for y* and compute the corresponding values of x.

$\left(\dfrac{1}{3}\right)^y = x$	
x	**y**
1	0
$\dfrac{1}{3}$	1
$\dfrac{1}{9}$	2
3	-1
9	-2

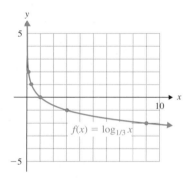

The domain of f is $(0, \infty)$, and the range is $(-\infty, \infty)$.

You Try 10

Graph $f(x) = \log_{1/4} x$.

The graphs in Examples 9 and 10 are typical of the graphs of logarithmic functions— Example 9 for functions where $a > 1$ and Example 10 for functions where $0 < a < 1$. Next is a summary of some characteristics of logarithmic functions.

Characteristics of a Logarithmic Function $f(x) = \log_a x$, where $a > 0$ and $a \neq 1$

1) If $f(x) = \log_a x$ where $a > 1$, the value of y increases as the value of x increases.

2) If $f(x) = \log_a x$ where $0 < a < 1$, the value of y decreases as the value of x increases.

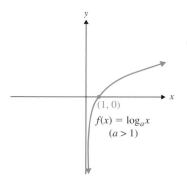

$f(x) = \log_a x$
$(a > 1)$

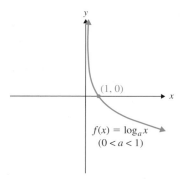

$(1, 0)$

$f(x) = \log_a x$
$(0 < a < 1)$

3) The function is one-to-one.

4) The x-intercept is $(1, 0)$.

5) The domain is $(0, \infty)$, and the range is $(-\infty, \infty)$.

6) The inverse of $f(x) = \log_a x$ is $f^{-1}(x) = a^x$.

Compare these characteristics of logarithmic functions to the characteristics of exponential functions on p. 866 in Section 13.2. The domain and range of logarithmic and exponential functions are interchanged since they are inverse functions.

9. Solve an Applied Problem Using a Given Logarithmic Equation

Example 11

A hospital has found that the function $A(t) = 50 + 8 \log_2(t + 2)$ approximates the number of people treated each year since 1990 for severe allergic reactions to peanuts. If $t = 0$ represents the year 1990, answer the following.

a) How many people were treated in 1990?

b) How many people were treated in 1996?

c) In what year were approximately 82 people treated for allergic reactions to peanuts?

Solution

a) The year 1990 corresponds to $t = 0$. Let $t = 0$, and find $A(0)$.

$$
\begin{aligned}
A(0) &= 50 + 8 \log_2(0 + 2) && \text{Substitute 0 for } t. \\
&= 50 + 8 \log_2 2 \\
&= 50 + 8(1)^2 && \log_2 2 = 1 \\
&= 58
\end{aligned}
$$

In 1990, 58 people were treated for peanut allergies.

b) The year 1996 corresponds to $t = 6$. Let $t = 6$, and find $A(6)$.

$$A(6) = 50 + 8 \log_2(6 + 2) \qquad \text{Substitute 6 for } t.$$
$$= 50 + 8 \log_2 8$$
$$= 50 + 8(3) \qquad\qquad \log_2 8 = 3$$
$$= 50 + 24$$
$$= 74$$

In 1996, 74 people were treated for peanut allergies.

c) To determine in what year 82 people were treated, let $A(t) = 82$ and solve for t.

$$82 = 50 + 8 \log_2(t + 2) \qquad \text{Substitute 82 for } A(t).$$

To solve for t, we first need to get the term containing the logarithm on a side by itself. Subtract 50 from each side.

$$32 = 8 \log_2(t + 2) \qquad \text{Subtract 50.}$$
$$4 = \log_2(t + 2) \qquad \text{Divide by 8.}$$
$$2^4 = t + 2 \qquad\qquad \text{Write in exponential form.}$$
$$16 = t + 2$$
$$14 = t$$

$t = 14$ corresponds to the year 2004. (Add 14 to the year 1990.)

82 people were treated for peanut allergies in 2004.

 You Try 11

The amount of garbage (in millions of pounds) collected in a certain town each year since 1985 can be approximated by $G(t) = 6 + \log_2(t + 1)$, where $t = 0$ represents the year 1985.

a) How much garbage was collected in 1985?

b) How much garbage was collected in 1992?

c) In what year would it be expected that 11,000,000 pounds of garbage will be collected? [Hint: Let $G(t) = 11$.]

Answers to You Try Exercises

1) a) $3^4 = 81$ b) $5^{-2} = \dfrac{1}{25}$ c) $64^{1/2} = 8$ d) $13^1 = 13$

2) a) $\log_7 49 = 2$ b) $\log_5 \dfrac{1}{625} = -4$ c) $\log_{19} 1 = 0$ d) $\log_{144} 12 = \dfrac{1}{2}$

3) a) $\{32\}$ b) $\{38\}$ c) $\{13\}$ d) $\{2\}$ e) $\left\{ \dfrac{1}{10} \right\}$

4) a) 2   b) 4   c) -1 d) $\dfrac{1}{2}$ 5) 3   6) $\{7\}$ 7) a) 1   b) 0   c) 1   8) $f^{-1}(x) = \log_6 x$

9)

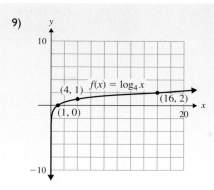

10)

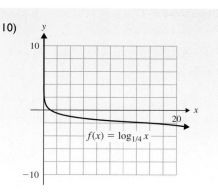

11) a) 6,000,000 lb   b) 9,000,000 lb   c) 2016

13.3 Exercises

Objectives 1 and 2

1) In $y = \log_a x$, a must be what kind of number?

2) In $y = \log_a x$, x must be what kind of number?

3) What is the base of $y = \log x$?

4) A base 10 logarithm is called a _____ logarithm.

Write in exponential form.

5) $\log_7 49 = 2$

6) $\log_{11} 121 = 2$

7) $\log_2 8 = 3$

8) $\log_2 32 = 5$

 9) $\log_9 \dfrac{1}{81} = -2$

10) $\log_8 \dfrac{1}{64} = -2$

11) $\log 1{,}000{,}000 = 6$

12) $\log 10{,}000 = 4$

13) $\log_{25} 5 = \dfrac{1}{2}$

14) $\log_{64} 4 = \dfrac{1}{3}$

15) $\log_{13} 13 = 1$

16) $\log_9 1 = 0$

Objective 3

Write in logarithmic form.

17) $9^2 = 81$

18) $12^2 = 144$

19) $10^2 = 100$

20) $10^3 = 1000$

21) $3^{-4} = \dfrac{1}{81}$

22) $2^{-5} = \dfrac{1}{32}$

23) $10^0 = 1$

24) $10^1 = 10$

25) $169^{1/2} = 13$

26) $27^{1/3} = 3$

27) $\sqrt{9} = 3$

28) $\sqrt{64} = 8$

video 29) $\sqrt[3]{64} = 4$

30) $\sqrt[4]{81} = 3$

Objectives 4 and 6

31) Explain how to solve a logarithmic equation of the form $\log_a b = c$.

32) A student solves $\log_x 9 = 2$ and gets the solution set $\{-3, 3\}$. Is this correct? Why or why not?

Solve each logarithmic equation.

33) $\log_{11} x = 2$

34) $\log_5 k = 3$

35) $\log_4 r = 3$

36) $\log_2 y = 4$

37) $\log p = 5$

38) $\log w = 2$

39) $\log_m 49 = 2$

40) $\log_x 4 = 2$

41) $\log_6 h = -2$

42) $\log_4 b = -3$

43) $\log_2(a + 2) = 4$

44) $\log_6(5y + 1) = 2$

video 45) $\log_3(4t - 3) = 3$

46) $\log_2(3n + 7) = 5$

47) $\log_{81} \sqrt[4]{9} = x$

48) $\log_{49} \sqrt[3]{7} = d$

49) $\log_{125} \sqrt{5} = c$

50) $\log_{16} \sqrt[5]{4} = k$

51) $\log_{144} w = \dfrac{1}{2}$

52) $\log_{64} p = \dfrac{1}{3}$

53) $\log_8 x = \dfrac{2}{3}$

54) $\log_{16} t = \dfrac{3}{4}$

55) $\log_{(3m-1)} 25 = 2$

56) $\log_{(y-1)} 4 = 2$

Objectives 5–7

Evaluate each logarithm.

57) $\log_5 25$

58) $\log_9 81$

59) $\log_2 32$

60) $\log_4 64$

61) $\log 100$

62) $\log 1000$

63) $\log_{49} 7$

64) $\log_{36} 6$

65) $\log_8 \dfrac{1}{8}$

66) $\log_3 \dfrac{1}{3}$

67) $\log_5 5$

68) $\log_2 1$

 69) $\log_{1/4} 16$

70) $\log_{1/3} 27$

Objective 8

71) Explain how to graph a logarithmic function of the form $f(x) = \log_a x$.

72) What are the domain and range of $f(x) = \log_a x$?

Graph each logarithmic function.

73) $f(x) = \log_3 x$

74) $f(x) = \log_4 x$

75) $f(x) = \log_2 x$

76) $f(x) = \log_5 x$

77) $f(x) = \log_{1/2} x$

78) $f(x) = \log_{1/3} x$

79) $f(x) = \log_{1/4} x$

80) $f(x) = \log_{1/5} x$

Find the inverse of each function.

81) $f(x) = 3^x$

82) $f(x) = 4^x$

83) $f(x) = \log_2 x$

84) $f(x) = \log_5 x$

Objective 9

Solve each problem.

85) The function $L(t) = 1800 + 68 \log_3(t + 3)$ approximates the number of dog licenses issued by a city each year since 1980. If $t = 0$ represents the year 1980, answer the following.

   a) How many dog licenses were issued in 1980?

   b) How many were issued in 2004?

   c) In what year would it be expected that 2072 dog licenses were issued?

86) Until the 1980s, Rock Glen was a rural community outside of a large city. In 1984, subdivisions of homes began to be built. The number of houses in Rock Glen t years after 1984 can be approximated by

$$H(t) = 142 + 58 \log_2(t + 1)$$

where $t = 0$ represents 1984.

   a) Determine the number of homes in Rock Glen in 1984.

   b) Determine the number of homes in Rock Glen in 1987.

   c) In what year were there approximately 374 homes?

87) A company plans to introduce a new type of cookie to the market. The company predicts that its sales over the next 24 months can be approximated by

$$S(t) = 14 \log_3(2t + 1)$$

where t is the number of months after the product is introduced, and $S(t)$ is in thousands of boxes of cookies.

   a) How many boxes of cookies were sold after they were on the market for 1 month?

   b) How many boxes were sold after they were on the market for 4 months?

   c) After 13 months, sales were approximately 43,000. Does this number fall short of, meet, or exceed the number of sales predicted by the formula?

88) Based on previous data, city planners have calculated that the number of tourists (in millions) to their city each year can be approximated by

$$N(t) = 10 + 1.2 \log_2(t + 2)$$

where t is the number of years after 1990.

   a) How many tourists visited the city in 1990?

   b) How many tourists visited the city in 1996?

   c) In 2004, actual data puts the number of tourists at 14,720,000. How does this number compare to the number predicted by the formula?

Section 13.4 Properties of Logarithms

Objectives

1. Use the Product Rule for Logarithms

2. Use the Quotient Rule for Logarithms

3. Use the Power Rule for Logarithms

4. Use the Properties $\log_a a^x = x$ and $a^{\log_a x} = x$

5. Combine the Properties of Logarithms to Rewrite Logarithmic Expressions

Logarithms have properties that are very useful in applications and in higher mathematics. In Section 13.3, we stated the following two properties.

If $a > 0$ and $a \neq 1$, then

1) $\log_a a = 1$

2) $\log_a 1 = 0$

In this section, we will learn more properties of logarithms, and we will practice using them because being able to use the properties of logarithms can make some very difficult mathematical calculations much easier. *The properties of logarithms come from the properties of exponents.*

1. Use the Product Rule for Logarithms

The product rule for logarithms can be derived from the product rule for exponents.

The Product Rule for Logarithms

Let $x, y,$ and a be positive real numbers where $a \neq 1$. Then,

$$\log_a xy = \log_a x + \log_a y$$

The logarithm of a product, xy, is the same as the sum of the logarithms of each factor, x and y.

 BE CAREFUL $\log_a xy \neq (\log_a x)(\log_a y)$

Example 1

Rewrite as the sum of logarithms and simplify, if possible. Assume the variables represent positive real numbers.

a) $\log_6(4 \cdot 7)$ b) $\log_4 16t$ c) $\log_8 y^3$ d) $\log 10pq$

Solution

a) The logarithm of a product equals the *sum* of the logs of the factors. Therefore,

$$\log_6(4 \cdot 7) = \log_6 4 + \log_6 7 \qquad \text{Product rule}$$

b) $\log_4 16t = \log_4 16 + \log_4 t \qquad$ Product rule
$ = 2 + \log_4 t \qquad \log_4 16 = 2$

Evaluate logarithms, like $\log_4 16$, when possible.

c) $\log_8 y^3 = \log_8(y \cdot y \cdot y)$ Write y^3 as $y \cdot y \cdot y$.

          $= \log_8 y + \log_8 y + \log_8 y$ Product rule

          $= 3 \log_8 y$

d) Recall that if no base is written, then it is assumed to be 10.

$$\log 10pq = \log 10 + \log p + \log q \quad \text{Product rule}$$
$$= 1 + \log p + \log q \quad \log 10 = 1$$

You Try 1

Rewrite as the sum of logarithms and simplify, if possible. Assume the variables represent positive real numbers.

a) $\log_9(2 \cdot 5)$ b) $\log_2 32k$ c) $\log_6 c^4$ d) $\log 100yz$

We can use the product rule for exponents in the "opposite" direction, too. That is, given the sum of logarithms we can write a single logarithm.

Example 2

Write as a single logarithm. Assume the variables represent positive real numbers.

a) $\log_8 5 + \log_8 3$ b) $\log 7 + \log r$ c) $\log_3 x + \log_3(x + 4)$

Solution

a) $\log_8 5 + \log_8 3 = \log_8(5 \cdot 3)$ Product rule

             $= \log_8 15$ $5 \cdot 3 = 15$

b) $\log 7 + \log r = \log 7r$ Product rule

c) $\log_3 x + \log_3(x + 4) = \log_3 x(x + 4)$ Product rule

                        $= \log_3(x^2 + 4x)$ Distribute.

BE CAREFUL $\log_a(x + y) \neq \log_a x + \log_a y$. Therefore, $\log_3(x^2 + 4x)$ does *not* equal $\log_3 x^2 + \log_3 4x$.

You Try 2

Write as a single logarithm. Assume the variables represent positive real numbers.

a) $\log_5 9 + \log_5 4$ b) $\log_6 13 + \log_6 c$ c) $\log y + \log(y - 6)$

2. Use the Quotient Rule for Logarithms

The quotient rule for logarithms can be derived from the quotient rule for exponents.

The Quotient Rule for Logarithms

Let x, y, and a be positive real numbers where $a \neq 1$. Then,

$$\log_a \frac{x}{y} = \log_a x - \log_a y$$

The logarithm of a quotient, $\frac{x}{y}$, is the same as the logarithm of the numerator *minus* the logarithm of the denominator.

BE CAREFUL

$$\log_a \frac{x}{y} \neq \frac{\log_a x}{\log_a y}.$$

 Example 3

Write as the difference of logarithms and simplify, if possible. Assume $w > 0$.

a) $\log_7 \dfrac{3}{10}$ b) $\log_3 \dfrac{81}{w}$

Solution

a) $\log_7 \dfrac{3}{10} = \log_7 3 - \log_7 10$ Quotient rule

b) $\log_3 \dfrac{81}{w} = \log_3 81 - \log_3 w$ Quotient rule

$\qquad\qquad = 4 - \log_3 w$ $\log_3 81 = 4$

 You Try 3

Write as the difference of logarithms and simplify, if possible. Assume $n > 0$.

a) $\log_6 \dfrac{2}{9}$ b) $\log_5 \dfrac{n}{25}$

Example 4

Write as a single logarithm. Assume the variable is defined so that the expressions are positive.

a) $\log_2 18 - \log_2 6$ b) $\log_4(z - 5) - \log_4(z^2 + 9)$

Solution

a) $\log_2 18 - \log_2 6 = \log_2 \dfrac{18}{6}$ Quotient rule

$\qquad\qquad\qquad = \log_2 3$ $\dfrac{18}{6} = 3$

b) $\log_4(z - 5) - \log_4(z^2 + 9) = \log_4 \dfrac{z - 5}{z^2 + 9}$ Quotient rule

BE CAREFUL $\log_a(x - y) \neq \log_a x - \log_a y$

You Try 4

Write as a single logarithm. Assume the variable is defined so that the expressions are positive.

a) $\log_4 36 - \log_4 3$ b) $\log_5(c^2 - 2) - \log_5(c + 1)$

3. Use the Power Rule for Logarithms

In Example 1c) we saw that $\log_8 y^3 = 3 \log_8 y$ since

$$\begin{aligned}
\log_8 y^3 &= \log_8(y \cdot y \cdot y) \\
&= \log_8 y + \log_8 y + \log_8 y \\
&= 3 \log_8 y
\end{aligned}$$

This result can be generalized as the next property and comes from the power rule for exponents.

The Power Rule for Logarithms

Let x and a be positive real numbers, where $a \neq 1$, and let r be any real number. Then,

$$\log_a x^r = r \log_a x$$

BE CAREFUL The rule applies to $\log_a x^r$ *not* $(\log_a x)^r$. Be sure you can distinguish between the two expressions.

Example 5

Rewrite each expression using the power rule and simplify, if possible. Assume the variables represent positive real numbers and that the variable bases are positive real numbers not equal to 1.

a) $\log_9 y^4$ b) $\log_2 8^5$ c) $\log_a \sqrt{3}$ d) $\log_w \dfrac{1}{w}$

Solution

a) $\log_9 y^4 = 4 \log_9 y$ Power rule

b) $\log_2 8^5 = 5 \log_2 8$ Power rule

           $= 5(3)$ $\log_2 8 = 3$

           $= 15$ Multiply.

c) *It is common practice to rewrite radicals as fractional exponents when applying the properties of logarithms.* This will be our first step.

$$\log_a \sqrt{3} = \log_a 3^{1/2} \qquad \text{Rewrite as a fractional exponent.}$$
$$= \frac{1}{2} \log_a 3 \qquad \text{Power rule}$$

d) Rewrite $\dfrac{1}{w}$ as w^{-1}.

$$\log_w \frac{1}{w} = \log_w w^{-1} \qquad \frac{1}{w} = w^{-1}$$
$$= -1 \log_w w \qquad \text{Power rule}$$
$$= -1(1) \qquad \log_w w = 1$$
$$= -1 \qquad \text{Multiply.}$$

You Try 5

Rewrite each expression using the power rule and simplify, if possible. Assume the variables represent positive real numbers and that the variable bases are positive real numbers not equal to 1.

a) $\log_8 t^9$ b) $\log_3 9^7$ c) $\log_a \sqrt[3]{5}$ d) $\log_m \dfrac{1}{m^8}$

The next properties we will look at can be derived from the power rule and from the fact that $f(x) = a^x$ and $g(x) = \log_a x$ are inverse functions.

4. Use the Properties $\log_a a^x = x$ and $a^{\log_a x} = x$

Other Properties of Logarithms

Let a be a positive real number such that $a \neq 1$. Then,

1) $\log_a a^x = x$ for any real number x.

2) $a^{\log_a x} = x$ for $x > 0$.

Example 6

Evaluate each expression.

a) $\log_6 6^7$ b) $\log 10^8$ c) $5^{\log_5 3}$

Solution

a) $\log_6 6^7 = 7$ $\log_a a^x = x$

b) $\log 10^8 = 8$ The base of the log is 10.

c) $5^{\log_5 3} = 3$ $a^{\log_a x} = x$

 You Try 6

Evaluate each expression.

a) $\log_3 3^{10}$ b) $\log 10^{-6}$ c) $7^{\log_7 9}$

Next is a summary of the properties of logarithms. The properties presented in Section 13.3 are included as well.

Properties of Logarithms

Let x, y, and a be positive real numbers where $a \neq 1$, and let r be any real number. Then,

1) $\log_a a = 1$

2) $\log_a 1 = 0$

3) $\log_a xy = \log_a x + \log_a y$ Product rule

4) $\log_a \dfrac{x}{y} = \log_a x - \log_a y$ Quotient rule

5) $\log_a x^r = r \log_a x$ Power rule

6) $\log_a a^x = x$ for any real number x

7) $a^{\log_a x} = x$

Many students make the same mistakes when working with logarithms. Keep in mind the following to avoid these common errors.

 BE CAREFUL

1) $\log_a xy \neq (\log_a x)(\log_a y)$

2) $\log_a(x + y) \neq \log_a x + \log_a y$

3) $\log_a \dfrac{x}{y} \neq \dfrac{\log_a x}{\log_a y}$

4) $\log_a(x - y) \neq \log_a x - \log_a y$

5) $(\log_a x)^r \neq r \log_a x$

5. Combine the Properties of Logarithms to Rewrite Logarithmic Expressions

Not only can the properties of logarithms simplify some very complicated computations, they are also needed for solving some types of logarithmic equations. The properties of logarithms are also used in calculus and many areas of science.

Next, we will see how to use different properties of logarithms together to rewrite logarithmic expressions.

Example 7

Write each expression as the sum or difference of logarithms in simplest form. Assume all variables represent positive real numbers and that the variable bases are positive real numbers not equal to 1.

a) $\log_8 r^5 t$ b) $\log_3 \dfrac{27}{ab^2}$ c) $\log_7 \sqrt{7p}$ d) $\log_a(4a + 5)$

Solution

a) $\log_8 r^5 t = \log_8 r^5 + \log_8 t$ Product rule
    $= 5 \log_8 r + \log_8 t$ Power rule

b) $\log_3 \dfrac{27}{ab^2} = \log_3 27 - \log_3 ab^2$ Quotient rule

    $= 3 - (\log_3 a + \log_3 b^2)$ $\log_3 27 = 3$; product rule
    $= 3 - (\log_3 a + 2 \log_3 b)$ Power rule
    $= 3 - \log_3 a - 2 \log_3 b$ Distribute.

c) $\log_7 \sqrt{7p} = \log_7(7p)^{1/2}$ Rewrite radical as fractional exponent.

    $= \dfrac{1}{2} \log_7 7p$ Power rule

    $= \dfrac{1}{2}(\log_7 7 + \log_7 p)$ Product rule

    $= \dfrac{1}{2}(1 + \log_7 p)$ $\log_7 7 = 1$

    $= \dfrac{1}{2} + \dfrac{1}{2} \log_7 p$ Distribute.

d) $\log_a(4a + 5)$ cannot be rewritten using any properties of logarithms. [Recall that $\log_a(x + y) \neq \log_a x + \log_a y$.]

You Try 7

Write each expression as the sum or difference of logarithms in simplest form. Assume all variables represent positive real numbers and that the variable bases are positive real numbers not equal to 1.

a) $\log_2 8s^2t^5$ b) $\log_a \dfrac{4c^2}{b^3}$ c) $\log_5 \sqrt[3]{\dfrac{25}{n}}$ d) $\dfrac{\log_4 k}{\log_4 m}$

Example 8

Write each as a single logarithm in simplest form. Assume the variable represents a positive real number.

a) $2 \log_7 5 + 3 \log_7 2$ b) $\dfrac{1}{2} \log_6 s - 3 \log_6(s^2 + 1)$

Solution

a) $\begin{aligned}
2 \log_7 5 + 3 \log_7 2 &= \log_7 5^2 + \log_7 2^3 && \text{Power rule} \\
&= \log_7 25 + \log_7 8 && 5^2 = 25;\ 2^3 = 8 \\
&= \log_7 (25 \cdot 8) && \text{Product rule} \\
&= \log_7 200 && \text{Multiply.}
\end{aligned}$

b) $\begin{aligned}
\dfrac{1}{2} \log_6 s - 3 \log_6(s^2 + 1) &= \log_6 s^{1/2} - \log_6(s^2 + 1)^3 && \text{Power rule} \\
&= \log_6 \sqrt{s} - \log_6(s^2 + 1)^3 && \text{Write in radical form.} \\
&= \log_6 \dfrac{\sqrt{s}}{(s^2 + 1)^3} && \text{Quotient rule}
\end{aligned}$

You Try 8

Write each as a single logarithm in simplest form. Assume the variables are defined so that the expressions are positive.

a) $2 \log 4 + \log 5$ b) $\dfrac{2}{3} \log_5 c + \dfrac{1}{3} \log_5 d - 2 \log_5(c - 6)$

Given the values of logarithms, we can compute the values of other logarithms using the properties we have learned in this section.

Example 9

Given that $\log 6 \approx 0.7782$ and $\log 4 \approx 0.6021$, use the properties of logarithms to approximate the following.

a) $\log 24$ b) $\log \sqrt{6}$

Solution

a)   To find the value of $\log 24$, we must determine how to write 24 in terms of 6 or 4 or some combination of the two.

$$24 = 6 \cdot 4$$
$$\begin{aligned}
\log 24 &= \log(6 \cdot 4) && 24 = 6 \cdot 4 \\
&= \log 6 + \log 4 && \text{Product rule} \\
&\approx 0.7782 + 0.6021 && \text{Substitute.} \\
&= 1.3803 && \text{Add.}
\end{aligned}$$

b) We can write $\sqrt{6}$ as $6^{1/2}$.

$$\log \sqrt{6} = \log 6^{1/2} \qquad \sqrt{6} = 6^{1/2}$$
$$= \frac{1}{2} \log 6 \qquad \text{Power rule}$$
$$\approx \frac{1}{2}(0.7782) \qquad \log 6 \approx 0.7782$$
$$= 0.3891 \qquad \text{Multiply.}$$

You Try 9

Using the values given in Example 9, approximate each of the following.

a) $\log 16$ b) $\log \dfrac{6}{4}$ c) $\log \sqrt[3]{4}$ d) $\log \dfrac{1}{6}$

Answers to You Try Exercises

1) a) $\log_9 2 + \log_9 5$ b) $5 + \log_2 k$ c) $4 \log_6 c$ d) $2 + \log y + \log z$ 2) a) $\log_5 36$
b) $\log_6 13c$ c) $\log(y^2 - 6y)$ 3) a) $\log_6 2 - \log_6 9$ b) $\log_5 n - 2$ 4) a) $\log_4 12$
b) $\log_5 \dfrac{c^2 - 2}{c + 1}$ 5) a) $9 \log_8 t$ b) 14 c) $\dfrac{1}{3} \log_a 5$ d) -8 6) a) 10 b) -6 c) 9
7) a) $3 + 2 \log_2 s + 5 \log_2 t$ b) $\log_a 4 + 2 \log_a c - 3 \log_a b$ c) $\dfrac{2}{3} - \dfrac{1}{3} \log_5 n$
d) cannot be simplified   8) a) $\log 80$ b) $\log_5 \dfrac{\sqrt[3]{c^2 d}}{(c - 6)^2}$ 9) a) 1.2042 b) 0.1761
c) 0.2007 d) -0.7782

13.4 Exercises

Objectives 1–5

Decide whether each statement is true or false.

1) $\log_6 8c = \log_6 8 + \log_6 c$

2) $\log_5 \dfrac{m}{3} = \log_5 m - \log_5 3$

3) $\log_9 \dfrac{7}{2} = \dfrac{\log_9 7}{\log_9 2}$

4) $\log 1000 = 3$

5) $(\log_4 k)^2 = 2 \log_4 k$

6) $\log_2(x^2 + 8) = \log_2 x^2 + \log_2 8$

7) $5^{\log_5 4} = 4$

8) $\log_3 4^5 = 5 \log_3 4$

Write as the sum or difference of logarithms and simplify, if possible. Assume all variables represent positive real numbers.

9) $\log_8(3 \cdot 10)$

10) $\log_2(6 \cdot 5)$

11) $\log_7 5d$

12) $\log_4 6w$

13) $\log_9 \dfrac{4}{7}$

14) $\log_5 \dfrac{20}{17}$

15) $\log_5 2^3$

16) $\log_8 10^4$

17) $\log p^8$

18) $\log_3 z^5$

 19) $\log_3 \sqrt{7}$

20) $\log_7 \sqrt[3]{4}$

21) $\log_5 25t$

22) $\log_2 16p$

23) $\log_2 \dfrac{8}{k}$

24) $\log_3 \dfrac{x}{9}$

25) $\log_7 49^3$

26) $\log_8 64^{12}$

27) $\log 1000b$

28) $\log_3 27m$

29) $\log_2 32^7$

30) $\log_2 2^9$

31) $\log_5 \sqrt{5}$

32) $\log \sqrt[3]{10}$

33) $\log \sqrt[3]{100}$

34) $\log_2 \sqrt{8}$

35) $\log_6 w^4 z^3$

36) $\log_5 x^2 y$

37) $\log_7 \dfrac{a^2}{b^5}$

38) $\log_4 \dfrac{s^4}{t^6}$

39) $\log \dfrac{\sqrt[5]{11}}{y^2}$

40) $\log_3 \dfrac{\sqrt{x}}{y^4}$

41) $\log_2 \dfrac{4\sqrt{n}}{m^3}$

42) $\log_9 \dfrac{gf^2}{h^3}$

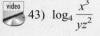 43) $\log_4 \dfrac{x^3}{yz^2}$

44) $\log \dfrac{3}{ab^2}$

45) $\log_5 \sqrt{5c}$

46) $\log_8 \sqrt[3]{\dfrac{z}{8}}$

47) $\log k(k-6)$

48) $\log_2 \dfrac{m^5}{m^2 + 3}$

Write as a single logarithm. Assume the variables are defined so that the variable expressions are positive and so that the bases are positive real numbers not equal to 1.

49) $\log_a m + \log_a n$

50) $\log_4 7 + \log_4 x$

51) $\log_7 d - \log_7 3$

52) $\log_p r - \log_p s$

53) $4 \log_3 f + \log_3 g$

54) $5 \log_y m + 2 \log_y n$

55) $\log_8 t + 2 \log_8 u - 3 \log_8 v$

56) $3 \log a + 4 \log c - 6 \log b$

57) $\log(r^2 + 3) - 2 \log(r^2 - 3)$

58) $2 \log_2 t - 3 \log_2(5t + 1)$

59) $3 \log_n 2 + \dfrac{1}{2} \log_n k$

60) $2 \log_z 9 + \dfrac{1}{3} \log_z w$

61) $\dfrac{1}{3} \log_d 5 - 2 \log_d z$

62) $\dfrac{1}{2} \log_5 a - 4 \log_5 b$

 63) $\log_6 y - \log_6 3 - 3 \log_6 z$

64) $\log_7 8 - 4 \log_7 x - \log_7 y$

65) $4 \log_3 t - 2 \log_3 6 - 2 \log_3 u$

66) $2 \log_9 m - 4 \log_9 2 - 4 \log_9 n$

67) $\dfrac{1}{2} \log_b (c + 4) - 2 \log_b(c + 3)$

68) $\dfrac{1}{2} \log_a r + \dfrac{1}{2} \log_a(r - 2) - \log_a(r + 2)$

Given that $\log 5 \approx 0.6990$ and $\log 9 \approx 0.9542$, use the properties of logarithms to approximate the following. **Do not use a calculator.**

 69) $\log 45$

70) $\log 25$

71) $\log 81$

72) $\log \dfrac{9}{5}$

73) $\log \dfrac{5}{9}$

74) $\log \sqrt{5}$

75) $\log 3$

76) $\log \dfrac{1}{9}$

77) $\log \dfrac{1}{5}$

78) $\log 5^8$

79) $\log \dfrac{1}{81}$

80) $\log 90$

81) $\log 50$

82) $\log \dfrac{25}{9}$

Section 13.5 Common and Natural Logarithms and Change of Base

In this section, we will focus our attention on two widely used logarithmic bases—base 10 and base e.

Common Logarithms

1. Evaluate Common Logarithms Without a Calculator

In Section 13.3, we said that a base 10 logarithm is called a **common logarithm**. It is often written as $\log x$.

$$\log x \text{ means } \log_{10} x$$

We can evaluate many logarithms without the use of a calculator because we can write them in terms of a base of 10.

Example 1

Evaluate.

a) $\log 1000$ b) $\log \dfrac{1}{100}$

Solution

a) $\log 1000 = 3$ since $10^3 = 1000$

b) $\log \dfrac{1}{100} = \log 10^{-2}$ $\dfrac{1}{100} = 10^{-2}$

$= \log_{10} 10^{-2}$

$= -2$ $\log_a a^x = x$

You Try 1

Evaluate.

a) $\log 100{,}000$ b) $\log \dfrac{1}{10}$

2. Evaluate Common Logarithms Using a Calculator

Common logarithms are used throughout mathematics and other fields to make calculations easier to solve applications. Often, however, we need a calculator to evaluate the logarithms. Next we will learn how to use a calculator to find the value of a base 10 logarithm. *We will approximate the value to four decimal places.*

Example 2

Find log 12.

Solution

Enter $\boxed{12}$ $\boxed{\text{LOG}}$ or $\boxed{\text{LOG}}$ $\boxed{12}$ $\boxed{\text{ENTER}}$ into your calculator.

$$\log 12 \approx 1.0792$$

(Note that $10^{1.0792} \approx 12$. Press $\boxed{10}$ $\boxed{y^x}$ $\boxed{1.0792}$ $\boxed{=}$ to evaluate $10^{1.0792}$.)

 You Try 2

Find log 3.

We can solve logarithmic equations with or without the use of a calculator.

3. Solve an Equation Containing a Common Logarithm

Example 3

Solve log $x = -3$.

Solution

Change to exponential form, and solve for x.

$$\log x = -3 \quad \text{means} \quad \log_{10} x = -3$$
$$10^{-3} = x \qquad \text{Exponential form}$$
$$\frac{1}{1000} = x$$

The solution set is $\left\{ \dfrac{1}{1000} \right\}$. This is the *exact* solution.

 You Try 3

Solve log $x = 2$.

For the equation in Example 4, we will give an exact solution *and* a solution that is approximated to four decimal places. This will give us an idea of the size of the exact solution.

Example 4

Solve $\log x = 2.4$. Give an exact solution and a solution that is approximated to four decimal places.

Solution

Change to exponential form, and solve for x.

$$\log x = 2.4 \quad \text{means} \quad \log_{10} x = 2.4$$
$$10^{2.4} = x \qquad \text{Exponential form}$$
$$251.1886 \approx x \qquad \text{Approximation}$$

The exact solution is $\{10^{2.4}\}$. This is approximately $\{251.1886\}$.

You Try 4

Solve $\log x = 0.7$. Give an exact solution and a solution that is approximated to four decimal places.

4. Solve an Applied Problem Given an Equation Containing a Common Logarithm

Example 5

The loudness of sound, $L(I)$ in decibels (dB), is given by

$$L(I) = 10 \log \frac{I}{10^{-12}}$$

where I is the intensity of sound in watts per square meter (W/m^2). Fifty meters from the stage at a concert, the intensity of sound is $0.01 \ \text{W/m}^2$. Find the loudness of the music at the concert 50 m from the stage.

Solution

Substitute 0.01 for I and find $L(0.01)$.

$$L(0.01) = 10 \log \frac{0.01}{10^{-12}}$$
$$= 10 \log \frac{10^{-2}}{10^{-12}} \qquad 0.01 = 10^{-2}$$
$$= 10 \log 10^{10} \qquad \text{Quotient rule for exponents}$$
$$= 10(10) \qquad \log 10^{10} = 10$$
$$= 100$$

The sound level of the music 50 m from the stage is 100 dB. (To put this in perspective, a normal conversation has a loudness of about 50 dB.)

 **You Try 5**

The intensity of sound from a thunderstorm is about 0.001 W/m². Find the loudness of the storm, in decibels.

Natural Logarithms

5. Define a Natural Logarithm

Another base that is often used for logarithms is the number e. In Section 13.2, we said that e, like π, is an irrational number. To four decimal places, $e \approx 2.7183$.

A base e logarithm is called a **natural logarithm** or **natural log.** The notation used for a base e logarithm is $\ln x$ (read as "*the natural log of x*" or "*ln of x*"). Since it is a base e logarithm, it is important to remember that

$$\ln x \text{ means } \log_e x$$

We can graph $y = \ln x$ the same way we graph other logarithmic functions.

$$y = \ln x \quad \text{means} \quad \begin{array}{l} y = \log_e x \\ x = e^y \end{array} \quad \text{Write } y = \log_e x \text{ in exponential form.}$$

Make a table of values, choose values for x, use a calculator to approximate the values of x, plot the points, and sketch the curve.

$x = e^y$	
x	**y**
1	0
2.72	1
7.39	2
0.37	−1
0.14	−2

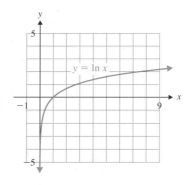

Additionally, the inverse of $y = \ln x$ is $y = e^x$. Verify this by following the procedure for finding the inverse of a function.

6. Evaluate Natural Logarithms Without a Calculator

Using the properties $\log_a a^x = x$ and $\log_a a = 1$, we can find the value of some natural logarithms without using a calculator.

Example 6

Evaluate.

a) $\ln e$ b) $\ln e^2$

Solution

a)   To evaluate ln e, remember that

$$\ln e = \log_e e = 1$$
$$\text{since} \quad \log_a a = 1$$

ln e = 1

This is a value you should remember. We will use this in Section 13.6 to solve exponential equations with base e.

b) $\ln e^2 = \log_e e^2 \qquad \log_a a^x = x$
$$= 2$$
$$\ln e^2 = 2$$

You Try 6

Evaluate.

a) 5 ln e b) ln e^8

Next we will learn how to use a calculator to *approximate natural logarithms to four decimal places.*

7. Evaluate Natural Logarithms Using a Calculator

Example 7

Find ln 5.

Solution

Enter $\boxed{5}\ \boxed{\text{LN}}$ or $\boxed{\text{LN}}\ \boxed{5}\ \boxed{\text{ENTER}}$ into your calculator.

$$\ln 5 \approx 1.6094$$

(Note that $e^{1.6094} \approx 5$.)

You Try 7

Find ln 9.

8. Solve an Equation Containing a Natural Logarithm

To solve an equation containing a natural logarithm like ln $x = 4$, we change to exponential form and solve for the variable. We can give an exact solution and a solution that is approximated to four decimal places.

Example 8

Solve each equation. Give an exact solution and a solution that is approximated to four decimal places.

a) $\ln x = 4$ b) $\ln(2x + 5) = 3.8$

Solution

a) $\ln x = 4$ means $\log_e x = 4$

$$e^4 = x \qquad \text{Exponential form}$$
$$54.5982 \approx x \qquad \text{Approximation}$$

The exact solution is $\{e^4\}$. This is approximately $\{54.5982\}$.

b) $\ln(2x + 5) = 3.8$ means $\log_e(2x + 5) = 3.8$

$$e^{3.8} = 2x + 5 \qquad \text{Exponential form}$$
$$e^{3.8} - 5 = 2x \qquad \text{Subtract 5.}$$
$$\frac{e^{3.8} - 5}{2} = x \qquad \text{Divide by 2.}$$
$$19.8506 \approx x \qquad \text{Approximation}$$

The exact solution is $\left\{\dfrac{e^{3.8} - 5}{2}\right\}$. This is approximately $\{19.8506\}$.

You Try 8

Solve each equation. Give an exact solution and a solution that is approximated to four decimal places.

a) $\ln y = 2.7$ b) $\ln(3a - 1) = 0.5$

9. Solve Applied Problems Using Exponential Functions

One of the most practical applications of exponential functions is for compound interest.

Definition

Compound Interest: The amount of money, *A*, in dollars, in an account after *t* years is given by

$$A = P\left(1 + \frac{r}{n}\right)^{nt}$$

where *P* (the principal) is the amount of money (in dollars) deposited in the account, *r* is the annual interest rate, and *n* is the number of times the interest is compounded (paid) per year.

We can also think of this formula in terms of the amount of money owed, *A*, after *t* yr when *P* is the amount of money loaned.

Example 9

If $2000 is deposited in an account paying 4% per year, find the total amount in the account after 5 yr if the interest is compounded

a) quarterly. b) monthly.

(We assume no withdrawals or additional deposits are made.)

Solution

a) If interest compounds quarterly, then interest is paid four times per year. Use

$$A = P\left(1 + \frac{r}{n}\right)^{nt}$$

with $P = 2000$, $r = 0.04$, $t = 5$, $n = 4$.

$$A = 2000\left(1 + \frac{0.04}{4}\right)^{4(5)}$$
$$= 2000(1.01)^{20}$$
$$\approx 2440.3801$$

Since A is an amount of money, round to the nearest cent.
The account will contain $2440.38 after 5 yr.

b) If interest is compounded monthly, then interest is paid 12 times per year. Use

$$A = P\left(1 + \frac{r}{n}\right)^{nt}$$

with $P = 2000$, $r = 0.04$, $t = 5$, $n = 12$

$$A = 2000\left(1 + \frac{0.04}{12}\right)^{12(5)}$$
$$\approx 2441.9932$$

Round A to the nearest cent.
The account will contain $2441.99 after 5 yr.

You Try 9

If $1500 is deposited in an account paying 5% per year, find the total amount in the account after 8 yr if the interest is compounded

a) monthly. b) weekly.

In Example 9 we saw that the account contained more money after 5 yr when the interest compounded monthly (12 times per year) versus quarterly (four times per year). This will always be true. The more often interest is compounded each year, the more money that accumulates in the account.

If interest *compounds continuously* we obtain the formula for *continuous compounding, $A = Pe^{rt}$.*

Definition

Continuous Compounding: If *P* dollars is deposited in an account earning interest rate *r* compounded continuously, then the amount of money, *A* (in dollars), in the account after *t* years is given by

$$A = Pe^{rt}$$

Example 10

Determine the amount of money in an account after 5 yr if $2000 was initially invested at 4% compounded continuously.

Solution

Use $A = Pe^{rt}$ with $P = 2000$, $r = 0.04$, and $t = 5$.

$$
\begin{aligned}
A &= 2000e^{0.04(5)} &&\text{Substitute values.}\\
&= 2000e^{0.20} &&\text{Multiply } (0.04)(5).\\
&\approx 2442.8055 &&\text{Evaluate using a calculator.}
\end{aligned}
$$

Round *A* to the nearest cent.

The account will contain $2442.81 after 5 yr. Note that, as expected, this is more than the amounts obtained in Example 9 when the same amount was deposited for 5 yr at 4% but the interest was compounded quarterly and monthly.

 You Try 10

Determine the amount of money in an account after 8 yr if $1500 was initially invested at 5% compounded continuously.

---

10. Use the Change-of-Base Formula

Sometimes, we need to find the value of a logarithm with a base other than 10 or *e*—like $\log_3 7$. Some calculators, however, do not calculate logarithms other than common logs (base 10) and natural logs (base *e*). In such cases, we can use the change-of-base formula to evaluate logarithms with bases other than 10 or *e*.

Definition

Change-of-Base Formula: If *a*, *b*, and *x* are positive real numbers and $a \neq 1$ and $b \neq 1$, then

$$\log_a x = \frac{\log_b x}{\log_b a}.$$

We can choose any positive real number not equal to 1 for *b*, but it is most convenient to choose 10 or *e* since these will give us common logarithms and natural logarithms, respectively.

Example 11

Find the value of $\log_3 7$ to four decimal places using

a) common logarithms.   b) natural logarithms.

Solution

a) The base we will use to evaluate $\log_3 7$ is 10—this is the base of a common logarithm. Then,

$$\log_3 7 = \frac{\log_{10} 7}{\log_{10} 3} \qquad \text{Change of base formula}$$
$$\approx 1.7712 \qquad \text{Use a calculator.}$$

b) The base of a natural logarithm is e. Then,

$$\log_3 7 = \frac{\log_e 7}{\log_e 3}$$
$$= \frac{\ln 7}{\ln 3}$$
$$\approx 1.7712 \qquad \text{Use a calculator.}$$

Using either base 10 or base e gives us the same result.

You Try 11

Find the value of $\log_5 38$ to four decimal places using

a) common logarithms.   b) natural logarithms.

Using Technology

Here is the graph of $y = \ln x$. Let's look at how we can use the properties of logarithms to predict what the graphs of other natural log functions will look like.

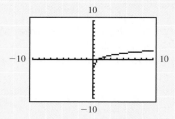

1. Use the properties of logarithms to rewrite $\ln(ex)$.

2. Use the properties of logarithms to rewrite $\ln\left(\dfrac{x}{e}\right)$.

3. Based on the result in 1, how should the graph of $y = \ln(ex)$ compare to the graph of $y = \ln x$?

4. Based on the result in 2, how should the graph of $y = \ln\left(\dfrac{x}{e}\right)$ compare to the graph of $y = \ln x$?

5. On the same screen, graph $y = \ln x, y = \ln(ex)$, and $y = \ln\left(\dfrac{x}{e}\right)$ to verify your conclusions in 3 and 4.

Answers to You Try Exercises

1) a) 5   b) −1   2) 0.4771   3) {100}   4) {$10^{0.7}$}; {5.0119}   5) 90 dB

6) a) 5   b) 8   c) $\dfrac{1}{2}$ d) −3   7) 2.1972   8) a) {$e^{2.7}$}; {14.8797}   b) $\left\{\dfrac{e^{0.5}+1}{3}\right\}$; {0.8829}

9) a) $2235.88   b) $2237.31   10) $2237.74   11) a) 2.2602   b) 2.2602

Answers to Technology Exercises

1. $\ln(ex) = 1 + \ln x$ 2. $\ln\left(\dfrac{x}{e}\right) = \ln(x) - 1$ 3. The graph of $y = \ln(ex)$ will be the same shape as the graph of $y = \ln x$ but $y = \ln(ex)$ will be shifted up 1 unit.   4. The graph of $y = \ln\left(\dfrac{x}{e}\right)$ will be the same shape as the graph of $y = \ln x$ but $y = \ln\left(\dfrac{x}{e}\right)$ will be shifted down 1 unit.

5.

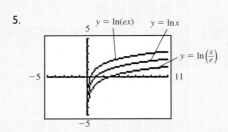

13.5 Exercises

Objectives 1, 5, and 6

1) What is the base of $\ln x$?

2) What is the base of $\log x$?

Evaluate each logarithm. Do *not* use a calculator.

3) $\log 100$

4) $\log 10,000$

5) $\log \dfrac{1}{1000}$

6) $\log \dfrac{1}{100,000}$

7) $\log 0.1$

8) $\log 0.01$

9) $\log 10^9$

10) $\log 10^7$

11) $\log \sqrt[4]{10}$

12) $\log \sqrt[5]{10}$

13) $\ln e^6$

14) $\ln e^{10}$

15) $\ln \sqrt{e}$

16) $\ln \sqrt[3]{e}$

17) $\ln \dfrac{1}{e^5}$

18) $\ln \dfrac{1}{e^2}$

19) $\ln 1$

20) $\log 1$

Objectives 2 and 7

Use a calculator to find the approximate value of each logarithm to four decimal places.

21) $\log 16$

22) $\log 23$

23) $\log 0.5$

24) $\log 627$

25) $\ln 3$

26) $\ln 6$

27) $\ln 1.31$

28) $\ln 0.218$

Objective 3

Solve each equation. Do *not* use a calculator.

29) $\log x = 3$

30) $\log z = 5$

31) $\log k = -1$

32) $\log c = -2$

33) $\log(4a) = 2$

34) $\log(5w) = 1$

35) $\log(3t + 4) = 1$

36) $\log(2p + 12) = 2$

Objectives 3 and 8

Solve each equation. Give an exact solution and a solution that is approximated to four decimal places.

37) $\log a = 1.5$

38) $\log y = 1.8$

39) $\log r = 0.8$

40) $\log c = 0.3$

41) $\ln x = 1.6$

42) $\ln p = 1.1$

43) $\ln t = -2$

44) $\ln z = 0.25$

45) $\ln(3q) = 2.1$

46) $\ln\left(\frac{1}{4}m\right) = 3$

47) $\log\left(\frac{1}{2}c\right) = 0.47$

48) $\log(6k) = -1$

49) $\log(5y - 3) = 3.8$

50) $\log(8x + 15) = 2.7$

51) $\ln(10w + 19) = 1.85$

52) $\ln(7a - 4) = 0.6$

53) $\ln(2d - 5) = 0$

54) $\log(3t + 14) = 2.4$

Objective 10

Use the change-of-base formula with either base 10 or base e to approximate each logarithm to four decimal places.

55) $\log_2 13$

56) $\log_6 25$

57) $\log_9 70$

58) $\log_3 52$

59) $\log_{1/3} 16$

60) $\log_{1/2} 23$

61) $\log_5 3$

62) $\log_7 4$

Objectives 4 and 9

For Exercises 63–66, use the formula

$$L(I) = 10 \log \frac{I}{10^{-12}}$$

where I is the intensity of sound, in watts per square meter, and $L(I)$ is the loudness of sound in decibels. Do *not* use a calculator.

63) The intensity of sound from fireworks is about 0.1 W/m². Find the loudness of the fireworks, in decibels.

64) The intensity of sound from a dishwasher is about 0.000001 W/m². Find the loudness of the dishwasher, in decibels.

65) The intensity of sound from a refrigerator is about 0.00000001 W/m². Find the loudness of the refrigerator, in decibels.

66) The intensity of sound from the takeoff of a space shuttle is $1,000,000$ W/m². Find the loudness of the sound made by the space shuttle at take-off, in decibels.

Use the formula $A = P\left(1 + \frac{r}{n}\right)^{nt}$ to solve each problem. See Example 9.

67) Isabel deposits $3000 in an account earning 5% per year compounded monthly. How much will be in the account after 3 yr?

68) How much money will Pavel have in his account after 8 yr if he initially deposited $6000 at 4% interest compounded quarterly?

69) Find the amount Christopher owes at the end of 5 yr if he borrows $4000 at a rate of 6.5% compounded quarterly.

70) How much will Anna owe at the end of 4 yr if she borrows $5000 at a rate of 7.2% compounded weekly?

Use the formula $A = Pe^{rt}$ to solve each problem. See Example 10.

71) If $3000 is deposited in an account earning 5% compounded continuously, how much will be in the account after 3 yr?

72) If \$6000 is deposited in an account earning 4% compounded continuously, how much will be in the account after 8 yr?

73) How much will Cyrus owe at the end of 6 yr if he borrows \$10,000 at a rate of 7.5% compounded continuously?

74) Find the amount Nadia owes at the end of 5 yr if she borrows \$4500 at a rate of 6.8% compounded continuously.

 75) The number of bacteria, $N(t)$, in a culture t hr after the bacteria is placed in a dish is given by

$$N(t) = 5000e^{0.0617t}$$

a) How many bacteria were originally in the culture?

b) How many bacteria are present after 8 hr?

 76) The number of bacteria, $N(t)$, in a culture t hr after the bacteria is placed in a dish is given by

$$N(t) = 8000e^{0.0342t}$$

a) How many bacteria were originally in the culture?

b) How many bacteria are present after 10 hr?

 77) The function $N(t) = 10{,}000e^{0.0492t}$ describes the number of bacteria in a culture t hr after 10,000 bacteria were placed in the culture. How many bacteria are in the culture after 1 day?

 78) How many bacteria are present 2 days after 6000 bacteria are placed in a culture if the number of bacteria in the culture is

$$N(t) = 6000e^{0.0285t}$$

t hr after the bacteria is placed in a dish?

In chemistry, the pH of a solution is given by

$$\text{pH} = -\log[\text{H}^+]$$

where $[\text{H}^+]$ is the molar concentration of the hydronium ion. A neutral solution has pH $= 7$. *Acidic solutions* have pH < 7, and *basic solutions* have pH > 7.

For Exercises 79–82, the hydronium ion concentrations, $[\text{H}^+]$, are given for some common substances. Find the pH of each substance (to the tenths place), and determine whether each substance is acidic or basic.

 79) Cola: $[\text{H}^+] = 2 \times 10^{-3}$

 80) Tomatoes: $[\text{H}^+] = 1 \times 10^{-4}$

81) Ammonia: $[\text{H}^+] = 6 \times 10^{-12}$

82) Egg white: $[\text{H}^+] = 2 \times 10^{-8}$

Extension

83) Show that the inverse of $y = \ln x$ is $y = e^x$.

Section 13.6 Solving Exponential and Logarithmic Equations

Objectives

1. Solve an Exponential Equation by Taking the Natural Logarithm of Both Sides

2. Solve Logarithmic Equations Using the Properties of Logarithms

3. Solve Applied Problems Involving Exponential Functions Using a Calculator

4. Solve an Applied Problem Involving Exponential Growth or Decay

In this section, we will learn another property of logarithms that will allow us to solve additional types of exponential and logarithmic equations.

Properties for Solving Exponential and Logarithmic Equations

Let a, x, and y be positive, real numbers, where $a \neq 1$.

1) If $x = y$, then $\log_a x = \log_a y$.

2) If $\log_a x = \log_a y$, then $x = y$.

For example, 1) tells us that if $x = 3$, then $\log_a x = \log_a 3$. Likewise, 2) tells us that if $\log_a 5 = \log_a y$, then $5 = y$.

We can use the properties above to solve exponential and logarithmic equations that we could not solve previously.

1. Solve an Exponential Equation by Taking the Natural Logarithm of Both Sides

We will look at two types of exponential equations—equations where both sides *can* be expressed with the same base and equations where both sides *cannot* be expressed with the same base.

If the two sides of an exponential equation *cannot* be expressed with the same base, we will use logarithms to solve the equation.

Example 1

Solve.

a) $2^x = 8$ b) $2^x = 12$

Solution

a)   Since 8 is a power of 2, we can solve $2^x = 8$ by expressing each side of the equation with the same base and setting the exponents equal to each other.

$$2^x = 8$$
$$2^x = 2^3 \qquad 8 = 2^3$$
$$x = 3 \qquad \text{Set the exponents equal.}$$

The solution set is $\{3\}$.

b)   Can we express both sides of $2^x = 12$ with the same base? No. *We will use property 1) to solve* $2^x = 12$ *by taking the logarithm of each side.*

We can use a logarithm of *any* base. It is most convenient to use base 10 (common logarithm) or base e (natural logarithm) because this is what we can find most easily on our calculators. *We will take the natural log of both sides.*

$$2^x = 12$$
$$\ln 2^x = \ln 12 \qquad \text{Take the natural log of each side.}$$
$$x \ln 2 = \ln 12 \qquad \log_a x^r = r \log_a x$$
$$x = \frac{\ln 12}{\ln 2} \qquad \text{Divide by } \ln 2.$$

The exact solution is $\left\{ \dfrac{\ln 12}{\ln 2} \right\}$. Use a calculator to get an approximation to four decimal places.

$$x \approx 3.5850$$

The approximation is $\{3.5850\}$. We can verify the solution by substituting it for x in $2^x = 12$.

$$2^{3.5850} \approx 12$$

BE CAREFUL

$$\frac{\ln 12}{\ln 2} \neq \ln 6$$

Solving an Exponential Equation

Begin by asking yourself, *"Can I express each side with the same base?"*

1) If the answer is **yes**, then write each side of the equation with the same base, set the exponents equal, and solve for the variable.

2) If the answer is **no**, then take the natural logarithm of each side, use the properties of logarithms, and solve for the variable.

You Try 1

Solve.

a) $3^{a-5} = 9$ b) $3^t = 24$

Example 2

Solve $5^{x-2} = 16$.

Solution

Ask yourself, *"Can I express each side with the same base?"* **No**. Therefore, take the natural log of each side.

$$5^{x-2} = 16$$
$$\ln 5^{x-2} = \ln 16 \qquad \text{Take the natural log of each side.}$$
$$(x-2)\ln 5 = \ln 16 \qquad \log_a x^r = r \log_a x$$

$(x - 2)$ *must* be in parentheses since it contains two terms.

$$x \ln 5 - 2 \ln 5 = \ln 16 \qquad \text{Distribute.}$$
$$x \ln 5 = \ln 16 + 2 \ln 5 \qquad \text{Add 2 ln 5 to get the } x\text{-term by itself.}$$
$$x = \frac{\ln 16 + 2 \ln 5}{\ln 5} \qquad \text{Divide by ln 5.}$$

The exact solution is $\left\{ \dfrac{\ln 16 + 2 \ln 5}{\ln 5} \right\}$. This is approximately $\{3.7227\}$.

You Try 2

Solve $9^{k+4} = 2$.

Recall that $\ln e = 1$. This property is the reason it is convenient to take the *natural logarithm* of both sides of an equation when a base is e.

Example 3

Solve $e^{5n} = 4$.

Solution

Begin by taking the natural log of each side.

$$e^{5n} = 4$$
$$\ln e^{5n} = \ln 4 \qquad \text{Take the natural log of each side.}$$
$$5n \ln e = \ln 4 \qquad \log_a x^r = r \log_a x$$
$$5n(1) = \ln 4 \qquad \ln e = 1$$
$$5n = \ln 4$$
$$n = \frac{\ln 4}{5} \qquad \text{Divide by 5.}$$

The exact solution is $\left\{ \dfrac{\ln 4}{5} \right\}$. The approximation is $\{0.2773\}$.

You Try 3

Solve $e^{6c} = 2$.

2. Solve Logarithmic Equations Using the Properties of Logarithms

We learned earlier that to solve a logarithmic equation like $\log_2(t + 5) = 4$ we write the equation in exponential form and solve for the variable.

$$\log_2(t + 5) = 4$$
$$2^4 = t + 5 \qquad \text{Write in exponential form.}$$
$$16 = t + 5 \qquad 2^4 = 16$$
$$11 = t \qquad \text{Subtract 5.}$$

In this section, we will learn how to solve other types of logarithmic equations as well. We will look at equations where

1) each term in the equation contains a logarithm.

2) one term in the equation does *not* contain a logarithm.

To Solve an Equation Where Each Term Contains a Logarithm

1) Use the properties of logarithms to write the equation in the form $\log_a x = \log_a y$.

2) Set $x = y$ and solve for the variable.

3) Check the proposed solution(s) in the original equation to be sure the values satisfy the equation.

Example 4

Solve.

a) $\log_5(m - 4) = \log_5 9$ b) $\log x + \log(x + 6) = \log 16$

Solution

a)   To solve $\log_5(m - 4) = \log_5 9$, use the property that states if $\log_a x = \log_a y$, then $x = y$.

$$\log_5(m - 4) = \log_5 9$$
$$m - 4 = 9$$
$$m = 13 \qquad \text{Add 4.}$$

Check to be sure that $m = 13$ satisfies the original equation.

$$\log_5(13 - 4) \overset{?}{=} \log_5 9$$
$$\log_5 9 = \log_5 9 \quad ✓$$

The solution set is $\{13\}$.

b)   To solve $\log x + \log(x + 6) = \log 16$, we must begin by using the product rule for logarithms to obtain one logarithm on the left side.

$$\log x + \log(x + 6) = \log 16$$
$$\log x(x + 6) = \log 16 \qquad \text{Product rule}$$
$$x(x + 6) = 16 \qquad \text{If } \log_a x = \log_a y, \text{ then } x = y.$$
$$x^2 + 6x = 16 \qquad \text{Distribute.}$$
$$x^2 + 6x - 16 = 0 \qquad \text{Subtract 16.}$$
$$(x + 8)(x - 2) = 0 \qquad \text{Factor.}$$
$$x + 8 = 0 \quad \text{or} \quad x - 2 = 0 \qquad \text{Set each factor equal to 0.}$$
$$x = -8 \quad \text{or} \qquad x = 2 \qquad \text{Solve.}$$

Check to be sure that $x = -8$ and $x = 2$ satisfy the original equation.

Check $x = -8$:
$$\log x + \log(x + 6) = \log 16$$
$$\log(-8) + \log(-8 + 6) \overset{?}{=} \log 16$$

We reject $x = -8$ as a solution because it leads to $\log(-8)$, which is undefined.

Check $x = 2$:
$$\log x + \log(x + 6) = \log 16$$
$$\log 2 + \log(2 + 6) \overset{?}{=} \log 16$$
$$\log 2 + \log 8 \overset{?}{=} \log 16$$
$$\log(2 \cdot 8) \overset{?}{=} \log 16$$
$$\log 16 = \log 16$$

$x = 2$ satisfies the original equation.

The solution set is $\{2\}$.

BE CAREFUL   Just because a proposed solution is a negative number does *not* mean it should be rejected. You *must* check it in the original equation; it may satisfy the equation.

You Try 4

Solve.

a) $\log_8(z + 3) = \log_8 5$ b) $\log_3 c + \log_3(c - 1) = \log_3 12$

To Solve an Equation Where One Term Does *Not* Contain a Logarithm

1) Use the properties of logarithms to get one logarithm on one side of the equation and a constant on the other side. That is, write the equation in the form $\log_a x = y$.

2) Write $\log_a x = y$ in exponential form, $a^y = x$, and solve for the variable.

3) Check the proposed solution(s) in the original equation to be sure the values satisfy the equation.

Example 5

Solve $\log_2 3w - \log_2(w - 5) = 3$.

Solution

Notice that one term in the equation $\log_2 3w - \log_2(w - 5) = 3$ does *not* contain a logarithm. Therefore, we want to use the properties of logarithms to get *one* logarithm on the left. Then, write the equation in exponential form and solve.

$$\log_2 3w - \log_2(w - 5) = 3$$

$$\log_2 \frac{3w}{w - 5} = 3 \qquad \text{Quotient rule}$$

$$2^3 = \frac{3w}{w - 5} \qquad \text{Write in exponential form.}$$

$$8 = \frac{3w}{w - 5} \qquad 2^3 = 8$$

$$8(w - 5) = 3w \qquad \text{Multiply by } w - 5.$$

$$8w - 40 = 3w \qquad \text{Distribute.}$$

$$-40 = -5w \qquad \text{Subtract } 8w.$$

$$8 = w \qquad \text{Divide by } -5.$$

Verify that $w = 8$ satisfies the original equation.

The solution set is $\{8\}$.

You Try 5

Solve.

a) $\log_4(7p + 1) = 3$ b) $\log_3 2x - \log_3(x - 14) = 2$

Let's look at the two types of equations we have discussed side by side. Notice the difference between the two equations.

Solve each equation

1) $\log_3 x + \log_3(2x + 5) = \log_3 12$

Use the properties of logarithms to get one log on the left.

$$\log_3 x(2x + 5) = \log_3 12$$

Since both terms contain logarithms, use the property that states if $\log_a x = \log_a y$, then $x = y$.

$$x(2x + 5) = 12$$
$$2x^2 + 5x = 12$$
$$2x^2 + 5x - 12 = 0$$
$$(2x - 3)(x + 4) = 0$$
$$2x - 3 = 0 \quad \text{or} \quad x + 4 = 0$$
$$x = \frac{3}{2} \quad \text{or} \quad x = -4$$

Reject -4 as a solution. The solution set is $\left\{\dfrac{3}{2}\right\}$.

---

2) $\log_3 x + \log_3(2x + 5) = 1$

Use the properties of logarithms to get one log on the left.

$$\log_3 x(2x + 5) = 1$$

The term on the right does *not* contain a logarithm. Write the equation in exponential form and solve.

$$3^1 = x(2x + 5)$$
$$3 = 2x^2 + 5x$$
$$0 = 2x^2 + 5x - 3$$
$$0 = (2x - 1)(x + 3)$$
$$2x - 1 = 0 \quad \text{or} \quad x + 3 = 0$$
$$x = \frac{1}{2} \quad \text{or} \quad x = -3$$

Reject $x = -3$ as a solution. The solution set is $\left\{\dfrac{1}{2}\right\}$.

3. Solve Applied Problems Involving Exponential Functions Using a Calculator

Recall that $A = Pe^{rt}$ is the formula for continuous compound interest where P (the principal) is the amount invested, r is the interest rate, and A is the amount (in dollars) in the account after t yr.

Here we will look at how we can use the formula to solve a different problem from the kind we solved in Section 13.5.

Example 6

If $3000 is invested at 5% interest compounded continuously, how long would it take for the investment to grow to $4000?

Solution

In this problem, we are asked to find t, the amount of *time* it will take for $3000 to grow to $4000 when invested at 5% compounded continuously.

Use $A = Pe^{rt}$ with $P = 3000$, $A = 4000$, and $r = 0.05$.

$$A = Pe^{rt}$$
$$4000 = 3000e^{0.05t} \qquad \text{Substitute the values.}$$
$$\frac{4}{3} = e^{0.05t} \qquad \text{Divide by 3000.}$$
$$\ln \frac{4}{3} = \ln e^{0.05t} \qquad \text{Take the natural log of both sides.}$$
$$\ln \frac{4}{3} = 0.05t \ln e \qquad \log_a x^r = r \log_a x$$
$$\ln \frac{4}{3} = 0.05t(1) \qquad \ln e = 1$$
$$\ln \frac{4}{3} = 0.05t$$
$$\frac{\ln \frac{4}{3}}{0.05} = t \qquad \text{Divide by 0.05.}$$
$$5.75 \approx t \qquad \text{Use a calculator to get the approximation.}$$

It would take about 5.75 yr for $3000 to grow to $4000.

You Try 6

If $4500 is invested at 6% interest compounded continuously, how long would it take for the investment to grow to $5000?

The amount of time it takes for a quantity to double in size are called the *doubling time*. We can use this in many types of applications.

Example 7

The number of bacteria, $N(t)$, in a culture t hr after the bacteria are placed in a dish is given by

$$N(t) = 5000e^{0.0462t}$$

where 5000 bacteria are initially present. How long will it take for the number of bacteria to double?

Solution

If there are 5000 bacteria present initially, there will be $2(5000) = 10{,}000$ bacteria when the number doubles.

Find t when $N(t) = 10,000$.

$$N(t) = 5000e^{0.0462t}$$

$10,000 = 5000e^{0.0462t}$	Substitute 10,000 for $N(t)$.
$2 = e^{0.0462t}$	Divide by 5000.
$\ln 2 = \ln e^{0.0462t}$	Take the natural log of both sides.
$\ln 2 = 0.0462t \ln e$	$\log_a x^r = r \log_a x$
$\ln 2 = 0.0462t(1)$	$\ln e = 1$
$\ln 2 = 0.0462t$	
$\dfrac{\ln 2}{0.0462} = t$	Divide by 0.0462.
$15 \approx t$	

It will take about 15 hr for the number of bacteria to double.

You Try 7

The number of bacteria, $N(t)$, in a culture t hr after the bacteria are placed in a dish is given by

$$N(t) = 12{,}000e^{0.0385t}$$

where 12,000 bacteria are initially present. How long will it take for the number of bacteria to double?

4. Solve an Applied Problem Involving Exponential Growth or Decay

We can generalize the formulas used in Examples 6 and 7 with a formula widely used to model situations that grow or decay exponentially. That formula is

$$y = y_0 e^{kt}$$

where y_0 is the initial amount or quantity at time $t = 0$, y is the amount present after time t, and k is a constant. If k is positive, it is called a *growth constant* because the quantity will *increase* over time. If k is negative, it is called a *decay constant* because the quantity will *decrease* over time.

Example 8

In April 1986, an accident at the Chernobyl nuclear power plant released many radioactive substances into the environment. One such substance was cesium-137. Cesium-137 decays according to the equation

$$y = y_0 e^{-0.0230t}$$

where y_0 is the initial amount present at time $t = 0$ and y is the amount present after t yr. If a sample of soil contains 10 g of cesium-137 immediately after the accident,

a) How many grams will remain after 15 yr?

b) How long would it take for the initial amount of cesium-137 to decay to 2 g?

c) The **half-life** of a substance is the amount of time it takes for a substance to decay to half its original amount. What is the half-life of cesium-137?

Solution

a) The initial amount of cesium-137 is 10 g, so $y_0 = 10$. We must find y when $y_0 = 10$ and $t = 15$.

$$y = y_0 e^{-0.0230t}$$
$$= 10e^{-0.0230(15)} \qquad \text{Substitute the values.}$$
$$\approx 7.08 \qquad \text{Use a calculator to get the approximation.}$$

There will be about 7.08 g of cesium-137 remaining after 15 yr.

b) The initial amount of cesium-137 is $y_0 = 10$. To determine how long it will take to decay to 2 g, let $y = 2$ and solve for t.

$$y = y_0 e^{-0.0230t}$$
$$2 = 10e^{-0.0230t} \qquad \text{Substitute 2 for } y.$$
$$0.2 = e^{-0.0230t} \qquad \text{Divide by 10.}$$
$$\ln 0.2 = \ln e^{-0.0230t} \qquad \text{Take the natural log of both sides.}$$
$$\ln 0.2 = -0.0230t \ln e \qquad \log_a x^r = r \log_a x$$
$$\ln 0.2 = -0.0230t \qquad \ln e = 1$$
$$\frac{\ln 0.2}{-0.0230} = t \qquad \text{Divide by } -0.0230.$$
$$69.98 \approx t \qquad \text{Use a calculator to get the approximation.}$$

It will take about 69.98 yr for 10 g of cesium-137 to decay to 2 g.

c) Since there are 10 g of cesium-137 in the original sample, to determine the half-life we will determine how long it will take for the 10 g to decay to 5 g because $\frac{1}{2}(10) = 5$.

Let $y_0 = 10$, $y = 5$, and solve for t.

$$y = y_0 e^{-0.0230t}$$
$$5 = 10e^{-0.0230t} \qquad \text{Substitute the values.}$$
$$0.5 = e^{-0.0230t} \qquad \text{Divide by 10.}$$
$$\ln 0.5 = \ln e^{-0.0230t} \qquad \text{Take the natural log of both sides.}$$
$$\ln 0.5 = -0.0230t \ln e \qquad \ln_a x^r = r \log_a x$$
$$\ln 0.5 = -0.0230t \qquad \ln e = 1$$
$$\frac{\ln 0.5}{-0.0230} = t \qquad \text{Divide by } -0.0230.$$
$$30.14 \approx t \qquad \text{Use a calculator to get the approximation.}$$

The half-life of cesium-137 is about 30.14 yr. This means that it would take about 30.14 yr for any quantity of cesium-137 to decay to half of its original amount.

You Try 8

Radioactive strontium-90 decays according to the equation

$$y = y_0 e^{-0.0244t}$$

where t is in years. If a sample contains 40 g of strontium-90,

a) How many grams will remain after 8 yr?

b) How long would it take for the initial amount of strontium-90 to decay to 30 g?

c) What is the half-life of strontium-90?

Using Technology

We can solve exponential and logarithmic equations in the same way that we solved other equations—by graphing both sides of the equation and finding where the graphs intersect.

In Example 2 of this section, we learned how to solve $5^{x-2} = 16$. Because the right side of the equation is 16 the graph will have to go at least as high as 16. So set the Y_{max} to be 20, enter the left side of the equation as Y_1 and the right side as Y_2 and press $\boxed{\text{GRAPH}}$:

 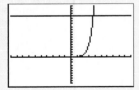

Recall that the x-coordinate of the point of intersection is the solution to the equation. To find the point of intersection, press $\boxed{\text{2nd}}$ $\boxed{\text{TRACE}}$ and then highlight 5:intersect and press $\boxed{\text{ENTER}}$. Press $\boxed{\text{ENTER}}$ three more times to see that the x-coordinate of the point of intersection is approximately 3.723.

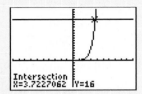

Remember, while the calculator can sometimes save you time, it will often give an approximate answer and not an exact solution.

Use a graphing calculator to solve each equation. Round your answer to the nearest thousandth.

1. $7^x = 49$

2. $6^{2b+1} = 13$

3. $5^{4a+7} = 8^{2a}$

4. $\ln x = 1.2$

5. $\log(k + 9) = \log 11$

6. $\ln(x + 3) = \ln(x - 2)$

Answers to You Try Exercises

1) a) $\{7\}$ b) $\left\{\dfrac{\ln 24}{\ln 3}\right\}; \{2.8928\}$ 2) $\left\{\dfrac{\ln 2 - 4\ln 9}{\ln 9}\right\}; \{-3.6845\}$ 3) $\left\{\dfrac{\ln 2}{6}\right\}; \{0.1155\}$

4) a) $\{2\}$ b) $\{4\}$ 5) a) $\{9\}$ b) $\{18\}$ 6) 1.76 yr   7) 18 hr

8) a) 32.91 g   b) 11.79 yr   c) 28.41 yr

Answers to Technology Exercises

1. $\{2\}$ 2. $\{0.216\}$ 3. $\{-4.944\}$ 4. $\{3.320\}$ 5. $\{2\}$ 6. $\varnothing$

13.6 Exercises

Boost your grade at mathzone.com! MathZone

> Practice Problems > NetTutor > Self-Test > e-Professors > Videos

Objective I

Solve each equation. Give the exact solution. If the answer contains a logarithm, approximate the solution to four decimal places.

1) $7^x = 49$

2) $5^c = 125$

3) $7^n = 15$

4) $5^a = 38$

5) $8^z = 3$

6) $4^y = 9$

 7) $6^{5p} = 36$

8) $2^{3t} = 32$

9) $4^{6k} = 2.7$

10) $3^{2x} = 7.8$

11) $2^{4n+1} = 5$

12) $6^{2b+1} = 13$

13) $5^{3a-2} = 8$

14) $3^{2x-3} = 14$

15) $4^{2c+7} = 64^{3c-1}$

16) $27^{5m-2} = 3^{m+6}$

17) $9^{5d-2} = 4^{3d}$

18) $5^{4a+7} = 8^{2a}$

 Solve each equation. Give the exact solution and the approximation to four decimal places.

19) $e^y = 12.5$

20) $e^t = 0.36$

21) $e^{-4x} = 9$

22) $e^{3p} = 4$

23) $e^{0.01r} = 2$

24) $e^{-0.08k} = 10$

25) $e^{0.006t} = 3$

26) $e^{0.04a} = 12$

27) $e^{-0.4y} = 5$

28) $e^{-0.005c} = 16$

Objective 2

Solve each equation.

29) $\log_6(k + 9) = \log_6 11$

30) $\log_5(d - 4) = \log_5 2$

31) $\log_7(3p - 1) = \log_7 9$

32) $\log_4(5y + 2) = \log_4 10$

33) $\log x + \log(x - 2) = \log 15$

34) $\log_9 r + \log_9(r + 7) = \log_9 18$

35) $\log_3 n + \log_3(12 - n) = \log_3 20$

36) $\log m + \log(11 - m) = \log 24$

37) $\log_2(-z) + \log_2(z - 8) = \log_2 15$

38) $\log_5 8y - \log_5(3y - 4) = \log_5 2$

39) $\log_6(5b - 4) = 2$

40) $\log_3(4c + 5) = 3$

41) $\log(3p + 4) = 1$

42) $\log(7n - 11) = 1$

43) $\log_3 y + \log_3(y - 8) = 2$

44) $\log_4 k + \log_4(k - 6) = 2$

 45) $\log_2 r + \log_2(r + 2) = 3$

46) $\log_9(z + 8) + \log_9 z = 1$

47) $\log_4 20c - \log_4(c + 1) = 2$

48) $\log_6 40x - \log_6(1 + x) = 2$

49) $\log_2 8d - \log_2(2d - 1) = 4$

50) $\log_6(13 - x) + \log_6 x = 2$

Objectives 3 and 4

Use the formula $A = Pe^{rt}$ to solve 51–58.

51) If $2000 is invested at 6% interest compounded continuously, how long would it take

a) for the investment to grow to $2500?

b) for the initial investment to double?

52) If $5000 is invested at 7% interest compounded continuously, how long would it take

a) for the investment to grow to $6000?

b) for the initial investment to double?

53) How long would it take for an investment of $7000 to earn $800 in interest if it is invested at 7.5% compounded continuously?

54) How long would it take for an investment of $4000 to earn $600 in interest if it is invested at 6.8% compounded continuously?

55) Cynthia wants to invest some money now so that she will have $5000 in the account in 10 yr. How much should she invest in an account earning 8% compounded continuously?

56) How much should Leroy invest now at 7.2% compounded continuously so that the account contains $8000 in 12 yr?

57) Raj wants to invest $3000 now so that it grows to $4000 in 4 yr. What interest rate should he look for? (Round to the nearest tenth of a percent.)

58) Marisol wants to invest $12,000 now so that it grows to $20,000 in 7 yr. What interest rate should she look for? (Round to the nearest tenth of a percent.)

59) The number of bacteria, $N(t)$, in a culture t hr after the bacteria is placed in a dish is given by

$$N(t) = 4000e^{0.0374t}$$

where 4000 bacteria are initially present.

a) After how many hours will there be 5000 bacteria in the culture?

b) How long will it take for the number of bacteria to double?

60) The number of bacteria, $N(t)$, in a culture t hr after the bacteria is placed in a dish is given by

$$N(t) = 10,000e^{0.0418t}$$

where 10,000 bacteria are initially present.

a) After how many hours will there be 15,000 bacteria in the culture?

b) How long will it take for the number of bacteria to double?

61) The population of an Atlanta suburb is growing at a rate of 3.6% per year. If 21,000 people lived in the suburb in 2004, determine how many people will live in the town in 2012. Use $y = y_0e^{0.036t}$.

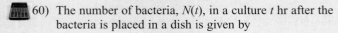

62) The population of a Seattle suburb is growing at a rate of 3.2% per year. If 30,000 people lived in the suburb in 2003, determine how many people will live in the town in 2010. Use $y = y_0e^{0.032t}$.

63) A rural town in South Dakota is losing residents at a rate of 1.3% per year. The population of the town was 2470 in 1990. Use $y = y_0e^{-0.013t}$ to answer the following questions.

a) What was the population of the town in 2005?

b) In what year would it be expected that the population of the town is 1600?

64) In 1995, the population of a rural town in Kansas was 1682. The population is decreasing at a rate of 0.8% per year. Use $y = y_0e^{-0.008t}$ to answer the following questions.

a) What was the population of the town in 2000?

b) In what year would it be expected that the population of the town is 1000?

65) Radioactive carbon-14 is a substance found in all living organisms. After the organism dies, the carbon-14 decays according to the equation

$$y = y_0e^{-0.000121t}$$

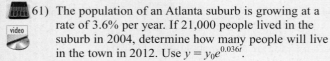

where t is in years, y_0 is the initial amount present at time $t = 0$, and y is the amount present after t yr.

a) If a sample initially contains 15 g of carbon-14, how many grams will be present after 2000 yr?

b) How long would it take for the initial amount to decay to 10 g?

c) What is the half-life of carbon-14?

66) Plutonium-239 decays according to the equation

$$y = y_0e^{-0.0000287t}$$

where t is in years, y_0 is the initial amount present at time $t = 0$, and y is the amount present after t yr.

a) If a sample initially contains 8 g of plutonium-239, how many grams will be present after 5000 yr?

b) How long would it take for the initial amount to decay to 5 g?

c) What is the half-life of plutonium-239?

67) Radioactive iodine-131 is used in the diagnosis and treatment of some thyroid-related illnesses. The concentration of the iodine in a patient's system is given by

$$y = 0.4e^{-0.086t}$$

where t is in days and y is in the appropriate units.

a) How much iodine-131 is given to the patient?

b) How much iodine-131 remains in the patient's system 7 days after treatment?

68) The amount of cobalt-60 in a sample is given by

$$y = 30e^{-0.131t}$$

where t is in years and y is in grams.

a) How much cobalt-60 is originally in the sample?

b) How long would it take for the initial amount to decay to 10 g?

Definition/Procedure	Example	Reference

Inverse Functions

13.1

One-to-One Function

In order for a function to be a *one-to-one function,* each x-value corresponds to exactly one y-value, and each y-value corresponds to exactly one x-value.

Determine whether each function is one-to-one.

a) $f = \{(-2, 9), (1, 3), (3, -1), (7, -9)\}$ *is* one-to-one.

b) $g = \{(0, 9), (2, 1), (4, 1), (5, 4)\}$ is *not* one-to-one since the y-value 1 corresponds to two different x-values.

c)

p. 852

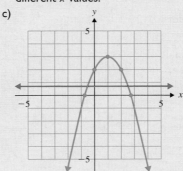

The *horizontal line test* tells us how we can determine if a graph represents a one-to-one function:

If every horizontal line that could be drawn through a function would intersect the graph at most once, then the function is one-to-one.

No. It fails the horizontal line test.

Inverse Function

Let f be a one-to-one function. The *inverse* of f, denoted by f^{-1}, is a one-to-one function that contains the set of all ordered pairs (y, x) where (x, y) belongs to f.

How to Find an Equation of the Inverse of $y = f(x)$

Step 1: Replace $f(x)$ with y.
Step 2: Interchange x and y.
Step 3: Solve for y.
Step 4: Replace y with the inverse notation, $f^{-1}(x)$.

The graphs of $f(x)$ and $f^{-1}(x)$ are symmetric with respect to the line $y = x$.

Find an equation of the inverse of $f(x) = 2x - 4$.

Step 1: $y = 2x - 4$ Replace $f(x)$ with y.
Step 2: $x = 2y - 4$ Interchange x and y.
Step 3: Solve for y.
$$x + 4 = 2y \qquad \text{Add 4.}$$
$$\frac{x + 4}{2} = y \qquad \text{Divide by 2.}$$
$$\frac{1}{2}x + 2 = y \qquad \text{Simplify.}$$

Step 4: $f^{-1}(x) = \dfrac{1}{2}x + 2$ Replace y with $f^{-1}(x)$.

p. 855

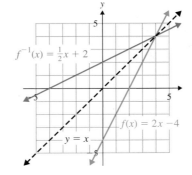

Definition/Procedure	Example	Reference

Exponential Functions | | **13.2**

An *exponential function* is a function of the form

$$f(x) = a^x$$

where $a > 0$, $a \neq 1$, and x is a real number.

$f(x) = 3^x$ | **p. 864**

Characteristics of an Exponential Function

$$f(x) = a^x$$

1) If $f(x) = a^x$, where $a > 1$, the value of y increases as the value of x increases.

2) If $f(x) = a^x$ where $0 < a < 1$, the value of y decreases as the value of x increases.

3) The function is one-to-one.

4) The y-intercept is $(0, 1)$.

5) The domain is $(-\infty, \infty)$, and the range is $(0, \infty)$.

1)

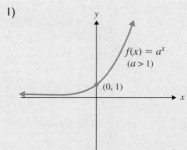

2)

p. 866

Solving an Exponential Equation

Step 1: *If possible, express each side of the equation with the same base. If it is not possible to get the same base, a method in 13.6 can be used.*

Step 2: *Use the rules of exponents to simplify the exponents.*

Step 3: *Set the exponents equal and solve for the variable.*

Solve $5^{4x-1} = 25^{3x+4}$.

Step 1: $5^{4x-1} = (5^2)^{(3x+4)}$ Both sides are powers of 5.

Step 2: $5^{4x-1} = 5^{2(3x+4)}$ Power rule for exponents

$5^{4x-1} = 5^{6x+8}$ Distribute.

Step 3: $4x - 1 = 6x + 8$ The bases are the same. Set the exponents equal.

$-2x = 9$ Subtract $6x$; add 1.

$x = -\dfrac{9}{2}$ Divide by -2.

The solution set is $\left\{-\dfrac{9}{2}\right\}$.

p. 869

Definition/Procedure	Example	Reference

Logarithmic Functions

13.3

Definition of Logarithm

If $a > 0$, $a \neq 1$, and $x > 0$, then for every real number y, $y = \log_a x$ means $x = a^y$.

Write $\log_5 125 = 3$ in exponential form.

$$\log_5 125 = 3 \text{ means } 5^3 = 125$$

p. 875

A *logarithmic equation* is an equation in which at least one term contains a logarithm.

To solve a logarithmic equation of the form

$$\log_a b = c$$

write the equation in exponential form ($a^c = b$) and solve for the variable.

Solve $\log_2 k = 3$.
Write the equation in exponential form and solve for k.

$\log_2 k = 3$ means $2^3 = k$.
$$8 = k$$

The solution set is $\{8\}$.

p. 877

To evaluate $\log_a b$ means *to find the power to which we raise a to get b.*

Evaluate $\log_7 49$.

$$\log_7 49 = 2 \text{ since } 7^2 = 49$$

p. 879

A base 10 logarithm is called a *common logarithm.* A base 10 logarithm is often written without the base.

$\log x$ means $\log_{10} x$.

p. 880

Characteristics of a Logarithmic Function

$f(x) = \log_a x$, **where $a > 0$ and $a \neq 1$**

1) If $f(x) = \log_a x$, where $a > 1$, the value of y increases as the value of x increases.

2) If $f(x) = \log_a x$, where $0 < a < 1$, the value of y decreases as the value of x increases.

3) The function is one-to-one.

4) The x-intercept is $(1, 0)$.

5) The domain is $(0, \infty)$, and the range is $(-\infty, \infty)$.

6) The inverse of $f(x) = \log_a x$ is $f^{-1}(x) = a^x$.

1)

$f(x) = \log_a x$
$(a > 1)$

2)

$f(x) = \log_a x$
$(0 < a < 1)$

p. 885

Definition/Procedure	Example	Reference

Properties of Logarithms

13.4

Let $x, y,$ and a be positive real numbers where $a \neq 1$ and let r be any real number. Then,

1) $\log_a a = 1$
2) $\log_a 1 = 0$
3) $\log_a xy = \log_a x + \log_a y$ Product rule
4) $\log_a \dfrac{x}{y} = \log_a x - \log_a y$ Quotient rule
5) $\log_a x^r = r \log_a x$ Power rule
6) $\log_a a^x = x$ for any real number x.
7) $a^{\log_a x} = x$

Write $\log_4 \dfrac{c^5}{d^2}$ as the sum or difference of logarithms in simplest form. Assume c and d represent positive real numbers.

$$\log_4 \frac{c^5}{d^2} = \log_4 c^5 - \log_4 d^2 \qquad \text{Quotient rule}$$
$$= 5 \log_4 c - 2 \log_4 d \qquad \text{Power rule}$$

p. 889

Common and Natural Logarithms and Change of Base

13.5

A base 10 logarithm is usually written without a base, $\log x$. It is called a *common logarithm*.

$$\log x \text{ means } \log_{10} x$$

Find the value of each.

a) $\log 100$ b) $\log 53$

a) $\log 100 = \log_{10} 100 = \log_{10} 10^2 = 2$
b) Using a calculator, we get $\log 53 \approx 1.7243$.

p. 899

The number e is approximately equal to 2.7183. A base e logarithm is called a *natural logarithm*. The notation used for a natural logarithm is $\ln x$.

$$f(x) = \ln x \quad \text{means} \quad f(x) = \log_e x$$

p. 902

ln e = 1 since $\ln e = 1$ means $\log_e e = 1$.

We can find the values of some natural logarithms using the properties of logarithms. We can approximate the values of other natural logarithms using a calculator.

Find the value of each.
a) $\ln e^{12}$ b) $\ln 18$

a) $\ln e^{12} = 12 \ln e$ Power rule
$\qquad\quad = 12(1)$ $\ln e = 1$
$\qquad\quad = 12$

b) Using a calculator, we get $\ln 18 \approx 2.8904$.

p. 902

To solve an equation such as $\ln x = 1.6$, change to exponential form and solve for the variable.

Solve $\ln x = 1.6$.
$\ln x = 1.6$ means $\log_e x = 1.6$.

$$\log_e x = 1.6.$$
$$e^{1.6} = x \qquad \text{Exponential form}$$
$$4.9530 \approx x \qquad \text{Approximation}$$

The exact solution is $\{e^{1.6}\}$. The approximation is $\{4.9530\}$.

p. 904

Applications of Exponential Functions

Continuous Compounding

If P dollars is deposited in an account earning interest rate r compounded continuously, then the amount of money, A (in dollars), in the account after t years is given by $A = Pe^{rt}$.

Determine the amount of money in an account after 6 yr if $3000 was initially invested at 5% compounded continuously.

$A = Pe^{rt}$
$\quad = 3000e^{0.05(6)}$ Substitute values.
$\quad = 3000e^{0.30}$ Multiply (0.05)(6).
$\quad \approx 4049.5764$ Evaluate using a calculator.
$\quad \approx \$4049.58$ Round to the nearest cent.

p. 904

Definition/Procedure	Example	Reference

Change-of-Base Formula

If a, b, and x are positive real numbers and $a \neq 1$ and $b \neq 1$, then

$$\log_a x = \frac{\log_b x}{\log_b a}$$

Find $\log_2 75$ to four decimal places.

$$\log_2 75 = \frac{\log_{10} 75}{\log_{10} 2} \approx 6.2288$$

p. 906

Solving Exponential and Logarithmic Equations

13.6

Let a, x, and y be positive real numbers, where $a \neq 1$.

1) If $x = y$, then $\log_a x = \log_a y$.

2) If $\log_a x = \log_a y$, then $x = y$.

Solving an Exponential Equation

Begin by asking yourself, "*Can I express each side with the same base?*"

1) If the answer is *yes*, then write each side of the equation with the same base, set the exponents equal, and solve for the variable.

2) If the answer is *no*, then take the natural logarithm of each side, use the properties of logarithms, and solve for the variable.

Solve each equation.

a) $4^x = 64$.

Ask yourself, "*Can I express both sides with the same base?*" Yes.

$$4^x = 64$$
$$4^x = 4^3$$
$$x = 3 \qquad \text{Set the exponents equal.}$$

The solution set is $\{3\}$.

b) $4^x = 9$.

Ask yourself, "*Can I express both sides with the same base?*" No. Take the natural logarithm of each side.

$$4^x = 9$$
$$\ln 4^x = \ln 9 \qquad \text{Take the natural log of each side.}$$
$$x \ln 4 = \ln 9 \qquad \log_a x^r = r \log_a x$$
$$x = \frac{\ln 9}{\ln 4} \qquad \text{Divide by } \ln 4.$$
$$x \approx 1.5850 \qquad \text{Use a calculator to get the approximation.}$$

The exact solution is $\left\{ \dfrac{\ln 9}{\ln 4} \right\}$. The approximation is $\{1.5850\}$.

p. 910

Definition/Procedure	Example	Reference
Solve an exponential equation with base e by taking the natural logarithm of each side.	Solve $e^y = 35.8$. $\ln e^y = \ln 35.8$ Take the natural log of each side. $y \ln e = \ln 35.8$ $\log_a x^r = r \log_a x$ $y(1) = \ln 35.8$ $\ln e = 1$ $y = \ln 35.8$ $y \approx 3.5779$ Approximation The exact solution is $\{\ln 35.8\}$. The approximation is $\{3.5779\}$.	**p. 913**
Solving Logarithmic Equations Sometimes we must use the properties of logarithms to solve logarithmic equations.	Solve $\log x + \log(x - 3) = \log 28$. $\log x + \log(x - 3) = \log 28$ $\log x(x - 3) = \log 28$ Product rule $x(x - 3) = 28$ If $\log_a x = \log_a y$, then $x = y$. $x^2 - 3x = 28$ Distribute. $x^2 - 3x - 28 = 0$ Subtract 28. $(x - 7)(x + 4) = 0$ Factor. $x - 7 = 0$ or $x + 4 = 0$ Set each factor equal to 0. $x = 7$ or $x = -4$ Solve. Verify that only 7 satisfies the original equation. The solution set is $\{7\}$.	**p. 913**

(13.1) **Determine whether each function is one-to-one. If it is one-to-one, find its inverse.**

1) $f = \{(-7, -4), (-2, 1), (1, 5), (6, 11)\}$

2) $g = \{(1, 4), (3, 7), (6, 4), (10, 9)\}$

Determine whether each function is one-to-one. If it is one-to-one, graph its inverse.

3)

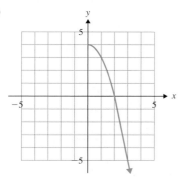

4)

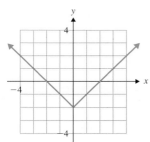

Find the inverse of each one-to-one function. Graph each function and its inverse on the same axes.

5) $f(x) = x + 4$

6) $g(x) = 2x - 10$

7) $h(x) = \dfrac{1}{3}x - 1$

8) $f(x) = \sqrt[3]{x} + 2$

Given each one-to-one function $f(x)$, find the following function values *without* finding an equation of $f^{-1}(x)$. Find the value in a) before b).

9) $f(x) = 6x - 1$
   a) $f(2)$
   b) $f^{-1}(11)$

10) $f(x) = \sqrt[3]{x + 5}$
   a) $f(-13)$
   b) $f^{-1}(-2)$

(13.2) Graph each exponential function.

11) $f(x) = 2^x$

12) $g(x) = 3^x$

13) $h(x) = \left(\dfrac{1}{3}\right)^x$

14) $f(x) = 3^{x-2}$

Solve each exponential equation.

15) $2^c = 64$

16) $7^{m+5} = 49$

17) $16^{3z} = 32^{2z-1}$

18) $9^y = \dfrac{1}{81}$

19) $\left(\dfrac{3}{2}\right)^{x+4} = \left(\dfrac{4}{9}\right)^{x-3}$

20) The value, $V(t)$, in dollars, of a luxury car t yr after it is purchased is given by $V(t) = 38{,}200(0.816)^t$.

   a) What was the purchase price of the car?

   b) What will the car be worth 4 yr after purchase?

(13.3)

21) In $y = \log_a x$, x must be what kind of number?

22) In $y = \log_a x$, a must be what kind of number?

Write in exponential form.

23) $\log_5 125 = 3$

24) $\log_{16} \dfrac{1}{4} = -\dfrac{1}{2}$

25) $\log 100 = 2$

26) $\log 1 = 0$

Write in logarithmic form.

27) $3^4 = 81$

28) $\left(\dfrac{2}{3}\right)^{-2} = \dfrac{9}{4}$

29) $10^3 = 1000$

30) $\sqrt{121} = 11$

Solve.

31) $\log_2 x = 3$

32) $\log_9(4x + 1) = 2$

33) $\log_{32} 16 = x$

34) $\log(2x + 5) = 1$

Evaluate.

35) $\log_8 64$

36) $\log_3 27$

37) $\log 1000$

38) $\log 1$

39) $\log_{1/2} 16$

40) $\log_{1/5} \dfrac{1}{25}$

Graph each logarithmic function.

41) $f(x) = \log_2 x$

42) $g(x) = \log_3 x$

43) $h(x) = \log_{1/3} x$

44) $f(x) = \log_{1/4} x$

Find the inverse of each function.

45) $f(x) = 5^x$

46) $g(x) = 3^x$

47) $h(x) = \log_6 x$

Solve.

48) A company plans to test market its new dog food in a large metropolitan area before taking it nationwide. The company predicts that its sales over the next 12 months can be approximated by

$$S(t) = 10 \log_3(2t + 1)$$

where t is the number of months after the dog food is introduced, and $S(t)$ is in thousands of bags of dog food.

a) How many bags of dog food were sold after 1 month on the market?

b) How many bags of dog food were sold after 4 months on the market?

(13.4) Decide whether each statement is true or false.

49) $\log_5(x + 4) = \log_5 x + \log_5 4$

50) $\log_2 \dfrac{k}{6} = \log_2 k - \log_2 6$

Write as the sum or difference of logarithms and simplify, if possible. Assume all variables represent positive real numbers.

51) $\log_8 3z$

52) $\log_7 \dfrac{49}{t}$

53) $\log_4 \sqrt{64}$

54) $\log \dfrac{1}{100}$

55) $\log_5 c^4 d^3$

56) $\log_4 m\sqrt{n}$

57) $\log_a \dfrac{xy}{z^3}$

58) $\log_4 \dfrac{a^2}{bc^4}$

59) $\log p(p + 8)$

60) $\log_6 \dfrac{r^3}{r^2 - 5}$

Write as a single logarithm. Assume the variables are defined so that the variable expressions are positive and so that the bases are positive real numbers not equal to 1.

61) $\log c + \log d$

62) $\log_4 n - \log_4 7$

63) $9 \log_2 a + 3 \log_2 b$

64) $\log_5 r - 2 \log_5 t$

65) $\log_3 5 + 4 \log_3 m - 2 \log_3 n$

66) $\dfrac{1}{2} \log_z a - \log_z b$

67) $3 \log_5 c - \log_5 d - 2 \log_5 f$

68) $2 \log_6 x + \dfrac{1}{3} \log_6(x - 4)$

Given that log 7 ≈ 0.8451 and log 9 ≈ 0.9542, use the properties of logarithms to approximate the following. Do NOT use a calculator.

69) $\log 49$

70) $\log 63$

71) $\log \dfrac{7}{9}$

72) $\log \dfrac{1}{7}$

(13.5)

73) What is the base of $\ln x$?

74) Evaluate $\ln e$.

Evaluate each logarithm. Do not use a calculator.

75) $\log 10$

76) $\log 100$

77) $\log \sqrt{10}$

78) $\log \dfrac{1}{100}$

79) $\log 0.001$

80) $\ln e^4$

81) $\ln 1$

82) $\ln \sqrt[3]{e}$

Use a calculator to find the approximate value of each logarithm to four decimal places.

83) $\log 8$

84) $\log 0.3$

85) $\ln 1.75$

86) $\ln 0.924$

Solve each equation. Do not use a calculator.

87) $\log p = 2$

88) $\log(5n) = 3$

89) $\log\left(\dfrac{1}{2}c\right) = -1$

90) $\log(6z - 5) = 1$

Solve each equation. Give an exact solution and a solution that is approximated to four decimal places.

91) $\log x = 2.1$

92) $\log k = -1.4$

93) $\ln y = 2$

94) $\ln c = -0.5$

95) $\log(4t) = 1.75$

96) $\ln(2a - 3) = 1$

Use the change-of-base formula with either base 10 or base e to approximate each logarithm to four decimal places.

97) $\log_4 19$

98) $\log_9 42$

99) $\log_{1/2} 38$

100) $\log_6 0.82$

For Exercises 101 and 102, use the formula $L(I) = 10 \log \dfrac{I}{10^{-12}}$, where I is the intensity of sound, in watts per square meter, and $L(I)$ is the loudness of sound in decibels. Do *not* use a calculator.

101) The intensity of sound from the crowd at a college basketball game reached 0.1 W/m². Find the loudness of the crowd, in decibels.

102) Find the intensity of the sound of a jet taking off if the noise level can reach 140 dB 25 m from the jet.

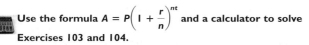
Use the formula $A = P\left(1 + \dfrac{r}{n}\right)^{nt}$ and a calculator to solve Exercises 103 and 104.

103) Pedro deposits $2500 in an account earning 6% interest compounded quarterly. How much will be in the account after 5 yr?

104) Find the amount Chelsea owes at the end of 6 yr if she borrows $18,000 at a rate of 7% compounded monthly.

Use the formula $A = Pe^{rt}$ and a calculator to solve Exercises 105 and 106.

105) Find the amount Liang will owe at the end of 4 yr if he borrows $9000 at a rate of 6.2% compounded continuously.

106) If $4000 is deposited in an account earning 5.8% compounded continuously, how much will be in the account after 7 yr?

107) The number of bacteria, $N(t)$, in a culture t hr after the bacteria is placed in a dish is given by

$$N(t) = 6000e^{0.0514t}$$

a) How many bacteria were originally in the culture?

b) How many bacteria are present after 12 hr?

108) The pH of a solution is given by pH $= -\log[H^+]$, where $[H^+]$ is the molar concentration of the hydronium ion. Find the ideal pH of blood if $[H^+] = 3.98 \times 10^{-8}$.

(13.6) Solve each equation. Give the exact solution. If the answer contains a logarithm, approximate the solution to four decimal places. *Some of these exercises require the use of a calculator to obtain a decimal approximation.*

109) $2^y = 16$

110) $3^n = 7$

111) $9^{4k} = 2$

112) $125^{m-4} = 25^{1-m}$

113) $6^{2c} = 8^{c-5}$

114) $e^z = 22$

115) $e^{5p} = 8$

116) $e^{0.03t} = 19$

Solve each logarithmic equation.

117) $\log_3(5w + 3) = 2$

118) $\log(3n - 5) = 3$

119) $\log_2 x + \log_2(x + 2) = \log_2 24$

120) $\log_4 y + \log_4(y + 1) = \log_4 30$

121) $\log 5r - \log(r + 6) = \log 2$

122) $\log_7 10p - \log_7(p - 8) = \log_7 6$

123) $\log_4 k + \log_4(k - 12) = 3$

124) $\log_3 12m - \log_3(1 + m) = 2$

Use the formula $A = Pe^{rt}$ to solve Exercises 125 and 126.

125) Jamar wants to invest some money now so that he will have $10,000 in the account in 6 yr. How much should he invest in an account earning 6.5% compounded continuously?

126) Samira wants to invest $6000 now so that it grows to $9000 in 5 yr. What interest rate (compounded continuously) should she look for? (Round to the nearest tenth of a percent.)

127) The population of a suburb is growing at a rate of 1.6% per year. The population of the suburb was 16,410 in 1990. Use $y = y_0e^{0.016t}$ to answer the following questions.

a) What was the population of the town in 1995?

b) In what year would it be expected that the population of the town is 23,000?

128) Radium-226 decays according to the equation

$$y = y_0e^{-0.000436t}$$

where t is in years, y_0 is the initial amount present at time $t = 0$, and y is the amount present after t yr.

a) If a sample initially contains 80 g of radium-239, how many grams will be present after 500 yr?

b) How long would it take for the initial amount to decay to 20 g?

c) What is the half-life of radium-226?

Use a calculator only where indicated.

Determine whether each function is one-to-one. If it is one-to-one, find its inverse.

1) $f = \{(-4, 5), (-2, 7), (0, 3), (6, 5)\}$

2) $g = \left\{ (2, 4), (6, 6), \left(9, \dfrac{15}{2}\right), (14, 10) \right\}$

3) Is this function one-to-one? If it is one-to-one, graph its inverse.

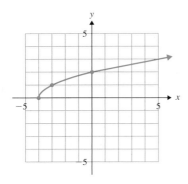

4) Find an equation of the inverse of $f(x) = -3x + 12$.

Use $f(x) = 2^x$ and $g(x) = \log_2 x$ for Problems 5–8.

5) Graph $f(x)$.

6) Graph $g(x)$.

7) a) What is the domain of $g(x)$?

   b) What is the range of $g(x)$?

8) How are the functions $f(x)$ and $g(x)$ related?

9) Write $3^{-2} = \dfrac{1}{9}$ in logarithmic form.

Solve each equation.

10) $9^{4x} = 81$

11) $125^{2c} = 25^{c-4}$

12) $\log_5 y = 3$

13) $\log(3r + 13) = 2$

14) $\log_6(2m) + \log_6(2m - 3) = \log_6 40$

15) Evaluate.

   a) $\log_2 16$

   b) $\log_7 \sqrt{7}$

16) Find $\ln e$.

Write as the sum or difference of logarithms and simplify, if possible. Assume all variables represent positive real numbers.

17) $\log_8 5n$

18) $\log_3 \dfrac{9a^4}{b^5}$

19) Write as a single logarithm.
   $2 \log x - 3 \log (x + 1)$

Use a calculator for the rest of the problems.

Solve each equation. Give an exact solution and a solution that is approximated to four decimal places.

20) $\log w = 0.8$

21) $e^{0.3t} = 5$

22) $\ln x = -0.25$

23) $4^{4a+3} = 9$

24) Approximate $\log_5 17$ to four decimal places.

25) If \$6000 is deposited in an account earning 7.4% interest compounded continuously, how much will be in the account after 5 yr? Use $A = Pe^{rt}$.

26) Polonium-210 decays according to the equation

$$y = y_0 e^{-0.00495t}$$

where t is in days, y_0 is the initial amount present at time $t = 0$, and y is the amount present after t days.

   a) If a sample initially contains 100 g of polonium-210, how many grams will be present after 30 days?

   b) How long would it take for the initial amount to decay to 20 g?

   c) What is the half-life of polonium-210?

1) Evaluate $40 + 8 \div 2 - 3^2$.

2) Evaluate $\dfrac{5}{6} - \dfrac{14}{15} \cdot \dfrac{10}{7}$.

Simplify. The answer should not contain any negative exponents.

3) $(-5a^2)(3a^4)$

4) $\dfrac{40z^3}{10z^{-5}}$

5) $\left(\dfrac{2c^{10}}{d^3}\right)^{-3}$

6) Write 0.00009231 in scientific notation.

7) *Write an equation and solve.*
   A watch is on sale for $38.40. This is 20% off of the regular price. What was the regular price of the watch?

8) Solve $-4x + 7 < 13$. Graph the solution set and write the answer in interval notation.

9) Solve using the elimination method.
   $x + 4y = -2$
   $-2x + 3y = 15$

10) Solve using the substitution method.
    $6x + 5y = -8$
    $3x - y = 3$

11) Write the equation of a line containing the points $(-2, 5)$ and $(2, -1)$. Write it in slope-intercept form.

12) Divide $(6c^3 - 7c^2 - 22c + 5) \div (2c - 5)$.

Factor completely.

13) $4w^2 + w - 18$

14) $3p^3 + 2p^2 - 3p - 2$

15) Solve $x^2 + 14x = -48$.

16) Subtract $\dfrac{r}{r^2 - 49} - \dfrac{3}{r^2 - 2r - 63}$.

17) Solve $\dfrac{9}{y + 6} + \dfrac{4}{y - 6} = \dfrac{-4}{y^2 - 36}$.

18) Graph the compound inequality $x + 2y \geq 6$ and $y - x \leq -2$.

Simplify. Assume all variables represent positive real numbers.

19) $\sqrt{120}$

20) $\sqrt{45t^9}$

21) $\sqrt{\dfrac{36a^5}{a^3}}$

22) $(27)^{2/3}$

23) Solve $\sqrt{h^2 + 2h - 7} = h - 3$.

24) Multiply and simplify $(2 - 7i)(3 + i)$.

25) Solve by completing the square $k^2 - 8k + 4 = 0$.

Solve.

26) $r^2 + 5r = -2$

27) $t^2 = 10t - 41$

28) $4m^4 + 4 = 17m^2$

29) Find the domain of $f(x) = \dfrac{4}{3x - 2}$.

30) Graph $f(x) = |x| - 4$ and identify the domain and range.

31) Graph $g(x) = 2x^2 + 4x + 4$.

32) Let $f(x) = x^2 - 6x + 2$ and $g(x) = x - 3$.
    a) Find $f(-1)$.
    b) Find $(f \circ g)(x)$.
    c) Find x so that $g(x) = -7$.

33) Solve $16^y = \dfrac{1}{64}$.

34) Solve $\log_4(5x + 1) = 2$.

35) Write as a single logarithm.
    $\log a + 2 \log b - 5 \log c$

36) Solve $e^{-0.04t} = 6$. Give an exact solution and an approximation to four decimal places.

CHAPTER 14

Conic Sections, Nonlinear Inequalities, and Nonlinear Systems

Algebra at Work: Forensics

We will look at one more application of mathematics to forensic science. Conic sections are used to help solve crimes.

Vanessa is a forensics expert and is called to the scene of a shooting. Blood is spattered everywhere. She uses algebra and trigonometry to help her analyze the blood stain patterns to determine how far the shooter was standing from the victim and the angle at which the victim was shot. The location and angle help police determine the height of the gun when the shots were fired.

Forensics experts perform blood stain pattern analysis. Pictures are taken of the crime scene, and scientists measure the distance between each of the blood stains and the pool of blood. The individual blood stains are roughly elliptical in shape with tails at the ends. On the pictures, scientists draw a best-fit ellipse into each blood stain, measure the longest axis of the ellipse (the major axis) and the shortest axis of the ellipse (the minor axis) and do calculations that reveal where the shooter was standing at the time of the shooting.

In this chapter, we will learn about the ellipse and other conic sections.

Section 14.1 The Circle

Objectives

1. Define a Conic Section
2. Graph a Circle Given in the Form $(x - h)^2 + (y - k)^2 = r^2$
3. Graph a Circle Given in the Form $Ax^2 + Ay^2 + Cx + Dy + E = 0$

1. Define a Conic Section

In this chapter, we will study the *conic sections*. When a right circular cone is intersected by a plane, the result is a **conic section**. The conic sections are parabolas, circles, ellipses, and hyperbolas. The following figures show how each conic section is obtained from the intersection of a cone and a plane.

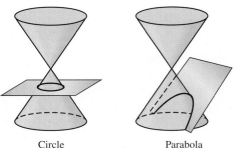 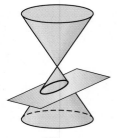 

| Circle | Parabola | Ellipse | Hyperbola |

We learned in Chapter 12 that the graph of a quadratic function, $f(x) = ax^2 + bx + c$, is a *parabola* that opens vertically. Another form this function may take is $f(x) = a(x - h)^2 + k$. The graph of a quadratic equation of the form $x = ay^2 + by + c$, or $x = a(y - k)^2 + h$, is a *parabola* that opens horizontally. The next conic section we will discuss is the circle.

2. Graph a Circle Given in the Form $(x - h)^2 + (y - k)^2 = r^2$

A **circle** is defined as the set of all points in a plane equidistant (the same distance) from a fixed point. The fixed point is the **center** of the circle. The distance from the center to a point on the circle is the **radius** of the circle. We can derive the equation of a circle from the formula for the distance between two points, which was presented in Section 11.2.

Let the center of a circle have coordinates (h, k) and let (x, y) represent any point on the circle. Let r represent the distance between these two points. r is the radius of the circle.

We will use the distance formula to find the distance between the center, (h, k), and the point (x, y) on the circle.

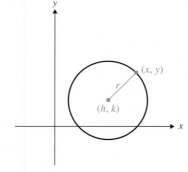

$$d = \sqrt{(x_2 - x_1)^2 + (y_2 - y_1)^2}$$ Distance formula

Substitute (x, y) for (x_2, y_2), (h, k) for (x_1, y_1), and r for d.

$$r = \sqrt{(x - h)^2 + (y - k)^2}$$
$$r^2 = (x - h)^2 + (y - k)^2$$ Square both sides.

This is the **standard form** for the equation of a circle.

Definition

Standard Form for the Equation of a Circle: The standard form for the equation of a circle with center (h, k) and radius r is

$$(x - h)^2 + (y - k)^2 = r^2$$

Example 1

Graph $(x - 2)^2 + (y + 1)^2 = 9$.

Solution

Standard form is $(x - h)^2 + (y - k)^2 = r^2$.
Our equation is $(x - 2)^2 + (y + 1)^2 = 9$.

$$h = 2 \qquad k = -1 \qquad r = \sqrt{9} = 3$$

The center is $(2, -1)$. The radius is 3.
   To graph the circle, first plot the center $(2, -1)$.
Use the radius to locate four points on the circle.
From the center, move 3 units up, down, left, and
right. Draw a circle through the four points.

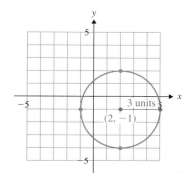

You Try 1

Graph $(x + 3)^2 + (y - 1)^2 = 16$.

Example 2

Graph $x^2 + y^2 = 1$.

Solution

Standard form is $(x - h)^2 + (y - k)^2 = r^2$.
Our equation is $x^2 \quad + \quad y^2 \quad = 1$.

$$h = 0 \qquad k = 0 \qquad r = \sqrt{1} = 1$$

The center is $(0, 0)$. The radius is 1. Plot $(0, 0)$, then
use the radius to locate four points on the circle.
From the center, move 1 unit up, down, left, and
right. Draw a circle through the four points.
   The circle $x^2 + y^2 = 1$ is used often in other areas
of mathematics such as trigonometry. $x^2 + y^2 = 1$ is called the **unit circle**.

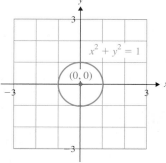

You Try 2

Graph $x^2 + y^2 = 25$.

If we are told the center and radius of a circle, we can write its equation.

Example 3

Find an equation of the circle with center $(-5, 0)$ and radius $\sqrt{7}$.

Solution

The x-coordinate of the center is h. $h = -5$
The y-coordinate of the center is k. $k = 0$

$$r = \sqrt{7}$$

Substitute these values into $(x - h)^2 + (y - k)^2 = r^2$.

$$(x - (-5))^2 + (y - 0)^2 = (\sqrt{7})^2 \qquad \text{Substitute } -5 \text{ for } x, 0 \text{ for } k, \text{ and } \sqrt{7} \text{ for } r.$$
$$(x + 5)^2 + y^2 = 7$$

You Try 3

Find an equation of the circle with center $(4, 7)$ and radius 5.

3. Graph a Circle Given in the Form $Ax^2 + Ay^2 + Cx + Dy + E = 0$

The equation of a circle can take another form—general form.

Definition

General Form for the Equation of a Circle: An equation of the form $Ax^2 + Ay^2 + Cx + Dy + E = 0$, where A, C, D, and E are real numbers, is the **general form** for the equation of a circle.

The coefficients of x^2 and y^2 must be the same in order for this to be the equation of a circle.

To graph a circle given in this form, we complete the square on x and on y to put it into standard form.

After having learned *all* of the conic sections, it is very important that we understand how to identify each one. To do this we will usually look at the coefficients of the square terms.

Example 4

Graph $x^2 + y^2 + 6x + 2y + 6 = 0$.

Solution

The coefficients of x^2 and y^2 are each 1. Therefore, this is the equation of a circle.

Our goal is to write the given equation in standard form, $(x - h)^2 + (y - k)^2 = r^2$, so that we can identify its center and radius. To do this we will group x^2 and $6x$ together, group y^2 and $2y$ together, then complete the square on each group of terms.

$$x^2 + y^2 + 6x + 2y + 6 = 0 \qquad \text{Group } x^2 \text{ and } 6x \text{ together.}$$
$$(x^2 + 6x) + (y^2 + 2y) = -6 \qquad \begin{array}{l} \text{Group } y^2 \text{ and } 2y \text{ together.} \\ \text{Move the constant to the other side.} \end{array}$$

Complete the square for each group of terms.

$$(x^2 + 6x + 9) + (y^2 + 2y + 1) = -6 + 9 + 1$$

$$(x + 3)^2 + (y + 1)^2 = 4$$

Since 9 and 1 are added on the left, they must also be added on the right.

Factor; add.

The center of the circle is $(-3, -1)$. The radius is 2.

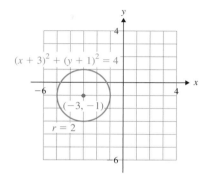

You Try 4

Graph $x^2 + y^2 + 10x - 4y + 20 = 0$.

If we rewrite $Ax^2 + Ay^2 + Cx + Dy + E = 0$ in standard form and get $(x - h)^2 + (y - k)^2 = 0$, then the graph is just the point (h, k). If the constant on the right side of the standard form equation is a negative number then the equation has no graph.

Using Technology

Recall that the equation of a circle is not a function. However, if we want to graph an equation on a graphing calculator, it must be entered as a function or a pair of functions. Therefore, to graph a circle we must solve the equation for y in terms of x.

Let's discuss how to graph $x^2 + y^2 = 4$ on a graphing calculator.

We must solve the equation for y.

$$x^2 + y^2 = 4$$

$$y^2 = 4 - x^2$$

$$y = \pm\sqrt{4 - x^2}$$

Now the equation of the circle $x^2 + y^2 = 4$ is rewritten so that y is in terms of x. In the graphing calculator, enter $y = \sqrt{4 - x^2}$ as Y_1. This represents the top half of the circle since the y-values are positive above the x-axis. Enter $y = -\sqrt{4 - x^2}$ as Y_2. This represents the bottom half of the circle since the y-values are negative below the x-axis. Here we have the window set from -3 to 3 in both the x- and y-directions. Press GRAPH.

The graph is distorted and does not actually look like a circle! This is because the screen is rectangular, and the graph is longer in the x-direction. We can "fix" this by squaring the window.

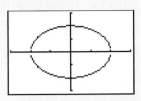

To square the window and get a better representation of the graph of $x^2 + y^2 = 4$, press ZOOM and choose 5:ZSquare. The graph reappears on a "squared" window and now looks like a circle.

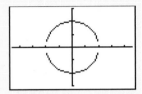

Identify the center and radius of each circle. Then, rewrite each equation for y in terms of x, and graph each circle on a graphing calculator. These problems come from the homework exercises so that the graphs can be found in the Answers to Selected Exercises appendix at the back of the book.

1. $x^2 + y^2 = 36$; Exercise 11

2. $x^2 + y^2 = 9$; Exercise 13

3. $(x + 3)^2 + y^2 = 4$; Exercise 7

4. $x^2 + (y - 1)^2 = 25$; Exercise 15

Answers to You Try Exercises

1)

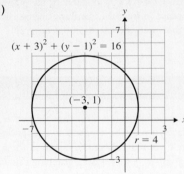

2)

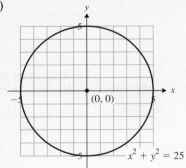

3) $(x - 4)^2 + (y - 7)^2 = 25$

4)

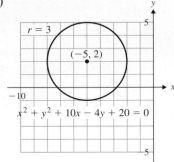

Answers to Technology Exercises

1. Center $(0, 0)$; radius $= 6$; $Y_1 = \sqrt{36 - x^2}$, $Y_2 = -\sqrt{36 - x^2}$
2. Center $(0, 0)$; radius $= 3$; $Y_1 = \sqrt{9 - x^2}$, $Y_2 = -\sqrt{9 - x^2}$
3. Center $(-3, 0)$; radius $= 2$; $Y_1 = \sqrt{4 - (x + 3)^2}$, $Y_2 = -\sqrt{4 - (x + 3)^2}$
4. Center $(0, 1)$; radius $= 5$; $Y_1 = 1 + \sqrt{25 - x^2}$, $Y_2 = 1 - \sqrt{25 - x^2}$

14.1 Exercises

 Boost your grade at mathzone.com! MathZone > **Practice Problems** > **NetTutor** > **Self-Test** > **e-Professors** > **Videos**

Objective 1

1) Is the equation of a circle a function? Explain your answer.

2) The standard form for the equation of a circle is
$$(x - h)^2 + (y - k)^2 = r^2$$
Identify the center and the radius.

Objective 2

Identify the center and radius of each circle and graph.

3) $(x + 2)^2 + (y - 4)^2 = 9$

4) $(x + 1)^2 + (y + 3)^2 = 25$

5) $(x - 5)^2 + (y - 3)^2 = 1$

6) $x^2 + (y - 5)^2 = 9$ 7) $(x + 3)^2 + y^2 = 4$

8) $(x - 2)^2 + (y - 2)^2 = 36$

9) $(x - 6)^2 + (y + 3)^2 = 16$

10) $(x + 8)^2 + (y - 4)^2 = 4$

11) $x^2 + y^2 = 36$ 12) $x^2 + y^2 = 16$

13) $x^2 + y^2 = 9$ 14) $x^2 + y^2 = 25$

15) $x^2 + (y - 1)^2 = 25$ 16) $(x + 3)^2 + y^2 = 1$

Find an equation of the circle with the given center and radius.

17) Center $(4, 1)$; radius $= 5$

18) Center $(3, 5)$; radius $= 2$

19) Center $(-3, 2)$; radius $= 1$

20) Center $(4, -6)$; radius $= 3$

21) Center $(-1, -5)$; radius $= \sqrt{3}$

22) Center $(-2, -1)$; radius $= \sqrt{5}$

23) Center $(0, 0)$; radius $= \sqrt{10}$

24) Center $(0, 0)$; radius $= \sqrt{6}$

25) Center $(6, 0)$; radius $= 4$

26) Center $(0, -3)$; radius $= 5$

27) Center $(0, -4)$; radius $= 2\sqrt{2}$

28) Center $(1, 0)$; radius $= 3\sqrt{2}$

Objective 3

Put the equation of each circle in the form $(x - h)^2 + (y - k)^2 = r^2$, identify the center and the radius, and graph.

29) $x^2 + y^2 + 2x + 10y + 17 = 0$

30) $x^2 + y^2 - 4x - 6y + 9 = 0$

31) $x^2 + y^2 + 8x - 2y - 8 = 0$

32) $x^2 + y^2 - 6x + 8y + 24 = 0$

33) $x^2 + y^2 - 10x - 14y + 73 = 0$

34) $x^2 + y^2 + 12x + 12y + 63 = 0$

35) $x^2 + y^2 + 6y + 5 = 0$

36) $x^2 + y^2 + 2x - 24 = 0$

37) $x^2 + y^2 - 4x - 1 = 0$

38) $x^2 + y^2 - 10y + 22 = 0$

39) $x^2 + y^2 - 8x + 8y - 4 = 0$

40) $x^2 + y^2 - 6x + 2y - 6 = 0$

41) $4x^2 + 4y^2 - 12x - 4y - 6 = 0$
(Hint: Begin by dividing the equation by 4.)

42) $16x^2 + 16y^2 + 16x - 24y - 3 = 0$
(Hint: Begin by dividing the equation by 16.)

Section 14.2 The Ellipse and the Hyperbola

Objectives

1. Graph an Ellipse
2. Graph a Hyperbola

The Ellipse

1. Graph an Ellipse

The next conic section we will study is the *ellipse*. An **ellipse** is the set of all points in a plane such that the *sum* of the distances from a point on the ellipse to two fixed points is constant. Each fixed point is called a **focus** (plural: **foci**). The point halfway between the foci is the **center** of the ellipse.

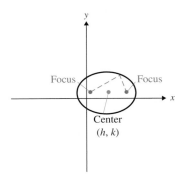

The orbits of planets around the sun as well as satellites around the Earth are elliptical. Statuary Hall in the U.S. Capitol building is an ellipse. If a person stands at one focus of this ellipse and whispers, a person standing across the room on the other focus can clearly hear what was said. Properties of the ellipse are used in medicine as well. One procedure for treating kidney stones involves immersing the patient in an elliptical tub of water. The kidney stone is at one focus, while at the other focus, high energy shock waves are produced, which destroy the kidney stone.

Definition | **Standard Form for the Equation of an Ellipse:** The standard form for the equation of an ellipse is

$$\frac{(x-h)^2}{a^2} + \frac{(y-k)^2}{b^2} = 1$$

The center of the ellipse is (h, k).

It is important to remember that the terms on the left are *both* positive quantities.

Example 1

Graph $\dfrac{(x-3)^2}{16} + \dfrac{(y-1)^2}{4} = 1$.

Solution

Standard form is $\dfrac{(x-h)^2}{a^2} + \dfrac{(y-k)^2}{b^2} = 1$.

Our equation is $\dfrac{(x-3)^2}{16} + \dfrac{(y-1)^2}{4} = 1$.

$$h = 3 \qquad k = 1$$
$$a = \sqrt{16} = 4 \qquad b = \sqrt{4} = 2$$

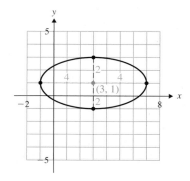

The center is (3, 1).

To graph the ellipse, first plot the center (3, 1). Since $a = 4$ and a^2 is under the squared quantity containing the **x**, move 4 units each way in the x-direction from the center. These are two points on the ellipse.

Since $b = 2$ and b^2 is under the squared quantity containing the **y**, move 2 units each way in the y-direction from the center. These are two more points on the ellipse. Sketch the ellipse through the four points.

 You Try 1

Graph $\dfrac{(x+2)^2}{25} + \dfrac{(y-3)^2}{16} = 1$.

Example 2

Graph $\dfrac{x^2}{9} + \dfrac{y^2}{25} = 1$.

Solution

Standard form is $\dfrac{(x-h)^2}{a^2} + \dfrac{(y-k)^2}{b^2} = 1$.

Our equation is $\dfrac{x^2}{9} + \dfrac{y^2}{25} = 1$.

$$h = 0 \qquad k = 0$$
$$a = \sqrt{9} = 3 \qquad b = \sqrt{25} = 5$$

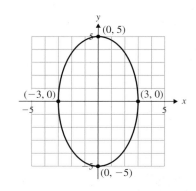

The center is (0, 0).

Plot the center (0, 0). Since $a = 3$ and a^2 is under the x^2, move 3 units each way in the x-direction from the center. These are two points on the ellipse.

Since $b = 5$ and b^2 is under the y^2, move 5 units each way in the y-direction from the center. These are two more points on the ellipse. Sketch the ellipse through the four points.

You Try 2

Graph $\dfrac{x^2}{36} + \dfrac{y^2}{9} = 1$.

In Example 2, note that the *origin*, (0, 0), is the center of the ellipse. $a = 3$ and the x-intercepts are (3, 0) and (−3, 0). $b = 5$ and the y-intercepts are (0, 5) and (0, −5). We can generalize these relationships as follows.

Definition

Equation of an Ellipse with Center at Origin: The graph of $\dfrac{x^2}{a^2} + \dfrac{y^2}{b^2} = 1$ is an ellipse with center at the origin, x-intercepts $(a, 0)$ and $(−a, 0)$, and y-intercepts $(0, b)$ and $(0, −b)$.

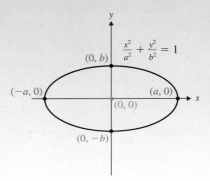

Looking at Examples 1 and 2 we can make another interesting observation.

Example 1	**Example 2**

$$\dfrac{(x-3)^2}{16} + \dfrac{(y-1)^2}{4} = 1$$
$$a^2 = 16 \quad b^2 = 4$$
$$a^2 > b^2$$

$$\dfrac{x^2}{9} + \dfrac{y^2}{25} = 1$$
$$a^2 = 9 \quad b^2 = 25$$
$$b^2 > a^2$$

The number under $(x - 3)^2$ is greater than the number under $(y - 1)^2$. *The ellipse is longer in the x-direction.*

The number under y^2 is greater than the number under x^2. *The ellipse is longer in the y-direction.*

This relationship between a^2 and b^2 will always produce the same result.

The equation of an ellipse can take other forms.

Example 3

Graph $4x^2 + 25y^2 = 100$.

Solution

How can we tell if this is a circle or an ellipse? We look at the coefficients of x^2 and y^2. Both of the coefficients are positive, *and* they are different. *This is an ellipse.* (If this was a circle, the coefficients would be the same.)

Since the standard form for the equation of an ellipse has a 1 on one side of the $=$ sign, divide both sides of $4x^2 + 25y^2 = 100$ by 100 to obtain a 1 on the right.

$$4x^2 + 25y^2 = 100$$

$$\frac{4x^2}{100} + \frac{25y^2}{100} = \frac{100}{100} \qquad \text{Divide both sides by 100.}$$

$$\frac{x^2}{25} + \frac{y^2}{4} = 1 \qquad \text{Simplify.}$$

The center is $(0, 0)$. $a = \sqrt{25} = 5$ and $b = \sqrt{4} = 2$. Plot $(0, 0)$. Move 5 units each way from the center in the x-direction. Move 2 units each way from the center in the y-direction.

Notice that the x-intercepts are $(5, 0)$ and $(-5, 0)$. The y-intercepts are $(2, 0)$ and $(-2, 0)$.

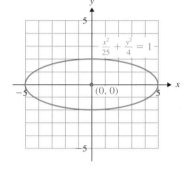

You Try 3

Graph $x^2 + 4y^2 = 4$.

You may have noticed that if $a^2 = b^2$, then the ellipse is a circle.

The Hyperbola

2. Graph a Hyperbola

The last of the conic sections is the *hyperbola*. A **hyperbola** is the set of all points in a plane such that the absolute value of the *difference* of the distances from two fixed points is constant. Each fixed point is called a **focus**. The point halfway between the foci is the **center** of the hyperbola.

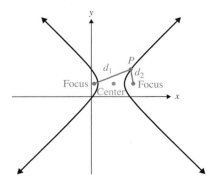

Some navigation systems used by ships are based on the properties of hyperbolas. Additionally, many telescopes use hyperbolic lenses.

A hyperbola is a graph consisting of two branches. The hyperbolas we will consider will have branches that open either in the x-direction or in the y-direction.

Standard Form for the Equation of a Hyperbola

1) A hyperbola with center (h, k) with branches that open in the *x-direction* has equation

$$\frac{(x - h)^2}{a^2} - \frac{(y - k)^2}{b^2} = 1$$

Its graph is to the right.

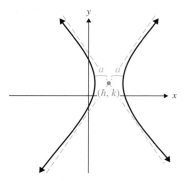

2) A hyperbola with center (h, k) with branches that open in the *y-direction* has equation

$$\frac{(y - k)^2}{b^2} - \frac{(x - h)^2}{a^2} = 1.$$

Its graph is to the right.

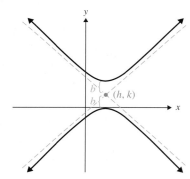

Notice in 1) that $\dfrac{(x - h)^2}{a^2}$ is the positive quantity, and the branches open in the x-direction.

In 2), the positive quantity is $\dfrac{(y - k)^2}{b^2}$, and the branches open in the y-direction.

In 1) and 2) notice how the branches of the hyperbola get closer to the dotted lines as the branches continue indefinitely. These dotted lines are called **asymptotes**. They are not an actual part of the graph of the hyperbola, but we can use them to help us obtain the hyperbola.

Example 4

Graph $\dfrac{(x+2)^2}{9} - \dfrac{(y-1)^2}{4} = 1.$

Solution

How do we know that this is a hyperbola and not an ellipse? *It is a hyperbola because there is a subtraction sign between the two quantities on the left.* If it was addition, it would be an ellipse.

Standard form is $\dfrac{(x-h)^2}{a^2} - \dfrac{(y-k)^2}{b^2} = 1.$

Our equation is $\dfrac{(x+2)^2}{9} - \dfrac{(y-1)^2}{4} = 1.$

$$h = -2 \qquad k = 1$$
$$a = \sqrt{9} = 3 \qquad b = \sqrt{4} = 2$$

The center is $(-2, 1)$. *Since the quantity* $\dfrac{(x-h)^2}{a^2}$ *is the* positive *quantity, the branches of the hyperbola will open in the x-direction.*

We will use the center, $a = 3$, and $b = 2$ to draw a *reference rectangle*. The diagonals of this rectangle are the asymptotes of the hyperbola.

First, plot the center $(-2, 1)$. Since $a = 3$ and a^2 is under the squared quantity containing the *x*, move 3 units each way in the *x*-direction from the center. These are two points on the rectangle.

Since $b = 2$ and b^2 is under the squared quantity containing the *y*, move 2 units each way in the *y*-direction from the center. These are two more points on the rectangle.

Draw the rectangle containing these four points, then draw the diagonals of the rectangle as dotted lines. These are the asymptotes of the hyperbola.

Sketch the branches of the hyperbola opening in the *x*-direction with the branches approaching the asymptotes.

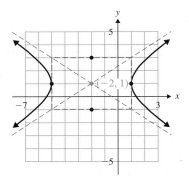

You Try 4

Graph $\dfrac{(x+1)^2}{9} - \dfrac{(y+1)^2}{16} = 1.$

Example 5

Graph $\dfrac{y^2}{4} - \dfrac{x^2}{25} = 1$.

Solution

Standard form is $\dfrac{(y-k)^2}{b^2} - \dfrac{(x-h)^2}{a^2} = 1$.

Our equation is $\dfrac{y^2}{4} - \dfrac{x^2}{25} = 1$.

$$k = 0 \qquad h = 0$$
$$b = \sqrt{4} = 2 \qquad a = \sqrt{25} = 5$$

The center is (0, 0). *Since the quantity* $\dfrac{y^2}{4}$ *is the* positive *quantity, the branches of the hyperbola will open in the y-direction.*

Use the center, $a = 5$, and $b = 2$ to draw the reference rectangle and its diagonals.

Plot the center (0, 0). Since $a = 5$ and a^2 is under the x^2, move 5 units each way in the x-direction from the center to get two points on the rectangle.

Since $b = 2$ and b^2 is under the y^2, move 2 units each way in the y-direction from the center to get two more points on the rectangle.

Draw the rectangle and its diagonals as dotted lines. These are the asymptotes of the hyperbola.

Sketch the branches of the hyperbola opening in the y-direction approaching the asymptotes.

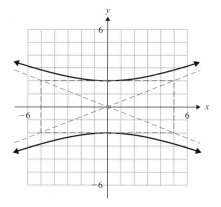

You Try 5

Graph $\dfrac{y^2}{16} - \dfrac{x^2}{16} = 1$.

Equation of a Hyperbola with Center at the Origin

1) The graph of $\dfrac{x^2}{a^2} - \dfrac{y^2}{b^2} = 1$ is a hyperbola with

   center $(0, 0)$ and x-intercepts $(a, 0)$ and $(-a, 0)$.

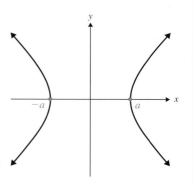

2) The graph $\dfrac{y^2}{b^2} - \dfrac{x^2}{a^2} = 1$ is a hyperbola with

   center $(0, 0)$ and y-intercepts $(0, b)$ and $(0, -b)$.

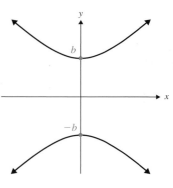

We will look at one more form of the hyperbola.

Example 6

Graph $y^2 - 9x^2 = 9$.

Solution

This is a hyperbola since there is a subtraction sign between the two terms.

   Since the standard form for the equation of the hyperbola has a 1 on one side of the $=$ sign, divide both sides of $y^2 - 9x^2 = 9$ by 9 to obtain a 1 on the right.

$$y^2 - 9x^2 = 9$$
$$\frac{y^2}{9} - \frac{9x^2}{9} = \frac{9}{9} \qquad \text{Divide both sides by 9.}$$
$$\frac{y^2}{9} - x^2 = 1 \qquad \text{Simplify.}$$

The center is $(0, 0)$. *The branches of the hyperbola will open in the y-direction since* $\dfrac{y^2}{9}$ *is a positive quantity.*

x^2 is the same as $\dfrac{x^2}{1}$, so $a = \sqrt{1} = 1$ and $b = \sqrt{9} = 3$.

Plot the center at the origin. Move 1 unit each way in the *x*-direction from the center and 3 units each way in the *y*-direction. Draw the rectangle and the asymptotes.

Sketch the branches of the hyperbola opening in the *y*-direction approaching the asymptotes.

The *y*-intercepts are (0, 3) and (0, −3). There are no *x*-intercepts.

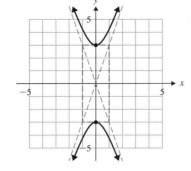

 You Try 6

Graph $4x^2 - 9y^2 = 36$.

More about the conic sections and their characteristics are studied in later mathematics courses.

 Using Technology

We graph ellipses and hyperbolas on a graphing calculator in the same way that we graphed circles: solve the equation for *y* in terms of *x*, enter both values of *y*, and graph both equations.

Let's graph the ellipse given in Example 3: $4x^2 + 25y^2 = 100$.

Solve for *y*.

$$4x^2 + 25y^2 = 100$$
$$25y^2 = 100 - 4x^2$$
$$y^2 = \frac{100 - 4x^2}{25}$$
$$y^2 = 4 - \frac{4x^2}{25}$$
$$y = \pm\sqrt{4 - \frac{4x^2}{25}}$$

Enter $y = \sqrt{4 - \dfrac{4x^2}{25}}$ as Y_1. This represents the top half of the ellipse since the *y*-values are

positive above the *x*-axis. Enter $y = -\sqrt{4 - \dfrac{4x^2}{25}}$ as Y_2. This represents the bottom half of

the ellipse since the *y*-values are negative below the *x*-axis. Set an appropriate window, and press ⎢GRAPH⎥.

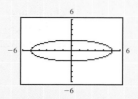

Use the same technique to graph a hyperbola on a graphing calculator.

Identify each conic section as either an ellipse or a hyperbola. Identify the center of each, rewrite each equation for y in terms of x, and graph each equation on a graphing calculator. These problems come from the homework exercises so that the graphs can be found in the Answers to Selected Exercises appendix at the back of the book.

1. $4x^2 + 9y^2 = 36$; Exercise 21

2. $x^2 + \dfrac{y^2}{4} = 1$; Exercise 15

3. $\dfrac{x^2}{25} + (y + 4)^2 = 1$; Exercise 17

4. $9x^2 - y^2 = 36$; Exercise 37

5. $\dfrac{y^2}{16} - \dfrac{x^2}{4} = 1$; Exercise 27

6. $y^2 - \dfrac{(x - 1)^2}{9} = 1$; Exercise 33

Answers to You Try Exercises

1)

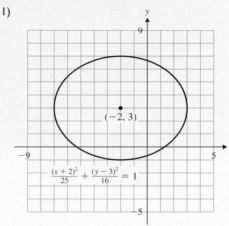

$$\dfrac{(x + 2)^2}{25} + \dfrac{(y - 3)^2}{16} = 1$$

2)

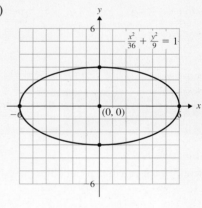

$$\dfrac{x^2}{36} + \dfrac{y^2}{9} = 1$$

3)

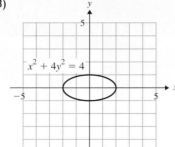

$$x^2 + 4y^2 = 4$$

4)

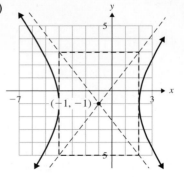

5)

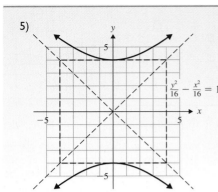

$$\frac{y^2}{16} - \frac{x^2}{16} = 1$$

6)

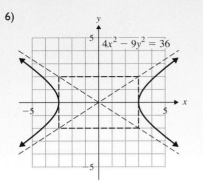

$$4x^2 - 9y^2 = 36$$

Answers to Technology Exercises

1. ellipse with center $(0, 0)$; $Y_1 = \sqrt{4 - \frac{4x^2}{9}}$, $Y_2 = -\sqrt{4 - \frac{4x^2}{9}}$

2. ellipse with center $(0, 0)$; $Y_1 = \sqrt{4 - 4x^2}$, $Y_2 = -\sqrt{4 - 4x^2}$

3. ellipse with center $(0, -4)$; $Y_1 = -4 + \sqrt{1 - \frac{x^2}{25}}$, $Y_2 = -4 - \sqrt{1 - \frac{x^2}{25}}$

4. hyperbola with center $(0, 0)$; $Y_1 = \sqrt{9x^2 - 36}$; $Y_2 = -\sqrt{9x^2 - 36}$

5. hyperbola with center $(0, 0)$; $Y_1 = \sqrt{16 + 4x^2}$; $Y_2 = -\sqrt{16 + 4x^2}$

6. hyperbola with center $(1, 0)$; $Y_1 = \sqrt{1 + \frac{(x - 1)^2}{9}}$; $Y_2 = -\sqrt{1 + \frac{(x - 1)^2}{9}}$

14.2 Exercises

Boost your grade at mathzone.com! MathZone

> **Practice Problems** > **NetTutor** > **e-Professors** > **Videos**
> **Self-Test**

Objective 1

Identify each equation as an ellipse or a hyperbola.

1) $\frac{x^2}{36} + \frac{y^2}{4} = 1$

2) $\frac{x^2}{9} - \frac{y^2}{25} = 1$

3) $\frac{(y - 3)^2}{4} - \frac{(x + 5)^2}{9} = 1$

4) $\frac{(x - 4)^2}{16} + \frac{(y - 1)^2}{9} = 1$

5) $16x^2 - y^2 = 16$

6) $4x^2 + 25y^2 = 100$

7) $\frac{x^2}{25} + y^2 = 1$

8) $-\frac{(x + 6)^2}{4} + \frac{y^2}{10} = 1$

10) $\frac{(x - 4)^2}{4} + \frac{(y - 3)^2}{16} = 1$

11) $\frac{(x - 3)^2}{9} + \frac{(y + 2)^2}{16} = 1$

12) $\frac{(x + 4)^2}{25} + \frac{(y - 5)^2}{16} = 1$

13) $\frac{x^2}{36} + \frac{y^2}{16} = 1$

14) $\frac{x^2}{36} + \frac{y^2}{4} = 1$

15) $x^2 + \frac{y^2}{4} = 1$

16) $\frac{x^2}{9} + y^2 = 1$

17) $\frac{x^2}{25} + (y + 4)^2 = 1$

Identify the center of each ellipse and graph the equation.

 9) $\frac{(x + 2)^2}{9} + \frac{(y - 1)^2}{4} = 1$

18) $(x + 3)^2 + \frac{(y + 4)^2}{9} = 1$

19) $\dfrac{(x+1)^2}{4} + \dfrac{(y+3)^2}{9} = 1$

20) $\dfrac{(x-2)^2}{16} + \dfrac{y^2}{25} = 1$ 21) $4x^2 + 9y^2 = 36$

22) $x^2 + 4y^2 = 16$ 23) $25x^2 + y^2 = 25$

24) $9x^2 + y^2 = 36$

Objective 2

Identify the center of each hyperbola and graph the equation.

25) $\dfrac{x^2}{9} - \dfrac{y^2}{25} = 1$ 26) $\dfrac{x^2}{9} - \dfrac{y^2}{4} = 1$

27) $\dfrac{y^2}{16} - \dfrac{x^2}{4} = 1$ 28) $\dfrac{y^2}{4} - \dfrac{x^2}{4} = 1$

29) $\dfrac{(x-2)^2}{9} - \dfrac{(y+3)^2}{16} = 1$

30) $\dfrac{(x+3)^2}{4} - \dfrac{(y+1)^2}{16} = 1$

31) $\dfrac{(y+1)^2}{25} - \dfrac{(x+4)^2}{4} = 1$

32) $\dfrac{(y-1)^2}{36} - \dfrac{(x+1)^2}{9} = 1$

33) $y^2 - \dfrac{(x-1)^2}{9} = 1$ 34) $\dfrac{(y+4)^2}{4} - x^2 = 1$

35) $\dfrac{(x-1)^2}{25} - \dfrac{(y-2)^2}{25} = 1$

36) $\dfrac{(x-2)^2}{16} - \dfrac{(y-3)^2}{9} = 1$

37) $9x^2 - y^2 = 36$ 38) $4y^2 - x^2 = 16$

39) $y^2 - x^2 = 1$ 40) $x^2 - y^2 = 25$

Mid-Chapter Summary

Objective	1. Learn How to Distinguish Between the Different Types of Conic Sections, and Graph Them

Objective

1. Learn How to Distinguish Between the Different Types of Conic Sections, and Graph Them

1. Learn How to Distinguish Between the Different Types of Conic Sections, and Graph Them

Sometimes the most difficult part of graphing a conic section is identifying which type of graph will result from the given equation. In this section, we will discuss how to look at an equation and decide what type of conic section it represents.

Example 1

Graph $x^2 + y^2 + 4x - 6y + 9 = 0$.

Solution

First, notice that this equation has two squared terms. Therefore, its graph cannot be a parabola since the equation of a parabola contains only one squared term. Next, observe that the coefficients of x^2 and y^2 are each 1. Since the coefficients are the same, *this is the equation of a circle.*

Write the equation in the form $(x - h)^2 + (y - k)^2 = r^2$ by completing the square on the x-terms and on the y-terms.

$$x^2 + y^2 + 4x - 6y + 9 = 0$$
$$(x^2 + 4x) + (y^2 - 6y) = -9$$
$$(x^2 + 4x + 4) + (y^2 - 6y + 9) = -9 + 4 + 9$$

$$(x + 2)^2 + (y - 3)^2 = 4$$

Group the x-terms together and group the y-terms together. Move the constant to the other side.
Complete the square for each group of terms.
Factor; add.

The center of the circle is $(-2, 3)$. The radius is 2.

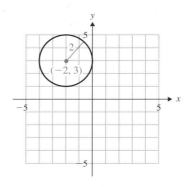

Example 2

Graph $x = y^2 + 4y + 3$.

Solution

This equation contains only one squared term. Therefore, *this is the equation of a parabola.* Since the squared term is y^2 and $a = 1$, the parabola will open to the right.

Use the formula $y = -\dfrac{b}{2a}$ to find the y-coordinate of the vertex.

$$a = 1 \quad b = 4 \quad c = 3$$
$$y = -\dfrac{4}{2(1)} = -2$$
$$x = (-2)^2 + 4(-2) + 3 = -1$$

The vertex is $(-1, -2)$. Make a table of values to find other points on the parabola, and use the axis of symmetry to find more points.

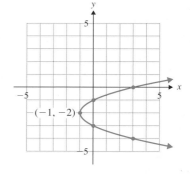

x	y
0	-1
3	0

Plot the points in the table. Locate the points $(0, -3)$ and $(3, -4)$ using the axis of symmetry, $y = -2$.

Example 3

Graph $\dfrac{(y - 1)^2}{9} - \dfrac{(x - 3)^2}{4} = 1$.

Solution

In this equation we see the *difference* of two squares. *The graph of this equation is a hyperbola. The branches of the hyperbola will open in the y-direction since the quantity containing the variable y,* $\dfrac{(y - 1)^2}{9}$, *is the positive, squared quantity.*

The center is $(3, 1)$. $a = \sqrt{4} = 2$ and $b = \sqrt{9} = 3$. Draw the reference rectangle and its diagonals, the *asymptotes* of the graph.

The branches of the hyperbola will approach the asymptotes.

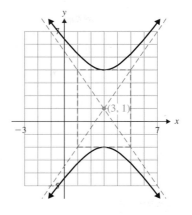

Example 4

Graph $x^2 + 9y^2 = 36$.

Solution

This equation contains the *sum* of two squares with *different* coefficients. *This is the equation of an ellipse.* (If the coefficients were the same, the graph would be a circle.)

Divide both sides of the equation by 36 to get 1 on the right side of the $=$ sign.

$$x^2 + 9y^2 = 36$$
$$\frac{x^2}{36} + \frac{9y^2}{36} = \frac{36}{36} \qquad \text{Divide both sides by 36.}$$
$$\frac{x^2}{36} + \frac{y^2}{4} = 1 \qquad \text{Simplify.}$$

The center is $(0, 0)$. $a = 6$ and $b = 2$.

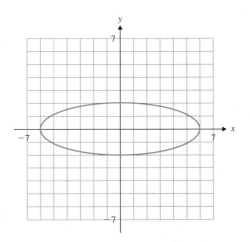

You Try I

Determine whether the graph of each equation will be a parabola, circle, ellipse, or hyperbola. Then, graph each equation.

a) $4x^2 - 25y^2 = 100$

b) $x^2 + y^2 - 6x - 12y + 9 = 0$

c) $y = -x^2 - 2x + 4$

d) $x^2 + \dfrac{(y+4)^2}{9} = 1$

Answers to You Try Exercises

1) a) hyperbola

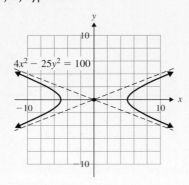

$4x^2 - 25y^2 = 100$

b) circle

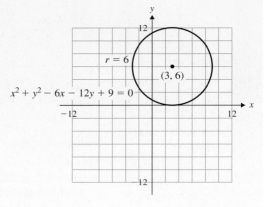

$r = 6$

$(3, 6)$

$x^2 + y^2 - 6x - 12y + 9 = 0$

c) parabola

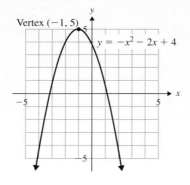

Vertex $(-1, 5)$

$y = -x^2 - 2x + 4$

d) ellipse

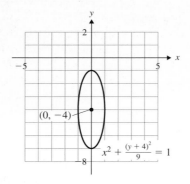

$(0, -4)$

$x^2 + \dfrac{(y+4)^2}{9} = 1$

Mid-Chapter Summary Exercises

Boost your grade at mathzone.com! MathZone

> Practice Problems
> NetTutor
> Self-Test
> e-Professors
> Videos

Objective 1

Determine whether the graph of each equation will be a parabola, circle, ellipse, or hyperbola. Then, graph each equation.

1) $y = x^2 + 4x + 8$

2) $(x + 5)^2 + (y - 3)^2 = 25$

 3) $\dfrac{(y+4)^2}{9} - \dfrac{(x+1)^2}{4} = 1$

4) $x = (y - 1)^2 + 8$ 5) $16x^2 + 9y^2 = 144$

6) $x^2 - 4y^2 = 36$

 7) $x^2 + y^2 + 8x - 6y - 11 = 0$

8) $\dfrac{(x-2)^2}{25} + \dfrac{(y-2)^2}{36} = 1$

9) $(x-1)^2 + \dfrac{y^2}{16} = 1$

10) $x^2 + y^2 + 8y + 7 = 0$

11) $x = -(y+4)^2 - 3$

12) $\dfrac{(y+4)^2}{4} - (x+2)^2 = 1$

13) $25x^2 - 4y^2 = 100$

14) $4x^2 + y^2 = 16$

15) $(x-3)^2 + y^2 = 16$

16) $y = -x^2 + 6x - 7$

17) $x = \dfrac{1}{2}y^2 + 2y + 3$

18) $\dfrac{(x-3)^2}{16} + y^2 = 1$

19) $(x-2)^2 - (y+1)^2 = 9$

20) $x^2 + y^2 + 6x - 8y + 9 = 0$

Section 14.3 Nonlinear Systems of Equations

Objectives

1. Define a Nonlinear System of Equations

2. Solve a Nonlinear System by Substitution

3. Solve a Nonlinear System Using the Elimination Method

1. Define a Nonlinear System of Equations

Recall from Chapter 5 that a system of linear equations in two variables is a system such as

$$3x - 2y = 8$$
$$x + 4y = -5$$

We learned to solve such a system by graphing, substitution, and the elimination method. We can use these same techniques for solving a *nonlinear* system of equations in two variables. A **nonlinear system of equations** is a system in which at least one of the equations is not linear.

Solving a nonlinear system by graphing is not very practical since it would be very difficult (if not impossible) to accurately read the points of intersection. Therefore, we will solve the systems using substitution and the elimination method. We will graph the equations, however, so that we can visualize the solution(s) as the point(s) of intersection of the graphs.

We will be interested only in real number solutions. If a system has imaginary solutions, then the graphs of the equations do not intersect in the real number plane.

2. Solve a Nonlinear System by Substitution

When one of the equations in a system is linear, it is often best to use the substitution method to solve the system.

Example 1

Solve the system $x^2 - 2y = 2$ (1)
$ -x + y = 3$ (2)

Solution

The graph of equation (1) is a parabola, and the graph of equation (2) is a line. Let's begin by thinking about the number of possible points of intersection the graphs can have.

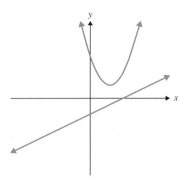

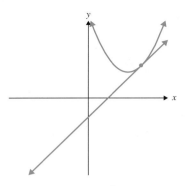

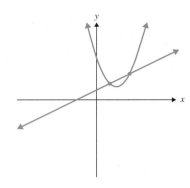

No points of intersection
The system has no solution.

One point of intersection
The system has one solution.

Two points of intersection
The system has two solutions.

Solve the linear equation for one of the variables.

$$-x + y = 3$$
$$y = x + 3 \quad (3) \qquad \text{Solve for } y.$$

Substitute $x + 3$ for y in equation (1).

$$x^2 - 2y = 2 \quad (1)$$
$$x^2 - 2(x + 3) = 2 \qquad \text{Substitute.}$$
$$x^2 - 2x - 6 = 2 \qquad \text{Distribute.}$$
$$x^2 - 2x - 8 = 0 \qquad \text{Subtract 2.}$$
$$(x - 4)(x + 2) = 0 \qquad \text{Factor.}$$
$$x = 4 \text{ or } x = -2 \qquad \text{Solve.}$$

To find the corresponding value of y for each value of x, we can substitute $x = 4$ and then $x = -2$ into *either* equation (1), (2), or (3). No matter which equation you choose, you should always check the solutions in *both* of the original equations. We will substitute the values into equation (3) because this is just an alternative form of equation (2), and it is already solved for y.

Substitute each value into equation (3) to find y.

$$x = 4: \; y = x + 3 \qquad\qquad x = -2: \; y = x + 3$$
$$y = 4 + 3 \qquad\qquad\qquad\qquad y = -2 + 3$$
$$y = 7 \qquad\qquad\qquad\qquad\qquad y = 1$$

The proposed solutions are $(4, 7)$ and $(-2, 1)$. Verify that they solve the system by checking them in equation (1).

The solution set is $\{(4, 7), (-2, 1)\}$.

Graphing each equation we can see that these are the points of intersection.

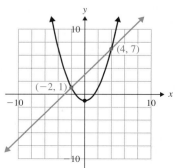

You Try 1

Solve the system $x^2 + 3y = 6$
$$x + y = 2$$

Another system that can be solved with substitution is shown in Example 2.

Example 2

Solve the system $x^2 + y^2 = 1$ (1)
$$x + 2y = -1 \quad (2)$$

Solution

The graph of equation (1) is a circle, and the graph of equation (2) is a line. These graphs can intersect at zero, one, or two points. Therefore, this system will have zero, one, or two solutions.

We will not solve equation (1) for a variable because doing so would give us a radical in the expression. It will be easiest to solve equation (2) for x because its coefficient is 1.

$$x + 2y = -1 \qquad (2)$$
$$x = -2y - 1 \qquad (3) \qquad \text{Solve for } x.$$

Substitute $-2y - 1$ for x in equation (1).

$$x^2 + y^2 = 1 \quad (1)$$
$$(-2y - 1)^2 + y^2 = 1 \qquad \text{Substitute.}$$
$$4y^2 + 4y + 1 + y^2 = 1 \qquad \text{Expand } (-2y - 1)^2.$$
$$5y^2 + 4y = 0 \qquad \text{Combine like terms; subtract 1.}$$
$$y(5y + 4) = 0 \qquad \text{Factor.}$$
$$y = 0 \quad \text{or} \quad 5y + 4 = 0 \qquad \text{Set each factor equal to zero.}$$
$$5y = -4 \qquad \text{Subtract 4.}$$
$$y = -\frac{4}{5} \qquad \text{Divide by 5.}$$

Substitute $y = 0$ and then $y = -\dfrac{4}{5}$ into (3) to find their corresponding values of x.

$$y = 0: x = -2y - 1$$
$$x = -2(0) - 1$$
$$x = -1$$

$$y = -\frac{4}{5}: x = -2y - 1$$
$$x = -2\left(-\frac{4}{5}\right) - 1$$
$$x = \frac{8}{5} - 1$$
$$x = \frac{3}{5}$$

The proposed solutions are $(-1, 0)$ and $\left(\dfrac{3}{5}, -\dfrac{4}{5}\right)$. Verify that they solve the system by checking them in equation (1).

The solution set is

$$\left\{ (-1, 0), \left(\frac{3}{5}, -\frac{4}{5} \right) \right\}.$$

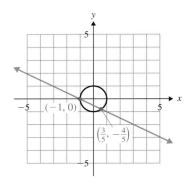

 You Try 2

Solve the system $x^2 + y^2 = 25$
$\qquad\qquad\qquad\quad x - y = 7$

We must be certain the proposed solutions satisfy *each* equation in the system. If we had substituted $y = 0$ and $y = -\dfrac{4}{5}$ into $x^2 + y^2 = 1$ in Example 2, then for each value of y we would get *two* corresponding values for x.

$$y = 0: \quad \begin{aligned} x^2 + y^2 &= 1 \\ x^2 + (0)^2 &= 1 \\ x^2 &= 1 \\ x &= \pm\sqrt{1} = \pm 1 \end{aligned} \qquad\qquad y = -\frac{4}{5}: \quad \begin{aligned} x^2 + y^2 &= 1 \\ x^2 + \left(-\frac{4}{5} \right)^2 &= 1 \\ x^2 + \frac{16}{25} &= 1 \\ x^2 &= \frac{9}{25} \\ x &= \pm\sqrt{\frac{9}{25}} = \pm\frac{3}{5} \end{aligned}$$

This gives the proposed solutions $(1, 0)$, $(-1, 0)$, $\left(\dfrac{3}{5}, -\dfrac{4}{5} \right)$, and $\left(-\dfrac{3}{5}, -\dfrac{4}{5} \right)$. If we check $(1, 0)$ and $\left(-\dfrac{3}{5}, -\dfrac{4}{5} \right)$ in $x + 2y = -1$, however, we see that they do not satisfy that equation.

$$\text{Check: } (1, 0): \quad \begin{aligned} x + 2y &= -1 \\ 1 + 2(0) &\overset{?}{=} -1 \\ 1 &= -1 \end{aligned}$$
$$\qquad\qquad\qquad\qquad \text{False}$$

$$\text{Check } \left(-\frac{3}{5}, -\frac{4}{5} \right): \quad \begin{aligned} x + 2y &= -1 \\ \left(-\frac{3}{5} \right) + 2\left(-\frac{4}{5} \right) &\overset{?}{=} -1 \\ -\frac{3}{5} - \frac{8}{5} &\overset{?}{=} -1 \\ -\frac{11}{5} &= -1 \end{aligned}$$
$$\qquad\qquad\qquad\qquad\qquad\qquad \text{False}$$

Once the value(s) for one of the variables is (are) found, it is best to substitute them into the *linear* equation to avoid getting extraneous solutions to the system.

3. Solve a Nonlinear System Using the Elimination Method

The elimination method can be used to solve a system when both equations are second-degree equations.

| Example 3 |

Solve the system $5x^2 + 3y^2 = 21$ (1)
$4x^2 - y^2 = 10$ (2)

Solution

Each equation is a second-degree equation. The first is an ellipse and the second is a hyperbola. They can have zero, one, two, three, or four points of intersection. Multiply equation (2) by 3. Then adding the two equations will eliminate the y^2-terms.

$$3(4x^2 - y^2) = 3(10) \qquad \text{Multiply equation (2) by 3.}$$
$$12x^2 - 3y^2 = 30$$

Original System		**Rewrite the System**
$5x^2 + 3y^2 = 21$	$\rightarrow$	$5x^2 + 3y^2 = 21$
$4x^2 - y^2 = 10$		$12x^2 - 3y^2 = 30$

Add the equations to eliminate y^2.

$$\begin{aligned} 5x^2 + 3y^2 &= 21 \\ +\underline{12x^2 - 3y^2} &= \underline{30} \\ 17x^2 &= 51 \\ x^2 &= 3 \\ x &= \pm\sqrt{3} \end{aligned}$$

Find the corresponding values of y for $x = \sqrt{3}$ and $x = -\sqrt{3}$. We will substitute the values into equation (2).

$x = \sqrt{3}$:
$$\begin{aligned} 4x^2 - y^2 &= 10 \\ 4(\sqrt{3})^2 - y^2 &= 10 \\ 12 - y^2 &= 10 \\ -y^2 &= -2 \\ y^2 &= 2 \\ y &= \pm\sqrt{2} \end{aligned}$$

$x = -\sqrt{3}$:
$$\begin{aligned} 4x^2 - y^2 &= 10 \\ 4(-\sqrt{3})^2 - y^2 &= 10 \\ 12 - y^2 &= 10 \\ -y^2 &= -2 \\ y^2 &= 2 \\ y &= \pm\sqrt{2} \end{aligned}$$

This gives us $(\sqrt{3}, \sqrt{2})$ and $(\sqrt{3}, -\sqrt{2})$.

This gives us $(-\sqrt{3}, \sqrt{2})$ and $(-\sqrt{3}, -\sqrt{2})$.

Check the proposed solutions in equation (1) to verify that they satisfy that equation as well.

The solution set is $\{(\sqrt{3}, \sqrt{2}), (\sqrt{3}, -\sqrt{2}), (-\sqrt{3}, \sqrt{2}), (-\sqrt{3}, -\sqrt{2})\}$.

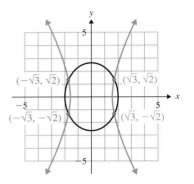

 You Try 3

Solve the system $2x^2 - 13y^2 = 20$
$-x^2 + 10y^2 = 4$

For solving some systems, using *either* substitution or the elimination method works well. Look carefully at each system to decide which method to use.

We will see in Example 4 that not all systems have solutions.

Example 4

Solve the system $y = \sqrt{x}$ (1)
$y^2 - 4x^2 = 4$ (2)

Solution

The graph of the square root function $y = \sqrt{x}$ is half of a parabola. The graph of equation (2) is a hyperbola. Solve this system by substitution. Replace y in equation (2) with $\sqrt{x}$ from equation (1).

$$y^2 - 4x^2 = 4 \quad (2)$$
$$(\sqrt{x})^2 - 4x^2 = 4 \qquad \text{Substitute } y = \sqrt{x} \text{ into equation (2).}$$
$$x - 4x^2 = 4$$
$$0 = 4x^2 - x + 4$$

Since the right-hand side of the equation does not factor, we will solve it using the quadratic formula.

$$4x^2 - x + 4 = 0$$
$$a = 4 \qquad b = -1 \qquad c = 4$$
$$x = \frac{-(-1) \pm \sqrt{(-1)^2 - 4(4)(4)}}{2(4)} = \frac{1 \pm \sqrt{1 - 64}}{8} = \frac{1 \pm \sqrt{-63}}{8}$$

Since $\sqrt{-63}$ is not a real number, there are no real number values for x. The system has no solution, or $\varnothing$.

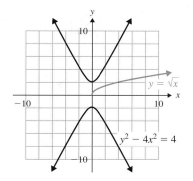

You Try 4

Solve the system $4x^2 + y^2 = 4$
$x - y = 3$

Using Technology

We can solve systems of nonlinear equations on the graphing calculator just like we solved systems of linear equations in Chapter 5—graph the equations and find their points of intersection.

Let's look at Example 3:

$$5x^2 + 3y^2 = 21$$
$$4x^2 - y^2 = 10$$

Solve each equation for y and enter them into the calculator.

Solve $5x^2 + 3y^2 = 21$ for y: Solve $4x^2 - y^2 = 10$ for y:

$$y = \pm\sqrt{7 - \frac{5}{3}x^2}$$ $$y = \pm\sqrt{4x^2 - 10}$$

Enter $\sqrt{7 - \dfrac{5}{3}x^2}$ as Y_1.   Enter $\sqrt{4x^2 - 10}$ as Y_3.

Enter $-\sqrt{7 - \dfrac{5}{3}x^2}$ as Y_2.   Enter $-\sqrt{4x^2 - 10}$ as Y_4.

After entering the equations, press $\boxed{\text{GRAPH}}$.

   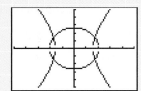

There are four real solutions to the system since the graphs have four points of intersection. We can use the INTERSECT option to find the solutions. Since we graphed four functions, we must tell the calculator which point of intersection we want to find. Note that the point where the graphs intersect in the first quadrant comes from the intersection of equations Y_1 and Y_3. Press 2nd TRACE and choose 5:intersect and you will see this screen:

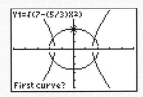

Notice that the top left of the screen displays the function Y_1. Since we want to find the intersection of Y_1 and Y_3, press ENTER when Y_1 is displayed. Now Y_2 appears at the top left, but we do not need this function. Press the down arrow to see the equation for Y_3 and be sure that the cursor is close to the intersection point in quadrant I. Press ENTER twice. You will see the approximate solution (1.732, 1.414):

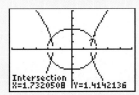

In Example 3 we found all of the exact solutions algebraically. The solution of (1.732, 1.414) found using the calculator is an approximation of the exact solution ($\sqrt{3}$, $\sqrt{2}$).

The other solutions of the system can be found in the same way.

Use the graphing calculator to find all real number solutions of each system. These are taken from the examples in the section and from the Chapter Summary.

1. $x^2 - 2y = 2$
   $-x + y = 3$
2. $x^2 + y^2 = 1$
   $x + 2y = -1$

3. $x - y^2 = 3$
   $x - 2y = 6$
4. $y = \sqrt{x}$
   $y^2 - 4x^2 = 4$

Answers to You Try Exercises

1) $\{(0, 2), (3, -1)\}$ 2) $\{(4, -3), (3, -4)\}$ 3) $\{(6, 2), (6, -2), (-6, 2), (-6, -2)\}$ 4) $\varnothing$

Answers to Technology Exercises

1. $\{(4, 7), (-2, 1)\}$ 2. $\{(-1, 0), (0.6, -0.8)\}$ 3. $\{(12, 3), (4, -1)\}$ 4. $\varnothing$

14.3 Exercises

Boost your grade at mathzone.com! MathZone

> Practice Problems   > NetTutor   > e-Professors   > Videos
> Self-Test

Objective I

If a nonlinear system consists of equations with the following graphs,

- a) sketch the different ways in which the graphs can intersect.

- b) make a sketch in which the graphs do not intersect.

- c) how many possible solutions can each system have?

1) circle and line

2) parabola and line

3) parabola and ellipse

4) ellipse and hyperbola

5) parabola and hyperbola

6) circle and ellipse

Objectives 2 and 3

Solve each system.

7) $x^2 + 4y = 8$
   $x + 2y = -8$

8) $x^2 + y = 1$
   $-x + y = -5$

 9) $x + 2y = 5$
   $x^2 + y^2 = 10$

10) $y = 2$
    $x^2 + y^2 = 8$

11) $y = x^2 - 6x + 10$
    $y = 2x - 6$

12) $y = x^2 - 10x + 22$
    $y = 4x - 27$

13) $x^2 + 2y^2 = 11$
    $x^2 - y^2 = 8$

14) $2x^2 - y^2 = 7$
    $2y^2 - 3x^2 = 2$

15) $x^2 + y^2 = 6$
    $2x^2 + 5y^2 = 18$

16) $5x^2 - y^2 = 16$
    $x^2 + y^2 = 14$

17) $3x^2 + 4y = -1$
    $x^2 + 3y = -12$

18) $2x^2 + y = 9$
    $y = 3x^2 + 4$

19) $y = 6x^2 - 1$
    $2x^2 + 5y = -5$

20) $x^2 + 2y = 5$
    $-3x^2 + 2y = 5$

21) $x^2 + y^2 = 4$
    $-2x^2 + 3y = 6$

22) $x^2 + y^2 = 49$
    $x - 2y^2 = 7$

 23) $x^2 + y^2 = 3$
   $x + y = 4$

24) $y - x = 1$
    $4y^2 - 16x^2 = 64$

25) $x = \sqrt{y}$
    $x^2 - 9y^2 = 9$

26) $x = \sqrt{y}$
    $x^2 - y^2 = 4$

27) $9x^2 + y^2 = 9$
    $x^2 + y^2 = 5$

28) $x^2 + y^2 = 6$
    $5x^2 + y^2 = 10$

29) $y = -x^2 - 2$
    $x^2 + y^2 = 4$

30) $x^2 + y^2 = 1$
    $y = x^2 + 1$

Write a system of equations and solve.

31) Find two numbers whose product is 40 and whose sum is 13.

32) Find two numbers whose product is 28 and whose sum is 11.

 33) The perimeter of a rectangular computer screen is 38 in. Its area is 88 in². Find the dimensions of the screen.

34) The area of a rectangular bulletin board is 180 in², and its perimeter is 54 in. Find the dimensions of the bulletin board.

35) A sporting goods company estimates that the cost y, in dollars, to manufacture x thousands of basketballs is given by

$$y = 6x^2 + 33x + 12$$

The revenue y, in dollars, from the sale of x thousands of basketballs is given by

$$y = 15x^2$$

The company breaks even on the sale of basketballs when revenue equals cost. The point, (x, y), at which this occurs is called the *break-even point*. Find the break-even point for the manufacture and sale of the basketballs.

36) A backpack manufacturer estimates that the cost y, in dollars, to make x thousands of backpacks is given by

$$y = 9x^2 + 30x + 18$$

The revenue y, in dollars, from the sale of x thousands of backpacks is given by

$$y = 21x^2$$

Find the break-even point for the manufacture and sale of the backpacks. (See problem 35 for explanation.)

Section 14.4 Quadratic and Rational Inequalities

In Chapter 3, we learned how to solve *linear* inequalities such as $3x - 5 \le 16$. In this section, we will discuss how to solve *quadratic* and *rational* inequalities.

Quadratic Inequalities

> **Definition**
>
> A **quadratic inequality** can be written in the form
>
> $$ax^2 + bx + c \le 0 \quad \text{or} \quad ax^2 + bx + c \ge 0$$
>
> where a, b, and c are real numbers and $a \ne 0$. ($<$ and $>$ may be substituted for $\le$ and $\ge$.)

1. Understand How to Solve a Quadratic Inequality by Looking at the Graph of a Quadratic Function

To understand how to solve a quadratic inequality, let's look at the graph of a quadratic function.

██   **Example 1**

a)   Graph $y = x^2 - 2x - 3$.

b)   Solve $x^2 - 2x - 3 < 0$.

c)   Solve $x^2 - 2x - 3 \ge 0$.

Solution

a)   The graph of the quadratic function $y = x^2 - 2x - 3$ is a parabola that opens upward. Use the vertex formula to confirm that the vertex is $(1, -4)$.

To find the y-intercept, let $x = 0$ and solve for y.

$$y = 0^2 - 2(0) - 3$$
$$y = -3$$

The y-intercept is $(0, -3)$.

To find the x-intercepts, let $y = 0$ and solve for x.

$$0 = x^2 - 2x - 3$$
$$0 = (x - 3)(x + 1) \qquad \text{Factor.}$$
$$x - 3 = 0 \quad \text{or} \quad x + 1 = 0 \qquad \text{Set each factor equal to 0.}$$
$$x = 3 \quad \text{or} \qquad x = -1 \qquad \text{Solve.}$$

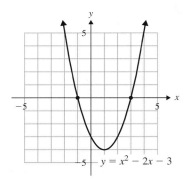

$y = x^2 - 2x - 3$

b) We will use the graph of $y = x^2 - 2x - 3$ to solve the inequality $x^2 - 2x - 3 < 0$. That is, to solve $x^2 - 2x - 3 < 0$ we must ask ourselves, "Where are the *y-values* of the function *less than* zero?"

The *y*-values of the function are less than zero when the *x*-values are greater than -1 and less than 3.

The solution set of $x^2 - 2x - 3 < 0$ (in interval notation) is $(-1, 3)$.

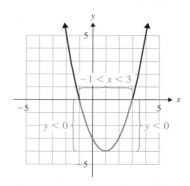

c) To solve $x^2 - 2x - 3 \geq 0$ means to find the *x*-values for which the *y-values* of the function $y = x^2 - 2x - 3$ are *greater than or equal to* zero. (Recall that the *x*-intercepts are where the function equals zero.)

The *y*-values of the function are greater than or equal to zero when $x \leq -1$ or when $x \geq 3$.

The solution set of $x^2 - 2x - 3 \geq 0$ (in interval notation) is $(-\infty, -1] \cup [3, \infty)$.

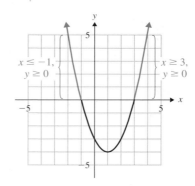

When $x \leq -1$ or $x \geq 3$, the *y*-values are greater than or equal to 0.

You Try 1

a) Graph $y = x^2 + 6x + 5$.   b) Solve $x^2 + 6x + 5 \leq 0$.   c) Solve $x^2 + 6x + 5 > 0$.

2. Solve a Quadratic Inequality Algebraically

Example 1 illustrates how the *x*-intercepts of $y = x^2 - 2x - 3$ break up the *x*-axis into the three separate intervals $x < -1$, $-1 < x < 3$, and $x > 3$. We can use this idea of intervals to solve a quadratic inequality without graphing.

Example 2

Solve $x^2 - 2x - 3 < 0$.

Solution

Begin by solving the equation $x^2 - 2x - 3 = 0$.

$$
\begin{aligned}
x^2 - 2x - 3 &= 0 \\
(x - 3)(x + 1) &= 0 \qquad &\text{Factor.} \\
x - 3 = 0 \quad \text{or} \quad x + 1 &= 0 \qquad &\text{Set each factor equal to 0.} \\
x = 3 \quad \text{or} \quad x &= -1 \qquad &\text{Solve.}
\end{aligned}
$$

(These are the x-intercepts of $y = x^2 - 2x - 3$.)

> The $<$ indicates that we want to find the values of x that will make $x^2 - 2x - 3 < 0$ —
> or—find the values of x that make $x^2 - 2x - 3$ a *negative* number.

Put $x = 3$ and $x = -1$ on a number line with the smaller number on the left. This breaks up the number line into three intervals: $x < -1$, $-1 < x < 3$, and $x > 3$.

Choose a test number in each interval and substitute it into $x^2 - 2x - 3$ to determine if that value makes $x^2 - 2x - 3$ positive or negative. (If one number in the interval makes $x^2 - 2x - 3$ positive, then *all* numbers in that interval will make $x^2 - 2x - 3$ positive.) Indicate the result on the number line.

Interval A: ($x < -1$) As a test number, choose any number less than -1. We will choose -2. Evaluate $x^2 - 2x - 3$ for $x = -2$.

$$
\begin{aligned}
x^2 - 2x - 3 &= (-2)^2 - 2(-2) - 3 \qquad &\text{Substitute } -2 \text{ for } x. \\
&= 4 + 4 - 3 \\
&= 8 - 3 = 5
\end{aligned}
$$

When $x = -2$, $x^2 - 2x - 3$ is *positive*. Therefore, $x^2 - 2x - 3$ will be positive for all values of x in this interval. Indicate this on the number line.

Interval B: ($-1 < x < 3$) As a test number, choose any number between -1 and 3. We will choose 0. Evaluate $x^2 - 2x - 3$ for $x = 0$.

$$
\begin{aligned}
x^2 - 2x - 3 &= (0)^2 - 2(0) - 3 \qquad &\text{Substitute 0 for } x. \\
&= 0 - 0 - 3 = -3
\end{aligned}
$$

When $x = 0$, $x^2 - 2x - 3$ is negative. Therefore, $x^2 - 2x - 3$ will be negative for all values of x in this interval. Indicate this on the number line.

Interval C: $(x > 3)$ As a test number, choose any number greater than 3. We will choose 4. Evaluate $x^2 - 2x - 3$ for $x = 4$.

$$x^2 - 2x - 3 = (4)^2 - 2(4) - 3 \qquad \text{Substitute 4 for } x.$$
$$= 16 - 8 - 3$$
$$= 8 - 3 = 5$$

When $x = 4$, $x^2 - 2x - 3$ is *positive*. Therefore, $x^2 - 2x - 3$ will be positive for all values of x in this interval. Indicate this on the number line.

Look at the number line. The solution set of $x^2 - 2x - 3 < 0$ consists of the interval(s) where $x^2 - 2x - 3$ is *negative*. This is in Interval B, $(-1, 3)$.

The graph of the solution set is

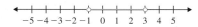

$$-5 \;-4 \;-3 \;-2 \;-1 \quad 0 \quad 1 \quad 2 \quad 3 \quad 4 \quad 5$$

The solution set is $(-1, 3)$.

This is the same as the result we obtained in Example 1 by graphing.

You Try 2

Solve $x^2 + 5x + 4 \leq 0$. Graph the solution set and write the solution in interval notation.

Next we will summarize how to solve a quadratic inequality.

How to Solve a Quadratic Inequality

Step 1: **Write the inequality in the form $ax^2 + bx + c \leq 0$ or $ax^2 + bx + c \leq 0$.**
($<$ and $>$ may be substituted for $\leq$ and ≥ 0.) If the inequality symbol is $<$ or $\leq$, we are looking for a *negative* quantity in the interval on the number line. If the inequality symbol is $>$ or $\geq$, we are looking for a *positive* quantity in the interval.

Step 2: **Solve the equation $ax^2 + bx + c = 0$.**

Step 3: **Put the solutions of $ax^2 + bx + c = 0$ on a number line.** These values break up the number line into intervals.

Step 4: **Choose a test number in each interval to determine whether $ax^2 + bx + c$ is positive or negative in each interval.** Indicate this on the number line.

Step 5: **If the inequality is in the form $ax^2 + bx + c \leq 0$ or $ax^2 + bx + c < 0$, then the solution set contains the numbers in the interval where $ax^2 + bx + c$ is negative. If the inequality is in the form $ax^2 + bx + c \geq 0$ or $ax^2 + bx + c > 0$, then the solution set contains the numbers in the interval where $ax^2 + bx + c$ is positive.**

Step 6: **If the inequality symbol is $\leq$ or $\geq$, then the endpoints of the interval(s) (the numbers found in step 3) are included in the solution set.** Indicate this with brackets in the interval notation.
If the inequality symbol is $<$ or $>$, then the endpoints of the interval(s) are not included in the solution set. Indicate this with parentheses in interval notation.

Example 3

Solve $2p^2 - 11p \geq -5$.

Solution

Step 1:   Add 5 to each side so that there is a zero on one side of the inequality symbol.

$$2p^2 - 11p + 5 \geq 0$$

Since the inequality symbol is $\geq$, the solution set will contain the interval(s) where the quantity $2p^2 - 11p + 5$ is *positive*.

Step 2:   Solve $2p^2 - 11p + 5 = 0$.

$$
\begin{array}{lll}
(2p - 1)(p - 5) = 0 & & \text{Factor.} \\
2p - 1 = 0 \quad \text{or} \quad p - 5 = 0 & & \text{Set each factor equal to 0.} \\
p = \dfrac{1}{2} \quad \text{or} \qquad p = 5 & & \text{Solve.}
\end{array}
$$

Step 3:   Put $p = \dfrac{1}{2}$ and $p = 5$ on a number line.

The solution set will contain the interval(s) where $2p^2 - 11p + 5$ is positive.

Step 4:   Choose a test number in each interval to determine the sign of $2p^2 - 11p + 5$.

	Interval A	Interval B	Interval C
Test number	$p = 0$	$p = 1$	$p = 6$
Evaluate $2p^2 - 11p + 5$	$2(0)^2 - 11(0) + 5 = 5$	$2(1)^2 - 11(1) + 5 = -4$	$2(6)^2 - 11(6) + 5 = 11$
	Positive	Negative	Positive

Step 5:   The solution set will contain the numbers in the intervals where $2p^2 - 11p + 5$ is positive. These are intervals A and C.

Step 6:   The endpoints of the intervals are included since the inequality is $\geq$.

The graph of the solution set is

The solution set of $2p^2 - 11p \geq -5$ is $\left(-\infty, \dfrac{1}{2}\right] \cup [5, \infty)$.

You Try 3

Solve $3n^2 - 10n < 8$. Graph the solution set and write the solution in interval notation.

3. Solve a Quadratic Inequality with No Solution or with a Solution Set of $(-\infty, \infty)$

We should look carefully at the inequality before trying to solve it. Sometimes, it is not necessary to go through all of the steps.

Example 4

Solve.

a) $(y + 4)^2 \geq -5$ b) $(t - 8)^2 < -3$

Solution

a) $(y + 4)^2 \geq -5$ says that a squared quantity, $(y + 4)^2$, is greater than or equal to a *negative* number, -5. *This is always true.* (A squared quantity will *always* be greater than or equal to zero.) Any real number, y, will satisfy the inequality.

The solution set is (∞, ∞).

b) $(t - 8)^2 < -3$ says that a squared quantity, $(t - 8)^2$, is less than a *negative* number, -3. *There is no real number value for t so that $(t - 8)^2 < -3$.*

The solution set is $\varnothing$.

You Try 4

Solve.

a) $(k + 2)^2 \leq -4$ b) $(z - 9)^2 > -1$

4. Solve an Inequality of Higher Degree

Other polynomial inequalities in factored form can be solved in the same way that we solve quadratic inequalities.

Example 5

Solve $(c - 2)(c + 5)(c - 4) < 0$.

Solution

This is the factored form of a third-degree polynomial. Since the inequality is $<$, the solution set will contain the intervals where $(c - 2)(c + 5)(c - 4)$ is *negative*.

Solve $(c - 2)(c + 5)(c - 4) = 0$.

$c - 2 = 0$ or $c + 5 = 0$ or $c - 4 = 0$ Set each factor equal to 0.
$c = 2$ or $c = -5$ or $c = 4$ Solve.

Put $c = 2$, $c = -5$, and $c = 4$ on a number line.

	Interval A	Interval B	Interval C	Interval D
	Negative	Positive	Negative	Positive
$(c - 2)(c + 5)(c - 4)$		-5	2	4

The solution set will contain the interval(s) where $(c - 2)(c + 5)(c - 4)$ is negative.

	Interval A	Interval B
Test number	$c = -6$	$c = 0$
Evaluate $(c - 2)(c + 5)(c - 4)$	$(-6 - 2)(-6 + 5)(-6 - 4)$ $= (-8)(-1)(-10)$ $= -80$	$(0 - 2)(0 + 5)(0 - 4)$ $= (-2)(5)(-4)$ $= 40$
	Negative	Positive

	Interval C	Interval D
Test number	$c = 3$	$c = 5$
Evaluate $(c - 2)(c + 5)(c - 4)$	$(3 - 2)(3 + 5)(3 - 4)$ $= (1)(8)(-1)$ $= -8$	$(5 - 2)(5 + 5)(5 - 4)$ $= (3)(10)(1)$ $= 30$
	Negative	Positive

The solution set will contain the intervals where $(c - 2)(c + 5)(c - 4)$ is negative. These are intervals A and C, $(-\infty, -5)$ and $(2, 4)$. The endpoints are not included since the inequality is $<$.

The graph of the solution set is

The solution set of $(c - 2)(c + 5)(c - 4) < 0$ is $(-\infty, -5) \cup (2, 4)$.

You Try 5

Solve $(y + 3)(y - 1)(y + 1) \geq 0$. Graph the solution set and write the solution in interval notation.

5. Solve a Rational Inequality

An inequality containing a rational expression, $\dfrac{p}{q}$, where p and q are polynomials is called a **rational inequality**. The way we solve rational inequalities is very similar to how we solve quadratic inequalities.

How to Solve a Rational Inequality

Step 1: **Write the inequality so that there is a 0 on one side and only one rational expression on the other side.** If the inequality symbol is $<$ or $\leq$, we are looking for a *negative* quantity in the interval on the number line. If the inequality symbol is $>$ or $\geq$, we are looking for a *positive* quantity in the interval.

Step 2: **Find the numbers that make the numerator equal 0 and any numbers that make the denominator equal 0.**

Step 3: **Put the numbers found in step 2 on a number line.** These values break up the number line into intervals.

Step 4: **Choose a test number in each interval to determine whether the rational inequality is positive or negative in each interval.** Indicate this on the number line.

Step 5: **If the inequality is in the form $\dfrac{p}{q} \leq 0$ or $\dfrac{p}{q} < 0$, then the solution set contains the numbers in the interval where $\dfrac{p}{q}$ is negative.**

If the inequality is in the form $\dfrac{p}{q} \geq 0$ or $\dfrac{p}{q} > 0$, then the solution set contains the numbers in the interval where $\dfrac{p}{q}$ is positive.

Step 6: **Determine whether the endpoints of the intervals are included in or excluded from the solution set.** Do not include any values that make the denominator equal 0.

Example 6

Solve $\dfrac{5}{x + 3} > 0$.

Solution

Step 1: The inequality is in the correct form—zero on one side and only one rational expression on the other side. Since the inequality symbol is > 0, the solution set will contain the interval(s) where $\dfrac{5}{x + 3}$ is *positive*.

Step 2: Find the numbers that make the numerator equal 0 and any numbers that make the denominator equal 0.

Numerator: 5	Denominator: $x + 3$
The numerator is a constant, 5, so it cannot equal 0.	Set $x + 3 = 0$ and solve for x. $x + 3 = 0$ $x = -3$

Step 3: Put -3 on a number line to break it up into intervals.

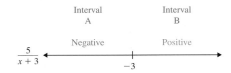

The solution set will contain the interval(s) where $\dfrac{5}{x + 3}$ is positive.

Step 4: Choose a test number in each interval to determine whether $\dfrac{5}{x + 3}$ is positive or negative in each interval.

	Interval A	Interval B
Test number	$x = -4$	$x = 0$
Evaluate $\dfrac{5}{x + 3}$	$\dfrac{5}{-4 + 3} = \dfrac{5}{-1} = -5$	$\dfrac{5}{0 + 3} = \dfrac{5}{3}$
	Negative	Positive

Step 5: The solution set of $\dfrac{5}{x + 3} > 0$ will contain the numbers in the interval where $\dfrac{5}{x + 3}$ is positive. This is interval B.

Step 6: Since the inequality symbol is $>$, the endpoint of the interval, -3, will not be included in the solution set.

The graph of the solution set is

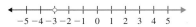

The solution set of $\dfrac{5}{x + 3} > 0$ is $(-3, \infty)$.

You Try 6

Solve $\dfrac{2}{y - 6} < 0$. Graph the solution set and write the solution in interval notation.

Example 7

Solve $\dfrac{7}{a + 2} \leq 3$.

Solution

Step 1: Get a zero on one side and only one rational expression on the other side.

$$\frac{7}{a + 2} \leq 3$$

$$\frac{7}{a + 2} - 3 \leq 0 \qquad \text{Subtract 3.}$$

$$\frac{7}{a + 2} - \frac{3(a + 2)}{a + 2} \leq 0 \qquad \text{Get a common denominator.}$$

$$\frac{7}{a + 2} - \frac{3a + 6}{a + 2} \leq 0 \qquad \text{Distribute.}$$

$$\frac{7 - 3a - 6}{a + 2} \leq 0 \qquad \text{Combine numerators.}$$

$$\frac{1 - 3a}{a + 2} \leq 0 \qquad \text{Combine like terms.}$$

From this point forward we will work with the inequality $\dfrac{1 - 3a}{a + 2} \leq 0$. It is equivalent to the original inequality. Since the inequality symbol is $\leq$ in $\dfrac{1 - 3a}{a + 2} \leq 0$, the solution set will contain the interval(s) where $\dfrac{1 - 3a}{a + 2}$ is *negative*.

Step 2: Find the numbers that make the numerator equal 0 and any numbers that make the denominator equal 0.

Numerator	Denominator
$1 - 3a = 0$	$a + 2 = 0$
$-3a = -1$	$a = -2$
$a = \dfrac{1}{3}$	

Step 3: Put $\dfrac{1}{3}$ and -2 on a number line to break it up into intervals.

$$\dfrac{1 - 3a}{a + 2}$$

	Interval A	Interval B	Interval C
	Negative	Positive	Negative
	-2	$\dfrac{1}{3}$	

The solution set will contain the interval(s) where $\dfrac{1 - 3a}{a + 2}$ is negative.

Step 4: Choose a test number in each interval to determine whether $\dfrac{1 - 3a}{a + 2}$ is positive or negative in each interval.

	Interval A	**Interval B**	**Interval C**
Test number	$a = -3$	$a = 0$	$a = 1$
Evaluate $\dfrac{1 - 3a}{a + 2}$	$\dfrac{1 - 3(-3)}{-3 + 2} = \dfrac{10}{-1} = -10$	$\dfrac{1 - 3(0)}{0 + 2} = \dfrac{1}{2}$	$\dfrac{1 - 3(1)}{1 + 2} = -\dfrac{2}{3}$
	Negative	Positive	Negative

Step 5: The solution set of $\dfrac{1 - 3a}{a + 2} \leq 0$ $\left(\text{and therefore } \dfrac{7}{a + 2} \leq 3\right)$ will contain the numbers in the interval where $\dfrac{1 - 3a}{a + 2}$ is negative. These are intervals A and C.

Step 6: Determine if the endpoints of the intervals, -2 and $\dfrac{1}{3}$, are included in the solution set. $\dfrac{1}{3}$ *will be included since it does not make the denominator equal 0. But -2 is not included because it makes the denominator equal 0.*

The graph of the solution set of $\dfrac{7}{a + 2} \leq 3$ is

The solution set is $(-\infty, -2) \cup \left[\dfrac{1}{3}, \infty\right)$.

BE CAREFUL Although an inequality symbol may be $\leq$ or $\geq$, an endpoint cannot be included in the solution set if it makes the denominator equal 0.

You Try 7

Solve $\dfrac{3}{z + 4} \geq 2$. Graph the solution set and write the solution in interval notation.

Answers to You Try Exercises

1) a)

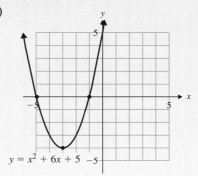

$y = x^2 + 6x + 5$

b) $[-5, -1]$ c) $(-\infty, -5) \cup (-1, \infty)$

2) ![number line with closed dots at -4 and -1] $-5\ -4\ -3\ -2\ -1\ \ 0\ \ 1\ \ 2\ \ 3\ \ 4\ \ 5$; $[-4, -1]$

3) ![number line with open circles at -2/3 and 4] $-\frac{2}{3}$ $-5\ -4\ -3\ -2\ -1\ \ 0\ \ 1\ \ 2\ \ 3\ \ 4\ \ 5$; $\left(-\dfrac{2}{3}, 4\right)$

4) a) $\varnothing$ b) $(-\infty, \infty)$

5) ![number line] $-5\ -4\ -3\ -2\ -1\ \ 0\ \ 1\ \ 2\ \ 3\ \ 4\ \ 5$; $[-3, -1] \cup [1, \infty)$

6) ![number line with open circle at 6] $-2\ -1\ \ 0\ \ 1\ \ 2\ \ 3\ \ 4\ \ 5\ \ 6\ \ 7\ \ 8$; $(-\infty, 6)$

7) ![number line with open circle at -4 and closed dot at -5/2] $-\frac{5}{2}$ $-5\ -4\ -3\ -2\ -1\ \ 0\ \ 1\ \ 2\ \ 3\ \ 4\ \ 5$; $\left(-4, -\dfrac{5}{2}\right]$

1) When solving a quadratic inequality, how do you know when to include and when to exclude the endpoints in the solution set?

2) If a rational inequality contains a $\leq$ or $\geq$ symbol, will the endpoints of the solution set always be included? Explain your answer.

Objective 1

For Exercises 3–6, use the graph of the function to solve each inequality.

3) $y = x^2 + 4x - 5$

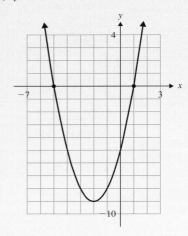

a) $x^2 + 4x - 5 \leq 0$

b) $x^2 + 4x - 5 > 0$

4) $y = x^2 - 6x + 8$

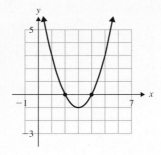

a) $x^2 - 6x + 8 > 0$

b) $x^2 - 6x + 8 \leq 0$

5) $y = -\dfrac{1}{2}x^2 + x + \dfrac{3}{2}$

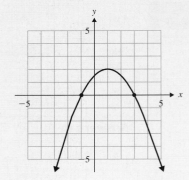

a) $-\dfrac{1}{2}x^2 + x + \dfrac{3}{2} \geq 0$

b) $-\dfrac{1}{2}x^2 + x + \dfrac{3}{2} < 0$

6) $y = -x^2 - 8x - 12$

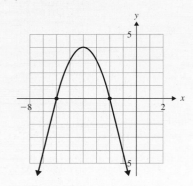

a) $-x^2 - 8x - 12 < 0$

b) $-x^2 - 8x - 12 \geq 0$

Objective 2

Solve each quadratic inequality. Graph the solution set and write the solution in interval notation.

7) $x^2 + 6x - 7 \geq 0$

8) $m^2 - 2m - 24 > 0$

9) $c^2 + 5c < 36$

10) $t^2 - 36 \leq 9t$

 11) $r^2 - 13r > -42$

12) $v^2 + 10v < -16$

13) $3z^2 + 14z - 24 \leq 0$

14) $5k^2 + 36k + 7 \geq 0$

15) $7p^2 - 4 > 12p$

16) $4w^2 - 19w < 30$

17) $b^2 - 9b > 0$

18) $c^2 + 12c \le 0$

19) $4y^2 \le -5y$

20) $2a^2 \ge 7a$

21) $m^2 - 64 < 0$

22) $p^2 - 144 > 0$

23) $121 - h^2 \le 0$

24) $1 - d^2 > 0$

 25) $144 \ge 9s^2$

26) $81 \le 25q^2$

Objective 3

Solve each inequality.

27) $(k + 7)^2 \ge -9$

28) $(h + 5)^2 \ge -2$

29) $(3v - 11)^2 > -20$

30) $(r + 4)^2 < -3$

31) $(2y - 1)^2 < -8$

32) $(4d - 3)^2 > -1$

33) $(n + 3)^2 \le -10$

34) $(5s - 2)^2 \le -9$

Objective 4

Solve each inequality. Graph the solution set and write the solution in interval notation.

35) $(r + 2)(r - 5)(r - 1) \le 0$

36) $(b + 2)(b - 3)(b - 12) > 0$

37) $(j - 7)(j - 5)(j + 9) \ge 0$

38) $(m + 4)(m - 7)(m + 1) \le 0$

39) $(6c + 1)(c + 7)(4c - 3) < 0$

40) $(t + 2)(4t - 7)(5t - 1) \ge 0$

Objective 5

Solve each rational inequality. Graph the solution set and write the solution in interval notation.

41) $\dfrac{7}{p + 6} > 0$

42) $\dfrac{3}{v - 2} < 0$

43) $\dfrac{5}{z + 3} \le 0$

44) $\dfrac{9}{m - 4} \ge 0$

45) $\dfrac{x - 4}{x - 3} > 0$

46) $\dfrac{a - 2}{a + 1} < 0$

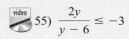 47) $\dfrac{h - 9}{3h + 1} \le 0$

48) $\dfrac{2c + 1}{c + 4} \ge 0$

49) $\dfrac{k}{k + 3} \le 0$

50) $\dfrac{r}{r - 7} \ge 0$

51) $\dfrac{7}{t + 6} < 3$

52) $\dfrac{3}{x + 7} < -2$

53) $\dfrac{3}{a + 7} \ge 1$

54) $\dfrac{5}{w - 3} \le 1$

55) $\dfrac{2y}{y - 6} \le -3$

56) $\dfrac{3z}{z + 4} \ge 2$

57) $\dfrac{3w}{w + 2} > -4$

58) $\dfrac{4h}{h + 3} < 1$

59) $\dfrac{(6d + 1)^2}{d - 2} \le 0$

60) $\dfrac{(x + 2)^2}{x + 7} \ge 0$

61) $\dfrac{(4t - 3)^2}{t - 5} > 0$

62) $\dfrac{(2y + 3)^2}{y + 3} \le 0$

63) $\dfrac{n + 6}{n^2 + 4} < 0$

64) $\dfrac{b - 3}{b^2 + 2} > 0$

65) $\dfrac{m + 1}{m^2 + 3} \ge 0$

66) $\dfrac{w - 7}{w^2 + 8} \le 0$

67) $\dfrac{s^2 + 2}{s - 4} \le 0$

68) $\dfrac{z^2 + 10}{z + 6} \le 0$

Definition/Procedure	Example	Reference

The Circle

14.1

Parabolas, circles, ellipses, and hyperbolas are called *conic sections*.

The *standard form for the equation of a circle* with center (h, k) and radius r is

$$(x - h)^2 + (y - k)^2 = r^2$$

Graph $(x + 3)^2 + y^2 = 4$.

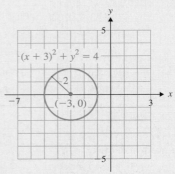

The center is $(-3, 0)$. The radius is $\sqrt{4} = 2$.

p. 936

The *general form for the equation of a circle* is

$$Ax^2 + Ay^2 + Cx + Dy + E = 0$$

where A, C, D, and E are real numbers.

To rewrite the equation in the form $(x - h)^2 + (y - k)^2 = r^2$, divide the equation by A so that the coefficient of each squared term is 1, then complete the square on x and on y to put it into standard form.

Write $x^2 + y^2 - 16x + 4y + 67 = 0$ in the form $(x - h)^2 + (y - k)^2 = r^2$.

Group the x-terms together and group the y-terms together.

$$(x^2 - 16x) + (y^2 + 4y) = -67$$

Complete the square for each group of terms.

$$(x^2 - 16x + 64) + (y^2 + 4y + 4) =$$
$$-67 + 64 + 4$$
$$(x - 8)^2 + (y + 2)^2 = 1$$

p. 938

The Ellipse and the Hyperbola

14.2

The *standard form for the equation of an ellipse* is

$$\frac{(x - h)^2}{a^2} + \frac{(y - k)^2}{b^2} = 1$$

The center of the ellipse is (h, k).

Graph $\dfrac{(x - 1)^2}{9} + \dfrac{(y - 2)^2}{4} = 1$.

The center is $(1, 2)$.

$$a = \sqrt{9} = 3 \qquad b = \sqrt{4} = 2$$

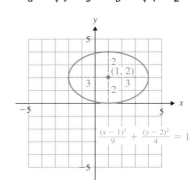

$$\frac{(x-1)^2}{9} + \frac{(y-2)^2}{4} = 1$$

p. 942

Definition/Procedure	Example	Reference

Standard Form for the Equation of a Hyperbola

1) A hyperbola with center (h, k) with branches that open in the *x-direction* has equation

$$\frac{(x - h)^2}{a^2} - \frac{(y - k)^2}{b^2} = 1$$

2) A hyperbola with center (h, k) with branches that open in the *y-direction* has equation

$$\frac{(y - k)^2}{b^2} - \frac{(x - h)^2}{a^2} = 1$$

Notice in 1) that $\dfrac{(x - h)^2}{a^2}$ is the positive quantity, and the branches open in the *x-direction*.

In 2), the positive quantity is $\dfrac{(y - k)^2}{b^2}$, and the branches open in the *y-direction*.

Graph $\dfrac{(y - 1)^2}{9} - \dfrac{(x - 4)^2}{4} = 1$.

The center is $(4, 1)$.

$$a = \sqrt{4} = 2 \qquad b = \sqrt{9} = 3$$

Use the center, $a = 2$, and $b = 3$ to draw the reference rectangle. The diagonals of the rectangle are the asymptotes of the hyperbola.

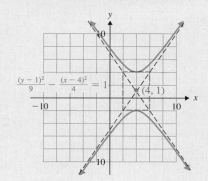

p. 946

Nonlinear Systems of Equations

14.3

A *nonlinear system of equations* is a system in which at least one of the equations is not linear. We can solve nonlinear systems by substitution or the elimination method.

Solve $\quad x - y^2 = 3 \qquad$ (1)
$\qquad\quad x - 2y = 6 \qquad$ (2)

$\quad x - y^2 = 3 \qquad$ (1) $\qquad$ Solve equation
$\qquad x = y^2 + 3 \qquad$ (3) $\qquad$ (1) for x.

Substitute $x = y^2 + 3$ into equation (2).

$$(y^2 + 3) - 2y = 6$$
$$y^2 - 2y - 3 = 0 \qquad \text{Subtract 6.}$$
$$(y - 3)(y + 1) = 0 \qquad \text{Factor.}$$
$y - 3 = 0 \quad$ or $\quad y + 1 = 0 \qquad$ Set each factor
$\qquad\qquad\qquad\qquad\qquad\qquad$ equal to 0.
$\quad y = 3 \quad$ or $\qquad y = -1 \quad$ Solve.

Substitute each value into equation (3).

$y = 3$: $\ x = y^2 + 3 \quad | \quad y = -1$: $\ x = y^2 + 3$
$\qquad\quad x = (3)^2 + 3 \quad | \qquad\qquad x = (-1)^2 + 3$
$\qquad\quad x = 12 \qquad\quad | \qquad\qquad x = 4$

The proposed solutions are $(12, 3)$ and $(4, -1)$.
Verify that they also satisfy (2).
The solution set is $\{(12, 3), (4, -1)\}$.

p. 957

Definition/Procedure	**Example**	**Reference**

Quadratic and Rational Inequalities

14.4

A *quadratic inequality* can be written in the form

$$ax^2 + bx + c \leq 0 \quad \text{or} \quad ax^2 + bx + c \geq 0$$

where a, b, and c are real numbers and $a \neq 0$.
($<$ and $>$ may be substituted for $\leq$ and $\geq$.)

An inequality containing a rational expression, like $\dfrac{c - 5}{c + 1} \leq 0$, is called a *rational inequality*.

Solve $r^2 - 4r \geq 12$.

Step 1: $r^2 - 4r - 12 \geq 0$ Subtract 12.
Since the inequality symbol is $\geq$, the solution set will contain the interval(s) where the quantity $r^2 - 4r - 12$ is *positive*.

Step 2: Solve $r^2 - 4r - 12 = 0$.

$$(r - 6)(r + 2) = 0 \quad \text{Factor.}$$
$$r - 6 = 0 \quad \text{or} \quad r + 2 = 0$$
$$r = 6 \quad \text{or} \quad r = -2$$

Step 3: Put $r = 6$ and $r = -2$ on a number line.

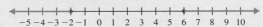

The solution set will contain the intervals where $r^2 - 4r - 12$ is positive.

Step 4: Choose a test number in each interval to determine the sign of $r^2 - 4r - 12$.

Step 5: The solution set will contain the numbers in the intervals where $r^2 - 4r - 12$ is positive.

Step 6: The endpoints of the intervals are included since the inequality is $\geq$.
The graph of the solution set is

The solution set of $r^2 - 4r - 12$ is $(-\infty, -2] \cup [6, \infty)$.

p. 966

(14.1) Identify the center and radius of each circle and graph.

1) $(x + 3)^2 + (y - 5)^2 = 36$

2) $x^2 + (y + 4)^2 = 9$

3) $x^2 + y^2 - 10x - 4y + 13 = 0$

4) $x^2 + y^2 + 4x + 16y + 52 = 0$

Find an equation of the circle with the given center and radius.

5) Center $(3, 0)$; radius $= 4$

6) Center $(1, -5)$; radius $= \sqrt{7}$

(14.2)

7) How can you distinguish between the equation of an ellipse and the equation of a hyperbola?

8) When is an ellipse also a circle?

Identify the center of the ellipse and graph the equation.

9) $\dfrac{x^2}{25} + \dfrac{y^2}{36} = 1$

10) $\dfrac{(x + 3)^2}{9} + \dfrac{(y - 3)^2}{4} = 1$

11) $(x - 4)^2 + \dfrac{(y - 2)^2}{16} = 1$

12) $25x^2 + 4y^2 = 100$

Identify the center of the hyperbola and graph the equation.

13) $\dfrac{y^2}{9} - \dfrac{x^2}{25} = 1$

14) $\dfrac{(y - 3)^2}{4} - \dfrac{(x + 2)^2}{9} = 1$

15) $\dfrac{(x + 1)^2}{4} - \dfrac{(y + 2)^2}{4} = 1$

16) $16x^2 - y^2 = 16$

(14.1–14.2) Determine whether the graph of each equation will be a parabola, circle, ellipse, or hyperbola. Then, graph each equation.

17) $x^2 + 9y^2 = 9$

18) $x^2 + y^2 = 25$

19) $x = -y^2 + 6y - 5$

20) $x^2 - y = 3$

21) $\dfrac{(x - 3)^2}{16} - \dfrac{(y - 4)^2}{25} = 1$

22) $\dfrac{(x + 3)^2}{16} + \dfrac{(y + 1)^2}{25} = 1$

23) $x^2 + y^2 - 2x + 2y - 2 = 0$

24) $4y^2 - 9x^2 = 36$

25) $y = \dfrac{1}{2}(x + 2)^2 + 1$

26) $x^2 + y^2 - 6x - 8y + 16 = 0$

(14.3)

27) If a nonlinear system of equations consists of an ellipse and a hyperbola, how many possible solutions can the system have?

28) If a nonlinear system of equations consists of a line and a circle, how many possible solutions can the system have?

Solve each system.

29) $-4x^2 + 3y^2 = 3$
    $7x^2 - 5y^2 = 7$

30) $y - x^2 = 7$
    $3x^2 + 4y = 28$

31) $y = 3 - x^2$
    $x - y = -1$

32) $x^2 + y^2 = 9$
    $8x + y^2 = 21$

33) $4x^2 + 9y^2 = 36$
    $y = \dfrac{1}{3}x - 5$

34) $4x + 3y = 0$
    $4x^2 + 4y^2 = 25$

Write a system of equations and solve.

35) Find two numbers whose product is 36 and whose sum is 13.

36) The perimeter of a rectangular window is 78 in., and its area is 378 in². Find the dimensions of the window.

(14.4) Solve each inequality. Graph the solution set and write the solution in interval notation.

37) $a^2 + 2a - 3 < 0$

38) $4m^2 + 8m \geq 21$

39) $6h^2 + 7h > 0$

40) $64v^2 \geq 25$

41) $36 - r^2 > 0$

42) $(5c + 2)(c - 4)(3c - 1) < 0$

43) $(6x - 5)^2 \geq -2$

44) $(p - 6)^2 \leq -5$

45) $\dfrac{t + 7}{2t - 3} > 0$

46) $\dfrac{6}{g - 7} \leq 0$

47) $\dfrac{4w + 3}{5w - 6} \leq 0$

48) $\dfrac{z}{z - 2} \leq 3$

49) $\dfrac{1}{n - 4} > -3$

50) $\dfrac{(3y - 4)^2}{y - 1} > 0$

51) $\dfrac{r^2 + 4}{r - 7} \geq 0$

52) $\dfrac{2k + 1}{k^2 + 5} \leq 0$

Determine whether the graph of each equation will be a parabola, circle, ellipse, or hyperbola. Then, graph each equation.

1) $\dfrac{(x-2)^2}{25} + \dfrac{(y+3)^2}{4} = 1$

2) $y = -2x^2 + 6$

3) $y^2 - 4x^2 = 16$ 4) $x^2 + (y-1)^2 = 9$

5) Write $x^2 + y^2 + 2x - 6y - 6 = 0$ in the form $(x-h)^2 + (y-k)^2 = r^2$. Identify the center, radius, and graph the equation.

6) Write an equation of the circle with center $(5, 2)$ and radius $\sqrt{11}$.

7) If a nonlinear system consists of the equation of a parabola and a circle,

   a)   sketch the different ways in which the graphs can intersect.

   b)   make a sketch in which the graphs do not intersect.

   c)   how many possible solutions can the system have?

Solve each system.

8) $x - 2y^2 = -1$
    $x + 4y = -1$

9) $2x^2 + 3y^2 = 21$
    $-x^2 + 12y^2 = 3$

10) $x^2 + y^2 = 7$
     $3x - 2y^2 = 0$

11) The perimeter of a rectangular picture frame is 44 in. The area is 112 in^2. Find the dimensions of the frame.

Solve each inequality. Graph the solution set and write the solution in interval notation.

12) $y^2 + 4y - 45 \geq 0$

13) $2w^2 + 11w < -12$

14) $49 - 9p^2 \leq 0$

15) $\dfrac{m-5}{m+3} \geq 0$

16) $\dfrac{6}{n-2} > 2$

Perform the indicated operations and simplify.

1) $\dfrac{1}{6} - \dfrac{11}{12}$

2) $16 + 20 \div 4 - (5 - 2)^2$

Find the area and perimeter of each figure.

3)

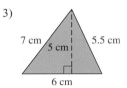

4)

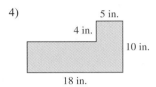

Evaluate.

5) $(-1)^5$

6) 2^4

7) Simplify $\left(\dfrac{2a^8 b}{a^2 b^{-4}} \right)^{-3}$

8) Solve $\dfrac{3}{8} k + 11 = -4$.

9) Solve for n. $an + z = c$

10) Solve $8 - 5p \le 28$

11) *Write an equation and solve.*
The sum of three consecutive odd integers is 13 more than twice the largest integer. Find the numbers.

12) Find the slope of the line containing the points $(-6, 4)$ and $(2, -4)$.

13) What is the slope of the line with equation $y = 3$?

14) Graph $y = -2x + 5$

15) Write the slope-intercept form of the line containing the points $(-4, 7)$ and $(4, 1)$.

16) Solve the system $\begin{array}{l} 3x + 4y = 3 \\ 5x + 6y = 4 \end{array}$

17) *Write a system of equations and solve.*
How many milliliters of an 8% alcohol solution and how many milliliters of a 16% alcohol solution must be mixed to make 20 mL of a 14% alcohol solution?

18) Subtract $5p^2 - 8p + 4$ from $2p^2 - p + 10$.

19) Multiply and simplify. $(4w - 3)(2w^2 + 9w - 5)$

20) Divide. $(x^3 - 7x - 36) \div (x - 4)$

Factor completely.

21) $6c^2 - 14c + 8$

22) $m^3 - 8$

23) Solve $(x + 1)(x + 2) = 2(x + 7) + 5x$.

24) Multiply and simplify. $\dfrac{a^2 + 3a - 54}{4a + 36} \cdot \dfrac{10}{36 - a^2}$

25) Simplify $\dfrac{\frac{t^2 - 9}{4}}{\frac{t - 3}{24}}$.

26) Solve $|3n + 11| = 7$.

27) Solve $|5r + 3| > 12$.

Simplify. Assume all variables represent nonnegative real numbers.

28) $\sqrt{75}$

29) $\sqrt[3]{48}$

30) $\sqrt[3]{27a^5 b^{13}}$

31) $(16)^{-3/4}$

32) $\dfrac{18}{\sqrt{12}}$

33) Rationalize the denominator of $\dfrac{5}{\sqrt{3} + 4}$.

Solve.

34) $(2p - 1)^2 + 16 = 0$

35) $y^2 = -7y - 3$

36) Given the relation $\{(4, 0), (3, 1), (3, -1), (0, 2)\}$,

a) What is the domain?

b) What is the range?

c) Is the relation a function?

37) Graph $f(x) = \sqrt{x}$ and $g(x) = \sqrt{x + 3}$ on the same axes.

38) $f(x) = 3x - 7$ and $g(x) = 2x + 5$.

a) Find $g(-4)$.

b) Find $(g \circ f)(x)$.

c) Find x so that $f(x) = 11$.

39) Given the function
$f = \{(-5, 9), (-2, 11), (3, 14), (7, 9)\}$,

a) Is f one-to-one?

b) Does f have an inverse?

40) Find an equation of the inverse of $f(x) = \dfrac{1}{3} x + 4$.

Solve.

41) $8^{5t} = 4^{t-3}$

42) $\log_3 (4n - 11) = 2$

43) Evaluate $\log 100$.

44) Graph $f(x) = \log_2 x$

45) Solve $e^{3k} = 8$. Give an exact solution and an approximation to four decimal places.

46) Graph $\dfrac{y^2}{4} - \dfrac{x^2}{9} = 1$.

47) Graph $x^2 + y^2 - 2x + 6y - 6 = 0$.

48) Solve the system. $\begin{aligned} y - 5x^2 &= 3 \\ x^2 + 2y &= 6 \end{aligned}$

49) Solve $25p^2 \leq 144$.

50) Solve $\dfrac{t-3}{2t+5} > 0$.

Appendix: Beginning Algebra Review

Section A1 The Real Number System and Geometry

Objectives

1. Multiply, Divide, Add, and Subtract Fractions
2. The Order of Operations
3. Review Concepts from Geometry
4. Define and Identify Sets of Numbers
5. Define Absolute Value and Perform Operations on Real Numbers
6. Learn the Vocabulary for Algebraic Expressions
7. Learn the Properties of Real Numbers

1. Multiply, Divide, Add, and Subtract Fractions

To **multiply** fractions, we can divide out the common factors then multiply numerators and denominators. To **divide** fractions, multiply the first fraction by the reciprocal of the second fraction. To **add** or **subtract** fractions, they must have a common denominator.

Example 1

Perform the operations and simplify. Write the answer in lowest terms.

a) $\dfrac{5}{9} \div \dfrac{4}{7}$ b) $\dfrac{3}{8} + \dfrac{1}{6}$

Solution

a) $\dfrac{5}{9} \div \dfrac{4}{7} = \dfrac{5}{9} \cdot \dfrac{7}{4}$ Multiply $\dfrac{5}{9}$ by the reciprocal of $\dfrac{4}{7}$.

$\qquad\qquad = \dfrac{35}{36}$ Multiply.

b) $\dfrac{3}{8} + \dfrac{1}{6}$ Identify the least common denominator (LCD): LCD = 24.

$\dfrac{3}{8} \cdot \dfrac{3}{3} = \dfrac{9}{24} \qquad \dfrac{1}{6} \cdot \dfrac{4}{4} = \dfrac{4}{24} \qquad \dfrac{3}{8} + \dfrac{1}{6} = \dfrac{9}{24} + \dfrac{4}{24} = \dfrac{13}{24}$

2. The Order of Operations

We evaluate expressions using the following rules called the **order of operations**.

The Order of Operations—simplify expressions in the following order:

1) If parentheses or other grouping symbols appear in an expression, simplify what is in these grouping symbols first.
2) Simplify expressions with exponents.
3) Perform multiplication and division from left to right.
4) Perform addition and subtraction from left to right.

Example 2

Evaluate $15 \div 3 + (5 - 2)^3$.

Solution

$$
\begin{aligned}
15 \div 3 + (5 - 2)^3 &= 15 \div 3 + 3^3 && \text{Perform operations in parentheses.} \\
&= 15 \div 3 + 27 && \text{Evaluate } 3^3. \\
&= 5 + 27 && \text{Perform the division.} \\
&= 32 && \text{Subtract.}
\end{aligned}
$$

A good way to remember the order of operations is to remember the sentence, "**P**lease **E**xcuse **M**y **D**ear **A**unt **S**ally." (**P**arentheses, **E**xponents, **M**ultiplication, **D**ivision, **A**ddition, **S**ubtraction)

3. Review Concepts from Geometry

Recall that two angles are **complementary** if the sum of their measures is 90°. Two angles are **supplementary** if their measures add up to 180°.

Perimeter and Area

Students should be familiar with the following area and perimeter formulas.

Figure		Perimeter	Area
Rectangle:		$P = 2l + 2w$	$A = lw$
Square:		$P = 4s$	$A = s^2$
Triangle: $h = $ height		$P = a + b + c$	$A = \dfrac{1}{2}bh$
Parallelogram: $h = $ height		$P = 2a + 2b$	$A = bh$
Trapezoid: $h = $ height		$P = a + c + b_1 + b_2$	$A = \dfrac{1}{2}h(b_1 + b_2)$
Circle:		Circumference $C = 2\pi r$	Area $A = \pi r^2$

The **radius, *r*** is the distance from the center of the circle to a point on the circle. A line segment that passes through the center of the circle and has its endpoints on the circle is called a **diameter**.

"π" is the ratio of the circumference of any circle to its diameter. $\pi \approx 3.14159265\ldots$, but we will use 3.14 as an approximation for π.

Example 3

For the rectangle 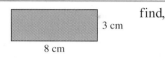 find,

3 cm

8 cm

a)   the perimeter.   b)   the area.

Solution

a) $P = 2l + 2w$
$\quad\quad = 2(8 \text{ cm}) + 2(3 \text{ cm})$
$\quad\quad = 16 \text{ cm} + 6 \text{ cm}$
$\quad\quad = 22 \text{ cm}$

b) $A = lw$
$\quad\quad = (8 \text{ cm})(3 \text{ cm})$
$\quad\quad = 24 \text{ cm}^2$

4. Define and Identify Sets of Numbers

We can define the following sets of numbers:

Natural numbers: $\{1, 2, 3, \ldots\}$ Whole numbers: $\{0, 1, 2, 3, \ldots\}$
Integers: $\{\ldots, -3, -2, -1, 0, 1, 2, 3, \ldots\}$

A **rational number** is any number of the form $\dfrac{p}{q}$, where p and q are integers and $q \neq 0$.

The set of rational numbers includes terminating decimals and repeating decimals. The set of numbers that *cannot* be written as the quotient of two integers is the set of **irrational numbers**. Written in decimal form, an irrational number is a nonrepeating, nonterminating decimal.

The set of **real numbers** consists of the rational and irrational numbers.

Example 4

Given this set of real numbers $\left\{ -8, 0, 3, \sqrt{19}, 1.4, \dfrac{5}{9}, 0.\overline{26}, 7.68412\ldots \right\}$, list the

a)   natural numbers   b)   integers   c)   rational numbers
d)   irrational numbers

Solution

a)   3   b) $-8, 0, 3$ c) $-8, 0, 3, 1.4, \dfrac{5}{9}, 0.\overline{26}$

d) $\sqrt{19}, 7.68412\ldots$

5. Define Absolute Value and Perform Operations on Real Numbers

Recall that on a number line, positive numbers are to the right of zero, and negative numbers are to the left of zero.

The **absolute value** of a number is the distance between that number and 0 on the number line. The absolute value of a number is always positive or zero.

If a is any real number, then the *absolute value* of a is denoted by $|a|$. For example, $|2| = 2$ since 2 is two units from 0. It is also true that $|-2| = 2$, since -2 is two units from 0.

Example 5

Perform the operations.

a) $-6 + (-4)$ b) $5 - 14$ c) $4 \cdot (-3) = -12$

d) $-42 \div (-6) = 7$

Solution

a) To add two numbers with the same sign, find the absolute value of each number and add them. The sum will have the same sign as the numbers being added.
$-6 + (-4) = -(|-6| + |-4|) = -(6 + 4) = -10$

b) To subtract two numbers, $a - b$, change subtraction to addition and add the additive inverse of b. $5 - 14 = 5 + (-14) = -9$

c) The product or quotient of two real numbers with *different signs* is *negative*.
$4 \cdot (-3) = -12$

d) The product or quotient of two real numbers with the *same sign* is positive.
$-42 \div (-6) = 7$

6. Learn the Vocabulary for Algebraic Expressions

An **algebraic expression** is a collection of numbers, variables, and grouping symbols connected by operations symbols such as $+$, $-$, $\times$, and $\div$.

Example 6

List the terms and coefficients of the expression $t^3 - t^2 - 4t + 7$. What is the constant term?

Solution

Term	Coefficient
t^3	1
$-t^2$	-1
$-4t$	-4
7	7

The constant term is 7.

We can **evaluate** an algebraic expression by substituting value(s) for the variable(s) and simplifying.

7. Learn the Properties of Real Numbers

Next, we will summarize some properties of real numbers.

Properties of Real Numbers

Assume a, b, and c are real numbers. Then the following properties are true for a, b, and c:

1) Commutative properties: $a + b = b + a$; $ab = ba$

2) Associative properties: $(a + b) + c = a + (b + c)$; $(ab)c = a(bc)$

3) Identity properties: $a + 0 = a$; $a \cdot 1 = a$

4) Inverse properties: $a + (-a) = 0$; $b \cdot \dfrac{1}{b} = 1$ $(b \neq 0)$

5) Distributive properties: $a(b + c) = ab + ac$
$$a(b - c) = ab - ac$$

Example 7

Rewrite each expression using the indicated property.

a) $5 + (8 + 1)$; associative   b) $3(n + 7)$; distributive

Solution

a) $5 + (8 + 1) = (5 + 8) + 1$

b) $3(n + 7) = 3n + 3(7)$
$$= 3n + 21$$

A1 Exercises

Evaluate.

1) $-|6|$

2) $-|-1.4|$

3) 2^5

4) -7^2

Decide if each statement is true or false.

5) $\dfrac{2}{3} < \dfrac{8}{9}$

6) $0.06 > 0.6$

Graph the numbers on a number line. Label each.

7) $3, -4, \dfrac{2}{3}, 1.5, -2\dfrac{1}{4}$

8) $5, \dfrac{3}{4}, -2, -3.2, 2\dfrac{1}{5}$

Perform the indicated operations and simplify.

9) $\dfrac{5}{8} \cdot \dfrac{6}{7}$

10) $8 - (-9)$

11) $\dfrac{2}{7} \div \dfrac{4}{21}$

12) $-14.6 - (-21.4)$

13) $\dfrac{9 \cdot 2 - 7 \cdot 4}{\sqrt{81} + (7 - 4)^4}$

14) $\dfrac{-48}{-3}$

15) $\dfrac{5}{9} \cdot \dfrac{9}{5}$

16) $-5\dfrac{1}{3} \div 1\dfrac{3}{7}$

17) $2\dfrac{1}{2} + \dfrac{1}{6}$

18) $\dfrac{20 - 2^3 + 9}{2 + 5 \cdot 4 - 10}$

19) $16 - 30$

20) $-7(4.3)$

21) $8 \cdot 7 - (1 + 3)^3 + 6 \cdot 2$

22) $\dfrac{2}{5} + \dfrac{3}{4} - \dfrac{3}{20}$

23) $5 - 16 + 2 - 24 - (-11)$

24) $\sqrt{16} + |8 - 11| - 45 \div 5$

25) $-\dfrac{2}{3} \cdot \dfrac{4}{5}$

26) $-\dfrac{3}{4} + \dfrac{2}{9}$

For Exercises 27–30, write a mathematical expression and simplify.

27) 10 less than 16

28) The quotient of -24 and 3

29) Twice the sum of -19 and 4

30) 9 less than the product of 7 and 6

For Exercises 31–34, use this set to list the indicated numbers.

$$A = \left\{ \dfrac{8}{11}, -14, 3.7, 5, \sqrt{7}, 0, -1\dfrac{1}{2}, 6.\overline{2}, 2.8193\ldots \right\}$$

31) The integers in set A

32) The whole numbers in set A

33) The rational numbers in set A

34) The irrational numbers in set A

35) The supplement of $41°$ is _____.

36) The complement of $32°$ is _____.

For Exercises 37 and 38, find the measure of the missing angle, and classify each triangle as acute, obtuse, or right.

37)

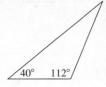

38)

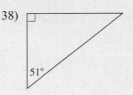

For Exercises 39 and 40, find the area of each figure. Include the correct units.

39)

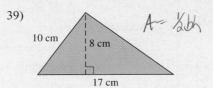

40)

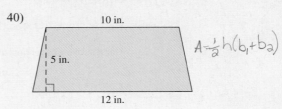

For Exercises 41 and 42, find the area and perimeter of each figure. Include the correct units.

41)

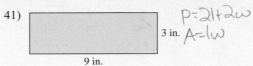

42)

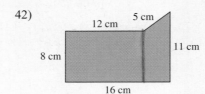

43) Find the a) area and b) circumference of the circle. Give an exact answer for each and give an approximation using 3.14 for π. Include the correct units.

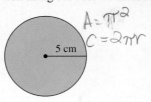

List the terms and their coefficients. Also, identify the constant.

44) $7z^3 - 3z^2 + z + 2.8$

45) Evaluate $5p^2 - 4p + 2$ when

a) $p = 3$

b) $p = -2$

46) Evaluate $\dfrac{a - b^2}{2a + 3b}$ when $a = 6$ and $b = -3$.

47) What is the multiplicative inverse of -8?

48) What is the additive inverse of 7?

Which property of real numbers is illustrated by each example? Choose from the commutative, associative, identity, inverse, or distributive properties.

49) $4(3 - 5) = 4 \cdot 3 - 4 \cdot 5$

50) $9 + 2 = 2 + 9$

51) $-18 + (6 + 1) = (-18 + 6) + 1$

52) $\left(\dfrac{6}{5}\right)\left(\dfrac{5}{6}\right) = 1$

53) Is $x - 4$ equivalent to $4 - x$? Why or why not?

54) Is $9 + 2w$ equivalent to $2w + 9$? Why or why not?

Rewrite each expression using the indicated property.

55) $8 + 3$; commutative

56) $(2 + 7) + 4$; associative

57) $-5p$; identity

58) $6t + 1$; commutative

Rewrite each expression using the distributive property.

59) $-7(2w + 1)$ 60) $-2(5m + 6)$

61) $6(5 - 7r)$ 62) $5(1 - 9h)$

63) $-8(3a - 4b - c)$ 64) $-(t - 8)$

Section A2 Variables and Exponents

Objectives

1. Combine Like Terms
2. Translate English Expressions to Algebraic Expressions
3. Use the Rules of Exponents
4. Use Scientific Notation

1. Combine Like Terms

Like terms contain the same variables with the same exponents.

We can add and subtract only those terms that are like terms. To simplify an expression containing parentheses, we use the distributive property to clear the parentheses, and then combine like terms.

Example 1

Simplify by combining like terms: $5(4x + 3y) - 3(x - 2y)$

Solution

$$
\begin{aligned}
5(4x + 3y) - 3(x - 2y) &= 20x + 15y - 3x + 6y && \text{Distributive property} \\
&= 17x + 21y && \text{Combine like terms.}
\end{aligned}
$$

2. Translate English Expressions to Algebraic Expressions

To translate from English to an algebraic expression, we can do the following:

Read the phrase carefully, choose a variable to represent the unknown quantity, and then translate the phrase to an algebraic expression.

Example 2

Write an algebraic expression for *five more than twice a number.*

Solution

After reading the phrase carefully, choose a variable to represent the unknown quantity.

Let $x =$ the number.

Slowly, break down the phrase to write the expression.

Five	more than	twice a number
5	+	$2x$

The expression is $5 + 2x$ or $2x + 5$. They are equivalent.

3. Use the Rules of Exponents

Exponential notation is used as a shorthand way to represent repeated multiplication.

$$2 \cdot 2 \cdot 2 \cdot 2 \cdot 2 = 2^5 \qquad x \cdot x \cdot x = x^3$$

Next, we will review some rules for working with expressions containing exponents.

Rule	Example
Product Rule: $a^m \cdot a^n = a^{m+n}$	$p^4 \cdot p^{11} = p^{4+11} = p^{15}$
Basic Power Rule: $(a^m)^n = a^{mn}$	$(c^8)^3 = c^{8 \cdot 3} = c^{24}$
Power Rule for a Product: $(ab)^n = a^n b^n$	$(3z)^4 = 3^4 \cdot z^4 = 81z^4$
Power Rule for a Quotient: $\left(\dfrac{a}{b}\right)^n = \dfrac{a^n}{b^n}$, where $b \neq 0$	$\left(\dfrac{w}{2}\right)^4 = \dfrac{w^4}{2^4} = \dfrac{w^4}{16}$
Zero Exponent: If $a \neq 0$, then $a^0 = 1$.	$(-3)^0 = 1$
Negative Exponent: For $a \neq 0$, $a^{-n} = \left(\dfrac{1}{a}\right)^n = \dfrac{1}{a^n}$. If $a \neq 0$ and $b \neq 0$, then $\dfrac{a^{-m}}{b^{-n}} = \dfrac{b^n}{a^m}$.	$5^{-3} = \left(\dfrac{1}{5}\right)^3 = \dfrac{1^3}{5^3} = \dfrac{1}{125}$ $\dfrac{r^{-6}}{s^{-3}} = \dfrac{s^3}{r^6}$
Quotient Rule: If $a \neq 0$, then $\dfrac{a^m}{a^n} = a^{m-n}$.	$\dfrac{m^7}{m^5} = m^{7-5} = m^2$

Sometimes, it is necessary to combine the rules of exponents to simplify a product or a quotient. We must also remember the order of operations.

Example 3

Simplify.

a) $(2t^4)^3(5t)^2$

b) $\dfrac{(4n^6)^3}{(10m^3)^2}$

c) $\left(\dfrac{7}{4}\right)^{-2}$

Solution

a) $(2t^4)^3(5t)^2 = (2t^4)^3 \cdot (5t)^2$

$\qquad\qquad\quad = (2^3)(t^4)^3 \cdot (5)^2(t)^2$ Power rule for products

$\qquad\qquad\quad = 8t^{12} \cdot 25t^2$ Basic power rule

$\qquad\qquad\quad = 200t^{14}$ Multiply coefficients and use the product rule.

b) $\dfrac{(4n^6)^3}{(10m^3)^2} = \dfrac{64n^{18}}{100m^6}$ Power rule for products

$= \dfrac{\overset{16}{\cancel{64}}n^{18}}{\underset{25}{\cancel{100}}\,m^6}$ Divide out the common factor of 4.

$= \dfrac{16n^{18}}{25m^6}$

BE CAREFUL $(ab)^n = a^n b^n$ is different from $(a + b)^n$. $(a + b)^n \neq a^n + b^n$. See Section A6 or Chapter 6.

c)   The reciprocal of $\dfrac{7}{4}$ is $\dfrac{4}{7}$, so

$$\left(\dfrac{7}{4}\right)^{-2} = \left(\dfrac{4}{7}\right)^2 = \dfrac{4^2}{7^2} = \dfrac{16}{49}$$

Next we will simplify more expressions containing negative exponents.

Example 4

Rewrite each expression with positive exponents. Assume the variables do not equal zero.

a) $5k^{-3}$ b) $\dfrac{7a^4 b^{-1}}{c^{-5} d^{-2}}$

Solution

a)   The base in $5k^{-3}$ is k. The 5 is not part of the base since it is not in parentheses.

$$5k^{-3} = 5 \cdot \left(\dfrac{1}{k}\right)^3 = 5 \cdot \dfrac{1}{k^3} = \dfrac{5}{k^3}$$

b) $\dfrac{7a^4 b^{-1}}{c^{-5} d^{-2}} = \dfrac{7a^4 c^5 d^2}{b}$ The 7 and the a^4 keep their positions within the expression. b^{-1}, c^{-5}, and d^{-2} switch their positions, and the exponents become positive.

Example 5

Simplify. Assume that the variables represent nonzero real numbers. The final answer should not contain negative exponents.

$$\left(\dfrac{c^4 d^{-2}}{10c^9 d^{-8}}\right)^{-3}$$

Solution

To begin, eliminate the negative exponent *outside* of the parentheses by taking the reciprocal of the base. Notice that we have *not* eliminated the negatives on the exponents *inside* the parentheses.

$$\left(\frac{c^4 d^{-2}}{10c^9 d^{-8}}\right)^{-3} = \left(\frac{10c^9 d^{-8}}{c^4 d^{-2}}\right)^{3}$$

$$= (10c^5 d^{-6})^3 \qquad \text{Subtract the exponents.}$$
$$= 1000c^{15} d^{-18} \qquad \text{Power rule}$$
$$= \frac{1000c^{15}}{d^{18}} \qquad \text{Write the answer using positive exponents.}$$

4. Use Scientific Notation

Scientific notation is a shorthand method for writing very large and very small numbers.

Definition   A number is in **scientific notation** if it is written in the form $a \times 10^n$, where $1 \le |a| < 10$ and n is an integer.

First, we will convert a number from scientific notation to a number without exponents.

Example 6

Write without exponents. 7.382×10^4

Solution

Since $10^4 = 10,000$, multiplying 7.382 by 10,000 will give us a result that is *larger* than 7.382. Move the decimal point in 7.382 four places to the right.

$$7.382 \times 10^4 = 7.3820 = 73,820$$

Four places to the right

To write a number in scientific notation, $a \times 10^n$, remember that a must have one number to the left of the decimal point.

Example 7

Write in scientific notation.   0.00000974

Solution

0.00000974

The decimal point
will be here.

$$0.00000974 = 9.74 \times 10^{-6} \qquad \text{Move the decimal point six places to the right.}$$

A2 Exercises   Boost your grade at mathzone.com!   MathZone   > Practice Problems   > NetTutor   > e-Professors   > Videos   > Self-Test

Determine if the following groups of terms are like terms.

1) $-6a^2, 5a^3, a$

2) $3p^2, -\dfrac{1}{4}p^2, -8p^2$

Simplify by combining like terms.

3) $5z^2 - 7z + 10 - 2z^2 - z + 3$

4) $6(2k + 1) + 5(k - 3)$

5) $10 - 4(3n - 2) + 7(2n - 1)$

6) $t^2 + t + 3 - (7t + 2) - 6t^2$

Write a mathematical expression for each phrase, and combine like terms if possible. Let x represent the unknown quantity.

7) Ten more than the sum of -7 and half of a number

8) Seven less than the sum of 4 and three times a number

Write in exponential form.

9) $y \cdot y \cdot y \cdot y$

10) $6 \cdot k \cdot k \cdot k$

Simplify using the rules of exponents. Assume the variables do not equal zero.

11) $2^2 \cdot 2^4$

12) $\left(\dfrac{1}{4}\right) \cdot \left(\dfrac{1}{4}\right)^2$

13) $4m^9 \cdot 3m$

14) $-6w^8 \cdot 9w^5 \cdot w^2$

15) $(-7k^5)^2$

16) $\left(\dfrac{1}{5}z^9\right)^3$

17) $\left(\dfrac{x}{y}\right)^9$

18) $\left(\dfrac{12}{n}\right)^2$

19) $\dfrac{w^{10}}{w^7}$

20) $\dfrac{m^9}{m^4}$

21) $\dfrac{2p^4q^7}{p^3q^2}$

22) $\dfrac{8a^6b^4}{3ab^2}$

Evaluate.

23) 6^0

24) $(-3)^0$

25) $\left(\dfrac{1}{5}\right)^{-3}$

26) $\left(\dfrac{3}{2}\right)^{-4}$

Rewrite each expression with only positive exponents. Assume the variables do not equal zero.

27) x^{-8}

28) h^{-5}

29) $\left(\dfrac{t}{4}\right)^{-3}$

30) $\left(\dfrac{10}{k}\right)^{-2}$

31) $\dfrac{a^{-4}}{7b^{-2}}$

32) $\dfrac{20x^{-1}}{y^{-6}}$

33) $r^{-3}s$

34) $\dfrac{8c^{-5}}{24d^2}$

Simplify. Assume the variables do not equal zero. The final answer should not contain negative exponents.

35) $\left(-\dfrac{8a^5b}{4c^3}\right)^3$

36) $\left(\dfrac{m^3n^8}{m^7n^2}\right)^2$

37) $(3c^5d^2)^3(-c^4d)^4$

38) $2(3a^6b^2)^3$

39) $\left(\dfrac{r^{-3}s}{r^{-2}s^6}\right)^3$

40) $(-3k^5)^2\left(\dfrac{2}{7}k^{-3}\right)^2\left(\dfrac{1}{6}k\right)$

41) $\left(\dfrac{v^{-3}}{v}\right)^{-5}$

42) $(-c^{-8}d^2)^3(2c^2d^{-1})^4$

43) $\left(\dfrac{14a^{-3}b}{7a^5b^{-6}}\right)^{-4}$

44) $\left(\dfrac{4x^5y^{-1}}{12x^2y^2}\right)^{-2}$

45) Find an algebraic expression for the area and perimeter of this rectangle.

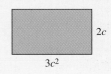

$2c$

$3c^2$

46) Find an algebraic expression for the area of this triangle.

$\dfrac{5}{4}t$

t

Write each number without an exponent.

47) 5.07×10^{-4}

48) 2.8×10^3

Write each number in scientific notation.

49) 94,000

50) 0.00000295

51) Divide. Write the final answer without an exponent.

$$\dfrac{3.6 \times 10^{11}}{9 \times 10^6}.$$

52) Suppose 1,000,000 atoms were lined up end-to-end. If the diameter of an atom is about 1×10^{-8} cm, find the length of the chain of atoms. Write the final answer without an exponent.

Section A3 Linear Equations and Inequalities

Objectives

1. Solve a Linear Equation
2. Solve Linear Equations with No Solution or an Infinite Number of Solutions
3. Solve Applied Problems
4. Solve an Equation for a Specific Variable
5. Solve a Linear Inequality in One Variable
6. Solve Compound Inequalities Containing *And* or *Or*

1. Solve a Linear Equation

> **Definition**   An **equation** is a mathematical statement that two expressions are equal.

An *equation* contains an equal ($=$) sign, and an *expression* does not.

We can **solve** equations, and we can **simplify** expressions.

There are many different types of equations. In this section, we will discuss how to solve a linear equation in one variable.

> **Definition**   A **linear equation in one variable** is an equation that can be written in the form
>
> $$ax + b = 0$$
>
> where a and b are real numbers and $a \neq 0$.

Examples of linear equations in one variable include $7t + 6 = 27$ and $4(m + 1) - 9 = 3m + 10$.

Notice that the exponent on each variable is 1. For this reason, these equations are also known as first-degree equations.

To **solve an equation** means to find the value or values of the variable that make the equation true.

We use the following properties of equality to help us solve equations.

> Let $a, b,$ and c be expressions representing real numbers. Then, the following properties hold.
>
> 1) *Addition Property of Equality*   If $a = b$, then $a + c = b + c$.
> 2) *Subtraction Property of Equality*   If $a = b$, then $a - c = b - c$.
> 3) *Multiplication Property of Equality*   If $a = b$, then $ac = bc$.
> 4) *Division Property of Equality*   If $a = b$, then $\dfrac{a}{c} = \dfrac{b}{c}$ ($c \neq 0$).

Example 1

Solve $7n - 9 = 5$.

Solution

Use the properties of equality to solve for the variable.

$$7n - 9 = 5$$
$$7n - 9 + 9 = 5 + 9 \qquad \text{Add 9 to each side.}$$
$$7n = 14$$
$$\frac{7n}{7} = \frac{14}{7} \qquad \text{Divide each side by 7.}$$
$$n = 2$$

The solution set is $\{2\}$. The check is left to the student.

When linear equations contain variables on both sides of the equal sign, we need to get the variables on one side of the equal sign and get the constants on the other side. With this in mind, we summarize the steps used to solve a linear equation in one variable.

How to Solve a Linear Equation

1) If there are terms in parentheses, begin by distributing.

2) Combine like terms on each side of the equal sign.

3) Get the terms with the variables on one side of the equal sign and the constants on the other side using the addition and subtraction properties of equality.

4) Solve for the variable using the multiplication or division property of equality.

5) Check the solution in the original equation.

Example 2

Solve $3a + 10 - 2(4a + 7) = 6(a - 1) - 5a$.

Solution

Follow the steps listed above.

$$3a + 10 - 2(4a + 7) = 6(a - 1) - 5a$$
$$3a + 10 - 8a - 14 = 6a - 6 - 5a \qquad \text{Distribute.}$$
$$-5a - 4 = a - 6 \qquad \text{Combine like terms.}$$
$$-5a - a - 4 = a - a - 6 \qquad \begin{array}{l}\text{Subtract } a \text{ from each side to get}\\ \text{the } a\text{-terms on the same side.}\end{array}$$
$$-6a - 4 = -6 \qquad \text{Combine like terms.}$$
$$-6a - 4 + 4 = -6 + 4 \qquad \text{Add 4 to each side.}$$
$$-6a = -2$$
$$\frac{-6a}{-6} = \frac{-2}{-6}$$
$$a = \frac{1}{3} \qquad \text{Simplify.}$$

The solution set is $\left\{\dfrac{1}{3}\right\}$. The check is left to the student.

Some equations contain several fractions or decimals. Before applying the steps for solving a linear equation, we can eliminate the fractions and decimals from the equation.

To eliminate the fractions, determine the least common denominator (LCD) for all of the fractions in the equation. Then, multiply both sides of the equation by the LCD. To eliminate decimals, multiply both sides of the equation by the appropriate power of 10.

2. Solve Linear Equations with No Solution or an Infinite Number of Solutions

Some equations have no solution while others have an infinite number of solutions.

| **Example 3** |

Solve each equation.

a) $3(2k - 1) = 6k + 5$ b) $z + 8 - 5z = -4(z - 2)$

Solution

a)
$$3(2k - 1) = 6k + 5$$
$$6k - 3 = 6k + 5 \qquad \text{Distribute.}$$
$$6k - 6k - 3 = 6k - 6k + 5 \qquad \text{Subtract } 6k \text{ from each side.}$$
$$-3 = 5 \qquad \text{False}$$

There is no solution to this equation. An equation that has no solution is called a **contradiction**. We say that the solution set is the **empty set**, denoted by $\varnothing$.

b)
$$z + 8 - 5z = -4(z - 2)$$
$$-4z + 8 = -4z + 8 \qquad \text{Distribute.}$$
$$-4z + 4z + 8 = -4z + 4z + 8 \qquad \text{Add } 4z \text{ to each side.}$$
$$8 = 8 \qquad \text{True}$$

The variable was eliminated, and we are left with a true statement. This means that any real number we substitute for z will make the original equation true. An equation that has all real numbers in its solution set is called an **identity**. This equation has an **infinite number of solutions**, and the solution set is {all real numbers}.

3. Solve Applied Problems

Next we will discuss how to translate information presented in English into an algebraic equation. The following approach is suggested to help in the problem-solving process.

Steps for Solving Applied Problems

Step 1: Read the problem carefully. Then read it again. Draw a picture, if applicable.

Step 2: Identify what you are being asked to find. *Define the variable;* that is, assign a variable to unknown quantity. Also,
- If there are other unknown quantities, define them in terms of the variable.
- Label the picture with the variable and other unknowns as well as with any given information in the problem.

Step 3: Translate from English to math. Some suggestions for doing so are
- Restate the problem in your own words.
- Read and think of the problem in "small parts."
- Make a chart to separate these "small parts" of the problem to help you translate to mathematical terms.
- Write an equation in English, then translate it to an algebraic equation.

Step 4: Solve the equation.

Step 5: Interpret the meaning of the solution as it relates to the problem. If there are other unknowns, find them. State the answer in a complete sentence. Be sure your answer makes sense in the context of the problem.

Example 4

Write an equation and solve.

Roy advises 23 more students than Tobi. Together, they advise a total of 459 students. How many students are assigned to each advisor?

Solution

We will follow the steps listed above to solve this problem.

Step 1: Read the problem carefully, then read it again.

Step 2: Define the unknown quantities.

$$x = \text{the number of Tobi's advisees}$$
$$x + 23 = \text{the number of Roy's advisees}$$

Step 3: We can think of an equation in English and then translate it to an algebraic equation.

$$\begin{array}{ccccc} \text{Number of} & + & \text{Number of} & = & \text{Total number} \\ \text{Tobi's advisees} & & \text{Roy's advisees} & & \text{of advisees} \\ x & + & (x + 23) & = & 459 \end{array}$$

The equation is $x + (x + 23) = 459$.

Step 4:   Solve the equation.

$$x + (x + 23) = 459$$
$$2x + 23 = 459 \qquad \text{Combine like terms.}$$
$$2x = 436 \qquad \text{Subtract 23 from each side.}$$
$$x = 218 \qquad \text{Divide by 2.}$$

Step 5:   Find the other unknown. Since $x = 218$, $x + 23 = 241$.

Tobi advises 218 students, and Roy advises 241 students. The answer makes sense since $218 + 241 = 459$, the total number of students both Tobi and Roy advise.

Investment Problems

There are different ways to calculate the amount of interest earned from an investment. Here we will discuss **simple interest**.

The initial amount of money deposited in an account is called the *principal*. The formula used to calculate simple interest, I, is $I = PRT$, where

$$I = \text{interest earned (simple)}$$
$$P = \text{principal (initial amount invested)}$$
$$R = \text{annual interest rate (expressed as a decimal)}$$
$$T = \text{amount of time the money is invested (in years)}$$

We can use this to solve an application.

Example 5

Write an equation and solve.

When Francisco received an $8000 bonus at work, he decided to invest some of it in an account paying 4% simple interest, and he invested the rest of it in a certificate of deposit that paid 7% simple interest. He earned a total of $410 in interest after 1 yr. How much did Francisco invest in each account?

Solution

Step 1:   Read the problem carefully, twice.

Step 2:   Find the amount Francisco invested in the 4% account and the amount he invested in the certificate of deposit that paid 7% simple interest. Define the unknowns.

$$x = \text{amount invested in the 4\% account}$$
$$8000 - x = \text{amount invested in the CD paying 7\% simple interest}$$

Step 3:   Translate from English to an algebraic equation.

Total interest earned = Interest from 4% account + Interest from 7% account

$$
\underset{}{410} \quad = \quad \underset{P \quad r \quad t}{x(0.04)(1)} \quad + \quad \underset{P \qquad r \quad t}{(8000 - x)(0.07)(1)}
$$

The equation can be rewritten as

$$410 = 0.04x + 0.07(8000 - x)$$

Step 4: Solve the equation.

Begin by multiplying both sides of the equation by 100 to eliminate the decimals.

$$410 = 0.04x + 0.07(8000 - x)$$
$$100(410) = 100[0.04x + 0.07(8000 - x)] \qquad \text{Multiply by 100.}$$
$$41{,}000 = 4x + 7(8000 - x)$$
$$41{,}000 = 4x + 56{,}000 - 7x \qquad \text{Distribute.}$$
$$41{,}000 = -3x + 56{,}000 \qquad \text{Combine like terms.}$$
$$-15{,}000 = -3x$$
$$5{,}000 = x$$

Step 5: $x = 5000$ and $8000 - x = 3000$

Francisco invested \$5000 in the account earning 4% interest and \$3000 in the certificate of deposit earning 7% interest. The check is left to the student.

4. Solve an Equation for a Specific Variable

Formulas are widely used not only in mathematics but also in disciplines such as business, economics, and the sciences. Often, it is necessary to solve a formula for a particular variable.

Example 6

Solve $A = P + PRT$ for T.

Solution

We will put a box around the T to remind us that this is the variable for which we are solving. The goal is to get the T on a side by itself.

$$A = P + PR\,\boxed{T}$$
$$A - P = P - P + PR\,\boxed{T} \qquad \text{Subtract } P \text{ from each side.}$$
$$A - P = PR\,\boxed{T} \qquad \text{Simplify.}$$
$$\frac{A - P}{PR} = \frac{PR\,\boxed{T}}{PR} \qquad \text{Since } T \text{ is being multiplied by } PR, \text{ divide both sides by } PR.$$
$$\frac{A - P}{PR} = T \qquad \text{Simplify.}$$

5. Solve a Linear Inequality in One Variable

While an equation states that two expressions are equal, an **inequality** states that two expressions are not necessarily equal.

Definition

A **linear inequality in one variable** can be written in the form $ax + b < c$, $ax + b \le c$, $ax + b > c$, or $ax + b \ge c$, where a, b, and c are real numbers and $a \ne 0$.

We solve linear inequalities in very much the same way we solve linear equations *except when we multiply or divide by a negative number we must reverse the direction of the inequality symbol.* To represent the solution to an inequality we can:

1) Graph the solution set.

2) Write the answer in set notation.

3) Write the answer in interval notation.

Example 7

Solve each inequality. Graph the solution set and write the answer in both set and interval notations.

a) $2 - y \leq 5$

b) $-2 \leq \dfrac{3}{4}c - 5 < 1$

Solution

a) $2 - y \leq 5$

   $\quad -y \leq 3$ Subtract 2 from each side.

   $\quad\ \ y \geq -3$ Divide by -1 and reverse the inquality symbol.

The graph of the solution set is

$$-4\ -3\ -2\ -1\quad 0\quad 1\quad 2\quad 3\quad 4$$

In set notation, we represent the solution as $\{y \mid y \geq -3\}$.

In interval notation, we represent the solution as $[-3, \infty)$.

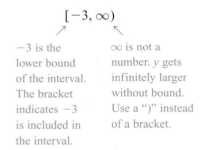

$[-3, \infty)$

-3 is the lower bound of the interval. The bracket indicates -3 is included in the interval.

∞ is not a number. y gets infinitely larger without bound. Use a ")" instead of a bracket.

b) $-2 \leq \dfrac{3}{4}c - 5 < 1$

This is a *compound inequality.* A **compound inequality** contains more than one inequality symbol. To solve this type of inequality we must remember that *whatever operation we perform on one part of the inequality we must perform on all parts of the inequality.*

$$-2 \leq \frac{3}{4}c - 5 < 1$$

$$-2 + 5 \leq \frac{3}{4}c - 5 + 5 < 1 + 5 \qquad \text{Add 5 to each part of the inequality.}$$

$$3 \leq \frac{3}{4}c < 6 \qquad\qquad \text{Combine like terms.}$$

$$\frac{4}{3} \cdot 3 \leq \frac{4}{3} \cdot \frac{3}{4}c < \frac{4}{3} \cdot 6 \qquad \text{Multiply each part by } \frac{4}{3}.$$

$$4 \leq c < 8 \qquad\qquad\qquad \text{Simplify.}$$

The graph of the solution set is

In set notation, we write the solution as $\{c \mid 4 \le c < 8\}$.
In interval notation, we write the solution as $[4, 8)$

In Example 7b), we said that a compound inequality contains more than one inequality symbol. Next we will look at compound inequalities containing the words *and* or *or*.

6. Solve Compound Inequalities Containing *And* or *Or*

The solution set of a compound inequality joined by *and* is the *intersection* of the solution sets of the individual inequalities. The solution set of a compound inequality joined by *or* is the *union* of the solution sets of the individual inequalities.

Example 8

Solve each compound inequality, and write the answer in interval notation.

a) $3x < 12$ and $x + 6 > 4$ b) $\dfrac{3}{2}w - 1 \ge 2$ or $2w - 5 \le -5$

Solution

a) **Step 1:** These two inequalities are connected by "**and**." That means the solution set will consist of the values of x that make *both* inequalities true. The solution set will be the **intersection** of the solution sets of $3x < 12$ and $x + 6 > 4$.

    Step 2: Solve each inequality separately.

$$3x < 12 \quad \text{and} \quad x + 6 > 4$$

Divide by 3. $x < 4 \quad \text{and} \quad x > -2$ Subtract 6.

    Step 3: Graph the solution set to each inequality on its own number line even if the problem does not require you to graph the solution set. This will help you to visualize the solution set of the compound inequality.

$x < 4$: $x > -2$:

    Step 4: The *intersection* of the two solution sets is the region where their graphs intersect. Ask yourself, *"If I were to put the number lines on top of each other, where would they intersect?"* They would intersect between -2 and 4.

    The graph of $x < 4$ and $x > -2$ is

    Step 5: In interval notation, we can represent the shaded region above as $(-2, 4)$. Every real number in this interval will satisfy each inequality.

b) ***Step 1:*** The solution to the compound inequality $\frac{3}{2}w - 1 \geq 2$ or

$2w - 5 \leq -5$ is the **union** of the solution sets of the individual inequalities.

Step 2: Solve each inequality separately.

$$\frac{3}{2}w - 1 \geq 2 \quad \text{or} \quad 2w - 5 \leq -5$$

Add 1. $\quad\quad \frac{3}{2}w \geq 3 \quad \text{or} \quad\quad 2w \leq 0 \quad$ Add 5.

Multiply by $\frac{2}{3}$. $\quad w \geq 2 \quad \text{or} \quad\quad\quad w \leq 0 \quad$ Divide by 2.

Step 3: Graph the solution set to each inequality on its own number line.

$w \geq 2$: $\qquad\qquad\qquad\qquad\qquad w \leq 0$:

Step 4: The *union* of the two solution sets consists of the *total* of what would be shaded if the number lines above were placed on top of each other.

The graph of $w \geq 2$ or $w \leq 0$ is

Step 5: In interval notation, we can represent the shaded region above as $(-\infty, 0] \cup [2, \infty)$. (Use the union symbol, $\cup$, for **or**.)

A3 Exercises

Determine if the given value is a solution to the equation.

1) $8b + 5 = 1; b = -\frac{1}{2}$

2) $3 - 2(t + 1) = 1 + t; t = 3$

Solve and check each equation.

3) $y + 6 = 10$

4) $n - 7 = -2$

5) $-\frac{3}{4}x = 3$

6) $-16 = -\frac{4}{9}c$

7) $7b + 2 = 23$

8) $3v - 8 = 16$

9) $4 - \frac{1}{3}h = 6$

10) $0.2q + 7 = 8$

11) $3y + 4 - 9y + 7 = 14$

12) $11 + 2m + 6m + 3 = 12$

13) $9n - 4 = 3n + 14$

14) $10b + 9 = 2b + 25$

15) $2(4z - 3) - 5z = z - 6$

16) $8 - 3d + 9d - 1 = 4(d - 2) + 3$

17) $4(k + 1) + 3k = 7k + 9$

18) $x - 3(2x - 7) = 5(4 - x) + 1$

19) $\frac{2}{3}r - 1 = \frac{2}{5}r + \frac{3}{5}$

20) $\frac{1}{6}t + \frac{3}{8} = \frac{1}{8}t + \frac{1}{12}$

21) $0.4(p + 5) + 0.2(p - 3) = 0.8$

22) $0.12(v - 2) - 0.03(2v - 5) = 0.09$

Solve using the five "Steps for Solving Applied Problems."

23) Eleven less than a number is 15. Find the number.

24) An electrician must cut a 60-ft cable into two pieces so that one piece is twice as long as the other. Find the length of each piece.

25) In the 2006 FIFA World Cup, Sweden received two fewer yellow cards than Mexico. The players on the two teams received a total of 20 yellow cards. How many yellow cards did each team receive?

26) The sum of three consecutive even integers is 102. Find the numbers.

27) Gonzalo inherited $8000 and invested some of it in an account earning 6% simple interest and the rest of it in an account earning 7% simple interest. After 1 yr, he earned $500 in interest. How much did Gonzalo deposit in each account?

28) A pair of running shoes is on sale for $60.80. This is 20% off of the original price. Find the original price of the shoes.

29) Find the measures of angles A and B.

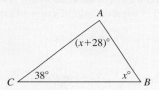

30) How many milliliters of a 12% acid solution and how many milliliters of a 4% acid solution must be mixed to make 100 mL of a 10% acid solution?

31) A collection of coins contains twice as many quarters as dimes. The coins are worth $18.60. How many quarters are in the collection?

32) Nick and Jessica leave home going in opposite directions. Nick is walking at 3 mph, and Jessica is jogging at 6 mph. After how long will they be 3 mi apart?

Solve each formula for the indicated variable.

33) $V = \dfrac{AH}{3}$ for H 34) $A = P + PRT$ for R

35) $A = \dfrac{1}{2}(b_1 + b_2)$ for b_2

36) $A = \pi(R^2 - r^2)$ for R^2

Solve each inequality. Graph the solution set and write the answer in interval notation.

37) $r - 9 \geq -3$ ✦ 38) $4t < 12$

39) $-5n + 7 \geq 22$ 40) $8 - a \leq 11$

41) $3(2z - 1) - z > 2z - 15$

42) $\dfrac{2}{3}(k - 2) \geq \dfrac{5}{4}(2k + 1) - \dfrac{5}{3}$

43) $-1 \leq 4 - 3p < 10$

44) $-3 \leq \dfrac{1 - w}{2} \leq 1$

Solve each compound inequality. Graph the solution set, and write the answer in interval notation.

45) $6z + 4 > 1$ and $z - 3 < -1$

46) $5 - r \leq 3$ and $2r + 5 \geq 1$

47) $2c - 5 > -3$ or $-4c \geq 12$

48) $t + 7 \leq 7$ or $3t + 4 \geq 10$

Section A4 Linear Equations in Two Variables and Functions

Objectives

1. Define a Linear Equation in Two Variables
2. Graph a Line by Plotting Points and Finding Intercepts
3. Find the Slope of a Line
4. Graph a Line Given in Slope-Intercept Form
5. Write an Equation of a Line
6. Determine if a Relation Is a Function, and Find the Domain and Range
7. Given an Equation, Determine Whether y Is a Function of x and Find the Domain
8. Use Function Notation and Find Function Values

1. Define a Linear Equation in Two Variables

Definition	A **linear equation in two variables** can be written in the form $Ax + By = C$ where A, B, and C are real numbers and where both A and B do not equal zero.

Two examples of such equations are $5x + 4y = 8$ and $y = -2x + 1$.

A **solution** to a linear equation in two variables is an ordered pair, (x, y), that satisfies the equation.

2. Graph a Line by Plotting Points and Finding Intercepts

> The graph of a linear equation in two variables, $Ax + By = C$, is a straight line. Each point on the line is a solution to the equation, and every linear equation has an infinite number of solutions.

We can graph a line by making a table of values, finding the intercepts, or using the slope and y-intercept of the line. We will look at a couple of examples here.

Example 1

Graph $y = \dfrac{2}{3}x - 2$ by finding the intercepts and one other point.

Solution

We will begin by finding the intercepts.

x-intercept: Let $y = 0$, and solve for x.   *y-intercept:* Let $x = 0$, and solve for y.

$$y = \frac{2}{3}x - 2 \qquad\qquad\qquad y = \frac{2}{3}x - 2$$

$$0 = \frac{2}{3}x - 2 \qquad\qquad\qquad y = \frac{2}{3}(0) - 2$$

$$2 = \frac{2}{3}x \qquad\qquad\qquad\qquad y = 0 - 2$$

$$\qquad\qquad\qquad\qquad\qquad\qquad y = -2$$

$$3 = x \qquad\qquad\qquad\qquad \text{The } y\text{-intercept is } (0, -2).$$

The x-intercept is $(3, 0)$.

We must find another point. The coefficient of x is $\dfrac{2}{3}$. If we choose a value for x that is a multiple of 3 (the denominator of the fraction), then $\dfrac{2}{3}x$ will not be a fraction.

Let $x = -3$.
$$y = \frac{2}{3}x - 2$$
$$= \frac{2}{3}(-3) - 2$$
$$= -2 - 2$$
$$= -4$$

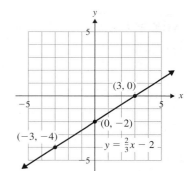

The third point is $(-3, -4)$.
Plot the points, and draw the line through them.

To graph a vertical or horizontal line, follow these rules:

If c is a constant, then the graph of $x = c$ is a *vertical line* going through the point $(c, 0)$.

If d is a constant, then the graph of $y = d$ is a *horizontal line* going through the point $(0, d)$.

3. Find the Slope of a Line

The slope of a line measures its steepness. It is the ratio of the vertical change (the change in y) to the horizontal change (the change in x). Slope is denoted by m.

For example, if a line has a slope of $\frac{2}{5}$, then the rate of change between two points on the line is a vertical change (change in y) of 2 units for every horizontal change (change in x) of 5 units.

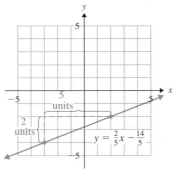

If a line has a *positive slope*, then it slopes *upward* from left to right. This means that as the value of x increases, the value of y increases as well. If a line has a *negative slope*, then it slopes *downward* from left to right. This means that as the value of x increases, the value of y decreases.

If we know two points on a line, we can find its slope.

The slope, m, of a line containing the points (x_1, y_1) and (x_2, y_2) is given by

$$m = \frac{\text{vertical change}}{\text{horizontal change}} = \frac{y_2 - y_1}{x_2 - x_1}$$

We can also think of slope as: $\dfrac{\text{rise}}{\text{run}}$ or $\dfrac{\text{change in } y}{\text{change in } x}$.

Example 2

Find the slope of the line containing $(-2, 3)$ and $(1, 9)$.

Solution

The slope formula is $m = \dfrac{y_2 - y_1}{x_2 - x_1}$.

Let $(x_1, y_1) = (-2, 3)$ and $(x_2, y_2) = (1, 9)$. Then,

$$m = \frac{9 - 3}{1 - (-2)} = \frac{6}{3} = 2$$

The slope is 2.

We have graphed lines by finding the intercepts and plotting another point. We can also use the slope to help us graph a line.

4. Graph a Line Given in Slope-Intercept Form

If we require that A, B, and C are integers and that A is positive, then $Ax + By = C$ is called the **standard form** of the equation of a line. Lines can take other forms, too.

Definition

The **slope-intercept form of a line** is $y = mx + b$, where m is the slope and $(0, b)$ is the y-intercept.

Example 3

Graph $y = 2x - 4$.

Solution

The equation is in slope-intercept form. Identify the slope and y-intercept.

$$y = 2x - 4$$

The slope $= m = 2$, and the y-intercept is $(0, -4)$.

Graph the line by first plotting the y-intercept and then by using the slope to locate another point on the line. Think of the slope as $m = \dfrac{2}{1} = \dfrac{\text{change in } y}{\text{change in } x}$. To get from the point $(0, -4)$ to another point on the line, move up 2 units and right 1 unit.

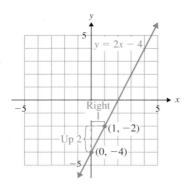

5. Write an Equation of a Line

In addition to knowing how to graph a line, it is important that we know how to write the equation of a line given certain information.

Here are some guidelines to follow when we want to write the equation of a line:

Writing Equations of Lines

If you are given

1) **The slope and y-intercept of the line**, use $y = mx + b$ and substitute those values into the equation.

2) **The slope of the line and a point on the line**, use the point-slope formula:

$$y - y_1 = m(x - x_1)$$

Substitute the slope for m and the point you are given for (x_1, y_1). Write your answer in slope-intercept or standard form.

3) **Two points on the line**, find the slope of the line and then use the slope and *either one* of the points in the point-slope formula. Write your answer in slope-intercept or standard form.

The equation of a *horizontal line* containing the point (c, d) is $y = d$.

The equation of a *vertical line* containing the point (c, d) is $x = c$.

Example 4

Find an equation of the line containing the point $(5, 1)$ with slope $= -3$. Express the answer in slope-intercept form.

Solution

First, ask yourself, "*What kind of information am I given?*" Since the problem tells us the slope of the line and a point on the line, we will use the point-slope formula.

Use $y - y_1 = m(x - x_1)$.

Substitute -3 for m. Substitute $(5, 1)$ for (x_1, y_1).

$$\begin{aligned} y - 1 &= -3(x - 5) && \text{Substitute } -3 \text{ for } m, 5 \text{ for } x_1, \text{ and } 1 \text{ for } y_1. \\ y - 1 &= -3x + 15 && \text{Distribute.} \\ y &= -3x + 16 && \text{Solve for } y. \end{aligned}$$

The equation is $y = -3x + 16$.

Often we need to write the equations of parallel or perpendicular lines. Remember,

---

Parallel lines have the same slopes and different y-intercepts.
Two lines are *perpendicular* if their slopes are negative reciprocals of each other.

6. Determine if a Relation Is a Function, and Find the Domain and Range

A set of ordered pairs like $\{(1, 40), (2.5, 100), (3, 120)\}$ is a *relation*.

Definition | A **relation** is any set of ordered pairs.

The **domain** of a relation is the set of all values of the independent variable (the first coordinates in the set of ordered pairs). The **range** of a relation is the set of all values of the dependent variable (the second coordinates in the set of ordered pairs).

The relation $\{(1, 40), (2.5, 100), (3, 120)\}$ is also a *function*.

Definition | A **function** is a special type of relation. If each element of the domain corresponds to *exactly one* element of the range, then the relation is a function.

Example 5

Identify the domain and range of each relation, and determine whether each relation is a function.

a) $\{(-5, 4), (-2, 1), (0, -2), (6, -7)\}$

b)

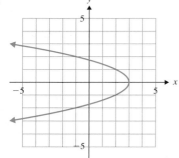

Solution

a) The *domain* is the set of first coordinates, $\{-5, -2, 0, 6\}$.

The *range* is the set of second coordinates, $\{4, 1, -2, -7\}$.

To determine whether $\{(-5, 4), (-2, 1), (0, -2), (6, -7)\}$ is a function ask yourself, "Does every first coordinate correspond to *exactly one* second coordinate?" *Yes.* So, this relation is a function.

b) The domain is $(-\infty, 3]$. The range is $(-\infty, \infty)$.

To determine if this graph represents a function, recall that we can use the *vertical line test*. The **vertical line test** says that if there is no vertical line that can be drawn through a graph so that it intersects the graph more than once, then the graph represents a function.

This graph fails the vertical line test because we can draw a vertical line through the graph that intersects it more than once. *This graph does* not *represent a function.*

7. Given an Equation, Determine Whether y Is a Function of x and Find the Domain

In an equation like $y = 5x$, we say that x is the independent variable and y is the dependent variable. That is, the value of y depends on the value of x.

If a relation is written as an equation so that y is in terms of x, then the **domain** is the set of all real numbers that can be substituted for the independent variable, x. The resulting set of real numbers obtained for y, the dependent variable, is the **range**.

The domain of a relation that is written as an equation, where y is in terms of x, is the set of all real numbers that can be substituted for the independent variable, x. When determining the domain of a relation, it can be helpful to keep these tips in mind.

1) Ask yourself, "Is there any number that *cannot* be substituted for x?"

2) If x is in the denominator of a fraction, determine what value of x will make the denominator equal 0 by setting the denominator equal to zero. Solve for x. This x-value is *not* in the domain.

Example 6

Determine whether $y = \dfrac{8}{x + 5}$ describes y as a function of x, and determine its domain.

Solution

Ask yourself, "Is there any number that *cannot* be substituted for x in $y = \dfrac{8}{x + 5}$?"

Look at the denominator. When will it equal 0? Set the denominator equal to 0 and solve for x to determine what value of x will make the denominator equal 0.

$$x + 5 = 0 \qquad \text{Set the denominator} = 0.$$
$$x = -5 \qquad \text{Solve.}$$

When $x = -5$, the denominator of $y = \dfrac{8}{x + 5}$ equals zero. The domain contains all real numbers *except* -5. Write the domain in interval notation as $(-\infty, -5) \cup (-5, \infty)$.

$y = \dfrac{8}{x + 5}$ *is a function.* For every value that can be substituted for x there is only one corresponding value of y.

8. Use Function Notation and Find Function Values

If y is a function of x, then we can use *function notation* to represent this relationship.

Definition

$y = f(x)$ is called **function notation**, and it is read as, "y equals f of x."
$y = f(x)$ means that y is a function of x (that is, y depends on x).

If such a relation is a function, then $f(x)$ can be used in place of y. $f(x)$ *is the same as* y. One special type of function is a *linear function*.

Definition	A **linear function** has the form $f(x) = mx + b$, where m and b are real numbers, m is the *slope* of the line, and $(0, b)$ is the *y-intercept*.

We graph linear functions just like we graph lines in the form $y = mx + b$, and we can evaluate a linear function for values of the variable. We call this **finding function values**. (We can find function values for any type of function not just linear functions.)

Example 7

Let $f(x) = -2x + 5$. Find $f(3)$.

Solution

To find $f(3)$ (read as "f of 3") means to find the value of the function when $x = 3$.

$$f(x) = -2x + 5$$
$$f(3) = -2(3) + 5 \qquad \text{Substitute 3 for } x.$$
$$f(3) = -1$$

We can also say that the ordered pair $(3, -1)$ satisfies $f(x) = -2x + 5$, where the ordered pair represents $(x, f(x))$.

A4 Exercises

Boost your grade at mathzone.com! MathZone > Practice Problems > NetTutor > Self-Test > e-Professors > Videos

---

Determine if each ordered pair is a solution of the equation $3x - 5y = 7$.

1) $(-1, -2)$ 2) $(2, 3)$

Make a table of values, and graph each equation.

3) $y = x - 4$ 4) $2x - 3y = 9$

5) $x = -3$ 6) $y = 2$

Graph each equation by finding the intercepts and at least one other point.

7) $y = -\dfrac{3}{2}x + 3$ 8) $y = x - 4$

9) $3x - y = 2$ 10) $3x + 4y = 4$

Use the slope formula to find the slope of the line containing each pair of points.

11) $(5, 2)$ and $(-2, -3)$ 12) $(-6, 2)$ and $(4, -3)$

Graph the line containing the given point and with the given slope.

13) $(-2, -5); m = \dfrac{1}{4}$ 14) $(0, 1); m = -\dfrac{2}{5}$

15) $(2, 3)$; slope is undefined

16) $(-3, -2); m = 0$

Each of the following equations is in slope-intercept form. Identify the slope and the y-intercept, then graph each line using this information.

17) $y = -4x + 1$ 18) $y = -x + 3$

19) $y = -x$ 20) $y = \dfrac{2}{3}x$

Graph each line using any method.

21) $x - y = 5$ 22) $3x + 5y = 10$

23) $7x + 2y = 6$ 24) $y - 2x = 3$

25) $y = \dfrac{5}{3}x - 2$

26) $y = x$

Write the slope-intercept form of the equation of the line, if possible, given the following information. $m = \dfrac{y_2 - y_1}{x_2 - x_1}$

27) $m = \dfrac{3}{8}$ and y-intercept $(0, 4)$

28) $m = -2$ and contains $(-3, 5)$

29) contains $(-1, 6)$ and $(7, 2)$

30) contains $(-8, -1)$ and $(-2, -11)$

31) $m = 0$ and contains $(4, 7)$

32) slope undefined and contains $(-3, 2)$

Write the standard form of the equation of the line given the following information. $Ax + By = C$

33) $m = -6$ and contains $(0, 2)$

34) $m = 1$ and contains $(3, 8)$

35) contains $(-4, -1)$ and $(-1, 8)$

36) contains $(0, -3)$ and $(2, 7)$

Write the slope-intercept form, if possible, of the equation of the line meeting the given conditions.

37) perpendicular to $y = \dfrac{3}{2}x - 6$ containing $(9, -1)$

38) parallel to $y = \dfrac{1}{4}x - 5$ containing $(8, 2)$

39) parallel to $x = -4$ containing $(-1, 4)$

40) perpendicular to $y = 3$ containing, $(5, -2)$

41) Naresh works in sales, and his income is a combination of salary and commission. He earns $24,000 per year plus 10% of his total sales. The equation $I = 0.10s + 24,000$ represents his total income I, in dollars, when his sales total s dollars.

Naresh's Income

a) What is the I-intercept? What does it mean in the context of the problem?

b) What is the slope? What does it mean in the context of the problem?

c) Use the graph to find Naresh's income if his total sales are $100,000. Confirm your answer using the equation.

Identify the domain and range of each relation, and determine whether each relation is a function.

42) $\{(-4, 16), (-1, 7), (1, 1), (3, -5)\}$

43)

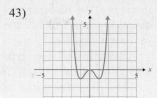

44)

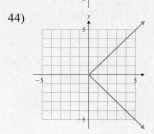

Determine whether each relation describes y as a function of x, and determine the domain of the relation.

45) $y = x + 8$

46) $y = \dfrac{7}{x}$

47) $y^2 = x$

48) $y = -\dfrac{2}{5x + 6}$

Let $f(x) = 2x - 9$ and $g(x) = x^2 - 6x - 4$. Find the following function values.

49) $f(3)$

50) $g(-3)$

51) $f(z)$

52) $g(t)$

53) $f(c + 3)$

54) $f(m - 4)$

55) $f(x) = -3x + 5$. Find x so that $f(x) = -7$.

56) $k(x) = 4x - 3$. Find x so that $k(x) = 9$.

Graph each function by making a table of values and plotting points.

57) $f(x) = x - 3$

58) $g(x) = 3x - 2$

Graph each function by finding the x- and y-intercepts and one other point.

59) $f(x) = -x + 3$

60) $g(x) = -2x + 3$

Graph each function using the slope and y-intercept.

61) $h(x) = -2x - 1$

62) $g(x) = 4x - 5$

63) A plane travels at a constant speed of 420 mph. The distance D (in miles) that the plane travels after t hours can be defined by the function

$$D(t) = 420t$$

a) How far will the plane travel after 2 hr?

b) How long does it take the plane to travel 1890 mi?

c) Graph the function.

Section A5 Solving Systems of Linear Equations

Objectives

1. Solve a System of Equations by Graphing

2. Solve a System of Equations by Substitution

3. Solve a System of Equations Using the Elimination Method

4. Solve an Applied Problem Using a System of Two Equations in Two Variables

5. Solve a System of Linear Equations in Three Variables

A **system of linear equations** consists of two or more linear equations with the same variables. We begin by learning how to solve systems of two equations in two variables. Later in this section, we will discuss how to solve a system of three equations in three variables.

A **solution of a system** of two equations in two variables is an ordered pair that is a solution of each equation in the system.

We will review three methods for solving a system of linear equations:

1. Graphing
2. Substitution
3. Elimination

1. Solve a System of Equations by Graphing

When solving a system of linear equations by graphing, *the point of intersection of the two lines is the solution of the system.* A system that has one solution is called a **consistent** system.

 Example 1

Solve the system by graphing.

$$2x + y = 1$$
$$y = x + 4$$

Solution

Carefully graph each line on the same axes. As you can see, the lines intersect at the point $(-1, 3)$. Therefore, the solution of the system is $(-1, 3)$. You can verify by substituting the ordered pair into each equation to see that it satisfies each equation.

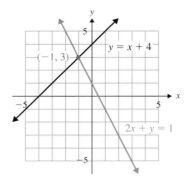

The lines in Example 1 intersect at one point, so the system has one solution. Some systems, however, have no solution while others have an infinite number of solutions.

> When solving a system of equations by graphing, if the lines are parallel then the system has **no solution**. We write this as $\varnothing$. Furthermore, we say that a system that has no solution is an **inconsistent system**.
>
> When solving a system of equations by graphing, if the graph of each equation is the same line then the system has an **infinite number of solutions**. Furthermore, we say that the system is **dependent**.

2. Solve a System of Equations by Substitution

Another method we can use to solve a system of equations is substitution. This method is especially good when one of the variables has a coefficient of 1 or -1.

Here are the steps we can follow to solve a system by substitution:

Solving a System of Equations by Substitution

Step 1: Solve one of the equations for one of the variables. If possible, solve for a variable that has a coefficient of 1 or -1.

Step 2: Substitute the expression in *step 1* into the *other* equation. The equation you obtain should contain only one variable.

Step 3: Solve the equation in one variable from *step 2*.

Step 4: Substitute the value found in *step 3* into either of the equations to obtain the value of the other variable.

Step 5: Check the values in each of the original equations, and write the solution as an ordered pair.

Example 2

Solve the system by substitution.

$$x + 3y = -5 \qquad (1)$$
$$3x + 2y = 6 \qquad (2)$$

Solution

We number the equations to make the process easier to follow. Let's follow the steps listed above.

Step 1: For which variable should we solve? The x in the first equation is the only variable with a coefficient of 1 or -1. Therefore, we will solve the first equation for x.

$$x + 3y = -5 \qquad (1) \qquad \text{First equation}$$
$$x = -5 - 3y \qquad \text{Subtract } 3y.$$

Step 2: Substitute $-5 - 3y$ for the x in equation (2).

$$3x + 2y = 6 \qquad (2)$$
$$3(-5 - 3y) + 2y = 6 \qquad \text{Substitute.}$$

Step 3: Solve the equation above for y.

$$3(-5 - 3y) + 2y = 6$$
$$-15 - 9y + 2y = 6 \qquad \text{Distribute.}$$
$$-7y - 15 = 6$$
$$-7y = 21$$
$$y = -3$$

Step 4:   To determine the value of x we can substitute -3 for y in either equation. We will substitute it in equation (1).

$$x + 3y = -5 \qquad (1)$$
$$x + 3(-3) = -5 \qquad\qquad \text{Substitute.}$$
$$x - 9 = -5$$
$$x = 4$$

Step 5:   The check is left to the reader. The solution of the system is $(4, -3)$.

When we were solving systems by graphing, we learned that some systems are inconsistent (have no solution) and some systems are dependent (have an infinite number of solutions).

   If we are trying to solve a system by substitution and it is either inconsistent or dependent, then somewhere in the process of solving the system, the variables are eliminated.

   When we are solving a system of equations and the variables are eliminated:

1)   If we get a **false statement**, like $3 = 5$, then the system has **no solution**.

2)   If we get a **true statement**, like $-4 = -4$, then the system has an **infinite number of solutions**.

3. Solve a System of Equations Using the Elimination Method

The **elimination method** (or **addition method**) is another technique we can use to solve a system of equations. It is based on the addition property of equality which says that we can add the same quantity to each side of an equation and preserve the equality.

Addition Property of Equality: If $a = b$, then $a + c = b + c$.

We can extend this idea by saying that we can add *equal* quantities to each side of an equation and still preserve the equality.

If $a = b$ and $c = d$, then $a + c = b + d$.

The object of the elimination method is to add the equations (or multiples of one or both of the equations) so that one variable is eliminated. Then, we can solve for the remaining variable.

Solving a System of Two Linear Equations by the Elimination Method

Step 1:   Write each equation in the form $Ax + By = C$.

Step 2:   Look at the equations carefully to determine which variable to eliminate. If necessary, multiply one or both of the equations by a number so that the coefficients of the variable to be eliminated are negatives of one another.

Step 3:   Add the equations, and solve for the remaining variable.

Step 4:   Substitute the value found in step 3 into either of the original equations to find the value of the other variable.

Step 5:   Check the solution in each of the original equations.

Example 3

Solve the system using the elimination method.

$$-3x + 4y = -8 \quad (1)$$
$$2x + 5y = 13 \quad (2)$$

Solution

Step 1: *Write each equation in the form $Ax + By = C$.*
Each equation is written in the form $Ax + By = C$.

Step 2: *Determine which variable to eliminate from equations (1) and (2). Often, it is easier to eliminate the variable with the smaller coefficients. Therefore, we will eliminate x.*

In equation (1) the coefficient of x is -3, and in equation (2) the coefficient of x is 2. *Multiply equation (1) by 2 and equation (2) by 3.* The x-coefficient in equation (1) will be -6, and the x-coefficient in equation (2) will be 6. When we add the equations, the x will be eliminated.

$$2(-3x + 4y) = 2(-8) \qquad \text{2 times equation (1)} \qquad \longrightarrow \qquad -6x + 8y = -16$$
$$3(2x + 5y) = 3(13) \qquad \text{3 times equation (2)} \qquad\qquad\qquad 6x + 15y = 39$$

Step 3: *Add the resulting equations to eliminate x. Solve for y.*

$$\begin{array}{rcr} -6x + 8y &=& -16 \\ + \quad 6x + 15y &=& 39 \\ \hline 23y &=& 23 \\ y &=& 1 \end{array}$$

Step 4: *Substitute $y = 1$ into equation (1) and solve for x.*

$$\begin{array}{rl} -3x + 4y = -8 & \quad (1) \\ -3x + 4(1) = -8 & \quad \text{Substitute 1 for } y. \\ -3x + 4 = -8 & \\ -3x = -12 & \quad \text{Subtract 4.} \\ x = 4 & \quad \text{Divide by } -3. \end{array}$$

Step 5: *Check* to verify that (4, 1) satisfies each of the original equations. The solution is (4, 1).

When solving a system by graphing and by substitution, we learned that some systems have no solution and some have an infinite number of solutions. The same can be true when solving using the elimination method.

4. Solve an Applied Problem Using a System of Two Equations in Two Variables

In Section A3 (and in Chapter 3) we introduced the five Steps for Solving Applied Problems. We defined unknown quantities in terms of *one* variable to write a linear equation to solve a problem.

Sometimes, it is easier to use *two* variables and a system of *two* equations to solve an applied problem. Here are some steps you can follow to solve an applied problem using a system of **two equations in two variables**. These steps can be applied to solving applied problems involving a system of three equations as well.

Solving an Applied Problem Using a System of Equations

Step 1: **Read the problem carefully, twice.** Draw a picture if applicable.

Step 2: Identify what you are being asked to find. **Define the unknowns; that is, write down** what each variable represents. If applicable, label a picture with the variables.

Step 3: **Write a system of equations relating the variables.** It may be helpful to begin by writing the equations in words.

Step 4: **Solve the system.**

Step 5: **Interpret the meaning of the solution as it relates to the problem.** Be sure your answer makes sense.

Example 4

Write a system of equations and solve.

On her iPod, Shelby has 27 more hip-hop songs than reggae songs. If she has a total of 85 hip-hop and reggae tunes on her iPod, how many of each type does she have?

Solution

Step 1: Read the problem carefully, twice.

Step 2: We must find the number of hip-hop and reggae songs on Shelby's iPod. Define the unknowns; assign a variable to represent each unknown quantity.

$$x = \text{number of hip-hop songs}$$
$$y = \text{number of reggae songs}$$

Step 3: Write a system of equations relating the variables.

We must write two equations. Let's think of the equations in English first. Then, we can translate them to algebraic equations.

i) *To get one equation,* use the information that says Shelby has 27 more hip-hop songs than reggae songs.

Number of hip-hop songs	is	27 more than	Number of reggae songs
↓	↓	↓	↓
x	$=$	$27+$	y

One equation is $x = 27 + y$.

ii) *To get the second equation,* use the information that says Shelby has a total of 85 hip-hop and reggae songs on her iPod. Write an equation in words, then translate it to an algebraic equation.

Number of hip-hop songs	+	Number of reggae songs	=	Number of hip-hop *and* reggae songs
x	$+$	y	$=$	85

The second equation is $x + y = 85$.
The system of equations is $x = 27 + y$.
$$x + y = 85$$

Step 4: Solve the system.

Use substitution.

$$x + y = 85 \qquad \text{Second equation}$$
$$(27 + y) + y = 85 \qquad \text{Substitute } 27 + y \text{ for } x.$$
$$27 + 2y = 85 \qquad \text{Combine like terms.}$$
$$2y = 58 \qquad \text{Subtract 27 from each side.}$$
$$y = 29$$

Find x by substituting $y = 29$ into $x = 27 + y$.

$$x = 27 + 29$$
$$x = 56$$

The solution to the system is $(56, 29)$.

Step 5: Interpret the meaning of the solution as it relates to the problem.

Shelby has 56 hip-hop songs and 29 reggae songs on her iPod.

Does the answer make sense? Yes. The total number of hip-hop and reggae songs is 85 and $56 + 29 = 85$, and the number of hip-hop songs is 27 more than the number of reggae songs: $56 = 27 + 29$.

5. Solve a System of Linear Equations in Three Variables

We will extend our study of solving a system of two equations in two variables to solving a system of three equations in three variables.

> **Definition** A **linear equation in three variables** is an equation of the form $Ax + By + Cz = D$, where A, B, and C are not all zero and where A, B, C, and D are real numbers. Solutions to this type of an equation are **ordered triples** of the form (x, y, z).

An example of a linear equation in three variables is

$$x + 2y - 3z = -8$$

One solution to the equation is $(2, 1, 4)$, since $2 + 2(1) - 3(4) = -8$

Like systems of linear equations in two variables, systems of linear equations in *three* variables can have *one solution* (the system is consistent), *no solution* (the system is inconsistent), or *infinitely many solutions* (the system is dependent).

Example 5

Solve. ① $-x + 2y + z = 3$
② $3x + y - 4z = 13$
③ $2x - 2y + 3z = -5$

Solution

Step 1: **Label** the equations ①, ②, and ③.

Step 2: **Choose a variable to eliminate.**

We will eliminate y from *two* sets of *two* equations.

a) *Equations* ① *and* ③. Add the equations to eliminate y. Label the resulting equation $\boxed{A}$.

$$
\begin{array}{ll}
① & \quad -x + 2y + z = 3 \\
③ & +\ \underline{2x - 2y + 3z = -5} \\
\boxed{A} & \quad x + 4z = -2
\end{array}
$$

b) *Equations* ② *and* ③. To eliminate y, multiply equation ② by 2 and add it to equation ③. Label the resulting equation $\boxed{B}$.

$$
\begin{array}{ll}
2 \times ② & \quad 6x + 2y - 8z = 26 \\
③ & +\ \underline{2x - 2y + 3z = -5} \\
\boxed{B} & \quad 8x - 5z = 21
\end{array}
$$

> Equations $\boxed{A}$ and $\boxed{B}$ contain only *two* variables, and they are the *same* variables, x and z.

Step 3: **Use the elimination method to eliminate a variable from equations $\boxed{A}$ and $\boxed{B}$.**

We will eliminate x from equations $\boxed{A}$ and $\boxed{B}$. Multiply equation $\boxed{A}$ by -8 and add it to equation $\boxed{B}$.

$$
\begin{array}{ll}
-8 \times \boxed{A} & \quad -8x - 32z = 16 \\
\boxed{B} & +\ \underline{8x - 5z = 21} \\
& \qquad\qquad -37z = 37 \\
& \qquad\qquad\ \boxed{z = -1} \qquad \text{Divide by } -37.
\end{array}
$$

Step 4: **Find the value of another variable** by substituting $z = -1$ into either equation $\boxed{A}$ or $\boxed{B}$.

We will substitute $z = -1$ into equation $\boxed{A}$.

$$
\begin{array}{ll}
\boxed{A} & \quad x + 4z = -2 \\
& \quad x + 4(-1) = -2 \qquad \text{Substitute } -1 \text{ for } z. \\
& \qquad\quad x - 4 = -2 \qquad \text{Multiply.} \\
& \qquad\qquad \boxed{x = 2} \qquad\quad \text{Add 4.}
\end{array}
$$

Step 5: **Find the value of the third variable** by substituting $x = 2$ and $z = -1$ into either equation ①, ②, or ③.

We will substitute $x = 2$ and $z = -1$ into equation ① to solve for y.

$$
\begin{array}{ll}
① & \quad -x + 2y + z = 3 \\
& \quad -(2) + 2y + (-1) = 3 \qquad \text{Substitute 2 for } x \text{ and } -1 \text{ for } z. \\
& \qquad\quad -2 + 2y - 1 = 3 \\
& \qquad\qquad\quad 2y - 3 = 3 \qquad \text{Combine like terms.} \\
& \qquad\qquad\qquad\ 2y = 6 \qquad \text{Add 3.} \\
& \qquad\qquad\qquad\ \boxed{y = 3} \qquad \text{Divide by 2.}
\end{array}
$$

Step 6: **Check** the solution, $(2, 3, -1)$ in each of the original equations. This will be left to the student.

The solution is $(2, 3, -1)$.

When one or more of the equations in the system contain only two variables, we can modify the steps we have just used to solve the system.

To solve applications involving a system of three linear equations in three variables, we use the same process that is outlined for systems of two equations on p. A-34 of this section.

A5 Exercises

Determine if the ordered pair is a solution of the system of equations.

1) $2x - 5y = 4$
   $x + y = 9$
   $(7, 2)$

2) $-x + 4y = 7$
   $2x - y = -5$
   $(-3, 1)$

Solve each system of equations by graphing. If the system is inconsistent or dependent, identify it as such.

3) $y = \dfrac{2}{3}x - 1$
   $x + y = 4$

4) $x + 2y = -2$
   $y = -2x + 2$

5) $2y - 2x = -3$
   $x = y - 2$

6) $2x - 6y = 5$
   $18y = 6x - 15$

Solve each system by substitution.

7) $y = 2x - 15$
   $-x + 3y = -5$

8) $4x + 5y = -13$
   $x = 3y + 1$

9) $x = 2 - 5y$
   $-4x - 20y = -8$

10) $4x - y = 7$
    $-8x + 2y = 9$

Solve each system using the elimination method.

11) $12x + 5y = 2$
    $-2x + 3y = 15$

12) $7x - 2y = -8$
    $5x + 4y = -22$

13) $-8x + 12y = 3$
    $6x - 9y = 1$

14) $\dfrac{1}{4}x + \dfrac{3}{2}y = -\dfrac{1}{2}$
    $\dfrac{1}{6}x + \dfrac{2}{3}y = \dfrac{1}{3}$

Solve each system using any method.

15) $\dfrac{1}{2}x = -\dfrac{1}{3}y - \dfrac{4}{3}$
    $\dfrac{3}{4}x + \dfrac{1}{2}y = \dfrac{1}{2}$

16) $3x + 2y = 6$
    $3x - 8y = 1$

17) $7(x + 1) + 5y = 6(x + 2) - 4$
    $2x + 3(3y + 2) = 2(y + 1)$

18) $y + 4 = 2(2y - x) + 1$
    $7x - 2(2y + 5) = 2(x - 4) + 1$

19) $y = \dfrac{2}{5}x$
    $x - 5y = 15$

20) $4y - 3(x - 1) = 2(y + 2x)$
    $6(x + y) + 4x = x + 2y - 6$

21) $0.04x - 0.07y = 0.22$
    $0.6x + 0.5y = 0.2$

22) $6x - 3(2y + 4) = 3(y + 2)$
    $6y = 4(x - 3)$

Write a system of equations and solve.

23) The tray table on the back of a seat on an airplane is 7 in. longer than it is wide. Find the length and width of the tray table if its perimeter is 54 in.

24) Find the measures of angles x and y if the measure of angle y is 34° less than the measure of angle x and if the angles are related according to the figure below.

25) How many ounces of an 18% alcohol solution and how many ounces of an 8% solution should be mixed to obtain 60 oz of a 12% alcohol solution?

26) On a round-trip flight between Chicago and Denver, two economy-class tickets and one business-class ticket cost $1050 while one economy ticket and two business tickets cost $1500. Find the cost of an economy-class ticket and the cost of a business-class ticket.

27) A passenger train and a freight train leave cities that are 300 mi apart and travel toward each other. The passenger train is traveling 15 mph faster than the freight train. Find the speed of each train if they pass each other after 4 hr.

28) Lydia inherited $12,000 and invested some of it in an account earning 5% simple interest and put the rest of it in an account earning 8% simple interest. If she earned a total of $690 in interest after 1 yr, how much did she invest in each account?

Solve each system.

29) $\begin{aligned} x + 4y + 2z &= 12 \\ 3x - y + z &= -2 \\ -2x + 3y - z &= 2 \end{aligned}$

30) $\begin{aligned} 2x + 6y - z &= -2 \\ -x + 9y + 2z &= -1 \\ 2x - 3y + z &= 12 \end{aligned}$

31) $\begin{aligned} a - 4b + 2c &= 1 \\ -2a + 3b - c &= -8 \\ -a + 2b + c &= -7 \end{aligned}$

32) $\begin{aligned} -3a + b + 2c &= 1 \\ 4a - b - 6c &= 5 \\ 6a - 2b - 4c &= 3 \end{aligned}$

33) $\begin{aligned} -3a + 2b + c &= 5 \\ 5a - b - 2c &= -6 \\ 4a - c &= 1 \end{aligned}$

34) $\begin{aligned} 2x + 9y &= -7 \\ 3y - 4z &= 11 \\ 2x - z &= -14 \end{aligned}$

Write a system of three equations and solve.

35) During their senior year of high school, Ryan, Seth, and Summer applied to colleges. Summer applied to three more schools than Seth while Seth applied to five fewer schools than Ryan. All together, the friends sent out 11 applications. To how many schools did each one apply?

36) The measure of the smallest angle of a triangle is half the measure of the largest angle. The third angle measures 10° less than the largest angle. Find the measures of each angle of the triangle. (Hint: Recall that the sum of the measures of the angles of a triangle is 180°.)

Section A6 Polynomials

Objectives

1. Learn the Vocabulary Associated with Polynomials

2. Add and Subtract Polynomials

3. Define a Polynomial Function and Find Function Values

4. Multiply Polynomials

5. Divide Polynomials

1. Learn the Vocabulary Associated with Polynomials

Definition

A **polynomial in x** is the sum of a finite number of terms of the form ax^n, where n is a whole number and a is a real number. (The exponents must be whole numbers.)

An example of a polynomial is $7x^3 + x^2 - 8x + \dfrac{2}{3}$. Let's look closely at some characteristics of this polynomial.

1) The polynomial is written in **descending powers of x**, since the powers of x decrease from left to right. Generally, we write polynomials in descending powers of the variable.

2) The **degree of a term** equals the exponent on its variable. (If a term has more than one variable, the degree equals the *sum* of the exponents on the variables.)

Term	Degree
$7x^3$	3
x^2	2
$-8x$	1
$\dfrac{2}{3}$	0

$\cdot$ $\dfrac{2}{3} = \dfrac{2}{3}x^0$ since $x^0 = 1$.

3) The **degree of the polynomial** equals the highest degree of any nonzero term. The degree of $7x^3 + x^2 - 8x + \dfrac{2}{3}$ is 3. Or, we say that this is a *third-degree polynomial*.

A **monomial** is a polynomial which consists of one term. A **binomial** is a polynomial that consists of exactly two terms. A polynomial that consists of exactly three terms is called a **trinomial**.

2. Add and Subtract Polynomials

To add polynomials, add like terms. Polynomials can be added horizontally or vertically.

Example 1

Add the polynomials $(3a^2b^2 + 8a^2b - 10ab - 9) + (2a^2b^2 + 3ab - 1)$.

Solution

We will add these polynomials vertically. Line up like terms in columns and add.

$$
\begin{array}{r}
3a^2b^2 + 8a^2b - 10ab - 9 \\
+\,2a^2b^2 \qquad\quad +\ 3ab - 1 \\
\hline
5a^2b^2 + 8a^2b - 7ab - 10
\end{array}
$$

To subtract two polynomials, change the sign of each term in the second polynomial. Then, add the polynomials.

3. Define a Polynomial Function and Find Function Values

We can use function notation to represent a polynomial like $x^2 - 12x - 7$ since each value substituted for the variable produces *only one value* of the expression.

$f(x) = x^2 - 12x - 7$ is a **polynomial function** since $x^2 - 12x - 7$ is a polynomial.

Finding $f(-3)$ when $f(x) = x^2 - 12x - 7$ is the same as evaluating $x^2 - 12x - 7$ when $x = -3$.

Example 2

If $f(x) = x^2 - 12x - 7$, find $f(-3)$.

Solution

Substitute -3 for x.

$$
\begin{aligned}
f(x) &= x^2 - 12x - 7 \\
f(-3) &= (-3)^2 - 12(-3) - 7 \\
&= 9 + 36 - 7 \\
&= 45 - 7 \\
&= 38
\end{aligned}
$$

4. Multiply Polynomials

Whenever we multiply polynomials, we use the distributive property. The *way* in which we use it, however, may vary depending on what types of polynomials are in the product.

Multiplying a Monomial and a Polynomial

To find the product of a monomial and a larger polynomial, multiply each term of the polynomial by the monomial.

Example 3

Multiply $7z^3(z^2 - 5z + 3)$.

Solution

$$7z^3(z^2 - 5z + 3) = (7z^3)(z^2) + (7z^3)(-5z) + (7z^3)(3) \qquad \text{Distribute.}$$
$$= 7z^5 - 35z^4 + 21z^3 \qquad \text{Multiply.}$$

Multiplying Two Polynomials

> To multiply two polynomials, multiply each term in the second polynomial by each term in the first polynomial. Then, combine like terms. The answer should be written in descending powers.

Example 4

Multiply $(4c - 5)(3c^2 + 8c - 2)$.

Solution

We will multiply each term in the second polynomial by the $4c$ in the first polynomial. Then, multiply each term in $(3c^2 + 8c - 2)$ by the -5 in $(4c - 5)$. Then, add like terms.

$$(4c - 5)(3c^2 + 8c - 2)$$
$$= (4c)(3c^2) + (4c)(8c) + (4c)(-2) + (-5)(3c^2) \qquad \text{Distribute.}$$
$$\quad + (-5)(8c) + (-5)(-2)$$
$$= 12c^3 + 32c^2 - 8c - 15c^2 - 40c + 10 \qquad \text{Multiply.}$$
$$= 12c^3 + 17c^2 - 48c + 10 \qquad \text{Combine like terms.}$$

Multiplying Two Binomials

We can use the distributive property as we did above to multiply two binomials such as $(x + 6)(x + 2)$.

$$(x + 6)(x + 2) = (x)(x) + (x)(2) + (6)(x) + (6)(2) \qquad \text{Distribute.}$$
$$= x^2 + 2x + 6x + 12 \qquad \text{Multiply.}$$
$$= x^2 + 8x + 12 \qquad \text{Combine like terms.}$$

Or, we can apply the distributive property in a different way—we can use FOIL. **FOIL** stands for **First Outside Inner Last**.

The letters in the word FOIL tell us how to multiply the terms in the binomials. Then, we add like terms.

Example 5

Use FOIL to multiply $(x + 6)(x + 2)$.

Solution

$$(x + 6)(x + 2) = (x + 6)(x + 2)$$

First, Outer, Inner, Last

F: Multiply the **First** terms of the binomials: $(x)(x) = x^2$
O: Multiply the **Outer** terms of the binomials: $(x)(2) = 2x$
I: Multiply the **Inner** terms of the binomials: $(6)(x) = 6x$
L: Multiply the **Last** terms of the binomials: $(6)(2) = 12$

After multiplying, add like terms.

$$(x + 6)(x + 2) = x^2 + 2x + 6x + 12$$
$$= x^2 + 8x + 12$$

Special Products

Binomial multiplication is very common in algebra, and there are three types of special products we explain here. The first special product we will review is

$$(a + b)(a - b) = a^2 - b^2$$

Example 6

Multiply $(q + 7)(q - 7)$.

Solution

$(q + 7)(q - 7)$ is in the form $(a + b)(a - b)$, where $a = q$ and $b = 7$.

$$(q + 7)(q - 7) = q^2 - 7^2$$
$$= q^2 - 49$$

Two more special products involve squaring a binomial like $(a + b)^2$. We have the following formulas for the square of binomials:

$$(a + b)^2 = a^2 + 2ab + b^2$$
$$(a - b)^2 = a^2 - 2ab + b^2$$

We can think of the formulas in words as:

> To square a binomial, you square the first term, square the second term, then multiply 2 times the first term times the second term and add.

Finding the product of a binomial raised to a power is also called *expanding* the binomial.

Example 7

Expand $(k + 8)^2$.

Solution

Using the formula $(a + b)^2 = a^2 + 2ab + b^2$ with $a = k$ and $b = 8$, we get

$$(k + 8)^2 = \underset{\uparrow}{k^2} + \underset{\uparrow}{2(k)(8)} + \underset{\uparrow}{8^2} = k^2 + 16k + 64$$

Square the first term.	Two times first term times second term	Square the second term.

5. Divide Polynomials

The last operation we will discuss is division of polynomials. We will break up this topic into two parts, division by a monomial and division by a polynomial containing two or more terms.

Dividing a Polynomial by a Monomial

> To divide a polynomial by a monomial, divide each term in the polynomial by the monomial and simplify.
>
> $$\frac{a + b}{c} = \frac{a}{c} + \frac{b}{c} \quad (c \neq 0)$$

Example 8

Divide $\dfrac{12x^3 - 8x^2 + 2x}{2x}$.

Solution

Since the polynomial in the numerator is being divided by a *monomial*, we will divide each term in the numerator by $2x$.

$$\frac{12x^3 - 8x^2 + 2x}{2x} = \frac{12x^3}{2x} - \frac{8x^2}{2x} + \frac{2x}{2x}$$
$$= 6x^2 - 4x + 1 \qquad \text{Simplify.}$$

We can check our answer the same way we can check any division problem—we can multiply the quotient by the divisor, and the result should be the dividend.

Check: $2x(6x^2 - 4x + 1) = 12x^3 - 8x^2 + 2x$ ✓

Dividing a Polynomial by a Polynomial

When dividing a polynomial by a polynomial containing two or more terms, we use *long division of polynomials.* When we perform long division, the polynomials must be written so that the exponents are in descending order.

Example 9

Divide $\dfrac{2x^2 + 17x + 30}{x + 6}$.

Solution

First, notice that we are dividing by more than one term. That tells us to use long division of polynomials.

$$
\begin{array}{r}
2x \\
x + 6 \overline{)\; 2x^2 + 17x + 30} \\
-(2x^2 + 12x) \quad \downarrow \\
\hline
5x + 30
\end{array}
$$

1) By what do we multiply x to get $2x^2$? $2x$
    Line up terms in the quotient according to exponents, so write $2x$ above $17x$.

2) Multiply $2x$ by $(x + 6)$. $2x(x + 6) = 2x^2 + 12x$

3) Subtract: $(2x^2 + 17x) - (2x^2 + 12x) = 5x$.

4) Bring down the $+30$.

Start the process again.

$$
\begin{array}{r}
2x + 5 \\
x + 6 \overline{)\; 2x^2 + 17x + 30} \\
-(2x^2 + 12x) \\
\hline
5x + 30 \\
-(5x + 30) \\
\hline
0
\end{array}
$$

1) By what do we multiply x to get $5x$? 5
    Write $+5$ above $+30$.

2) Multiply 5 by $(x + 6)$. $5(x + 6) = 5x + 30$

3) Subtract: $(5x + 30) - (5x + 30) = 0$.

$$\frac{2x^2 + 17x + 30}{x + 6} = 2x + 5 \quad \text{(There is no remainder.)}$$

Check: $(x + 6)(2x + 5) = 2x^2 + 5x + 12x + 30 = 2x^2 + 17x + 30$ ✓

When we are dividing polynomials, we must watch out for missing terms. If a dividend is missing one or more terms, we put them into the dividend with coefficients of zero.

Synthetic Division

When we divide a polynomial by a binomial of the form $x - c$, another method called *synthetic division* can be used. Synthetic division uses only the numerical coefficients of the variables to find the quotient.

Example 10

Use synthetic division to divide $(2x^3 - x^2 - 16x + 8)$ by $(x - 3)$.

Solution

Remember, in order to be able to use synthetic division, the divisor must be in the form $x - c$. $x - 3$ is in the form $x - c$, and $c = 3$.

Write the 3 in the open box, and then write the coefficients of the dividend. Skip a line and draw a horizontal line under the first coefficient. Bring down the 2. Then, multiply the 2 by 3 to get 6. Add the -1 and 6 to get 5.

$$
\begin{array}{r|rrrr}
3 & 2 & -1 & -16 & 8 \\
  &   & 6 &   &   \\
\hline
  & 2 & 5 &   &
\end{array}
\qquad 3 \cdot 2 = 6; \quad -1 + 6 = 5
$$

We continue working in this way until we get

$$
\begin{array}{r|rrrr}
 & \multicolumn{4}{c}{\overbrace{2x^3 \; -x^2 \; -16x \; + 8}^{\text{Dividend}}} \\
3 & 2 & -1 & -16 & 8 \\
  &   & 6 & 15 & -3 \\
\hline
  & 2 & 5 & -1 & 5 \;\rightarrow \text{Remainder}
\end{array}
$$

$$\underbrace{2x^2 \; +5x \; -1}_{\text{Quotient}}$$

The numbers in the last row represent the quotient and the remainder. The last number is the remainder. The numbers before it are the coefficients of the quotient. *The degree of the quotient is one less than the degree of the dividend.*

$$(2x^3 - x^2 - 16x + 8) \div (x - 3) = 2x^2 + 5x - 1 + \frac{5}{x - 3}$$

A6 Exercises

Boost your grade at mathzone.com! MathZone

> Practice Problems   > NetTutor   > e-Professors   > Videos
> Self-Test

Evaluate.

1) $(-2)^3 \cdot (-2)^2$

2) $\dfrac{5^{11}}{5^{14}}$

3) $\left(\dfrac{3}{4}\right)^{-3}$

Simplify. Assume all variables represent nonzero real numbers. The answer should not contain negative exponents.

4) $(w^3)^5$

5) $(9z^8)^2$

6) $\left(-\dfrac{3}{2}xy^7\right)^3$

7) $\dfrac{40t^{11}}{8t}$

8) $\dfrac{9a^4}{18a^{-5}}$

9) $\left(\dfrac{5n^6}{m^2}\right)^{-3}$

10) Identify each term in the polynomial, the coefficient and degree of each term, and the degree of the polynomial.

$$5x^3 - x^2 - 8x + 6$$

11) Evaluate $-c^2 + 7c + 10$ for $c = -2$.

Perform the operations and simplify.

12) $(8y^2 - 6y + 5) + (3y^2 + 2y - 3)$

13) $(7a^2b^2 - a^2b + 9ab + 14)$
   $\qquad - (3a^2b^2 + 14a^2b - 5ab^2 - 8ab + 20)$

14) Subtract $\left(\dfrac{4}{15}w^2 + \dfrac{8}{9}w - \dfrac{3}{4}\right)$ from $\left(\dfrac{4}{5}w^2 + \dfrac{1}{6}w + \dfrac{11}{4}\right)$.

15) If $f(x) = x^2 + 8x - 11$, find

   a) $f(5)$ 　　　　　　 b) $f(-3)$

Multiply and simplify.

16) $3x(2x + 1)$

17) $-7q^3(2q^4 - 9q^2 - 6)$

18) $0.4h^2(12h^4 + 8h^2 - 4h - 9)$

19) $(5b + 2)(3b^3 - 2b^2 - b + 9)$

20) $(6n^3 - 9n + 5)(2n^2 - 1)$

21) $(-5d^2 + 12d + 1)(6d^4 - d - 2)$

22) $(t - 11)(t - 2)$ 23) $(3r + 7)(2r + 1)$

24) $(5a + 2b)(a - b)$ 25) $3d^2(2d - 5)(4d + 1)$

26) $(c + 12)(c - 12)$ 27) $(3a + 4)(3a - 4)$

28) $\left(t + \dfrac{1}{4}\right)\left(t - \dfrac{1}{4}\right)$ 29) $\left(h^2 - \dfrac{3}{8}\right)\left(h^2 + \dfrac{3}{8}\right)$

Expand.

30) $(x + 9)^2$ 31) $(r - 10)^2$

32) $(11j + 2)^2$ 33) $\left(\dfrac{3}{4}c - 6\right)^2$

34) $(a - 2)^3$ 35) $(v + 5)^3$

Divide.

36) $\dfrac{14n^4 - 28n^3 - 35n^2}{7n^2}$

37) $(-110h^5 + 40h^4 + 10h^3) \div 10h^3$

38) $\dfrac{30t^5 - 15t^4 + 20t^3 - 6t}{6t^4}$

39) $\dfrac{3a^2b^2 + ab^2 - 27ab}{-3a^2b}$

40) $\dfrac{c^2 + 10c + 16}{c + 8}$ 41) $\dfrac{2a^2 + a - 15}{2a - 5}$

42) $(35t^3 + 31t^2 + 16t + 6) \div (5t + 3)$

43) $(6x^2 + 11x - 25) \div (2x + 3)$

44) $\dfrac{11w - 19w^2 - 8 + 25w^4}{5w - 2}$

45) $\dfrac{b^2 - 9 + 20b^3}{4b - 3}$ 46) $\dfrac{c^3 - 64}{c - 4}$

47) $(12h^4 + 8h^3 - 23h^2 + 4h - 6) \div (2h^2 + 3h - 1)$

For each figure, find a polynomial that represents a) its perimeter and b) its area.

48)

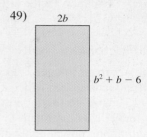

$k + 5$; $k - 2$

49)

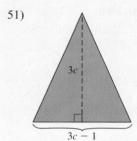

$2b$; $b^2 + b - 6$

Find a polynomial that represents the area of each triangle.

50)

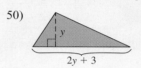

y ; $2y + 3$

51)

$3c$; $3c - 1$

52) Find a polynomial that represents the length of the rectangle if its area is given by $4x^2 + 7x - 15$ and its width is $x + 3$.

$x + 3$

53) Find a polynomial that represents the base of the triangle if the area is given by $3a^3 - a^2 - 7a$ and the height is $2a$.

$2a$

Section A7 Factoring Polynomials

Objectives

1. Factor Out the Greatest Common Factor
2. Factor by Grouping
3. Factor Trinomials of the Form $x^2 + bx + c$
4. Factor Trinomials of the Form $ax^2 + bx + c$ $(a \neq 1)$
5. Factor a Perfect Square Trinomial
6. Factor the Difference of Two Squares
7. Factor the Sum and Difference of Two Cubes
8. Solve a Quadratic Equation by Factoring
9. Use the Pythagorean Theorem to Solve an Applied Problem

In this section, we will review different techniques for factoring polynomials, and then we will discuss how to solve equations by factoring.

Recall that the **greatest common factor (GCF)** of a group of monomials is the *largest* common factor of the terms in the group. For example, the greatest common factor of $24t^5$, $56t^3$, and $40t^8$ is $8t^3$.

1. Factor Out the Greatest Common Factor

To **factor an integer** is to write it as the product of two or more integers. For example, a factorization of 15 is $3 \cdot 5$ since $15 = 3 \cdot 5$.

Likewise, to **factor a polynomial** is to write it as a product of two or more polynomials. It is important to understand that factoring a polynomial is the opposite of multiplying polynomials. Example 1 shows how these procedures are related.

Example 1

Factor out the GCF from $3p^2 + 12p$.

Solution

Use the distributive property to factor out the greatest common factor from $3p^2 + 12p$.

$$GCF = 3p$$
$$3p^2 + 12p = (3p)(p) + (3p)(4)$$
$$= 3p(p + 4) \qquad \text{Distributive property}$$

We can check our result by multiplying. $3p(p + 4) = 3p^2 + 12p$ ✓

Sometimes we can take out a binomial factor. This leads us to our next method of factoring a polynomial—factoring by grouping.

2. Factor by Grouping

When we are asked to factor a polynomial containing four terms, we often try to **factor by grouping**.

> The first step in factoring *any* polynomial is to ask yourself, *"Can I factor out a GCF?"* If you can, then factor it out.

Example 2

Factor completely. $5z^5 - 10z^4 + 15z^3 - 30z^2$

Solution

Notice that this polynomial has four terms. This is a clue for us to try factoring by grouping.

The first step in factoring this polynomial is to factor out $5z^2$.

$$5z^5 - 10z^4 + 15z^3 - 30z^2 = 5z^2(z^3 - 2z^2 + 3z - 6) \qquad \text{Factor out the GCF, } 5z^2.$$

The polynomial in parentheses has four terms. Try to factor it by grouping.

$$5z^2(\underbrace{z^3 - 2z^2}_{\text{Group 1}} + \underbrace{3z - 6}_{\text{Group 2}})$$

$$= 5z^2[z^2(z - 2) + 3(z - 2)] \qquad \text{Take out the common factor in each group.}$$
$$= 5z^2(z - 2)(z^2 + 3) \qquad\qquad \text{Factor out } (z - 2) \text{ using the distributive property.}$$

3. Factor Trinomials of the Form $x^2 + bx + c$

One of the factoring problems encountered most often in algebra is the factoring of trinomials. Next we will discuss how to factor a trinomial of the form $x^2 + bx + c$—notice that the coefficient of the squared term is 1.

Understanding that factoring is the opposite of multiplying will help us understand how to factor this type of trinomial. Consider the following multiplication problem:

$$(x + 3)(x + 5) = x^2 + 5x + 3x + 3 \cdot 5$$
$$= x^2 + (5 + 3)x + 15$$
$$= x^2 + 8x + 15$$

So, if we were asked to *factor* $x^2 + 8x + 15$, we need to think of two integers whose *product* is 15 and whose *sum* is 8. Those numbers are 3 and 5. The *factored form* of $x^2 + 8x + 15$ is $(x + 3)(x + 5)$.

Example 3

Factor completely.

a) $n^2 + 9n + 18$ b) $2k^3 - 8k^2 - 24k$

Solution

a) $n^2 + 9n + 18$

Begin by asking yourself, *"Can I factor out a GCF?" No.* We must find the two integers whose *product* is 18 and whose *sum* is 9. Both integers will be positive. Those numbers are 3 and 6.

$$n^2 + 9n + 18 = (n + 3)(n + 6)$$

Check: $(n + 3)(n + 6) = n^2 + 6n + 3n + 18$
$$= n^2 + 9n + 18 \quad \checkmark$$

b) $2k^3 - 8k^2 - 24k$

Ask yourself, *"Can I factor out a GCF?" Yes.* The GCF is $2k$.

$$2k^3 - 8k^2 - 24k = 2k(k^2 - 4k - 12)$$

Look at the trinomial and ask yourself, *"Can I factor again?" Yes.* The integers whose product is -12 and whose sum is -4 are -6 and 2. Therefore,

$$2k^3 - 8k^2 - 24k = 2k(k^2 - 4k - 12)$$
$$= 2k(k - 6)(k + 2)$$

We cannot factor again. The check is left to the student.
The completely factored form of $2k^3 - 8k^2 - 24k$ is $2k(k - 6)(k + 2)$.

> After performing one factorization you should always ask yourself, *"Can I factor again?"* If you can, then factor the polynomial again. If not, then you know that the polynomial has been completely factored.

4. Factor Trinomials of the Form $ax^2 + bx + c$ $(a \neq 1)$

We will discuss two methods for factoring a trinomial of the form $ax^2 + bx + c$ when $a \neq 1$ and when we cannot factor out the leading coefficient of a.

Factoring $ax^2 + bx + c$ $(a \neq 1)$ by Grouping

Example 4

Factor $4t^2 - 7t - 2$ completely.

Solution

Ask yourself, *"Can I factor out a GCF?"* No. Multiply the 4 and -2 to get -8.

$$4t^2 - 7t - 2$$
$$\text{Product: } 4(-2) = -8$$

Find two integers whose *product* is -8 and whose *sum* is -7. The numbers are -8 and 1. Rewrite $-7t$ as $-8t + 1t$.

$$4t^2 - 7t - 2 = 4t^2 - 8t + 1t - 2$$
$$\underbrace{\qquad}_{\text{Group 1}} \quad \underbrace{\qquad}_{\text{Group 2}}$$
$$= 4t(t - 2) + 1(t - 2) \qquad \text{Take out the common factor from each group.}$$
$$= (t - 2)(4t + 1) \qquad \text{Factor out } t - 2.$$
$$4t^2 - 7t - 2 = (t - 2)(4t + 1)$$

The check is left to the student.

Factoring $ax^2 + bx + c$ $(a \neq 1)$ by Trial and Error

Example 5

Factor $3r^2 - 29r + 18$ completely.

Solution

Can we factor out a GCF? *No.* To get a product of $3r^2$, we will use $3r$ and r.

$$3r^2 - 29r + 18 = (3r \qquad)(r \qquad)$$
$$\underbrace{\qquad}_{3r^2}$$

Since the last term is positive and the middle term is negative, we want pairs of *negative* integers that multiply to 18. The pairs are -1 and -18, -2 and -9, and -3 and -6.

When we try -2 and -9, we get $\qquad 3r^2 - 29r + 18 \overset{?}{=} (3r - 2)(r - 9)$

$$-2r$$
$$+ (-27r)$$
$$-29r \quad \text{Correct!}$$

$$3r^2 - 29r + 18 = (3r - 2)(r - 9)$$

5. Factor a Perfect Square Trinomial

We can use the following formulas to factor perfect square trinomials:

$$a^2 + 2ab + b^2 = (a + b)^2 \qquad a^2 - 2ab + b^2 = (a - b)^2$$

In order for a trinomial to be a perfect square, two of its terms must be perfect squares.

Example 6

Factor $16c^2 - 24c + 9$ completely.

Solution

We cannot take out a GCF. Let's see if this trinomial fits the pattern of a perfect square trinomial.

$$16c^2 - 24c + 9$$
$$\downarrow \qquad\qquad \downarrow$$

What do you square to get $16c^2$? $4c$ $\qquad (4c)^2 \qquad (3)^2 \qquad$ What do you square to get 9? 3

Does the middle term equal $2 \cdot 4c \cdot 3$? Yes.

$$2 \cdot 4c \cdot 3 = 24c$$

Therefore, $16c^2 - 24c + 9 = (4c)^2 - 2(4c)(3) + (3)^2$

$$= (4c - 3)^2$$

Check by expanding $(4c - 3)^2$.

6. Factor the Difference of Two Squares

Another common type of factoring problem is a **difference of two squares**. Some examples of these types of binomials are

$$x^2 - 25 \qquad 4t^2 - 49u^2 \qquad 100 - n^2 \qquad z^4 - 1$$

Notice that in each binomial, the terms are being *subtracted*, and each term is a perfect square. To factor the difference of two squares we use the following formula:

$$a^2 - b^2 = (a + b)(a - b)$$

Example 7

Factor $x^2 - 25$ completely.

Solution

Since each term is a perfect square, we can use the formula
$a^2 - b^2 = (a + b)(a - b)$.

Identify a and b.

$$x^2 - 25$$
$$\downarrow \qquad \downarrow$$

What do you square $(x)^2 \qquad (5)^2$ What do you square
to get x^2? x to get 25? 5

Then, $a = x$ and $b = 5$.

$$x^2 - 25 = (x + 5)(x - 5)$$

Check by multiplying: $(x + 5)(x - 5) = x^2 - 25$ ✓

7. Factor the Sum and Difference of Two Cubes

A binomial that is either the sum of two cubes or the difference of two cubes can be factored using the following formulas:

> **Factoring the Sum and Difference of Two Cubes:** $a^3 + b^3 = (a + b)(a^2 - ab + b^2)$
> $$a^3 - b^3 = (a - b)(a^2 + ab + b^2)$$

Notice that each factorization is the product of a binomial and a trinomial.

Example 8

Factor $r^3 + 27$ completely.

Solution

Identify a and b: $a = r$ and $b = 3$
    Using $a^3 + b^3 = (a + b)(a^2 - ab + b^2)$ we get

$$r^3 + 27 = (r + 3)[r^2 - (r)(3) + 3^2] \qquad \text{Let } a = r \text{ and } b = 3.$$
$$= (r + 3)(r^2 - 3r + 9)$$

In addition to learning the different factoring techniques, it is important to remember these two things:

1. The first thing you should do when factoring is ask yourself, *"Can I factor out a GCF?"*

2. The last thing you should do when factoring is look at the result and ask yourself, *"Can I factor again?"*

8. Solve a Quadratic Equation by Factoring

A **quadratic equation** can be written in the form $ax^2 + bx + c = 0$, where a, b, and c are real numbers and $a \neq 0$.

There are many different ways to solve quadratic equations. Here, we will review how to solve them by factoring.

Solving an equation by factoring is based on the **zero product rule**, which states:

$$\text{If } ab = 0, \text{ then } a = 0 \text{ or } b = 0.$$

Here are the steps to use to solve a quadratic equation by factoring:

Solving a Quadratic Equation by Factoring

1) Write the equation in the form $ax^2 + bx + c = 0$ so that all terms are on one side of the equal sign and zero is on the other side.

2) Factor the expression.

3) Set each factor equal to zero, and solve for the variable.

4) Check the answer(s).

Example 9

Solve by factoring: $3w^2 + 7w = 6$

Solution

Begin by writing $3w^2 + 7w = 6$ in standard form, $ax^2 + bx + c = 0$.

$$3w^2 + 7w - 6 = 0 \qquad \text{Standard form}$$
$$(3w - 2)(w + 3) = 0 \qquad \text{Factor.}$$

$$3w - 2 = 0 \quad \text{or} \quad w + 3 = 0 \qquad \text{Set each factor equal to zero.}$$
$$3w = 2$$
$$w = \frac{2}{3} \quad \text{or} \qquad w = -3 \qquad \text{Solve.}$$

Check the solutions by substituting them back into the original equation. The solution set is $\left\{ -3, \dfrac{2}{3} \right\}$.

9. Use the Pythagorean Theorem to Solve an Applied Problem

A **right triangle** is a triangle that contains a 90° (*right*) angle. We label a right triangle as follows.

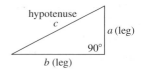

The side opposite the 90° angle is the longest side of the triangle and is called the **hypotenuse**. The other two sides are called the **legs**. The Pythagorean theorem states a relationship between the lengths of the sides of a right triangle. This is a very important relationship in mathematics and is one that is used in many different ways.

Definition

Pythagorean Theorem: Given a right triangle with legs of length a and b and hypotenuse of length c,

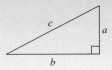

the Pythagorean theorem states that $a^2 + b^2 = c^2$ [or $(\text{leg})^2 + (\text{leg})^2 = (\text{hypotenuse})^2$].

Example 10

Write an equation and solve.

A garden situated between two walls of a house will be in the shape of a right triangle with a fence on the third side. The side with the fence will be 4 ft longer than the shortest side, and the other side will be 2 ft longer than the shortest side. How long is the fence?

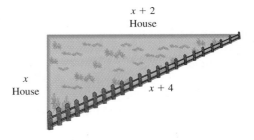

Solution

Step 1:   Read the problem carefully, twice.

Step 2:   Find the length of the fence. Define the unknowns.

$$x = \text{length of shortest side}$$
$$x + 2 = \text{length of the other side along the house}$$
$$x + 4 = \text{length of the fence}$$

Label the picture.

Step 3:   Translate from English to math.

The garden is in the shape of a right triangle. The hypotenuse is $x + 4$ since it is the side across from the right angle. The legs are x and $x + 2$. From the Pythagorean theorem we get

$$a^2 + b^2 = c^2 \qquad \text{Pythagorean theorem}$$
$$x^2 + (x + 2)^2 = (x + 4)^2 \qquad \text{Substitute.}$$

Step 4:   Solve the equation.

$$x^2 + (x + 2)^2 = (x + 4)^2$$
$$x^2 + x^2 + 4x + 4 = x^2 + 8x + 16 \quad \text{Expand.}$$
$$2x^2 + 4x + 4 = x^2 + 8x + 16$$
$$x^2 - 4x - 12 = 0 \quad \text{Write in standard form.}$$
$$(x - 6)(x + 2) = 0 \quad \text{Factor.}$$

$$x - 6 = 0 \quad \text{or} \quad x + 2 = 0 \quad \text{Set each factor equal to 0.}$$
$$x = 6 \quad \text{or} \quad x = -2 \quad \text{Solve.}$$

Step 5:   Interpret the meaning of the solution as it relates to the problem.

x represents the length of the shortest side, so x cannot equal -2.
Therefore, the length of the shortest side must be 6 ft.
The length of the fence $= x + 4 = 6 + 4 = 10$ ft.
The length of the fence is 10 ft.
The check is left to the student.

A7 Exercises

Factor out the greatest common factor.

1) $28k + 8$
2) $12p^3 - 6p$
3) $26y^3z^3 + 8y^3z^2 - 12yz^3$
4) $a(b + 7) + 4(b + 7)$

Factor by grouping.

5) $vw + 3v + 12w + 36$
6) $rs - 6r + 10s - 60$
7) $21ab - 18a - 56b + 48$
8) $18pq - 66p + 15q - 55$

Completely factor the following trinomials.

9) $x^2 + 7x + 10$
10) $z^2 + 4z - 32$
11) $2d^4 - 2d^3 - 24d^2$
12) $x^2 + 10xy + 24y^2$
13) $3n^2 + 10n + 8$
14) $2b^2 + 17b + 21$
15) $8z^2 - 14z - 9$
16) $7m^2 + 4m - 6$
17) $12c^2 - 28c + 16$
18) $h^2 + 4h + 4$
19) $x^2 - 8xy + 16y^2$
20) $100w^2 + 40w + 4$

Completely factor each binomial.

21) $m^2 - 9$
22) $\dfrac{1}{4} - a^2$
23) $25q^6 - 9q^4$
24) $x^4 - 1$
25) $u^2 + 25$
26) $h^3 - 64$
27) $64s^3 - 27t^3$
28) $5a^4 - 40a$

Factor completely. These exercises consist of all of the different types of polynomials presented in this section.

29) $45m - 54$
30) $a^2 - 10a + 21$
31) $4t^2 - 81$
32) $m^4n + 15m^3n + 36m^2n$
33) $w^2 + 13w - 48$
34) $n^2 + \dfrac{2}{3}n + \dfrac{1}{9}$
35) $2k^4 - 10k^3 + 8k^2$
36) $h^2 + 9$
37) $10r^2 - 43r + 12$
38) $9k^2 - 30k + 25$
39) $cd - c + d - 1$
40) $28 - 7a^2$
41) $m^2 + 5m - 50$
42) $r^2 - 7rs - 30s^2$

Solve each equation.

43) $(y + 8)(y - 3) = 0$
44) $(5m - 2)(m - 1) = 0$
45) $r^2 + 7r + 12 = 0$
46) $x^2 + 15x + 54 = 0$
47) $4w^2 = 5w$
48) $10a - 8 = 3a^2$
49) $k^2 - 169 = 0$
50) $18c^2 = 8$
51) $(y + 4)(y + 8) = 5$
52) $a(a - 1) = 56$

53) $6q^2 = -24q$

54) $2(d + 8) = (d + 9)^2 - 5$

55) $2(h - 5)^2 - 11 = 2(15 - h^2)$

56) $16q^3 = 4q$ 57) $z^3 + 7z^2 + 6z = 0$

58) $(7r - 3)(r^2 + 12r + 36) = 0$

Use the Pythagorean Theorem to find the length of the missing side.

59)

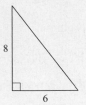

60)

Write an equation and solve.

61) A wire is attached to the top of a pole. The pole is 2 ft shorter than the wire, and the distance from the wire on the ground to the bottom of the pole is 9 ft less

than the length of the wire. Find the length of the wire and the height of the pole.

62) A 15-ft board is leaning against a wall. The distance from the top of the board to the bottom of the wall is 3 ft more than the distance from the bottom of the board to the wall. Find the distance from the bottom of the board to the wall.

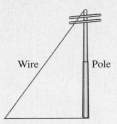

Section A8 Rational Expressions

Objectives

1. Evaluate a Rational Expression, and Determine Where It Equals Zero and Where It Is Undefined

2. Define a Rational Function and Determine the Domain of a Rational Function

3. Write a Rational Expression in Lowest Terms

4. Multiply and Divide Rational Expressions

5. Add and Subtract Rational Expressions

6. Simplify Complex Fractions

7. Solve Rational Equations

8. Solve an Equation for a Specific Variable

1. Evaluate a Rational Expression, and Determine Where It Equals Zero and Where It Is Undefined

We begin by defining a rational expression.

> **Definition**
>
> A **rational expression** is an expression of the form $\dfrac{P}{Q}$, where P and Q are polynomials and where $Q \neq 0$.

We can *evaluate* rational expressions for given values of the variable(s), but just as a fraction like $\dfrac{9}{0}$ is undefined because its denominator equals zero, a rational expression is undefined when its denominator equals zero.

> 1) A fraction (rational expression) equals zero when its *numerator* equals zero.
> 2) A fraction (rational expression) is *undefined* when its denominator equals zero.

Example 1

Given the rational expression $\dfrac{3w + 4}{w + 5}$, answer the following.

a)   Evaluate the expression for $w = -2$.

b)   For what value of the variable does the expression equal zero?

c)   For what value of the variable is the expression undefined?

Solution

a)   Substitute -2 for w in the expression and simplify.

$$\frac{3(-2) + 4}{-2 + 5} = \frac{-6 + 4}{3} = \frac{-2}{3} = -\frac{2}{3}$$

b) $\dfrac{3w + 4}{w + 5} = 0$ when its *numerator* equals zero. Set the numerator equal to zero, and solve for w.

$$3w + 4 = 0$$
$$3w = -4 \qquad \text{Subtract 4 from each side.}$$
$$w = -\frac{4}{3} \qquad \text{Divide by 3.}$$

$\dfrac{3w + 4}{w + 5} = 0$ when $w = -\dfrac{4}{3}$.

c) $\dfrac{3w + 4}{w + 5}$ is *undefined* when its *denominator* equals zero. Set the denominator equal to zero, and solve for *w.*

$$w + 5 = 0$$
$$w = -5$$

$\dfrac{3w + 4}{w + 5}$ is *undefined* when $w = -5$. So, *w* cannot equal -5 in the expression.

2. Define a Rational Function and Determine the Domain of a Rational Function

$f(x) = \dfrac{x + 10}{x - 3}$ is an example of a **rational function** since $\dfrac{x + 10}{x - 3}$ is a rational expression and since each value that can be substituted for *x* will produce *only one* value for the expression.

The domain of a rational function consists of all real numbers except the value(s) of the variable that make the denominator equal zero.

Therefore, to determine the domain of a rational function we set the denominator equal to zero and solve for the variable. The value(s) that make the denominator equal to zero are *not* in the domain of the function.

To determine the domain of a rational function, sometimes it is helpful to ask yourself, "Is there any number that *cannot* be substituted for the variable?"

Example 2

Determine the domain of $f(x) = \dfrac{6x - 1}{x^2 + 12x + 32}$.

Solution

To determine the domain of $f(x) = \dfrac{6x - 1}{x^2 + 12x + 32}$, ask yourself, "Is there any number that *cannot* be substituted for *x*? Yes, $f(x)$ is *undefined* when its *denominator* equals zero. Set the denominator equal to zero and solve for *x.*

$x^2 + 12x + 32 = 0$	Set the denominator $= 0$.
$(x + 8)(x + 4) = 0$	Factor.
$x + 8 = 0$ or $x + 4 = 0$	Set each factor equal to 0.
$x = -8$ or $x = -4$	Solve.

When $x = -8$ or $x = -4$, the denominator of $f(x) = \dfrac{6x - 1}{x^2 + 12x + 32}$ equals zero.

The domain contains all real numbers *except* -8 and -4.

Write the domain in interval notation as $(-\infty, -8) \cup (-8, -4) \cup (-4, \infty)$

3. Write a Rational Expression in Lowest Terms

A rational expression is in lowest terms when its numerator and denominator contain no common factors except 1. We can use the fundamental property of rational expressions to write a rational expression in lowest terms.

Definition

Fundamental Property of Rational Expressions: If P, Q, and C are polynomials such that $Q \neq 0$ and $C \neq 0$, then

$$\frac{PC}{QC} = \frac{P}{Q}$$

Writing a Rational Expression in Lowest Terms

1) Completely *factor* the numerator and denominator.

2) *Divide* the numerator and denominator by the greatest common factor.

Example 3

Write $\dfrac{d^2 + 3d - 18}{5d^2 - 15d}$ in lowest terms.

Solution

$$\frac{d^2 + 3d - 18}{5d^2 - 15d} = \frac{(d - 3)(d + 6)}{5d(d - 3)} \qquad \text{Factor.}$$

$$= \frac{(\cancel{d - 3})(d + 6)}{5d(\cancel{d - 3})} \qquad \text{Divide out the common factor, } d - 3.$$

$$= \frac{(d + 6)}{5d}$$

4. Multiply and Divide Rational Expressions

We multiply and divide rational expressions the same way that we multiply and divide rational numbers.

Definition

Multiplying Rational Expressions:

If $\dfrac{P}{Q}$ and $\dfrac{R}{T}$ are rational expressions, then

$$\frac{P}{Q} \cdot \frac{R}{T} = \frac{PR}{QT}$$

To multiply two rational expressions, multiply their numerators, multiply their denominators, and simplify.

Use the following steps to multiply two rational expressions:

1) Factor.

2) Divide out common factors and multiply.

All products must be written in lowest terms.

Example 4

Multiply $\dfrac{k^2 - 7k + 12}{3k^2 - 8k - 3} \cdot \dfrac{3k^2 + k}{2k^2 - 32}$.

Solution

$$\dfrac{k^2 - 7k + 12}{3k^2 - 8k - 3} \cdot \dfrac{3k^2 + k}{2k^2 - 32} = \dfrac{(k - 3)(k - 4)}{(3k + 1)(k - 3)} \cdot \dfrac{3k(k + 1)}{2(k + 4)(k - 4)} \quad \text{Factor.}$$

$$= \dfrac{3k(k + 1)}{2(3k + 1)(k + 4)} \quad \text{Divide out common factors and multiply.}$$

To divide rational expressions we multiply the first rational expression by the reciprocal of the second expression.

5. Add and Subtract Rational Expressions

In order to add or subtract fractions, they must have a common denominator. The same is true for rational expressions.

To find the least common denominator (LCD) of a group of rational expressions, begin by factoring the denominators. The LCD will contain each unique factor the *greatest* number of times it appears in any single factorization. The LCD is the product of these factors.

Once we have identified the least common denominator for a group of rational expressions, we must be able to rewrite each expression with this LCD.

Writing Rational Expressions as Equivalent Expressions with the Least Common Denominator

Step 1: Identify and write down the LCD.

Step 2: Look at each rational expression (with its denominator in factored form) and compare its denominator with the LCD. Ask yourself, *"What factors are missing?"*

Step 3: Multiply the numerator and denominator by the "missing" factors to obtain an equivalent rational expression with the desired LCD. Multiply the terms in the numerator, but leave the denominator as the product of factors.

Example 5

Identify the least common denominator of $\dfrac{8}{5t^2 - 10t}$ and $\dfrac{3t}{t^2 - 6t + 8}$, and rewrite each as an equivalent fraction with the LCD as its denominator.

Solution

Follow the steps.

Step 1:   Identify and write down the LCD of $\dfrac{8}{5t^2 - 10t}$ and $\dfrac{3t}{t^2 - 6t + 8}$. First, we must factor the denominators.

$$\dfrac{8}{5t^2 - 10t} = \dfrac{8}{5t(t - 2)}, \quad \dfrac{3t}{t^2 - 6t + 8} = \dfrac{3t}{(t - 2)(t - 4)}$$

We will work with the factored forms of the expressions.

$$\text{LCD} = 5t(t - 2)(t - 4)$$

Step 2: Compare the denominators of $\dfrac{8}{5t(t - 2)}$ and $\dfrac{3t}{(t - 2)(t - 4)}$ to the LCD and ask yourself, *"What's missing from each denominator?"*

$\dfrac{8}{5t(t - 2)}$: $5t(t - 2)$ is "missing" the factor $t - 4$.

$\dfrac{3t}{(t - 2)(t - 4)}$: $(t - 2)(t - 4)$ is "missing" $5t$.

Step 3: Multiply the numerator and denominator by $t - 4$.

$$\frac{8}{5t(t - 2)} \cdot \frac{t - 4}{t - 4} = \frac{8(t - 4)}{5t(t - 2)(t - 4)}$$
$$= \frac{8t - 32}{5t(t - 2)(t - 4)}$$

Multiply the numerator and denominator by $5t$.

$$\frac{3t}{(t - 2)(t - 4)} \cdot \frac{5t}{5t} = \frac{15t^2}{5t(t - 2)(t - 4)}$$

Multiply the factors in the numerators.

$$\frac{8}{5t(t - 2)} = \frac{8t - 32}{5t(t - 2)(t - 4)} \quad \text{and} \quad \frac{3t}{(t - 2)(t - 4)} = \frac{15t^2}{5t(t - 2)(t - 4)}$$

Now that we have reviewed how to rewrite rational expressions with a least common denominator, we summarize the steps we can use to add and subtract rational expressions.

Steps for Adding and Subtracting Rational Expressions with Different Denominators

Step 1: Factor the denominators.

Step 2: Write down the LCD.

Step 3: Rewrite each rational expression as an equivalent rational expression with the LCD.

Step 4: Add or subtract the numerators and keep the common denominator in factored form.

Step 5: After combining like terms in the numerator ask yourself, *"Can I factor it?"* If so, factor.

Step 6: Reduce the rational expression, if possible.

Example 6

Add $\dfrac{8x - 24}{x^2 - 36} + \dfrac{x}{x + 6}$.

Solution

Step 1: Factor the denominator of $\dfrac{8x - 24}{x^2 - 36}$.

$$\frac{8x - 24}{x^2 - 36} = \frac{8x - 24}{(x + 6)(x - 6)}$$

Step 2: Identify the LCD of $\dfrac{8x - 24}{(x + 6)(x - 6)}$ and $\dfrac{x}{x + 6}$.

$$\text{LCD} = (x + 6)(x - 6)$$

Step 3: Rewrite $\dfrac{x}{x + 6}$ with the LCD.

$$\frac{x}{x + 6} \cdot \frac{x - 6}{x - 6} = \frac{x(x - 6)}{(x + 6)(x - 6)}$$

Step 4: $\dfrac{8x - 24}{x^2 - 36} + \dfrac{x}{x + 6} = \dfrac{8x - 24}{(x + 6)(x - 6)} + \dfrac{x}{x + 6}$ Factor the denominator.

$$= \frac{8x - 24}{(x + 6)(x - 6)} + \frac{x(x - 6)}{(x + 6)(x - 6)}$$ Write each expression with the LCD.

$$= \frac{8x - 24 + x(x - 6)}{(x + 6)(x - 6)}$$ Add the expressions.

$$= \frac{8x - 24 + x^2 - 6x}{(x + 6)(x - 6)}$$ Distribute.

$$= \frac{x^2 + 2x - 24}{(x + 6)(x - 6)}$$ Combine like terms.

Steps 5 & 6: Ask yourself, *"Can I factor the numerator?"* Yes.

$$\frac{x^2 + 2x - 24}{(x + 6)(x - 6)} = \frac{\cancel{(x + 6)}(x - 4)}{\cancel{(x + 6)}(x - 6)}$$ Factor.

$$= \frac{x - 4}{x - 6}$$ Reduce.

6. Simplify Complex Fractions

A **complex fraction** is a rational expression that contains one or more fractions in its numerator, its denominator, or both.

A complex fraction is not considered to be an expression in simplest form. Let's review how to simplify complex fractions.

Example 7

Simplify each complex fraction.

a) $\dfrac{\dfrac{2a + 12}{9}}{\dfrac{a + 6}{5a}}$

b) $\dfrac{1 - \dfrac{3}{4}}{\dfrac{1}{2} + \dfrac{1}{3}}$

Solution

a)   We can think of $\dfrac{\frac{2a+12}{9}}{\frac{a+6}{5a}}$ as a division problem.

$$\dfrac{\frac{2a+12}{9}}{\frac{a+6}{5a}} = \frac{2a+12}{9} \div \frac{a+6}{5a} \qquad \text{Rewrite the complex fraction as a division problem.}$$

$$= \frac{2a+12}{9} \cdot \frac{5a}{a+6} \qquad \text{Change division to multiplication by the reciprocal of } \frac{a+6}{5a}.$$

$$= \frac{2\cancel{(a+6)}}{9} \cdot \frac{5a}{\cancel{a+6}} \qquad \text{Factor and divide numerator and denominator by } a+6 \text{ to simplify.}$$

$$= \frac{10a}{9} \qquad \text{Multiply.}$$

If a complex fraction contains one term in the numerator and one term in the denominator, it can be simplified by rewriting it as a division problem and then performing the division.

b)   We can simplify the complex fraction $\dfrac{1-\frac{3}{4}}{\frac{1}{2}+\frac{1}{3}}$ in two different ways. We can

combine the terms in the numerator, combine the terms in the denominator, and then proceed as in part a). Or, we can follow the steps below.

Step 1:   Look at all of the fractions in the complex fraction. They are $\dfrac{3}{4}, \dfrac{1}{2},$

and $\dfrac{1}{3}$. Write down their LCD: LCD $= 12$

Step 2:   Multiply the numerator and denominator of the complex fraction by the LCD, 12.

$$\dfrac{1-\frac{3}{4}}{\frac{1}{2}+\frac{1}{3}} = \dfrac{12\left(1-\frac{3}{4}\right)}{12\left(\frac{1}{2}+\frac{1}{3}\right)}$$

Step 3:   Simplify.

$$\dfrac{12\left(1-\frac{3}{4}\right)}{12\left(\frac{1}{2}+\frac{1}{3}\right)} = \dfrac{12\cdot 1 - 12\cdot\frac{3}{4}}{12\cdot\frac{1}{2} + 12\cdot\frac{1}{3}} \qquad \text{Distribute.}$$

$$= \dfrac{12-9}{6+4} \qquad \text{Multiply.}$$

$$= \dfrac{3}{10} \qquad \text{Simplify.}$$

If a complex fraction contains more than one term in the numerator and/or denominator, we can multiply the numerator and denominator by the LCD of all of the fractions in the expression and simplify.

7. Solve Rational Equations

When we add and subtract rational expressions, we must write each of them with the least common denominator. When we solve rational equations, however, we must *multiply* the entire equation by the LCD to eliminate the denominators. Here is a summary of the steps we use to solve rational equations:

How to Solve a Rational Equation

1) If possible, factor all denominators.

2) Write down the LCD of all of the expressions.

3) Multiply both sides of the equation by the LCD to *eliminate* the denominators.

4) Solve the equation.

5) Check the solution(s) in the original equation. If a proposed solution makes a denominator equal 0, then it is rejected as a solution.

Example 8

Solve $\dfrac{m}{m-2} - \dfrac{1}{4} = \dfrac{5}{m-2}$.

Solution

We do not need to factor the denominators, so identify the LCD of the expressions. Then, multiply both sides of the equation by the LCD to eliminate the denominators.

$$\text{LCD} = 4(m-2)$$

$$4(m-2) \cdot \left(\frac{m}{m-2} - \frac{1}{4} \right) = 4(m-2) \cdot \frac{5}{m-2} \qquad \text{Multiply both sides of the equation by the LCD, } 4(m-2).$$

$$4(m-2) \cdot \left(\frac{m}{m-2} \right) - 4(m-2) \cdot \left(\frac{1}{4} \right) = 4(m-2) \cdot \frac{5}{m-2} \qquad \text{Distribute and divide out common factors.}$$

$$4m - (m-2) = 20 \qquad \text{Multiply.}$$
$$4m - m + 2 = 20 \qquad \text{Distribute.}$$
$$3m + 2 = 20 \qquad \text{Combine like terms.}$$
$$3m = 18 \qquad \text{Subtract 2 from each side.}$$
$$m = 6 \qquad \text{Divide by 3.}$$

$$\text{Check: } \frac{6}{6-2} - \frac{1}{4} \overset{?}{=} \frac{5}{6-2} \qquad \text{Substitute 6 for } m \text{ in the original equation.}$$

$$\frac{6}{4} - \frac{1}{4} \overset{?}{=} \frac{5}{4}$$

$$\frac{5}{4} = \frac{5}{4} \quad \checkmark$$

The solution is $\{6\}$.

BE CAREFUL   **Always** check what *appears* to be the solution or solutions to an equation containing rational expressions. If a value makes a denominator zero, then it *cannot* be a solution to the equation.

Example 9 shows how to solve a special type of rational equation.

Example 9

Solve $\dfrac{15}{c + 3} = \dfrac{6}{c}$.

Solution

This equation is a *proportion*. A **proportion** is a statement that two ratios are equal. We can solve this proportion as we have solved the other equations in this section, by multiplying both sides of the equation by the LCD. Or, *we can solve a proportion by setting the cross products equal to each other.*

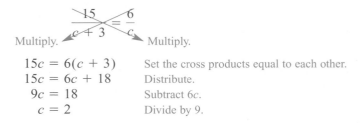

Multiply.   Multiply.

$$15c = 6(c + 3)$$ Set the cross products equal to each other.
$$15c = 6c + 18$$ Distribute.
$$9c = 18$$ Subtract $6c$.
$$c = 2$$ Divide by 9.

The check is left to the student.
The solution is $\{2\}$.

8. Solve an Equation for a Specific Variable

When an equation contains more than one letter and we are asked to solve for a specific variable, we use the same ideas that we used in the previous examples.

Example 10

Solve $n = \dfrac{3k - t}{r}$ for r.

Solution

Since we are asked to solve for r, put the r in a box to indicate that it is the variable for which we must solve. Multiply both sides of the equation by r to eliminate the denominator.

$$n = \dfrac{3k - t}{\boxed{r}}$$ Put r in a box.

$$\boxed{r}(n) = \boxed{r}\left(\dfrac{3k - t}{\boxed{r}}\right)$$ Multiply both sides by r to eliminate the denominator.

$$\boxed{r}n = 3k - t$$ Multiply.

$$r = \dfrac{3k - t}{n}$$ Divide by n.

Evaluate for a) $x = 5$ and b) $x = -2$.

1) $\dfrac{x^2 - 4}{3x - 1}$

Determine the value(s) of the variable for which a) the expression equals zero and b) the expression is undefined.

2) $\dfrac{h - 4}{3h + 10}$

3) $\dfrac{x^2 - 9}{7}$

4) $\dfrac{8}{t - 7}$

Determine the domain of each rational function.

5) $f(x) = \dfrac{x - 1}{x + 9}$

6) $g(c) = \dfrac{10}{2c - 3}$

7) $h(t) = \dfrac{t}{t^2 - 5t - 6}$

8) $k(n) = \dfrac{3n + 7}{n^2 + 1}$

Write each rational expression in lowest terms.

9) $\dfrac{40z^3}{48z}$

10) $\dfrac{56m^{11}}{7m}$

11) $\dfrac{12k + 48}{7k + 28}$

12) $\dfrac{c^2 + 4}{c + 2}$

13) $\dfrac{a^2 + 2a - 8}{a^2 + 11a + 28}$

14) $\dfrac{14 - 18q}{9q^2 - 16q + 7}$

Multiply or divide as indicated.

15) $\dfrac{9t^3 - 3t^2 + t}{t + 2} \div \dfrac{27t^3 + 1}{9t^2 - 1}$

16) $\dfrac{z^2 + 8z + 12}{z^3} \cdot \dfrac{3z^2 + 4z}{z^2 + 12z + 36}$

17) $\dfrac{16 - r^2}{12} \div \dfrac{2r^2 - 9r + 4}{6r - 3}$

18) $\dfrac{2a^2 - 162}{ab - 5a + 9b - 45} \cdot \dfrac{b^3 - 125}{4a - 36}$

Identify the least common denominator of each pair of fractions, and rewrite each as an equivalent fraction with the LCD as its denominator.

19) $\dfrac{3}{8w^3}, \dfrac{1}{6w}$

20) $\dfrac{c}{c - 2}, \dfrac{5}{c + 3}$

Add or subtract as indicated.

21) $\dfrac{5}{4j} + \dfrac{11}{12j}$

22) $\dfrac{3}{8z^4} - \dfrac{1}{12z}$

23) $\dfrac{7}{w} + \dfrac{4}{w - 2}$

24) $\dfrac{6}{x + 5} - \dfrac{1}{x}$

25) $\dfrac{a}{4a - 3} + \dfrac{6}{a + 2}$

26) $\dfrac{5}{c - 6} + \dfrac{2c}{c + 8}$

27) $\dfrac{k^2 + 21}{k^2 - 49} + \dfrac{5}{7 - k}$

28) $\dfrac{n + 4}{3n^2 - 5n - 2} + \dfrac{4}{3n - 1} - \dfrac{7n}{9n^2 - 1}$

Perform the indicated operations and simplify.

29) $\dfrac{3r}{4r^2 - 12r + 9} \div \dfrac{9r + 9}{2r^2 - r - 3}$

30) $\dfrac{5}{d - 6} - \dfrac{3}{d}$

31) $\dfrac{w - 4}{w^2 - 2w - 8} \cdot \dfrac{w^2 - 16}{2w^2 + 5w - 12}$

32) $\dfrac{2a - 3}{a^2 + 4a - 5} - \dfrac{a}{a^2 - 2a + 1}$

33) $\dfrac{x^2 - y^2}{x + y} \cdot \dfrac{2x^2 - 8xy}{x^2 - 5xy + 4y^2}$

34) $\dfrac{2}{5z + 4} + \dfrac{z - 1}{3z - 2}$

35) $\dfrac{9 - m^2}{5m^2 + 15m} \div (m^3 - 27)$

36) $\dfrac{h + 1}{4h^3 + h^2} - \dfrac{2}{12h + 3}$

37) $\dfrac{6x}{x - 8} + \dfrac{1}{8 - x}$

38) $\dfrac{6}{2p - 1} - \dfrac{5}{1 - 4p^2}$

Simplify completely.

39) $\dfrac{\dfrac{7}{9}}{\dfrac{5}{6}}$

40) $\dfrac{\dfrac{3s^4 t}{7}}{\dfrac{9s^2 t^4}{28}}$

41) $\dfrac{\dfrac{r - 9}{r}}{\dfrac{r - 9}{8}}$

42) $\dfrac{\dfrac{1}{8} + \dfrac{2}{5}}{\dfrac{3}{4} - \dfrac{3}{10}}$

43) $\dfrac{m + \dfrac{5}{m}}{2 + \dfrac{3}{m}}$

44) $\dfrac{\dfrac{9}{c} + 1}{\dfrac{7}{c} - c}$

45) $\dfrac{\dfrac{w^2 - 1}{4}}{\dfrac{3w + 3}{8}}$

46) $\dfrac{\dfrac{1}{h + 1} - \dfrac{h}{h^2 - 1}}{\dfrac{4}{h^2 - 1} - \dfrac{1}{h - 1}}$

Solve each equation.

47) $\dfrac{r}{2} + \dfrac{1}{3} = -\dfrac{7}{6}$

48) $\dfrac{2}{5}m - \dfrac{4}{15} = \dfrac{1}{3}$

49) $2 + \dfrac{z}{z - 6} = \dfrac{3z}{4}$

50) $\dfrac{x}{x + 9} + \dfrac{x}{2} = -2$

51) $\dfrac{2}{a} + \dfrac{1}{2} = \dfrac{6}{a}$

52) $\dfrac{8w + 15}{18w - 6} - \dfrac{1}{3w - 1} = \dfrac{2w}{3}$

53) $\dfrac{11}{3z - 15} + \dfrac{z + 3}{z - 5} = \dfrac{z}{3}$

54) $\dfrac{k}{k + 6} + \dfrac{4}{k - 3} = \dfrac{11k}{k^2 + 3k - 18}$

55) $\dfrac{x - 1}{2x^2 - 3x} + \dfrac{3}{2x^2 + 3x} = \dfrac{2x + 1}{4x^2 - 9}$

56) $\dfrac{b}{2b^2 + 7b + 3} - \dfrac{b - 1}{b^2 - 9} = \dfrac{1}{2b^2 - 5b - 3}$

Solve for the indicated variable.

57) $\dfrac{y - b}{x} = m$ for y

58) $V = \dfrac{nRT}{P}$ for P

59) $C = \dfrac{ab}{d - t}$ for d

60) $\dfrac{1}{R_1} + \dfrac{1}{R_2} = \dfrac{1}{R_3}$ for R_1

For each rectangle, find a rational expression in simplest form to represent its a) area and b) perimeter.

61)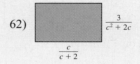

$\dfrac{a - 9}{4}$

$\dfrac{20}{a - 2}$

62)

$\dfrac{3}{c^2 + 2c}$

$\dfrac{c}{c + 2}$

Write an equation for each and solve.

63) The ratio of jazz music to reggae in Consuela's music library is 5 to 3. If there are 18 fewer reggae songs than jazz songs, how many songs of each type are in her music library?

64) A cold water faucet can fill a sink in 4 min, while it takes a hot water faucet 6 min. How long would it take to fill the skin if both faucets are on?

Answers to Selected Exercises

Chapter 1

Section 1.1

1) a) $\dfrac{2}{5}$ b) $\dfrac{2}{3}$ c) 1

3) $\dfrac{1}{2}$

5) a) 1, 2, 3, 6, 9, 18 b) 1, 2, 4, 5, 8, 10, 20, 40 c) 1, 23

7) a) composite b) composite c) prime

9) Composite. It is divisible by 2 and has other factors as well.

11) a) $2 \times 3 \times 3$ b) $2 \times 3 \times 3 \times 3$ c) $2 \times 3 \times 7$
    d) $2 \times 3 \times 5 \times 5$

13) a) $\dfrac{3}{4}$ b) $\dfrac{3}{4}$ c) $\dfrac{12}{5}$ or $2\dfrac{2}{5}$ d) $\dfrac{3}{7}$

15) a) $\dfrac{6}{35}$ b) $\dfrac{10}{39}$ c) $\dfrac{7}{15}$ d) $\dfrac{12}{25}$ e) $\dfrac{1}{2}$ f) $\dfrac{7}{4}$ or $1\dfrac{3}{4}$

17) She multiplied the whole numbers and multiplied the fractions. She should have converted the mixed numbers to improper fractions before multiplying. Correct answer: $\dfrac{77}{6}$ or $12\dfrac{5}{6}$.

19) a) $\dfrac{1}{12}$ b) $\dfrac{15}{44}$ c) $\dfrac{4}{7}$ d) 7 e) $\dfrac{24}{7}$ or $3\dfrac{3}{7}$ f) $\dfrac{1}{14}$

21) 30 23) a) 30 b) 24 c) 36

25) a) $\dfrac{8}{11}$ b) $\dfrac{3}{5}$ c) $\dfrac{3}{5}$ d) $\dfrac{7}{18}$ e) $\dfrac{29}{30}$ f) $\dfrac{1}{18}$
    g) $\dfrac{71}{63}$ or $1\dfrac{8}{63}$ h) $\dfrac{7}{12}$ i) $\dfrac{7}{5}$ or $1\dfrac{2}{5}$ j) $\dfrac{41}{54}$

27) a) $14\dfrac{7}{11}$ b) $11\dfrac{2}{5}$ c) $6\dfrac{1}{2}$ d) $5\dfrac{9}{20}$ e) $1\dfrac{2}{5}$ f) $4\dfrac{13}{40}$
    g) $11\dfrac{5}{28}$ h) $8\dfrac{9}{20}$

29) four bears; $\dfrac{1}{3}$ yd remaining 31) 36

33) $16\dfrac{1}{2}$ in. by $22\dfrac{5}{8}$ in. 35) $3\dfrac{5}{12}$ cups 37) $5\dfrac{3}{20}$ gal

39) 35 41) $7\dfrac{23}{24}$ in. 43) 240

Section 1.2

1) a) base: 6; exponent: 4 b) base: 2; exponent: 3
   c) base: $\dfrac{9}{8}$; exponent: 5

3) a) 9^4 b) 2^8 c) $\left(\dfrac{1}{4}\right)^3$

5) a) 64 b) 121 c) 16 d) 125 e) 81 f) 144 g) 1
   h) $\dfrac{9}{100}$ i) $\dfrac{1}{64}$ j) 0.09

7) $(0.5)^2 = 0.5 \cdot 0.5 = 0.25$ or $(0.5)^2 = \left(\dfrac{1}{2}\right)^2 = \dfrac{1}{4}$

9) Answers may vary.

11) 19 13) 20 15) 23 17) 17 19) $\dfrac{7}{24}$ 21) $\dfrac{19}{18}$ or $1\dfrac{1}{18}$

23) 4 25) 3 27) 19 29) 9 31) 28 33) 8 35) $\dfrac{5}{6}$

Section 1.3

1) right 3) obtuse 5) supplementary; complementary
7) 66° 9) 45° 11) 80° 13) 109°
15) $m\angle A = m\angle C = 151°, m\angle B = 29°$ 17) 180
19) right; 49° 21) acute; 77°
23) isosceles 25) equilateral 27) true
29) $A = 81$ ft^2; $P = 36$ ft 31) $A = 9$ mm^2; $P = 14.6$ mm
33) $A = 6\dfrac{9}{16}$ mi^2; $P = 10\dfrac{3}{4}$ mi 35) $A = 30$ cm^2; $P = 30$ cm
37) a) $A = 25\pi$ in^2; $A \approx 78.5$ in^2
    b) $C = 10\pi$ in; $C \approx 31.4$ in.
39) a) $A = 100\pi$ cm^2; $A \approx 314$ cm^2
    b) $C = 20\pi$ cm; 62.8 cm
41) $A = 51.84\pi$ ft^2; $C = 14.4\pi$ ft
43) $A = 64\pi$ yd^2; $C = 16\pi$ yd
45) a) $A = 58$ ft^2; $P = 36$ ft b) $A = 112$ cm^2; $P = 48$ cm
    c) $A = 154$ in^2; $P = 60.6$ in.
47) 66 in^2 49) 43.99 m^2 51) 70 m^3 53) 36π in^3
55) $\dfrac{256}{3}\pi$ ft^3 57) 52.2π cm^3
59) a) 270 ft^2 b) No, it would cost $621 to use this carpet.
61) 423.9 cm^3 63) a) 43.96 in. b) 153.86 in^2
65) 426,360 gal 67) Yes. It would cost $1221 per month.
69) 29.5 in. 71) $44.80 73) 25.12 ft^3

Section 1.4

1) a) 6, 0 b) $-14, 6, 0$ c) $\sqrt{19}$ d) 6
   e) $-14, 6, \dfrac{2}{5}, 0, 3.\overline{28}, -1\dfrac{3}{7}, 0.95$
   f) $-14, 6, \dfrac{2}{5}, \sqrt{19}, 0, 3.\overline{28}, -1\dfrac{3}{7}, 0.95$

3) true 5) false 7) true

9)

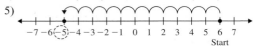

11)

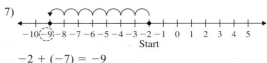

13)

$-6.8 \ -4\frac{1}{3}$ $-\frac{3}{8} 0.2 \ 1\frac{8}{9}$

15) 13　17) $\dfrac{3}{2}$　19) -10　21) -19　23) -11　25) 7

27) 4.2　29) $-10, -2, 0, \dfrac{9}{10}, 3.8, 7$

31) $-9, -4\dfrac{1}{2}, -0.3, \dfrac{1}{4}, \dfrac{5}{8}, 1$　33) true　35) true　37) false

39) false　41) true　43) -27　45) 1.1%

47) -4371　49) 3000

Section 1.5

1) Answers may vary.　3) Answers may vary.

5)

$6 - 11 = -5$

7)

$-2 + (-7) = -9$

9) -4　11) -14　13) 13　15) 18　17) -1451　19) $\dfrac{1}{2}$

21) $-\dfrac{31}{24}$ or $-1\dfrac{7}{24}$　23) $-\dfrac{13}{36}$　25) -3.9　27) 7.31

29) 15　31) -11　33) -2　35) -19　37) 23

39) $-\dfrac{19}{18}$ or $-1\dfrac{1}{18}$　41) $\dfrac{5}{24}$　43) 12　45) 11　47) -7

49) true　51) false　53) true

55) $29{,}028 - (-36{,}201) = 65{,}229$. There is a 65,229-ft difference between Mt. Everest and the Mariana Trench.

57) $51{,}700 - 51{,}081 = -619$. The median income for a male with a bachelor's degree rose by \$619 from 2004 to 2005.

59) $-79.8 + 213.8 = 134$. The highest temperature on record in the United States is 134°F.

61) $7 + 4 + 1 + 6 - 10 = 8$. The Patriots' net yardage on this offensive drive was 8 yd.

63) a) 1900　b) -1200　c) 1300　d) -2100

65) 65) $5 + 7; 12$　67) $10 - 16; -6$　69) $9 - (-8); 17$

71) $-21 + 13; -8$　73) $-20 + 30; 10$　75) $23 - 19; 4$

77) $(-5 + 11) - 18; -12$

Section 1.6

1) positive　3) negative　5) -45　7) 42　9) 30　11) $-\dfrac{14}{15}$

13) -0.3　15) -64　17) 0　19) negative　21) positive

23) 36　25) -125　27) 9　29) -49　31) -32　33) 7

35) -8　37) -4　39) $\dfrac{10}{13}$　41) 0　43) $-\dfrac{9}{7}$ or $-1\dfrac{2}{7}$

45) -33　47) 43　49) 16　51) 16　53) $\dfrac{1}{4}$　55) $-12 \cdot 6; -72$

57) $9 + (-7)(-5); 44$　59) $\dfrac{63}{-9} + 7; 0$　61) $(-4)(-8) - 19; 13$

63) $\dfrac{-100}{4} - (-7 + 2); -20$　65) $2[18 + (-31)]; -26$

67) $\dfrac{2}{3}(-27); -18$　69) $12(-5) + \dfrac{1}{2}(36); -42$

Section 1.7

1)

Term	Coeff.
$7p^2$	7
$-6p$	-6
4	4

The constant is 4.

3)

Term	Coeff.
x^2y^2	1
$2xy$	2
$-y$	-1
11	11

The constant is 11.

5)

Term	Coeff.
$-2g^5$	-2
$\dfrac{g^4}{5}$	$\dfrac{1}{5}$
$3.8g^2$	3.8
g	1
-1	-1

The constant is -1.

7) a) 11　b) -17　9) a) 37　b) 28　11) -9　13) 0

15) -19　17) -1　19) $-\dfrac{9}{5}$　21) 0　23) $\dfrac{1}{6}$　25) associative

27) commutative　29) associative　31) distributive

33) identity　35) distributive　37) $7u + 7v$　39) $4 + k$

41) m　43) $-5 + (3 + 6)$

45) No. Subtraction is not commutative.

47) $5 \cdot 4 + 5 \cdot 3 = 20 + 15 = 35$

49) $(-2) \cdot 5 + (-2) \cdot 7 = -10 + (-14) = -24$

51) $4 \cdot 11 + 4 \cdot (-3) = 44 + (-12) = 32$

53) $-9 + 5 = -4$　55) $8 \cdot 4 + (-2) \cdot 4 = 32 + (-8) = 24$

57) $2 \cdot (-6) + 2 \cdot 5 + 2 \cdot 3 = -12 + 10 + 6 = 4$

59) $9g + 9 \cdot 6 = 9g + 54$　61) $4t + 4(-5) = 4t - 20$

63) $-5z + (-5) \cdot 3 = -5z - 15$

65) $-8u + (-8) \cdot (-4) = -8u + 32$ 67) $-v + 6$

69) $10m + 10 \cdot 5n + 10 \cdot (-3) = 10m + 50n - 30$

71) $8c - 9d + 14$

Chapter 1 Review Exercises

1) 1, 2, 4, 8, 16   3) $2 \cdot 2 \cdot 7$ 5) $\dfrac{2}{5}$ 7) $\dfrac{12}{55}$ 9) $\dfrac{25}{12}$ or $2\dfrac{1}{12}$

11) $\dfrac{21}{4}$ or $5\dfrac{1}{4}$ 13) $\dfrac{2}{3}$ 15) $\dfrac{53}{80}$ 17) $\dfrac{2}{5}$ 19) $8\dfrac{43}{72}$ 21) $3\dfrac{3}{4}$ yd

23) 64   25) 0.36   27) 29   29) 39°   31) acute, 56°

33) $A = 45$ in^2; $P = 38.8$ in.   35) $A = 376$ m^2; $P = 86$ m

37) a) $A = 6.25\pi$ m^2; $A \approx 19.625$ m^2

   b) $C = 5\pi$ m; $C \approx 15.7$ m

39) 936 in^3 41) $\dfrac{400}{3}\pi$ ft^3 43) 2.25π in^3

45)

47) -14 49) $-\dfrac{31}{24}$ or $-1\dfrac{7}{24}$ 51) $-17°$F   53) -10 55) -30

57) -9 59) $-\dfrac{19}{22}$ 61) $\dfrac{2}{9}$ 63) 25   65) -1 67) -2

69) -40 71) $\dfrac{-120}{-3}$; 40   73) $(-4) \cdot 7 - 15$; -43

75)

Term	Coeff.
c^4	1
$12c^3$	12
$-c^2$	-1
$-3.8c$	-3.8
11	11

77) $-\dfrac{29}{9}$ 79) associative   81) commutative

83) $3 \cdot 10 - 3 \cdot 6 = 30 - 18 = 12$ 85) $-12 - 5 = -17$

Chapter 1 Test

1) $2 \cdot 3 \cdot 5 \cdot 7$ 3) $\dfrac{3}{16}$ 5) $3\dfrac{1}{12}$ 7) $-\dfrac{11}{42}$ 9) 28   11) 48

13) 0   15) 14,787 ft   17) -81 19) -10 21) 26°; obtuse

23) a) $A = 40$ in^2; $P = 28$ in.   b) $A = 78$ yd^2; $P = 40$ yd

25) a) 3π in.   b) 4.5π in^3

27)

29)

Term	Coeff.
$5a^2b^2$	5
$2a^2b$	2
$\dfrac{8}{9}ab$	$\dfrac{8}{9}$
a	1
$-\dfrac{b}{7}$	$-\dfrac{1}{7}$
-14	-14

31) a) distributive   b) inverse   c) commutative

Chapter 2

Section 2.1

1) No; the exponents are different.   3) Yes; both are x^4y^3-terms.

5) $24p + 7$ 7) $-19y^2 + 30$ 9) $\dfrac{16}{5}r - \dfrac{2}{9}$ 11) $7w + 10$

13) 0   15) $-5g + 2$ 17) $-3t$ 19) $26x + 37$

21) $\dfrac{11}{10}z + \dfrac{13}{2}$ 23) $\dfrac{65}{8}t - \dfrac{19}{16}$ 25) $-1.1x - 19.6$

27) $18 + x$ 29) $x - 6$ 31) $x - 3$ 33) $12 + 2x$

35) $(3 + 2x) - 7$; $2x - 4$ 37) $(x + 15) - 5$; $x + 10$

Section 2.2a

1) 9^4 3) $\left(\dfrac{1}{7}\right)^5$ 5) $(-5)^7$ 7) $(-3y)^8$ 9) base: 6; exponent: 8

11) base: 0.05; exponent: 7   13) base: -8; exponent: 5

15) base: $9x$; exponent: 8   17) base: $-11a$; exponent: 2

19) base: p; exponent: 4   21) base: y; exponent: 2

23) $(3 + 4)^2 = 49$, $3^2 + 4^2 = 25$. They are not equivalent
   because when evaluating $(3 + 4)^2$, first add $3 + 4$ to get 7,
   then square the 7.

25) Answers may vary.

27) No. $3t^4 = 3 \cdot t^4$; $(3t)^4 = 3^4 \cdot t^4 = 81t^4$

29) 32   31) 121   33) 16   35) -81 37) -8 39) $\dfrac{1}{125}$

41) 64   43) 81   45) 200   47) $\dfrac{1}{32}$ 49) 8^{12} 51) 7^{10}

53) $(-4)^7$ 55) a^5 57) k^6 59) $8y^5$ 61) $54m^{15}$ 63) $-42r^5$

65) $28t^{16}$ 67) $-40x^6$ 69) $8b^{15}$ 71) x^{12} 73) t^{42} 75) 64

77) $(-5)^6$ 79) $\dfrac{1}{32}$ 81) $\dfrac{64}{y^3}$ 83) $\dfrac{d^8}{c^8}$ 85) $125z^3$ 87) $81p^4$

89) $-64a^3b^3$ 91) $6x^3y^3$ 93) $-9t^4u^4$

95) a) $A = 3w^2$ sq units; $P = 8w$ units
   b) $A = 5k^5$ sq units; $P = 10k^3 + 2k^2$ units

97) $\dfrac{3}{8}x^2$ sq units

Section 2.2b

1) operations   3) k^{24} 5) $200z^{26}$ 7) $-9p^{51}q^{16}$ 9) 64

11) $-64t^{18}u^{26}$ 13) $288k^{14}t^4$ 15) $\dfrac{3}{4g^{15}}$ 17) $\dfrac{49}{4}n^{22}$

19) $900h^{28}$ 21) $-147w^{45}$ 23) $\dfrac{36x^6}{25y^{10}}$ 25) $\dfrac{d^{18}}{4c^{30}}$ 27) $\dfrac{3a^{40}b^{15}}{4c^3}$

29) $\dfrac{r^{39}}{242t^5}$ 31) $\dfrac{2}{3}x^{24}y^{14}$ 33) $-\dfrac{1}{10}c^{29}d^{18}$ 35) $\dfrac{125x^{15}y^6}{z^{12}}$

37) $\dfrac{81t^{16}u^{36}}{16v^{28}}$ 39) $\dfrac{9w^{10}}{x^6y^{12}}$ 41) a) $20l^2$ units   b) $25l^4$ sq units

43) a) $\dfrac{3}{8}x^2$ sq units   b) $\dfrac{11}{4}x$ units

Section 2.3a

1) false   3) true   5) 1   7) -1 9) 0   11) 2   13) $\dfrac{1}{36}$

15) $\dfrac{1}{16}$ 17) $\dfrac{1}{125}$ 19) 64   21) 32   23) $\dfrac{27}{64}$ 25) $\dfrac{49}{81}$ 27) -64

29) $\dfrac{64}{9}$ 31) $-\dfrac{1}{64}$ 33) -1 35) $\dfrac{1}{16}$ 37) $\dfrac{13}{36}$ 39) $\dfrac{83}{81}$

Section 2.3b

1) a) w b) n c) $2p$ d) c 3) 1   5) -2 7) 2   9) $\dfrac{1}{y^4}$

11) $\dfrac{1}{p}$ 13) $\dfrac{b^3}{a^{10}}$ 15) $\dfrac{x^5}{y^8}$ 17) $\dfrac{x^4y^5}{10}$ 19) $\dfrac{9a^4}{b^3}$ 21) c^5d^8

23) $\dfrac{8a^6c^{10}}{5bd}$ 25) $2x^7y^6z^4$ 27) $\dfrac{36}{a^2}$ 29) $\dfrac{q^5}{32n^5}$ 31) $\dfrac{c^2d^2}{144b^2}$

33) $-\dfrac{9}{k^2}$ 35) $\dfrac{3}{t^3}$ 37) $-\dfrac{1}{m^9}$ 39) z^{10} 41) j 43) $5n^2$ 45) cd^3

Section 2.4

1) d^7 3) m^2 5) $9t^5$ 7) 36   9) 81   11) $\dfrac{1}{16}$ 13) $\dfrac{1}{125}$

15) $10k^3$ 17) $\dfrac{2}{3}c^5$ 19) $\dfrac{1}{z^6}$ 21) $\dfrac{1}{x^9}$ 23) $\dfrac{1}{r^2}$ 25) $\dfrac{1}{a^{10}}$

27) t^3 29) $\dfrac{15}{w^8}$ 31) $-\dfrac{6}{k^3}$ 33) a^3b^7 35) $\dfrac{m^4}{3n^8}$ 37) $\dfrac{10}{x^2y^8}$

39) $\dfrac{w^6}{9v^3}$ 41) $\dfrac{3}{8}c^4d$ 43) $(x+y)^4$ 45) $(c+d)^6$

Chapter 2 Mid-Chapter Summary

1) $\dfrac{16}{81}$ 3) 1   5) $\dfrac{9}{100}$ 7) 25   9) $\dfrac{1}{81}$ 11) $\dfrac{1}{81}$ 13) $-\dfrac{125}{27}$

15) $-\dfrac{1}{48}$ 17) $270g^{12}$ 19) $\dfrac{23}{t^{10}}$ 21) $\dfrac{8}{27}x^{30}y^{18}$ 23) $\dfrac{s^{16}}{49r^6}$

25) $-k^{12}$ 27) $-32m^{25}n^{10}$ 29) $-\dfrac{3}{2}z^4$ 31) $\dfrac{b^{24}}{a^{30}}$ 33) $a^{14}b^3c^7$

35) $\dfrac{27u^{30}}{64v^{21}}$ 37) $81t^8u^{20}$ 39) $\dfrac{1}{h^{21}}$ 41) $\dfrac{h^5}{32}$ 43) $56c^{14}$ 45) $\dfrac{3}{a^5}$

47) $\dfrac{6}{55}d^{11}$ 49) $\dfrac{9}{64n^6}$ 51) p^{10n} 53) y^{11m} 55) $\dfrac{1}{x^{3a}}$ 57) $\dfrac{3}{5c^{6x}}$

Section 2.5

1) yes   3) no   5) no   7) yes   9) Answers may vary.
11) Answers may vary.   13) -0.000068 15) $-52,600$
17) 0.000008   19) 602,196.7   21) 3,000,000   23) -0.000744
25) 2.1105×10^3 27) 9.6×10^{-5} 29) -7×10^6
31) 3.4×10^3 33) 8×10^{-4} 35) -7.6×10^{-2}
37) 6×10^3 39) 3.808×10^8 kg   41) 1×10^{-8} cm
43) 30,000   45) 690,000   47) -1200 49) -0.06
51) -0.0005 53) 160,000   55) 0.0001239
57) 5,256,000,000 particles   59) 17,000 lb/cow
61) 26,400,000 droplets

Chapter 2 Review Exercises

1) $17y^2 - 3y - 3$ 3) $\dfrac{31}{4}n - \dfrac{9}{2}$ 5) $x - 10$

7) $(x+8) - 3; x + 5$ 9) a) 8^6 b) $(-7)^4$

11) a) 32   b) $\dfrac{1}{27}$ c) 7^{12} d) k^{30}

13) a) $125y^3$ b) $-14m^{16}$ c) $\dfrac{a^6}{b^6}$ d) $6x^2y^2$ e) $\dfrac{25}{3}c^8$

15) a) z^{22} b) $-18c^{10}d^{16}$ c) 125   d) $\dfrac{25t^6}{2u^{21}}$

17) a) 1   b) -1 c) $\dfrac{1}{9}$ d) $-\dfrac{5}{36}$ e) $\dfrac{125}{64}$

19) a) $\dfrac{1}{v^9}$ b) $\dfrac{c^2}{81}$ c) y^8 d) $-\dfrac{7}{k^9}$ e) $\dfrac{19a}{z^4}$ f) $\dfrac{20n^5}{m^6}$ g) $\dfrac{k^5}{32j^5}$

21) a) 9   b) r^8 c) $\dfrac{3}{2t^5}$ d) $\dfrac{3x^7}{5y}$

23) a) $81s^{16}t^{20}$ b) $2a^{16}$ c) $\dfrac{y^{18}}{z^{24}}$ d) $-36x^{11}y^{11}$ e) $\dfrac{d^{25}}{c^{35}}$

f) $8m^3n^{12}$ g) $\dfrac{125t^9}{27k^{18}}$ h) 14

25) a) y^{10k} b) x^{10p} c) z^{7c} d) $\dfrac{1}{t^{5d}}$ 27) -418.5 29) 0.00067

31) 20,000   33) 5.75×10^{-5} 35) 3.2×10^7 37) 1.78×10^5
39) 9.315×10^{-4} 41) 0.0000004   43) 3.6   45) 7500
47) 30,000 quills   49) 0.00000000000000299 g

Chapter 2 Test

1) a) $-6k^2 + 4k - 14$ b) $6c - \dfrac{49}{6}$ 3) 81   5) $\dfrac{1}{32}$ 7) $-\dfrac{27}{64}$

9) $125n^{18}$ 11) m^6 13) $-\dfrac{t^{33}}{27u^{27}}$ 15) t^{13k} 17) 1.65×10^{-4}

19) 21,800,000

Cumulative Review for Chapters 1–2

1) $\dfrac{3}{5}$ 3) $\dfrac{7}{25}$ 5) -28 7) 42   9) 62   11) $-3t^2 + 37t - 25$

13) $-28z^8$ 15) $-\dfrac{32b^5}{a^{30}}$ 17) 58,280

Chapter 3

Section 3.1

1) equation   3) expression   5) No, it is an expression.

7) b, d   9) no   11) yes   13) yes   15) $\{17\}$ 17) $\{-6\}$

19) $\{-4\}$ 21) $\left\{-\dfrac{1}{8}\right\}$ 23) $\{10\}$ 25) $\{-7\}$ 27) $\{8\}$

29) $\{48\}$ 31) $\{-48\}$ 33) $\{-15\}$ 35) $\{18\}$ 37) $\left\{\dfrac{11}{5}\right\}$

39) $\{12\}$ 41) $\{7\}$ 43) $\{0\}$ 45) $\{8\}$ 47) $\{2\}$ 49) $\{0\}$

51) $\left\{-\dfrac{2}{5}\right\}$ 53) $\{-1\}$ 55) $\{27\}$ 57) $\left\{-\dfrac{7}{5}\right\}$ 59) $\left\{-\dfrac{3}{2}\right\}$

61) $\{6\}$ 63) $\{-3\}$ 65) $\{19\}$ 67) $\left\{-\dfrac{5}{4}\right\}$ 69) $\{-3\}$

71) $\{0\}$

Section 3.2

1) $\{8\}$ 3) $\{3\}$ 5) $\{-3\}$ 7) $\left\{\dfrac{7}{3}\right\}$ 9) $\{4\}$ 11) $\varnothing$

13) {all real numbers}   15) {all real numbers}   17) $\{0\}$

19) $\left\{\dfrac{11}{2}\right\}$ 21) $\{5\}$ 23) $\{3\}$ 25) $\left\{-\dfrac{15}{2}\right\}$ 27) $\{-8\}$

29) $\{5\}$ 31) $\{4\}$ 33) $\{300\}$ 35) 11   37) 29   39) -8

41) 15   43) 14   45) 16   47) -14 49) 3   51) 6   53) -11

Section 3.3

1) $c + 5$ 3) $p - 31$ 5) $3w$ 7) $14 - x$

9) Pepsi = 9.8 tsp, Gatorade = 3.3 tsp

11) Greece: 16, Thailand: 8

13) Columbia: 1240 mi, Ohio: 1310 mi   15) 11 in. and 25 in.

17) 12 ft and 6 ft   19) 64, 65, 66   21) 12, 14

23) $-15, -13, -11$ 25) 172, 173   27) $49   29) 64 mi

31) 124 mi   33) ride: 4 mi; walk: 3 mi   35) 72, 74, 76

37) *Shrek:* $267.7 million; *Harry Potter:* $309.7 million

39) 148 mi

41) Avril Lavigne: 4.1 million
Nelly: 4.9 million
Eminem: 7.6 million

43) 13 ft, 19 ft, and 26 ft

Section 3.4

1) $63.75   3) $11.55   5) $11.60   7) $140.00   9) $10.95

11) $32.50   13) 80   15) 425   17) $32   19) $6890   21) $380

23) $9000 at 6% and $6000 at 7%

25) $1400 at 6% and $1600 at 5%

27) $2800 at 9.5% and $4200 at 7%   29) $8.75   31) 8500

33) CD: $1500; IRA: $3000; Mutual Fund: $2500

35) $79.00   37) $38,600   39) $6000 at 5% and $14,000 at 9%

Section 3.5

1) 25 ft   3) 6 in.   5) 415 ft^2 7) 12 ft   9) 6 in.

11) $m\angle A = 26°$, $m\angle B = 52°$

13) $m\angle A = 37°$, $m\angle B = 55°$, $m\angle C = 88°$

15) 68°, 68°   17) 150°, 150°   19) 133°, 47°   21) 79°, 101°

23) $180 - x$ 25) 17°   27) angle: 20°, comp.: 70°, supp.: 160°

29) 35°   31) 40°   33) 2   35) 4   37) 18   39) 2   41) 7

43) 20   45) a) $x = 23$ b) $x = p - n$ c) $x = v - q$

47) a) $n = 6$ b) $n = \dfrac{c}{y}$ c) $n = \dfrac{d}{w}$

49) a) $c = 21$ b) $c = ur$ c) $c = xt$

51) a) $d = 3$ b) $d = \dfrac{z + a}{k}$ 53) a) $z = -\dfrac{5}{2}$ b) $z = \dfrac{w - t}{y}$

55) $m = \dfrac{F}{a}$ 57) $c = nv$ 59) $\sigma = \dfrac{E}{T^4}$ 61) $h = \dfrac{3V}{\pi r^2}$

63) $E = IR$ 65) $R = \dfrac{I}{PT}$ 67) $I = \dfrac{P - 2w}{2}$ 69) $N = \dfrac{2.5H}{D^2}$

71) $b_2 = \dfrac{2A}{h} - b_1$ or $b_2 = \dfrac{2A - hb_1}{h}$ 73) $h^2 = \dfrac{S}{\pi} - \dfrac{c^2}{4}$

75) a) $w = \dfrac{P - 2l}{2}$ b) 3 cm

77) a) $F = \dfrac{9}{5}C + 32$ b) 77°F

Section 3.6

1) $\dfrac{3}{5}$ 3) $\dfrac{4}{3}$

5) A ratio is a quotient of two quantities. A proportion is a statement that two ratios are equal.

7) false   9) true   11) 2   13) $\dfrac{48}{5}$ 15) -1 17) -2

19) 30   21) 82.5 mg   23) 168   25) $x = 10$

27) $x = 13$ 29) $x = 63$ 31) a) $.80   b) 80¢

33) a) $2.17   b) 217¢   35) a) $2.95   b) 295¢

37) a) $0.25q$ b) $25q$ 39) a) $0.10d$ b) $10d$

41) a) $0.01p + 0.25q$ b) $p + 25q$ 43) 9 nickels, 17 quarters

45) 11 $5 bills, 14 $1 bills

47) 38 adult tickets, 19 children's tickets   49) 2 oz   51) 7.6 mL

53) 16 oz of the 4% acid solution, 8 oz of the 10% acid solution

55) $2\dfrac{1}{2}$ L   57) 2 lb of Aztec and 3 lb of Cinnamon

61) 4:30 P.M.   63) 48 mph   65) 23 dimes, 16 quarters

67) 1560 calories   69) jet: 400 mph, small plane: 200 mph

71) 8 cc of the 0.08% solution and 12 cc of the 0.03% solution

73) $1\dfrac{1}{5}$ gal

Section 3.7

1)
   a) $\{x|x \ge 3\}$ b) $[3, \infty)$

3)
   a) $\{c|c < -1\}$ b) $(-\infty, -1)$

5)
   a) $\left\{w\middle|w > -\dfrac{11}{3}\right\}$ b) $\left(-\dfrac{11}{3}, \infty\right)$

7)
   a) $\{n|1 \le n \le 4\}$ b) $[1, 4]$

9)
   a) $\{a|-2 < a < 1\}$ b) $(-2, 1)$

11)
   a) $\left\{z\middle|\dfrac{1}{2} < z \le 3\right\}$ b) $\left(\dfrac{1}{2}, 3\right]$

13) You use parentheses when there is a $<$ or $>$ symbol or when you use ∞ or $-\infty$.

15)
   a) $\{n|n \le 5\}$ b) $(-\infty, 5]$

17)
   a) $\{y|y \ge -4\}$ b) $[-4, \infty)$

19)
   a) $\{c|c > 4\}$ b) $(4, \infty)$

21)
   a) $\left\{k\middle|k < -\dfrac{11}{3}\right\}$ b) $\left(-\infty, -\dfrac{11}{3}\right)$

23)
   a) $\{b|b \ge -8\}$ b) $[-8, \infty)$

25)
   a) $\{w|w < 3\}$ b) $(-\infty, 3)$

27)
   a) $\{x|x < -6\}$ b) $(-\infty, -6)$

29)
   a) $\{p|p \le -10\}$ b) $(-\infty, -10]$

31)
   $(-1, \infty)$

33)
   $\left(-\infty, -\dfrac{3}{7}\right]$

35)
   $(-3, \infty)$

37)
   $(-\infty, -2]$

39)
   $(-\infty, 4)$

41)
   $(0, \infty)$

43)
   $(-\infty, 5]$

45)
   $[-3, 1]$

47)
   $\left(\dfrac{3}{2}, 3\right)$

49)
   $[-4, -1]$

51)
   $\left[\dfrac{7}{4}, 3\right)$

53)
   $(-8, 4)$

55)
   $\left[-1, -\dfrac{2}{5}\right]$

57)
   $(1, 3]$

59)
   $\left[-1, \dfrac{4}{3}\right)$

61) He can rent the truck for at most 2 hr 45 min.

63) at most 8 mi   65) 89 or higher

Section 3.8

1) $A \cap B$ means "A intersect B." $A \cap B$ is the set of all numbers which are in set A *and* in set B.

3) $\{8\}$ 5) $\{2, 4, 5, 6, 7, 8, 9, 10\}$ 7) $\varnothing$

9) $\{1, 2, 3, 4, 5, 6, 8, 10\}$

11)
$[-3, 2]$

13)
$(-1, 3)$

15)
$[3, \infty)$

17)
$\varnothing$

19)
$[2, 5]$

21)
$(-2, 3)$

23)
$(-3, 4]$

25)
$\varnothing$

27)
$(3, \infty)$

29)
$[-4, 1]$

31)
$(-\infty, -1) \cup (5, \infty)$

33)
$\left(-\infty, \dfrac{5}{3}\right] \cup (4, \infty)$

35)
$(1, \infty)$

37)
$(-\infty, \infty)$

39)
$(-\infty, -1) \cup (3, \infty)$

41)
$\left(-\infty, \dfrac{7}{2}\right] \cup (6, \infty)$

43)
$(-5, \infty)$

45)
$(-\infty, -6) \cup [-3, \infty)$

47)
$(-\infty, \infty)$

49)
$(-\infty, -2]$

Chapter 3 Review Exercises

1) yes   3) $\left\{-\dfrac{10}{3}\right\}$ 5) $\{19\}$ 7) $\left\{\dfrac{45}{14}\right\}$ 9) $\{35\}$ 11) $\{8\}$

13) $\{-2\}$ 15) $\{2\}$ 17) $\{5\}$ 19) $\{-2\}$

21) $\{$all real numbers$\}$ 23) $\{3\}$ 25) 17   27) 10   29) $26 - c$

31) Kelly Clarkson: 297,000 copies; Clay Aiken: 613,000 copies

33) 125 ft   35) $6724   37) $19.9 million   39) 12 cm

41) 79°, 101°   43) 48°   45) $R = \dfrac{pV}{nT}$ 47) $\{20\}$ 49) 17

51) 58 dimes, 33 quarters   53) 45 min

55)
$\left(-\infty, -\dfrac{5}{2}\right)$

57)
$(-2, 3]$

59) 88 or higher

61)
$[1, 3]$

63)
$(-1, \infty)$

Chapter 3 Test

1) $\left\{-\dfrac{5}{3}\right\}$ 3) -3 5) $\varnothing$ 7) 21

9) 3.5 qt of regular oil, 1.5 qt of synthetic oil

11) 70 ft   13) $h = \dfrac{S - 2\pi r^2}{2\pi r}$

15)
$\left(\dfrac{3}{4}, \infty\right)$

17) $[0, 3]$

Cumulative Review for Chapters 1–3

1) $-\dfrac{13}{36}$ 3) 64   5) $0, 8, \sqrt{81}, -2$ 7) $\dfrac{2}{3}k^7$ 9) $-\dfrac{5}{3}m^2$

11) 2.79×10^8

13)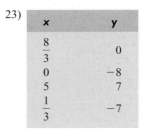

$\left(-4, -\dfrac{7}{6}\right)$

15) $R = \dfrac{A - P}{PT}$ 17) $\dfrac{1}{2}$ hr

Chapter 4

Section 4.1

1) Answers may vary.   3) yes   5) yes   7) no   9) yes

11) no   13) -3 15) $-\dfrac{7}{2}$ 17) -4

19)

x	y
0	4
1	1
2	-2
-1	7

21)

x	y
0	10
$-\dfrac{1}{2}$	4
-1	-2
3	46

23)

x	y
$\dfrac{8}{3}$	0
0	-8
5	7
$\dfrac{1}{3}$	-7

25)

x	y
5	0
5	4
5	-1
5	-8

27) Answers may vary.

29) A: (5, 1); quadrant I
   B: (2, −3); quadrant IV
   C: (−2, 4); quadrant II
   D: (−3, −4); quadrant III
   E: (3, 0); no quadrant
   F: (0, −2); no quadrant

31–33)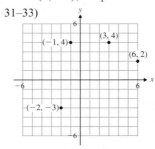

35–37)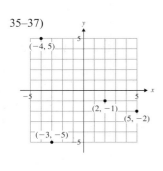

39–41)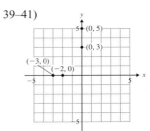

43–45)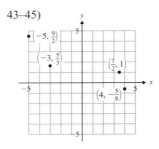

47)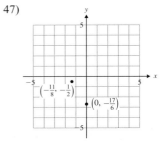

49)

x	y
0	3
$\dfrac{3}{4}$	0
2	-5
-1	7

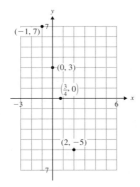

51)

x	y
0	0
4	4
-2	-2
-3	-3

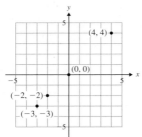

53)

x	y
0	-3
2	0
4	3
-3	$-\dfrac{15}{2}$

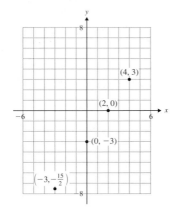

55)

x	y
−6	0
−6	−5
−6	2
−6	4

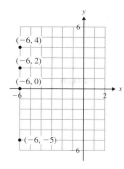

57)

x	y
0	2
4	−3
8	−8
1	$\frac{3}{4}$

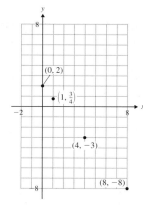

59) a) $(3, -5), (6, -3), (-3, -9)$

b) $\left(1, -\frac{19}{3}\right), \left(5, -\frac{11}{3}\right), \left(-2, -\frac{25}{3}\right)$

c) The x-values in part a) are multiples of the denominator of $\frac{2}{3}$. So, when you multiply $\frac{2}{3}$ by a multiple of 3 the fraction is eliminated.

61) positive   63) positive   65) negative   67) zero

69) a) $(16, 1300), (17, 1300), (18, 1600), (19, 1500)$

b)

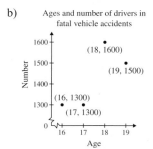

c) There were 1600 18-year-old drivers involved in fatal motor vehicle accidents in 2002.

71) a)

x	y
1	120
3	160
4	180
6	220

$(1, 120), (3, 160), (4, 180), (6, 220)$

b)

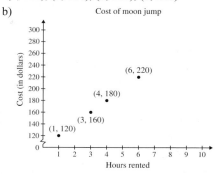

c) The cost of renting the moon jump for 4 hr is $180.

d) Yes, they lie on a straight line.   e) 9 hr

Section 4.2

1) an infinite number

3)

x	y
0	−1
1	2
2	5
−1	−4

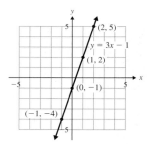

5)

x	y
0	4
−3	6
3	2
6	0

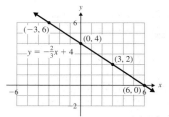

7)

x	y
0	$\frac{3}{2}$
−3	0
5	4
−1	1

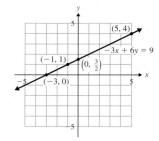

9)

x	y
0	−4
−3	−4
−1	−4
2	−4

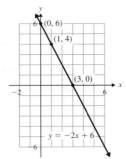

25) (5, 0), (5, 2), (5, −1)
Answers may vary.

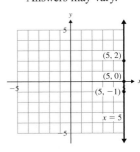

27) (0, 0), (1, 0), (−2, 0)
Answers may vary.

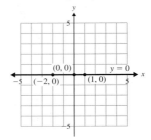

11) Let $x = 0$ in the equation, and solve for y.

13) (3, 0), (0, 6), (1, 4)

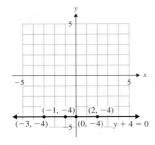

15) $(4, 0), (0, -3), \left(2, -\dfrac{3}{2}\right)$

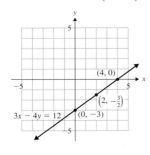

29) (0, −3), (1, −3), (−3, −3)
Answers may vary.

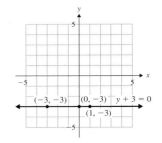

17) (−1, 0), (0, 4), (1, 8)

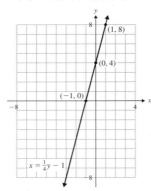

19) $(-3, 0), (0, -2), \left(1, -\dfrac{8}{3}\right)$

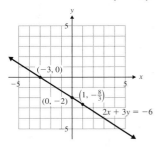

31) $(8, 0), \left(0, \dfrac{8}{3}\right), (2, 2)$

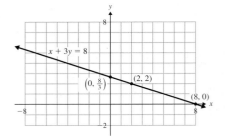

21) (0, 0), (1, −1), (−1, 1)

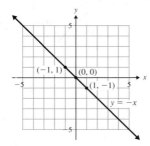

23) (0, 0), (5, 2), (−5, −2)

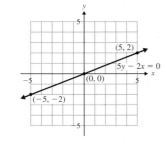

33) a)

x	y
0	0
10	15
20	30
60	90

(0, 0), (10, 15), (20, 30), (60, 90)

b) (0, 0): Before engineers began working ($x = 0$ days), the tower did not move toward vertical ($y = 0$).
(10, 15): After 10 days of working, the tower was moved 15 mm toward vertical.
(20, 30): After 20 days of working, the tower was moved 30 mm toward vertical.
(60, 90): After 60 days of working, the tower was moved 90 mm toward vertical.

c)

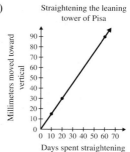

Straightening the leaning tower of Pisa

d) 300 days

17) $\dfrac{1}{2}$ 19) -1 21) $-\dfrac{2}{9}$ 23) 0

25) undefined   27) $\dfrac{14}{3}$ 29) -2

31) a) $\dfrac{5}{6}$ b) $\dfrac{2}{3}$ c) $m = \dfrac{1}{3}$; 4-12 pitch

33) a) \$22,000   b) negative

c) The value of the car is decreasing over time.

d) $m = -2000$; the value of the car is decreasing \$2000 per year.

35) a) Answers will vary.

b) 29.86 in.; 28.86 in.; 26.36 in.; 24.86 in.

c)

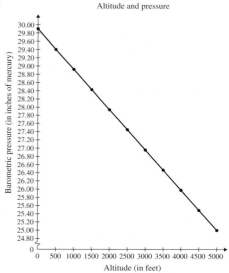

Altitude and pressure

d) No, because the problem states that the equation applies to altitudes 0 ft–5000 ft.

Section 4.3

1) The slope of a line is the ratio of vertical change to horizontal change. It is $\dfrac{\text{change in } y}{\text{change in } x}$ or $\dfrac{\text{rise}}{\text{run}}$ or $\dfrac{y_2 - y_1}{x_2 - x_1}$ where (x_1, y_1) and (x_2, y_2) are points on the line.

3) It slants downward from left to right.   5) undefined

7) $m = \dfrac{3}{4}$ 9) $m = -\dfrac{1}{3}$ 11) $m = -5$ 13) $m = 0$

15)

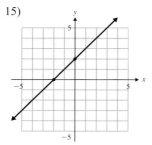

35)

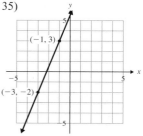

37)

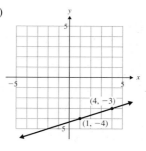

39)

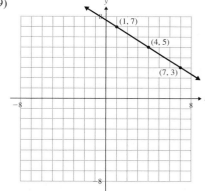

41)

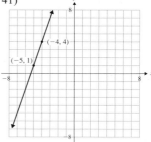

43)

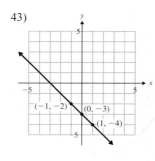

45)
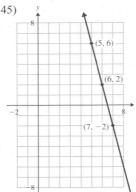

47)
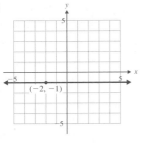

11) $m = -1$, y-int: $(0, 5)$ 13) $m = \dfrac{3}{2}$, y-int: $\left(0, \dfrac{1}{2}\right)$

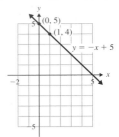

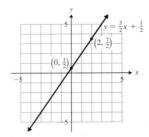

49)
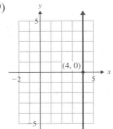

51)
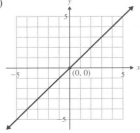

15) $m = 0$, y-int: $(0, -2)$ 17) $y = -\dfrac{1}{3}x - 2$

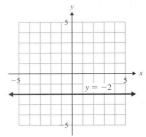

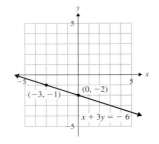

Section 4.4

1) The slope is m, and the y-intercept is $(0, b)$.

3) $m = \dfrac{2}{5}$, y-int: $(0, -6)$ 5) $m = -\dfrac{5}{3}$, y-int: $(0, 4)$

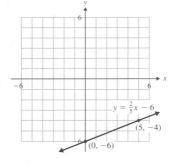

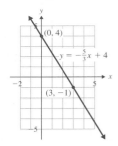

19) $y = \dfrac{3}{2}x - 4$

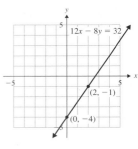

21) This cannot be written in slope-intercept form.

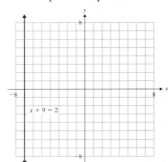

23) $y = \dfrac{5}{2}x + 3$ 25) $y = 0$

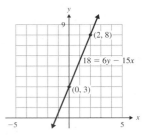

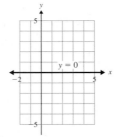

7) $m = \dfrac{3}{4}$, y-int: $(0, 1)$ 9) $m = 4$, y-int: $(0, -2)$

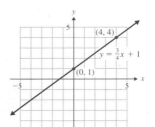

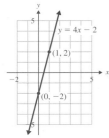

27) a) $(0, 34{,}000)$; if Dave has $0 in sales, his income is $34,000.

   b) $m = 0.05$; Dave earns $0.05 for every $1 in sales.

   c) $38,000

29) a) In 1945, 40.53 gal of whole milk were consumed per person, per year.

   b) $m = -0.59$; since 1945, Americans have been consuming 0.59 fewer gallons of whole milk each year.

   c) Estimate from the graph: 7.5 gal. Consumption from the equation: 8.08 gal.

31) $y = -4x + 7$ 33) $y = \frac{8}{5}x - 6$ 35) $y = \frac{1}{3}x + 5$

37) $y = -x$ 39) $y = -2$

Section 4.5

1) $8x - y = -3$ 3) $x + 2y = -11$ 5) $4x - 5y = 5$

7) $6x + 4y = 1$

9) Substitute the slope and y-intercept into $y = mx + b$.

11) $y = -7x + 2$ 13) $x - y = 3$ 15) $x + 3y = -12$

17) $y = x$ 19) a) $y - y_1 = m(x - x_1)$

   b) Substitute the slope and point into the point-slope formula.

21) $y = 5x + 1$ 23) $y = -x - 5$ 25) $4x - y = -7$

27) $2x + 3y = 33$ 29) $y = \frac{1}{6}x - \frac{13}{3}$ 31) $5x + 9y = 30$

33) Find the slope and use it and one of the points in the point-slope formula.

35) $y = x + 1$ 37) $y = -\frac{1}{3}x + \frac{10}{3}$ 39) $3x - 4y = 17$

41) $2x + y = 9$ 43) $y = 5.0x - 8.3$ 45) $x + 8y = -6$

47) $y = \frac{3}{2}x - 4$ 49) $y = -x - 2$ 51) $y = 5$

53) $y = 4x + 15$ 55) $y = \frac{8}{3}x - 9$ 57) $y = -\frac{3}{4}x - \frac{11}{4}$

59) $x = 3$ 61) $y = -8$ 63) $y = -\frac{1}{2}x - 3$

65) $y = -\frac{1}{3}x + 2$ 67) $y = x$

69) a) $y = 8700x + 1{,}257{,}900$

   b) The population of Maine is increasing by 8700 people per year.

   c) 1,257,900; 1,292,700   d) 2018

71) a) $y = -6.4x + 124$

   b) The number of farms with milk cows is decreasing by 6.4 thousand (6400) per year.

   c) 79.2 thousand (79,200)

73) a) $y = 12{,}318.7x + 6479$

   b) The number of registered hybrid vehicles is increasing by 12,318.7 per year.

   c) 31,116.4; this is slightly lower than the actual value.

   d) 129,666

Section 4.6

1) Answers may vary.   3) perpendicular   5) parallel

7) neither   9) perpendicular   11) neither   13) parallel

15) parallel   17) perpendicular   19) parallel

21) perpendicular   23) neither   25) parallel   27) perpendicular

29) $y = 4x + 2$ 31) $x - 2y = -6$ 33) $4x + 3y = -24$

35) $y = -\frac{1}{5}x + 10$ 37) $y = -\frac{3}{2}x + 6$ 39) $x - 5y = 10$

41) $y = x$ 43) $5x + 8y = 24$ 45) $y = -3x + 8$

47) $y = 2x - 5$ 49) $x = -1$ 51) $x = 2$

53) $y = -\frac{2}{7}x + \frac{1}{7}$ 55) $y = -\frac{5}{2}$

Section 4.7

1) a) any set of ordered pairs

   b) Answers may vary.

   c) Answers may vary.

3) domain: $\{-8, -2, 1, 5\}$ 5) domain: $\{1, 9, 25\}$
   range: $\{-3, 4, 6, 13\}$ range: $\{-3, -1, 1, 5, 7\}$
   function   not a function

7) domain: $\{-2, 1, 2, 5\}$ 9) domain: $\{-2, 1, 5, 8\}$
   range: $\{6, 9, 30\}$ range: $\{-7, -3, 12, 19\}$
   function   not a function

11) domain: $(-\infty, \infty)$;   13) domain: $[-\infty, 4]$;
   range: $(-\infty, \infty)$ range: $(-\infty, \infty)$
   function   not a function

15) domain: $(-\infty, \infty)$;
   range: $(-\infty, 6]$
   function

17) yes   19) yes   21) no   23) no   25) yes

27) $(-\infty, \infty)$; function   29) $(-\infty, \infty)$; function

31) $[0, \infty)$; not a function   33) $(-\infty, 0) \cup (0, \infty)$; function

35) $(-\infty, -4) \cup (-4, \infty)$; function

37) $(-\infty, 5) \cup (5, \infty)$; function

39) $(-\infty, -8) \cup (-8, \infty)$; function

41) $\left(-\infty, \frac{3}{5}\right) \cup \left(\frac{3}{5}, \infty\right)$; function

43) $\left(-\infty, -\frac{4}{3}\right) \cup \left(-\frac{4}{3}, \infty\right)$; function

45) $(-\infty, 0) \cup (0, \infty)$; function

47) $\left(-\infty, \frac{1}{2}\right) \cup \left(-\frac{1}{2}, \infty\right)$; function

49) $(-\infty, 3) \cup (3, \infty)$; function

Section 4.8

1) y is a function, and y is a function of x.

3) a) $y = 7$ b) $f(3) = 7$ 5) -13 7) 7   9) 50

11) -10 13) $-\frac{25}{4}$ 15) $f(-1) = 10, f(4) = -5$

17) $f(-1) = 6, f(4) = 2$ 19) -4 21) 6

23) a) $f(c) = -7c + 2$ b) $f(t) = -7t + 2$

   c) $f(a + 4) = -7a - 26$ d) $f(z - 9) = -7z + 65$

   e) $g(k) = k^2 - 5k + 12$ f) $g(m) = m^2 - 5m + 12$

25)

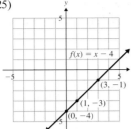

27)

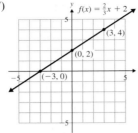

45)

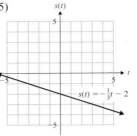

47)

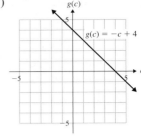

29)

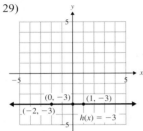

31) x-int: (−1, 0); y-int: (0, 3)

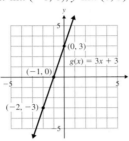

49)

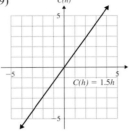

51) a) 108 mi   b) 216 mi   c) 2.5 hr
   d)

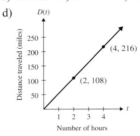

53) a) $E(10) = 75$; when Jenelle works for 10 hr, she earns $75.00.

   b) $E(15) = 112.5$; when Jenelle works 15 hr, she earns $112.50.

   c) $t = 28$; for Jenelle to earn $210.00, she must work 28 hr.

55) a) 253.56 MB   b) 1267.8 MB   c) 20 sec

   d)

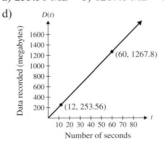

33) x-int: (4, 0); y-int: (0, 2)

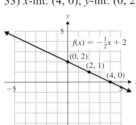

35) intercept: (0, 0)

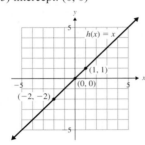

57) a) 60,000   b) 3.5 sec

59) a) $S(50) = 2205$; after 50 sec, the CD player reads 2,205,000 samples of sound.

   b) $S(180) = 7938$; after 180 sec (or 3 min), the CD player reads 7,938,000 samples of sound.

   c) $t = 60$; the CD player reads 2,646,000 samples of sound in 60 seconds (or 1 min).

37) $m = -4$; y-int: (0, −1)

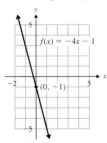

39) $m = \dfrac{3}{5}$; y-int: (0, −2)

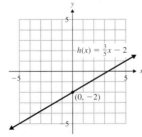

41) $m = 2$; y-int: $\left(0, \dfrac{1}{2}\right)$

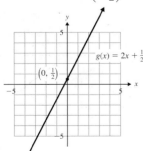

43) $m = -\dfrac{5}{2}$; y-int: (0, 4)

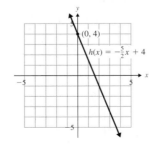

Chapter 4 Review Exercises

1) yes   3) yes   5) 28   7) −8

9)

x	y
0	−11
3	−8
−1	−12
−5	−16

11)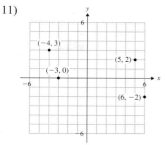

13) a)

x	y
1	0.10
2	0.20
7	0.70
10	1.00

(1, 0.10), (2, 0.20), (7, 0.70), (10, 1.00)

b)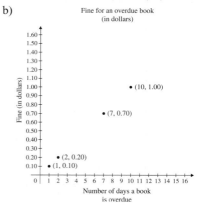

c) If a book is 14 days overdue, the fine is $1.40.

15)

x	y
0	3
1	1
2	−1
−2	7

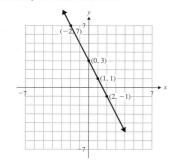

17) (6, 0), (0, −3);
    (2, −2) may vary.

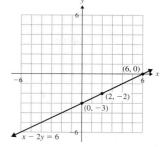

19) (24, 0), (0, 4);
    (12, 2) may vary.

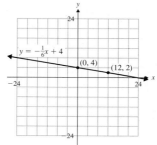

21) (5, 0), (5, 1) may vary. (5, 2) may vary.

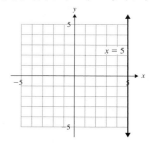

23) $-\dfrac{2}{5}$ 25) 1   27) $-\dfrac{13}{5}$ 29) −2   31) undefined

33) a) In 2002, one share of the stock was worth $32.

b) The slope is positive, so the value of the stock is increasing over time.

c) $m = 3$; the value of one share of stock is increasing by $3.00 per year.

35)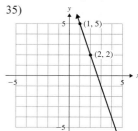

37)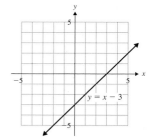

39) $m = 1$, y-int: (0, −3)

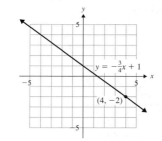

41) $m = -\dfrac{3}{4}$, y-int: (0, 1)

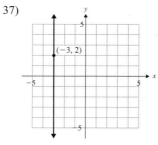

43) $m = \dfrac{1}{3}$, y-int: $(0, 2)$ 45) $m = -1$, y-int: $(0, 0)$

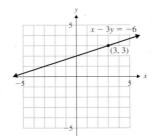

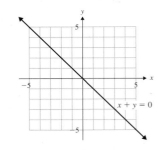

47) a) $(0, 5920.1)$; in 1998 the amount of money spent for personal consumption was \$5920.1 billion.

   b) It has been increasing by \$371.5 billion per year.

   c) estimate from the graph: \$7400 billion; number from the equation: \$7406.1 billion.

49) $y = 7x - 9$ 51) $y = -\dfrac{4}{9}x + 2$ 53) $y = -\dfrac{1}{3}x - 5$

55) $y = 9$ 57) $x - 2y = -6$ 59) $3x + y = 5$

61) $6x - y = 0$ 63) $3x - 4y = -1$

65) a) $y = 186.2x + 944.2$

   b) The number of worldwide wireless subscribers is increasing by 186.2 million per year.

   c) 1316.6 million; this is slightly less than the number given on the chart.

67) parallel   69) perpendicular   71) neither   73) $y = 5x + 6$

75) $2x + y = -7$ 77) $y = \dfrac{5}{3}x + \dfrac{4}{3}$ 79) $y = 2x - 7$

81) $y = -\dfrac{11}{2}x + 4$ 83) $x + 4y = -20$ 85) $x = 2$

87) $x = -1$

89) domain: $\{-4, 0, 2, 5\}$ 91) domain: $\{-6, 2, 5\}$
   range: $\{-9, 3, 9, 18\}$ range: $\{0, 1, 8, 13\}$
   function   not a function

93) domain: $(-\infty, \infty)$
   range: $(-\infty, \infty)$
   function

95) $(-\infty, \infty)$; function   97) $(-\infty, 0) \cup (0, \infty)$; function

99) $(-\infty, \infty)$; function   101) $f(3) = -14, f(-2) = -5$

103) $f(3) = -2, f(-2) = 1$

105) a) 8   b) -27 c) 32   d) 5   e) $5a - 12$
    f) $t^2 + 6t + 5$ g) $5k + 28$ h) $5c - 22$

107) 6

109)

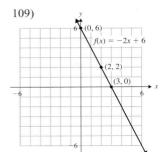

111)

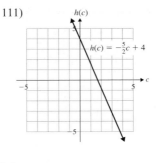

113) a) 960 MB; 2880 MB   b) 2.5 sec

Chapter 4 Test

1) yes

3) negative; positive

5)

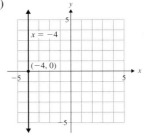

7)

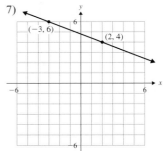

9) $y = 3x - 4$)

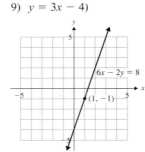

11) $x - 2y = -13$ 13) a) $y = \dfrac{1}{3}x - 9$ b) $y = \dfrac{5}{2}x - 6$

15) domain: $\{-2, 1, 3, 8\}$
   range: $\{-5, -1, 1, 4\}$
   function

17) a) $(-\infty, \infty)$ b) yes   19) -3 21) -22 23) $t^2 - 3t + 7$

25)

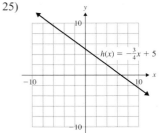

Cumulative Review for Chapters 1–4

1) $\dfrac{3}{10}$ 3) -64 5) $\dfrac{13}{5}$ 7) $-21t^{14}$ 9) $\{-3\}$

11) 300 calories   13) Lynette's age $= 41$; daughter's age $= 16$

15)

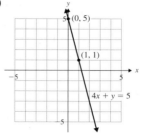

17) $y = -3x$ 19) -37 21) $8t + 19$

Chapter 5

Section 5.1

1) yes   3) no   5) yes   7) no   9) The lines are parallel.

11) $(3, 1)$

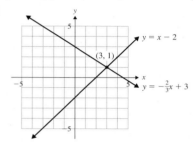

13) $(-1, -1)$

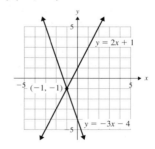

15) $(4, -5)$

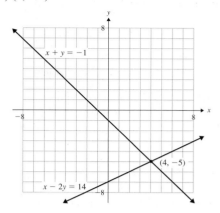

17) $\varnothing$; inconsistent system

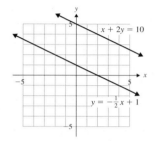

19) infinite number of solutions of the form
$\{(x, y)\,|\,6x - 3y = 12\}$; dependent system

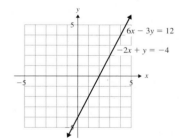

21) $\varnothing$; inconsistent system   23) $(4, 5)$

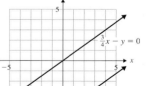

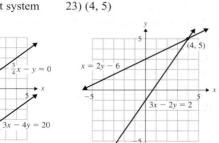

25) infinite number of solutions of the form $\{(x, y)|y = -3x + 1\}$; dependent system

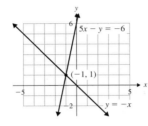

27) $(-1, -3)$

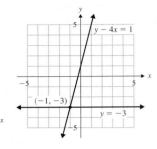

29) $(-1, 1)$

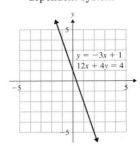

31) $\varnothing$; inconsistent system

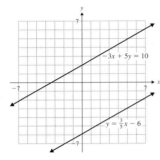

33–37) Answers may vary.   39) D. $(5, -3)$ is in quadrant IV.

41) C. $(0, -5)$ is on the y-axis not the x-axis.

43) The slopes are different.

45) infinite number of solutions   47) one solution

49) no solution   51) one solution   53) no solution

55) a) after 2001   b) 2001; 5.3 million

c) 1999–2001   d) 1999–2001

57) 4   59) -1

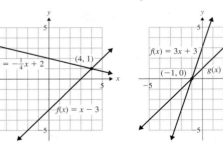

61) 0

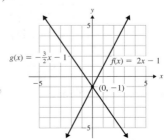

63) $(3, -2)$ 65) $(-5, -1)$ 67) $(-0.5, -1.25)$

Section 5.2

1) It is the only variable with a coefficient of 1.

3) $(2, 10)$ 5) $(-1, -7)$ 7) $(4, 0)$ 9) $\varnothing$

11) infinite number of solutions of the form $\{(x, y)|6x + y = -6\}$

13) $\left(-\dfrac{4}{5}, 3\right)$ 15) $(8, -1)$ 17) $\left(-4, -\dfrac{2}{3}\right)$ 19) $\varnothing$

21) $(0, -6)$ 23) $(3, 2)$ 25) $\left(-5, \dfrac{4}{3}\right)$

27) infinite number of solutions of the form $\{(x, y)|4y - 10x = -8\}$

29) Multiply the equation by the LCD of the fractions to eliminate the fractions.

31) $(6, 1)$ 33) $(3, -2)$ 35) $(0, 2)$ 37) $(-4, -4)$

39) $(3, 5)$ 41) $(-6, 2)$ 43) $(20, 10)$ 45) $(-3, 3)$

47) $(5, 0)$ 49) $\left(\dfrac{1}{2}, 6\right)$

51) a) Rent-for-Less: \$24; Frugal: \$30

b) Rent-for-Less: \$64; Frugal: \$60

c) $(120, 48)$; if the car is driven 120 mi, the cost would be the same from each company: \$48.

d) If a car is driven less than 120 mi, it is cheaper to rent from Rent-for-Less. If a car is driven more than 120 mi, it is cheaper to rent from Frugal Rentals. If a car is driven exactly 120 mi, the cost is the same from each company.

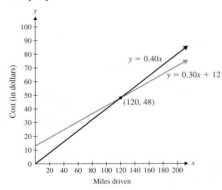

Section 5.3

1) $(4, 1)$ 3) $(0, -2)$ 5) $(-3, 4)$

7) infinite number of solutions of the form $\{(x, y)|3x - y = 4\}$

9) $(-1, -3)$ 11) $(8, 1)$ 13) $\varnothing$ 15) $\left(\dfrac{3}{2}, -1\right)$

17) $\left(5, -\dfrac{3}{2}\right)$ 19) $\varnothing$

21) infinite number of solutions of the form $\{(x, y)|7x + 2y = 12\}$

23) $(0, 2)$

25) Eliminate the decimals. Multiply the first equation by 10, and multiply the second equation by 100.

27) $(-6, 1)$ 29) $(1, 2)$ 31) $(7, 4)$ 33) $(12, -1)$ 35) $(-5, -2)$

37) $(3, -4)$ 39) $\left(\dfrac{1}{3}, 9\right)$ 41) $(1, 1)$ 43) $\left(-\dfrac{123}{17}, \dfrac{78}{17}\right)$

45) $\left(\dfrac{85}{46}, \dfrac{45}{23}\right)$ 47) $\left(-\dfrac{39}{14}, -\dfrac{9}{28}\right)$

49) a) 8   b) c can be any real number except 8.

51) a) 4   b) a can be any real number except 4.   53) 5

Chapter 5 Mid-Chapter Summary

1) Elimination method; none of the coefficients is 1 or -1; $(5, 6)$

3) Since the coefficient of x in the first equation is 1, you can solve for x and use substitution. Or, multiply the first equation by -3 and use the elimination method. Either method will work well; $(-4, -1)$

5) Substitution; the second equation is solved for x and does not contain any fractions; $(1, -7)$

7) $(2, 5)$ 9) $\left(7, -\dfrac{1}{2}\right)$ 11) $\varnothing$

13) $(-6, 1)$ 15) $(3, -6)$ 17) $\left(\dfrac{1}{2}, 2\right)$

19) infinite number of solutions of the form
$\{(x, y) | y = -6x + 5\}$

21) $(10, 9)$ 23) $\left(\dfrac{3}{4}, 0\right)$ 25) $(-4, 0)$

27) $(2, 2)$ 29) $(-2, -1)$

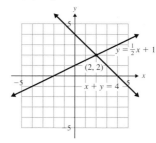

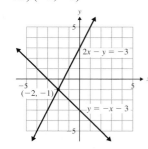

31) $(5, -3)$

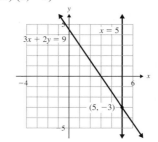

Section 5.4

1) 17 and 19   3) *The Aviator*, 11; *Finding Neverland*, 7

5) IHOP: 1156; Waffle House: 1470

7) 1939: 5500; 2004: 44,468

9) beef: 63.4 lb; chicken: 57.1 lb

11) length: 26 in.; width: 13 in.   13) width: 30 in.; height: 80 in.

15) length: 9 cm; width: 5 cm   17) $m\angle x = 72°$; $m\angle y = 108°$

19) Marc Anthony: $86; Santana: $66.50

21) two-item: $5.19; three-item: $6.39

23) key chain: $2.50; postcard: $0.50

25) cantaloupe: $1.50; watermelon: $3.00

27) hamburger: $0.47; small fry: $1.09

29) 9%: 3oz; 17%: 9 oz   31) peanuts: 7 lb; cashews: 3 lb

33) 3%: $2800; 5%: $1200   35) 52 quarters, 58 dimes

37) 4 L of pure acid, 8 L of 10% solution

39) car: 60 mph; truck: 50 mph

41) walking: 4 mph; biking: 11 mph

43) Nick: 14 mph; Scott: 12 mph

Section 5.5

1) yes   3) no   5) $(-2, 0, 5)$ 7) $(1, -1, 4)$ 9) $\left(2, -\dfrac{1}{2}, \dfrac{5}{2}\right)$

11) $\varnothing$ 13) $\{(x, y, z) | 5x + y - 3z = -1\}$

15) $\{(a, b, c) | -a + 4b - 3c = -1\}$ 17) $(2, 5, -5)$

19) $\left(-4, \dfrac{3}{5}, 4\right)$ 21) $(0, -7, 6)$ 23) $(1, 5, 2)$

25) $\left(-\dfrac{1}{4}, -5, 3\right)$ 27) $\varnothing$ 29) $\left(4, -\dfrac{3}{2}, 0\right)$ 31) $(4, 4, 4)$

33) $\{(x, y, z) | -4x + 6y + 3z = 3\}$ 35) $\left(1, -7, \dfrac{1}{3}\right)$

37) $(-3, -1, 1)$

39) hot dog: $2.00, fries: $1.50, soda: $2.00

41) Clif Bar: 11g, Balance Bar: 15 g, PowerBar: 24 g

43) Knicks: $160 million, Lakers: $149 million, Bulls: $119 million

45) value: $15; regular: $28; prime: $38

47) 104°, 52°, 24°   49) 80°, 64°, 36°   51) 12 cm, 10 cm, 7 cm

Chapter 5 Review Exercises

1) no   3) yes

5) $(1, 1)$ 7) $(-2, 3)$

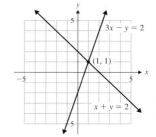

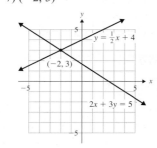

9) one solution   11) infinite number of solutions

13) a) during the 4th quarter of 2003

b) During the 3rd quarter of 2003 approximately 89.1% of flights left on time.

15) infinite number of solutions of the form
$$\left\{(x, y)\,\middle|\, y = \frac{5}{6}x - 2\right\}$$

17) $\left(\frac{1}{3}, 5\right)$ 19) $\varnothing$ 21) $(4, 7)$ 23) $(-5, 3)$ 25) $(1, 4)$

27) $(1, -6)$ 29) $(2, 1)$

31) $\left(\frac{83}{23}, -\frac{50}{23}\right)$ 33) 34 dogs, 17 cats

35) hot dog: \$5.00; soda: \$3.25

37) width: 15 in; length: 18 in.

39) gummi bears: 4 lb; jelly beans: 6 lb

41) car: 60 mph; bus: 48 mph

43) no   45) $(3, -1, 4)$ 47) $\left(-1, 2, \frac{1}{2}\right)$ 49) $\left(3, \frac{2}{3}, -\frac{1}{2}\right)$

51) $\varnothing$ 53) $\{(a, b, c)\,|\,3a - 2b + c = 2\}$ 55) $(1, 0, 3)$

57) $\left(\frac{3}{4}, -2, 1\right)$

59) Propel: 35 mg; Powerade: 55 mg; Gatorade Xtreme: 110 mg

61) Verizon: 32.4 million; Cingular: 21.9 million; T-Mobile: 10.0 million

63) ice cream cone: \$1.50; shake: \$2.50; sundae: \$3.00

65) 92°, 66°, 22°

Chapter 5 Test

1) yes

3) $\varnothing$

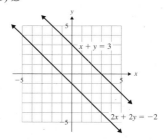

5) infinite number of solutions of the form
$\{(x, y)\,|-9x + 12y = 21\}$

7) $(4, 0)$ 9) $(9, 1)$ 11) $(-2, 7)$ 13) screws: \$4; nails: \$3

Cumulative Review for Chapters 1–5

1) $-\dfrac{1}{6}$ 3) -29 5) $-24x^2 + 8x + 56$ 7) $\dfrac{15}{c^6}$

9) $8.319 \cdot 10^{-5}$ 11) $\varnothing$ 13) 2,100,000

15)

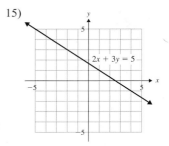

17) $y = -\dfrac{7}{8}x - \dfrac{17}{8}$ 19) $(7, -8)$ 21) $\varnothing$

23) Aretha Franklin: 16; Whitney Houston: 6; Christina Aguilera: 3

Chapter 6

Section 6.1

1) 64   3) -64 5) $\dfrac{1}{6}$ 7) 81   9) $\dfrac{16}{81}$ 11) -31 13) $\dfrac{1}{64}$

15) t^{13} 17) $-16c^9$ 19) z^{24} 21) $125p^{30}$ 23) $-\dfrac{8}{27}a^{21}b^3$

25) f^4 27) $7v$ 29) $\dfrac{d^4}{6}$ 31) $\dfrac{1}{x^6}$ 33) $\dfrac{1}{m}$ 35) $\dfrac{3}{2k^4}$

37) $20m^8n^{14}$ 39) $24y^8$ 41) $\dfrac{1}{49a^8b^2}$ 43) $\dfrac{b^5}{a^3}$

45) $\dfrac{y^{15}}{x^5}$ 47) $\dfrac{64c^6}{a^6b^3}$ 49) $\dfrac{9}{h^8k^8}$ 51) $\dfrac{c^6}{27d^{18}}$

53) $A = 10x^2$ sq units, $P = 14x$ units

55) $A = \dfrac{13}{16}p^2$ sq units, $P = 2p$ units   57) k^{6a} 59) g^{8x} 61) x^{3b}

Section 6.2

1) Yes; the coefficients are real numbers and the exponents are whole numbers.

3) No; one of the exponents is a negative number.

5) No; two of the exponents are fractions.

7) binomial   9) trinomial   11) monomial

13) It is the same as the degree of the term in the polynomial with the highest degree.

15) Add the exponents on the variables.

17)

Term	Coeff.	Degree
$7y^3$	7	3
$10y^2$	10	2
$-y$	-1	1
2	2	0

Degree is 3.

19)

Term	Coeff.	Degree
$-9r^3s^2$	-9	5
$-r^2s^2$	-1	4
$\dfrac{1}{2}rs$	$\dfrac{1}{2}$	2
$6s$	6	1

Degree is 5.

21) a) 22 b) 2   23) 13   25) -55 27) 0   29) $5w^2 + 7w - 6$

31) $-8p + 7$ 33) $-5a^4 + 9a^3 - a^2 - \dfrac{3}{2}a + \dfrac{5}{8}$

35) $\dfrac{25}{8}x^3 + \dfrac{1}{12}$ 37) $2.6k^3 + 8.7k^2 - 7.3k - 4.4$

39) $17x - 8$ 41) $12r^2 + 6r + 11$

43) $1.5q^3 - q^2 + 4.6q + 15$ 45) $-7a^4 - 12a^2 + 14$

47) $8j^2 + 12j$ 49) $9s^5 - 3s^4 - 3s^2 + 2$

51) $-b^4 - 10b^3 + b + 20$ 53) $-\dfrac{5}{14}r^2 + r - \dfrac{7}{6}$

55) $15v - 6$ 57) $-b^2 - 12b + 7$

59) $3a^4 - 11a^3 + 7a^2 - 7a + 8$ 61) Answers may vary.

63) No. If the coefficients of the like terms are opposite in sign, their sum will be zero. Example:
$(3x^2 + 4x + 5) + (2x^2 - 4x + 1) = 5x^2 + 6$

65) $-9b^2 - 4$ 67) $\dfrac{5}{4}n^3 - \dfrac{3}{2}n^2 + 3n - \dfrac{1}{8}$

69) $5u^3 - 10u^2 - u - 6$ 71) $-\dfrac{5}{8}k^2 - 9k + \dfrac{1}{2}$

73) $3t^3 - 7t^2 - t - 3$ 75) $9a^3 - 8a + 13$ 77) $3a + 8b$

79) $-m + \dfrac{11}{6}n - \dfrac{1}{4}$ 81) $5y^2z^2 + 7y^2z - 25yz^2 + 1$

83) $10x^3y^2 - 11x^2y^2 + 6x^2y - 12$ 85) $5v^2 + 3v - 8$

87) $4g^2 + 10g - 10$ 89) $-8n^2 + 11n$ 91) $8x + 14$ units

93) $6w^2 - 2w - 6$ units   95) a) 16 b) 4   97) a) 29 b) 5

99) $-\dfrac{1}{2}$ 101) 40

Section 6.3

1) Answers may vary.   3) $14k^6$ 5) $-24t^9$ 7) $\dfrac{7}{4}d^{11}$

9) $28y^2 - 63y$ 11) $-36b^2 - 32b$ 13) $6v^5 - 24v^4 - 12v^3$

15) $-27t^5 + 18t^4 + 12t^3 + 21t^2$

17) $2x^4y^3 + 16x^4y^2 - 22x^3y^2 + 4x^3y$

19) $-15t^7 - 6t^6 + \dfrac{15}{4}t^5$ 21) $18g^3 + 10g^2 + 14g + 28$

23) $-r^3 - 39r^2 - 31r - 9$

25) $-47a^3b^3 - 39a^3b^2 + 108a^2b - 51$ 27) $5q^3 - 36q + 27$

29) $2p^3 - 9p^2 - 23p + 30$

31) $15y^4 - 23y^3 + 28y^2 - 29y - 4$

33) $6k^4 + \dfrac{5}{2}k^3 + 31k^2 + 15k - 30$

35) $a^4 + 3a^3 - 3a^2 + 14a - 6$

37) $-24v^5 + 26v^4 - 22v^3 + 27v^2 - 5v + 10$

39) $8x^3 - 22x^2 + 19x - 6$ 41) First, Outer, Inner, Last

43) $w^2 + 15w + 56$ 45) $k^2 + 4k - 45$ 47) $y^2 - 7y + 6$

49) $4p^2 - 7p - 15$ 51) $24n^2 + 41n + 12$

53) $5y^2 - 16y + 12$ 55) $-m^2 - 4m + 21$

57) $12a^2 + ab - 20b^2$ 59) $60p^2 + 68pq + 15q^2$

61) $a^2 + \dfrac{21}{20}a + \dfrac{1}{5}$

63) a) $4y + 8$ units   b) $y^2 + 4y - 12$ sq units

65) a) $2a^2 + 4a + 16$ units   b) $3a^3 - 3a^2 + 24a$ sq units

67) Both are correct.   69) $15y^2 + 54y - 24$

71) $-54a^2 - 90a + 36$ 73) $-7r^4 + 77r^3 - 126r^2$

75) $c^3 + 6c^2 + 5c - 12$ 77) $2p^3 - 15p^2 + 27p - 10$

79) $5n^5 + 55n^3 + 150n$ 81) $x^2 - 36$ 83) $t^2 - 9$

85) $4 - r^2$ 87) $n^2 - \dfrac{1}{4}$ 89) $\dfrac{4}{9} - k^2$ 91) $9m^2 - 4$

93) $b^2 - 36a^2$ 95) $y^2 + 16y + 64$ 97) $t^2 - 22t + 121$

99) $k^2 - 4k + 4$ 101) $16w^2 + 8w + 1$

103) $4d^2 - 20d + 25$

105) No. The order to operations tells us to perform exponents, $(t + 3)^2$, before multiplying by 4.

107) $6x^2 + 12x + 6$ 109) $2a^3 + 12a^2 + 18a$

111) $-3m^2 + 6m - 3$ 113) $r^3 + 15r^2 + 75r + 125$

115) $s^3 - 6s^2 + 12s - 8$ 117) $c^4 - 18c^2 + 81$

119) $\dfrac{4}{9}n^2 + \dfrac{16}{3}n + 16$ 121) $y^4 + 8y^3 + 24y^2 + 32y + 16$

123) No; $(x + 5)^2 = x^2 + 10x + 25$

125) $h^3 + 6h^2 + 12h + 8$ cubic units

127) $9x^2 + 33x + 14$ sq units

Section 6.4

1) dividend: $12c^3 + 20c^2 - 4c$; divisor: $4c$; quotient: $3c^2 + 5c - 1$

3) Answers may vary.   5) $2a^2 - 5a + 3$

7) $2u^5 + 2u^3 + 5u^2 - 8$ 9) $-5d^3 + 1$

11) $\dfrac{3}{2}w^2 + 7w - 1 + \dfrac{1}{2w}$ 13) $\dfrac{5}{2}v^3 - 9v - \dfrac{11}{2} - \dfrac{5}{4v^2} + \dfrac{1}{4v^4}$

15) $9a^3b + 6a^2b - 4a^2 + 10a$

17) $-t^4u^2 + 7t^3u^2 + 12t^2u^2 - \dfrac{1}{9}t^2$

19) The answer is incorrect. When you divide $4t$ by $4t$, you get 1. The quotient should be $4t^2 - 9t + 1$.

21) $g + 4$ 23) $p + 6$ 25) $k - 5$ 27) $h + 8$

29) $2a^2 - 7a - 3$ 31) $3p^2 + 7p - 1$ 33) $6t + 23 + \dfrac{119}{t - 5}$

35) $4z^2 + 8z + 7 - \dfrac{72}{3z + 5}$ 37) $w^2 - 4w + 16$

39) $2r^2 + 8r + 3$

41) Use synthetic division when the divisor is in the form $x - c$.

43) No. The divisor, $2x + 5$, is not in the form $x - c$.

45) $t + 9$ 47) $5n + 6 + \dfrac{2}{n + 3}$ 49) $2y^2 - 3y + 5 - \dfrac{4}{y + 5}$

51) $3p^2 - p + 1$ 53) $5x^3 + 2x^2 - 3x - 5 + \dfrac{7}{x + 1}$

55) $r^2 - r - 2$ 57) $m^3 + 3m^2 + 9m + 27$

59) $2c^4 + c^3 + 2c^2 + 4c - 3 - \dfrac{6}{c - 2}$ 61) $2x^2 + 8x - 12$

63) $x^2y^2 + 5x^2y - \dfrac{1}{6} + \dfrac{1}{2xy}$ 65) $-2g^3 - 5g^2 + g - 4$

67) $6t + 5 + \dfrac{20}{t - 8}$ 69) $4n^2 + 10n + 25$ 71) $5x^2 - 7x + 3$

73) $4a^2 - 3a + 7$ 75) $5h^2 - 3h - 2 + \dfrac{1}{2h^2 - 9}$

77) $3d^2 - d - 8$ 79) $9c^2 + 8c + 3 + \dfrac{7c + 4}{c^2 - 10c + 4}$

81) $k^2 - 9$ 83) $-\dfrac{5}{7}a^3 - 7a + 2 + \dfrac{15}{7a}$ 85) $4y + 1$

87) $2a^2 - 5a + 1$ 89) $12h^2 + 6h + 2$

Chapter 6 Review Exercises

1) 81   3) $\dfrac{64}{125}$ 5) z^{18} 7) $7r^5$ 9) $-54t^7$ 11) $\dfrac{1}{k^8}$

13) $-\dfrac{40b^4}{a^6}$ 15) $\dfrac{4q^{30}}{9p^6}$ 17) $\dfrac{1}{s^2t^2}$

19) $A = 8x^2$ sq units, $P = 12x$ units   21) x^{8t} 23) r^{6a} 25) y^{6p}

27)

Term	Coeff.	Degree
$4r^3$	4	3
$-7r^2$	-7	2
r	1	1
5	5	0

Degree is 3.

29) 5   31) a) 32   b) -7 33) $-2t^2 + 10t + 5$

35) $7.9p^3 + 5.1p^2 + 4.8p - 3.6$ 37) $\dfrac{5}{4}k^2 - \dfrac{2}{3}k + 7$

39) $-2a^2b^2 + 17a^2b - 4ab + 5$ 41) $-m + 15n - 5$

43) $5s^2 - 18s - 26$ 45) $4d^2 - 4d + 14$ units

47) $28r^2 - 60r$ 49) $-24w^4 - 48w^3 + 26w^2 - 4w + 15$

51) $y^2 + y - 56$ 53) $6n^2 - 41n + 63$ 55) $-a^2 - a + 132$

57) $-56u^6v^3 + 49u^5v^4 + 84u^5v^2 - 21u^4v^2$

59) $6x^2 - 13xy - 8y^2$ 61) $a^2b^2 + 11ab + 30$

63) $12x^6 + 48x^5 - 139x^4 - 28x^3 + 86x^2 + 36x - 99$

65) $24c^5 - 78c^4 + 60c^3$ 67) $z^3 + 10z^2 + 29z + 20$

69) $\dfrac{1}{5}m^2 - \dfrac{26}{15}m - 8$ 71) $b^2 + 14b + 49$

73) $25q^2 - 20q + 4$ 75) $x^3 - 6x^2 + 12x - 8$

77) $z^2 - 81$ 79) $\dfrac{1}{25}n^2 - 4$ 81) $\dfrac{49}{64} - r^4$

83) $-18c^2 + 48c - 32$

85) a) $4m^2 - 7m - 15$ sq units   b) $10m + 4$ units

87) $4t^2 - 7t - 10$ 89) $c + 10$ 91) $4r^2 - 7r + 3$

93) $3a + 8 - \dfrac{3}{2a} + \dfrac{2}{a^2}$ 95) $-\dfrac{5}{2}x^2y^3 + 7xy^3 + 1 - \dfrac{5}{3x^2}$

97) $2q - 4 - \dfrac{7}{3q + 7}$ 99) $3a + 7 + \dfrac{21}{5a - 4}$

101) $3m^2 + m - 4$ 103) $b^2 + 4b + 16$

105) $8w^2 - 6w - 7 - \dfrac{2}{4w + 3}$ 107) $7u^2 + u + 4$

109) $5y + 4$

Chapter 6 Test

1) $\dfrac{27}{125}$ 3) $-27d^8$ 5) $\dfrac{a^4}{b^{12}}$ 7) a) -1 b) 3   9) -1

11) $14r^3s^2 - 2r^2s^2 - 5rs + 8$ 13) $-12n^3 - 15n^2 - 9$

15) $6y^2 + 13y + 5$ 17) $6a^2 - 13ab - 5b^2$

19) $3x^3 + 24x^2 + 48x$ 21) $s^3 - 12s^2 + 48s - 64$

23) $2t^2 - 5t + 1 - \dfrac{2}{3t}$ 25) $x^2 + 2x + 4$ 27) $3k + 7$

Cumulative Review for Chapters 1–6

1) a) 43, 0   b) $-14, 43, 0$

c) $\dfrac{6}{11}, -14, 2.7, 43, 0.\overline{65}, 0$

3) $\dfrac{12}{5}$ or $2\dfrac{2}{5}$ 5) $\dfrac{8}{n^{18}}$ 7) $\left\{ -\dfrac{45}{4} \right\}$ 9) $(-\infty, 1]$

11) x-int: $(2, 0)$, y-int: $(0, -5)$

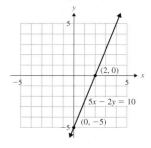

13) $x + y = 3$ 15) width: 15 cm; length: 23 cm

17) $k^2 - 3k - 54$ 19) $ab - \dfrac{5}{2b} + \dfrac{7}{a^2b^2} + \dfrac{1}{a^3b^2}$

21) $2p^2 - 3p + 1 + \dfrac{5}{p + 4}$ 23) $2c^3 - 17c^2 + 48c - 45$

Chapter 7

Section 7.1

1) $5m^2$ 3) $6k^5$ 5) $9x^2y$ 7) $4uv^3$ 9) s^2t 11) $(n - 7)$

13) Answers may vary.   15) $6(5s + 3)$ 17) $4(6z - 1)$

19) $3d(d - 2)$ 21) $7y^2(6 + 5y)$ 23) $t^4(t - 1)$

25) $\frac{1}{2}c(c + 5)$ 27) $5n^3(2n^2 - n + 8)$

29) $2v^5(v^3 - 9v^2 - 12v + 1)$ 31) does not factor

33) $a^3b^2(a + 4b)$ 35) $10x^2y(5xy^2 - 7xy + 4)$

37) $(n - 12)(m + 8)$ 39) $(8r - 3)(p - q)$

41) $(z + 11)(y + 1)$ 43) $(3r + 4)(2k^2 - 1)$

45) $-8(8m + 5)$ 47) $-5t^2(t - 2)$ 49) $-a(3a^2 - 7a + 1)$

51) $-1(b - 8)$ 53) $(t + 3)(k + 8)$ 55) $(g - 7)(f + 4)$

57) $(s - 3)(2r + 5)$ 59) $(3x - 2)(y + 9)$

61) $(2b + 5c)(4b + c^2)$ 63) $(a^2 - 3b)(4a + b)$

65) $(k + 7)(t - 5)$ 67) $(n - 8)(m - 10)$

69) $(g - 1)(d + 1)$ 71) $(5u + 6)(t - 1)$

73) $(12g^3 + h)(3g - 8h)$ 75) Answers may vary.

77) $2(b + 4)(a + 3)$ 79) $8s(s - 5)(t + 2)$

81) $(d + 4)(7c + 3)$ 83) $(7k - 3d)(6k^2 - 5d)$

85) $9fj(f + 1)(j + 5)$ 87) $2x^3(2x - 1)(y + 7)$

Section 7.2

1) a) $10, 3$ b) $7, -4$ c) $-8, 1$ d) $-8, -7$

3) They are negative.   5) Can I factor out a GCF?

7) Can I factor again? If so, factor again.

9) $n + 3$ 11) $c - 9$ 13) $x + 8$ 15) $(g + 6)(g + 2)$

17) $(w + 7)(w + 6)$ 19) $(c - 4)(c - 9)$

21) $(b - 4)(b + 2)$ 23) $(u + 12)(u - 11)$

25) $(q - 5)(q - 3)$ 27) prime   29) $(w + 5)(w - 1)$

31) $(p - 10)(p - 10)$ or $(p - 10)^2$ 33) $(d + 12)(d + 2)$

35) $-(a + 8)(a + 2)$ 37) $-(h - 5)(h + 3)$

39) $-(k - 7)(k - 4)$ 41) $-(x + 10)(x - 9)$

43) $-(n + 7)(n + 7)$ or $-(n + 7)^2$ 45) $(a + 5b)(a + b)$

47) $(m - 3n)(m + 7n)$ 49) $(x - 12y)(x - 3y)$

51) $(f + g)(f - 11g)$ 53) $(c - 5d)(c + 11d)$

55) $2(r + 3)(r + 1)$ 57) $4q(q - 3)(q - 4)$

59) $m^2n(m - 4n)(m + 11n)$ 61) $pq(p - 7q)(p - 10q)$

63) $(r - 9)(r - 2)$ 65) $7cd^2(c - 2)(c + 1)$

67) prime   69) $2r^2(r + 6)(r + 7)$

71) $8x^2y^3(xy - 4)(xy + 2)$ 73) $(a + b)(k + 9)(k - 2)$

75) $(x + y)(t + 3)(t - 7)$

77) No; from $(2x + 8)$ you can factor out a 2. The correct answer is $2(x + 4)(x + 5)$.

Section 7.3

1) a) $10, -5$ b) $-27, -1$ c) $6, 2$ d) $-12, 6$

3) $(2k + 9)(k + 5)$ 5) $(7y - 6)(y - 1)$

7) $(4a - 7b)(2a + 3b)$ 9) Can I factor out a GCF?

11) because 2 can be factored out of $2x - 4$, but 2 cannot be factored out of $2x^2 + 13x - 24$

13) $a + 2$ 15) $2k + 3$ 17) $9x - 4y$ 19) $(2r + 5)(r + 3)$

21) $(5p - 1)(p - 4)$ 23) $(11m + 4)(m - 2)$

25) $(3v + 7)(2v - 1)$ 27) $(5c + 2)(2c + 3)$

29) $(9x - 4y)(x - y)$ 31) $(5w + 6)(w + 1)$

33) $(3u - 5)(u - 6)$ 35) $(7k - 6)(k + 3)$

37) $(4r + 3)(2r + 5)$ 39) $(6v - 7)(v - 2)$

41) $(7d + 2)(3d - 4)$ 43) $(12v + 1)(4v + 5)$

45) $(5a - 4b)(2a - b)$

47) $(3t + 4)(2t - 1)$; the answer is the same.

49) $(2y - 3)(y - 8)$ 51) $3c(4c - 3)(c + 2)$

53) $(2t - 5)(6t + 1)$ 55) $3(5h + 3)(3h + 2)$ 57) prime

59) $(13t - 9)(t + 2)$ 61) $(5c + 3d)(c + 4d)$

63) $2(d + 5)(d - 4)$ 65) $4c^2d(2d - 5)(d + 3)$

67) $(6a - 1)^2$ 69) $(y + 6)^2(3x + 4)(x - 5)$

71) $(z - 10)^3(9y + 4)(y + 8)$

73) $(v + 8)(2u - 11)(4u + 3)$ 75) $-(h + 9)(h - 6)$

77) $-(5z - 2)(2z - 3)$ 79) $-3(7v + 3)(v - 3)$

81) $-2j(j + 10)(j + 6)$ 83) $-(8y - 3)(2y + 5)$

85) $-3cd(2c - d)(c - 4d)$

Section 7.4

1) a) 36 b) 100 c) 16 d) 121 e) 9 f) 64 g) 144

h) $\frac{1}{4}$ i) $\frac{9}{25}$

3) a) n^2 b) $5t$ c) $7k$ d) $4p^2$ e) $\frac{1}{3}$ f) $\frac{5}{2}$

5) $z^2 + 18z + 81$

7) The middle term does not equal $2(3c)(-4)$. It would have to equal $-24c$ to be a perfect square trinomial.

9) $(t + 8)^2$ 11) $(g - 9)^2$ 13) $(2y + 3)^2$ 15) $(3k - 4)^2$

17) $\left(a + \frac{1}{3}\right)^2$ 19) $\left(v - \frac{3}{2}\right)^2$ 21) $(x + 3y)^2$

23) $(6t - 5u)^2$ 25) $4(f + 3)^2$ 27) $2p^2(p - 6)^2$

29) $-2(3d + 5)^2$ 31) $3c(4c^2 + c + 9)$

33) a) $x^2 - 16$ b) $16 - x^2$ 35) $(x + 3)(x - 3)$

37) $(n + 11)(n - 11)$ 39) prime

41) $\left(y + \frac{1}{5}\right)\left(y - \frac{1}{5}\right)$ 43) $\left(c + \frac{3}{4}\right)\left(c - \frac{3}{4}\right)$

45) $(6 + h)(6 - h)$ 47) $(13 + a)(13 - a)$

49) $\left(\frac{7}{8} + j\right)\left(\frac{7}{8} - j\right)$ 51) $(10m + 7)(10m - 7)$

53) $(4p + 9)(4p - 9)$ 55) prime   57) $\left(\frac{1}{2}k + \frac{2}{3}\right)\left(\frac{1}{2}k - \frac{2}{3}\right)$

59) $(b^2 + 8)(b^2 - 8)$ 61) $(12m + n^2)(12m - n^2)$

63) $(r^2 + 1)(r + 1)(r - 1)$

65) $(4h^2 + g^2)(2h + g)(2h - g)$

67) $4(a + 5)(a - 5)$ 69) $2(m + 8)(m - 8)$

71) $5r^2(3r + 1)(3r - 1)$

73) a) 64 b) 1 c) 1000 d) 27 e) 125 f) 8

75) a) y b) $2c$ c) $5r$ d) x^2 77) $(d + 1)(d^2 - d + 1)$

79) $(p - 3)(p^2 + 3p + 9)$ 81) $(k + 4)(k^2 - 4k + 16)$

83) $(3m - 5)(9m^2 + 15m + 25)$

85) $(5y - 2)(25y^2 + 10y + 4)$

87) $(10c - d)(100c^2 + 10cd + d^2)$

89) $(2j + 3k)(4j^2 - 6jk + 9k^2)$

91) $(4x + 5y)(16x^2 - 20xy + 25y^2)$

93) $6(c + 2)(c^2 - 2c + 4)$

95) $7(v - 10w)(v^2 + 10vw + 100w^2)$

97) $(h + 2)(h - 2)(h^2 - 2h + 4)(h^2 + 2h + 4)$

99) $7(2x + 3)$ 101) $(3p + 7)(p - 1)$

103) $(t + 7)(t^2 + 8t + 19)$ 105) $(k - 10)(k^2 - 17k + 73)$

Chapter 7 Mid-Chapter Summary Exercises

1) $(m + 10)(m + 6)$ 3) $(u + 9)(v + 6)$

5) $(3k - 2)(k - 4)$ 7) $8d^4(2d^2 + d + 9)$

9) $10w(3w + 5)(2w - 1)$ 11) $(t + 10)(t^2 - 10t + 100)$

13) $(7 + p)(7 - p)$ 15) $(2x + y)^2$ 17) $3z^2(z - 8)(z + 1)$

19) prime   21) $5(2x - 3)(4x^2 + 6x + 9)$

23) $\left(c + \dfrac{1}{2}\right)\left(c - \dfrac{1}{2}\right)$ 25) $(3s + 1)(3s - 1)(5t - 4)$

27) $(k + 3m)(k + 6m)$ 29) $(z - 11)(z + 8)$

31) $5(4y - 1)^2$ 33) $2(10c + 3d)(c + d)$

35) $(n^2 + 4m^2)(n + 2m)(n - 2m)$

37) $2(a - 9)(a + 4)$ 39) $\left(r - \dfrac{1}{2}\right)^2$

41) $(4g - 9)(7h + 4)$ 43) $(4b + 3)(2b - 5)$

45) $5a^2b(11a^4b^2 + 7a^3b^2 - 2a^2 - 4)$

47) prime   49) $(3p - 4q)^2$ 51) $(6y - 1)(5y + 7)$

53) $10(2a - 3b)(4a^2 + 6ab + 9b^2)$

55) $(r - 1)(t - 1)$ 57) $4(g + 1)(g - 1)$

59) $3(c - 4)^2$ 61) $(12k + 11)(12k - 11)$

63) $-4(6g + 1)(2g + 3)$ 65) $(q + 1)(q^2 - q + 1)$

67) $(9u^2 + v^2)(3u + v)(3u - v)$ 69) $(11f + 3)(f + 3)$

71) $j^3(2j^8 - 1)$ 73) $(w - 8)(w + 6)$ 75) prime

77) $(m + 2)^2$ 79) $(3t + 8)(3t - 8)$

81) $(2z + 1)(y + 11)(y - 5)$ 83) $r(r + 3)$

85) $-3x(x - 2y)$ 87) $(n + p + 6)(n - p + 6)$

Section 7.5

1) It says that if the product of two quantities equals 0, then one or both of the quantities must be zero.

3) $\{-9, 8\}$ 5) $\{4, 7\}$ 7) $\left\{-\dfrac{3}{4}, 9\right\}$ 9) $\{-15, 0\}$

11) $\left\{\dfrac{5}{6}\right\}$ 13) $\left\{-3, -\dfrac{7}{4}\right\}$ 15) $\left\{-\dfrac{3}{2}, \dfrac{1}{4}\right\}$ 17) $\{0, 2.5\}$

19) $\{-8, -7\}$ 21) $\{-15, 3\}$ 23) $\left\{-\dfrac{5}{3}, 2\right\}$ 25) $\left\{-\dfrac{4}{7}, 0\right\}$

27) $\{6, 9\}$ 29) $\{-7, 7\}$ 31) $\left\{-\dfrac{6}{5}, \dfrac{6}{5}\right\}$ 33) $\{-12, 5\}$

35) $\left\{-2, -\dfrac{3}{4}\right\}$ 37) $\{0, 11\}$ 39) $\left\{-\dfrac{1}{2}, 3\right\}$ 41) $\{-8, 12\}$

43) $\left\{\dfrac{7}{2}, \dfrac{9}{2}\right\}$ 45) $\{-9, 5\}$ 47) $\{1, 6\}$ 49) $\left\{-3, -\dfrac{4}{5}\right\}$

51) $\left\{\dfrac{3}{2}\right\}$ 53) $\{-11, -3\}$ 55) $\left\{-\dfrac{7}{2}, \dfrac{3}{4}\right\}$ 57) $\left\{\dfrac{2}{3}, 8\right\}$

59) $\left\{-4, 0, \dfrac{1}{2}\right\}$ 61) $\left\{-1, \dfrac{2}{9}, 11\right\}$ 63) $\left\{\dfrac{5}{2}, 3\right\}$

65) $\{0, -8, 8\}$ 67) $\{-4, 0, 9\}$ 69) $\{-12, 0, 5\}$

71) $\left\{0, -\dfrac{3}{2}, \dfrac{3}{2}\right\}$ 73) $\left\{-5, -\dfrac{2}{3}, \dfrac{7}{2}\right\}$ 75) $\left\{-\dfrac{3}{4}, \dfrac{2}{5}, \dfrac{1}{2}\right\}$

77) $-7, -3$ 79) $\dfrac{5}{2}, 4$ 81) $-4, 4$ 83) $0, 1, 4$

Section 7.6

1) length = 12 in.; width = 3 in.

3) base = 3 cm; height = 8 cm

5) base = 6 in.; height = 3 in.

7) length = 10 in.; width = 6 in.

9) length = 9 ft; width = 5 ft

11) length = 9 in.; width = 6 in.

13) width = 12 in.; height = 6 in.

15) height = 10 cm; base = 7 cm

17) 5 and 7 or -1 and 1   19) 0, 2, 4, or 2, 4, 6

21) 6, 7, 8   23) Answers may vary.

25) 9   27) 15   29) 10   31) 8 in.   33) 5 ft   35) 5 mi

37) a) 144 ft   b) after 2 sec   c) 3 sec

39) a) 288 ft   b) 117 ft   c) 324 ft   d) 176 ft   e) 648 ft
   f) 213 ft   g) 1089 ft   h) 586 ft   i) 360 ft   j) 765 ft
   k) The 10-in. shell would need to be 410 ft farther horizontally from the point of explosion than the 3-in. shell.

41) a) $4000   b) $4375   c) $30

Chapter 7 Review Exercises

1) 9   3) $11p^4q^3$ 5) $12(4y + 7)$ 7) $7n^3(n^2 - 3n + 1)$

9) $(b + 6)(a - 2)$ 11) $(n + 2)(m + 5)$

13) $(r - 2)(5q - 6)$ 15) $-4x(2x^2 + 3x - 1)$

17) $(p + 8)(p + 5)$ 19) $(x + 5y)(x - 4y)$

21) $3(c - 6)(c - 2)$ 23) $(5y + 6)(y + 1)$

25) $(2m - 5)(2m - 3)$ 27) $4a(7a + 4)(2a - 1)$

29) $(3s - t)(s + 4t)$ 31) $(n + 5)(n - 5)$

33) prime   35) $10(q + 9)(q - 9)$ 37) $(a + 8)^2$

39) $(h + 2)(h^2 - 2h + 4)$

41) $(3p - 4q)(9p^2 + 12pq + 16q^2)$ 43) $(7r - 6)(r + 2)$

45) $\left(\dfrac{3}{5} + x\right)\left(\dfrac{3}{5} - x\right)$ 47) $(s - 8)(t - 5)$

49) $w^2(w - 1)(w^2 + w + 1)$ 51) prime   53) $-4ab$

55) $\left\{0, \dfrac{1}{2}\right\}$ 57) $\left\{-1, \dfrac{2}{3}\right\}$ 59) $\{-3, 15\}$ 61) $\left\{-\dfrac{6}{7}, \dfrac{6}{7}\right\}$

63) $\{-4, 8\}$ 65) $\left\{\dfrac{4}{5}, 1\right\}$ 67) $\left\{0, -\dfrac{3}{2}, 2\right\}$ 69) $\{-8, 9\}$

71) $\left\{\dfrac{1}{6}, 3, 7\right\}$ 73) base $= 5$ in.; height $= 6$ in.

75) height $= 3$ in.; length $= 8$ in.   77) 12

79) length $= 6$ ft; width $= 2.5$ ft

81) $-1, 0, 1$ or $8, 9, 10$ 83) 5 in.

85) a) 0 ft   b) after 2 sec and after 4 sec   c) 144 ft
   d) after 6 sec

Chapter 7 Test

1) See if you can factor out a GCF.   3) $(4 + b)(4 - b)$

5) $7p^2q^3(8p^4q^3 - 11p^2q + 1)$ 7) $2d(d + 9)(d - 2)$

9) $(3h + 4)^2$ 11) $(s - 7t)(s + 4t)$

13) $(x + 3)^2(y + 7)(y + 8)$ 15) $m^9(m + 1)(m^2 - m + 1)$

17) $\{0, -5, 5\}$ 19) $\{-4, 7\}$ 21) $\left\{\dfrac{5}{3}, 2\right\}$ 23) 5, 7, 9

25) length $= 16$ ft; width $= 6$ ft

Cumulative Review for Chapters 1–7

1) $\dfrac{1}{8}$ 3) $\dfrac{3t^4}{2u^6}$ 5) 481,300   7) $R = \dfrac{A - P}{PT}$

9)

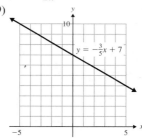

$y = -\dfrac{3}{5}x + 7$

11) $(-2, -2)$ 13) $8p^3 - 50p^2 + 95p - 56$

15) $3a^2b^2 - 7a^2b - 5ab^2 + 19ab - 8$

17) $3r + 1 - \dfrac{5}{2r} + \dfrac{3}{4r^2}$ 19) $6(3q - 7)(3q - 1)$

21) $(t^2 + 9)(t + 3)(t - 3)$ 23) $\{-8, 5\}$

Chapter 8

Section 8.1

1) when its numerator equals zero

3) Set the denominator equal to zero and solve for the variable. That value cannot be substituted into the expression because it will make the denominator equal to zero.

5) a) 3   b) $\dfrac{3}{2}$ 7) a) undefined   b) $\dfrac{32}{21}$ 9) a) 0   b) 7

11) a) -2 b) 0   13) a) $-\dfrac{1}{3}$ b) $\dfrac{9}{2}$ 15) a) 0, 10   b) $\dfrac{4}{3}$

17) a) never equals 0   b) 0   19) a) 0   b) $-12, 3$

21) a) -10 b) $-\dfrac{3}{2}, 4$ 23) a) $-5, -3$ b) 0

25) a) 0
   b) never undefined—any real number can be substituted
      for y

27) $\dfrac{2y}{3}$ 29) $\dfrac{2}{5d^3}$ 31) $\dfrac{3}{5}$ 33) $-\dfrac{6}{5}$ 35) $\dfrac{13}{10}$ 37) $b - 7$

39) $\dfrac{1}{r + 4}$ 41) $\dfrac{3k + 4}{k + 2}$ 43) $\dfrac{p - 4}{2p - 1}$ 45) $\dfrac{w + 5}{5}$

47) $\dfrac{8}{c^2 - 3c + 9}$ 49) $\dfrac{4(m - 5)}{11}$ 51) $\dfrac{x + y}{x^2 + xy + y^2}$ 53) -1

55) -1 57) $-k - 7$ 59) $-\dfrac{5}{x + 2}$ 61) $-4(b + 2)$

63) $-\dfrac{4t^2 + 6t + 9}{2t + 3}$

65) Possible answers:
   $\dfrac{-b - 7}{b - 2}, \dfrac{-(b + 7)}{b - 2}, \dfrac{b + 7}{2 - b}, \dfrac{b + 7}{-(b - 2)}, \dfrac{b + 7}{-b + 2}$

67) Possible answers:
   $\dfrac{-9 + 5t}{2t - 3}, \dfrac{5t - 9}{2t - 3}, \dfrac{-(9 - 5t)}{2t - 3},$
   $\dfrac{9 - 5t}{-2t + 3}, \dfrac{9 - 5t}{3 - 2t}, \dfrac{9 - 5t}{-(2t - 3)}$

69) Possible answers:
   $\dfrac{w - 6}{4w - 7}, \dfrac{6 - w}{4w - 7}, \dfrac{w - 6}{7 - 4w}, -\dfrac{6 - w}{-4w + 7}$

71) $2x - 7$ 73) $9t^2 + 6t + 4$ 75) $5x + 3$ 77) $c - 2$

79) $(-\infty, 7) \cup (7, \infty)$ 81) $\left(-\infty, -\dfrac{2}{5}\right) \cup \left(-\dfrac{2}{5}, \infty\right)$

83) $(-\infty, 1) \cup (1, 8) \cup (8, \infty)$

85) $(-\infty, -9) \cup (-9, 9) \cup (9, \infty)$ 87) $(-\infty, \infty)$

Section 8.2

1) $\dfrac{7}{24}$ 3) $\dfrac{3}{4}$ 5) $\dfrac{11r^2}{3}$ 7) $\dfrac{8v^4}{3u^3}$ 9) $-\dfrac{4x^6}{9}$ 11) $\dfrac{a + 1}{5(a - 6)}$

13) $\dfrac{1}{2t(3t - 2)}$ 15) $\dfrac{2(2p - 1)}{9}$ 17) $\dfrac{4v - 1}{4}$ 19) $\dfrac{3}{10}$

21) $\dfrac{2}{7}$ 23) $\dfrac{49k^4}{10}$ 25) $\dfrac{1}{6c^6}$ 27) $\dfrac{5(x - 7)(x + 8)}{3}$

29) $\dfrac{3a^2}{2a - 1}$ 31) $\dfrac{5}{(2y + 5)}$ 33) $\dfrac{1}{7}$ 35) $\dfrac{c + 1}{6c}$ 37) $\dfrac{7}{6}$

39) $\dfrac{1}{12}$ 41) 3   43) $\dfrac{4}{3r^2}$ 45) $\dfrac{4z^4}{z + 5}$ 47) $\dfrac{3}{4a(3a - 2)}$

49) $\dfrac{(d - 7)(d + 1)}{4}$ 51) $\dfrac{1}{4}$ 53) $\dfrac{5(2n - 1)}{3n^2(n - 4)}$ 55) $-\dfrac{a}{6}$

57) $\dfrac{2(t + 3)}{2t - 1}$ 59) $\dfrac{2u + 3}{5(u + 2)}$ 61) $\dfrac{4x^2}{x^2 + 5}$

63) $\dfrac{8}{9(a^2 - ab + b^2)}$ 65) $\dfrac{w - 8}{6w}$ 67) $\dfrac{1}{a + 10}$

69) $\dfrac{4t(3t - 2)}{5t + 1}$

Section 8.3

1) 40   3) 84   5) c^4 7) $36p^8$ 9) $21w^7$ 11) $12k^5$

13) $24a^3b^4$ 15) $2(n + 4)$ 17) $w(2w + 1)$ 19) $18(k + 5)$

21) $4a^3(3a - 1)$ 23) $(r + 7)(r - 2)$

25) $(x - 8)(x - 2)(x + 3)$ 27) $(w - 5)(w + 2)(w + 3)$

29) $b - 4$ or $4 - b$ 31) $u - v$ or $v - u$

33) Answers may vary.   35) $\dfrac{8p}{6p^5}$ 37) $\dfrac{66c^2d}{24c^3d^3}$

39) $\dfrac{5m}{m(m - 9)}$ 41) $\dfrac{2a^2}{16a(3a + 1)}$ 43) $\dfrac{3b^2 + 15b}{(b + 2)(b + 5)}$

45) $\dfrac{w^2 - 5w - 6}{(4w - 3)(w - 6)}$ 47) $-\dfrac{8}{n - 5}$ 49) $\dfrac{a}{2 - 3a}$

51) $\dfrac{3}{t} = \dfrac{3t^2}{t^3}; \dfrac{8}{t^3} = \dfrac{8}{t^3}$ 53) $\dfrac{9}{8n^6} = \dfrac{27}{24n^6}; \dfrac{2}{3n^2} = \dfrac{16n^4}{24n^6}$

55) $\dfrac{1}{x^3y} = \dfrac{5y^4}{5x^3y^5}; \dfrac{6}{5xy^5} = \dfrac{6x^2}{5x^3y^5}$

57) $\dfrac{t}{5t - 6} = \dfrac{7t}{7(5t - 6)}; \dfrac{10}{7} = \dfrac{50t - 60}{7(5t - 6)}$

59) $\dfrac{3}{c} = \dfrac{3c + 3}{c(c + 1)}; \dfrac{2}{c + 1} = \dfrac{2c}{c(c + 1)}$

61) $\dfrac{z}{z - 9} = \dfrac{z^2}{z(z - 9)}; \dfrac{4}{z} = \dfrac{4z - 36}{z(z - 9)}$

63) $\dfrac{a}{24a + 36} = \dfrac{3a}{36(2a + 3)}; \dfrac{1}{18a + 27} = \dfrac{4}{36(2a + 3)}$

65) $\dfrac{4}{h + 5} = \dfrac{4h - 12}{(h + 5)(h - 3)}; \dfrac{7h}{h - 3} = \dfrac{7h^2 + 35h}{(h + 5)(h - 3)}$

67) $\dfrac{b}{3b - 2} = \dfrac{b^2 - 9b}{(3b - 2)(b - 9)}; \dfrac{1}{b - 9} = \dfrac{3b - 2}{(3b - 2)(b - 9)}$

69) $\dfrac{9y}{y^2 - y - 42} = \dfrac{18y^2}{2y(y + 6)(y - 7)};$

$\dfrac{3}{2y^2 + 12y} = \dfrac{3y - 21}{2y(y + 6)(y - 7)}$

71) $\dfrac{z}{z^2 - 10z + 25} = \dfrac{z^2 + 3z}{(z - 5)^2(z + 3)};$

$\dfrac{15z}{z^2 - 2z - 15} = \dfrac{15z^2 - 75z}{(z - 5)^2(z + 3)}$

73) $\dfrac{11}{g - 3} = \dfrac{11g + 33}{(g + 3)(g - 3)}; \dfrac{4}{9 - g^2} = -\dfrac{4}{(g + 3)(g - 3)}$

75) $\dfrac{10}{4k - 1} = \dfrac{40k + 10}{(4k + 1)(4k - 1)};$

$\dfrac{k}{1 - 16k^2} = -\dfrac{k}{(4k + 1)(4k - 1)}$

77) $\dfrac{4}{w^2 - 4w} = \dfrac{28w - 112}{7w(w - 4)^2}; \dfrac{6}{7w^2 - 28w} = \dfrac{6w - 24}{7w(w - 4)^2};$

$\dfrac{11}{w^2 - 8w + 16} = \dfrac{77w}{7w(w - 4)^2}$

79) $-\dfrac{1}{a + 4} = -\dfrac{a^2 - 3a - 4}{(a + 4)(a - 4)(a + 1)};$

$\dfrac{a}{a^2 - 16} = \dfrac{a^2 + a}{(a + 4)(a - 4)(a + 1)};$

$\dfrac{3}{a^2 + 5a + 4} = \dfrac{3a - 12}{(a + 4)(a - 4)(a + 1)}$

Section 8.4

1) $\dfrac{5}{11}$ 3) $\dfrac{4}{5}$ 5) $\dfrac{10}{a}$ 7) $\dfrac{8}{3k^2}$ 9) $\dfrac{n - 9}{n + 6}$

11) 2   13) $\dfrac{6}{w}$ 15) $\dfrac{r - 3}{r + 2}$

17) a) $x(x - 3)$

b) Multiply the numerator and denominator of $\dfrac{8}{x - 3}$ by x,

and the numerator and denominator of $\dfrac{2}{x}$ by $x - 3$.

c) $\dfrac{8}{x - 3} = \dfrac{8x}{x(x - 3)}, \dfrac{2}{x} = \dfrac{2x - 6}{x(x - 3)}$

19) $\dfrac{19}{24}$ 21) $\dfrac{3x}{20}$ 23) $\dfrac{15u - 16}{24u^2}$ 25) $\dfrac{3(7a + 4)}{14a^2}$

27) $\dfrac{11k + 20}{k(k + 10)}$ 29) $\dfrac{11d + 32}{d(d - 8)}$ 31) $\dfrac{5z + 26}{(z + 6)(z + 2)}$

33) $\dfrac{x^2 - x - 3}{(2x + 1)(x + 5)}$ 35) $\dfrac{t - 3}{t - 7}$

37) $\dfrac{b^2 + b - 40}{(b + 4)(b - 4)(b - 9)}$ 39) $\dfrac{(c - 5)(c + 2)}{(c + 6)(c - 2)(c - 4)}$

41) $\dfrac{(4m + 21)(m - 1)}{(m + 3)(m - 2)(m + 6)}$ 43) $\dfrac{4b^2 + 28b + 3}{3(b - 4)(b + 3)}$

45) No; if the sum is rewritten as $\dfrac{5}{x - 7} - \dfrac{2}{x - 7}$ then the

LCD $= x - 7$. If the sum is rewritten as $\dfrac{-5}{7 - x} + \dfrac{2}{7 - x}$,

then the LCD is $7 - x$.

47) $\dfrac{7}{z - 6}$ or $-\dfrac{7}{6 - z}$ 49) $\dfrac{8}{c - d}$ or $-\dfrac{8}{d - c}$

51) $\dfrac{-m - 1}{m - 3}$ or $\dfrac{m + 1}{3 - m}$ 53) 1   55) $\dfrac{2c + 13}{12b - 7c}$ or $-\dfrac{2c + 13}{7c - 12b}$

57) $-\dfrac{5(t + 6)}{(t + 8)(t - 8)}$ 59) $\dfrac{3(3a + 4)}{(2a + 3)(2a - 3)}$

61) $\dfrac{6j^2 - j - 2}{3j(j + 8)}$ 63) $\dfrac{13k^2 + 15k - 109}{k(k - 4)(2k - 7)}$

65) $\dfrac{c^2 - 2c + 20}{(c - 4)^2(c + 3)}$ 67) $-\dfrac{3y}{(x + y)(x - y)}$

69) $\dfrac{-n^2 + 33n + 1}{(4n - 1)(3n + 1)(n + 2)}$ 71) $\dfrac{3y^2 + 6y + 26}{y(y - 4)(2y - 5)}$

73) a) $\dfrac{2(x + 1)}{x - 3}$ b) $\dfrac{x^2 - 2x + 5}{x - 3}$

75) a) $\dfrac{w}{(w + 2)^2(w - 2)}$ b) $\dfrac{2(w - 1)^2}{(w + 2)(w - 2)}$

Chapter 8 Mid-Chapter Summary

1) a) $\dfrac{2}{7}$ b) undefined   3) a) $-\dfrac{8}{3}$ b) 9   5) a) $-3, 3$ b) 0

7) a) $-4, 2$ b) $\dfrac{3}{5}$

9) a) never undefined—any real number can be substituted
for t b) 6

11) $\dfrac{4}{3n^3}$ 13) $\dfrac{1}{j-4}$ 15) $-\dfrac{3}{n+2}$ 17) $\dfrac{3f-16}{f(f+8)}$

19) $\dfrac{4a^2b}{9}$ 21) $\dfrac{8q^2-37q+21}{(q-5)(q+4)(q+7)}$ 23) $-\dfrac{m+7}{8}$

25) $\dfrac{9}{r-8}$ 27) $\dfrac{4}{d^3}$ 29) $\dfrac{5}{3}$ 31) $\dfrac{34}{15z}$ 33) $\dfrac{2a^3(a^2+5)}{5}$

35) $\dfrac{2(7x-1)}{(x-8)(x+3)}$ 37) -1 39) $\dfrac{5u^2+37u-19}{u(3u-2)(u+1)}$

41) 3 43) $\dfrac{x^2+2x+12}{(2x+1)^2(x-4)}$ 45) $\dfrac{5(c^2+16)}{3(c-2)}$ 47) $-\dfrac{11}{5w}$

49) a) $\dfrac{x(x-3)}{8}$ b) $\dfrac{3(x-1)}{2}$

Section 8.5

1) Method 1: Rewrite it as a division problem, then simplify.

$$\dfrac{2}{9} \div \dfrac{5}{18} = \dfrac{2}{\underset{1}{9}} \cdot \dfrac{\overset{2}{18}}{5} = \dfrac{4}{5}$$

Method 2: Multiply the numerator and denominator by 18, the LCD of $\dfrac{2}{9}$ and $\dfrac{5}{18}$. Then, simplify.

$$\dfrac{\overset{2}{18}\left(\dfrac{2}{9}\right)}{\underset{1}{18}\left(\dfrac{5}{18}\right)} = \dfrac{4}{5}$$

3) $\dfrac{14}{25}$ 5) ab^2 7) $\dfrac{1}{st^2}$ 9) $\dfrac{2m^4}{15n^2}$ 11) $\dfrac{t}{5}$ 13) $\dfrac{4}{3(y-8)}$

15) $\dfrac{5}{6w^4}$ 17) $x(x-3)$ 19) $\dfrac{3}{2}$ 21) $\dfrac{7d+2c}{d(c-5)}$ 23) $\dfrac{4z+7}{5z-7}$

25) $\dfrac{8}{y}$ 27) $\dfrac{x^2-7}{x^2-11}$ 29) $-\dfrac{52}{15}$ 31) $\dfrac{2ab}{a+b}$ 33) $\dfrac{r(r^2+s)}{s^2(sr+1)}$

35) $\dfrac{t-5}{t+4}$ 37) $\dfrac{b^2+1}{b^2-3}$ 39) $\dfrac{1}{m^3n}$ 41) $\dfrac{2(x-9)(x+2)}{3(x+3)(x+1)}$

43) $\dfrac{r}{20}$ 45) $\dfrac{a}{12}$ 47) $\dfrac{25}{18}$ 49) $\dfrac{2(n+3)^2}{4n+7}$

Section 8.6

1) Eliminate the denominators.   3) sum; $\dfrac{3m-14}{8}$

5) equation; $\{-9\}$ 7) difference; $\dfrac{z^2-4z+24}{z(z-6)}$

9) equation; $\{3\}$ 11) $0, -10$ 13) $0, 9, -9$ 15) $4, 9$

17) $\{2\}$ 19) $\{-2\}$ 21) $\{-21\}$ 23) $\{4\}$ 25) $\left\{-\dfrac{5}{2}\right\}$

27) $\{3\}$ 29) $\left\{\dfrac{1}{2}\right\}$ 31) $\varnothing$ 33) $\{6\}$ 35) $\{-4\}$ 37) $\{5\}$

39) $\{12\}$ 41) $\{-11\}$ 43) $\varnothing$ 45) $\{-3\}$ 47) $\{0, 6\}$

49) $\{-20\}$ 51) $\{2, 9\}$ 53) $\{1, 3\}$ 55) $\{-3\}$ 57) $\{-4, -2\}$

59) $\{-10\}$ 61) $\{-6\}$ 63) $\{2, 8\}$ 65) $\{-8, 1\}$ 67) $\varnothing$

69) $P = \dfrac{nRT}{V}$ 71) $z = \dfrac{kx}{y}$ 73) $x = \dfrac{t+u}{3B}$

75) $b = \dfrac{a-zc}{z}$ 77) $t = \dfrac{Aq-4r}{A}$ 79) $c = \dfrac{na-wb}{wk}$

81) $r = \dfrac{st}{s+t}$ 83) $C = \dfrac{AB}{3A-2B}$

Section 8.7

1) $\{35\}$ 3) $\{48\}$ 5) 111

7) 3 cups of tapioca flour and 6 cups of potato-starch flour

9) length: 48 ft; width: 30 ft

11) stocks: $8000; bonds: $12,000   13) 1355

15) a) 7 mph   b) 13 mph   17) a) $x+30$ mph   b) $x-30$ mph

19) 20 mph   21) 4 mph   23) 260 mph   25) 2 mph

27) $\dfrac{1}{4}$ job/hr   29) $\dfrac{1}{t}$ job/hr   31) $1\dfrac{1}{5}$ hr   33) $3\dfrac{1}{13}$ hr

35) 20 min   37) 3 hr   39) 3 hr

Chapter 8 Review Exercises

1) a) $\dfrac{18}{7}$ b) undefined   3) a) -5 b) 1 5) a) $-\dfrac{1}{2}, 6$ b) $0, 3$

7) a) 2

b) never undefined—any real number can be substituted for d

9) $\dfrac{7}{a^9}$ 11) $\dfrac{5}{11}$ 13) $\dfrac{1}{3z+1}$ 15) $-\dfrac{1}{x+10}$

17) Possible answers:

$\dfrac{-u+6}{u+2}, \dfrac{6-u}{u+2}, \dfrac{u-6}{-u-2}, \dfrac{-(u-6)}{u+2}, \dfrac{u-6}{-(u+2)}$

19) $2l+1$ 21) $(-\infty, 5) \cup (5, \infty)$

23) $(-\infty, -4) \cup (-4, 6) \cup (6, \infty)$

25) $\dfrac{4}{15}$ 27) $\dfrac{36k^2}{m}$ 29) $\dfrac{1}{4w}$ 31) $-\dfrac{a+5}{4(a^2+5a+25)}$ 33) $\dfrac{10}{27}$

35) $\dfrac{4m}{3}$ 37) 18 39) k^2 41) $m(m+5)$

43) $3d(2d-1)$ 45) $b-2$ or $2-b$

47) $(c+4)(c+6)(c-7)$ 49) $3x(x+8)^2$

51) $(c+d)(c-d)$ or $(c+d)(d-c)$

53) $\dfrac{24r^2}{20r^3}$ 55) $\dfrac{8z}{z(3z+4)}$ 57) $\dfrac{t^2+2t-15}{(2t+1)(t+5)}$

59) $\dfrac{4}{9z^3} = \dfrac{16z^2}{36z^5}; \dfrac{7}{12z^5} = \dfrac{21}{36z^5}$

61) $\dfrac{8}{p+7} = \dfrac{8p}{p(p+7)}; \dfrac{2}{p} = \dfrac{2p+14}{p(p+7)}$

63) $\dfrac{1}{g-10} = \dfrac{1}{g-10}; \dfrac{3}{10-g} = -\dfrac{3}{g-10}$

65) $-\dfrac{11}{36}$ 67) $\dfrac{4m-5}{m-3}$ 69) $\dfrac{11t+12}{t(t+4)}$ 71) $\dfrac{27-y}{(y-2)(y+3)}$

73) $\dfrac{10p^2-67p-53}{4(p+1)(p-7)}$ 75) $\dfrac{17}{11-m}$ or $-\dfrac{17}{m-11}$

77) $-\dfrac{1}{r+8}$ 79) $\dfrac{15w^2+2w+54}{5w(w+7)}$ 81) $\dfrac{6d^2+7d-32}{d(d+2)(5d-3)}$

83) a) $\dfrac{3x}{2(x-4)}$ b) $\dfrac{x^2-4x+96}{4(x-4)}$

85) $\dfrac{6}{7}$ 87) $\dfrac{p^2 + 6}{p^2 + 8}$ 89) $\dfrac{2}{3n}$ 91) $\dfrac{(y + 3)(y - 10)}{(y - 9)(y + 5)}$

93) $\dfrac{c - 1}{c - 2}$ 95) $\left\{\dfrac{2}{5}\right\}$ 97) $\{10\}$ 99) $\{-2\}$ 101) $\{1, 20\}$

103) $\varnothing$ 105) $\{-3, 5\}$ 107) $c = \dfrac{2p}{A}$ 109) $a = \dfrac{t - nb}{n}$

111) $s = \dfrac{rt}{t - r}$ 113) 12 g 115) 280 mph

Chapter 8 Test

1) $-\dfrac{1}{5}$ 3) a) $-8, 6$ b) never equals zero 5) $\dfrac{7v - 1}{v - 8}$

7) $b(b + 8)$ 9) $-\dfrac{7n^2}{4m^4}$ 11) $\dfrac{r^2 + 11r + 3}{(2r + 1)(r + 5)}$

13) $\dfrac{11}{15 - c}$ or $-\dfrac{11}{c - 15}$ 15) $-\dfrac{1}{d}$ 17) $\{4\}$ 19) $\{-1, 5\}$

21) $1\dfrac{7}{8}$ hr

Cumulative Review for Chapters 1–8

1) area $= 16$ cm^2; perimeter $= 20$ cm

3) $81y^8$ 5) $\left[-\dfrac{7}{2}, 1\right]$

7) x-int: $\left(\dfrac{10}{3}, 0\right)$; y-int: $(0, 2)$

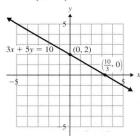

9) $\left(-\dfrac{3}{2}, 4\right)$ 11) $m^2 - 14m + 49$

13) $\dfrac{9}{2}c^2 + 7c - 1 + \dfrac{3}{4c}$ 15) $(2h + 5)(4h^2 - 10h + 25)$

17) $(a + 3)(b - 2)$ 19) a) $0, 4$ b) $-\dfrac{1}{9}$ 21) $-\dfrac{14d^3}{2d + 1}$

23) $\dfrac{11 - x}{x - 4}$ 25) $\{-20, 3\}$

Chapter 9

Section 9.1

1) Answers may vary. 3) $\{-6, 6\}$ 5) $\{2, 8\}$ 7) $\left\{-\dfrac{1}{2}, 3\right\}$

9) $\left\{-\dfrac{1}{2}, -\dfrac{1}{3}\right\}$ 11) $\left\{-1, \dfrac{5}{4}\right\}$ 13) $\{-24, 15\}$

15) $\left\{-\dfrac{10}{3}, \dfrac{50}{3}\right\}$ 17) $\left\{\dfrac{28}{15}, \dfrac{52}{15}\right\}$ 19) $\varnothing$ 21) $\left\{-\dfrac{1}{5}\right\}$

23) $\{-10, 22\}$ 25) $\{-5, 0\}$ 27) $\{-14\}$ 29) $\varnothing$

31) $\left\{-\dfrac{16}{5}, 2\right\}$ 33) $\varnothing$ 35) $\left\{-\dfrac{14}{3}, 4\right\}$ 37) $\left\{\dfrac{1}{2}, 4\right\}$

39) $\left\{\dfrac{2}{5}, 2\right\}$ 41) $\{-4, -1\}$ 43) $\{10\}$ 45) $|x| = 9$, may vary

47) $|x| = \dfrac{1}{2}$, may vary

Section 9.2

1) $[-1, 5]$

3) $(-\infty, 2) \cup (9, \infty)$

5) $\left(-\infty, -\dfrac{9}{2}\right] \cup \left[\dfrac{3}{5}, \infty\right)$

7) $\left(4, \dfrac{17}{3}\right)$

9) $[-7, 7]$

11) $(-4, 4)$

13) $(-2, 6)$

15) $\left[-\dfrac{14}{3}, -2\right]$

17) $\left[\dfrac{2}{3}, \dfrac{5}{3}\right]$

19) $\varnothing$

21) $\varnothing$

23) $\left(-\dfrac{1}{4}, 1\right)$

25) $[-12, 4]$

27) $(-\infty, -7] \cup [7, \infty)$

29) $\left(-\infty, -\dfrac{2}{5}\right) \cup \left(\dfrac{2}{5}, \infty\right)$

31) $(-\infty, -14] \cup [-6, \infty)$

33) $\left(-\infty, -\dfrac{3}{2}\right] \cup [3, \infty)$

35) $(-\infty, 2) \cup \left(\dfrac{11}{3}, \infty\right)$

37) $(-\infty, \infty)$

39) $(-\infty, \infty)$

41) $(-\infty, 0) \cup (1, \infty)$

43) $\left(-\infty, -\dfrac{27}{5}\right] \cup \left[\dfrac{21}{5}, \infty\right)$

45) The absolute value of a quantity is always 0 or positive; it cannot be less than 0.

47) The absolute value of a quantity is always 0 or positive, so for any real number, x, the quantity $|2x + 1|$ will be greater than -3.

49) $(-\infty, -6) \cup (-3, \infty)$ 51) $\left\{-2, -\dfrac{1}{2}\right\}$

53) $\left(-\infty, -\dfrac{1}{4}\right] \cup [2, \infty)$ 55) $(-3, \infty)$ 57) $[3, 13]$

59) $\varnothing$ 61) $\{-21, -3\}$ 63) $(-\infty, 0) \cup \left(\dfrac{16}{3}, \infty\right)$ 65) $\varnothing$

67) $\left(-\infty, -\dfrac{1}{25}\right]$ 69) $(-\infty, \infty)$ 71) $[-15, -1]$

73) $\left(-\infty, \dfrac{1}{5}\right) \cup (3, \infty)$ 75) $\{-4, 20\}$ 77) $\left[-\dfrac{11}{5}, 1\right]$

79) $|a - 128| \le 0.75$; $127.25 \le a \le 128.75$; there is between 127.25 oz and 128.75 oz of milk in the container.

81) $|b - 38| \le 5$; $33 \le b \le 43$; he will spend between \$33 and \$43 on his daughter's gift.

Section 9.3

1) Answers may vary.   3) Answers may vary.

5) Answers may vary.

7)

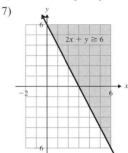

9)

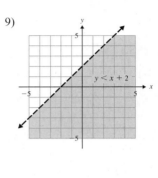

11)

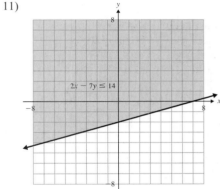

13)

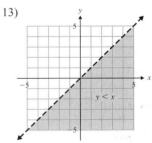

15)

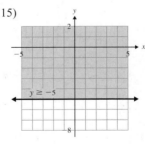

17)

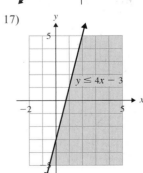

19)

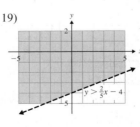

21)
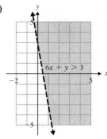
$6x + y > 3$

23)
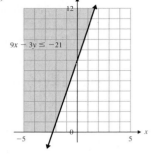
$9x - 3y \leq -21$

45)
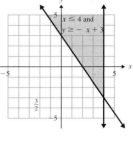
$x \leq 4$ and $y \geq -x + 3$

25)
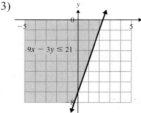
$x > 2y$

27) Answers may vary.

47)
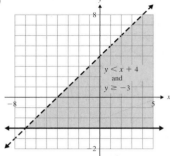
$y < x + 4$ and $y \geq -3$

29)
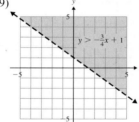
$y > -\frac{3}{4}x + 1$

31)
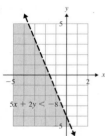
$5x + 2y < -8$

49)
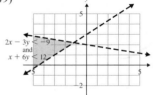
$2x - 3y < -9$ and $x + 6y < 12$

33)
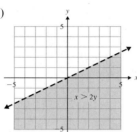
$9x - 3y \leq 21$

35)
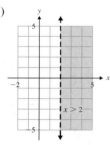
$x > 2$

51)
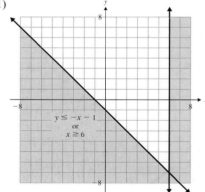
$y \leq -x - 1$ or $x \geq 6$

37)
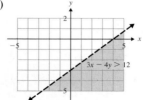
$3x - 4y > 12$

53)
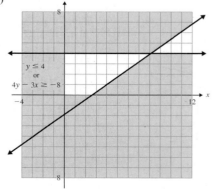
$y \leq 4$ or $4y - 3x \geq -8$

39) Answers may vary. 41) Answers may vary.

43) No; (3, 5) satisfies $x - y \geq -6$ but not $2x + y < 7$. Since the inequality contains *and*, it must satisfy *both* inequalities.

55)
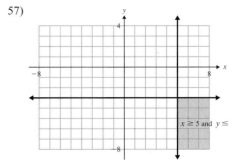

$y > -\frac{2}{3}x + 1$
or
$-2x + 5y \le 0$

57)
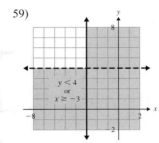

$x \ge 5$ and $y \le$

59)
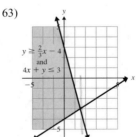

$y < 4$
or
$x \ge -3$

61)
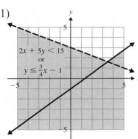

$2x + 5y < 15$
or
$y \le \frac{3}{4}x - 1$

63)
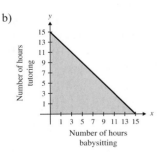

$y \ge \frac{2}{3}x - 4$
and
$4x + y \le 3$

65) a) $x \ge 0$
          $y \ge 0$
          $x + y \le 15$
   c) Answers may vary.
   d) Answers may vary.

b)
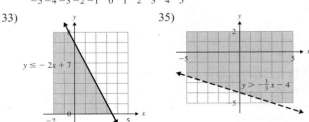

Number of hours tutoring

Number of hours babysitting

Section 9.4

1) $\begin{bmatrix} 1 & -7 & | & 15 \\ 4 & 3 & | & -1 \end{bmatrix}$ 3) $\begin{bmatrix} 1 & 6 & -1 & | & -2 \\ 3 & 1 & 4 & | & 7 \\ -1 & -2 & 3 & | & 8 \end{bmatrix}$

5) $3x + 10y = -4$ 7) $x - 6y = 8$
   $x - 2y = 5$ $y = -2$

9) $x - 3y + 2z = 7$ 11) $x + 5y + 2z = 14$
     $4x - y + 3z = 0$ $y - 8z = 2$
     $-2x + 2y - 3z = -9$ $z = -3$

13) $(3, -1)$ 15) $(-6, -5)$ 17) $(0, -2)$ 19) $(-1, 4, 8)$

21) $(10, 1, -4)$ 23) $(0, 1, 8)$ 25) $\varnothing$

Chapter 9 Review Exercises

1) $\{-9, 9\}$ 3) $\left\{-1, \frac{1}{7}\right\}$ 5) $\left\{-\frac{15}{8}, -\frac{7}{8}\right\}$ 7) $\left\{\frac{11}{5}, \frac{13}{5}\right\}$

9) $\left\{-8, \frac{4}{15}\right\}$ 11) $\varnothing$ 13) $\left\{-\frac{4}{9}\right\}$ 15) $|a| = 4$, may vary

17) $[-3, 3]$

-5 -4 -3 -2 -1 0 1 2 3 4 5

19) $(-\infty, -2) \cup (2, \infty)$

-5 -4 -3 -2 -1 0 1 2 3 4 5

21) $(-\infty, -1] \cup \left[\frac{1}{6}, \infty\right)$

$\frac{1}{6}$
-5 -4 -3 -2 -1 0 1 2 3 4 5

23) $(-5, 13)$

0 2 4 5 6 8 10 13

25) $\left[-\frac{15}{4}, -\frac{3}{4}\right]$

$-\frac{15}{4}$ $-\frac{3}{4}$
-5 -4 -3 -2 -1 0 1 2 3 4 5

27) $\left(-\infty, -\frac{19}{5}\right] \cup [-1, \infty)$

$-\frac{19}{5}$
-5 -4 -3 -2 -1 0 1 2 3 4 5

29) $(-\infty, \infty)$

-5 -4 -3 -2 -1 0 1 2 3 4 5

31) $\left\{-\frac{1}{12}\right\}$

$-\frac{1}{12}$
-5 -4 -3 -2 -1 0 1 2 3 4 5

33)

$y \le -2x + 7$

35)

$y > -\frac{1}{3}x - 4$

37)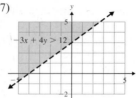

39)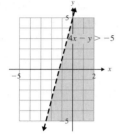

41)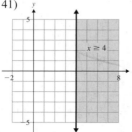

43)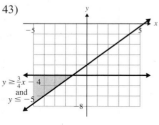

45)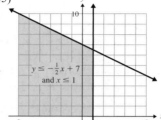

47)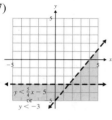

49)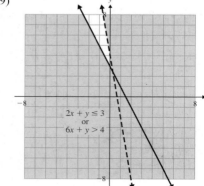

51)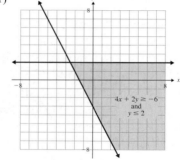

53) $(-9, 2)$ 55) $(1, 0)$ 57) $(5, -2, 6)$

Chapter 9 Test

1) $\left\{-\dfrac{1}{2}, 5\right\}$ 3) $\left\{-8, \dfrac{3}{2}\right\}$ 5) $|x| = 8$, may vary

7) $[-1, 8]$

9) $|w - 168| \le 0.75$; $167.25 \le w \le 168.75$; Thanh's weight is between 167.25 lb and 168.75 lb.

11)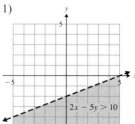

13)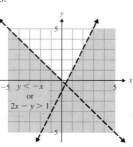

15) $(1, -1, 1)$

Cumulative Review for Chapters 1–9

1) 26   3) 81   5) $\dfrac{1}{64}$ 7) 9.14×10^{-6} 9) $(-\infty, -21]$

11) $-\dfrac{1}{5}$ 13) $y = \dfrac{1}{3}x - \dfrac{1}{3}$ 15) $-12p^4 + 28^3 + 4p^2$

17) $t^2 + 16t + 64$ 19) $(3m + 11)(3m - 11)$

21) $\{-3\}$ 23) $\dfrac{-r^2 + 2r + 17}{2(r + 5)(r - 5)}$ 25) $\{16, 40\}$

27)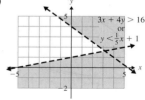

Chapter 10

Section 10.1

1) False; the $\sqrt{}$ symbol means to find only the positive square root of 121. $\sqrt{121} = 11$

3) true

5) False; the even root of a negative number is not a real number.

7) 7 and -7 9) 1 and -1 11) 20 and -20

13) 50 and -50 15) $\dfrac{2}{3}$ and $-\dfrac{2}{3}$ 17) $\dfrac{1}{9}$ and $-\dfrac{1}{9}$ 19) 7

21) 1   23) 13   25) not real   27) $\dfrac{9}{5}$ 29) $\dfrac{7}{8}$ 31) -6 33) $-\dfrac{1}{11}$

35) 3.3

37) 6.8

39) 4.1

$\sqrt{17}$

0 1 2 3 4 5 6 7 8 9

41) 2.2

$\sqrt{5}$

0 1 2 3 4 5 6 7 8 9

43) 7.8

$\sqrt{61}$

0 1 2 3 4 5 6 7 8 9

45) $\sqrt[3]{64}$ is the number you cube to get 64. $\sqrt[3]{64} = 4$

47) No; the even root of a negative number is not a real number.

49) 2   51) 5   53) -1 55) 3   57) not real   59) -2 61) -2

63) 3   65) not real   67) $\dfrac{2}{5}$ 69) 7   71) 5   73) not real

Section 10.2

1) The denominator of 2 becomes the index of the radical.
$25^{1/2} = \sqrt{25}$

3) 3   5) 10   7) 2   9) -5 11) $\dfrac{2}{11}$ 13) $\dfrac{5}{4}$ 15) $-\dfrac{6}{13}$

17) The denominator of 4 becomes the index of the radical. The numerator of 3 is the power to which we raise the radical expression. $16^{3/4} = (\sqrt[4]{16})^3$

19) 16   21) 25   23) 32   25) -81 27) $\dfrac{8}{27}$ 29) $-\dfrac{100}{9}$

31) False; the negative exponent does not make the result negative. $81^{-1/2} = \dfrac{1}{9}$

33) $\dfrac{1}{7}$ 35) $\dfrac{1}{10}$ 37) 3   39) -4 41) $\dfrac{1}{32}$ 43) $\dfrac{1}{25}$ 45) $\dfrac{8}{125}$

47) $\dfrac{25}{16}$ 49) 8   51) 3   53) $8^{4/5}$ 55) 16   57) $\dfrac{1}{125}$

59) 6   61) $7^{1/3}$ 63) $k^{5/2}$ 65) $\dfrac{1}{j^{3/10}}$ 67) $-72v^{11/8}$

69) $a^{1/9}$ 71) $\dfrac{5}{18c^{3/2}}$ 73) q^8 75) $\dfrac{1}{x^{2/3}}$ 77) $z^{2/15}$

79) $27u^2v^3$ 81) $8r^{1/5}s^{4/15}$ 83) $\dfrac{f^{2/7}g^{5/9}}{3}$ 85) $x^{10}w^9$

87) $\dfrac{1}{y^{2/3}}$ 89) $\dfrac{a^{12/5}}{4b^{2/5}}$ 91) 7   93) 10   95) 5   97) 3   99) 12

101) 15   103) x^4 105) $\sqrt[3]{k}$ 107) $\sqrt{z}$ 109) d^2

Section 10.3

1) $\sqrt{21}$ 3) $\sqrt{30}$ 5) $\sqrt{6y}$

7) False; 20 contains a factor of 4 which is a perfect square.

9) True; 42 does not have any factors (other than 1) that are perfect squares.

11) $2\sqrt{5}$ 13) $3\sqrt{6}$ 15) simplified   17) $2\sqrt{2}$ 19) $4\sqrt{5}$

21) $7\sqrt{2}$ 23) simplified   25) 20   27) $\dfrac{12}{5}$ 29) $\dfrac{2}{7}$ 31) 2

33) 3   35) $2\sqrt{3}$ 37) $2\sqrt{5}$ 39) $\sqrt{15}$ 41) $\dfrac{\sqrt{6}}{7}$ 43) $\dfrac{3\sqrt{5}}{4}$

45) x^4 47) m^5 49) w^7 51) $10c$ 53) $8k^3$ 55) $2r^2\sqrt{7}$

57) $10q^{11}\sqrt{3}$ 59) $\dfrac{9}{c^3}$ 61) $\dfrac{2\sqrt{10}}{t^4}$ 63) $\dfrac{5\sqrt{3}}{y^6}$ 65) $a^2\sqrt{a}$

67) $g^6\sqrt{g}$ 69) $b^{12}\sqrt{b}$ 71) $6x\sqrt{2x}$ 73) $q^3\sqrt{13q}$

75) $5t^5\sqrt{3t}$ 77) c^4d 79) $a^2b\sqrt{b}$ 81) $u^2v^3\sqrt{uv}$

83) $6m^4n^2\sqrt{m}$ 85) $2x^6y^2\sqrt{11y}$ 87) $4t^2u^3\sqrt{2tu}$

89) $\dfrac{a^3\sqrt{a}}{9b^3}$ 91) $\dfrac{r^4\sqrt{3r}}{s}$ 93) $5\sqrt{2}$ 95) $3\sqrt{7}$ 97) w^3

99) $n^3\sqrt{n}$ 101) $4k^3$ 103) $2x^4y^2\sqrt{3xy}$ 105) $2c^5d^4\sqrt{10d}$

107) $3k^4$ 109) $2h^3\sqrt{10}$ 111) $a^4\sqrt{10a}$

Section 10.4

1) Answers may vary.

3) i) Its radicand will not contain any factors that are perfect cubes.

ii) The radicand will not contain fractions.

iii) There will be no radical in the denominator of a fraction.

5) $\sqrt[3]{20}$ 7) $\sqrt[5]{9m^2}$ 9) $\sqrt[3]{a^2b}$ 11) $2\sqrt[3]{3}$ 13) $2\sqrt[4]{4}$

15) $3\sqrt[3]{2}$ 17) $10\sqrt[3]{2}$ 19) $2\sqrt[5]{2}$ 21) $\dfrac{1}{2}$ 23) -3 25) $2\sqrt[3]{3}$

27) $2\sqrt[4]{5}$ 29) d^2 31) n^5 33) xy^3 35) $w^4\sqrt[3]{w^2}$ 37) $y^2\sqrt[4]{y}$

39) $d\sqrt[3]{d^2}$ 41) $u^3v^5\sqrt[3]{u}$ 43) $b^5c\sqrt[3]{bc^2}$ 45) $n^4\sqrt[4]{m^3n^2}$

47) $2x^3y^4\sqrt[3]{3x}$ 49) $5wx^5\sqrt[3]{2wx}$ 51) $\dfrac{m^2}{3}$ 53) $\dfrac{2a^4\sqrt[5]{a^3}}{b^3}$

55) $\dfrac{t^2\sqrt[4]{t}}{3s^6}$ 57) $\dfrac{u^9\sqrt[3]{u}}{v}$ 59) $2\sqrt[3]{3}$ 61) $3\sqrt[4]{4}$ 63) $2\sqrt[3]{10}$

65) m^3 67) k^4 69) $r^3\sqrt[3]{r^2}$ 71) $p^4\sqrt[5]{p^3}$ 73) $3z^6\sqrt[3]{z}$

75) h^4 77) $c^2\sqrt[3]{c}$ 79) $3d^4\sqrt[4]{d^3}$ 81) $\sqrt[6]{p^5}$ 83) $n\sqrt[4]{n}$

85) $c\sqrt[15]{c^4}$ 87) $\sqrt[4]{w}$ 89) $\sqrt[15]{t^2}$

Section 10.5

1) They have the same index and the same radicand.

3) $14\sqrt{2}$ 5) $-5\sqrt{3}$ 7) $15\sqrt[3]{4}$ 9) $11 - 3\sqrt{13}$

11) $-5\sqrt[3]{z^2}$ 13) $-9\sqrt[3]{n^2} + 10\sqrt[5]{n^2}$ 15) $2\sqrt{5c} - 2\sqrt{6c}$

17) $4\sqrt{3}$ 19) $5\sqrt{3}$ 21) $14\sqrt{5}$ 23) $41\sqrt{2}$ 25) $-2\sqrt{2}$

27) $6\sqrt{3}$ 29) $-8\sqrt{5}$ 31) $10\sqrt[3]{9}$ 33) $-8\sqrt[3]{3}$ 35) $-\sqrt[3]{6}$

37) $13q\sqrt{q}$ 39) $-20d^2\sqrt{d}$ 41) $5n^2\sqrt{n}$ 43) $4t^3\sqrt[3]{t}$

45) $6a^2\sqrt[4]{a^3}$ 47) $15\sqrt{5z}$ 49) $-2\sqrt{2p}$ 51) $25a\sqrt[3]{3a^2}$

53) $-18c^2\sqrt[3]{4c}$ 55) $4y\sqrt{xy}$ 57) $3c^2d\sqrt{2d}$ 59) $17m\sqrt{3mn}$

61) $20a^5\sqrt[3]{7a^2b}$ 63) $14cd\sqrt[4]{9cd}$ 65) $\sqrt{m}(m^2 + n)$

67) $\sqrt[3]{b}(a^3 - b^2)$ 69) $\sqrt[3]{u^2}(v^2 + 1)$

Section 10.6

1) $3x + 15$ 3) $7\sqrt{6} + 14$ 5) $\sqrt{30} - \sqrt{10}$ 7) $-30\sqrt{2}$

9) $4\sqrt{5}$ 11) $-\sqrt{30}$ 13) $4\sqrt{3} + 3\sqrt{2}$ 15) $2\sqrt{42} + \sqrt{35}$

17) $t - 9\sqrt{tu}$ 19) $a\sqrt{5b} + 3b\sqrt{3a}$

21) Both are examples of multiplication of two binomials. They can be multiplied using FOIL.

23) $(a + b)(a - b) = a^2 - b^2$ 25) $p^2 + 13p + 42$

27) $19 + 8\sqrt{7}$ 29) $-22 + 5\sqrt{2}$ 31) $22 - 9\sqrt{15}$

33) $86 - 14\sqrt{3}$ 35) $5\sqrt{7} + 5\sqrt{2} + 2\sqrt{21} + 2\sqrt{6}$

37) $x + 6\sqrt{2xy} + 10y$ 39) $-2\sqrt{6pq} + 30p - 16q$

41) $25y^2 - 40y + 16$ 43) $4 + 2\sqrt{3}$ 45) $16 - 2\sqrt{55}$

47) $22 + 4\sqrt{30}$ 49) $98 - 16\sqrt{3}$ 51) $h + 2\sqrt{7h} + 7$

53) $x - 2\sqrt{xy} + y$ 55) $c^2 - 81$ 57) -7 59) 31 61) 46

63) -64 65) $c - d$ 67) $25 - t$ 69) $64f - g$

71) $\sqrt[3]{4} - 9$ 73) 41 75) $7 + 6\sqrt{2}$

Section 10.7

1) Eliminate the radical from the denominator.

3) $\dfrac{\sqrt{5}}{5}$ 5) $\dfrac{3\sqrt{2}}{2}$ 7) $\dfrac{3\sqrt{6}}{2}$ 9) $-5\sqrt{2}$ 11) $\dfrac{\sqrt{21}}{14}$ 13) $\dfrac{\sqrt{3}}{3}$

15) $\dfrac{3\sqrt{13}}{13}$ 17) $\dfrac{\sqrt{42}}{6}$ 19) $\dfrac{\sqrt{30}}{3}$ 21) $\dfrac{\sqrt{15}}{10}$ 23) $\sqrt{2}$

25) $\dfrac{8\sqrt{y}}{y}$ 27) $\dfrac{\sqrt{5t}}{t}$ 29) $\dfrac{f\sqrt{10fg}}{g}$ 31) $\dfrac{8v^3\sqrt{5vw}}{5w}$

33) $\dfrac{a\sqrt{3b}}{3b}$ 35) $-\dfrac{5\sqrt{3b}}{b^2}$ 37) $\dfrac{\sqrt{13j}}{j^3}$ 39) 2^2 or 4 41) 3

43) c^2 45) 2^3 or 8 47) m 49) $\dfrac{4\sqrt[3]{9}}{3}$ 51) $6\sqrt[3]{4}$ 53) $\dfrac{9\sqrt[5]{5}}{5}$

55) $\dfrac{\sqrt[4]{45}}{3}$ 57) $\dfrac{\sqrt[5]{12}}{2}$ 59) $\dfrac{\sqrt[4]{18}}{3}$ 61) $\dfrac{10\sqrt[3]{z^2}}{z}$ 63) $\dfrac{\sqrt[3]{3n}}{n}$

65) $\dfrac{\sqrt[3]{28k}}{2k}$ 67) $\dfrac{9\sqrt[5]{a^2}}{a}$ 69) $\dfrac{\sqrt[4]{cd}}{d}$ 71) $\dfrac{\sqrt[4]{40m^3}}{2m}$

73) Change the sign between the two terms.

75) $(5 - \sqrt{2}); 23$ 77) $(\sqrt{2} - \sqrt{6}); -4$

79) $(\sqrt{t} + 8); t - 64$ 81) $6 - 3\sqrt{3}$ 83) $\dfrac{90 + 10\sqrt{2}}{79}$

85) $2\sqrt{6} - 4$ 87) $\dfrac{\sqrt{30} - 5\sqrt{2} + 3 - \sqrt{15}}{7}$

89) $\dfrac{m - \sqrt{mn}}{m - n}$ 91) $\sqrt{b} + 5$ 93) $\dfrac{x + 2\sqrt{xy} + y}{x - y}$

95) $1 + 2\sqrt{3}$ 97) $\dfrac{15 - 9\sqrt{5}}{2}$ 99) $\dfrac{\sqrt{5} + 2}{3}$ 101) $-2 - \sqrt{2}$

Section 10.8

1) Sometimes there are extraneous solutions.

3) $\{49\}$ 5) $\left\{\dfrac{4}{9}\right\}$ 7) $\varnothing$ 9) $\{125\}$ 11) $\{-64\}$ 13) $\{20\}$

15) $\varnothing$ 17) $\left\{\dfrac{2}{3}\right\}$ 19) $\left\{-\dfrac{3}{2}\right\}$ 21) $\{1\}$ 23) $\{-1\}$

25) $\{-3\}$ 27) $\{10\}$ 29) $x^2 + 6x + 9$ 31) $\{2\}$ 33) $\{-3\}$

35) $\{10\}$ 37) $\varnothing$ 39) $\{1, 3\}$ 41) $\{4\}$ 43) $x + 10\sqrt{x} + 25$

45) $85 - 18\sqrt{a + 4} + a$ 47) $12n + 28\sqrt{3n - 1} + 45$

49) $\{5, 13\}$ 51) $\{2\}$ 53) $\{1, 5\}$ 55) $\varnothing$ 57) $\{2, 11\}$

59) $\left\{\dfrac{1}{4}\right\}$ 61) $\{9\}$ 63) $\{-1\}$ 65) $E = \dfrac{mv^2}{2}$

67) $b^2 = c^2 - a^2$ 69) $\sigma = \dfrac{E}{T^4}$

71) a) 320 m/sec   b) 340 m/sec

c) The speed of sound increases.   d) $T = \dfrac{V_s^2}{400} - 273$

73) a) 2 in.   b) $V = \pi r^2 h$

Chapter 10 Review Exercises

1) 5 3) -9 5) 4 7) -1 9) not real

11) 5.8

13) The denominator of the fractional exponent becomes the index on the radical. The numerator is the power to which we raise the radical expression. $8^{2/3} = (\sqrt[3]{8})^2$

15) 6 17) $\dfrac{3}{5}$ 19) -2 21) 25 23) $\dfrac{16}{9}$ 25) $\dfrac{1}{9}$ 27) $\dfrac{1}{27}$

29) $\dfrac{100}{9}$ 31) 9 33) 64 35) 1 37) $32a^{10/3}b^{10}$ 39) $\dfrac{2c}{3d^{7/4}}$

41) 3 43) 7 45) k^7 47) w^3 49) $10\sqrt{10}$ 51) $\dfrac{3\sqrt{2}}{7}$

53) k^6 55) $x^4\sqrt{x}$ 57) $3t\sqrt{5}$ 59) $6x^3y^6\sqrt{2xy}$ 61) $\sqrt{15}$

63) $2\sqrt{6}$ 65) $11x^6\sqrt{x}$ 67) $10k^8$ 69) $2\sqrt[3]{2}$ 71) $2\sqrt[4]{3}$

73) z^6 75) $a^6\sqrt[3]{a^2}$ 77) $2z^5\sqrt[3]{2}$ 79) $\dfrac{h^3}{3}$ 81) $\sqrt[3]{21}$

83) $2t^4\sqrt[4]{2t}$ 85) $\sqrt[6]{n^5}$ 87) $11\sqrt{5}$ 89) $6\sqrt{5} - 4\sqrt{3}$

91) $-4p\sqrt{p}$ 93) $-12d^2\sqrt{2d}$ 95) $6k\sqrt{5} + 3\sqrt{2k}$

97) $23\sqrt{2rs} + 8r + 15s$ 99) $2 + 2\sqrt{y + 1} + y$

101) $\dfrac{14\sqrt{3}}{3}$ 103) $\dfrac{3\sqrt{2kn}}{n}$ 105) $\dfrac{7\sqrt[3]{4}}{2}$ 107) $\dfrac{\sqrt[3]{x^2y^2}}{y}$

109) $\dfrac{3 - \sqrt{3}}{3}$ 111) $1 - 3\sqrt{2}$ 113) $\{1\}$ 115) $\varnothing$

117) $\{-4\}$ 119) $\{2, 6\}$ 121) $V = \dfrac{1}{3}\pi r^2 h$

Chapter 10 Test

1) 12 3) -3 5) 2 7) $\dfrac{1}{7}$ 9) $m^{5/8}$ 11) $\dfrac{y^2}{32x^{3/2}}$ 13) $2\sqrt[3]{6}$

15) y^3 17) $t^4\sqrt{t}$ 19) $c^7\sqrt[3]{c^2}$ 21) 6 23) $2w^5\sqrt{15w}$

25) $3\sqrt{2} - 4\sqrt{3}$ 27) $2\sqrt{3} - 5\sqrt{6}$ 29) 4

31) $2t - 2\sqrt{3tu}$

33) $12 - 4\sqrt{7}$ 35) $\dfrac{5\sqrt[3]{3}}{3}$ 37) $\{1\}$ 39) $\{1, 5\}$

Cumulative Review for Chapters 1–10

1) $\frac{10}{3}x - 2y + 8$ 3) $\left\{-\frac{3}{4}\right\}$ 5) $y = \frac{5}{4}x - \frac{13}{4}$

7) $15p^4 - 20p^3 - 11p^2 + 8p + 2$ 9) $(4w - 3)(w + 2)$

11) $\{6, 9\}$ 13) $\frac{2a^2 + 2a + 3}{a(a + 4)}$ 15) $\{-8, -4\}$

17) $(4, -1, 2)$ 19) a) 9   b) 16   c) $\frac{1}{3}$

21) a) $\frac{\sqrt{10}}{5}$ b) $3\sqrt[3]{4}$ c) $\frac{x\sqrt[3]{y}}{y}$ d) $\frac{a - 2 - \sqrt{a}}{1 - a}$

Chapter 11

Section 11.1

1) $\{-7, 6\}$ 3) $\{-11, -4\}$ 5) $\{-7, 8\}$ 7) $\left\{-\frac{1}{10}, \frac{1}{10}\right\}$

9) $\left\{\frac{2}{5}, 4\right\}$ 11) $\left\{-\frac{10}{3}, -\frac{1}{2}\right\}$ 13) $\left\{0, \frac{2}{7}\right\}$

15) quadratic   17) linear   19) quadratic

21) linear   23) $\left\{\frac{1}{2}, 6\right\}$ 25) $\{-5, 2\}$ 27) $\left\{\frac{9}{2}\right\}$

29) $\left\{-\frac{5}{3}, -1, 0\right\}$ 31) $\left\{-\frac{1}{2}, 0\right\}$ 33) $\left\{\frac{9}{4}\right\}$ 35) $\{-4, 2\}$

37) $\left\{-\frac{1}{2}\right\}$ 39) $\left\{-2, -\frac{3}{5}\right\}$ 41) $\{-7, -2, 2\}$

43) width = 2 in., length = 7 in.

45) width = 5 cm, length = 9 cm

47) base = 9 in.; height = 4 in.

49) base = 6 cm; height = 12 cm

51) legs = 5, 12; hypotenuse = 13

53) legs = 6, 8; hypotenuse = 10

Section 11.2

1) factoring and the square root property; $\{-4, 4\}$

3) $\{-6, 6\}$ 5) no real number solution   7) $\{-3\sqrt{3}, 3\sqrt{3}\}$

9) $\left\{-\frac{2}{3}, \frac{2}{3}\right\}$ 11) $\{-\sqrt{14}, \sqrt{14}\}$ 13) $\{-5, 5\}$

15) $\{-2\sqrt{5}, 2\sqrt{5}\}$ 17) $\{-1, 1\}$ 19) no real number solution

21) $\{-12, -8\}$ 23) $\{6, 8\}$ 25) $\{-1 - \sqrt{22}, -1 + \sqrt{22}\}$

27) $\{-4 - 3\sqrt{2}, -4 + 3\sqrt{2}\}$ 29) $\left\{-\frac{2}{5}, \frac{6}{5}\right\}$

31) $\left\{\frac{-1 - 2\sqrt{5}}{2}, \frac{-1 + 2\sqrt{5}}{2}\right\}$

33) $\left\{\frac{10 - \sqrt{14}}{3}, \frac{10 + \sqrt{14}}{3}\right\}$

35) no real number solution   37) $\left\{\frac{-7 - 2\sqrt{6}}{10}, \frac{-7 + 2\sqrt{6}}{10}\right\}$

39) $\left\{8, \frac{40}{3}\right\}$ 41) $\left\{-\frac{6}{5}, -\frac{2}{5}\right\}$ 43) 6   45) $\sqrt{29}$

47) 6 in.   49) $\sqrt{57}$ cm   51) 12 ft   53) 5   55) $\sqrt{13}$

57) 6   59) $\sqrt{61}$ 61) $2\sqrt{5}$

Section 11.3

1) false   3) true   5) $9i$ 7) $5i$ 9) $i\sqrt{6}$

11) $3i\sqrt{3}$ 13) $2i\sqrt{15}$

15) Write each radical in terms of i *before* multiplying.
$$\sqrt{-5} \cdot \sqrt{-10} = i\sqrt{5} \cdot i\sqrt{10}$$
$$= i^2\sqrt{50}$$
$$= (-1)\sqrt{25} \cdot \sqrt{2}$$
$$= -5\sqrt{2}$$

17) $-\sqrt{5}$ 19) -6 21) 2   23) -13

25) Add the real parts and add the imaginary parts.   27) -1

29) $3 + 11i$ 31) $4 - 9i$ 33) $-\frac{1}{4} - \frac{5}{6}i$ 35) $7i$

37) $24 - 15i$ 39) $-6 + \frac{4}{3}i$ 41) $-36 + 30i$ 43) $-28 + 17i$

45) $14 + 18i$ 47) $36 - 42i$ 49) $\frac{3}{20} + \frac{9}{20}i$

51) conjugate: $11 - 4i$ 53) conjugate: $-3 + 7i$
    product: 137   product: 58

55) conjugate: $-6 - 4i$, product: 52

57) $\frac{8}{13} + \frac{12}{13}i$ 59) $\frac{8}{17} + \frac{32}{17}i$ 61) $\frac{7}{29} - \frac{3}{29}i$

63) $-\frac{74}{85} + \frac{27}{85}i$ 65) $-\frac{8}{61} + \frac{27}{61}i$ 67) $-9i$ 69) $\{-2i, 2i\}$

71) $\{-i\sqrt{3}, i\sqrt{3}\}$ 73) $\{-2i\sqrt{3}, 2i\sqrt{3}\}$ 75) $\{-3, 3\}$

77) $\{-3 - 5i, -3 + 5i\}$ 79) $\{2 - i\sqrt{14}, 2 + i\sqrt{14}\}$

81) $\{9 - 2i\sqrt{2}, 9 + 2i\sqrt{2}\}$

83) $\left\{\frac{5}{4} - \frac{\sqrt{30}}{4}i, \frac{5}{4} + \frac{\sqrt{30}}{4}i\right\}$

85) $\left\{-\frac{11}{6} - \frac{7}{6}i, -\frac{11}{6} + \frac{7}{6}i\right\}$ 87) $\{0, -7i, 7i\}$

89) $\{0, -3i\sqrt{6}, 3i\sqrt{6}\}$

Section 11.4

1) A trinomial whose factored form is the square of a binomial; examples may vary.

3) No, because the coefficient of y^2 is not 1.

5) $a^2 + 12a + 36; (a + 6)^2$ 7) $c^2 - 18c + 81; (c - 9)^2$

9) $r^2 + 3r + \frac{9}{4}; \left(r + \frac{3}{2}\right)^2$ 11) $b^2 - 9b + \frac{81}{4}; \left(b - \frac{9}{2}\right)^2$

13) $x^2 + \frac{1}{3}x + \frac{1}{36}; \left(x + \frac{1}{6}\right)^2$ 15) Answers may vary.

17) $\{-4, -2\}$

19) $\{3, 5\}$ 21) $\{-5 - \sqrt{15}, -5 + \sqrt{15}\}$

23) $\{1 - 2i\sqrt{2}, 1 + 2i\sqrt{2}\}$ 25) $\{-2 - 2i, -2 + 2i\}$

27) $\{6 - \sqrt{33}, 6 + \sqrt{33}\}$ 29) $\{-8, 5\}$ 31) $\{3, 4\}$

33) $\left\{\frac{1}{2} - \frac{\sqrt{13}}{2}, \frac{1}{2} + \frac{\sqrt{13}}{2}\right\}$

35) $\left\{-\dfrac{5}{2} - \dfrac{\sqrt{3}}{2}i, -\dfrac{5}{2} + \dfrac{\sqrt{3}}{2}i\right\}$ 37) $\{1 - i\sqrt{3}, 1 + i\sqrt{3}\}$

39) $\{-3 - \sqrt{11}, -3 + \sqrt{11}\}$

43) $\left\{\dfrac{5}{4} - \dfrac{\sqrt{39}}{4}i, \dfrac{5}{4} + \dfrac{\sqrt{39}}{4}i\right\}$ 45) $\left\{\dfrac{3}{4}, 1\right\}$

47) $\{-1 - \sqrt{21}, -1 + \sqrt{21}\}$ 49) $\left\{\dfrac{5}{4} - \dfrac{\sqrt{7}}{4}i, \dfrac{5}{4} + \dfrac{\sqrt{7}}{4}i\right\}$

51) width = 9 in., length = 17 in.

Section 11.5

1) The fraction bar should also be under $-b$:

$$x = \frac{-b \pm \sqrt{b^2 - 4ac}}{2a}$$

3) You cannot divide only the -2 by 2.

$$\frac{-2 \pm 6\sqrt{11}}{2} = \frac{2(-1 \pm 3\sqrt{11})}{2} = -1 \pm 3\sqrt{11}$$

5) $\{-3, -1\}$ 7) $\left\{-2, \dfrac{5}{3}\right\}$ 9) $\left\{\dfrac{5 - \sqrt{17}}{2}, \dfrac{5 + \sqrt{17}}{2}\right\}$

11) $\{4 - 3i, 4 + 3i\}$ 13) $\left\{\dfrac{1}{5} - \dfrac{\sqrt{14}}{5}i, \dfrac{1}{5} + \dfrac{\sqrt{14}}{5}i\right\}$

15) $\{-7, 0\}$ 17) $\left\{\dfrac{-1 - \sqrt{13}}{3}, \dfrac{-1 + \sqrt{13}}{3}\right\}$ 19) $\left\{\dfrac{7}{2}, 4\right\}$

21) $\{-4 - \sqrt{31}, -4 + \sqrt{31}\}$ 23) $\left\{-\dfrac{2}{3} - \dfrac{1}{3}i, -\dfrac{2}{3} + \dfrac{1}{3}i\right\}$

25) $\{-10, 4\}$ 27) $\left\{-\dfrac{3}{2}i, \dfrac{3}{2}i\right\}$ 29) $\{-3 - 5i, -3 + 5i\}$

31) $\{-1 - \sqrt{10}, -1 + \sqrt{10}\}$ 33) $\left\{\dfrac{3}{2}\right\}$

35) $\left\{-\dfrac{3}{8} - \dfrac{\sqrt{7}}{8}i, -\dfrac{3}{8} + \dfrac{\sqrt{7}}{8}i\right\}$

37) $\left\{\dfrac{5 - \sqrt{19}}{2}, \dfrac{5 + \sqrt{19}}{2}\right\}$ 39) -39; two complex solutions

41) 0; one rational solution 43) 16; two rational solutions

45) 56; two irrational solutions 47) -8 or 8

49) 9 51) 4 53) 2 in., 5 in.

55) a) 2 sec b) $\dfrac{3 + \sqrt{33}}{4}$ sec or about 2.2 sec

Chapter 11 Mid-Chapter Summary Exercises

1) $\{-5\sqrt{2}, 5\sqrt{2}\}$ 3) $\{-5, 4\}$

5) $\left\{\dfrac{-7 - \sqrt{13}}{2}, \dfrac{-7 + \sqrt{13}}{2}\right\}$ 7) $\left\{-3, \dfrac{1}{4}\right\}$

9) $\{-7 - i\sqrt{11}, -7 + i\sqrt{11}\}$

11) $\left\{\dfrac{1}{3} - \dfrac{\sqrt{6}}{3}i, \dfrac{1}{3} + \dfrac{\sqrt{6}}{3}i\right\}$ 13) $\{-2, 6\}$

15) $\{2 - \sqrt{7}, 2 + \sqrt{7}\}$ 17) $\left\{\dfrac{3}{2}, 2\right\}$ 19) $\{3, 7\}$

21) $\left\{\dfrac{1}{3} - \dfrac{\sqrt{2}}{3}i, \dfrac{1}{3} + \dfrac{\sqrt{2}}{3}i\right\}$ 23) $\{5 - 4i, 5 + 4i\}$

25) $\{0, 3\}$ 27) $\left\{-\dfrac{3}{2}, 0, \dfrac{3}{2}\right\}$

29) $\left\{-\dfrac{5}{4} - \dfrac{\sqrt{39}}{4}i, -\dfrac{5}{4} + \dfrac{\sqrt{39}}{4}i\right\}$

Section 11.6

1) $\{-4, 12\}$ 3) $\left\{-2, \dfrac{4}{5}\right\}$ 5) $\{4 - \sqrt{6}, 4 + \sqrt{6}\}$

7) $\left\{\dfrac{1 - \sqrt{7}}{3}, \dfrac{1 + \sqrt{7}}{3}\right\}$ 9) $\left\{-\dfrac{9}{5}, 1\right\}$

11) $\left\{\dfrac{11 - \sqrt{21}}{10}, \dfrac{11 + \sqrt{21}}{10}\right\}$ 13) $\{5\}$ 15) $\left\{\dfrac{4}{5}, 2\right\}$

17) $\{9\}$ 19) $\{1, 16\}$ 21) $\{5\}$ 23) $\{0, 2\}$ 25) yes

27) yes 29) no 31) yes 33) no 35) $\{-3, -1, 1, 3\}$

37) $\{-\sqrt{7}, -2, 2, \sqrt{7}\}$ 39) $\{-i\sqrt{7}, -i\sqrt{5}, i\sqrt{5}, i\sqrt{7}\}$

41) $\{-8, -1\}$ 43) $\{-64, 1000\}$ 45) $\left\{-\dfrac{27}{8}, 125\right\}$

47) $\{4, 36\}$ 49) $\{49\}$ 51) $\{16\}$

53) $\left\{-\dfrac{2\sqrt{3}}{3}i, -\dfrac{\sqrt{3}}{3}i, \dfrac{\sqrt{3}}{3}i, \dfrac{2\sqrt{3}}{3}i\right\}$

55) $\{-\sqrt{5}, \sqrt{5}, -i\sqrt{3}, i\sqrt{3}\}$

57) $\{-\sqrt{3 + \sqrt{7}}, \sqrt{3 + \sqrt{7}}, -\sqrt{3 - \sqrt{7}}, \sqrt{3 - \sqrt{7}}\}$

59) $\left\{-\dfrac{\sqrt{7 + \sqrt{41}}}{2}, \dfrac{\sqrt{7 + \sqrt{41}}}{2},\right.$ $\left.-\dfrac{\sqrt{7 - \sqrt{41}}}{2}, \dfrac{\sqrt{7 - \sqrt{41}}}{2}\right\}$

61) $\left\{-\dfrac{1}{2}, \dfrac{1}{6}\right\}$ 63) $\left\{\dfrac{1}{4}, 3\right\}$ 65) $\{-6, -1\}$ 67) $\left\{-\dfrac{1}{2}, 0\right\}$

69) $\left\{-\dfrac{2}{5}, \dfrac{2}{5}\right\}$ 71) $\left\{-11, -\dfrac{20}{3}\right\}$ 73) $\left\{\dfrac{3}{2}\right\}$

75) $\{2 - \sqrt{2}, 2 + \sqrt{2}\}$

Section 11.7

1) $r = \dfrac{\pm\sqrt{A\pi}}{\pi}$ 3) $v = \pm\sqrt{ar}$ 5) $d = \dfrac{\pm\sqrt{IE}}{E}$

7) $r = \dfrac{\pm\sqrt{kq_1q_2F}}{F}$ 9) $A = \dfrac{1}{4}\pi d^2$ 11) $l = \dfrac{gT_p^2}{4\pi^2}$

13) $g = \dfrac{4\pi^2 l}{T_p^2}$

15) a) Both are written in the standard form for a quadratic equation, $ax^2 + bx + c = 0$.

  b) Use the quadratic formula.

17) $x = \dfrac{5 \pm \sqrt{25 - 4rs}}{2r}$ 19) $z = \dfrac{-r \pm \sqrt{r^2 + 4pq}}{2p}$

21) $a = \dfrac{h \pm \sqrt{h^2 + 4dk}}{2d}$ 23) $t = \dfrac{-v \pm \sqrt{v^2 + 2gs}}{g}$

25) length = 12 in., width = 9 in. 27) 2 ft

29) base = 8 ft, height = 15 ft   31) 10 in.

33) a) 0.75 sec on the way up, 3 sec on the way down.

b) $\dfrac{15 + \sqrt{241}}{8}$ sec or about 3.8 sec

35) a) 9.5 million   b) 1999   37) $2.40

Chapter 11 Review Exercises

1) $\{-6, 9\}$ 3) $\left\{-\dfrac{1}{2}, \dfrac{3}{2}\right\}$ 5) $\{-4, -3, 4\}$

7) width = 8 cm, length = 12 cm   9) $\{-12, 12\}$

11) $\{-2i, 2i\}$ 13) $\{-4, 10\}$ 15) $\left\{-\dfrac{\sqrt{10}}{3}, \dfrac{\sqrt{10}}{3}\right\}$ 17) 3

19) $\sqrt{29}$ 21) 5   23) $7i$ 25) -4 27) $12 - 3i$

29) $\dfrac{3}{10} - \dfrac{4}{3}i$ 31) $-30 + 35i$ 33) $-36 - 21i$

35) $-24 - 42i$ 37) conjugate: $2 + 7i$, product: 53

39) $\dfrac{12}{29} - \dfrac{30}{29}i$ 41) $-8i$ 43) $\dfrac{58}{37} - \dfrac{15}{37}i$

45) $r^2 + 10r + 25; (r + 5)^2$ 47) $c^2 - 5c + \dfrac{25}{4}; \left(c - \dfrac{5}{2}\right)^2$

49) $a^2 + \dfrac{2}{3}a + \dfrac{1}{9}; \left(a + \dfrac{1}{3}\right)^2$ 51) $\{-2, 8\}$

53) $\{-5 - \sqrt{31}, -5 + \sqrt{31}\}$

55) $\left\{-\dfrac{3}{2} - \dfrac{\sqrt{5}}{2}, -\dfrac{3}{2} + \dfrac{\sqrt{5}}{2}\right\}$

57) $\left\{\dfrac{7}{6} - \dfrac{\sqrt{95}}{6}i, \dfrac{7}{6} + \dfrac{\sqrt{95}}{6}i\right\}$ 59) $\{-6, 2\}$

61) $\left\{\dfrac{5 - \sqrt{15}}{2}, \dfrac{5 + \sqrt{15}}{2}\right\}$ 63) $\{1 - i\sqrt{3}, 1 + i\sqrt{3}\}$

65) $\left\{-\dfrac{2}{3}, \dfrac{1}{2}\right\}$ 67) 64; two rational solutions

69) -15; two complex solutions   71) -12 or 12

73) $\{-5, 3\}$ 75) $\left\{\dfrac{4}{3}, 2\right\}$ 77) $\{25\}$

79) $\{-\sqrt{2}, \sqrt{2}, -i\sqrt{7}, i\sqrt{7}\}$ 81) $\{1, 4\}$ 83) $\left\{-\dfrac{7}{2}, -1\right\}$

85) $v = \dfrac{\pm\sqrt{Frm}}{m}$ 87) $A = \pi r^2$ 89) $n = \dfrac{l \pm \sqrt{l^2 + 4km}}{2k}$

91) 3 in.   93) $8.00

Chapter 11 Test

1) $\sqrt{101}$ 3) $3i\sqrt{2}$ 5) $5 + 3i$ 7) $\dfrac{10}{3} + \dfrac{3}{2}i$ 9) $\dfrac{48}{37} - \dfrac{8}{37}i$

11) $\{-2 - \sqrt{11}, -2 + \sqrt{11}\}$ 13) $\{4 - i, 4 + i\}$

15) $\left\{-2, \dfrac{4}{3}\right\}$ 17) $\left\{\dfrac{7 - \sqrt{17}}{4}, \dfrac{7 + \sqrt{17}}{4}\right\}$

19) $\left\{-\dfrac{3}{2}, -1\right\}$ 21) $\sqrt{19}$ 23) $V = \dfrac{1}{3}\pi r^2 h$

25) width = 13 in., length = 19 in.

Cumulative Review for Chapters 1–11

1) $-\dfrac{1}{6}$ 3) area: 238 cm², perimeter: 82 cm   5) $\dfrac{45x^6}{y^4}$ 7) 90

9) x-int: $(4, 0)$; y-int: $\left(0, -\dfrac{8}{5}\right)$

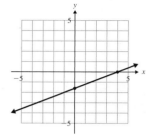

11) chips: $0.80; soda: $0.75   13) $3r^2 - 30r + 75$

15) $(a + 5)(a^2 - 5a + 25)$ 17) $\dfrac{c(c + 3)}{1 - 4c}$

19) $(0, 3, -1)$ 21) $2\sqrt[3]{5}$ 23) 16   25) $34 - 77i$

27) $\left\{\dfrac{1}{6} - \dfrac{\sqrt{11}}{6}i, \dfrac{1}{6} + \dfrac{\sqrt{11}}{6}i\right\}$ 29) $\{-9, 3\}$

Chapter 12

Section 12.1

1) It is a special type of relation in which each element of the domain corresponds to exactly one element in the range.

3) domain: $\{5, 6, 14\}$ 5) domain: $\{-2, 2, 5, 8\}$
range: $\{-3, 0, 1, 3\}$ range: $\{4, 25, 64\}$
not a function   is a function

7) domain: $(-\infty, \infty)$
range: $[-4, \infty)$
is a function

9) yes   11) yes   13) no   15) no

17) False; it is read as "f of x."

19)

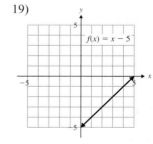

21)

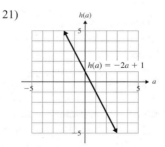

23)

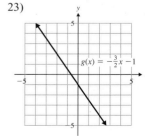

25)

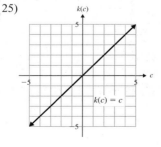

27) 11   29) -12 31) $3a - 7$

33) $d^2 - 4d - 9$ 35) $3c + 5$

37) $t^2 - 13$

39) $h^2 - 6h - 4$ 41) -18

43) -4 45) $15k + 2$

47) $25t^2 + 35t + 2$ 49) $-5b - 3$

51) $r^2 + 15r + 46$

53) 5   55) $-\dfrac{5}{2}$ 57) -2 or 8

59) $(-\infty, \infty)$

61) Set the denominator equal to 0 and solve for the variable. The domain consists of all real numbers *except* the values that make the denominator equal to 0.

63) $(-\infty, \infty)$ 65) $(-\infty, \infty)$

67) $(-\infty, -8) \cup (-8, \infty)$

69) $(-\infty, 0) \cup (0, \infty)$

71) $\left(-\infty, \dfrac{1}{2}\right) \cup \left(\dfrac{1}{2}, \infty\right)$

73) $\left(-\infty, -\dfrac{3}{7}\right) \cup \left(-\dfrac{3}{7}, \infty\right)$

75) $(-\infty, \infty)$

77) $(-\infty, -8) \cup (-8, -3) \cup (-3, \infty)$

79) $(-\infty, -4) \cup (-4, 9) \cup (9, \infty)$

81) $[0, \infty)$

83) $[-2, \infty)$ 85) $[8, \infty)$

87) $\left[\dfrac{5}{2}, \infty\right)$ 89) $(-\infty, 0]$

91) $(-\infty, 9]$ 93) (∞, ∞)

95) a) \$440   b) \$1232   c) 35 yd^2

d)
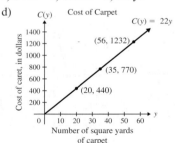
Cost of Carpet
$C(y) = 22y$
(56, 1232)
(35, 770)
(20, 440)

97) a) $L(1) = 90$. The labor charge for a 1-hr repair job is \$90.
 b) $L(1.5) = 115$. The labor charge for a 1.5-hr repair job is \$115.
 c) $h = 2.5$. If the labor charge is \$165, the repair job took 2.5 hr.

99) a) $A(r) = \pi r^2$
 b) $A(3) = 9\pi$. When the radius of the circle is 3 cm, the area of the circle is 9π cm^2.

c) $A(5) = 25\pi$. When the radius of the circle is 5 in., the area of the circle is 25π in^2.
 d) $r = 8$ in.

Section 12.2

1) domain: $(-\infty, \infty)$;
 range: $[3, \infty)$
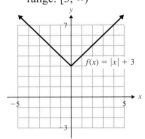
$f(x) = |x| + 3$

3) domain: $(-\infty, \infty)$;
 range: $[0, \infty)$
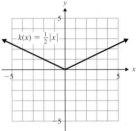
$k(x) = \frac{1}{2}|x|$

5) domain: $(-\infty, \infty)$;
 range: $[-4, \infty)$
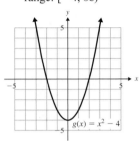
$g(x) = x^2 - 4$

7) domain: $(-\infty, \infty)$;
 range: $(-\infty, -1]$
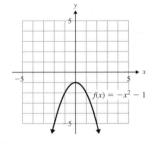
$f(x) = -x^2 - 1$

9) domain: $[-3, \infty)$;
 range: $[0, \infty)$
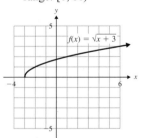
$f(x) = \sqrt{x + 3}$

11) domain: $[0, \infty)$;
 range: $[0, \infty)$
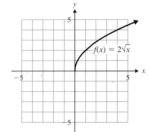
$f(x) = 2\sqrt{x}$

13) The graph of $g(x)$ is the same shape as $f(x)$, but g is shifted down 2 units.

15) The graph of $g(x)$ is the same shape as $f(x)$, but g is shifted left 2 units.

17) The graph of $g(x)$ is the reflection of $f(x)$ about the x-axis.

19)
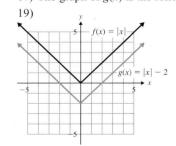
$f(x) = |x|$
$g(x) = |x| - 2$

21)
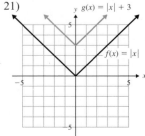
$g(x) = |x| + 3$
$f(x) = |x|$

23)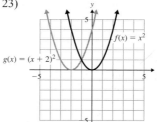

25)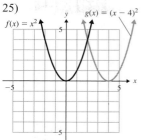

41)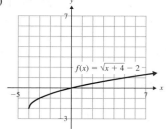

27)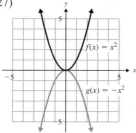

29)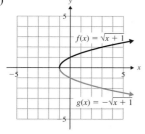

43)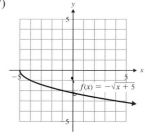

45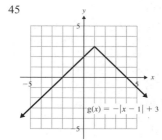

31)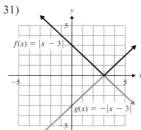

33)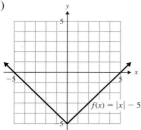

47)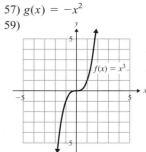

35)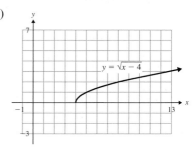

49) $h(x)$ b) $f(x)$ c) $g(x)$ d) $k(x)$ 51) $g(x) = \sqrt{x + 5}$

53) $g(x) = |x + 2| - 1$ 55) $g(x) = (x + 3)^2 + \dfrac{1}{2}$

57) $g(x) = -x^2$

59) 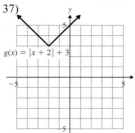   a)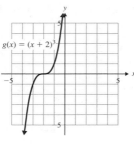

37) 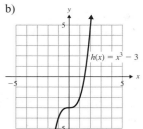   39)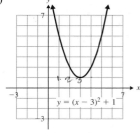

b) 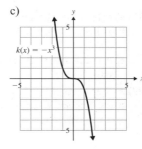   c)

d)

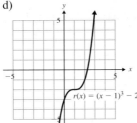

61)

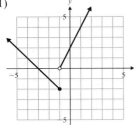

89)

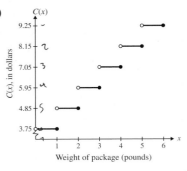

63)

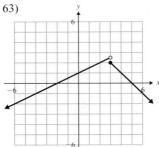

65)

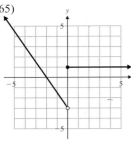

91)

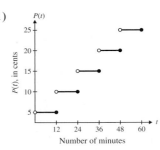

67)

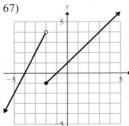

69)

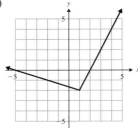

71) 3   73) 9   75) 8   77) −7   79) −9

81)

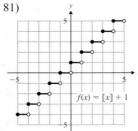

83)

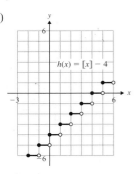

Section 12.3

1) (h, k) 3) a is positive.   5) $|a| > 1$

7) $V(-1, -4)$; $x = -1$; x-ints: $(-3, 0)$, $(1, 0)$; y-int: $(0, -3)$

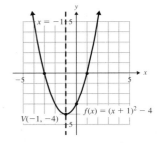

9) $V(2, 3)$; $x = 2$; x-ints: none; y-int: $(0, 7)$

11) $V(4, -2)$; $x = 4$; x-ints: $(4 - \sqrt{2}, 0)$, $(4 + \sqrt{2}, 0)$; y-int: $(0, 14)$

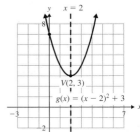

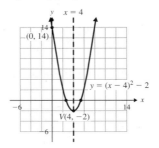

85)

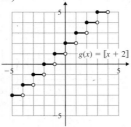

87)

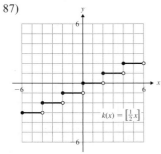

13) $V(-3, 6)$; $x = -3$; x-ints: $(-3 - \sqrt{6}, 0)$, $(-3 + \sqrt{6}, 0)$;
y-int: $(0, -3)$

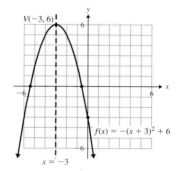

23) $V(-4, 3)$; $x = -4$; x-ints: $(-7, 0)$, $(-1, 0)$; y-int: $\left(0, -\frac{7}{3}\right)$

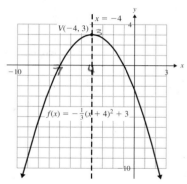

15) $V(-1, -5)$; $x = -1$;
x-ints: none; y-int: $(0, -6)$

17) $V(1, -8)$; $x = 1$;
x-ints: $(-1, 0)$,
$(3, 0)$; y-int: $(0, -6)$

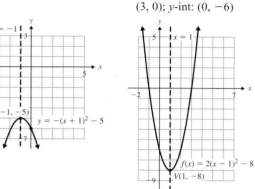

25) $V(-2, 5)$; $x = -2$;
x-int: none; y-int: $(0, 17)$

27) $f(x) = (x - 1)^2 - 4$;
x-ints: $(-1, 0)$, $(3, 0)$;
y-int: $(0, -3)$

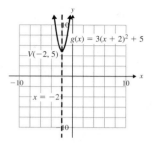
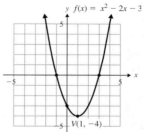

29) $y = (x + 3)^2 - 2$;
x-ints: $(-3 - \sqrt{2}, 0)$,
$(-3 + \sqrt{2}, 0)$;
y-int: $(0, 7)$

31) $g(x) = (x + 2)^2 - 4$;
x-ints: $(-4, 0)$, $(0, 0)$;
y-int: $(0, 0)$

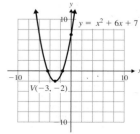
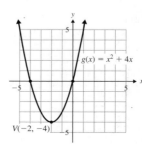

19) $V(-4, 0)$; $x = -4$; x-int: $(-4, 0)$; y-int: $(0, 8)$

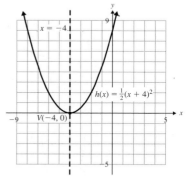

33) $h(x) = -(x + 2)^2 + 9$;
x-ints: $(-5, 0)$, $(1, 0)$;
y-int: $(0, 5)$

35) $y = -(x - 3)^2 - 1$;
x-ints: none;
y-int: $(0, -10)$

21) $V(0, 5)$; $x = 0$; x-ints: $(-\sqrt{5}, 0)$, $(\sqrt{5}, 0)$; y-int: $(0, 5)$

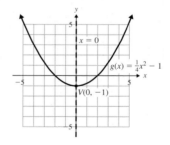

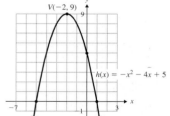
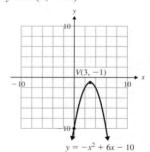

37) $f(x) = 2(x - 2)^2 - 4$;
x-ints: $(2 - \sqrt{2}, 0)$,
$(2 + \sqrt{2}, 0)$; y-int: $(0, 4)$

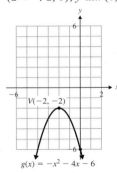

$g(x) = -x^2 - 4x - 6$

39) $g(x) = -\dfrac{1}{3}(x + 3)^2 - 6$;
x-ints: none;
y-int: $(0, -9)$

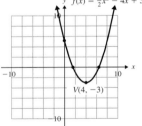

$g(x) = -\frac{1}{3}x^2 - 2x - 9$

41) $y = \left(x - \dfrac{3}{2}\right)^2 - \dfrac{1}{4}$;
x-ints: $(1, 0)$, $(2, 0)$;
y-int: $(0, 2)$

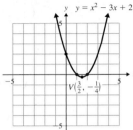

$y = x^2 - 3x + 2$

$V\left(\frac{3}{2}, -\frac{1}{4}\right)$

43) $V(-1, -4)$; x-ints: $(-3, 0)$,
$(1, 0)$; y-int: $(0, -3)$

$y = x^2 + 2x - 3$

$V(-1, -4)$

45) $V(-4, 3)$; x-ints: $(-4 - \sqrt{3}, 0)$, $(-4 + \sqrt{3}, 0)$;
y-int: $(0, -13)$

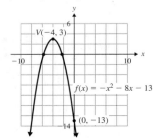

$V(-4, 3)$

$f(x) = -x^2 - 8x - 13$

$(0, -13)$

47) $V(1, 2)$; x-int: none;
y-int: $(0, 4)$

49) $V(1, 4)$; x-ints: $\left(1 + \dfrac{2\sqrt{3}}{3}, 0\right)$;
$\left(1 - \dfrac{2\sqrt{3}}{3}, 0\right)$, y-int: $(0, 1)$

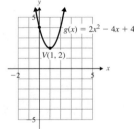

$g(x) = 2x^2 - 4x + 4$

$V(1, 2)$

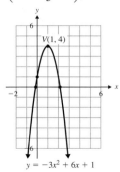

$V(1, 4)$

$y = -3x^2 + 6x + 1$

51) $V(4, -3)$; x-ints:
$(4 - \sqrt{6}, 0)$, $(4 + \sqrt{6}, 0)$;
y-int: $(0, 5)$

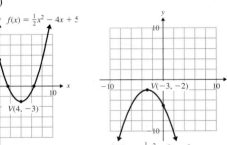

$f(x) = \frac{1}{2}x^2 - 4x + 5$

$V(4, -3)$

53) $V(-3, -2)$; x-int: none;
y-int: $(0, -5)$

$V(-3, -2)$

$h(x) = -\frac{1}{3}x^2 - 2x - 5$

Section 12.4

1) If a is positive the graph opens upward, so the y-coordinate of the vertex is the minimum value of the function. If a is negative the graph opens downward, so the y-coordinate of the vertex is the maximum value of the function.

3) a) minimum   b) $(-3, 0)$ c) 0

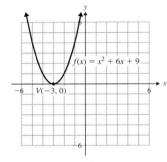

$f(x) = x^2 + 6x + 9$

$V(-3, 0)$

5) a) maximum   b) $(4, 2)$ c) 2

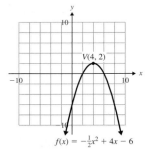

$V(4, 2)$

$f(x) = -\frac{1}{2}x^2 + 4x - 6$

7) a) 10 sec   b) 1600 ft   c) 20 sec   9) July; 480 people
11) 1991; 531,000   13) 625 ft²   15) 12 ft × 24 ft   17) 9 and 9
19) 6 and −6   21) (h, k) 23) to the left

25) $V(-4, 1)$; $y = 1$;
   x-int: $(-3, 0)$;
   y-ints: $(0, -1)$, $(0, 3)$

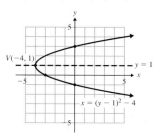

27) $V(2, 0)$; $y = 0$;
   x-int: $(2, 0)$; y-int: none

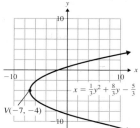

39) $x = \dfrac{1}{3}(y + 4)^2 - 7$

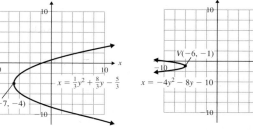

41) $x = -4(y + 1)^2 - 6$

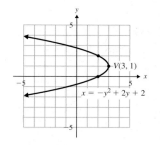

29) $V(5, 4)$; $y = 4$; x-int: $(-11, 0)$;
   y-ints: $(0, 4 - \sqrt{5})$, $(0, 4 + \sqrt{5})$

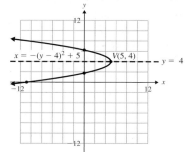

43) $V(-1, 2)$; x-int: $(3, 0)$;
   y-ints: $(0, 1)$, $(0, 3)$

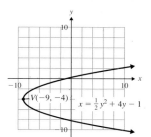

45) $V(3, 1)$; x-int: $(2, 0)$; y-ints:
   $(0, 1 - \sqrt{3})$, $(0, 1 + \sqrt{3})$

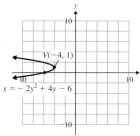

31) $V(-9, 2)$; $y = 2$; x-int:
   $(-17, 0)$; y-int: none

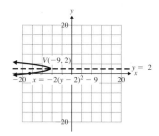

33) $V(0, -2)$; $y = -2$;
   x-int: $(1, 0)$; y-int: $(0, -2)$

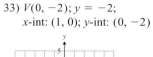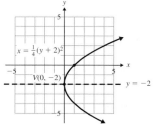

47) $V(-4, 1)$; x-int: $(-6, 0)$;
   y-int: none

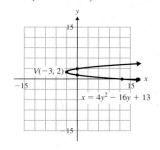

49) $V(-3, 2)$; x-int: $(13, 0)$;
   y-int: $\left(0, 2 - \dfrac{\sqrt{3}}{2}\right)$,
   $\left(0, 2 + \dfrac{\sqrt{3}}{2}\right)$

35) $x = (y - 2)^2 + 1$

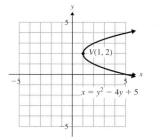

37) $x = -(y - 3)^2 + 15$

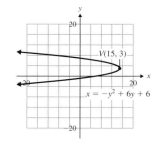

51) $V(6, 1)$; x-int: $\left(\dfrac{25}{4}, 0\right)$; y-int: none

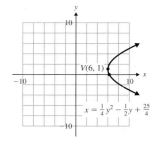

Section 12.5

1) a) $-x - 10$ b) -15 c) $-5x + 12$ d) 2

3) a) $5x^2 - 4x - 7$ b) 98   c) $3x^2 - 10x + 5$ d) -3

5) a) $-x^2 + 5x$ b) -24

7) a) $6x^2 + 11x + 3$ b) 24

9) a) $\dfrac{6x + 9}{x + 4}, x \neq 4$ b) $-\dfrac{3}{2}$

11) a) $x + 3, x \neq 8$ b) 1

13) a) $x + 4, x \neq -\dfrac{2}{3}$ b) 2

15) a) $P(x) = 4x - 2000$ b) \$4000

17) a) $P(x) = 3x - 2400$ b) \$0

19) a) $5x + 31$ b) $5x + 3$ c) 46

21) a) $-6x + 26$ b) $-6x + 11$ c) 32

23) a) $x^2 - 6x + 7$ b) $x^2 - 14x + 51$ c) 11

25) a) $18x^2 - 51x + 25$ b) $6x^2 + 9x - 35$ c) -8

27) a) $-x^2 - 13x - 48$ b) $-x^2 + 3x$ c) 0

29) a) $r(5) = 20$. The radius of the spill 5 min after the ship started leaking was 20 ft.

b) $A(20) = 400\pi$. The area of the oil slick is 400π ft^2 when its radius is 20 ft.

c) $A(r(t)) = 16\pi t^2$. This is the area of the oil slick in terms of t, the number of minutes after the leak began.

d) $A(r(5)) = 400\pi$. The area of the oil slick 5 min after the ship began leaking was 400π ft^2.

Section 12.6

1) increases   3) direct   5) inverse

7) combined   9) $M = kn$

11) $h = \dfrac{k}{j}$ 13) $T = \dfrac{k}{c^2}$

15) $s = krt$ 17) $Q = \dfrac{k\sqrt{z}}{m}$

19) a) 9   b) $z = 9x$ c) 54

21) a) 48   b) $N = \dfrac{48}{y}$ c) 16

23) a) 5   b) $Q = \dfrac{5r^2}{w}$ c) 45

25) 56   27) 18   29) 70

31) \$500.00   33) 12 hr

35) 180 watts   37) 162,000 J

39) 200 cycles/sec

41) 3 ohms   43) 320 lb

Chapter 12 Review Exercises

1) domain: $\{-7, -5, 2, 4\}$ 3) domain: $(-\infty, \infty)$
   range: $\{-4, -1, 3, 5, 9\}$ range: $[0, \infty)$
   not a function   is a function

5) yes   7) no   9) yes   11) yes

13)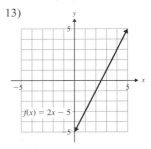

15)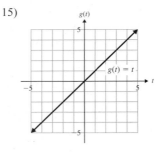

17) a) -37 b) 35   c) -22 d) 18   e) $-8c + 3$
    f) $r^2 + 7r - 12$ g) $-8p + 27$ h) $t^2 + 15t + 32$

19) 12   21) 2 or 6   23) $(-\infty, 5) \cup (5, \infty)$ 25) $(-\infty, \infty)$

27) $\left[\dfrac{7}{5}, \infty\right)$ 29) $(-\infty, 0) \cup (0, \infty)$ 31) $(-\infty, \infty)$

33) $(-\infty, -1) \cup (-1, 8) \cup (8, \infty)$

35) a) \$32   b) \$46   c) 150 mi   d) 80 mi

37) domain: $[0, \infty)$;   39) domain: $(-\infty, \infty)$;
    range: $[0, \infty)$ range: $[0, \infty)$

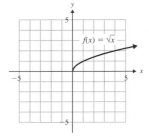
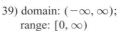

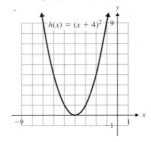
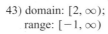

41) domain: $(-\infty, \infty)$;   43) domain: $[2, \infty)$;
    range: $(-\infty, 5]$ range: $[-1, \infty)$

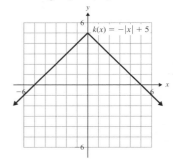

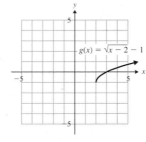

45)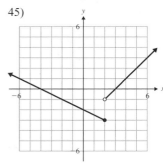

47) 7   49) −9   51) 0

53)

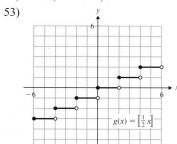

55) $g(x) = |x - 5|$

57) a) (h, k) b) $x = h$

c) If a is positive, the parabola opens upward. If a is negative, the parabola opens downward.

59) a) (h, k) b) $y = k$

c) If a is positive, the parabola opens to the right. If a is negative, the parabola opens to the left.

61) $V(-2, -1)$; $x = -2$;   63) $V(-1, 0)$; $y = 0$;
    x-ints: $(-3, 0), (-1, 0)$; x-int: $(-1, 0)$; y-ints: none
    y-int: $(0, 3)$

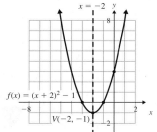

65) $V(11, 3)$; $y = 3$; x-int: $(2, 0)$;
    y-ints: $(0, 3 - \sqrt{11}), (0, 3 + \sqrt{11})$

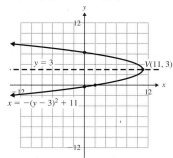

67) $x = (y + 4)^2 - 9$; x-int: $(7, 0)$; y-ints: $(0, -1), (0, -7)$

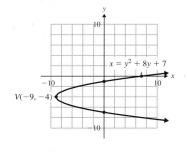

69) $y = \frac{1}{2}(x - 4)^2 + 1$;   71) $V(1, -5)$; x-ints:
    x-ints: none; y-int: $(0, 9)$ $(1 - \sqrt{5}, 0), (1 + \sqrt{5}, 0)$;
    y-int: $(0, -4)$

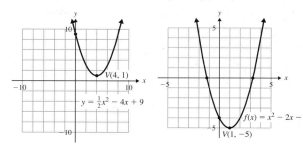

73) $V(2, -3)$; x-int: $\left(-\frac{5}{2}, 0\right)$; y-ints: $(0, -5), (0, -1)$

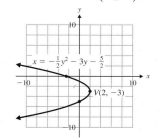

75) a) 1 sec   b) 256 ft   c) 5 sec

77) $4x + 6$ 79) −3

81) $-5x^2 + 18x + 8$

83) a) $\frac{6x - 5}{x + 4}, x \neq -4$ b) $\frac{13}{7}$

85) a) $P(x) = 6x - 400$ b) $800

87) a) $4x^2 + 17$ b) $16x^2 - 24x + 14$ c) 126

89) a) $(N \circ G)(h) = 9.6h$. This is Antoine's net pay in terms of how many hours he has worked.

b) $(N \circ G)(30) = 288$. When Antoine works 30 hr, his net pay is $288.   c) $384

91) 72   93) 3.24 lb

Chapter 12 Test

1) It is a special type of relation in which each element of the domain corresponds to exactly one element of the range.

3) a) $\left[-\frac{7}{3}, \infty\right)$ b) yes

5) $\left(-\infty, \frac{8}{7}\right) \cup \left(\frac{8}{7}, \infty\right)$ 7) $4c + 3$

9) $k^2 + 4k + 5$

11) a) $C(3) = 210$. The cost of delivering 3 yd³ of cedar mulch is $210.

b) 6 yd³

13) domain: $[-3, \infty)$;
    range: $[0, \infty)$

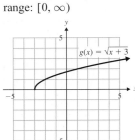

15)

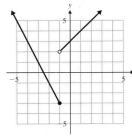

17)

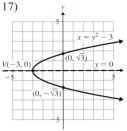

19)

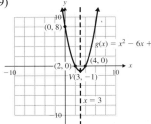

21) $x^2 + 3x - 10$ 23) $2x^2 + 10x + 1$ 25) 300

Cumulative Review for Chapters 1–12

1) inverse   3) $\dfrac{1}{32}$ 5) 1

7) $(-\infty, -2] \cup [3, \infty)$

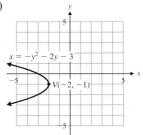

9) $(3, 2)$ 11) $\dfrac{3}{2}r - 10 + \dfrac{3r}{2} + \dfrac{1}{r^2}$

13) $(10 + 3m)(10 - 3m)$ 15) $\dfrac{c + 4}{3(2c + 3)}$ 17) $\varnothing$

19) $2\sqrt{15}$ 21) $3c^3d^5\sqrt{2d}$ 23) $7\sqrt{3}$

25) $\dfrac{2}{13} - \dfrac{16}{13}i$ 27) $\left\{-\dfrac{5}{2}, \dfrac{1}{2}\right\}$

29) a) 8   b) $x + 3, x \neq -1$ c) 4   d) $x^2 + 4x + 4$

31)

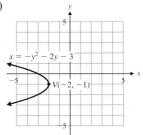

Chapter 13

Section 13.1

1) no   3) yes; $h^{-1} = \{(-16, -5), (-4, -1), (8, 3)\}$
5) yes; $g^{-1} = \{(1, 2), (2, 5), (14, 7), (19, 10)\}$

7) yes   9) No; only one-to-one functions have inverses.
11) False; it is read "f inverse of x."   13) True
15) False; they are symmetric with respect to $y = x$.
17) a) yes
    b)

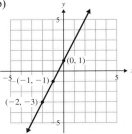

19) no
21) a) yes
    b)

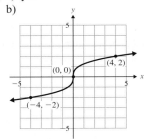

23) $g^{-1}(x) = x + 6$

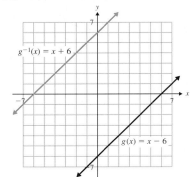

25) $f^{-1}(x) = -\dfrac{1}{2}x + \dfrac{5}{2}$ 27) $g^{-1}(x) = 2x$

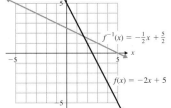
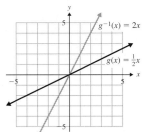

29) $f^{-1}(x) = \sqrt[3]{x}$

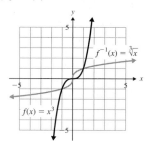

31) $f^{-1}(x) = \dfrac{1}{2}x + 3$ 33) $h^{-1}(x) = -\dfrac{2}{3}x + \dfrac{8}{3}$

35) $g^{-1}(x) = x^3 - 2$ 37) $f^{-1}(x) = x^2, x \ge 0$

39) a) 3 b) 1 41) a) 2 b) 9 43) a) 10 b) -7

45) a) 8 b) 3 47–54) Answers will vary.

Section 13.2

1) Choose values for the variable that will give positive numbers, negative numbers, and zero in the exponent.

3)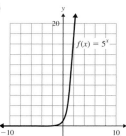

5)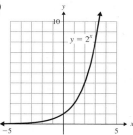

7)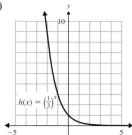

9) $(-\infty, \infty)$

11)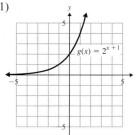

13)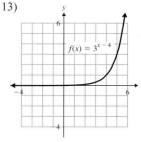

15)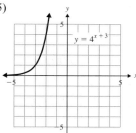

17)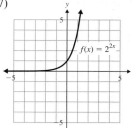

19)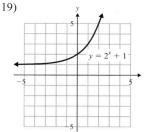

21)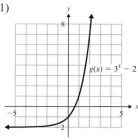

23)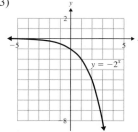

25) 2.7183 27) B 29) D 31) $\{2\}$ 33) $\left\{\dfrac{3}{4}\right\}$

35) $\{8\}$ 37) $\{-2\}$ 39) $\{1\}$ 41) $\left\{\dfrac{23}{4}\right\}$ 43) $\{-3\}$

45) $\{-2\}$ 47) $\{-3\}$ 49) $\left\{-\dfrac{3}{2}\right\}$ 51) $\left\{\dfrac{3}{2}\right\}$ 53) $\{-1\}$

55) a) $32,700 b) $17,507.17 57) a) $16,800 b) $4504.04

59) a) $185,200 b) $227,772.64 61) $90,036.92

63) $59,134.40 65) 40.8 mg

Section 13.3

1) a must be a positive real number that is not equal to 1.

3) 10 5) $7^2 = 49$ 7) $2^3 = 8$ 9) $9^{-2} = \dfrac{1}{81}$

11) $10^6 = 1,000,000$ 13) $25^{1/2} = 5$ 15) $13^1 = 13$

17) $\log_9 81 = 2$ 19) $\log_{10} 100 = 2$ 21) $\log_3 \dfrac{1}{81} = -4$

23) $\log_{10} 1 = 0$ 25) $\log_{169} 13 = \dfrac{1}{2}$ 27) $\log_9 3 = \dfrac{1}{2}$

29) $\log_{64} 4 = \dfrac{1}{3}$

31) Write the equation in exponential form, then solve for the variable.

33) $\{121\}$ 35) $\{64\}$ 37) $\{100,000\}$ 39) $\{7\}$ 41) $\left\{\dfrac{1}{36}\right\}$

43) $\{14\}$ 45) $\left\{\dfrac{15}{2}\right\}$ 47) $\left\{\dfrac{1}{8}\right\}$ 49) $\left\{\dfrac{1}{6}\right\}$ 51) $\{12\}$

53) $\{4\}$ 55) $\{2\}$ 57) $\{2\}$ 59) $\{5\}$ 61) $\{2\}$ 63) $\left\{\dfrac{1}{2}\right\}$

65) $\{-1\}$ 67) $\{1\}$ 69) $\{-2\}$

71) Replace $f(x)$ with y, write $y = \log_a x$ in exponential form, make a table of values, then plot the points and draw the curve.

73)
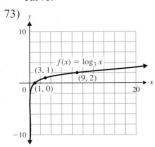

75)
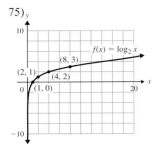

77)
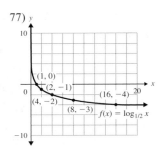

79)
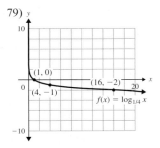

81) $f^{-1}(x) = \log_3 x$ 83) $f^{-1}(x) = 2^x$

85) a) 1868   b) 2004   c) 2058

87) a) 14,000   b) 28,000

   c) It is 1000 more than what was predicted by the formula.

Section 13.4

1) true   3) false   5) false   7) true   9) $\log_8 3 + \log_8 10$

11) $\log_7 5 + \log_7 d$ 13) $\log_9 4 - \log_9 7$ 15) $3 \log_5 2$

17) $8 \log p$ 19) $\dfrac{1}{2}\log_3 7$ 21) $2 + \log_5 t$ 23) $3 - \log_2 k$

25) 6   27) $3 + \log b$ 29) 35   31) $\dfrac{1}{2}$ 33) $\dfrac{2}{3}$

35) $4 \log_6 w + 3 \log_6 z$ 37) $2 \log_7 a - 5 \log_7 b$

39) $\dfrac{1}{5}\log 11 - 2 \log y$ 41) $2 + \dfrac{1}{2}\log_2 n - 3 \log_2 m$

43) $3 \log_4 x - \log_4 y - 2 \log_4 z$ 45) $\dfrac{1}{2} + \dfrac{1}{2}\log_5 c$

47) $\log k + \log(k - 6)$ 49) $\log_a(mn)$ 51) $\log_7 \dfrac{d}{3}$

53) $\log_3(f^4 g)$ 55) $\log_8 \dfrac{tu^2}{v^3}$ 57) $\log \dfrac{r^2 + 3}{(r^2 - 3)^2}$

59) $\log_n(8\sqrt{k})$ 61) $\log_d \dfrac{\sqrt[3]{5}}{z^2}$ 63) $\log_6 \dfrac{y}{3z^3}$ 65) $\log_3 \dfrac{t^4}{36u^2}$

67) $\log_b \dfrac{\sqrt{c + 4}}{(c + 3)^2}$ 69) 1.6532   71) 1.9084   73) -0.2552

75) 0.4771   77) -0.6990 79) -1.9084 81) 1.6990

Section 13.5

1) e 3) 2   5) -3 7) -1 9) 9   11) $\dfrac{1}{4}$ 13) 6   15) $\dfrac{1}{2}$

17) -5 19) 0   21) 1.2041   23) -0.3010 25) 1.0986

27) 0.2700   29) $\{1000\}$ 31) $\left\{\dfrac{1}{10}\right\}$ 33) $\{25\}$ 35) $\{2\}$

37) $\{10^{1.5}\}; \{31.6228\}$ 39) $\{10^{0.8}\}; \{6.3096\}$

41) $\{e^{1.6}\}; \{4.9530\}$ 43) $\left\{\dfrac{1}{e^2}\right\}; \{0.1353\}$

45) $\left\{\dfrac{e^{2.1}}{3}\right\}; \{2.7221\}$ 47) $\{2.10^{0.47}\}; \{5.9024\}$

49) $\left\{\dfrac{3 + 10^{3.8}}{5}\right\}; \{1262.5147\}$

51) $\left\{\dfrac{e^{1.85} - 19}{10}\right\}; \{-1.2640\}$ 53) $\{3\}$ 55) 3.7004

57) 1.9336   59) -2.5237 61) 0.6826   63) 110 dB

65) 40 dB   67) \$3484.42   69) \$5521.68   71) \$3485.50

73) \$15,683.12   75) a) 5000   b) 8191   77) 32,570

79) 2.7; acidic   81) 11.2; basic   83) Answers will vary.

Section 13.6

1) $\{2\}$ 3) $\left\{\dfrac{\ln 15}{\ln 7}\right\}; \{1.3917\}$ 5) $\left\{\dfrac{\ln 3}{\ln 8}\right\}; \{0.5283\}$

7) $\left\{\dfrac{2}{5}\right\}$ 9) $\left\{\dfrac{\ln 2.7}{6 \ln 4}\right\}; \{0.1194\}$

11) $\left\{\dfrac{\ln 5 - \ln 2}{4 \ln 2}\right\}; \{0.3305\}$

13) $\left\{\dfrac{\ln 8 + 2 \ln 5}{3 \ln 5}\right\}; \{1.0973\}$ 15) $\left\{\dfrac{10}{7}\right\}$

17) $\left\{\dfrac{2 \ln 9}{5 \ln 9 - 3 \ln 4}\right\}; \{0.6437\}$ 19) $\{\ln 12.5\}; \{2.5257\}$

21) $\left\{-\dfrac{\ln 9}{4}\right\}; \{-0.5493\}$ 23) $\left\{\dfrac{\ln 2}{0.01}\right\}; \{69.3147\}$

25) $\left\{\dfrac{\ln 3}{0.006}\right\}; \{183.1021\}$ 27) $\left\{-\dfrac{\ln 5}{0.4}\right\}; \{-4.0236\}$

29) $\{2\}$ 31) $\left\{\dfrac{10}{3}\right\}$ 33) $\{5\}$ 35) $\{2, 10\}$ 37) $\varnothing$ 39) $\{8\}$

41) $\{2\}$ 43) $\{9\}$ 45) $\{2\}$ 47) $\{4\}$ 49) $\left\{\dfrac{2}{3}\right\}$

51) a) 3.72 yr   b) 11.55 yr   53) 1.44 yr   55) $2246.64
57) .2%   59) a) 6 hr   b) 18.5 hr   61) 28,009
63) a) 2032   b) 2023   65) a) 11.78 g   b) 3351 yr   c) 5728 yr
67) a) 0.4 units   b) 0.22 units

Chapter 13 Review Exercises

1) yes; $\{(-4, -7), (1, -2), (5, 1), (11, 6)\}$
3) yes

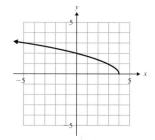

5) $f^{-1}(x) = x - 4$ 7) $h^{-1}(x) = 3x + 3$

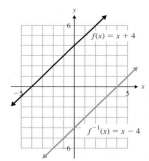   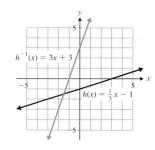

9) a) 11   b) 2

11)   13)

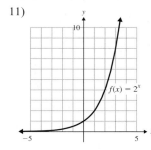   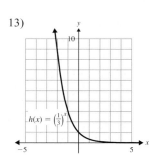

15) $\{6\}$ 17) $\left\{-\dfrac{5}{2}\right\}$ 19) $\left\{\dfrac{2}{3}\right\}$

21) x must be a positive number.   23) $5^3 = 125$ 25) $10^2 = 100$

27) $\log_3 81 = 4$ 29) $\log 1000 = 3$ 31) $\{8\}$ 33) $\left\{\dfrac{4}{5}\right\}$

35) $\{2\}$ 37) $\{3\}$ 39) $\{-4\}$

41)   43)

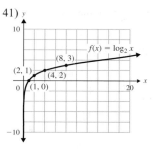   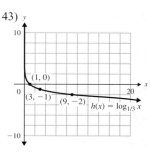

45) $f^{-1}(x) = \log_5 x$ 47) $h^{-1}(x) = 6^x$ 49) false

51) $\log_8 3 + \log_8 z$ 53) $\dfrac{3}{2}$ 55) $4 \log_5 c + 3 \log_5 d$

57) $\log_a x + \log_a y - 3 \log_a z$ 59) $\log p + \log(p + 8)$

61) $\log(cd)$ 63) $\log_2 a^9 b^3$ 65) $\log_3 \dfrac{5m^4}{n^2}$ 67) $\log_5 \dfrac{c^3}{df^2}$

69) 1.6902   71) -0.1091 73) e 75) 1   77) $\dfrac{1}{2}$ 79) -3

81) 0   83) 0.9031   85) 0.5596   87) $\{100\}$ 89) $\left\{\dfrac{1}{5}\right\}$

91) $\{10^{2.1}\}$; $\{125.8925\}$ 93) $\{e^2\}$; $\{7.3891\}$

95) $\left\{\dfrac{10^{1.75}}{4}\right\}$; $\{14.0585\}$ 97) $\{2.1240\}$ 99) $\{-5.2479\}$

101) 110 dB   103) $3367.14   105) $11,533.14
107) a) 6000   b) 11,118   109) $\{4\}$

111) $\left\{\dfrac{\ln 2}{4 \ln 9}\right\}$; $\{0.0789\}$

113) $\left\{\dfrac{5 \ln 8}{\ln 8 - 2 \ln 6}\right\}$; $\{-6.9127\}$ 115) $\left\{\dfrac{\ln 8}{5}\right\}$; $\{0.4159\}$

117) $\left\{\dfrac{6}{5}\right\}$ 119) $\{4\}$ 121) $\{4\}$ 123) $\{16\}$ 125) $6770.57

127) a) 17,777   b) 2011

Chapter 13 Test

1) no

3) yes   5)

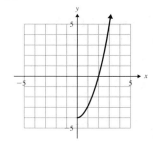   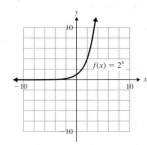

7) a) $(0, \infty)$ b) $(-\infty, \infty)$ 9) $\log_3 \dfrac{1}{9} = -2$ 11) $\{-2\}$

13) $\{29\}$ 15) a) $\{4\}$ b) $\left\{\dfrac{1}{2}\right\}$ 17) $\log_8 5 + \log_8 n$

19) $\log \dfrac{x^2}{(x+1)^3}$ 21) $\left\{\dfrac{\ln 5}{0.3}\right\}; \{5.3648\}$

23) $\left\{\dfrac{\ln 9 - 3\ln 4}{4\ln 4}\right\}; \{-0.3538\}$ 25) $8686.41

Cumulative Review for Chapters 1–13

1) 35   3) $-15a^6$ 5) $\dfrac{d^9}{8c^{30}}$ 7) $48.00   9) $(-6, 1)$

11) $y = -\dfrac{3}{2}x + 2$ 13) $(4w+9)(w-2)$ 15) $\{-8, -6\}$

17) $\{2\}$ 19) $2\sqrt{30}$ 21) $6a$ 23) $\varnothing$

25) $\{4 - 2\sqrt{3}, 4 + 2\sqrt{3}\}$ 27) $\{5 + 4i, 5 - 4i\}$

29) $\left(-\infty, \dfrac{2}{3}\right) \cup \left(\dfrac{2}{3}, \infty\right)$

31)

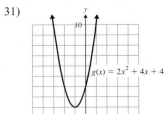

$g(x) = 2x^2 + 4x + 4$

33) $\left\{-\dfrac{3}{2}\right\}$ 35) $\log\dfrac{ab^2}{c^5}$

Chapter 14

Section 14.1

1) No; there are values in the domain that give more than one value in the range. The graph fails the vertical line test.

3) center: $(-2, 4)$; $r = 3$

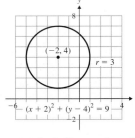

$(x+2)^2 + (y-4)^2 = 9$

5) center: $(5, 3)$; $r = 1$

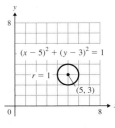

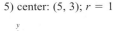

$(x-5)^2 + (y-3)^2 = 1$

7) center: $(-3, 0)$; $r = 2$

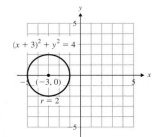

$(x+3)^2 + y^2 = 4$

9) center: $(6, -3)$; $r = 4$

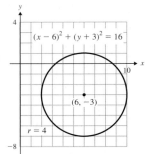

$(x-6)^2 + (y+3)^2 = 16$

11) center: $(0, 0)$; $r = 6$

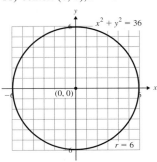

$x^2 + y^2 = 36$

13) center: $(0, 0)$; $r = 3$

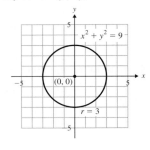

$x^2 + y^2 = 9$

15) center: $(0, 1)$; $r = 5$

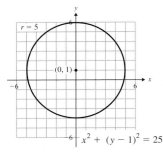

$x^2 + (y-1)^2 = 25$

17) $(x-4)^2 + (y-1)^2 = 25$

19) $(x+3)^2 + (y-2)^2 = 1$

21) $(x+1)^2 + (y+5)^2 = 3$

23) $x^2 + y^2 = 10$

25) $(x-6)^2 + y^2 = 16$

27) $x^2 + (y+4)^2 = 8$

29) $(x+1)^2 + (y+5)^2 = 9$;
center: $(-1, -5)$; $r = 3$

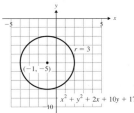

$x^2 + y^2 + 2x + 10y + 17 = 0$

31) $(x+4)^2 + (y-1)^2 = 25$;
center: $(-4, 1)$; $r = 5$

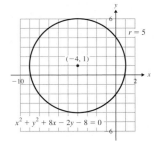

$x^2 + y^2 + 8x - 2y - 8 = 0$

33) $(x-5)^2 + (y-7)^2 = 1$;
center: $(5, 7)$; $r = 1$

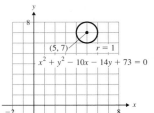

$x^2 + y^2 - 10x - 14y + 73 = 0$

35) $x^2 + (y+3)^2 = 4$;
center: $(0, -3)$; $r = 2$

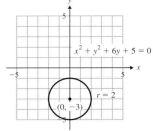

$x^2 + y^2 + 6y + 5 = 0$

37) $x^2 + y^2 - 4x - 1 = 0$; $(x - 2)^2 + y^2 = 5$;
center: $(2, 0)$; $r = \sqrt{5}$

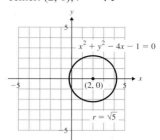

39) $(x - 4)^2 + (y + 4)^2 = 36$; center: $(4, -4)$; $r = 6$

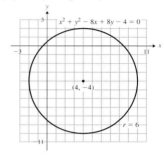

41) $\left(x - \dfrac{3}{2}\right)^2 + \left(y - \dfrac{1}{2}\right)^2 = 4$; center: $\left(\dfrac{3}{2}, \dfrac{1}{2}\right)$; $r = 2$

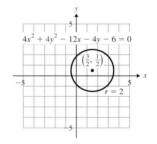

Section 14.2

1) ellipse   3) hyperbola   5) hyperbola   7) ellipse

9) center: $(-2, 1)$ 11) center: $(3, -2)$

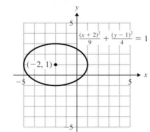
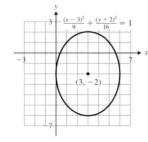

13) center: $(0, 0)$ 15) center: $(0, 0)$

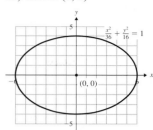
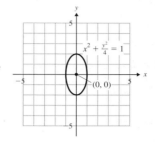

17) center: $(0, -4)$ 19) center: $(-1, -3)$

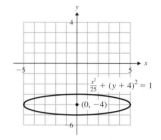
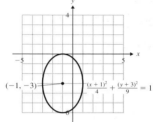

21) center: $(0, 0)$ 23) center: $(0, 0)$

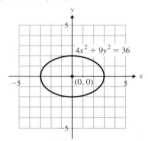
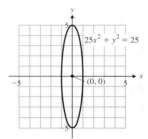

25) center: $(0, 0)$ 27) center: $(0, 0)$

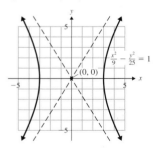
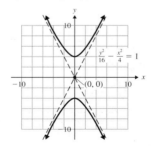

29) center: $(2, -3)$ 31) center: $(-4, -1)$

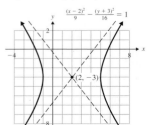
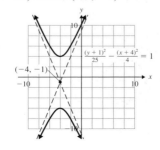

33) center: (1, 0)

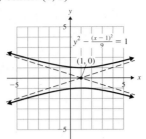

35) center: (1, 2)

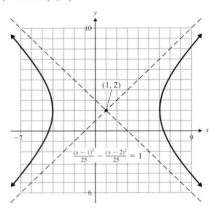

37) center: (0, 0)   39) center: (0, 0)

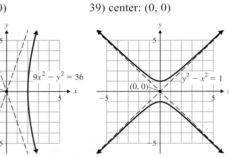

Chapter 14 Mid-Chapter Summary

1) parabola   3) hyperbola

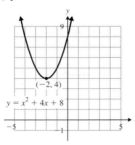   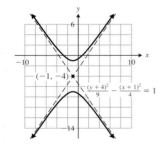

5) ellipse   7) circle

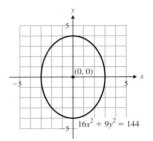   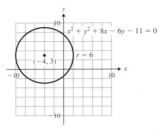

9) ellipse   11) parabola

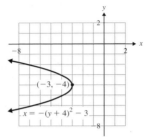   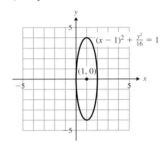

13) hyperbola   15) circle

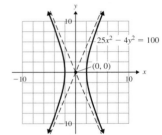   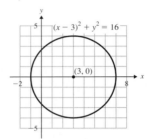

17) parabola   19) hyperbola

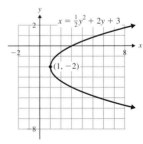   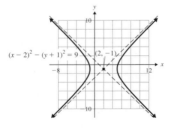

Section 14.3

1) c) 0, 1, or 2

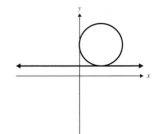

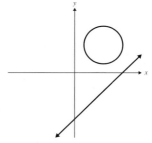

3) c) 0, 1, 2, 3, or 4

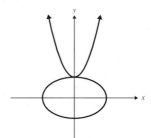

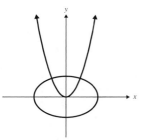

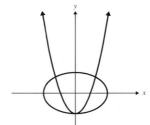

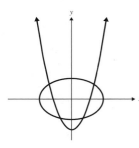

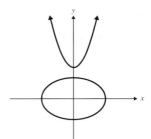

5) c) 0, 1, 2, 3, or 4

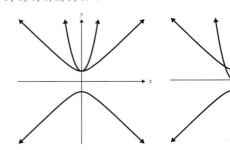

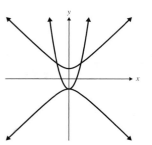

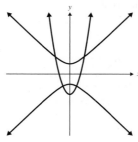

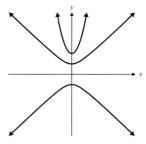

7) $\{(-4, -2), (6, -7)\}$

9) $\{(-1, 3), (3, 1)\}$ 11) $\{(4, 2)\}$

13) $\{(3, 1), (3, -1), (-3, 1), (-3, -1)\}$

15) $\{(2, \sqrt{2}), (2, -\sqrt{2}), (-2, \sqrt{2}), (-2, -\sqrt{2})\}$

17) $\{(3, -7), (-3, -7)\}$ 19) $\{(0, -1)\}$

21) $\{(0, 2)\}$ 23) $\varnothing$ 25) $\varnothing$

27) $\left\{\left(\dfrac{\sqrt{2}}{2}, \dfrac{3\sqrt{2}}{2}\right), \left(\dfrac{\sqrt{2}}{2}, -\dfrac{3\sqrt{2}}{2}\right),\right.$
$\left.\left(-\dfrac{\sqrt{2}}{2}, \dfrac{3\sqrt{2}}{2}\right), \left(-\dfrac{\sqrt{2}}{2}, -\dfrac{3\sqrt{2}}{2}\right)\right\}$

29) $\{(0, -2)\}$ 31) 8 and 5   33) 8 in. $\times$ 11 in.

35) 4000 basketballs, \$240

Section 14.4

1) The endpoints are included when the inequality symbol is $\leq$ or $\geq$. The endpoints are not included when the symbol is $<$ or $>$.

3) a) $[-5, 1]$ b) $(-\infty, -5) \cup (1, \infty)$

5) a) $[-1, 3]$ b) $(-\infty, -1) \cup (3, \infty)$

7) $(-\infty, -7] \cup [1, \infty)$

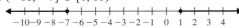

9) $(-9, 4)$

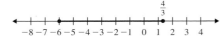

11) $(-\infty, 6) \cup (7, \infty)$

13) $\left[-6, \dfrac{4}{3}\right]$

15) $\left(-\infty, -\dfrac{2}{7}\right) \cup (2, \infty)$

17) $(-\infty, 0) \cup (9, \infty)$

19) $\left[-\dfrac{5}{4}, 0\right]$

21) $(-8, 8)$

23) $(-\infty, -11] \cup [11, \infty)$

25) $[-4, 4]$

27) $(-\infty, \infty)$ 29) $(-\infty, \infty)$ 31) $\varnothing$ 33) $\varnothing$

35) $(-\infty, -2] \cup [1, 5]$

37) $[-9, 5] \cup [7, \infty)$

39) $(-\infty, -7) \cup \left(-\dfrac{1}{6}, \dfrac{3}{4}\right)$

41) $(-6, \infty)$

43) $(-\infty, -3)$

45) $(-\infty, 3) \cup (4, \infty)$

47) $\left(-\dfrac{1}{3}, 9\right]$

49) $(-3, 0]$

51) $(-\infty, -6) \cup \left(-\dfrac{11}{3}, \infty\right)$

53) $(-7, -4]$

55) $\left[\dfrac{18}{5}, 6\right)$

57) $(-\infty, -2) \cup \left(-\dfrac{8}{7}, \infty\right)$

59) $(-\infty, 2)$

61) $(5, \infty)$

63) $(-\infty, -6)$

65) $[-1, \infty)$

67) $(-\infty, 4)$

Chapter 14 Review Exercises

1) center: $(-3, 5)$; $r = 6$

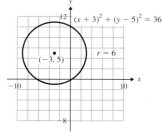

3) center: $(5, 2)$; $r = 4$

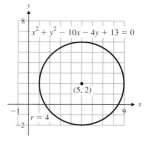

5) $(x - 3)^2 + y^2 = 16$

7) The equation of an ellipse contains the sum of two squares, $\dfrac{(x - h)^2}{a^2} + \dfrac{(y - k)^2}{b^2} = 1$, but the equation of a hyperbola contains the difference of two squares, $\dfrac{(x - h)^2}{a^2} - \dfrac{(y - k)^2}{b^2} = 1$ or $\dfrac{(y - k)^2}{b^2} - \dfrac{(x - h)^2}{a^2} = 1$.

9) center (0, 0)

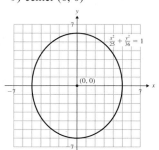

11) center (4, 2)

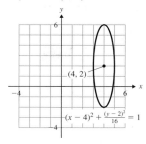

27) 0, 1, 2, 3, or 4   29) $\{(6, 7), (6, -7), (-6, 7), (-6, -7)\}$

31) $\{(1, 2), (-2, 1)\}$ 33) $\varnothing$ 35) 9 and 4

37) $(-3, 1)$

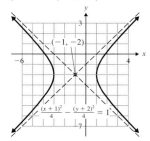

39) $\left(-\infty, -\dfrac{7}{6}\right) \cup (0, \infty)$

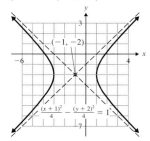

41) $(-6, 6)$

43) $(-\infty, \infty)$

45) $(-\infty, -7) \cup \left(\dfrac{3}{2}, \infty\right)$

47) $\left[-\dfrac{3}{4}, \dfrac{6}{5}\right)$

49) $\left(-\infty, \dfrac{11}{3}\right) \cup (4, \infty)$

51) $(7, \infty)$

13) center (0, 0)

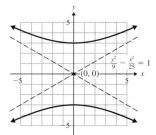

15) center: $(-1, -2)$

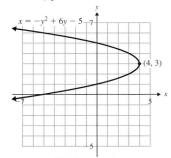

17) ellipse

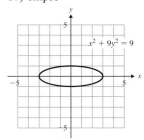

19) parabola

21) hyperbola

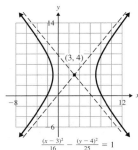

23) circle

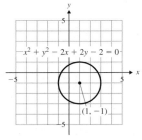

25) parabola

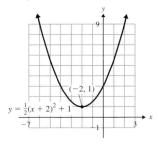

Chapter 14 Test

1) ellipse

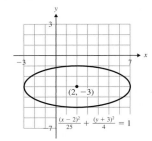

3) hyperbola

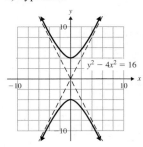

5) $(x + 1)^2 + (y - 3)^2 = 16$; center $(-1, 3)$; $r = 4$

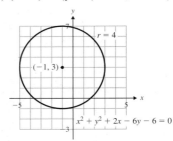

7) a)

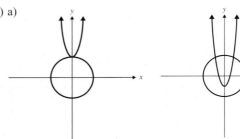

b)

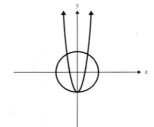

9) $\{(3, 1), (3, -1), (-3, 1), (-3, -1)\}$ 11) 8 in. $\times$ 14 in.

13) $\left(-4, -\dfrac{3}{2}\right)$

$$-\dfrac{3}{2}$$
$-5\ -4\ -3\ -2\ -1\ \ 0\ \ 1\ \ 2\ \ 3\ \ 4\ \ 5$

15) $(-\infty, -3) \cup [5, \infty)$

$-8\ -7\ -6\ -5\ -4\ -3\ -2\ -1\ \ 0\ \ 1\ \ 2\ \ 3\ \ 4\ \ 5\ \ 6\ \ 7\ \ 8$

Cumulative Review for Chapters 1–14

1) $-\dfrac{3}{4}$ 3) $A = 15$ cm^2; $P = 18.5$ cm

5) -1 7) $\dfrac{1}{8a^{18}b^{15}}$ 9) $n = \dfrac{c - z}{a}$ 11) 15, 17, 19   13) 0

15) $y = -\dfrac{3}{4}x + 4$ 17) 5 mL of 8%, 15 mL of 16%

19) $8w^3 + 30w^2 - 47w + 15$ 21) $2(3c - 4)(c - 1)$

23) $\{-2, 6\}$ 25) $6(t + 3)$ 27) $(-\infty, -3) \cup \left(\dfrac{9}{5}, \infty\right)$

29) $2\sqrt[3]{6}$ 31) $\dfrac{1}{8}$ 33) $\dfrac{20 - 5\sqrt{3}}{13}$

35) $\left\{-\dfrac{7}{2} + \dfrac{\sqrt{37}}{2}, -\dfrac{7}{2} - \dfrac{\sqrt{37}}{2}\right\}$

37)

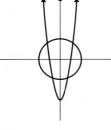

39) a) no   b) no   41) $\left\{-\dfrac{6}{13}\right\}$ 43) 2   45) $\left\{\dfrac{\ln 8}{3}\right\}$; 0.6931

47)   49) $\left[-\dfrac{12}{5}, \dfrac{12}{5}\right]$

Appendix

Section A1

1) -6 3) 32   5) true

7)
$-2\dfrac{1}{4}$ $\dfrac{2}{3}$ 1.5
$-4\ -3\ -2\ -1\ \ 0\ \ 1\ \ 2\ \ 3\ \ 4$

9) $\dfrac{15}{28}$ 11) $\dfrac{3}{2}$ 13) $-\dfrac{1}{9}$ 15) 1   17) $2\dfrac{2}{3}$ 19) -14 21) 4

23) -22 25) $-\dfrac{8}{15}$ 27) $16 - 10$; 6   29) $2(-19 + 4)$; -30

31) $-14, 0, 5$ 33) $\dfrac{8}{11}, -14, 3.7, 5, 0, -1\dfrac{1}{2}, 6.\overline{2}$

35) 139°   37) 28°; obtuse   39) 68 cm^2

41) $A = 27$ in^2; $P = 24$ in.

43) a) $A = 25\pi$ cm^2; $A \approx 78.5$ cm^2
   b) $C = 10\pi$ cm; $C \approx 31.4$ cm

45) a) 35 b) 30 47) $-\dfrac{1}{8}$ 49) distributive 51) associative

53) No; subtraction is not commutative. 55) $3 + 8$ 57) $-5p$

59) $-14w - 7$ 61) $30 - 42r$ 63) $-24a + 32b + 8c$

Section A2

1) no 3) $3z^2 - 8z + 13$ 5) $2n + 11$

7) $10 + \left(-7 + \dfrac{1}{2}x\right)$; $\dfrac{1}{2}x + 3$ 9) y^4 11) 64 13) $12m^{10}$

15) $49k^{10}$ 17) $\dfrac{x^9}{y^9}$ 19) w^3 21) $2pq^5$ 23) 1 25) 125

27) $\dfrac{1}{x^8}$ 29) $\dfrac{64}{t^3}$ 31) $\dfrac{b^2}{7a^4}$ 33) $\dfrac{s}{r^3}$ 35) $-\dfrac{8a^{15}b^3}{c^9}$

37) $27c^{31}d^{10}$ 39) $\dfrac{1}{r^3 s^{15}}$ 41) v^{20} 43) $\dfrac{a^{32}}{16b^{28}}$

45) $A = 6c^3$ square units; $P = 6c^2 + 4c$ units 47) 0.000507

49) 9.4×10^4 51) 40,000

Section A3

1) yes 3) $\{4\}$ 5) $\{-4\}$ 7) $\{3\}$ 9) $\{-6\}$ 11) $\left\{-\dfrac{1}{2}\right\}$

13) $\{3\}$ 15) $\{0\}$ 17) $\varnothing$ 19) $\{6\}$ 21) $\{-1\}$ 23) 26

25) Sweden: 9; Mexico: 11 27) $6000 at 6% and $2000 at 7%

29) $m\angle A = 85°$, $m\angle B = 57°$ 31) 62 33) $H = \dfrac{3V}{A}$

35) $b_2 = \dfrac{2A}{h} - b_1$ or $b_2 = \dfrac{2A - hb_1}{h}$

37) $[6, \infty)$

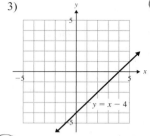

39) $(-\infty, -3]$

40)

41) $(-4, \infty)$

43) $\left(-2, \dfrac{5}{3}\right]$

45) $\left(-\dfrac{1}{2}, 2\right)$

47) $(-\infty, -3] \cup (1, \infty)$

Section A4

1) yes

3)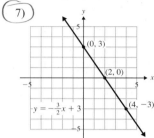

5)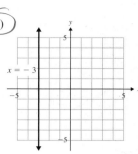

7)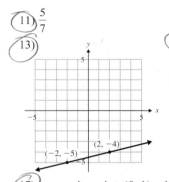

9)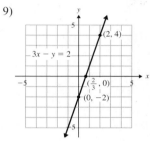

11) $\dfrac{5}{7}$

13)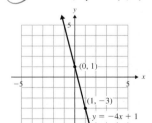

15)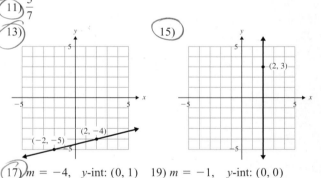

17) $m = -4$, y-int: $(0, 1)$ 19) $m = -1$, y-int: $(0, 0)$

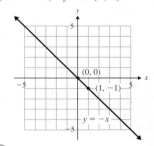

19) $m = -1$, y-int: $(0, 0)$ 21)

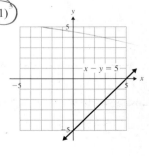

23)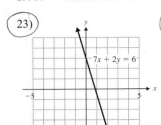

25)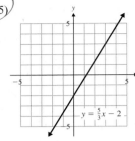

27) $y = \dfrac{3}{8}x + 4$ 29) $y = -\dfrac{1}{2}x + \dfrac{11}{2}$ 31) $y = 7$

33) $6x + y = 2$ 35) $3x - y = -11$

37) $y = -\dfrac{2}{3}x + 5$ 39) $x = -1$

41) a) $(0, 24000)$; If Naresh has $0 in sales, his income is
$24,000.

b) $m = 0.10$; Naresh earns $0.10 for $1.00 in sales

c) $34,000

43) domain: $(-\infty, \infty)$ 45) function; $(-\infty, \infty)$
range: $[-1, \infty)$
function

47) not a function; $[0, \infty)$ 49) -3 51) $2z - 9$

53) $2c - 3$ 55) 4

57)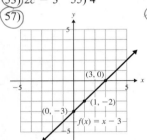

59)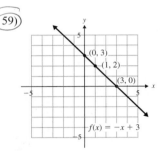

61) $m = -2$, y − int: $(0, -1)$ 63) a) 840 mi   b) 4.5 hours

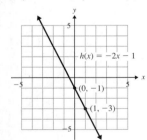

c)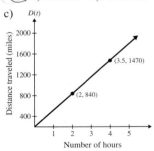

Section A5

1) yes

3) $(3, 1)$ 5) $\varnothing$

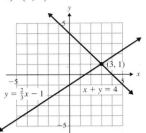

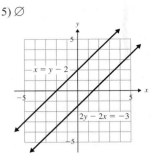

7) $(8, 1)$

9) infinite number of solutions of the form
$\{(x, y)\mid x = 2 - 5y\}$

11) $\left(-\dfrac{3}{2}, 4\right)$ 13) $\varnothing$ 15) $\varnothing$ 17) $(-9, 2)$ 19) $(-15, -6)$

21) $(2, -2)$ 23) length = 17 in. width = 10 in.

25) 24 oz of 18% solution and 36 oz of 8% solution

27) passenger train: 45 mph; freight train: 30 mph

29) $(-2, 1, 5)$ 31) $(5, 0, -2)$ 33) $(2, 2, 7)$

35) Ryan: 6; Seth: 1; Summer: 4

Section A6

1) -32 3) $\dfrac{64}{27}$ 5) $81z^{16}$ 7) $5t^{10}$ 9) $\dfrac{m^6}{125n^{18}}$ 11) -8

13) $4a^2b^2 - 15a^2b + 5ab^2 + 17ab - 6$ 15) a) 54   b) -26

17) $-14q^7 + 63q^5 + 42q^3$

19) $15b^4 - 4b^3 - 9b^2 + 43b + 18$

21) $-30d^6 + 72d^5 + 6d^4 - 2d^2 - 25d - 2$

23) $6r^2 + 17r + 7$ 25) $24d^4 - 54d^3 - 15d^2$ 27) $9a^2 - 16$

29) $h^4 - \dfrac{9}{64}$ 31) $r^2 - 20r + 100$ 33) $\dfrac{9}{16}c^2 - 9c + 36$

35) $v^3 + 15v^2 + 75v + 125$ 37) $-11h^2 + 4h + 1$

39) $-b - \dfrac{b}{3a} + \dfrac{9}{a}$ 41) $a + 3$ 43) $3x + 1 - \dfrac{28}{2x + 3}$

45) $5b^2 + 4b + 3$ 47) $6h^2 - 5h - 1 + \dfrac{2h - 7}{2h^2 + 3h - 1}$

49) a) $2b^2 + 6b - 12$ units   b) $2b^3 + 2b^2 - 12b$ square units

51) $\dfrac{9}{2}c^2 - \dfrac{3}{2}c$ square units   53) $3a^2 - a - 7$ units

Section A7

1) $4(7k + 2)$ 3) $2yz^2(13y^2z + 4y^2 - 6z)$

5) $(v + 12)(w + 3)$ 7) $(3a - 8)(7b - 6)$

9) $(x + 5)(x + 2)$ 11) $2d^2(d - 4)(d + 3)$

13) $(3n + 4)(n + 2)$ 15) $(2z + 1)(4z - 9)$

17) $4(3c - 4)(c - 1)$ 19) $(x - 4y)^2$

21) $(m + 3)(m - 3)$ 23) $q^4(5q + 3)(5q - 3)$

25) prime   27) $(4s - 3t)(16s^2 + 12st + 9t^2)$

29) $9(5m - 6)$ 31) $(2t + 9)(2t - 9)$

33) $(w + 16)(w - 3)$ 35) $2k^2(k - 4)(k - 1)$

37) $(10r - 3)(r - 4)$ 39) $(c + 1)(d - 1)$

41) $(m + 10)(m - 5)$ 43) $\{-8, 3\}$ 45) $\{-4, -3\}$

47) $\left\{0, \dfrac{5}{4}\right\}$ 49) $\{-13, 13\}$ 51) $\{-9, -3\}$ 53) $\{-4, 0\}$

55) $\left\{\dfrac{1}{2}, \dfrac{9}{2}\right\}$ 57) $\{-6, -1, 0\}$ 59) 10

61) length of wire = 17 ft; length of pole = 15 ft

Section A8

1) a) $\dfrac{3}{2}$ b) 0

3) a) -3 or 3

b) never undefined—any real number may be substituted for x.

5) $(-\infty, -9) \cup (-9, \infty)$

7) $(-\infty, -1) \cup (-1, 6) \cup (6, \infty)$ 9) $\dfrac{5z^2}{6}$ 11) $\dfrac{12}{7}$

13) $\dfrac{a - 2}{a + 7}$ 15) $\dfrac{t(3t - 1)}{t + 2}$ 17) $-\dfrac{r + 4}{4}$

19) $\dfrac{3}{8w^3} = \dfrac{9}{24w^3}; \dfrac{1}{6w} = \dfrac{4w^2}{24w^3}$ 21) $\dfrac{13}{6j}$ 23) $\dfrac{11w - 14}{w(w - 2)}$

25) $\dfrac{a^2 + 26a - 18}{(4a - 3)(a + 2)}$ 27) $\dfrac{k + 2}{k + 7}$ 29) $\dfrac{r}{3(2r - 3)}$

31) $\dfrac{w - 4}{(2w - 3)(w + 2)}$ 33) $2x$ 35) $-\dfrac{1}{5m(m^2 + 3m + 9)}$

37) $\dfrac{6x - 1}{x - 8}$ 39) $\dfrac{14}{15}$ 41) $\dfrac{8}{r}$ 43) $\dfrac{m^2 + 5}{2m + 3}$ 45) $\dfrac{2(w - 1)}{3}$

47) $\{-3\}$ 49) $\{2, 8\}$ 51) $\{8\}$ 53) $\{-2, 10\}$ 55) $\{2\}$

57) $y = mx + b$ 59) $d = \dfrac{ab + ct}{c}$

61) a) $\dfrac{5(a - 9)}{a - 2}$ square units   b) $\dfrac{a^2 - 11a + 98}{2(a - 2)}$ units

63) 45 jazz and 27 reggae

Photo Credits

Index

Figure		Perimeter	Area
Rectangle:	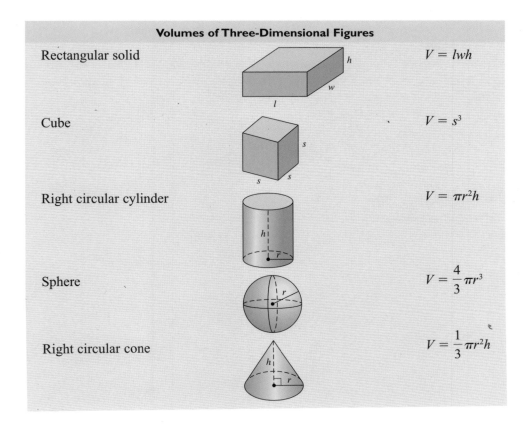	$P = 2l + 2w$	$A = lw$
Square:		$P = 4s$	$A = s^2$
Triangle: $h = $ height		$P = a + b + c$	$A = \dfrac{1}{2}bh$
Parallelogram: $h = $ height		$P = 2a + 2b$	$A = bh$
Trapezoid: $h = $ height		$P = a + c + b_1 + b_2$	$A = \dfrac{1}{2}h(b_1 + b_2)$
Circle:		**Circumference** $C = 2\pi r$	**Area** $A = \pi r^2$

Volumes of Three-Dimensional Figures

Rectangular solid		$V = lwh$
Cube		$V = s^3$
Right circular cylinder		$V = \pi r^2 h$
Sphere		$V = \dfrac{4}{3}\pi r^3$
Right circular cone		$V = \dfrac{1}{3}\pi r^2 h$

Angles

An **acute angle** is an angle whose measure is greater than 0° and less than 90°.

A **right angle** is an angle whose measure is 90°.

An **obtuse angle** is an angle whose measure is greater than 90° and less than 180°.

A **straight angle** is an angle whose measure is 180°.

Acute angle   Right angle   Obtuse angle   Straight angle

Two angles are **complementary** if their measures add to 90°.

Two angles are **supplementary** if their measures add to 180°.

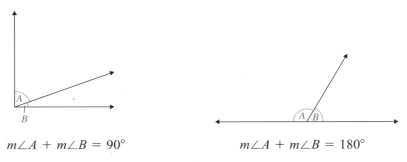

$$m\angle A + m\angle B = 90°$$ $$m\angle A + m\angle B = 180°$$

Vertical Angles

The pair of opposite angles are called **vertical angles**. Angles A and C are *vertical angles,* and angles B and D are *vertical angles. The measures of vertical angles are equal.* Therefore, $m\angle A = m\angle C$ and $m\angle B = m\angle D$.

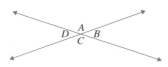

Parallel and Perpendicular Lines

Parallel lines are lines in the same plane that do not intersect.
Perpendicular lines are lines that intersect at right angles.

Perpendicular lines

Parallel lines

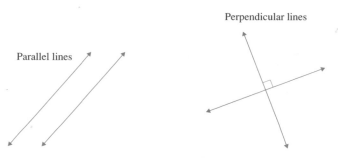